Managing Air Quality and Energy Systems

Environmental Management Handbook, Second Edition

Edited by
Brian D. Fath and Sven E. Jørgensen

Volume 1
Managing Global Resources and Universal Processes

Volume 2
Managing Biological and Ecological Systems

Volume 3
Managing Soils and Terrestrial Systems

Volume 4
Managing Water Resources and Hydrological Systems

Volume 5
Managing Air Quality and Energy Systems

Volume 6
Managing Human and Social Systems

Managing Air Quality and Energy Systems

Second Edition

Edited by
Brian D. Fath and Sven E. Jørgensen

Assistant to Editor
Megan Cole

CRC Press
Taylor & Francis Group
Boca Raton London New York

CRC Press is an imprint of the
Taylor & Francis Group, an **informa** business

Cover photo: Hounslow, United Kingdom, N. Fath

Second edition published 2021
by CRC Press
6000 Broken Sound Parkway NW, Suite 300, Boca Raton, FL 33487-2742

and by CRC Press
2 Park Square, Milton Park, Abingdon, Oxon, OX14 4RN

First edition published by CRC Press 2013

CRC Press is an imprint of Taylor & Francis Group, LLC

ISBN: 978-1-138-34267-5 (hbk)
ISBN: 978-1-003-04346-1 (ebk)

Typeset in Minion
by codeMantra

Contents

Preface

Given the current state of the world as compiled in the massive Millennium Ecosystem Assessment Report, humans have changed ecosystems more rapidly and extensively during the past 50 years than in any other time in human history. These are unprecedented changes that need certain action. As a result, it is imperative that we have a good scientific understanding of how these systems function and good strategies on how to manage them.

In a very practical way, this multi-volume *Environmental Management Handbook* provides a comprehensive reference to demonstrate the key processes and provisions for enhancing environmental management. The experience, evidence, methods, and models relevant for studying environmental management are presented here in six stand-alone thematic volumes, as follows:

VOLUME 1 – Managing Global Resources and Universal Processes
VOLUME 2 – Managing Biological and Ecological Systems
VOLUME 3 – Managing Soils and Terrestrial Systems
VOLUME 4 – Managing Water Resources and Hydrological Systems
VOLUME 5 – Managing Air Quality and Energy Systems
VOLUME 6 – Managing Human and Social Systems

In this manner, the handbook introduces in the first volume the general concepts and processes used in environmental management. The next four volumes deal with each of the four spheres of nature (biosphere, geosphere, hydrosphere, and atmosphere). The last volume ties the material together in its application to human and social systems. These are very important chapters for a wide spectrum of students and professionals to understand and implement environmental management. In particular, the features include the following:

- The first handbook that demonstrates the key processes and provisions for enhancing environmental management.
- Addresses new and cutting-edge topics on ecosystem services, resilience, sustainability, food–energy–water nexus, socio-ecological systems, etc.
- Provides an excellent basic knowledge on environmental systems, explains how these systems function, and gives strategies on how to manage them.
- Written by an outstanding group of environmental experts.

Since the handbook covers such a wide range of materials from basic processes, to tools, technologies, case studies, and legislative actions, each handbook entry is further classified into the following categories:

APC: Anthropogenic chemicals—the chapters cover human-manufactured chemicals and activities
COV: Indicates that the chapters give comparative overviews of important topics for environmental management

CSS: The chapters give a case study of a particular environmental management example
DIA: Means that the chapters are about diagnostic tools—monitoring, ecological modeling, ecological indicators, and ecological services
ELE: Focuses on the use of legislation or policy to address environmental problems
ENT: Addresses environmental management using environmental technologies
NEC: Natural elements and chemicals—the chapters cover basic elements and chemicals found in nature
PRO: The chapters cover basic environmental processes.

Overall, these volumes will be a valuable resource for all libraries supporting programs in environmental science and studies, earth science, geography, and policy.

In this volume, #5, the focus is on managing air quality and the closely related topic of energy systems, as represented in over 50 entries. Energy basics and physics for conventional and alternative sources are considered. Specific impacts such as global climate change, acid rain, and ozone are covered. New entries include specific tools to measure road traffic emissions, the importance of managing micropower grids, and the role of individual and household behavior in emission scenarios. Case studies look at energy conversion and the impact of wind farm noise. This volume contains a number of entries on air pollution control strategies.

Brian D. Fath
Brno, Czech Republic
December 2019

Editors

Brian D. Fath is Professor in the Department of Biological Sciences at Towson University (Maryland, USA) and Senior Research Scholar at the International Institute for Applied Systems Analysis (Laxenburg, Austria). He has published over 180 research papers, reports, and book chapters on environmental systems modeling, specifically in the areas of network analysis, urban metabolism, and sustainability. He co-authored the books *A New Ecology: Systems Perspective* (2020), *Foundations for Sustainability: A Coherent Framework of Life–Environment Relations* (2019), and *Flourishing Within Limits to Growth: Following Nature's Way* (2015). He is also Editor-in-Chief for the journal *Ecological Modelling* and Co-Editor-in-Chief for *Current Research in Environmental Sustainability.* Dr. Fath was the 2016 recipient of the Prigogine Medal for outstanding work in systems ecology and twice a Fulbright Distinguished Chair (Parthenope University, Naples, Italy, in 2012 and Masaryk University, Czech Republic, in 2019). In addition, he has served as Secretary General of the International Society for Ecological Modelling, Co-Chair of the Ecosystem Dynamics Focus Research Group in the Community Surface Modeling Dynamics System, and member and past Chair of Baltimore County Commission on Environmental Quality.

Sven E. Jørgensen (1934–2016) was Professor of environmental chemistry at Copenhagen University. He received a doctorate of engineering in environmental technology and a doctorate of science in ecological modeling. He was an honorable doctor of science at Coimbra University (Portugal) and at Dar es Salaam (Tanzania). He was Editor-in-Chief of *Ecological Modelling* from the journal inception in 1975 until 2009. He was Editor-in-Chief for the *Encyclopedia of Environmental Management* (2013) and *Encyclopedia of Ecology* (2008). In 2004, Dr. Jørgensen was awarded the Stockholm Water Prize and the Prigogine Medal. He was awarded the Einstein Professorship by the Chinese Academy of Sciences in 2005. In 2007, he received the Pascal Medal and was elected a member of the European Academy of Sciences. He had published over 350 papers, and has edited or written over 70 books. Dr. Jørgensen gave popular and well-received lectures and courses in ecological modeling, ecosystem theory, and ecological engineering worldwide.

Contributors

Kalyan Annamalai
Paul Pepper Professor of Mechanical Engineering
Texas A&M University
College Station, Texas

Senthil Arumugam
Enerquip, Inc.
Medford, Wisconsin

Muhammad Asif
School of the Built and Natural Environment
Glasgow Caledonian University
Glasgow, United Kingdom

Lu Aye
Renewable Energy and Energy Efficiency Group
Department of Infrastructure Engineering
Melbourne School of Engineering
University of Melbourne
Melbourne, Victoria, Australia

Udo Becker
Chair of Transport Ecology
Faculty of Transportation Sciences "Friedrich List"
Institute of Transport Planning and Road Traffic
Dresden University of Technology
Dresden, Germany

John O. Blackburn
Professor Emeritus of Economics
Duke University
Maitland, Florida

Pascal Boeckx
Faculty of Agricultural and Applied Biological Sciences
University of Ghent
Ghent, Belgium

Jens Borken-Kleefeld
Air Quality and Greenhouse Gases
International Institute for Applied Systems Analysis
Laxenburg, Austria

Felix Creutzig
Mercator Research Institute on Global Commons and Climate Change (MCC)
and
Technical University of Berlin
Berlin, Germany

Bernd Delakowitz
Faculty of Mathematics and Natural Sciences
University of Applied Sciences
Zittau, Germany

Harvey E. Diamond
Energy Management International
Conroe, Texas

Ibrahim Dincer
Faculty of Engineering and Applied Science
University of Ontario Institute of Technology (UOIT)
Oshawa, Ontario, Canada

Svetlana Drozdova
Institute of Chemical Technologies and Analytics
Vienna University of Technology
Vienna, Austria

Sarina J. Ergas
Department of Civil and Environmental Engineering
University of Massachusetts
Amherst, Massachusetts

Sara Evangelisti
Interuniversity Research Center for Sustainable Development (CIRPS)
Sapienza University of Rome
Rome, Italy

Andrew Fang
Humphrey School of Public Affairs
University of Minnesota
Minneapolis, Minnesota

Stefan Fraenzle
Department of Biological and Environmental Sciences
Research Group of Environmental Chemistry
International Graduate School Zittau
Zittau, Germany

Regine Gerike
Chair of Integrated Transport Planning and Traffic Engineering
Faculty of Transportation Sciences "Friedrich List"
Institute of Transport Planning and Road Traffic
Dresden University of Technology
Dresden, Germany

Fredric S. Goldner
Energy Management & Research Associates
East Meadow, New York

Dan Golomb
Department of Environmental, Earth and Atmospheric Sciences
University of Massachusetts—Lowell
Lowell, Massachusetts

Alex E.S. Green
Professor Emeritus
University of Florida
Gainesville, Florida

Eliane Tigre Guimarães
Experimental Air Pollution Laboratory
Department of Pathology
School of Medicine
University of Sao Paulo
Sao Paulo, Brazil

Ryan Hanna
School of Global Policy and Strategy
University of California San Diego
La Jolla, California

Chris Hanning
Sleep Medicine
University Hospitals of Leicester
Leicester, United Kingdom

Fabian Heidegger
Chair of Integrated Transport Planning and Traffic Engineering
Faculty of Transportation Sciences "Friedrich List"
Institute of Transport Planning and Road Traffic
Dresden University of Technology
Dresden, Germany

Steven D. Heinz
Good Steward Software
State College, Pennsylvania

Arif Hepbasli
Faculty of Engineering
Department of Energy Systems Engineering
Yaşar University
Bornova, Izmir, Turkey

Nathan E. Hultman
University of Maryland
College Park, Maryland

Keith E. Idso
Center for the Study of Carbon Dioxide and Global Change
Tempe, Arizona

Sherwood Idso
Center for the Study of Carbon Dioxide and Global Change
Tempe, Arizona

Kui Jiao
Department of Mechanical Engineering
University of Waterloo
Waterloo, Ontario, Canada

Sven Erik Jørgensen
Institute A, Section of Environmental Chemistry
Copenhagen University
Copenhagen, Denmark

Gholamreza Karimi
Department of Mechanical Engineering
University of Waterloo
Waterloo, Ontario, Canada

Janey Kaster
Yamas Controls West
San Francisco, California

Ronald L. Klaus
VAST Power Systems
Elkhart, Indiana

Ewa Klugmann-Radziemska
Chemical Faculty
Gdansk University of Technology
Gdansk, Poland

Milivoje M. Kostic
Department of Mechanical Engineering
Northern Illinois University
DeKalb, Illinois

Umesh Kulshrestha
School of Environmental Sciences
Jawaharlal Nehru University
New Delhi, India

Xianguo Li
Department of Mechanical Engineering
University of Waterloo
Waterloo, Ontario, Canada

Bernd Markert
Environmental Institute of Scientific Networks (EISN)
Haren-Erika, Germany

Rudolf Marloth
San Diego State University
San Diego, California

Andrea Micangeli
Interuniversity Research Center for Sustainable Development (CIRPS)
Sapienza University of Rome
Rome, Italy

Adnan Midili
Department of Mechanical Engineering
Faculty of Engineering
Nigde University
Nigde, Turkey

Luisa T. Molina
Massachusetts Institute of Technology
Cambridge, Massachusetts

Martin A. Mozzo, Jr.
M and A Associates Inc.
Robbinsville, New Jersey

Tariq Muneer
School of Engineering
Napier University
Edinburgh, United Kingdom

Leila Niamir
Mercator Research Institute on Global Commons and Climate Change (MCC)
Berlin, Germany

and

University of Twente
Enschede, The Netherlands

K.E. Ohrn
Cypress Digital Ltd.
Vancouver, British Columbia, Canada

Bohdan W. Oppenheim
U.S. Department of Energy Industrial Assessment Center
Loyola Marymount University
Los Angeles, California

Andrea Nunes Vaz Pedroso
Nucleus Research in Ecology
Institute of Botany
Sao Paulo, Brazil

Soyuz Priyadarsan
Texas A&M University
College Station, Texas

Anu Ramaswami
Humphrey School of Public Affairs
University of Minnesota
Minneapolis, Minnesota

Marc A. Rosen
Faculty of Engineering and Applied Science
University of Ontario Institute of Technology
Oshawa, Ontario, Canada

Erwin Rosenberg
Institute of Chemical Technologies and Analytics
Vienna University of Technology
Vienna, Austria

Danilo Sbordone
Interuniversity Research Center for Sustainable Development (CIRPS)
Sapienza University of Rome
Rome, Italy

Wolfram Schmidt
Chair of Transport Ecology
Faculty of Transportation Sciences "Friedrich List"
Institute of Transport Planning and Road Traffic
Dresden University of Technology
Dresden, Germany

Daniel Shepherd
Auckland University of Technology
Auckland, New Zealand

Leslie A. Solmes
LAS and Associates
Mill Valley

Amanda Staudt
Climate Scientist
National Wildlife Federation
Reston, Virginia

Therese Stovall
Oak Ridge National Laboratory
Oak Ridge, Tennessee

John M. Sweeten
Texas A&M University
Amarillo, Texas

Bob Thorne
Massey University
Palmerston North, New Zealand

Jill S. Tietjen
Technically Speaking, Inc.
Greenwood Village, Colorado

Wayne C. Turner
Industrial Engineering and Management
Oklahoma State University
Stillwater, Oklahoma

Robert E. Uhrig
Department of Nuclear Engineering
University of Tennessee
Knoxville, Tennessee

Oswald Van Cleemput
Faculty of Agricultural and Applied Biological Sciences
University of Ghent
Ghent, Belgium

George F. Vance
Department of Ecosystem Sciences and Management
University of Wyoming
Laramie, Wyoming

Waldemar Wardencki
Department of Chemistry
Gdansk University of Technology
Gdansk, Poland

Eric A. Woodroof
Profitable Green Solutions
Plano, Texas

Simone Wuenschmann
Environmental Institute of Scientific Networks (EISN)
Haren-Erika, Germany

Kazuyuki Yagi
National Institute for Agro-Environmental Sciences
Ibaraki, Japan

Andrew R. Zimmerman
Department of Geological Sciences
University of Florida
Gainesville, Florida

I

APC: Anthropogenic Chemicals and Activities

1

Genotoxicity and Air Pollutions

Eliane Tigre Guimarães and Andrea Nunes Vaz Pedroso

Air Pollutants

Air pollution can be generated by natural and anthropogenic sources. The natural sources, such as electrical discharge, decomposition of organic matter, volcano eruption, and natural fires, do not depend on human actions and emit large amounts of pollution, usually in restricted and sparsely populated areas.

The anthropogenic sources can be stationary or mobile. Stationary sources are mainly industries that cause local problems of air contamination. Their pollution emissions are determined by the characteristics of the manufacturing processes, which include the sort of raw materials and fuels used and the products furnished, as well as by the efficiency of the industrial processes and the control measurements adopted.

Mobile sources consist of automotive vehicles, trains, airplanes, ships, and motorboats, which release pollutants into the atmosphere due to incomplete burning of fossil fuels. However, the automotive vehicles are the main mobile sources.

The atmosphere in large cities is usually contaminated by a range of pollutants from stationary and mobile sources. The pollutant emissions of mobile sources are difficult to be controlled, mainly because of the increasing number of automotive vehicles in the last 50 years. This number increased tenfold during this period.[1] The emissions of air pollutants in urban centers have been causing great concerns all over the world and have been causing harmful effects on living organisms.

The main pollutants in urban centers are gases such as carbon, nitrogen, and sulfur oxides, and organic compounds such as hydrocarbons, volatile organic compounds (VOCs), and particulate matter. They will be described below.

The gas carbon monoxide is an odorless, colorless, and tasteless gas formed during the incomplete combustion of carbon-containing fuels.[2]

The main source of sulfur dioxide (SO_2) is the combustion of fuels. Fossil fuels have 1%–5% sulfur in their composition. During combustion, the sulfur is converted to SO_2. Nowadays, in developed countries, a large quantity of the sulfur is removed from motor fuels during the refining process and gas emission from chimneys. However, in developing countries, unabated burning of coal and the use of fuel oils and automotive diesel with higher sulfur content are major sources of SO_2.[2]

The nitrogen derived from the combustion process of fossil fuels is converted to nitrogen oxides[2] such as nitrogen monoxide (NO) and nitrogen dioxide (NO_2). They are considered the precursors of tropospheric ozone (O_3) formation. Nitrogen dioxide diffuses into the atmosphere, where it is usually oxidized and can react with water to form acid rain, causing corrosion in materials and damage to human beings.[3]

The nitrogen oxides and VOCs are considered precursors and produce by photochemical reactions many secondary pollutants, among them O_3 and peroxyacetyl nitrate,[1,3] which compose so-called photochemical smog.

When O_3 is formed in an atmosphere without pollutants, it is consumed within minutes by the photostationary equilibrium between NO and NO_2. Nevertheless, in a polluted atmosphere, NO is converted to NO_2 and can be consumed by RO_2 (organic radical), and, as a consequence, O_3 is accumulated.[3,4] The O_3 is considered one of the most damaging gaseous pollutants to human health and plants, because it forms the reactive oxygen species (ROS) such as superoxide, hydrogen peroxide, and hydroxyl, among others. Reactive oxygen species are oxidative and affect lipids, proteins, and nucleic acids; the cell membranes, composed by polyunsaturated fatty acids, represent the initial target of ROS, changing their permeability and triggering lipid peroxidation,[5] amino acid oxidation, and inactivation of enzymes.[6]

The particulate matter (PM) is a mixture of solid or liquid particles suspended in the air, including smoke, fumes, soot, and other combustion by-products, besides natural particles such as wind-blown dust, sea salt, pollen, and spores.[7] These components can be characterized by their size and composition.[1] Based on the aerodynamic diameter, which ranges from 0.002 to 100 µm, the particulate matter is classified into three categories: 1) coarse particles, ranging from 2.5 to 100 µm; 2) fine particulate matter, below 2.5 µm; 3) ultrafine particles, below 0.1 µm.[8]

Air quality is now regulated by standard concentrations established by laws, based on experiments on humans and/or animals and epidemiological investigations. The standards for air pollutants in Europe are proposed by the European Commission and in the United States by the Environmental Protection Agency (EPA). In Brazil, the standard values for air quality control are defined by a resolution proposed by the National Council of Environment (CONAMA; Table 1).[9]

Although these standards are often revised in order to protect the human health, the World Health Organization states that around 2.4 million people still die each year due to causes related to air pollution.

TABLE 1 Standard Values for Pollutants Established by Environmental Agencies (European Commission, EPA, and CONAMA)

Pollutant	Sampling Time	European Commission	EPA	CONAMA
CO (carbon monoxide)	1 hr	26 ppm	35 ppm	35 ppm
	8 hr	10 mg/m³	–	–
NO_2	8 hr	–	–	9 ppm
	1 hr	200 µg/m³	100 ppb	320 µg/m³
PM_{10} (particulate matter)	Annual	40 µg/m³	53 ppb	100 µg/m³
	24 hr	50 µg/m³	150 µg/m³	150 µg/m³
$PM_{2,5}$ (particulate matter)	Annual	25 µg/m³	150 µg/m³	50 µg/m³
	24 hr	–	35 µg/m³	–
O_3	8 hr	120 µg/m³	80 ppb	120 µg/m³
	24 hr	–	120 ppb	125 µg/m³
SO_2	24 hr	125 µg/m³	140 ppb	125 µg/m³
	Annual	–	75 ppb	80 µg/m³

Source: Adapted from "Air Quality Standards,"[16] "National Ambient Air Quality Standards,"[17] and "Qualidade do Ar."[18]
–, Limit not defined.

About 1.5 million deaths are attributable to indoor air pollution (estimated deaths).[2] Epidemiological studies suggest that Americans and Europeans have high rates of deaths from cardiopulmonary diseases arising from air pollution.[10] Worldwide, the number of deaths per year caused by pollution is greater than that caused by car accidents.[11] The individual response to air pollutants depends of the type of pollutant, the degree of exposure, the health conditions and the individual genetics[12] and still, socioeconomic profile.[13,14]

Air pollutants affect the vital molecules of human beings, such as nucleic acids, causing genotoxic effects, among numerous other health problems. Thus, this will be the main focus of this entry from here on. The genotoxic effects most commonly reported in the literature and bioassays proposed for prognosis of genotoxic risks will be reported. Finally, other effects to human health will be mentioned at the end of this entry.

Genotoxic Effects of Air Pollutants

Genotoxicity is defined as every alteration occurring in genetic material that causes loss of cellular integrity.[15]

The literature defines that mutations can modify the amino acid sequence of the gene encoding the protein or damage in the DNA molecule. They may occur over the life span of living beings. This process is of extreme importance for the evolution of species. Among gene mutations, we may find the following cases: 1) when a single nucleotide base is replaced by another; and 2) when extra base pairs are added or deleted from the DNA. These are also referred to as insertions and deletions, respectively. These mutations can be devastating, because the messenger RNA is translated into new groups of three nucleotides and the protein produced can cause serious damage.

The mutations that encompass larger portions of DNA are called macrolesions, which change the structure of chromosomes, resulting in damage of the genotype and phenotype of the organism, such as translocations, inversions, deletions, and duplications. They occur most frequently during meiosis.[19]

The mutation may result from exposure to different environmental or chemical agents. Throughout the numerous and successive divisions, the cell can accumulate a large number of mutations and trigger the loss of division control and contribute to the initial stage of tumor development.[20] Therefore, the genotoxic agents are considered mutagenic or carcinogenic,[15] and there is a slight difference between these classifications: 1) mutagenic effect is an alteration in the genetic material of the cell of a living organism that is more or less permanent and that can be transmitted to the cell's descendants; and 2) carcinogenic effects are caused by genotoxic substances that can produce tumors, abnormal tissue growths caused by a loss of control in cell replication. Nevertheless, repairing mechanisms of DNA are present in all organisms, and their complexity is directly proportional to the complexity of the organism.[21] The mechanisms can be classified into direct reversal, damage excision, or recombination.[22]

In human beings, most of the pollutants studied are particulate matter.[7,23,24] In the previous section, we said that smaller fractions of PM, i.e., <2.5 μm, are more harmful, and this was confirmed by Rossner and colleagues,[25] who analyzed the organic fractions of PM of polluted sites and found DNA damage. A similar result was found by Coronas and colleagues;[26] they employed cells of lymphocytes and buccal mucosa cells to assess the genotoxic potential of PM in people living and working near refinery oil. The authors used two genotoxicity assays—comet assay and micronucleus assay—and pointed out that the comet assay was more sensitive. These tests will be detailed in the next section.

Other studies show organic extracts of particulate matter to cause genotoxicity. Roubicek and colleagues[27] found that in regions polluted in Mexico City, organic extracts of PM containing Cd and PAHs induce micronucleus formation in human epithelial cells. In several European cities were collected organic extracts mixtures of PAHs and particulate matter, confirmed this genotoxic, using HepG2 cells have the metabolic capacity for PAHs similar to human hepatocytes and represent the best in vitro model for investigating the genotoxic potential of complex mixtures containing PAHs.[28]

Another study found DNA damage in human lung cells when exposed to particulate matter.[29] Still, Gilli and colleagues[30] obtained positive correlations statistically between PM_{10} and mutagenicity, bioavailable iron, sulfates, and nitrates.

Genotoxicity Tests

Currently, there are numerous protocols with prokaryotic and eukaryotic organisms, which evaluate the mutagenic effects of different substances in order to identify risks that living organisms are exposed to. The genotoxic tests most used to detect genotoxicity of air pollutants will be mentioned below.

The Ames test is also known as the *Salmonella* mutagenicity test. This test was developed by Bruce Ames and colleagues and aimed to evaluate the carcinogenic potential of different substances using mutant strains of auxotrophic *Salmonella typhimurium* with respect to histidine. It detects mutagens that cause the displacement of the reading frame (frameshift) or substitution of base pairs of DNA.[31,32]

Other tests use eukaryotic organisms, aiming to evaluate the mutagenic potential of different substances by means of numerical and/or structural chromosomal abnormalities involving at least 10 million base pairs (10 Mb). Among them, we may include the in vitro cytogenetic test in mammalian cells (mouse lymphoma assay) that quantifies the genetic changes that affect the expression of the TK gene of the enzyme thymidine kinase (tk) cells in cultures of L5178Y tk+/tk− lymphoma mice. Although the mammalian cells present locus heterozygosity, only one gene copy is functional in this mouse strain. Loss of locus heterozygosity of the enzyme thymidine kinase, when the gene is affected by a mutagen, causes the resistance of cells mutated to supplementation of medium with trifluorothymidine (TFT). The TFT causes inhibition of metabolism, preventing cell division and leading to death of cells that have the entire way of nucleotides recovery. This test was developed by Clive and colleagues in 1979[33] and modified by Cole and colleagues in 1990.[34]

The comet assay is also a well-known test. It is used to detect not chromosomal mutations but genomic lesions. The test is based on the technique of gel electrophoresis, which detects DNA damage. Since DNA is negatively charged, the electric current causes migration of small broken pieces through the gel, faster and farther than larger pieces. As a result, the damaged cell looks like a comet, with the pieces of damaged DNA forming the tail. The smaller the pieces of DNA, the more they migrate from the cell body. Therefore, a longer tail with smaller pieces implies a greater genetic damage. The content and fragment length of the tail are directly proportional to the amount of DNA damage.

Among mutagenic tests, the micronucleus test is widely used because it is applicable with different eukaryotic organisms. By definition, the micronucleus is a small nucleus, regarded as a product of breakage of genomic DNA of eukaryotic cells. During cell division, genetic material is duplicated and distributed equally between two daughter cells. Radiation and chemicals can cause chromosomal breakage or damage, affecting the distribution of genetic material between daughter cells. Parts or fragments of chromosomes resulting from this damage can be distributed to any of the daughter cells. It is not incorporated into the new core; they may be presented in the form of micronuclei clearly observable on optical microscope.[35]

The micronucleus test in erythrocytes of bone marrow of rodents was developed by Matter and Schmid[36] and modified by Heddle and Salamone[37] in the following years and more recently by Mavournin and colleagues.[38] The test is based on the fact that the effect of genotoxic agent is observed in polychromatic erythrocytes. In addition, the micronuclei are easily observed, and the frequency of micronuclei is dependent on the sampling time.

The micronucleus test in human peripheral blood lymphocytes is also used to detect the mutagenic potential of substances. Nonetheless, some technical problems occur due to the fragility of the cell and the variability in the process of mitotic lymphocytes. These problems were solved by cell hypotonization[39] and radioactive labeling with cytochalasin B to identify the cells that suffered mitosis by inhibiting cytokinesis without blocking mitosis.[40]

Although studies on animals used to detect the effects of air pollution are successful,[41] simple and more efficient analyses to investigate the environmental risks and to determine the genotoxicity induced by pollution are needed. The bioassays with plants are generally more sensitive than other systems for this purpose.[42] Several studies with genus *Tradescantia* have been considered since 1960 as effective biomarkers for determining the genotoxic potential of air pollutants.[43–50]

The *Tradescantia* micronucleus bioassay is the quantification of micronuclei formed in meiotic prophase I, better seen in the young tetrad stage.[44] Among the genus *Tradescantia*, we may highlight the 4430 clone (hybrid between *Tradescantia hirsutiflora* and *T. subacaulis*).[44,48] A cultivar of *T. pallida* Purpurea from Mexico and Honduras has also been used for the micronucleus test since 1999.[47,49]

Nonetheless, in urban areas, gaseous and particulate pollutants interact with each other, thereby enhancing the genotoxic effects on the living organisms. Some studies with *Tradescantia* showed a significant increase in genotoxic potential in plants exposed in the urban environment compared with the rural environment,[49,51,52] also to detect the genotoxic potential of water-soluble fraction of PM_{10}[47] and a dose of 60 ppb ozone in fumigation chambers.[53]

Furthermore, other tests are also conducted with micronuclei in different plant species such as *Allium cepa* and *Vicia faba*; however, the micronuclei are formed from errors in mitotic division in those species.[50]

Another test used to evaluate the risks caused by mutagens also using the genus *Tradescantia* is the *Tradescantia* stamen hair bioassay.[54] It was developed by Arnold H. S. Sparrow based on the fact that the stamen hair cells of plants are heterozygous for color, making it possible to detect mutations based on the change in pigmentation from blue (dominant) to pink (recessive).[55] A pink mutant cell can continue to divide, giving rise to a series of contiguous pink cells, representing a single mutation event. Two mutant cells separated by blue cells are considered two mutation events.[54,56]

In addition to the genotoxic tests using plant species, there is another test known as pollen abortion, which has the ability to detect lethal mutations in haploid cells (microsporous) that end up affecting the development of pollen grains.[57,58] This was confirmed by Micieta and Murín,[59] who evaluated approximately 40 species of native vegetation in Slovakia subjected to industrial pollution, and they observed a positive relationship, i.e., high rate of pollen abortion in polluted area.

Other plant species used in genotoxic tests, like *Nicotiana tabacum*, in different degrees of polluted environments show the largest amount of DNA damage in high pollutant concentrations.[52]

The genotoxic effects caused by air pollutants are studied with more emphasis on the respiratory system in human beings. Effects as changes in pulmonary functions, modification of biochemical and cellular functions, or secretions could happen in the respiratory epithelium. Pacini and colleagues[60] observed a higher amount of DNA damage in people living in the polluted region in Florence, when compared with those living in less polluted areas in Sardinia, both in Italy. In Suwon, Korea, the genotoxic potential of organic extract of $PM_{2,5}$ in lung bronchial epithelial cells was also detected.[61] In addition, the same genotoxic effect can be observed when human beings were exposed to different concentrations of NO_2.[62] Additionally, Tova-lin and colleagues[63] noted the severity of DNA damage in workers in a large urban center, due to the combination of air pollutants VOC, PM_{25}, and O_3.

A considerable amount of tests to detect the mutagenicity of different substances is available, although only the most used were described. A review on urban air mutagenicity and experimental systems reported that 50% of the studies apply the *Salmonella* assay (Ames test); about 30% apply the plant systems (micronucleus tests, chromosomal aberrations, among others); and the other 20% of the studies used other bioassays (such as damage in DNA), animals, and other combinations of studies.[64]

Air Pollutants and Other Health Effects

Historically, the harmful effects of air pollution on human health have become evident. The association between high levels of pollution and diseases manifested by the population exposed has been long detected. Perhaps the most known episode occurred in London in December 1952, where a thermal inversion was responsible for 4000 deaths in 2 weeks.[65] Other similar episodic events confirm this evidence, for example, Meuse Valley in Belgium in 1930,[66] Donora in Pennsylvania in 1948.[67] and St. Louis in 1985.[68]

There is evidence to support the concept that particulate matter causes human mortality, morbidity,[69,70] and genotoxic effects.[71] There is an association between particulate matter and alteration in the respiratory system with restricted activity and severe breathing conditions (acute bronchitis and asthma), resulting in difficulty in breathing and insomnia in adults,[72] emergency room visits, hospital admissions,[73] and pulmonary vasoconstriction.[74,75] It affects lung growth in rats after chronic exposure[76] and significant association between lung cancer and long-term exposure to fine particles, reinforcing the role of fine particulates in the pathogenesis of lung cancer.[77]

Still, in the circulatory system, the fine particulate matter reaches the alveolar regions, transposes the alveolar capillary barrier, and, as a consequence, intensifies the risk of functional abnormalities,[78] such as acute vascular dysfunction, increases thrombus formation,[79,80] arrhythmia, and sudden death.[81]

The toxicity of carbon monoxide has been widely investigated and is well known. Studies show that a major change in humans is the formation of a stable complex between CO and hemoglobin, called carboxyhemoglobin. It decreases the release transport of oxygen to the tissues via blood.[82] In relation to nitrogen dioxide, epidemiological studies have shown that it affects the respiratory system of humans when inhaled[83] and, in high levels of concentration, can be correlated with increased symptoms of asthmatic bronchitis and reduced lung function in children.[2]

In humans, high concentrations of O3 are associated with reduced forced expiratory volume in 1 sec and forced expiratory flow at 50% and 75% of forced vital capacity.[83] Regions with higher ozone concentrations present a higher incidence of asthmatic patients.[83,84] Additionally, according to Pereira et al.,[85] this pollutant showed a positive correlation with the incidence of lung and larynx cancer. Animal studies suggest that O_3 may damage the ciliate cells of the epithelium with changes in the air–blood barrier permeability, causing an inflammatory response.[86]

The complex mixtures of pollutants may affect the human circulatory system, with changes in the levels of fibrinogen, increases erythrocyte count, and plasma viscosity.[87–89] Some studies evidenced deleterious effects on lung defense mechanism, causing inflammatory changes in the airway and distal lung parenchyma.[90–92] Others confirm the carcinogenic effect resulting from air pollution, which acts as a promoter and/or initiator of pulmonary tumor in mice.[93,94]

Over the past few years, some evidence focused particularly on male fertility and pregnancy, showing the negative effects of urban air pollutants on reproductive health in humans.[95–97] Moreover, few studies have been able to demonstrate an association between air pollution and changes in fertility in women, probably due to multiple factors involved in female reproductive function.[98] Male and female mice exposed to urban pollution in São Paulo, Brazil, show changes in the genus distribution in their offspring, suggesting that air pollution can change the proportion of XY sperm in exposed animals.[99]

Previous studies showed that air pollution has a significant impact on female reproductive function in mice. Exposure to fine particulate matter has been implicated in disruption of the pattern of segregation of inner cell mass and trophectoderm cell lineages at the blastocyst stage,[100,101] an important marker of embryo viability and development potential.[102] A retrospective epidemiological study confirmed the increased risk of early pregnancy loss, which was already observed in experimental studies in women exposed to air pollution. In addition, an association between brief exposure to high levels of environmental particles during pre-conception and early pregnancy loss was found, independently from conception method (natural or after in vitro fertilization treatment), and the risk of miscarriage increased 2.6-fold.[100] Furthermore, a positive association between air pollution and intrauterine mortality was found in a study conducted in São Paulo, southeastern Brazil, suggesting that pollution in São Paulo may promote adverse effects on fetuses.[103]

Conclusion and Remarks

Air pollution is a reality in megacities, and it is intensified mainly by a huge number of vehicles circulating. According to the literature, the major air pollutants that human populations are exposed to in urban areas have a number of substances with carcinogenic activity. These substances can cause mutations and trigger neoplasias.

Therefore, the high levels of vehicular emissions require complex demands and strict actions from the government. These solutions can be taken individually with proper maintenance of vehicles; indeed, need the improvement of public transport, which means efficient and rapid tubes, trains, trams, and buses. Alternative measurements such as rodizio have been implemented in large cities such as São Paulo, Mexico City, and Beijing, aiming to reduce the daily and enormous traffic and, consequently, the direct emissions from sources.

Although not emphasized in this entry, industrial air pollution also should not be underestimated. However, the emissions of industries are more easily controlled and must be applied, followed by periodic monitoring, employment of cleaner technologies, and cleaning up of filters on the chimneys, which should be constructed with enough height to guarantee adequate dispersion of pollutants. Furthermore, it is important to choose raw materials that produce waste with lower pollution potential at the end of the production process. It is mandatory to apply more resources to develop new technologies, which can be developed at universities, research institutes, and private companies. The use of cleaner energy sources, such as wind and biogas, is desirable.

Nevertheless, the pollutants produced locally may not stay in the same place and its surrounding area. It can also damage remote areas of the planet due to long-term transportation. Each country has its own formulation of public policies that impose limits on environment pollution levels. Finally, not only signed agreements among governments will bring significant changes to reduce emissions of pollutants. Individual and local actions organized by nongovernmental organizations, schools, and media are equally important to preserve life.

References

1. Fenger, J. Air pollution in the last 50 years—From local to global. Atmos. Environ. **2009**, *43*, 13–22.
2. WHO—World Health Organization Air quality guidelines; Copenhagen, Denmark, 2005.
3. Freedman, B. *Environmental Ecology;* Academic Press: San Diego, 1995.
4. Kley, D.; Kleinmann, M.; Sanderman, H.; Kruppa, S. Photochemical oxidants: State of the science. Environ. Pollut. **1999**, *100*, 19–42.
5. Nohl, H.; Kozlov, A.V.; Gille; L.; Staniek, K. Cell respiration and formation of reactive oxygen species: Facts and artefacts. Biochem. Soc. Trans. **2003**, *31* (6), 1306–1311.
6. Kovacic, P.; Somanathan, R. Biomechanisms of nanoparticles (toxicants, antioxidants and therapeutics): Electron transfer and reactive oxygen species. J. Nanosci. Nanotechnol. **2010**, *10*, 7919–7930.
7. Dockery, D.W. Health effects of particulate air pollution. Ann. Epidemiol. **2009**, *19*(4), 257–263.
8. Seinfeld, J.H.; Pandis, S.N. *Atmospheric Chemistry and Physics—From Air Pollution to Climate Changes*, 1st Ed; John Wiley and Sons, Inc.: New York, 1949.
9. CETESB—Companhia de Tecnologia de Saneamento Ambiental, Relatório de qualidade do ar do Estado de São Paulo de 2008. Série Relatórios, São Paulo, 2009.
10. Zanobetti, A.; Woodhead, M. Air pollution and pneumonia: The "old man" has a new "friend." Am. J.Respir. Crit. Care Med. **2010**, *181*, 5–6.
11. Enander, R.T.; Gagnon. R.N.; Hanumara, R.C.; Park, E.; Armstrong, T.; Gute, D.M. Environmental health practice: Statistically based performance measurement. Am. J. Pub. Health **2007**, *97*, 819–824.

12. Kaplan, C. Indoor air pollution from unprocessed solid fuels in developing countries. Rev. Environ. Health **2010**, *25*, 221–242.
13. Martins, M.C.H.; Fatigati, F.L.; Véspoli, T.C.; Martins, L.C.; Pereira, L.A.A.; Martins, M.A.; Saldiva, P.H.N.; Braga, A.L.F. Influence of socioeconomic conditions on air pollution adverse health effects in elderly people: An analysis of six regions in São Paulo, Brazil. J. Epidemiol. Commun. Health **2004**, *58*, 41–46.
14. Peled, R. Air pollution exposure: Who is at high risk? Atmos. Environ. **2011**, *45* (12), 1781–1785.
15. Weisburger, J.H. Carcinogenicity and mutagenicity testing, then and now. Mutat. Res. **1999**, *437* (2), 105–112.
16. Air Quality Standards, available at http://ec.europa.eu/envi-ronment/air/quality/standards.htm, (accessed May 3, 2011).
17. National Ambient Air Quality Standards, available at http://www.epa.gov/air/criteria.html, (accessed May 3, 2011).
18. Qualidade do Ar, available at http://www.cetesb.sp.gov.br/ar/qualidade-do-ar/31-publicacoes-e-relatorios, (accessed May 3, 2011).
19. Alberts, B.; Alexander, J.; Lewis, L.; Raff, M.; Roberts, K.; Walter, P. *Molecular Biology of the Cell*, Garland Science: New York, 2002.
20. Cohen, S.M.; Arnold, L.L. Chemical carcinogenesis. Toxicol. Sci. **2011**, *120*, 76–92.
21. Evans, M.D.; Dizdaroglu, M.; Cooke, M. S. Oxidative DNA damage and disease: Induction, repair and significance Mutat. Res. **2004**, *567*, 1–61.
22. Oga, S.; Camargo, M.M.A.; Batistuzzo, J.A.O. *Fundamentos da Toxicologia*, 3rd Ed; Atheneu Editora: São Paulo, 2008.
23. Oh, S.M.; Kim, H.R.; Park Y.J.; Lee, S.Y.; Chung, K.H. Organic extracts of urban air pollution particulate matter (PM2.5)-induced genotoxicity and oxidative stress in human lung bronchial epithelial cells (BEAS-2B cells). Mutat. Res. **2011**, *723*, 142–151.
24. Risom, L.; Møller, P.; Loft, S. Oxidative stress-induced DNA damage by particulate air pollution. Mutat. Res. **2005**, *592*, 119–137.
25. Rossner, P.; Topinka, J.; Hovorka, J.; Milcova, A.; Schmuczerova, J.; Krouzek, J.; Sram, R.J. An acellular assay to assess the genotoxicity of complex mixtures of organic pollutants bound on size segregated aerosol. Part II: Oxidative damage to DNA. Toxicol. Lett. **2010**, *198* (3), 312–316.
26. Coronas, M.V.; Pereira, T.S.; Rocha, J.A.V.; Lemos, A.T.; Fachel, J.M.G.; Salvadori, D.M.F.; Vargas, V.M.F. Genetic biomonitoring of an urban population exposed to mutagenic airborne pollutants. Environ. Int. **2009**, *35*, 1023–1029.
27. Roubicek, D.A.; Gutierrez-Castillo, M.E.; Sordo,M.; Cebrian- Garcia, M.E.; Ostrosky-Wegman, P. Micronuclei induced by airborne particulate matter from Mexico City. Mutat. Res., Genet. Toxicol. Environ. Mutagen. **2007**, *631* (1), 9–15.
28. Sevastyanova, O.; Binkova, B.; Topinka, J.; Sram, R.J.; Kalina, I.; Popov, T.; Novakova, Z.; Farmer, P.B. In vitro genotoxicity of PAH mixtures and organic extract from urban air particles—Part II: Human cell lines. Mutat. Res., Fundam. Mol. Mech. Mutagen. **2007**, *620*, 123–134
29. Bonetta, Sa.; Gianotti, V.; Bonetta, Si.; Gosetti, F.; Oddone, M.; Gennaro, M.C.; Carraro, E. DNA damage in A549 cells exposed to different extracts of $PM_{2.5}$ from industrial, urban and highway sites. Chemosphere **2009**, *77*, 1030–1034.
30. Gilli, G.; Traversi, D.; Rovere, R.; Pignata, C.; Schilirò, T. Chemical characteristics and mutagenic activity of PM_{10} in Torino, a Northern Italian City. Sci. Total Environ. **2007**, *385*, 97–107.
31. Ames, B.N.; Yamasaki, E. The detection of chemical mutagens with enteric bacteria. In *Chemical Mutagens: Principles and Methods for their Detection*; Hollander, A., Ed.; Plenum Press: New York, 1971.
32. Mortelmans, K.; Zeiger, E. The Ames *Salmonella/micro-* some mutagenicity assay. Mutat. Res. **2000**, *455*, 29–60.

33. Clive, D.; Johnson, K.O.; Spector, J.F.S.; Batson, A.G.; Brown, M.M.M. Validation and characterization of the L5178Y/TK+/– mouse lymphoma mutagen assay system. Mutat. Res. **1979**, *589*, 61–108.
34. Cole, J.; McGregor, D.B.; Fox, M.; Thacker, J.; Garner, R.C. Gene mutation assays in cultured mammalian cells. In *Mutagenicity Tests*; Kirkland, D.J., Ed.; Ed Basic, Cambridge Press: Cambridge, U.K., 1990.
35. Ribeiro, L.R.; Salvadori, D.M.F.; Marques, E.K. *Mutagênese ambiental*, 1st Ed; Editora ULBRA: Canoas, Brasil, 2003.
36. Matter, B.; Schmid, W. Treminon-induced chromosomal damage in bone narrow cells of six mammalian species evaluated by the micronucleus test. Mutat. Res. **1971**, *12*, 417–425.
37. Heddle, J.A.; Salamone, M.F. The micronucleus assay in vivo. In Proceedings of the international workshop on short-term tests for chemical carcinogens; Stich, H., San, R.H., Eds.; Springer-Verlag: New York, 1981.
38. Mavournin, K.H.; Blakey, D.H.; Cimino, M.C.; Salamone, M.F.; Heddle, J.A. The in vivo micronucleus assay in mammalian bone marrow and peripheral blood. A report of the U.S. Environmental Protection Agency Gene-Tox Program. Mutat. Res. **1990**, *239*, 29–80.
39. Iskander, O. An improved method for the detection of micronuclei in human lymphocytes. Stain Technol. **1989**, *54*, 221–223.
40. Fenech, M.; Morley, A.A. Measurement of micronuclei in lymphocytes. Mutat. Res. **1985**, *147*, 29–36.
41. Saldiva, P.H.N.; Böhm, G.M. Animal indicator of adverse effects associated with air pollution. Ecosyst. Health **1998**, *4*, 230–235.
42. EPA. Current status of bioassays in genetic genotoxicology (Gene-Tox), 1980.
43. Sparrow, A.H.; Underbrink, A.G.; Rossi, H.H. Mutation induced in *Tradescantia* by small doses of x-rays and neutrons: Analysis of dose–response curves. Science **1972**, *176*, 916–918.
44. Ma, T.H. *Tradescantia* micronuclei (Trad-MCN) test for environmental clastogens. In *In Vitro Toxicity Testing of Environmental Agent. Current and Future Possibilities. Part A: Survey of Test Systems;* Kolber, A.R., Wong, T.K., Grant, L.D., DeWoskin, R.S., Hughes, T.J., Eds.; Plenum Press: New York, 1983; 191–214.
45. Rodrigues, G.S.; Ma, T.H.; Pimentel, D.; Weinstein, L.H. *Tradescantia* bioassays as monitoring systems for environmental mutagenesis—Review. Crit. Rev. Plant Sci. **1997**, *16* (4), 325–359.
46. Arutyunyan, R.M.; Pogosyan, V.S.; Simonyan, E.H.; Atoyants, A.L.; Djigardjian, E.M. *In situ* monitoring of the ambient air around the chloroprene rubber industrial plant using the *Tradescantia-stamen-hair* mutation assay. Mutat. Res. **1999**, *426* (2), 117–120.
47. Batalha, J.R.F.; Guimarães, E.T.; Lobo, D.J.A.; Lichtenfels, A.J.F.C.; Deur, T.; Carvalho, H.A.; Alves, E.S.; Domingos, M.; Rodrigues, G.S.; Saldiva, P.H.N. Exploring the clastogenic effects of air pollutants in São Paulo (Brazil) using the *Tradescantia* micronuclei assay. Mutat. Res. **1999**, *426*, 229–232.
48. Monarca, S.; Feretti, D.; Zanardini, A.; Falistocco, E.; Nardi, G. Monitoring of mutagens in urban air samples. Mutat. Res. **1999**, *426*, 189–192.
49. Guimarães, E.T.; Domingos, M.; Alves, E.S.; Caldini Jr., N.; Lobo, D.J.A.; Lichtenfels, A.J.F.C.; Saldiva, P.H.N. Detection of the genotoxicity of air pollutants in and around the city of São Paulo (Brazil) with the *Tradescantia*-micronucleus (Trad-MCN) assay. Environ. Exp. Bot. **2000**, *44*, 1–8.
50. Misík, M.; Ma, T.H.; Nersesyan, A.; Monarca, S.; Kim, J.K..; Knasmueller, S. Micronucleus assays with *Tradescantia* pollen tetrads: an update. Mutagenesis **2011**, *26* (1), 215–221.
51. Prajapati, S.K.; Tripathi, B.D. Assessing the genotoxicity of urban air pollutants in Varanasi City using *Tradescantia* micronucleus (Trad-MCN) bioassay. Environ. Int. **2008**, *34* (8), 1092–1096.

52. Villarini, M.; Fatigoni, C.; Dominici, L.; Maestri, S.; Ederli, L.; Pasqualini, S.; Monarca, S.; Moretti, M. Assessing the genotoxicity of urban air pollutants using two in situ plant bioassays. Environ. Pollut. **2009,** *157* (12), 3354–3356.
53. Lima, E.D.; de Souza, S.R.; Domingos, M. Sensitivity of *Tradescantia pallida* (Rose) Hunt. 'Purpurea' Boom to genotoxicity induced by ozone. Mutat. Res. **2009,** *675* (1–2), 41–45.
54. Ichikawa, S. *Tradescantia* stamen-hair system as an excellent botanical tester of mutagenicity: Its responses to ionizing radiation and chemical mutagens, and some synergistic effects found. Mutat. Res. **1992,** *270* (1), 3–22.
55. Ma, T.H.; Cabrera, G.L.; Cebulska-Wasilewska, A.; Chen, R.; Loarca, F.; Vandenberg, A.L.; Salamone, M.F. *Tradescantia* stamen hair mutation bioassay. Mutat. Res. **1994,** *310* (2), 211–220.
56. Takahashi, C.S. Teste dos pêlos estaminais de *Tradescantia.* In *Mutagênese, Carcinogênese e Teratogênese: Métodos e critérios de avaliação;* Rabelo-Gay, M.N., Rodrigues, M.A.L.R., Monteleone-Neto, R., Eds.; Rev. Brasileira de Genética—Ribeirão Preto, Sociedade Brasileira de Genética, 1991; 61–66.
57. Malallah, G.; Afzal, M.; Murin, G.; Murin, A.; Abraham, D. Genotoxicity of oil pollution on some species of Kuwaiti flora. Biologia **1997,** *52* (1), 61–70.
58. Micieta, K.; Kunová, K. Phytoindication of genotoxic deterioration of polluted environment. Biológia **2000,** *55,* 75–79.
59. (56) Micieta, K.; Murín, G. Microspore analysis for genotoxicity of a polluted environment. Environ. Exp. Bot. **1996,** *36,* 21–27.
60. Pacini, S.; Giovanelli, L.; Gulisano, M.; Peruzzi, B.; Polli, G.; Boddi, V.; Ruggiero, M.; Bozzo, C.; Stomeo, F.; Fenu, G.; Pezzatini, S.; Pitozzi, V.; Dolara, P. Association between atmospheric ozone levels and damage to human nasal mucosa in Florence, Italy. Environ. Mol. Mutagen. **2003,** *42* (3), 127–135.
61. Oh, S.M.; Kim, H.R.; Park, Y.P.; Lee, S.Y.; Chung, K.H. Organic extracts of urban air pollution particulate matter ($PM_{2.5}$)-induced genotoxicity and oxidative stress in human lung bronchial epithelial cells (BEAS-2B cells). Mutat. Res., Genet. Toxicol. Environ. Mutagen. **2011,** doi:10.106/j.mrgentox.2011.04.003.
62. Koehler, C.; Ginzkey, C.; Friehs, G.; Hackenberg, S.; Froelich, K.; Scherzed, A.; Burghartz, M. Aspects of nitrogen dioxide toxicity in environmental urban concentrations in human nasal epithelium. Toxicol. Appl. Pharmacol. **2010,** *245* (2), 219–225.
63. Tovalin, H.; Valverde, M.; Morandi, M.T.; Blanco, S.; Whitehead, L.; Rojas, E. DNA damage in outdoor workers occupationally exposed to environmental air pollutants. Oc-cup. Environ. Med. **2006,** *63,* 230–236.
64. Claxton, L.D.; Woodall, G.M.. A review of the rodent carcinogenicity and mutagenicity of ambient air. Mutat. Res. **2007,** *636* (1–3), 36–94.
65. Logan, W.P.D. Mortality in the London fog incident, 1952. Lancet **1953,** *1,* 336–338.
66. Firket, J. Fog along the Meuse Valley. Trans. Faraday Soc. **1936,** *32,* 1192–1197.
67. Helfand, W.H.; Lazarus, J.; Theerman, P. Dondora, Pennsylvania: An environmental disaster of the 20th century. Am. J. Pub. Health **2001,** *91* (4), 553.
68. Dockery, D.W.; Schwartz, J.; Spengler, J.D. Air pollution and daily mortality: Associations with particulates and acid aerosol. Environ. Res. **1992,** *59* (2), 362–373.
69. Dockery, D.W. Epidemiologic evidence of cardiovascular effects of particulate air pollution. Environ. Health Perspect. **2001,** *109* (4), 483–486.
70. Lin, C.A.; Pereira, L.A.A.; Conceição, G.M.S.; Kishi, H.S.; Milani, R., Jr.; Braga, A.L.F.; Saldiva, P.H.N. Association between air pollution and ischemic cardiovascular emergency room visits. Environ. Res. **2003,** *92* (1), 57–63.
71. Avogbe, P.H.; Ayi-Fanou, L.; Autrup, H.; Loft, S.; Fayomi, B; Sanni, A.; Vinzents, P.; Møller, P. Ultrafine particulate matter and high-level of benzene urban air pollution in relation to oxidative DNA damage. Carcinogenesis **2005,** *26* (3), 613–620.

72. Ostro, B.D.; Rothschild, S. Air pollution and acute respiratory morbidity: An observational study of multiple pollutants. Environ. Res. **1989**, *50* (2), 238–247.
73. Ostro, B. The association of air pollution and mortality: Examining the case for inference. Arch. Environ. Health **1993**, *48(5)*, 336–342.
74. Batalha, J.R.F.; Saldiva, P.H.N.; Clarke, R.W.; Coull, B.A.; Stearns, R.C.; Lawrence, J.; Krishna Murthy, G.G.; Koutrakis, P.; Godleski, J.J. Concentrated ambient air particles induce vasoconstriction of small pulmonary arteries in rats. Environ. Health Perspect. **2002**, *110*, 1191–1197.
75. Rivero, D.H.R.F.; Soares, S.R.C.; Lorenzi-Filho, G.; Saiki, M.; Godleski J.J.; Antonangelo, L.; Dolhnikoff; M.; Saldiva, P.H.N. Acute cardiopulmonary alterations induced by fine particulate matter of São Paulo, Brazil. Toxicol. Sci. **2005**, *85*, 898–905.
76. Mauad, T.; Rivero, D.H.R.F.; Oliveira, R.C.; Lichtenfels, A.J.F.C.; Guimarães, E.T.; Andre, P.A.; Kasahara, D.I.; Bueno, H.M.S.; Saldiva, P.H.N. Chronic exposure to ambient levels of urban particles affects mouse lung development. Am. J.Respir. Crit. Care Med. **2008**, *178*, 721–728.
77. Pope, C.A. III; Burnett, R.T.; Thun, M.J.; Calle, E.E.; Krewski, D.; Ito, K.; Thurston, G.D. Lung cancer, cardiopulmonary mortality, and long-term exposure to fine particulate air pollution. J. Am. Med. Assoc. **2002**, *287*, 1132–1141.
78. Schwartz, J. Air pollution and blood markers of cardiovascular risk. Environ. Health Perspect. **2001**, *109* (3), 405–409.
79. Lucking, A.J.; Lundback, M.; Mills, N.L.; Faratian, D.; Barath, S.L.; Pourazar, J.; Cassee, F.R.; Donaldson, K.; Boon, N.A.; Badimon, J.J.; Sandstrom, T.; Blomberg, A.; Newby, D.E. Diesel exhaust inhalation increases thrombus formation in man. Eur. Heart J. **2008**, *29* (24), 3043–3051.
80. Törnqvist, H.; Mills, N.L.; Gonzalez, M.; Miller, M.R.; Robinson, S.D.; Megson, I.L.; Macnee, W.; Donaldson, K.; Söderberg, S.; Newby, D.E.; Sandström, T.; Blomberg, A. Persistent endothelial dysfunction in humans after diesel exhaust inhalation. Am. J. Respir. Crit. Care Med. **2007**, *176* (4), 395–400.
81. Stone, P.H.; Godleski, J.J. First steps toward understanding the pathophysiologic link between air pollution and cardiac mortality. Am. Heart J. **1999**, *138*, 804–807.
82. Roughton, F.J.W.; Darling, R.C. The effect of carbon monoxide on the oxyhemoglobin dissociation curve. Am. J. Physiol. **1944**, *141*, 17–31.
83. Schmitzberger, R.; Rhomberg, K.; Buchele, H.; Puchegger, R.; Schmitzberger-Natzmer, D.; Kemmler, G.; Panosch, B. Effects of air pollution on the respiratory tract of children. Pediatr. Pulmonol. **1993**, *15* (2), 68–74.
84. Sousa, S.I.V.; Ferraz, C.; Alvim-Ferraz, M.C.M.; Martins, F.G.; Vaz, L.G.; Pereira, M.C. Spirometric tests to assess the prevalence of childhood asthma at Portuguese rural areas: Influence of exposure to high ozone levels. Environ. Int. **2011**, *37* (2), 474–478.
85. Pereira, F.A.C.; Assunção, J.V.; Saldiva, P.H.N.; Pereira. L.A.A.; Mirra, A.P.; Braga, A.L.F. Influence of air pollution on the incidence of respiratory tract neoplasm. J. Air Waste Manage. Assoc. **2005**, *55* (1), 83–87.
86. Yang, W.; Omaye, S.T. Air pollutants, oxidative stress and human health. Mutat. Res. **2011**, *674*, 45–54.
87. Baskurt, O.K.; Levi, E.; Caglayan, S.; Dikmenoglu, N.; Kutman, M.N. Hematological and hemorheological effects of air pollution. Arch. Environ. Health **1990**, *45* (4), 224–228.
88. Seaton, A.; Soutar, A.; Crawford, V.; Elton, R.; McNerlan, S.; Cherrie, J.; Watt, M.; Agius, R.; Stout, R. Particulate air pollution and the blood. Thorax **1999**, *54*, 1027–1032.
89. Gardner, S.Y.; Lehmann, J.R.; Costa, D.L. Oil fly ash-induced elevation of plasma fibrinogen levels in rats. Toxicol. Sci. **2000**, *56*, 175–180.
90. Böhm, G.M.; Saldiva, P.H.N.; Pasqualucci, C.A.G.; Massad, E.; Martin, M.A.; Zin, W.A.; Cardoso, W.V.; Criado, P.M.P.; Komatsuzaki, M.; Sakae, R.S.; Negri, E.M.; Lemos, M.; Capelozzi, V.D.; Crestana, C.; Silva, R. Biological effects of air pollution in São Paulo and Cubatao. Environ. Res. **1989**, *49*, 208–216.

91. Saldiva, P.H.N.; King, M.; Delmonte, V.L.C.; Macchione, M.; Parada, M.A.C.; Daliberto, M.L.; Sakae, R.S.; Criado, P.M.P.; Silveira, P.L.P.; Zin, W.A.; Böhm, G.M. Respiratory alterations due to urban air pollution: An experimental study in rats. Environ. Res. **1992,** *57* (1), 19–33.
92. Lemos, M.; Lichtenfels, A.J.F.C.; Amaro Jr, E.; Macchione, M.; Martins, M.A.; King, M.; Böhm, G.M.; Saldiva, P.H.N. Quantitative pathology of nasal passages in rats exposed to urban levels of air pollution. Environ. Res. **1994,** *66* (1), 87–95.
93. Reymão, M.S.F.; Cury, P.M.; Lichtenfels, A.J.F.C.; Lemos, M.; Battlehner, C.N.; Conceição, G.M.S.; Capelozzi, V.L.; Montes, G.S.; Júnior, M.F.; Martins, M.A.; Böhm, G.M.; Saldiva, P.H.N. Urban air pollution enhances the formation of urethane-induced lung tumors in mice. Environ. Res. **1997,** *74* (2), 150–158.
94. Cury, P.M.; Lichtenfels, A.J.F.C.; Reymão, M.S.F.; Con- ceição, G.M.S.; Capelozzi, V.L.; Saldiva, P.H.N. Urban levels of air pollution modifies the progression of urethane- induced lung tumours in mice. Pathol. Res. Pract. **2000,** *196,* 627–633.
95. Lee, B.E.; Ha, E.H.; Park, H.S.; Kim, Y.J.; Hong, Y.C.; Kim, H.; Lee, J.T. Exposure to air pollution during different gestational phases contributes to risks of low birth weight. Hum. Reprod. **2003,** *18* (3), 638–643.
96. Rubes, J.; Selevan, S.G.; Evenson, D.P.; Zudova, D.; Vozdova, M.; Zudova, Z.; Robbins, W.A.; Perreaut, S.D. Episodic air pollution is associated with increased DNA fragmentation in human sperm without other changes in semen quality. Hum. Reprod. **2005,** *20* (10), 2776–2783.
97. Silva, I.R.R.; Lichtenfels, A.J.F.C.; Pereira, L.A.A.; Saldiva, P.H.N. Effects of ambient levels of air pollution generated by traffic on birth and placental weights in mice. Fertil. Steril. **2008,** *90* (5), 1921–1924.
98. Tomei, G.; Ciarrocca, M.; Fortunato, B.R.; Capozzella, A.; Rosati, M.V.; Cerratti, D.; Tomao, E.; Anzelmo, V.; Monti, C.; Tomei, F. Exposure to traffic pollutants and effects on 17-beta-estradiol (E2) in female workers. Int. Arch. Occup. Environ. Health **2006,** *80* (1), 70–77.
99. Lichtenfels, A.J.F.C.; Gomes. J.B.; Pieri, P.C.; Miraglia, S.G.E.K.; Hallak, J.; Saldiva, P.H.N. Increased levels of air pollution and a decrease in the human and mouse male-to-female ratio in São Paulo, Brazil. Fertil. Steril. **2007,** *87,* 230–232.
100. Perin, P.M.; Maluf, M.; Januário, D.N.; Saldiva, P.H. Effects of short-term exposure of female mice to diesel exhaust particles on in vitro fertilization and embryo development. Fertil. Steril. **2008,** *90* (1), S206.
101. Maluf, M.; Perin, P.M.; Januário, D.A.N.F.; Saldiva, P.H.N. In vitro fertilization, embryo development, and cell lineage segregation after pre- and/or postnatal exposure of female mice to ambient fine particulate matter. Fertil. Steril. **2009,** *92* (5), 1725–1735.
102. Kovacic, B.; Vlaisavljevic, V.; Reljic, M.; Cizek-Sajko, M. Developmental capacity of different morphological types of day 5 human morulae and blastocysts. Reprod. BioMed. **2004,** *8* (6), 687–694.
103. Pereira, L.A.A.; Loomis, D.; Conceição, G.M.S.; Braga, A.L.F.; Arcas, R.M.; Kishi, H.S.; Singer, J.M.; Böhm, G.M.; Saldiva, P.H.N. Association between air pollution and intrauterine mortality in São Paulo, Brazil. Environ. Health Perspect. **1998,** *106* (6), 325–329.

2

Methane Emissions: Rice

Kazuyuki Yagi

Introduction

The atmospheric concentration of methane (CH_4) has increased rapidly in recent years. Because it is a radiative trace gas and takes part in atmospheric chemistry, the rapid increase could be of significant environmental consequence. Of the wide variety of sources, rice fields are considered an important source of atmospheric CH_4, because the harvest area of rice has increased by about 70% during last 50 years and it is likely that CH_4 emission has increased proportionally. Recent estimates suggest that global emission rates of CH_4 from rice fields account for about 4%–19% of the emission from all sources.[1] Due to the large amount of the global emission from rice cultivation, reduction of CH_4 emission from this source is very important in order to stabilize atmospheric concentration. In addition, because of the possibility of controlling the emission by agronomic practices, rice cultivation must be one of the most hopeful sources for mitigating CH_4 emission.

Processes Controlling CH_4 Emissions from Rice Fields

Table 1 provides a summary of measured methane emissions at a number of specific research sites around the world.[2] It should be noted that methane fluxes from rice fields show pronounced diel and seasonal variations and vary substantially with different climate, soil properties, agronomic practices, and rice cultivars.

Processes involved in CH_4 emission from rice fields are illustrated in Figure. 1. Like other biogenic sources, CH_4 is produced by the activity of CH_4 producing bacteria, or methanogens, as one of the terminal products in the anaerobic food web in paddy soils. Methanogens are known as strict anaerobes that require highly reducing conditions. After soil is flooded, the redox potential of soil decreases rapidly by sequential biochemical reactions.

Flooded paddy soils have a high potential to produce CH_4, but part of CH_4 produced is consumed by CH_4 oxidizing bacteria, or methanotrophs. In rice fields, it is possible that a proportion of CH_4 produced in the anaerobic soil layer is oxidized in the aerobic layers, such as the surface soil–water interface and the rhizosphere of rice plants.

The emission pathways of CH_4 that is accumulated in flooded paddy soils is: diffusion into the flood water, loss through ebullition, and transport through the aerenchyma system of rice plants. In the

TABLE 1 Methane Emission from Rice Fields in Various World Locations[a]

Country	Daily Average (g/m^2 day)	Flooding Period (days)	Season Total Average (g/m^2)	Season Total Range (g/m^2)
China	0.19–1.39	75–150	13	10–22
India	0.04–0.46	60	10	5–15
Italy	0.10–0.68	130	36	17–54
Japan	0.01–0.39	110–130	11	3–19
Spain	0.10	120	12	
Thailand	0.04–0.77	80–110	16	4–40
U.S.A.	0.05–0.48	80–100	25	15–35

[a] The data are for the fields without organic fertilizer.

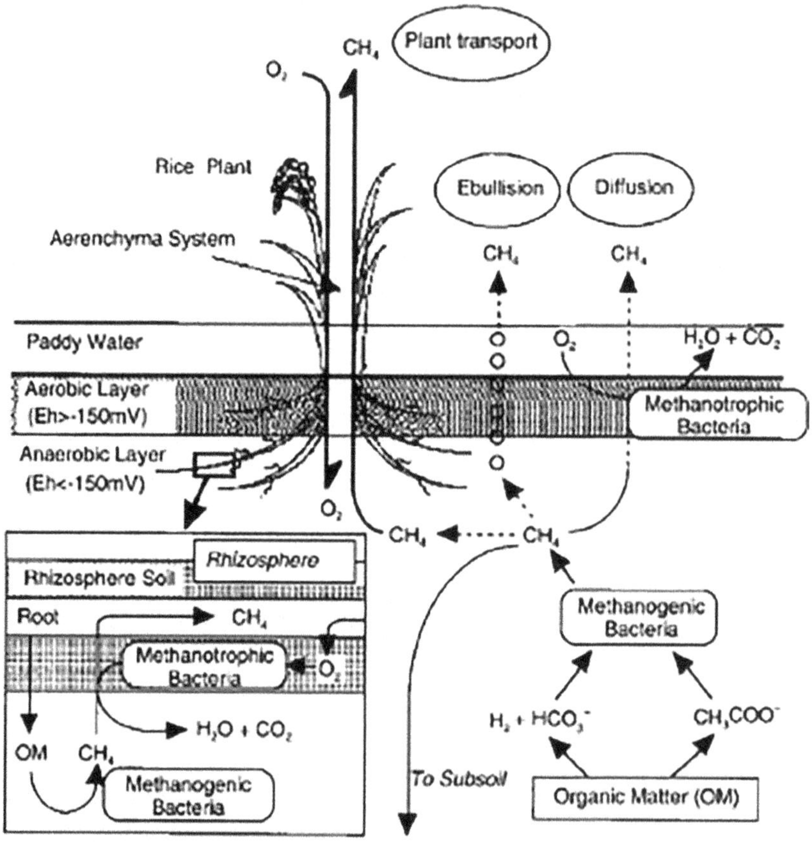

FIGURE 1 Production, oxidation and emission of CH4 in rice paddy fields.
Source: Conrad[3] and Knowles.[4]

temperate rice fields, more than 90% of CH_4 is emitted through plants,[5] while significant amounts of CH_4 may evolve by ebullition, in particular during the early part of the season in the tropical rice fields.[6] Therefore, it is concluded that possible strategies for mitigating CH_4 emission from rice cultivation can be made by controlling either production, oxidation, or transport processes.

Options for Mitigating CH_4 Emission

Water Management

Mid-season drainage (aeration)in flooded rice fields supplies oxygen into soil, resulting in a reduction of CH_4 production and a possible enhancement of CH_4 oxidation in soil.[7,8] A study using an automated sampling and analyzing system clearly showed that short-term drainage had a strong effect on CH_4 emission, as shown in Figure. 2. Total emission rates of CH_4 during the cultivation period were reduced by 42%–45% by short-term drainage practices compared with continuously flooded treatment.[9] These results indicate that improvement in water management can be one of the most promising mitigation strategies for CH_4 emission from rice fields. Increasing the rate of water percolation in rice fields by installing underground pipe drainage may also have an influence on CH_4 production and emission.

Soil Amendments and Mineral Fertilizers

The progress of soil reduction can be retarded by adding one of several electron acceptors in the sequential soil redox reactions. Sulfate is one of the most promising candidates for this strategy because it is commonly used as a component of mineral fertilizer and soil amendment. Field measurements have shown that CH_4 emission rate decreased by at most 55%–70% by application of ammonium sulfate or gypsum.[10,11]

Additions of other oxidants, such as nitrate and iron-containing materials, may influence CH_4 emission from rice fields. As well as adding oxidants, dressing paddy fields with other soils that contain a large amount of free iron and manganese may decrease CH_4 emission. Other chemical candidates are nitrification inhibitors and acetylene releasing materials.

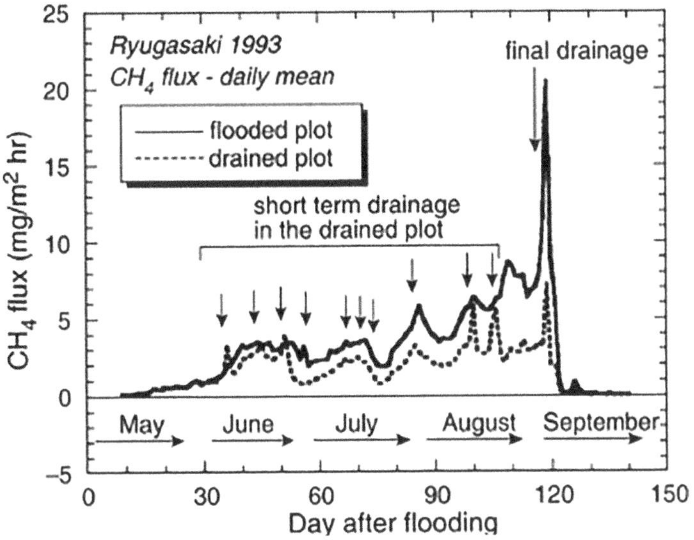

FIGURE 2 Effect of water management on CH_4 emission from a rice paddy field. The arrows indicate period of midseason drainage in the intermittent irrigation plot and the timing of final drainage in both of the plots.

Organic Matter Management

In rice cultivation, fresh organic matter and animal wastes are often applied as fertilizers. In the fields, a proportion of the biomass of previous crops and weeds remains in soils at the start of rice cultivation. Such organic matter is decomposed in soils and acts as a substrate for fermentation reactions. Many researchers have demonstrated that incorporation of rice straw and green manure into rice paddy soils dramatically increases CH_4 emission.[7,10,12] The impact of organic amendments on CH_4 emissions can be described by a dose–response curve which adopts correction factors for composted and fermented organic matter.[6] Mitigation of CH_4 emission requires that the quantities of organic amendments be minimized.

Field experiments also indicated that composted or fermented organic matter increased CH_4 emission much less than fresh organic matter, due to a lower content of easily decomposable carbon.[6,7] Therefore, stimulation of composting organic amendments appears to be a promising mitigation option. Plowing the fields during the fallow period and promoting aerobic degradation of organic matter is also likely to reduce CH_4 emission.

Others

Different tillage and cropping practices change the physical, chemical, and microbiological properties of the plow layer soil and may reduce CH_4 emission. These include deep tillage, no tillage, and flooded rice-upland crop rotation.

Selecting and breeding rice cultivars that emit lower CH_4 is a desirable approach because it is easy to adopt. There are four points to consider for selecting cultivars: 1) they should exude low levels carbon from their roots; 2) they should have a low level of CH_4 transport and a high level of CH_4 oxidation in the rhizosphere; 3) they should have a higher harvest index, in order to reduce organic matter input into soil after harvest; and 4) they should be suitable and have a high productivity when other mitigation options are performed.

Problems and Feasibility of the Options

If the above mitigation options could be applied to world's rice cultivation, global CH_4 emission from rice fields could decrease significantly. However, there are several formidable obstacles to adopting the mitigation options into local rice farming. Table 2 summarizes the problems and feasibility of the individual mitigation options along with the efficiency of the options.

Application of some options is limited to specific types of rice fields. In particular, altering water management practices may be limited to rice paddy fields where the irrigation system is well equipped. Long midseason drainage and short flooding may cause possible negative effects on grain yield and soil fertility. Improving percolation by underground pipe drainage requires laborious engineering work. The increased water requirement is another problem in the water management options because water is a scarce commodity in many regions.

Cost and labor are serious obstacles for applying each option to local farmers. Most of the mitigation options will decrease profitability and the farmer net returns in the short run. To overcome these obstacles, an effort to maximize net returns by joining CH_4 mitigation and increased rice production will be needed, as well as political support.

It is recognized that the mitigation options should not have any significant trade-off effects, such as decreased rice yield, a decline in soil fertility, or increased environmental impact by nitrogen compounds. The development of anaerobic conditions in soil by flooding decreases decomposition rates of soil organic matter compared with aerobic soils, resulting in soil fertility being sustained for a long time. Flooded rice cultivation shows very little growth retardation by continuous cropping. Some mitigation options may reduce these advantages of rice fields. Application of sulfate-containing fertilizer may cause a reduction in rice yield due to the toxicity of hydrogen sulfide. Mid-season aeration and soil amendments may induce nitrogen transformation resulting in enhanced N_2O emissions.[16,17]

TABLE 2 Evaluation of the Mitigation Options for Methane Emission from Fields

		Problem for Application							
		Applicability		Economy		Effects On			
	CH_4 Mitigation Efficiency	Irrigated	Rain-fed	Cost	Labor	Yield	Fertility	Time Span	Other Trade-off Effects
Water management									
Midseason drainage	□	o	•	~	↑	+	~	o	May promote N_2O emission
Short flooding	□	o	•	~	~	–	–	o	May promote N_2O emission
High percolation	□	o	•	↑	↑	+	~	o	May promote nitrate leaching
Soil amendments									
Sulfate fertilizer	□	o	o	↑	~	Δ	–	o	May cause H_2S injury
Oxidants	□	o	o	↑	↑	Δ	–	o	
Soil dressing	o	o	o	↑	↑	–	–	o	
Organic matter									
Composting	□	o	o	↑	↑	+	+	o	
Aerobic	□	o	o	~	↑	~	~	o	
decomposition	□								
Burning	o	o	o	~	↑	~	~	o	Causing atmospheric pollution
Others									
Deep tillage	o	o	o	↑	↑	–	–	o	
No tillage	?	o	o	~	↓	–	~	o	
Rotation	o	o	Δ	~	↑	–	–	o	
Cultivar	o	o	o	~	~	~	~	•	

Source: Ranganathan et al.,[13] Neue et al.,[14] and Yagi et al.[15]

Key:
□ Very effective
o Effective/applicable Δ Case by case
• Not applicable/require long time
? No information
↑ Increase
↓ Decrease
~ About equal to previous situation
+ Positive
– Negative

References

1. Prather, M.; Derwent, R.; Ehhalt, D.; Fraser, P.; Sanhueza, E.; Zhou, X. Other trace gases and atmospheric chemistry. In *Climate Change 1994, Radiative Forcing of Climate Change and an Evaluation of the IPCCIS92 Emission Scenarios*; Houghton, J.T., Meira Filho, L.G., Bruce, J., Lee, H., Callander, B.A., Haites, E., Harris, N., Maskell, K., Eds.; Cambridge University Press: Cambridge, England, 1995; 73–126.
2. International panel on climate change. *Greenhouse Gas Inventory Reference Manual*; IPCC Guidelines for National Greenhouse Gas Inventories, OECD: Paris, France, 1997; Vol. 3, 46–60.
3. Conrad, R. Control of methane production in terrestrial ecosystems. In *Exchange of Trace Gases Between Terrestrial Ecosystems and the Atmosphere*; Andreae, M.O., Schimel, D.S., Eds.; John Wiley and Sons Ltd.: New York, 1989; 39–58.

4. Knowles, R. Processes of production and consumption. In *Agricultural Ecosystem Effects on Trace Gases and Global Climate Change*; Harper, L.A., Mosier, A.R., Duxbury, J.M., Rolston, D.E., Eds.; American Society of Agronomy: Madison, WI, 1993; 145–156.
5. Cicerone, R.J.; Shetter, J.D. Sources of atmospheric methane: measurements in rice paddies and a discussion. J. Geo-phys. Res. **1981**, *86*, 7203–7209.
6. Denier van der Gon, H.A.C.; Neue, H.-U. Influence of organic matter incorporation on the methane emission from a wetland rice field. Global Biogeochem. Cycles **1995**, *9*, 11–22.
7. Yagi, K.; Minami, K. Effect of organic matter application on methane emission from some Japanese paddy fields. Soil Sci. Plant Nutr. **1990**, *36*, 599–610.
8. Sass, R.L.; Fisher, F.M.; Wang, Y.B.; Turner, F.T.; Jund, M.F. Methane emission from rice fields: the effect of flood-water management. Global Biogeochem. Cycles **1992**, *6*, 249–262.
9. Yagi, K.; Tsuruta, H.; Kanda, K.; Minami, K. Effect of water management on methane emission from a Japanese rice paddy field: automated methane monitoring. Global Biogeochem. Cycles **1996**, *10*, 255–267.
10. Schütz, H.; Holzapfel-Pschorn, A.; Conrad, R.; Rennenberg, H.; Seiler, W. A 3years continuous record on the influence of daytime, season, and fertilizer treatment on methane emission rates from an Italian rice paddy. J. Geophys. Res. **1989**, *94*, 16405–16416.
11. Denier van der Gon, H.A.C.; Neue, H.-U. Impact of gypsum application on methane emission from a Wetland rice field. Global Biogeochem. Cycles **1994**, *8*, 127–134.
12. Sass, R.L.; Fisher, F.M.; Harcombe, P.A.; Turner, F.T. Mitigation of methane emission from rice fields: possible adverse effects of incorporated rice straw. Global Biogeochem. Cycles **1991**, *5*, 275–287.
13. Ranganathan, R.; Neue, H.-U.; Pingali, P.L. Global climate change: role of rice in methane emission and prospects for mitigation. In *Climate Change and Rice;* Peng, S., Ingram, K.T., Neue, H.-U., Ziska, L.H., Eds.; Springer-Verlag: Berlin, Germany, 1995; 122–135.
14. Neue, H.-U.; Wassmann, R.; Lantin, R.S. Mitigation options for methane emissions from rice fields. In *Climate Change and Rice;* Peng, S., Ingram, K.T., Neue, H.-U., Ziska, L.H., Eds.; Springer-Verlag: Berlin, Germany, 1995; 137–144.
15. Yagi, K.; Tsuruta, H.; Minami, K. Possible options for mitigating methane emission from rice cultivation. Nutr. Cycling Agro-Ecosys. **1997**, *49*, 213–220.
16. Cai, Z.; Xing, G.; Yan, X.; Xu, H.; Tsuruta, H.; Yagi, K.; Minami, K. Methane and nitrous oxide emissions from rice paddy fields as affected by nitrogen fertilizers and water management. Plant Soil **1997**, *196*, 7–14.
17. Bronson, K.F.; Neue, H.-U.; Singh, U.; Abao, E.B., Jr. Automated chamber measurements of methane and nitrous oxide flux in a flooded rice soil: I. Residue, nitrogen, and water management. Soil Sci. Soc. Am. J. **1997**, *61*, 981–987.

3

Petroleum: Hydrocarbon Contamination

Svetlana Drozdova and Erwin Rosenberg

Introduction

Historically, environmental analyses focused on monitoring compounds that pose a threat to humans and their environment. Petroleum hydrocarbon compounds are among them. Contamination of water, soil, and sediment samples by petroleum hydrocarbons is a common and severe environmental problem caused by improper handling, storage, transport, or use of petrochemical products or raw materials.

Petroleum products are the major source of energy for industry and daily life. Leaks and accidental spills occur regularly during exploration, production, refining, transport, and storage. In addition, natural processes can result in seepage of crude oil from geologic formations below the seafloor. The total input of crude oil and petroleum into the environment is estimated to be 1.3 million tons per year. To understand the potential effect of petroleum contaminations on the environment, it is important to understand the nature and distribution of sources and their inputs. Petroleum poses a range of environmental risks when released into the environment. Catastrophic and large-scale spills have a very severe physical impact in addition to the chemical pollution that they cause; chronic discharges and small releases can damage and eventually kill the exposed flora and fauna due to toxicity of many of the individual compounds contained in petroleum. Oil contamination in the environment is primarily assessed by measuring the chemical concentrations of petroleum products in the affected environmental compartment (e.g., sediment, biota, water).

This entry provides a discussion of the environmental relevance of petroleum hydrocarbons; the principal sources of petroleum contaminations in the environment; and the nature and composition of crude oil and petroleum products derived from it. The fate of petroleum hydrocarbons in the environment, possible effects from exposure to them, and their toxicity are discussed as well. The entry is concluded by an overview of analytical methods for determination of petroleum hydrocarbon contamination.

Petroleum Hydrocarbons and Their Environmental Relevance

Oil and gas resources are organic compounds, formed by the effects of heat and pressure on sediments trapped beneath the earth's surface over millions of years. The remains of animals and plants that lived millions of years ago in a marine environment were covered by layers of sand and silt over the years. Heat and pressure from these layers helped the remains turn into crude oil or petroleum.[1] The word "petroleum" means "rock oil" (from Greek: petra [rock] + Latin: oleum [oil])[2] or "oil from the earth." While ancient societies made some use of these resources, the modern petroleum age began less than a century and a half ago, when in 1859, Colonel Drake discovered oil in Oil Creek in Titusville, Philadelphia, United States. From that time on, the world's demand for fossil fuel and the production of oil have continuously increased. From the 1980s, in particular, after the second oil crisis of 1979, the petroleum business has developed into a high-technology industry. Advances in technology have greatly improved the ability to find and extract oil and gas and to convert them to efficient fuels and useful consumer products. About 100 countries produce crude oil. Russia, Saudi Arabia, the United States, Iran, and China are the top five producing countries in 2009 (Table 1).[3] In the United States, the oil and gas industry employs 1.4 million people and generates about 4% of U.S. economic activity. It is larger than the domestic automobile industry and larger than education and social services, the computer industry, and the steel industry combined.[4] At a refinery, different fractions of the crude oil are separated into useable petroleum products. Various sources of information provide a good overview of the different processes in petroleum refining.[5–7] Petroleum products are used worldwide for energy production, as fuel for transport, and as a raw material for many chemical processes. The United States is the biggest consumer of oil in the world (Table 1). Although there exist well-developed alternatives to the use of oil (particularly for energy production and transportation), our societies are still strongly dependent on oil, which is an environmental burden, an economic problem, and a political hazard. However, at the current time, the economic situation still favors the use of petroleum and petroleum products for these applications rather than its alternatives, which at the moment are not competitive from an economic point of view.

Petroleum poses a range of environmental risks when it is released into the environment (whether by catastrophic spills or through chronic discharges). In addition to the physical impact of large spills, the toxicity of many of the individual compounds contained in crude oils or petroleum products is significant. Information on how petroleum hydrocarbons enter and diffuse in the environment is abundant.[8,9] The sources of petroleum input to the environment, particularly to the sea, are diverse. They can be categorized effectively into four major groups, namely, natural seeps, petroleum extraction, petroleum transportation, and petroleum consumption.

Natural seeps are frequently encountered phenomena that occur when crude oil seeps from the geologic strata beneath the seafloor to the overlying water column as a natural process.[10] Recognized by geologists for decades as indicating the existence of potentially exploitable reserves of petroleum, these seeps release vast amounts of crude oil annually. Yet these large volumes are released at a rate low enough that the surrounding ecosystem can adapt and even thrive in their presence; which is not true in case of the catastrophic and accidental impact of a tanker or oil well spill. Natural processes are, therefore, responsible for over 45% of the petroleum entering the marine environment worldwide (Table 2).[11]

As result of human activities, about 700,000 tons of petroleum is released annually into the sea worldwide. Processes such as petroleum extraction, transportation, and consumption can cause soil and groundwater contamination in case of equipment failure or operation errors and other reasons. Petroleum extraction can result in release of both crude oil and refined products as a result of human activities associated with efforts to explore and produce petroleum. The nature and size of these releases are highly variable—see Table 3 for the largest oil spills observed until 2010[12]—and can include accidental spills of crude oil from platforms and blowouts such as that of the oil rig Deepwater Horizon in the Gulf of Mexico in April 2010 or slow chronic releases of water produced from oil- or gas- bearing formations during extraction. Under current industry practices, this "produced water" is treated to separate from crude oil and either injected back into the reservoir or discharged overboard. Produced water is

TABLE 1 Annual Production and Consumption of Oil by the Top 10 Industrial Nations and by the Top 10 Countries in the European Union

Oil production by country					Oil consumption by country in the world				
Rank	Country	Amount bbl/day	Date	Percentage %	Rank	Countries	Amount bbl/d	Date	Percentage %
1	Russia	10,120,000	2010	11.9	1	United States	18,690,000	2009	22.6
2	Saudi Arabia	9,764,000	2009	11.5	2	China	8,200,000	2009	9.9
3	United States	9,056,000	2009	10.7	3	Japan	4,363,000	2009	5.3
4	Iran	4,172,000	2009	4.9	4	India	2,980,000	2009	3.6
5	China	3,991,000	2009	4.7	5	Russia	2,740,000	2010	3.3
6	Canada	3,289,000	2009	3.9	6	Brazil	2,460,000	2009	3.0
7	Mexico	3,001,000	2009	3.5	7	Germany	2,437,000	2009	2.9
8	United Arab Emirates	2,798,000	2009	3.3	8	Saudi Arabia	2,430,000	2009	2.9
9	Brazil	2,572,000	2009	3.0	9	Korea, South	2,185,000	2010	2.6
10	Kuwait	2,494,000	2009	2.9	10	Canada	2,151,000	2009	2.6
Total		84,764,555			**Total**		82,769,370		
Oil production by EU member states					Oil consumption by EU member states				
Rank	Countries	Amount bbl/d	Date	Percentage %	Rank	Countries	Amount bbl/d	Date	Percentage %
1	United Kingdom	1,502,000	2009	60.4	1	Germany	2,437,000	2009	16.2
2	Denmark	262,100	2009	10.5	2	France	1,875,000	2009	12.5
3	Germany	156,800	2009	6.3	3	United Kingdom	1,669,000	2009	11.1
4	Italy	146,500	2009	5.9	4	Italy	1,537,000	2009	10.2
5	Romania	117,000	2009	4.7	5	Spain	1,482,000	2009	9.9
6	France	70,820	2009	2.8	6	Hungary	1,373,000	2009	9.1
7	Netherlands	57,190	2009	2.3	7	Netherlands	922,800	2009	6.1
8	Poland	34,140	2009	1.4	8	Belgium	608,200	2009	4.1
9	Spain	27,230	2009	1.1	9	Poland	545,400	2009	3.6
10	Austria	21,880	2009	0.9	10	Greece	414,400	2009	2.8
Total (EU, 27 countries):		2,485,550	2009		**Total (EU, 27 countries):**		15,012,050		
	Norway	2,350,000	2009			Norway	204,100		
	Turkey	52,980	2009			Turkey	579,500		

Source: Adapted from Energy Statistics: Oil-Production (Most Recent) by Country.[3]
bbl, barrel; EU, European Union. 1 bbl ≈ ca. 159 L.

TABLE 2 Petroleum Input to the Sea

	North America		Worldwide	
Source of Input	Tons	%	Tons	%
Natural seeps	160,000	61	600,000	46
Petroleum extraction	3,000	1	38,000	3
Petroleum transportation	9,100	4	150,000	12
Petroleum consumption	84,000	32	480,000	37
Other	3,900	2	32,000	2
Total input:	260,000 tons		1,300,000 tons	

Source: Adapted from *Oil in the Sea III Inputs, Fates, and Effects.*[11]

TABLE 3 Top 10 Oil Spills in the World as of 2010

	Incident	Location	Year	Type of Incident	Magnitude of Oil Spill (gallons)
1	Gulf War	Kuwait	1991	Oil spill due to war action and sabotage of oil drilling stations and pipelines, encompassing also the dumping of the charge of several oil tankers into the Persian Gulf by Iraqi troops during the Gulf War.	520,000,000
2	Deepwater Horizon	Gulf of Mexico	2010	Oil spill as a consequence of a methane blowout (which could not be prevented due to a technical problem) at the oil rig Deepwater Horizon, which caused an explosion and and the subsequent loss of the oil drilling platform. The well continued to leak for over 100 days.	172,000,000
3	Ixtoc I	Mexico	1979	After an unexpected blowout at the offshore oil rig Ixtoc 1 in the Gulf of Mexico, the platform exploded and collapsed. Oil escaped freely from the well for almost 1 year until the well could be capped.	138,000,000
4	Atlantic Empress/ Aegean Captain	Trinidad and Tobago	1979	Collision of two ships, the Aegean Captain and the supertanker Atlantic Empress, during a heavy storm in the Caribbean Sea. The Atlantic Empress exploded, sank, and lost its freight.	90,000,000
5	Fergana Valley/ Mingbulak	Russia	1992	Technical failure of an oil well in the Fergana Valley located between Kyrgyzstan and Uzbekistan from which oil blew out for a period of 8 months.	88,000,000
6	Nowruz Oil Field	Persian Gulf	1983	Collision of an oil tanker with an oil platform at the Nowruz Oil Field during the Iran–Iraq War. After the oil drilling platform collapsed, the wellhead was destroyed and leaked oil into the Persian Gulf for more than 6 months before being capped. A similar event at the same oilfield resulted directly form war action.	80,000,000
7	Castillo de Bellver	South Africa	1983	A fire at the tanker Castillo de Bellver caused the ship to drift and then break into two separate pieces. Relatively little damage was done to the South African coastline since the oil may have sunk into the sea or burned during the fire.	79,000,000
8	The Amoco Cadiz	France	1978	The crude oil carrier Amoco Cadiz ran aground off the French Atlantic coast and finally spilt into halves, whereby it lost its complete freight, which contaminated 200 km of the French coastline.	69,000,000
9	ABT Summer	Angola	1991	Following a fire aboard the oil tanker *ABT Summer*, it sank and all its freight either leaked to the sea or sank to the ground about 900 miles from the coast of Angola.	51,000,000
10	The MT Haven	Genova, Italy	1991	After unloading the oil tanker MT Haven, a fire broke out, followed by explosions after which the ship sank and continued to leak oil for 12 years.	45,000,000

Source: Adapted from Top 10 Worst Oil Spills.[12]

the largest single wastewater stream in oil and gas production. The amount of produced water from a reservoir varies widely and increases over time as the reservoir is depleted. Petroleum transportation can result also in releases of dramatically varying sizes of petroleum products (not just crude oil) from major incidents (mostly from tankers, such as the one in 1979 off the coast of Tobago, when two tankers collided and one of these, the Atlantic Empress, sank, losing all its freight) to relatively small operational releases that occur regularly, such as those from pipelines.

Releases that occur during the consumption of petroleum, whether by individual car and boat owners, non-tank vessels, or runoff from urban or industrial areas, are typically small but frequent and widespread and are responsible for the vast majority (70%) of petroleum introduced to the environment through human activity.

Because crude oil and petroleum products are a complex and highly variable mixture of hundreds to thousands of individual hydrocarbon compounds, characterizing the risks posed by petroleum-contaminated soil and water has proven to be difficult and inexact. It is very important to have an understanding of the toxicology, analytical science, environmental fate and behavior, risk, and technological implications of petroleum hydrocarbons in order to interpret, evaluate the risk of, and make decisions about the hazardous effect to and ensure the appropriate protection of the environment.

General Chemical Composition Features of Crude Oils and Petroleum Products

Crude oil is an extremely complex mixture of several thousands of different compounds; its compositions and physical properties vary widely depending on the source from which the oils are produced, the geologic environment, and location in which they migrated and from which they are extracted. The nature of the refining processes has an effect on crude oil compositions as well. As indicated in Table 4, petroleum and petroleum products contain primarily hydrocarbons, heteroatom compounds, and relatively small concentrations of (organo)metallic constituents.[13,14] The complexity of petroleum and petroleum products increases with carbon number of its constituents, so it is impossible to identify all components. Petroleum and petroleum products are typically characterized in terms of boiling range and approximate carbon number. Raw petroleum is usually dark brown or almost black, although some fields deliver a greenish or sometimes yellow petroleum. Depending upon the oil field and the way the petroleum composition was formed, the crude oil will also differ in viscosity. The composition of crude oil impacts certain physical properties of the oil, and it is these physical properties (e.g., density or viscosity) by which crude oils are generally characterized, classified, and traded. These physical properties can be used to classify crude oils as light, medium, or heavy. The American Petroleum Institute (API) gravity[15] is a measure of the specific gravity of a petroleum liquid compared with water (API = 10).

TABLE 4 Main Constituents of Petroleum Hydrocarbons and Representative Examples

Petroleum Hydrocarbon Compounds						
Aliphatics/Alicyclics				Aromatics		
Saturated hydrocarbons		Unsaturated hydrocarbons				
Alkanes(paraffins)	Cycloalkanes	Alkenes (olefins)	Alkynes (acetylenes)	Benzene and alkylbenzenes (BTEX)	Polynuclear aromatics (PAH)	Heterocyclic compounds
Single carbon bonds, straight and branched structure	Straight and cyclic structure	One or more double carbon bonds, straight, branched, or cyclic	One or more triple carbon bonds, straight, branched, or cyclic	Single aromatic ring or with attached functional group	Two or more aromatic rings fused together, can be with attached functional group	Aromatic ring structures with one or more heteroatom (N, S, O) in the ring
C_nH_{2n+2} 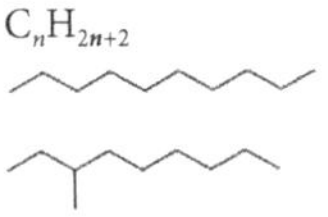	C_nH_{2n} 	C_nH_{2n} 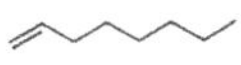	C_nH_{2n-2} 	 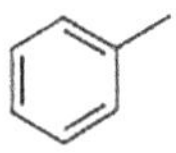	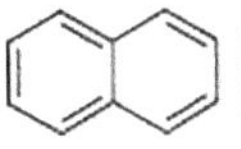	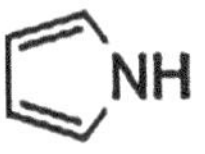
n-Decane 3-Methylnonane	Cyclohexane	1-Octene	1-Hexyne	Benzol Toluene	Naphthalene	Pyrrole

Light oils are defined as having an API < 22.3, heavy oils are those with API > 31.1, and medium oils have an API gravity between 22.3 and 31.1.

Regardless of the complexity, petroleum compounds can be separated into two major categories: hydrocarbons and non-hydrocarbons. Hydrocarbons (compounds composed solely of carbon and hydrogen) comprise the majority of the components in most petroleum products and are the compounds that are primarily (but not always) measured as total petroleum hydrocarbons (TPH).[16] The nonhydrocarbon components are heterocyclic hydrocarbons (compounds containing heteroatoms such as sulfur, nitrogen, or oxygen in addition to carbon and hydrogen). These heterocyclic hydrocarbons are typically present in oils at relatively low concentrations and can be found in most refined motor fuels as they are concentrated in the heavier fractions and residues during refining. Most organic nitrogen hydrocarbons in crude oils are present as alkylated aromatic heterocycles, mostly with a pyrrolic structure. Crude oils also contain small amounts of organometallic compounds (of nickel, vanadium, and other metals up to atomic number 42, with the exception of rubidium and niobium) and inorganic salts. Although, depending on the analytical method, sulfur-, oxygen-, and nitrogen-containing compounds are sometimes included in the value reported as TPH concentration, they do not fall under the definition of petroleum hydrocarbons in the strict sense.[16]

Depending on the structure of petroleum hydrocarbons, the individual compounds are grouped into aliphatic(saturated and unsaturated) hydrocarbons and aromatics. Saturated hydrocarbons are the major class of compounds found in crude oil. The common names of these types of compounds are alkanes and isoalkanes or, as used in petroleum industry, paraffins and isoparaffins, respectively. Unsaturated hydrocarbons have at least one multiple bond (double bond [alkenes] or triple bond [alkynes]), and they are typically not present in crude oil but can be formed during the cracking process. Aromatic hydrocarbons are based on the benzene ring structure and are further categorized depending on the number of rings. Benzene rings are very stable and therefore persistent in the environment, and particularly, the mono- and polycyclic aromatic compounds can have toxic effects on organisms. Aromatic hydrocarbons with one benzene ring and with one or more side chains are alkyl benzenes and include benzene; toluene; ethylbenzene; and o-, p-, and m-xylenes (BTEX). This class of compounds has significant water solubility and is more mobile in the environment. Polycyclic aromatic hydrocarbons (PAHs) are aromatic compounds with two or more fused aromatic rings. Occurrence of PAH compounds in oils is dominated almost completely by the C1- to C4-alkylated homologues of the parent PAH, in particular, for naphthalene, phenanthrene, dibenzothiophene (a sulfur-containing aromatic heterocycle), fluorine, and chrysene. These alkylated PAH homologues form the basis of chemical characterization and identification of oil spills.[17,18] A typical crude oil may contain 0.2% to more than 7% total PAHs. Of the hydrocarbon compounds common in petroleum, PAHs appear to pose the greatest toxicity to the environment.

Different crude oil sources usually have a unique hydrocarbon composition.[19,20] The actual overall properties of each different petroleum source are defined by the percentage of the main hydrocarbons found within petroleum as part of the petroleum composition. The percentages for these hydrocarbons can vary greatly. It gives the crude oil a quite-specific compound personality depending on geographic region. The typical percentage of hydrocarbons (although covering very wide ranges) is as follows: paraffins (15%–60%), naphthenes (30%–60%), aromatics (3%–30%), and asphaltenes making up the remainder. Furthermore, due to differences in refining technologies and refinery operating conditions, each refining process has a distinct impact on the hydrocarbon composition of the product.

Refined petroleum products are primarily produced through distillation processes that separate fractions from crude oil according to their boiling ranges. Production processes may also be directed to increase the yield of low-molecular-weight fractions, reduce the concentration of undesirable sulfur and nitrogen components, and incorporate performance-enhancing additives. Therefore, each petroleum product has its unique, product-specific hydrocarbon pattern. The petroleum products are composed of both aliphatic and aromatic hydrocarbons in a range of molecules that include C6 and greater. The different classes of compounds contained in various petroleum products are summarized in Table 5.[20,21] The main products are gasoline (benzene), naphtha/solvents, jet fuels, kerosene, diesel fuel, and lubricating (motor) oils. Due to the variety of components in petroleum, they are typically characterized using the boiling range of the mixture and the carbon number rather than individual components. For example, diesel is a fraction with boiling points between 200°C and 325°C and is represented as C10–C22.

While a physical property such as boiling range may establish the initial product specification, other finer specifications define their ultimate use in certain applications. A lighter, less dense, raw petroleum composition with a composition that contains higher percentages of hydrocarbons is much more profitable as a fuel source. On the other hand, other denser petroleum compositions with a less flammable level of hydrocarbons and containing higher levels of sulfur are expensive to refine into a fuel and are therefore more suitable for plastics manufacturing and other uses. In contrast to the ever-increasing demand, the world's reserves of light petroleum (light crude oil) are severely depleted, and refineries are forced to refine and process more and more heavy crude oil and bitumen.

Petroleum fractions are among the most complex samples an analyst can face in terms of the number of compounds present. The characterization of petroleum fractions is typically done by gas chromatography (GC). As can be seen in Figure 1, the petroleum products contain such a large number of hydrocarbon constituents that complete chromatographic separation is not possible. Even then, GC remains the most informative analytical technique, providing both quantitative information (deduced from the total signal recorded in a chromatogram) and qualitative information, which derives from the fact that the retention times in the chromatograms can be correlated with the boiling points of the compounds contained in the petroleum. To illustrate the complexity of chemical composition of petroleum products, Figure 2 shows the chromatograms for six different petroleum products, including a crude oil with API of 18.7 and the BAM (Bundesanstalt für Materialprüfung, Berlin) petroleum hydrocarbon standard. The BAM standard K-010 is a certified reference material for the determination of mineral hydrocarbons, which is a synthetic mixture of a diesel and a lubricating oil. It is evident that these six samples are very different according to their carbon ranges. The difference is clearly seen from the comparison of their chromatograms. The volatile fuel with a content of hydrocarbons with less than 10 C atoms (benzene and premium gasoline) has the majority of its constituents at the beginning of chromatogram (Figure 2c). The peaks in the chromatogram of diesel are shifted to the retention time window where hydrocarbons from C10 to C22 are eluted (Figure 2d). In turn, the chromatogram of motor oil shows a characteristic "bump" (because the fraction of saturated alkanes is very small) situated in the

TABLE 5 Overview of Petroleum Products with Respect to Boiling Point Ranges, Approximate Carbon Number, and Average Percentage Amount of Aliphatic and Aromatic Compounds

		Fractions				Hydrocarbons	
	Boiling Range	<C7 (% w/w)	C7–C10 (% w/w)	C10–C40 (% w/w)	>C40 (% w/w)	Aliphatic	Aromatic
Statfjord C (39.1)[a]		11.6	18.1	56.6	13.7		
Crude oil (API = 18.7)		0.9	3.0	63.2	32.8		
Grane[a]							
Normal benzene[b]	40–200°C	~100 (C5–C12)				~70%	20%–50%
Jet fuel[b]	150–300°C		~100 (C6–C14, C16)			80%–90%	10–20%
Kerosene[b]	150–300°C		~100 (C6, C9–C16)			60%–80%	5%–20%
Diesel[b]	200–325°C		~100 (C10–C22)			60%–90%	30%–40%
Light heating oil[b]	200–325°C		~100 (C10–C22)				
Lubricant or motor oil[b]	325–600°C				~100 (C20–C40)	70%–90%	10%–30%
Heavy heating oil[b]	325–600°C				~100 (C20–C50)		

[a]**Source:** Data from Crude Oil Assays.[20]
[b]**Source:** Data from Statoil Web site.[21]

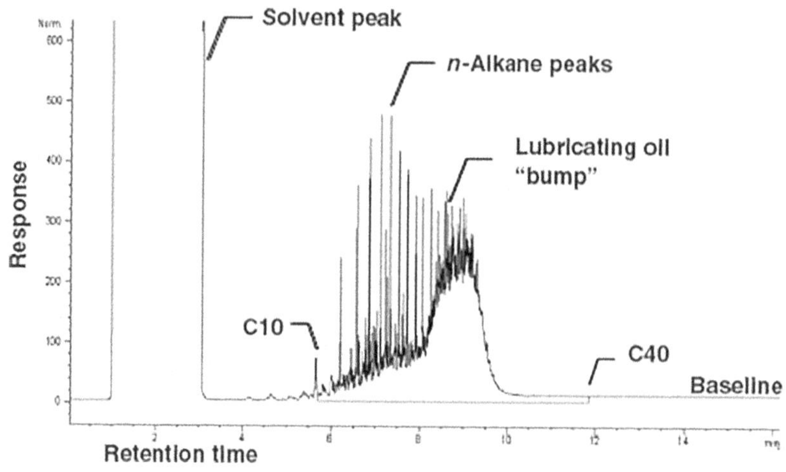

FIGURE 1 Chromatogram of a mixture of petroleum products (diesel and lubricating oil, 1:1), obtained by GC with FID.

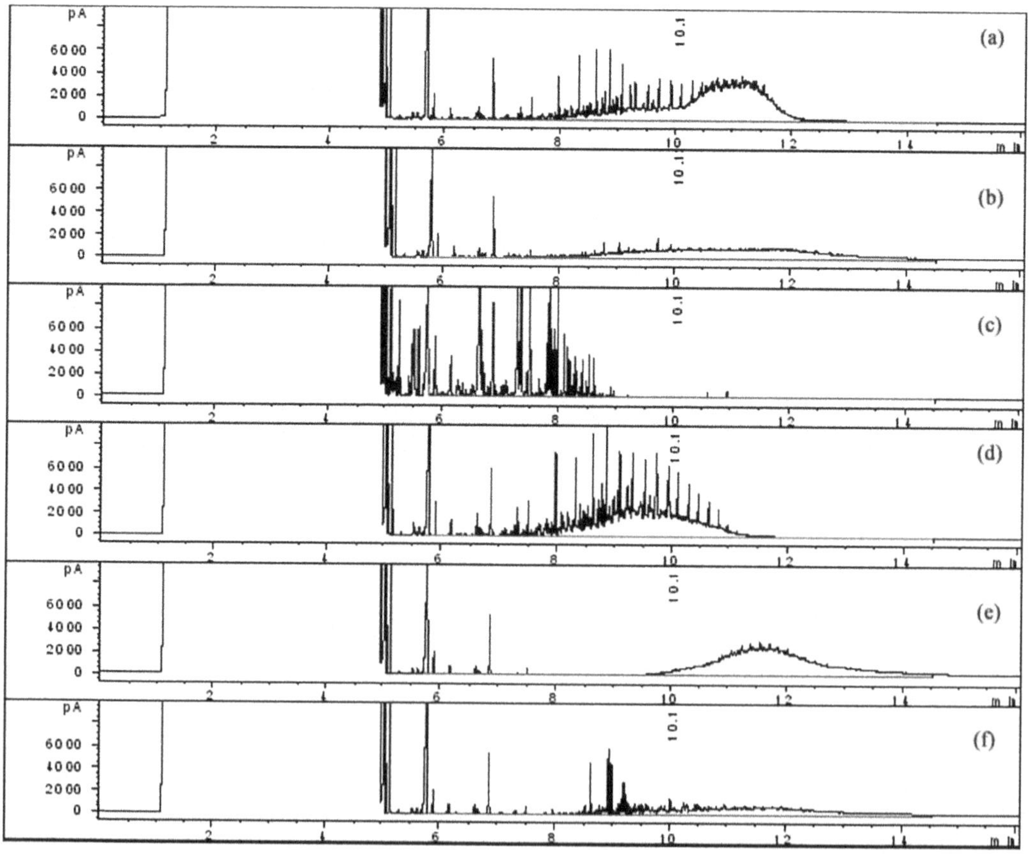

FIGURE 2 Comparison of chromatograms of different oil samples: (a) BAM; (b) crude oil (API = 18.7); (c) gasoline; (d) diesel; (e) motor oil; and (f) heavy heating oil with the same concentration (20 ppm oil in water) obtained by GC-FID method

region where heavier hydrocarbons C20–C40 are eluted (Figure 2e). Thus, GC- based methods provide important qualitative information, which in the ideal case even allows the assignment of the source of contamination. This is proven by the comparison of chromatograms of different oil samples (Figure 2).

Fate of Petroleum Hydrocarbons in the Environment

The effects of petroleum hydrocarbons entering the environment are a complex function of the magnitude and the rate of release; the nature of the released petroleum (its physicochemical properties and, in particular, the amount of toxic compounds it may contain); and the affected geographical, hydrogeological, and biological ecosystem. The fate of petroleum-type pollutants in the environment has been investigated in many studies.[22] Complex transformation and degradation processes of oil in the environment start from its first contact with the atmosphere, seawater, and soil. They depend on the physical properties (volatility, solubility, etc.), as well as on the chemical properties (chemical composition) of the oil. While the former are responsible for transport, or diffusion of the petroleum hydrocarbons in the environment, the latter are responsible for their chemical, photo-, and microbial degradation. The main processes affecting the environmental fate of petroleum hydrocarbons after their release to the environment are thus their volatilization, dissolution/dispersion and emulsification in water, adsorption to soil, oxidation, destruction, and biodegradation.[23,24] In addition to the parameters that characterize the oil's composition, reactivity, and toxicity, the environmental conditions, i.e., the meteorological and hydrological factors, also play an important role in the fate of petroleum hydrocarbons.

When petroleum hydrocarbons are released to the water column, certain fractions will float on top and form thin surface films. This process is controlled by the viscosity of the oil and the surface tension of water. A spill of 1 ton of oil can disperse over a radius of 50 m in 10 min, forming a slick 10 mm thick. Later, it spreads, gets thinner, and covers an area of up to 12 km^2.[11] It should be pointed out that much of the environmental and ecological damage caused by oil spills actually is due to this oil film that covers the surface of the sea, or the coastline, thus physically impairing birds and other animals and causing suffocation of fish as oxygen will not permeate the oil layers to a sufficient degree anymore. In the first days after the spill, the volatile compounds from oil evaporate. Only a small proportion of the hydrocarbon constituents of petroleum products are significantly soluble in water. Dissolution takes more time compared with evaporation, considering that most oil components are soluble in water only to a limited degree (although the degradation products typically are more polar and thus more soluble). Other heavier fractions (up to 10%–30%) will accumulate in the sediment at the bottom of the water, which may affect bottom-feeding fish and organisms. This happens mainly in the narrow coastal zone and shallow waters, where water is intensively mixing.

Crude oil released to the soil may percolate and reach the groundwater. Because petroleum has a lower specific gravity than water, free (undissolved) product and most dissolved contamination are usually concentrated near the top of the groundwater.[25] This may then lead to a fractionation of the original complex mixture, depending on the chemical properties of the compound. Some of these compounds will evaporate, while others will dissolve into the groundwater and be diffused from the release area. Other compounds will adsorb to soil or sediments and will remain there for a long period of time, while others will be metabolized by organisms found in the soil.[26,27]

While evaporation and dissolution redistribute the oil, photochemical oxidation and bacterial degradation transform it. Where crude oil is exposed to sunlight and oxygen in the environment, both photooxidation and aerobic microbial oxidation take place. The photochemical oxidation of hydrocarbons is dependent upon ultraviolet (UV) radiation and will therefore occur only in the upper surface layers. The aromatic hydrocarbons absorb UV radiation with high efficiency and are transformed mainly into hydrogen peroxides. Alkanes are much less efficient in absorbing UV radiation, and only small quantities are transformed by this process. The final products of oxidation (hydroperoxides, phenols, carboxylic acids, ketones, aldehydes, and others) usually have increased water solubility and toxicity. Where oxygen and sunlight are excluded in anoxic environments, anaerobic microbial oxidation takes place.[28,29]

Generally, saturated alkanes are more quickly degraded by microorganisms than aromatic compounds; alkanes and smaller-sized aromatics are degraded before branched alkanes, multiring and substituted aromatics, and cyclic compounds.[30,31] Polar petroleum compounds such as sulfur- and nitrogen-containing species are the most resistant to microbial degradation. Complex structures (e.g., branched methyl groups) and the stability of hydrocarbons decrease the rates of mineralization, which are likely a consequence of the greater stability of carbon–carbon bonds in aromatic rings than in straight-chain compounds. Emulsification also provides greater surface area for microorganisms to attach.

It has been shown in experiments that *n*-alkanes are among the most biodegradable hydrocarbons, and therefore, they are easily broken down and preferentially depleted from soil samples.[32] Also, it has been proven in simulation experiments of the biodegradation of two different samples of crude petroleum (paraffinic and naphthenic type) that microbial cultures that were isolated as dominant microorganisms from the surface of a wastewater canal of an oil refinery (most abundant species: *Phormidium foveolarum*, filamentous Cyanobacteria [blue-green algae] and *Achnanthes minutissima*, diatoms, algae) show a strongly differentiated degradation behavior with clear preference for the degradation of *n*-alkanes and isoprenoid aliphatic alkanes.[33] As can be seen in Figure 3, the largest degree of biodegradation was achieved in a medium containing the base nutrients $Ca(NO_3)_2.4H_2O$, $K_2HPO_4.7H_2O$, KCl, $FeCl_2$, and K_2SO_4, at $pH \approx 8$ and exposed to light. Biodegradation activity is somewhat lower with the same medium in the dark. With a medium containing not only the nutrient broth but also organic compounds (tryptone, yeast extract, glucose, at $pH \approx 7$), degradation occurs at a much lower rate, especially without light.

When crude oil or petroleum products are accidentally released to the environment, they are immediately subjected to a variety of weathering processes that lead to compositional changes and to the depletion of certain hydrocarbon compounds. Weathering processes include all previously mentioned physicochemical processes, such as dissolution, evaporation, photooxidation, polymerization, adsorptive interactions between hydrocarbons and the soil, and some biological factors. Furthermore, due to the fact that the degree of biodegradation is different for different types of petroleum hydrocarbons and varies depending on their nature, the weathering rate also depends on the type of petroleum contaminant. If we thus observe in the analysis of petroleum hydrocarbon contaminants changing patterns of hydrocarbons with time, this may be either due to the segregation of the oil according to the physical properties or due to the action of bacteria and microorganisms. As these are able to degrade only certain classes of compounds, or at least they exhibit a strong preference for some over other compounds, characteristic changes of the hydrocarbon pattern will result, as observed by GC (Figure 3).

Possible Toxic Effects from Exposure to Petroleum Hydrocarbons

As it was discussed earlier, crude oil and petroleum products are complex mixtures of groups of compounds. Many of the compounds are apparently benign, but many others are known to have toxic effects. Due to petroleum hydrocarbon toxicity, spilled hydrocarbons pose a threat that affects not only the sea and land but also the lakes, rivers, and groundwater and can be harmful for animals and human health.

Much of what is known about the impacts of petroleum hydrocarbons comes from studies of catastrophic oil spills and chronic seeps. Large oil spills usually receive considerable public attention because of the obvious environmental damage, oil-coated shorelines, and dead or moribund wildlife, including, in particular, oiled seabirds and marine animals. The acute toxicity of petroleum hydrocarbons to marine organisms is dependent on the persistence and bioavailability of specific hydrocarbons. The exposure to them may alter an organism's chances for survival and reproduction in the environment, and the narcotic effects of hydrocarbons on nerve transmission are a major biological factor in determining the ecologic impact of any release. Marine birds and mammals may be especially vulnerable to oil spills. In addition to acute effects such as high mortality, chronic, low-level exposure to hydrocarbons may affect reproductive performance and physiological impairment of seabirds and some marine mammals as well.[11] Petroleum contamination may also cause unfavorable impacts on nearby plants and animals. Plants growing in contaminated soils or water may die or appear distressed.

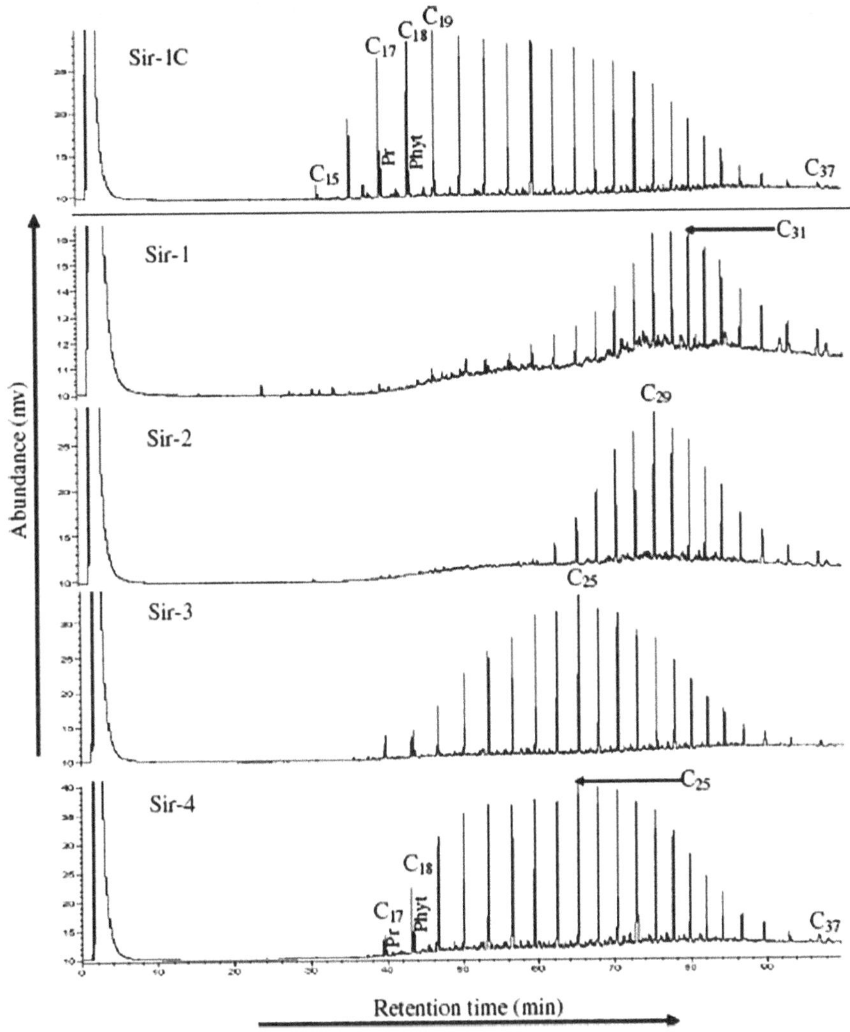

FIGURE 3 Gas chromatograms of the alkane fractions derived from crude oil Sirakovo (Sir, paraffinic type) after 90 days of simulated biodegradation with *Phormidium foveolarum* and *Achnanthes minutissima* with inorganic medium in the light (Sir-1), with inorganic medium in the dark (Sir-2), with organic medium in the light (Sir-3), and with organic medium in the dark (Sir-4), together with chromatogram of alkane fraction typical for the control experiments (Sir-1C), pristane (Pr), phytane (Phyt).
Source: Adapted from Antić et al.[33]

In turn, natural seeps, leaking pipelines, and production discharges release small amounts of oil over long periods of time, resulting in chronic exposure of organisms to oil and oil chemical compounds. The lower-molecular-weight compounds are usually the more water-soluble components of a product, and hence, attention has also been paid to the water-soluble fractions of petroleum and related products. Concentrations in the environment are usually comparatively low, and chronic effects are usually more significant.[26] The persistence of some compounds such as PAH in sediments, especially in urban areas, is also an example of chronic pollution and toxicity.

Nowadays, humans can be exposed to petroleum hydrocarbons through ingestion of contaminated drinking water and soil residues; inhalation of vapors and airborne soils; and contact of contaminants with skin (dermal exposure) from many sources, including gasoline fumes at the pump, spilled crankcase oil on pavement, chemicals used at home or work, or certain pesticides that contain petroleum hydrocarbon components as solvents. Most petroleum hydrocarbon constituents will enter the bloodstream rapidly when inhaled or ingested. Incorporated petroleum hydrocarbons are widely distributed by the blood throughout the body and quickly are metabolized into less harmful compounds. Others may be degraded into more harmful chemicals. Even other compounds are distributed by the blood to other parts of the body and do not readily break down but are accumulated instead in fat tissue. The resorption of petroleum compounds through dermal tissue is slower; that is why direct exposure of the skin to petroleum hydrocarbons is generally harmless when exposure is only occasional and of short duration.

Studies on animals have shown effects on the lungs, central nervous system, liver, kidney, developing fetus, and reproductive system from exposure to petroleum compounds, generally after breathing or swallowing the compounds. Health impacts of exposure to petroleum contamination may include lung irritation, headaches, dizziness, fatigue, diarrhea, cramps, and nervous system effects. Benzene and other chemicals found in petroleum products have been determined to be carcinogenic (cause cancer). More information regarding toxicity of petroleum chemicals is available, for example, from the Agency for Toxic Substances and Diseases Registry (ATSDR), an agency of the U.S. Department of Health and Human Services,[34] or from the European Chemicals Agency.[35]

Oil products are complex mixtures of hundreds of chemicals, with each compound having its own toxicity characteristics. There are many difficulties associated with assessing the health effects of such complex mixtures with regard to hazardous waste site remediation. This means that the traditional approach of evaluating individual components is largely inappropriate. Toxicity information is in the best case available for the pure product; however, once a petroleum product is released to the environment, it changes its composition as a result of weathering. These compositional changes may be reflected in changes in the toxicity of the product.

One approach for assessing the toxicity of oil products is to use toxicity information from studies conducted on the whole product. A second approach is to identify and quantify all components and then consider their toxicities. This approach produces data that theoretically could be compared with the known toxicity of each compound. The impracticality of this approach stems from its high analytical cost and the lack of toxicity data for many of the component chemicals found in hydrocarbon mixtures. A third approach is to consider a series of hydrocarbon fractions and determine appropriate tolerable concentrations and toxicity specific for those fractions. A number of groups have examined such an approach, but the most widely accepted and internationally used are the ones developed by the Total Petroleum Hydrocarbons Criteria Working Group (TPHCWG) and the Massachusetts Department of Environmental Protection (MA DEP) in the United States, although they have been subject to adjustments in many cases. For example, in the United Kingdom, the TPHCWG approach is modified and extended to consider heavier hydrocarbon fractions. It has been developed as part of the Environment Agency's[36] environment sciences program and published in documents related to petroleum hydro- carbons.[37]

The MA DEP introduced in 1994 the concept of petroleum hydrocarbon size-based fractions for use in evaluating the human health effects of exposure to complex mixtures of hydrocarbons[38,39] and provided oral toxicity values for each of the fractions. The toxicity value assigned for each fraction is used in dose–response evaluations. Cancer risks or hazard amounts are subsequently summed across the fractions to get the total values. The TPHCWG has developed and published a series of five monographs[16,40–42] detailing the data on petroleum hydrocarbons and, in addition, has developed tolerable intakes for a series of total hydrocarbon fractions. The TPHCWG independently identified largely similar groupings of hydrocarbon fractions with somewhat different toxicity values in 1997. Of the 250 individual compounds identified in petroleum by the TPHCWG, toxicity data were available for only 95. Of these 95, the TPHCWG concluded that there were sufficient data to develop toxicity criteria for only 25.

As there are differences in toxicity between different hydrocarbon compounds, it is impossible to accurately predict toxic effects of contamination for which only total hydrocarbon data are available. Health assessors often select surrogate or reference compounds (or combinations of compounds) to represent TPH so that toxicity and environmental fate can be evaluated. Correspondence dates relating the toxicologically derived hydrocarbon fractions and their toxicity values to the analytically defined reporting fractions (by MA DEP) are contained in Table 6 for ingestion and inhalation exposure. Inhaled or ingested volatile hydrocarbons have both general and specific effects. The toxicity values are represented as a reference dose (RfD), which is the U.S. Environmental Protection Agency's (EPA's) maximum acceptable oral dose of a toxic substance. Significant efforts have been undertaken by MA DEP to describe an approach for the evaluation of human health risks from ingestion exposure to complex petroleum hydrocarbon mixtures. The methods offered by MA DEP for determination of air-phase (APH), volatile (VPH), and extractable (EPH) petroleum hydrocarbons[43–45] are designed to complement and support the toxicological approach. The ranges of quantified hydrocarbons within each method and their reporting limits are shown in Table 7.

The components of petroleum can be generally divided into broad chemical classes: alkanes, cycloalkanes, alkenes, and aromatics. A review of Table 6 shows that a U.S. EPA RfD is available for only one alkane, *n*-hexane. In general terms, alkanes have relatively low acute toxicity, but alkanes having carbon numbers in the range of C5–C12 have narcotic properties, particularly following inhalation exposure to high concentrations, because of their relatively high volatility and low solubility in water. Repeated exposure to high

TABLE 6 Oral and Inhalation Toxicity Values by MA DEP for Petroleum Hydrocarbon Fractions and Individual Compounds Present in Petroleum Products

		Toxicity Value, RfD		
Carbon range	Compound	Inhalation mg/m^3	Oral mg/kg/day	Critical Effect
Aliphatic				
C5–C8		0.2	0.04	Neurotoxicity
	n-Hexane	0.2	0.06	
C9–C18		0.2	0.1	Neurotoxicity, hepatic, and hematological effects
C19–C32		NA	2	Liver granuloma
Aromatic				
C6–C8		Use individual RfCs for compounds in this range		
	Benzene	NA	0.03	
	Toluene	0.4	0.2	
BTEX	Ethylbenzene	1.0	0.1	
	Styrene	1.0	0.2	
	Xylene (o-, p-, m-)	NA	2	
C9-C18		0.05		Body weight reduction; hepatic, renal, and developmental effects
	Isopropylbenzene	0.4	0.1	
	Naphthalene	0.003	0.02	
	Acenaphthene	NA	0.06	
	Biphenyl	NA	0.05	
	Fluorene	NA	0.04	
	Anthracene	NA	0.3	
	Fluoranthene	NA	0.04	
	Pyrene	NA	0.03	
C9–C32			0.3	Neurotoxicity
C19–C32		NA		

Source: Adapted from The U.K. Approach for Evaluating Human Health Risks from Petroleum Hydrocarbons in Soil[37] and Interim Final Petroleum Report Development of Health-Based Alternative to the Total Petroleum Hydrocarbon TPH Parameter.[38]
NA, not applicable.

TABLE 7 The Ranges of Hydrocarbons Quantified within the Methods for Determination of APH, VPH, and EPH by MA DEP and Their Reporting Limits

	APH 28°C–218°C		VPH 36°C–220°C		EPH 150°C–265°C		
Aliphatic	C5–C8	C9–C12	1C5–C8	C9–C12 C9–C10	C9–C18	C19–C36	
Aromatic	C9–C10		***Reporting limits***		C11–C22		PAH
For the individual target analytes							
In air phase	2–5 g/m³						
In soil			0.05–0.25 mg/kg		20 mg/kg		0.2–1 mg/kg
In water			1-5 μ g/L		100μ g/L		2–5 μg/L
For the collective hydrocarbon ranges							
In air phase	10–12 g/m³						
In soil			5–10 mg/kg		20 mg/kg		
In water			100–150 μg/L		100 μg/L		

Source: Adapted from *Interim Final Petroleum Report Development of Health-Based Alternative to the Total Petroleum Hydrocarbon TPH Parameter.*[38]

concentrations, for example, of *n*-hexane (RfD, 0.06 mg/kg/day) may lead to irreversible effects on the nervous system. Hexane is considered to be the most toxic compound in the C5–C8 aliphatic fraction. No RfDs are available for other alkanes, nor for any cycloalkane or alkene. Alkenes exhibit little toxicity other than weak anesthetic properties. Alkanes and cycloalkanes are treated similarly and have similar toxic effects.

Aromatic compounds with less than nine carbon atoms (such as BTEX) are evaluated separately because the toxicity values for each are well supported and these compounds have a wide range of toxicity. However, most of the smaller aromatic compounds have low toxicity, with the exception of benzene, which is a known human carcinogen (RfD, 0.029 mg/kg/day). Most petroleum hydrocarbon mixtures contain very low concentrations of PAHs. The major concern regarding PAHs is the potential carcinogenicity of some of these. Benzo(*a*)pyrene and benz(*a*)anthracene are classified as probable human carcinogens. Benzo(*a*)pyrene is normally considered to be the most potent carcinogenic PAH, but the carcinogenic potency of most PAHs is not well characterized. In case of spills of petroleum products affecting water, PAHs are not usually a specific concern; however, this concern becomes more specific if these compounds are released into the soil due to a bioaccumulation of PAH in soil.

Different regulations and guidelines to protect public health have been developed. These public health statements tell as well about petroleum hydrocarbons and the effects of exposure. The U.S. EPA[46] identifies the most serious hazardous waste sites in the United States. The EPA lists certain wastes containing petroleum hydrocarbons as hazardous. It regulates certain petroleum fractions, products, and some individual petroleum compounds. General health and safety data are as well discussed by the Energy Institute,[7] which is the main professional organization for the energy industry within the United Kingdom that promotes the safe, environmentally responsible, and efficient supply and use of energy in all its forms and applications. The Occupational Safety and Health Administration and the Food and Drug Administration are other agencies that develop regulations for toxic substances in the United States. The information provided by all of them is regularly updated as more information becomes available. The Dutch National Institute for Public Health and the Environment (RIVM), has been involved in a number of studies on risk assessment for petroleum hydrocarbons which were commissioned by the Dutch government and the European Commission.[47] Also the U.K. Environment Agency, mentioned before, is the leading public body protecting and improving the environment in the United Kingdom, including protection from petroleum contaminations.

Total Petroleum Hydrocarbons and Analytical Methods for Determination of Petroleum Hydrocarbons in Environmental Media

Due to the compositional complexity of petroleum products, it is impossible to assess the extent of petroleum hydrocarbon contamination by directly measuring the concentration of each hydrocarbon contaminant. For this reason, at the present time, no single analytical method is capable of providing comprehensive chemical information on petroleum contaminants. Total petroleum hydrocarbon is one parameter and definition that is currently widely used for expressing the total concentration of nonpolar petroleum hydrocarbons in soil, water, or other investigated samples. In the United States, for example, there are no federal regulations or guidelines for TPH in general. Many states have standards for controlling the concentrations of petroleum hydrocarbons or components of petroleum products. These are designed to protect the public from the possible harmful health effects of these chemicals. Analytical methods are specified as well, many of which are considered to be methods for TPH. These generate basic information that is a surrogate for contamination, such as a single TPH concentration. Such data are not suitable for risk assessment. However, they are relatively quick and easy to obtain and can offer useful preliminary information.

The term TPH is widely used, but it is rarely well defined. In essence, TPH is defined by the analytical method—in other words, estimates of TPH concentration often vary depending on the analytical method used to measure it. Thus, the ATSDR defines the TPH as a term used to describe a broad family of several hundred chemical compounds that originally come from crude oil. In this sense, TPH is really a mixture of chemicals. As per the TPHCWG, TPH, also called "hydrocarbon index," refers sometimes to mineral oil, hydrocarbon oil, extractable hydrocarbon, oil, and grease. The TPHCWG also says that the TPH measurement is the total concentration of the hydrocarbons extracted and measured by a particular method, and it depends on the analytical method used for determination. According to the MA DEP, the TPH is also a loosely defined parameter, which can be quantified using a number of different analyses, and this parameter is an estimate of the total concentration of petroleum hydrocarbons in a sample. Again, depending on the analytical method used to quantify TPH, the TPH concentration may represent the entire range of petroleum hydrocarbons from C9 to C36 or the sum of concentrations of a number of single compounds (for instance, BTEX) and groups of compounds (fractions, e.g., primarily aliphatics C9–C18, C19-C36, and aromatics C11–C20). Great improvements in the definition and analysis of TPH were finally introduced by the International Organization for Standardization (ISO)[48] in 2000, when it published the standard method ISO 9377-2:2000[49] for the quality control of water in which a method for the determination of the hydrocarbon oil index within the C10–C40 range in waters by means of GC is specified. The definition of "hydrocarbon oil index by GC-FID" was introduced, which defines the fraction of compounds extractable with a hydrocarbon solvent, boiling point between 36°C and 69°C, not adsorbed on Florisil, and which may be chromatographed with retention times between those of *n*-decane ($C_{10}H_{22}$) and *n*-tetracontane ($C_{40}H_{82}$). (Substances complying with this definition are long-chain or branched aliphatic, alicyclic, aromatic, or alkyl -substituted aromatic hydrocarbons.)

The TPHCWG and MA DEP evaluated the risk implications and arrived at the conclusion that TPH concentration data cannot be used for a quantitative estimation of the human health risk. The same concentration of TPH may represent very different compositions and very different risks to human health and the environment because the TPH parameter includes a number of compounds of differing toxicities and the health effects associated with exposure to particular concentrations of TPH cannot be determined. For example, two sites may have the same amount of TPH, but constituents at one site may include carcinogenic compounds while these compounds may be absent at the other site. If TPH data indicate that there may be significant contamination of environmental media, then fractionated measurements and the separate determination of BTEX compounds and PAHs are necessary so that potential risk to human health can be quantitatively assessed.[50] The hydrocarbon index is thus a good indicator of the (magnitude of the) relative contamination of oil; however, it will not be suitable to give a true representation of the actual concentration of TPH in the investigated sample. There are several reasons why TPH data do not provide the ideal information for investigated samples and do not establish target cleanup criteria. This is due to many factors including the complex nature of petroleum hydrocarbons, their interaction with the environment over time, and the non-specificity of some of the methods used. The scope of the methods used for TPH determination varies greatly. There are few, if any, methods that are capable of quantifying all hydrocarbons without interference from non-hydrocarbons. All methods are subject to interferences from non-hydrocarbons, some to a greater extent than others.

There are numerous established analytical methods that are available for detecting, measuring, or monitoring TPH and its metabolites. Analytical methods used for analysis of petroleum hydrocarbons in environmental media should provide a sufficient degree of robustness. At the current time, however, the correctness and precision of results for the petroleum hydrocarbon determination strongly depend on the proper choice of method and measurement parameters whose correct selection is left to the judgment of the analyst. Besides methods that measure the TPH concentration, two other types of methods can be distinguished. These are methods that measure the concentration of a group or fraction of petroleum compounds and methods that measure individual petroleum constituent concentrations. For product identification, the results of analyses of the petroleum groups or fractions can be useful because they separate and quantify different categories of hydrocarbons. Individual constituent methods quantify concentrations of specific compounds that might be present in petroleum-contaminated samples, such as BTEX and PAHs, which can be used to evaluate human health risk.

There are several basic steps related to the separation of analytes of interest from a sample matrix prior to their measurement, such as extraction, concentration, and cleanup. These steps are common to the analytical processes for all methods, irrespective of the method type or the environmental matrix. Each of these steps together with the sampling, which is also an important step in performing petroleum analyses, affects the final result and has a certain impact onto the measurement uncertainty.[51,52]

Sample taking and sample handling have been recognized as probably the most significant factors that contribute random errors and uncertainties in the analysis of offshore oil in produced water. There are some general guidelines available through a number of studies that have been carried out on this subject. To separate the analytes from the matrix, extraction is performed using one of the many available extraction methods. Heating of the sample or purging with an inert gas can be used in the analysis of volatile compounds; solid-phase extraction or extraction into a solvent is usually applied for water samples, the latter extraction method also being used for soil samples. For some types of solid samples, the extraction efficiency depends on the extraction method and time. However, ultrasonication and extraction by shaking are equally used for this purpose. It was demonstrated by some studies that extraction and cleanup are the most crucial steps in sample preparation procedures. According to the results. the most critical factors affecting TPH recovery are extraction solvent and type of cosolvent, extraction time, adsorbent and its mass, and the TPH concentration.[53] The results of a study where the occurrence of matrix effects in the gas chromatographic determination of petroleum hydrocarbons in soil was evaluated indicate that solid-phase extraction does not appear to be effective enough in removing interfering matrix components from the extract.[54]

Most of the methods for the determination of TPH involve a cleanup step using Florisil (a particular form of magnesium silicate) and sodium sulfate (anhydrous), which essentially aims at removing the polar, non-petroleum hydrocarbons of biological origin and remaining traces of water. It appears that the found hydrocarbon concentration strongly depends on the used cleanup technique. The efficiency of the cleanup procedure for removing polar compounds is not limited to heteroatomic substances like O-, N-, or Cl-containing compounds. Also, some hydrocarbons have a tendency to adsorb on Florisil, e.g., aromatic compounds with p- electrons or alkyl aromatics. The TPH recoveries after a cleanup procedure might depend on the composition of the oil investigated. Lower TPH recoveries may be expected for oils containing high concentrations of unsaturated hydrocarbons or PAHs. Also, lubricating oils often contain different amounts and types of (non-petrogenic) additives that may behave differently from the other compounds during the cleanup procedure.[55] The results demonstrate also that the ratio of Florisil amount and extract volume are of importance for the recovery of the purified extracts.[56]

The three most commonly used TPH testing methods include GC,[49,57–60] infrared absorption (IR),[61,62] and gravimetric analysis.[63–65] Conventional TPH methods are summarized in Table 8.

Methods based on solvent extraction followed by quantitative IR measurement (at a frequency of 2930 cm^{-1}, which corresponds to the stretching vibration of aliphatic CH2 groups) have been widely used in the past for TPH measurement because they are simple, quick, and inexpensive. However, the use of these methods has been discontinued, since the sale and use of Freons (required for the extraction of hydrocarbons from the sample) is no longer allowed, and Freons are generally phased out worldwide due to their ozone layer–destructing potential. Recently, a new IR-based method was introduced, based on Freon- free extraction. This method defines oil and grease in water and wastewater as the fraction that is extractable with a cyclic aliphatic hydrocarbon (for example, cyclohexane) and measured by IR absorption in the narrow spectral region of 1370–1380 cm^{-1} (which corresponds to the excitation frequency of the symmetrical deformation vibration of CH_3 groups) using mid-IR quantum cascade lasers.[62] The method also considers the volatile fraction of petroleum hydrocarbons, which is lost by gravimetric methods that require solvent evaporation prior to weighing, as well as by solventless IR methods that require drying of the employed solid-phase material prior to measurement. Similarly, a more complete fraction of extracted petroleum hydrocarbon is accessible by this method as compared with GC methods that use a time window for quantification, as petroleum hydrocarbons eluting outside these windows are also quantified. On the other hand, IR-based methods hardly provide any information on the chemical composition of the oil or the presence or absence of other relevant compounds (aromatics, PAHs). In contrast, they even detect compounds that are not typically considered as TPH, such as surfactants, which also may absorb IR radiation due to the presence of CH bonds. However, this statement is only partially true, since it depends mainly on the cleanup whether the IR method determines also compounds other than the TPH.

Gravimetric-based methods are also simple, quick, and inexpensive; they measure anything that is extractable by a solvent, not removed during solvent evaporation, and capable of being weighed. Consequently, they do not offer any selectivity or information on the type of oil detected. Gravimetric-based methods may be useful for oily sludges and wastewaters at high(er) concentrations but are not suitable for measurement of light hydrocarbons (less than C15), which will be lost by evaporation below 70–85°C.

Gas chromatography–based methods are currently the preferred laboratory methods for TPH measurement because they detect a broad range of hydrocarbons, they provide both sensitivity and selectivity, and they can be used for TPH identification as well as quantification. The potential of GC for producing information on the product-specific hydrocarbon pattern has been long recognized by researchers in the field of petroleum hydro-carbon analysis. [66–68]

Currently, there are several standard methodologies based on GC for different types of samples (water, soil, wastes). The ISO has published the standard ISO 93772:2000 for the quality control of water and specifies a method for the determination of the hydrocarbon oil index within the C10–C40 range in waters by means of GC. The method is suitable for surface water, wastewater, and water from sewage treatment plants and allows the determination of the hydrocarbon oil index in concentrations above 0.1 mg/L. Due to systematic differences, which became evident between the results from the DIN ISO method and those from the IR-based method, the GC-based method was subsequently modified.[69,70] As a result, the modified version of DIN ISO 9377-2:2000, the OSPAR (Oslo–Paris commission) reference method,[58] was published in 2005 and taken into force as a reference method in the field of petroleum production in January 2007. The OSPAR reference method is applicable for the determination of dispersed oil content in produced water and other types of wastewater discharged from gas, condensate, and oil platforms. It also allows the determination of the dispersed mineral oil content in concentrations above 0.1 mg/L and includes the determination of certain hydrocarbons within the C7–C10 range, with the TEX (toluene, ethylbenzene, and *o-/p-/m*-xylene) compounds being reported separately.

TABLE 8 Summary of Common TPH Methods

Analytical Method	Method Name	Matrix	Scope of Method	Carbon Range	Approximate Detection Limits	Advantages	Limitations	Reference
GC based	DIN ISO 9377-2:2000	Water	Solvent (hydrocarbon) extraction, cleanup using Florisil, evaporation, 1 μL injection, GC-FID	C10-C40	0.1 mg/L	Detects broad range of hydrocarbons; provides information (e.g., a chromatogram) for identification	Does not quantify below C10; chlorinated compounds can be quantified as TPH	[49]
	OSPAR (2007)	Water	n-Pentane extraction, cleanup using Florisil, 50 μL injection, GC-FID	C7–C40 + TEX compounds	0.1 mg/L	Does not need preconcentration step; detect broad range of hydrocarbons and polar hydrocarbons; provide information for identification	Does not quantify below C7	[57]
	DIN ISO 16703:2005–12	Soil	Acetone/n-heptane extraction, cleanup using Florisil, evaporation, GC-FID	C10–C40	10 mg/kg	Detects broad range of hydrocarbons; provide information (e.g., a chromatogram) for identification	Does not quantify below C10; chlorinated compounds can be quantified as TPH	[58]
	DIN EN 14039:2004	Wastes	Acetone/n-heptane extraction, cleanup using Florisil, evaporation, GC-FID	C10–C40	10 mg/kg			[59]
IR based	EPA 418.1 (1991/1992)	Water, soil	Freon extraction, silica gel treatment to remove polar compounds	Most hydrocarbons with exception of volatile and very high hydrocarbons	1 mg/mL in water, 10 mg/kg in soil	Technique is simple, quick, and inexpensive	Freon is banned now; low sensitivity; lack of specificity; prone to interference; provides quantitation only	[62]
	ASTM D7678-11 (2011)	Water, wastewater	Solvent (cyclic aliphatic hydrocarbon) extraction, cleanup using Florisil, IR Absorption in the region of 1370–1380 cm^{-1} (7.25–7.30 mm)	Most hydrocarbons with volatile	0.5 mg/mL	Technique is simple, very quick; a more complete fraction of extracted petroleum hydrocarbon is accessible		[61]

(*Continued*)

TABLE 8 (*Continued*) Summary of Common TPH Methods

Analytical Method	Method Name	Matrix	Scope of Method	Carbon Range	Approximate Detection Limits	Advantages	Limitations	Reference
Gravimetry	EPA 413.1 (1979) ASTM D4281–95(2005)el	Most appropriate for wastewater, sludge, sediment	Freon extraction, solvent evaporation	Anything that is extractable (with exception of volatiles which are lost)	5 mg/mL in water, 50 mg/kg in soil	Technique is simple, quick, and inexpensive	Freon is banned now; low sensitivity; lack of specificity not suitable for low boiling fractions; prone to interference (organic acids, phenols, and other polar hydrocarbons); provides quantitation only	[63,65]
	EPA 1664 (1999)	Most appropriate for water and wastewater	n-Hexane extraction, silica gel treatment to remove polar compounds, solvent evaporation	Anything that is extractable (with exception of volatiles which are lost)	5 mg/mL	Technique is simple, quick, and inexpensive	Low sensitivity; lack of specificity not suitable for low boiling fractions; prone to interference; provides quantitation only	[64]

Gas chromatography–based methods are based on the extraction of water samples with a nonpolar (hydrocarbon) solvent, the removal of polar substances by cleanup with Florisil, and capillary GC measurements using a nonpolar column and a flame ionization detector (FID), cumulating the total peak area of compounds eluted between *n*- decane (C10H22) and *n*-tetracontane ($C_{40}H_{82}$) for the DIN ISO 9377-2:2000 standard method and for the DIN ISO 16703:2005–12[59] standard method for soil samples. The OSPAR method was modified in order to include the determination of certain hydrocarbons with a boiling point between 98°C and 174°C (that is, from *n*-heptane to *n*-decane), with the TEX compounds being determined separately by integration and subtraction of their peak areas from the total integrated area. The GC-based methods usually cannot quantitatively detect compounds with a lower boiling point than *n*-heptane because these compounds are highly volatile and are interfered by the solvent peak. Furthermore, the EPA method 8240,[61] which is used to determine volatile organic compounds in a variety of waste matrices by GC/mass spectrometry (MS), exists. It can be used to quantitate most volatile organic compounds that have boiling points below 20°C and that are insoluble or slightly soluble in water. The estimated quantitation limit of the EPA 8240 method for an individual compound is approximately 5 μg/kg (wet weight) for soil/sediment samples, 0.5 mg/kg (wet weight) for wastes, and 5 μg/L for groundwater.

Gas chromatography–based methods are suitable for surface water, wastewater, and other types of wastewater discharged from gas, concentrate, and oil platforms and allow the determination of hydrocarbon oil concentration above 0.1 mg/L. To reach the required detection limit, the method according to DIN ISO 9377-2:2000 foresees preconcentration of the extracts by solvent evaporation, which bears the risk of losing the more volatile constituents of the sample. In contrast to this, the OSPAR method does not allow for any external apparatus for preconcentration, for which reason the GC must be equipped with an injection system that allows the injection of a volume of up to 100 μL of the extract. This is most easily realized with programmed-temperature vaporizer large-volume injectors. This technique can reduce the loss of volatile analytes, can increase sensitivity, and is a viable, fast, and automated alternative to an external preconcentration procedure.[71–73]

Petroleum products easily contain thousands of different compounds. Classical capillary GC cannot resolve such mixtures up to the level of individual compounds. A powerful analytical tool for separation of complex mixtures, such as petroleum hydrocarbons, is comprehensive two-dimensional GC (GC×GC or 2D-GC).[74–76] The use of 2D-GC with MS detection (GC×GC/MS) is expected to not only allow the separation of the various constituents of complex TPH samples but also to identify them based on MS detection (Figure 4b). It is known that a certain class of chemical compounds (a series of "homologues") forms a very distinct, clearly identifiable pattern in the two-dimensional space of the GC×GC separation. The diesel total ion (TIC) GC×GC/MS chromatogram, illustrated in Figure 4a, is characterized by very typical group-type patterns: saturated hydrocarbons, which present low second-dimension retention times, are followed by monocyclic and dicyclic aromatics; tri-and tetracyclic aromatics are the most retained on the secondary polar column.[77] Moreover, partial overlapping between chemical groups occurs, the monocyclic aromatics are situated in a rather narrow band, and the tri-and tetracyclic aromatics are hardly visible in the two-dimensional chromatogram. The analytical potential of such a two-dimensional system is great.

Conclusion

Due to the importance and widespread use of petroleum hydrocarbons for energy production, for transport, and as a raw material in the chemical industries, there are many routes for their inadvertent or accidental release into the environment. Thus, they do represent one of the most important sources of large-scale environmental pollution. While petroleum hydrocarbons also are introduced into the oceans from natural seeps, these continuous emissions of comparatively low intensity represent a less significant environmental problem since the resident flora and fauna have adapted to this continuous input of hydrocarbons and effects are limited to local scale. Large oil spills in contrast exceed the self-cleaning capacity

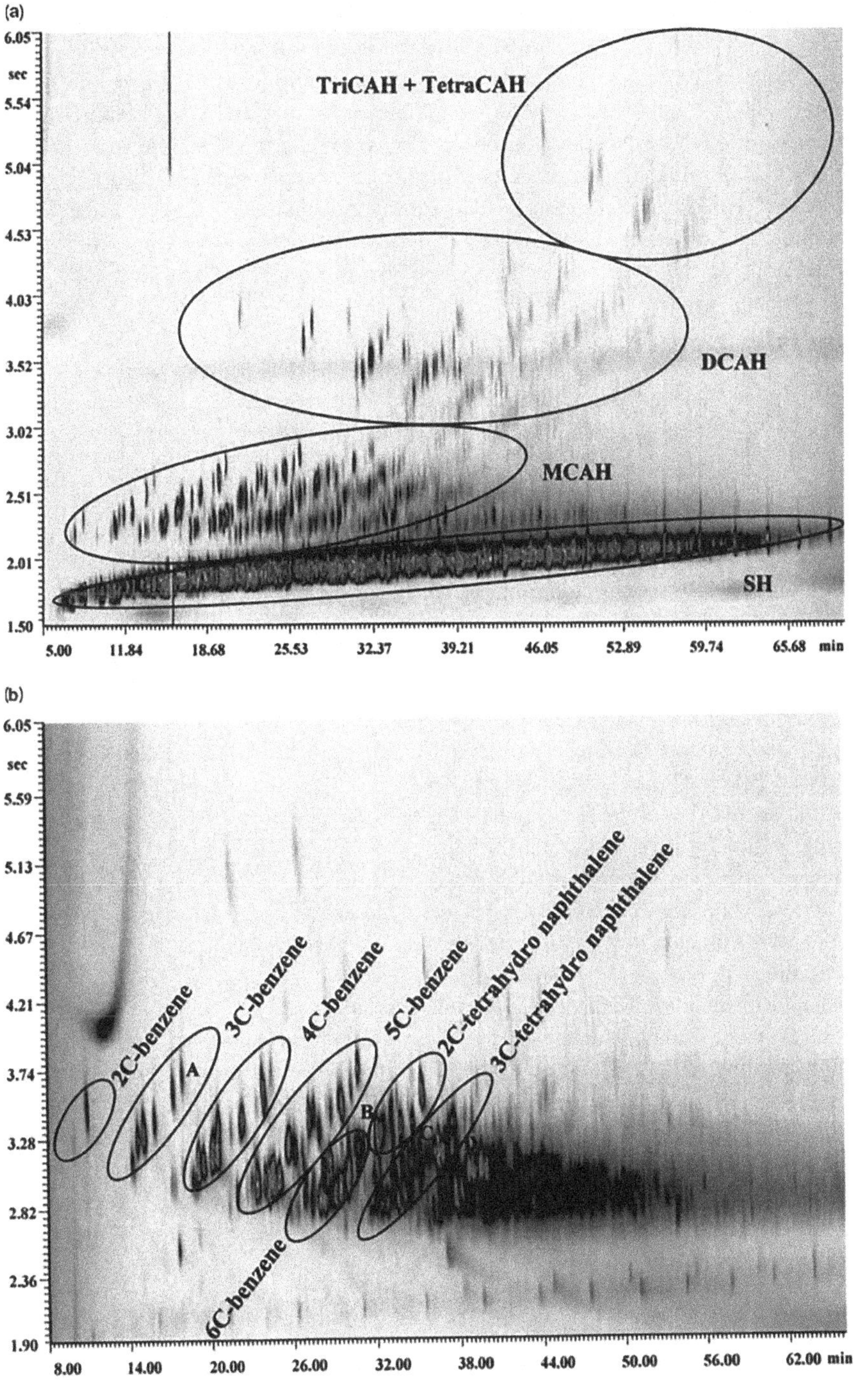

FIGURE 4 (a) TIC GC×GC-qMS (quadruple-mass spectrometry) chromatogram of diesel oil. SH, saturated hydrocarbons; MCAH, monocyclic aromatics; DCAH, dicyclic aromatics; TriCAH, tricyclic aromatics; TetraCAH, tetracyclic aromatics. (b) TIC LC-GC×GC- qMS chromatogram of the monocyclic aromatic fraction of diesel oil. A) Indane, B) 1,2,3,4-Tetrahydro-2,7-dimethyl naphthalene, C) 1-Cyclohexyl 3-methyl benzene, D) 1,2,3,4-Tetrahydro-2,5,8-trimethyl naphthalene.
Source: Adapted from Sciarrone et al.[77]

of the ecosystem, which cannot regenerate without human intervention to both physically and chemically immobilize, bind, and remove oil from the affected region. Although such techniques are available, large-scale oil spills always have caused severe damage to the environment, with the affected ecosystems recovering only slowly. Analytical methods are available for the qualitative and quantitative determination of the composition of oil samples and the assessment of pollution levels in various environmental compartments. Gas chromatographic techniques mostly have supplanted the former analytical standard method based on Freon extraction and mid-IR determination, but there is further research and development going on to develop either more powerful analytical methods—such as two-dimensional GC—or alternative detection methods, such as the ones based on mid-IR lasers as light sources.

Acknowledgments

This report was compiled within the frame of project 818084-16604 SCK/KUG of the Austrian Science Foundation (FFG), whose financial support is gratefully acknowledged.

References

1. Tissot, B.P.; Welte, D.H. *Petroleum Formation and Occurrence;* Springer-Verlag: Berlin, 1984.
2. Availableathttp://www.oxforddictionaries.com/view/entry/m_en_gb0623910#m_en_gb0623910 (accessed September 2011).
3. Available at http://www.nationmaster.com/graph/ene_oil_pro-energy-oil-production (accessed September 2011).
4. Available at http://www.eia.gov/petroleum/ (accessed September 2011).
5. Leffler, W.L. *Petroleum Refining in Nontechnical Language*, 4th Ed.; PennWell Corporation: Tulsa, OK, 2008.
6. Fahimm, M.A.; Al-Sahhaff, T.A.; Lababidii, H.M.S.; Elkilanii A. *Fundamentals of Petroleum Refining*, 1st Ed.; Elsevier: Oxford, U.K., 2010.
7. Available at http://www.energyinst.org (accessed September 2011).
8. Nancarrow, D.J.; Adams, A.L.; Slade, N.J.; Steeds, J.E. Land *Contamination: Technical Guidance on Special Sites: Petroleum Refineries*; R&D Technical Report P5–042/ TR/05; Environment Agency: Bristol, 2001.
9. *Toxicological Profile for Total Petroleum Hydrocarbons (TPH)*; U.S. Department of Health and Human Services, Public Health Service Agency for Toxic Substances and Disease Registry: Atlanta, GA, 1999.
10. Leifer, I.; Kamerling, M.J.; Luyendyk, B.P.; Douglas, S.W. Geologic control of natural marine hydrocarbon seep emissions, Coal Oil Point seep field, California. Geo-Mar. Lett. **2010,** *30,* 331–338.
11. *Oil in the Sea III Inputs, Fates, and Effects;* The National Academies Press: Washington, DC, 2003.
12. Available at http://www.toptenz.net/top-10-worst-oil-spills. php (accessed September 2011).
13. Weggen, K.; Pusch, G.; Rischmüller, H. Oil and gas. In *Ullmann's Encyclopedia of Industrial Chemistry;* Wiley-VCH: Weinheim, 2000.
14. Sanin, P.I. Petroleum hydrocarbons. Russ. Chem. Rev. **1976,** *45* (8), 684–700.
15. Available at http://api-ep.api.org/ (accessed September 2011).
16. Analysis of petroleum hydrocarbons in environmental media. In *Total Petroleum Hydrocarbon Criteria Working Group Series;* Weisman. W., Ed.; Amherst Scientific Publishers: Amherst, MA, 1998; vol. 1.
17. Wang, Z.; Fingas, M.F. Development of oil hydrocarbon fingerprinting and identification techniques. Mar. Pollut. Bull. **2003,** *47,* 423–452.
18. Wang, Z.; Fingas, M.; Pag.e, D.S. Oil spill identification. J. Chromatogr. A **1999,** *843,* 369–411.
19. Available at http://www.etc-cte.ec.gc.ca/databases/OilProperties/oil_prop_e.html (accessed September 2011).

20. Availableat http://www.statoil.com/en/ouroperations/tradingproducts/crudeoil/crudeoilassays/pages/default.aspx (accessed September 2011).
21. Availableathttp://www.castrol.com/liveassets/bp_internet/castrol/castrol_switzerland/STAGING/local_assets/ downloads/a/ABC_D_Mai_2009.pdf (accessed September 2011).
22. *Spills of Emulsified Fuels: Risks and Responses;* The National Academy of Sciences: Washington, DC, 2001.
23. *Petroleum Products in Drinking-Water;* Background document for development of WHO Guidelines for Drinking-water Quality, WHO/SDE/WSH/05.08/123; World Health Organization: Geneva, 2005.
24. *Fate of Spilled Oil In Marine Waters: Where Does It Go? What Does It Do? How Do Dispersants Affect It?;* An Information Booklet for Decision Makers, Publication 4691; American Petroleum Institute: Virginia, 1999.
25. Afifi, S.M. Petroleum hydrocarbon contamination of groundwater in Suez: causes severe fire risk. Proceedings 24th AGU Hydrology Days, March 10–12, 2004, pp. 1–9. Colorado State University (2004).
26. Vaajasaar, K.; Joutti, A.; Schultz, E.; Selonen, S.; Westerholm, H. Comparisons of terrestrial and aquatic bioassays for oil-contaminated soil toxicity. J. Soils Sediments **2002,** *2* (4), 194–202.
27. *Guidelines for Assessing and Managing Petroleum Hydrocarbon Contaminated Sites in New Zealand;* Module 2—Hydrocarbon contamination fundamentals; Ministry for Environment: New Zealand, 1999.
28. Das, N.; Chandran, P. Microbial degradation of petroleum hydrocarbon contaminants: An overview. Biotechnol. Res. Int. **2011,** Article ID 941810, 13 pages.
29. Atlas, R.M. Microbial degradation of petroleum hydrocarbons: An environmental perspective. Microbiol. Rev. **1981,** *45,* 180–209.
30. Van Hamme, J.D.; Singh, A.; Ward, O.P. Recent advances in petroleum microbiology. Microbiol. Mol. Biol. Rev.**2003,***67* (4), 503–549.
31. Kaplan, I.R.; Galperin, Y.; Lu, S.T.; Lee, R.P. Forensic environmental geochemistry: Differentiation of fuel-types, their sources and release time. Org. Geochem. **1997,** *2,* 289–317.
32. Šepieč, E.; LeskovŠek, H.; Trier, C. Aerobic bacterial degradation of selected polyaromatic compounds and n-alkanes found in petroleum. J. Chromatogr. A **1995,** *697,* 515–523.
33. Antić, M.P.; Jovaneičević, B.S.; Ilić, M.; Vrvić, M.M.; Schwarzbauer, J. Petroleum pollutant degradation by surface water microorganisms. Environ. Sci. Pollut. Res. **2006,** *13* (5), 320–327.
34. Available at http://www.atsdr.cdc.gov/ (accessed September 2011).
35. Available at http://echa.europa.eu/home_en.asp (accessed September 2011).
36. Available at http://www.environment-agency.gov.uk/ (accessed September 2011).
37. *The U.K. Approach for Evaluating Human Health Risks from Petroleum Hydrocarbons in Soil;* Science report P5–080/TR3; Environment Agency: Bristol, U.K., 2005.
38. MA DEP 1994. *Interim Final Petroleum Report Development of Health-Based Alternative to the Total Petroleum Hydrocarbon TPH Parameter*; Massachusetts Department of Environmental Protection: Boston, Massachusetts, 1994.
39. MA DEP 2003. *Updated Petroleum Hydrocarbon Fraction Toxicity Values for the VPH/EPH/APH*; Massachusetts Department of Environmental Protection: Boston, Massachusetts, 2003.
40. Edwards, D.A.; Andriot, M.D.; Amoruso, M.A.; Tummey, A.C.; Tveit, A.; Bevan, C.J.; Hayes, L.A.; Youngren, S.H.; Nakles, D.V. Development of fraction specific reference doses (RfDs) and reference concentrations (RfCs) for total petroleum hydrocarbons. In *Total Petroleum Hydrocarbon Criteria Working Group Series*; Amherst Scientific Publishers: Amherst, Massachusetts, 1997; Vol. 4.
41. Potter, T.L.; Simmons, K.E. Composition of petroleum mixtures. In *Total Petroleum Hydrocarbon Criteria Working Group Series;* Amherst Scientific Publishers: Amherst, Massachusetts, 1998; Vol. 2.
42. Vorhees, D.J.; Weisman, W.H.; Gustafson, J.B. Human health risk-based evaluation of petroleum release sites: implementing the working group approach. In *Total Petroleum Hydrocarbon Criteria Working Group Series*; Amherst Scientific Publishers: Amherst, Massachusetts, 1999; Vol. 5.

43. MA DEP 2003. *Method for the Determination of Air-Phase Petroleum Hydrocarbons (APH)*; Massachusetts Department of Environmental Protection: Boston, Massachusetts, 2009.
44. MA DEP 2004. *Method for the Determination of Volatile Petroleum Hydrocarbons (VPH)*; Massachusetts Department of Environmental Protection: Boston, Massachusetts,2004.
45. MA DEP 2004. *Method for the Determination of Extract-able Petroleum Hydrocarbons (EPH)*; Massachusetts Department of Environmental Protection: Boston, Massachusetts, 2004.
46. Available at http://www.epa.gov/ (accessed September 2011).
47. Verbruggen, E.M.J. *Environmental Risk Limits for Mineral Oil (Total Petroleum Hydrocarbons)*; RIVM report 601501021; National Institute for Public Health and the Environment: Bilthoven, the Netherlands, 2004.
48. Available a3t http://www.iso.org/iso/home.html (accessed September 2011).
49. DIN ISO 9377-2:2000. *Water Quality—Part 2, Method Using Solvent Extraction and Gas Chromatography*; International Organization for Standardisation: Geneva, 2000.
50. Pollard, S.J.T.; Duarte-Davidson, R.; Askari, K.; Stutt, E. Managing the risk from petroleum hydrocarbons at contaminated sites achievements and future research directions. Land Contam. Reclam. **2005,** *13* (2), 115–122.
51. Saari, E.; Perämäki, P.; Jalonen, J. Measurement uncertainty in the determination of total petroleum hydrocarbons (**TPH**) in soil by GC-FID. Chemom. Intell. Lab. Syst. **2008,** *92* (1), 3–12.
52. Becker, R.; Buge, H.G.; Bremser, W.; Nehls, I. Mineral oil content in sediments and soils: Comparability, traceability and a certified reference material for quality assurance. Anal. Bioanal. Chem. **2006,** *385* (3), 645–651.
53. Saari, E.; Perämäki, P.; Jalonen, J. Evaluating the impact of extraction and cleanup parameters on the yield of total petroleum hydrocarbons in soil. Anal. Bioanal. Chem. **2008,** *392* (6), 1231–1240.
54. Saari, E.; Perämäki, P.; Jalonen, J. Effect of sample matrix on the determination of total petroleum hydrocarbons (**TPH**) in soil by gas chromatography–flame ionization detection. Microchem. J. **2007,** *87* (2), 113–118.
55. Muijs, B.; Jonker, M.T.O. Evaluation of clean-up agents for total petroleum hydrocarbon analysis in biota and sediments. J. Chromatogr. A **2009,** *1216* (27), 5182–5189.
56. Koch, M.; Liebich, A.; Win, T.; Nehls, I. *Certified Reference Materials for the Determination of Mineral Oil Hydrocarbons in Water, Soil and Waste;* Forschungsbericht 272; Bundesanstalt für Materialforschung und - prüfung (BAM): Berlin, 2005.
57. OSPAR. *Reference Method of Analysis for Determination of the Dispersed Oil Content in Produced Water*; OSPAR Commission, ref. no. 2005–15: Malahide, published in 2005, *taken into force in 2007.*
58. DIN ISO 16703:2005–12. *Soil Quality—Determination of Content of Hydrocarbon in the Range C10 to C40 by Gas Chromatography*; International Organization for Standardisation: Brussels, 2005.
59. DIN EN 14039:2004. *Characterization of Waste—Determination of Hydrocarbon Content in the Range of C10 to C40 by Gas Chromatography*; German version; International Organization for Standardisation: Brussels, 2004.
60. EPA Method 8240. *Gas Chromatography/Mass Spectrometry for Volatile Organics, Test Methods for Evaluating Solid Wastes*; US Environmental Protection Agency: Washington, 1986, Vol. 1B.
61. ASTM Standard D7678111. *Standard Test Method for Total Petroleum Hydrocarbons (TPH) in Water and Wastewater with Solvent Extraction using Mid-IR Laser Spectroscopy*; ASTM International: West Conshohocken, PA, 2011; DOI: 10.1520/D7678-11.
62. EPA Method 418.1. *Total Recoverable Petroleum Hydrocarbons by IR, Groundwater Analytical Technical Bulletin*; Groundwater Analytical Inc.: Buzzards Bay, MA, 1991/1992.
63. EPA method 413.1. *Standard Test Method for Oil and Grease Using Gravimetric Determination*; issued in 1974, editorial revision in 1978 (withdrawn).
64. EPA method 1664. *Revisiow A: w-Hexawe Extractable Material (HEM; Oil awd Grease) awd Silica Gel Treated n-Hexawe Extractable Material (SGT-HEM: Now-polar Material) by Extractiow awd Gravimetry*; US Environmental Protection Agency: Washington, 1999.

65. ASTM D4281-95(2005)e1. *Standard Test Method for Oil awd Grease (Fluorocarbow Extractable Substawces) by Gravimetric Determiwatiow;* ASTM International: West Conshohocken, PA, 2005.
66. Blomberg, J.; Schoenmakers, P.J.; Brinkman, U.A.T. GC methods for oil analysis. J. Chromatogr. A **2002,** *972* (2), 137–173.
67. Beens, J.; Brinkman, U.A.T. The role of GC in compositional analyses in the petroleum industry. Trends Anal. Chem. **2000,** *19* (4), 260–275.
68. Saari, E.; Perämäki, P.; Jalonen, J. Evaluating the impact of GC operating settings on GC–FID performance for total petroleum hydrocarbon (TPH) determination. Microchem. J. **2010,** *94* (1), 73–78.
69. Thomey, N.; Bratberg, D.; Kalisz, C. A comparison of methods for measuring total petroleum hydrocarbons in soil. In Proceedings of the Petroleum Hydrocarbons and Organic Chemicals in Groundwater: Prevention, Detection and Restoration, November 15–17, 1989; National Water Well Association: Houston, Texas, 1989.
70. Xie, G.; Barcelona, M.J.; Fang, J. Quantification and interpretation of total petroleum hydrocarbons in sediment samples by a GC/MS method and comparison with EPA 418.1 and a rapid field method. Anal. Chem. **1999,** *71* (9), 1899–1904.
71. Hoh, E.; Mastovska, K. Large volume injection techniques in capillary gas chromatography. J. Chromatogr. A **2008,** *1186,* 2–15.
72. Miñones Vázqiez, M.; Vázquez Blanco, M.E.; Muniategui Lorenzo, S.; Loópez Mahía, P.; Fernández-Fernández, E.; Prada Rodríguez, D. Application of programmed-temperature split/splitless injection to the trace analysis of aliphatic hydrocarbons by gas chromatography. J. Chromatogr. A **2001,** *919,* 363–371.
73. Dellavedova, P.; Vitelli, M.; Ferraro, V.; Di Toro, M.; Santoro, M. *Applicatiow of enhawced large volume injectiow; an approach to the analysis* of *petroleum hydrocarbons iw water.* Chromatographia **2006,** *63,* 73–76.
74. Van De Weghe, H.; Vanermen, G.; Gemoets, J.; Lookman, R.; Bertels, D. Application of comprehensive two-dimensional gas chromatography for the assessment of oil contaminated soils. J. Chromatogr. A **2006,** *1137,* 91–100.
75. von Mühlen, C.; Alcaraz Zini, C.; Bastos Caramão, E.; Marriott, P. Applications of comprehensive two-dimensional gas chromatography to the characterization of petrochemical and related samples. J. Chromatogr. A **2006,** *1105,* 39–50.
76. van Deursen, M.M.; Beens, J.; Reijenga, J.C.; Lipman, P.J.L.; Camers, C.A.M.G.; Blomberg, J. Group-type identification of oil samples using comprehensive two-dimensional gas chromatography coupled to a time-of-flight mass spectrometer (GCxGC-TOF). J. High Resolut. Chromatogr. **2000,** *23* (7–8), 507–510.
77. Sciarrone, D.; Tranchida, P.Q.; Costa, R.; Donato, P.; Ragonese, P.; Dugo, P.; Dugo, G.; Mondello, L. Offline LC-GCxGC in combination with rapid-scanning quadrupole mass spectrometry. J. Sep. Sci. **2008,** *31,* 3329–3336.

4

Road-Traffic Emissions

Fabian Heidegger,
Regine Gerike,
Wolfram Schmidt,
Udo Becker, and
Jens Borken-Kleefeld

Introduction

The transport sector is a major contributor of air pollution and greenhouse gas emissions worldwide. Governments around the world are under pressure to deal with the consequences of climate change while also striving to meet growing transport demands. The prime challenge is to find and apply measures that reduce the environmental impact of transport, i.e., emissions (including air pollution, greenhouse gases, noise and vibration), land use, separation effects, and effects on landscape.

This chapter focuses solely on the topic of emissions, which includes energy consumption, air pollutants and greenhouse gases of road traffic and their measurements, as well as the possibility to model and influence the reduction of emissions via traffic planning measures, traffic management, or economic policy and user-related measures. Off-road emissions from vehicles are not included in this article, nor are upstream or downstream processes (e.g., well-to-tank emissions). Noise also contributes to road-traffic emissions but is not discussed here; however, it is important to note that the impact of noise can have major consequences on humans and animals.

Particularly in emerging economies, the aviation and road transport sectors have contributed considerably to the increase in emissions worldwide (although the latter sector has a high potential for emission reduction). The problems caused by road-traffic emissions can be separated into three categories:

- Climate Impact (mainly CO_2 emissions): In 2016, the transport sector produced around 8.0 $GtCO_2$ (25% of the total emissions that year), which measured 71% higher than those in 1990 (International Energy Agency 2018); 74% of those emissions were produced by road transport (International Energy Agency 2018).

- Major Share in Oil Consumption: With 92% of transport-final energy demand being composed of oil products, the transport sector is the least diversified energy end-use sector (International Energy Agency 2017). Hence, limited fossil fuel resources ensure a large share of mobility needs, especially in rural but also in urban areas today. Despite future energy efficiency gains and a wider application of electric cars, fuel needs will be higher in 2040 than they are today as a direct result of an increase in car use (International Energy Agency 2018).
- Air Pollutant Emissions: Many substances directly impact human health; in particular, there is high emission density in urban areas, which drives cities to develop solutions such as anti-pollution schemes.

Emissions are released in gaseous or particulate form into the atmosphere and are subject to physical and chemical transformation processes as well as to the influence of meteorological parameters during the transmission process (from the source to the place with pollution damage). Depending on the location, pollution damage has an effect on human health, vegetation, soil, water, building materials, and the climate. These substances can be grouped by the following impact categories (VDI 2019):

- Climate relevant substances (e.g., CO_2, CH_4, N_2O, black carbon)
- Substances contributing to ozone formation (e.g., NO_x, non-methane hydrocarbons [NMHC], carbon monoxide [CO])
- Eutrophying substances (e.g., NO_x, SO_2)
- Acidifying substances (e.g., NH_3, NO_x)
- Toxic substances (e.g., black carbon, heavy metals, polycyclic aromatic hydrocarbons, CO)
- Carcinogenic substances (e.g., black carbon, heavy metals, lead [Pb], aromatic hydrocarbons, e.g., benzene [C_6H_6], polycyclic aromatic hydrocarbons).

In the context of emissions and pollutants, the substances PM_{10} and NO_x (NO and NO_2) present the biggest problems in transportation. Particulate matter (PM) includes particles <10 μm in diameter (PM_{10}) as well as particles <2.5 μm in diameter ($PM_{2.5}$) (Pfäfflin 2018). PM occurs in incomplete combustion processes or due to abrasion and resuspension. PM may cause respiratory and cardiovascular diseases, and the smaller the size of the particle, the more dangerous it is to human health. NO_x is formed during high-temperature and high-pressure combustion processes and can lead to respiratory disorders, inflammation, and bronchitis. With long-term exposure, the number of heart attacks and fatalities as a result of respiratory disease significantly increases (Pfäfflin 2018).

Table 1 illustrates thresholds for nitrogen dioxide (NO_2), PM_{10}, and $PM_{2.5}$ in Europe. It is clearly visible that the guideline values of the WHO (World Health Organization) are substantially lower than the legally effective values of the European Union.

The European Union legislation directed toward creating a framework for the compliance of motor vehicles and their components limits the mass and number of particles, NO_x, and additional air pollutants, as well as greenhouse gases for vehicles (DIRECTIVE 2007/46/EC 2007), (REGULATION No

TABLE 1 Thresholds for Nitrogen Dioxide (NO_2), PM_{10}, and $PM_{2.5}$

	NO_2 [μg/m$_3$]		PM_{10} [μg/m$_3$]		$PM_{2.5}$ [μg/m$_3$]	
EU	40	Annual mean of hourly measurements	40	Annual mean of hourly measurements	25	Annual mean of hourly measurements
	–	–	50	24-hour mean with 35 permitted exceedances	–	–
WHO	40	Annual mean of hourly measurements	20	Annual mean of hourly measurements	10	Annual mean of hourly measurements
	200	1-hour mean	50	24-hour mean	25	24-hour mean

Source: According to (DIRECTIVE 2008/50/EC 2008) and (WHO 2006).

715/2007/EC 2007). Mainly CO_2 emissions are relevant regarding the direct climate impact of road transportation. As greenhouse gases do not cause local environmental damages or injuries to health, there is no need for local greenhouse gas thresholds. In 2009, the European Union introduced mandatory targets for average CO_2 emissions of 130 g CO_2/km for 2015 and 95 g CO_2/km for 2021 to be applied to each newly manufactured vehicle fleet (REGULATION No 333/2014/EU 2014).

Road vehicles are certified according to exhaust emission standards. In the United States, the Environmental Protection Agency (EPA) manages emission standards nationally. The European Union has its own set of emission standards, which all new vehicles must meet: The latest *Euro 6* targets passenger cars, light commercial vehicles, and motorcycles, and the *Euro VI* is for large goods vehicles and coaches (DIRECTIVE 2007/46/EC 2007), (REGULATION No 715/2007/EC 2007). Many Asian countries adopted these European emission standards; however, since the specified driving cycles, e.g., the WLTC (Worldwide Harmonized Light Vehicles Test Cycle), do not adequately represent real driving performance, the official limit values for air pollutants[1] and greenhouse gases (CO_2) are below actual driving emissions (although the vehicles fulfill the requirements of the test) (Hooftman et al. 2018). Therefore, the official exhaust thresholds cannot be used for a true and reliable emission calculation. In recent years, the topic of air pollution has received remarkable attention from the public, as well as from environmental and consumer organizations. Legal actions have been enforced upon vehicle manufacturers (e.g., Volkswagen, Daimler, Fiat, and Chrysler) for emission frauds (i.e., not complying with the limit thresholds of their vehicles) and upon cities and governments due to noncompliance with air quality standards.

Reliable emission projections/inventories combined with an integrated and systematic reduction of emissions—preferably in the form of pro-active prevention—are greatly needed. Substantial energy use and high levels of exhaust emissions are an unwanted outcome of transport demand; these are a consequence of many individual and non-individual interconnected processes. Mode choice and route choice, for example, are behaviors which could be adjusted rather quickly toward more sustainable options, while the implementation of change on the density of transportation networks, transportation planning, land use, and other economic factors would require an extensive amount of time and planning. In addition, emission factors (how much each vehicle emits per km [g/km or g/s]) and fuel use factors [l/km] of the vehicles, depending on, for example, vehicle type choice, topography, and driving behavior, are influencing coefficients for exhaust emissions.

This chapter is organized as follows: "Modeling of Road-Traffic Emissions" section gives an overview of energy consumption, vehicle propulsion technologies, default parameters of emission calculation, and types of emissions; "Potential Influences on Emission Reduction and Their Impacts" section introduces the modeling of road-traffic emissions by presenting different approaches and model types, followed by an example of an emission model and validation measures of emissions; and "Conclusion" section is dedicated to the potential influences for emission reduction and their impact, which can also be found in anti-air pollution schemes and climate action plans of cities and countries.

Energy Consumption and Types of Emissions

Default Parameters for Emission Calculation

Figure 1 summarizes the default parameters (all boxes on top) for determining road-traffic emissions. Colored (straight) arrows indicate effects to the type of emissions (all boxes in the middle); for example, the roughness of road surface will have an impact on abrasion and resuspension emissions. Moreover, vehicle fleet composition and traffic mileage affect the sum of all traffic emissions (box at the bottom). Resultant effects are shown as black (arcuated) arrows which close the circuit in a mutually reinforcing manner: An increase in the release of emissions directly affects climate change and meteorological conditions, just as

[1] True notably for NO_x (from diesel cars), but not necessarily for other regulated pollutants.

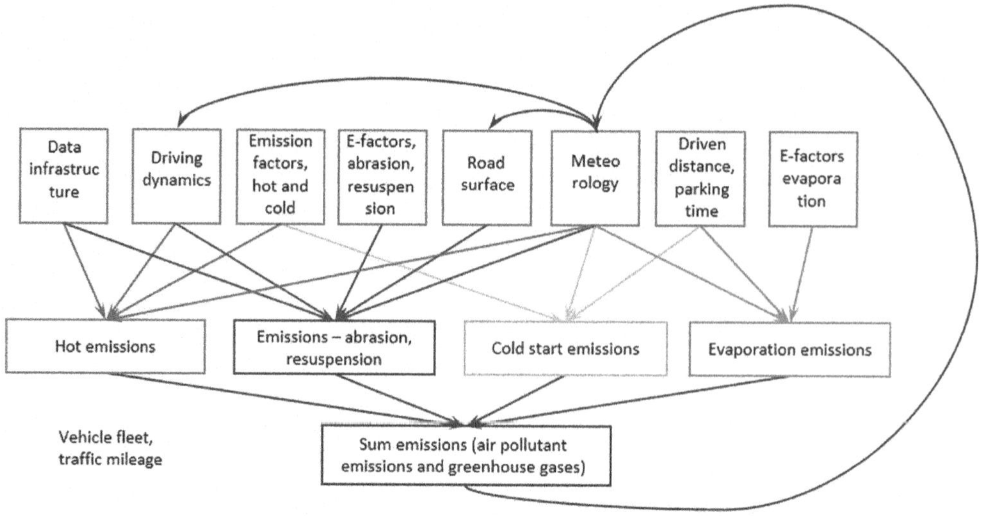

FIGURE 1 Default parameters (all boxes on top) for determining road-traffic emissions with resultant effects.

weather conditions affect driving dynamics (driving behavior) and road surfaces. Though not shown in Figure 1, emissions from trains, ships, airplanes, power stations and industry, domestic fuels and small businesses, mobile sources, agriculture, and biogenic sources are also part of the circuit.

Basics of Driving Dynamics

A vehicle moves or accelerates by converting externally added energy (fuel, electricity) into kinetic energy. When braking, kinetic energy is transferred to thermal energy. Some vehicle concepts are also able to recuperate and store electric energy for subsequent vehicle movement.

The level of energy consumption of a vehicle is dependent on the kinetic energy demand (to overcome driving resistances) as well as on the degree of efficiency of the engine and the use of auxiliary units such as air conditioning or lighting.

Kinetic energy demand arises from vehicle parameters and traffic situations (Pfäfflin 2018). Vehicle parameters include

- Acceleration Resistance: Overcoming vehicle mass inertia (including loading) and rotating masses (wheels, gearboxes, engine)
- Air Resistance: Force of displacement depending on vehicle size, drag coefficient, and speed of the vehicle
- Rolling Resistance: A result of tire contact with the road, the inner frictional resistance, and the weight force
- Gradient Resistances: Dependent on vehicle mass and topography.

During low speeds or stop-and-go traffic situations, the energy demand for the acceleration of vehicle mass is the decisive parameter. With a constant driving speed, rolling resistance and air resistance—especially at higher speeds—are the dominating factor for energy demand.

Energy Conversion in the Vehicle

As previously mentioned, kinetic energy is generated in either a combustion or an electric engine. Electric engines have a higher degree of efficiency than combustion engines. A hybrid vehicle obtains its energy from an electric engine and a fuel-based engine. Combustion engines are characterized by a cyclic, non-stationary, and non-optimal combustion process (Pfäfflin 2018). For a comparable vehicle type with similar engine power, the diesel engine has an advantage in terms of fuel consumption as compared to the gasoline engine. The spark ignition of the diesel leads to a more energy-efficient yield of fuel: This combustion process results in less fuel consumption and lowers CO_2 emissions; however, rising combustion temperatures lead to greater NO_x emissions.

Types of Emissions

The following is a list of the main types of emissions, divided by their components, as well as the chemical reactions in which they are released:

- Ammoniac (NH_3) and Dinitrogen (N_2O)—Released through catalytic converters
- Lead (Pb)—If the fuel contains lead
- Carbon Monoxide (CO) and Carbon Dioxide (CO_2)—During the combustion of fuel
- Hydrocarbons (HC)—During the combustion and evaporation of fuel
- PM (PM_{10}/$PM_{2.5}$)—In incomplete combustion or abrasion and resuspension
- Sulphur Dioxide (SO_2)—During the combustion of sulfurous fuel
- Nitrogen Oxide (NO_x)—During combustion process, especially from diesel engines.

There exist further emissions in road traffic, e.g., heavy metals, however, their effects on humans and the environment have yet to be sufficiently researched.

The following is a list of the main types of emissions, divided by point of origin:

- **Hot/Warm**

  These are emissions (CO, CO_2, NO_x, PM, HC, etc.) released when the engine and the exhaust reduction systems are hot, i.e., have reached their ideal operating temperature. Furthermore, hot emissions depend on traffic situations. Except for PM, hot/warm emissions have the highest share of total emissions on national inventories. Hot/warm emissions are given in units of [g/km].

- **Cold Start**

  When a vehicle has been parked for a minimum of 12 hours before the start of the engine, engine and exhaust reduction systems have reached cold start/ambient temperature; these technical systems are not at an ideal operating temperature. Upon start, an optimal operating temperature is first attained after a few kilometers. This produces cold-start supplement emissions, especially carbon monoxide (CO) and hydrocarbons (HC). Cold-start emissions are given as excess emissions (in [g/start]). A cool start arises in standstill times between 0.5 and 12 hours.

- **Abrasion and Resuspension**

  Abrasion from tires, brakes, clutch, and road surfaces are sources for non-exhaust particles (PMs). In addition, resuspension of generated and deposited particles from road surfaces occurs as a result of vehicle-induced airflow and wind. Technical improvements have reduced engine-based particles; thus, the emissions from abrasion and resuspension can exceed exhaust emissions.

- **Evaporation**

  Evaporative emissions in the form of HCs can be distinguished as follows:

  - Hot/warm soak emissions, given in [g/stop]—These emissions occur after switching off the engine when the engine is still hot or warm.
  - Diurnal evaporative emissions per vehicle given in [g/day].
  - Running losses—Generated as a result of vapor in the fuel tank during vehicle operation, given in [g/km].

Modeling of Road-Traffic Emissions

Overview of Approaches

In general, there are two different approaches for calculating road-traffic emissions for air pollutants and greenhouse gases: the top-down approach and the bottom-up approach. The selection of the suitable approach depends on the acceptable tolerance limit.

In the top-down approach, emissions are calculated based on "fuel sold" or "fuel used" (not including fuel exports to other countries) (European Environment Agency 2016). This approach is a simple means for calculating the (national) total quantity of emissions and is well suited for large-scale areas (in which the volume of fuel sold/fuel used within the country is known and inventoried), without, however, the use of temporal or spatial differentiation parameters or vehicle data. Moreover, the fuel-based approach is exact for CO_2 emissions (calculated from fuel quantity) but only approximate for some air pollutants.

All emission types for specific street sections as well as for partial or complete road networks on the meso- and microlevels can be calculated only with a bottom-up approach for all temporal and spatial differentiations on the basis of vehicle data, i.e., by means of (agent-based) traffic and emission models or by modeled vehicles. These two bottom-up approaches try to imitate or model driving behavior. Therefore, all the following remarks relate to the bottom-up approaches based on vehicle data only.

The bottom-up quantification of emissions can be implemented in different degrees. Potential spatial and temporal resolutions depend on the detail level of (available) input variables. Ericsson (2000) identifies cause–effect relationships (partly also interacting with each other) that influence variability of driving patterns: driver factors, vehicle factors, weather factors, traffic factors, street environment factors, and travel behavior factors (see Figure 2). Figure 2 allows a more detailed perspective into the top boxes of Figure 1 that can be used in a bottom-up approach.

Fontaras et al. (2017) add to Figure 2 driving style and vehicle maintenance (for driver factors), wind (for weather factors), aerodynamic and rolling resistance and auxiliary systems (for vehicle factors), road grade (for street environment factors), and none for travel behavior factors and traffic factors.

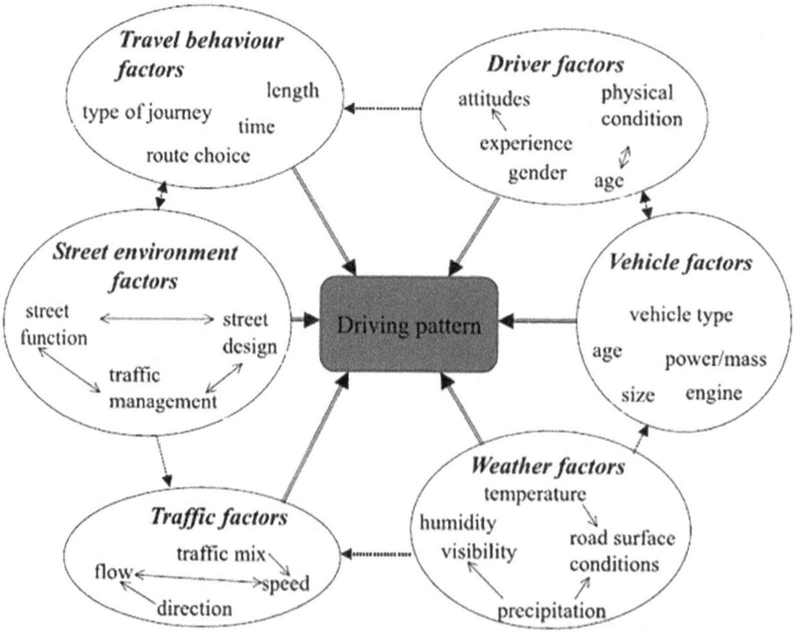

FIGURE 2 Cause–effect model of variability in driving patterns. [According to Ericsson (2000).]

Two technical possibilities for measuring emissions and one for modeling emissions are known. Emissions can be measured as a function of time or distance (known as emission factors):

- Either by a Portable Emission Measurement System (PEMS) at the exhaust of a test car that moves in traffic (exact, but costly)
- Or emissions can be determined in the laboratory with engine test facilities or chassis dynamometers. The necessary engine power to overcome driving resistances and the engine speed (according to a gearshift model) generate emission maps which allow for the calculation of emissions.

Emission models provide emission factors representative of vehicle classes/layers under specific driving situations. The model approach is ideal for the emission determination of large traffic networks, as it is less complex than individual measurements on each street. A traffic-supply model, a traffic-demand model, and an emission model are coupled. Data components for each model are discussed in the following sections.

Traffic-Supply Model

A traffic-supply model includes all infrastructure data of the road network (e.g., length of road sections, number of lanes, speed limits, gradients, traffic lights, and road surfaces). The result is a geo-referenced static traffic model allowing for the mapping of emissions back to their place of origin.

Traffic-Demand Model

The factors for the specification of driving dynamics (see Figure 2) are bundled in a traffic (demand) model. Traffic volume, traffic mileage (traffic volume multiplied by road section length), traffic flow, traffic composition, and traffic quality all serve as input variables for the energy demand of vehicles and emissions. Traffic volume is modeled either as Average Annual Daily Traffic (AADT) based on an average day of the year or in a higher temporal resolution with traffic flow by means of time series of a day, a week, or a year. Traffic flow is the number of vehicles which cross a specific road section over a certain period of time.

Traffic factors have to be quantified by indicators to make them useful for emission calculations in the study area. Ericsson (2000) identifies three parameter categories, which represent changes in driving patterns and have a high impact on fuel use and emissions:

- Level Indicators: Level measures describe, e.g., (average) speed, vehicle kilometers traveled, and number of vehicles. However, no information is given about variation or frequency of variation.
- Oscillation Indicators: Oscillation measures describe frequency, e.g., relative positive acceleration (RPA). RPA is defined as (see Eq. 1)

$$RPA = 1/x \int v * a^{+} \, dt \tag{1}$$

where x = total distance, v = speed, a^+ = (positive) acceleration, and t = time.

- Distribution Indicators: Distribution measures indicate the distribution of a variable in a certain section, e.g., the percentage time share of acceleration ($a > 1.5\,m/s^2$).

According to Ericsson (2000), the five most important parameters for fuel consumption are

- Relative positive acceleration
- Number of accelerations per 100 m
- Percentage of time with $a > 1.5\,m/s^2$
- Percentage of time with $-2.5\,m/s^2 < a < -1.5\,m/s^2$
- Percentage of time with $v < 15$ km/h.

Data sources for quantifying traffic factors can be traffic detectors, radar, video detection, Floating car data (FCD), short-time counting, traffic messages, and data from police and traffic models with origin–destination matrices and assignment processes.

Traffic-Emission Model Types

The literature encompasses two general bottom-up approaches for calculating vehicle emissions by means of a traffic model: the macroscopic modeling approach with aggregated network parameters, which has rather low accuracy but enables faster computation of emission estimates, and the microscopic modeling approach with instantaneous vehicle emission rates, using either vehicle engine or vehicle speed/acceleration data (Jiang et al. 2018). According to Smit et al. (2010), there are five different types of traffic-emission models (see Table 2) with different input data and different interfaces between traffic volume and emission factors.

The first step for creating the model is to allocate road infrastructure data and traffic activity data into different categories (e.g., point sources, line sources, area sources) and into different temporal resolutions which refer back to each road section by means of geographical (interpolation) approaches and by assignment processes in traffic models. The second step is to select the emission factors of the model type that correspond as accurately as possible to the merged data of the road section. Subsequently, activity data (*A*) per road section are multiplied with specific emission factors (*EF*) per vehicle category, where E = emission (see Eq. 2):

$$E = A * EF \tag{2}$$

The total amount of emissions of a road network results in the sum of all mentioned vehicle emissions that are released on the road sections.

An emission model implies classed emission factors per substance, country, and year for all vehicle categories. These vehicle categories exist for the bottom-up approach (VDI 2019):

- Cars
- Light duty vehicles (≤ 3.5 tons permissible total weight, e.g. small buses, camper vans)
- Heavy duty vehicles (> 3.5 tons permissible total weight, e.g., trucks, truck trains, semitrailer trucks)
- Buses (regular buses, coaches)
- Motorized two-wheelers (motorcycles and mopeds)
- Per vehicle layer/segment: Vehicle category with the same propulsion technique and same size (mass or engine capacity) with the same emission performance to make emission calculation more manageable.

TABLE 2 Different Types of Traffic-Emission Models

Model Type	Input Data	Source	Example Models
Average speed models	Mean traveling speed	Traffic models or field measurements	COPERT, MOBILE or EMFAC
Traffic situation models	Vehicle kilometers traveled per traffic situation	Traffic models	ARTEMIS or HBEFA
Traffic variable models	Traffic-flow variables	Macroscopic and microscopic traffic models	TEE or Matzoros
Cycle variable models	Function of driving cycle variables at high resolution (seconds to minutes)	Obtained from vehicle movements (equipped with GPS) or microscopic traffic models	MEASURE or VERSIT+
Modal models	Function of driving cycle variables	Obtained via engine or vehicle operating models at the highest resolution (one to several seconds)	PHEM, CMEM, or VeTESS

Source: Own Representation Based on Smit, Ntziachristos, and Boulter (2010).

Vehicle-specific components depend predominantly on the following:

- Vehicle size (weight and motorization)
- Drive concept (induced engine [spark ignition engine], or autoignition [diesel engine], natural gas, hybrid)
- In-Engine technology-based reduction measures (e.g., exhaust gas recirculation [EGR]) or exhaust gas after-treatment systems (particulate traps, selective catalytic reduction [SCR], lean NO_x trap, three-way catalytic converter)
- Gear ratio
- Air resistance
- Rolling resistance of tires
- Vehicle conditions (age, mileage, maintenance).

Operational-specific components depend predominantly on the following:

- Engine and exhaust gas temperature
- Gear selection
- Operation of accessories (e.g., air conditioner, alternator)
- Cargo load
- Fuel quality
- Gradient of road
- Ambient temperature
- Height above sea level
- Driving behavior.

To manage the multiple variability of emission calculations, vehicles with similar technical attributes are summarized into the abovementioned vehicle categories with similar consumption and emission behavior, and then, emission factors are attached. The lowest level of detail are vehicle layers. Each vehicle is considered homogenous with respect to its emission characteristics and reasonably distinct from any other layer. Layer criteria are the abovementioned vehicle-specific components. The share of each vehicle layer in the vehicle category changes over time and place due to fleet modernization; therefore, the emission factors are updated yearly for the whole vehicle fleet. Traffic-mileage-related emission calculations need fleet compositions which are weighted accordingly, as average yearly traffic mileage may vary within the vehicle layers (e.g., cars with diesel engines have a higher traffic mileage than cars with gasoline engines).

The HBEFA Traffic Situation Model

The Handbook of Emission Factors for Road Transport (HBEFA) provides a traffic situation model which can be used for large-scale emission calculations while also taking driving dynamics into consideration.

HBEFA maps a comprehensive range of real driving specifications through 1190 classed driving cycles, simulated with the emission model Passenger Car and Heavy Duty Emission Model (PHEM). Kinematic parameters (velocity, percentage of standstill time, and RPA) are considered by traffic situations. They are differentiated by

- Area (urban area/rural area)
- Road type (motorway, arterial road, collecting roads, etc.)
- Speed limit
- Traffic conditions/levels of service (LOS): Fluent, dense, saturated, stop and go, etc.

HBEFA model generates emission factors for all current vehicle categories and concepts for different countries (Germany, Austria, Switzerland, Sweden, France, and Norway).

Figure 3 shows NOx emission factors for a HBEFA traffic situation with examples given for different vehicle categories, such as diesel cars, gasoline cars, light duty vehicles, heavy goods vehicles, and

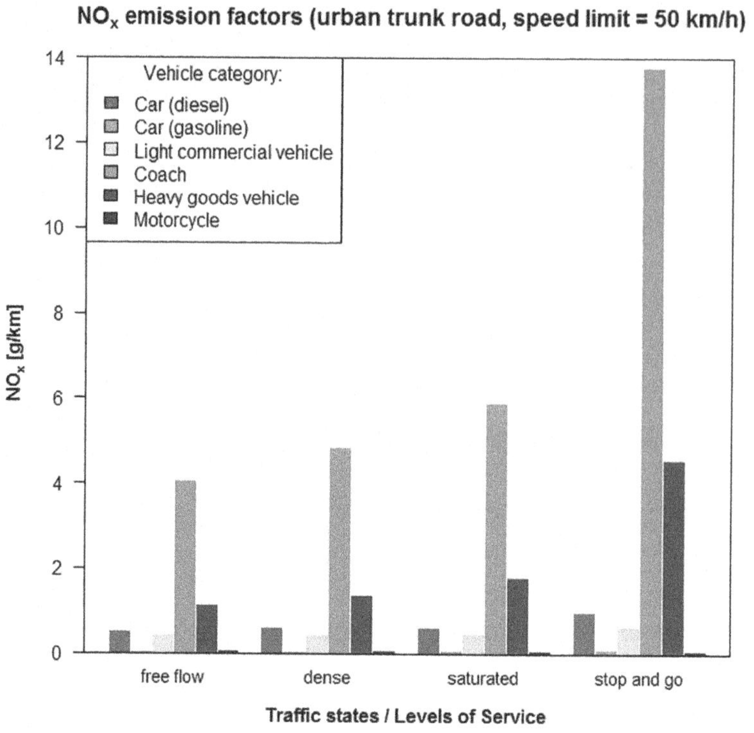

FIGURE 3 Emission factors produced by the HBEFA traffic situation model (urban area/speed limit = 50 km/h on a trunk road) for an average German car in 2019. (INFRAS 2017.)

motorcycles (all from the 2019 German vehicle fleet; ordered from left to right in figure 3), on an urban trunk road with speed limit = 50 km/h without gradient for different LOS (INFRAS 2017).

Validation of Emissions

According to Smit et al. (2010), the calculated emissions can be validated at different spatial scales (local, road, journey, area) by means of six different methods:

1. In *laboratory measurements*, driving conditions can be imitated under controlled parameters
2. *Onboard measurements* are able to test real driving conditions of a vehicle.
3. *Tunnel validation measures* at the tunnel entrance and exit, linked with tunnel features, e.g., tunnel airflow, cross-sectional areas, road length, and traffic density, measure emissions from a large sample of the on-road fleet.
4. Compared to the tunnel validation with relatively "controlled" conditions, *ambient concentration measurements* require results from combined emission and dispersion modeling (including background concentrations from households, industry, agriculture, and natural resources).
5. In the *remote sensing method*, an infrared/ultraviolet beam is directed from a pylon across the road to measure instantaneous ratios of various pollutants of a large sample of the fleet.
6. *Ambient mass balance studies* specify emission fluxes (kg/h) through the measurement of pollutant concentrations upwind and downwind of certain areas at different heights to compare these data with area-wide emission prognoses by the model for the same period.

In addition to the above methods, Poehler et al. (2019) use the *plume chasing method* to measure real vehicle emissions of a vehicle with a testing vehicle driven behind. The ratio of a pollutant (e.g., NO_x) to

CO_2 is independent from the dilution of the exhaust plume after correction of the background concentration. The ratio enables the calculation of the emission values of the vehicle.

Potential Influences on Emission Reduction and Their Impacts

As mentioned in "Energy Consumption and Types of Emissions" section, traffic demand and the resulting environmental impacts are embedded in complex interdisciplinary relationships. Therefore, systematic emission-reduction measurements, preferably implemented as pro-active prevention approaches, are needed—for example, by covering the dimensions of the four "E's". These include

- Engineering (e.g., technical reduction possibilities, traffic planning measures)
- Encouragement/Economy (e.g., pricing measures, incentives to change traffic behavior)
- Education (e.g., pedagogical and communicative measures, such as the promotion of environmentally-friendly modes)
- Enforcement (e.g., legislative measures, control, and monitoring).

The four "E's" merge in a bundle of vehicle-related, infrastructural, traffic-planning-related, traffic-management-related, and user-related measures, in which the impacts increase as more measures are implemented together (Schlag 1998).

Vehicle-Related Measures

Undesirable secondary outcomes emerge from combustion processes while driving. Their amount can be reduced by using high-quality fuel and in-engine technology as well as through the utilization of five different after-treatment systems of exhaust gases, as follows:

1. A *three-way catalytic converter* (for suction pipe gasoline engines) oxidizes the air pollutants carbon monoxide (CO), hydrocarbons (HC), and nitrogen oxides (NO_x) to CO_2 and inert N_2, if the exhaust temperature is sufficiently high.
2. An *oxidation catalytic converter* (for diesel engines and gasoline engines with direct injection) oxidizes HC, CO, and soluble particles.
3. *SCR* reduces nitrogen oxides with ammoniac (NH_3)—based on an aqueous urea solution—to nitrogen (N_2) and water (H_2O).
4. In *particulate reduction systems*, the exhaust gases flow over porous surfaces in open or closed filter systems.
5. *EGR* mixes exhaust gases back to the combustion chamber to reduce thermic NO_x (Pfäfflin 2018).

Infrastructural, Traffic Planning, Traffic Management-Related Measures

Traffic-planning measures—often coupled with traffic-management measures—reveal steering effects on traffic flows, which usually include, in addition to primary local (desired) impacts, secondary (unwanted) effects on the transportation system (e.g., induced traffic[2] [increasing traffic], spatial shift of traffic flows, etc.). Thus, traffic-planning measures are typically valued on a case-by-case basis. Often, the effectiveness of a planning measure is also dependent on framework conditions and the type of chosen implementation. Pfäfflin (2018) has outlined seven typical planning measures as follows:

1. *Measures of improvement of traffic efficiency/decongestion* include, for example, capacity expansions or construction of roundabouts (consider maximum thresholds of traffic volume) from an infrastructure perspective. From a traffic management perspective, traffic control centers can implement environmentally oriented traffic management based on the following:

[2] Hills (1996) explains the usage of this term in the context of transportation.

- Traffic models for an interpolation or prognosis of traffic scenarios and strategies
- Data archives to recognize traffic behavior patterns and causal relations
- Strategy management systems to run predefined strategies.

Traffic control centers use Intelligent Transportation Systems (ITS); these include sensor systems for traffic monitoring, devices for transfer systems for traffic incidents (e.g., road works, accidents, events), devices to control traffic flow (e.g., [virtual] variable message signs), and further information devices for transportation users. Further traffic management scenarios are optimization of traffic lights (green waves), ramp metering, dedicated lanes for vehicle with high occupancy (HOV lane), or high-occupancy toll lanes (HOT lanes), in which high-occupancy vehicles are without charge, and other vehicles are required to pay a variable, demand-adjusted fee.

All these measures lead—under the assumption of same boundary conditions—to an improvement in local traffic quality with less stop and go and, thus, less emissions; however, rebound effects must be considered. Lower travel times induce traffic (either from changed-route choice or from modal shift from public transport) which may lead to an increase in the total emissions in the network than before the implementation of the traffic planning measure.

2. *Construction of new bypass roads* leads to traffic shifts and/or induced traffic with a shift or increase of emissions. Dismantling of main through-roads reduces the risk of the abovementioned rebound effects.
3. *Access and transit prohibitions* restrict temporarily or permanently an entry of certain vehicle categories or vehicle groups to specified roads or areas, e.g., implemented as environmental zones or city-center tolls. An environmental zone can lead to an acceleration of vehicle-fleet modernization with less specific emissions—also outside of the environmental zone. In contrast, traffic shifts and a rise in vehicle kilometers traveled (due to longer trips to detour around the environmental zone) can lead to increased emissions.
4. *Parking management* includes the dismantling or creation of parking spaces as well as the establishment of parking guidance systems. Parking guidance systems reduce the emissions created by traffic in search of parking; yet, at the same time, the attractiveness of parking spaces has the counter effect of induced traffic (increasing traffic). Consistent parking-space management zones of sufficient size with homogenous prices will reduce variable attractiveness between parking options and, thus, avoid extended searches for parking.
5. The *reduction of permissible maximum speed* is a measure that does not directly affect emissions but rather driving behavior. A maximal reduction in emissions can be seen in the mid-speed range of 60–80 km/h. Emissions increase at faster speeds due to higher engine power demand; they also increase at speeds below 50 km/h as a result of external factors such as traffic control, traffic signal settings, and gradient. Therefore, an overarching statement concerning permissible maximum speed is not possible.
6. The *attraction of environmentally-friendly modes* includes modes of transportation with zero or less emissions than motorized traffic, e.g., walking, cycling, public transportation, or trains. Measures that increase attractiveness of these modes include a safe infrastructure for non-motorized traffic, an increase in network density and frequency, barrier-free vehicles, public relations (information and communication, marketing, consultations), educational measures (e.g., biking safety classes), carsharing, ridesharing, and intermodal mobility stations (mobility as a service). For the most part, reducing environmental impacts can only be achieved when pull measures are combined with push measures of motorized transportation (reduction of parking spaces, increase of parking prices, etc.).
7. *Measures without secondary effects to traffic quality and traffic volumes* include construction of noise-protection barriers for lowering emissions, planting vegetation, or installing photocatalytic surface coatings for the reduction of NO_2 concentrations. The reduction potential is very low compared to other measures.

User-Related Measures: Internalizing External Costs of Environmental Impact, Mobility Management, and Land Use Planning

The abovementioned traffic-planning measures have a predominantly local or regional impact on transportation. International policies must set incentives for all emission sectors to develop and implement a common cooperation strategy for reducing and eliminating emissions. This can be achieved through the realization of a climate-friendly legal framework which follows the *true-cost principle* (i.e., usage-related charges with no external effects on other regions, people, or time frames) combined with municipal and operational *mobility management* (a target-oriented approach for influencing individual mobility behavior by perceiving, revealing, and assessing mobility options for traffic prevention). A climate-friendly transportation policy and legislation also encourages the development of livable, green, and healthy cities / (small) towns composed of the following:

- Diversity: Buildings and public spaces serve multiple purposes, such as for employment, shopping, or recreation.
- Density: Critical for short distances and the (public) transit system, density concentrates trip origins or destinations, allowing for ride sharing to become practical as well as economical.
- Design: This provides an accessible, barrier-free, safe, and rewarding infrastructure with an easy-to-understand orientation system. Streets would not only have link functions (which minimize travel times) but also have place functions (which maximize the time traffic participants spend for social interaction in public spaces) (Cervero and Kockelman 1997).

A successive realization of the true-cost principle in transport with social compensation measures leads to traffic behavior changes, sustainability, and a shift toward eco-conscious transportation.

Conclusion

Transport emissions are a significant contributor to total emissions worldwide; in particular, road-traffic emissions have been increasing, and this trend will continue in the future. The substances PM_{10} and NO_x (NO and NO_2) present the greatest problems in transport, with the resultant CO_2 contributing directly to global warming. Fuel-based vehicles are the majority, but electric vehicles gain market shares. Emissions arise from combustion processes, abrasion, resuspension, and evaporation. Either the emissions can be measured with test vehicles or the emissions can be modeled with traffic and emission models for large networks. Several methods exist on different spatial scales to validate emissions. Emissions can be reduced most effectively with an integrated approach consisting of vehicle-related, infrastructural, traffic-planning-related, traffic-management-related, and user-related measures. (Rebound effects must also be considered.) Traffic and emission models can be used for environmentally oriented traffic management. A realization of the true-cost principle will support a shift toward eco-conscious transportation.

References

Cervero, Robert, and Kara Kockelman. 1997. "Travel demand and the 3Ds: Density, diversity, and design." *Transportation Research Part D*, 9: 199–219.

DIRECTIVE 2007/46/EC. 2007. DIRECTIVE 2007/46/EC OF THE EUROPEAN PARLIAMENT AND OF THE COUNCIL of 5 September 2007 establishing a framework for the approval of motor vehicles and their trailers, and of systems, components and separate technical units intended for such vehicles. https://eur-lex.europa.eu/legal-content/EN/TXT/PDF/?uri=CELEX:32007L0046&qid=1561907271004&from=EN.

DIRECTIVE 2008/50/EC. 2008. DIRECTIVE 2008/50/EC OF THE EUROPEAN PARLIAMENT AND OF THE COUNCIL of 21 May 2008 on ambient air quality and cleaner air for Europe. https://eur-lex.europa.eu/legal-content/EN/TXT/PDF/?uri=CELEX:32008L0050&from=EN.

Ericsson, Eva. 2000. "Variability in urban driving patterns." *Transportation Research Part D*, 5: 337–354.
European Environment Agency. 2016. EMEP/EEA air pollutant emission inventory guidebook 2016 – Technical guidance to prepare national emission inventories, Vol. 21.
Fontaras, Georgios, Nikiforos-Georgios Zacharof, and Biagio Ciuffo. 2017. "Fuel consumption and CO2 emissions from passenger cars in Europe – Laboratory versus real-world emissions." *Progress in Energy and Combustion Science*, 60: 97–131.
Hills, Peter J. 1996. 'What is induced traffic?" *Transportation*, 2: 5–16. https://link.springer.com/article/10.1007/BF00166216.
Hooftman, Nils, Maarten Messagie, Joeri Van Mierlo, and Thierry Coosemans. 2018. "A review of the European passenger car regulations – Real driving emissions vs local air quality." *Renewable and Sustainable Energy Reviews*, 86: 1–21. doi:10.1016/j.rser.2018.01.012.
INFRAS. 2017. Handbook of Emission Factors for Road Transport, Version 3.3. www.hbefa.net.
International Energy Agency. 2018. CO2 emissions from fuel combustion – Highlights. Comp. Fatih Birol.
International Energy Agency. 2017. Energy Technology Perspectives 2017 – Catalysing Energy Technology Transformations. Comp. Fatih Briol.
International Energy Agency. 2018. World Energy Outlook 2018. Comp. Fatih Birol.
Jiang, Yan-Qun, Pei-Jie Ma, and Shu-Guang Zhou. 2018. "Macroscopic modeling approach to estimate traffic-related emissions in urban areas." *Transporation Research Part D*, 5: 41–55.
Pfäfflin, Florian. 2018. Hinweise zu Energie, luftbezogenen Emissionen und Immissionen im Straßenverkehr H EEIS. Translation of Title: Information on energy, air-related emissions and pollution in road traffic. Arbeitsgruppe Straßenentwurf (Working Group Road Design). Cologne, Germany: Road and Transportation Research Association (FGSV).
Poehler, Denis, Tobias Engel, Uli Roth, Joscha Reber, Martin Horbanski, Johannes Lampel, and Ulrich Platt. 2019. "Messung realer Fahrzeugemissionen mit dem 'Plume Chasing'-Verfahren [Measurement of Real Vehicle Emissions using the 'Plume Chasing' Method]." *Kolloquium Luftqualitaet an Straßen* 2019 (*Colloquium Air Quality at Roads* 2019). Bergisch Gladbach, Germany: Federal Highway Research Institute (bast), 63–78.
REGULATION No 333/2014/EU. 2014. REGULATION (EU) No 333/2014 OF THE EUROPEAN PARLIAMENT AND OF THE COUNCIL of 11 March 2014 amending Regulation (EC) No 443/2009 to define the modalities for reaching the 2020 target to reduce CO2 emissions from new passenger cars. https://eur-lex.europa.eu/legal-content/EN/TXT/PDF/?uri=CELEX:32014R0333&from=EN.
REGULATION No 715/2007/EC. 2007. REGULATION (EC) OF THE EUROPEAN PARLIAMENT AND OF THE COUNCIL of 20 June 2007 on type approval of motor vehicles with respect to emissions from light passenger and commercial vehicles (Euro 5+6) and on access to vehicle repair and maintenance information.
Schlag, Bernhard. 1998. "Zur Akzeptanz von Straßenbenutzungsgebühren, Translation of Title: Acceptance of road pricing." *Internationales Verkehrswesen*, 50 (7/8): 308–312.
Smit, Robin, Leonidas Ntziachristos, and Paul Boulter. 2010. "Validation of road vehicle and traffic emission models – A review and meta-analysis." *Atmospheric Environment*, 44: 2943–2953.
VDI. 2019. "Umweltmeteorologie – Kfz-Emissionsbestimmung – Luftbeimengungen [Environmental meteorology – Determination of the emission from motor vehicles – Air quality]." VDI-Richtlinien (VDI Guidelines), Fachbereich Umweltmeteorologie (Department Environmental Meteorology), Verein Deutscher Ingenieure (The Association of German Engineers).
WHO. 2006. WHO Air quality guidelines for particulate matter, ozone, nitrogen dioxide and sulfur dioxide – Global update 2005 – Summary of risk assessment. World Health Organization.

II

COV: Comparative Overviews of Important Topics for Environmental Management

5

Alternative Energy

Bernd Markert, Simone Wuenschmann, Stefan Fraenzle, and Bernd Delakowitz

Introduction

Regenerative resources like wind and solar energies, running and falling water (hydropower), biomass processed in some way and geothermic energy are in a position to replace crude oil, natural gas, coal, and uranium step by step depending on both technological innovations and political decisions. Here, technological innovations are not meant to require fundamental inventions to render some renewable resource useful for energy (electric current) "harvesting." Rather, this is about making existing technologies cheaper and overcoming specific material problems, like device corrosion and fast "blocking" of underground heat exchange pathways with geothermic energy, storage in an easy-to-handle form (methanol?) for both solar energy and biomass, or using less toxic and less brittle semiconductors in much thinner layers in photovoltaics [thin-film cells based on either copper indium dichalcogenides (chalcopyrites) or organic semiconductors].

Given the relativeness of time and the notorious "difficulty to make predictions which refer to the future," what does it say about classical fossil resources running out in the foreseeable future? Putting this into proper context means to distinguish between "reserve" and "resource": "reserve" just encompasses those deposits of energy carriers that are actually known and can be really accessed following both technical and economic criteria. In the pre-1973 world of %2.70 per barrel of crude oil (some $19 per ton), it would have been considered a fancy to try to extract oil from shales or sand and drill deep below the seafloor (now always performed in the Atlantic Ocean and the Mexican Gulf, off Brazil and Angola) or in remote arctic regions. Currently, the meaning of underground hard coal production or making access to very deep natural gas deposits is doubtful. While oil production from oil sands (Alberta Province, Canada) is now economically viable—but still an ecological disaster in a sensitive surrounding—other methods of accessing certain fossil resources will probably never be viable: it simply takes more energy to extract and process traces of ^{235}U dissolved in seawater than can be obtained from its fission afterwards. As the term "reserves" by definition ("share of total potential which can be mined and exploited economically reasonable by currently available technical means") includes both the present level of technology and current pricing, reserves are subject to changes other than due to ongoing prospection and exploitation. For example, when disregarding the environmental and safety issues associated with either kind of fossil energy carrier for this moment, both oil sands and shale gas are close to the lower limit of economic feasibility given the current crude and natural gas prices. However, available amounts are not settled with shale gas, not permitting to include it into the reserves, while oil sands—although a blueprint for ecological disaster—are economically and technically feasible to produce and hence are part of (Canada's and global) crude reserves. Matters are different once again if the technology to actually obtain energy from some fossil or far-spread reservoir around is not yet at hand or it is doubtful whether energy required to mine and concentrate that particular source might even exceed the energetic payback obtained thereafter: consider amounts/concentrations of deuterium, ^{6}Li, and ^{235}U in seawater. The former two (^{6}Li being a precursor to fusion "fuel" tritium) would combine to an inexhaustible source of energy if "only" net-energy-yield nuclear fusion would be at hand already, while there are just some 15 ng (!) of ^{235}U that can be actually extracted from 1 L of ocean water, providing a few hundred joules ($\approx 10^{-4}$ kWh) of electric energy in nuclear (now, fission) power plants thereafter. Even though the corresponding extraction was already demonstrated in Japan, using ion-exchange resins, the energy for producing the resins and running the device is so large that ocean-derived uranium is not a viable resource either and thus cannot be counted among the reserves.

"Resources," on the other hand, are those energy carriers either already discovered (not a single "elephant field" of crude oil was spotted after the 1960s anywhere in the world!) or *reasonably believed* to exist from geological arguments in Earth's crust, but cannot be exploited right now for either technological or economic reasons. Due to the difficulties and risks associated with drilling either below some 9 km underground or into active magma regions, most of the huge amounts of geothermic energy can never be actually used. Hence, for fossil resources as well, we are left with what is in the crust or the ocean (floor). Given this distinction, the residual economic lifetimes for traditional forms of energy carriers such as oil, natural gas, brown and hard coal, or uranium given in Figure 1 are obtained.

Energy Depletion of Fossil Fuels

Counting from year 2000 onwards, the German Federal Institute for Geosciences and Natural Resources estimated in 2002 that reserves of oil, natural, gas and uranium will last for just another 40 to 65 years. Reserves of both hard and brown coal will still last up to 200 years, whereas the resources of coal and gas and uranium are to last more than 200 years. Things are more critical with crude oil, reserves of which will be gone within 60 years while resources are estimated to last for 160 years at best. Data and predictions in Figure 1 do not cover and include the energy consumption of current growth regions [BRICS states like Brazil, Russia, India, China (PR), and other Latin American countries], leaving us with the

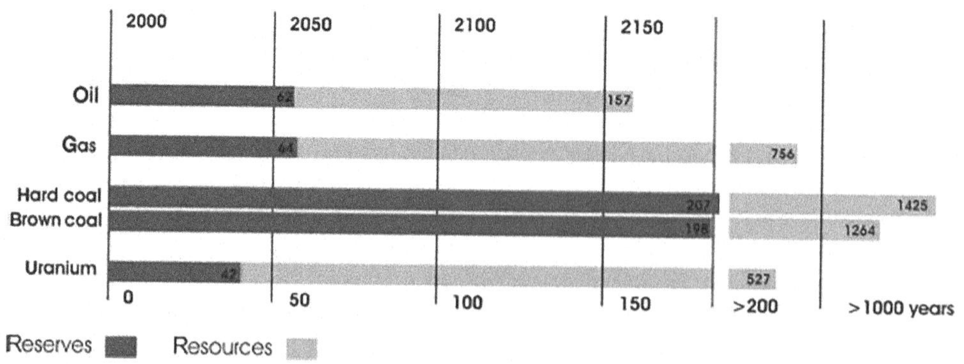

FIGURE 1 Predicted reserves/resources of fossil energy carriers from year 2000 onwards.
Source: German Federal Institute for Geosciences and Natural Resources, 2002.

conclusion that we are left with much less time to change our bases of energy supply altogether.[1–4] The resources of crude oil are much disputed. Apparently, most of the additional stockpiles—beyond established reserves still considered a few years ago—simply do not exist. Speculation on oil term in markets is considerably influenced by this insecurity of affairs while the big oil companies obviously produce an exaggerated picture of resources rather than make speculation go its way.

Another matter is the regional distribution of these reserves/resources all around the globe (Figure 2; Federal Institute for Geosciences and Raw Material Research.[5]): the problem is obvious with crude oil, and everybody is aware of it, but there are also biases/imbalances with hard coal. Considering the total energy stored in it, some 60% of it rests with brown or hard coal; the smaller part is included in liquid and gaseous hydrocarbons. Note that efficiencies of power plants differ considerably among these energy carriers, with hard coal and gas steam plants being superior to others ($\eta_{el} > 55\%$). Crude oil, which may be produced by current technologies, reasonably amounts to 6682 EJ, a little less than with natural gas (7136 EJ). 1000 EJ (exajoule) = 10^{21} J. Standard heats of formation of the compounds/mixture/combustion products involved are as follows:

CH_4	–50 kJ/mol
C (graphite)	zero [by definition (standard state of an element)]
"CH_2" (fraction of crude oil)	≈ –20 kJ/mol
CO_2	–394 kJ/mol
H_2O	–237 kJ/mol

Thus, 1 mol (44 g) of CO_2 produced from combustion of natural gas (CH_4, essentially), crude oil, and coal yields 818 kJ, 611 kJ, and 394 kJ, respectively. The annual global anthropogenic CO_2 output is of the order of gigatons (1015 g, Pg), with the atmosphere containing some 770 Pg of it now, considerably more than which is tied up in living biota (about 610 Pg). One gigaton of CO2 from hard coal, crude oil, and natural gas natural translates into somewhat less than 9, 13.9, and 18.6 EJ of thermal energy, respectively. Thus, actually combusting the above-estimated resources would leave us with some 50,000 gigatons, that is, 65 times (!) the present CO2 inventory of the atmosphere.

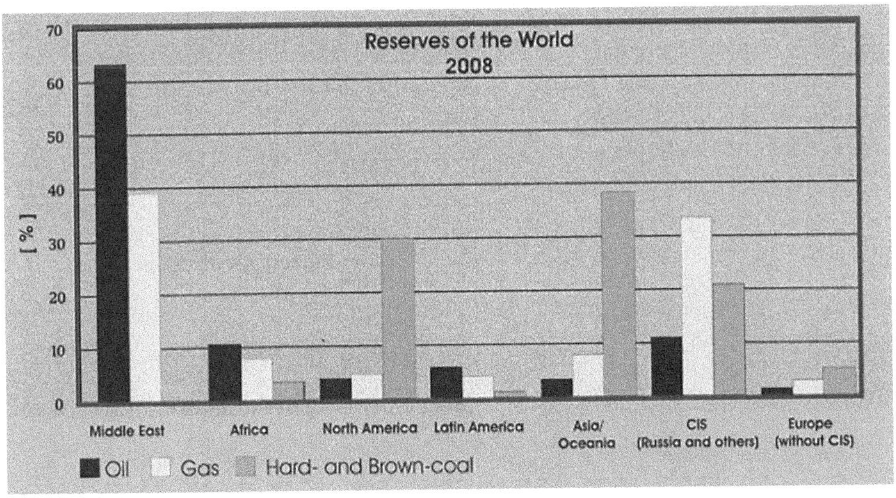

FIGURE 2 Estimate of the regional/continental distribution of oil/gas/coal reserves.[5] As for the Commonwealth of Independent States (former USSR except of Baltic states and Georgia), the biggest share is with just three of them: Russia (the current biggest oil producer in the world), Azerbaijan (both oil and gas), and Turkmenistan (almost gas only).

The total energy from conventional fossil energy carriers (resources combined) would be some 448,289 EJ, 97% of these resources being hard or brown coal. Besides the above-mentioned conventional energy carriers, there are substantial amounts of other energy carriers such as oil sands, oil shales, tars, and other kinds of the heaviest, most condensed oils, natural gas from dense storage sites, carbon deposits (adsorbed), or aquifer waters, plus gas (CH_4) hydrates on deep shelf (below some 350 m of seawater or in the uppermost sediments down there) reserves and resources that correspond to an energy equivalent of 2368 EJ and 116,270 EJ, respectively.[6]

Most industrial activities now depend on oil, as do almost all traffic systems, be it airplanes, ships, or cars. Oil getting scarce thus does not only cause prices to rise; there will also be ramifications on workforce in petrochemical branches and political implications. The largest share of oil (2005), some 742 billion barrels, is located in Central and Southern Asia (Bangladesh, Bhutan, India, Maldives, Nepal, Pakistan, Afghanistan, Sri Lanka, and parts of Iran).

Considering the global oil reserves according to British Petrol in 2005, this translates into 62% of global stockpiles, whereas North America commands just 5%, being one of the meta-regions scarcest in oil besides Asia/Pacific and Europe.

With oil getting scarce, a one-sided economic dependence on the Middle East poses increasing political risks, causing everybody to consider oil and gas resources located elsewhere and how they and additional geological goods might be obtained. Thus, there is a recent growing geopolitical interest in the Arctic region, which will become void of drifting ice during the next 20 or 30 years due to climate change. According to the U.S. Geological Survey Institute,[7] some 30% of natural gas and 13% of crude oil are located there. However, most of these are located far offshore [the remainder already exploited for decades in Russia (Taimyr Peninsula) and Alaska], with most of the gas belonging to the Russian Federation while oil is scattered among Canada, Alaska, and Greenland (still partly governed by Denmark). Though very large, experts say[7] that these deposits are not enough to consider relocation of most production activities from the Middle East. In addition, long transport distances (thus, costs) and still adverse weather and climate conditions pose grave problems. Everybody is still aware of what can happen with deep sea-based oil production considering the Deepwater Horizon catastrophe of April 2010, environmental concerns translating into larger political obstacles, higher insurance fees, and eventually less consumer acceptance.

That the so-called peak-oil level where global oil production has reached its maximum ever was already achieved in the beginning of the 21st century is evident from a depiction of development of oil production from 1930 until (predicted) 2050 (Figure 3): scarce oil means there will be no more cheap oil. Until the beginning of the 21st century, one barrel of oil commanded between $20 and a maximum

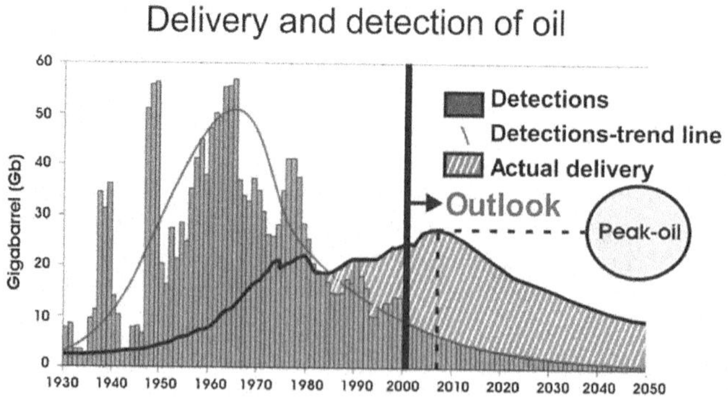

FIGURE 3 Oil findings and delivery rates from 1930 to 2050 with the outlook after Campbell (2006).[8] The delivery rate per year is calculated by the Association for Peak-Oil and Gas Studies, Germany. Data from ExxonMobil (2002). Peak oil: point of time when the maximum rate of global petroleum extraction is reached.
Source: Figure modified from Blum (2005).[9]

of $40. The financial crisis made it (Brent) rise up to $147 after 2007. Declining somewhat, it now (2011) stabilizes around $100. In the long term, it is more likely to increase again, having severe drawbacks on economic conditions in industrialized countries especially.

Figure 1 tells us that hard and brown coal reserves are to last another about 200 years, rendering coal-fired power plants most attractive if it were not for coal being one of the most polluting sources of energy. Combustion of coal produces plenty of CO2, and therefore, it contributes to the anthropogenic part of greenhouse effect warming of our atmosphere.

Climate Protection

CO_2 is one of the most prominent greenhouse gases in the atmosphere, contributing to heating the atmosphere and thus the Earth's surface.[10,11] There is both a natural and an anthropogenic (share of) greenhouse effect. CO_2, produced by animal respiration, wildfires, and volcanoes, is a natural (<300 ppm) component of the atmosphere, heating our planet from a radiation equilibrium ("blackbody") value of some −18°C to a global average of +15°C, that is, by a considerable 33°C.

A global increase of average atmospheric temperatures is now seen for decades, with Figure 4 displaying the distribution of this effect for 2000–2009 as compared to the reference time frame of 1951–1980. Satellite measurements by the NASA Earth Observatory showed the largest increases of the average temperature in the Arctic parts of the Northern Hemisphere, besides some parts of Antarctica. Between the beginning and the end of the 20th century, the average increase was 0.74 ± 0.18°C.

Time is imminent for rethinking energy production with a focus on sustainability in all ecological, economic terms and social acceptance. Man-made radiative greenhouse forcing is established for more than 20 years now. It causes enhanced glacier meltdown rates in the Arctic (Greenland actually turns green again), the Alps, and other mountainous regions, as well as sea level rising (both due to meltdown waters and to thermal expansion of warmed surface water), with more frequent droughts causing limnetic waters to evaporate often completely. We also would be urged and obliged to seriously think about what we are doing if, as some scientists still maintain to argue, this would be a

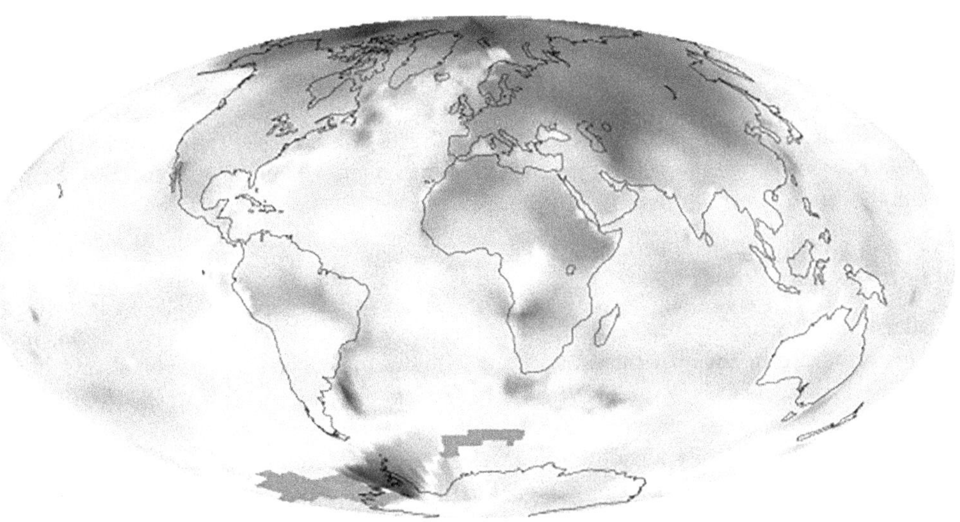

FIGURE 4 Regional distribution of relative greenhouse effect warming. The map shows the 10 years average (2000–2009) global mean temperature anomaly relative to the 1951–1980 mean. The largest temperature increases are in the Arctic and the Antarctic Peninsula. The darker the gray shades in the picture of NASA,[14] the larger the increase in local average temperatures.
Source: NASA Earth observatory.

normal climate excursion as occurred in the later Middle Ages. It does not take much proficiency in the natural sciences to acknowledge the problems associated with the industry emitting greenhouse gases[12] while striving to maintain and enhance our common well-being. The destruction of the environment concomitant with the exploitation of the developing Third World countries has already been an issue some 40 years ago, prompting the first Conference on the Human Environment at Stockholm (Sweden) on June 16, 1972, initiated and organized by the United Nations. Twenty years after, in 1992, the most comprehensive global conference on such issues ever held, the United Nations Conference on Environment and Development (UNCED), took place in Rio de Janeiro. One hundred years ago, Wilhelm Ostwald mentioned in his approach "Die Mühle des Lebens,"[13] "Do not waste energy, rather use it wisely," This is in opposite to the Kantian imperative at this time, and the first energetic step to a so-called sustainable chemistry of today.[15]

A total of 178 countries dedicated themselves to Agenda 21 to oblige the 118 more developed countries to embrace environmental restoration, preservation, and social development. Their aims are to meet the challenge of global warming, pollution, and biodiversity and to solve the interrelated social problems of poverty, health, and population. The Agenda 21 program furthermore encompasses constructing a network of "new and equitable global partnership through the creation of new levels of cooperation among States, key sectors of societies and people, while working towards international agreements which respect the interests of all and protect the integrity of the global environmental and developmental system."

In 1997, the World Climate Summit was held at Kyoto, Japan, focusing on the future climate protection policy and measures. The 158 states that had so far signed and ratified the Framework Convention on Climate Protection and 6 "observer states" had sent a total of almost 2300 delegates, plus another 3900 observers from non-governmental and other international organizations and 3700 journalists from global media, making up a total of close to 10,000.[16] The Kyoto process unleashed by this huge meeting now describes the attempts to maintain climate protection agreements beyond the Kyoto protocol running out in 2012 and to do so in a way that still is safeguarded by agreements of international law.

The industrialized states listed in Appendix 1 of the Kyoto Protocol agreed to reduce their collective greenhouse gas emissions by 5.2% from the 1990 levels by the year 2012. The common (purported) aim of the international community is to limit the increase of global average temperature within 2°C. By now, it has become most doubtful whether this "two-degree aim" can be kept at all. In addition, it was proven that even an aggressive forestation policy could not cope by photosynthetic absorption (assimilation) with the present upsurge of CO_2 produced. The progress made so far is modest at best, with real successes still to be waited for. The subsequent UN Climate Conference at Copenhagen (Denmark) in late 2009 saw just a communiqué of minimal consent without any mandatory aims in CO2 reduction although the "two-degree aim" was once again acknowledged to be worthwhile.

Role of Nuclear Power

The Fukushima (Japan) sequence of catastrophic events began on March 11, 2011 [the strongest earthquake ever recorded in Japan (Richter magnitude 9.1)], causing coolant pipes to break and starting a sequence of reactor core meltdowns in at least three adjacent nuclear power plants (NPPs). The area was then hit by a tsunami; any semblance of control of the preceding events was lost. Finally, the uncooled reactor systems exploded one after the other, releasing large amounts of radioactivity and visibly destroying reactor block 3. People reconsidered the risks of nuclear energy all over the world, although with grossly differing political consequences.

Now introduced into IAEA (International Atomic Energy Agency) accident level VII, the events at Fukushima were set equal to the Chernobyl disaster [April 1986, Ukraine (officially: Ukrainian Soviet Socialist Republic)] and considered worse than the explosion of a nuclear fission waste storage tank at Kyshtym in 1957 [near Chelyabinsk, Urals, then RSFSR (Russian Soviet Federative Socialist Republic), USSR (Union of Soviet Socialist Republic); level VI accident], which delivered very large amounts of

^{90}Sr [several EBq (exabecquerel: 10^{18} Bq)] that an area of some 15,000 km^2 had to be abandoned for any human use up to now. Japan is known to be tectonically most active, prompting the authorities and engineers to consider earthquakes up to Richter magnitude 7.9 in design. Two matters demand further consideration here:

a. What causes, e.g., main coolant pipes to break is acceleration and the mere amplitude of dislocation combined with inertia, that is, parameters rather covered by the "old-fashioned" Mercalli scale of earthquake intensity, with Mercalli 13 denoting accelerations larger than that of the Earth's gravity. Such quakes actually happened, e.g., the Easter quake of 1964 in Central Alaska (there is a very impressive movie showing cars and humans and even some homes losing contact with the ground during the quake). In contrast, Richter scaling gives the energy unleashed by a quake [or a landslide or an underground explosion (be it chemical or nuclear in origin)] in a logarithmic way: Depending on the depth of the epicenter, i.e., the amount of matter located above the vibrating or breaking sample of crustal matter, accelerations due to a given amount of tectonic energy can be quite different: the "fairly moderate" Haiti quake of February 2010 (Richter 7.1) became a real killer not because of its energy but since its epicenter was located just a few kilometers below the surface, producing a massive acceleration, thus aggravating the effects of non-adapted architecture. Inertia also is a matter of dislocation of the ground, which provides a kind of reference frame, physically speaking, to which any device that can vibrate, etc., will respond: this dislocation was about 2.5 m in Japan on March 11, 2011, that is, much larger than in the even stronger (more energetic) Chile 1960 (Richter 9.5) and Sumatra Christmas 2004 (Richter 9.3) earthquakes.
b. At least two NPPs, one of them in Japan, had already experienced earthquakes exceeding these limits of design: at Niigata (NW Japan) and at then Leninakan in Armenian Soviet Socialist Republic (SSR) in 1988. Hence, the design was obviously short of reasonable expectations [Central Europe, German, Swiss, and French NPPs next to the Upper Rhine Rift Valley do not take account of the fact that Basle City was almost completely destroyed by a quake in historic times (in 1356)].

The Fukushima earthquake was far stronger than this. The tragedy was worsened by a tsunami caused by this quake; it took thousands of lives in cities and villages near the coast, but most likely, the crucial damage to the Fukushima NPPs was caused by the earthquake, though the NPP area was inundated soon after. Certain radionuclides, mainly highly volatile ones were released; that core meltdowns had occurred was first contested and then confirmed by Japanese authorities only 2 1/2 mo later. These highly volatile radionuclides include ^{131}I, $^{134;137}Cs$ and some noble gases, and traces of ^{132}Te but not Sr, Ba, and rare earth elements. Apparently, the temperatures during core meltdown did not yet suffice to vaporize alkaline earths or rare earth element compounds (cesium becomes highly volatile as a hydroxide, CsOH molecules sublimating as easily as NaOH and KOH do), neither did volatile chlorides form after NaCl from seawater pumped inside to cool the reactor cores solidified. Including those still missing, their whereabouts unaccounted for but unlikely still to be alive, there were some 23,000 fatalities due to the quake and tsunami, and several workers received massive radiation damages during cleanup, with a nuclear catastrophe still looming ahead. The number of people forced to abandon their homes and jobs (permanently) at Fukushima and Iitate provinces [40 km NW Fukushima (the NPP site, not the town of Fukushima)] is more than 80,000 now, that is, close to that of people who became homeless due to the Chernobyl accident. Responses from different countries were different: while politicians and citizens in certain countries started to abandon energetic uses of nuclear fission (Germany and Switzerland) or reaffirmed prior decisions of this kind [New Zealand, referendum in Italy with a 94% (2011) turnout, Belgium], others chose to keep a pronuclear course or even kept up decisions to build national first-ever NPPs (Poland).

On a global scale, the issue of nuclear energy is controversial in all developing countries, classical industrialized states, and "official" and "unofficial" nuclear-arms-possessing states. While the United States plans to commence building the first new NPPs after a moratorium of some third of a century,[17]

Italy officially revoked the schedule to abandon nuclear energy adopted by a 1994 referendum (but only so in the old Berlusconi regime, the Monti administration being undecided on this issue), and the spring 2012 presidential election campaign in France triggered a broad discussion on the future of NPPs there for the first time. Several developing states extend their present NPP program apparently as a pretext to obtain more fissionable material for nuclear arms, like India and Iran, regardless of whether they signed the Non-Proliferation treaty (Iran), did not sign (India and Israel), or left the protocol some time later (North Korea). In "established" nuclear powers, on the other hand, NPP extension also is a means to get rid of excess amounts of weapons-grade or "special" nuclear fuels (United States, Russia, U.K.). The most interesting case is some 2 tons of ^{233}U left over from the thorium/uranium breeding cycle, which is/was used only in India for energy production and formerly in the United States for making arms. Now, the United States is left with a stockpile of this nuclide, a nuclide that actually would be usable by very-low-technology groups or states to produce "efficient" and reliable-yield warheads (different from that with any plutonium isotope [-mixture]!) and correspondingly must safeguard it to a level that would not be necessary with any other isotope—unless they "burn" it in reactors.

Hence, the ambiguity of nuclear technologies continues to influence decisions on introducing or maintaining it as an energy source, far beyond, and competing with, considerations of supply reliability (the number of states that provide/export significant amounts of uranium is considerably smaller than those that provide all natural gas, crude oil, and hard coal). Political decisions on increased or renewed use of nuclear energy in certain countries hence cannot be considered a signal that is globally significant, except for their drawbacks to the issue of proliferation.

The classical argument in favor of nuclear energy production is their releasing much less CO_2 as compared to oil, coal, or natural gas power plants, enabling economies to meet Kyoto ends more easily. For certain countries that produce nuclear fuels but command few other natural resources, there is the bonus of apparent independence; this includes Japan. In the EU, the Czech Republic presently is the only state to mine domestic uranium ores, globally large producers being differently organized and reliable in political terms [Australia, Canada, the United States, Kazakhstan, Niger, Gabon (Central Africa)].

Nuclear energy is rather "compact," a GW-class power plant taking a few hectares at most, which is important in countries as densely populated as Japan (or India, or Belgium), with Japan covering a total area hardly larger than that of Germany (372,000 km^2). Next to the United States (104) and France (59), Japan has the largest number of NPPs [53, from which 6 at Fukushima and 3 at another highly earthquake-prone site (scheduled for shutdown) must be subtracted; thus, effectively it is only 44] due to an experience from the 1970s: the OPEC oil crisis in 1973, rapidly increasing the prices of crude, of course hit all the already then industrialized countries except for those producing sufficient fossil fuels of their own (Canada, Australia) or relying on other sources mainly for different reasons (Norway, New Zealand, South Africa) but worst so in Japan: in fact, there were large-scale electricity shutdowns badly hurting Japan's economy. Hence, Japan—once again in its history—became eager for independent supply, planning to build yet more NPPs before Fukushima happened. By now, unlike France, Belgium, the United States, or formerly Lithuania, which obtains 60% or more of its electricity from NPPs, Japan's rate is 30%. During the period of Soviet dominance, Lithuania even had the largest NPP in the world, a 2400 MW_{el} plant at Ignalina (now Visaginas) of the notorious (Chernobyl) RbmK (*reaktor bolshoy moshchnosti kanalniy:* high-power channel-type reactor) construction type. Relying on nuclear energy by >70% shortly after regaining independence, Lithuania had to close down this plant as a precondition for joining the EU, as Bulgaria had with Kozloduy. Now, Lithuanians plan to erect a new, similarly huge NPP at the same site next to the border triangle with Latvia and Belarus to supply all of the three Baltic states even though Latvia is now the EU's champion in regenerative energy supply, producing almost 40% of its (admittedly rather limited) demand mainly from hydropower, wind, and some biomass use, and suffers from financial crisis (much like Ireland and Greece but less perceived to do so since they still maintain their own currency, the Lat). Figure 5 shows the global pattern of distribution of NPPs that is almost complete [note that there are 1) considerable differences in electric outputs of individual plants and 2) some plants were primarily meant to produce radionuclides rather than electricity, although they

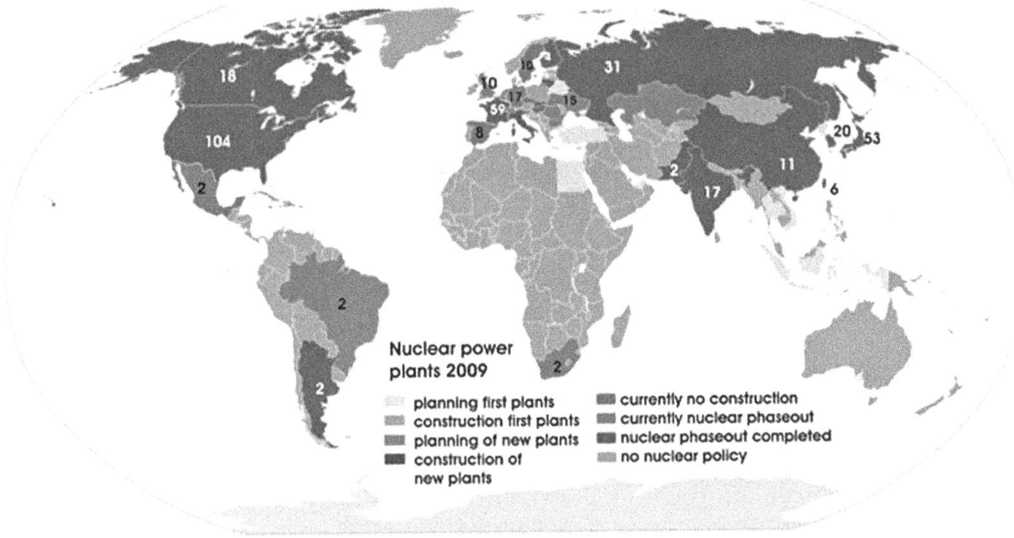

FIGURE 5 Nuclear power worldwide in 2009 (the figure was transferred from Wikipedia; the number of nuclear power plants has been added by collecting information separately by Wikipedia). The number of nuclear power plants is not complete.

are connected to the public current grids (e.g., Dimona in Southern Israel)]. The total number is about 450 NPPs.

As shown above (Figure 1), uranium reserves are to last for just another about 30 years from now (2011), adding another problem to safety concerns and disposal of radioactive wastes (see below) associated with nuclear (fission) energy. The increase in prices during the last 5 years even relatively surpassed that of oil, the process probably going on as demands for uranium did not yet reduce considerably on a global scale after Fukushima. Figure 6 shows a map indicating the 10 most prolific uranium-mining countries (by 2008).

Let us turn to the third issue associated with nuclear energy: radioactive wastes, which are produced by all nuclear fission, then located in "used" (irradiated) fuel rods, neutron impact on non-fissionable ^{238}U, and on construction materials from Zircaloy to concrete, and their first decay products, sometimes dissolved in activated wastewaters including dissolution residues (nitric acid) in nuclear fuel reprocessing (now banned in both the United States and Germany), must be disposed of and stored until radioactive towards stable products is essentially complete. A fuel rod in an NPP is a metal (Zr mainly) tube filled with cylindrical sintered pellets of UO2 or some mixture of UO2 and PuO2; metal alloys or other compounds (carbide UC2, hydride UH3) are used in minireactors only. After a while, a fuel rod is "spent," reducing the ^{235}U content from the original 3.5% to some 1.3% while 1–1.5% of plutonium—most of it fissionable also—were produced from ^{238}U in situ. Although use could be continued by further pulling out the control rods that absorb excess neutrons, it would be no longer safe to work with such rods. Thus, they are replaced, usually a third of them every year or so. The "spent" rods are stored for several decades, usually next to "their" reactor to get rid of the highly active short-lived isotopes of high yields, e.g., $^{141;144}$Ce and $^{103;106}$Ru and ^{91}Y, then either processed or put to a final depository (if there is one, by now only in Finland).

The case is not at all settled; there are no operating final deposit sites but only such ones meant to contain (withhold) the waste for a few decades. Obviously, radioactive waste solutions were discarded into the open sea or rivers both after accidents and routinely during nuclear reprocessing [Windscale/Sellafield (U.K.) and La Hague (France), Mayak near Chelyabinsk (Russian Federation)], not to mention simply dumping entire discarded reactors from nuclear-powered submarines into the sea, with or without the rest of the vessels ... What can be done responsibly with nuclear waste instead?

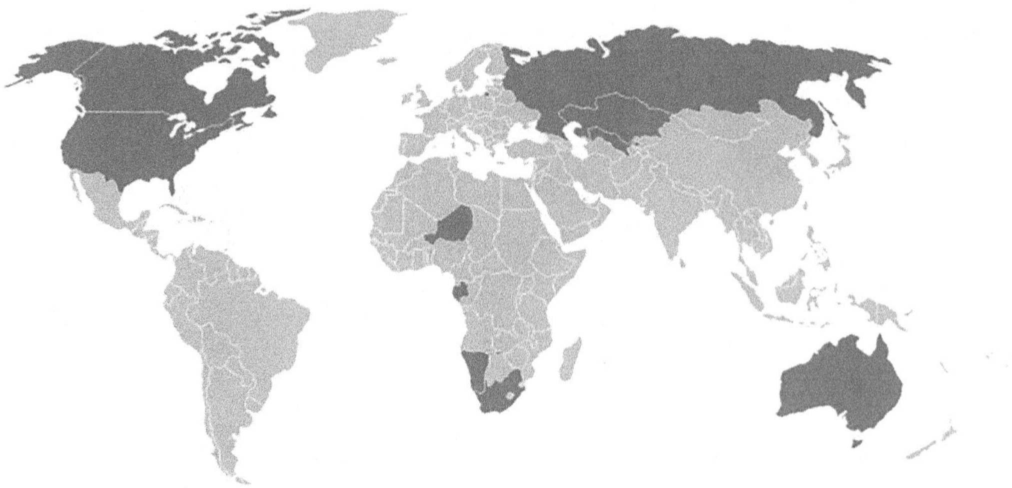

FIGURE 6 The 10 states (dark gray) that produced the largest amounts of uranium ores (in 2008). Besides the spatially largest states outside of South America and except for China and India, most are located in Western Africa, including Gabon (Central Africa) whose ultra-high-grade (>60% U) deposits at Oklo and neighboring sites gave rise to natural reactors some 2 billion years ago (map from Wikipedia.de).

The periods of time over which radionuclides from fission reactors must be stored and safeguarded are outright unimaginable and far beyond any other time frame of political or economic planning: there are nuclides with half-lives of several million years (^{94}Nb, ^{129}I, ^{237}Np); thus, they will create a danger for 10^7 year or the like. With ^{237}Np (tons of which currently exist) and ^{243}Am (americium, tens of kilograms of which exist), there is an additional problem: during very long storage, enriched samples of either nuclide will spontaneously turn into fissionable materials (^{233}U and ^{239}Pu, respectively), causing heat and neutron release to increase after millennia.

Here are some examples:

Uranium	^{238}U	4.468 billion years
Uranium	^{235}U	704 million years
Iodine	^{129}I	15 million years
Neptunium	^{237}Np	2.144 million years
Plutonium	^{239}Pu	24,110 years

While actinides [from the third, protactinium (Pa, Z = 91) onwards] generally are extremely chemotoxic, this does only matter for long-lived nuclides like the natural uranium isotopes, ^{237}Np, $^{242;244}$Pu, or $^{247;248}$Cm. In the other cases, say at $T_{1/2} \leq 10^5$ years for α emitters with negligible spontaneous fission shares, radiotoxicity, i.e., the effects caused by particles of ionizing radiation emitted during decay, prevails even against this chemotoxicity. Among these nuclides, $^{239;240}$Pu (which are produced together in a reactor given there is substantial irradiation of a uranium sample) are peculiar in their radiotoxicity but are rather long lived as they do reside in the body at very sensitive points: the marrow and mucosa around or in bones and liver, while radioactive ("hot") particles may be inhaled and reside in the lung, exposing it to radiation. Note that some 6 tons (!) of ^{239}Pu that escaped fission during nuclear bomb tests were spread up into the stratosphere as an aerosol that still keeps on being deposited ("fallout") apart from some other transuranic nuclides: from analyses of fallout composition, 1100 kg of ^{240}Pu and substantial amounts of $^{241;243}$Am (produced in nuclear weapons containing ^{241}Pu or ^{242m}Am as fissionable nuclides, which have much smaller critical masses with fast neutrons than all $^{233;235}$U and ^{239}Pu) are determined. Estimated amounts of the latter are 25 kg ^{241}Pu, 70 kg of its decay daughter ^{241}Am, and 2.5 kg ^{238}Pu [most of it from breakup of a RIG (radioisotope thermoelectric generator) during re-entry in a 1960s space probe mission; data calculated from Breban et al.[18] Besides this, there are natural contributions of plutonium in the biosphere: some 5 kg of ^{244}Pu is still left over from the origins of the solar system owing to its 83-milion-year half-life and trace amounts (<1 g) of ^{238}Pu from the double-β-decay branch of ^{238}U. Most fallout radioactivity is still due to ^{241}Pu ($T_{1/2}$ = 14 years), but this is less destructive in radiotoxicological terms because it does undergo β-decay, rather than α-decay, followed by each comparable activities of $^{239;240}$Pu and ^{241}Am ($T_{1/2}$ = 433 years). This compromises sensitive tissues.

Cancer rates, e.g., bone skin sarcomas and their metastases, will thus increase after a Super-GAU (German: *größter anzunehmender Unfall* means worst case, meltdown) if plutonium is released, like in the 1951 Mayak accident (Russian Federation) or after a Pu-based nuclear fission bomb was destroyed 1) in an airplane crash at Thule (Greenland) in 1968 or 2) by mis-ignition ("fizzle") in the *Hardtack Quince* test (explosion yield but 0.02 kt TNT) at Runit Island, Enewetak Atoll (now some part of Republic of Marshall Islands) in 1958. Both latter events spread several kilograms each of ^{239}Pu over a very restricted area, which made effectively cleaning it impossible. The doubtful results of (2 out of 5) nuclear bomb tests in Pakistan (1998) and the first one in North Korea (2006)—neither yet precluding burst of Pu and fission products to the surface though these were underground test "fizzles"—must be added to this list. Runit Island (Pacific Ocean) is now used as a dumpsite for highly contaminated materials from nearby test craters, the hole capped by a concrete dome of some 100 m diameter. This is the gravest possible kind of catastrophe from an ecological and ecotoxicological point of view, going beyond even what happened at Fukushima or Chernobyl. The environment, including groundwater, is so polluted that access is strongly impeded for centuries or even longer.

As with Chernobyl and the Bikini Atoll test site, animals and plants apparently adapted to the harsh radiological conditions there. Starting soon after the Chernobyl accident, researchers noted that diverse animals (including wolves, foxes, lynx, moose, hares, and many kinds of birds)—and some humans—returned to the off-limits area around Chernobyl, including the so-called Red Forest and the decontamination lake. Among these, there are species that try to avoid man, prompting them to invade an area where there are (almost) no humans left behind. In addition, biodiversity is lower than in comparable areas, brain sizes tend to be reduced in both mammals and birds, and there are other malformations, tumors, and evidence of genetic alterations.[19] Likewise, there are lots of aquatic life in Bikini Atoll [except for the crater basin of the largest ever U.S. test, *Castle Bravo* (1954, some 15 MT)], but once again, there is reduced biodiversity.

At least since 9/11 some 10 years ago, but actually ever since the first threat to attack a NPP by a flying passenger airplane abducted before (in 1972 over Oak Ridge, Tennessee) citizens and public authorities of countries which make use of nuclear fission power: while older NPPs—all over the world—must be considered to be improperly protected against even the impact of a smaller airliner there is virtually none anywhere to withstand the perpendicular impact of a fighter plane. The jet engine compressor axle in a military plane is about 3–3.5 m long, made of dense metals such as niobium alloys, then hitting a concrete ($\rho \approx 2.8$ g/cm^3) shell at about the speed of sound. This will suffice to penetrate some 10 m (!) of concrete, as compared to an actual containment thickness of less than 2 m. In addition, global warming and eutrophication combine to increase the likelihood that river or ocean water cooling can no longer be taken for granted in summer at least: in 2003, rivers became so warm all over Central Europe (probably related to global warming) that NPPs from France to Lithuania had to be shut down for many weeks. In the same summer, cooling water entries at the Russian Sosnovy Bor plant were blocked by algae excessively growing in the adjacent Bight of Helsinki. Either problem is likely to occur more often in the future.

Transmutation[20,21] is discussed as a means of faster disposal of fissiogenic radionuclides, exposing the nuclide mixture obtained by fuel reprocessing to a high-energy proton beam. Fission products that are distinguished by a considerable excess of neutrons in their nuclei hence are brought closer to stability while actinides will either undergo fission directly or produce at least nuclides of much shorter lifetimes. It is completely unsettled and doubtful whether this can be done on a scale of tons of material.

Rethinking in the Way for Ecological Economics

The previous entry was mainly concerned with the problems of conventional and nuclear energy sources from an economic, ecological, and political point of view, which will challenge us more and more in the near future. There are several quite different—and mutually independent— reasons to switch to alternative,[4,22–25] that is, regenerative, resources of energy, including resources of fossil fuels becoming scarce(r), constraining greenhouse gas emissions, and striving for geopolitical risk management (spreading suppliers among most diverse regions and political systems, if you are not in a position to produce the materials related to your energy demands domestically). Although there are numerous incentives, this transformation will take several decades not only on a global scale but also in national dimensions, and it depends on both economic interest positions and technical innovations. While wind energy is, at present, technically "ripe," the size of rotors being no longer limited by mechanical problems with generators, etc., exposed to vibrations and bending along a horizontal axis, but by the size of available cranes required to erect them (maximum: some 200 m rotor diameter, 7–9 MW peak power), and photovoltaic devices that reasonably benefit from technologies of semiconductor processing—which is now advanced much beyond anything that would ever be required in solar energy conversion—becoming a large-scale-business as well [the combined area of highly integrated chips that make it to the market in computers and numerous other devices (some 70 of them alone in a common car) annually is in the square kilometer region also], extracting energy from the oceans actually still takes an engineer's ingenuity to make it reasonably work. For solar energy, the "surviving" technical innovation challenges are restricted to thin-layer and organic semiconductor systems and to conversion/storage in a convenient chemical storage form, like methanol.

The following part of this entry will deal with both those energy forms already present on the global market for energy conversion devices, plus a discussion of how energy can be harvested from the open sea, keeping in mind which will be the challenges for Germany or Italy who both decided either to abandon (D) the use of nuclear energy until 2022 or not to restart it (I). This prompts the question whether redesigning the energy supply of a medium-sized highly industrialized country, including consumers which are really demanding, within some 10 years, merits further consideration which is added, too.

Globally View of Renewable Energy

Available energy resources are "renewable" when they are sustainable and environmentally benign (which need not coincide) if

- They renew themselves in the short term by itself (e.g., biomass).
- Their use does not contribute to the depletion of the source (e.g., wind, sun, water).
- They do not have an impact on the environment.

Sustainability in that sense means a triad related to ecology, economy, and social affairs. Figure 7 gives an overview of different energy sources like sun, wind, water, geothermal energy, and biomass for the establishment of technical supported renewable energies.[4,22,24,26,27]

In 2010, the Global Status Report on Renewable Energies[28] was published, parts of which are now to be quoted. In its preface, El Ashry (UN Foundation) already noted more than a hundred states to have political agendas or strategies in 2010 in favor of spreading and widening the use of renewable energy sources, about a doubling from just 55 five years before. In 2009, the largest increases were observed with wind power and photovoltaics, investing some $150 billion into extending capacities and producing and implementing renewable energies vs. only $30 billion in 2004. Except sometimes for advancement of nuclear energy, more money is allocated into capacities and growth of renewable energy sources than into fossil types. By reaching substantial shares of energy input (particularly in some less-populated countries like Denmark, Latvia, or Mongolia), renewable energies did have a turning point, which renders them significant, as they also address the problems of climate change. As implied before, acceptance and broad

FIGURE 7 Renewable energy sources: onshore and offshore wind power, biomass, solar power, geothermal, hydropower. (a) Onshore wind turbines located outside of Palm Springs, California (photograph: Wikipedia, Tim1337); (b) sugar cane is a major supplier of biomass, which is used either as food or as energy supplier (photograph: Wikipedia, Culture sugar cane, Avaré, Sao Paulo. José Reynaldo da Fonseca, 2006); (c) newly constructed offshore windmills on the Thornton Bank, 28 km offshore, on the Belgian part of the North Sea (photograph: Wikipedia, Hans Hillewaert, 2008); (d) photovoltaic array near Freiberg (Germany) (photograph: Wikipedia, I, Eclipse.sx); (e) Krafla Geothermal Station (2006), North Iceland (photograph: Wikipedia, Mike Schiraldi); (f) Grand Coulee Dam is a hydroelectric gravity dam on the Columbia River in the U.S. state of Washington. It is the largest electric-power-producing facility in the United States (photograph: Wikipedia, (http://users.owt.com/chubbard/gcdam/html/photos/exteriors.html) U.S. Bureau of Reclamation). Larger dams are located in Russia, Brazil, and China.

application of renewables are not (no longer) restricted to highly developed industrialized countries but extend to developing nations now displaying more than half of implemented and operating renewable energy supply systems. However, it remains to be seen as to how far this is actually in favor of domestic development, better situations especially in rural areas, or larger autonomy towards globalized oil and coal markets or is another blueprint for export of goods to the North: growing sugar cane and oil palms for biofuel exports in Brazil or Indonesia threatens local primary forests if not even the food supply there. The Desertec blueprint (http://www.desertec.org) for producing lots of hydrogen from solar sources in Northern Africa revives existing supplier roles with countries like Algeria once again.

Figure 8 gives a more precise idea of the present relevance of renewables, their share being some 19% of total in 2008. Share of final energy means: at the point of end use, as electricity, heat, and directly used fuels. This method counts all forms of electricity equally, regardless of origin. The European Commission adopted this method in 2007 when setting the EU target of a 20% renewables share of energy by 2020. Thus, it could be called the "EC method."[29] This includes all traditional sources of biomass energy carriers (biogas, ethanol, wood, etc.), large hydropower plants, and the "novel" kinds of renewables such as small-scale hydropower plants, modern biomass sources, wind and solar energy, geothermics, and biofuels. To be exact, it should be noted that some of the "modern" ones actually are rather old, from small-scale water mills also delivering electricity up to biofuels [Rudolf Diesel demonstrated his engine to be run with peanut oil fuel already at the 1900 World Fair (EXPO) at Paris].

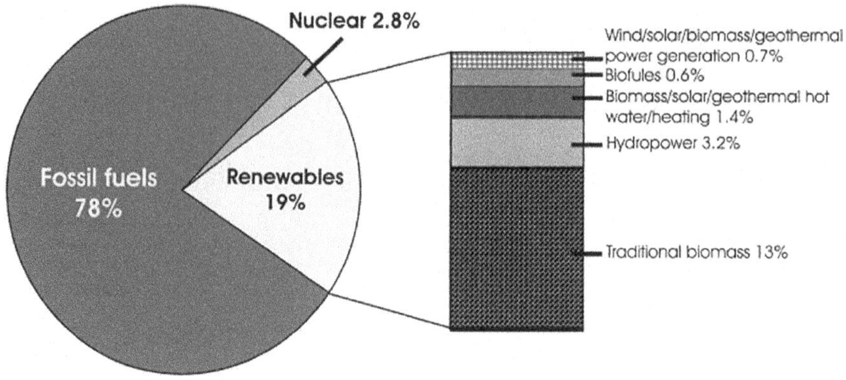

FIGURE 8 Global final energy consumption of conventional and renewable energy in 2008 after REN 21 (2010). Also consider footnote 327 of REN 21 (2010).[28]

Most of these 19% (13%, i.e., relatively 70% among the renewables) rests with traditional biomass used for cooking and heating. This, however, is subject to some change and even cutbacks, especially because there is virtually no possible firing wood (not even small branches) left over around (and this is to say, up to 50–70 km from the outer edge of "informal suburban settlements"!) megacities[30] all over developing and threshold countries, causing to replace with more advanced (e.g., solar cooking devices) or at least more efficiently use these resources. Hydropower represents 3.2% (17% out of renewables) and is growing modestly but from an already advanced level. The other renewable energies contribute 2.7% (relative share: some 14%) and undergo very fast growth in both industrialized and certain developing countries.

In developing countries, renewable energies have a particular role in national support and corresponding policies as well, contributing to their present share of renewable energy capacities of more than 50% [but compare this to both their share of global population, which, including the BRICS (Brazil, Russia, India, China) states except Russia, is >80%, on one hand, and their share of global energy consumption, on the other hand]. China now leads in several indicators of market growth. Among suppliers of wind energy, India now ranks No. 5 globally, owing to its substantial expansion of biogas and photovoltaic capacities especially in the countryside (India also plans to install another 15 NPPs owing to its being populated as densely as, e.g., Belgium or the Netherlands in total and citing growing industrial energy demands). Brazil is now a key supplier of ethanol from sugar cane, adding to other kinds of biomass and wind power in its renewable energy portfolio. Concerning the complete array of renewable energy sources, Argentina, Costa Rica, Egypt, Indonesia, Kenya, Tanzania, Thailand, Tunisia, Uruguay, and others boast high growth rates, often even surpassing their sometimes impressive gross economic growth rates.

In a nutshell, the geographical distribution and political significance of renewable energies have considerably been altered in favor of near-global distribution and application. As compared to quite a few countries in the 1990s, there are at least 82 countries presently operating wind power plants. As for manufacturing these devices, production was relocated from Europe to Asia, with the largest shares in China, India, and South Korea, which additionally become more devoted to renewable energy applications. China not only is the "production yard" of the world for conventional items including consumer electronics but also made, in 2009, 40% of PV devices (from solar cells to complete panels with control and DC/AC converter units), 30% of the world's wind power plants, and even 77% of solar-driven water-heating collectors. All over Latin America, countries like Argentina, Brazil, Colombia, Ecuador, and Peru increase their production, including exports of biofuels, additionally investing in other kinds and technologies of renewable energies. More than 20 countries in the "Solar Belt" of the Middle, Northern, and sub-Saharan Africa are involved in renewable energy markets.

Large economic gains and considerable further technological changes are to be achieved from this state of affairs and developments beyond the principal players of the highly developed world such as the European Union (EU), the United States, Australia, Canada, or Japan. With renewable energies gaining a truly global footage and application area, there is internationally growing confidence into renewable energies being less susceptible towards perturbations by either political turnovers or market changes such as financial crises than "classical" energy carriers (which readily, like other raw materials, become an item of speculation undergoing tremendous price level oscillations in such conditions).

Another impetus to renewable energy development—as to every other kind of large-scale technical innovation—is its inherent potential to create entire new industries (or at least create vast uncharted fields of opportunities for existing technical branches) and thus millions of new workplaces. In photovoltaics, much larger amounts of semiconductor-grade to medium-purity silicon, SiH_x, and GaAs are used than in integrated (chip) solid-state microelectronics (e.g., GaAs chips are applied in cellphones), whereas technologies required to obtain control of a large rotating propeller in changing conditions, originally developed for helicopters, now increase reliability and output of wind power systems. Creating new workplaces is, by the way, the positive, friendly side of the "dual-use" problem: both engineers and blue-collar technicians who were mainly concerned with development and production of arms systems had to strive to save and "humanize" their workplaces after the Cold War came to an unanticipatedly happy end around 1990; these highly skilled metal workers and engineers at Kiel harbor (Germany) then started to focus on wind power plants (with rotor techniques borrowed from military helicopters) and integrated current-heat support systems (using diesel engines originally designed for tank propulsion to consume and convert plant and waste oils to produce some 1 MW each of electricity and heat). Hence, although not creating environmentally benign branches from scratch, this "Arbeitskreis für Alternative Produktion" (workgroup planning alternative production) and similar endeavors in U.S., British, Australian, and French arms enterprises effectively increased the array of possible customers much beyond the state (i.e., department of defense and sometimes police forces) and in the same turn reduced political-military dependences even though most of the respective employers did not really like the idea of employees considering what should (better) be produced on their own.

This is part of the ethical issues and bonuses associated with alternative energy production: there is a considerable bonus in terms of both workplace safety and numbers of workers required to install and run 1 GWel of alternative energies as compared to fossil and nuclear types; besides, there are fewer risks associated with making PV devices than with coal or uranium mining for the miners themselves, counting and comparing, for example, mine accidents and cancer fatalities per MWy. In 2009, there were an estimated three million workforce directly related and devoted to renewable energies, about half of them concerned with biofuels, and many more than this in branches indirectly connected with renewables (Table 1).

What are the recent performances of renewable energies as of 2009 (Figure 9, source: REN 21[28])? For the second consecutive year, in 2009, in both the United States and the EU, the newly installed renewable energy capacities exceeded those of combined conventional fossil energies and nuclear power. Renewables accounted for 60% of newly installed power capacity in Europe in 2009, and nearly 20% of annual power production.

Christopher Flavin (Worldwatch Institute) pointed out in his entry "Renewable Energy at the Tipping Point" within the REN 21[28] report that China's recent leader role in producing wind rotors and photovoltaic devices just gives proof of the political prerogatives in favor of renewable energy exploitation, including both laws and funding, to be successful. Although there were initial problems, the important reforms in China starting with the national legislation on renewable energies of 2005 caused fast and efficient development there. China since then increased its efforts to become a leading innovative power as well as key producer of renewable energy technologies.

Table 2 shows the five countries that are the most important players concerning renewable energies, as of 2009.

TABLE 1 Jobs Worldwide from Renewable Energy (REN 21[28] with Added Information from the United Nations Environment Programme Report 2010)

Industry	Estimated Jobs Worldwide	Selected National Estimates
Biofuels	>1,500,000	Brazil, 730,000 for sugar cane and ethanol production
Wind power	>500,000	Germany, 100,000; United States, 85,000; Spain, 42,000; Denmark, 22,000; India, 10,000
Solar hot water	~300,000	China, 250,000
Solar PV	~300,000	Germany, 70,000; Spain, 26,000; United States, 7,000
Biomass power	–	Germany, 110,000; Unites States, 66,000; Spain, 5,000
Hydropower	–	Europe, 20,000; United States, 8,000; Spain, 7,000
Geothermal	–	Germany, 9,000; United States, 9,000
Solar thermal power	~2,000	Spain, 1,000; United States, 1,000
Total	>3,000,000	

Note: Further information about the evaluation of the data are reported in REN 21,[28] p. 75, note 226. The table is incomplete.

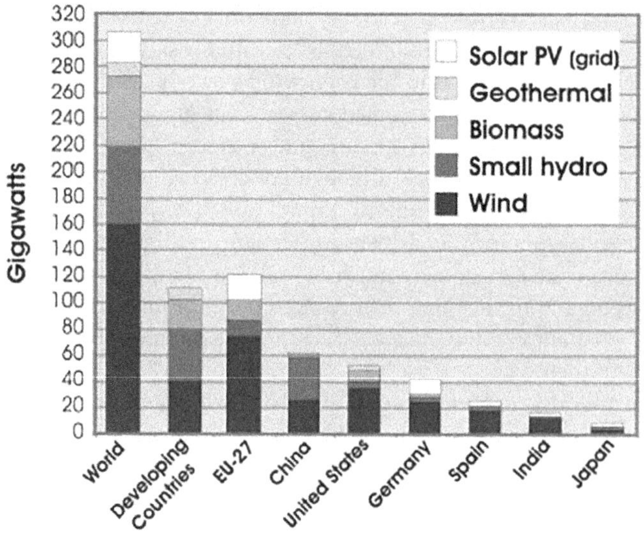

FIGURE 9 Renewable power capacities in 2009 without inclusion of large-scale hydropower: Developing World, European Union, and Top Six Countries (REN 21 2010).[28]
Note: Only includes small hydropower <10MW.

Owing to the well-known variabilities of renewable energy output, which are due to weather, time of day, and longer-period (tidal power plants) periodic or aperiodic changes, there are much larger theoretical (peak power output) renewable energy technical potentials than average yields. The present (as of end of 2009) capacity of a global 1.23 TW (1230 GW) that now constitutes just more than 25% of total electric generating capacity worldwide thus is considerably larger than the actual share/contribution of produced electricity.

What about Ocean-Related Energy [Waves, Ocean Currents, Tidal Power Plants, Osmotic Energy Conversion, Ocean Thermal Energy Conversion (OTEC)]?

The power associated with flowing water is impressive and has motivated people to use it many centuries ago in mills located at running creeks and rivers. It was an obvious idea to extend this technique to tapping ocean currents, like the Gulf current, as well as tidal water flows[31] that can reach speeds much above those seen in most rivers [e.g., some 6 m/sec (11 knots/hr) along the Welsh coast of the Atlantic

TABLE 2 The Five Countries That Are the Most Important Players concerning Renewable Energies [as of 2009[28]]

	#1	#2	#3	#4	#5
Annual amounts in 2009					
New capacity investment	Germany	China	United States	Italy	Spain
Wind power added	China	United States	Spain	Germany	India
Solar PV added (grid connected)	Germany	Italy	Japan	United States	Czech Republic
Ethanol production	United States	Brazil	China	Canada	France
Biodiesel production	France/Germany		United States	Brazil	Argentina
Existing capacity as of end of 2009					
Renewables' power capacity (including only small hydro)	China	United States	Germany	Spain	India
Renewables' power capacity (including all hydro)	China	United States	Canada	Brazil	Japan
Wind power	United States	China	Germany	Spain	India
Biomass power	United States	Brazil	Germany	China	Sweden
Geothermal power	United States	Philippines	Indonesia	Mexico	Italy
Solar PV (grid connected)	Germany	Spain	Japan	United States	Italy

Ocean (rather than perpendicular to it)], providing concentrated energy as flow speeds are close to those in air (wind), with water being 800 times as dense; thus, a rotor of equal diameter exposed to a water flow of equal speed delivers 800 times as much power [or the same power at $\sqrt[3]{(1/800)}$. this speed, i.e., some 10.8%]. Basins that are filled with water at maximum level differences of a few meters are commonplace in electric storage (in German: Pumpspeicherwerk), with the same being offered by tidal changes of ocean and estuary water levels [the largest tides are seen in river mounds, e.g., River Severn (Wales) or River Rance (France, Normandy) at some 10 m], and not just after pumping water into them by electrical power obtained otherwise, but for free twice a day (actually even 4 times daily using differences of either levels).

Concerning periodical filling and uncharging of storage basins connected to flowing water turbines, there is but one large power plant (240 MW_{el}) in the world still connected to the French grid back in 1967. It is located at La Rance in the mounding of River Rance, next to the famous island Mont St. Michel (Figure 10).

FIGURE 10 Aerial photograph of the world's biggest tidal power station. The Rance Tidal Power Station is located at La Rance (NW France) in the mounding of River Rance, next to the famous island Mont St. Michel.
Source: Photograph: Wikipedia.

Here, the tidal water level differences are far larger than the global average of about 3.5 m (isolated ocean basins, such as the Black Sea and the Baltic Sea, and even the Mediterranean Sea, tend to lack any significant tides, e.g., average tides in the Baltic Sea are some 20 cm); the minimum tidal heights required for a meaningful operation of a tidal power plant are estimated to be some 5 m.[32] Obviously, the gain of energy from a basin of given size interacting with the tide flows increases by the square of tidal water level changes: the amount of water flowing in and out is proportional to tidal height and so is the energy gain from a given mass of water flowing through the turbines. Now being operated for more than 40 years (it was connected to the French grid in 1967), effects from this plant and its overall performance can be well evaluated:

- Local tidal height decreased from some 14 m (!) to <8 m due to deposition and relocation of sediments.
- Corrosion is an issue, requiring both to avoid combination of different alloys in the construction to exclude galvanic effects in seawater, and active electrochemical protection measures.
- There were some ecological side effects concerning distributions of limnetic and marine fishes in and on either side of the basin, with the power plant shifting the limits of marine and limnetic populations of fishes somewhat downstream into the estuary. The turbines rotate so slowly that both fishes and squids can pass through them without being hurt, much unlike classical running-water power plants.
- Actual average output to the grid is some 540 GWh/yr, which is an average of 62 MW, about 25% of nominal power production.

Nevertheless, the La Rance plant remained unique, being the only large-scale plant existing, although sites having even higher tidal heights such as Bay of Fundy (Canada, up to 19 m) were used for yet smaller installations.

Waves[33] may be destructive, but their energy can also be technically exploited, e.g., with transducers based on either bending of some part of the device [hydraulically or by piezoelectric (piezoelectricity means electricity resulting from applying pressure, squeezing) devices]. Other devices employ either systems operating on periodically compressing and expanding air by by-passing volumes in a volume (some hollow metal or concrete bunker) under which the waves pass through and pass this air over a rotating turbine or using the alternating flow directions of water or directly converting the water level oscillation into electricity by electromagnetic induction.

It is estimated that one can gain some 15 kW_{el} (130,000 kWh/yr) from a single meter of coastline around the North Sea, which is comparable to the gain earned by covering about 1 km of land behind this shoreline by photovoltaics in the same climate (this sounds stunning if not paradoxical but consider the width of the seas where the waves could gain energy before from wind rather efficiently converting solar energy). So far, wave energy converters were constructed at Western ocean coasts in the Northern Hemisphere, including SW Norway (Toftestallen, near Bergen), the northernmost Scotch coast, open to the Atlantic Ocean and Oregon. A most straightforward way to convert mechanical movements into electricity, besides piezoelectric techniques, is by magnetic induction. A buoy is towed to the seafloor, with a magnetic shaft tethered to it, while the floating part of the buoy contains some induction coil that hence moves up and down around the shaft within the magnetic field produced by the latter, directly producing an alternating current even though its voltage and frequency still have to be adjusted.

Of course, such interferences with natural water flows may alter and affect the ecological situation in estuaries especially; in the best cases, using both tidal and wave energy reduces coastal erosion.

Less obvious sources of energy associated with ocean and shoreline,[34] respectively, include the temperature difference between surface waters and the deep sea (particularly in the tropics, except for the Red Sea and its annexes like the Gulf of Aqaba/Eilat), some 25–30°C there and the gradient in salt content and, thus, the osmotic pressure difference of some 26 bar between the ocean and freshwater from rivers and creeks running into it (the latter being identical to have this water falling through a Pelton turbine from about 260 m of height).

At a surface temperature of some 300–305°K in tropical waters, the Carnot (most efficient heat engine) maximum efficiency would be about 8%–10%. Usually, in OTEC plants, ammonia is used as a heat-transferring working medium; actual efficiencies are about 3% due to long hoses required to exchange hot and cold waters. Rather than steep cliffs selected earlier, OTEC plants are now located on either ships or floating constructions much like those employed in crude oil production. Given the problems associated with coral bleaching due to surface waters exceeding 30°C, some larger use of the OTEC technology should even provide a real ecological bonus while mixing nutrients between deep and surface waters might cause problems of eutrophication.

The osmotic pressure of some solution is about the pressure that would be exerted if the same concentration of dissolved entities (i.e., ions from, say, $MgCl_2$ solutions, to be counted individually, hence producing 3 times the osmotic pressure in water or acetonitrile than if dissolved in non-ionizing solvent like a long-chain alcohol); 1 M/kg of solvent (here water) hence is equivalent to a gas compressed to 1 M/L, which is tantamount to a pressure of some 24 bar at 20°C. A unimolar $MgCl_2$ solution in water would thus produce an osmotic pressure of more than 70 bar, while the corresponding value for seawater—which is 0.55 M NaCl solution to a first approximation—is about 26 bar [the actual value for Norway is a little smaller because of 1) the water being colder and 2) some dilution by the very freshwater discharges into the Fjord site used, which is rather remote from the open ocean].

Before dreaming of what the latter source could deliver when passing but substantial parts of the freshwater flows of giant rivers like River Amazon, Mississippi, Yangtzekiang (China's largest river), Kongo, or the Siberian rivers like Yenisei through such devices, be reminded of some problems that are not yet solved. The only demonstration plant existing so far is located at a creek mounding into a fjord in Norway, producing some 3–4 kW of electric power since November 2009. There are three quite different kinds of converting osmotic potentials into electricity:

- Using a water-permeable yet pressure-proof membrane (this is done in Norway). The saltwater content of this closed membrane volume is going to absorb pure water from the freshwater passed along outside (i.e., in the mounding of the river). Due to the osmotic pressure, the water level inside will increase considerably, allowing to pass it over a falling-water turbine and produce electricity eventually. This process is a somewhat periodical one: as the saltwater gets diluted during the process, you need to discard it into the ocean sooner or later and refill the chamber with "pure" (3.5% salt) ocean water. By now, electric power output is about 1 W/m^2 of membrane interface area; the aim is to achieve 5 W/m^2 soon.
- Electrochemical settings using the diffusion potential: concentration differences leveled off by diffusion create electric potentials even though no redox reactions are involved in charge transfer. Likewise, second-type electrochemical cells draw upon concentration potentials: an electrode made of combined silver and AgCl (both solid and mixed among each other) will adjust its potential according to the concentration level of chloride ions. This can be used both in analytical chemistry and for producing electric currents from chloride solutions that differ in concentration (freshwater typical values being 1–2 mM/L as opposed to 0.55 M/L in seawater, giving an open-cell voltage of 150–170 mV).
- Theoretically speaking, the zeta (electrokinetic) potential could also be used in osmotic energy conversion: by osmotic pressure differences forcing water through some membrane, or a column filled with a packed solid, it will produce an electrical potential difference between either side of the interface. This potential is due to selective adsorption of cations or anions onto a typically charged particle, charging being caused by oxide/ hydroxide particles (say, wet alumina) behaving as an acid (adsorbing hydroxide) or base (adsorbing protons) depending on local pH, co-sorbents, and material (point of zero zeta potential). Then, the other, non-adsorbed ions will be passed through along the solvent, and a potential of typically half a V forms. The effect is reversible (electrokinetic water pumping).

The problems are with membrane stability and, more generally, also affecting electrochemical systems, clogging of the interfaces by biomass (mainly phytoplankton) or even mineral concretions. The Norwegian success terminated a history of decades of failed experiments on osmotic power production.

Things became a little different—and better, rather more advanced—because of OTEC. The first plant ever of this kind was constructed by French Georges Claude (1870–1960) at the coast of Cuba back in the late 19th century. OTEC devices are simple thermal power plants using rather small T differences, more like a steam engine compared to turbines, coal-fired plants, or internal combustion engines.

The global distribution of chances for this way of harvesting is solar energy indirectly (having hot water from insolation in the tropics while cooling is provided by Arctic or Antarctic undercurrents at some 1000–2000 m of depth).

Eventually, there can be integrated offshore energy parks making use of almost all the energy sources discussed above combined on, or beneath, a tethered floating island.

Geothermal Energy

For many decades, *geothermal* energy[35,36] has been well established for heating purposes in countries like Iceland, New Zealand, and several developing countries in Central America such as El Salvador. Although electrical uses, with geothermal (fumarole) vapors directly run through a steam turbine, were first tried in Italy more than a century ago (at Larderello in 1904, delivering about 200 W), corrosion and clogging problems remain severe until this day. The obvious reason is the "contamination" of fumaroles with both clogging agents like boric acid and hydrolyzable volatile metal chlorides, besides the large shares of corrosive gaseous acid precursors like SO_2, HCl, and HF. Thus, one has to create a primary heat exchange cycle directly exposed to these corrosive items, as well as a secondary one linked to the heat/mechanical/electrical conversion systems, much like in NPPs and mainly for the same reasons (if not even worse here), and worse, due to the rather small heat difference, this decreases total efficiency of conversion. When obtaining the vapor from underground wells drilled several kilometers into the Earth's crust rather than operating close to active volcanoes, clogging of drilling holes or rock fractures required to circulate some operating medium also remains critical. Hence, it is safe to predict that geothermics will remain more concerned with heating (houses, swimming pools, etc.,) than with electricity production in the near future as well.

Wind Power

Wind power now is an established source of energy, with average production costs per energy unit (e.g., cent per kilowatt-hour) coming close to those by conventional (fossil) energy sources.[37–40] In certain countries, the share of wind power in total electricity production exceeds 20% (Denmark, Mongolia), and there is a broad international consensus that state subsidies are no longer required nor given to enhance the rate of implementation. Rather, as with all kinds of renewable energies that are subject to considerable periodic (sunlight) or non-periodic changes of supply, the optimum strategy of storage becomes imminent. Hence, electrolysis of water (and possibly secondary production of methane or methanol) by "excess" wind power (excesses being produced by mismatches with the grid also), storage of H_2 or CH_3OH, and use of the latter energy carriers in either vehicles or stationary fuel cells connected to the electric grid are gaining importance.

The size of individual plants is limited by the necessity to erect them and thereby place 100-ton items more than 100 m above the ground within millimeter precision, that is, by size of the available cranes. The largest wind power systems thus now have 200 m rotor diameter and deliver some 8 MW while arrays of them ("wind parks") can produce outputs in the size of classical power plants and NPPs both on- and offshore. However, in either case, interactions with local fauna may become significant.

Solar

Solar energy is the key source of almost all the biological and meteorological processes operating on Earth.[41,42] Semiconductor solid-state devices allow for a remarkably efficient exploitation of this source, with an additional role for thermal processes that latter rely on focusing and, thus, on non-scattered sunlight, that is, on clear skies. These thermal processes include production of fuels by cycles involving zinc or cerium oxides as well as metallurgical transformations.

Solar thermal power plants, augmented by natural gas or biogas combustion during night and other dark times, are now realized in a scale of hundreds of megawatts, while photovoltaics in 2011 first yielded an all-year average of more than 2000 MW (2 GW) electrical output in Germany (some 3% of total current). The price breakdown in production and processing of semiconductors (which need not be that pure or advanced than with electronic microdevices) supports the "boom" furthermore, regardless of fast cuts in state subsidies paid for supplying PV current to the public grid in all the EU member countries now.

When considering very large plants such as in the Desertec initiative, the increase of radiation absorption however becomes likely to influence the performance of solar parks by itself: large volumes of heated air will rise right above the plants, causing an increased dust advection to the panels as well as clouds to form on their top (both sailplane pilots and birds of prey look for typical kinds of clouds to spot regions of upwind over hotter surface areas!). High-yield photovoltaics by thin-layer solar cells [a few micrometers' total thickness, unlike 0.3–0.5 mm with polycrystalline SiH_x ("blue silicon")] demands rare (e.g., In) and/or highly toxic (Cd, As, Se, Te) elements, causing problems in all mining, processing them and eventually abandoning old PV devices when their performance sharply decreases after several decades.

Biomass

Biomass can be used in a variety of ways as a source of energy, with combustion of wood and vegetable or animal oils for both heating and illumination purposes dating back as far as the Stone Age.[43,44] More recently, plant oils were introduced into internal combustion engines (first with peanut oil; diesel, 1900), while other engines were fed with either wood distillates (containing mainly CH_3OH and acetone besides H_2 and CO) or ethanol produced by microbial activity, much after steam engines had been powered by either wood, wood-processing residues (e.g., sawdust), or peat to replace hard coal (which is biogenic in itself, of course).

Conversion methods of brown coal—lignite—by hydrogenation, gasification (steam gas process), and liquefaction can also be readily applied to biomass, including less obvious representatives of biomass such as sewage sludge (chiefly containing heterotrophic bacteria), and then mostly even take less vigorous conditions in terms of all temperature, H_2 pressure (3–10 bar rather than hundreds of bars), and needed catalysts. Finally, motivated by the fact that biomasses, especially scrap biomasses, became an item of fuel production (and waste treatment/compaction) once again, the very former coal liquefaction plants in, e.g., South Africa are now used for this purpose. Regionally, in Germany, success and economic performance were poorer, however.

Using scrap or digestible waste fractions relieves an ethical problem from the competition among food/fodder and "energy plants" for the same agrarian areas, but even after avoiding this, one must bear in mind the poor area productivity of photosynthesis—the only economically viable source of biomass energy carriers in a large scale—which typically is 0.5% or a few kilograms of reduced $C/m^{2}*a$, i.e., far short of photovoltaics. While there will be a role in waste processing, a large-scale use of biomass for energy purposes such as in Brazil poses a lot of difficult problems, including ecological ones associated with monocultures, possible fertilization/eutrophication, and high water requirements.

Renewable Energy in Germany and the Planned Nuclear Exit

Concerning its gross domestic product (GDP), Germany is the largest national economy in Europe and No. 4 in the world. In 2009, it was second in export and third in import values. Like with the GDP, Germany ranks No. 4 in energy consumption [measured in fossil fuel (hard coal) mass equivalents (BTU: British thermal unit)] but just No. 21 among the energy producers in the world).[46] The intention of the Federal Government of Germany in fulfilment of Kyoto Protocol obligations is a reduction of greenhouse gas production of about 40% by 2020 and up to about 80%–95% reduction by 2050. Quite recently, the greenhouse gas issue and the risk of NPPs (disaster of Fukushima in 2011) were aggravated by the decision to abandon nuclear energy use in Germany in the early 2020s. Figure 11 shows the fuel mix in 2010 and the aimed fuel mix in 2050 in Germany, which mainly will be supplied by wind and solar power.

By steadily replacing fossil energy sources with renewables, the share of the latter will increase, even allowing for a slight increase in energy consumption (which, in Germany, like most other highly developed countries, is rather constant for decades now, notwithstanding a slight decrease in population happening soon). Hydropower is fully established now except for reactivation of very small local plants, many of which had been in operation since the early 20th century. Hence, the present 3% share will remain almost the same; 35% from photovoltaics corresponds to an average output of some 23 GW, which, in our climates, is tantamount to an introduced peak power of 130–150 GW, more than twice that what now is funneled into the entire grid by all kinds of power plants. Dealing with this excess energy on sunny summer afternoons, possibly by chemical storage (water electrolysis, then linked to fuel cells), remains to be figured out. The total area required to produce this amount of PV electricity is about 1000 km^2 [<0.3% of Germany's total area (356,000 km^2) and <10% of the fields on which "energy plants" are now grown (>12,000 km^2 are covered with rape alone)], even assuming no further improvement in today's Si hydride polycrystallinic or $CuIn(S; Se)_2$ thin-layer solar cells (some 13% efficiency).

Growth and Booming Region Ems-Axis, Lower Saxony (NW Germany)

The previous entries dealt with the global-scale relevance of renewable energies. Besides the environmental issue, there are both economic and social surpluses produced by creating novel workplaces.

This can turn a formerly "just" agrarian region into some diversified boom area as will be shown by the example of the so-called River Ems-Axis (Lower Saxony, Northwestern Germany). For more than a decade now, the region keeps increasing its workforce by 3% per year—no "Mc jobs" but fully qualified jobs that produce social security and modest earnings, with >10,000 enterprises that keep expanding and creating new jobs one year after another. The regional motto reads: "Powerful, innovative and ready to achieve by unconventional solutions—these are our region's benchmarks."

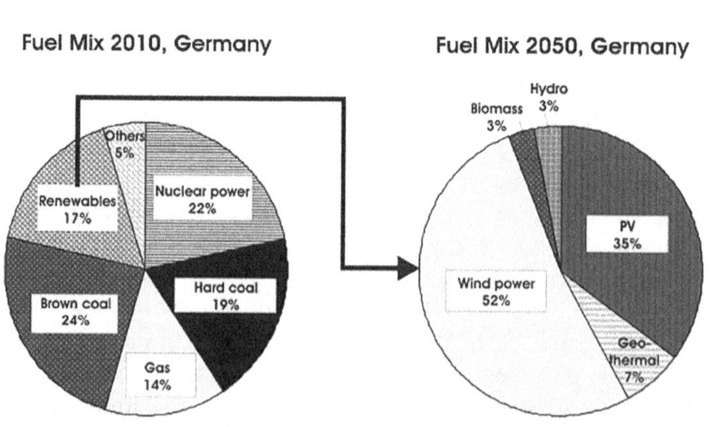

FIGURE 11 Energy mix contributions in Germany in 2010 and the probable future in 2050.[45]

As the Ems-Axis is located next to the North Sea, maritime-related activities are prominent by locating shipyards, shipping companies, and wind power plant producers, among suppliers of other renewable energies. This model region was created through a combination of prudent political support, improvement of infrastructures, synergy among regionally active enterprises, and finally the support of the public. It is located near the Dutch border in Central Europe, making use of already existing East-West connections, and, in addition, links the North Sea shores to the German megalopolis Ruhr district, which is the most populated part of the most populated and economically prolific *Bundesland* of Germany. The Ems-Axis includes the counties Wittmund, Aurich, Leer, Emsland, Grafschaft Bentheim, and Emden City with its large harbor (see Figure 12).

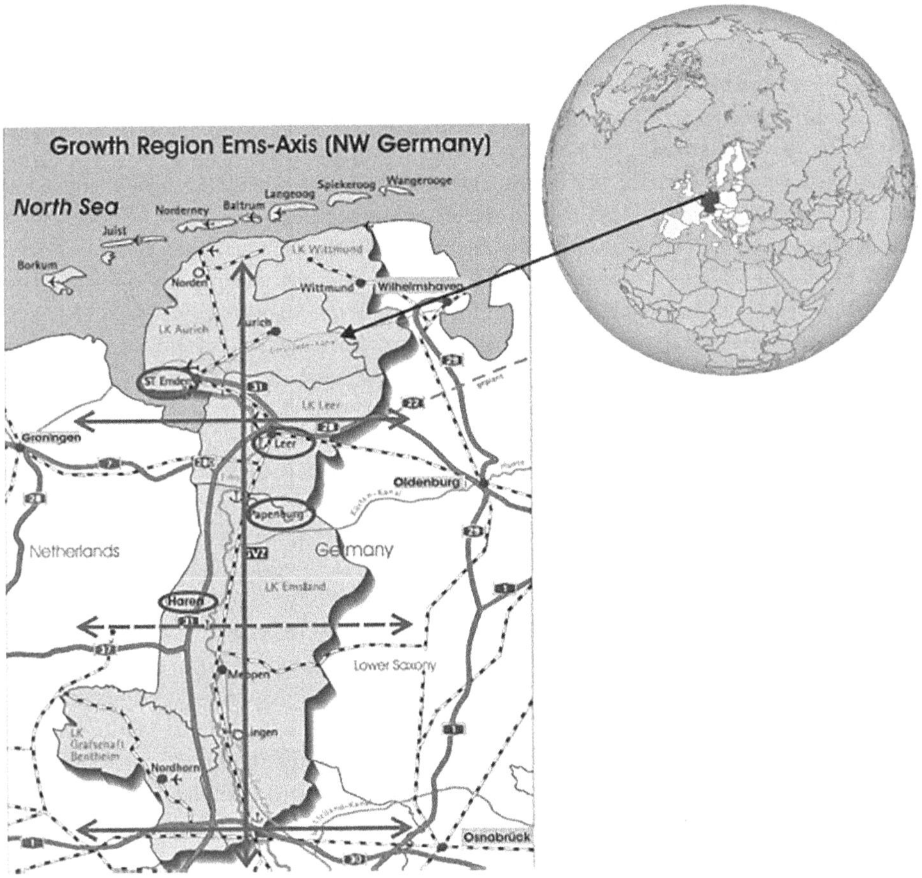

FIGURE 12 The growth region Ems-Axis in Northwestern Germany. This figure shows the excellent infrastructure that will imminently cause new enterprises to settle and expand here. The East-West and South-North highways (motorways) are marked with Nos. 7, 28, 30, 31, and 37, while rivers and channels for ship travel are marked medium gray. These are the River Ems, which is deep and wide enough to permit economically meaningful transportation by ship, and the Dortmund–Ems channel, which extends almost parallel to it. Railway tracks are outlined in black and white. Framed: the cities of Leer and Haren/Ems are among the most important locations for ship owners all over Germany. In Papenburg, there is the Meyer shipyards, among the largest in Europe and moreover the one producing the biggest ships (passenger and cruise ships). Other notable shipyards located next to the shore at Emden recently rather switched to producing wind power plants. The Ems-Axis is distinguished by intense economic activities covering all energy supply, integrated maritime activities, agriculture, and tourism; processing plastics and metals; building vehicles and machines; and providing logistical infrastructure.
Source: The main figure of the Ems-Axis is modified after http://www.emsachse.de. The figure on the right is from Wikipedia, TUBS.

There are six permanent workgroups concerned with energy, integrated maritime economy, tourism, production of plastics items, vehicles and machines, and finally logistics to initiate and run projects. It is the aim of these workgroups and the economic region to make Ems-Axis an independent axis along which economic, travel, and transport activities will organize. This implies strengthening economy-related infrastructure and creating networks for regional economy.

The cities of Haren and Leer combine to be the second-largest shipowner's site in Germany. A total of 750 ships are run from here, making these two special and significant players in running ship travel and dockyards and providing additional maritime goods and items, together with Papenburg. The existing travel infrastructure allows to process materials inshore and, using local logistics, build huge ocean liners such as that for Disney Cruise Line (340 m long and 37 m wide, 128,000 tons, can accommodate 2500 passengers) at Papenburg's Meyer dockyards (Figure 13). The latter commands the world's most advanced instrumentation and facilities for building ships, its workforce being about 2500.

One should mention that there is minimal required bureaucracy used to acquire these infrastructures. This enabled Motorway 31—a crucial North-South connection—to be completed years ahead of planning, with the region providing the required funds itself by joint and coordinated action. Another ambitious project was Euroharbour Emsland at Haren (operated jointly with nearby Meppen city), the construction of which began in 2007. In August 2011, construction of the plant of ENERCON wind power devices began here. ENERCON is the manufacturer of the most advanced wind rotors (the actual propellers), producing blades that are aimed to deliver 3 MW per unit at the Emsland Euroharbour site. The principal administrative person *(Landrat)* of the largest of the involved counties and cities of Ems-Axis, Emsland itself, uses to call this a "pro-climate climate," stressing that currently, an impressive 82% of the energy consumed in the county are derived from renewable energy. The location of Euroharbour, the 24,000-population town of Haren, even boasts a 100% renewable electric current production. Among the renewables, wind is most important for the Ems-Axis region. With the shore nearby and little terrain roughness, it is most suited to create onshore wind plants; thus, NW Germany outcompetes the southern parts of the country in this respect.

Yet, there are also offshore wind power parks in the region now. In 2010, the first one in German domestic waters, "Alpha Ventus," was erected and connected to the grid. As for crucial parts of wind power technology, BARD Energy at Emden both produces rotor blades specifically designed for offshore application (there are special criteria to withstand salt corrosion, impact of water drops on the fastly moving blades, etc.,) and likewise constructs entire power plants at offshore sites (Figure 14).

Suffice this to show features of the booming Ems-Axis economic region, which additionally sports, for example, the Transrapid (maglev) testbed at Lathen, and notably a big plant at Werlte, which will be the first in the world to convert excess wind power energy via hydrogen and hydrogenation of CO_2 into methane for energy storage purposes (to be combined with natural gas CH_4 and biogas). With regard to issues of energy use efficiency and extending the amount of renewable energy supply, the Ems-Axis consortium stated in May 2011:

FIGURE 13 Meyer shipyards at Papenburg (NW Germany). The largest dock is incredibly 504 m long. Right figure: Norwegian Jewel in front of the 70 m tall Meyerwerft Hall.
Source: Pictures courtesy of Wikipedia: left, C. Walther; right, satermedia.de, C. Brinkmann.

FIGURE 14 BARD Emden Energy GmbH & Co. KG produces rotor parts, etc., for offshore wind power plants, then mounting them at sea also. Located at Emden, it belongs to the economic region Ems-Axis. Photos: (Left) German special crane ship for the setups of offshore wind farms called Wind Lift I (BARD) in the harbor of Emden. **Source:** Wikipedia, photographer Carschten. (Right) BARD offshore 1 (Mai, 2011). Courtesy of the BARD Group.

"Partners in growing region Ems-Axis consider big chances for local and regional economy to be obtained from making energy supply a cornerstone of economical politics. Simultaneously they respond to their environmental responsibility by making energy use more efficient and increasing the share of renewable energy sources.

Growing region Ems-Axis is capable of becoming a model (blueprint) energy supply region for the future. Concerning Germany, this region both has the largest concentration of wind power plants and is the site of globally active producers of wind power devices. In addition, renewable energy is earned here from all biomass, sun and geothermal resources. So there is a bandwidth of competence in energy supply which yields new impetus to the region by enhanced cooperation and thus advantages in competition which in turn once more improves the economic performance of the local enterprises".

More pieces of information on the Ems-Axis region, including pertinent enterprises, can be obtained via http://www.emsachse.de.

Conclusion

The present mix of renewable resources used in both thermal and electrical energy delivery represents a superposition of both technical problems still to be overcome (the less so) and political decisions, many of which are made in favor of protecting the respective domestic industries for both producing energies and the very power plants required to obtain and convert them: this partly is a quite reasonable and, to some extent, even responsible industrial policy. Now, there are "old" energy sources, exploitation of which has become so costly that it is worthwhile only in certain most simple conditions, including hard coal and, in another way, oil sands. This statement refers to all economic costs of exploitation, ecological side effects (as well as cultural ones such as destruction of villages and first-nation settlements in favor of open pits), and risk production causes to the workers. The renewables make it to the market step by step with their increasing ability to compete economically and the perspective to relieve old dependences, in addition to avoiding the above risks by offering genuine technical alternatives.

Of course, this might produce problems for countries that have virtually nothing else to offer to today's global markets than their fossil energy carriers, including uranium, but not to some of the "big shots" in fossil fuel mining—highly industrialized countries such as the Unites States, Canada, Australia, and Russia. Apparently, however, there is no convincing perspective of sustainable development by which the common population might benefit from exploitation of fossil energy carriers alone for countries such as Niger in Western Africa (uranium) or Yemen in the Middle East (oil). Other large uranium suppliers like Gabon (West Central Africa) or Kazakhstan (Central Asia/ Eastern Europe) have a more diversified supply portfolio. Several of the Arab oil-producing countries are very aware of what might happen to them, their regimes, their population, and their common welfare (which is often truly restricted to some indigenous minorities) when oil continues to get scarce, and there are cautionary economic examples of countries, societies, and national economies running out of the single, principal minable resource the entire economy was based on, such as the tiny South Pacific Republic of Nauru (phosphate) and Bolivia in Central South America (tin, silver).

Nevertheless, the exchange of our joint economic basis for energy production appears feasible globally within some 50 years from now. It remains to be seen whether this is fast enough both to control climate effects from fossil combustion within acceptable limits and to reorganize completely our strategies of personal transportation while avoiding yet more catastrophes like those in Chernobyl or Fukushima [as well as the failure of a hydropower plant in Longarone (Friaul, NE Italy) which took some 2000 lives in 1963]. Besides this, nuclear power plants—like other technical systems—can run into operation states where they almost or entirely escape control. If a catastrophic accident then can be avoided due to self-regulation or simply luck, it is by no means satisfying or consoling that, e.g., nuclear reactors arrived at states that were not even known to their own operators for extended periods of time (like in Forsmark, Sweden, in 2006), let alone these people would be able to influence it anymore.

The future awaits us but is notoriously hard to predict, but we should take chances, even severe ones, if we decide either way, and we should be aware that doing nothing is tantamount not only to taking chances but also pursuing ways that we know for sure to be not sustainable, not even in the shorter term.

Acknowledgments

We are deeply thankful to the thousands of comments given by colleagues, students, friends, and especially opposite thinking people during attractive and highly motivating discussions during the past decades. For intensive support during the preparation of this entry, we would like to thank Prof. Michael Tomaschek, University of Applied Sciences Emden/Leer.

This entry corresponds in parts with chapter 4.4 (Energy—One of the Biggest Challenges of the 21st Century) of the textbook by Fraenzle, S., Markert, B., and Wuen-schmann, S. (2012) on an "Introduction to Environmental Engineering" published by Wiley/VCH, Weinheim.

References

1. Shafiee, S.; Topal, E. When will fossil fuel reserves be diminished? Energy Policy **2009**, *37* (1), 181–189.
2. Kutz, M. *Environmentally Conscious Fossil Energy Production;* John Wiley and Sons, Hoboken, NJ, 2010.
3. Kreith, F.; Goswami, D. *Principles of Sustainable Energy;* CRC Press, Boca Raton, FL, 2011.
4. Fraenzle, S.; Markert, B.; Wuenschmann, S. *Introduction to Environmental Engineering—Innovative Technologies for Soil, Air and (Ground)water Remediation and Pollution Control;* Wiley-VCH: Weinheim, 2012.
5. Federal Institute for Geosciences and Raw Material Research. *Reserven und Verfügbarkeit von Energierohstoffen (Kurzstudie);* Bundesanstalt für Geowissenschaften und Rohstoffe: Hannover, 2009.
6. Federal Ministry of Economics and Technology. *Energie in Deutschland (Energy in Germany);* BMWI: Berlin, 2010.
7. U.S. Geological Survey. *Circum-Arctic Resource Appraisal: Estimates of Undiscovered Oil and Gas North of the Arctic Circle;* Washington, DC, 2008.
8. Campbell, C. The Rimini Protocol: An oil depletion protocol. Heading off economic chaos and political conflict during the second half of the age of oil. Energy Policy **2006**, *34* (12), 1319–1325.

9. Blum, A. Die finale Ölkrise—fossile Brennstoffe, vor allem Erdöl, sind endlich und eine Verknappung ist absehbar, 2005, available at http://www.raize.ch/Geologie/erdoel/oil.html (accessed on April 11, 2005).
10. Kondratyev, K.; Krapivin, V.; Varostos, C. *Global Carbon Cycle and Climate Change;* Springer: Berlin, 2003.
11. Leroux, M.; Comby, J. *Global Warming. Myth or Reality;* Springer: Berlin, 2005.
12. Hoel, M.; Kverndokk, S. Depletion of fossil fuels and the impacts of global warming. Resour. Energy Econ. **1996**, *18* (2), 115–136.
13. Ostwald, W. *Die Mühle des Lebens;* Theod. Thomas: Leipzig, 1911.
14. Voiland, A. *2009: Second Warmest Year on Record; End of Warmest Decade;* NASA Goddard Institute for Space Studies (accessed January 22, 2010).
15. Reschetilowski, W. Vom energetischen Imperativ zur nachhaltigen Chemie. Nachr. Chem. **2012**, 134–136.
16. United Nations. United Nations Framework Convention on Climate Change: FCCC/CP/1997/INF. 5, List of participants (COP 3), available at http://en.wikipedia.org/wiki/United_Nations_Framework_Convention_on_Climate_ Change (accessed January 21, 2011).
17. Romberg, B. Reaktorneubau: Amerika wagt Renaissance der Kernenergie. Süddeutsche Zeitung, Febr. 13th, 2012.
18. Breban, D.C.; Moreno, J.; Mocanu, N. Activities of Pu radionuclides and ^{241}Am in soil samples from an alpine pasture in Romania. J. Radioanal. Nucl. Chem. **2003**, *258*, 613–617.
19. Gill, V. Chernobyl zone shows decline in biodiversity. BBC News online, Science and Environment, 2010, available at http://www.bbc.co.uk/news/science-environment-10819027 (accessed January 20, 2010).
20. Etspüler, M. *Transmutation: Die zauberhafte Entschärfung des Atommülls*; Frankfurter Allgemeine Zeitung: Frankfurt, 2011.
21. De Bruyn, D. European fast neutron transmutation reactor projects (MYRRHA/XT-ADS). IAEA Review Paper, 2009, available at http://www-pub.iaea.org/MTCD/publications/PDF/P1433_CD/datasets/summaries/Sum_SM-ADS.pdf (accessed November 2011).
22. Hoffert, M.; Caldeira, K.; Benford, G.; Criswell, D.; Green, C.; Herzog, H.; Jain, A.; Kheshgi, H.; Lackner, K.; Lewis, J.; Lightfoot, D.; Manheimer, W.; Mankins, J.; Mauel, M.; Perkins, J.; Schlesinger, M.; Volk, T.; Wigley, T. Advanced technology paths to global climate stability: Energy for a greenhouse planet. Science **2002**, *298* (5595), 981–987.
23. Turner, J. A realizable renewable energy future. Science **1999**, *285* (5428), 687–689.
24. Kaltschmitt, M.; Streicher, W.; Wiese, A. *Renewable Energy: Technology, Economics and Environment;* Springer: Berlin Heidelberg, 2010.
25. Pimentel, D., Ed. *Biofuels, Solar and Wind as Renewable Energy System. Benefits and Risks;* Springer, Science+Business Media B.V., Dordrecht, 2008.
26. Olah, G.; Goeppert, A.; Prakash, C.K. *Beyond Oil and Gas: The Methanol Economy*; Wiley-VCH: Weinheim, 2006.
27. REN 21. Renewables 2010 Global Status Report (Paris: REN 21 Secretariat). Deutsche Gesellschaft für Technische
28. Zusammenarbeit (GTZ) GmbH, 2010, http://www.ren21. net (accessed January 22, 2010).
29. REN 21. Renewables 2007 Global Status Report (Paris: REN 21 Secretariat and Washington, DC: Worldwatch Institute). Deutsche Gesellschaft für Technische Zusammenarbeit (GTZ) GmbH, 2008; 21, available at http://www. ren21.net (accessed January 22, 2010).
30. Markert, B.; Wuenschmann, S.; Fraenzle, S.; Figueiredo, A.; Ribeiro, A.P.; Wang, M. Bioindication of trace metals—With special reference to megacities. Environ. Pollut. **2011**, *159*, 1991–1995.
31. Charlier, R.; Finkl, C. *Ocean Energy: Tide and Tidal Power;* Springer: Berlin, Heidelberg, 2009.
32. Hoffmann, V. Energie aus Sonne, Wind und Meer. Harri Deutsch Thun (SUI): Frankfurt/Main, 1990.

33. Cruz, J. *Ocean Wave Energy. Current Status and Future Perspectives;* Springer: Heidelberg, 2008.
34. Multon, B. *Marine Renewable Handbook*; Wiley and Sons, Weinheim, 2011.
35. Glassley, W. *Geothermal Energy. Renewable Energy and the Environment*; CRC Press, Taylor and Francis Group, Boca Raton, FL, 2010.
36. Ghosh, T.; Prelas, M. Geothermal energy. In *Energy Resources and Systems;* Gosh, T., Prelas, M., Eds.; Springer, Weinheim, 2011; 217–266.
37. European Wind Energy Association. *Wind Energy—The Facts: A Guide to the Technology, Economics and Future of Wind Power;* Taylor and Francis Ltd., Oxford, 2009.
38. Nelson, V. *Wind Energy: Renewable Energy and the Environment;* CRC Press, Boca Raton, Florida, 2009.
39. Hau, E. *Wind Turbines: Fundamentals, Technologies, Application,* Economics, 3rd Ed.; Springer: Berlin, 2012.
40. Maki, K.; Sbragio, R.; Vlahopoulos, N. System design of a wind turbine using a multi-level optimization approach. Renewable Energy, **2012**, *43,* 101–110.
41. Goetzberger, A.; Hoffmann, V. *Photovoltaic Solar Energy Generation;* Springer: Berlin, Heidelberg, 2005.
42. Foster, R.; Ghassemi, M.; Cota, A. *Solar Energy: Renewable Energy and the Environment*; CRC Press, Taylor and Francis Group, Boca Raton, FL, 2009.
43. Goldemberg, J.; Coelho, S. Renewable energy—Traditional biomass vs. modern biomass. Energy Policy, **2004**, 32 (6), 711–714.
44. Fraenzle, S.; Markert, B. Metals in biomass: From the biological system of elements to reasons of fractionation and element use. Environ. Sci. Pollut. Res. **2007**, (6), 404–413.
45. Burkhardt, M.; Weigand, T. Der Strom-Mix in Deutschland (the fuel mix in Germany). German television ZDF, 2011, available at http://www.heute.de/ZDFheute/inhalt/24/ 0,3672,8233016,00.html (accessed January 21, 2011).
46. U.S. Energy Information Administration. Total Primary Energy. Statistics on Germany, available at http://www.eia.gov/countries/country-data.cfm?fips=GM (accessed January 22, 2010).
47. Flavin, C. Last word: Renewable energy at the tipping point. In *Renewable Energy Policy Network for the 21st Century (REN 21) (2010) Renewables 2010 Global Status Report (Paris: REN 21 Secretariat)*; Deutsche Gesellschaft für Technische Zusammenarbeit (GTZ) GmbH, 2010; 52–53.

6

Energy and Environmental Security

Muhammad Asif

Introduction

The economies of all countries, and particularly of the developed countries, are dependent on secure supplies of energy. Energy security means consistent availability of sufficient energy in various forms at affordable prices. These conditions must prevail over the long term if energy is to contribute to sustainable development. Owing to the pivotal role of energy in the modern age, energy security is at the heart of national and international energy policies across the world.

The global environmental scene has changed dramatically over the last century. Global warming and its consequent climatic changes driven by human activities, in particular the production of greenhouse gases (GHGs), directly impact the environment. The environmental security to a certain extent is linked with the energy security as there is an intimate relationship between energy and environment. The production and use of all energy sources results in undesirable environmental effects, which vary based on the health of the existing ecosystem, the size and health of the human population, energy production and consumption technology, and chemical properties of the energy source or conversion device. A shorthand equation for the environmental impacts of energy production and use has been provided by Solomon:[1]

$$I = PAT$$

where I is the environmental impact, P is the size of the human population, A is the affluence of the population (e.g., per capita income and/or energy use), and T is the technology (e.g., energy efficiency, emission rate of air and water pollution). A comparison of different types of energy systems in terms of CO_2 emissions is provided in Table 1.[2]

The energy resources, presently being consumed in the world, can be broadly classified into three groups: fossil fuels, nuclear power, and renewable energy. Fossil fuels contribute to more than 80%

TABLE 1 Comparison of CO_2 Emissions from Different Energy Systems

Type of Power Plant	Fuel/Type of Energy	CO_2 (kg/kWh)
Steam power plant	Lignite	1.04–1.16
Steam power plant	Hard coal	0.83
Gas power plant	Pit coal	0.79
Thermal power plant	Fuel oil (heavy)	0.76
Gas turbine power plant	Natural gas	0.58
Nuclear power plant (pressurized water)	Uranium	0.025
Thermal power plant	Natural gas	0.45
Solar thermal power plant	Solar energy	0.1–0.15
Photovoltaic power plant	Solar energy	0.1–0.2
Wind power plant	Solar/wind energy	0.02
Hydroelectric power plant	Hydropower	0.004

of world's total energy supplies and are mainly classified into the following three types: coal, oil, and natural gas. It is the set of fossil fuels that primarily bond the environmental security with energy security. Global warming—the predominant threat to environmental security—to a great extent is a consequence of fossil fuel consumption. As the two issues go side by side, their potential solutions would also be closely related.

Energy and Sustainable Development

Energy is at the heart of the existence of present-day societies. The accomplishments of civilization have largely been achieved through the increasingly efficient and extensive harnessing of various forms of energy to extend human capabilities and ingenuity. Providing adequate and affordable energy is essential for eradicating poverty, improving human welfare, and raising living standards worldwide. The per capita energy consumption is an index used to measure the socioeconomic prosperity in any society—the human development index (HDI) of a country has strong relationship with its energy prosperity.[3] A direct correlation between the access to electricity and the economic well-being in a range of countries, for example, is indicated in Figure 1.[4] Throughout the course of history, with the evolution of civilizations, the human demand for energy has continuously swallowed. Of present, key factors driving the growth in energy demand include increasing human population, modernization, and urbanization.

Poverty, hunger, disease, illiteracy, and environmental degradation are among the most important challenges facing the world. Poor and inadequate access to secure and affordable energy is one of the crucial factors behind these issues. Electricity, for example, is vital for providing basic social services such as education and health, water supply and purification, sanitation, and refrigeration of essential medicines. Electricity can also be helpful in supporting a wide range of income-generating opportunities. Access to electricity remains to be a serious issue in developing countries as around 13% of the world's population—do not have access to it yet.[5] As of 2017, around 3 billion people rely on traditional biomass, including wood, agricultural residues, and dung, for cooking and heating. More than 99% of people without electricity live in developing regions, and four out of five live in rural areas of South Asia and sub-Saharan Africa.[6,7]

There is a global consensus that the provision of secure, affordable, and socially acceptable energy services is a prerequisite for eradicating poverty in order to achieve the Millennium Development Goals (MDGs). The Earth Summit 2002 strongly urged the nations to "Take joint actions and improve efforts to work together at all levels to improve access to reliable and affordable energy services for sustainable development sufficient to facilitate the achievement of the MDGs, including the goal of halving the proportion of people in poverty by 2015, and as a means to generate other important services that mitigate poverty, bearing in mind that access to energy facilitates the eradication of poverty." United

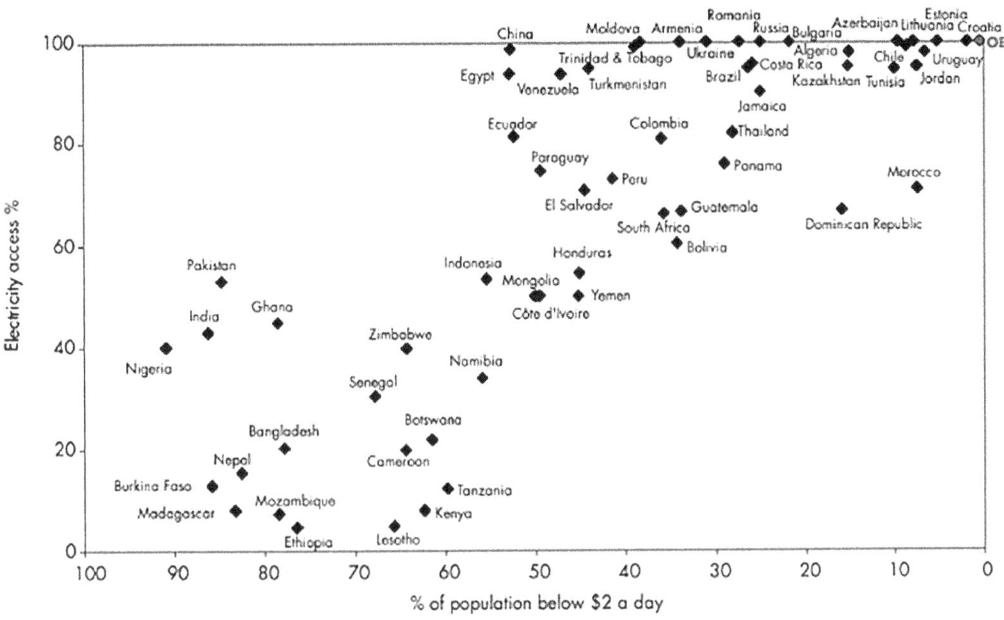

FIGURE 1 Relationship between economic prosperity and availability of electricity.

Nations also acknowledged that "without increased investment in the energy sector, the MDGs will not be achieved in the poorest countries."

With the growing world population and people's innate aspirations for improved life, a central and collective global issue in the new century is to sustain socioeconomic growth within the constraints of the Earth's limited natural resources while at the same time preserving the environment. This target—sustainable development—can only be met by ensuring energy sustainability.

Energy Security

Energy has attained the status of an indispensable strategic commodity, and ensuring its availability is one of the important responsibilities of governments across the world. Failure to ensure robust provision of interruption-free and affordable energy may result into serious financial and social problems. Breach of energy security is also prone to lead to national instability. A great number of countries in the world, particularly the industrialized mature economies, have placed energy security on top of their national policies. The high degree of multidisciplinary interdependency among nations makes energy security vital from both national and international perspectives. German Foreign Minister Frank-Walter Steinmeier emphasizes this fact by stating that maintaining global security in the 21st century will "inseparably also be linked to energy security."[8] At their summit in Brussels in March 2007, the European Union (EU) leaders adopted a road map, the highlight of which was "Energy is what makes Europe tick. It is essential, then, for the European Union to address the major energy challenges facing us today and in future." Similarly, the 2006 G8 summit held in St. Petersburg had energy security on top of its agenda ahead of issues such as education, health, trade, environment, and terrorism.[9] The seventh sustainable development goal (SDG) by the United Nations—ensure access to affordable, reliable, sustainable, and modern energy—is all about energy security.[5]

Owing to a great degree of dependency upon oil and gas, the international energy market generally gauges the energy security in terms of secure supplies of these two commodities. The balance of evidence suggests that the present international energy market is vulnerable to different types of security risks. Energy security risks can be broadly classified into two categories: first, man-made— international geopolitical and geostrategic conflicts resulting into military attacks and wars, and sabotage and terrorist activities on relatively smaller scale; and second, natural— depletion of fossil fuel reserves, floods, fires, and earthquakes.

Risk Assessment

As with the supply of any commodity, the security of energy supply depends on the availability, size, and location of the energy reserves. The availability and the size of the reserves and their location then determine the supply line: the exploitation, transportation, and utilization arrangements. Conversely, the security of the exploitation, transportation, and utilization arrangements then determines the security of energy supplies. Using a very simple model, the major components of the risk can be identified:

- The number of potential disruptive events
- The probability that, given a potential disruptive event, the supply line will be successfully disrupted
- The potential consequences of such a disruption.

The contingency strategy could have three possible prongs. In order to minimize the risk, the number of potential disruptive events must be made as low as possible. Similarly, the probability that such an event will result in disruption must also be minimized. Finally, contingencies must be in place to successfully survive such a disruption.

The effects of human errors can be minimized by proper education and training of the operators, and by ergonomically suitable design. Energy companies have always strived to achieve this but failed spectacularly on a number of occasions, such as Windscale in the U.K., Chernobyl in the Ukraine, or Three Mile Island in the United States, and most recently Fukushima in Japan. However, based on past experience, the probability of human error can be estimated with relative accuracy. Determining the probability of sabotage or a military attack is difficult, since it depends upon various political and economic considerations. Nevertheless, it can be observed that such events are more frequent in unstable countries with many unresolved internal and external conflicts. Finally, the probability and severity of the natural events can be, once again, estimated for various locations, based on geographical, geological, and historical data.

In order to minimize the subsequent disruptions, or the effect of the potentially disruptive events, defense systems or protection and safeguards systems are employed. These systems have been well developed, and in order to maximize their effectiveness, they use the principles of defense in depth, redundancy, and diversity. All energy companies are geared to deal with human errors and natural events, but problems can be encountered when dealing with sabotage or military attacks. To minimize the impact of sabotage and military attacks, security services must be called for assistance. These services are generally able to deal with these problems domestically in countries without serious internal and external conflicts. However, in countries with serious internal and external conflicts, either the internal security forces can be ineffective, particularly in remote locations, or their loyalty can be divided. In these cases, the governments sometimes require security support from other countries, which, while improving the defenses, may make the internal conflicts worse and lead to an increasing frequency of attacks. Hence, stable countries, where the probability of an initial attack is already low, can, generally, protect their installations, but countries, with significant internal and external conflicts, where the probability of an initial attack is already high, cannot, generally, protect their installations.

All developed countries have contingency plans in place to deal with the effects of supply disruptions in the form of strategic sock reserves. These plans used to be based on the diversity of the energy supplies and the availability of strategic stocks. Due to the reliance on oil and natural gas by the developed countries, the principle of diversity of energy supplies holds less and less. All developed countries now deal with the potential problems primarily by having strategic reserves of oil supplies. By definition, the strategic reserves would be only available for certain activities and a certain period of time. This selectivity and time limit would provide a serious blow to the economies of the developed countries, if the reserves had to be used for a significant period of time.

Many of the world's leading oil-producing countries such as Iraq and Nigeria are politically unstable. The Middle East region as a whole has quite a volatile geopolitical situation and has experienced frequent conflicts over the last 100 years. There are serious reservations regarding security of oil; production and supply channels of many countries are regarded as the legitimate targets of radical elements because of various internal and external conflicts.

Getting oil from the well to the refinery and from there to the service station involves a complex transportation and storage system. Millions of barrels of oil are transported every day in tankers, pipelines, and trucks. This transportation system has always been a possible weakness of the oil industry, but it has become even more so in the present volatile geopolitical situation, especially in the Middle East region. The threats of global terrorism have made the equation more complex. Tankers and pipelines are quite vulnerable targets. There are approximately 4000 tankers employed, and each of them can be attacked in the high seas and more seriously while passing through narrow straits in hazardous areas. Pipelines, through which about 40% of the world's oil flows, are no less vulnerable, and due to their length, they are very difficult to protect. This makes pipelines potential targets for terrorists. In recent years, there have been an increased number of pipeline sabotages in different countries particularly Nigeria and Iraq, sending shockwaves in international energy markets.

Depleting Oil Reserves

A combination of constrained production capacity and growing fears of a rapid depletion of oil reserves in the world is also an important factor that has been playing its behind-the-scene role in pushing oil prices. In recent years, global oil infrastructure particularly with regard to extraction and refining capacity has been stretched to its limits. The production capacity of various oil-rich countries in the world, such as Iraq and Venezuela, has also been curtailed. The aspect of depleting oil reserves, despite its critical role, for various reasons, is not being publicly accepted by the market forces.

The world's ultimate conventional oil reserves are estimated at 2,000 billion barrels. This is the amount of production that would have been produced when production eventually ceases. The demand for oil has grown rapidly over the last few decades as shown in Table 2.[10] The surging demand for oil has already stretched the production to its limits—in mid-2002, there were more than 6 million barrels per day of excess production capacity, but by mid-2003, the daily excess capacity was below 2 million barrels, which further skewed to less than 1 million barrels by 2006.[11]

Different countries are at different stages of their reserve depletion curves. Some, such as the United States, are past their midpoint and are in terminal decline, whereas others are close to midpoint such as U.K. and Norway. However, the five major Gulf producers—Saudi Arabia, Iraq, Iran, Kuwait, and United Arab Emirates—are at an early stage of depletion and can exert a swing role, making up the difference between world demand and what others can supply.

The expert consensus is that the world's midpoint of reserve depletion will be reached when 1,000 billion barrels of oil have been produced—that is to say, half the ultimate reserves of 2,000 billion barrels. It is estimated that around 1,000 billion barrels have already been consumed and 1,000 billion barrels of proven oil reserves are left in the world.[12] According to BP statistical review of oil reserves in 2017, the reserve-to-production ratio for North America, South and Central America, Europe, Asia Pacific, Middle East, and Africa was 12, 48, 13, 32, 120, and 62, respectively. The reserve-to-production ratio for the whole world is reported to be equal to 52 years.[13] In the backdrop of a continuous growth in oil demand—as according to the U.S. Department of Energy, by 2025 the global daily demand could be as much as 110 barrels/day—the global reserves are actually going to run out much quicker than 42 years.[14]

A growing number of opinions among energy experts suggest that global oil production will probably peak sometime during this decade as indicated in Table 3.[15] A siren call regarding energy

TABLE 2 Growth in World Oil Demand

Year	World Population (million)	Average Daily Oil Demand (million barrels/day)	World Average per Capita Consumption (barrels/yr)
1965	3310	31.23	3.65
1968	3520	39.04	4.05
1971	3750	51.76	5.04
1974	3990	59.39	5.44
1977	4200	63.66	5.53
1980	4410	64.14	5.31
1983	4650	58.05	4.56
1986	4890	61.76	4.60
1989	5150	65.88	4.67
1992	5400	66.95	4.52
1995	5610	69.88	4.54
1998	5870	72.92	4.51
2001	6140	75.99	4.53
2004	6400	82.35	4.67
2007	6610	86.30	4.76
2010	6958	86.41	4.53
2013	7213	91.82	4.64
2016	7466	96.15	4.70

TABLE 3 Various Projections of the Global Oil Reserves and the Peak Year (billions of barrels)

Author	Affiliation	Year	Estimated Ultimate Reserves	Peak Year
Hubert	Shell	1969	2100	2000
Bookout	Shell	1989	2000	2010
Mackenzie	Researcher	1996	2600	2007–2019
Appleby	BP	1996		2010
Ivanhoe	Consultant	1996		2010
Edwards	University of Colorado	1997	2836	2020
Campbell	Consultant	1997	1800–2000	2010
Bernabe	ENI	1998		2005
Schollenberger	Amoco	1988		2015–2035
IEA	OECD	1998	2800	2010–2020
EIA	DOE	1998	4700	2030
Laherrere	Consultant	1999	2700	2010
USGS	International Department	2000	3270	
Salameh	Consultant	2000	2000	2004–2005
Deffeyes	Princeton University	2001	1800–2100	2004

security was also raised in a recent speech made by the British Ambassador to the United States, Sir David Manning. He eloquently puts forward the case thus: "The International Energy Agency predicts that, if we do nothing, the global oil demand will reach 121 million barrels per day by 2030, up from 85 million barrels today. That will require increasing production by 37 million barrels per day over the next 25 years, of which 25 million barrels per day has yet to be discovered. That is, we'll have to find four petroleum systems that are each the size of the North Sea. Production from existing fields is dropping at about 5% per year." Only one barrel of oil is now being discovered for every

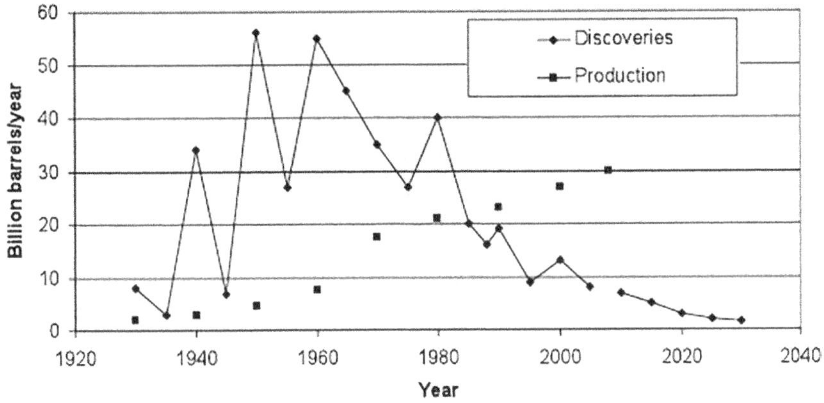

FIGURE 2 An overview of global oil discoveries and production.

four consumed as highlighted in Figure 2. Globally, the discovery rate of untapped oil peaked in the late 1960s. Over the past decade, oil production has been falling in 33 of the world's 48 largest oil-producing countries, including 6 of the 13 members of OPEC (Organization of the Petroleum Exporting Countries).[16]

Concerns over Security of Supplies from the Middle East

Oil and gas are jointly contributing to around 56% of the present global energy requirements. In terms of existence, both oil and gas are extremely localized by their very nature. Vast majority of their resources are found in relatively confined regions. For example, almost 88% of the world oil reserves exist within ten countries. Middle East is the oil headquarters of the world, holding almost 48% of the global oil as highlighted in Figure 3.[13] The situation with natural gas is even more intense. Statistics indicate that of the remaining known gas reserves in the world, equivalent to 6260 trillion cubic feet (tcf), around 57% are shared by only three countries: Russia, Iran, and Qatar, as shown in Figure 4.[17]

Middle East has been under the limelight for its vast oil reserves for nearly a century now. By the 1950s, it had established its standing as the oil-rich region of the world—its oil reserves were equal to the rest of the world's combined and double the U.S. reserves.[18] For the last six decades, Middle East

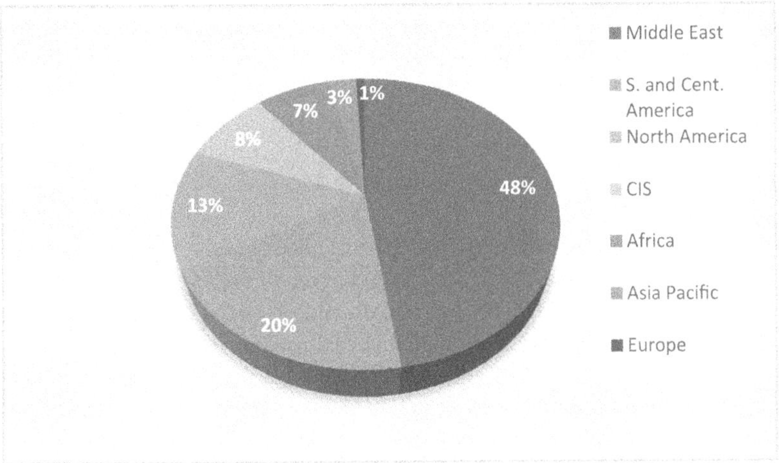

FIGURE 3 Remaining oil reserves in the world, 2017.

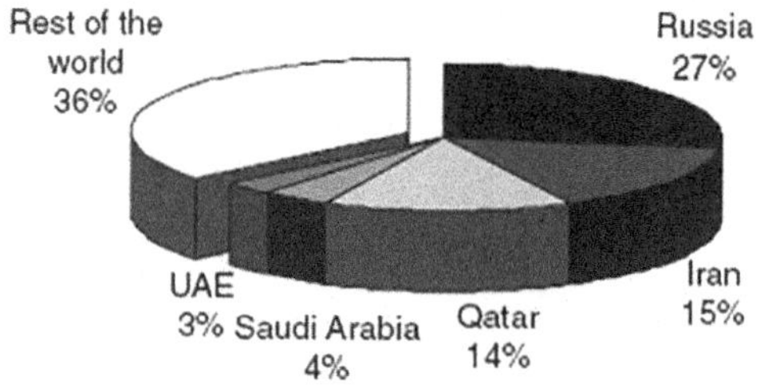

FIGURE 4 Remaining natural gas reserves in the world, 2009.

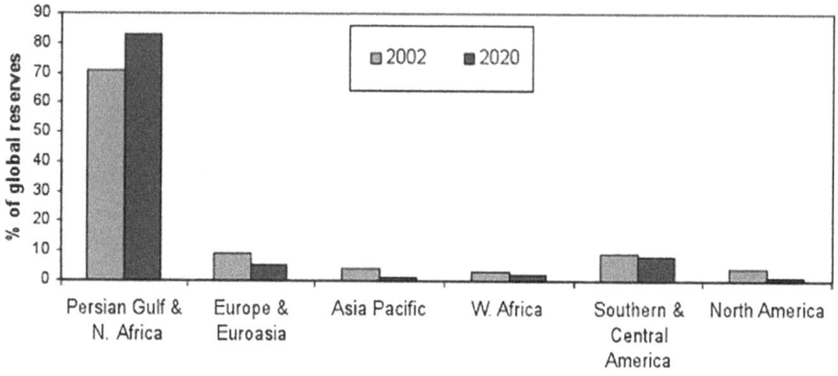

FIGURE 5 Growing share of Middle Eastern oil reserves.

has been the source of abundant and cheap oil that propelled the economic growth of the world over this period. In 2006, it produced about 28% of the world's total oil. During the same year, the region exported 18.2 million barrels of oil per day. Besides oil, the region also has huge reserves (2509 tcf) of natural gas, accounting for 41% of total proven world gas reserves.[19]

The importance of the Arabian Gulf countries is several-fold. The oil and gas reserves in non-Middle East countries are being depleted more rapidly than those of Middle East producers. If production continues at the present rate, many of the largest, non-Middle Eastern producers in 2002, such as Russia, Mexico, United States, Norway, and Brazil, will cease to be relevant players in the oil market in less than two decades. At that point, the Middle East will be the only major reservoir of abundant crude oil—within 20 years or so, about four-fifths of oil reserves could be in the hands of the Middle Eastern countries as shown in Figure 5.[11] The role of the region in future oil supplies of the world is thus going to become even more prominent.

Global Warming

Since the industrial revolution, human activities have resulted in a sharp rise in the concentration of GHGs in the Earth's atmosphere. The GHGs such as carbon dioxide (CO_2), nitrous oxide (N_2O), methane (CH_4), ozone (O_3), sulfur hexafluoride (SF_6), hydrofluorocarbons (HFCs), and perfluorocarbons (PFCs) are widely considered to be the root cause of a rapid increase in the atmospheric temperature, a phenomenon recognized as global warming. As a consequence of global warming, the climatic patterns of the earth are experiencing wide-ranging changes that are having numerous and far-reaching implications for the planet. The concentration levels of GHGs in the atmosphere are usually described in terms of carbon dioxide equivalent (CO_2e). Estimates suggest that since the advent of the industrial revolution, the level of GHGs in the atmosphere has grown from 280 parts per million (ppm) CO_2e to 430 ppm CO_2e.

Scientific circles believe that in order to avoid catastrophic implications, the level should be limited to 450–550 ppm CO_2e.[20] The growing concentration of GHGs in the atmosphere has increased the global average temperature. Some estimates suggest that the global temperature has increased by 0.8°C since the industrial revolution. The rise in the atmospheric temperature leads to the melting of glaciers and ice sheets, which, in turn, increases the sea level. According to the Intergovernmental Panel on Climate Change (IPCC), over the last century, the global sea level has increased at a rate of around 2 mm/yr. Estimates also indicate that the sea level rose twice as much between 1993 and 2003 as in the previous three decades.[21]

Global warming and climate change lead to a pattern of more frequent and more intense weather events such as floods, storms, droughts, heat waves, diseases, and loss of habitat. These events affect people across the world and result into huge casualties. According to the World Health Organization (WHO), as many as 160,000 people die each year from the side effects of climate change, and the numbers could almost double by 2020. These side effects range from malaria to malnutrition and diarrhea that follow in the wake of floods, droughts, and warmer temperatures.[22] Evidence suggests that the weather-related disasters have quadrupled over the last two decades. According to Oxfam, from an average of 120 disasters a year in the early 1980s, there are now as many as 500, with the rise to be attributed to unpredictable weather conditions caused by global warming. The year 2007 saw floods in South Asia, across the breadth of Africa and Mexico, which affected more than 250 million people. Devastating floods in 2010 affected a similar magnitude of people in China and Pakistan alone.[23] Statistics also suggest that over the last 25 years, the number of people affected by disasters has risen by 68%, from an average of 174 million a year from 1985 to 1994 to 254 million a year from 1995 to 2004.[24]

Climate change is responsible for huge economic consequences. Between the 1960s and the 1990s, the number of significant natural catastrophes such as floods and storms rose nine-fold, and the associated economic losses rose by a factor of nine. Figures indicate that the economic losses as a direct result of natural catastrophes over 5 years between 1954 and 1959 were $35 billion, while between 1995 and 1999, these losses were around $340 billion.[25] Natural catastrophes associated with global warming killed more than 190,000 people in 2004, twice as many as in 2003, with an economic cost of $145 billion. The August 2005 Hurricane Katrina was responsible for taking more than 1,000 human lives. Hurricane Katrina caused at least $125 billion in economic damage and could cost the insurance industry up to $60 billion in claims. That is significantly higher than the previous record-setting storm, Hurricane Andrew in 1992, which caused nearly $21 billion in insured losses in today's dollars.[26] Katrina shut down large portions of oil and gas production in the Gulf of Mexico at a time when worldwide energy output was already stretched thin. While the storm's impact was most acute in the United States, it also sent fuel costs higher around the globe, squeezing consumers in Europe and Asia.[27] It has been reported that since the advent of the 20th century, natural disasters such as floods, storms, earthquakes, and bushfires have resulted in an estimated loss of nearly 8 million lives and over $7 trillion of economic loss.[27]

The scale and intensity of global warming are set to increase in the 21st century. Global mean temperature is forecast to rise by between 1°C and 4.5°C by 2100, with best estimates somewhere between 2°C and 3°C. All projections produce rates of warming that are greater than those experienced in the last 10,000 years. Sea level is projected to rise by about 50 cm by 2100 (with a range of 20–90 cm).[11,28] Temperature and sea level changes will not be globally uniform. Land areas, particularly at high latitudes, will warm faster than the oceans, with a more vigorous hydrological cycle potentially affecting the rate and scale of various extreme events such as drought, flood, and rainfall. Impacts on natural and semi-natural ecosystems, agriculture, water resources, human infrastructure, and human health are subject to many uncertainties, but all will be subject to stresses that will exacerbate stresses from other sources such as land degradation, pollution, population growth and migration, and rising per capita exploitation of natural resources. Global warming is also set to have colossal ramifications on biodiversity—Stern Report suggests that a warming of even only 2°C could leave 15–40% species facing extinction.[20]

Global Warming Threats for Developing Countries

Although global warming is a threat to the whole planet, its intensity is not uniformly distributed. It is the low-lying and small island countries that are being hit harder. Most of them are actually poor and developing countries. It is feared that several hundred million people in densely populated coastal regions—particularly river deltas in Asia—are threatened by rising sea levels and the increasing risk of flooding. It is estimated that more than one-sixth of the world's population live in areas affected by water sources from glaciers and snow pack that will very likely disappear.[29] The number of people at risk of flooding by coastal storm surges is projected to increase from the current 75 million to 200 million by 2080, when sea levels may have risen by more than 1 ft. In the United Nation's list of countries under severe threat by global warming, Bangladesh is at the top. Being a low-lying and densely populated country, Bangladesh would be worst hit by any rise in the sea level. Coastal areas would experience erosion and inundation due to intensification of tidal action. A rise in seawater would enable saline water to intrude further inland during high tides. Destruction of agricultural land and loss of sweet water fauna and flora could also occur. The shoreline would retreat inland, causing changes in the coastal boundary and coastal configuration. The process will also shrink the land area of Bangladesh. Worst scenarios suggest that by the year 2050, one-third of the country could be under water, making more than 70 million people homeless.[30] For small island developing states (SIDSs), global warming poses an enormous set of challenges for their livelihood, safety, and security. Since most of the infrastructure in these countries is on the coast, the damage from consequent erosion and flooding is likely to be hugely burdensome for their already fragile economies. Owing to their smaller land area compared to other countries, they cannot afford to lose land due to surging sea level. For some, for example, Maldives, it threatens their very existence.

The health-related implications of global warming are also expected to be more formidable for developing countries. The WHO estimates that global warming is already causing about 5 million extra cases of severe illness a year. By 2030, however, the number of climate-related diseases is likely to more than double, with a dramatic increase in heat-related deaths caused by heart failure, respiratory disorders, the spread of infectious diseases, and malnutrition from crop failures. Countries with coastlines along the Indian and Pacific oceans and sub-Saharan Africa would suffer a disproportionate share of the extra health burden. According to WHO experts, many of the most important diseases in poor countries, such as diarrhea and malnutrition, are highly sensitive to climate. Also, that the health sector is already struggling to control these diseases and climate change threatens to undermine these efforts.[31]

One of the significant heartbreaks of global warming is that developed and industrialized countries are acutely responsible for the phenomenon, but the heavier price is to be paid by the poor and developing nations. For example, the average value of the per capita energy consumption—an index to measure the contribution towards global warming—in industrialized and developed countries is almost six times greater than that in developing countries. In 2007, the former U.S. Vice President Al Gore, in his famous work on global warming, heavily criticized the United States for turning a blind eye on the issue of global warming. According to him, the U.S. stance on the Kyoto Protocol is unfavorable despite the fact that it alone is responsible for more than 30% of the world GHG emissions. He concludes that a U.S. citizen emits nearly six times greater amount of carbon as compared to the world average emission as indicated in Figure 6, whereas compared to African or South Asian countries such as Pakistan, India, and Bangladesh, the per capita carbon emission in the United States is around 22 times higher.[32] Figure 7 provides a relationship between carbon dioxide emissions and gross national income (GNI) of a range of developing and developed countries. It can be seen that China has a substantially higher emission rate compared to other countries. Australia, Canada, and the United States also have a significantly higher emission rate in comparison with other developed countries. According to the World Wildlife Fund (WWF), in terms of eco-footprint, an index of sustainability, the excessive consumption of natural resources by the developed countries is imposing serious implications on the ecosystem of the planet.[33] These pieces of evidence indicate that the responsibility of the disturbance in the global ecosystem, also

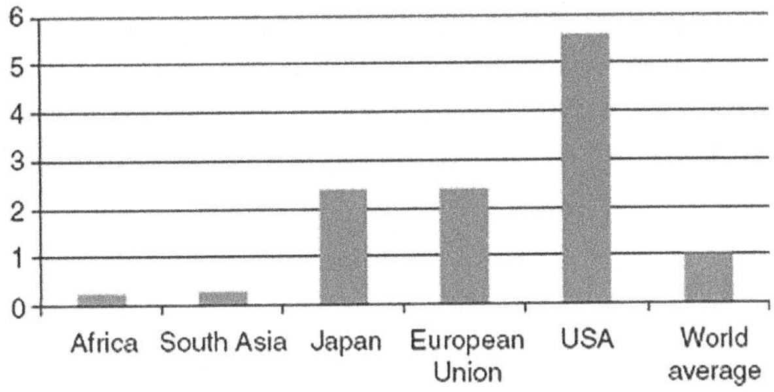

FIGURE 6 Annual per capita carbon emission in tons.

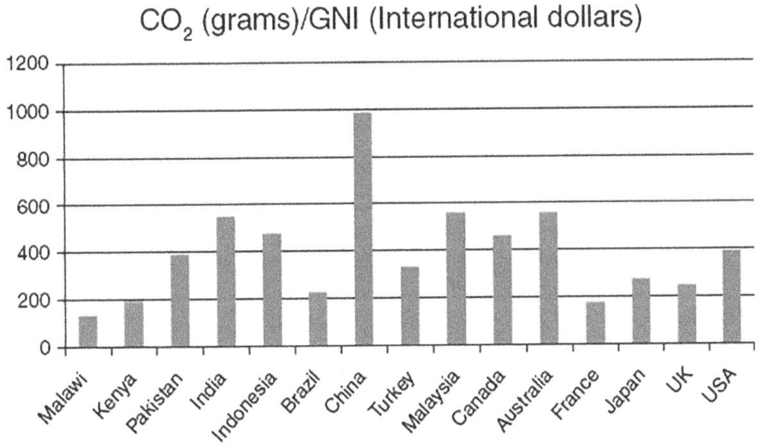

FIGURE 7 The per capita values of grams of carbon dioxide emitted for each international dollar of GNI in various countries.

leading to the phenomenon of global warming and climate change, to a great extent rests on the shoulders of the developed and industrialized countries of the world.

In the backdrop of the trajectory of the issue, the Kyoto Protocol urges the industrialized countries to reduce their collective emissions of GHGs by 5.2% compared to the year 1990. However, compared to the emission levels that would be expected by 2010 without the Protocol, this target represents a 29% cut. The goal is to lower overall emissions of six GHGs—carbon dioxide, methane, nitrous oxide, sulfur hexafluoride, HFCs, and PFCs—calculated as an average over the 5-year period of 2008–2012. National targets range from 8% reductions for the EU and some others to 7% for the United States and 6% for Japan. The developing countries of the world have been spared any emission reduction obligations not only in the Kyoto Protocol but also in its predecessor accords such as Agenda 21. According to Professor Jonathan Patz, it is incumbent on those countries bearing the greatest responsibility for climate change to show moral leadership. He goes on to say: "Those least able to cope and least responsible for the GHGs that cause global warming are most affected. Herein exists an enormous global ethical challenge."[34] On similar lines, Sir Nicholas Stern acknowledges that the developed countries should realize their responsibility towards addressing global warming. He concludes that climate change is the greatest and widest-ranging challenge mankind has ever faced. All countries will be affected by climate change, but the poorest countries will suffer earliest and most. Loss of biodiversity is another inevitable consequence—more than 40% of the species are likely to face extinction. He further urges that climate change should be fully integrated into development policy, and rich countries should honor pledges to increase support through overseas development assistance.[20]

To tackle global warming, a radical change in human attitude towards the environment and consumption of natural resources is required. A major shift in energy consumption practices—from the reliance on currently employed environmentally dangerous resources to the environmentally friendly ones—would be imperative to attain sustainable development. To safeguard the future of coming generations, the world thus has to move towards low-carbon energy systems.

Renewable Energy

Renewable energy as the name implies is the energy obtained from natural resources such as wind power, solar energy, hydropower, biomass energy, and geothermal energy. Renewable energy resources have also been important for humans since the beginning of civilization. Biomass, for example, has been used for heating, cooking, and steam production; wind has been used for moving ships; both hydropower and wind have been used for powering mills to grind grains. Renewable energy resources are abundant in nature and have the potential to provide energy services with zero or almost zero emissions. Renewable energy is acknowledged as a vital and plentiful source of energy. Technically, renewable energy resources have enormous potential and can meet many times the present world energy demand. However, due to their intermittent nature, they have to be used in conjunction with other energy resources. They can enhance diversity in energy supply markets, secure long-term sustainable energy supplies, and reduce local and global atmospheric emissions. They can also provide commercially attractive options to meet specific needs for energy services (particularly in developing countries and rural areas), create new employment opportunities, and offer possibilities for local manufacturing of equipment.

Renewable energy's favorable dimensions in terms of resources availability, reducing cost trends, and environmental friendliness are propelling its rapid growth.[35,36] Technological advancements and supportive policy frameworks have also played a vital role in the rapid growth of renewable energy application around the world.[37–39] In 2017, renewable energy technologies received a cumulative investment of $280 billion. At the end of the year, the global installed capacity of renewables stood at 2,195 GW. Renewables contributed to over 26% of the world's total electricity supplies.[40] Overall, renewables accounted for an estimated 70% of net additions to global power capacity in 2017. Solar photovoltaic (PV) made up led the renewables march, making up nearly 55% of the newly installed renewable power capacity in the year. Wind power and hydropower with respective shares of 29% and 11% account for most of the remaining capacity additions. Renewable energy technologies are becoming competitive with conventional power systems—in 2017, solar PV and wind power projects received bids as low as $30/MWh. Since 2010, the global weighted average levelized cost of electricity (LCOE) of utility-scale solar PV has dropped by over 73%.[35]

Conclusions

The world faces stringent energy and environmental challenges. Fossil fuels, contributing to more than 80% of global energy supplies, interlink the security of energy and environment. The present energy scenario has a number of concerns including depletion of fossil fuel reserves, surging energy prices, geopolitical conflicts in fossil-fuel-rich regions, and emission of GHGs. The emission of GHGs leads to environmental insecurity by contributing to global warming. The intensity of the global warming concerns can be gauged from the fact that the average atmospheric temperature is forecasted to rise by as much as 4.5°C during the 21st century compared to a 0.8°C increment since the industrial revolution. Global warming and the consequent sea level rise—which could be up to 50 cm over the same period—would have catastrophic implications for natural and seminatural ecosystems, agriculture, water resources, human infrastructure, and human health. Even the very existence of some of the small island countries is under threat. Energy and environmental security is thus absolutely vital for sustainable development. The world needs to shift to energy resources that are more reliable, affordable, secure, and environmentally friendly. Renewable energy is one possible solution that can significantly help address the energy and environmental security concern.

References

1. Solomon, B. Economic geography of energy. In Cleveland, C., *Encyclopedia of Energy*; Amsterdam: Elsevier, 2004; Vol. 2.
2. Asif, M. Energy Crisis in Pakistan: Origins, Challenges and Sustainable Solutions; Karachi: Oxford University Press, 2011.
3. BP Energy Outlook, 2019 Edition, British Petroleum, UK.
4. Weynand, G. Energy Sector Assessment for US Aid/Pakistan. United States Agency for International Development, June 2007.
5. Sustainable Development Goals, United Nations, 2019, https://www.un.org/sustainabledevelopment/energy/ (accessed on 25 March 2019).
6. Poverty, Energy and Society, Energy Forum. The Baker Institute, Rice University.
7. Energy Services for the Millennium Development Goals, UNDP/World Bank, 2006.
8. Muller-Kraenner, S. *Energy Security*; London: Earthscan, 2008.
9. Issues and Instruments, G8 Summit 2006, St Petersburg, 15–17 July 2006.
10. McKillop, A. Oil Shock and Energy Transition. POGEE Conference, Karachi, May 2008.
11. Asif, M.; Muneer, T. Energy Supply, its Demand and Security Issues for Developed and Emerging Economies. *Renewable and Sustainable Energy Reviews* 2007, 11 (7): 1388–1413.
12. Asif, M.; Currie, J.; Muneer, T. The Role of Renewable and Non-renewable Sources for Meeting Future U.K. Energy Needs. International Journal *of* Nuclear Governance, Economy *and* Ecology 2007, 1 (4): 372–383.
13. BP Statistical Review of World Energy 2018, British Petroleum, UK.
14. International Energy Outlook 2006.
15. Salameh, M. Oil Crises: Historic Perspective. In Cleveland, C., *Encyclopaedia of Energy*; Amsterdam: Elsevier, 2004; Vol. 4.
16. British Ambassador Sir David Manning Discusses Energy: A Burning Issue for Foreign Policy at Stanford University. British Embassy, Washington, DC, March 2006. https://fsi.stanford.edu/multimedia/energy-burning-issue-foreign-policy (accessed 15 May 2020).
17. International Energy Agency Statistics 2009, International Energy Agency, Washington DC, USA.
18. Chomsky, N. The Fateful Triangle: The United States Israel and The Palestinians; London: Pluto Press, 1999.
19. Persian Gulf Region, Energy Information Administration, EIA, 2008, available at http://www.eia.doe.gov/emeu/cabs/Persian_Gub7Background.html (accessed June 2011).
20. Stern Report: The Key Points. *The Guardian*, 30 Hilary Osborne, October 2006.
21. Dodds, F.; Higham, A.; Sherman, R. *Climate Change and Energy Insecurity*; UK: Earthscan, 2009.
22. Asif, M.; Muneer, T. Energy Supply, its Demand and Security Issues for Developed and Emerging Economies. Renewable and Sustainable Energy Reviews 2007, 11 (7): 1388–1413.
23. Available at http://en.wikipedia.org/wiki/2010_China_floods (accessed June 2011). China Flood, Wikipedia, Available at http://en.wikipedia.org/wiki/2010_China_floods (accessed 15 May 2020).
24. John Sinnott, Disasters quadruple over last 20 years: Oxfam, Reuters, 25 November 2007, https://www.reuters.com/article/us-britain-climate-oxfam-idUSL2518480220071125 (accessed: 15 May 2020)
25. Muneer, T.; Asif, M. Generation and Transmission Prospects for Solar Electricity: U.K. and Global Markets. *Energy Conversion and Management* 2003, 44: 35–52.
26. Available at http://www.usatoday.com/money/economy/2005-09-09-katrina-damage_x.htmS (accessed June 2011).
27. Khan, H.; Asif, M. Impact of Green Roof and Orientation on the Energy Performance of Buildings: A Case Study from Saudi Arabia. *Sustainability* 2017, 9(4): 640. doi:10.3390/su9040640
28. Intergovernmental Panel on Climate Change, IPCC Working Group I (scenario IS92a), available at http://www.ipcc.ch/ipccreports/sres/regional/140.htm (accessed June 2011).

29. IPCC,2007:Climate Change 2007: Impacts, Adaptation and Vulnerability. Contribution of Working Group II to the Fourth Assessment Report of the Intergovernmental Panel on Climate Change, M.L. Parry O.F. Canziani, J.P. Palutikof, P.J. van der Linden and C.E. Hanson, Eds., Cambridge University Press, Cambridge, UK, 976.
30. UNDP. 2007/8. Human Development Report 2007/8: Fighting climate change - Human solidarity in a divided world. New York.
31. Connor, S. Climate Change Map Reveals Countries Most Under Threat, The Independent, 17 November 2005.
32. Guggenheim, D. An Inconvenient Truth, Documentary on Former U.S. Vice President Al Gore's Campaign on Environment, 2006.
33. Living Planet Report 2004, WWF. Living Planet Report 2004, WWF. http://wwf.panda.org/knowledge_hub/all_publications/living_planet_report_timeline/lpr_2004/ (accessed 15 May 2020).
34. Jonathan A. Patz, Diarmid Campbell-Lendrum, Tracey Holloway & Jonathan A. Foley, Impact of regional climate change on human health, Nature volume 438, 310–317, 2005.
35. Asif, M. Growth and Sustainability Trends in the GCC Countries with Particular Reference to KSA and UAE. *Renewable & Sustainable Energy Reviews Journal* 2016, 55: 1267–1273.
36. Dehwah, A.; Asif, M. Assessment of Net Energy Contribution to Buildings by Rooftop PV Systems in Hot-Humid Climates. *Renewable Energy* 2019, 131: 1288–1299, https://doi.org/10.1016/j.renene.2018.08.031
37. Asif, M.; Hassanain, M.; Nahiduzzaman, Kh.; Swalha, H. Techno-Economic Assessment of Application of Solar PV in Building Sector-A Case Study from Saudi Arabia. *Smart and Sustainable Built Environment* 2019, 8 (1): 34–52.
38. Dehwah, A.; Asif, M.; Tauhidurrahman, M. Prospects of PV Application in Unregulated Building Rooftops in Developing Countries: A Perspective from Saudi Arabia. *Energy and Buildings* 2018, 171: 76–87.
39. Khan, M.; Asif, M.; Stach, S. Rooftop PV Potential in the Residential Sector of the Kingdom of Saudi Arabia. *Buildings* 2017, 7(2): 46. doi:10.3390/buildings7020046.
40. Renewables 2018 Global Status Report, Renewable Energy Policy Network for the 21st Century, Paris.

7

Energy Commissioning: New Buildings

Janey Kaster

Introduction

This entry provides an overview of commissioning—the processes one employs to optimize the performance characteristics of a new facility being constructed. Commissioning is important to achieve customer satisfaction, optimal performance of building systems, cost containment, and energy efficiency, and it should be understood by contractors and owners.

After providing an overview of commissioning and its history and prevalence, this entry discusses what systems should be part of the commissioning process, the benefits of commissioning, how commissioning is conducted, and the individuals and teams critical for successful commissioning. Then the entry provides a detailed discussion of each of the different phases of a successful commissioning process, followed by a discussion of the common mistakes to avoid and how one can measure the success of a commissioning effort, together with a cost-benefit analysis tool.

The purpose of this entry will be realized if its readers decide that successful commissioning is one of the most important aspects of construction projects and that commissioning should be managed carefully and deliberately throughout any project, from predesign to postacceptance. As an introduction to those unfamiliar with the process and as a refresher for those who are, the following section provides an overview of commissioning, how it developed, and its current prevalence today.

Overview of Commissioning

Commissioning Defined

Commissioning is the methodology for bringing to light design errors, equipment malfunctions, and improper control strategies at the most cost-effective time to implement corrective action. Commissioning facilitates a thorough understanding of a facility's intended use and ensures that the design meets the intent through coordination, communication, and cooperation of the design and installation team. Commissioning ensures that individual components function as a cohesive system. For these reasons, commissioning is best when it begins in the predesign phase of a construction project and can in one sense be viewed as the most important form of quality assurance for construction projects.

Unfortunately, there are many misconceptions associated with commissioning, and perhaps for this reason, commissioning has been executed with varying degrees of success, depending on the level of understanding of what constitutes a "commissioned" project. American Society of Heating, Refrigerating and Air-Conditioning Engineers (AHSRAE) guidelines define commissioning as: the process of ensuring that systems are designed, installed, functionally tested, and capable of being operated and maintained to perform conformity with the design intent … [which] begins with planning and includes design, construction, startup, acceptance, and training, and is applied throughout the life of the building.[4] However, for many contractors and owners, this definition is simplified into the process of system startup and checkout or completing punch-list items.

Of course, a system startup and checkout process carried out by a qualified contractor is one important aspect of commissioning. Likewise, construction inspection and the generation and completion of punch-list items by a construction manager are other important aspects of commissioning. However, it takes much more than these standard installation activities to have a truly "commissioned" system. Commissioning is a comprehensive and methodical approach to the design and implementation of a cohesive system that culminates in the successful turnover of the facility to maintenance staff trained in the optimal operation of those systems.

Without commissioning, a contractor starts up the equipment but doesn't look beyond the startup to system operation. Assessing system operation requires the contractor to think about how the equipment will be used under different conditions. As one easily comprehended example, commissioning requires the contractor to think about how the equipment will operate as the seasons change. Analysis of the equipment and building systems under different load conditions due to seasonal conditions at the time of system startup will almost certainly result in some adjustments to the installed equipment for all but the most benign climates. However, addressing this common requirement of varying load due to seasonal changes most likely will not occur without commissioning. Instead, the maintenance staff is simply handed a building with minimal training and left to figure out how to achieve optimal operation on their own. In this seasonal example, one can just imagine how pleased the maintenance staff would be with the contractor when a varying load leads to equipment or system failure—often under very hot or very cold conditions!

Thus, the primary goal of commissioning is to achieve optimal building systems performance. For heating, ventilation, and air-conditioning (HVAC) systems, optimal performance can be measured by thermal comfort, indoor air quality, and energy savings. Energy savings, however, can result simply from successful commissioning targeted at achieving thermal comfort and excellent indoor air quality. Proper commissioning will prevent HVAC system malfunction—such as simultaneous heating and cooling, and overheating or overcooling—and successful malfunction prevention translates directly into energy savings. Accordingly, energy savings rise with increasing comprehensiveness of the commissioning plan. Commissioning enhances energy performance (savings) by ensuring and maximizing the performance of specific energy efficiency measures and correcting problems causing excessive energy use.[3] Commissioning, then, is the most cost-effective means of improving energy efficiency in commercial buildings. In the next section, the two main types of commissioning in use today—acceptance-based and processed-based—are compared and contrasted.

Acceptance-Based vs. Process-Based Commissioning

Given the varied nature of construction projects, contractors, owners, buildings, and the needs of the diverse participants in any building projection, commissioning can of course take a variety of forms. Generally, however, there are two types of commissioning: acceptance-based and process-based. Process-based commissioning is a comprehensive process that begins in the predesign phase and continues through postacceptance, while acceptance-based commissioning, which is perceived to be the cheaper method, basically examines whether an installation is compliant with the design and accordingly achieves more limited results.

Acceptance-based commissioning is the most prevalent type due to budget constraints and the lack of hard cost/benefit data to justify the more extensive process-based commissioning. Acceptance-based commissioning does not involve the contractor in the design process but simply constitutes a process to ensure that the installation matches the design. In acceptance-based commissioning, confrontational relationships are more likely to develop between the commissioning agent and the contractor because the commissioning agent and the contractor, having been excluded from the design phase, have not "bought in" to the design and thus may be more likely to disagree in their interpretation of the design intent.

Because the acceptance-based commissioning process simply validates that the installation matches the design, installation issues are identified later in the cycle. Construction inspection and regular commissioning meetings do not occur until late in the construction/installation phase with acceptance-based commissioning. As a result, there is no early opportunity to spot errors and omissions in the design, when remedial measures are less costly to undertake and less likely to cause embarrassment to the designer and additional costs to the contractor. As most contractors will readily agree, addressing issues spotted in the design or submittal stages of construction is typically much less costly than addressing them after installation, when correction often means tearing out work completed and typically delays the completion date.

Acceptance-based commissioning is cheaper, however, at least on its face, being approximately 80% of the cost of process-based commissioning.[2] If only the initial cost of commissioning services is considered, many owners will conclude that this is the most cost-effective commissioning approach. However, this 20% cost differential does not take into account the cost of correcting defects after the fact that process-based commissioning could have identified and corrected at earlier stages of the project. One need encounter only a single, expensive-to-correct project to become a devotee of process-based commissioning.

Process-based commissioning involves the commissioning agent in the predesign through the construction, functional testing, and owner training. The main purpose is quality assurance—assurance that the design intent is properly defined and followed through in all phases of the facility life cycle. It includes ensuring that the budget matches the standards that have been set forth for the project so that last-minute "value engineering" does not undermine the design intent, that the products furnished and installed meet the performance requirements and expectation compliant with the design intent, and that the training and documentation provided to the facility staff equip them to maintain facility systems true to the design intent.

As the reader will no doubt already appreciate, the author believes that process-based commissioning is far more valuable to contractors and owners than acceptance-based commissioning. Accordingly, the remainder of this entry will focus on process-based commissioning, after a brief review of the history of commissioning from inception to date, which demonstrates that our current, actively evolving construction market demands contractors and contracting professionals intimately familiar with and expert in conducting process-based commissioning.

History of Commissioning

Commissioning originated in the early 1980s in response to a large increase in construction litigation. Owners were dissatisfied with the results of their construction projects and had recourse only to the courts and litigation to resolve disputes that could not be resolved by meeting directly with their contractors. While litigation attorneys no doubt found this satisfactory approach to resolving construction project issues, owners did not, and they actively began looking for other means to gain assurance that they were receiving systems compliant with the design intent and with the performance characteristics and quality specified. Commissioning was the result.

While commissioning enjoyed early favor and wide acceptance, the recession of the mid-1980s placed increasing market pressure on costs, and by the mid-to late 1980s it forced building professionals to reduce fees and streamline services. As a result, acceptance-based commissioning became the norm, and process-based commissioning became very rare. This situation exists in most markets today; however, the increasing cost of energy, the growing awareness of the global threat of climate change and the need to reduce CO2 emissions as a result, and the legal and regulatory changes resulting from both are creating a completely new market in which process-based commissioning will become ever more important, as discussed in the following section.

Prevalence of Commissioning Today

There are varying degrees of market acceptance of commissioning from state to state. Commissioning is in wide use in California and Texas, for example, but it is much less widely used in many other states. The factors that impact the level of market acceptance depend upon

- The availability of commissioning service providers
- State codes and regulations
- Tax credits
- Strength of the state's economy[1]

State and federal policies with regard to commissioning are changing rapidly to increase the demand for commissioning. Also, technical assistance and funding are increasingly available for projects that can serve as demonstration projects for energy advocacy groups. The owner should investigate how each of these factors could benefit the decision to adopt commissioning in future construction projects.

Some of the major initiatives driving the growing market acceptance of commissioning are:

- Federal government's U.S. Energy Policy Act of 1992 and Executive Order 12902, mandating that federal agencies develop commissioning plans
- Portland Energy Conservation, Inc.; National Strategy for Building Commissioning; and their annual conferences
- ASHRAE HVAC Commissioning Guidelines (1989)
- Utilities establishing commissioning incentive programs
- Energy Star building program
- Leadership in Energy Environmental Design (LEED) certification for new construction
- Building codes
- State energy commission research programs

Currently, the LEED is having the largest impact in broadening the acceptance of commissioning. The Green Building Council is the sponsor of LEED and is focused on sustainable design—design and construction practices that significantly reduce or eliminate the cradle-to-grave negative impacts of buildings on the environment and building occupants. Leadership in energy efficient design encourages sustainable site planning, conservation of water and water efficiency, energy efficiency and renewable energy, conservation of materials and resources, and indoor environmental quality.

With this background on commissioning, the various components of the commissioning process can be explored, beginning with an evaluation of what building systems should be subject to the commissioning process.[5]

Commissioning Process

Systems to Include in the Commissioning Process

The general rule for including a system in the commissioning process is: the more complicated the system is the more compelling is the need to include it in the commissioning process. Systems that are required to integrate or interact with other systems should be included. Systems that require specialized trades working independently to create a cohesive system should be included, as well as systems that are critical to the operation of the building. Without a commissioning plan on the design and construction of these systems, installation deficiencies are likely to create improper interaction and operation of system components.

For example, in designing a lab, the doors should be included in the commissioning process because determining the amount of leakage through the doorways could prove critical to the ability to maintain critical room pressures to ensure proper containment of hazardous material. Another common example is an energy retrofit project. Such projects generally incorporate commissioning as part of the measurement and verification plan to ensure that energy savings result from the retrofit process.

For any project, the owner must be able to answer the question of why commissioning is important.

Why Commissioning?

A strong commissioning plan provides quality assurance, prevents disputes, and ensures contract compliance to deliver the intended system performance. Commissioning is especially important for HVAC systems that are present in virtually all buildings because commissioned HVAC systems are more energy efficient.

The infusion of electronics into almost every aspect of modern building systems creates increasingly complex systems requiring many specialty contractors. Commissioning ensures that these complex subsystems will interact as a cohesive system.

Commissioning identifies design or construction issues and, if done correctly, identifies them at the earliest stage in which they can be addressed most cost effectively. The number of deficiencies in new construction exceeds existing building retrofit by a factor of 3.[3] Common issues that can be identified by commissioning that might otherwise be overlooked in the construction and acceptance phase are: air distribution problems (these occur frequently in new buildings due to design capacities, change of space utilization, or improper installation), energy problems, and moisture problems.

Despite the advantages of commissioning, the current marketplace still exhibits many barriers to adopting commissioning in its most comprehensive and valuable forms.

Barriers to Commissioning

The general misperception that creates a barrier to the adoption of commissioning is that it adds extra, unjustified costs to a construction project. Until recently, this has been a difficult perception to combat because there are no energy-use baselines for assessing the efficiency of a new building. As the cost of energy continues to rise, however, it becomes increasingly less difficult to convince owners that commissioning is cost effective. Likewise, many owners and contractors do not appreciate that commissioning can reduce the number and cost of change orders through early problem identification. However, once the contractor and owner have a basis on which to compare the benefit of resolving a construction issue earlier as opposed to later, in the construction process, commissioning becomes easier to sell as a win–win proposal.

Finding qualified commissioning service providers can also be a barrier, especially in states where commissioning is not prevalent today. The references cited in this entry provide a variety of sources for identifying associations promulgating commissioning that can provide referrals to qualified commissioning agents.

For any owner adopting commissioning, it is critical to ensure acceptance of commissioning by all of the design construction team members. Enthusiastic acceptance of commissioning by the design team will have a very positive influence on the cost and success of your project. An objective of this entry is to provide a source of information to help gain such acceptance by design construction team members and the participants in the construction market.

Selecting the Commissioning Agent

Contracting an independent agent to act on behalf of the owner to perform the commissioning process is the best way to ensure successful commissioning. Most equipment vendors are not qualified and are likely to be biased against discovering design and installation problems—a critical function of the commissioning agent—with potentially costly remedies. Likewise, systems integrators have the background in control systems and data exchange required for commissioning but may not be strong in mechanical design, which is an important skill for the commissioning agent. Fortunately, most large mechanical consulting firms offer comprehensive commissioning services, although the desire to be competitive in the selection processes sometimes forces these firms to streamline their scope on commissioning.

Owners need to look closely at the commissioning scope being offered. An owner may want to solicit commissioning services independently from the selection of the architect/mechanical/electrical/plumbing design team or, minimally, to request specific details on the design team's approach to commissioning. If an owner chooses the same mechanical, electrical, and plumbing (MEP) firm for design and commissioning, the owner should ensure that there is physical separation between the designer and commissioner to ensure that objectivity is maintained in the design review stages. An owner should consider taking on the role of the commissioning agent directly, especially if qualified personnel exist in-house. This approach can be very cost effective. The largest obstacles to success with an in-house commissioning agent are the required qualifications and the need to dedicate a valuable resource to the commissioning effort. Many times, other priorities may interfere with the execution of the commissioning process by an in-house owner's agent.

There are three basic approaches to selecting the commissioning agent:

Negotiated—best approach for ensuring a true partnership
Selective bid list—preapproved list of bidders
Competitive—open bid list

Regardless of the approach, the owner should clearly define the responsibilities of the commissioning agent at the start of the selection process. Fixed-cost budgets should be provided by the commissioning agent to the owner for the predesign and design phases of the project, with not-to-exceed budgets submitted for the construction and acceptance phases. Firm service fees should be agreed upon as the design is finalized.

Skills of a Qualified Commissioning Agent

A commissioning agent needs to be a good communicator, both in writing and verbally. Writing skills are important because documentation is critical to the success of the commissioning plan. Likewise, oral communication skills are important because communicating issues uncovered in a factual and nonaccusatory manner is most likely to resolve those issues efficiently and effectively. The commissioning agent should have practical field experience in MEP controls design and startup to be able to identify potential issues early. The commissioning agent likewise needs a thorough understanding of how building structural design impacts building systems. The commissioning agent must be an effective facilitator and must be able to decrease the stress in stressful situations. In sum, the commissioning agent is the cornerstone of the commissioning team and the primary determinant of success in the commissioning process.

At least ten organizations offer certifications for commissioning agents. However, there currently is no industry standard for certifying a commissioning agent. Regardless of certification, the owner should carefully evaluate the individuals to be performing the work from the commissioning firm selected. Individual experience and reputation should be investigated. References for the lead commissioning agent are far more valuable than references for the executive members of a commissioning firm in evaluating potential commissioning agents. The commissioning agent selected will, however, only be one member of a commissioning team, and the membership of the commissioning team is critical to successful commissioning.

Commissioning Team

The commissioning team is composed of representatives from all members of the project delivery team: the commissioning agent, representatives of the owner's maintenance team, the architect, the MEP designer, the construction manager, and systems contractors. Each team member is responsible for a particular area of expertise, and one important function of the commissioning agent is to act as a facilitator of intrateam communication.

The maintenance team representatives bring to the commissioning team the knowledge of current operations, and they should be involved in the commissioning process at the earliest stage, defining the design intent in the predesign phase, as described below. Early involvement of maintenance team representatives ensures a smooth transition from construction to a fully operational facility, and aids in the acceptance and full use of the technologies and strategies that have been developed during the commissioning process. Involvement of the maintenance team representatives also shortens the building turnover transition period.

The other members of the commissioning team have defined and important functions. The architect leads the development of the design intent document (DID). The MEP designer's responsibilities are to develop the mechanical systems that support the design intent of the facility and comply with the owner's current operating standards. The MEP schematic design is the basis for the systems installed and is discussed further below. The construction manager ensures that the project installation meets the criteria defined in the specifications, the budget requirements, and the predefined schedule. The systems contractors' responsibilities are to furnish and install a fully functional system that meets the design specifications. There are generally several contractors whose work must be coordinated to ensure that the end product is a cohesive system.

Once the commissioning team is in place, commissioning can take place, and it occurs in defined and delineated phases—the subject of the following section.

Commissioning Phases

The commissioning process occurs over a variety of clearly delineated phases. The commission plan is the set of documents and events that defines the commissioning process over all phases. The commissioning plan needs to reflect a systematic, proactive approach that facilitates communication and cooperation of the entire design and construction team.

The phases of the commissioning process are:

Predesign
Design
Construction/installation
Acceptance
Postacceptance

These phases and the commissioning activities associated with them are described in the following sections.

Predesign Phase

The predesign phase is the phase in which the design intent is established in the form of the DID. In this phase of a construction project, the role of commissioning in the project is established if process-based commissioning is followed. Initiation of the commissioning process in the predesign phase increases acceptance of the commissioning process by all design team members. Predesign discussions about commissioning allow all team members involved in the project to assess and accept the importance of commissioning to a successful project. In addition, these discussions give team members more time to assimilate the impact of commissioning on their individual roles and responsibilities in the project. A successful project is more likely to result when the predesign phase is built around the concept of commissioning instead of commissioning's being imposed on a project after it has been designed.

Once an owner has decided to adopt commissioning as an integral part of the design and construction of a project, the owner should be urged to follow the LEED certification process, as discussed above. The commissioning agent can assist in the documentation preparation required for the LEED certification, which occurs in the postacceptance phase.

The predesign phase is the ideal time for an owner to select and retain the commissioning agent. The design team member should, if possible, be involved in the selection of the commissioning agent because that member's involvement will typically ensure a more cohesive commissioning team. Once the commissioning agent is selected and retained, the commissioning-approach outline is developed. The commissioning-approach outline defines the scope and depth of the commissioning process to be employed for the project. Critical commissioning questions are addressed in this outline. The outline will include, for most projects, answers to the following questions:

What equipment is to be included?
What procedures are to be followed?
What is the budget for the process?

As the above questions suggest, the commissioning budget is developed from the choices made in this phase. Also, if the owner has a commissioning policy, it needs to be applied to the specifics of the particular project in this phase.

The key event in the predesign phase is the creation of the DID, which defines the technical criteria for meeting the requirements of the intended use of the facilities. The DID document is often created based in part upon the information received from interviews with the intended building occupants and maintenance staff. Critical information—such as the hours of operation, occupancy levels, special environmental considerations (such as pressure and humidity), applicable codes, and budgetary considerations and limitations—is identified in this document. The owner's preference, if any, for certain equipment or contactors should also be identified at this time. Together, the answers to the critical questions above and the information in the DID are used to develop the commissioning approach outline. A thorough review of the DID by the commissioning agent ensures that the commissioning-approach outline will be aligned with the design intent.

With the commissioning agent selected, the DID document created, and the commissioning approach outline in place, the design phase is ready to commence.

Design Phase

The design phase is the phase in which the schematics and specifications for all components of a project are prepared. One key schematic and set of specifications relevant to the commissioning plan is the MEP schematic design, which specifies installation requirements for the MEP systems. As noted, the DID is the basis for creating the commissioning approach outline in the predesign phase. The DID also serves as the basis for creating the MEP schematic design in the design phase. The DID provides the MEP designer with the key concepts from which the MEP schematic design is developed.

The completed MEP schematic design is reviewed by the commissioning agent for completeness and conformance to the DID. At this stage, the commissioning agent and the other design team members should consider what current technologies, particularly those for energy efficiency, could be profitably included in the design. Many of the design enhancements currently incorporated into existing buildings during energy retrofitting for operational optimization are often not considered in new building construction. This can result in significant lost opportunity, so these design enhancements should be reviewed for incorporation into the base design during this phase of the commissioning process. This point illustrates the important principle that technologies important to retrocommissioning should be applied to new building construction—a point that is surprisingly often overlooked in the industry today.

For example, the following design improvements and technologies should always be considered for applicability to a particular project:

- Variable-speed fan and pumps installed
- Chilled water cooling (instead of DX cooling)
- Utility meters for gas, electric, hot water, chilled water, and steam at both the building and system level
- CO_2 implementation for minimum indoor air requirements

This list of design improvements is not exhaustive; the skilled commissioning agent will create and expand personalized lists as experience warrants and as the demands of particular projects suggest.

In addition to assisting in the evaluation of potential design improvements, the commissioning agent further inspects the MEP schematic design for:

- Proper sizing of equipment capacities
- Clearly defined and optimized operating sequences
- Equipment accessibility for ease of servicing

Once the commissioning agent's review is complete, the feedback is discussed with the design team to determine whether its incorporation into the MEP schematic design is warranted. The agreed-upon changes or enhancements are incorporated, thus completing the MEP schematic design.

The completed MEP schematic design serves as the basis on which the commissioning agent will transform the commissioning-approach outline into the commissioning specification.

The commissioning specification is the mechanism for binding contractually the contractors to the commissioning process. Expectations are clearly defined, including:

- Responsibilities of each contractor
- Site meeting requirements
- List of the equipment, systems, and interfaces
- Preliminary verification checklists
- Preliminary functional-performance testing checklists
- Training requirements and who is to participate
- Documentation requirements

- Postconstruction documentation requirements
- Commissioning schedule
- Definition for system acceptance
- Impact of failed results

Completion of the commissioning specification is required to select the systems contractor in a competitive solicitation. Alternatively, however, owners with strong, preexisting relationships with systems contractors may enter into a negotiated bid with those contractors, who can then be instrumental in finalizing the commissioning specification.

Owners frequently select systems contractors early in the design cycle to ensure that the contractors are involved in the design process. As noted above, if there are strong, preexisting relationships with systems contractors, early selection without a competitive selection process (described in the following paragraph) can be very beneficial. However, if there is no competitive selection process, steps should be taken to ensure that the owner gets the best value. For example, unit pricing should be negotiated in advance to ensure that the owner is getting fair and reasonable pricing. The commissioning agent and the MEP designer can be good sources for validating the unit pricing. The final contract price should be justified with the unit pricing information.

If the system selection process is competitive, technical proposals should be requested with the submission of the bid price. The systems contractors need to demonstrate a complete understanding of the project requirements to ensure that major components have not been overlooked. Information such as the project schedule and manpower loading for the project provide a good basis from which to measure the contractor's level of understanding. If the solicitation does not have a preselected list of contractors, the technical proposal should include the contractor's financial information, capabilities, and reference lists. As in the negotiated process described above, unit pricing should be requested to ensure the proper pricing of project additions and deletions. The review of the technical proposals should be included in the commissioning agent's scope of work.

A mandatory prebid conference should be held to walk the potential contractors through the requirements and to reinforce expectations. This conference should be held regardless of the approach— negotiated or competitive— used for contractor selection. The contractor who is to bear the financial burden for failed verification tests and subsequent functional-performance tests should be reminded of these responsibilities to reinforce their importance in the prebid meeting. The prebid conference sets the tone of the project and emphasizes the importance of the commissioning process to a successful project.

Once the MEP schematic design and commissioning specification are complete, and the systems contractors have been selected, the construction/installation phase begins.

Construction/Installation Phase

Coordination, communication, and cooperation are the keys to success in the construction and installation phase. The commissioning agent is the catalyst for ensuring that these critical activities occur throughout the construction and installation phase.

Frequently, value engineering options are proposed by the contractors prior to commencing the installation. The commissioning agent should be actively involved in the assessment of any options proposed. Many times, what appears to be a good idea in construction can have a disastrous effect on a facility's long-term operation. For example, automatic controls are often value engineered out of the design, yet the cost of their inclusion is incurred many times over in the labor required to perform their function manually over the life of the building. The commissioning agent can ensure that the design intent is preserved, the life-cycle costs are considered, and the impact on all systems of any value engineering modification proposed is thoroughly evaluated.

Once the design aspects are complete and value engineering ideas have been incorporated or rejected, the submittals, including verification checklists, need to be finalized. The submittals documentation is prepared by the systems contractors and reviewed by the commissioning agent. There are two types of submittals: technical submittals and commissioning submittals. Both types of submittals are discussed below.

Technical submittals are provided to document the systems contractors' interpretation of the design documents. The commissioning agent reviews the technical submittals for compliance and completeness. It is in this submittal review process that potential issues are identified prior to installation, reducing the need for rework and minimizing schedule delays. The technical submittals should include:

- Detailed schematics
- Equipment data sheets
- Sequence of operation
- Bill of material

A key technical submittal is the testing, adjusting, and balancing submittal (TAB). The TAB should include:

- TAB procedures
- Instrumentation
- Format for results
- Data sheets with equipment design parameters
- Operational readiness requirements
- Schedule

In addition to the TAB, other technical submittals, such as building automation control submittals, will be obtained from the systems contractors and reviewed by the commissioning agent.

The commissioning submittal generally follows the technical submittal in time and includes:

- Verification checklists
- Startup requirements
- Test and balance plan
- Training plan

The commissioning information in the commissioning submittal is customized for each element of the system.

These submittals, together with the commissioning specification, are incorporated into the commissioning plan, which becomes a living document codifying the results of the construction commissioning activities. This plan should be inspected in regular site meetings. Emphasis on the documentation aspect of the commissioning process early in the construction phase increases the contractors' awareness of the importance of commissioning to a successful project.

In addition to the submittals, the contractors are responsible for updating the design documents with submitted and approved equipment data and field changes on an ongoing basis. This update design document should be utilized during the testing and acceptance phase.

The commissioning agent also performs periodic site visits during the installation to observe the quality of workmanship and compliance with the specifications. Observed deficiencies should be discussed with the contractor and documented to ensure future compliance. Further inspections should be conducted to ensure that appropriate corrective action has been taken.

The best way to ensure that the items discussed above are addressed in a timely manner is to hold regularly scheduled commissioning meetings that require the participation of all systems contractors. This is the mechanism for ensuring that communication occurs. Meeting minutes prepared by the commissioning agent document the discussions and decisions reached. Commissioning meetings should be coordinated with the regular project meetings because many participants in a construction project need to attend both meetings.

Typical elements of a commissioning meeting include:

- Discussing field installation issues to facilitate rapid response to field questions
- Updating design documents with field changes
- Reviewing the commissioning agent's field observations
- Reviewing progress against schedule
- Coordinating multicontractor activities

Once familiar with the meeting process, an agenda will be helpful but not necessary. Meeting minutes should be kept and distributed to all participants.

With approved technical and commissioning plan submittals, as installation progresses, the contractor is ready to begin the system verification testing. The systems contractor generally executes the system verification independently of the commissioning agent. Contractor system verification includes:

- Point-to-point wiring checked out
- Sensor accuracy validated
- Control loops exercised

Each of the activities should be documented for each control or system element, and signed and dated by the verification technician.

The documentation expected from these activities should be clearly defined in the commissioning specification to ensure its availability to the commissioning agent for inspection of the verification process. The commissioning agent's role in the system verification testing is to ensure that the tests are completed and that the results reflect that the system is ready for the functional-performance tests. Because the commissioning agent is typically not present during the verification testing, the documentation controls how successfully the commissioning agent performs this aspect of commissioning.

In addition to system verification testing, equipment startup is an important activity during this phase. Equipment startup occurs at different time frames relative to the system verification testing, depending on the equipment and system involved. There may be instances when the system verification needs to occur prior to equipment startup to prevent a catastrophic event that could lead to equipment failure. The commissioning agent reviews the startup procedures prior to the startup to ensure that equipment startup is coordinated properly with the system verification. Unlike in verification testing, the commissioning agent should be present during HVAC equipment startup to document the results. These results are memorialized in the final commissioning report, so their documentation ultimately is the responsibility of the commissioning agent.

Once system verification testing and equipment startup have been completed, the acceptance phase begins.

Acceptance Phase

The acceptance phase of the project is the phase in which the owner accepts the project as complete and delivered in accordance with the specifications, and concludes with acceptance of the project in its entirety. An effective commissioning process during the installation phase should reduce the time and labor associated with the functional-performance tests of the acceptance phase.

Statistical sampling is often used instead of 100% functional-performance testing to make the process more efficient. A 20% random sample with a failure rate less than 1% indicates that the entire system was properly installed. If the failure rate exceeds 1%, a complete testing of every system may need to be completed to correct inadequacies in the initial checkout and verification testing. This random sampling statistical approach holds the contractor accountable for the initial checkout and test, with the performance testing serving only to confirm the quality and thoroughness of the installation. This approach saves time and money for all involved. It is critical, however, that the ramifications of not meeting the desired results of the random tests are clearly defined in the commissioning specifications.

The commissioning agent witnesses and documents the results of the functional-performance tests, using specific forms and procedures developed for the system being tested. These forms are created with the input of the contractor in the installation phase. Involvement of the maintenance staff in the functional-performance testing is important. The maintenance team is often not included in the design process, so they may not fully understand the design intent. The functional-performance testing can provide the maintenance team an opportunity to learn and appreciate the design intent. If the design intent is to be preserved, the maintenance team must fully understand the design intent. This involvement of the maintenance team increases their knowledge of the system going into the training and will increase the effectiveness of the training.

Training of the maintenance team is critical to a successful operational handover once a facility is ready for occupancy. This training should include:

- Operations and maintenance (O&M) manual overview
- Hardware component review
- Software component review
- Operations review
- Interdependencies discussion
- Limitations discussion
- Maintenance review
- Troubleshooting procedures review
- Emergency shutdown procedures review

The support level purchased from the systems contractor determines the areas of most importance in the training and therefore should be determined prior to the training process. Training should be videotaped for later use by new maintenance team members and in refresher courses, and for general reference by the existing maintenance team. Using the O&M manuals as a training manual increases the maintenance team's awareness of the information contained in them, making the O&M manuals more likely to be referenced when appropriate in the future.

The O&M manuals should be prepared by the contractor in an organized and easy-to-use manner. The commissioning agent is sometimes engaged to organize them all into an easily referenced set of documents. The manuals should be provided in both hard-copy and electronic formats, and should include:

- System diagrams
- Input/output lists
- Sequence of operations
- Alarm points list
- Trend points list
- Testing documentation
- Emergency procedures

These services—including functional-performance testing, training, and preparing O&M manuals—should be included in the commissioning plan to ensure the project's successful acceptance. The long-term success of the project, however, is determined by the activities that occur in the postacceptance phase.

Postacceptance Phase

The postacceptance phase is the phase in which the owner takes beneficial occupancy and forms an opinion about future work with the design team, contractors, and the commissioning agent who completed the project. This is also the phase in which LEED certification, if adopted, is completed. Activities that usually occur in the acceptance phase should instead occur in the postacceptance phase. This is due to constraints that are not controllable by the contractor or owner. For example, seasonal changes may make functional-performance testing of some HVAC systems impractical during the acceptance phase for certain load conditions. This generally means that in locations that experience significant seasonal climate change, some of the functional-performance testing is deferred until suitable weather conditions exist. The commissioning agent determines which functional-performance tests need to be deferred and hence carried out in the postacceptance phase.

During the postacceptance phase, the commissioning agent prepares a final commissioning report that is provided to the owner and design team. The executive summary of this report provides an overall assessment of the design intent conformance. The report details whether the commissioned equipment and systems meet the commissioning requirements. Problems encountered and corrective actions taken are documented in this report. The report also includes the signed and dated startup and functional-performance testing checklists.

The final commissioning report can be used profitably as the basis of a "lessons learned" meeting involving the design team so that the commissioning process can be continuously improved and adaptations can be made to the owner's commissioning policy for future projects. The owner should use the experience of the first commissioned project to develop the protocols and standards for future projects. The documentation of this experience is the owner's commissioning policy. Providing this policy and the information it contains to the design and construction team for the next project can help the owner reduce budget overruns by eliminating any need to reinvent protocols and standards and by setting the right expectations earlier in the process.

Commissioning therefore should not be viewed as a onetime event but should instead be viewed as an operational philosophy. A recommissioning or continuous commissioning plan should be adopted for any building to sustain the benefits delivered from a commissioning plan. The commissioning agent can add great value to the creation of the recommissioning plan and can do so most effectively in the postacceptance phase of the project.

Figure 1 depicts the information development that occurs in the evolution of a commissioning plan and summarizes the information presented in the preceding sections by outlining the various phases of the commissioning process.

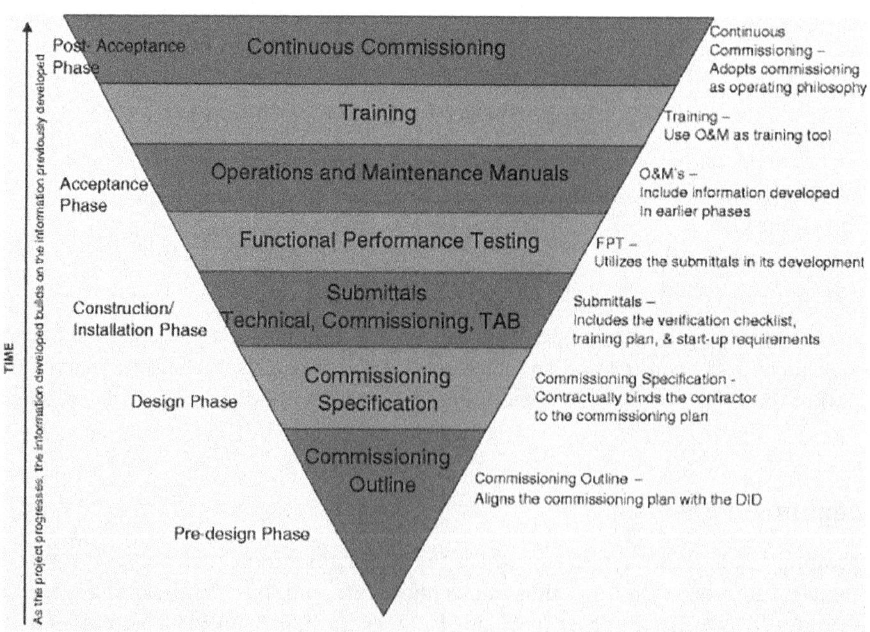

FIGURE 1 The commissioning plan.

With this background, the reader is better positioned for success in future commissioning projects and better prepared to learn the key success factors in commissioning and how to avoid common mistakes in the commissioning process.

Commissioning Success Factors

Ultimately, the owner will be the sole judge of whether a commissioning process has been successful. Thus, second only to the need for a competent, professional commissioning agent, keeping the owner or the owner's senior representative actively involved in and informed at all steps of the commissioning process is a key success factor. The commissioning agent should report directly to the owner or the owner's most senior representative on the project, not only to ensure that this involvement and information transfer occur, but also to ensure the objective implementation of the commissioning plan—a third key success factor.

Another key success factor is an owner appreciation— which can be enhanced by the commissioning agent—that commissioning must be an ongoing process to get full benefit. For example, major systems should undergo periodic modified functional testing to ensure that the original design intent is being maintained or to make system modification if the design intent has changed. If an owner appreciates that commissioning is a continuous process that lasts for the entire life of the facility, the commissioning process will be a success.

Most owners will agree that the commissioning process is successful if success can be measured in a cost/benefit analysis. Cost/benefit or return on equity is the most widely used approach to judge the success of any project. Unfortunately, the misapplication of cost/benefit analyses has been the single largest barrier to the widespread adoption of commissioning. For example, because new construction does not have an energy baseline from which to judge energy savings, improper application of a cost/benefit analysis can lead to failure to include energy savings technologies—technologies the commissioning agent can identify—in the construction process. Similarly, unless one can appreciate how commissioning can prevent schedule delays and rework by spotting issues and resolving them early in the construction process, one cannot properly offset the costs of commissioning with the benefits.

Fortunately, there are now extensive studies analyzing the cost/benefit of commissioning that justify its application. A study performed jointly by Lawrence Berkeley National Laboratory; Portland Energy Conservation, Inc.; and the Energy Systems Laboratory at Texas A&M University provides compelling analytical data on the cost/benefit of commissioning. The study defines the median commissioning cost for new construction as $1 per square foot or 0.6% of the total construction cost. The median simple payback for new construction projects utilizing commissioning is 4.8 years. This simple payback calculation does not take into account the quantified nonenergy impacts, such as the reduction in the cost and frequency of change orders or premature equipment failure due to improper installation practices. The study quantifies the median nonenergy benefits for new construction at $1.24 per square foot per year.[3]

While the primary cost component of assessing the cost/benefit of commissioning lies in whether there was a successful negotiation of the cost of services with the commissioning service provider, the more important aspect of the analysis relates to the outcomes of the process. For example, after a commissioning process is complete, what are the answers to these questions?

Are the systems functioning to the design intent?
Has the owner's staff been trained to operate the facility?
How many of the systems are operated manually a year after installation?

Positive answers to these and similar questions will ensure that any cost/benefit analysis will demonstrate the value of commissioning.

To ensure that a commissioning process is successful, one must avoid common mistakes. A commissioning plan is a customized approach to ensuring that all the systems operate in the most effective and efficient manner. A poor commissioning plan will deliver poor results. A common mistake is to use an existing commissioning plan and simply insert it into a specification to address commissioning. Each commissioning plan should be specifically tailored to the project to be commissioned.

Also, perhaps due to ill-conceived budget constraints, commissioning is implemented only in the construction phase. Such constraints are ill conceived because the cost of early involvement of the commissioning agent in the design phases is insignificant compared with the cost of correcting design defects in the construction phase. Significant cost savings can arise from identifying design issues prior to construction. Studies have shown that 80% of the cost of commissioning occurs in the construction phase.[2] Also, the later the commissioning process starts, the more confrontational commissioning becomes, making it more expensive to implement later in the process.[2] Therefore, adopting commissioning early in the project is a key success factor.

Value engineering often results in ill-informed, last-minute design changes that have an adverse and unintended impact on the overall building performance and energy use.[3] By ensuring that the commissioning process includes careful evaluation of all value engineering proposals, the commissioning agent and owner can avoid such costly mistakes.

Finally, the commissioning agent's incentive structure should not be tied to the number of issues brought to light during the commissioning process, as this can create an antagonistic environment that may create more problems than it solves. Instead, the incentive structure should be outcome based and the questions outlined above regarding compliance with design intent, training results, and postacceptance performance provide excellent bases for a positive incentive structure.

Conclusion

Commissioning should be performed on all but the most simplistic of new construction projects. The benefits of commissioning include:

- Optimization of building performance
  —Enhanced operation of building systems
  —Better-prepared maintenance staff
  —Comprehensive documentation of systems
  —Increased energy efficiency
  —Improved quality of construction
- Reduced facility life-cycle cost
  —Reduced impact of design changes
  —Fewer change orders
  —Fewer project delays
  —Less rework or postconstruction corrective work
  —Reduced energy costs
- Increased occupant satisfaction
  —Shortened turnover transition period
  —Improved system operation
  —Improved system reliability

With these benefits, owners and contractors alike should adopt the commissioning process as the best way to ensure cost-efficient construction and the surest way to a successful construction project.

References

1. Quantum Consulting, Inc. *Commissioning in Public Buildings: Market Progress Evaluation Report*, June 2005. Available at http://www.nwalliance.org/resources/reports/141.pdf.
2. ACG AABC Commissioning Group. ACG Commissioning Guidelines 2005. Available at http://www.commissioning.org/commissioningguideline/ACGCommissioningGuide-line.pdf.

3. Mills, E.; Friedman, H.; Powell, T. et al. *The Cost-Effectiveness of Commercial-Buildings Commissioning*, December 2004. Available at http://eetd.lbl.gov/Emills/PUBS/Cx-Costs-Benefits.html.
4. American Society of Heating. *Refrigerating and Air-Conditioning Engineers,* ASHRAE Guidelines 1–1996.
5. Green Building Council Web site. Available at http://www.usgbc.org/leed (accessed October 2005).

Bibliography

1. SBW Consulting, Inc. *Cost-Benefit Analysis for the Commissioning in Public Buildings Project,* May 2004. Available at http://www.nwalliance.org/resources/documents/CPBRe-port.pdf (accessed).
2. Turner, W.C. *Energy Management Handbook,* 5th Ed.; Fairmont Press: Lilburn, GA, 2005.
3. Research News Berkeley Lab. *Berkeley Lab Will Develop Energy-Efficient Building Operation Curriculum for Community Colleges,* December 2004. Available at http://www.lbl.gov/Science-Articles/Archive/EETD-college-curriculum.html.
4. Interview with Richard Holman. *Director of Commissioning and Field Services for Affiliated Engineers, Inc.* Walnut Creek, CA.

8

Energy Sources: Renewable versus Non-Renewable

Marc A. Rosen

Introduction

Energy can exist in many forms and be converted from form to form by energy-conversion technologies. Society and the people within it use energy carriers (often simply referred to as energy), which are produced from energy sources, in all aspects of living.

A basic understanding of renewable energy and non-renewable energy is provided in this entry. Renewable energy includes the energy received directly and indirectly from the sun as well as energy derived from other natural forces. Non-renewable energy includes non-renewable energy resources as well as energy forms that do not exist naturally but are produced by people. The main types of renewable energy are listed in Table 1, and the main types of non-renewable energy are listed in Table 2. All of these have been studied extensively.[1–10]

From the breakdown of renewable and non-renewable energy in Tables 1 and 2, it is clear that energy resources are often categorized into two groups: (1) those generally acknowledged to be finite and non-renewable and therefore not sustainable over the long term (e.g., fossil fuels, peat, uranium) and (2) those generally considered renewable and therefore sustainable over the relatively longer term (e.g., sunlight, wind, tides, falling water). Wastes (convertible to useful energy forms through, for example, waste-to-energy incineration facilities) and biomass fuels are also sometimes viewed as renewable energy sources.

In this entry, energy forms, sources, and carriers are explained. Renewable energy and non-renewable energy are discussed, and energy-conversion technologies are described. Then, energy use and factors

TABLE 1 Types of Renewable Energy

Direct solar radiation		
Solar-related energy*		
	Water based	
		Hydraulic energy (falling and running water, including large and small hydro)
		Wave energy
		Ocean thermal energy (from temperature difference between surface and deep waters of the ocean)
	Air based	
		Wind energy
	Land based	
		Biomass (where the rate of use does not exceed the rate of replenishment)
		Geothermal energy (ambient)
	Non-solar-related energy	
	Geothermal energy (internal heat of the earth)	
	Tidal energy (from gravitational forces of the sun and moon and the rotation of the earth)	

*As explained in the text, fossil fuels are originally solar energy, but with a very long time lag for their transformation. So technically, they are solar-related energies. But for practical purposes, fossil fuels are non-renewable, so they are not listed in this table.

TABLE 2 Types of Non-Renewable Energy

Energy sources			
	Fossil fuels		
		Conventional	
			Coal
			Oil
			Natural gas
		Alternative	
			Oil shales
			Tar sands
			Peat
	Non-fossil fuels		
		Uranium	
		Fusion material (e.g., deuterium)	
		Wastes (which can be used as energy forms or converted to more useful energy forms)	
Energy currencies			
	Work		
	Electricity		
	Thermal energy		
		Heat (or a heated medium such as hot air, steam, exhaust gases)	
		Cold (or a cooled medium such as cold brine, ice)	
	Secondary chemical fuels		
		Conventional	
			Oil products (e.g., gasoline, diesel fuel, naphtha)
			Synthetic gaseous fuels (e.g., from coal gasification)
			Coal products (e.g., coke)
		Non-conventional	
			Methanol
			Ammonia
			Hydrogen

in energy selection are discussed. Finally, efficiencies for energy use are presented, along with measures to improve energy efficiency.

Forms, Sources, and Carriers of Energy

Energy comes in a variety of forms, including fossil fuels (e.g., coal, oil, natural gas), fossil fuel-based products (e.g., gasoline, diesel fuel), uranium, electricity, work (e.g., mechanical energy in a rotating engine shaft), heat, heated substances (e.g., steam, hot air), and light and other electromagnetic radiation.

Energy sources (sometimes called primary energy forms) are found in the natural environment. Some are available in finite quantities (e.g., fossil fuels, fossil fuel-containing substances such as oil sands, peat, and uranium). Some energy resources are renewable (or relatively renewable), including sunlight (or solar energy), falling water, wind, tides, geothermal heat, wood, and other biomass fuels (provided the growth rate exceeds or meets the rate of use). Energy sources are often processed from their raw forms prior to use.

Energy carriers (sometimes called energy currencies) are the energy forms that we transport and use, and include some energy sources (e.g., fossil fuels) and processed (or secondary) energy forms (e.g., gasoline, electricity, work, heat). The processed energy forms are not found in the environment.

The distinction between energy carriers and sources is important. Energy carriers can exist in a variety of forms and can be converted from one form to another, while energy sources are the original resource from which an energy carrier is produced. Misunderstanding sometimes results between energy sources and carriers because some energy sources are also energy carriers. For example, hydrogen is an energy carrier, not an energy source, and can be produced from a wide range of resources using various energy-conversion processes (e.g., water electrolysis, reforming of natural gas, coal gasification). In this way, hydrogen energy is analogous to electricity. Nevertheless, hydrogen is often erroneously referred to as an energy source, especially in discussions of its potential future role as a chemical energy carrier to replace fossil fuels.

Renewable Energy

Renewable energy includes the solar radiation incident on the earth and the energy forms that directly result from that radiation. Renewable energy also includes the energy supplied by other natural forces, such as gravitation and the rotation of the earth. The types of renewable energy are summarized in Table 1. It is this energy that makes possible the existence of ecosystems, human civilizations, and life itself.

Solar Energy

Direct solar radiation is the main type of renewable energy.[11] The daily energy output of the sun is 8.33×10^{25} kWh, of which the earth receives 4.14×10^{15} kWh. At any instant, the rate of solar energy reaching the earth is 1.75×10^{17} W, which is about 20,000 times greater than the total energy-use rate of the world. Solar radiation can be converted directly to electricity in photovoltaic devices. Also, solar energy can be collected as heat and used for thermal processes such as space and water heating or concentrated for use for in high-temperature heating and thermal electricity generation.

Most of the energy that enters the system of the earth and its atmosphere eventually exits to space. This concept can be demonstrated by considering the earth–sun energy balance (see Figure 1). A general energy balance

$$\text{Energy input} - \text{Energy output} = \text{Energy accumulation}$$

can be applied to the earth when

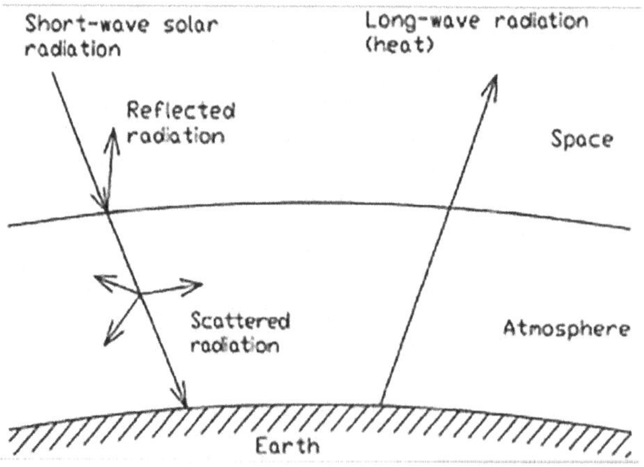

FIGURE 1 Earth–sun energy balance.

- The energy input is the short-wave solar radiation entering the atmosphere.
- The energy output is the long-wave radiation exiting the atmosphere to space.
- The energy accumulation term is the increase in energy of the earth and its atmosphere.

The main implication of this global energy balance is that since the average temperature of the earth is relatively constant, the energy accumulation term is zero. Therefore, the energy output is equal to the energy input for the planet.

Global warming is caused by a disruption in the earth–sun energy balance. The main cause of global warming is increased releases of atmospheric "greenhouse gases" that absorb radiation in the 8–20 μm region. When greenhouse gas concentrations increase in the atmosphere, energy output from the earth and its atmosphere (Figure 1) is reduced while energy input remains constant. Thus, the energy accumulation term becomes positive, leading to an increase in the average temperature of the earth. Eventually, if concentrations of greenhouse gases in the atmosphere stabilize at new levels, the energy balance is re-established but at some higher average planetary temperature.

Solar-Related Energy

Several types of renewable energy are a consequence of solar radiation. These can be loosely arranged into water, air, and land groupings. Some forms of solar-derived energy are more common, for reasons of practicality, economics, and environmental impact.

Water-Based Solar-Related Energy. The most common solar-related energy is hydraulic energy, which includes falling and running water in natural settings such as rivers and waterfalls. Large-scale hydroelectric generating installations are common. Most economically utilizable hydraulic resources have already been developed. Recently, interest in the potential uses of small-scale hydro has grown. Other forms of solar-related energy involve water. Wave energy systems that take advantage of the motion of waves have been proposed, although the potential contribution from wave energy is relatively small. Ocean thermal energy arises from the temperature difference between surface and deep waters of the ocean. This temperature difference can be utilized to drive a heat engine, and several ocean thermal energy-conversion (OTEC) devices have been tested.

Land-Based Solar-Related Energy. Biomass energy includes wood and other forms of plants and organic matter. Biomass can act as a fuel itself or can be converted into more desirable fuels. Several fast-growing trees have been identified as good candidates for biomass energy production. Biomass energy is only a renewable resource when the rate at which it is used does not exceed the rate at which it is replenished. Another form of solar-related energy that involves land is ambient geothermal energy. This is the energy at ambient temperatures in the ground near the surface, which is maintained predominantly by solar radiation and ambient air temperatures. This ambient energy can be used for heating and/or cooling, especially in conjunction with technologies like heat pumps.

Air-Based Solar-Related Energy. Wind energy is used extensively in some countries (e.g., Denmark) for electricity generation. Its use is growing in recent years, making it more widespread than even a decade ago.[12]

Non-Solar-Related Energy

The main types of renewable energy in this category are geothermal energy,[13] which exists as a consequence of the internal heat of the earth, and tidal energy, which is attributable to the gravitational forces of the sun and moon and the rotation of the earth. Both of these energy sources have been used in limited ways.

Non-Renewable Energy

Non-renewable energy, often called secondary energy, includes both energy resources that are available in limited quantities and not renewable and energy forms produced by humankind.[2–7,13] Different types of non-renewable energy are summarized in Table 2. Non-renewable energy is related to the level of technological development of a society.

Two main categories exist for non-renewable energy:

- Energy resources that are not renewable. The most common of these are fossil fuels, which are the basis of the economy for most industrialized countries. In addition to conventional fossil fuels, there exist alternative fossil fuels such as oil shales, tar sands, and peat. Other non-renewable energy resources include uranium and fusion material (e.g., deuterium).
- Energy currencies not existing naturally. This category of non-renewable energy encompasses energy currencies that do not exist naturally. They include such basic energy forms as work, electricity, and thermal energy. The latter can be either heat or a heated medium like steam or cold water.

At this point, a note on the non-renewability of fossil fuels is necessary for clarity and comprehensiveness. Fossil fuels are originally solar energy, but with a very long time lag for their transformation. Thus, considering the origin of fossil fuels, they could in theory be listed in Table 1 along with other solar-related energies. The non-renewability of fossil fuels deals with the lengthy time lag as well as the unique conditions on Earth that permit the creation of biomass at a much faster rate than its decomposition. Thus, fossil fuels are for practical purposes non-renewable, so they are not listed in Table 1.

Thermal energy in the form of heat or cold can be transported to users over long distances in district heating and/or cooling systems. District heating systems use centralized heating facilities to produce a heated medium that is transported to many users connected along a district heating network. For example, buildings in the cores of many cities are often connected by pipes through which hot water or steam flows to provide space and water heating. Similarly, district cooling involves the central production of a cold medium, which is transported to users through a piping network to provide cooling. Many cities and industrial parks utilize such district energy systems.

Wastes, which include recovered materials and energy that would otherwise be discarded, are also sometimes considered as an energy source in the category of non-renewable energy. Wastes can be used directly as energy forms or converted to more useful forms. Waste materials and waste heat can be recovered for utilization both within a facility and in other facilities where they are needed. For example, waste heat from hot gases (e.g., stack gases) and liquids (e.g., cooling-water discharges) can sometimes be recovered. Also, material wastes can be used in waste-to-energy incineration facilities, which burn garbage to provide heat and to generate electricity. Utilizing such wastes offsets the need for further supplies of external energy.

Non-renewable energy also includes secondary chemical fuels. Some conventional ones include oil-derived products such as gasoline, diesel fuel, and naphtha, as well as synthetic gaseous fuels (e.g., from coal gasification) and coal products (e.g., coke). The types of non-conventional chemical fuels proposed are numerous and include methanol, ammonia, and hydrogen.

Life Cycle Considerations

As many types of non-renewable energy can be produced from energy resources or converted from other types of energy, it is important to understand and consider all the steps in the entire life cycle of an energy product. The following life stages are usually included in assessments:

- Extraction or collection of raw energy resources
- Manufacturing and processing of the desired energy form(s)
- Transportation and distribution of the energy to users
- Energy storage
- Use of the energy to provide services and tasks
- Recovery and reuse of output energy that would otherwise be wasted (e.g., waste heat recovery)
- Recycling of wastes from any of the above steps
- Disposal of final wastes (e.g., materials such as stack gases and solid wastes including ash).

For example, the life cycle of a general energy form may involve the following chain of events:

Raw resource → Finished resource → Energy product → Waste → Waste disposal

Two or more types of energy can be simultaneously produced in some systems. For example, cogeneration is a process that usually refers to the combined generation of electricity (or work) and heat (or a heated medium). Trigeneration refers to an extended cogeneration process in which cooling is provided as a third product.

Some examples of the different life cycles for electricity generation methods from a range of energy sources are presented in Table 3. In that table, methods based on fossil fuels and non-fossil resources are considered. In addition, electricity generation from different energy sources via a less conventional technology, fuel cells, is considered in Table 4. All processes in Tables 3 and 4 include equipment disposal as a final step in the life cycle, although that step is not shown.

TABLE 3 Life Cycles for Selected Energy Sources and Methods for Electricity Generation

Method and Energy Source	Energy-conversion Processes in the Life Cycle
Fossil fuel-based methods	
Natural gas	Extraction via gas well → Transport → Processing → Natural gas-fired power plant
Coal	Coal mining → Transport → Processing → Coal-fired power plant
Oil	Extraction via oil well → Transport (tanker or pipeline) → Refining → Oil-fired power plant
Non-fossil-based methods	
Nuclear	Uranium mining → Transport → Processing → Nuclear power plant
Hydro	Hydraulic turbine
Wind	Wind generator
Solar (thermal)	Solar energy collection → Thermal power plant
Solar (non-thermal)	Solar photovoltaic panels
Geothermal	Well to geothermal source → Geothermal power plant
Ocean (thermal)	OTEC power plant

TABLE 4 Life Cycles for Electricity Generation from Various Energy Sources Using Fuel Cells

Energy Source	Energy-conversion Processes in the Life Cycle
Fossil fuels	Fossil-based hydrogen production → Fuel cell
Non-fossil energy sources	Hydrogen production → Fuel cell
High-temperature heat (e.g., solar)	Hydrogen production via thermochemical cycle → Fuel cell

Energy Use

Energy use involves the production of useful energy forms through energy-conversion processes, as well as energy transportation, distribution and storage, and the utilization of energy resources and processed forms of energy to provide services and perform tasks.

Energy-Conversion Technologies

Desired energy carriers are produced from energy sources, or converted from one form to another, using energy-conversion technologies. The energy-conversion technology appropriate in a given instance depends on the initial (or source) and final (or desired) energy forms. Conventional energy-conversion technologies include hydroelectric, fossil fuel, and nuclear generating stations; oil refineries; engines and motors; and heaters.[14–16] Some examples of less conventional energy-conversion technologies include fuel cells, solar photovoltaics, and high-efficiency and/or clean technologies for fossil fuels (e.g., combined-cycle systems).[4,17,18] Some energy-conversion technologies yield more than one product (e.g., cogeneration). Various technologies for electricity generation are listed in Tables 3 and 4.

A major energy-conversion process is combustion, which drives furnaces, engines, transportation vehicles, etc., and which can be expressed for a hydrocarbon fuel as

$$\text{Hydrocarbon fuel} + \text{Air}\ (\text{mainly } O_2 \text{ and } N_2) \rightarrow CO_2 + H_2O + N_2 + \text{Other substances}$$

The other substances are leftover reactants and other reaction products. The reaction for stoichiometric combustion (i.e., complete combustion with a balanced amount of reactants and no reactions that yield other products) of a general hydrocarbon (C_mH_n) in air (treated as only nitrogen and oxygen) can be written as

$$C_mH_n + (m + n/4)(O_2 + 3.76N_2) \rightarrow m\ CO_2\ + n/2H_2O + 3.76\ (m + \ n/4)N_2$$

Here, m and n are variables that take on different values for different hydrocarbons (e.g., in approximate terms, $m = 1$ and $n = 4$ for natural gas, which is mainly methane, $m = 1$ and $n = 2$ for oil, $m = 1$ and $n = 1$ for typical coal, and $m = 1$ and $n = 0$ for pure carbon). Several important points stem from these expressions. First, carbon dioxide is an inherent product of the combustion of any carbon-containing fuel. Generally, the only way to avoid carbon dioxide emissions is to eliminate the use of carbon-based fuels (e.g., by using hydrogen as a fuel). Second, if insufficient oxygen is available (or if mixing is poor or reaction time short), the carbon will often only partially combust, yielding CO. Third, excess air (over stoichiometric) is usually used to improve fuel burnup, often lowering combustor efficiencies since some of the fuel energy must go towards heating the excess air. Fourth, often the fuel is not a pure hydrocarbon and contains other substances, e.g., sulfur. Sulfur combusts, yielding heat, and reacts to sulfur dioxide.

Energy Selection

Energy selection involves choosing both energy carriers and sources. The selection often depends on the energy service to be provided and the energy-conversion technologies available. Some energy selections involve energy or fuel substitution (e.g., heating with natural gas rather than electricity).

To prevent environmental impact, energy sources are usually preferred, which are renewable and cause relatively lower environmental impacts.[19–23] By extension, preference is also placed on energy sources and carriers that can be used with higher efficiency and more environmentally benign energy-conversion technologies (e.g., boilers having low emissions of nitrogen oxides, NO_x).[9,24,25]

Renewable energy sources (Table 1), usually being derived from sunlight or solar-derived sources (e.g., wind, waves), are sustainable. A major barrier to renewable energy sources is that they are usually more costly to use than non-renewable energy sources such as fossil fuels, although this observation is not true in some niche applications.

Environmental Considerations

The use of fossil fuels leads to combustion emissions. Problematic pollutants are generally lower for fuels having higher hydrogen-to-carbon atomic ratios, so natural gas is more benign than oil, which is more benign than coal. There are no emissions during normal operation of nuclear power facilities except the spent fuel, which remains radioactive for many years. Although falling and running water, derived from solar energy, drive hydroelectric generation, flooding of lands sometimes occurs. The use of renewable energy is relatively benign, but resources are required to build the relevant energy-conversion technologies and large land tracts are often needed (e.g., for solar collectors or wind generators). In all assessments of environmental impact, adoption of a life cycle perspective is critical.

Energy Efficiency and Other Measures of Merit for Energy Use

Many efficiencies can be defined for energy use. Other measures are also often used to assess the merit of energy use relative to other criteria. One important measure is energy intensity, which reflects the energy use per some unit of output, e.g., energy consumption per monetary unit of gross domestic product for a country or region. Since they deal with raw materials, primary processing industries (e.g., petroleum/coal production, chemicals production, primary metals) generally have higher energy intensities than intermediate and secondary processing industries, which, although they come in a wide variety of forms, tend to deal with more finished products.

Numerous methods and technologies exist for improving energy efficiency:

Use of High-Efficiency Devices. Energy efficiency can generally be increased via high-efficiency versions of energy intensive devices (e.g., home appliances, furnaces, air conditioners, motors, boilers). New lighting fixtures (including bulbs, reflectors, diffusers, and ballasts), for instance, have significantly higher efficiencies and longer lives than older equipment. Lighting efficiency can also be increased by task lighting (i.e., directing light where it is needed), reducing lighting to levels adequate for the human eye and the nature of the facility, and using timers, dimmers, and occupancy sensors.

Energy Leak and Loss Prevention. Energy losses are associated with energy leaks. Approaches to avoid such energy leaks include (1) applying leak-prevention technologies and methods (e.g., sealing leaks in storage vessels and pipelines and insulating to reduce undesired heat flows) and (2) inspecting periodically to detect leaks and initiate appropriate actions.

Application of Advanced and Integrated Energy Systems. Many advanced energy systems feature increased efficiencies compared to conventional technologies. Efficiency can also be increased by advantageously integrating energy systems so that wastes become inputs to other processes. Waste-recovery and waste-to-energy plants normally increase efficiency by offsetting some of the need for external energy. Cogeneration and trigeneration are often more efficient than separate processes for the individual products. Compared to the alternative of each facility having its own heating and/ or cooling plant, district heating and/or cooling often provide increased efficiency and reduced environmental impact because many efficiency-improvement and environmental control measures are possible in centralized, large-scale facilities. Attempts have been made to optimize regional heat markets and systems.[26]

Improved Monitoring, Control, and Maintenance. Energy efficiency can be improved through better monitoring, control, and maintenance of operations, so that performance degradation from design specifications is avoided. Computer systems allow automated acquisition of frequent and widespread readings and diagnostic evaluations. The life span of energy equipment can be increased through diligent maintenance (e.g., cleaning, lubricating, calibrating, tuning, regular testing, periodic overhauling, replacing consumable items regularly).

Improved Matching of Energy Supplies and Demands. Rather than supply an energy form of a level that greatly exceeds that required for a specific energy demand, it is usually more efficient to supply an energy form of a better matched level. For instance, better matching can often be achieved for heat-transfer flow temperatures (e.g., for space heating at about 22°C, a heat supply of perhaps 40°C could be sufficient, rather than furnace combustion gases at hundreds of degrees). Thus, industrial waste heat and low-temperature cogenerated heat can satisfy some heating needs.

Energy Storage. Sometimes, available energy supply exceeds demand, and the energy is not utilized.[27,28] At times, the supply is not controllable (e.g., sunlight may be available for heating during the day whether or not it is needed), while in other cases, the supply may not be easily alterable (e.g., the waste heat from a factory). Energy storage can improve system efficiency by storing energy between times when it is available and needed.

Improved Building Envelopes. The energy efficiency of a building can be improved by increasing insulation to reduce heat infiltration in summer and heat loss in winter, applying weather stripping and caulking to reduce air leakages, and utilizing advanced and high-efficiency windows. The latter reduce heat losses by using multiple glazings that are sometimes separated by insulating gases or a vacuum, electronic and photosensitive windows that automatically reflect or absorb excessive sunlight, low-emissivity window coatings that increase a window's resistance to heat loss, and efficient window shades that adjust themselves using photosensitive sensors to block excessive sunlight while permitting natural room lighting.

Use of Passive Strategies to Reduce Energy Use. Passive, as opposed to active, methods can be used to reduce energy use. The following are some examples: adjusting design temperatures such as those used for space heating in winter and space cooling in summer; shutting energy devices during periods of non-use and using timers to control operating hours; using daylight harvesting to reduce artificial lighting needs; applying "free cooling" by using cool outdoor air when it is available to cool warm indoor spaces, rather than active air conditioning; using solar radiation to heat buildings by exploiting the thermal storage capacity of buildings; and locating trees, windows, and window shades so as to keep buildings cool during summer.

Use of Exergy Analysis. Exergy analysis is used to improve and optimize the efficiency and performance of energy and other systems and processes.[1,13,29–31]

Energy Sustainability

Achieving appropriate energy technologies, resources, and efficiencies can help in efforts to make energy systems more sustainable, in terms of environmental protection as well as economic and social development.[32] Energy sustainability requires appropriate energy sources and carriers, efficiencies, and environmental protection, in addition to other needs.[33,34] Efforts to achieve energy sustainability greatly support efforts to achieve overall sustainability,[35] and contribute significantly to attaining the United Nations Sustainable Development Goals for 2015–2030.[36]

Example

Many examples can illustrate the methods and technologies described in the section "Energy Efficiency and Other Measures of Merit for Energy Use" for improving energy efficiency. Here, we focus on the use of high-efficiency lighting. New fixtures have significantly higher efficiencies and longer lives than older equipment. Several operating parameters for various types of lighting are shown in Table 5. For example, the efficiencies for some light sources (in lumens of light delivered per watt of electricity consumed) are as follows: incandescent (10–30), mercury (20–55), fluorescent (20–60), high-pressure sodium (50–130), and low-pressure sodium (80–155). Thus, the same amount of light can be delivered with high-efficiency lighting using much less than 20% of the electricity required for an incandescent bulb. Lighting using

TABLE 5 Operating Parameters for Various Types of Lighting

Lighting Type	Normalized Light Output (lumens/W) Mean	Lower Range	Upper Range	Normalized Electricity Use for Lighting (W/lumens) Mean	Upper Range	Lower Range	Increase in Normalized Light Output (%)[a]	Decrease in Normalized Electrical Use (%)[a]
Incandescent	20	10	30	0.050	0.100	0.033	0	0
Mercury	38	20	55	0.026	0.050	0.018	90	48
Fluorescent	40	20	60	0.025	0.050	0.017	100	50
High-pressure sodium	90	50	130	0.011	0.020	0.0080	350	78
Low-pressure sodium	118	80	155	0.0085	0.013	0.0065	490	83

[a] Changes are relative to incandescent lighting and based on mean values.

light-emitting diode (LED) technology is capable today of over 100 lumens per watt, and some prototypes have demonstrated the ability to achieve 200 and even 300 lumens per watt.

On a regional or national level, the savings can be significant. Since lighting is a significant energy consumer, accounting for about 20% of U.S. electrical energy use, the use of high-efficiency lighting across the United States could reduce that country's electricity consumption by 18% (i.e., 90% of the 20% of electricity use attributable to lighting). Of course, cost and other considerations limit the implementation of high-efficiency lighting.

Note that other measures to increase lighting efficiency could reduce the electricity used for lighting further. If, for instance, electricity use for lighting is reduced by an additional 5% through application of (1) timers, dimmers, and occupancy sensors to turn lights off when rooms are unoccupied; (2) reduced lighting intensities; and (3) task lighting, then electricity use in the United States could be reduced by 19% (i.e., 95% of the 20% of electricity use attributable to lighting).

Conclusion

Categorizing energy forms as renewable energy and non-renewable energy allows for improved understanding of energy systems. Renewable energy and non-renewable energy are important in societies, particularly since they affect many facets of life and living standards. These energy categories and their applications are described in this entry, as are energy forms, sources, carriers, and technologies; energy use; energy selection and efficiencies; and efficiency-improvement measures.

Glossary

m number of carbon atoms in a hydrocarbon
n number of hydrogen atoms in a hydrocarbon

Acknowledgments

The author is grateful for the support provided for this work by the Natural Sciences and Engineering Research Council of Canada.

References

1. Dincer, I., Rosen, M.A.; Ahmadi, P. *Optimization of Energy Systems*; Wiley: London, 2017.
2. Organization for Economic Cooperation and Development. *Energy: The Next Fifty Years*; OECD Publishing: Washington, DC, 1999.
3. Organization for Economic Cooperation and Development. *World Energy Outlook 2010*; OECD Publishing: Washington, DC, 2010.

4. Asdrubali, F.; Desideri, U., Eds. *Handbook of Energy Efficiency in Buildings: A Life Cycle Approach*; Elsevier: Oxford, 2019.
5. Niele, F. *Energy: Engine of Evolution*; Elsevier: Oxford, 2005.
6. Roosa, S.A.; Doty, S.; Turner, W.C. *Energy Management Handbook*, 9th Ed.; Fairmont Press: Lilburn, GA, 2018.
7. Goldemberg, J.; Johansson, T.B.; Reddy, A.K.N.; Williams, R.H. *Energy for a Sustainable World*; Wiley: New York, 1988.
8. Asif, M.; Muneer, T. Energy supply, its demand and security issues for developed and emerging economies. *Renewable and Sustainable Energy Reviews* **2007**, *11*, 1388–1413.
9. Ramanathan, R. A multi-factor efficiency perspective to the relationships among world GDP, energy consumption and carbon dioxide emissions. *Technological Forecasting and Social Change* **2006**, *73*, 483–494.
10. Birol, F. World energy prospects and challenges. *Asia Pacific Business Review* **2007**, *14*, 1–12.
11. Moharamian, A.; Soltani, S.; Rosen, M.A.; Mahmoudi, S.M.S.; Bhattacharya, T. Modified exergy and modified exergoeconomic analyses of a solar based biomass co-fired cycle with hydrogen production. *Energy* **2019**, *167*, 715–729.
12. Ehyaei, M.A.; Ahmadi, A.; Rosen, M.A. Energy, exergy, economic and advanced and extended exergy analyses of a wind turbine. *Energy Conversion and Management* **2019**, *183*, 369–381.
13. Rosen, M.A.; Koohi-Fayegh, S. *Geothermal Energy: Sustainable Heating and Cooling Using the Ground*; Wiley: London, 2017.
14. International Energy Agency. *Energy Technology Transitions for Industry: Strategies for the Next Industrial Revolution*; International Energy Agency: Paris, 2009.
15. International Energy Agency. *Transport, Energy and CO_2: Moving Toward Sustainability*; International Energy Agency: Paris, 2009.
16. Organisation for Economic Co-cooperation and Development. *Nuclear Energy Outlook 2008*; Organisation for Economic Co-operation and Development Nuclear Energy Agency, OECD Publishing: Paris, 2008.
17. Gnanapragasam, N.V.; Reddy, B.V.; Rosen, M.A. Feasibility of an energy conversion system in Canada involving large-scale integrated hydrogen production using solid fuels. *The International Journal of Hydrogen Energy* **2010**, *35*, 4788–4807.
18. Energy Technology Perspectives 2010: Scenarios & Strategies to 2030; International Energy Agency: Paris, 2010.
19. Shogren, J. (Ed.) *Encyclopedia of Energy, Natural Resource, and Environmental Economics*; Elsevier: Oxford, 2013.
20. Singh, D.P.; Kothari, R.; Tyagi, V.V., Eds. *Emerging Energy Alternatives for Sustainable Environment*; CRC Press: Boca Raton, FL, 2019.
21. Michaelides, E.E. *Energy, the Environment, and Sustainability*; CRC Press: Boca Raton, FL, 2018.
22. Lund, H.; Mathiesen, B.V. Energy system analysis of 100% renewable energy systems—The case of Denmark in years 2030 and 2050. *Energy* **2009**, *34*, 524–531.
23. Rosen, M.A. Combating global warming via non-fossil fuel energy options. *The International Journal of Global Warming* **2009**, *1*, 2–28.
24. Kreith, F; Goswami, D.Y., Eds. *Energy Management and Conservation Handbook*, 2nd Ed.; CRC Press: Boca Raton, FL, 2016.
25. Herring, H. Energy efficiency—A critical view. *Energy* **2006**, *31*, 10–20.
26. Karlsson, M.; Gebremedhin, A.; Klugman, S.; Henning, D.; Moshfegh, B. Regional energy system optimization—Potential for a regional heat market. *Applied Energy* **2009**, *86*, 441–451.
27. Ibrahim, H.; Ilinca, A.; Perron, J. Energy storage systems—Characteristics and comparisons. *Renewable and Sustainable Energy Reviews* **2008**, *12*, 1221–1250.
28. Dincer, I.; Rosen, M.A. *Thermal Energy Storage: Systems and Applications*, 2nd Ed.; Wiley: London, 2011.

29. Kotas, T.J. *The Exergy Method of Thermal Plant Analysis*, Reprint Ed.; Krieger: Melbourne, FL, 1995.
30. Dincer, I.; Rosen, M.A. *Exergy: Energy, Environment and Sustainable Development*, 2nd Ed.; Elsevier: Oxford, 2013.
31. Rosen, M.A. *Economics and Exergy: An Enhanced Approach to Energy Economics*; Nova Science Publishers: Hauppauge, NY, 2011.
32. Krewitt, W.; Simon, S.; Graus, W.; Teske, S.; Zervos, A.; Schafer, O. The 2°C scenario—A sustainable world energy perspective. *Energy Policy* **2007**, *35*, 4969–4980.
33. Rosen, M.A. Energy sustainability: A pragmatic approach and illustrations. *Sustainability* **2009**, *1*, 55–80.
34. Rosen, M.A. Towards energy sustainability: A quest of global proportions. *Forum on Public Policy Online: A Journal of the Oxford Round Table*, **2008**, *2008* (2), 1–20. Available at http://forumonpublicpolicy.com/Summer08papers/archivesummer08/rosen.pdf (accessed May 5, 2012).
35. Rosen, M.A. Issues, concepts and applications for sustainability. *Glocalism: Journal of Culture, Politics and Innovation* **2018**, *3*, 1–21.
36. Rosen, M.A. How can we achieve the UN sustainable development goals? *European Journal of Sustainable Development Research* **2017**, *1*(2), 6.

9

Energy: Physics

Milivoje M. Kostic

Introduction: From Work to Heat to General Energy Concept

Energy is a fundamental concept indivisible from matter and space, and energy exchanges or transfers are associated with all processes (or changes), thus indivisible from time. Actually, energy is "the building block" and fundamental property of matter and space, thus a fundamental property of existence. Energy transfer is needed to produce a process to change other system properties. Also, among all properties, energy is the only one that is directly related to mass and vice versa: $E = mc^2$ (known in some literature as *mass energy,* or mass-energy equivalence; the c is the speed of light in a vacuum); thus, mass and energy are interrelated. Mass and energy are a manifestation of each other and are equivalent; they have a holistic meaning of *mass* energy.[1,2]

Energy moves cars and trains, and boats and planes. It bakes food and keeps it frozen for storage. Energy lights our homes and plays our music. Energy makes our bodies alive and grow, and allows our minds to think. Through centuries, people have learned how to use energy in different forms in order to do work more easily and live more comfortably. No wonder that energy is often defined as *ability to perform work,* i.e., as a potential for energy transfer in a specific direction (displacement in force direction), thus achieving a purposeful process, as opposed to dissipative (less-purposeful) energy transfer in the form of heat. The above definition could be generalized as: "energy is providing existence, and if exchanged, it has the ability to perform change."[3,4]

Any and every material system in nature possesses energy. The structure of any matter and field is "energetic," meaning active; i.e., photon waves are traveling in space, electrons are orbiting around an atom nucleus or flowing through a conductor, atoms and molecules are in constant rotation, vibration,

TABLE 1 Material System Structure and Related Forces and Energies

Particles	Forces	Energies
Photons	Electromagnetic	Electromagnetic
Atom nucleus	Strong and weak	Nuclear
Electron shell, or electron flow	Electromagnetic	Electrical, magnetic, electromagnetic
Atoms/Molecules	Interatomic/molecular	Chemical
Molecules	Inertial kinetic due to random collision and potential Intermolecular	Sensible thermal
Molecules	Potential Intermolecular	Latent thermal
Molecules	Potential Intermolecular	Mechanical elastic
System mass	Inertial kinetic and gravitational	Mechanical kinetic and gravitational potential

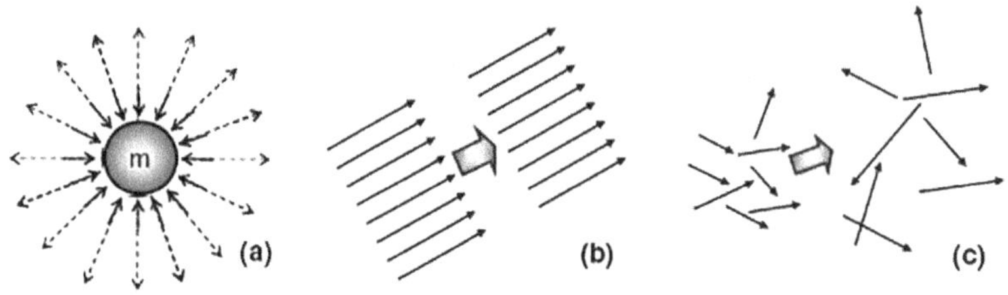

FIGURE 1 Different types of energy. (a) Potential gravitational and electromagnetic radiation. (b) Organized energy as work transfer. (c) Disorganized thermal energy as heat transfer.

or random thermal motion, etc., (see Table 1 and Figure 1). Thus, energy is a property of a material system (simply referred here as *system*) and, together with other properties, defines the system equilibrium state or existence in space and time.

Energy in transfer ($E_{transfer}$) is manifested as work (W) or heat (Q) when it is exchanged or transferred from one system to another, as explained next (see Figure 2).

Work

Work is a mode of energy transfer from one acting body (or system) to another resisting body (or system), with an acting force (or its component) in the direction of motion, along a path or displacement. A body that is acting (forcing) in motion-direction in time is doing work on another body that is resisting the motion (displacement rate) by an equal resistance force, including inertial force, in opposite direction of action. Energy transfer is a forced-displacement interaction between inertial material systems, in effect accelerating the reacting mass-energy sink system (i.e., its structure) at the expense of decelerating

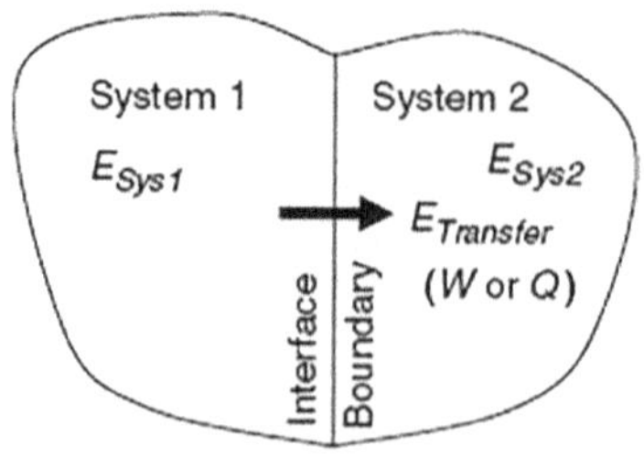

FIGURE 2 Energy as material system property and energy transfer from one system to another.

an acting-source system structure. The acting body (energy source) is imparting (transferring away) its energy to the resisting body (energy sink), and the amount of energy transfer is the work done by the acting onto the resisting body, equal to the product of the force component in the motion direction multiplied with the corresponding displacement, or vice versa (force multiplied by displacement component in the force direction, see Figure 3). If the force ($\vec{F}$) and displacement vectors ($d\vec{s} = d\vec{r}$) are not constant, then integration of differential work transfers from initial (1) to final state (2), defined by the corresponding position vectors $\vec{r}$, will be necessary (see Figure 4).

Work is a directional energy transfer. However, it is a scalar quantity like energy and is distinctive from another energy transfer in the form of *heat,* which is due to random motion (chaotic or random in all directions) and collisions of system molecules and their structural components.

Work transfer cannot occur without existence of a resisting body or system, or without finite displacement in the force direction. This may not always be obvious. For example, if we are holding a heavy weight or pushing hard against a stationary wall, there will be no work done against the weight or wall (neglecting their small deformations). However, we will be doing work internally due to the contraction and expansion of our muscles (thus force with displacement), thus converting (spending) a lot of chemical energy via muscle work and then dissipating it into thermal energy and heat transfer (sweating and getting tired).

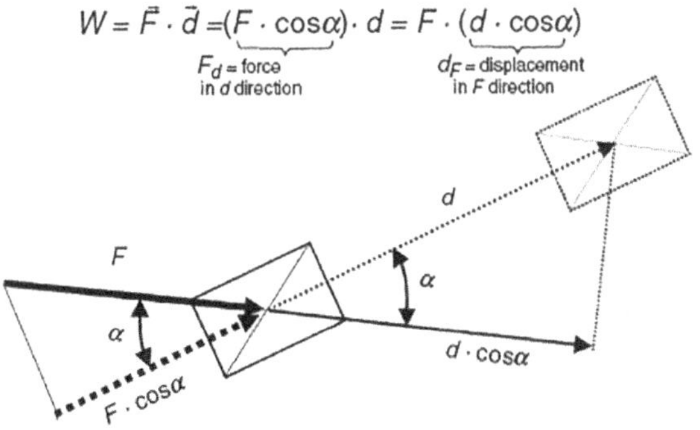

FIGURE 3 Work, force, and displacement.

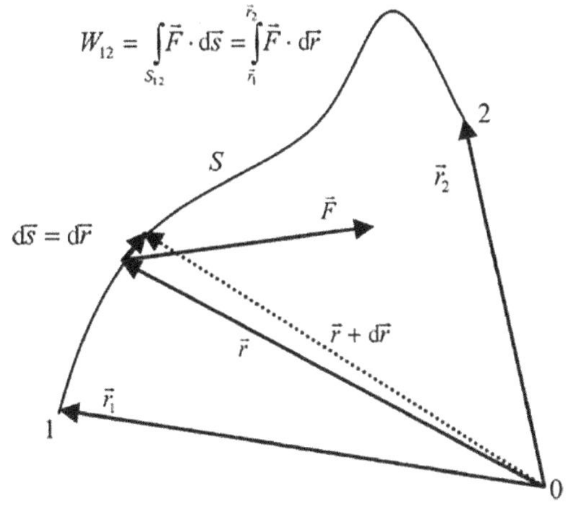

FIGURE 4 Work along arbitrary path.

Heat

Heat is another mode of energy transfer from one system at a higher temperature to another at a lower temperature due to their temperature difference. Fire was civilization's first great invention, long before people could read and write, and wood was the main heat source for cooking and heating for ages. However, true physical understanding of the nature of heat was discovered rather recently, in the middle of the 19th century, thanks to the development of the kinetic theory of gases. Thermal energy and its transfer as heat are defined as the energy associated with the random motion of atoms and molecules. The prior concept of heat was based on the *caloric* theory proposed by the French chemist Antoine Lavoisier in 1789. The caloric theory defines heat as a massless, colorless, odorless, and tasteless, fluid-like substance called the *caloric* that can be transferred or "poured" from one body into another. When caloric was added to a body, its temperature increased and vice versa. The caloric theory came under attack soon after its introduction. It maintained that heat is a substance that could not be created or destroyed. Yet it was known that heat can be generated indefinitely by rubbing hands together or from mechanical energy during friction, like mixing or similar. Finally, careful experiments by James P. Joule (1843) quantified correlation between mechanical work and heat, and thus put the caloric theory to rest by convincing the skeptics that heat was not the caloric substance after all. Although the caloric theory was totally abandoned in the middle of the 19th century, it contributed greatly to the development of thermodynamics and heat transfer. Actually, if conversion of all other energy types to heat (thermal energy "generation" from all types of "phlogistons") is included, i.e., conservation of energy in general, then caloric theory will be valid in general as stated by Clausius.[5]

Heat may be transferred by three distinctive mechanisms: conduction, convection, and thermal radiation (see Figure 5).[6] Heat conduction is the transfer of thermal energy due to interaction between the more energetic particles of a substance, like atoms and molecules (thus at higher temperature), and the adjacent, less energetic ones (thus at lower temperature). Heat convection is the transfer of thermal energy between a solid surface and the adjacent moving fluid, and it involves the combined effects of conduction and fluid motion. Thermal radiation is the transfer of thermal energy due to the emission of electromagnetic waves (or photons), which are products of random interactions between energetic electron shells of substance particles (thus related to the temperature). During those interactions, the electron shell energy level is changed, thus causing emission of photons, i.e., electromagnetic thermal radiation.

Joule's experiments of establishing equivalency between work and heat paved the way to establish the concept of internal thermal energy, to generalize the concept of energy, and to formulate the general law of energy conservation. The total internal energy includes all other possible but mechanical energy types or forms, including thermal, chemical, and nuclear energies. This allowed extension of

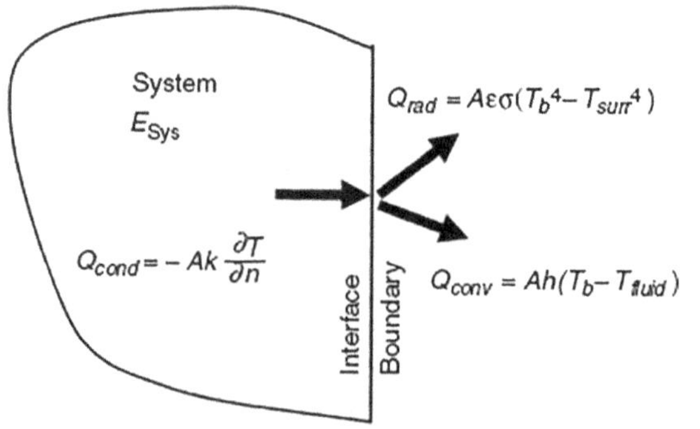

FIGURE 5 Heat as energy transfer by (a) conduction, (b) convection, and (c) radiation is due to difference in temperature.

the well-established law of mechanical energy conservation to the general law of energy conservation, known as the *first law of thermodynamics,* which includes all possible energy forms that a system could possess, and heat and all types of work as all possible energy transfers between the systems. The *law of energy conservation* will be elaborated later.

Energy

Energy is fundamental property of a physical system and refers to its potential to maintain a system identity or structure and to influence changes with other systems (via forced interaction) by imparting work (forced directional displacement) or heat (forced chaotic displacement/motion of a system's molecular or related structures). Energy exists in many forms: electromagnetic (including light), electrical, magnetic, nuclear, chemical, thermal, and mechanical (including kinetic, elastic, gravitational, and sound), where, for example, electromechanical energy may be kinetic or potential, while thermal energy represents overall potential and chaotic motion energy of molecules and/or related microstructure.[3,7,8]

Energy, Work, and Heat Units

Energy is manifested via work and heat transfer, with corresponding Force × Length dimension for work (N·m, kg_f·m, and lb_f·ft, in SI, metric, and English system of units, respectively), and the caloric units, in kilocalorie (kcal) or British thermal unit (Btu), the last two defined as heat needed to increase a unit mass of water (at reference state: specified pressure and temperature) for one degree of temperature in their respective units. Therefore, the water specific heat at reference state is 1 kcal/(kg °C) = 1 Btu/(lb °F) by definition, in metric and English system of units, respectively. It was demonstrated by Joule that 4187 Nm of work, when dissipated in heat, is equivalent to 1 kcal. In his honor, 1 N·m of work is named after him as 1 joule, or 1 J, the SI energy unit, also equal to electrical work of 1 W·s = 1 V·A·s. The SI unit for power, or work rate, is watt, i.e., 1 J/s = 1 W, and also corresponding units in other system of units, like Btu/hr, etc. Horsepower is defined as 1 HP = 550 lb_f·ft/sec = 745.7 *W*. Other common units for energy, work, and heat are given in Table 2.

Energy Forms and Classifications: Energy Transfer versus Energy Property

Any and all changes or processes (happening in space and time) are caused by energy exchanges or transfers from one substance *(system* or *subsystem)* to another (see Figure 2). A part of a system may be considered as a subsystem if energy transfer within a system is taking place, and inversely, a group of interacting systems may be considered as a larger isolated system, if they do not interact with the rest of the surroundings. Energy transfer may be in organized form (different types of work transfer due to different force actions) or in chaotic disorganized form (heat transfer due to temperature difference).

TABLE 2 Typical Energy Units with Conversion Factors

Energy Units	J	kWh	Btu
1 joule (J)	1	2.78×10^{-7}	9.49×10^{-4}
1 kilowatt hour (kWh)	3.6×10^{6}	1	3.412×10^{3}
1 kilocalorie (kcal = Cal = 1000 cal)	4187	1.19×10^{-3}	3.968
1 British thermal unit (Btu)	1055	2.93×10^{-4}	1
1 pound-force foot ($lb_f \cdot ft$)	1.36	3.78×10^{-7}	1.29×10^{-3}
1 electron volt (eV)	1.6×10^{-19}	4.45×10^{-26}	1.52×10^{-22}
1 horsepower × second (HP · s)	745.7	2.071×10^{-4}	0.707

Energy transfer into a system builds up energy potential or generalized force (called simply potential for short, like pressure, temperature, voltage, etc.) over energy displacement (like volume, entropy, charge, etc.). Conversely, if energy is transferred from a system, its energy potential is decreased. That is why net energy is transferred from a higher to a lower energy potential until the two equalize, due to virtually infinite probability of equipartition of energy over system microstructure, causing system equilibrium; otherwise, virtually impossible singularity of energy potential at infinite magnitude would result.[4]

There are many forms and classifications of energy (see Table 3), all of which could be classified as *microscopic* (or internal within a system microstructure) and/or *macroscopic* (or external as related to the system mass as a whole with reference to other systems or fields). Furthermore, energy may be "quasi-potential" (associated with a system equilibrium state and structure, i.e., system property) or "quasi-kinetic" (energy in transfer from one system or one structure to another, in the form of *work* or *heat*).

Every material system state is an equilibrium potential "held" by forces (space force fields); i.e., the forces "define" the potential and state—action and reaction; otherwise, a system will undergo dynamic change (in time) or quasi-kinetic energy exchange towards another stable equilibrium. Atoms (mass) are held by atomic and electromagnetic forces in small scale and by gravity in large scale (see Figure 1a); otherwise, mass would disintegrate ("evaporate" or radiate into energy) like partly in nuclear reactions—*nuclear* energy and/or electromagnetic radiation. Molecules are held by electrochemical bonding (e.g., valence) forces (chemical reactions—*chemical* energy). Liquids are held by latent intermolecular forces (gas condensation, when kinetic energy is reduced by cooling—*latent thermal* energy). Solids are held by "firm" intermolecular forces (freezing/fusing/solidification when energy is further reduced by cooling—*latent thermal* energy again). *Sensible thermal* energy represents intermolecular potential energy and energy of random molecular motion and is related to the temperature of a system. "Holding" potential forces may be "broken" by energy transfer (e.g., radiation, heating, high-energy particle interactions, etc.). States and potentials are often "hooked" (i.e., stable) and thus need to be "unhooked" (or to "be broken") to overcome existing "threshold" or equilibrium, like in igniting combustion, starting nuclear reaction, etc.

As stated above, energy transfer can be directional (purposeful or organized) and chaotic (dissipative or disorganized). For example, mass in motion (*mechanical kinetic* energy) and electron in motion (*electrical kinetic* energy) are organized kinetic energies (Figure 1b), while *thermal* energy is disorganized chaotic energy of motion of molecules and atoms (Figure 1c). System energy may be defined with reference to position in a force field, like *elastic potential* (stress) energy, *gravitational potential* energy, or *electromagnetic* field energy. There are many different energy forms and types (see Table 3). We are usually not interested in (absolute) energy level, but in the change of energy (during a process) from an initial state (*i*) to a final state (*f*), and thus zero reference values for different energy forms are irrelevant, and often taken arbitrarily for convenience. The following are basic correlations for energy changes of several typical energy forms, often encountered in practice: motion kinetic energy ($E_K = KE$) as a function of system velocity (V); potential elastic deformational, e.g., pressure or spring elastic energy ($E_{\text{Pdeff}} = E_{\text{Pp}} = E_{\text{Ps}}$), as a function of spring deformation displacement (x); gravitational potential energy ($E_{\text{Pg}} = PE_g$) as a function of gravitational elevation (z); and sensible thermal energy ($E_U = U$) as a function of system temperature (T):

$$\Delta E_k = \frac{1}{2} m\left(V_f^2 - V_i^2\right) \qquad \Delta E_{\text{Ps}} = \frac{1}{2} k\left(x_f^2 - x_i^2\right) \tag{1}$$

$$\Delta E_{\text{Pg}} = mg\left(z_f - z_i\right) \qquad \Delta E_U = mc_v\left(T_f - T_i\right)$$

If the reference energy values are taken to be zero when the above initial (i) variables are zero, then the above equations will represent the energy values for the final values (f) of the corresponding variables. If the corresponding parameters, spring constant k, gravity g, or constant-volume specific heat c_v, are not constant, then integration of differential energy changes from initial to final state will be necessary.

TABLE 3 Energy Forms and Classifications

Energy									
Scale							Type		Transfer (release)
							Kinetic (Motion)		Kinetic (Motion)
MACRO/External (mass Based)	MICRO/Internal (structure Based)	Energy Form (energy Storage)		Energy Process	Potential (state of field)	Directional[a]	Chaotic dissipative	Work[b] Directional[a]	Heat Dissipative
		Mechanical							
x		—Kinetic	$mV^2/2$	Acceleration			X		X
x		—Gravitational[c]	mgz	Elevation		X			X
	x	—Elastic	$kx^2/2$ or $PV = mP/\rho$	Deformation		X			X
		Thermal	U_{th}						
	x	—Sensible	$U_{th} = mc_{avg}T$	Heating				X	X
	x	—Latent	$U_{th} = H_{latent}$	Melting Evaporation		X			X
	x	Chemical	U_{ch}	Chemical reaction		X			X
	x	Nuclear	U_{nucl}	Nuclear reaction		X			X
		Electrical	E_{el}						
	x	—Electro kinetic	$V(It)$ or $LI^2/2$	Electro-current flow			X		X
	x	—Electrostatic	$(It)^2/(2C)$	Electro-charging		X			X
		Magnetic	E_{magn}	Magnetization		X			X
	x[d]	Electromagnetic	E_{el_mag}	Radiation			X[d]	X[e]	X

[a] Electromechanical kinetic energy type (directional/organized, the highest energy quality) is preferable since it may be converted to any other energy form/type with high efficiency.
[b] All processes (involve energy transfer) are to some degree irreversible (i.e., dissipative or chaotic/disorganized).
[c] Due to mass position in a gravitational field.
[d] Electromagnetic form of energy is the smallest known scale of energy.
[e] Thermal radiation.

Energy transfer via work *W* (net-out), and heat transfer *Q* (net-in), may be expressed for reversible processes as product of related energy potentials (pressure *P*, or temperature *T*) and corresponding energy displacements (change of volume *V* and entropy *S*, respectively), i.e.:

$$W_{12} = \vec{F} \cdot \vec{d} = \left[\underbrace{(P \cdot A\vec{n}).\vec{d}}_{\Delta V} \right] = P \cdot V_{12} = \left| \underbrace{\int_{V1}^{V2} P \cdot dV}_{P \neq \text{const}} \right. \tag{2}$$

$$Q_{12} = T \cdot \Delta S_{12|T \neq const} = \int_{S1}^{S2} T \cdot dS \tag{3}$$

Note, in Eq. 2, that force cannot act at a point but is distributed as pressure (*P*) over some area *A* (with orthogonal unit vector $\vec{n}$), which when displaced will cause the volume change ΔV. Also, note that it is custom in some references to denote heat transfer "in" and work transfer "out" as positive (as they appear in the above equations and a heat engine). In general, "in" (means "net-in") is negative "out" (means "net-out") and vice versa.

In general, energy transfer between systems is taking place at the system boundary interface and is equal to the product of energy potential or generalized force and the corresponding generalized displacement:[7–9]

$$\begin{aligned}\delta E_{\text{Transfer}} &= \delta Q_{\text{netIN}} - \left[\sum \delta W_{\text{netOUT}}\right] \\ &= \delta Q_{\text{netIN}} + \left[\sum \delta W_{\text{netIN}}\right] \\ &= T ds + \left[\underbrace{-PdV}_{\text{COMPR.}} + \underbrace{\sigma dA}_{\text{STRETCHING.}} + \underbrace{\tau d(A\cdot s)}_{\text{SHEARING.}}\right] \\ &\quad + \underbrace{Vdq}_{\text{CHARGING}} + \underbrace{\vec{E}\cdot d(V\vec{P})}_{\text{POLARIZATION}} \\ &\quad + \underbrace{\mu_\circ \vec{H}\cdot d(V\vec{P})}_{\text{MAGNETIZATION}} + \underbrace{\cdots}_{\text{ETC.}}\end{aligned} \tag{4}$$

where the quantities after the last equal sign are temperature and entropy; pressure and volume; surface tension and area; tangential stress and area with tangential displacement, voltage, and electrical charge; electric field strength and volume-electric dipole moment per unit volume product; and permeability of free space, magnetic field strength, and volume-magnetic dipole moment per unit volume product, respectively.

The total system energy stored within the system, as energy property, is:

$$E_{\text{Sys}} = \underbrace{E_K + E_{Pg} + E_{\text{Pdeff.}}}_{E_{\text{Mechanical}}} + \underbrace{\underbrace{E_{\text{Uth}}}_{\text{Thermal}} + E_{\text{Ch}} + E_{\text{Nucl}} + E_{\text{El}} + E_{\text{Magn}} + \underbrace{\cdots}_{\text{Etc.}}}_{\text{Internal(total)}} \tag{5}$$

where the quantities after the equal sign are kinetic, potential gravitational, potential elastic deformational, thermal, chemical, nuclear, electric, magnetic energies, etc.

First Law of Energy Conservation: Work-Heat-Energy Principle

Newton formulated the general theory of motion of objects due to applied forces (1687). This provided for concepts of mechanical work, kinetic and potential energies, and development of solid-body mechanics. Furthermore, in absence of non-mechanical energy interactions, excluding friction and other dissipation effects, it is straightforward to derive (and thus prove) energy conservation, i.e.:

$$\underbrace{\int_{s1}^{s2} F_s ds}_{W_{Fs}} = \int_{s1}^{s2} \left(\underbrace{\frac{dV_s}{dt}}_{a_s}\right) ds = \int_{s1}^{s2} \left(\underbrace{m\frac{dV_s}{ds}\left\{\frac{ds}{dt}\right\}}_{V_s}\right) ds$$

$$= \int_{V_1}^{V_2} m V_s \mathrm{d}V_s = \underbrace{\frac{1}{2} m \left(V_2^2 - V_1^2\right)}_{KE_2 - KE_1}$$

$$\underbrace{\left(E_{\text{Transfer}} = W_{Fs}\right)}_{\text{work-energy transfer}} = \underbrace{\left(KE_2 - KE_1 = E_2 - E_1\right)}_{\text{energy transfer}} \tag{6}$$

The above correlation is known as the *work-energy principle.* The work-energy principle could be easily expended to include work of gravity force and gravitational potential energy as well as elastic spring force and potential elastic spring energy.

During a free gravity fall (or a free bounce) without air friction, for example, the potential energy is being converted to kinetic energy of the falling body (or vice versa for free bounce) and at any time the total mechanical energy (sum of kinetic and potential mechanical energies) is conserved, i.e., stays the same (see Figure 6). The mechanical energy is also conserved if a mass freely vibrates on an ideally elastic spring, or if a pendulum oscillates around its pivot, both in the absence of dissipative effects, like friction or non-elastic deformation. In general, for work of conservative forces only, the mechanical energy, E_{mech}, for N isolated systems is conserved since there is no dissipative conversion into thermal energy and thus no heat transfer, i.e.:

$$E_{\text{mech}} = E_K + E_{pg} + E_{Ps} = \sum_{i=1}^{N} \left(\frac{1}{2} m v^2 + m g z + \frac{1}{2} k x^2 \right) = \text{const} \tag{7}$$

The mechanical work-energy concept could also be expended to fluid motion by inclusion of elastic-pressure force and potential elastic-pressure energy (referred to in some references as flow work; however, note that elastic- pressure energy is a system property while flow work is related energy transfer; see the *Bernoulli* equation below). For flowing or stationary fluid without frictional effects, the mechanical energy is conserved, including fluid elastic- pressure energy, $PV = mP/\rho$ (where V is volume, whereas v is used for velocity here), as expressed by the Bernoulli or hydrostatic equations below (see also Figure 7).[3]

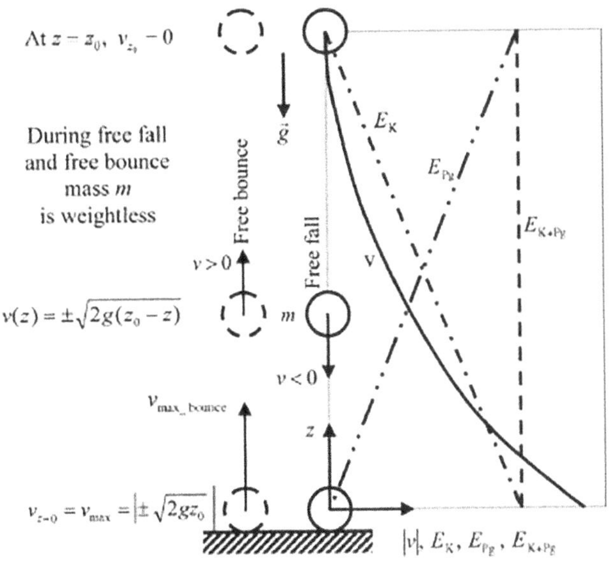

FIGURE 6 Energy and work due to conservative gravity force.

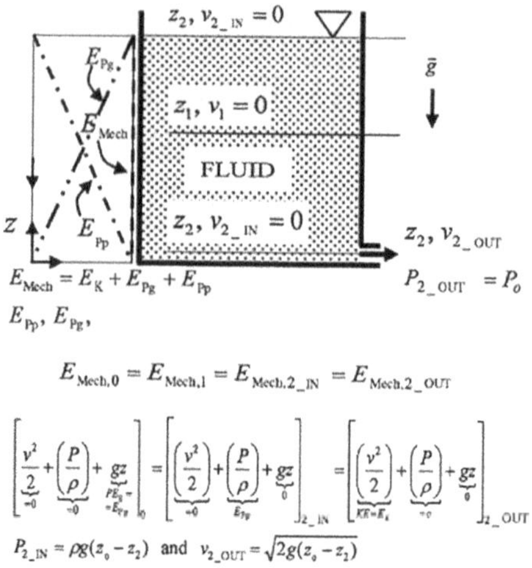

FIGURE 7 Conservation of fluid mechanical energy: Bernoulli equation, hydrostatic equation, and Torricelli's orifice velocity.

$$\frac{E_{\text{Mech}}}{m} = \frac{1}{m}\left(E_K + E_{Ps} + E_{Pg}\right) = \underbrace{\frac{v^2}{2} + \frac{P}{\rho} + gz}_{\substack{\text{Bernoulli} \\ \text{equation}}}\Bigg|_{v=0} = \underbrace{\frac{P}{\rho} + gz}_{\substack{\text{Hydrostatic} \\ \text{equation } (v=0)}} = \text{const.} \tag{8}$$

Work against inertial and/or conservative forces (also known as internal, or volumetric, or space potential field) is path independent, and during such a process, the mechanical energy is conserved. However, work of nonconservative, dissipative forces is process path dependent and part of the mechanical energy is converted (dissipated) to thermal energy (see Figure 8a).

When work of non-conservative (dissipative) forces, W_{nc}, is exchanged between N isolated systems, from an initial (i) to final state (f), then the total mechanical energy of all systems is reduced by that work amount, i.e.:

$$W_{nc,\,i\to f} = \left(\sum_{i=1}^{N} E_{\text{mech},\,i}\right)_i - \left(\sum_{i=1}^{N} E_{\text{mech},\,i}\right)_f \tag{9}$$

Regardless of the traveled path (or displacement), the work against conservative forces (like gravity or elastic spring in above cases) in absence of any dissipative forces will depend on the final and initial position (or state) only. However, the work of non-conservative, dissipative forces (W_{nc}) depends on the traveled path since the energy is dissipated during the force displacement, and mechanical energy will not be conserved but in part converted (via dissipation and heat transfer) into thermal energy. This should not be misunderstood with the total energy conservation, which is always the case, and it must include both work and heat transfer.

As already stated, there are many different types of work transfer into (or out of) a system that will change the corresponding energy form stored in the system (system properties). In addition to work, energy may be transferred as heat caused by temperature difference and associated with change of the thermal energy of a system. Furthermore, different forms of stored energy are often coupled so that one type of energy transfer may change more than one form of stored energy, particularly due to inevitable dissipative conversion of work to heat, i.e., to internal thermal energy. Conversely, heat and thermal

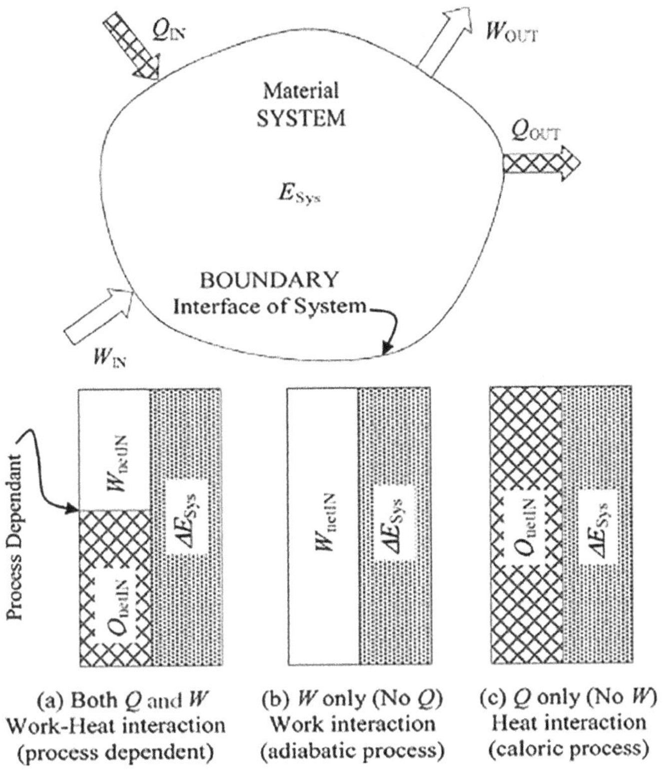

FIGURE 8 System energy and energy boundary interactions (transfers) for (a) arbitrary, (b) adiabatic, and (c) caloric processes.

energy may be converted into other energy forms. In the absence of a powerful nuclear reaction (conversion of mass into energy, $E = mc^2$, is virtually negligible), we then may assume that the mass and energy are conserved separately for an isolated system, a group of isolated systems, or for the Universe. Since the material system structure is of particulate form, then systems' interactions (collisions at different scale sizes) will exchange energy during the forced displacement— and similarly to the mechanical energy conservation—the totality of all forms of energy will be conserved[7,8] (also see Figure 8), which could be expressed as:

$$\underbrace{\sum_{\text{All } i\text{'s}} E_{i,\,\text{Trans.}}}_{\text{BOUNDARY}} = \underbrace{\Delta E_{\text{Change}}}_{\text{SYSTEM}} \text{ or}$$

$$\underbrace{\sum_{\text{All } j\text{'s}} W_{j,\,\text{netIN}} = \sum_{\text{All } k\text{'s}} Q_{k,\,\text{netIN}}}_{\text{BOUNDARY}} \tag{10}$$

$$= \underbrace{\Delta E_{\text{netIncrease}}}_{\text{SYSTEM}} = \Delta E_{\text{Sys}}$$

Energy interactions or transfers across a system boundary, in the form of work, $W_{\text{netIN}} = \sum W_{\text{IN}} - \sum W_{\text{OUT}} = -\left(\sum W_{\text{OUT}} - \sum W_{\text{IN}}\right) = -\sum W_{\text{netOUT}}$, and heat, $Q_{\text{netIN}} = \sum Q_{\text{IN}} - \sum Q_{\text{OUT}}$, will change the system total energy, $\Delta E_{\text{Sys}} = E_{\text{Sys},2} - E_{\text{Sys},1}$. Different energy-type transfers are process (or process path) dependent, for the same ΔE_{Sys} change, except for special two cases: for adiabatic processes with work interaction only (no heat transfer, but fully irreversible, all work to heat generation) or for caloric processes with heat interaction only (no work transfer) (see Figure 8a–c). For the latter caloric processes without work interactions (no volumetric expansion or other mechanical energy changes), the internal thermal energy is conserved by being transferred from one system to another via heat transfer only, known as *caloric*. This demonstrates the value of the caloric theory of heat that was established by Lavoisier and Laplace (1789), the great minds of the 18th century. Ironically, the caloric theory was creatively used by Sadi Carnot (1824) to develop the concept of reversible cycles for conversion of caloric heat to mechanical work as it "flows" from high- to low-temperature reservoirs that later helped in dismantling the caloric theory. The caloric theory was discredited by establishing the "heat equivalent of work, e.g., mechanical equivalent of heat" by Mayer (1842) and experimentally confirmed by Joule (1843), which paved the way for establishing the *first law of energy conservation* and new science of *thermodynamics* (Clausius, Rankine, and Kelvin, 1850 and later). Prejudging the caloric theory now as a "failure" is unfair and unjustified since it made great contributions in calorimetry and heat transfer, and it is "valid" for caloric processes (without work interactions).[5] The coupling of work-heat interactions and conversion between thermal and mechanical energy are outside of the caloric theory domain and are further developed within the first and second laws of thermodynamics.

The first law of energy conservation for the control volume (CV, with boundary surface BS) flow process (see Figure 9) is:

$$\underbrace{\frac{d}{dt}E_{\text{CV}}}_{\substack{\text{RATE OF ENERGY}\\\text{CHANGE IN CV}}} = \underbrace{\sum_{\text{BS}} \dot{Q}_{\text{netIN},i}}_{\substack{\text{BS TRANSFER}\\\text{RATE OF HEAT}}} - \underbrace{\sum_{\text{BS}} \dot{W}_{\text{netOUT},i}}_{\substack{\text{BS TRANSFER}\\\text{RATE OF WORK}}} + \underbrace{\sum_{\text{IN}} \dot{m}_j (e+Pv)_j}_{\substack{\text{ENERGY TRANSPORT}\\\text{RATE WITH MASS IN}}} - \underbrace{\sum_{\text{OUT}} \dot{m}_K (e+Pv)_K}_{\substack{\text{ENERGY TRANSPORT}\\\text{RATE WITH MASS OUT}}} \tag{11}$$

The first law of energy conservation equation for a material system in a differential volume per unit of volume around a point (x, y, z) in a flowing fluid is:[8]

$$\underbrace{\rho\frac{De}{Dt}}_{\text{energy change in time}} = \underbrace{-\vec{V}.(\nabla P)}_{\substack{\text{work rate}\\\text{of normal pressure stresses}}} + \underbrace{\nabla.\left(\vec{V}.\tau_{ij}\right)}_{\substack{\text{work rate}\\\text{of shearing stresses}}} + \underbrace{\nabla\cdot(k\nabla T)}_{\substack{\text{heat rate}\\\text{via thermal conduction}}} \tag{12}$$

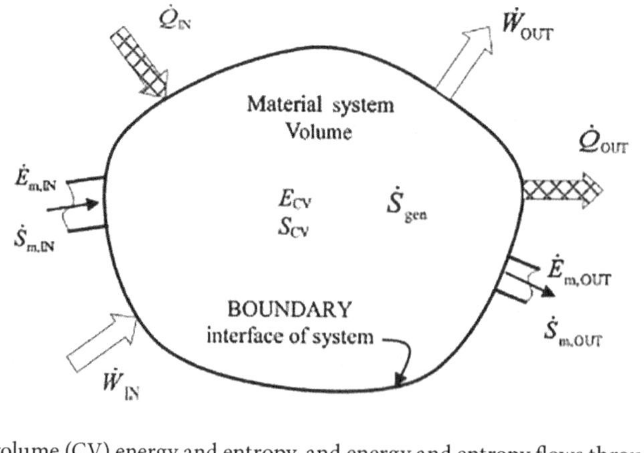

FIGURE 9 Control volume (CV) energy and entropy, and energy and entropy flows through the boundary interface of the CV.

Where $e = \hat{u} + \frac{\vec{V}^2}{2} + gz$.

Eq. 12 reduces after substitution, $\nabla \cdot \left(\vec{V} \cdot \tau_{ij}\right) = \vec{V}\left(\nabla \cdot \tau_{ij}\right) + \Phi$, and using the momentum equation, to:

$$\rho \frac{\mathrm{D}\hat{u}}{\mathrm{D}t} = -p\left(\nabla.\vec{V}\right) + \Phi_k + \Phi + \nabla \cdot \left(k\nabla T\right) \tag{13}$$

where $\Phi_k = k\left(\nabla \cdot \vec{V}\right)^2$ is the bulk-viscosity dissipation and Φ is the shear-viscosity dissipation function, which is the rate of mechanical work conversion to internal thermal energy for a differential volume per unit of volume, with [W/m³] unit, given for Newtonian fluid as:

$$\begin{aligned}\Phi = \frac{\mathrm{d}\dot{W}_\Phi}{\mathrm{d}V} = &\left[\left(\frac{\partial u}{\partial x}\right)^2\left(\frac{\partial v}{\partial y}\right)^2\left(\frac{\partial w}{\partial z}\right)^2\right] \\ &+\mu\left[\left(\frac{\partial v}{\partial x}+\frac{\partial u}{\partial y}\right)^2+\left(\frac{\partial w}{\partial y}+\frac{\partial v}{\partial z}\right)^2\right. \\ &\left.+\left(\frac{\partial u}{\partial z}+\frac{\partial w}{\partial x}\right)^2\right]-\frac{2}{3}\mu\left(\nabla\cdot\vec{V}\right)^2\end{aligned} \tag{14}$$

The power or work rate of viscous dissipation (irreversible conversion of mechanical to thermal energy) in a control volume V is:

$$\dot{W}_\Phi = \int_V \Phi dV \tag{15}$$

Second Law of Energy Degradation: Entropy and Exergy

Every organized or directional kinetic energy will, in part or in whole (and ultimately in whole), disorganize/dissipate in all directions within the microstructure of a system (in time over its mass and space) into disorganized thermal energy. Entropy represents the measure of energy disorganization or random energy redistribution to smaller-scale structure and/or space, per absolute temperature level. Contrary to energy and mass, which are conserved in the universe, entropy is continuously generated (increased) due to continuous redistribution and disorganization of energy in transfer and thus degradation of the quality of energy ("spreading" of energy towards and over lower potentials in time until equipartitioned over system structure and space). Often, we want to extract energy from one system in order to purposefully change another system, thus to transfer energy in organized form (as work, the ultimate energy quality). No wonder that energy is defined as *ability to perform work*, and a special quantity, *exergy*, is defined as the maximum possible work that may be obtained from a system by bringing it to the equilibrium in a process with reference surroundings (called reference *dead state*). The maximum possible work will be obtained if we prevent energy disorganization, thus with limiting *reversible* processes where the existing non-equilibrium is conserved (but rearranged and could be reversed) within interacting systems. Since the energy is conserved during any process, only in ideal reversible processes will entropy (measure of energy disorganization or degradation) and exergy (maximum possible work with regard to the reference surroundings) be conserved, while in real irreversible processes, entropy will be generated and exergy will be partly (or even fully) destroyed or permanently lost. Therefore, conversion of other energy types in thermal energy or heat generation and transfer are universal manifestations of all natural and artificial (man-made) processes, where all organized potential and/or quasi-kinetic energies are disorganized or dissipated in the form of thermal energy, in irreversible and spontaneous processes.

Reversibility and Irreversibility: Energy Transfer and Disorganization, and Entropy Generation

Energy transfer (when energy moves from one system or subsystem to another) through a system boundary and in time is of kinetic nature and may be directionally organized as work or directionally chaotic and disorganized as heat. However, the net-energy transfer is in one direction only, from a higher to a lower energy potential, and the process cannot be reversed. Thus, *all processes are irreversible* in the direction of decreasing energy potential (like pressure and temperature) and increasing energy displacement (like volume and entropy) as a consequence of energy and mass conservation in the universe. This implies that the universe (as isolated and unrestricted system) is expending in space with entropy generation (or increase) as a measure of continuous energy degradation, i.e., energy redistribution and disorganization per unit of absolute temperature. It is possible in the limit to have an energy transfer process with infinitesimal potential difference (still from higher to infinitesimally lower potential, P). Then, if infinitesimal change of potential difference direction is reversed ($P + \mathrm{d}P \rightarrow P - \mathrm{d}P$, with infinitesimally small external energy, since $\mathrm{d}P \rightarrow 0$), the process will be reversed too, which is characterized by infinitesimal entropy generation, thus in the limit, without energy degradation (no further energy disorganization) and no entropy generation—thus achieving a limiting, ideal *reversible process*. Such processes at infinitesimal potential differences, called quasi-equilibrium processes, maintain the system space equilibrium at any instant but with incremental changes in time. The quasi-equilibrium processes are reversible and vice versa. In effect, the quasi-equilibrium reversible processes are infinitely slow processes at infinitesimally small potential differences, but they could be reversed to any previous state and forwarded to any future equilibrium state, without any "permanent" change to the rest of the surroundings. Therefore, the changes are "fully" reversible and, along with their rate of change and time, completely irrelevant, as if nothing is effectively changing—time is irrelevant as if it does not exist since it could be reversed (no permanent change and relativity of time). Since the real time cannot be reversed, it is a measure of permanent changes, i.e., irreversibility, which is, in turn, measured by entropy generation. In this regard, time and irreversible entropy generation are related.[2]

Entropy and the Second Law of Thermodynamics

Entropy is also a system property, which, together with energy, defines its equilibrium state and actually represents the system energy displacement or random energy disorganization (dissipation) per absolute temperature level over its mass and the space it occupies. Therefore, entropy as property of a system, for a given system state, is the same regardless whether it is reached by reversible heat transfer (Eq. 16) or irreversible heat or irreversible work transfer (caloric or adiabatic processes, Eq. 17). For example, an ideal gas system entropy increase will be the same during a reversible isothermal heat transfer and reversible expansion to a lower pressure (heat-in equal to expansion work-out), as during an irreversible adiabatic unrestricted expansion (no heat transfer and no expansion work) to the same pressure and volume, as illustrated in Fig. 10a and b, respectively.

If heat or work at higher potential (temperature or pressure) than necessary is transferred to a system, the energy at excess potential will dissipate spontaneously to a lower potential (if left alone) before a new equilibrium state is reached, with entropy generation, i.e., increase of entropy (energy degradation per absolute temperature level). A system will "accept" energy if it is transferred at minimum necessary (infinitesimally higher) or higher potential with regard to the system energy potential. Furthermore, the higher potential energy will dissipate and entropy increase will be the same as with minimum necessary potential, like in reversible heating process, i.e.

$$dS = \frac{\delta Q}{T} \text{ or } S = \int \frac{\delta Q}{T} + S_{\text{ref}} \tag{16}$$

However, source entropy will decrease to a smaller extent over higher potential, thus resulting in overall entropy generation for the two (or all) interacting systems, which may be considered as a combined isolated system (no energy exchange with the rest of the surroundings). The same is true for energy exchange between different system parts (could be considered as subsystems) at different energy potentials (non-uniform, not at equilibrium at a given time). Energy at higher potential (say, close to boundary within a system) will dissipate ("mix") to parts at lower energy potential with larger entropy increase than decrease at higher potential, resulting in internal irreversibility and entropy generation, i.e., energy "expansion" from higher potential over more mass and/or space with lower potential. Entropy is not only displacement for heat as often stated but also displacement for any energy dissipation (energy disorganization) and the measure of irreversibility. Examples are spontaneous heat transfer or throttling (unrestricted expansion) where "work potential" is lost ($W_{\text{pot, LOSS}}$), i.e., spontaneously converted (dissipated) in thermal energy, known as heat generation ($Q_{\text{gen}} = W_{\text{pot, LOSS}}$), resulting in entropy generation:

$$S_{\text{ref}} = \frac{Q_{\text{gen}}}{T} \geq \text{ where } Q_{\text{gen}} = W_{\text{pot, LOSS}} \tag{17}$$

Eq. 17 may refer to the entropy generation rate at any instant within infinitesimal system or for the entropy generated during a larger process duration for an integral system. Therefore, entropy generation is the fundamental measure of irreversibility or "permanent change" that cannot be reversed, and therefore cannot be negative. This is true in general for closed and open processes and without any exception, thus representing the most general statement of the second law of energy degradation.[10]

However, there are two classical statements of the second law (both negative, about impossibility; see Figure 11).[7–9] One is the *Kelvin-Planck statement,* which expresses the never-violated fact that it is impossible to obtain work in a continuous cyclic process *(perpetual mobile)* from a single thermal reservoir (100% efficiency impossible, since it is not possible to spontaneously create non-equilibrium within the single reservoir in equilibrium), and the second is the *Clausius statement,* which refers to the direction of heat transfer, expressing the never-violated fact that it is impossible for heat transfer to take place spontaneously by itself (without any work input) from a lower- to a higher-temperature thermal reservoir: it is impossible to spontaneously create non-equilibrium. Actually, the two statements imply each other and thus are the same, as well as imply that all reversible cycles between the two temperature reservoirs (or all reversible processes between the two states) are the most (and equally) efficient with regard to extracting the maximum work out of a system or requiring the minimum possible work into the system (thus conserving the existing non-equilibrium).

Even though directionally organized energy transfer as work does not possess or generate any entropy (no energy disorganization, Figure 1b), it is possible to obtain work from the equal amount of disorganized thermal energy or heat if such process is reversible. There are two typical reversible processes where disorganized heat or thermal energy could be entirely converted into organized work, and vice versa. Namely, they are the following:[2]

1. Reversible expansion at constant thermal energy, e.g., isothermal-heating ideal-gas expansion ($\delta W = \delta Q$) (Figure 10a)
2. Reversible adiabatic expansion ($\delta W = -dU$)

During a reversible isothermal heat transfer and expansion of an ideal-gas system (S) for example (Figure 10a), the heat transferred from a thermal reservoir (R) will reduce its entropy for ΔS_R

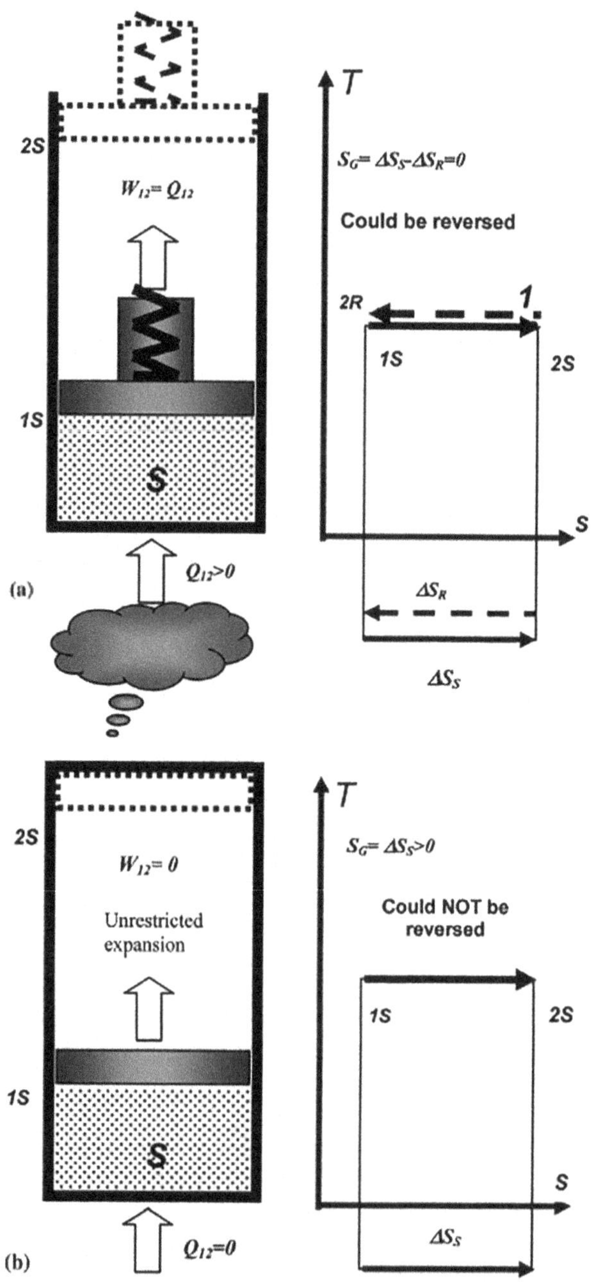

FIGURE 10 (a) Isothermal reversible heat transfer and restricted reversible expansion. (b) Adiabatic unrestricted irreversible expansion of the same initial system to the same final state (thus the same system entropy change).

magnitude, while ideal gas expansion in space (larger volume and lower pressure) will further disorganize its internal thermal energy and increase the gas entropy for ΔS_S, and in the process, an organized expansion work, equivalent to the heat transferred, will be obtained ($W_{12} = Q_{12}$). The process could be reversed, and thus, it is a reversible process with zero total entropy generation ($S_G = \Delta S_S - \Delta S_r = 0$). On the other hand, if the same initial system (ideal gas) is adiabatically expanded without any restriction (Figure 10b, thus zero expansion work) to the same final state, but without heat transfer, the system internal energy will remain the same but more disorganized over the larger volume, resulting in the same entropy increase as during the reversible isothermal heating and restricted expansion. However, this process cannot be autonomously reversed, since no work was obtained to compress back the system, and indeed the system entropy increase represents the total entropy generation ($S_G = \Delta S_S > 0$). Similarly, during reversible adiabatic expansion, the system internal thermal energy will be reduced and transferred in organized expansion work with no change of system entropy (isentropic process), since the reduction of disorganized internal energy and potential reduction of entropy will be compensated with equal increase of disorder and entropy in expanded volume. The process could be reversed back and forth (like elastic compression-expansion oscillations of a system) without energy degradation and entropy change, thus a reversible isentropic process. In reversible processes, energy is exchanged at minimum needed, not higher-than-needed potential, and isolated interacting systems as a whole do not undergo any energy-potential- related degradation/disorganization and with total conservation of entropy. The total non-equilibrium is conserved by reversible energy transfer within interacting systems, i.e., during reversible processes.

Heat Engines

Heat engines are devices undergoing thermomechanical cycles (converting thermal into mechanical energy), similar to the one in Figure 12, with mechanical expansion and compression network ($W = Q_h - Q_c$), obtained as the difference between the heat transferred to the engine from the

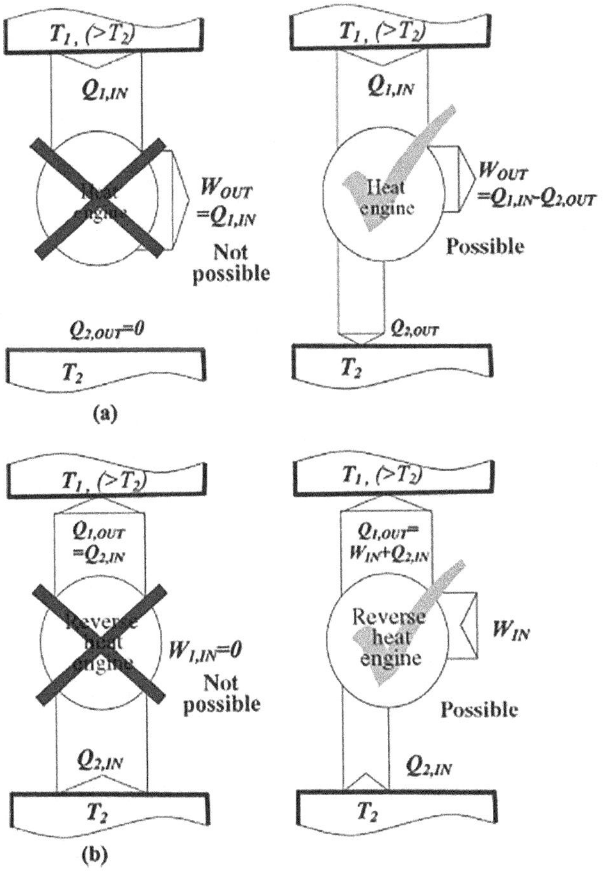

FIGURE 11 The second law: (a) Kelvin-Planck and (b) Clausius statements.

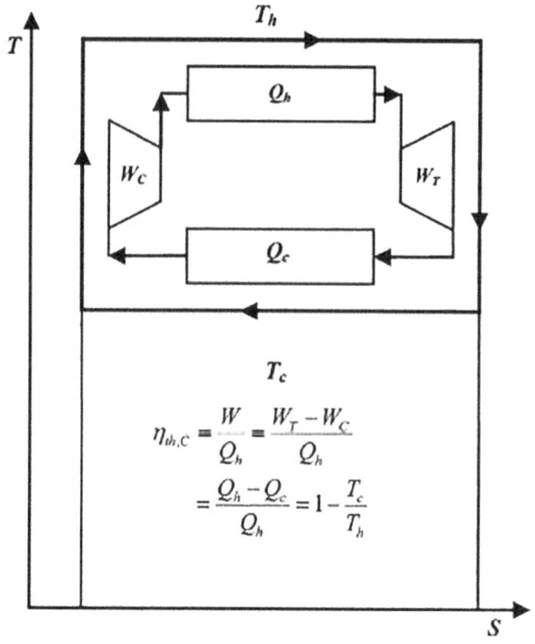

FIGURE 12 Heat engine ideal Carnot cycle.

high-temperature heat reservoir (at T_h) and rejected to a low-temperature (cold) heat reservoir (at T_c), thus converting part of the thermal energy into mechanical work. In a close-system cycle, the net-work-out is due to net-work of thermal expansion and thermal compression. Therefore, heat engine cycle cannot be accomplished without two thermal reservoirs at different temperatures, one at higher temperature to accomplish thermal expansion and work out, and another at lower temperature to accomplish thermal compression to initial volume and thus complete the cycle.[10]

The combustion process itself is an irreversible one, where chemical energy (electrochemical energy binding atoms in reactants' molecules) is chaotically released during combustion, i.e., converted in random thermal energy of products' molecules and cannot be fully converted into directional work energy. The *second law of thermodynamics* limits the maximum amount of work that could be obtained from thermal energy between any two thermal reservoirs at different temperatures, hot T_h and cold T_c, by using the ideal, reversible *Carnot cycle* (see Figure 12), with thermal efficiency given by Eq. 18. As an example, consider $T_c = 293$ K and $T_h = 2273$ K, then:

$$\eta_{\text{th, C\_ad}} = \frac{W}{Q_h} = \frac{Q_h - Q_c}{Q_h} = 1 - \frac{T_c}{T_h}\bigg|_{T_h = T_{\text{ad}} = 2273K,\ T_c = 293K} = 1 - \frac{293}{2273} = 87.1\% \tag{18}$$

where $W = W_T - W_C$ is the net-work of expansion, usually turbine (W_T) and compression (W_C) difference. The maximum efficiency is achieved if heat is supplied at the highest possible temperature, T_h, and released at the lowest possible temperature, T_c. However, both temperatures are limited by the fact that fuel combustion is performed using oxygen with ambient air, resulting in the maximum so-called adiabatic, stoichiometric combustion temperature, T_{ad}, which is for most fuels about 2000°C or 2273 K. A part of the heat supplied at high temperature, T_h, must be released to the surroundings at a low temperature of about $T_c = 20°C = 293$ K, which results in a Carnot efficiency of 87.1% (see Eq. 18 and Figure 12). However, the fuel heating value energy, $Q_{\text{HV}} = Q_{\text{ad\_var}}$, is not all available at the adiabatic temperature of the products but is distributed over their variable temperature range from an initial surrounding temperature before combustion, T_c, to a final adiabatic temperature, T_{ad}. For such a variable heat reservoir, a large number (infinite in the limit) of ideal Carnot engines operating at different temperatures (with $dW = dQ$) must be employed to achieve a reversible cycle, resulting in the variable- temperature Carnot cycle with the maximum possible combustion-products-to-work conversion efficiency:[1]

$$\eta_{\text{th, Cvar max}} = \left.\frac{\ln(T_{ad}/T_c)}{(T_{ad}/T_c)-1}\right|_{T_{ad}=2273K,\ T_c=293K} = 69.7\% \tag{19}$$

Eq. 19 is valid if the cyclic medium has constant specific heat; otherwise, integration will be required. Due to engine material property limitations and other unavoidable irreversibilities, it is impossible to reach the ideal Carnot efficiency. Different actual heat engines undergo similar but different cycles, depending on the system design. For example, internal combustion engines undergo the *Otto* cycle with gasoline fuel and the *Diesel* cycle with diesel fuel, while the steam and gas turbine power plants undergo the *Rankine* and *Brayton* cycles, respectively.[7,8] However, with improvements in material properties, effective component cooling and regeneration, and combining gas and steam turbine systems, the efficiencies above 50% are being achieved, which is a substantial improvement over the usual 30%–35% efficiency. The ideal Carnot cycle is an important reference to guide researchers and engineers to better understand limits and possibilities for new concepts and practical performance improvements of real heat engines.

The reversible Carnot cycle efficiency (Eq. 18) between the two thermal reservoirs does not depend on the cycling system, but only on the ratio of the absolute temperatures of the two reservoirs ($T = T_h$ and $T_{ref} = T_c$), known as Carnot cycle efficiency. Then, we may deduce the following ratio equality, where $Q = Q_h$ and $Q_{ref} = Q_c$, i.e.:

$$\frac{T}{T_{ref}} = \left.\frac{Q}{Q_{ref}}\right|_{\text{rev cycle}} \tag{20}$$

The above correlation, also known as the Carnot ratio equality, is much more important than what it appears at first. Actually, it is probably the most important equation in thermodynamics and among the most important equations in natural sciences.[10] Not only does Eq. 20 define the thermodynamic (absolute) temperature scale, which is independent of the substance of a thermometer, but it was also used by Clausius to deduce a fundamental property of matter, the entropy S, as well as the related process-equilibrium functions and correlations, and to quantify irreversibility (loss of work potential) and the far-reaching second law of thermodynamics.

The entropy balance equation for the control-volume flow process, complementing the related energy balance (Eq. 11, see Figure 9), is:[7–9]

$$\underbrace{\frac{d}{dt}S_{CV}}_{\substack{\text{RATE OF ENTROPY}\\ \text{CHANGE IN CV}}} = \underbrace{\sum_{CS}\frac{\dot{Q}_i}{T_i}}_{\substack{\text{CS TRANSFER}\\ \text{RATE WITH Q}}} + \underbrace{\sum_{IN}\dot{m}_j S_j}_{\substack{\text{TRANSPORT}\\ \text{RATE WITH}\\ \text{MASS IN}}} - \underbrace{\sum_{OUT}\dot{m}_k S_k}_{\substack{\text{TRANSPORT}\\ \text{RATE WITH}\\ \text{MASS OUT}}} + \underbrace{\dot{S}_{gen}}_{\substack{\text{IRREVERSIBLE}\\ \text{GENERATION}\\ \text{RATE}}} \tag{21}$$

Exergy and the Second-Law Efficiency

Exergy, E_X, as described above, is defined as the maximum possible work (i.e., the work potential) that may be obtained from a system by bringing it to the equilibrium in a process with reference surroundings (called reference *dead state*). The maximum possible work will be obtained if we prevent energy disorganization, thus with limiting *reversible* processes where the existing non-equilibrium is conserved (but rearranged and could be reversed) within interacting systems. The specific exergy per unit of system mass, e_X, with reference to the surroundings with T_o, P_o, $V_o = 0$, and $z_o = 0$, is:[7]

$$e_X = \left(\breve{u} - \breve{u}_o\right) + P_o\left(\breve{v} - \breve{v}_o\right) - T_o\left(s - s_o\right) + \frac{1}{2}V^2 + gz \tag{22}$$

The specific exergy of a flowing fluid, used for control- volume analysis, in addition to Eq. 22 includes the so- called flow exergy, $(P - P_o)$, $\breve{v}$ resulting in:

$$e_{xf} = \left(\breve{h} - \breve{h}_o\right) - T_o\left(s - s_o\right) + \frac{1}{2}V^2 + gz \tag{23}$$

where system state is defined by its specific properties (internal energy $\breve{u}$, enthalpy $\breve{h}$, specific volume $\breve{v}$, entropy s, velocity V, and elevation z, while 0th subscripted properties refer to the corresponding system properties in equilibrium with the surroundings (at T_o and P_o).

The exergy balance equation for the control-volume flow process, complementing the related energy and entropy balance (Eqs. 11 and 21, see Figure 9), is:[7,–9]

$$\frac{d}{dt}E_x = \sum\left(1 - \frac{T_o}{T_i}\right)\dot{Q}_{IN,\,i} - \left(\dot{W}_{out} - P_o\frac{d}{dt}\dot{V}_{CV}\right) + \sum_{in}\dot{m}e_{xf} - \sum_{out}\dot{m}e_{xf} - \dot{E}_{X,\,LOSS} \tag{24}$$

The destroyed or lost exergy ($E_{X,\,LOSS}$), or irreversibility (I), is a measure of energy degradation causing entropy generation (S_{gen}), or vice versa (see Eq. 17), and represents the lost work potential ($W_{pot,\,LOSS}$) with regard to reference or "dead" state (surroundings at T_o absolute temperature), as expressed by the following correlation:

$$I = W_{pot,\,LOSS} = E_{X,\,LOSS} = T_o S_{gen} \geq 0 \tag{25}$$

The second-law efficiency is defined by comparing the real irreversible processes or cycles with the corresponding ideal reversible processes or cycles:

$$\eta_{II} = \frac{E_{X,\,out}}{E_{X,\,in}} = 1 - \frac{E_{X,\,LOSS}}{E_{X,\,in}}, \text{ where } E_{X,\,LOSS} = E_{X,\,in} - E_{X,\,out} \tag{26}$$

The second-law efficiency is true efficiency related to the "goodness" or quality of energy processes. For example, the ideal Carnot cycle has the second-law efficiency 100% as the best possible heat-engine conversion process, and the efficiency (quality) of the rejected heat to the surrounding is 0%; i.e., the rejected heat in an ideal cycle has no value at all; therefore, it is not a "waste energy" as usually presented but is a useful and necessary one for accomplishing the heat engine cycle. Many other misconceptions based on the first-law analysis are clarified using the second-law or exergy analysis.

Concluding Remarks: Energy Provides Existence and Is Cause for Change

Energy is the fundamental property of a physical system and refers to its potential to maintain a system identity or structure and to influence changes (via forced-displacement interactions) with other systems by imparting work (forced directional displacement) or heat (forced chaotic displacement/motion of a system's molecular or related structures). Energy exists in many forms: electromagnetic (including light), electrical, magnetic, nuclear, chemical, thermal, and mechanical (including kinetic, elastic, gravitational, and sound). Note that if all energy is literally expelled from a confined space, then nothing will be left (empty space). As long as any matter is left, it will contain energy—even at zero absolute temperature, the electrons will be orbiting around very energetic nucleus. Matter is and must be energetic, $E = mc^2$; thus, literally, "energy is everything"—no energy, nothing in the space. Mass and energy are a manifestation of each other and are equivalent; they have a holistic meaning of *mass energy*.

The philosophical and practical aspects of energy and entropy, including reversibility and irreversibility, could be summarized as follows:

- Energy is a fundamental concept indivisible from matter and space, and energy exchanges or transfers are associated with all processes (or changes), thus indivisible from time.
- Energy is "the building block" and fundamental property of matter and space, thus a fundamental property of existence. Any and every material system in nature possesses energy.
- For a given system state (structure and phase), addition of energy will spontaneously tend to randomly redistribute (disorganize, degrade, dissipate) over the system's finer microstructure and the space it occupies, called thermal energy, equalizing the thermal energy potential (temperature) and increasing the energy randomization per absolute temperature (entropy), and vice versa.
- Entropy is an integral measure of (random) thermal energy redistribution (due to heat transfer and/or irreversible heat generation due to energy degradation) within a system mass and/or space (during system expansion), per absolute temperature level. Entropy is increasing from a perfectly ordered (singular and unique) crystalline structure at zero absolute temperature (zero reference) during reversible heating (entropy transfer) and entropy generation during irreversible energy conversion (loss of work-potential to degraded thermal energy), i.e., energy degradation or random equipartition within system material structure and space per absolute temperature level. Energy and mass are conserved within interacting systems (all of which may be considered as a combined isolated system not interacting with its surrounding systems), and energy transfer (in time) is irreversible (in one direction) from a higher to a lower energy potential only, which then results in continuous generation (increase) of energy displacement, called entropy generation, which is a fundamental measure of irreversibility or permanent changes.
- Exergy refers to the maximum system work potential if it is reversibly brought to the equilibrium with reference surroundings; i.e., exergy is a measure of a system's non-equilibrium with regard to the reference system. Exergy is not conserved but is destroyed or lost during real irreversible processes commensurate to the related entropy generation.
- Reversible energy transfer is only possible as a limiting case of irreversible energy transfer at infinitesimally small energy potential differences, thus in quasiequilibrium processes, with conservation of entropy (no entropy generation). Since such changes are reversible, they are not permanent and, along with time, irrelevant.
- Non-equilibrium, i.e., non-uniform distribution of mass energy in space, tends in time to spontaneously and irreversibly redistribute over space towards a common equilibrium; thus, non-equilibrium cannot be spontaneously created. Therefore, all natural spontaneous or autonomous processes (proceeding by itself and without interaction with the rest of the surroundings) between systems in non-equilibrium have irreversible tendency towards a common equilibrium—and thus irreversible loss of the original work potential (measure of nonequilibrium), by converting other energy forms into the thermal energy accompanied with increase of entropy (randomized equipartition of energy per absolute temperature level). The spontaneous forced tendency of mass-energy transfer is due to a difference or nonequilibrium in space of the mass-energy space density or mass-energy potential. As mass energy is transferred from a higher to a lower potential, and thus conserved, the lower mass-energy potential is increased on the expense of the higher potential until the two equalize, i.e., until a lasting equilibrium is established. This explains a process tendency towards the common equilibrium and impossibility of otherwise (impossibility of spontaneous creation of non-equilibrium).[10]

In summary, energy provides existence and is cause for change. Energy is possessed (thus equilibrium property) by material systems and redistributed (transferred) between and within systems, due to systems' non-equilibrium, via forced- displacement interactions (process) towards the equilibrium (equipartition of energy over mass and space); thus, energy is conserved (the first law) but degraded (the second law). Finally, we may unify the two laws of thermodynamics as one *grand law of nature:* The universe consists of local material (mass energy) structures in forced equilibrium and their interactions via force fields. The forces are balanced at any time (including inertial—process rate forces), thus conserving momentum, while charges/mass and energy are transferred and conserved during forced displacement in space all the time, but energy is degraded (dissipated) as it is redistributed (transferred) from a higher to a lower non-equilibrium potential towards equilibrium (equipartition of energy).

Glossary

Energy: It is a fundamental property of a physical system and refers to its potential to maintain a material system identity or structure (force field in space) and to influence changes (via forced-displacement interactions, i.e., systems' restructuring) with other systems by imparting work (forced directional displacement) or heat (forced chaotic displacement/motion of a system's molecular or related structures). Energy exists in many forms: electromagnetic (including light), electrical, magnetic, nuclear, chemical, thermal, and mechanical (including kinetic, elastic, gravitational, and sound).

Energy conservation: It may refer to the fundamental law of nature that energy and mass are conserved, i.e., cannot be created or destroyed but only transferred from one form or one system to another. Another meaning of energy conservation is improvement of efficiency of energy processes so that they could be accomplished with minimal use of energy sources and minimal impact on the environment.

Energy conversion: A process of transformation of one form of energy to another, like conversion of chemical to thermal energy during combustion of fuels, or thermal to mechanical energy in heat engines, etc.

Energy efficiency: Ratio between useful (or minimally necessary) energy to complete a process and actual energy used to accomplish that process. Efficiency may also be defined as the ratio between energy used in an ideal energy-consuming process versus energy used in the corresponding real process, or vice versa, for an energy- producing process. Energy, as per the conservation law, cannot be lost (destroyed), but part of energy input that is not converted into "useful energy" is customarily referred to as "energy loss."

Entropy: It is an integral measure of (random) thermal energy redistribution (due to heat transfer or irreversible heat generation) within a system mass and/or space (during system expansion), per absolute temperature level. Entropy is increasing from a perfectly ordered (singular and unique) crystalline structure at zero absolute temperature (zero reference) during reversible heating (entropy transfer) and entropy generation during irreversible energy conversion (loss of work potential to thermal energy), i.e., energy degradation or random equipartition within system material structure and space per absolute temperature level.

Exergy: It is the maximum system work potential if it is reversibly brought to the equilibrium with reference surroundings; i.e., exergy is a measure of a system's nonequilibrium with regard to the reference system.

Heat: It is inevitable (spontaneous) energy transfer due to temperature differences (from a higher to a lower level), to a larger or smaller degree without control (dissipative) via chaotic (in all directions, non-purposeful) displacement/motion of system molecules and related microstructure, as opposed to controlled (purposeful and directional) energy transfer referred to as *work*.

Heat engine: It is a device undergoing thermomechanical cycle that partially converts thermal energy into mechanical work and is limited by the ideal *Carnot cycle* efficiency. The cycle mechanical expansion and compression net-work is obtained due to difference between heat transferred to the engine from a high-temperature heat reservoir and rejected to a low temperature reservoir, thus converting part of thermal energy into mechanical work.

Mechanical energy: It is defined as the energy associated with ordered motion of moving matter at a large scale (kinetic) and ordered elastic potential energy within the material structure (potential elastic), as well as potential energy in gravitational field (potential gravitational).

Power: It is the energy rate per unit of time and is related to work or heat transfer processes (different work power or heating power).

System (also *Particle* or *Body* or *Object*) refers to any arbitrarily chosen but fixed physical or material system in space (from a single particle to a system of particles), often called "closed system," which is subject to observation and analysis. System occupies a so-called system volume within its own enclosure interface or system boundary and thus separates itself from its surroundings, i.e., other surrounding systems.

Temperature: It refers to the average kinetic energy during thermal interaction of disordered microscopic motion of molecules and atoms. The concept of temperature is complicated by the particle internal degrees of freedom like molecular rotation and vibration and by the existence of internal interactions in solid materials, which can include the so-called collective molecular or atomic behavior. All of these phenomena could contribute to the kinetic energy during particle (thermal) interactions. When two objects are in thermal contact (i.e., interaction of random motion of their particles), the one that tends to spontaneously give away (lose) energy is at the higher temperature. In general, temperature is a measure of the tendency of an object to spontaneously exchange thermal energy with another object until their temperatures equalize, that is, until they achieve interacting particle kinetic energy equipartition (statistically equalize).

Thermal energy: It refers to the energy associated with the random, disordered motion of molecules and potential energy of intermolecular forces, as opposed to the macroscopic ordered energy associated with ordered "bulk" motion of system structure at a large scale, and excluding internal binding energy within atoms (nuclear) and within molecules (chemical).

Total internal energy: It refers to the energy associated with the random, disordered motion of molecules and intermolecular potential energy (thermal), potential energy associated with chemical molecular structure (chemical), and atomic nuclear structure (nuclear), as well as with other structural potentials in force fields (electrical, magnetic, elastic, etc.). It refers to the "invisible" microscopic energy on the subatomic, atomic, and molecular scale as opposed to "visible" mechanical, bulk energy.

Work: It is a type of controlled energy transfer when one system is exerting force in a specific direction and thus making a purposeful change (forced displacement) in the other systems. It is inevitably (spontaneously) accompanied, to a larger or smaller degree (negligible in ideal processes), by dissipative (without control) energy transfer referred to as *heat* (see above).

References

1. Kostic, M. Work, power, and energy. In *Encyclopedia of Energy*; Cleveland, C.J., Ed.; Elsevier Inc.: Oxford, U.K., 2004; Vol. 6, 527–538.
2. Kostic, M. Irreversibility and reversible heat transfer: The quest and nature of energy and entropy, IMECE2004. In ASME Proceedings, ASME: New York, 2004; 6 pp.
3. Kostic, M. Treatise with Reasoning Proof of the First Law of Energy Conservation, Copyrighted Manuscript, Northern Illinois University: DeKalb, IL, 2006.
4. Kostic, M. Treatise with Reasoning Proof of the Second Law of Energy Degradation, Copyrighted Manuscript, Northern Illinois University: DeKalb, IL, 2006.

5. Clausius, R. Über die bewegende Kraft der Warme. Annalen der Physik **1850**, *79*, 368–397, 500–524. English Translation: "On the moving force of heat, and the laws regarding the nature of heat itself which are deducible therefrom." Phil. Mag. **1957**, *2* (1–21), 102–119.
6. Cengel, Y.A.; Ghajar, A.J. *Heat and Mass Transfer, Fundamentals and Applications,* 4th Ed.; McGraw-Hill: New York, 2011.
7. Cengel, Y.A.; Boles, M.A. *Thermodynamics, An Engineering Approach,* 5th Ed.; WCB McGraw-Hill: Boston, 2006.
8. Bejan, A. *Advanced Engineering Thermodynamics*; Wiley: New York, 1988.
9. Zemansky, M.W.; Dittman, R.H. *Heat and Thermodynamics*; McGraw-Hill: New York, 1982.
10. Kostic, M.; Revisiting the 2nd Law of Energy Degradation and Entropy Generation: From Sadi Carnot's Ingenious Reasoning to Holistic Generalization. AIP Conf. Proc. 2011, 1411, 327, doi: 10.1063/1.3665247.

10

Energy: Renewable

John O. Blackburn

Introduction

Renewable energy is a category of energy supply that is receiving increasing attention in the United States. The term refers, of course, to energy sources that can be fully utilized without diminishing future supplies. The obvious contrast is with fossil fuels which, once used, are no longer available.

Renewable energy is certainly nothing new. Mankind lived on it almost exclusively until the industrial revolution in the 1700s. Falling water and wind provided mechanical energy for tasks like pumping water or grinding grain, while ships were propelled by wind. Wood fires provided heat for dwellings or for tasks like smelting metals.

This entry examines the various renewable energy resources in the United States and the status of conversion technologies. Because the largest two sources are also intermittent, integration into reliable supply systems is also examined. This turns out to be a less formidable task than it first appears to be.

Present energy end uses are examined in order to see how renewable sources might be deployed in a highly renewable energy future.

Renewable Energy Forms

Renewable energy comes in many forms, though most sources come directly or indirectly from solar radiation. Solar energy may be used directly as a source of heat as, for example, in heating water. It may also produce electricity, either directly in photovoltaic cells or by using concentrating collectors to produce steam for turbines. Wind energy, which is derived from solar energy, is now used almost exclusively to produce electricity, as is falling water. Winds over ocean waters produce waves, which are beginning to be harnessed as a source of electricity. Ocean currents or temperature differences between surface and deep waters also offer opportunities to extract useful energy.

Trees, crops, and other plant matter (referred to as "biomass" in the renewable energy literature) likewise represent stored solar energy. Half the world still uses wood as a fuel, while forest product industries in the industrial world use wood waste to produce process heat, steam, and electricity. Energy crops, trees, or even grasses could be grown. Fuel alcohols are now produced mostly from sugarcane and corn, and they are beginning to be produced from cheaper, more abundant cellulosic biomass materials, such as wheat straw or other waste products. Other biomass sources include animal manures and sewage, which can produce methane through anaerobic digestion. There are a few renewable sources of energy that do not derive from solar sources.

The moon's gravitational pull on the earth creates tidal movements of ocean waters. In some locations, the rise and fall of water levels is sufficiently great to generate electricity.

Heat generated from nuclear processes deep in the earth sometimes comes close enough to the surface to be captured as useful energy. When underground water is present as well, steam or hot water may be available for electricity generation.

The principal difference between age-old uses of renewable energy and the present possibilities is the availability of modern technologies with conversion efficiencies much higher than those previously available. Most importantly, all forms of renewable energy can be converted into electricity—the most versatile form of energy for human purposes.

Potential Renewable Resources in the United States

Solar radiation falling on the lower forty-eight states amounts to an estimated 44,000 quads (quadrillion Btu) annually—440 times the annual energy use for the nation.[1] Few other resources are found in such abundance. Windpower, the next most abundant renewable energy source, has an estimated electricity-generating potential of 10,000 billion kilowatt-hours in the United States, almost three times present annual electricity consumption.[2] This estimate is for on-shore wind turbines only; the additional offshore potential is large, but has not been as carefully estimated. Moreover, recent studies find the U.S. potential to be underestimated in quantity and in the number of areas with economic wind-generating potential.[3]

Hydroelectric facilities already generate 7%–8% of the nation's electricity, with another 3%–4% that might still be developed.[4] Geothermal resources that are suitable for electricity generation are found only in the western third of the United States, but they amount to 25,000 megawatts— enough to provide 25%–30% of that area's present electricity use. Geothermal heat is available to more of the nation, with a much larger potential.[5] The wave-generating potential has not yet been fully assessed in the United States. It appears to be substantial along the Pacific coast and the North Atlantic coast. Tidal power possibilities in the United States are limited to Alaska, Maine, and Washington.

As for biomass sources, U.S. forests and farms now produce materials with an energy content of some 14 quads. With energy crops, this could increase fourfold, according to estimates of the National Renewable Energy Laboratory.[6] Other recent estimates give a near-term potential of 19–20 quads over and above materials taken for food, feed, and forest products, given relatively minor modifications in farm and forestry practices.[7] The author gives a figure of 1 cent per kwh in Japan, but this calculation does not include amortization of the equipment or annual maintenance costs.

There is clearly no shortage of renewable energy. It comes in many forms and, with respect to solar radiation and windpower, in great abundance.

In order to be useful in human activities, renewable energy must be captured and converted into forms which are consistent with energy demands. Time gaps between availability and need must also be bridged. For these tasks, equipment that is reliable, durable, and affordable must be available.

Conversion Efficiencies

Conversion efficiencies for various forms of renewable energy vary considerably. The growth of trees and other plant life generally convert only 0.5%–1% of solar radiation into stored biomass energy, which can be used commercially. The fastest growing species in areas with long growing seasons (sugarcane, for example) convert 2%–3% of solar energy into recoverable energy.[8]

Other renewable technologies, fortunately, have much higher conversion ratios. Solar panels for heating water convert 40%–55% of the solar energy that falls on their surfaces to useful energy.[9] Crystalline silicon photovoltaic cells now on the market convert 11%–14% of solar radiation into electricity, lab models reach 33%, and nanotechnologies hold some promise of yet-higher figures.[10] It is these conversion efficiencies that facilitate the use of large quantities of renewable energy collected in a relatively small portion of a nation's total area.

State of Technologies: Costs

Technologies Now Fully Developed and Cost Competitive

Hydroelectricity has been competitive since the earliest days of grid electricity. Expansion in the United States is constrained by environmental considerations, though output can be expanded by repowering existing facilities or installing generators at small existing dams. Hydro capacity becomes extremely valuable in grids with high components of intermittent power from wind and solar sources because it enables them to operate over a considerable range without further backup.

Wind energy is competitive with conventional fuels in areas with good wind resources and its costs continue to fall.

Geothermal Electricity

This technology is now well established in suitable sites around the world. Initial capital costs are high, both for drilling into steam or hot water sources and for generating equipment. California has 2200 megawatts in operation, and new facilities are planned or under construction in several Western states.[11]

Biomass

About three quads are already used annually for energy in the United States. The forest products industries use 1.8 quads of wood and wood wastes for process heat and cogenerated electricity. Firewood heats three million homes and supplies some occasional heat to another 20 million. These uses are fully competitive with fossil fuels, as is methane collected from landfills.[12] Other biomass possibilities are listed in the following section.

Ocean Sources

A small number of tidal plants have been installed around the world, but none have been installed as of yet in the United States. Other technologies for using ocean energy, except for waves, are not under active development at this time.[13]

Technologies That Are Well Developed but Not Yet Fully Cost Competitive

Solar Water Heating

This is a relatively simple technology consisting basically of water pipes in a box covered with glass and painted black. Today's versions are, to be sure, much more sophisticated, with freeze-protection features to make the devices useable nationwide. Making costs competitive with present energy sources will be possible upon mass production and mass installation of the equipment, a feat long since accomplished in Israel.

Photovoltaic Electricity

Fifty years of research and development efforts have brought higher efficiencies and greatly reduced costs. Costs are still too high, by a factor of two or three, for photovoltaic electricity to be fully competitive on customer facilities in the United States. They are already competitive for users remote from a power grid or for utilities at peak summer hours.

Cost reductions in photovoltaic systems manufacturing are proceeding more slowly than anticipated in the 1980s and 1990s, while conventional generation, with which these systems compete, has shown declining inflation- adjusted cost, as well. Authoritative analyses in the 1980s projected cost reductions of photovoltaic modules and other system costs to about one dollar per peak watt (two dollars in 2005 prices), which implied Photovoltaic (PV) electricity costs of six to seven cents per kilowatt-hour.[14] Module prices have indeed declined to about $3.50 for volume purchases in 2005 (in current dollars).[15] Costs are not likely to fall significantly until some alternatives to the (now) standard silicon wafer are developed for volume production. Nonetheless, manufacturing costs have fallen below three dollars for modules.[16] With rapid market and manufacturing development, Japan can produce photovoltaic electricity at about 15 cents per kwh, competitive with (expensive) Japanese-delivered grid electricity.[17]

Photovoltaic equipment benefits from its potential location at the point of consumption, thus avoiding most of the costs of transmission and distribution. PV electricity, therefore, can be priced higher than electricity from central generating plants. Because there is already enough rooftop space to accommodate nearly any conceivable volume of photovoltaic power, site costs are avoided as well.

Other Biomass Sources

Subsidized fuel ethanol is made mostly from corn, but more abundant cellulose sources are beginning to be utilized. Scale economies and cost decreases with production experience should make this technology competitive in the next few years. Iogen, the Canadian producer of ethanol from wheat straw, estimates a price of $1.30 per gallon from a proposed plant in Idaho.[18] Economically feasible processes and equipment to produce methane from animal wastes are already available, but the technology is still not widely used.

Technologies for Which Further Development Is Required

Solar Industrial Process Heat

Development of equipment for these tasks received some attention in the late 1970s, but was abandoned after that as fuel prices declined. Industrial process heat requirements and insolation overlap considerably in the United States, so their potential remains to be exploited.[19] Further development awaits higher fossil fuel costs and some renewed research attention.

Wave Energy

Several small demonstration facilities now operate. The first full-scale commercial generation facilities are about to be installed in Portugal by British and Portuguese firms using British designs.[20]

Ocean Currents and Thermal Differences between Surface and Deep Ocean Waters

There is very little activity at present.

In summary, hydroelectric, wind, and geothermal electricity are fully cost-competitive with fossil and nuclear-generated electricity in areas with good resources. These areas are, respectively, three-quarters of the states for wind, all but three or four states for hydroelectricity, and the western third of the nation for geothermal. Biomass-based heat and electricity are likewise well-established sources and they are available in every state. Widespread market penetration of solar hot water systems awaits economies of mass production and mass installation. Biomass-based liquid fuel production is set to expand rapidly. Sales of photovoltaic equipment are rising rapidly, but still with subsidies and incentives. Cost reduction proceeds apace at about 4%–5% per year, while several promising new and potentially much cheaper versions are moving into production.

Solar industrial process heat is receiving little attention, though it probably will as oil and gas prices trend upward in coming decades.

Dealing with Intermittent Sources

Solar radiation and wind are the most abundant renewable energy sources, but they are also intermittent in nature. This feature raises issues of complementarity and energy storage. Utilities think of generating capacity in terms of baseload, intermediate, and peaking capacities. Because intermittent sources do not fit into any of these categories, they are likely to be relegated to marginal roles.

Inquiries that pose the question of usefulness in another way have produced much more favorable conclusions. Utilities are already accustomed to demand patterns that are subject to seasonal, daily, weather-related, and random fluctuations. Their supply sources, in contrast, can be turned on or off as they wish. The more fruitful approach in considering intermittent sources is to regard them as "negative demand"—the side which is already subject to variations. Models that start with actual hour-by-hour demand loads through the year and then subtract out wind or solar contributions, hour-by-hour, consistently show sizeable potential contributions from intermittent renewables without further backup. Needless to say, the presence of hydroelectric facilities or pumped storage units supports the ability to accommodate intermittent sources.

A simulation of the British electricity system showed that windpower, though not steadily available and not correlated with demand patterns, could still meet 25%–45% of system demand without additional backup, provided that the conventional components could be reconfigured to accommodate wind patterns.[21] A U.S. simulation found that in a system with demand patterns similar to those of most utilities, a grid with a substantial hydroelectric component could accommodate 50% of its outputs from intermittent renewable sources (i.e., wind and solar) and at costs no higher than those for conventional generation.[22] These are certainly counter-intuitive results, and they would need to be supplemented by modeling studies of other utilities. The findings, though, are consistent with experiences in Denmark and North Germany. Denmark now obtains 21% of its electricity from wind turbines, without evident problems of integrating this output into the national grid. German reports of strains on the grid seem to have more to do with transmission bottlenecks than generation sources.

Solar electricity has the advantage, for most utilities, of coinciding in availability with summer peaks in air-conditioning use. A recent New Jersey study found that photovoltaic generation would not only supply peak power (and thereby displace fuel generation) but 40%–70% of PV capacity would displace the capital costs of conventional generation, fractions which rose quickly in the presence of grid storage capacity.[23]

These results are so much at variance with traditional thinking in utility managements that time, experience, and external pressures will be required to bring solar and wind technologies into widespread use, even when they are cost-competitive. Therein lies the case for incentives and regulatory requirements, such as the renewable portfolio standards now found in 18 states. Expanding markets enable the solar and wind equipment manufacturers to scale up and further reduce costs, while the utilities gain actual operating experience with intermittent sources.

Moving toward an All-Renewable Energy Future

A consideration of energy futures in which renewable sources contribute most or all energy needs begins with an assessment of the quantity of energy that is likely to be needed, and its appropriate forms. As a starting point, one examines present energy sources and end uses in the economy, and then allows for increasing efficiencies. One must show that renewable resources are sufficient in quantity, quality and cost to provide for each of the end uses.

A word of caution is in order. It is much too easy to focus on energy supply, while neglecting the rich possibilities for improving energy efficiency in all end uses. One thus can miss the most economical opportunities for meeting energy demands because efficiency measures usually are less expensive than any new supplies, renewable or conventional. The reader is referred to the many entries in this encyclopedia that deal with the significant potential of increasing the efficiency with which energy is used.

TABLE 1 Fuel and Energy Use by Sector, United States, 2003 (Quadrillion Btus)

Sector	Fuel	Electricity	Total End Energy Use	Electricity Losses	Total Primary Energy	Fuel End Uses in Sector
Residential	7.2	4.4	11.5	8.7	21.3	Space heat
Commercial	4.3	4.1	8.4	9.2	17.6	Water heat Space heat
Transportation	26.8	—a	26.8	0.1	26.9	Water heat Liquid fuel
Industry						
Manufacturing						
Agriculture, Mining	15.8	3.4	19.2	7.7	26.9	Process heat
Construction	3.0	—a	3.0		3.0	
Feed stocks	3.0	—	3.0		3.0	
Total industry	21.8	3.4	25.2	7.7	32.9	
Total all sectors	60.1	11.9	72.0	26.7	98.7	

[a]Less than 0.1 quad.

Source: Adapted from Energy Information Administration (additional detail for industry from International Energy Agency, IEA Energy Statistics, 2003, Energy Balance Table for the United States).[25]

Present U.S. Energy Use

U.S. primary energy use is now running at about 100 quads per year and growing at about the same rate as population, unlike the growth pattern prior to 1973. Data for 2003 are shown in Table 1.

As the table indicates, primary energy use in the United States in 2003 was 98.7 quads. About 40% of this primary energy was used to generate electricity, and two-thirds (26.7 quads) of that was lost as waste heat in generation and transmission. End-use energy was thus 72 quads. Transportation and industry each take about one third of end-use energy, with the rest going to the residential and commercial sectors. In U.S. energy statistics, electricity losses are allocated to each of the sectors in order to see the primary energy demand occasioned by the activities of each sector.

Table 1 also lists the kinds of energy end-uses provided by fuels because these would have to be supplied from renewable sources in the absence of fuels. (Sector end- uses for electricity are not shown, since electricity is readily produced from renewable sources.) In the residential sector, for example, fuels are used mostly to heat space and water or, put differently, to provide low temperature (under 100 degrees centigrade) heat. A much smaller use of fuel is for cooking, which utilizes temperatures in the 100–250 degree centigrade range. Fuel use in the commercial sector is applied for similar purposes. Eighty percent of industrial energy use is in manufacturing, while 20% goes to mining, agriculture, and construction or is used for fuel-based feedstocks. Manufacturing end-use energy is 19 quads of which three to four quads is for refining petroleum.

Energy Efficiency Gains

All of these uses can be reduced with gains in energy efficiency, and some of them can be reduced significantly. Indeed, if the best practices already existing in building design and construction, lights, appliances, heating and cooling systems, vehicle mileage, and industrial processes were uniformly applied in the whole economy, end-use energy demands use would be 50%–70% lower, or in the 22–36 quad range—not the 72 quads shown in Table 1.[24] These potential reductions in energy use make renewable energy futures much more readily conceivable for high-income, high-output nations. Any energy supply system in a world of diminishing oil and gas output would be hard pressed to meet the energy demands of our present inefficient economy.

Adequacy of Renewable Resources

The questions for renewable energies then become:

1. In the United States, can renewable energy sources provide 22–36 quads, of which 6 to 9 or so would meet efficiency-reduced electricity demands, 3 to 5 quads would be used for heat in buildings, 7 to 10 quads or so would power the transportation sector, and 6 to 10 quads would be composed of fuels for industry?
2. If that can be shown, then can these sources provide for any continuing growth in the economy?
3. What technological developments are needed for this potential to be realized?

The adequacy of renewable energy sources, considered in gross, is easy to establish. Wind and solar resources alone provide these amounts of energy many times over. In addition, the "nonintermittent" renewable sources (biomass, hydroelectric, and geothermal resources) sum to another 17 quads at a minimum.

As to end-uses, 6 to 9 quads of electricity fall well under available sources, with storage capability being required only in areas lacking hydroelectric resources.

Space and water heat can be provided in part from direct solar sources, with the rest coming from biomass or perhaps electricity.

For those who envision a hydrogen-fueled transportation system, intermittent sources of renewable electricity are tailor-made. Hydrogen can be produced by electrolysis whenever solar or wind sources exceed grid demands. Liquid fuels from biomass are likely to be available, as well, especially with emerging technologies to produce liquid fuels from cellulosic materials.

Industrial process heat would come primarily from biomass sources, with some assistance from solar equipment or perhaps hydrogen produced from renewable electricity.

As to growth over time in energy demands over time, one can observe that economic growth in the United States, at least in per capita terms, is directed toward sectors that generally do not have high energy contents (services vs. basic commodities). If the United States is ever to reach a measure of sustainability with respect to its physical resource demands, its population must level off, as must that of any other nation seeking sustainability.

There are two major technological developments that remain to be achieved in order to make an all-renewable future attainable. The first is cost reductions in the manufacture and installation of photovoltaic systems and the second is the rapid commercialization and cost reduction in processes which make fuel ethanol from cellulosic materials, rather than foods. These developments are already well underway.

Space does not permit an extended analysis of these matters; this brief discussion is intended to establish, in general terms, that a largely or wholly renewable energy future is possible, given several decades to adjust. This view is not now widely accepted, yet it is certainly the most hopeful one for mankind in an age of diminishing fossil fuel resources.

References

1. Kendall, H.W.; Nadis, S.J. Eds.; Energy strategies: toward a solar future. In *A Report of the Union of Concerned Scientists*. Ballenger: Cambridge, MA, 1979.
2. Grubb, M.J.; Meyer, N.I. Wind energy: resources, systems, and regional strategies. In *Renewable Energy: Sources for Fuel and Electricity*; Johansson, T.B., Kelly, H., Reddy, A.K.N. et al., Eds.; Island Press: Washington, DC, 1992.
3. Archer, C.L.; Jacobson, M.Z. Spatial and temporal distributions of U.S. winds and windpower at 80 m derived from measurements. J. Geophys. Res. Atmos. **2003**, *108* (D9), 4289.
4. National Hydropower Association, Hydrofacts. http://www.hydro.org/hydrofacts/forecast/asp (accessed May 24, 2005).
5. Wright, P.M. Status and potential of the U. S. geothermal industry. AAPG Bull. **1993**, *77* (8).

6. National Renewable Energy Laboratory, Biomass Energy, Potential, http://www.nrel.gov/documents/biomass_energy.html (accessed May 21, 2005).
7. Perlack, R.D.; Wright, L.L.; Turhollow, A.; Graham, R.L. *Biomass as Feedstock for a Bioenergy and Bioproducts Industry: The Technical Feasibility of a Billion-Ton Annual Supply;* Oak Ridge National Laboratory: Oak Ridge, TN, April, 2005.
8. Hall, D.O.; Rosillo-Calle, F.; Williams, R.H.; Woods, J. Biomass for energy: supply prospects. In *Renewable Energy: Sources for Fuels and Electricity*; Johansson, T.B., Kelly, H., Reddy, A.K.N., Williams, R.H., Eds.; Island Press: Washington, DC, 1992.
9. Florida Solar Energy Center. http://www.fsec.ucf.edu/solar/testcert/collectr/tpndhw.htm (accessed May 23, 2005).
10. http://www.oja-services.nl/iea-pvps/pv/materials.htm (accessed May 24, 2005).
11. Renewable Energy Policy Project http://www.repp.org./geothermal_brief_power_technologyand-generation.html (accessed May 24, 2005).
12. http://www.nrel.gov/documents/biomass_energy.html (viewed May 21, 2005).
13. http://www.poemsinc.org/FAQtidal.html (accessed May 25, 2005).
14. Maycock, P.D. *Photovoltaics: Sunlight to Electricity in One Step*; Brick House Publishing Co.: Andover, MA, 1981.
15. http://www.solarbuzz.com Price Survey, May 2005. (accessed May 20, 2005).
16. http://www.nrel.gov/docs/fy04osti/32578.pdf (viewed May 21, 2005).
17. Sawin, J. L. *Mainstreaming renewable energy in the 21st century.* Worldwatch Paper 169, Worldwatch Institute: Washington, DC, 2004.
18. Governors seek wider ethanol use. Wall Street Journal, April 12, 2005.
19. *A New Prosperity; Building a Sustainable Energy Future. The SERI Solar/Conservation Study.* Brick House Publishing: Andover, MA, 1981.
20. http://www.renewableenergyaccess.com (accessed May 20, 2005).
21. Grubb, M.J.; Meyer, N.I. Wind energy: resources, systems, and regional strategies. In *Renewable Energy: Sources for Fuels and Electricity*; Johanssen, T.B., Kelly, H., Reddy, A.K.N., Williams, R.H. Eds.; Island Press: Washington, DC, 1992.
22. Kelly, H.; Weinberg, C. Utility strategies for using renewables. In *Renewable Energy: Sources for Fuels and Electricity;* Johanssen, T.B., Kelly, H., Reddy, A.K.N., Williams, R.H. Eds.; Island Press: Washington, DC, 1992.
23. Perez, R. Determination of Photovoltaic Effective Capacity for New Jersey. http://cleanpower.com/research/capacityvaluation/ELCC_New_Jersey.pdf (accessed May 24, 2005).
24. Blackburn, J.O. Possible efficiency gains identified for the U.S. as of the mid-1980's are examined. *The Renewable Energy Alternative*; Duke University Press: Durham, NC, 1987.
25. Energy Information Administration, Monthly Energy Review, March 2005.

11

Energy: Storage

Rudolf Marloth

Introduction

Although specific types of energy storage are introduced below, note that this encyclopedia has separate and detailed entries on batteries, compressed air storage, and pumped hydroelectricity. First, some definitions are given:

- Primary storage methods are designed for a single use. A regular flashlight cell is used once and discarded (NOT into landfill).
- Secondary storage methods allow energy to be stored and recovered repeatedly. Rechargeable flashlight cells are available.
- Renewable energy is endlessly available in nature. Wind and solar power are currently in the spotlight.

The greater part of the entry concerns secondary storage. Renewables and primary storage are discussed briefly. A glossary is included at the end of the entry.

Need for Storage

Depending upon the use, energy is stored for varying periods. The energy stored in the gas tank of an automobile is used over a period of about a week. Part of that energy is stored in the car's battery, to be used two or several times in a day.

Power companies use large-scale energy storage[1] to defer investment in new generation, transmission, and distribution facilities. It is useful for integration with renewables, for voltage and frequency regulation, for demand leveling, and in lieu of spinning reserves. Long-term energy storage might depend upon availability and the cost of transporting it, such as the pile of coal at a generating plant.

Rate payers can use energy storage for load leveling and to protect themselves from the vagaries of the grid. Large quantities of electric energy generated at night when utility rates are cheap are stored in order to be released in daytime when rates are high. Smaller quantities are stored indefinitely to keep critical or sensitive equipment running during power failures or surges, sometimes only for seconds or minutes, until backup generators can come on line. A small and isolated user might want to store irregular energy supplied by wind or sun rather than build a road, pipeline, or power line for access to conventional sources.

Example 1. Consider electric rates that are \$10.00/kW-mo and \$0.10/kWh during weekdays from noon to 6:00 P.M. and \$5.00/kW-mo and \$0.05/kWh during weeknights from midnight to 6:00 A.M. and on weekends. Making reasonable assumptions about on-peak and off-peak hours, calculate a benefit of this use of energy storage.

Solution. Using an average of 22 weekdays and 8 weekend days per month gives 132 hr/mo on-peak hours and 324 hr/mo off-peak hours. Storing energy over all the offpeak hours will reduce the demand contribution to the stored energy cost. Assuming 75% round-trip efficiency apportioned evenly between storage and recovery (86.6% each way) and spreading the demand cost cover each hour of use, the blended cost of stored energy is

$$\frac{\$5.00/\text{kW}-\text{mo}/324\text{hr/mo}+\$0.05/\text{kWh}}{0.866}=\$0.076/\text{kWh} \tag{1}$$

while the blended cost of recovered energy is

$$\frac{\$10.00/\text{kW}-\text{mo}/132\text{hr/mo}+\$0.10/\text{kWh}}{0.866}=\$0.20/\text{kWh} \tag{2}$$

Valuation Criteria

Energy storage methods can be evaluated by many different criteria, with importance depending upon the application. Possible criteria are first cost in \$/kWh, lifetime in cycles or hours, maintenance cost, energy density in kWh/lb, allowable depth of discharge, footprint in kWh/ft^2 and area in ft^2, reliability, and safety.

Storing Electricity

A journal entry[2] whose title implies the possibility of storing electric energy seems to allow in its content that, technically, one cannot. The methods cited require conversion and reconversion, with storage in any of several non-electric forms. For storage in batteries, electricity is converted to chemical energy. For flywheel storage, electricity is converted to rotational kinetic energy. In pumped hydroelectric storage, electricity is converted to gravitational potential energy. Compressed air and energy stored in springs are mechanical potential energy. Superconducting magnetic energy storage (SMES) is magnetic energy. The energy stored in a capacitor is in an electric field, so at least this can be called stored electric energy.

Renewable Energy

A distinction is commonly made between stored energy and renewable (natural) energy. Some might say that energy is "stored" in the wind or ocean waves where it is available to be converted into a more useful form. Others contend that at least some processing is required for energy to qualify as stored, for instance, oil from a well must be refined, transported, and stored in a tank. For another example of primary stored energy, sulfur, charcoal, and saltpeter are combined to make gunpowder. Other examples of renewable energy are waterfalls, sunlight, trees and other biomass (saw grass, bagasse, corn), and geothermal energy. The case of water behind a dam is not so simple.

However, once some of these resources are harvested, processed, transported, and deposited at the consuming site, they can be considered as primary stored energy. Yet another category of natural energy becomes available at unpredictable times, such as that in volcanoes, hurricanes and tornadoes, and falling trees. A fascinating example of this type[3] is the energy in rocks tumbling down the San Gabriel Mountains into the backyards of suburban Los Angeles.

Hydrocarbons (coal, oil, and gas) are not considered renewable because they require eons to form, and indeed, they are being used up much faster than they form.

Primary Storage

Primary storage involves extraction or manufacture and a storage site or vessel, for example: 1) coal mined, transported, and stored at a power plant; 2) gasoline in the fuel tank of an automobile; and 3) uranium ore mined and processed into pellets of fuel. Examples of manufactured products containing primary stored energy are gunpowder, candles, lamp oil, and, indeed, non-rechargeable flashlight batteries (cells). Taking this one step further, the energy used in manufactured durable goods is stored when the goods are inventoried. This has the advantage of a very low dissipation rate but the disadvantage of frozen capital.

Secondary Methods

Storing on the Grid

References to storage on the electric grid conjure visions of electrons running about on transmission lines, looking for a place to light. In fact, the term merely means storage facilities connected to the grid, and it is in vogue because the lack of such storage is an impediment to the development of renewable power.[4] The problem is the need for a stable supply of power combined with the erratic availability of renewables. The solution is threefold: 1) an extended grid from the sources to the users; 2) increased storage facilities to receive energy when available and to supply it when needed; and 3) improved controls to gain access to stored energy and/or to limit the demand by users.

Capacitors

The energy, E, stored in a capacitor is equal to

$$E = \frac{1}{2} C V^2 \tag{3}$$

where the capacitance, C, of a plate capacitor is equal to

$$C = \frac{\varepsilon A}{d} \tag{4}$$

where ε is the permittivity of the dielectric between the plates, V is the voltage across the plates, A is the area of each plate, and d is the distance between the plates.

Evidently then, a high-energy capacitor requires a high voltage, a high permittivity, and a large area-to-separation ratio.

Conventional electrostatic capacitors have two metal plates separated by a dielectric. Electrolytic capacitors[5] are made up of two aluminum foils. The anode foil is coated with aluminum oxide, Al_2O_3, and the cathode is the other foil with a spacer saturated with electrolyte facing the oxide layer. Many types of both aqueous and non-aqueous electrolytes have been used.

Electric double-layer capacitors (EDLCs), also known hyperbolically as supercapacitors or ultracapacitors, have, between the electrodes, two layers of a sponge-like material with a separator that takes the place of plates. The material is typically activated carbon, although carbon nanotubes are envisioned for the purpose. These materials have thousands as much area per volume as conventional dielectrics and the distance between charges is submicroscopic, thus making their capacitance and energy density that much greater than that of electrolytic capacitors. Because of the small spacing, the voltage is only 2–3 V.

A major use of EDLCs is in combination with batteries. While batteries store more energy, EDLCs can charge and discharge much faster. Thus the EDLC is used in cameras to supply the flash and in electric cars to assist in hill climbing and acceleration or to absorb braking energy.

Chemical Batteries

The lead–acid battery has several disadvantages. It is heavy, toxic, maintenance-intensive, and short-lived. However, the technology is well-developed, and the materials are inexpensive, and this trumps all the disadvantages. The positive electrode (cathode) is PbO_2, lead dioxide, the negative electrode (anode) is Pb, lead, and the electrolyte is sulfuric acid, H_2SO_4. The overall reaction is

$$PbO_2 + Pb + 2H_2SO_4 \leftrightarrow 2PbSO_4 + 2H_2O \tag{5}$$

moving to the right during discharge and to the left during charging. Thus both electrodes and the electrolyte are used up during discharge, forming lead sulfate and water. The reaction can be reversed by application of a reverse current at the electrodes. A half-reaction occurs at each electrode, with electrons being produced at the anode and absorbed at the cathode.

A large lead–acid battery test facility[6] was built by Southern California Edison in Chino, California, and went into operation in 1998. Four strings of series-wired cells were housed in each of two large buildings, with a smaller building in between to house the converter and controls. Each string consisted of 1032 two-volt cells tied in series to provide a nominal voltage of 2000 Vdc. The facility was programed for an 80% depth of discharge to provide 20.8 kAh, or 40 MWh plus the energy to run auxiliaries, primarily the air compressor and air handlers. Typical discharge cycles were 5 MW for 8 hr or 10 MW for 4 hr.

Compare these figures to those for an automotive SLI (starting, lighting, ignition) battery that has six 2V cells to provide 40 Ah or 480 Wh. The SLI battery is designed for high current for a short time and for shallow discharge, whereas a traction battery, e.g., a 6, 8, or 12V golf cart battery, is designed for deep discharge and power over long periods.

Compressed Air

In the compressed air energy storage (CAES) method,[7] air is stored at high pressure in an underground cavern for use in a gas turbine. Since a conventional gas turbine consumes about two-thirds of its fuel to compress air at the time of generation, the CAES method benefits by using electricity from the grid at off-peak times for use at on-peak times in lieu of the gas turbine compressor. One improvement to the basic method is a surface pond connected to the cavern that serves to keep the pressure in the cavern constant. When air is stored, water is displaced, and when air is withdrawn, water reenters. Second, between the compressor and the cavern is a thermal storage device to store the heat of compression. When the air is withdrawn, it is reheated.

Example 2. Compute the specific work required to compress air at 1 atm and 20°C to 50 atm in an adiabatic compressor having an isentropic efficiency of 80%. Assuming the air is stored at 20°C in a cavern of 200,000 m^3, what total work is required?

Solution. The specific work of compression is

$$w = \left(\frac{1}{\eta}\right)\left(\frac{k}{k-1}\right)RT_1\left[\left(\frac{P_2}{P_1}\right)^{\frac{k-1}{k}} - 1\right] = \left(\frac{1}{0.80}\right)\left(\frac{1.4}{0.4}\right)(0.287\text{kJ/kg}-\text{K})\left[(20+273)\text{K}\right]$$

$$\left[\left(\frac{50}{1}\right)^{\frac{1.4-1}{1.4}} - 1\right] = 758\text{kJ/kg} \tag{6}$$

$$m = \frac{P_2 V_2}{RT_2} = \frac{50(101.3\text{ kPa})(2\times10^5\text{ m}^3)}{(0.287\text{ kPa}-\text{m}^3/\text{kg}-\text{K})293\text{ K}} = 1.20\times10^7\text{ kg} \tag{7}$$

$$W = \frac{mw}{c_1} = \frac{(1.20\times10^7\text{ kg})(758\text{kJ/kg})}{3.6\times10^6\text{ kJ/MWh}} = 2530\text{ MWh} \tag{8}$$

Flow Batteries

These batteries are so named because the electrolyte flows between the cell and a storage tank for each electrode. In the best-known flow battery the electrolyte is a solution of vanadium and sulfuric acid and the reaction is an oxidation-reduction reaction, hence the name vanadium redox battery[8] (VRB). The cathode electrolyte carries V^{5+} and V^{4+} ions, while the anode electrolyte carries V^{2+} and V^{3+} ions. The reactions during discharge are oxidation at the cathode

$$V^{4+} + V^{5+} + e^{-} \tag{9}$$

and reduction at the anode

$$V^{3+} + e^{-} \rightarrow V^{2+} \tag{10}$$

These reactions are reversed during charging. The electrodes are separated by a permeable ionic membrane to allow hydrogen ions to pass through to complete the circuit.

The electrodes do not take part in the reactions, so they do not deteriorate as in a lead–acid battery. The electrolyte can be recycled indefinitely, eliminating the problem of disposal. The battery can be fully discharged without damage, it can be discharged at high voltage, and it could be "instantly" recharged by replacing the electrolyte.

Flywheels

Originally, flywheels were used to regulate the speed of rotating machinery. By virtue of its large rotational inertia, the flywheel tended to oppose any tendency in the machine to slow down or speed up. Now flywheels are being used as energy storage devices.

Modern energy storage flywheels are massive rotating cylinders supported on stators by magnetically levitated bearings to eliminate bearing wear. The flywheel is operated in a low vacuum environment to reduce drag. Some of the key features of flywheels are low maintenance, long life, and inert material.

The choice of solid steel vs. composite rims is based on the system cost, weight, and size. Steel is denser (heavier) but composites can withstand much higher tip speeds. Actual delivered energy depends on the speed range of the flywheel as it cannot deliver its rated power at very low speeds.

The energy stored in a flywheel can be calculated as follows: The energy of translation of a particle of mass m and velocity v is $1/2mv^2$, so the energy of rotation of a particle of mass m and rotational speed ω rotating about a center at distance r is $1/2m(r\omega)^2$. For a group of particles all at distance R and totaling mass M, the rotational energy $1/2MR^2\omega^2$, is or $1/2I\omega^2$ where $I = MR^2$ is the rotational inertia (or moment of inertia). Therefore to maximize the stored energy, a flywheel should be heavy (large M), big (large R), fast (large ω), and have the mass concentrated at the outer rim (mass at R).

Flywheels are used to provide power to uninterruptible power supplies.[9] They pair well with batteries in this application, since flywheels are tolerant to frequent cycling (which batteries are not), and batteries can provide power for longer periods. Comparative advantages of flywheels over batteries are longer life, smaller footprint, less maintenance, and less stringent housing conditions. The main disadvantage of a flywheel is the short period in which backup power can be supplied because of the continuously decreasing speed of rotation. This can be overcome by introducing special circuitry, but this adds cost and complexity.

Example 3. Consider a modest-sized flywheel of 40 kg mass and 0.3 m diameter rotating at 5000 rpm. Find the stored energy. Assume all the mass to be located at the extremity.

Solution. The rotational speed is (5000 rev/min) (2π)/(60 sec/min) = 523.6 rad/sec. The kinetic energy in the wheel is (1/2) (40 kg) $(0.3\text{ m})^2$ $(523.6\text{ rad/sec})^2$ = 493,500 (kg-m/sec^2)-m (= N-m = J = W-sec). If only 10% of this energy is usable until the flywheel slows to an unacceptable speed, 49,350 W-sec is available, or about 5 kW for 10 sec until a longer-term backup can be brought on line.

Fuel Cells

Fuel cells are not properly an energy storage device, but an energy conversion device, since they take a continuous flow of materials and combine them to produce energy and heat. However, they have a certain analogy with chemical batteries. Fuel cells convert hydrogen and oxygen into electricity and heat. A fuel cell, like a battery, is made up of an anode, a cathode, and an electrolyte. A current is produced when hydrogen ionizes at the anode with the aid of a catalyst. The hydrogen ions pass through the electrolyte to the cathode, where they combine with oxygen to produce water. The electrons released at the anode travel through an electric circuit to the cathode. The overall reaction is

$$2H_2 + O_2 \rightarrow 2H_2O \tag{11}$$

Notice that the reaction only goes to the right, i.e., it is not reversible as in an energy storage device. As in a lead-acid battery, the half-reactions produce electrons at the anode and absorb them at the cathode.

Fuel cells use a variety of electrolytes, including phosphoric acid, molten carbonate, solid oxide, and a proton exchange membrane.

Pumped Hydroelectricity

In this method, water is pumped into an elevated reservoir, with the energy recovered when the water flows back down to a lower reservoir through a turbine generator. The reservoirs may be man-made or natural bodies of water. Reversible turbine/pumps reduce the equipment required. Pumped hydro is particularly advantageous in conjunction with gas- or coal-fired plants, first by demand leveling, allowing the thermal plant to run at its most efficient load, and second by responding quickly to fluctuations in demand, where a thermal plant cannot. Pumped hydro also has been found to pair well with wind power and its variations in availability. Of course pumped hydro can be used on a regular schedule of pumping at night when rates are cheap and recovering by day when rates are high. The energy stored, E, is the potential energy of position, $E = mgh$, where m is the mass of water, g is the acceleration due to gravity, and h is the difference in height between reservoirs. Conversion efficiency is good, with losses depending on efficiency of the motor/generator and pump/turbine, pipe friction, and evaporation/seepage. Dam failures notwithstanding, this method is generally safe. The essential requirement is a site where high and low reservoirs are naturally available or can be constructed at a reasonable cost.

Springs

The force, F, required to stretch or compress a linear spring is $F = kx$, where k is the spring constant and x is the change in length. The energy, E, stored is the integral of force over the change in length, $E = 1/2kx^2$. The torque, τ, required to twist a linear torsion spring is $\tau = k\theta$, where k is the spring constant and θ is the change in angle. The energy, E, stored is the integral of torque over the change in angle, $E = 1/2k\theta^2$.

Torsion springs are typically used in instruments in spiral form. A torsion pendulum is a heavy disk that rotates at the end of a wire in torsion.

It is interesting to note that a bolt is a spring, and the clamping force (i.e., the stored energy) is the stretch in the bolt times its spring constant.

Superconducting Magnets

SMES is the storage of a current, *I*, in a cryogenically cooled inductor with energy *E* equal to

$$E = \frac{1}{2} L I^2 \tag{12}$$

where the inductance of a coil, *L*, is equal to

$$L = \frac{\mu N^2 A}{X} \tag{13}$$

where μ is the permeability of the core, *N* is the number of turns in the coil, *A* is the cross-sectional area of the coil, and *X* is the length of the coil.

Evidently then, a high-energy SMES requires high current, a highly permeable core, many turns, and a large cross-sectional area-to-length ratio.

The energy is maintained without dissipation by cryo-genically cooling the coil.[10] Helium is used as a coolant because its freezing temperature is 4.2°C above absolute zero. The salient characteristic of SMES is fast response with extremely high power. Thus a prominent use would be to stabilize short-term faults in grid power. Advantages cited for this method of energy storage are high conversion efficiency because of low resistance losses, and low maintenance because of few moving parts.

Thermal Storage

A wag has suggested that the thermos bottle is the greatest invention of the human mind because it keeps hot things hot and cold things cold. How does it know? Three examples of thermal storage are discussed here: chilled water, regenerators, and Trombe walls.

The energy consumed by a chiller can be stored in the form of chilled water or ice in an insulated tank, usually as a play on varying electric rates. Water is an excellent medium for this since it is inexpensive but also because it has a high specific heat, giving a good energy-to-volume ratio. For instance, an energy-efficient chiller will operate at 0.5 kW/T, (T=ton) so that over 6 hr, a 100T chiller will make

$$(100\text{T})(12{,}000\ \text{Btu/hr} - \text{T})(6\text{hr}) = 7.2\ \text{MMBtu} \tag{14}$$

of cooling for

$$(0.5\ \text{kW/T})(100\ \text{T})(6\ \text{hr.}) = 300\ \text{kWh of electrity}$$

This energy can be stored by chilling water from 70°F to 35°F, which will require a volume of water equal to

$$\frac{7.2\ \text{MM Btu}}{(1\ \text{Btu/1 bm} - {}^\circ\text{F})(70 - 35^\circ\text{F})(62.41\text{bm/ft}^3)} = 3300\text{ft}^3 \tag{15}$$

which is equivalent to a cubical tank about 15 ft on a side. The energy would typically be recovered by running the chilled water through an air handler to provide air conditioning. Good insulation, the small temperature difference between the medium and the surroundings of the tank, and the relatively short storage time would keep the storage loss low.

A regenerator is a regenerative heat exchanger, in which energy is stored in the material of the heat exchanger itself. A regenerator cycles such that half of it is heated by hot flue or exhaust gases while cool combustion air is drawn in and heated in the other half. In the second half of the cycle, cool air enters through the previously heated material of the regenerator. For example, older industrial furnaces have two chimneys, where combustion air enters in one while flue gases heat the other. Then the paths are switched so that new air is heated as it descends the first, heated chimney. A newer type of regenerator is a slowly rotating heat wheel, with half of it being heated by exiting flue gas and the other half heating the incoming air. The paths of the intake and exhaust remain constant while the heat-storing element moves.

A Trombe wall[11] (after the engineer Felix Trombe who used it extensively) is a form of passive solar energy storage. It is most commonly used in small buildings in combination with other passive energy-saving methods. The wall is a south-facing thick construction of brick or concrete with a layer of glass on the outside and an air space between. The greenhouse effect allows solar radiation to heat the wall but blocks the longer-wavelength, lower-temperature radiation from the wall. The wall stores heat during the day and releases it into the living space at night. Vents with one-way flaps pierce the wall at the top and bottom, whereby cool air from the room enters the air space at the bottom, is heated, rises to the top of the space, and reenters the room through the top vent. The flaps block reverse flow.

Conclusion

Stored energy is considered here as involving some human activity, ruling out natural energy contained in the winds or the tides or the Earth's crust. Primary storage is advantageous when the energy source is cheap or available. A disadvantage of acquiring in advance of need is that energy leaks from the medium.

Most secondary methods involve converting from and to electrical energy, so the electric grid becomes a primary factor. The practice of "buy low, sell high" leads to storing energy during off-peak hours and recovering it during on-peak hours. The availability and location of renewables causes that energy to be transported from the source to the point of use and stored until most valuable. Finally, the grid is not completely reliable, so energy is stored to carry critical services through outages.

Since energy is lost in every round trip to and from a storage medium, ongoing efforts to improve conversion efficiencies are of great economic importance.

Glossary

Bagasse	the solid residue of a plant, e.g., sugarcane, after the juice has been extracted
Battery	a collection of chemical cell
Btu	British Thermal Unit, 1055 J or 778 ft-lbf
CAES	See Compressed Air Energy Storage
Chemical Cell	a unit that stores chemical energy and converts it from and electric energy
Chemical Energy	energy stored or released in a chemical reaction
Chiller	a mechanical device that cools a refrigerant in a thermodynamic cycle
Combustion	a chemical reaction that oxidizes fuel and releases thermal energy
Compressed Air Energy storage	large-scale secondary storage, used in gas turbines
Demand Leveling	what the power company does to reduce the peak demand
EDLC	electric double-layer capacitor

Energy	the capacity to do work, usual units Btu or kWh
Ethane	C_2H_6, a minor component of natural gas
Ethanol	C_2H_6O, ethyl alcohol, sometimes used as a component of automotive fuel
Ethanol	ethyl alcohol, grain alcohol, C_2H_5OH
Flow battery	a battery in which the electrolyte flows between the cell and a storage tank
Flywheel	a device that spins to store kinetic energy
Fuel Cell	an energy conversion device that generates energy from hydrogen and oxygen
Gasoline	primarily octane
Geothermal Energy	energy associated with heat in the Earth's crust
Heat Energy	more properly called thermal energy
Kinetic	energy associated with motion
Latent Energy	energy associated with a change in phase, also called latent heat
Load Leveling	what the rate payer does to reduce the peak demand
Mechanical	Energy kinetic or potential energy
Methane	CH_4, the main component of natural gas, 1000 Btu/ft3
MMBtu	million Btu
MMBtu/hr	million Btu per hour
Natural Gas	a mixture of hydrocarbon gases, primarily methane
Natural Energy	renewable energy
Octane	C_8H_{18}, the main component of gasoline
Potential Energy	energy associated with the position of a system or its component parts
Regenerator	an energy-storing heat exchanger
Renewable Energy	energy endlessly available in nature
Saw Grass	a sedge, with leaves having edges set with small sharp teeth, a source of ethanol
Sensible Energy	energy associated with a change in temperature, also called sensible heat
SMES	See superconducting magnetic energy storage
Superconducting Magnetic Energy Storage	energy stored by energizing (inducing a magnetic current in) a magnet
T	ton of cooling, 12,000 Btu/hr
Thermal Energy	sensible plus latent internal energy
Trombe Wall	a building component that stores heat from the sun

References

1. Akhil, A.; Swaminathan, S.; Sen, R.K. Cost Analysis of Energy Storage Systems for Electric Utility Applications, available at http://infoserve.sandia.gov/cgi-bin/techlib/access-control.pl/1997/970443.pdf (accessed January 2009).
2. Zink, J.C. Who says you cant [sic] store electricity? *Power Engineering,* March 1997, available at http://pepei.pennnet.com/articles/article_display.cfm?article_id=44573 (accessed October 2008).
3. McPhee, J. *The Control of Nature,* Farrar, Strauss, and Giroux: New York, 1989.
4. Talbot, D. Lifeline for renewable power, *Technology Review,* January/February 2009.
5. Schindall, J. The Charge of the Ultra-Capacitors, available at http://www.spectrum.ieee.org/nov07/5636 (accessed February 2009).
6. Rodriguez, G.D. Operating Experience with the Chino 10 MW/40 MWh Battery Energy Storage Facility, available at http://ieeexplore.ieee.org/ielx5/852/2490/00074691.pdf? arnumber=74691 (accessed February 2009).
7. El-Wakil, M.M. *Powerplant Technology*; McGraw–Hill Primus: New York, 2002.
8. Miyake, S.; Tokuda, N. Vanadium Redox Flow Battery (VRB) for a Variety of Applications, available at http://www.electricitystorage.org/pubs/2001/IEEE_PES_Sum-mer2001/Miyake.pdf (accessed December 2008).

9. Flywheel Energy Storage, available at http://www1.eere.energy.gov/femp/pdfs/fta_flywheel.pdf (accessed February 2009).
10. Boyes, J.D. Technologies for Energy Storage, available at http://www.osti.gov/bridge/servlets/purl/756082-Q5CBAv/webviewable/756082.pdf (accessed August 2009).
11. Cengel, Y.A. *Heat and Mass Transfer*, 3rd Ed.; McGraw Hill: New York, 2007.

12

Fossil Fuel Combustion: Air Pollution and Global Warming

Dan Golomb

Introduction

Currently, fossil fuels supply about 86% of global primary energy consumption (39% oil, 24% coal, and 23% natural gas), providing energy for transportation, electricity generation, and industrial, commercial and residential uses. The air emissions of fossil fuel combustion are transported by winds and dispersed by atmospheric turbulence, eventually falling or migrating to the surface of the earth or ocean at various rates. While in the atmosphere, pollutants cause considerable harmful effects on human health, animals, vegetation, structures, and aesthetics.

In recent decades, it has become evident that rising levels of atmospheric carbon dioxide (and other greenhouse gases) have already warmed the earth's surface slightly. It is predicted that, with continuous and increasing use of fossil fuels, the warming trend will increase. Most of the increase in carbon dioxide levels is a direct consequence of fossil fuel combustion.

The goal of this entry is to describe the characteristics of fossil fuel-generated air pollutants, including carbon dioxide and other greenhouse gases, their transport and fate in the atmosphere, and their effects on human health, the environment, and global climate change.

Ambient and Emission Standards of Air Pollutants

Air pollutants, when they exceed certain concentrations, can cause acute or chronic diseases in humans, animals, and plants. They can impair visibility, cause climatic changes, and damage materials and structures. Regulatory agencies in most developed countries prescribe ambient standards (concentrations) for air pollutants. The standards are set at a level below which it is estimated that no harmful effect will ensue to human health and the environment. On the other hand, if the standards are exceeded, increased human mortality and morbidity, as well as environmental degradation, is expected. For example, in the United States, the Environmental Protection Agency (EPA) sets National Ambient Air Quality Standards (NAAQS) for six pollutants, the so-called criteria pollutants: carbon monoxide (CO), nitrogen dioxide (NO_2), ozone (O_3), sulfur dioxide (SO_2), particulate matter (PM), and lead (Pb). The standards are listed in Table 1. In the past, particulate matter was regulated regardless of size (the so-called Total Suspended Particles, TSP). Starting in 1978, particles were regulated with an aerodynamic diameter less than 10μm. This is called PM_{10}. In 1997, a new standard was introduced for particles with an aerodynamic diameter less than 2.5μm. This is called $PM_{2.5}$. The reason for regulating only small particles is that these particles can be lodged deep in the alveoli of the lungs, and hence are detrimental to our health, whereas the larger particles are filtered out in the upper respiratory tract. While there is a standard for lead concentrations in the air, this pollutant is no longer routinely monitored. With the phasing out of leaded gasoline in the 1970s, lead concentrations in the air steadily declined, and allegedly, in the United States, airborne lead no longer poses a health hazard. (Of course, lead in its other forms, such as in paint, pipes, groundwater, solder, and ores, is still a hazard.)

In addition, the EPA sets emission standards for stationary and mobile sources. Stationary sources include power plants, incinerators, steel, cement, paper and pulp factories, chemical manufacturers, refineries, and others. Mobile sources include automobiles, trucks, locomotives, ships, and aircraft. As an example, Table 2 lists the U.S. emission standards (New Source Performance Standards) for vfossil-fueled steam generators, which include large fossil-fueled power plants and industrial boilers that were constructed after 1970. The emission standards are set in terms of mass emitted per fuel heat input (g/GJ). However, in 1978, new regulations were implemented. Instead of numerical emission standards, EPA prescribed *emission control technologies*, the so-called best available control technologies (BACT), purporting to reduce emissions to a minimum with practical and economic pollution abatement devices. For example, current BACT for sulfur oxides is a "scrubber," usually employing a slurry of pulverized limestone ($CaCO_3$) counter-flowing to the flue gas that contains the sulfur oxides. For control of nitric oxides (NO_x), the current BACT is the low-NO_x-burner (LNB). Because the LNB reduces

TABLE 1 U.S. 2000 National Ambient Air Quality Standards (NAAQS)

Pollutant	Standard	Averaging Time
Carbon monoxide	9 ppm (10 mg/m^3)	8-h[a]
	35 ppm (40 mg/m^3)	1-h[a]
Lead	1.5 μg/m^3	Quarterly Average
Nitrogen dioxide	0.053 ppm (100 μg/m^3)	Annual (arithmetic mean)
Particulate matter (PM10)	50 μg/m^3	Annual[b] (arithmetic mean)
	150 μg/m^3	24-h[a]
Particulate matter (PM2.5)	15.0 pg/m^3	Annual[c] (arithmetic mean)
	65 μg/m^3	24-h[d]
Ozone	0.08 ppm	8-h[e]
Sulfur oxides	0.03 ppm	Annual (arithmetic mean)
	0.14 ppm	24-h[a]
	–	3-h[a]

[a]Not to be exceeded more than once per year.

[b]To attain this standard, the 3-year average of the weighted annual mean PM_{10} concentration at each monitor within an area must not exceed 50 μg/m^3.

[c]To attain this standard, the 3-year average of the weighted annual mean PM2.5 concentrations from single or multiple community-oriented monitors must not exceed 15.0 μg/m^3.

[d]To attain this standard, the 3-year average of the 98th percentile of 24-h concentrations at each population-oriented monitor within an area must not exceed 65 μg/m^3.

[e]To attain this standard, the 3-year average of the fourth-highest daily maximum 8-h average ozone concentrations measured at each monitor within an area over each year must not exceed 0.08 ppm.

TABLE 2 U.S. New Source Performance Standards (NSPS) for Fossil Fuel Steam Generators with Heat Input >73 MW

Pollutant	Fuel	Heat Input (g/GJ)
SO_2	Coal	516
	Oil	86
NO_2	Gas	86
	Coal (bituminous)	260
PM[a]	Coal (subbituminous)	210
	Oil	130
	Gas	86
	All	13

[a]PM, particulate matter. For PM emissions an opacity standard also applies, which allows a maximum obscuration of the background sky by 20% for a 6-min period.
Source: Data from EPA. Standards of performance for new stationary sources. Electric steam generating units. *Federal Register,* 45, February 1980, 8210–8213.

NO_x emissions only up to 50%, new sources may be required to employ more efficient NO_x-reducing technologies, such as selective catalytic reduction (SCR) or selective non-catalytic reduction (SNCR), based on ammonia or urea injection into the flue gas. For control of particulate matter, the current prescribed technology is the electrostatic precipitator (ESP). Because the ESP is not efficient in removing sub-micron and 1–2 pm particles, and in view of the new $PM_{2.5}$ standard, new sources may be required to install instead of an ESP a Fabric Filter (FF), also called a bag house. The FF consists of a porous fabric or fiberglass membrane that filters out efficiently the smallest particles, albeit at an increased cost and energy penalty compared to the ESP. The detailed description of the workings of these emission control technologies is beyond the scope of this entry; the reader is referred to the excellent handbooks on the subject.[1,2]

Table 3 lists the U.S. emission standards in units of mass emitted per length traveled (g/km) for passenger cars and light trucks for the different model years from 1968 to 1994. At present, the 1994 standards are still in effect. Notice that light truck standards are not as strict as those for light duty vehicles. This dichotomy is very controversial, because the light truck category includes sport utility vehicles (SUV), minivans, and pick-up trucks, which are mostly used for personal and not for cargo transport. Therefore, their emission standards ought to be equal to those for passenger cars. Until now, these vehicles captured the majority of sales in the United States. (With rising gasoline prices, this trend may be reversed.) The achievement of the emission standards relies on emission control technologies. For unleaded gasoline-fueled vehicles, the prevailing control technology is the three-way catalytic converter, which simultaneously reduces carbon monoxide (CO), nitric oxides (NO_x), and fragmentary hydrocarbon (HC) emissions. Unfortunately, the catalytic converter would not work on diesel-fueled vehicles, because the relatively high sulfur and particle emissions from diesel engines would poison the catalyst. Automobile and truck manufacturers are intensively investigating possible technologies that would reduce emissions from diesel-fueled autos and trucks.

In addition to the aforementioned criteria pollutants, one finds in the air a host of other gaseous and particulate pollutants, generally designated as hazardous air pollutants (HAP), or simply *air toxics.* The EPA has identified 189 HAPs. Of course, not all HAPs are related to fossil fuel usage. Examples of fossil fuel HAPs are products of incomplete combustion (PIC), volatile organic compounds (VOC), polycyclic aromatic hydrocarbons (PAH), toxic metals (e.g., mercury, cadmium, selenium, arsenic, vanadium, etc.) Many of the fossil fuel–related HAPs are found as condensed matter on particles emitted by stationary sources (e.g., fly ash from power plants) or mobile sources (e.g., exhaust smoke from trucks). While HAPs may be more harmful to our health than the criteria pollutants (some of them are carcinogens), it is difficult to establish a dose–response relationship. Therefore, instead of setting HAP emission standards,

TABLE 3 U.S. Federal Vehicle Emission Standards

	Light-duty Vehicles (Auto)				Light-duty Trucks (Gasoline)			
Model Year	HC (g/km)	CO (g/km)	NO (g/km)	PM (g/km)	HC (g/km)	CO (g/km)	NOx (g/km)	PM (g/km)
1968	2.00	20.50						
1971[a]	2.90	29.20	2.49					
1974	2.11	24.20	1.86					
1977	0.93	9.32	1.24					
1978	0.93	9.32	1.24		1.24	12.40	1.93	
1979	0.93	9.32	1.24		1.06	11.20	1.43	
1980	0.25	4.35	1.24		1.06	11.20	1.43	
1981	0.25	2.11	0.62		1.06	11.20	1.43	
1982	0.25	2.11	0.62	0.37	1.06	11.20	1.43	
1985	0.25	2.11	0.62	0.37	0.50	6.21	1.43	0.99
1987	0.25	2.11	0.62	0.12	0.50	6.21	1.43	1.62
1988	0.25	2.11	0.62	0.37	0.50	6.21	0.75	1.62
1994	0.25	2.11	0.25	0.05	0.50	3.42	0.60	0.06

[a]Test method changed in 1971.

the EPA mandates that specific control technologies be installed on major emitting sources. These are called maximum achievable control technologies (MACT). For example, for toxic volatile gases, the most often used control technologies are physical adsorption on porous adsorbents (usually activated carbon), or chemical absorption in solvents, including water, that have a large absorption capacity for these gases. When the concentration of toxic gases in the effluent is high, secondary incineration of the effluent may be warranted. For detailed descriptions of MACTs, the reader is also referred to the appropriate handbooks.[1,2]

Pollutant Transport and Dispersion

When air pollutants exit a smoke stack or exhaust pipe (called the *sources*), they are transported by winds and dispersed by turbulent diffusion. Winds blow from high pressure toward low-pressure cells at speeds that depend on the pressure gradient. Because of the Coriolis force, wind trajectories are curvilinear in reference to fixed earth coordinates, although within a relatively short (few to tens of km) distance, wind trajectories can be approximated as linear. Winds have a horizontal and vertical component. Over flat terrain, the horizontal component predominates; in mountainous and urban areas with tall buildings, the vertical component can be significant, as well at the land/sea interface.

In the bottom layer of the atmosphere, called the troposphere, the temperature usually declines with altitude. For a dry atmosphere, the temperature gradient is −9.6°C/km. This is called the *dry adiabatic lapse rate.* In a moist atmosphere, the temperature gradient is less. Under certain conditions (e.g., nocturnal radiative cooling of the earth surface), the temperature gradient may even become positive. This is called *inversion.* The global and temporal average temperature gradient in the troposphere is −6.5°C/km. When a negative temperature gradient exists, upper parcels of the air are denser than lower ones, so they tend to descend, while lower parcels buoy upward, giving rise to eddy or turbulent mixing. The more negative the temperature gradient, the stronger the turbulence, and the faster the dispersion of a pollutant introduced into the troposphere. Wind shears also cause turbulence. A wind shear exists when, in adjacent layers of the atmosphere, winds blow in different directions and speeds.

FIGURE 1 Los Angeles smog. Note the low inversion beneath which the smog accumulates. Above the inversion, the air is relatively clear. Photo by the South Coast Air Quality Management District.

During an inversion, there is little turbulence, and a pollutant will disperse very slowly. Inversions may also occur aloft, that is, a negative temperature gradient exists at the bottom, followed by a positive gradient above. The layer up to the altitude at which the inversion occurs is called the *mixing layer* and the altitude at the inversion is called the *mixing height.* The shallower the mixing layer, the greater chances for air pollution episodes to develop, because pollutants emitted near the ground are confined to the shallow mixing layer. This often occurs in cities such as Los Angeles, Houston, Atlanta, Salt Lake City, Denver, Mexico City, Sao Paulo, Athens, Madrid, Rome, and Istanbul. Some of these cities are surrounded by mountain chains. In the basin of the mountain chain, or in the valley, the mixing layer is shallow, and the winds in the layer are usually weak, leading to poor ventilation. During the morning rush hour, pollutants are emitted into the shallow mixing layer, where they are concentrated because of the "lid" imposed by the inversion aloft. Later in the day, when solar radiation breaks up the inversion, the pollutants disperse to higher altitude, and the pollutant concentration becomes more diluted. The photo in Figure 1 shows a pollution episode in Los Angeles. The mixing height extends only to the middle of the tall building. The lower part is obscured by the smog, the upper part, which is above the mixing layer, is in relatively clear air.

Air Quality Modeling

The estimation of ambient pollutant concentrations in space and time due to emissions of single or multiple sources is called air quality modeling (AQM), or source–receptor modeling (SRM). The basic ingredients of AQM are the emission strengths of the sources, meteorological conditions, and solar irradiation. Air quality models are of the trajectory-type, where the coordinate system moves together with the plume of pollutants, or grid-type, where the coordinate system is fixed over an area, and the emission strengths and meteorological variables are inserted in each grid of the model domain. Some models consider only conservative pollutants, where the emitted pollutant does not change en route to the receptor. Other models consider chemical transformation processes and "sinks," e.g., dry and wet deposition to the ground. Air quality models where transformation and deposition processes need to be considered are acid deposition, regional haze, and photo-oxidants.

Acid Deposition

While commonly called acid rain, acid deposition is a better term, because deposition can occur both in the wet and dry form. The ingredients of acid deposition are sulfuric and nitric acids. The primary pollutants are SO_2 and NO_x, which is the sum of NO and NO_2 molecules. Both SO_2 and NO_x result from fossil fuel combustion. Sulfur is a ubiquitous ingredient of coal and petroleum. When these fuels are burned in air, SO_2 is emitted from a smoke stack or the exhaust pipe of a vehicle. Coal and petroleum also contain nitrogen in their molecular make-up, resulting in NO_x emissions when these fuels are combusted in air. In addition, some NO_x is formed from the recombination of air oxygen and nitrogen at the high flame temperatures. So, even the combustion of natural gas, which has no nitrogen in its molecular make-up, produces some NO_x

Primary emitted SO_2 and NO_x are transformed in the atmosphere to sulfuric and nitric acids. The resulting acids can be either deposited in the dry phase on land or water, a process called dry deposition, or scavenged by falling rain drops or snow flakes, resulting in wet deposition.[3] Acid deposition modeling is successful in predicting the amount of acid deposition given the emission strength of the precursors SO_2 and NO_x.[4] From these models, it was concluded that acid deposition within a geographic domain is approximately linearly dependent on SO_2 and NO_x emission strength in that domain, so a certain percentage of precursor emission reduction results in a proportional deposition reduction of sulfuric or nitric acid. Indeed, in countries and continents where serious curtailments of precursor emissions have been made, a proportional reduction of acid deposition occurred. In the United States, as a result of reducing emissions of SO_2 by approximately one-half since the enactment of the Clean Air Act of 1990, sulfuric acid deposition has declined by approximately one-half.[5] Nitric acid deposition has not fallen appreciably, because the control of NO_x is much more difficult to accomplish than the control of SO_2, especially from dispersed sources such as commercial and residential boilers and furnaces, automobiles, and diesel trucks.

Regional Haze

Small particles (also called fine particles or aerosols), less than 1–2 μm in diameter, settle very slowly on the ground. Small particles can be either in the solid or liquid phase. For example, fly ash or smoke particles are solid; mist is liquid. Small particles can be of natural origin (e.g., volcanic dust, wind blown soil dust, forest and brush fires) or of anthropogenic origin (e.g., fly ash, diesel truck smoke). They can be emitted as primary particles or formed by transformation and condensation from primary emitted gases. They can travel hundreds to thousands of kilometers from their emitting sources. The particles can envelope vast areas, such as the northeastern or southwestern United States, southeastern Canada, western and central Europe, and southeastern Asia. Satellite photos often show continental areas covered with a blanket of haze. The haze may extend far out over the ocean. This phenomenon is called regional haze. Small particles are efficient scatterers of sunlight. Light scattering prevents distant objects from being seen. This is called visibility impairment. Figure 2 shows an encroaching haze episode on a mountain chain in Vermont. As the haze thickens, distant mountain peaks are no longer visible, and eventually neighboring peaks disappear. Increasing concentration of particles in urbanized parts of continents causes the loss of visibility of the starlit nocturnal sky. These days, small stars, less than the fifth order of magnitude, rarely can be seen from populated areas of the world.

The composition of fine particles varies from region to region, depending on the precursor emissions. In the northeastern United States, central Europe, and southeastern Asia, more than half of the composition is made up of sulfate particles, due to the combustion of high sulfur coal and oil. The rest is made up of nitrate particles, carbonaceous material (elemental and organic carbon), and crustal matter (fugitive particles from soil, clay, and rock erosion).

FIGURE 2 Regional haze in the Vermont green mountains. Note that as the pollution episode progresses, adjacent peaks are no longer visible. Photo by the Vermont agency of environmental conservation.

Photo-Oxidants

The family of photo-oxidants includes tropospheric ozone, O_3 (the bad ozone), ketones, aldehydes, and nitrated oxidants, such as peroxyacetylnitrate (PAN) and peroxybenzoylnitrate (PBN). The modeling of photo-oxidants is more complicated than that of acid deposition.[6] Here, the primary precursor is NO_x, which as mentioned previously, is emitted because of fossil fuel combustion. A part of NO_x is the NO_2 molecule, which splits (photo-dissociates) by solar ultraviolet and blue photons into NO and atomic oxygen. The photo-dissociation rate is dependent on solar irradiation, which, in turn, is dependent on latitude, season, time of day, and cloudiness. Atomic oxygen combines with molecular oxygen to form O_3. The NO that is formed in the photo-dissociation is quickly re-oxidized into NO_2 by peroxy radicals, RO_2, present in the polluted atmosphere. The peroxy radicals are formed from VOCs that are emitted as a consequence of incomplete combustion of fossil fuels, or from evaporation and leakage of liquid fossil fuels and solvents. The VOCs are oxidized in a complicated sequence of photo-chemical reactions to the peroxy radicals. The oxidation rates of VOCs are also dependent on solar irradiation and on the specific VOC molecule. Long-and branch-chained hydrocarbons (e.g., *n*-octane and *iso*octane) are more reactive then short-and straight-chained ones (e.g., methane and ethane); unsaturated hydrocarbons (e.g., ethene) are more reactive than saturated ones (e.g., ethane). Aromatic hydrocarbons (e.g., benzene) are more reactive than aliphatic ones (e.g., hexane), and so on.

Thus, photo-oxidants have two kinds of precursors, NO_x and VOCs, which make abatement of these secondary pollutants, as well as their modeling, so complicated. First of all, as mentioned previously, complete NO_x emission control is difficult to accomplish, because in addition to coming from large stationary sources, NO_x is emitted from a myriad of dispersed sources, such as home and commercial furnaces and boilers, automobiles, trucks, off-road vehicles, aircraft, locomotives, and ships. In principle, anthropogenic VOCs could be substantially controlled, for example, by ensuring complete combustion of the fossil fuel, or with the catalytic converter on automobiles. But, not all VOCs are of anthropogenic origin. Trees and vegetation emit copious quantities of VOCs, such as terpenes and pinenes, which are pleasant smelling, but they do participate in photo-chemical reactions that produce photo-oxidants. Even though great effort and expenses are being made in many developed countries to control precursors, photo-oxidant concentrations over urban-industrial continents have improved only slightly, if at all. In less developed countries that do not have the means of controlling photo-oxidant precursors, their concentrations are on a steady increase.

Global Warming

Of all environmental effects of fossil fuel usage, global warming, including its concomitant climate change, is the most perplexing, potentially most threatening, and arguably most intractable problem. It is caused by the ever-increasing accumulation in the atmosphere of carbon dioxide (CO_2) and other gases, such as methane (CH_4), nitrous oxide (N_2O), and chloro-fluoro-carbons (CFC), collectively called greenhouse gases (GHG). Atmospheric aerosols, natural as well as anthropogenic, also may contribute to global warming.

The term greenhouse effect is derived by analogy to a garden greenhouse. There, a glass-covered structure lets in the sun's radiation, warming the soil and plants that grow in it, while the glass cover restricts the escape of heat into the ambient surroundings by convection and radiation. Similarly, the earth atmosphere lets through most of the sun's radiation, which warms the earth surface, but the GHGs and some aerosols trap outgoing terrestrial infrared (IR) radiation, keeping the earth's surface warmer than if the GHGs and aerosols were absent from the atmosphere.

A schematic of the greenhouse effect is represented in Figure 3. To the left of the schematic is the atmospheric temperature structure, which defines the various spheres. In the troposphere, the temperature decreases on average by 6.5°C/km. This sphere is thoroughly mixed by thermal and mechanical turbulence, and this sphere contains most of the air pollutants mentioned in the previous sections. In the stratosphere, the temperature increases steadily. Because of the positive temperature gradient, this sphere is very stable with very little turbulence and mixing. At about 50–60 km height, the temperature declines again with altitude, giving rise to the mesosphere. Finally, above 90–100 km, the temperature gradient reverses itself, and becomes positive. The highest sphere is called the thermosphere. Its temperature can reach hundreds to one thousand degrees, depending on solar radiation intensity that directly heats this sphere.

The average solar radiation that impinges on the top of the atmosphere is about 343 W/m^2. This is the annual, diurnal, and spatial average irradiation. Of this irradiation, currently about 30% (103 W/m^2) is immediately reflected into space by land, ocean, icecaps, and clouds, and scattered into space by atmospheric molecules. The reflected and scattered sunlight is called albedo. The albedo may not remain constant over time. With increased melting of the ice caps, and increased cloud cover, in part due to anthropogenic influences, the albedo may change over time. The remaining 70% of solar irradiation heats the earth's surface, land, and oceans. Currently, the global average surface temperature is 288 K (about 15°C). A body (the so-called black body) that is heated to 288 K radiates 390 W/m^2. The earth's surface radiation occurs in the far IR. This is called earth shine. A part of the earth shine is reflected back to the earth's surface by clouds and aerosols; another part is first absorbed by certain gaseous molecules, and then re-radiated back to the surface. The absorption/re-radiation occurs by poly-atomic molecules, including water vapor and the GHGs: CO_2, CH_4, N_2O, O_3, CFCs, and others. The reflection and re-radiation to the earth's surface of the outgoing terrestrial IR radiation is causing the earth's surface to become warmer than it would be merely by solar irradiation. This is the greenhouse effect. With increasing concentrations of anthropogenic GHGs and aerosols, the earth's surface temperature will

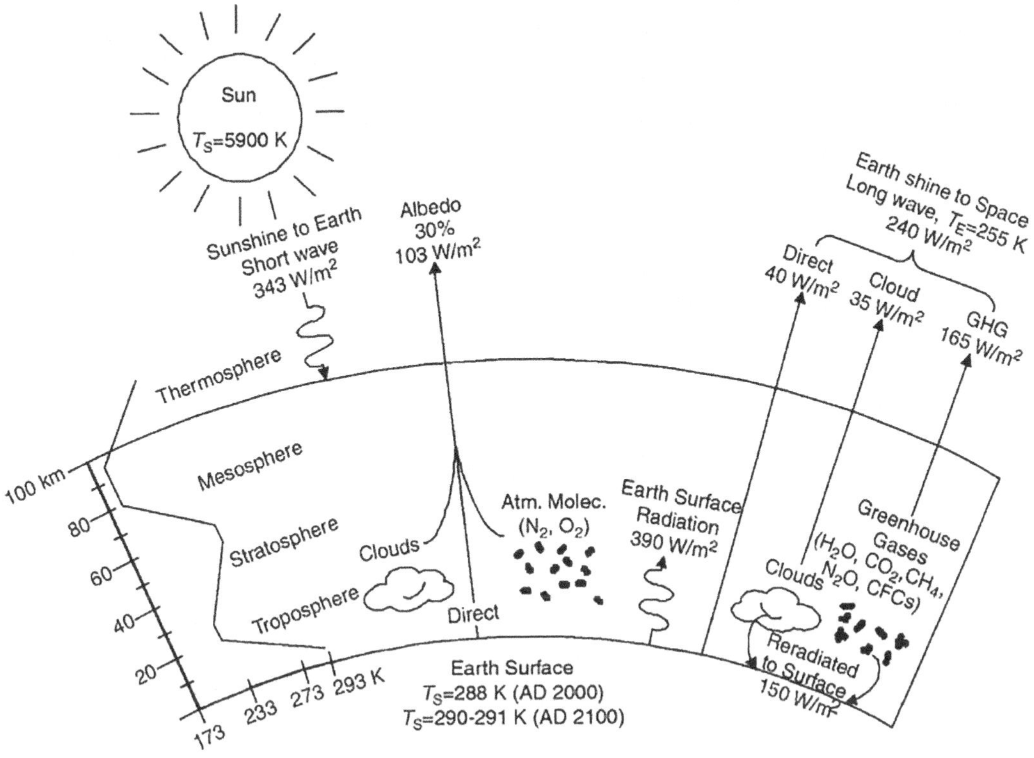

FIGURE 3 Schematic of the greenhouse effect.

increase. It should be emphasized that the greenhouse effect is not due to trapping of the incoming solar radiation, but outgoing terrestrial IR radiation. In fact, because of the trapping of earth shine by GHGs, the upper layers of the atmosphere (primarily the stratosphere) will become colder, not warmer.[7]

The extent of global warming can be predicted by radiative transfer models. These models include the radiative properties of GHGs and their distribution in the earth's atmosphere, as well as the temperature and pressure gradients in the atmosphere. There is general agreement among the models as to the extent of surface warming due to GHG absorption/re-radiation, called radiative forcing. Based on the models, it is predicted that the average earth surface temperature will increase as shown in Figure 4. The middle, "best," estimate predicts a rise of the earth's surface temperature by the end of the 21st century of about 2°C; the "optimistic" estimate predicts about 1°C, and the "pessimistic" estimate predicts about a 3°C rise. The optimistic estimate relies on the slowing of CO_2 and other GHG emissions; the pessimistic estimate relies on "business-as-usual," i.e., on the continuing rate of growth of CO_2 and other GHG emissions, and the "best" estimate is somewhere in between.[8] If the GHG concentrations increase in the atmosphere at their current rate, by the year 2100, CO_2 will contribute about 65% to global warming, CH_4 15%, N_2O 10%, and CFC about 5%–10%. (By international conventions, CFCs are being phased out entirely. But because of their very long lifetime in the atmosphere, they still will contribute to global warming by the year 2100.)

In addition to radiative forcing, global warming may be enhanced by the so-called feedback effects. For example, water vapor is a natural GHG. When the temperature of the ocean surface increases, the evaporation rate will increase. As a consequence, the average water vapor content of the atmosphere will increase. This causes more absorption of the outgoing infrared radiation and more global warming. Furthermore, increased evaporation may cause more cloud formation. Clouds and aerosols also can

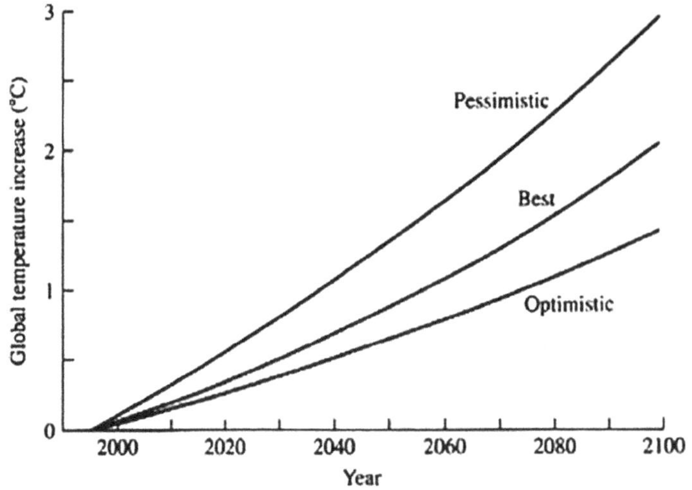

FIGURE 4 Projected trend of the earth surface temperature increase. Upper curve: pessimistic scenario with no emission curtailment; lower curve: optimistic scenario with significant emission curtailment; middle curve: in-between scenario.
Source: Cambridge University Press (see Intergovernmental Panel on Climate Change).[8]

trap outgoing terrestrial radiation, further increasing global warming. Melting ice caps and glaciers decrease the reflection of incoming solar radiation (reduced albedo), which also increases global warming. The prediction of the feedback effects is more un-certain than the prediction of radiative forcing, but generally, it is assumed that the feedback effects may double the surface temperature increases because of radiative forcing alone.

Has the surface temperature already increased due to anthropogenic activities? Figure 5 plots the global average surface temperature over the last century and a half. Even though there are large annual fluctuations, the smoothed curve through the data points indicates an upward trend of the temperature. From 1850, the start of the Industrial Revolution, to date, the global average surface temperature increased by about 0.5°C–1°C. This is in accordance with radiative models that predict such a trend, considering the increase of GHG concentrations over that period.

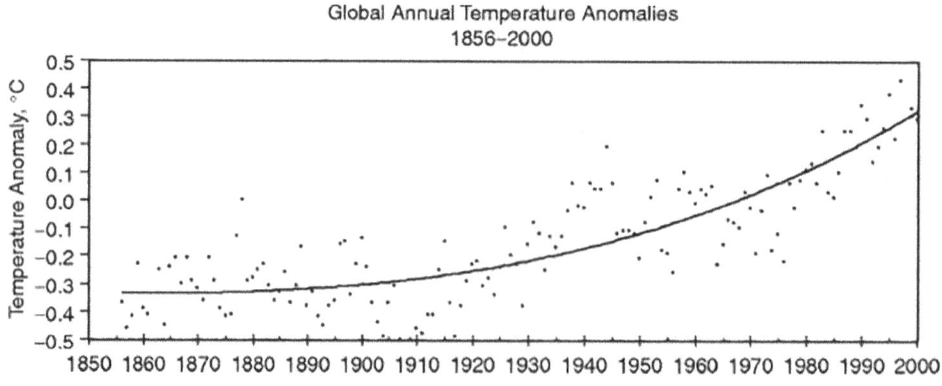

FIGURE 5 Global average surface temperature trend 1850–2000.
Source: Oak Ridge National Laboratory (see CO_2 Information Center[9]).

Other Effects of Global Warming

Because of increased GHG concentrations in the atmosphere, the earth's surface temperature may rise, as discussed in the previous section. The surface temperature rise may cause several ancillary effects on global climate and hydrogeology, which in turn will affect human habitat, welfare, and ecology.

Sea Level Rise

With increasing surface temperatures, the average sea level will rise because of three factors: melting of polar ice caps, receding of glaciers, and thermal expansion of the ocean surface waters. Combining all three factors, it is estimated that by the end of the next century, the average sea level may be 30–50 cm higher than it is today. This can seriously affect low-lying coastal areas, such as the Netherlands in Europe, Bangladesh in Asia, and low-lying islands in the Pacific and other oceans.[8]

Climate Changes

Predicting global and regional climatic changes because of average surface temperature rise is extremely difficult and fraught with uncertainties. It is expected that regional temperatures, prevailing winds, and storm and precipitation patterns will change, but where and when, and to what extent changes will occur, is a subject of intensive investigation and modeling on the largest available computers, the so-called supercomputers. Climate is not only influenced by surface temperature changes, but also by biological and hydrological processes, and by the response of ocean circulation, which are all coupled to temperature changes. It is expected that temperate climates will extend to higher latitudes, probably enabling the cultivation of grain crops further toward the north than at present. But crops need water. On the average, the global evaporation and precipitation balance will not change much, although at any instant, more water vapor (humidity) may be locked up in the atmosphere. However, precipitation patterns may alter, and the amount of rainfall in any episode may be larger than it is now. Consequently, the runoff (and soil erosion) may be enhanced, and areas of flooded watersheds may increase. Hurricanes and typhoons spawn in waters that are warmer than 27°C, in a band from 5 to 20° north and south latitude. As the surface waters become warmer, and the latitude band expands, it is very likely that the frequency and intensity of tropical storms will increase.

The sea level and climatic changes may cause a redistribution of agricultural and forestry resources, a considerable shift in population centers, and incalculable investments in habitat and property protection.

Greenhouse Gas Concentrations Trends

Currently, about 6.8 Gt/y of carbon (25 Gt/y CO_2) are emitted into the atmosphere by fossil fuel combustion. Another 1.5 ± 1 Gt/y are emitted due to deforestation and land use changes, mainly artificial burning of rain forests in the tropics, and logging of mature trees, which disrupts photosynthesis. Figure 6 plots the trend of atmospheric concentrations of CO_2, measured consistently at Mauna Loa, Hawaii since 1958 to date. At present, the average CO_2 concentration is about 375 parts per million by volume (ppmv). The plot shows seasonal variations due mainly to assimilation/respiration of CO_2 by plants, but there is a steady increase of the average concentration at a rate of approximately 0.4%/y. If that rate were to continue into the future, a doubling of the current CO_2 concentration would occur in about 175 years. However, if no measures are taken to reduce CO_2 emissions, then due to the population increase, and the concomitant enhancement of fossil fuel use, the rate of growth of CO_2 concentration will increase more than 0.4%/y, and the doubling time will be achieved sooner.

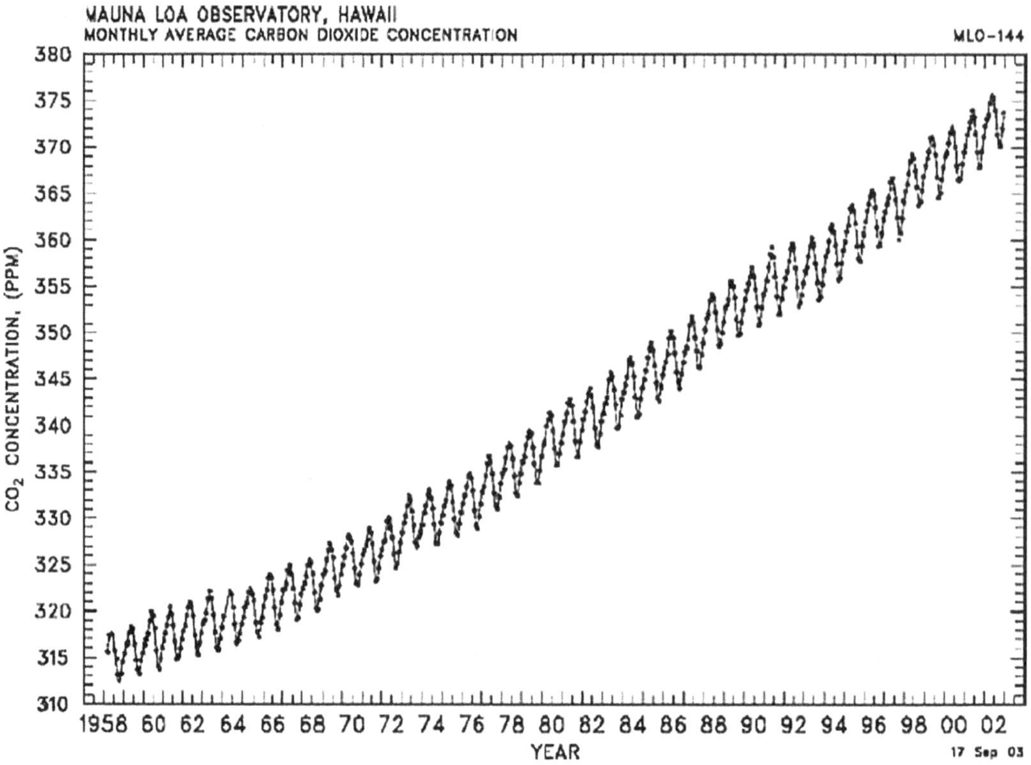

FIGURE 6 Carbon dioxide concentration trend 1958–2002.
Source: Oak Ridge National Laboratory (see CO_2 Information Center).[9]

Methane emissions are in part due to fossil energy usage, because CH_4 leaks from gas pipes, storage tanks, tankers, and coal mine shafts. Anthropogenic emissions of CH_4 from fossil fuel usage amount to about 100 Mt/y. However, CH_4 is also emitted from municipal waste landfills, sewage treatment, biomass burning, cultivated rice paddies, enteric fermentation of cattle, and other anthropogenic activities, so that the total amount of CH_4 emissions is about 400 Mt/y. Currently, the average atmospheric concentration of CH_4 is about 1.7 ppmv, growing at about 0.6%/y. Nitrous oxide (N_2O) is a minor product of combustion of fossil fuels. Currently, the concentration of N_2O is about 0.3 ppmv, growing by about 0.25%/y. CFCs are not directly associated with fossil fuel usage; however, they are emitted inadvertently from energy using devices, such as refrigerators, air conditioners, chillers, and heat pumps. Current concentrations of the various CFCs are about 0.5 parts per billion by volume (ppbv), and their concentrations are slowly declining due to the phase-out of production of CFCs.

What Can Be Done about Global Warming

Global warming could be lessened by reducing significantly the emissions of GHGs into the atmosphere. Most GHG emissions are a consequence of fossil fuel use. While it is important to reduce the emissions of all GHGs, the greatest preventative measure would come from reducing CO_2 emissions. CO_2 emission reductions can be accomplished by a combination of several of the following approaches.

Demand-Side Conservation and Efficiency Improvements

This includes less space heating and better insulation, less air conditioning, use of more fluorescent instead of incandescent lighting, more energy efficient appliances, process modification in industry, and very importantly, more fuel efficient automobiles. Some measures may even incur a negative cost, i.e., consumer savings by using less energy, or at least a rapid payback period for the investment in energy saving devices.

Supply-Side Efficiency Measures

Here we mean primarily increasing the efficiency of electricity production. Natural gas combined cycle power plants (NGCC) emit less CO_2 than single cycle coal fired power plants. First, because NG emits about one-half the amount of CO_2 per fuel heating value than coal, and second, because the thermal efficiency of combined cycle power plants is in the 45%–50% range vs. the 35%–38% range of single cycle plants. In the future, integrated coal gasification combined cycle power plants (IGCC) may come on-line with a thermal efficiency in the 40%–45% range reckoned on the basis of the coal heating value. Furthermore, IGCC may enable the capture of CO_2 at the gasification stage, with subsequent sequestration of the captured CO_2 in geologic and deep ocean repositories.

Shift to Non-Fossil Energy Sources

The choices here are agonizing, because the largest impact could be made by shifting to nuclear and hydro electricity, both presently very unpopular and fraught with environmental and health hazards. The shift to solar, wind, geothermal, and ocean energy is popular, but because of their limited availability and intermittency, and because of their larger cost compared to fossil energy, a substantial shift to these energy sources cannot be expected in the near future.

None of these options can prevent global warming by itself. They have to be taken in combination and on an incremental basis, starting with the least expensive ones and progressing to the more expensive ones. Even if the predictions of global climate change were to turn out exaggerated, the fact that fossil fuel usage entails many other environmental and health effects, and the certainty that fossil fuel resources are finite, makes it imperative that we curtail fossil energy usage as much as possible.

Summary and Conclusions

The use of fossil fuels to supply energy for the use of the world's population has resulted in the release to the atmosphere of troublesome chemical byproducts that present harm to humans and other natural species. These effects can be localized (near the emission source), can extend to large regional areas (involving sub-continents), and can even cover the globe, from pole to pole. A large portion of the human population is exposed to one or more of these environmental effects. The scientific understanding of how fossil fuel use causes these effects is well advanced, providing quantitative means for explaining what is currently observed and predicting what changes will occur in the future from projected future fuel consumption. These projections provide a basis for modifying the amount and character of future energy supply so as to lessen harmful environmental consequences. The technological systems that employ fossil fuel energy have been developed to lessen the amounts of harmful emissions, albeit at significant energy and economic cost. Further improvements can be expected, but at increasing marginal cost. The most severe emission control problem, in terms of economic and energy cost, is CO_2, a major contributor to global warming. The implementation of policies by national governments and international bodies to curtail the use of fossil energy and the concomitant emissions of CO_2 will become a growing task for humankind in this century.

References

1. Wark, K.; Warner, C.F.; Davis, W.T. *Air Pollution: Its Origin and Control,* 3rd Ed.; Addison-Wesley: Menlo Park, CA, 1998.
2. Cooper, C.D.; Alley, F.C. *Air Pollution Control: A Design Approach,* 3rd Ed.; Waveland Press, Prospects Heights: IL, 2002.
3. Seinfeld, J.H.; Pandis, S.N. *Atmospheric Chemistry and Physics: From Air Pollution to Climate Change*; Wiley: New York, 1998.
4. Fay, J.A.; Golomb, D.; Kumar, S. Source apportionment of wet sulfate deposition in Eastern North America. Atm. Environ. **1985**, 19, 1773–1782.
5. Butler, T.J.; Likens, G.E.; Stunder, B.J.B. Regional-scale impacts of phase I of the clean air act amendments in the U.S.A.: the relation between emissions and concentrations, both wet and dry. Atmos. Environ. **2000**, 35, 1015–1028.
6. National Research Council. *Rethinking the Ozone Problem in Urban and Regional Air Pollution*; National Academy Press: Washington, DC, 1991.
7. Mitchell, J.F.B. The greenhouse effect and climate change. Rev. Geophys. **1989**, 27, 115–139.
8. Intergovernmental Panel on Climate Change. *Climate Change: The Scientific Basis*; Cambridge University Press: Cambridge, U.K., 2001.
9. CO_2 Information Center. *Trends Online*; Oak Ridge National Laboratory: Oak Ridge, TN, 2000.

13

Geothermal Energy Resources

Ibrahim Dincer
and Arif Hepbasli

Introduction

With increasing awareness of the detrimental effects of the burning of fossil fuels on the environment, there has been an increasing interest worldwide in the use of clean and renewable energy sources. Although geothermal energy is categorized in international energy tables as one of the "new renewables," it is not a new energy source at all. People have used hot springs for bathing and washing clothes since the dawn of civilization in many parts of the world.[1] Heat is a form of energy, and *geothermal energy* is, literally, the heat contained within the Earth that generates geological phenomena on a planetary scale. "Geothermal energy" is often used nowadays, however, to indicate that part of the Earth's heat that can, or could, be recovered and exploited by man, and it is in this sense that we will use the term from now on.[2] Geothermal energy utilization is basically divided into two categories, i.e., electric energy production and direct uses. Direct use of geothermal energy is one of the oldest, most versatile, and also the most common form of utilization of geothermal energy.[3] The early history of geothermal direct use has been well documented for over 25 countries in the *Stories from a Heat Earth—Our Geothermal Heritage*[4] that documents geothermal use for over 2000 years.[5]

Geothermal energy has until recently had a considerable economic potential only in areas where thermal water or steam is found concentrated at depths less than 3 km in restricted volumes, analogous to oil in commercial oil reservoirs. The use of ground-source heat pumps has changed this. In this case, the earth is the heat source for the heating and/or the heat sink for cooling, depending on the season. This has made it possible for people in all countries to use the heat of the earth for heating and/or cooling, as appropriate. It should be stressed that the heat pumps can be used basically everywhere.[1]

Based on country update papers submitted for the World Geothermal Congress 2010 (WGC-2010), covering the period 2005 to 2009, an estimate of the installed thermal power for direct utilization at the end of 2009 is 50,583 MW_t, almost a 79% increase over the 2005 data, growing at a compound rate of 12.3% annually with a capacity factor of 0.27. The thermal energy used is 438,071 TJ/yr (121,696 GWh/yr), about a 60% increase over 2005, growing at a compound rate of 9.9% annually. The distribution of thermal energy used by category is approximately 49.0% for ground-source heat pumps, 24.9% for bathing and swimming (including balneology), 14.4% for space heating (of which 85% is for district heating), 5.3% for greenhouses and open ground heating, 2.7% for industrial process heating, 2.6% for aquaculture pond and raceway heating, 0.4% for agricultural drying, 0.5% for snow melting and cooling, and 0.2% for other uses. Energy savings amounted to 307.8 million barrels (46.2 million tons) of equivalent oil annually, preventing 46.6 million tons of carbon and 148.2 million tons of CO_2[5] being released to the atmosphere, which includes savings in geothermal heat pump cooling (compared to using fuel oil to generate electricity).

The main objective of this entry is to provide a comprehensive picture about geothermal energy resources and their use for various purposes, ranging from power production to hydrogen production. It first gives a brief historical development of geothermal energy resources and geothermal energy applications. The following sections delve the nature of geothermal resources, classification of geothermal resources by energy and exergy, Lindal diagram, and some energetic and exergetic parameters for practical applications.

Historical Background

Early humans probably used geothermal water that occurred in natural pools and hot springs for cooking, bathing, and hot water and heating purposes. Archeological evidence indicates that the Indians of the Americas occupied sites around these geothermal resources for over 10,000 years to recuperate from battle and take refuge. Many of their oral legends describe these places and other volcanic phenomena. Recorded history shows uses by Romans, Japanese, Turks, Icelanders, Central Europeans, and the Maori of New Zealand for bathing, cooking, hot water, and space heating. Baths in the Roman Empire, the middle kingdom of the Chinese, and the Turkish baths of the Ottomans were some of the early uses of balneology, where body health, hygiene, and discussions were the social custom of the day. This custom has been extended to geothermal spas in Japan, Germany, Iceland, and countries of the former Austro-Hungarian Empire, the Americas, and New Zealand. Early industrial applications include chemical extraction from the natural manifestations of steam, pools, and mineral deposits in the Larderello region of Italy, with boric acid being extracted commercially starting in the early 1800s. At Chaudes-Aigues in the heart of France, the world's first geothermal district heating system was started in the 14th century and is still in use. The oldest geothermal district heating project in the United States is on Warm Springs Avenue in Boise, Idaho, which came on line in 1892 and continues to provide space heating for up to 450 homes.[2,4]

As outlined in various literature (e.g., Dickson and Fanelli[2] and Cataldi[4]), a brief summary is given as follows. The first use of geothermal energy for electric power production was carried out in Italy with some experimental tests by Prince Gionori Conti between 1904 and 1905. The first commercial power plant with a capacity of 250 kW_e was commissioned in 1913 at Larderello, Italy. An experimental plant was installed in the Geysers in 1932 to power a local resort. Such initiatives were followed in Wairakei (New Zealand) in 1958, with an experimental plant in Pathe (Mexico) in 1959, and the first commercial plant at The Geysers in the United States in 1960. Japan followed with a capacity of 23 MW_e in Matsukawa in 1966. All of these early plants utilized steam directly from the earth (mainly dry steam fields), except the one in New Zealand, which was considered the first to use flashed or separated steam for running the turbines. The former USSR produced power from the first true binary power plant with a capacity of 680 kW_e using 81°C water in Paratunka on the Kamchatka Peninsula—the lowest temperature at that time. Iceland first produced power at Namafjall in the Northern part of the country by using a 3 MWe non-condensing turbine. These were later followed by various plants in El Salvador, China, Indonesia, Kenya, Turkey, Philippines, Portugal (Azores), Greece, and Nicaragua in the 1970s and 1980s. Some recent plants have been installed in Thailand, Argentina, Taiwan, Australia, Costa Rica, Austria, Guatemala, and Ethiopia, with the latest installations in Germany and Papua New Guinea.

More recently, renewable energy resources have received increasing interest for sustainable hydrogen production. In this regard, Balta et al.[6,7] have recently studied on geothermal based hydrogen production methods comprehensively and defined various potential methods using thermochemical and hybrid cycles driven essentially by heat, as considered appropriate for geothermal sources. This has further widened the scope of geothermal energy resources. Geothermal based hydrogen production offers potential advantages over other sources for a growing hydrogen economy. These advantages do not ensure that geothermal based hydrogen will be the only option, but it will compete with other hydrogen production options. In some countries, geothermal energy is considered a primary energy source of producing hydrogen since it provides reliable energy supply in an environmentally benign manner. Hydrogen can be produced from geothermal energy using five potential methods such as 1) direct production of hydrogen from geothermal steam; 2) water electrolysis using the electricity generated from a power plant; 3) high- temperature steam electrolysis; 4) thermochemical cycles driven by geothermal heat; and 5) hybrid cycles driven by geothermal heat and electricity. A comprehensive discussion of these methods is available elsewhere.[6,7]

Nature of Geothermal Resources

The *geothermal gradient* expresses the increase in temperature with depth in the Earth's crust. Down to the depths accessible by drilling with modern technology, i.e., over 10,000 m, the average geothermal gradient is about 2.5–3°C/100 m. For example, if the temperature within the first few meters below ground level, which on average corresponds to the mean annual temperature of the external air, is 15°C, then we can reasonably assume that the temperature becomes about 65–75°C at 2000 m depth, 90–105°C at 3000 m, and so on for a further few thousand meters. There are, however, vast areas in which the geothermal gradient is far from the average value. In areas in which the deep rock basement has undergone rapid sinking, and the basin is filled with geologically "very young" sediments, the geothermal gradient may be lower than 1°C/100 m. On the other hand, in some "geothermal areas," the gradient is more than 10 times the average value (further details are given in Dickson and Fanelli[2]).

Geothermal energy essentially comes from the natural heat of the earth as a result of the decay of the naturally radioactive isotopes of uranium, thorium, and potassium. Because of the internal heat, the Earth's surface heat flow averages 82 mW/m^2 with an equivalent total heat loss rate (the mean terrestrial heat flow of continents and oceans is 65 and 101 mW/m^2) of about 42 million MW.[2,8,9] The estimated total thermal energy above mean surface temperature to a depth of 10 km is 1.3×10^{27} J, which is equivalent to burning 3.0×10^{17} barrels of oil. Since the global energy consumptions for all types of energy are equivalent to use of about 100 million barrels of oil per day, the Earth's energy to a depth of 10 km could theoretically supply all of mankind's energy needs for about 6 million years (as further discussed in Wright[8] and Lund[9]).

A geothermal system essentially consists of three elements, namely, 1) a heat source; 2) a reservoir; and 3) a heat-carrying fluid (geothermal liquid) as illustrated in Figure 1. The heat source can be either at very high temperature greater than 600°C with magmatic intrusion at relatively shallow depths (5–10 km) or at relatively low temperatures. As mentioned above, the Earth's temperature increases with depth. The reservoir is a volume of hot permeable rocks from which the circulating fluids extract heat and connected to a surfacial recharge area through which the meteoric waters can replace or partly replace the fluids that escape from the reservoir through springs or are extracted by boreholes. The geothermal fluid is water (in the majority of cases, meteoric water) either in liquid or in vapor phase,

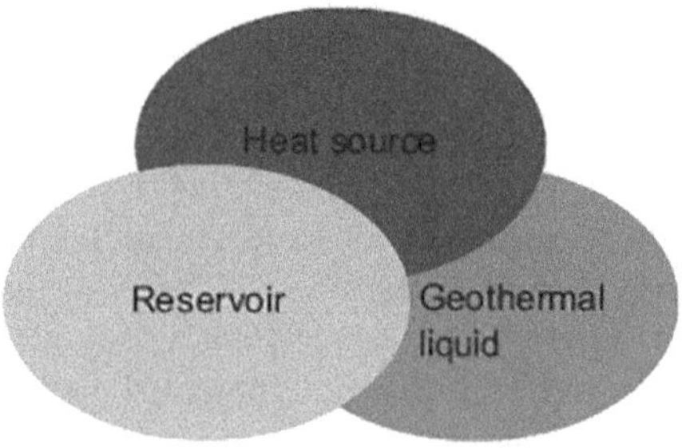

FIGURE 1 Geothermal system.

depending on its temperature and pressure. The geothermal water often carries with it some chemicals and gases such as CO_2 and H_2S.

Geothermal Energy and Environmental Impact

Problems with energy supply and use are related not only to global warming but also to such environmental concerns as air pollution to climate change, forest destruction, and emission of radioactive substances. These issues must be taken into consideration simultaneously if humanity is to achieve a bright energy future with minimal environmental impacts. Much evidence exists, which suggests that the future will be negatively impacted if humans keep degrading the environment.

One solution to both energy and environmental problems is to make much more use of renewable energy resources, particularly geothermal. This cause is sometimes espoused with a fervor that leads to extravagant and impossible claims being made. Engineering practicality, reliability, applicability, economy, scarcity of supply, and public acceptability should all be considered accordingly.

In order to achieve the energetic, economic, and environmental benefits that geothermal offers, the following integrated set of activities should be acted accordingly:

- *Research and Development:* Research and development priorities should be set in close consultation with industry to reflect their needs. Most research is conducted through cost-shared agreements and falls within the short-to- medium term. Partners in these activities should include a variety of stakeholders in the energy industry, such as private-sector firms, utilities across the country, provincial governments, and other federal departments.
- *Technology Assessment:* Appropriate technical data should be gathered in the laboratory and through field trials on factors such as cost benefit, reliability, environmental impact, safety, and opportunities for improvement. These data should also assist the preparation of technology status overviews and strategic plans for further research and development.
- *Standards Development:* The development of technical and safety standards is needed to encourage the acceptance of proven technologies in the marketplace. Standards development should be conducted in cooperation with national and international standards writing organizations, as well as other national and provincial regulatory bodies.
- *Technology Transfer:* Research and development results should be transferred through sponsorship of technical workshops, seminars, and conferences, as well as through the development of training manuals and design tools, Web tools, and the publication of technical reports.

Such activities will also encourage potential users to consider the benefits of adopting geothermal energy technologies. In support of developing near-term markets, a key technology transfer area is to accelerate the use of geothermal energy technologies, particularly for district heating applications and power generation.

Geothermal Energy and Sustainability

Sustainable development requires a sustainable supply of clean and affordable energy resources that do not cause negative societal impacts. Supplies of such energy resources as fossil fuels and uranium are finite. Energy resources such as geothermal are generally considered renewable and therefore sustainable over the relatively long term.

Sustainability often leads local and national authorities to incorporate environmental considerations into energy planning. The need to satisfy basic human needs and aspirations, combined with increasing world population, will make the need for successful implementation of sustainable development increasingly apparent. Various criteria that are essential to achieving sustainable development in a society follow:[10,11]

- Information about and public awareness of the benefits of sustainability investments
- Environmental education and training
- Appropriate energy and exergy strategies
- The availability of renewable energy sources and cleaner technologies
- A reasonable supply of financing
- Monitoring and evaluation tools

Geothermal as a renewable energy has an important role to play in meeting future energy needs in both rural and urban areas. The development and utilization of geothermal energy applications should be given a high priority, especially in the light of increased awareness of the adverse environmental impacts of fossil-based generation. The need for sustainable energy development is increasing rapidly in the world. Widespread use of geothermal energy is important for achieving sustainability in the energy sectors in both developing and industrialized countries.

Geothermal energy can be a key component for sustainable development for three main reasons:

1. They generally cause less environmental impact than other energy resources. The variety of geothermal energy applications provides a flexible array of options for their use.
2. They cannot be depleted. If used carefully in appropriate applications, renewable energy resources can provide a reliable and sustainable supply of energy almost indefinitely. In contrast, fossil fuel and uranium resources are diminished by extraction and consumption.
3. They favor both system centralization and decentralization and local solutions that are somewhat independent of the national network, thus enhancing the flexibility of the system and providing economic benefits to small isolated populations. Also, the small scale of the equipment often reduces the time required from initial design to operation, providing greater adaptability in responding to unpredictable growth and/or changes in energy demand.

Not all renewable energy resources are inherently clean in that they cause no burden on the environment in terms of waste emissions, resource extraction, or other environmental disruptions. Nevertheless, use of geothermal almost certainly can provide a cleaner and more sustainable energy system than increased controls on conventional energy systems.

To seize the opportunities, it is essential to establish a geothermal energy market and gradually build up the experience with the cutting-edge technologies. The barriers and constraints to the diffusion of geothermal energy use should be removed. The legal, administrative, and financing infrastructure should be established to facilitate planning and application of geothermal energy projects. Government could/should play a useful role in promoting geothermal energy technologies through fundings and incentives to encourage research and development as well as commercialization and implementation in both urban and rural areas.

Classification of Geothermal Resources

There is no standard international terminology in use throughout the geothermal community, which is unfortunate, as this would facilitate mutual comprehension.[2] In the following, some of the most common classifications in this discipline are presented in terms of energetic and exergetic aspects.

Classification of Geothermal Resources by Energy

The most common classification of geothermal resources is done based on the enthalpy values of the geothermal fluids, acting as the energy carrier for transporting the heat from deep hot rocks to the Earth's surface. In this regard, enthalpy and temperature levels are accordingly considered as given in Table 1. The temperatures of these geothermal resources range from the mean annual temperature of around 20°C to over 300°C. In general, resources above 150°C are used for electric power generation, although power has recently been generated at Chena Hot Springs Resort in Alaska using a 74°C geothermal resource.[12] Resources below 150°C are usually used in direct-use projects for individual and district heating and cooling applications. Furthermore, the ambient temperatures in the range of 5–30°C can be used with geothermal (ground source) heat pumps to provide both heating and cooling. The resources may also be categorized into three enthalpy levels (low, medium, and high) according to criteria that are generally based on the energy content of the fluids and their potential forms of utilization. Furthermore, Table 2 presents a summary of the types of geothermal resources, reservoirs temperatures, reservoirs fluids, applications, and systems.

Although numerous criteria can be utilized for classification, the reservoir temperature still remains as the primary criterion. Table 3 lists a possible scheme in which geothermal resources are classified into

TABLE 1 Geothermal Resource Types and Temperature Ranges

Resource Type		Temperature Range (°C)
Convective hydrothermal resources	Vapor dominated	≈240
	Hot water dominated	20–350
Other hydrothermal resources	Sedimentary basin	20–150
	Geopressured	90–200
	Radiogenic	30–150
Hot rock resources	Solidified (hot dry rock)	90–650
	Part still molten magma	>600

Source: Adapted from Lund[12] and Dickson and Fanelli.[13]

TABLE 2 Geothermal Resource Types, Applications, and System Types

Reservoir Temperature	Reservoir Fluid	Applications	Systems
High temperature (>220°C)	Water or steam	Power generation Direct use	• Flash steam • Combined (flash and binary) cycle • Direct fluid use • Heat exchangers • Heat pumps
Medium temperature (100–220°C)	Water	Power generation Direct use	• Binary cycle • Direct fluid use • Heat exchangers • Heat pumps
Low temperature (50–150°C)	Water	Direct use	• Direct fluid use • Heat exchangers • Heat pumps

Source: Adapted from World Bank, Geothermal Energy.[14]

TABLE 3 Classification of Different Schemes for Geothermal Resources

Class of Resource	Reservoir Temperature (°C)	Mobile Fluid Phase in Reservoir	Production Mechanism	Fluid State at Well Head	Well Productivity and Controlling Factors Other than Temperature	Applicable Power Conversion Technology	Unusual Development or Operational Problems
1. Non-electrical	<100	Liquid water	Artesian self-flowing wells; pumped wells	Liquid water	Well productivity dependent on reservoir flow capacity and static water level	Direct use	
2. Very low temperature	100 to <150	Liquid water	Pumped wells	Liquid water (for pumped wells); steam–water mixture (for self-flowing wells)	Typical well capacity 2 to 4 MW_e; dependent on reservoir flow capacity and gas content in water; well productivity often limited by pump capacity	Binary	
3. Low temperature	150 to <190	Liquid water	Pumped wells; self-flowing wells (only at the higher temperature end of the range)	Liquid water (for pumped wells); steam–water mixture (for self-flowing wells)	Typical well capacity 3 to 5 MW_e; dependent on reservoir pressures, reservoir flow capacity and gas content in water; productivity of pumped wells typically limited by pump capacity and pump parasitic power need; productivity of self-flowing wells strongly dependent on reservoir flow capacity	Binary; two-stage flash; hybrid	Calcite scaling in production wells and stibnite scaling in binary plant are occasional problems
4. Moderate temperature	190 to <230	Liquid water	Self-flowing wells	Steam–water mixture (enthalpy equal to that of saturated liquid at reservoir temperature)	Well productivity highly variable (3 to 12 MW_e); strongly dependent on reservoir flow capacity	Single-stage flash; two-stage flash; hybrid	Calcite scaling in production wells occasional problem; alumino-silicate scale in injection system a rare problem

(*Continued*)

TABLE 3 (*Continued*) Classification of Different Schemes for Geothermal Resources

Class of Resource	Reservoir Temperature (°C)	Mobile Fluid Phase in Reservoir	Production Mechanism	Fluid State at Well Head	Well Productivity and Controlling Factors Other than Temperature	Applicable Power Conversion Technology	Unusual Development or Operational Problems
5. High temperature	230 to <300	Liquid water; liquid-dominated two-phase	Self-flowing wells	Steam–water mixture (enthalpy equal to or higher than that of saturated liquid at reservoir temperature); saturated steam	Well productivity highly variable (up to 25 MW_e); dependent on reservoir flow capacity and steam saturation		
6. Ultrahigh temperature	300+	Liquid-dominated two-phase	Self-flowing wells	Steam–water mixture (enthalpy equal to or higher than that of saturated liquid at reservoir condition); saturated steam; superheated steam	Well productivity extremely variable (up to 50 MW_e); dependent on reservoir flow capacity and steam saturation		
7. Steam field	240 (33.5 bar-a pressure; 2800 kJ/kg enthalpy)	Steam	Self-flowing wells	Saturated or superheated steam	Well productivity extremely variable (up to 50 MW_e); dependent on reservoir flow capacity		

Source: Adapted from Sanyal.[15]

seven categories based on temperature, while they are explained in more detail in Sanyal.[15] This classification includes non-electrical grade (<100°C), very low temperature (100°C to <150°C), low temperature (150°C to 190°C), moderate temperature (190°C to <230°C), high temperature (230°C to <300°C), ultrahigh temperature (>300°C), and steam fields (approximately 240°C with steam as the only mobile phase).[16] In the first four classes of reservoirs, liquid water is the mobile phase; in the "high" and "ultra-high" temperature reservoirs, the mobile fluid phase is either liquid or a liquid–vapor mixture. Thus, this makes steam fraction a second criterion. Unfortunately, there is no readily available procedure to classify enhanced geothermal systems or hot dry rock resources other than perhaps considering temperature-based classification as discussed above.

Classification of Geothermal Resources by Exergy

Although many agree on the classification of geothermal resources based on temperature and enthalpy levels as part of the energy aspects under the first law of thermodynamics, there are still some ambiguities with respect to their quality that address the second law of thermodynamics through exergy. Therefore, geothermal resources should be classified by their exergy, a measure of their ability to do work[17,18]).

Lee[17] proposed a new parameter, the so-called specific exergy index (SExI) for better classification and evaluation as follows:

$$\text{SExI} = \frac{h_{\text{brine}} - 273.16\, S_{\text{brine}}}{1192} \tag{1}$$

which is a straight line on an $h-s$ plot of the Mollier diagram. Straight lines of SExI = 0.5 and SExI = 0.05 can therefore be drawn in this diagram and used as a map for classifying geothermal resources by taking into account the following criteria:

- SExI < 0.05 for low-quality geothermal resources
- 0.05 ≤ SExI < 0.5 for medium-quality geothermal resources
- SExI ≥ 0.5 for high-quality geothermal resources

Here, the demarcation limits for these indexes are exergies of saturated water and dry saturated steam at 1 bar absolute.

In order to map any geothermal field on the Mollier diagram as well as to determine the energy and exergy values of the geothermal brine, the average values for the enthalpy and entropy are then calculated from the following equations:[18]

$$h_{\text{brine}} = \frac{\sum_{i}^{n} \dot{m}_{w,i} h_{w,i}}{\sum_{i}^{n} \dot{m}_{w,i}} \tag{2}$$

$$S_{\text{brine}} = \frac{\sum_{i}^{n} \dot{m}_{w,i} S_{w,i}}{\sum_{i}^{n} \dot{m}_{w,i}} \tag{3}$$

The above-given exergetic classification has been used by some investigators in their studies. From these, Quijano[19] performed an exergy analysis of the Ahuachapán and Berlin geothermal fields. After plotting the thermodynamic states of both fields on the Mollier diagram and calculating the values of SExI, he classified these geothermal fields as the medium high exergy zone. Ozgener et al.[20] and Baba et al.[21] utilized this index (SExI) for the Balcova geothermal field located in the western part of Turkey, endowed with considerably rich geothermal resources. The number of the operating wells in this field varies depending on the heating days and operating strategies. Taking into account the eight operating wells and using Eqs. 1–3, the SExI is found to be 0.07 since it falls into the medium-quality geothermal resources according to the classification of Lee.[17] However, given the classification of the Balcova geothermal field with respect to its average reservoir as 113.9°C, the Balcova geothermal field becomes an intermediate-enthalpy resource. This clearly indicates that the classification with respect to the SExI is more meaningful as there is no need for a temperature range classification. Ozgener et al.[22,23] also determined the SExI values for the Gonen and Salihli geothermal fields in Turkey as 0.025 and 0.049, respectively. They concluded that both these fields fall into the low-quality geothermal resources according to the classification of Lee. Etemoglu and Can[24] studied geothermal resources in Turkey based on SExI. In their analysis, the triple point of water was selected as the dead-state conditions since the enthalpy and entropy values were zero. Geothermal resources with SExI values of 0.56 were classified as high-exergy resources; those whose values were <0.05 were low-exergy resources, and those whose values are between these two were medium-exergy resources.

The Lindal Diagram

There is another way of classifying the geothermal fields developed by Lindal[25] as illustrated in Figure 2. This diagram has frequently appeared in books, reports, and articles on direct uses of geothermal energy and shows the temperature required for both current and potential applications of geothermal energy. In other words, the minimum production temperatures in a geothermal field are generally required for the different types of use as shown in the diagram. This is now widely used by the geothermal community to depict temperature as the yardstick of direct applications. The boundaries, however, serve only as guidelines. Conventional electric power production is limited to fluid temperatures about 150°C, but considerably lower temperatures can be used with the application of binary fluids (outlet temperatures commonly at 100°C, while there are also some applications up to 74°C). The ideal inlet temperature into houses for space heating using radiators is about 80°C, but by using radiators of floor heating, or by applying heat pumps or auxiliary boilers, thermal waters with temperatures only a few degrees above the ambient can be used beneficially. It is a common misconception that direct use of geothermal is confined to low-temperature resources. High-temperature resources can, of course, also be used for heating and drying purposes even if the process is at a very low temperature. Refrigeration is, in fact, only possible with temperatures above about 120°C.[26]

Some Energetic and Exergetic Parameters

Coskun et al.[27] have introduced some geothermal energy systems-related parameters, namely, energetic renewability ratio, exergetic renewability ratio, energetic reinjection ratio, and exergetic reinjection ratio:

- *Energetic renewability ratio:* This is defined as the ratio of useful renewable energy obtained from the system to the total energy input (all renewable and non-renewable together) to the system.

$$R_{\mathrm{Ren}_E} = \frac{\dot{E}_{\mathrm{usf}}}{\dot{E}_{\mathrm{total}}} \tag{4}$$

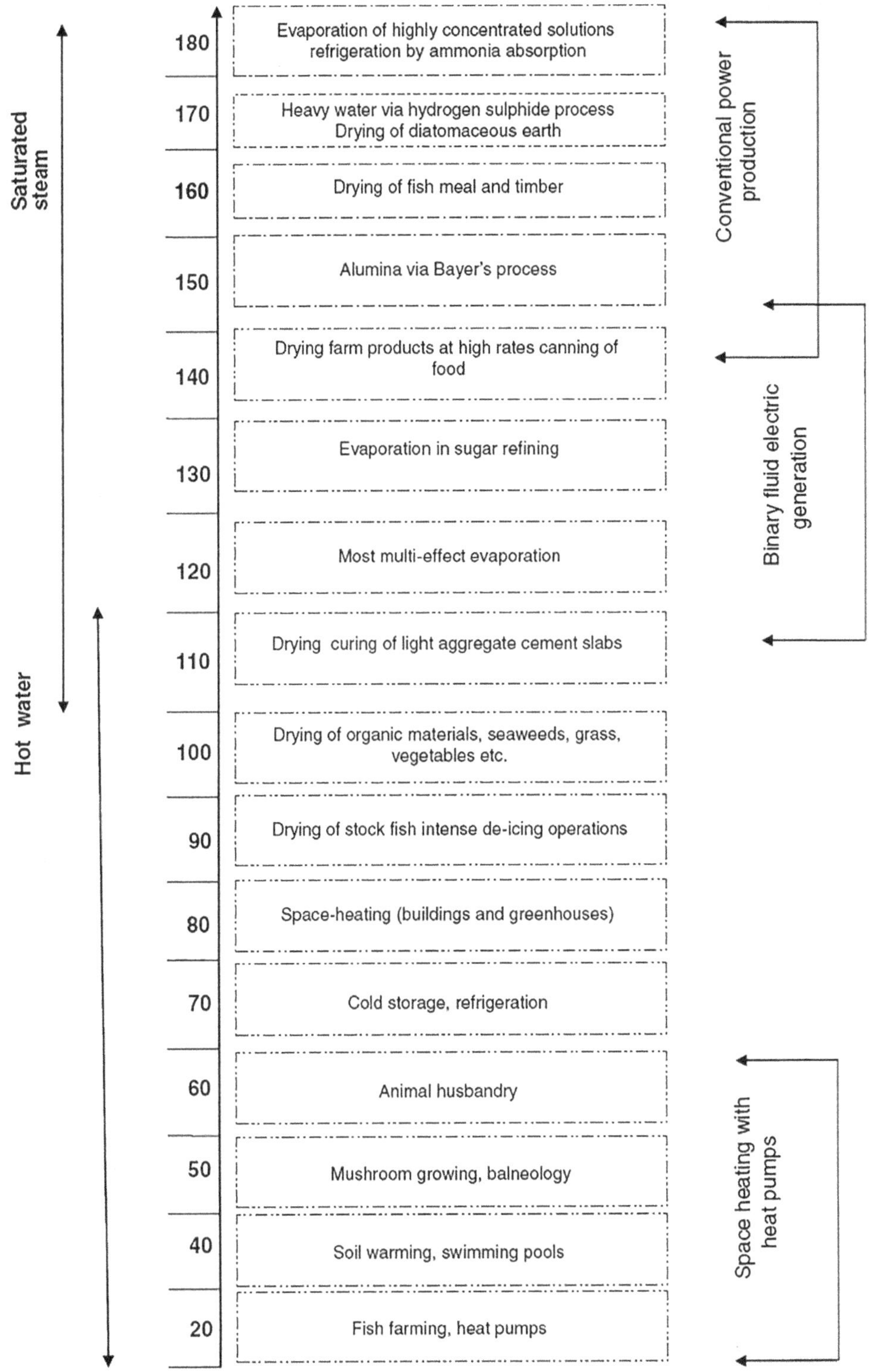

FIGURE 2 Lindal diagram indicating the applications for geothermal resources depending on the temperature. **Source:** Adapted from Hepbasli.[28]

- *Exergetic renewability ratio:* This is defined as the ratio of useful renewable exergy obtained from the system to the total exergy input (all renewable and non-renewable together) to the system:

$$R_{\text{Re in}_{\text{Ex}}} = \frac{\dot{E}x_{\text{usf}}}{\dot{E}x_{\text{total}}} \tag{5}$$

Here, total energy and exergy input terms include the wellhead geothermal water energy only in energy and exergy efficiencies. However, in the above equations for energetic and exergetic renewability ratios, all renewable and non-renewable energy and exergy inputs are considered for input.

- *Energetic reinjection ratio:* This is defined as the ratio of renewable energy discharged to environment or reinjected to the well from the system to the total geothermal energy supplied to the system:

$$R_{\text{Re in}_{\text{E}}} = \frac{\dot{E}_{\text{nd}}}{\dot{E}_{\text{gw}}} \tag{6}$$

- *Exergetic reinjection ratio:* This is defined as the ratio of renewable exergy discharged to environment or reinjected to the well from the system to the total geothermal exergy supplied to the system:

$$R_{\text{Re in}_{\text{Ex}}} = \frac{\dot{E}x_{\text{nd}}}{\dot{E}x_{\text{gw}}} \tag{7}$$

Here, the monthly variations of the above-listed parameters as studied by Coskun et al.[27] for the Edremit Geothermal District Heating System (GDHS) are illustrated in Figure 3. As seen in Figure 3a and b, both energetic and exergetic renewability ratios decrease with increasing temperature in the summer season. Here, one can extract the following:

- The energetic renewability ratio varies between 0.32 and 0.33 for the Edremit system during summer when heating is not required.
- The exergetic renewability ratio varies between 0.25 and 0.35 for the Edremit system during the summer when heating is not required.

It was observed that the outdoor reference ambient temperatures do not have a direct effect on both mass flow rate and energy/exergy losses as a result of no heating requirement for the summer season. Domestic hot water requirement is one important parameter for the summer season. Domestic hot water requirement increases for hot days. Because of the increase in domestic hot water requirement, pump energy consumption in total energy input increases. Moreover, calculations show that total exergy loss/destruction tended to increase in the summer season. For instant, pump exergy destruction increases in the summer season as a result of the increase in the required domestic mass flow rate. In the heating season, total "winter" heat demand is the sum of the domestic hot water + heating proper. The mass flow rates for the network water and geothermal water vary during the season dependent on the average useful energy demand. The demand is the highest in January, declines until May, and begins increasing until the end of December. It is obvious that based on the demand, the mass flow rates proportionally change. The energetic and exergetic renewability ratios become the lowest in January, increasing until May, and begin decreasing until the end of December. We have understood from these results that heat demand and renewability ratio values are opposite characteristic properties. The renewability values are given for the heating season as below:

- The energetic renewability ratio varies between 0.33 and 0.34 for the Edremit system.
- The exergetic renewability ratio varies between 0.53 and 0.63 for the Edremit system.

In addition, both energetic and exergetic reinjection ratios are calculated using Eqs. 6 and 7 for the Edremit GDHS annually as given in Figure 3c and d, respectively. These figures can be explained for summer and winter seasons as follows. The energetic reinjection ratio increases in higher temperatures with the effect of pump energy consumption. They vary between 0.59 and 0.62 for the Edremit system during the summer season when heating is not required (Figure 3c). Variation of the exergetic reinjection ratio depends on outdoor temperature and has similar characteristic with average annual temperature (see Figure 3d). Used geothermal water exergy rate decreases as a result of increase in reference ambient temperature in the summer season. Used geothermal exergy rate decreases through the middle of summer, at the highest temperature. The exergetic reinjection ratio varies between 0.10 and 0.20 for the Edremit system during the summer season (see Figure 3d).

Furthermore, Hepbasli[28] has recently proposed the following new exergetic parameters, based on the system given in Ozgener et al.[29] while their energetic approaches may be also considered using their exergetic relations:

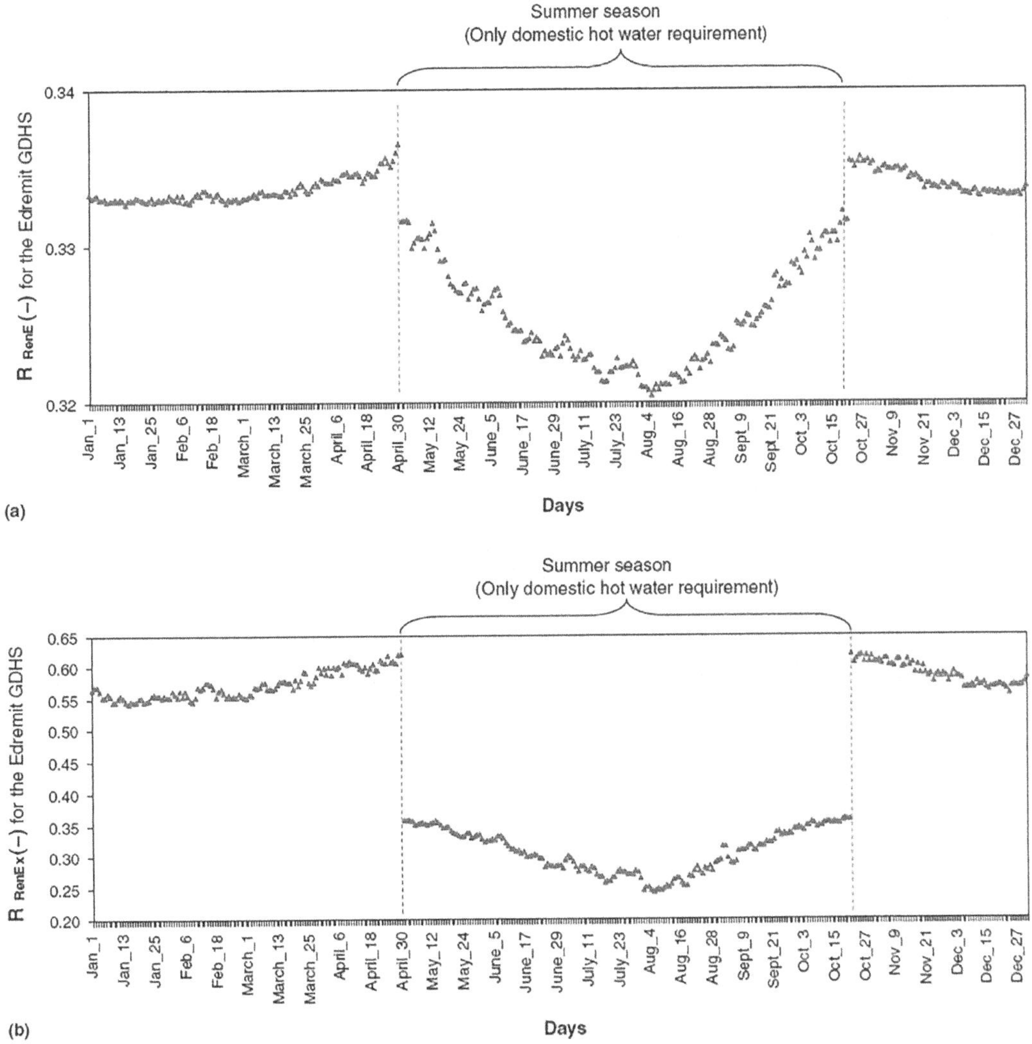

FIGURE 3 (a) Change of energetic renewability ratio for the Edremit GDHS annually. (b) Change of exergetic renewability ratio for the Edremit GDHS annually (modified from Coskun et al.[27]).

(Continued)

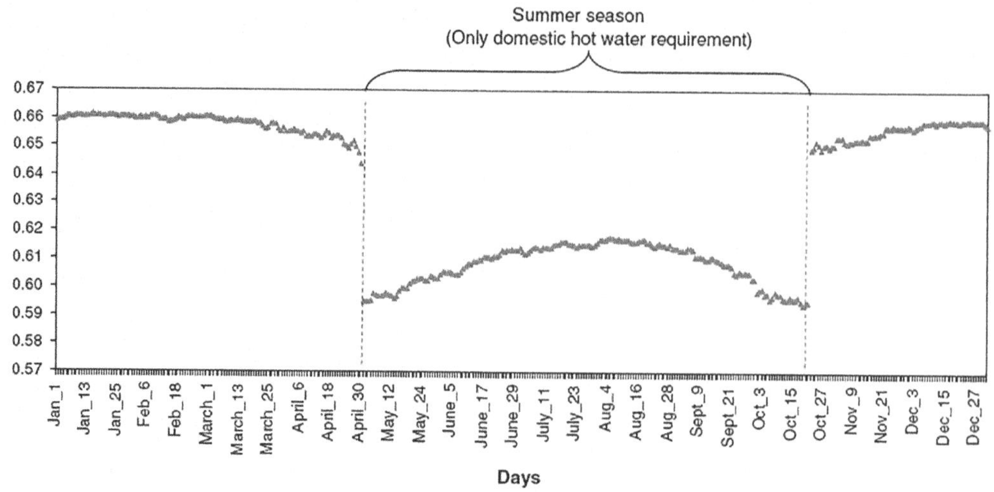

(c)

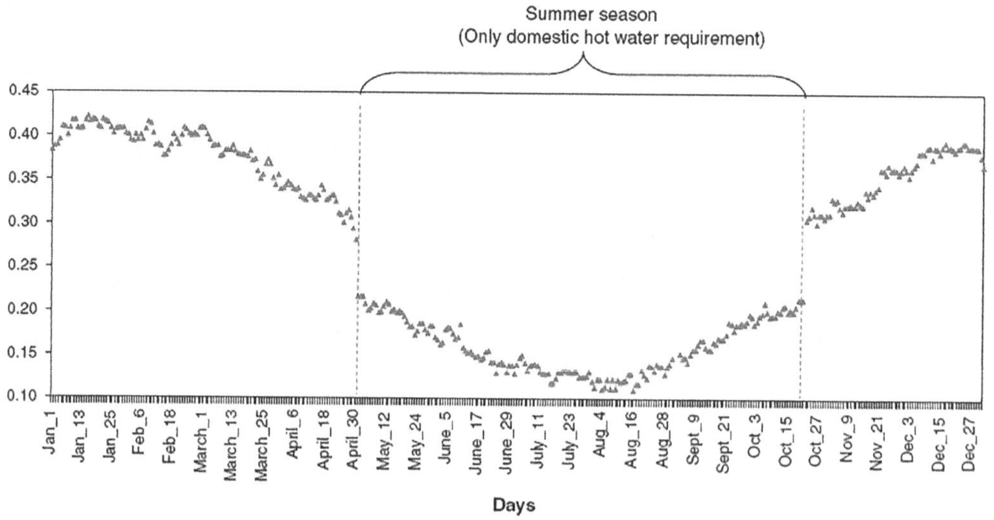

(d)

FIGURE 3 (CONTINUED) (c) Change of energetic reinjection ratio for Edremit GDHS annually. (d) Change of exergetic reinjection ratio for Edremit GDHS annually (modified from Coskun et al.[27])S

1. *Distribution cycle exergetic ratio* (R_{DCEx}): This is defined as the ratio of the exergetic capacity of the district heating water circulating through the district heating distribution network (MW) to the exergetic capacity of the geothermal reservoir in the field considered (MW):

$$R_{\text{DCEx}} = \frac{\dot{E}x_{\text{dhw}}}{\dot{E}x_{\text{gr}}} \tag{8}$$

2. *Energy consumptions circuit exergetic ratio* (R_{ECCEx}): This is defined as the ratio of the exergetic capacity of the fluid circulating through the energy consumptions circuit (building substations) district heating distribution network (MW) to the exergetic capacity of the geothermal reservoir in the field considered (MW):

$$R_{\text{ECCEx}} = \frac{\dot{E}x_{\text{dhw}}}{\dot{E}x_{\text{gr}}} \tag{9}$$

The average values for the enthalpy and entropy to be used in the calculations may be calculated from Eqs. 2 and 3.

3. *Reservoir specific exergy utilization index* (RSExUI): This is defined as the ratio of the exergetic capacity of the geothermal reservoir in the field considered (MJ) to the total heating surface area of all the buildings (m²) and given by

$$\text{RSExUI}\left(\text{MJ}/m^2\right) = \frac{\dot{E}x_{gr}\left(\text{MJ}\right)}{A_{\text{building}}\left(m^2\right)} \tag{10}$$

4. *Geothermal brine specific exergy utilization index* (GBSExUI): This is defined as the ratio of the geothermal brine exergy input value (MJ) to the total heating surface area of all the buildings (m²) and given by

$$\text{GBSxUI}\left(\text{MJ}/m^2\right) = \frac{\dot{E}x_{gb}\left(\text{MJ}\right)}{A_{\text{building}}\left(m^2\right)} \tag{11}$$

Some more details on the use of exergy for design and analysis of geothermal district heating systems are available elsewhere.[30]

Performance Evaluation

From the thermodynamics point of view, exergy is defined as the maximum amount of work that can be produced by a system or a flow of matter or energy as it comes to equilibrium with a reference environment. Exergy is a measure of the potential of the system or flow to cause change, as a consequence of not being completely in stable equilibrium relative to the reference environment. Unlike energy, exergy is not subject to a conservation law (except for ideal, or reversible, processes). Rather, exergy is consumed or destroyed, due to irreversibilities in any real process. The exergy consumption during a process is proportional to the entropy created due to irreversibilities associated with the process.

Exergy analysis is a technique that uses the conservation of mass and conservation of energy principles together with the second law of thermodynamics for the analysis, design, and improvement of geothermal energy systems as well as others. It is also useful for improving the efficiency of energy-resource use, for it quantifies the locations, types, and magnitudes of wastes and losses. In general, more meaningful efficiencies are evaluated with exergy analysis rather than energy analysis, since exergy efficiencies are always a measure of how nearly the efficiency of a process approaches the ideal. Therefore, exergy analysis identifies accurately the margin available to design more efficient energy systems by reducing inefficiencies. We can suggest that thermodynamic performance is best evaluated using exergy analysis because it provides more insights and is more useful in efficiency-improvement efforts than energy analysis. For exergy analysis, the characteristics of a reference environment must be specified. This is commonly done by specifying the temperature, pressure, and chemical composition of the reference environment. The results of exergy analyses, consequently, are relative to the specified reference environment, which, in most applications, is modeled after the actual local environment. The exergy of a system is zero when it is in equilibrium with the reference environment.

Energy and exergy balances for an unsteady-flow process in a system during a finite time interval can be written as:

$$\text{Energy input} - \text{Energy output} = \text{Energy accumulation} \tag{12}$$

$$\text{Exergy input} - \text{Exergy output} - \text{Exergy consumption} = \text{Energy accumulation} \tag{13}$$

For a general steady-state, steady-flow process, it is assumed that the accumulation terms in the above equations become zero. Therefore, input terms become equal to output terms.

For a steady-state, steady-flow process, the mass balance equation can be expressed as

$$\sum_{\dot{m}_{\text{in}}} = \sum_{\dot{m}_{\text{out}}} \tag{14}$$

For a general steady-state, steady-flow process, the general energy and exergy balance equations can also be written more explicitly as

$$\dot{Q} + \sum_{\dot{m}_{\text{in}}} h_{\text{in}} = \dot{W} + \sum_{\dot{m}_{\text{out}}} h_{\text{out}} \tag{15}$$

And

$$\sum\left(1 - \frac{T_0}{T_s}\right)\dot{Q}_s + \sum \dot{m}_{\text{in}}\left(h_{\text{in}} - T_0 s_{\text{in}}\right) - \dot{E}x_d \tag{16}$$

$$= \dot{W} + \sum \dot{m}_{\text{out}}\left(h_{\text{out}} - T_0 s_{\text{out}}\right)$$

We can generally define the energy and exergy efficiencies as follows:

$$\eta = \frac{\dot{E}_{\text{net}}}{\dot{E}_{\text{tot}}} \tag{17}$$

And

$$\psi = \frac{\dot{E}x_{\text{net}}}{\dot{E}x_{\text{tot}}} \tag{18}$$

For further details of both energy and exergy analyses and applications to various energy systems, see Dincer and Rosen.[31]

Note that in the calculations, the temperature T_0 and pressure P_0 of the environment are often taken as standard-state values, such as 1 atm and 25°C. However, these properties may be specified differently depending on the application. T_0 and P_0 might be taken as the average ambient temperature and pressure, respectively, for the location at which the system under consideration operates. Or, if the system uses atmospheric air, T_0 might be specified as the average air temperature. If both air and water from the natural surroundings are used, T_0 would be specified as the lower of the average temperatures for air and water.

Detailed energy and exergy as well as exergoeconomic analysis methodologies for geothermal energy systems are given elsewhere.[22,23]

Case Study

Here, we present a case study analysis of the Salihli Geothermal District Heating System (SGDHS), which has a maximal yield of 87 L/s at an average reservoir temperature of 95°C, with a minimal capacity of 838 MW.[30] The SGDHS was originally designed for 20,000 residences equivalence. Of these, 2400 residences equivalence are heated by geothermal energy as of February 2004. The outdoor and indoor

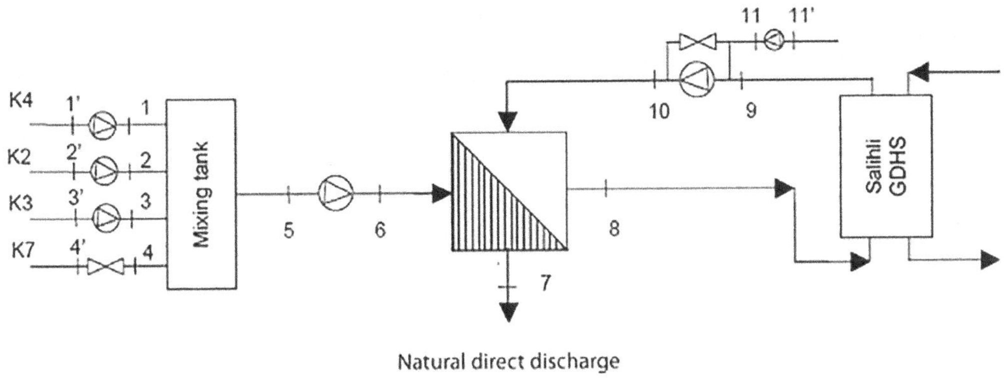

FIGURE 4 A schematic of the SGDHS.
Source: Adapted from Dincer et al.[30]

design temperatures for the system are 4°C and 22°C, respectively. Figure 4 illustrates a schematic of the SGDHS where two hospitals and official buildings heated by geothermal energy were also included. The SGDHS consists mainly of three cycles, such as 1) energy production cycle (geothermal well loop and geothermal heating center loop); 2) energy distribution cycle (district heating distribution network); and 3) energy consumption cycle (building substations). At the beginning of 2004, there were seven wells ranging in depth from 70 to 262 m in the SGDHS. Of these, five wells were in operation at the date studied and two wells (K5 and K6) were out of operation. Four wells (designated as K2, K3, K4, and K7) and one well (K1) are production and balneology wells, respectively. The well head temperatures of the production wells vary from 56°C to 115°C, while the volumetric flow rates of the wells range from 2 to 20 L/s. The geothermal fluid is basically sent to two primary plate-type heat exchangers and is cooled to about 45°C, as its heat is transferred to secondary fluid. The geothermal fluid[30] is discharged via natural direct discharge, with no recharge to Salihli geothermal field production, but reinjection studies are expected to be completed in the near future. The temperatures obtained during the operation of the SGDHS are, on average, 98°C/45°C for the district heating distribution network and 62°C/42°C for the building circuit. By using the control valves for flow rate and temperature at the building main station, the needed amount of water is sent to each housing unit and the heat balance of the system is achieved. Geothermal fluid, collected from the four production wells at an average well heat temperature of 95.5°C, is pumped to the inlet of the heat exchanger mixing tank later through a main collector (from four production wells) with a total mass flow rate of about 47.62 kg/s. Geothermal fluid of intermingling molecules of different species through molecular diffusion was neglected in this study. As a result, not only irreversibility of the mixing tank was assumed equal to zero but also heat losses from the tank and main collector pipeline through the mixing process were neglected. In the SGDHS investigated, the thermal data on pressures, temperatures, and flow rates of the fluids were measured using a number of devices and employed in the models.[22,23]

Results and Discussion

The thermodynamic data used in the case study are based on the actual data taken from the SGDHS on February 1, 2004. The number of the wells in operation in the Salihli geothermal field may normally vary depending on the heating days as well as operating strategy. Taking into account the four productive wells when this study was conducted, the specific exergy index (SExI) is found to be 0.049, which is very close to the limit of the medium-quality geothermal resources. This shows

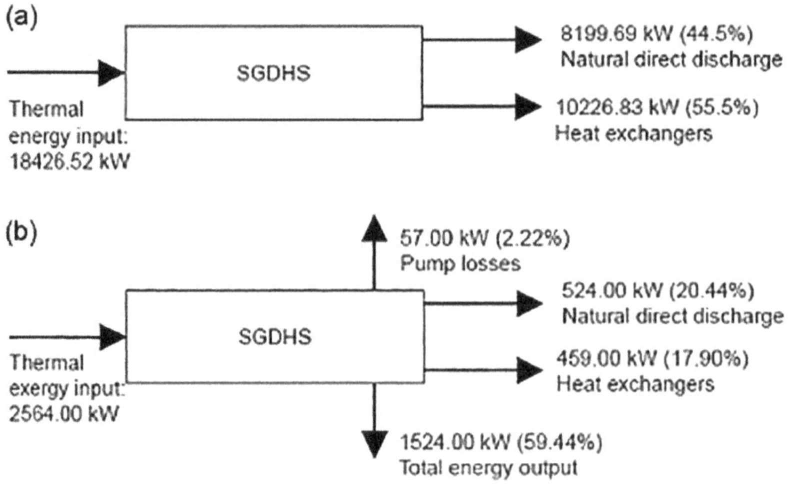

FIGURE 5 System flow diagrams for (a) energy and (b) exergy.
(**Source:** Adapted from Dincer et al.[30]

that the Salihli geothermal field falls into the low-quality geothermal resources according to the classification of Lee.[18]

The total mass flow rates of the geothermal fluid at the inlet of the heat exchanger (the total mass flow rate of the production wells) were measured to be 47.62 kg/s at an average temperature of 44.2°C on February 1, 2004. If hot water losses measured in the district heating distribution network are above 5 m^3/h, the water is generally added via a pump (using pressurized water tanks) to the network in order to cover the leaks taking place. However, the hot water losses were neglected in this study as these losses were below 5 m^3/h on the day of the study.

We present our calculation results, based on the actual data taken, given elsewhere.[30] These include all energetic and exergetic as well as other thermodynamic parameters along with the efficiencies. The energy and exergy efficiencies of the SGDHS are determined to be 55.5% and 59.4%, respectively. Here, the exergy efficiency is higher due to the recirculation and heat recovery processes.

We also present our energy and exergy flow diagrams in Figure 5. As illustrated in Figure 5a, the natural direct discharge of the system accounts for 44.5% of the total energy input. An investigation of the exergy flow diagram given in Figure 5b shows that 40.56% (accounting for about 1040 kW) of the exergy entering the system is lost, while the remaining 59.44% is utilized. The highest exergy loss of 20.44% occurs from the natural direct discharge in this study. The second largest exergy destruction occurs from the heat exchanger with 17.90% (about 459 kW) of the total exergy input. This is followed by the total exergy destruction associated with the pumps amounting to some 57 kW, which accounts for 2.22% of the exergy input to the system.

Conclusions

This entry has presented some historical background on geothermal energy resources and geothermal energy applications and classified geothermal resources by energy and exergy in detail along with some practical cases studies. It has discussed energetic, environmental, and sustainability aspects of geothermal energy and systems and proposed some energetic and exergetic parameters. Also, a procedure for performance assessment of geothermal energy systems is presented. A case study on how both energy and exergy analyses are necessary for better performance of the systems is likewise presented.

Nomenclature

$\dot{E}$	energy rate (kW)
$\dot{E}x$	exergy rate (kW)
$\dot{F}$	exergy rate of the fuel (kW)
h	specific enthalpy (kJ/kg)
IP	improvement potential rate (kW)
$\dot{m}$	mass flow rate (kg/s)
$\dot{P}$	product exergy rate (kW)
$\dot{Q}$	heat transfer rate (kW)
S	specific entropy (kJ/kg/K)
SExI	specific exergy index (dimensionless)
T	temperature (°C or K)
$\dot{W}$	work rate, power (kW)

Greek letters

β	exergetic factor (%)
η	energy efficiency (%)
ψ	exergy efficiency (%)
δ	fuel depletion rate (%)
ξ	productivity lack (%)
χ	relative irreversibility (%)

Subscripts

0	reference environment, dead state
d	natural direct discharge; destruction
HE	heat exchanger
in	inlet
s	source
out	outlet
tot	total

References

1. Fridleifsson, I.B. Status of geothermal energy amongst the world's energy sources. Geothermics **2003**, *32*, 379–388.
2. Dickson, M.H.; Fanelli, M. *What is geothermal energy?* Prepared on February 2004. Available at http://www. geothermal-energy.org/314,what_is_geothermal_energy.html (accessed June 2010).
3. Dickson, M.H.; Fanelli, M., Eds.; *Geothermal Energy: Utilization and Technology;* UNESCO Renewable Energy Series, UNESCO Publishing: Paris, 2003; 205 pp.
4. Cataldi, R.; Hodgson, S.F.; Lund, J.W., Eds.; *Stories from a Heated Earth—Our Geothermal Heritage;* Geothermal Resources Council and International Geothermal Association: Davis, CA, 1999; 569 pp.
5. Lund, J.W.; Freeston, D.H.; Boyd, T.L. Direct utilization of geothermal energy 2010 worldwide review. Geothermal Congress 2010, Bali, Indonesia, 25–29 April, 2010; 23 pp.

6. Balta, M.T.; Dincer, I.; Hepbasli, A. Geothermal-based hydrogen production using thermochemical and hybrid cycles: A review and analysis. Int. J. Energy Res., **2010**, *34*, 757–775.
7. Balta, M.T.; Dincer, I.; Hepbasli, A. Thermodynamic assessment of geothermal energy use in hydrogen production. Int. J. Hydrogen Energy, **2009**, *34* (7), 2925-2939.
8. Wright, M. Nature of geothermal resources. In *Geothermal Direct-Use Engineering and Design Guidebook;* Lund, J.W., Ed.; Geo-Heat Center: Klamath Falls, OR, 1998; 27–69.
9. Lund, J.W.; Bjelm, L.; Bloomquist, G.; Mortensen, A.K. Characteristics, development and utilization of geothermal resources—A Nordic perspective. Episodes **2008**, *31* (1), 140–147.
10. Dincer, I. Renewable energy and sustainable development: A crucial review. Renewable Sustainable Energy Rev. **2000**, *4* (2), 157–175.
11. Dincer, I.; Rosen, M.A. Thermodynamic aspects of renewables and sustainable development. Renewable Sustainable Energy Rev. **2005**, *9* (2), 169–189.
12. Lund, J.W. Chena Hot Springs. Geo-Heat Center Quarterly Bulletin **2006**, *27* (3) (September), Klamath Falls, OR, 2–4.
13. Dickson, M.H.; Fanelli, M. Geothermal energy and its utilization. In *Small Geothermal Resources*; Dickson, M.H., Fanelli, M., Eds.; UNITAR/UNDP Centre for Small Energy Resources: Rome, Italy, 1990; 1–29.
14. World Bank, Geothermal Energy, available at http://www.worldbank.org/html/fpd/energy/geothermal/technology.htm (accessed October 23, 2005).
15. Sanyal, S.K. Classification of geothermal systems—A possible scheme. Proceedings of Thirtieth Workshop on Geothermal Reservoir Engineering, SGP-TR-176, Stanford University: Stanford, California, January 31–February 2, 2005, 8 pp.
16. Sanyal, S.K.; Butler, S.J. National Assessment of U.S. Enhanced Geothermal Resources Base—A Perspective, Trans.; Geothermal Resources Council: Palm Springs, California, August-September, 2004; Vol. 28, 233–238.
17. Lee, K.C. Classification of geothermal resources by exergy. Geothermics **2001**, *30*, 431–442.
18. Lee, K.C. Classification of geothermal resources: An engineering approach. Proceedings, Twenty-First Workshop on Geothermal Reservoir Engineering; Stanford University, 1996; 5 pp.
19. Quijano J. Exergy analysis for the Ahuachapan and Berlin Geothermal fields, El Salvador. Proceedings World Geothermal Congress, May 28–June 10; Kyushu-Tohoku, Japan, 2000.
20. Ozgener, L.; Hepbasli, A.; Dincer, I. Thermo-mechanical exergy analysis of Balcova geothermal district heating system in Izmir, Turkey. ASME J. Energy Resour. Technol. **2004**, *126*, 293–301.
21. Baba, A.; Ozgener, L.; Hepbasli, A. Environmental and exergetic aspects of geothermal energy. Energy Sources **2006**, *28* (7), 597–609.
22. Ozgener, L.; Hepbasli, A.; Dincer, I. Energy and exergy analysis of the Gonen geothermal district heating system, Turkey. Geothermics **2005**, *34*, 632–645.
23. Ozgener, L.; Hepbasli, A.; Dincer, I. Exergy analysis of Salihli geothermal district heating system in Manisa, Turkey. Int. J. Energy Res. **2005**, *29* (5), 398–408.
24. Etemoglu, A.B.; Can, M. Classification of geothermal resources in Turkey by exergy analysis. Renewable Sustainable Energy Rev. **2007**, *11*, 1596–1606.
25. Lindal, B. Industrial and other applications of geothermal energy. In *Geothermal Energy*; Armstead, H.C.H., Ed.; Earth Science 12; UNESCO: Paris, 1973; 135–148.
26. Gudmundsson, J.S.; Freeston, D.H.; Lienau, P.J. The Lindal diagram. Trans. Geotherm. Resour. Counc. **1985**, *9*, Part 1, August.
27. Coskun, C.; Oktay, Z.; Dincer, I. New energy and exergy parameters for geothermal district heating systems. Appl. Therm. Eng. **2009**, *29* (11–12), 2235–2242.
28. Hepbasli, A. A review on energetic, exergetic and exergoeconomic aspects of geothermal district heating systems (GDHSs). Energy Convers. Manage. **2010**, *51* (10), 2041–2061.

29. Ozgener, L.; Hepbasli, A.; Dincer, I. A key review on performance improvement aspects of geothermal district heating systems and applications. Renewable Sustainable Energy Rev. **2007**, *11* (8), 1675–1697.
30. Dincer, I.; Hepbasli, A.; Ozgener, L. Geothermal energy resources. In *Encyclopedia of Energy Engineering and Technology;* Capehart, B.L., Ed.; CRC Press, Taylor and Francis Group, 2007; Vol. 1, 744–752.
31. Dincer, I.; Rosen, M.A., Exergy, Elsevier: New York, 2007.

14

Green Energy

Ibrahim Dincer
and Adnan Midili

Introduction

Green energy can be considered a catalyst for energy security, sustainable development, and social, technological, industrial, and governmental innovations in a country. An increase in the green energy consumption of a country provides a positive impact on the economic as well as the social development of the country. Moreover, the supply and utilization of low-priced green fuel are particularly significant for global stability and peace because energy plays a vital role in industrial and technological developments around the world.

Critical energy issues in the 21st century will likely include energy security for almost 7 billion people, the expected global population by the middle of the 21st century, and global warming, mainly caused by CO_{2s} emissions generated from the combustion of fossil fuels.[1,2] Fossil fuels have caused some major problems for human health and human welfare due to their extensive use in various industrial and nonindustrial sectors. In reality, the main source of these problems is the extensive use of fossil-based technologies and strategies used by humans to govern throughout the centuries. In recent decades, the world has struggled with shortages of fossil fuels, pollution, and increased global energy requirements due to fast population growth, fast technological developments, and higher living standards. These factors have led to world population transition, migration, hunger, environmental (especially air and water pollution) problems, deteriorating health and disease, terrorism, energy and natural resources concerns, and wars. Also, problems with energy supply and use are related not only to global unrest, but also to such environmental concerns as air pollution, acid precipitation, ozone depletion, forest destruction, and the emission of radioactive substances. These issues must be taken into consideration simultaneously if humanity is to achieve a bright energy future with minimal environmental impact.

Other environmental considerations have been given increasing attention by energy industries and the public. The concept that consumers share responsibility for pollution and its cost has been increasingly accepted. In some jurisdictions, the prices of many energy resources have increased over the last one to two decades, in part to account for environmental costs. Global demand for energy services is expected to increase by as much as an order of magnitude by 2050, while primary-energy demands are expected to increase by 1.5 ± 0.3 times.[2] Accordingly, humans have reached energy shortage and pollution levels that are not tolerable anymore. The urgent need in this regard is to develop green energy–based permanent solutions for a sustainable future without any negative environmental and societal impacts. As a consequence, investigations for green energy should be encouraged particularly for green energy-based sustainability and future world stability.

In this book contribution, the authors present key information on green energy–based sustainability and global stability in accordance with major considerations that are presented in the following subtitles. In addition, this presents some key parameters like the green energy impact ratio and the green energy–based sustainability ratio. Moreover, anticipated patterns of future energy use and consequent environmental impacts (focusing on acid precipitation, stratospheric ozone depletion, and the greenhouse effect) are comprehensively discussed. Also, potential solutions to current environmental problems are identified along with green energy technologies. The relationships between green energy and sustainable development are described. Throughout the entry, several issues relating to green energy, the environment, and sustainable development are examined from both current and future perspectives. The specific objectives of this entry can be enumerated as follows:

- To help understand the main concepts and issues about green energy use and sustainability aspects
- To develop relationships between green energy use and sustainability development
- To encourage the strategic use and conservation of green energy sources
- To provide methods for energy security, implementation, and development
- To increase motivation on the implementation of green energy strategies for better energy supply
- To provide solutions to reduce negative environmental impacts by considering the possible green energy strategies
- To discuss possible green energy strategies for sectoral use

Green Energy

The most important property of green energy sources is environmental compatibility. In line with this characteristic, green energy sources will likely become the most attractive energy sources in the near future and will be the most promising energy sources for the technological and environmental perspectives of the 21st century, particularly in the context of sustainable development for the future. Green energy is one of the main factors that must be considered in discussions of sustainable development and future world stability. Several definitions of sustainable development have been put forth, including the following common one: "development that meets the needs of the present without compromising the ability of future generations to meet their own needs."[3,4] There are many factors that can contribute to achieving sustainable development. One of them is the requirement for a supply of energy resources that is fully sustainable. A secure supply of green energy resources is generally agreed to be a necessary but not sufficient requirement for development within a society.

Furthermore, sustainable development within a society demands a sustainable supply of green energy sources and an effective and efficient utilization of green energy technologies. Green energy can be defined as an energy source that has zero or minimum environmental impact, is more environmentally benign and more sustainable, and is produced from solar, hydro, biomass, wind, and geothermal, etc., energies. These types of green energy reduce the negative effects of fossil energy resources and the overall emissions from electricity generation and decrease greenhouse gases. They give an opportunity to take an active role in improving the environment and meet clean energy demand for both industrial and nonindustrial applications. Considering the benefits of green energy, it can be said that the sustainability of green energy supply and progress is assumed to be a key element in the interactions between nature and society. In addition, sustainable development requires a supply of energy resources that is sustainably available at reasonable cost and causes no or minimal negative societal impacts. Clearly, energy resources such as fossil fuels are finite and thus lack the characteristics needed for sustainability while others such as green energy sources are sustainable over the relatively long term.[5] Particularly, low-priced green energy is the most essential means for increasing the sustainable technological development and industrial productivity as well as people's living standards in a society. Therefore, achieving solutions to the energy shortages and environmental problems presented today requires long-term potential actions for sustainable development. In this regard, green energy sources and technologies appear to be the one of the most efficient and effective solutions. It can be said that one solution to the impending energy shortage and environmental problems is to make much more use of green energy sources and technologies. Another important solution is to develop permanent and effective sustainable green energy strategies for the increase of sustainable global stability.[1,6] For these purposes, engineering practicality, reliability, applicability, economy, scarcity of supply, and public acceptability should also be considered.

Green Energy and Environmental Consequences

Some of the global problems affecting world stability are presented in Figure 1. Fossil fuel utilization effects such as global climate change, world energy conflicts, and energy source shortages have increasingly threatened world stability. These negative effects may be observed locally, regionally, and globally.

This concern arises due to world population growth, fast technological development, and increasing energy demand. In the past, fossil energy sources could be used to solve world energy problems. However, in this century, fossil fuels cannot continue indefinitely as the principal energy sources due to the rapid increase of world energy demand and energy consumption. Due to world population growth and the advance of technologies that depend on fossil fuels, the reserves of fossil fuels eventually will

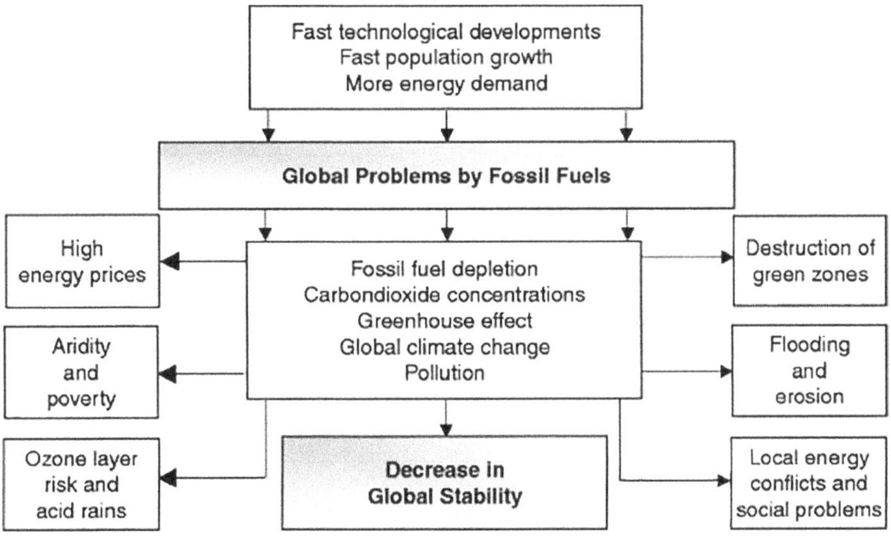

FIGURE 1 Problems affecting global stability.

not be able to meet world energy demand. Energy experts point out the fact that reserves account for less than 40 years for petroleum, 60 years for natural gas, and 250 years for coal.[1] The increase of the energy consumption and energy demand indicates our dependence on the fossil fuels. If the increase of fossil fuel utilization continues in this manner, it is likely that the world will be affected by many negative problems due to fossil fuels. As one outcome, global stability will probably decrease. This effect is presented as flow chart in Figure 2. For instance, shortages of fossil fuel reserves and global environmental problems will likely lead to global unrest throughout the world. As a result, local, regional, and world conflicts may appear across the world.

To further support these arguments, the observed and predicted consumptions of world primary energy, fossil fuels, and green energy from 1965 to 2050 are displayed in Figure 3 based on the data taken from literature.[7,8]

According to Figure 3, the quantities of world primary energy consumption are expected to reach 16502 Mtoe (Million tons of oil equivalent) by the year 2050. World population is now over 6 billion, double the population of 40 years ago, and it is likely to double again by the middle of the 21st century. The world's population is expected to rise to about 7.0 billion by 2010. Even if birth rates fall so that the world population becomes stable by 2050, the population will still be about 10 billion. Because the population is expected to increase drastically, conventional energy resource shortages are likely due to insufficient fossil fuel resource supplies. In the near future, therefore, green energy will become increasingly important to compensate for shortages of conventional resources. The correlations that have been applied here are as follows:

For world primary energy consumption:

$$\text{WPEC} = 143.57 \times Y - 277808 \left(R^2 = 0.9902\right) \tag{1}$$

For world fossil fuel consumption:

$$\text{WPEC} = 113.61 \times Y - 219092 \left(R^2 = 0.9968\right) \tag{2}$$

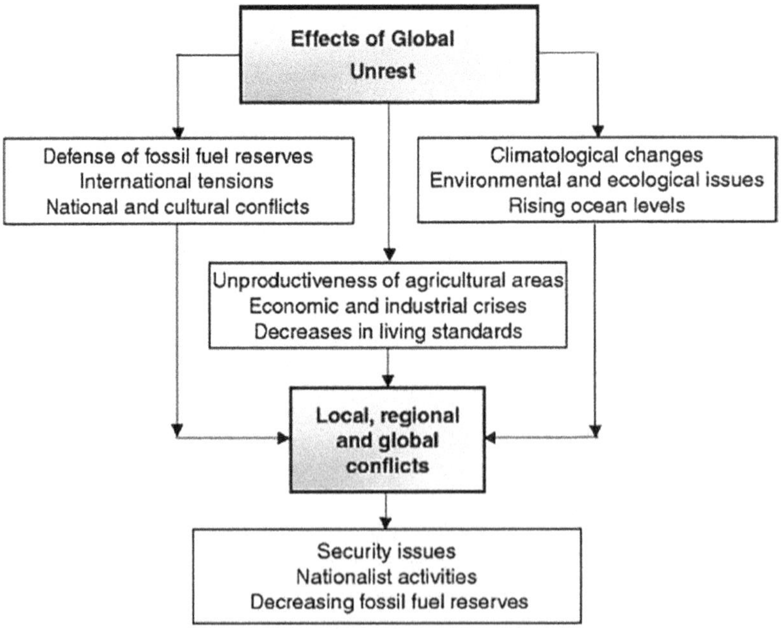

FIGURE 2 Some possible effects and results of the global unrest.

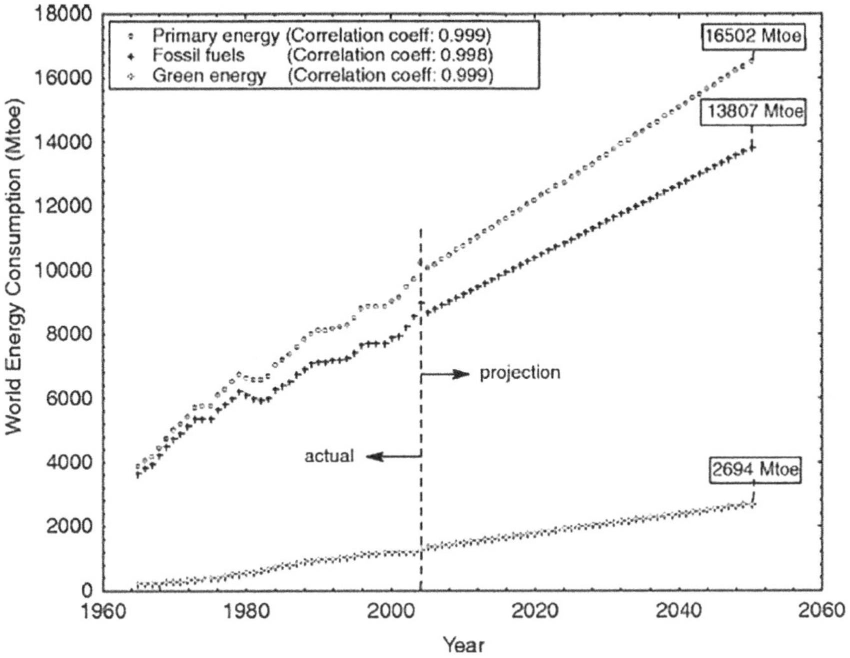

FIGURE 3 Observed and projected quantities of world primary energy, fossil fuels, and green energy consumptions (Mtoe: million tons of oil equivalent).

For world green energy consumption:

$$\text{WPEC} = 29.956 \times Y - 58715 \left(R^2 = 0.9985\right) \tag{3}$$

Here, WPEC is world primary energy consumption (Mtoe); WFFC, world fossil fuel consumption (Mtoe); WGEC, world green energy consumption; Y is time (years); and R is the correlation coefficient.

Energy shortages will accelerate the fluctuations of energy prices and economic recessions and decrease living standards and increase the unrest among countries. Decreased available fossil fuel reserves and increased fuel costs since the middle to late 1900s have led to variations in lifestyles and standards of life. These effects have, in some regions, decreased living standards of entire societies. Countries that need more energy resources have been purchasing cheaper energy sources. Countries that look after the future welfare of their societies have received the attention of countries that possess fossil fuel supplies, posing the potential for world conflict. Problems are often attributed to decreases in fossil fuel energy reserves. Those who seek a clean world must find appropriate alternatives to fossil fuels. Why not green energy? Green energy that is abundantly available all over the world can help:

- Reduce or stop conflicts among countries regarding energy reserves
- Facilitate or necessitate the development of new technologies
- Reduce air, water, and land pollution and the loss of forests
- Reduce illnesses and deaths caused by direct or indirect use of energy

Therefore, green energy and related technologies are needed to improve global stability by reducing the harmful effects of fossil-based energy consumption. Thus, the importance of green energy in reducing world problems and achieving a sustainable energy system should be emphasized and a transition to the green economy should be encouraged.

Energy, Environment, and Sustainability

Environmental concerns are an important factor in sustainable development. For a variety of reasons, activities that continually degrade the environment are not sustainable over time, e.g., the cumulative impact on the environment of such activities often leads over time to a variety of health, ecological, and other problems. A large portion of the environmental impact in a society is associated with its utilization of energy resources. Ideally, a society seeking sustainable development utilizes only energy resources that release no or minimal emissions to the environment and thus cause no or little environmental impact. However, because all energy resources lead to some environmental impact, it is reasonable to suggest that some (not all) of the concerns regarding the limitations imposed on sustainable development by environmental emissions and their negative impacts can be in part overcome through increased energy efficiency. Clearly, a strong relation exists between energy efficiency and environmental impact because for the same services or products less resource utilization and pollution is normally associated with increased energy efficiency. While not all green energy resources are inherently clean, there is such a diversity of choices that a shift to renewables carried out in the context of sustainable development could provide a far cleaner system than would be feasible by tightening controls on conventional energy.[9] Furthermore, being by nature site-specific, they favor power system decentralization and locally applicable solutions more or less independent of the national network. It enables citizens to perceive positive and negative externalities of energy consumption. Consequently, the small scale of the equipment often makes the time required from initial design to operation short, providing greater adaptability in responding to unpredictable growth and changes in energy demand. In this regard, sustainability often leads local and national authorities to incorporate environmental considerations into energy planning. The need to satisfy basic human needs and aspirations combined with the increasing world population will make the need for successful implementation of sustainable development increasingly apparent. Various criteria that are essential to achieving sustainable development in a society are as follows:[3,4,10]

- Information about and public awareness of the benefits of sustainability investments, environmental education, and training
- Appropriate energy strategies
- The availability of green energy resources and technologies
- A reasonable supply of financing
- Monitoring and evaluation tools

Environmental concerns are significantly linked to sustainable development. Activities which continually degrade the environment are not sustainable. For example, the cumulative impact on the environment of such activities often leads over time to a variety of health, ecological, and other problems. Clearly, a strong relation exists between efficiency and environmental impact because for the same services or products, less resource utilization and pollution is normally associated with increased efficiency.[3] Improved energy efficiency leads to reduced energy losses. Most efficiency improvements produce direct environmental benefits in two ways. First, operating energy input requirements are reduced per unit of output and pollutants generated are correspondingly reduced. Second, consideration of the entire life cycle for energy resources and technologies suggests that improved efficiency reduces environmental impact during most stages of the life cycle. In recent years, the increased acknowledgment of humankind's interdependence with the environment has been embraced in the concept of sustainable development. With energy constituting a basic necessity for maintaining and improving standards of living throughout the world, the widespread use of fossil fuels may have impacted the planet in ways far more significant than first thought. In addition to the manageable impacts of mining and drilling for fossil fuels and discharging wastes from processing and refining operations, the greenhouse gases created by burning these fuels is regarded as a major contributor to a global warming threat.

Global warming and large-scale climate change have implications for food chain disruption, flooding, and severe weather events.

Therefore, use of green energy sources can help reduce environmental damage and achieve sustainability. The attributes of green energy technologies (e.g., modularity, flexibility, low operating costs) differ considerably from those for traditional, fossil-fuel-based energy technologies (e.g., large capital investments, long implementation lead times, operating cost uncertainties regarding future fuel costs). Green energy technologies can provide cost-effective and environmentally beneficial alternatives to conventional energy systems. Some of the benefits of the green energy- based systems are as follow:[3,10]

- They are relatively independent of the cost of oil and other fossil fuels, which are projected to rise significantly over time. Thus, cost estimates can be made reliably for green energy systems and they can help reduce the depletion of the world's nongreen energy resources.
- Implementation is relatively straightforward.
- They normally cause minimum environmental degradation and so they can help resolve major environmental problems. Widespread use of green energy systems would certainly reduce pollution levels.
- They are often advantageous in developing countries. In fact, the market demand for renewable energy technologies in developing nations will likely grow as they seek a better standard of living.

Under these considerations, green energy resources have some characteristics that lead to problematic but often solvable technical and economic challenges:

- Generally diffuse
- Not fully accessible
- Sometimes intermittent
- Regionally variable

The overall benefits of green energy technologies are often not well understood, leading to such technologies often being assessed as less cost-effective than traditional technologies. For green energy technologies to be assessed comprehensively, all of their benefits must be considered. For example, many green energy technologies can provide, with short lead times, small incremental capacity additions to existing energy systems. Such power generation units usually provide more flexibility in incremental supply than large devices like nuclear power stations.

Green Energy and Sustainable Development

Sustainability has been called a key to the solution of current ecological, economic, and developmental problems by Dincer and Rosen[10] A secure supply of energy resources is generally agreed to be a necessary but not sufficient requirement for development within a society. Furthermore, sustainable development demands a sustainable supply of energy resources that, in the long term, is readily and sustainably available at reasonable cost and can be utilized for all required tasks without causing negative societal impacts. Supplies of such energy resources as fossil fuels (coal, oil, and natural gas) and uranium are generally acknowledged to be finite; other energy sources such as sunlight, wind and falling water are generally considered renewable and therefore sustainable over the relatively long term. Wastes (convertible to useful energy forms through, for example, waste-to-energy incineration facilities) and biomass fuels are also usually viewed as sustainable energy sources.

In general, the implications of these statements are numerous and depend on how sustainable is defined.[3] For sustainable development, green energy can play an important role for meeting energy requirements in both industrial and local applications. Therefore, development and utilization of green energy sources and technologies should be given a high priority for sustainable development in a country. The need for sustainable energy development is increasing rapidly in the world. Widespread use of green energy sources and technologies is important for achieving sustainability in the energy sectors in both developing and industrialized countries. Thus, green energy resources and technologies are a key component of sustainable development for three main reasons:

- They generally cause less environmental impact than other energy sources. The variety of green energy resources provides a flexible array of options for their use.
- They cannot be depleted. If used carefully in appropriate applications, green energy resources can provide a reliable and sustainable supply of energy almost indefinitely.
- They favor system decentralization and local solutions that are somewhat independent of the national network, thus enhancing the flexibility of the system and providing economic benefits to small isolated populations. Also, the small scale of the equipment often reduces the time required from initial design to operation, providing greater adaptability in responding to unpredictable growth and changes in energy demand.
- Major considerations involved in the development of green energy technologies for sustainable development as modified from[10] are presented in Figure 4.

As a consequence, it can be said that green energy and technologies are definitely needed for sustainable development that ensures a minimization of global unrest. The relation between green energy and sustainability is of great significance to the developed countries as well as developing or less developed countries. Moreover, examining the relations between green energy sources and sustainability makes it clear that green technology is directly related to sustainable development. Therefore, attaining

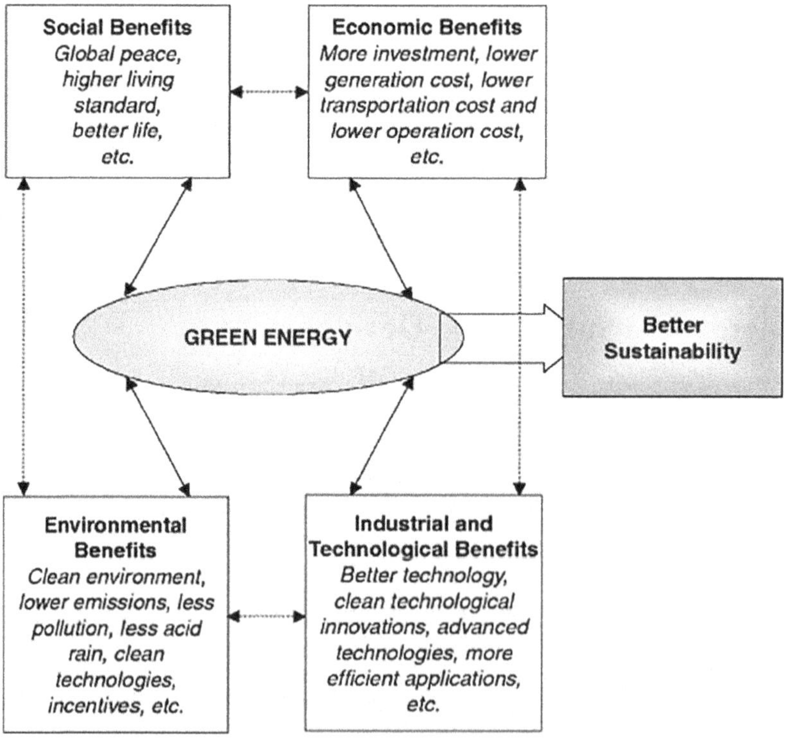

FIGURE 4 Major considerations involved in the development of green energy technologies for sustainable development.

sustainable development requires that green energy resources be also used, and is assisted if resources are used efficiently.[4,10] Thus, if sustainable green energy technologies are effectively put into practice, the countries may maximize the benefits from the green energy sources and technologies while minimizing the global unrest[1] associated with the use of fossil fuel energy sources. It is expected that this initiative can contribute to development over a longer period of time, i.e., to make development more sustainable.

Green Energy Resources and Technologies

Since the oil crises in the early 1970s, there has been active worldwide research and development in the field of green energy resources and technologies. During this time, energy conversion systems that were based on green energy technologies appeared to be most attractive because of facts such as the projected high cost of oil and the cost effectiveness estimates and easy implementation of green energy technologies. Furthermore, in more recent times, it has been realized that green energy sources and technologies can have a beneficial impact on essential technical, environmental, economic, and political issues of the world. As pointed out by Hartley,[3,9] green energy technologies produce marketable energy by converting natural phenomena into useful energy forms. These technologies use the energy inherent in sunlight and its direct and indirect impacts on the earth (photons, wind, falling water, heating effects, and plant growth), gravitational forces (the tides), and the heat of the earth's core (geothermal) as the resources from which they produce energy. These resources represent a massive energy potential which dwarfs that of equivalent fossil resources. Therefore, the magnitude of these is not a key constraint on energy production.

However, they are generally diffused and not fully accessible, some are intermittent, and all have distinct regional variabilities. Such aspects of their nature give rise to difficult, but solvable, technical, institutional, and economical challenges inherent in development and the use of green energy resources. Despite having such difficulties and challenges, the research and development on green energy resources and technologies has been expanded during the past two decades because of the facts listed above. Recently, significant progress has been made by (1) improving the collection and conversion efficiencies, (2) lowering the initial and maintenance costs, (3) increasing the reliability and applicability, and (4) understanding the phenomena of green energy technologies.

Green energy technologies become important as environmental concerns increase, utility (hydro) costs climb, and labor costs escalate.[3,9] The uncertain global economy is an additional factor. The situation may be turned around with an increase in research and development in the advanced technologies fields, some of which are closely associated with green energy technologies. This may lead to innovative products and job creation supported by the governments. The progress in other technologies, especially in advanced technologies, has induced some innovative ideas in green energy system designs. The ubiquitous computer has provided means for optimizing system performance, costs/benefits, and environmental impacts even before the engineering design was off the drawing board. The operating and financial attributes of green energy technologies, which include modularity and flexibility and low operating costs (suggesting relative cost certainty) are considerably different than those for traditional, fossil-based technologies, whose attributes include large capital investments, long implementation lead times, and operating cost uncertainties regarding future fuel costs. The overall benefits of green energy technologies are often not well understood and consequently they are often evaluated to be not as cost effective as traditional technologies. In order to comprehensively assess green energy technologies, however, some of their benefits that are often not considered must be accounted for. Green energy technologies, in general, are sometimes seen as direct substitutes for existing technologies so that their benefits and costs are conceived in terms of assessment methods developed for the existing technologies. Many government organizations and universities recognize the opportunity and support the efforts to exploit some commercial potential by:

- Analyzing opportunities for green energy and working in consultation with industry to identify research, development, and market strategies to meet technological goals.
- Conducting research and development in cooperation with industry to develop and commercialize technologies.
- Encouraging the application of green energy technologies to potential users, including utilities.
- Providing technical support and advice to industry associations and government programs that are encouraging the increased use of green energy. In order to realize the energy and the economic and environmental benefits that green energy sources offer, the following integrated set of activities should be acted on accordingly.[3,9,10]
- Conducting research and development. The priorities should be set in close consultation with industry to reflect their needs. Most research is conducted through cost- shared agreements and falls within the short-to-medium term. Partners in research and development should include a variety of stakeholders in the energy industry such as private sector firms, utilities across the country, provincial governments, and other federal departments.
- Assessing technology. Data should be gathered in the lab and through field trials on factors such as cost benefit, reliability, environmental impact, safety, and opportunities for improvement. This data should also assist the preparation of technology status overviews and strategic plans for research and development.
- Developing standards. The development of technical and safety standards is needed to encourage the acceptance of proven technologies in the marketplace. Standards development should be conducted in cooperation with national and international standards writing organizations as well as other national and provincial regulatory bodies.
- Transferring technology. Research and development results should be transferred through the sponsorship of technical workshops, seminars, and conferences, as well as through the development of training manuals and design tools and the publication of technical reports. Such activities will also encourage potential users to consider the benefits of adopting green energy technologies. In support of developing near-term markets, a key technology transfer area is to accelerate the use of green energy technologies in a country's remote communities.

Such activities will also encourage potential users to consider the benefits of adopting renewable energy technologies. In support of developing near-term markets, a key technology transfer area is to accelerate the use of green energy technologies in a country's remote communities.

Essential Factors for Sustainable Green Energy Technologies

There are various essential parameters, as outlined and detailed in Figure 5. These factors can help in identifying and achieving required green energy strategies and technologies for sustainable development. As shown in Figure 5, green energy technologies are largely shaped by broad and powerful trends that are rooted in basic human needs. In conjunction with this, the increasing world population requires the definition and successful implementation of green energy technologies. Briefly, the important parameters and their interrelations, as outlined in Figure 5, are definitely required to carry out the best green energy program and to select the most appropriate green energy technologies for sustainable development.

Exergetic Aspects of Green Energy Technologies

The impact of energy resource utilization on the environment and the achievement of increased resource-utilization efficiency are best addressed by considering exergy. The exergy of an energy form or a substance is a measure of its usefulness, quality, or potential to cause change and provide the basis for an effective measure of the potential of a substance or energy form to impact the environment. In practice,

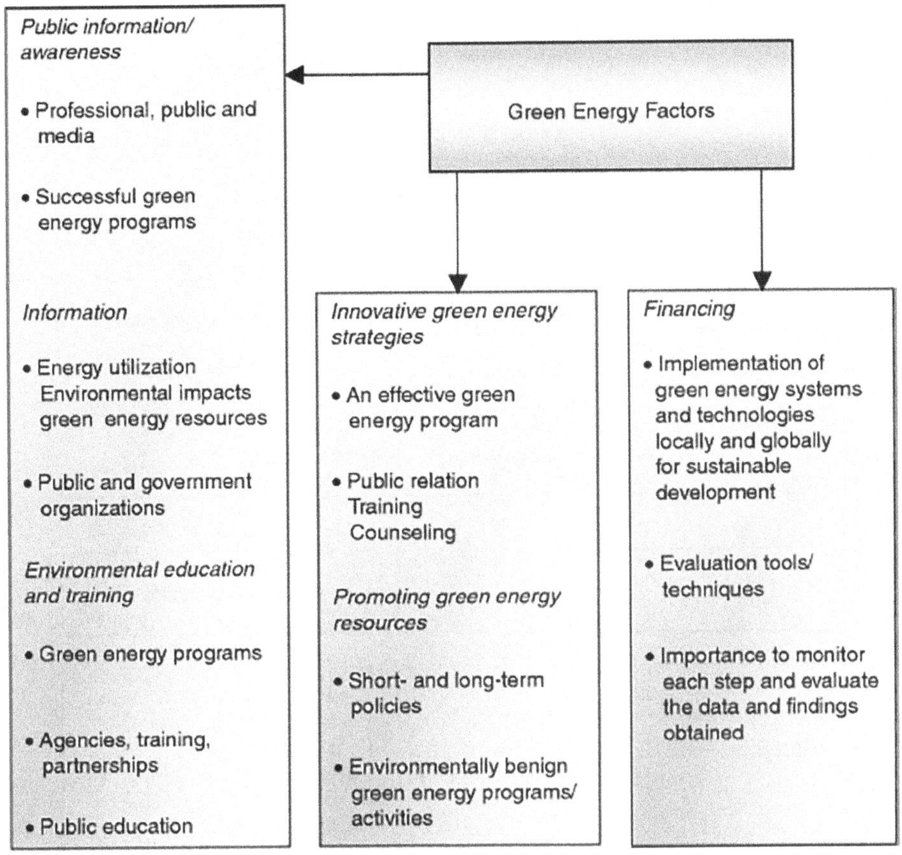

FIGURE 5 Essential factors for green energy technologies.

a thorough understanding of exergy and the insights it can provide into the efficiency, environmental impact, and sustainability of green energy technologies are helpful if not required for the engineer or scientist working in the area of green energy–based environmental systems.[4,11] For green energy technologies, applications of exergy methods can have numerous broad and comprehensive benefits:

- A better knowledge of the efficiencies and losses for the technologies and systems and how they behave and perform
- A clearer appreciation of the environmental impacts of green energy technologies as well as the mitigation of environmental that they can facilitate
- Better identification of the ways green energy technologies can contribute to sustainable development

Green Energy Applications

Green energy technologies are expected to play a key role in sustainable energy scenarios for the future. The foremost factor that will determine the specific role of green energy and technologies will likely be energy demand. Therefore, in order to compensate the energy requirement, it will be possible to produce green energy from renewable energy sources such as hydraulic, solar, wind, geothermal, wave and biomass, etc. If so, the green energy and technologies can be utilized for many application fields as shown in Figure 6.

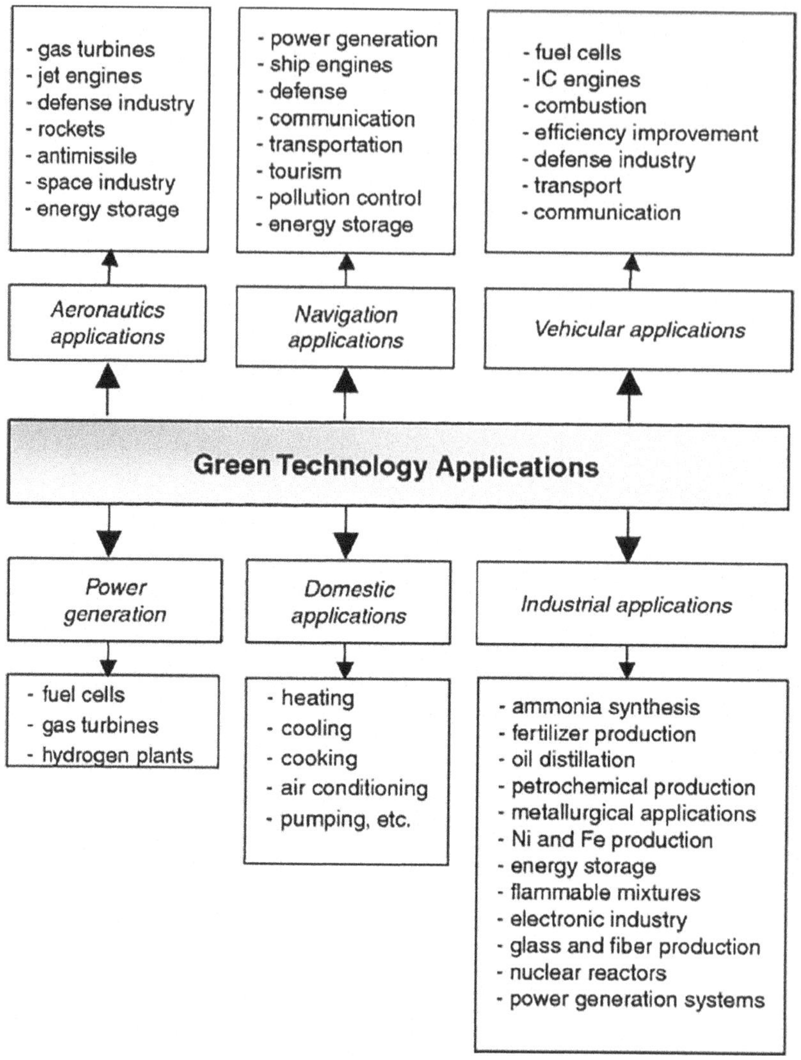

FIGURE 6 Possible application fields of green energy and technologies for sustainable

Thus, it can be said that green energy and technologies, which are abundantly available, can help:

- Provide more environmentally benign and more sustainable future
- Increase energy security
- Facilitate or necessitate the development of new, clean technologies
- Reduce air, water, and soil pollution and the loss of forests
- Reduce energy-related illnesses and deaths
- Reduce or stop conflicts among countries regarding energy reserves

Therefore, green energy and related technologies are needed to ensure global stability by reducing the harmful effects of fossil-based energy consumption. Thus, the importance of green energy in reducing the world problems and achieving a sustainable energy system should be emphasized considering the sustainable green energy strategies; and a transition to green energy economy should be encouraged and developed countries in particular should increase investments in green energy and technologies.

Green Energy Analysis

In light of the above major considerations, in order to accelerate the use of green energy sources and technologies and the implementation of green energy strategies, and in this regard, to describe the relationship between the global stability and sustainable development, some key steps are presented as follows:

- Key strategies
- Green energy based sustainability ratio
- Global unrest and peace

Green Energy-Based Sustainable Development Ratio

In order to discuss the key role of green energy for sustainable development and global stability, the general algebraic form of the equation that is the green energy-based sustainable development ratio should be presented based on the above strategies. For this purpose, the works that were early presented by Midilli et al.[1,2,6,12] and Dincer[3,10] were taken as the reference basis. The following important parameters, which are sectoral, technological, and practical application impact ratios, are taken into consideration to estimate the green-energy-based sustainable development ratio. The detailed derivations of these parameters are presented in the literature.[2] Briefly, the green energy–based sustainable development ratio is mainly based on the following parameters:

- *Sectoral impact ratio,* (R_{si}) (ranging from 1 to 1/3), is based on the provided financial support ($C_{p, si}$) of public, private, and media sectors for transition to green energy–based technologies, and depending on the total green energy financial budget (C_{geb}), as a reference parameter.[2]
- *Technological impact ratio,* (R_{ti}) (ranging from 1 to 1/3), is based on the provided financial support ($C_{p, ti}$) for research and development, security, and analysis of green energy–based technologies, and also depending on the total green energy financial budget (C_{geb}) as a reference parameter.[2]
- *Practical application impact ratio,* (R_{pai}) (ranging from 1 to 1/3), is based on the provided financial support ($C_{p, pai}$) for projection, production, conversion, marketing, distribution, management, and consumption of green fuel from green energy sources, and also depending on the total green energy financial budget (C_{geb}), as a reference parameter.[2] Here, it should be emphasized that these parameters can be defined as the ratio of the provided financial support to the total green energy financial budget in a country. In addition, it should be always remembered that it is assumed that the financial share of each parameter is equal to one-third of the total green energy financial budget in a country. The utilization ratio of green energy depends on the green energy utilization and the world primary energy quantity. Green energy–based global stability is a function of utilization ratios of coal, petroleum, and natural gas, and utilization ratios of green energy at certain utilization ratios of these fuels, and utilization ratio of green energy. Therefore, the green-energy-based sustainable development ratio is written in algebraic form as follows:[2]

$$R_{ges} = R_{gei} \times R_{geu} \tag{4}$$

where

$R_{gei} = \left(R_{si} + R_{ti} + R_{pai}\right)$ green energy impact ratio; $R_{si} = C_{p,si}/C_{geb}$ sectoral impact ratio, which is estimated based on the provided financial support of the public, private, and media sectors for transition to green energy ($C_{p, si}$) and also the green energy budget allocation of a country:(C_{geb});

$R_{ti} = C_{p,ti}/C_{geb}$ technological impact ratio, which is estimated based on the provided financial supports for research and development, security, and analysis of green- energy-based technologies ($C_{p, ti}$) and also the green energy budget allocation of a country (C_{geb});

$R_{pai} = C_{p,pai}/C_{geb}$ practical application impact ratio, which is estimated based on the provided financial supports for projection, production, conversion, marketing, distribution, management, and consumption of green fuel from green energy sources, and also the green energy budget allocation of a country (C_{geb});

$R_{geu} = 1 - Q_{wffc}/Q_{wpec}$ green energy utilization ratio, which is also defined as a function of fossil fuel utilization ratio (R_{ffu}).

Here Q_{wffc} explains world fossil fuel consumption (M) and Q_{wpec} world primary energy consumption (M).

Global Unrest and Peace

Fossil fuels such as petroleum, coal, and natural gas, which have been extensively utilized in industrial and domestic applications for a long time, have often been the cause of global destabilization and unrest. This problem is likely to increase in significance in the future and suggests the need for investigations of, among other factors, the role of hydrogen energy relative to future global unrest and global peace. In general, the global unrest arising from the use of fossil fuels is considered as a function of the usage ratios of these fossil fuels.[12] In order to estimate the level of global unrest, it is important to select a proper reference value. Therefore, it is assumed that the lowest value of global unrest and the highest value of global peace are equal to 1, which is a reference point to evaluate the interactions between global unrest and global peace.[12] Consequently, the general algebraic case form of global unrest expression is as follows[12]:

$$GU = \left\{1 - \left[\left(\frac{q_p}{q_{wpec}} + \frac{q_c}{q_{wpec}} + \frac{q_{ng}}{q_{wpec}}\right) - \left(r_{H_2-p} + r_{H_2-c} + r_{H_2-ng}\right)\right]\right\}^{-1} \quad (5)$$

where q_p, q_c and q_{ng} define the consumption quantities of petroleum, coal, and natural gas, respectively; q_{wpec} defines the quantity of world primary energy consumption; r_{H_2-p} is the utilization ratio of hydrogen from nonfossil fuels at a certain utilization ratio of petroleum; r_{H_2-c} is the utilization of hydrogen from nonfossil fuels at a certain utilization ratio of coal; r_{H_2-ng} is the utilization of hydrogen from nonfossil fuels at a certain utilization ratio of natural gas. The values of q_p, q_c, q_{ng}, and q_{wpec} can be taken from the literature.[8] The values for r_{H_2-p}, r_{H_2-c}, and r_{H_2-ng} can be taken depending on the utilization ratios of the petroleum, coal, and natural gas presented in the literature.

In order to estimate the level of global peace quantitatively, it is assumed that the highest value of global peace is equal to 1, which is a reference point to evaluate the interactions between global unrest and global peace.[12] Thus, the relationship between global unrest and global peace can be written as a function of the utilization ratio of hydrogen from nonfossil fuels. The general algebraic case form of global peace expression proposed is as follows:

$$GP = \frac{U_{H_2}}{Q_{wpec}} \times \frac{1}{(GU)} \quad (6)$$

where U_{H_2} defines the utilization of hydrogen from non-fossil fuels, which can be taken from the literature.

Case Study

The first case study is given to determine the green-energy- based sustainability ratio depending on the three subcases that are carried out based on the sectoral impact ratio, the technological impact ratio, and the practical application impact ratio by using actual data from the literature.[8] All cases are presented in Table 1. The results of the first case study are presented in Figures. 7 and 8a–c.

TABLE 1 The Cases Depending on Green Energy Impact Ratio for the Green Energy–Based Sustainability Ratio

Effect of Variable Parameters (entity)			10%	30%	50%	70%	90%
Cases		Percent of Total Financial Budget for Green Energy	Green Energy Impact Ratio (Rgei)				
I	2 constant parameters	n/3	0.699	0.766	0.833	0.899	0.966
	1 variable parameter	$k \times (1/3) \times \in$					
II	1 constant parameter	n/3	0.400	0.533	0.666	0.800	0.933
	2 variable parameters	$k \times (1/3) \times \in$					
III	3 variable parameters	$k \times (1/3) \times \in$	0.100	0.300	0.500	0.700	0.900

Here, n defines the number of constant parameters; k, the number of variable parameters.

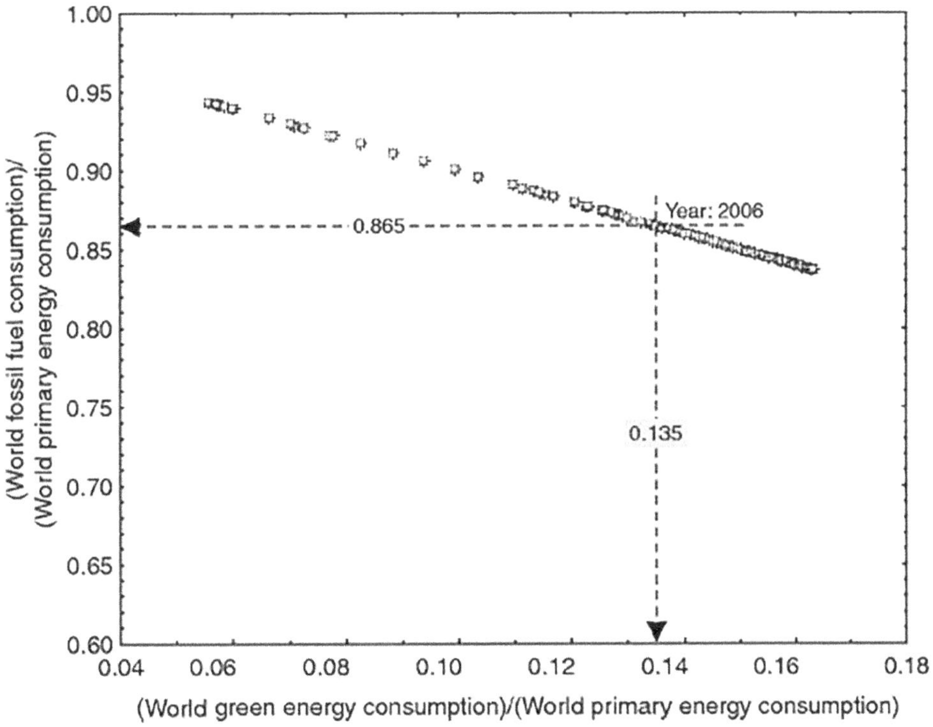

FIGURE 7 World fossil fuel consumption ratio as a function of green energy consumption ratio.

The second case study is performed to determine the global unrest and global peace level based on the predicted utilization ratios of hydrogen from nonfossil fuels. In this regard, two important empirical relations that describe the effects of fossil fuels on world peace and global unrest are taken into consideration for this case study. The results of this case study are presented in Figure 9.

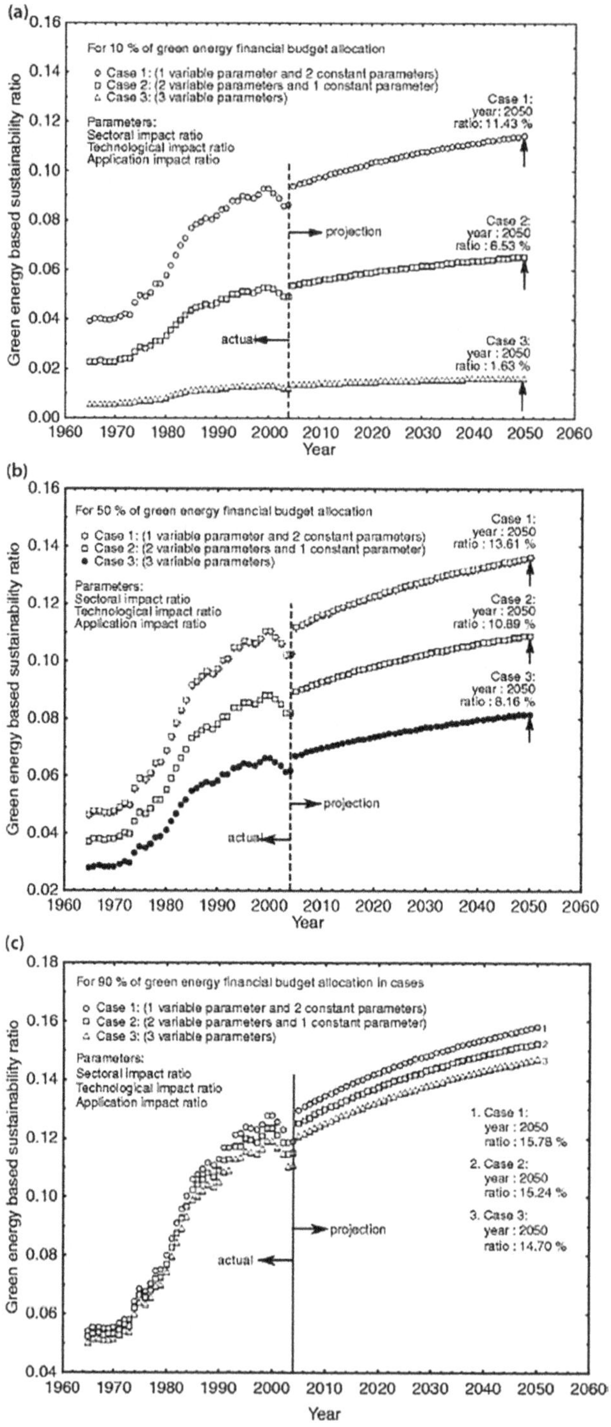

FIGURE 8 (a) Green energy-based sustainability ratio versus time depending on the actual and projected green energy consumption data for 10% of green energy financial budget allocation. (b) Green energy-based sustainability ratio versus time depending on the actual and projected green energy consumption data for 50% of green energy financial budget allocation. (c) Green energy-based sustainability ratio versus time depending on the actual and projected green energy consumption data for 90% of green energy financial budget allocation.

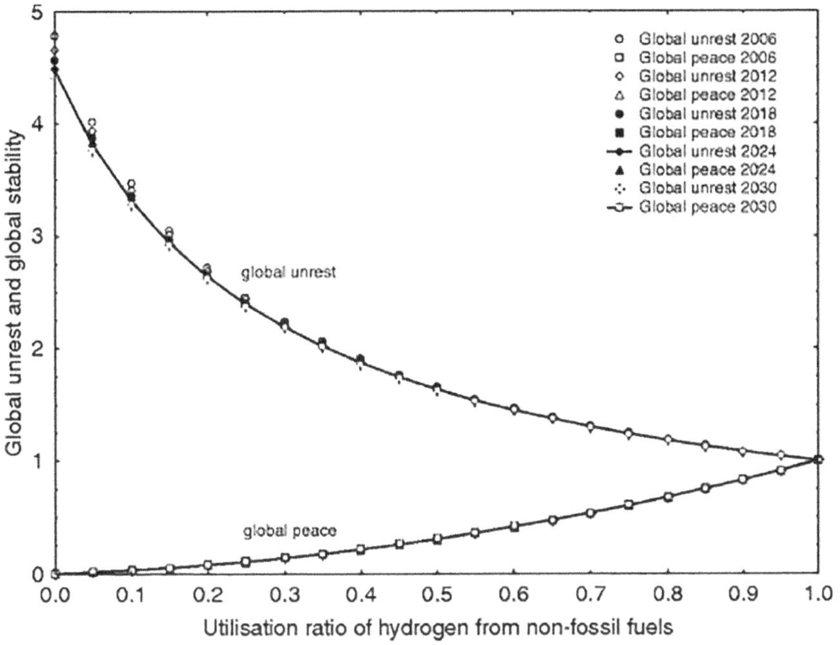

FIGURE 9 Comparison of levels of global unrest and global stability as a function of hydrogen utilization ratios from non-fossil fuels.

Results and Discussion

In accordance with the objective of this entry, the projection data is obtained using actual data of primary energy, fossil fuel, and green energy consumptions. Considering the technological impact ratio (max. value=1/3 of the green energy financial budget), the sectoral impact ratio (max. value=1/3 of the green energy financial budget allocation), and the practical application impact ratio (max. value=1/3 of the green energy financial budget allocation), three cases are analyzed and discussed in detail in this part.

Figure 7 shows the variations of the fossil fuel utilization ratio (world fossil fuel consumption/world primary energy consumption) as a function of the green energy utilization ratio (world green energy consumption/world primary energy consumption). It is found out from this figure that the fossil fuel utilization ratio decreases depending on the rise of the year while green energy utilization ratio increases. For example, the green energy utilization ratio was 5.71% in 1970, 8.25% in 1980, 11.67% in 1990, and 13.27% in 2000 while the fossil fuel utilization ratio was 94.29% in 1970, 91.75% in 1980, 88.32% in 1990, and 86.77% in 2000, based upon the actual data. However, it is observed that the green energy utilization ratio increased and reached 12.31% while fossil fuel utilization ratio decreased to 87.69% in 2004. Based on the projected data, it expected that green energy utilization ratio will reach almost 13.52% in 2006, 14.09% in 2012, 14.58% in 2018, 15.01% in 2024, and 15.38% in 2030. However, it is expected that the fossil fuel utilization ratio will decrease to almost 86.48% in 2006, 85.91% in 2012, 85.42% in 2018, 85.99% in 2024, and 84.62% in 2030. Thus, in order to increase the green energy utilization ratio and to reduce the harmful effects resulting from the fossil fuel consumption, the investments on green energy should be encouraged and the green energy strategies should be put into practice for sustainable development.

First, it should be stated that one or two of the parameters that are sectoral, technological, and practical application impact ratios can be selected as constant parameters. As shown in Table 1, a variable and two constant parameters are considered in Case 1, two variables and one constant parameter are considered in Case 2, and three variable parameters are considered in Case 3. When Case 1 is applied for green energy supply and progress, it is found that the green energy impact ratio changes between 0.699 and 0.966 depending on the percentages of the variable parameter. In the application of Case 2 it is obtained that the green energy impact ratio varies between 0.40 and 0.933 depending on the percentages of two variable parameters. In the application of Case 3, it is calculated that the green energy impact ratio changes from 0.1 to 0.9 depending on the percentages of three variable parameters. When the three Cases are compared to each other, it can be said that the highest values of green energy impact ratio are found by applying Case 1, and also that Case 3 gives the lowest green energy impact ratios. Thus, Case 1 should be selected to increase the green energy impact ratio and the green-energy-based sustainability ratio.

Figure 8a–c show a variation of the green energy–based sustainability ratio (R_{ges}) as a function of year by depending on the percentages of the green energy financial budget as 10%, 50%, and 90%, or the effect of the parameters in the Cases, respectively. The values of green-energy-based sustainability ratios were calculated using Figure 4.

As shown in these figures, the values of R_{ges} increase with time based on the cases. The highest values of R_{ges} are obtained when Case 1 is applied, as shown in Figure 8a–c. For example, the green-energy-based sustainability ratios are estimated to be 9.46% in 2006, 9.86% in 2012, 10.21% in 2018, 10.50% in 2024, and 10.76% in 2030 in case of 10% of green energy financial budget; 13.06% in 2006, 13.62% in 2012, 14.09% in 2018, 14.51% in 2024, and 14.86% in 2030 in case of 90% of green energy financial budget.

It is important to implement green energy strategies through green energy systems and applications for sustainable future. If so, the green energy–based sustainability ratio increases and green energy is more easily supplied, thus its technologies are more preferred and applied. Hence, as long as sustainable green energy strategies are increasingly applied and the green technologies are more utilized and encouraged, the negative effects stemming from the fossil fuel utilization will decrease and the green-energy-based sustainability ratio will increase.

Figure 9 compares the levels of global unrest and global peace as a function of the predicted utilization ratio of hydrogen from non-fossil fuels. Figure 4 indicates that there is an inversely proportional relationship between global peace and global unrest, depending on the utilization ratio of hydrogen from non-fossil fuels.

To better appreciate this figure, some of the key energy- related reasons for global unrest need to be understood. They include:

- Increases in fossil fuel prices
- Environmental effects of energy use, including pollution due to emissions, stratospheric ozone layer depletion, and global warming
- Decreases in the amount of fossil fuel available per capita and the associated decrease in living standards
- Increases in energy demand due to technological developments attributable to and based on fossil fuels
- Depletion of fossil fuel reserves
- Increases in conflicts for fossil fuel reserves throughout the world
- The lack of affordable and practical alternative energy sources to fossil fuels

It is found out from Figure 9 that an increase in hydrogen utilization accordingly decreases the reasons for global unrest, allowing the benefits of global peace to be realized over time; and the lowest levels of global unrest occur when hydrogen from non-fossil fuels is substituted completely for fossil fuels. In general, the level of global unrest is higher than 1 and the problems causing global unrest can be reduced by using hydrogen energy from non-fossil sources instead of fossil fuels. Figure 5 suggests that the utilization of hydrogen from non-fossil fuels at certain ratios of petroleum, coal, and natural gas decreases the amount of fossil fuel consumption and thus reduces the level of global unrest closer to 1. As shown in Figure 9, it is expected that, depending on the actual and projected fossil fuel consumption data, the levels of global unrest will be approximately 4.78 in 2006, 4.66 in 2012, 4.56 in 2018, 4.48 in 2024, and 4.41 in 2030 when the utilization ratio of hydrogen is zero. If the utilization ratio of hydrogen from non-fossil fuels is lower than 100%, the level of global peace is less than 1 and the reasons for global unrest increase. Therefore, it is beneficial to encourage the utilization of hydrogen from non-fossil fuels in place of fossil fuels. The highest level of global peace is attained when 100% of hydrogen from non-fossil fuels is used in place of fossil fuels. Some advantages of having the highest level of global peace are:

- Lifetimes of fossil fuel reserves are extended and real fossil fuel prices, consequently, can be held constant or reduced relative to present prices.
- Environmental effects from using fossil fuels are reduced or prevented because of the utilization of hydrogen from renewable energy sources and technologies.
- Technological developments based on hydrogen from nonfossil fuels increase and the requirement of technologies based on fossil fuels decrease.
- Living standards are probably higher than at present due in part to the increased consumption of the technologies related to hydrogen from sustainable green energy sources.
- Pressures to discover energy sources reduce because hydrogen can be abundantly produced and conflicts for energy supplies subside.

To increase global peace, the relationship of hydrogen to renewable energy sources needs to be understood, as does the importance of producing hydrogen from renewable energy sources.

Figure 10 describes routes using sustainable green energy sources for green power production. It is expected that the utilization of sustainable green energy sources will reduce the negative energy-related environmental effects such as global climate change and emissions of CO, CO_2, NO_x, SO_x, non-methane hydrocarbons, and particulate matter. In this regard, some potential solutions to decreasing the global unrest associated with the harmful pollutant emissions have evolved, including:[3]

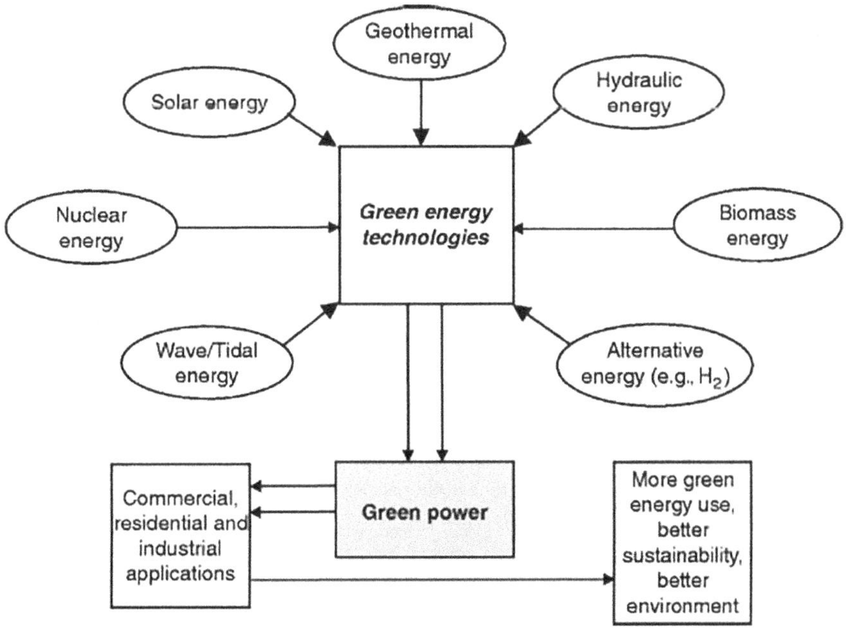

FIGURE 10 Routes for green power production from sustainable green energy sources.

- Education and training for a sustainable future
- Renewable energy technologies
- Energy conservation (efficient energy utilization)
- Cogeneration and district heating
- Energy storage technologies
- Alternative energy dimensions for transport
- Energy source switching from fossil fuels to environmentally benign energy forms
- Coal cleaning technologies
- Optimum monitoring and evaluation of energy indicators
- Policy integration
- Recycling
- Process change and sectoral shiftment
- Acceleration of forestation
- Carbon or fuel taxes
- Greener materials substitution
- Promoting public transport
- Changing life styles
- Increasing public awareness

Considering the above explanations, the following important remarks can be extracted. Fossil fuel consumption and green energy consumption are expected to reach 13807.2 and 2694.9 M, respectively, by the year 2050. This increase indicates that humans will still be dependent on fossil fuels. Based on the projected data, the green energy consumption ratio expects that the green energy utilization ratio will reach almost 16.33% and the fossil fuel utilization ratio will decrease to almost 83.67% in 2050. If the increase of fossil fuel consumption continues in this manner, it is likely that the world will be affected by many negative problems due to fossil fuels. More utilization of fossil fuels will harm world stability and increase local and global environmental problems, resulting in increasing global unrest. It is thus suggested that the utilization of fossil fuels should be reduced, and fossil-based technologies should be gradually converted to green-energy-based technologies.

Case 1 gives better results than Case 2 and Case 3. Therefore, for a higher green energy impact ratio in practice, Case 1 should be applied to increase the green energy sustainability ratio depending on the green energy strategies. Moreover, Case 1 gives the best results of the green energy–based sustainability ratio depending on the green energy impact ratio and the green energy utilization ratio.

The approximate quantified measures developed for level of global peace (ranging between 0 and 1) and level of global unrest (ranging between 1 and ∞)expressions can help understand and measure levels of global unrest and global peace. The highest level of global peace occurs when $GP=1$ and correspondingly, the lowest level of global unrest when $GU=1$, and efforts to increase global peace and stability should cause the values of GP and GU to shift towards these limiting cases. Hydrogen from non-fossil fuels can replace oil, coal, and natural gas to reduce the level of global unrest.

Sustainable green energy strategies are definitely required to ensure the global stability by reducing the harmful effects of fossil-based energy consumption. So, it is suggested that the importance of green energy and technologies that probably reduce world problems and achieve a sustainable energy system should be emphasized with consideration of the sustainable energy strategies. Moreover, a transition to a green energy–based economy should be encouraged and developed countries in particular should increase investments in green energy and technologies. Progress of green energy and technologies is based on sustainable green energy strategies for future green energy scenarios. The foremost factor that will determine the specific role of green energy and technologies will likely be energy demand. In order to balance the energy demand now and in the future, it is suggested that sustainable green energy sources and technologies be taken into consideration to increase the sustainable development in a country.

Conclusion

Green energy for sustainable development has been discussed and some key parameters to increase green energy-based sustainability and the global peace level have been presented. The effects of technological, sectoral, and practical application impact ratios on the green energy impact ratio and the green energy-based sustainability ratio are examined thoroughly. In addition, the key role of hydrogen—one of the most green energy carriers for the future—is discussed in terms of global unrest and peace. Accordingly, sustainable green energy and technologies are definitely required to ensure global stability by reducing the harmful effects of fossil-based energy consumption. The most important scenario that encourages the transition to green energy and technologies and promotes green energy-based technologies is to supply the required incentives and interactions among the countries, scientists, researchers, societies, and all. Therefore, the investments in green energy supply should be, for the future of world nations, encouraged by governments and other authoritative bodies who, for strategic reasons, wish to have a green alternative to fossil fuels.

References

1. Midilli, A.; Ay, M.; Dincer, I.; Rosen, M.A. On hydrogen and hydrogen energy strategies: I: current status and needs. Renewable Sustainable Energy Rev. **2005**, *9* (3), 255–271.
2. Midilli, A.; Dincer, I.; Ay, M. Green energy strategies for sustainable development. Energy Policy **2006**, *34* (18), 623–633.
3. Dincer, I. Renewable energy and sustainable development: a crucial review. Renewable Sustainable Energy Rev. **2000**, *4* (2), 157–175.
4. Dincer, I. The role of exergy in energy policy making. Energy Policy **2002**, *30*, 137–149.
5. Dincer, I.; Rosen, A.M. Exergy as a driver for Achieving Sustainability. Int. J. Green Energy **2004**, *1* (1), 1–19.
6. Midilli, A.; Ay, I.; Dincer, M.; Rosen, M. A.; *On Hydrogen Energy Strategy: The Key Role in This Century and Beyond*, Proceedings of The First Cappadocia International Mechanical Engineering Symposium (CMES-04), 1, 211–217, Cappadocia, Turkey, 2004.
7. World Energy Outlook, EO&W, Institute of Shipping Analysis, http://saiweb.sai.se, Available at http://www.world energyoutlook.org/weo/pubs/weo2002/WEO21.pdf, 2002.
8. Workbook, Statistical Review Full Report of World Energy 2005, Available at http://www.bp.com/centres/energy, 2005.
9. Dincer, I. Environmental impacts of energy. Energy Policy **1999**, *27*, 845–854.
10. Dincer, I.; Rosen, A.M. Thermodynamic aspects of renewables and sustainable development. Renewable Sustainable Energy Rev. **2005**, *9*, 169–189.
11. Rosen, M.A.; Scott, D.S. Entropy production and exergy destruction—part I: hierarchy of earth's major constituencies, and part II: illustrative technologies. Int. J. Hydrogen Energy **2003**, *28* (12), 1307–1323.
12. Midilli, A.; Ay, M.; Dincer, I.; Rosen, M.A. On hydrogen and hydrogen energy strategies II: future projections affecting global stability and unrest. Renewable Sustainable Energy Rev. **2005**, *9* (3), 273–287.

15

Ozone Layer

Luisa T. Molina

Introduction

The Earth's ozone layer shields all life from the sun's harmful ultraviolet (UV) radiation. It is mainly located in the lower stratosphere, between 12 and 30 km above Earth's surface. Ozone is a gas that is present naturally in the Earth's atmosphere; it is continuously being made by the action of solar radiation on molecular oxygen, predominantly in the upper stratosphere and at low latitudes; it is also continuously being destroyed throughout the atmosphere by a variety of chemical processes. The ozone abundances in the atmosphere are therefore determined by the balance between chemical production and destruction processes.

The average concentration of ozone in the atmosphere is about 300 parts per billion by volume (ppbv); most of it (~90%) is contained in the stratosphere. Even though it occurs in such small quantities, ozone plays a vital role in sustaining life on Earth by absorbing most of the biologically damaging UV sunlight. However, the fragile ozone layer is being depleted since the late 1970s as a consequence of the emission of human-made chemicals, chlorofluorocarbons (CFCs), to the atmosphere. The CFCs are industrial chemicals that have been used in the past as coolants for refrigerators and air conditioners, propellants for aerosol spray cans, foaming agents for plastics, and cleaning solvents for electronic components, among other uses. They are thought of as "miracle" compounds because they are non-flammable, noncorrosive, and unreactive with most other substances. Ironically, it is their chemical inertness that creates a global scale problem by enabling them to reach the stratosphere, where they decompose, releasing chlorine atoms that deplete the ozone layer.

Observations of the ozone layer itself showed that depletion was indeed occurring; the most dramatic loss was discovered over Antarctica—far from the emitted sources. In response to the likelihood of increasing depletion of ozone in the stratosphere, an international agreement, the Montreal Protocol on Substances that Protect the Ozone Layer, was signed by most national governments of the world calling for an orderly phase out of all ozone-depleting substances (ODSs). Thus, ozone, a trace constituent in the atmosphere, has become an issue of global prominence and the model for international cooperation to protect the environment from unintended consequences of human activities.

While the Montreal Protocol has made great strides in phasing out ODSs, new challenges are emerging. Changes in climate are expected to have an increasing influence on stratospheric ozone in the coming decades. International efforts to protect the ozone layer would require improved understanding of the complex linkages between stratospheric ozone and climate change. Moreover, some new ODSs replacements are extremely powerful global warming gases and represent a potential focal area within the overall climate change challenge. This entry describes the discovery of the ozone depletion phenomenon, the chemicals that cause ozone depletion in the stratosphere, polar ozone destruction processes, impacts of ozone depletion on human health and on ecosystems, international treaties that regulate the ODSs, and the linkage between stratospheric ozone and climate change.

Atmospheric Ozone

Ozone was discovered by the German chemist Christian Schoenbein in 1840 while observing an electrical discharge; he noted its distinctively pungent odor and named it "ozone, " which means "smell" in Greek. An ozone (O_3) molecule is made of three oxygen atoms, instead of two of the normal oxygen molecule (O_2), which makes up 21% of the air we breathe. Ozone is found mainly in two regions of the Earth's atmosphere (Figure 1):

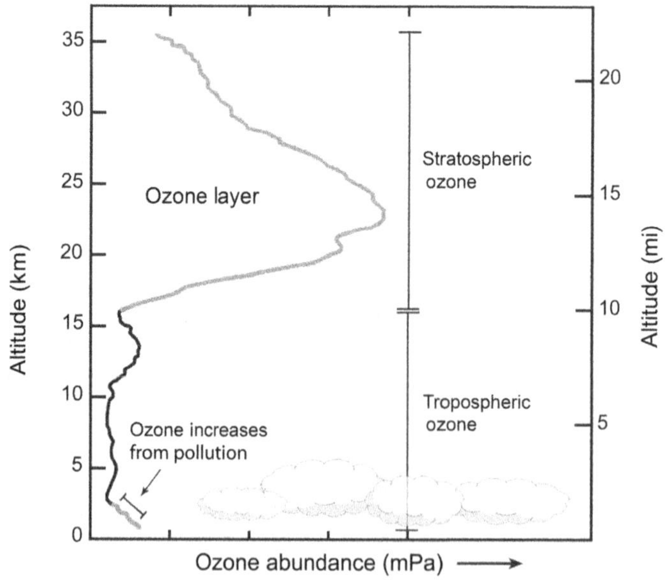

FIGURE 1 Typical atmospheric ozone profile. Ozone abundances are shown here as the pressure of ozone at each altitude using the unit "milli-Pascals" (mPa) (100 million mPa = atmospheric sea-level pressure).

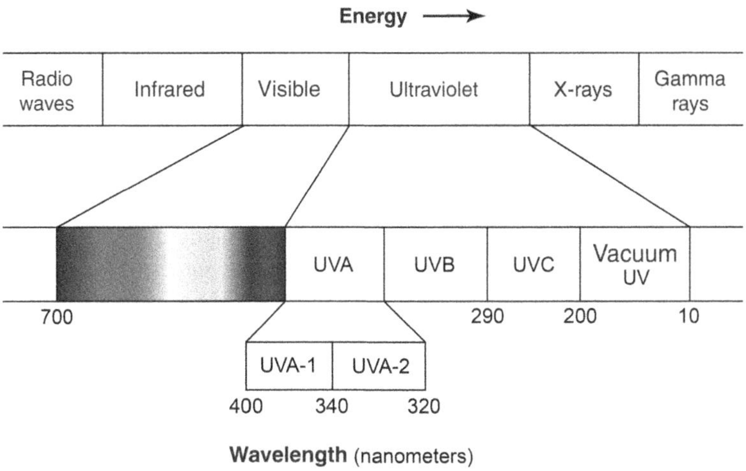

FIGURE 2 Spectrum of electromagnetic radiation.

1. Stratosphere, from about 10–16 km above Earth's surface up to 50 km. The ozone in this region is commonly known as the ozone layer.
2. Troposphere, the lowest region of the atmosphere, between Earth's surface and the stratosphere. Some of the tropospheric ozone is generated by atmospheric photochemical reactions (smog), and some is transported from the stratosphere.

The ozone molecules in the stratosphere and the troposphere are chemically identical; however, they have very different roles in the atmosphere and very different effects on humans and the biosphere. Stratospheric ozone—sometimes referred to as "good ozone"—plays a beneficial role by absorbing most of the biologically damaging UV sunlight (UV-B), allowing only a small amount to reach the Earth's surface (Figure 2). At the Earth's surface, ozone is a key component of urban smog and is harmful to human health, agriculture, and ecosystems; hence, it is often referred as "bad ozone." Furthermore, because ozone is a powerful greenhouse gas, increase in tropospheric ozone contributes to global warming.[1] Thus, depending on where ozone resides, it can have positive or negative impacts on human well-being and the environment.[2]

Origin of Ozone

The fundamental ozone formation–destruction mechanism consists of the following reactions, suggested initially by the British physicist Sidney Chapman[3] in the 1930s:

$$O_2 + \text{UV light} \rightarrow O + O \tag{1}$$

$$O + O_2 + M \rightarrow O_3 + M \tag{2}$$

$$O_3 + \text{UV light} \rightarrow O + O_2 \tag{3}$$

$$O + O_3 \rightarrow O_2 + O_2 \tag{4}$$

Molecular oxygen absorbs solar radiation at ⊠200 nm and releases oxygen atoms (Reaction 1), which rapidly combine with oxygen molecules to form ozone (Reaction 2). Ozone absorbs solar radiation very efficiently at wavelengths ⊠200 to 300 nm.[4] This absorption process leads to the decomposition of ozone, producing oxygen atoms (Reaction 3), which in turn regenerate the ozone molecule by Reaction 2. Thus, the net effect of Reactions 2 and 3 is the conversion of solar energy into heat, without the net loss of ozone. This process leads to an increase of temperature with altitude, which is the feature that gives rise to the stratosphere; the inverted temperature profile is responsible for its large stability toward vertical movements (Figure 3). In contrast, the troposphere is characterized by decreasing temperature with altitude because of the heating effects from the absorption of the sun's energy at the Earth's surface. Because hot air rises, this causes rapid vertical mixing so that chemical substances emitted on the ground can rise to the tropopause (transition zone between troposphere and stratosphere) in a matter of days; they are also dispersed horizontally throughout the troposphere on the time scale of weeks to months by winds and convection. However, once they reach the stratosphere, the transport time scales in the stratosphere are much slower and mixing in the stratosphere can take months to years. Movement of air between the troposphere and stratosphere is very slow compared to the movement of air within the troposphere itself. However, this small air exchange is an important source of ozone from the stratosphere to the troposphere.

Most of the time, oxygen atoms react with molecular oxygen to make ozone (Reaction 2), but occasionally they destroy ozone (Reaction 4). The overall amount of ozone is determined by a balance between the production and the removal processes. Models based only on the Chapman's mechanism were found to overpredict stratospheric ozone levels; thus, there are other reactions that contribute to the destruction of ozone.

In the early 1970s, Crutzen[5] suggested that trace amounts of nitrogen oxides ($NO_x = NO + NO_2$) formed in the stratosphere through the decomposition of nitrous oxide (N_2O), which originates from soil-borne microorganisms, control the ozone abundance through the following catalytic cycle:

$$NO + O_3 \rightarrow NO_2 + O_2 \tag{5}$$

$$NO_2 + O \rightarrow NO + O_2 \tag{6}$$

$$\frac{O_3 + \text{UV light} \rightarrow O + O_2 \left(\text{for Reaction 3}\right)}{\text{Net: } 2\,O_3 \rightarrow 3\,O_2}$$

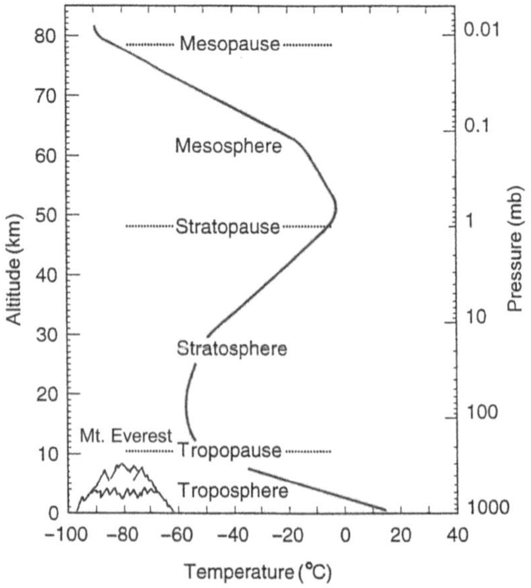

FIGURE 3 Typical variation of atmospheric temperatures and pressures with altitude.

The term "catalyst" refers to a compound that reacts with one or more reactants to form intermediates that subsequently give the final reaction product, in the process regenerating the catalyst. In the above cycle, the species NO (nitric oxide) and NO_2 (nitrogen dioxide) are still present after these three reactions have occurred, but two molecules of ozone have been destroyed. These species have an odd number of electrons; they are free radicals and are chemically very reactive. Although the concentration of NO and NO_2 is small (several ppbv), each radical pair can destroy thousands of ozone molecules before being temporarily removed, mainly by reaction with hydroxyl (OH) radical to form nitric acid:

$$OH + NO_2 \rightarrow HNO_3 \tag{7}$$

Independently, Johnston[6] suggested that the NO_x chain reaction can be initiated by the direct release of NO_x in the exhaust of supersonic transport (SST) aircraft and could disturb the delicate natural balance between ozone formation and destruction.

In the troposphere, NO_x, together with volatile organic compounds in the presence of sunlight, are the ingredients for the photochemical formation of ground-level ozone. Thus, NO_X plays a dual role, destroying or generating O_3 depending on the altitude.

Other free radicals that destroy stratospheric ozone are OH and HO_2, derived from the water molecule:

$$OH + O_3 \rightarrow HO_2 + O_2 \tag{8}$$

$$\frac{HO_2 + O_3 \rightarrow OH + 2O_2}{\text{Net: } 2\,O_3 \rightarrow 3\,O_2} \tag{9}$$

Chlorine atoms are also very efficient catalysts for ozone destruction and may proceed in a similar cycle[7–9] as will be discussed below. Small amounts of chlorine compounds of natural origin exist in the stratosphere; the most important source is methyl chloride (CH_3Cl), which is emitted mainly from oceanic and terrestrial ecosystems. Most of the CH_3Cl is destroyed in the troposphere, but a few percent reaches the stratosphere. There are also large natural sources of inorganic chlorine compounds at the Earth's surface, e.g., NaCl and HCl from the oceans; however, they are water soluble and are removed efficiently from the atmosphere by clouds and rainfall.

Measurements and Distribution of Stratospheric Ozone

In the 1920s, the British meteorologist G.M.B. Dobson developed the Dobson spectrophotometer that measures sunlight at two UV wavelengths: one that is strongly absorbed by ozone and one that is weakly absorbed. The difference in light intensity at the two wavelengths provides the total ozone above the location of the instrument.[10] In 1957, a global network of ground-based, total ozone observing stations was established as part of the International Geophysical Year; currently, there are about 100 sites distributed throughout the world. Ozone concentrations are also being routinely measured by a variety of instruments on balloons, aircraft, and satellites. One of the most commonly used units for measuring ozone concentration is called "Dobson unit" (DU), which is a measure of how much ozone is contained in a vertical column of air. The average amount of ozone in the atmosphere is about 300 DU, equivalent to a layer 3 mm thick. By comparison, if all of the air in a vertical column that extends from the ground up to space were collected and squeezed together at 0°C and 1 atm pressure, that column would be 8 km thick.

As shown in Figure 4, the distribution of total ozone over the globe varies with latitude, longitude, and season.[11] The variations are caused by large-scale movements of stratospheric air and the chemical production and destruction of ozone. In general, the total ozone is highest in the polar regions and lowest at the equator. In the Northern Hemisphere, the ozone layer is thicker during spring and thinner during autumn.

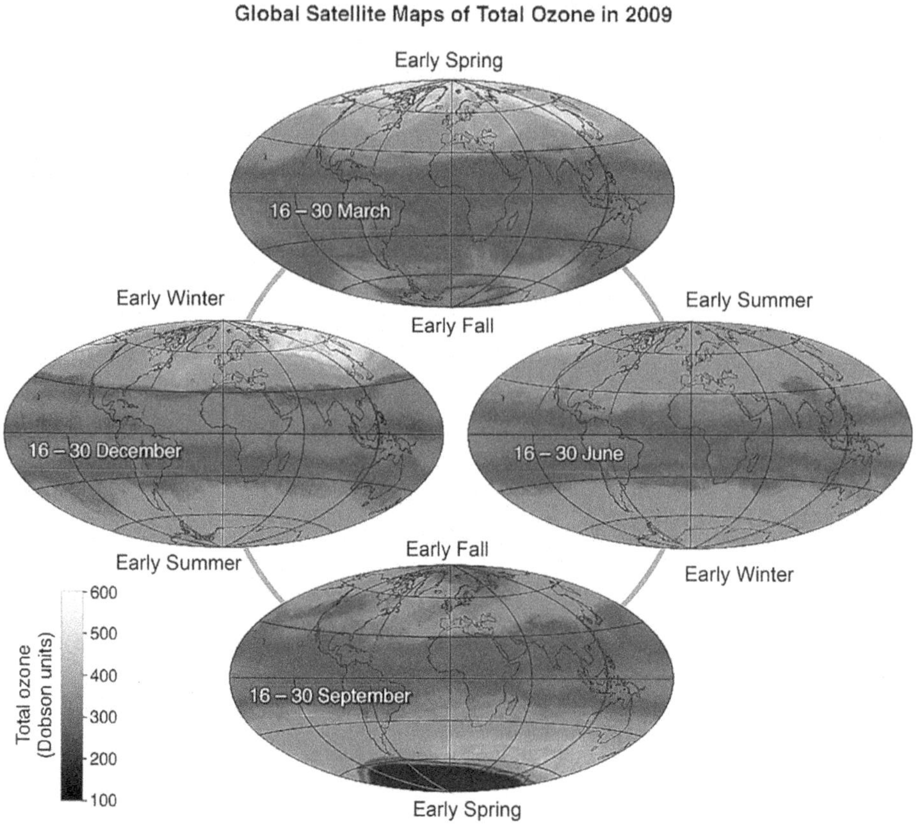

FIGURE 4 Global satellite maps of total ozone in 2009 showing the variation with latitude, longitude, and season. The variations are demonstrated here with 2-week averages of total ozone in 2009 as measured with a satellite instrument. Total ozone shows little variation in the tropics (20°N–20°S latitudes) over all seasons. Total ozone outside the tropics varies more strongly with time on a daily to seasonal basis as ozone-rich air is moved from the tropics and accumulates at higher latitudes. The low total ozone values over Antarctica in September constitute the "ozone hole" in 2009. Since the 1980s, the ozone hole in late winter and early spring represents the lowest values of total ozone that occur over all seasons and latitudes.
Source: Fahey and Hegglin).[11]

CFCS and the Ozone Layer

CFCs: The Miracle Compounds

In the 1930s, American mechanical engineer and chemist Thomas Midgley[12,13] invented the CFCs during a search for nontoxic and non-flammable substances that could be used as coolants in home refrigerators and air conditioners. The CFCs are compounds that contain only chlorine, fluorine, and carbon; they are also known under trademarks such as Freon (DuPont) and Genetron (Allied Signal).

The two important properties that make the CFCs commercially valuable are their volatility (they can be readily converted from a liquid to a vapor and vice versa) and their chemical inertness (they are stable, nontoxic, and nonflammable). The CFCs were thought of as miracle compounds and soon replaced the toxic ammonia and sulfur dioxide as the standard cooling fluids. Subsequently, the CFCs found uses as propellants for aerosol sprays, blowing agents for plastic foam, and cleaning agents for electronic components. All of these activities doubled the worldwide use of CFCs every 6 to 7 years and eventually reached over 700,000 metric tons annually by the early 1970s.

CFCs and the Destruction of Stratospheric Ozone

In the early 1970s, James Lovelock showed that trichlorofluoromethane (CCl_3F or CFC-11) was present in the air over Ireland using a newly developed electron capture detector.[14] Subsequently, Lovelock and co-workers[15] detected measurable levels of CCl_3F in the atmosphere over the South and North Atlantic and concluded that the CFCs were carried over by large-scale wind motions. Their interest in CFCs lies in the potential use of these compounds as inert tracers; they did not expect the CFCs to present any conceivable harm to the environment.

In 1973, Molina and Rowland[7,8,16–18] decided to investigate the ultimate fate of these new miracle compounds upon their release to the atmosphere. After carrying out a systematic search of chemical and physical processes that might destroy the CFCs in the lower atmosphere, they concluded that these compounds would break up in the middle stratosphere (~25–30 km) by solar UV radiation.

Because the CFCs are chemically inert and practically insoluble in water, they are not removed by the common cleansing mechanisms that operate in the lower atmosphere. Furthermore, the CFC molecules are transparent from 230 nm through the visible wavelengths; they are effectively protected below 25 km by the stratospheric ozone layer that shields the Earth's surface from UV light. Instead, they rise into the stratosphere, where they are eventually destroyed by the short-wavelength (~200 nm) solar UV radiation to yield radicals that can destroy stratospheric ozone through a catalytic process.[7] Because transport into the stratosphere is very slow, the atmospheric lifetime for the CFCs is about 50 to 100 years.

The destruction of CFCs by high-energy solar radiation leads to the release of chlorine atoms, as shown in Reaction 12 for CCl_3F:

$$CCl_3F + \text{UV light} \rightarrow CI + CCI_2F \tag{10}$$

The chlorine atoms attack ozone within a few seconds and are regenerated on a time scale of minutes; the net result is the conversion of one ozone molecule and one oxygen atom to two oxygen molecules.

$$CI + O_3 \rightarrow CIO + O_2 \tag{11}$$

$$\frac{CIO + O \rightarrow CI + O_2}{\text{Net: } O + O_3 \rightarrow O_2 + O_2} \tag{12}$$

In the above cycle, chlorine acts as a catalyst for ozone destruction because Cl and ClO react and are reformed, and ozone is simply removed. Oxygen atoms are formed when solar UV radiation reacts with ozone and oxygen molecule (Reactions 1 and 3). It is estimated that one Cl atom can convert 100,000 molecules of ozone into oxygen molecules before that chlorine becomes part of a less reactive compound, e.g., by reaction of ClO with HO_2 or NO_2 to produce hypochlorous acid (HOCl) or chlorine nitrate ($ClONO_2$), respectively, or by reaction of the Cl atom with methane (CH_4) to produce the relatively stable hydrogen chloride (HCl):

$$CIO + HO_2 \rightarrow HOCI + O_2 \tag{13}$$

$$CIO + NO_2 \rightarrow CINO_2 \tag{14}$$

$$CI + CH_4 \rightarrow HCI + CH_3 \tag{15}$$

The chlorine-containing product species HCl, $ClONO_2$, and HOCl function as temporary inert reservoirs: they are not directly involved in ozone depletion, but they are eventually broken down by reaction with other free radicals or by absorption of solar radiation, thus returning chlorine to its catalytically active free radical form. At low latitudes and in the upper stratosphere, where the

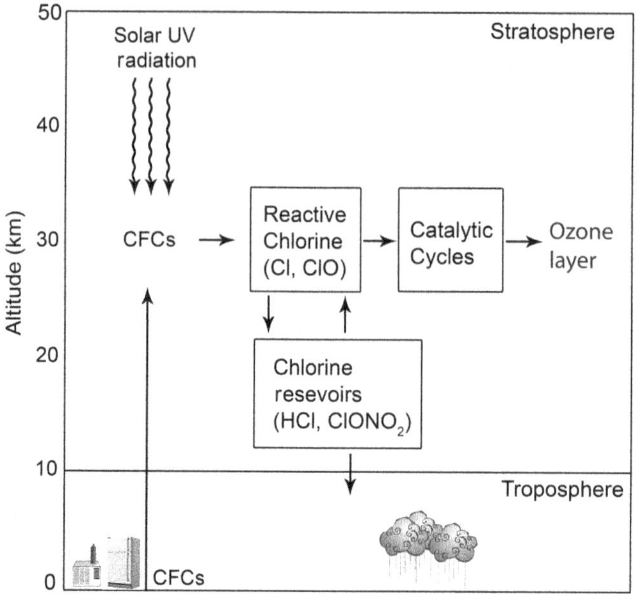

FIGURE 5 Schematic representation of the CFC-ozone depletion hypothesis.

formation of ozone is fastest, a few percent of the chlorine is in this active form; most of the chlorine is in the inert reservoir form, with HCl being the most abundant species. The temporary chlorine reservoirs remain in the stratosphere for several years before returning to the troposphere, where they are rapidly removed by rain or clouds. A schematic representation of these processes is presented in Figure 5.

Other ODSs

Besides CFCs, there are other chlorine substances from human activities that destroy ozone in the stratosphere. Carbon tetrachloride and methyl chloroform are important ODSs that are used as fire extinguisher, cleaning agents, and solvents.

Another category of ODSs contains bromine. There are industrial sources of brominated ODSs as well as natural ones; the most important are the halons and methyl bromide (CH_3Br). Halons are halocarbon gases containing carbon, bromine, fluorine, and (in some cases) chlorine; they are produced industrially as fire extinguishers. Methyl bromide is both natural and human-made; it is used as an agricultural fumigant. These sources release bromine to the stratosphere at pptv levels, compared with ppbv for chlorine. On the other hand, bromine atoms are about 60 times more efficient than chlorine atoms for ozone destruction on an atom-per-atom basis;[19–21] a large fraction of the bromine compounds is present as free radicals because the temporary reservoirs are less stable and are formed at considerably slower rates than the corresponding chlorine reservoirs.

Many of the ODSs contain fluorine atoms in addition to chlorine and bromine. However, in contrast to chlorine and bromine, fluorine atoms abstract hydrogen atoms very rapidly from methane and from water vapor, forming the stable hydrogen fluoride (HF), which serves as a permanent inert fluorine reservoir. Hence, fluorine free radicals are extremely scarce and the effect of fluorine on stratospheric ozone is negligible. Halogen source gases that contain fluorine and no other halogens are not classified as ODSs. An important category is the hydrofluorocarbons (HFCs) discussed in the "International Response: Montreal Protocol" section.

The publication of the 1974 Molina–Rowland article[7] stimulated numerous scientific studies, including laboratory studies, computer modeling, and field measurements, to understand the impacts of chlorine and bromine on stratospheric ozone.[20,21] The U.S. National Academy of Sciences issued two reports in 1976, verifying the Molina-Rowland findings.[22,23]

Stratospheric Ozone Depletion

Discovery of the Antarctic Ozone Hole

In 1985, Joseph Farman and coworkers[24] reported that the ozone concentrations recorded at the Halley Bay Observatory in Antarctica has dropped dramatically in the spring months starting in the early 1980s, compared to the data obtained since 1957, when the British Antarctic Survey began ozone measurements using a Dobson Spectrometer. The 1984 October monthly ozone averaged less than 200 DU, about 35% lower than the 300 DU levels recorded in 1957–1958 and on through the 1960s. Farman et al.'s findings were subsequently confirmed by satellite observation from the total ozone mapping spectrometer (TOMS).[25] Furthermore, satellite measurements confirmed that the bulk of the chlorine in the stratosphere is of human origin.[26] Additional measurements from ground-based Dobson instruments[27] and from satellites indicate that the extent of ozone depletion over Antarctica in the spring months continued to increase after 1985, with concentrations as low as 85 to 95 DU reported from some of the polar stations.

Measurements show that ozone was also being depleted in the Northern Hemisphere, particularly at high latitudes and in the winter and spring months. Examination of the ozone records shows that significant changes have also occurred in the lower stratosphere at mid- latitudes.[20,21] Figure 6 shows an acute drop in total atmospheric ozone during October in the early and mid-1980s measured by instruments from the ground and from a satellite.[28]

The discovery of the depletion of ozone over Antarctica—the ozone hole—was not predicted by the atmospheric scientists (see Figure 7). The large magnitude of the depletion suggests that the stratospheric

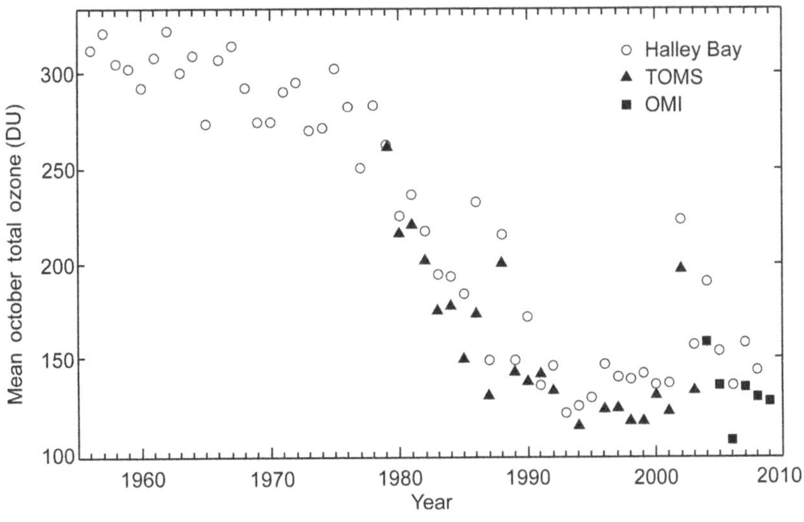

FIGURE 6 Average total amount of ozone measured in October over Antarctica. Instruments on the ground (at Halley Bay) and high above Antarctica (the TOMS and ozone monitoring instrument [OMI] measured an acute drop in total atmospheric ozone during October in the early and middle 1980s).
Source: Adapted from NASA Ozone Hole Watch.[28]

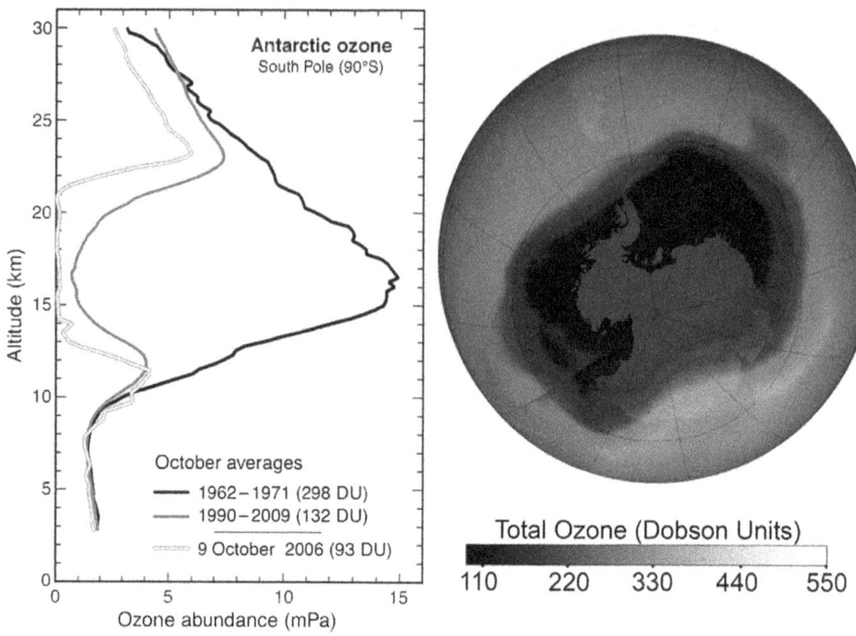

FIGURE 7 Antarctic ozone hole. Left panel: Vertical distributions of Antarctic ozone. A normal ozone layer was observed to be present between 1962 and 1971. In more recent years, as shown here for October 9, 2006, ozone is almost completely destroyed between 14 and 21 km in the Antarctic in spring (Source: Adapted from Fahey and Hegglin[11]). Right panel: The darker shaded regions over the Antarctic continent show the severe ozone depletion or ozone hole on October 9, 2006, measured by satellite instrument. The hole reached 26.2 million km^2, the greatest extent recorded in the Antarctic.
Source: NASA Ozone Hole Watch.[28]

ozone is influenced by processes that had not been considered previously. Researchers all over the world raced to develop plausible explanation; the cause of this depletion soon became very clear. Laboratory experiments, field measurements over Antarctica, and model calculations showed unambiguously that the ozone hole over Antarctica can indeed be traced to the industrial CFCs.[20,21]

Polar Ozone Chemistry

Characteristics of the Polar Regions

ODSs emitted at Earth's surface are transported over great distances to the stratosphere by atmospheric air motions and are present throughout the stratospheric ozone layer. Yet, the most dramatic ozone depletion was over Antarctica—the ozone hole—far from the emitted sources. A major reason why an Antarctic ozone hole of the observed extent could happen is because of the unique atmospheric and chemical conditions that exist there. The very low winter temperatures in the Antarctic stratosphere cause polar stratospheric clouds (PSCs) to form. Special reactions that occur on PSCs, combined with the relative isolation of polar stratospheric air, allow chlorine and bromine reactions to produce the ozone hole in the Antarctica when the sunlight returns in the springtime.

Both polar regions of the earth are cold, primarily because they receive far less solar radiation than the tropics and mid-latitudes do. Moreover, most of the sunlight that does shine on the polar regions is reflected by the bright white surface. Winter temperatures at the North Pole can range from about −45°C to −25°C and summer temperatures can average around the freezing point (0°C).

In comparison, the annual mean temperature at the South Pole is about −60°C in the winter and −28°C in the summer.

What makes the South Pole so much colder than the North Pole is that it sits on top of a very thick ice sheet, which itself sits on the continent of Antarctica. The surface of the ice sheet at the South Pole is more than 2700 m (9000 ft) above sea level; Antarctica is by far the highest continent on the earth. In contrast, the North Pole is at sea level in the middle of the Arctic Ocean, which also acts as an effective heat reservoir.

Stratospheric air in the polar regions is relatively isolated from other stratospheric regions for long periods in the winter months. The isolation comes about because of strong winds that encircle the poles, forming a polar vortex (or polar cyclone), which prevents substantial air masses into or out of the polar stratosphere. This cyclonic circulation strengthens in winter as stratospheric temperatures decrease.[11] All through the long, dark winter, chemical changes occur in polar regions from reactions on PSCs inside the vortex; the isolation preserves those changes until the spring sunlight strikes the stratosphere above the frozen continent in late August. The result is massive ozone destruction inside the vortex forming an ozone hole, as described below. The polar vortex diminishes when the continent and the air above it begin to warm up and ozone-rich air from outside the vortex flows in, replacing much of the ozone that was destroyed.

The Antarctic polar vortex is more pronounced and persistent than its Arctic counterpart with the result that the isolation of air inside the vortex is much more effective in the Antarctic than in the Arctic. In addition to being significantly warmer, the Northern Hemisphere also has numerous mountain ranges and a more active tropospheric meteorology, giving rise to enhanced planetary wave and a less stable Arctic vortex.

Polar Stratospheric Clouds

The conditions in the polar stratosphere are unique in several ways. Firstly, ozone is not generated there because the high-energy solar radiation that is absorbed by molecular oxygen is scarce over the poles. Secondly, the total ozone column abundance at high latitudes is large because ozone is transported toward the poles from higher altitudes and lower latitudes. Thirdly, the prevailing temperatures over the stratosphere above the poles in the winter and spring months are the lowest throughout the atmosphere, particularly over Antarctica. Typically, average daily minimum values are as low as −90°C in July and August over Antarctica and near −80°C in late December and January over the Arctic (Figure 8). Ozone is expected to be rather stable over the poles if one considers only gas-phase chemical and photochemical processes, because regeneration of ozone-destroying free radicals from the reservoir species would occur very slowly at those temperatures. However, another unique feature of the polar stratosphere is the seasonal presence of PSCs (Fig. 9). Different types of liquid and solid PSC particles form when stratospheric temperatures fall below −78°C. As a result, PSCs are often found over large areas of the winter polar regions and over significant altitude ranges. With a temperature threshold of −78°C, PSCs exist in larger areas and longer periods in the Antarctica than in the Arctic.

The stratosphere is normally very dry; water is present only at a level of a few ppmv, comparable to that of ozone itself. Over the poles, a somewhat larger amount of water is present, resulting from the oxidation of methane. Furthermore, the temperature can drop below −85°C over Antarctica in the winter and spring months, leading to the formation of ice clouds. The presence of trace amounts of nitric and sulfuric acids enables the formation of PSCs a few degrees above the frost point (the temperature at which ice can condense from the gas phase); these acids can form cloud particles consisting of crystalline hydrates.

Solomon et al.[29] first suggested that PSCs could play a major role in ozone depletion over Antarctica by promoting the release of photolytically active chlorine from its reservoir species (HCl, $ClONO_2$, and HOCl). This occurs mainly by the following reaction:

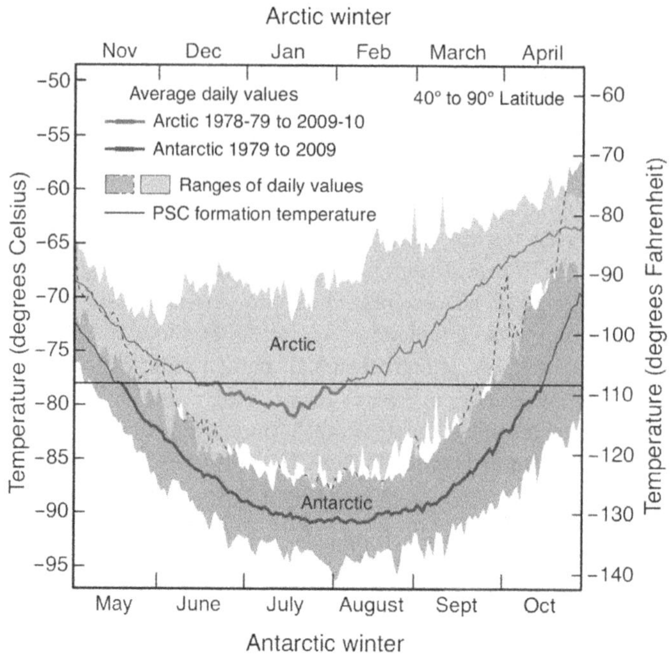

FIGURE 8 Arctic and Antarctic temperatures. Air temperatures in both polar regions reach minimum values in the lower stratosphere in the winter season. Typically, average daily minimum values are as low as −90°C in July and August in the Antarctic and near −80°C in late December and January in the Arctic. PSCs are formed in the polar ozone layer when winter minimum temperatures fall below the formation temperature of about −78°C. Note that the dashed black lines denote the upper limits of the Antarctic temperature range where they overlap with the Arctic temperature range.
Source: Fahey and Hegglin.[11]

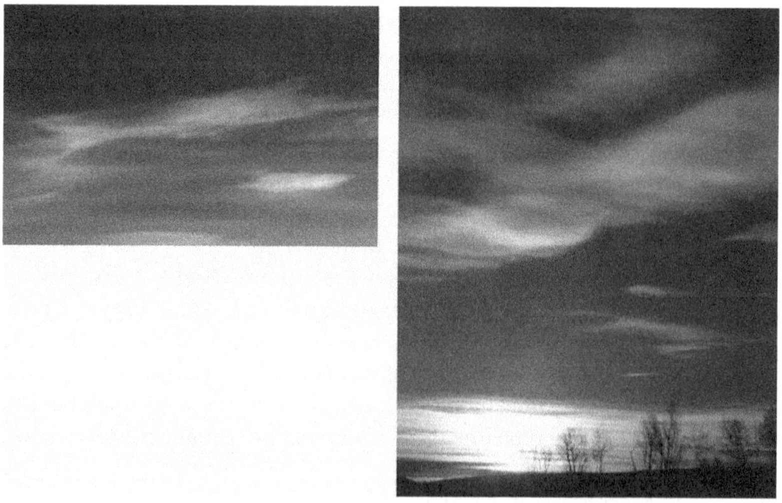

FIGURE 9 Polar stratospheric clouds. The frozen crystals that make up PSCs provide a surface for the reactions that free chlorine atoms in the Antarctic stratosphere (left panel[28]) and in the Arctic stratosphere (right panel[11]).

$$HCl + ClONO_2 \rightarrow Cl_2 + HNO_3 \quad (16)$$

$$Cl_2 + \text{sunlight} \rightarrow Cl + Cl \quad (17)$$

Indeed, laboratory studies have shown that this reaction occurs very slowly in the gas phase;[30] however, it proceeds with remarkable efficiency in the presence of ice surfaces.[31] The product, Cl_2, is immediately released to the gas phase while the other product, HNO_3, remains in the condensed phase. Cl_2 decomposes readily to Cl atoms even with the faint amount of sunlight present over Antarctica in the early spring:

When average temperatures begin increasing by late winter, PSCs form less frequently and their surface conversion reactions produce less ClO. Without continued ClO production, ClO amounts decrease and other chemical reactions re-form the reactive reservoirs, $ClONO_2$ and HCl. The most intense period of ozone depletion will end when PSC temperatures no longer occur.

An important feature for Reaction 16 in the presence of PSCs is the removal of NO_x from the gas phase; the source for these free radicals in the polar stratosphere is nitric acid, which condenses in the cloud particles. NO_x normally interfere with the catalytic ozone loss reactions by scavenging ClO to form chlorine nitrate (Reaction 14). In the absence of NO_x, the fraction of chlorine compounds in the form of free radical is larger and makes it possible for the ozone depletion reaction to occur. These experimental results have been corroborated by other studies.[32] Further laboratory studies[33,34] and theoretical calculations[35] indicate that HCl solvates readily on the ice surface, forming hydrochloric acid. Therefore, chlorine activation reactions on the surfaces of ice crystals proceed through ionic mechanisms analogous to those in aqueous solutions.

Another natural source of ozone depletion is the sulfate aerosols that come from volcanic eruptions. The most recent large eruption was that of Mt. Pinatubo in 1991, which ejected large amounts of sulfur dioxide into the stratosphere, resulting in up to a 10-fold increase in the number of particles available for surface reactions. These particles increased the global ozone depletion by 1%–2% for several years following the eruption.[11]

Polar Ozone Destruction

While the presence of the PSCs explains how chlorine can be released from the inactive reservoir chemicals, it remains unanswered how a catalytic cycle might be maintained to account for the large ozone destruction observed. The ClO_x cycles such as Reactions 11 and 12 are not efficient in the polar stratosphere because they require the presence of free oxygen atoms, which are scarce at high latitudes.

Several catalytic cycles have been suggested as occurring over Antarctica during winter and spring. One of the dominant cycles involving ClO dimer or chlorine peroxide (ClOOCl) was proposed by Molina and Molina.[36] The cycle is initiated by the combination of two ClO radicals forming ClOOCl, which then photolyzes to release molecular oxygen and free chlorine atoms:

$$ClO + ClO + M \rightarrow ClOOCl + M \quad (18)$$

$$ClOOCl + \text{sunlight} \rightarrow Cl + ClOO \quad (19)$$

$$ClOO + M \rightarrow Cl + O_2 + M \quad (20)$$

$$\frac{2\left[Cl + O_3 \rightarrow ClO + O_2\right] \ (\text{for Reaction 11})}{\text{Net: } 2\ O_3 \rightarrow 3\ O_2}$$

No free oxygen atoms are involved in this cycle. Visible sunlight is required to complete and maintain the cycle; thus, this cycle can occur only in late winter/early spring when sunlight returns to the polar region. Laboratory studies have shown that the photolysis products of ClOOCl are indeed Cl atoms.[37,38]

Another important cycle operating in the polar stratosphere involves the reaction of bromine monoxide (BrO) with ClO suggested by McElroy et al.:[39]

$$BrO + ClO + \text{sunlight} \rightarrow Cl + Br + O_2 \tag{21}$$

$$Br + O_3 \rightarrow BrO + O_2 \tag{22}$$

$$\frac{Cl + O_3 \rightarrow ClO + O_2 \ (\text{for Reaction 11})}{\text{Net: } 2\ O_3 \rightarrow 3\ O_2}$$

Field Measurements of Atmospheric Trace Species

Ground-based and aircraft expeditions were launched in the years following the ozone hole discovery to measure trace species in the stratosphere over Antarctica.[40] The results provided strong evidence for the crucial role played by industrial chlorine in the ozone depletion. One of the most convincing evidence was provided by the NASA ER-2 aircraft measurements in 1987, which flew into the Antarctic vortex. The flight data (Figure 10) showed an anticorrelation between ClO measured by Anderson et al.[41] with in situ ozone measurements monitored by Proffitt et al.[42] The results show that the ClO + ClO cycle accounts for about three-quarters of the observed ozone loss, with the BrO + ClO cycle accounting for the rest. Furthermore, NO*x* levels were found to be very low and nitric acid was shown to be present in the cloud particles, as expected from the laboratory studies.

Recent laboratory measurements of the dissociation cross section of ClO dimer[43] and analyses of observation from aircraft and satellites have reaffirmed that polar springtime ozone depletion is caused primarily by the ClO + ClO catalytic ozone destruction cycle, with substantial contributions from the BrO + ClO cycle.[44]

Field measurements have also been conducted in the Arctic stratosphere,[45] indicating that a large fraction of the chlorine is also activated there. Nevertheless, ozone depletion is less severe over the Arctic and is not as localized because the atmosphere above the Arctic is warmer than above the Antarctic; the active chlorine does not remain in contact with ozone long enough and at low enough temperatures to destroy it before the stratospheric air over the Arctic mixes with warmer air from lower latitudes. This warmer air also contains NO_2, which deactivates the chlorine. On the other hand, cold winters can lead to significant ozone depletion—30% or more—over large areas, as described below.[46]

Depletion of the Global Ozone Layer

Global total ozone levels are influenced not only by the concentrations of ODSs but also by atmospheric transport (winds), incoming solar radiation, aerosols (fine particles suspended in the air), and other natural compounds. Global total ozone has decreased beginning in the 1980s, reaching a maximum of about 5% in the early 1990s (Figure 11a). The lowest global total ozone values occurred in the years following the eruption of Mt. Pinatubo in 1991, which resulted in up to a 10-fold increase in the number of sulfuric-acid-containing particles available for surface reaction in the stratosphere, thereby increased global ozone depletion for several years before they were removed from the stratosphere by natural processes. The depletion has lessened since then; the average global ozone for 2005–2009 is about 3.5% below the 1964–1980 average.

Observed total ozone loss varies significantly with latitude on the globe. Figure 11b shows how the 2005–2009 ozone depletion varies with latitude. The ozone loss is very small near the equator and increases with latitude towards the poles. The largest decreases have occurred at high latitudes in both

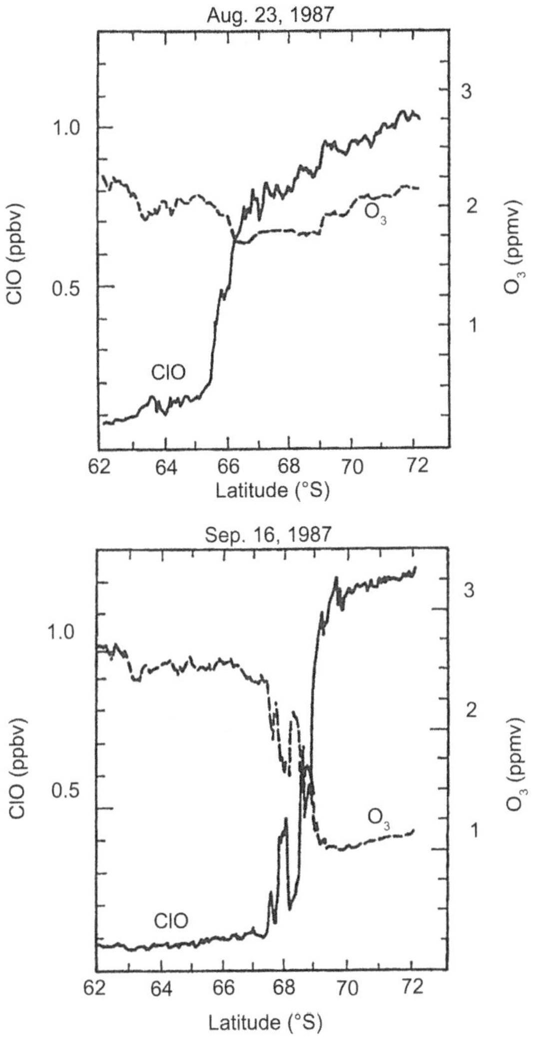

FIGURE 10 Aircraft measurements conducted on August 23 and September 16, 1987, of chlorine monoxide (ClO) by Anderson et al.[41] and of ozone (O_3) by Proffitt et al.[42]
Source: Adapted from Anderson.[41]

hemispheres because of the large winter/spring depletion in the polar regions; the losses are greater in the Southern Hemisphere because of the Antarctic ozone hole. Since the 1980s, Antarctic ozone loss in the springtime has been quite large, covering nearly the entire continent with virtually all of the ozone destroyed between 15 and 20 km. Ozone loss over the Arctic is smaller than its Antarctic counterpart; it is modulated strongly by variability in atmospheric dynamics, transport, and temperature. The degree of spring Arctic depletion is highly variable from year to year, but large Arctic depletion has also been observed recently with the most dramatic occurring in the spring of 2011. Observations over the Arctic region as well as from satellites show an unprecedented ozone column loss comparable to some Antarctic ozone holes. The formation of the "Arctic ozone hole" in 2011 was driven by an anomalously strong stratospheric polar vortex and an unusually long cold period, leading to persistent enhancement of active chlorine and severe ozone loss that exceeded 80% over 18–20 km altitude. This result raises the possibility of yet more severe depletion as lower stratospheric temperatures decrease. More acute Arctic

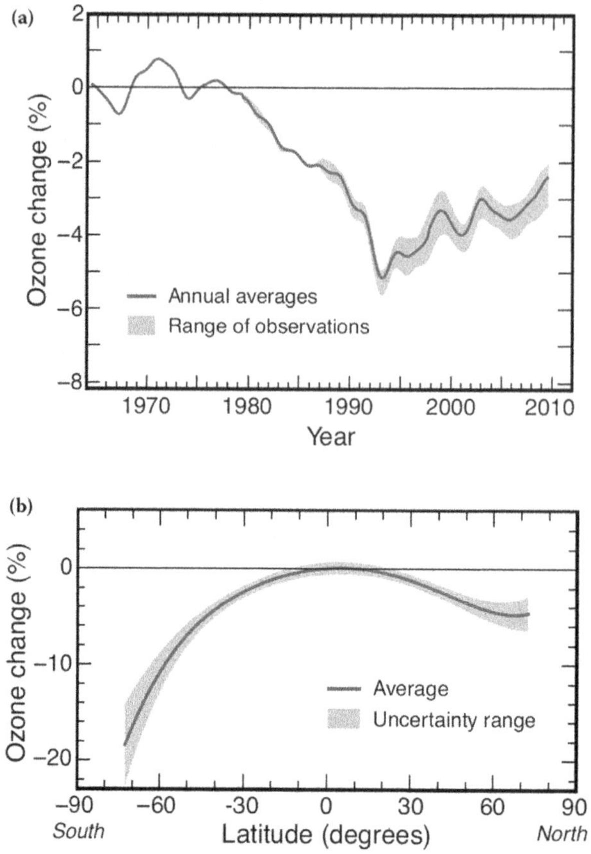

FIGURE 11 Global total ozone changes. Satellite observations show depletion of global total ozone beginning in the 1980s. Panel (a) compares annual averages of global ozone with the average from the period 1964 to 1980 before the ozone hole appeared. Seasonal and solar effects have been removed from the observational data set. On average, global ozone decreased each year between 1980 and 1990. The depletion worsened for a few years after 1991 following Mt. Pinatubo eruption. Panel (b) shows variation of the 2005–2009 depletion with latitude over the globe. The largest decreases have occurred at high latitudes in both hemispheres because of the large winter/spring depletion in polar regions; the losses are greater in the Southern Hemisphere because of the Antarctic ozone hole.
Source: Fahey and Hegglin.[11]

ozone loss could exacerbate biological risks from exposure to increased UV radiation, especially if the vortex shifted over densely populated mid-latitudes, as it did in April 2011.[46]

Ozone depletion is also observed at the mid-latitudes between equatorial and polar latitudes. Total ozone averaged for 2005–2009 is about 3.5% lower in northern mid-latitudes (35°N–60°N) and about 6% lower at southern mid-latitudes (35°S–60°S) compared with the 1964–1980 average (Figure 11b). Chemical destruction processes occurring at midlatitudes contributes to observed depletion in these regions, although it is much smaller than in the polar regions because the amounts of reactive halogen gases are lower and a seasonal increase of the most reactive halogen gases in Antarctic late winter does not occur in mid-latitude regions. Changes in mid-latitude ozone are also affected by changes occurring in the polar regions when the ozone-depleted air over both polar regions is dispersed away from the poles following the vortex breakdown, thus reducing average ozone concentrations at nonpolar latitudes.

Ozone Depletion and Biological Effects

Because ozone absorbs UV radiation from the sun, depletion of the stratospheric ozone layer is expected to lead to increases in the amount of solar UV radiation reaching the Earth's surface, predominantly in the wavelength range of 290 to 320 nanometers (UV-B radiation). UV-B radiation is also partially shielded by clouds, dust, and air pollutants. Large ozone losses in the Antarctica have produced a clear increase in surface UV radiation. Ground-based measurements show that the average spring erythemal (sunburning) irradiance for 1990–2006 is up to 85% greater than the modeled irradiance for 1963–1980, depending on site. The Antarctic spring erythemal irradiance is approximately twice that measured in the Arctic for the same season.

Analyses based on surface and satellite measurements show that erythemal UV irradiance over mid-latitudes has increased since the late 1970.[44] This is in qualitative agreement with the observed decrease in column ozone, although other factors (mainly clouds and aerosols) have influenced long-term changes in erythemal irradiance. Clear-sky UV observations from unpolluted sites in midlatitudes show that since the late 1990s, UV irradiance levels have been approximately constant, consistent with ozone column observations over this period.

The environmental and health effects of stratospheric ozone depletion have been summarized in several assessment reports.[47–49] UV-B radiation can induce acute skin damage in humans, such as sunburn, as well as eye diseases and infectious diseases.[50,51] Human epidemiological studies and animal experiments have established that UV-B radiation is a key risk factor for development of skin cancer, both melanoma and non-melanoma, especially in light-skinned population. Non-melanoma (squamous cell carcinoma and basal cell carcinoma) is the more common form of skin cancer and can be readily treated and is rarely fatal,[51] whereas melanoma is the most dangerous and the leading cause of death from skin cancer. Absorption of strong UV radiation damages the DNA molecule, eventually leading to faulty replication and mutation.[52] Numerous experiments have shown that the cornea and lens of the eye can also be damaged by UV-B radiation, and that chronic exposure to this radiation increased the likelihood of certain cataracts.[49] Studies in human subjects show that exposure to UV-B radiation can suppress proper functioning of the body's immune system. Animal experiments indicate that overexposure to UV-B radiation decreases the immune response to skin cancers and some infectious agents.[53,54]

Terrestrial plants can also be affected by UV-B radiation, although the response varies to a large extent among different species. In addition to plant growth, the changes induced by UV radiation can be indirect, for example, by affecting the timing of developmental phases or the allocation of biomass to the different parts of the plant.[55] Aquatic ecosystems can also be damaged by UV-B radiation; for example, there is evidence for impaired larval development and decreased reproductive capacity in some amphibians, shrimp, and fish.[56] There is also direct evidence of UV-B effects under the ozone hole on the productivity of natural phytoplankton communities in Antarctic waters.[57]

International Response: Montreal Protocol

Following the publication of the Molina–Rowland article,[7] the U.S. National Academy of Sciences issued two reports in 1976 stating that the atmospheric sequence outlined by Molina and Rowland was essentially correct.[22,23] The United States, Canada, Norway, and Sweden responded in late 1970s by banning the sale of aerosol spray cans containing CFCs; this caused a temporary halt in the growing demands for CFCs. However, worldwide use of the chemicals continued and the production rate began to rise again.

The discovery of the massive ozone losses in Antarctica in 1985 spurred a rush of scientific research activities, leading to improved understanding of stratospheric chemistry and the evolution of the ozone layer. These new scientific developments have provided the foundation for the critical policy decisions that followed.

In 1985, under the auspices of the United Nations Environment Programme (UNEP), 20 nations signed the Vienna Convention for the Protection of the Ozone Layer.[58] In September 1987, the recognition that CFC use was increasing and the mounting scientific evidence that this increase would cause large ozone depletions led to an international agreement limiting the production of CFCs.[58] This agreement, the Montreal Protocol on Substances that Deplete the Ozone Layer, initially called for a reduction of only 50% in the manufacture of CFCs by the end of the century. In view of the scientific evidence that emerged in the following years, the initial provisions were strengthened through the London (1990), Copenhagen (1992), Montreal (1997), and Beijing (1999) amendments as well as several adjustments by both controlling additional ODSs and by moving up the date by which already controlled substances must be phased out. At the 19th Meeting of the Parties to the Montreal Protocol in 2007, the Parties agreed to adjust their commitments related to the phase out of HCFCs.[58] Fig. 12 shows the projected abundance of ODSs in the stratosphere according to the provisions of the Montreal Protocol and subsequent amendments.[44]

The Montreal Protocol is now more than 20 years old and has been ratified by 196 countries, although not all the parties have ratified the subsequent amendments. It is widely regarded as one of the most effective multilateral environmental agreements in existence. The production of CFCs in industrialized countries was phased out at the end of 1995, and other compounds such as the halons, methyl bromide, carbon tetrachloride, and methyl chloroform (CH_3CCl_3) were also regulated. Developing countries were allowed to continue CFC production until 2010, to facilitate their transition to the newer CFC-free technologies. An important feature of the Montreal Protocol was the establishment of a funding mechanism to help these countries meet the costs of complying with the protocol and with its subsequent amendments.[58]

A significant fraction of the former CFC usage is being dealt with by conservation and recycling. Some of the former use of CFCs is being temporarily replaced by hydrochlorofluorocarbons (HCFCs)—compounds that have similar physical properties to the CFCs, but their molecules contain hydrogen atoms and are less stable in the atmosphere. A large fraction of the HCFCs released industrially reacts

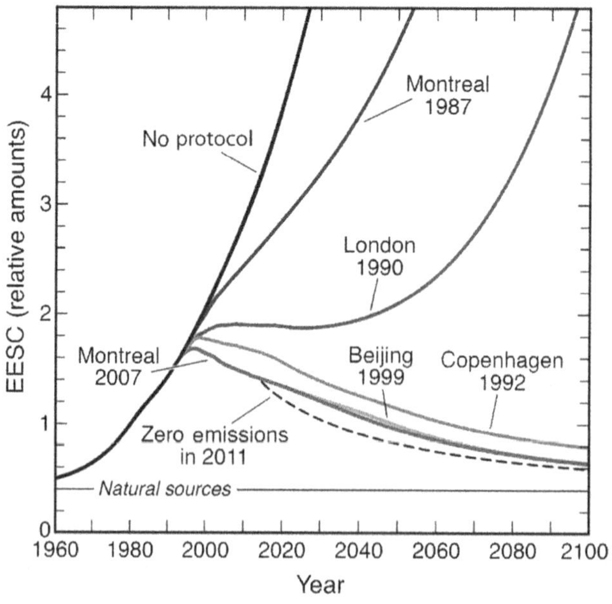

FIGURE 12 Measured and projected chlorine concentrations in the stratosphere according to the provisions of the Montreal Protocol and subsequent amendments. EESC is the equivalent effective stratospheric chlorine.
Source: Fahey and Hegglin.[11]

in the lower atmosphere with the OH radical before reaching the stratosphere, forming water and an organic free radical that rapidly photo-oxidizes to yield water-soluble products, which are then removed from the atmosphere mainly by rainfall. Although HCFCs are more ozone friendly than CFCs, they still destroy some ozone. They are now also regulated under the Montreal Protocol; the concentrations of HCFCs are projected to grow for another two decades before decreasing.

Some HFCs, which do not contain chlorine atoms, are now being used as CFC replacements; they are ozone friendly because fluorine forms stable compounds in the stratosphere. However, they have the potential to contribute to global warming. HFCs are now regulated under the Kyoto Protocol, an agreement under the United Nations Framework Convention on Climate Change.

About half of the CFC usage has been replaced by not-in-kind alternatives; for example, CFC-113—used extensively as a solvent to clean electronic components—has been phased out by CFC-free cleaning technologies such as soap and water or terpene-based solvents; there are also new technologies to manufacture clean electronic boards. Other examples include the use of stick or spray pump deodorants to replace CFC-12 aerosol deodorants and the use of mineral wool to replace CFC, HFC, or HCFC insulating foam.

Overall, the provisions of the Montreal Protocol have been successfully enforced. Atmospheric measurements indicate that the abundance of chlorine contained in the CFCs and other halocarbons has declined in response to the Montreal Protocol regulations.[44] On the other hand, because of the long lifetime of the CFCs in the atmosphere, relatively high chlorine levels in the stratosphere—with the consequent ozone depletion—are expected to continue well into the 21st century.

The total tropospheric abundance of chlorine from ODSs and methyl chloride had declined in 2008 to 3.4 parts per billion (ppb) from its peak of 3.7 ppb between 1992 and 1994. However, the rate of decline in total tropospheric chlorine was only two-thirds as fast as was expected because of the increase in the HCFC abundances. The rapid HCFC increases are coincident with increased production in the developing countries, particularly in East Asia. The rate of decline of total tropospheric bromine from controlled ODSs was close to that expected and was driven by changes in methyl bromide.[44]

By the middle of the 21st century, the amounts of halogens in the stratosphere are expected to be similar to those present in 1980 prior to the onset of the ozone hole. However, the influence of climate change could accelerate or delay ozone recovery.[44]

Stratospheric Ozone and Climate Change Linkage

Greenhouse Gases and Climate Change

The Earth absorbs energy from the Sun, and also radiates energy back into space. However, much of this energy going back to space is absorbed by greenhouse gases occurring naturally in the atmosphere, such as CO_2, CH_4, water vapor, N_2O, and ozone. Because the atmosphere then radiates most of this energy back to the Earth's surface, our planet is warmer than it would be if the atmosphere did not contain these gases. This is the natural "greenhouse effect," which maintains the surface of the Earth at a temperature that is suitable for life as we know it today. However, this natural greenhouse effect has intensified since the start of the industrial era as human activities emit more greenhouse gases to the atmosphere, including industrial compounds such as CFCs, HFCs, perfluorocarbons (PFCs), and sulfur hexafluoride (SF_6), resulting in a shift in the radiative balance of the Earth's atmosphere.

According to the Intergovernmental Panel on Climate Change, there is now visible and unequivocal evidence of climate change impacts, and there is a consensus that greenhouse gas emissions from human activities are the main drivers of change. The Earth's average temperature has been recorded to have increased by approximately 0.74°C over the past century.[1] Climate change is now recognized as a major global challenge that will have significant and long-lasting impacts on human well-being and development.[59]

One way of quantifying the contribution of greenhouse gases to climate change is through the standard metric known as radiative forcing, which is defined as positive if it results in a gain of energy for the Earth system (warming) and negative if it results in a loss (cooling).[1] The largest radiative forcing comes from CO_2, followed by CH_4, tropospheric ozone, halocarbons, and N_2O. All of these gases absorb infrared radiation emitted from the Earth's surface and re-emit it at a lower temperature, thus decreasing the outgoing radiation flux and producing a positive forcing, leading to warming. In contrast, stratospheric ozone depletion represents a small negative forcing, which leads to cooling of Earth's surface (Fig. 13); however, this contribution is expected to decrease as ODSs are gradually removed from the atmosphere.

Halocarbons and Climate Change

The understanding of the interaction between ozone depletion and climate change has been strengthened in recent years.[44,49] Stratospheric ozone and tropospheric ozone both absorb infrared radiation emitted by Earth's surface. Stratospheric ozone also significantly absorbs solar radiation. Therefore, changes in the stratospheric ozone and tropospheric ozone are directly linked to climate change. Stratospheric ozone and climate change are also indirectly linked because both ODSs and their replacements are greenhouse gases and represent an important contribution to the radiative forcing (see Fig. 13).

One approach of comparing the influence of individual halocarbons on ozone depletion and climate change is to use ozone depletion potentials (ODPs) and global warming potentials (GWPs). The ODP and GWP of a gas quantify its effectiveness in causing ozone depletion and climate forcing, respectively. GWP is defined as the total forcing attributed to a mass of emitted pollutant during a specified time after emissions (typically 100 years) as compared to the same mass of CO_2. ODP is the relative value that indicates the potential of a substance to destroy ozone as compared with the potential of CFC-11, which is assigned a reference value of 1.

Table 1 lists the atmospheric lifetime, global emissions, ODP, and GWP of some halogen source gases and the HFC replacement gases.[11,44] All ODSs and their substitutes shown here have a non-zero GWP,

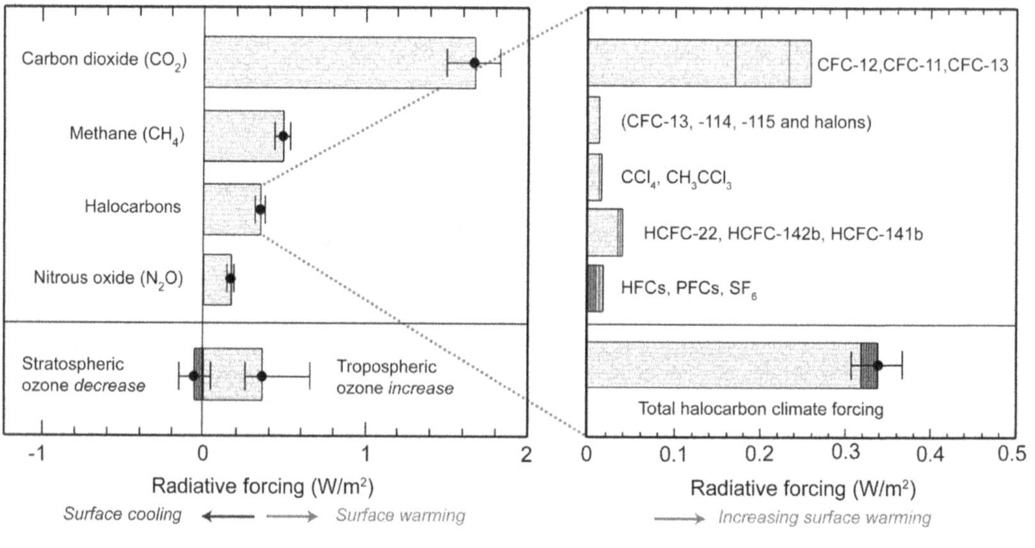

FIGURE 13 Radiative forcing of climate change. Left panel: radiative forcing of major greenhouse gases. Right panel: radiative forcing of halocarbons (darker shading designates Kyoto Protocol gases).
Source: Adapted from Fahey and Hegglin.[11]

TABLE 1 Atmospheric Lifetimes, Global Emissions, ODPs, and GWPs of Some Halogen Source Gases and HFC-Substituted Gases

Gas	Atmospheric Lifetime	Global Emissions (2008) (kt/yr)[a]	ODP	GWPb
Halogen source gases				
Chlorine gases				
CFC-11	45	52–91	1	4,750
CFC-12	100	41–99	0.82	10,900
CFC-113	85	3–8	0.85	6130
Carbon tetrachloride (CClq)	26	40–80	0.82	1400
HCFCs	1–17	385–481	0.01–0.12	77–2220
Methyl chloroform (CH_3CCl_3)	5	<10	0.16	146
Methyl chloride (CH_3Cl)	1	3600–4600	0.02	13
Bromine gases				
Halon-1301	65	1–3	15.9	7140
Halon-1211	16	4–7	7.9	1890
Methyl bromide (CH_3Br)	0.8	110–150	0.66	5
Very short-lived gases (e.g., $CHBr_3$)	<0.5	c	Very low[c]	Very low[c]
HFCs				
HFC-134	13.4	149±27	0	1370
HFC-23	222	12	0	14,200
HFC-143[a]	47.1	17	0	4180
HFC-125	28.2	22	0	3420
HFC-152[a]	1.5	50	0	133
HFC-32	5.2	8.9	0	716

[a]Includes both human activities (production and banks) and natural sources. Units are in kilotons per year.
[b]One-hundred-year GWPs. Values are calculated for emissions of an equal mass of each gas.
[c]Estimates are very uncertain for most species.
Source: Fahey and Hegglin[11] and WMO.[44]

with values spanning the wide range of 4 to 14,000; they are far more effective than equivalent amount of CO_2 in causing climate forcing. Therefore, the future selection of specific HFCs as replacement for ODS will have important consequences for climate change.

The CFCs, halons, and HCFCs are ODSs; they are controlled under the Montreal Protocol. Thus, the Montreal Protocol has provided collateral benefit of reducing the contributions of ODSs to climate change. In 2010, the decrease of annual ODS emissions under the Montreal Protocol is estimated to be about 10 gigatons of avoided CO_2-equivalent emissions per year, which is about 5 times larger than the annual emissions reduction target for the first commitment period (2008–2012) of the Kyoto Protocol.[44]

The HFCs, used as ODSs substitutes, do not destroy ozone (ODPs equal zero) and are considered ozone safe and have become the major replacement for CFCs and HCFCs. However, like the ODSs they replace, many HFCs have high GWP. HFCs, together with CO_2, CH_4, N_2O, PFCs, and SF_6, are controlled under the Kyoto Protocol. According to a new UNEP report, emissions of HFCs are growing at a rate of 8% per year due to growing demand in emerging economies and increasing populations. Without intervention, the increase in HFCs emissions is projected to offset much of the climate benefit achieved by the earlier reduction in ODS emissions. It is therefore important to select HFCs with low GWP potential and short lifetimes to minimize the climate impact while protecting the ozone layer.[60] This is one of the focal areas for the newly launched Climate and Clean Air Coalition on Short-lived Climate Pollutants.[61]

Impact of Climate Change on Ozone

The ODSs have declined as a result of the Montreal Protocol and its subsequent amendments; it is expected that this will lead to the recovery in the stratospheric ozone abundances. However, it is difficult to attribute ozone increases to the decreases in ODSs alone during the next few years because of natural variability, observational uncertainty, and confounding factors, such as changes in stratospheric temperature or water vapor. In contrast to the diminishing role of ODSs, changes in climate are expected to have an increasing influence on the recovery of ozone layer from the effects of ODSs.

Stratospheric ozone is influenced by changes in temperatures and winds in the stratosphere. For example, lower temperatures and stronger polar winds could both increase the extent and severity of winter polar ozone depletion. Observations show that the global-mean lower stratosphere has cooled by 1–2°C and the upper stratosphere cooled by 4–6°C between 1980 and 1995. There have been no significant long-term trends in global-mean lower stratospheric temperatures since about 1995. The two main reasons for the cooling of the stratosphere are depletion of stratospheric ozone and increase in atmospheric greenhouse gases. Ozone absorbs solar UV radiation, which heats the surrounding air in the stratosphere. Loss of ozone means that less UV light gets absorbed, resulting in cooling of the stratosphere. A significant portion of the observed stratospheric cooling is also due to human-emitted greenhouse gases. While the Earth's surface is expected to continue warming in response to the net positive radiative forcing from greenhouse gas increases, the stratosphere is expected to continue cooling.

A cooler stratosphere would extend the time period over which PSCs are present in winter and early spring, leading to increased polar ozone depletion. In the upper stratosphere at altitudes above PSC formation regions, a cooler stratosphere is expected to increase ozone amounts because lower temperatures decrease the effectiveness of ozone loss reactions.

Climate change may also affect the stratospheric circulation, which will significantly alter the distribution of ozone in the stratosphere. These changes tend to decrease total ozone in the tropics and increase total ozone at mid- and high latitudes. Changes in circulation induced by changes in ozone can also affect patterns of surface wind and rainfall. The projected changes in ozone and clouds may lead to large decreases in UV at high latitudes, where UV is already low, and to small increases at low latitudes, where it is already high. This could have important implications for human and ecosystem health. However, these projections depend strongly on changes in cloud cover, air pollutants, and aerosols, all of which are influenced by climate change. It is therefore important to improve our understanding of the processes involved and to continue monitoring ozone and surface UV spectral irradiances both from the surface and from satellites.[62]

Conclusions

The depletion of the stratospheric ozone layer exemplifies the global environmental challenges human face: it is an unintended consequence of human activity. Strong involvement and cooperation of stakeholders at all levels (scientists, technologists, economic and legal experts, environmentalists, and policy makers); strengthening of human and institutional capacities, coupled with suitable mechanisms for facilitating technological and financial flows; and changes in human behavior have been critical to the success in phasing out the ODSs.

The 1987 Montreal Protocol on Substances that Deplete the Ozone Layer is a landmark agreement that has successfully reduced the global production, consumption, and emissions of ODSs. By protecting the ozone layer from much higher levels of depletion, it has provided direct benefits to human health and agriculture, which in turn provide economic benefits by decreasing health costs and increasing crop production. Furthermore, because many ODSs are also potent greenhouse gases, the Montreal Protocol has provided substantial co-benefits to climate change.

On the other hand, demand for replacement substances such as HFCs has increased. Many of these substances are Kyoto gases. Additional climate benefits could be achieved by managing the emissions of replacement fluorocarbon gases and by implementing alternative gases with lower GWPs, as well as designing buildings that avoid the need for air conditioning.

Assuming full Montreal Protocol compliance, midlatitude ozone is expected to return to 1980 levels before mid-century. The recovery rate will be slower at high latitudes. Springtime ozone depletion is expected to continue to occur at polar latitudes, especially in Antarctica, in the next few decades. It is estimated that the ozone layer over the Antarctica will recover to pre-1980 levels between 2060 and 2075, and probably one or two decades earlier in the Arctic. However, effective control mechanisms for new chemicals threatening the ozone layer are essential; continued monitoring of the ozone layer is crucial to maintain momentum on recovering the ozone layer while simultaneously minimizing the influence on climate.

Changes in climate are expected to have an increasing influence on stratospheric ozone in the coming decades. An important scientific challenge is to project future ozone abundance based on an improved understanding of the complex linkages between stratospheric ozone and climate change.

Human activities will continue to change the composition of the atmosphere and new challenges that require international cooperation and collaboration will emerge. The ozone-hole phenomenon demonstrates the importance of long-term atmospheric monitoring and research, without which depletion of the ozone layer might not have been detected until more serious damage was evident. It is important for national and international agencies to continue their coordinated efforts on atmospheric monitoring, research, and assessment activities to provide sound scientific data needed to understand environmental changes on both regional and global scales.

Acknowledgments

The author gratefully acknowledges the use of the material and figures from "WMO Scientific Assessment of Ozone Depletion: 2010" and NASA Ozone Hole Watch presented in this entry.

References

1. IPCC (Intergovernmental Panel on Climate Change). Climate Change 2007: The Physical Science Basis, Contribution of Working Group I to the Fourth Assessment Report of the Intergovernmental Panel on Climate Change; Solomon, S., Qin, D., Manning, M., Chen, Z., Marquis, M., Avery, K.B., Tignor, M., Miller, H.L., Eds.; Cambridge University Press: Cambridge United Kingdom and New York, USA, 2007, 996 pp.
2. Molina M.J.; Molina L.T. Chlorofluorocarbons and destruction of the ozone layer. In *Environmental and Occupational Medicine,* 4th Ed.; Rom, W.N., Ed.; Lippincott, Williams and Wilkins: Philadelphia, 2007; 1605–1615.
3. Chapman, S. A theory of upper atmospheric ozone. Mem. R. Meteorol. Soc. **1930**, *3*, 103.
4. Molina, L.T.; Molina, M.J. Absolute absorption cross sections of ozone in the 185–350 nm wavelength range. J. Geophys. Res. **1986**, *91*, 14501–14508.
5. Crutzen, P.J. The influence of nitrogen oxides on atmosphere ozone content. Q. J. R. Meteorol. Soc. **1970**, *96*, 320–325.
6. Johnston H.S. Reduction of stratospheric ozone by nitrogen oxide catalysts from supersonic transport exhaust. Science **1971**, *173*, 517–522.
7. Molina, M.J.; Rowland, F.S. Stratospheric sink for chlorofluoromethanes: Chlorine-atom catalyzed destruction of ozone. Nature **1974**, *249*, 810–812.
8. Rowland, F.S.; Molina, M.J. Chlorofluoromethanes in the environment. Rev. Geophys. Space Phys. **1975**, *13*, 1–35.

9. Stolarski, R.S.; Cicerone, R. Stratospheric chlorine: A possible sink for ozone. Can. J. Chem. **1974**, *52*, 1610–1650.
10. Dobson, G.M.B.; Harrison, D.N. Measurement of the amount of ozone in the Earth's atmosphere and its relation to other geophysical conditions. Proc. R. Soc. London **1926**, *110*, 660–693.
11. Fahey, D.W.; Hegglin, M.I. Twenty Questions and Answers About the Ozone Layer: 2010 Update, Scientific Assessment of Ozone Depletion: 2010; World Meteorological Organization: Geneva, Switzerland, 2011; 72 pp.
12. Midgley, T. From the periodic table to production. Ind. Eng. Chem. **1937**, *29*, 241–244.
13. C&EN Special Issue **2008**, *86* (14), available at http://pubs.acs.org/cen/priestley/recipients/1941midgely.html (accessed May 29, 2011).
14. Lovelock, J.E. Atmospheric fluorine compounds as indicators of air movements. Nature **1971**, *230*, 379.
15. Lovelock, J.E.; Maggs, R.J.; Wade, R.J. Halogenated hydrocarbons in and over the Atlantic. Nature **1973**, *241*, 194–196.
16. Molina, M.J. Polar ozone depletion (Nobel lecture). Angew. Chem. Int. Ed. Engl. **1996**, *35*, 1778–1785.
17. Rowland, F.S. Stratospheric ozone depletion by chlorofluorocarbons (Nobel lecture). Angew. Chem. Int. Ed. Engl. **1996**, *35*, 1786–1798.
18. Rowland, F.S.; Molina, M.J. The CFC-Ozone Puzzle: Environmental Science in the Global Area. John H. Chafee Memorial Lecture on Science and the Environment. National Academy of Sciences, December 7, 2000.
19. DeMore, W.B.; Sander, S.D.; Golden, D.M. et al. Chemical kinetics and photochemical data for use in the stratospheric modeling. Evaluation no. 11, JPL publication no. *94–26*. NASA Jet Propulsion Laboratory: Pasadena, CA, 1994.
20. World Meteorological Organization. Scientific assessment of ozone depletion: 1994. WMO Global Ozone Research and Monitoring Project, report no. 37, Geneva: WMO, 1995.
21. World Meteorological Organization. Scientific assessment of ozone depletion: 1998. WMO Global Ozone Research and Monitoring Project, report no. 44, Geneva: WMO, 1999.
22. National Research Council. *Halocarbon: Environmental Effects of Chlorofluoromethane Release;* National Academy of Sciences: Washington, DC, 1976.
23. National Research Council. *Halocarbons: Effects on Stratospheric Ozone;* National Academy of Sciences: Washington, DC, 1976.
24. Farman, J.C.; Gardiner, B.G.; Shanklin, J.D. Large losses of total ozone in Antarctica reveal seasonal ClO_x/NO_x interactions. Nature **1985**, *315*, 207–210.
25. Stolarski, R.S.; Bloomfield, P.; McPeters, R.D.; Herman, J.R. Total ozone trends deduced from Nimbus 7 TOMS data. Geophys. Res. Lett. **1991**, *18*, 1015–1018.
26. Russell, J.M. III; Luo, M.; Cicerone, R.J.; Deaver, L.E. Satellite confirmation of the dominance of chlorofluorocarbons in the global stratospheric chlorine budget. Nature **1996**, *379*, 526–529.
27. Jones, A.E.; Shanklin, J.D. Continued decline of total ozone over Halley, Antarctica, since 1985. Nature **1995**, *376*, 409–411.
28. NASA Ozone Hole Watch, available at http://ozonewatch.gsfc.nasa.gov/facts/hole.html (accessed May 28, 2011).
29. Solomon S.; Garcia, R.R.; Rowland, F.S.; Wuebbles, D.J. On the depletion of Antarctic ozone. Nature **1986**, *321*, 755–758.
30. Molina, L.T.; Molina, M.J.; Stachnick, R.A.; Tom, R.D. An upper limit to the rate of the $HCl+ClONO_2$ reaction. J. Phys. Chem. **1985**, *89*, 3779–3781.
31. Molina, M.J.; Tso, T.-L.; Molina, L.T.; Wang, F.C.-Y. Antarctic stratospheric chemistry of chlorine nitrate, hydrogen chloride and ice. Release of active chlorine. Science **1987**, *238*, 1253–1260.
32. Tolbert, M.A.; Rossi, M.J.; Malhotra, R.; Golden, D.M. Reaction of chlorine nitrate with hydrogen chloride and water at Antarctic stratospheric temperatures. Science **1987**, *238*, 1258–1260.

33. Abbatt, J.P.D.; Beyer, K.D.; Fucaloro, A.F.; McMahon, J.R.; Wooldridge, P.J.; Zhang, R. Interaction of HCl vapor with water-ice: Implications for the stratosphere. J. Geophys. Res. **1992**, 97, 15819–15826.
34. Molina, M.J. The probable role of stratospheric 'ice' clouds: Heterogeneous chemistry of the ozone hole. In *Chemistry of the Atmosphere: The Impact of Global Change*; Calvert, J.G., Ed.; Blackwell Scientific: Oxford, U.K., 1994; 27–38.
35. Gertner, B.J.; Hynes, J.T. Molecular dynamics simulation of hydrochloric acid ionization at the surface of stratospheric ice. Science **1996**, *271,* 1563–1566.
36. Molina L.T.; Molina M.J. Production of Cl_2O_2 from the self-reaction of the ClO radical. J. Phys. Chem. **1987**, *91,* 433–436.
37. McElroy, M.B.; Salawitch, R.J.; Wofsy, S.C.; Logan, JA. Reduction of Antarctic ozone due to synergistic interactions of chlorine and bromine. Nature **1986**, *321,* 759–762.
38. Cox, R.A.; Hayman, G.D. The stability and photochemistry of dimers of the ClO radical and implications for Antarctic ozone depletion. Nature **1988**, *332,* 796–800.
39. Molina, M.J.; Colussi, A.J.; Molina, L.T.; Schindler, R.N.; Tso. T-L. Quantum yield of chlorine-atom formation in the photodissociation of chlorine peroxide (ClOOCl) at 308 nm. Chem. Phys. Lett. **1990**, *173,* 310–315.
40. Tuck, A.F.; Watson R.; Condon, E.P.; Margitan, J.J.; Toon, O.B. The planning and execution of ER-2 and DC-8 aircraft flights over Antarctica, August and September 1987. J. Geophys. Res. **1989**, *94,* 181–222.
41. Anderson, J.G., Toohey, D.W., Brune, W.H. Free radicals within the Antarctic Vortex: the role of CFCs in Antarctic ozone loss. Science **1991**, *251,* 39–46.
42. Proffitt, M.H.; Steinkamp, M.J.; Powell, J.A. et al. In situ ozone measurements within the 1987 Antarctic ozone hole from a high-altitude ER-2 aircraft. J. Geophys. Res. **1989**, 94, 547–555.
43. Chen, H.-Y.; Lien, C.Y.; Lin, W.-Y.; Lee, Y.T.; Lin, J.J. UV absorption cross sections of ClOOCl are consistent with ozone degradation models. Science **2009**, *324,* 781–784.
44. WMO (World Meteorological Organization). Scientific Assessment of Ozone Depletion: 2010, Global Ozone Research and Monitoring Project—Report No. 52, Geneva, Switzerland, 2011; 516 pp.
45. Turco, R.; Plumb, A.; Condon, E. The Airborne Arctic Stratospheric Expedition: Prologue. Geophys. Res. Lett. **1990**, *17,* 313–316.
46. Manney, G.L.; Santee, M.L.; Rex, M. et al. Unprecedented Arctic ozone loss in 2011. Nature **2011**, *478,* 469–475.
47. van der Leun, J.; Tang, X.; Tevini, M. Environmental effects of ozone depletion: 1994 assessment. Ambio **1995**, *24,* 138.
48. Biggs R.H.; Joyner, M.E.B., Eds. *Stratospheric Ozone Depletion/UV-B Radiation in the Biosphere;* Springer-Verlag: New York, 1994.
49. UNEP, Environmental Effects of Ozone Depletion and its Interaction with Climate Change: 2010 Assessment. United Nations Environment Programme, December 2010, Nairobi, Kenya, available at http://ozone.unep.org/Assessment_Panels/EEAP/eeap-report2010.pdf (accessed May 20, 2011).
50. Longstreth, J.D.; de Grujil, F.R.; Kripke, M.L.; Takizawa, Y.; van der Leun, J.C. Effects of solar radiation on human health. Ambio **1995**, *24,* 153–165.
51. Molina, M.J.; Molina L.T.; Fitzpatrick, T.B.; Nghiem, P.T. Ozone depletion and human health effects. In *Environmental Medicine;* Moller L., Ed.; Joint Industrial Safety Council Product 33: Sweden, 2000; 28–51.
52. Brash, D.E.; Rudolph, J.A.; Simon, J.A. et al. A role for sunlight in skin cancer: UV-induced p53 mutations in squamous cell carcinoma. Proc. Natl. Acad. Sci. U. S. A. **1991**, *88,* 124–128.
53. De Fabo, E.C.; Noonan, F.P. Mechanism of immune suppression by ultraviolet irradiation in vivo. I. Evidence for the existence of a unique photoreceptor in skin and its role in photo-immunology. J. Exp. Med. **1983**, *157,* 84–98.

54. Cooper, K.D.; Oberhelman, L.; Hamilton, T.A. et al. UV exposure reduces immunization rates and promotes tolerance to epicutaneous antigens in humans—Relationship to dose, CDla-DR+ epidermal macrophage induction and Langerhans cell depletion. Proc. Natl. Acad. Sci. U. S. A. **1992**, *89,* 8497–8501.
55. Caldwell, M.M.; Teramura, A.H.; Tevini, M.; Bomman, J.F.; Björn, L.O.; Kulandaivelu, G. Effects of increased solar ultraviolet radiation on terrestrial plants. Ambio **1995**, *24,* 166–173.
56. Häder, D.P.; Worrest, R.C.; Kumar, H.D.; Smith, R.C. Effects of increased solar ultraviolet radiation on aquatic ecosystems. Ambio **1995**, *24,* 174–180.
57. Smith, R.C.; Prezelin, B.B.; Baker, K.S. et al. Ozone depletion: Ultraviolet radiation and phytoplankton biology in Antarctic waters. Science **1992**, 255, 952–959.
58. UNEP (United Nations Environment Program). *Handbook for the Montreal Protocol on Substances that Deplete the Ozone Layer,* 8th Ed.; UNEP Ozone Secretariat, 2009.
59. UNEP (United Nations Environment Program). *Global Environmental Outlook (GEO-4): Environment for Development;* Progress Press Ltd.: Malta, **2007**.
60. UNEP (United Nations Environment Program). HFCs: A Critical Link in Protecting Climate and the Ozone Layer, November 2011; 36 pp.
61. The Climate and Clean Air Coalition to Reduce Short-Lived Climate Pollutants, Fact Sheet. http://www.state.gov/rZpa/prs/ps/2012/02/184055.htm (accessed February 16, 2012)
62. McKenzie, R.L.; Aucamp, P.J.; Bais, A.F.; Björn, L.O.; Ilyas, M.; Madronich, S. Ozone depletion and climate change: impacts on UV Radiation. Photochem. Photobiol. Sci. **2011**, *10,* 182–198.

16

Thermodynamics

Ronald L. Klaus

Introduction

Thermodynamics is concerned with the interaction between matter and energy on a macroscopic level. It is therefore the basic science that underlies the engineering of the use of energy in all its forms. It is governed by two basic laws with far-reaching implications. The first concerns the conservation of energy, and the second puts limits on the amount of energy that can be converted into work over and above what the first law might be thought to imply.

The actual performance of practical process equipment can be predicted by application of these two laws. That, in turn, depends on the ability to estimate the thermodynamic properties of the streams that enter and leave this equipment. The exact equations needed to calculate these properties can be derived from the formalism of thermodynamics. Certain simplifications, such as the ideal gas law or the assumption of an equation of state, are usually required to make numerical estimates of these properties.

First Law of Thermodynamics

Formulation

The first law of thermodynamics is a statement of the principle of conservation of energy for a closed system—namely, one in which there is no transfer of mass across the system boundaries.

$$\Delta U = Q - W_{\text{total}} \tag{1}$$

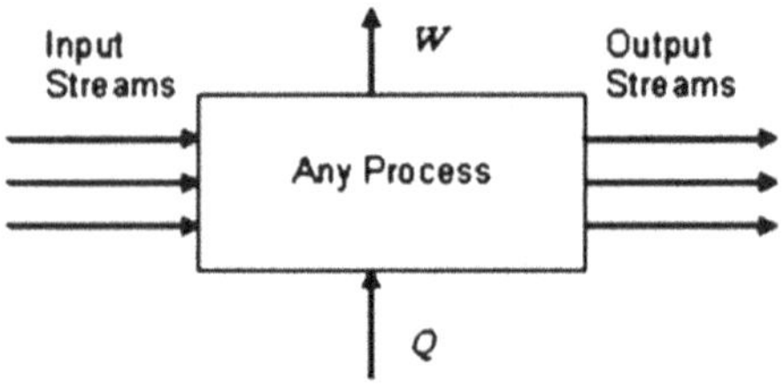

FIGURE 1 Schematic representation of any thermodynamic process.

This equation ignores certain effects such as differences in fluid velocities, which become important only at very high velocities, and differences in fluid elevations, which become important only when there are greatly varying elevations. *U* is a thermodynamic property called internal energy. It accounts for energy due to the motion of molecules in a fluid and the exchanges of energy between them. The Δs refer to differences in the property value between the final and initial states of the system. *U*, *Q*, and *W* all have units of energy. However, this equation can also have another interpretation for open flow systems at steady state. Such systems have input and outlet streams, and possibly an exchange of heat or work with the surroundings (Figure 1). However, the stream flows and property values throughout the system are all considered constant over time. In this case, the first law is a rate equation. ΔU is the difference between the sum of *U* times the mass flow rate for all the output streams minus that same sum for all the inlet streams. *Q* and *W* are rates of energy transfer. The arrows in Figure 1 indicate the direction of the flow of work, *W*, and heat, *Q*, when these quantities are positive.

When applying Eq. 1 to a flow process, W_{total} includes the work done in moving fluids into and out of the process. To eliminate these effects, another thermodynamic property, namely enthalpy, *H*, is defined,

$$H = U + PV \tag{2}$$

where *P* is pressure and *V* is volume.

For a flow process at steady state, this changes the first law to

$$\Delta H = Q - W \tag{3}$$

In Eq. 3, *W* is the work produced by the process apart from that involved in moving fluids into and out of the process. It is often called the shaft work.

In these equations, it is important to make a distinction between intrinsic and extrinsic properties. The former are properties per unit mass or per mole, whereas the latter are total values of the property for the mass flow of the given process stream. These and other thermodynamic relationships can be applied to either, provided that they are applied consistently.

U, *H*, *P*, and *V* are examples of thermodynamic state functions—that is, their intrinsic values depend only on the state of the fluid (usually its temperature, pressure, and composition) and not on the process used to bring the fluid to that state. On the other hand, quantities like *Q* and *W* are not state functions. When a fluid makes a transition from one state to another, the values of *Q* and *W* associated with that transition are dependent on the "path" over which that transition was made.

Enthalpies of Formation and Reaction

Enthalpy cannot be calculated in an absolute sense. It requires the definition of a reference state, at which it is taken to be zero. Enthalpies in all other states are referenced to this state. The most rigorous definition of a reference state is to take it as comprising the constituent atomic species in their normal molecular configurations at some temperature and pressure—typically 25°C and 1 atm. Normal molecular configuration means the normal configuration the atomic molecules take in nature, such as O_2, H_2, N_2, etc. Thus, these molecules would be taken to have zero enthalpy at 25°C and 1 atm. The phases of these elemental species also have to be stated. They are generally taken to be the normal phase of the element at the stated conditions. The reference states for gases are usually taken to be in their "ideal gas state" (ig). (The ideal gas state is difficult to describe briefly. It is a fluid with the same heat capacity as the real fluid's heat capacity at very low pressure. However, unlike the real fluid, it obeys the ideal gas law. For low-pressure calculations, the difference between the ideal gas state and the real fluid at low pressure is not important.)

With these ideas as a starting point, the enthalpies of compounds can be built up by considering real or imaginary reactions through which they are formed from their constituent elements. For example, the enthalpy of formation of methane would be the enthalpy change associated with the following reaction:

$$C(s) + 2H_2(ig) = CH_4(ig) \quad (25°C, 1\ atm)$$

where s indicates solid and ig indicates the ideal gas state. Such chemical reactions are usually accompanied by the absorption or release of a certain amount of heat. If the reaction is carried out at constant temperature and pressure, and if no work is done, according to the first law, this heat is the enthalpy difference of the reaction and defines the standard enthalpy (or heat) of formation of the resulting compound.

Standard enthalpies of formation are widely tabulated. An excellent source is the collection by the National Institute of Standards and Technology (NIST).[1] Enthalpies of formation for some common species are given in Table 1. They are given both in the units that appear in the NIST collection and in nondimensionalized form. The latter permits their use in any system of units simply by multiplying by RT_0 in the desired units. To is taken to be 0°C (= 273.15°K=459.67°*R*), and *R* is the Universal Gas Constant. Note that °*R* represents degrees Rankine and °K represents degrees Kelvin.

Many different units are used in energy calculations. The values of some useful conversion factors and the values of the universal gas constant appear in Table 2.

These enthalpies of formation can be combined to produce enthalpy changes for any chemical reaction whose species have known enthalpies of formation. For example, in the combustion reaction

$$CH_4 + 2O_2 = CO_2 + 2H_2O$$

the enthalpy change may be calculated as follows:

$$\Delta H = \Delta H_f(CO_2) + 2\Delta H_f(H_2O) - \Delta H_f(CH_4) - 2\Delta H_f(O_2)$$

$$= (-393.51) + 2(-285.830) - (-74.84) - 0 = -890.3 kj/gmole$$

TABLE 1 Enthalpies of Formation of Some Common Compounds

Chemical Specie	Formula	State	MW	ΔH_f (T=298.15°C)	ΔH_f (Non dim.)
Methane	CH_4	IG	16.04	−74.87	−32.968
Ethane	C_2H_6	IG	30.08	−83.8	−36.901
Propane	C_3H_8	IG	44.10	−104.7	−46.104
n-Octane	C_8H_{18}	IG	114.23	−208.4	−91.767
n-Octane	C_8H_{18}	L	114.23	−250.3	−110.217
Carbon dioxide	CO_2	IG	44.01	−393.51	−173.278
Carbon monoxide	CO	IG	28.01	−110.53	−48.671
Water	H_2O	IG	18.02	−241.826	−106.486
Water	H_2O	L	18.02	−285.830	−125.863

Note: Units of ΔH_f= kJ/gmole; Non-dimensionalization, $\Delta H_f/RT_0$ where T_0= 273.15°K.

TABLE 2 Some Useful Conversion Factors and Constants

Type	Value	Conversion
Length	1 m	3.28084 ft
		39.3701 in.
Mass	1 kg	2.20462 lbm
Force	1 N	1 kg m/s^2
		0.224809 lbf
Pressure	1 bar	10^5 N/m^2
		10^5 Pa
		0.986923 atm
		14.5038 psia
Pressure	1 atm	14.6960 psia
Volume	1 m^3	35.3147 ft^3
Density	1 g/cm^3	62.4278 lbm/ft^3
Energy	1 J	1 N m
		1 m^3 Pa
		10^{-5} m^3 bar
		10^{-3} kW s
		9.86923 cm^3 atm
		0.239006 cal
		5.12197×10^{-3} ft^3 psia
		0.737562 ft lbf
		9.47831×10^{-4} Btu
Power	1 kW	10^3 J/s
		239.006 cal/s
		737.562 ft lbf/s
		0.94783 Btu/s 1.34102 hp

Note: Values of the universal gas constant, R
8.314 J/gmole K = 8.314 m^3 Pa/(gmol K) = 83.14 cm^3 bar/(gmole K) 82.06 cm^3 atm/(gmole K)
1.987 cal/(gmole K) = 1.986 Btu/(lbmole R)
0.7302 ft^3 atm/(lbmole R) = 10.73 ft^3 psia/(lbmole R)
1545 ft lbf/(lbmole R)

If ΔH is positive, the reaction is called endothermic, whereas if it is negative, it is called exothermic. Because this particular reaction is a combustion reaction, the absolute value of its enthalpy change is also called the enthalpy change (or heat) of combustion.

Application to Combustion

The goal of a power cycle is to convert the chemical energy of a fuel into usable mechanical energy that can produce useful shaft work or drive a generator to produce electricity (Figure 2). What one hopes to do is to convert as much of the fuel's chemical energy into useful work. The hope for this ideal is reflected in what is called the heating value of the fuel. It is taken to be the negative of the enthalpy change of the combustion reaction through which a fuel is completely oxidized at the so-called ISO (International Organization for Standardization) conditions—namely, 15°C and 1 atm pressure. (ISO conditions for air also include a relative humidity of 60%.) Thus, it is the negative of the enthalpy change of the following kind of reaction:

$$\text{Fuel} + aO_2(\text{ig}) \rightarrow bCO_2(\text{ig}) + cH_2O(\text{ig or 1}) + \text{Any other products}(51°\text{C},1\ \text{atm})$$

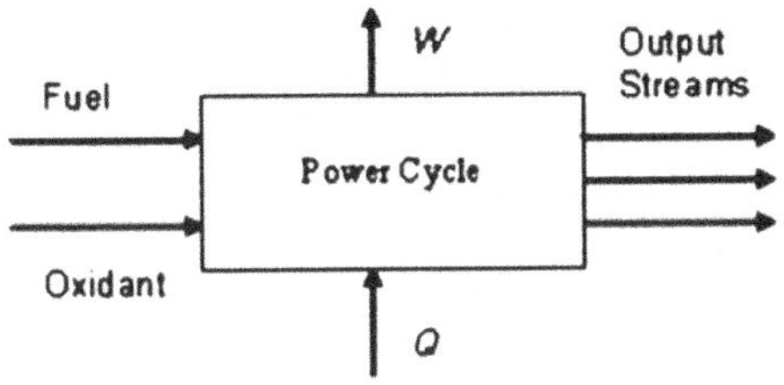

FIGURE 2 Schematic representation of a power cycle.

TABLE 3 Heating Values of Some Common Fuels

	HHV		LHV		LHV/
Fuel	Btu/lbm	Btu/gal	Btu/lbm	Btu/gal	HHV
No. 2 oil	19,580	142,031	18,421	133,623	0.9408
No. 4 oil	18,890	146,476	17,804	138,055	0.9425
No. 6 oil	18,270	150,808	17,312	142,901	0.9476
Diesel fuel	19,733		18,487		0.9368
Hydrogen	61,007		51,635		0.8464
Methane	23,876		21,518		0.9012
Typical natural gas	22,615		20,450		0.9019
Propane	21,653		19,922		0.9201
Butane	21,266		19,623		0.9227
Gasoline	19,657	121,808	18,434	114,235	0.9379
Reformulated gasoline	19,545	120,103	18,304	112,377	0.9365
Methanol	11,274	73,882	10,115	66,289	0.8972
Anthracite coal	14,661		14,317		0.9765
Bituminous coal	14,100		13,600		0.9645

Source: Based on calculations by the author and data from The Association of Energy Engineers.[2–4]

where *a*, *b*, and *c* are coefficients that depend on the particular fuel.

It is unfortunate that ISO conditions differ from the reference state at which enthalpies of formation are usually tabulated. Nevertheless, the latter can be adjusted to ISO conditions, and values for some common fuels appear in Table 3.

Heating Values, Heat Rates, and Power Cycle Efficiency

One area of confusion about heating values comes because of the choice of the phase of the water produced as a combustion product. If the product water is considered to be a liquid, its enthalpy will be lower than that if it is considered to be a vapor. The former choice leads to a higher heating value (HHV), whereas the latter leads to a lower heating value (LHV). If the chemical composition of the fuel is known, the heating value is the mole fraction average of the heating values of the individual constituents divided by the average molecular weight of the fuel. Where the composition of the fuel is not known, accurate heating values are determined experimentally.

The heat rate is one often-tabulated measure that is used to describe the effectiveness of a machine—such as an internal combustion engine or a gas turbine—used to convert the chemical energy of a fuel into usable mechanical or electrical energy. Several definitions go into it.

Heat Energy Input(Btu/s)

$$= \text{First Law Available Chemical Energy}$$

$$= \text{Heating value of a Fuel } (\text{Btu/lb}) \times \text{Fuel flow rate } (\text{lb/s})$$

Heat Rate (Btu/kWh)

$$= \text{Heat Energy Input } (\text{Btu/s}) \times 3600(\text{Btu/s})/\text{Power Produced } (\text{kW})$$

The heat rate is based on a particular choice of either HHV or LHV of the fuel. The higher the heat rate, the more fuel is being used per kW of power produced and therefore the less efficient the machine. The heat rate is directly related to the efficiency.

$$\text{Efficiency} = \frac{\text{Power Output}(\text{kW}) \times 9.47831 \times 10^{-4}\,\text{Btu/J}}{\text{Heat Energy Input } (\text{Btu/s})}$$

Comparison with the previous equations yields Efficiency

$$\text{Efficiency} = \text{Conversion Factor } (\text{Btu/kWh}) / \text{Heat Rate}(\text{Btu/kWh})$$

The conversion factor's role is strictly to convert the units. In the units shown it is equal to

$$\text{Conversion Factor} = 9.47831 \times 10^{-4}\,\text{Btu/J} \times 1000\text{J/kW} - \text{s} \times 3600\text{s/h} = 3412.19\text{Btu/kWh}$$

When the power-generating equipment is used to produce electricity, the efficiency is often called the electrical efficiency.

The efficiency calculated by the above equations is based on the hope of converting the entire heating value of the fuel into useful work. It is unfortunate that this has come to be the standard because there is a further limitation imposed by the second law of thermodynamics. It tells us that nature will not allow conversion of this much chemical energy into useful work, even if the power cycle were constructed as perfectly as possible under the most ideal conditions.

The Second Law of Thermodynamics

Reversibility

Before stating the second law specifically, it is necessary to introduce the concept of reversibility. Most processes produce changes in the substances on which they operate. Reversibility has to do with how easy it would be to undo the changes that are produced in a substance through some process. Specifically, reversibility means that a change that a process makes on a substance could theoretically be undone (reversed) with only an infinitesimal change in the conditions in the process.

For example, suppose that heat is exchanged by allowing heat to flow from a hotter substance to a cooler one. For this change to be reversible, temperatures at every point where heat transfer takes place must differ only infinitesimally. Thus, if we raised the temperature profile of the cooler substance by just an infinitesimal amount, the heat could be transferred back to the hotter substance.

As another example, suppose that a gas is compressed in a cylinder by moving a piston in such a way that it reduces its volume. For the change to be reversible, the pressure the piston exerts must be only differentially greater than the pressure of the gas. In this case, if the piston pressure were reduced only infinitesimally, the gas could be expanded back to its original state, and all the work of compression would be recovered. On the other hand, if a pressure were applied that differed substantially from that of the gas in the piston, the change brought about would be irreversible, because a differential change in the applied pressure could not reverse the effects of the compression.

Formulation

The second law of thermodynamics has a certain mystique to it because unlike most physical laws, it is not a conservation law (e.g., conservation of mass, energy, momentum, etc.). Instead, it puts certain limits on what nature allows in spite of the great conservation laws. It is also not as intuitive as the other great laws. Therefore, much has been written to try to explain it. However, it can be applied rigorously based on the following three-part formulation. This formulation refers to a given substance of constant mass that experiences changes from an initial to a final state through the possible exchange of heat and work with its surroundings.

Part 1: There is a thermodynamic state function, called entropy, S, which is defined by the following equation:

$$dS = \frac{dQ_{rev}}{T} \tag{4}$$

where dS is the differential change in entropy and dQ_{rev} refers to a differential amount of heat transferred to or from a substance in a reversible manner. That is, to calculate finite entropy differences, this equation needs to be integrated along a reversible path.

Stating that entropy is a state function is not a trivial claim. What it means is that no matter what reversible path is chosen, the net change in entropy will be the same, provided that the initial and final states are the same.

Part 2: No matter how a change of state is brought about in a given substance

$$\Delta S_{substance} + \Delta S_{surrounding} \geq 0 \tag{5}$$

The differences implied by the Δs are differences between the end state and the initial state of both the substance and surroundings. Notice that there is such a thing as a change in the entropy of the surroundings as well as the substance. It too can be calculated by Eq. 4.

Part 3: The inequality in the previous equation becomes an equality if, and only if, the process used to bring about the change is completely internally reversible and if the process also exchanges heat with the surroundings in a completely reversible manner.

Thus, this third part becomes a criterion of reversibility for a process or piece of equipment. Furthermore, the greater the inequality, the greater the irreversibility.

In general, the surroundings are considered to be at a constant ambient temperature, T_0. Thus,

$$\Delta S_{surroundings} = \frac{Q_{surroundings}}{T_0} = -\frac{Q_{substance}}{T_0} \tag{6}$$

where the two *Q*s are heat transfers to the surroundings and substance, respectively, which are the same in magnitude but opposite in sign. This equation can be combined with Eq. 5 to give

$$Q \leq T_0 \Delta S \tag{7}$$

Along with the first law for flow processes, this gives

$$W \leq T_0 \Delta S - \Delta H \tag{8}$$

In these equations, the subscripts have been dropped and all quantities refer to the substance, not the surroundings. Furthermore, although these equations were derived for changes in a substance of constant mass (closed system), as with the first law, they can also be applied to a steady flow system. In this case, the Δs refer to differences in the fluxes of the thermodynamic properties between the output and inlet streams.

The implications of these two equations are very significant. They mean that in a flow process, if the inlet and outlet conditions of the various streams are fixed, there are an upper bound to the amount of heat that can be transferred to the process and an upper bound on the amount of work it can produce. The latter claim has an impact on power generation systems because it imposes a limit on the amount of power that can be produced from a given amount of fuel, regardless of its energy content. Furthermore, because the entropy change of such processes is usually negative, another implication of Eq. 7 is that a certain amount of heat must be rejected to the surroundings regardless of how well the process is designed internally.

Carnot Engine

Some of the implications of the second law can be illustrated through the so-called Carnot engine. This is an imaginary reversible engine in which heat, Q_H, is transferred from a high-temperature source at T_H into a reversible engine, which produces work, W, and discards heat, Q_C, into a cold sink—possibly the environment—at T_C (Figure 3). From the first law,

$$W = Q_H - Q_C$$

From the second law,

$$\frac{Q_C}{T_C} - \frac{Q_H}{T_H} \geq 0$$

If the Carnot efficiency is defined as the amount of work that can be produced from the high-temperature heat (i.e., W/Q_H),

$$\frac{W}{Q_H} = \frac{Q_H - Q_C}{Q_H} = \frac{T_H - T_C}{T_H} \quad (9)$$

The last term on the right is of particular interest. It means that the Carnot efficiency depends on only the temperatures of the two heat reservoirs. All the heat in a high-temperature source can never be converted to work, even if the apparatus is perfectly reversible. The higher the temperature of the hot source relative to the cold sink, the higher the Carnot efficiency.

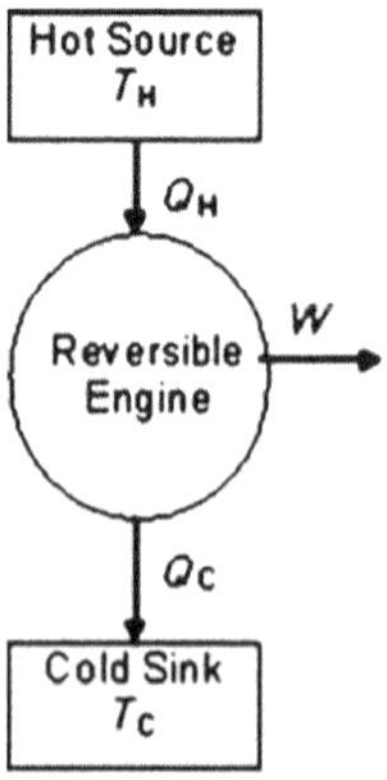

FIGURE 3 The Carnot engine.

Ideal Work, Lost Work

The second law gives a means to measure the irreversibility of a process or piece of equipment. That, in turn, is a measure of its departure from an ideal design—one which makes changes to substances in a way that most preserves their work-producing potential. This analysis begins by imagining a perfectly reversible process and defining W_{ideal} and Q_{ideal} to be the limiting values defined by Eqs. 7 and 8 (see Figure 4).

$$W_{ideal} = T_0 \Delta S - \Delta H \tag{10}$$

$$Q_{ideal} = T_0 \Delta S \tag{11}$$

These ideal values can be determined from the enthalpies and entropies of the various streams entering and leaving the process alone, without reference to any of the internal details of the process.

The level of irreversibility in a real process or in some part of it can be quantified through what is called lost work.[5] Because the actual work that can be obtained from any real process is always less than that what would have resulted from a completely reversible process, this leads to the following definition:

$$W_{lost} = W_{ideal} - W \tag{12}$$

By combining this with Eqs. 10 and 3, this becomes

$$W_{lost} = T_0 \Delta S - Q \tag{13}$$

A real process—or any subprocess—can be represented as shown in Figure 5. The W_{lost} stream is not an actual work output. However, showing it in this manner is a reminder that in any real process a certain amount of the potential to do work has been lost due to the irreversibilities of the process. Furthermore, the lost work of all the subprocesses within the overall process sum up to the lost work of the process as a whole,

$$W_{ideal} = W + \sum W_{lost} \tag{14}$$

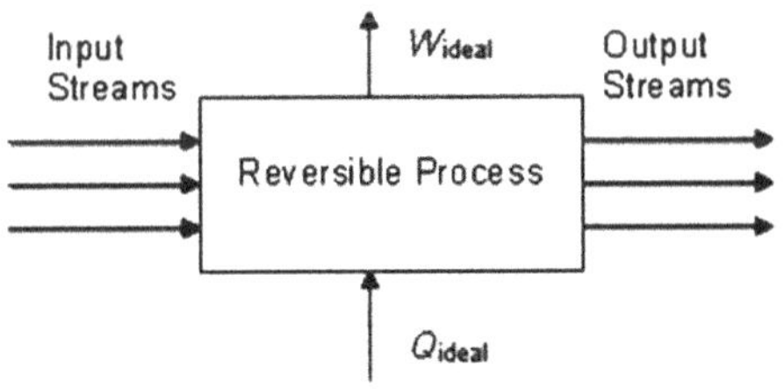

FIGURE 4 Schematic representation of a reversible thermodynamic process.

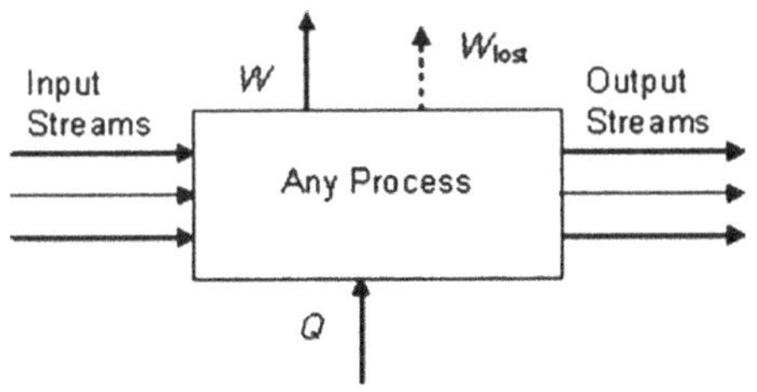

FIGURE 5 Schematic representation of any process showing lost work.

where the summation is carried out over each of the subprocesses. When analyzing a process, Eq. 13 can be applied to each piece of equipment. Then the W_{lost} for each of these can be summed according to Eq. 14. Such an analysis brings out where the greatest irreversibilities occur and therefore shows where the greatest opportunities are for improvement.

Exergy

Lost work is related to the modern concept of exergy. Lost work gives a quantitative measure of the irreversibility of a process or subprocesses within it. Furthermore, the term "lost work" seems quite descriptive. However, it is also possible to assign values—perhaps even monetary values—to streams as well as equipment. This can be done through the concept of exergy.

The exergy of a stream is the maximum amount of work that could theoretically be extracted from that stream if its temperature and pressure were reduced to that of the environment and if the concentrations of its chemical species were brought to that of the environment. It should be clear that from a power-production point of view, the economic value of a stream is related to its exergy because the production of work is usually more valuable than the production of heat alone.

Exergy analysis is based on the idea that it costs something in terms of equipment, power, adding costly streams, etc., to raise a stream's exergy—and hence its value. If what it costs to raise the value of a given stream is less than the increased value of the stream, it is a desirable thing to do.

Exergy is generally divided into four parts: physical, chemical, kinetic (having to do with the velocity of the fluid), and potential (having to do with its height above some datum level). The first two of these parts are of primary interest here. Physical exergy is the negative of the ideal amount of work that could be extracted from a stream if its temperature and pressure were reduced to ambient conditions. That is,

$$e^{Ph} = \Delta H - T_0 \Delta S \tag{15}$$

In Eq. 15 and the following equations, the Δs refer to differences between the present state of the fluid and its state at ambient conditions.

Chemical exergy, e^{Ch}, also has to be added because the exergy, say, of a fuel at ambient conditions ought to be higher than that of its eventual combustion products at those conditions because it has the potential of producing work in being transformed into its products. This can be accounted for in an equation similar to the one above, but this time involving properties of formation at ambient conditions.

$$e^{Ch} = \Delta H_f - T_0 \Delta S_f \tag{16}$$

Exergy is not quite a thermodynamic state function because it depends on environmental temperature. Also, unlike enthalpy or internal energy, it is not conserved in a process. Nevertheless, it is a useful measure of the power-producing value of a process stream. In any real processes, every process step results in a net loss of exergy.

The relationship between exergy and lost work is very straightforward—namely,

$$\text{Lost Work} = \sum \text{Exergies of incoming streams} - \sum \text{Exergies of outgoing streams} \tag{17}$$

where the summations are carried out over all incoming and outgoing streams, respectively.

Reversibility Revisited

The preceding analysis shows how the second law can be used to provide a measure of the irreversibility of a process or part of a process. Every irreversibility causes a loss in the potential to convert energy into work. In this era of diminishing supplies of cheap fuels and increasing demands for electric power, this is an important consideration. Furthermore, irreversibilities in one part of a process reduce the amount of work that can be produced—or else increase the amount of work that is needed—in the rest of the process, even though the source of the irreversibility may be far removed from the place in the process where work is produced or consumed.

The following is a list of some process steps that introduce irreversibility into processes, including power cycles:

- Heat transfer across a nonzero temperature difference
- Mixing of fluids of differing pressures, temperatures, or compositions
- Pressure drops that do not recover work
- Chemical reactions that do not take place at equilibrium conditions
- Flashing of liquids when pressure is reduced
- Two-phase contact and mass transfer between fluids whose various species have concentrations that differ from their equilibrium values
- Temperature and pressure shocks when fluids enter equipment at different temperatures or pressures from those that are present within the equipment

The most thermodynamically efficient designs—those that most preserve the work-producing potential of the fluids in the process—are those that strive for the greatest reversibilities in the various pieces of process equipment. However, there is another side to this question. The more reversible a piece of equipment is, the greater its capital cost is apt to be. Thus, in the practical design of equipment there is often a trade-off between reversibility and capital cost. Processes need to be evaluated not only for their thermodynamic efficiency, but also for their thermoeconomic effectiveness.

Calculation of Thermodynamic Properties

Rigorous Equations

Process equipment cannot be designed or evaluated apart from the ability to estimate the thermodynamic properties of the entering and exiting streams. The following are rigorous equations for the calculation of some of the important thermodynamic properties of pure or constant-composition substances (The equations can be found in many thermodynamics text books. A concise useful summary is found in Van Ness and Abott.)[6]:

$$dH = C_p dT + \left[V - T\left(\frac{\partial V}{\partial T_p}\right)\right] dP \quad (18)$$

$$dS = \left(\frac{C_p}{T}\right) dT - \left(\frac{\partial V}{\partial T}\right)_P dP \quad (19)$$

$$G = H - TS \quad (20)$$

C_p is the heat capacity at constant pressure, defined by

$$C_P = \left(\frac{\partial H}{\partial T}\right)_P \quad (21)$$

G is called the Gibbs free energy or Gibbs function and has importance in equilibrium calculations. Eq. 20 can be considered its definition. By differentiation and comparison with the other two, it can be shown that

$$dG = VdP - SdT \quad (22)$$

These are differential equations that emphasize that these properties need reference values from which differences can be calculated. To obtain values for the property differences for real substances, the pressure-volume-temperature (PVT) behavior of the fluid needs to be inserted into the equations, which would then be integrated to produce the final results.

Ideal Gas

One way of inserting PVT behavior is through equations of state. One of the simplest models for gases is that of the ideal gas, whose equation of state on a molar basis is

$$\mathrm{PV} = \mathrm{RT} \tag{23}$$

This equation reflects the fact that the molecules in the gas do not interact in any way, although they may have very complex energy interactions within a given molecule.

When this equation is inserted into the rigorous equations, one obtains

$$\mathrm{d}H^* = C_p^* \mathrm{d}T \tag{24}$$

$$\mathrm{d}S^* = \left(\frac{C_p^*}{T}\right)\mathrm{d}T + \left(\frac{R}{P}\right)\mathrm{d}P \tag{25}$$

$$\mathrm{d}G^* = RT\mathrm{d}\,\mathrm{In}\,P - S^*\mathrm{d}T \tag{26}$$

where the asterisk designates the properties as those of an ideal gas.

The integrated forms of Eqs. 24 and 25 are

$$\Delta H^*(P2,T2;P1,T1) = \int_{T1}^{T2} C_p^* \mathrm{d}T \tag{27}$$

$$\Delta S^*(P2,T2;P1,T1) = \int_{T1}^{T2} \left(\frac{C_p^*}{T}\right)\mathrm{d}T + R\,\mathrm{In}\left(\frac{P2}{P1}\right) \tag{28}$$

The heat capacity has been left behind the integral sign because even for most ideal gases, it varies with temperature. However, neither ideal gas heat capacities nor enthalpies vary with pressure, but ideal gas entropies do. Ideal gas heat capacity data is often presented in the form of an analytical equation. One common form used by NIST[7] is

$$C_p^* = A + BT + CT^2 + DT^2 + DT^3 + \frac{E}{T^2} \tag{29}$$

NIST tabulates the constants (*A*, *B*, etc.), for many compounds.

Real Fluids

Real fluids may be represented by the following general equation of state:

$$\mathrm{V} = \mathrm{Z}(\mathrm{P},\mathrm{T})\mathrm{RT}/\mathrm{P} \tag{30}$$

where Z is the compressibility factor, which can be a function of both P and T. It is common to report PVT data in the form of the compressibility factor. For an ideal gas, $Z=1$. Thus, the compressibility factor is a measure of the deviation from ideal gas behavior.

When Eq. 30 is substituted in Eqs. 18 and 19, one obtains

$$dH = C_P dT - \left[\frac{RT^2}{P}\left(\frac{\partial Z}{\partial T}\right)_P\right]dP \tag{31}$$

$$dS = \left(\frac{C_P}{T}\right)dT - R\left(Z + T\frac{\partial Z}{\partial T}\right)\frac{dP}{P} \tag{32}$$

One consequence of these equations is that thermodynamic properties of real fluids can be calculated from their ideal gas–heat capacity (as a function of temperature) and their PVT behavior. This follows from the fact that to get from one state to another, these equations can first be integrated to zero pressure (which requires only PVT data), then integrated at zero pressure to the final temperature (which requires only ideal gas heat capacity data), and then integrated back to the end-point pressure.

Several approaches have been used to apply these equations for practical calculations. A first approach is to take known ideal gas-heat capacity and PVT data, and use them directly in these equations. Such data are not often available, and even when they are, this is a tedious process. Nevertheless, it has been done for a number of well-studied chemical species, most notably water. This is the basis of the steam tables, which have been put into analytical form for easy computer calculations.[8]

A second approach is through generalized correlations. An early attempt to describe the PVT behavior of many fluids was begun by Pitzer[9] through the use of his so- called acentric factor. His correlations have been extended so that they can be used to predict the more important thermodynamic properties of real no-polar fluids in both vapor and liquid phases.[10]

A third approach is to insert an equation of state into Eqs. 31 and 32 so that they may be integrated analytically. One excellent equation of state for low-density gases is the virial equation

$$Z = 1 + \frac{B(T)}{V} + \frac{C(T)}{V^2} + \frac{D(T)}{V^3} + \cdots \tag{33}$$

where B, C, D, etc., are functions of temperature only and are called the second, third, fourth, etc., virial coefficients. There is a great deal of data for the second virial coefficients.[11] Where data are not available, correlations have been developed to estimate them.[12,13] Third virial coefficients are usually not available, but prediction methods have been proposed.[14] Very little information is available for higher-order coefficients. Other equations of state have also enjoyed success in predicting properties of real fluids, both in the vapor and liquid phases. Among the more successful are the Soave–Redlich–Kwong[15] and Peng–Robinson[16] equations.

Solutions

The properties of solutions—whether they are gas or liquid mixtures—are not merely the mole fraction averages of the properties of the pure components. For any thermodynamic property, M, the exact equation for the properties of mixtures is

$$(P,T,\bar{x}) = \sum_i x_i \overline{M_i}(P,T,\bar{x}) \tag{34}$$

where $\overline{M_i}$ is called the "partial molal property of i" and $\bar{x}$ is a vector containing all of the compositions of the various constituents of the solution. The summation is carried out over all constituents. $\overline{M_i}$ is the property of constituent i at pressure, P, and temperature, T, as it exists in solution of composition $\bar{x}$. In general, this is different from its value as a pure substance.

The most general equation through which partial molal properties may be evaluated is

$$\overline{M_i} = \left(\frac{\partial(nM)}{\partial_{ni}} \right)_{P,T,\mathrm{nj}} \tag{35}$$

where n is the total number of moles and n_i is the number of moles of specie "i".

The partial derivative implies that the extensive property, *nM,* is differentiated with respect to the particular specie of interest, *i*, with pressure, temperature, and all of the other constituent amounts held constant. If properties of the various possible mixtures are known or can be approximated analytically, this equation provides a means to calculate their partial molal properties.

An important approximation for solutions is that of so- called "ideal solutions," whose properties are calculated as follows,

$$V^{\mathrm{id}} = \sum_i x^i V^i \tag{36}$$

$$H^{\mathrm{id}} = \sum_i x_i H_i \tag{37}$$

$$S^{\mathrm{id}} = \sum_i x_i S_i - R \sum_i x_i \,\mathrm{In} x_i \tag{38}$$

$$G^{\mathrm{id}} = \sum_i x_i G_i + RT \sum_i x_i \,\mathrm{In} x_i \tag{39}$$

where the superscript, "id," designates these as the properties for ideal solutions. All ideal gases are ideal solutions. The ideal-solution approximation is useful in some situations, but most liquids depart from it significantly. Notice also that properties involving entropy have a mixing effect (second terms on the right side of Eqs. 38 and 39) even if they are ideal solutions. This reflects the fact that the entropy of a solution is always higher than the combined entropies of its constituents.

The true properties of solutions can be calculated from their PVT behavior. For humid air, for example, the mixing effect is often neglected. However, one recent study done by M. Conde Company[17] includes virial coefficient information, including information for mixtures, sufficient to produce very accurate psychrometric properties of air-water mixtures.

Phase Equilibrium

Phase equilibrium can also be predicted from thermodynamic properties. The way this is done goes beyond the scope of this entry. However, several ideas can be stated here. One aspect of phase equilibria—namely pure-component vapor pressures—has been widely measured and can be used as a starting point for multicomponent phase equilibrium estimates. Vapor pressure data is often correlated by the Antoine equation,

$$\log_{10}\left(P^V\right) = A - \left(\frac{B}{T+C} \right) \tag{40}$$

where *A*, *B*, and *C* are constants. Such constants for many compounds also appear in the NIST collection.[18]

One calculations is the two-phase equilibrium between (1) pure special case of interest in energy liquid water and (2) a vapor phase one of whose components is water vapor. If the vapor phase is considered to be an ideal gas (and thus an ideal mixture), and if certain other assumptions are made, the composition of water in the vapor phase can be estimated by

$$yH_2O = \frac{P^V}{P} \tag{41}$$

where the vapor pressure can be calculated from the Antoine equation. This is a useful approximation that can be refined if PVT data for the vapor mixture are available.

Conclusions

This entry attempts to give an introduction to some of the basic principles of thermodynamics and how they relate to various energy engineering applications. The two main laws of thermodynamics govern the performance of any energy-conversion process. Their application requires the ability to estimate the thermodynamic properties of the various process streams. Such estimates are also based on the formalism that thermodynamics provides. Thus, thermodynamics provides the basic framework within which the design and evaluation of all energy-conversion equipment and processes must be performed.

Acknowledgments

Support for this work by VAST Power Systems, Elkhart, Indiana, is gratefully acknowledged.

Nomenclature

All thermodynamic properties have the units of energy or energy per unit time except where otherwise specified.

C_p	Heat capacity at constant pressure (energy/temperature)
e^{Ch}	Chemical exergy
e^{Ph}	Physical exergy
G	Gibbs free energy, or Gibbs function
H	Enthalpy
M	General thermodynamic property
M_i	Partial molal property of specie "i"
n	total number of moles
n_i	number of moles of specie "i"
P	Pressure (force/area)
P^V	Vapor pressure (force/area)
Q	Heat or heat flux added to a process (energy or energy per unit time)
Q_{ideal}	Heat or heat flux added to a completely reversible process (energy or energy per unit time)
R	Universal gas constant (energy/temperature)
S	Entropy (energy/temperature)
T	Temperature
T_0	Reference temperature or temperature of the surroundings
U	Internal energy
V	Volume (length^3)
W	Work or power produced by a process. Usually does not include work done by or on fluids entering or leaving the process (energy or energy per unit time)
W_{ideal}	Work or power produced by a perfectly reversible process (energy or energy per unit time)
W_{lost}	Lost work or power (energy or energy per unit time)
W_{total}	Total work or power produced by the process (energy or energy per unit time)
X_i	mole fraction of specie "i"
x	vector containing all mole fractions

yH_2O	Mole fraction of water vapor in a vapor phase
Z	Compressibility factor
Δ	Difference in the value between two states or differences between the properties of outlet and input streams
ΔH_f	Enthalpy of formation
ΔS_f	Entropy of formation (energy/temperature)

Subscripts

f	Formation
o	Base or reference state
C	Cold sink
H	Hot source
rev	Reversible

Superscripts

*	Ideal gas or ideal gas state
id	Ideal solution

References

1. Available at http://webbook.nist.gov/chemistry/form-ser. html (accessed July 2005).
2. Capehart, B.L.; Salas, C.E. *The Theory and Practice of Distributed Generation and On-Site CHP Systems*; The Association of Energy Engineers: Atlanta, GA, June 2003.
3. Bosul, Ulf. *Well-to-Wheel Studies, Heating Values and the Energy Conversation Principle,* available at http://www. efce.com/reports/E10.pdf (accessed July 2005).
4. Available at http://www.chpcentermw.org/pdfs/toolkit/7c_ rules_thumb.pdf (accessed July 2005).
5. Van Ness, H.C.; Abbott, M.M. Thermodynamics. In *Perry's Chemical Engineers' Handbook,* 7th Ed.; Perry, R.H., Green, D.W., Eds.; McGraw Hill: New York, 1997; 4-1–4–36; However, note that because of a different sign convention their formulas for ideal work have the opposite signs from those in this entry.
6. Van Ness, H.C.; Abbott, M.M. In *Perry's Chemical Engineers' Handbook,* 7th Ed.; Perry, R.H., Green, D.W., Eds.; McGraw Hill: New York, 1997; 4-1–4–36.
7. Available at http://webbook.nist.gov/chemistry/form-ser. html (accessed July 2005).
8. Available at http://www.cheresources.com/iapwsif971.pdf (accessed July 2005). A free downloadable Excel add-in which embodies the equations is also available on this Web site.
9. Pitzer, K.S. *Thermodynamics,* 3rd Ed.; Mc-Graw Hill: New York, 1995; App. 3.
10. Lee, B.I.; Kessler, M.G. A generalized thermodynamic correlation based on three-parameter corresponding states. AIChE J. **1975,** *21* (3), 510–527.
11. Daubert, T.E.; Danner, R.P. *Physical and Thermodynamic Properties of Pure Chemicals Data Compilation;* Taylor and Francis: Bristol, PA, 1994.
12. Tsonopoulos, C. An empirical correlation of second virial coefficients. AIChE J. **1974,** *20* (2), 263–272.
13. Liley, P.E.; Thomson, G.H.; Daubert, T.E.; Buck, E. Physical and chemical data. In *Perry's Chemical Engineers' Handbook,* 7th Ed.; Perry, R.H., Green, D.W., Eds.; McGraw Hill: New York, 1997; 2-355–2-358.
14. De Santis, R.; Grande, B. An equation for predicting third virial coefficients of nonpolar gases. AIChE J. **1979,** *25* (6), 931–938.
15. Soave, G. Equilibrium constants from a modified Redlich- Kwong equation of state. Chem. Eng. Sci. **1972,** *27,* 1197–1203.

16. Peng, D.-Y.; Robinson, B.B. A new two-constant equation of state. Ind. Eng. Chem. Fund. **1976**, *15*, 59–64.
17. Available at http://www.mrc-eng.com/Downloads/Moist% 20Air%20Props%20English.pdf (accessed July 2005).
18. Available at http://webbook.nist.gov/chemistry/form-ser. html (accessed July 2005).

III

CSS: Case Studies of Environmental Management

17

Energy Conversion: Coal, Animal Waste, and Biomass Fuel

Kalyan Annamalai, Soyuz Priyadarsan, Senthil Arumugam, and John M. Sweeten

Introduction and Objectives

The overall objective of this entry is to provide the basics of energy conversion processes and to present thermochemical data for coal and biomass fuels. Energy represents the capacity for doing work. It can be converted from one form to another as long as the total energy remains the same. Common fuels like natural gas, gasoline, and coal possess energy as chemical energy (or bond energy) between atoms in molecules. In a reaction of the carbon and hydrogen in the fuel with oxygen, called an oxidation reaction (or more commonly called combustion), carbon dioxide (CO_2) and water (H_2O) are produced, releasing energy as heat measured in units of kJ or Btu (see Table 1 for energy units). Combustion processes are used to deliver (i) work, using external combustion (EC) systems by generating hot gases and producing steam to drive electric generators as in coal fired power plans, or internal combustion (IC) engines by using the hot gases directly as in automobiles or gas turbines; and (ii) thermal energy, for applications to manufacturing processes in metallurgical and chemical industries or agricultural product processing.

TABLE 1 Energy Units and Terminology

The section on energy Units and Conversion factors in Energy is condensed from Chapter 01 of Combustion Engineering by Annamalai and Puri [2005] and Tables.

Energy Units

1 Btu (British thermal unit) = 778.14 ft lbf = 1.0551 kJ, 1 kJ = 0.94782 Btu = 25,037 lbmft/s^2

1 mBtu = 1 k Btu = 1000 Btu, 1 mmBtu = 1000 k Btu = 106 Btu, 1 trillion Btu = 109 Btu or 1 giga Btu

1 quad = 1015 Btu or 1.05 × 1015 kJ or 2.93 × 10n kW h

1 Peta J = 1015 J = 1012 kJ»0.00095 Quads

1 kilowatt-hour of electricity = 3,412 BTU = 3.6 Mj

1 cal: 4.1868 J, One (food) calorie = 1000 cal or 1 Cal

1 kJ/kg = 0.43 Btu/lb, 1 Btu/lb = 2.326 kJ/kg

1 kg/GJ = 1 g/MJ = 2.326 lb/mmBtu; 1 lb/mmBtu = 0.430 kg/GJ = 0.430 g/MJ

1 Btu/SCF = 37 kJ/m^3

1 Therm = 105 Btu = 1.055 × 105 kJ

1 m^3/GJ = 37.2596 ft^3/mmBTU

1 hp = 0.7064 Btu/s = 0.7457 kW = 745.7 W = 550 lbf ft/s = 42.41 Btu/min

1 boiler HP = 33475 Btu/h, 1 Btu/h = 1.0551 kJ/h

1 barrel (42 gal) of crude oil = 5,800,000 Btu = 6120 MJ

1 gal of gasoline = 124,000 Btu = 131 MJ

1 gal of heating oil = 139,000 Btu = 146.7 MJ, 1 gal of diesel fuel = 139,000 Btu = 146.7 MJ

1 barrel of residual fuel oil = 6,287,000 Btu = 6633 MJ

1 cubic foot of natural gas = 1,026 Btu = 1.082 MJ, 1 Ton of Trash = 150 kWh

1 gal of propane = 91,000 Btu = 96 MJ, 1 short ton of coal = 20,681,000 Btu = 21821 MJ

Emission reporting for pollutants: (i) parts per million (ppm), (ii) normalized ppm, (iii) emission Index (EI) in g/kg fuel, (iv) g/GJ, v) mg/ m^3 of flue gas:

Conversions in emissions reporting: (ii) normalized ppm = ppm × (21-O_2% std)/(21-O_2% measured); (iii) EI of species k: C % by mass in fuel × mol Wt of k × ppm of species k × 10–3/{12.01(CO_2% + CO%)}, (iv) g/GJ = EI/ {HHV in GJ/kg}; (v) mg/m^3 = ppm of species k × Mol Wt of k/24.5

Volume of 1 kmol (SI) and 1 lb mole (English) of an ideal gas at STP conditions defined below: Pressure at 101.3 kPa (1 atm, 14.7 psia, 29.92 in.Hg, 760 Torr) fixed; T changes depending upon type of standard adopted

Scientific (or SATP, standard ambient T and P)	US standard (1976) or ISA (International standard atmosphere)	NTP (gas industry reference	Chemists-standard-atmosphere (CSA)
25°C (77°F)	15°C (60°F)	20°C(68°F), 101.3 kPa	0°C (32°F),
24.5 m^3/kmol (392 ft^3/lb mole); $\rho_{air,SATP}$ = 1.188 kg/m^3 = 0.0698 lbm/ft^3	23.7 m^3/kmol (375.6 ft^3/lb mole); $\rho_{air,ISA}$ = 1.229 kg/m^3 = 0.0767 lbm/ft^3	24.06 m^3/kmole or 385 ft^3/lb mole; $\rho_{air,NTP}$ = 1.208 kg/m^3 = 0.0754 lbm/ft^3	22.4 m^3/kmol (359.2 ft^3/lb mole), $\rho_{air,CSA}$ = 1.297 kg/m^3 = 0.0810 lbm/ft^3

Fuels can be naturally occurring (e.g., fossil fuels such as coal, oil, and gas, which are residues of ancient plant or animal deposits) or synthesized (e.g., synthetic fuels). Fuels are classified according to the phase or state in which they exist: as gaseous (e.g., natural gas), liquid (e.g., gasoline or ethanol), or solid (e.g., coal, wood, or plant residues). Gaseous fuels are used mainly in residential applications (such as water heaters, home heating, or kitchen ranges), in industrial furnaces, and in boilers. Liquid fuels are used in gas turbines, automotive engines, and oil burners. Solid fuels are used mainly in boilers and steelmaking furnaces.

During combustion of fossil fuels, nitrogen or sulfur in the fuel is released as NO, NO_2 (termed generally as NO_x) and SO_2 or SO_3 (termed as SO_x). They lead to acid rain (when SO_x or NO_x combine with H_2O and fall as rain) and ozone depletion. In addition, greenhouse gas emissions (CO_2, CH_4, N_2O,

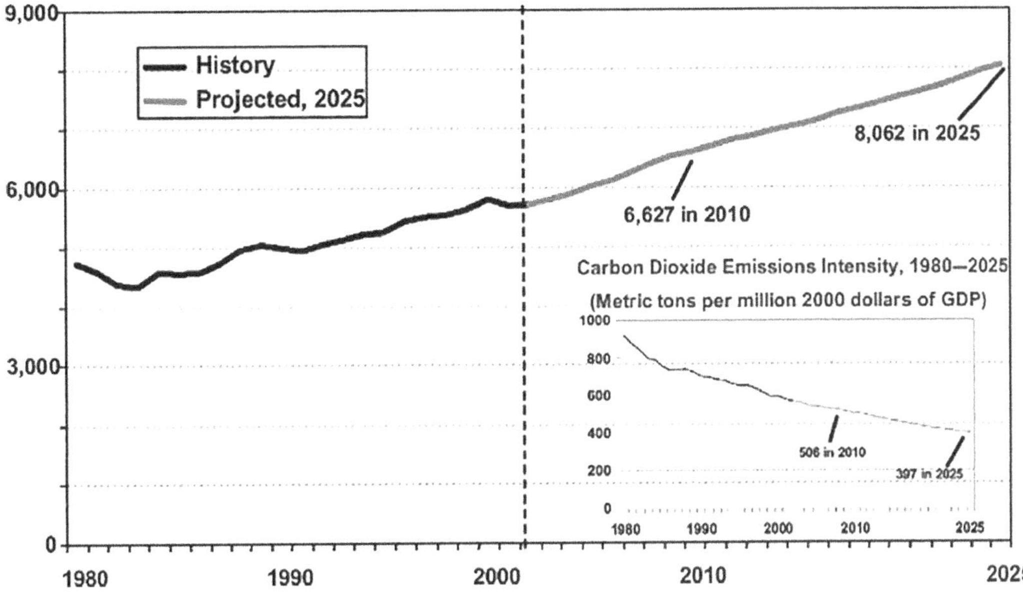

FIGURE 1 Total CO_2 emission in million metric tons per year: History and Projected 1980–2025.
Source: DOE-EIA.[1]

CFCs, SF_6, etc.,) are becoming a global concern due to warming of the atmosphere, as shown in Figure 1 for CO_2 emissions. Global surface temperature has increased by 0.6°C over the past 100 years. About 30%–40% of the world's CO_2 is from fossil fuels. The Kyoto protocol, signed by countries that account for 54% of the world's fossil based CO_2 emissions, calls for reduction of greenhouse gases by 5% from 1990 levels over the period from 2008 to 2012.

The total worldwide energy consumption is 421.5 quads of energy in 2003 and is projected to be 600 quads in 2020, while U.S. consumption in 2004 is about 100 quads and is projected to be 126 quads in 2020. The split is as follows: 40 quads for petroleum, 23 for natural gas, 23 for coal, 8 for nuclear power, and 6 for renewables (where energy is renewed or replaced using natural processes) and others sources. Currently, the United States relies on fossil fuels for 85% of its energy needs. Soon, the U.S. energy consumption rate which distributed as electrical power (40%), transportation (30%), and heat (30%), will outpace the growth in the energy production rate, increasing reliance on imported oil. The Hubbert peak theory (named after Marion King Hubbert, a geophysicist with Shell Research Lab in Houston, Texas) is based upon the rate of extraction and depletion of conventional fossil fuels, and predicts that fossil-based oil would peak at about 12.5 billion barrels per year worldwide some time around 2000. The power cost and percentage use of coal in various U.S. states varies from 10 cents (price per kWh) at 1% coal use for power generation in California to 48 cents at 94% use of coal in Utah.

Biomass is defined as "any organic material from living organisms that contains stored sunlight (solar energy) in the form of chemical energy."[1] These include agro-based materials (vegetation, trees, and plants); industrial wastes (sawdust, wood chips, and crop residues); municipal solid wastes (MSWs), which contain more than 70% biomass (including landfill gases, containing almost 50% CH_4); and animal waste. Biomass is a solid fuel in which hydrogen is locked with carbon atoms. Biomass production worldwide is 145 billion metric tons. Biomass now supplies 3% of U.S. energy, and it could be increased to as high as 20%. Renewable energy sources (RES) include biomass, wind, hydro, solar, flowing water or hydropower, anaerobic digestion, ocean thermal (20°C temperature difference), tidal energies, and geothermal (a nonsolar source of energy), and these supply 14% of the world demand. The RES constitute only 6%, while coal, petroleum, and natural gas account for 23%, 40%, and 24%, respectively. About 9%

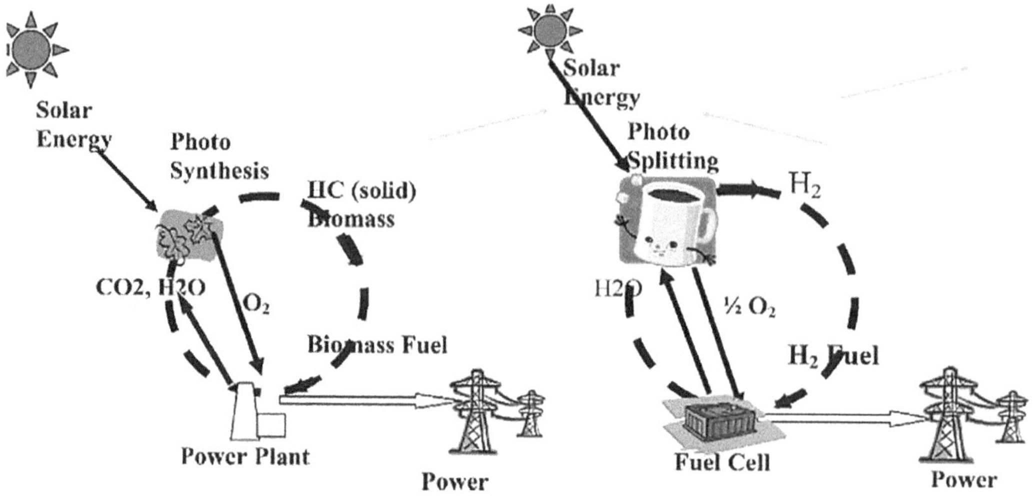

FIGURE 2 Comparison of biomass energy and H_2 energy cycles.

of the world's electricity is from RES, and 65% of the electricity contributed by biomass. About 97% of energy conversion from biomass is by combustion. Many U.S. states have encouraged the use of renewables by offering REC (Renewable Energy Credits). One REC = 1 MW/h = 3.412 mmBtu; hence the use of 1 REC is equivalent to replacing approximately 1500 lb of coal, reducing emission of NO_x and SO_x by 1.5 lb for every 1 REC, assuming that emissions of NO_x and SO_x are 0.45 lb per mmBtu generated by coal. Several emission-reporting methods and conversions are summarized in Table 1. Recently, H_2 is being promoted as a clean-burning, non-global-warming, and pollution-free fuel for both power generation and transportation.

Figure 2 shows a comparison between biomass and hydrogen energy cycles. In the biomass cycle, photosynthesis is used to split CO_2 into C and O_2, and H_2O into H_2 and O_2, producing Hydrocarbons (HC) fuel (e.g., leaves) and releasing O_2. The O_2 released is used to combust the HC and produce CO_2 and H_2O, which are returned to produce plant biomass (e.g., leaves) and O_2. On the other hand, in the hydrogen cycle, H_2O is disassociated using the photosplitting process to produce H_2 and O_2, which are then used for the combustion process. The hydrogen fuel can be used in fuel cells to obtain an efficient conversion. Photosynthesis is water intensive; most of the water supplied to plants evaporates through leaves into the atmosphere, where it re-enters the hydrology cycle.

This entry is organized in the following format: (i) coal and biosolid properties; (ii) coal and biosolid pyrolysis (a process of thermal decomposition in the absence of oxygen), combustion, and gasification; (iii) combustion by cofiring coal with biosolids; (iv) gasification of coal and biosolids (a process that includes pyrolysis, partial oxidation due to the presence of oxygen, and hydrogenation); and (v) reburn for NO_x reduction.

Fuel Properties

Fuel properties play a major role in the selection, design, and operation of energy conversion systems.

Solid Fuels

The primary solid fuel widely used in power plants is coal containing combustibles, moisture, and intrinsic mineral matter originating from dissolved salts in water. During the "coalification" process, lignite, the lowest rank of coal (low C/O ratio), is produced first from peat, followed by subbituminous (black lignite, typically low sulfur, noncaking), bituminous (soft coal that tends to stick when heated and is typically high in S), and finally anthracite (dense coal; has the highest carbon content, >90%, low volatile <15%) with a gradual increase in the coal C/O ratio. The older the coal, the higher its rank. Anthracite (almost carbon) is the highest-ranked coal, with a high heating value. To classify coals and ascertain the quality of coal, it is essential to perform proximate and ultimate analyses according to American Society of Testing Materials (ASTM) standards.

Proximate Analysis (ASTM D3172)

A solid fuel consists of combustibles, ash, and moisture. Combustibles together with ash are called the solid content of fuel. A proximate analysis provides the following information: surface moisture (SM) or dry loss (DL), i.e., moisture in air-dried coal; the inherent moisture in the coal (M); volatile matter (VM; produced by pyrolysis, a thermal decomposition process resulting in release of water, gases, oil and tar); fixed carbon (FC; skeletal matter left after release of volatiles); mineral matter (MM; inert collected with solid fuel); and heating value (HV). On combustion, the MM may be partially oxidized or reduced, and the material left after combustion of C and H in the fuel is called ash (CaO, $CaCO_3$, Fe_2O_3, FeO, etc.).

Table 2 shows comparative proximate analyses of coal, advanced feedlot biomass (FB, low-ash cattle manure; see "Coal and Bio-Solids Cofiring"), and litter biomass (LB, chicken manure).[2] Feedlot manure has higher moisture, nitrogen, chlorine, and ash content than coal. With aging or composting, the VM in manure decreases as a result of the gradual release of hydrocarbon gases or dehydrogenation, but fuel becomes more homogeneous.

Ultimate/Elemental Analysis (ASTM D3176)

Ultimate analysis is used to determine the chemical composition of fuels in terms of either the mass percent of their various elements or the number of atoms of each element. The elements of interest are C, H, N, O, S, Cl, P, and others. It can be expressed on an "as received" basis, on a dry basis (with the moisture in the solid fuel removed), or on a dry ash free (DAF) basis (also known as the moisture ash free basis MAF). Tables 3 and 4 show the ultimate analyses of various types of coal and biomass fuels.[3] While nitrogen is not normally present in natural gas, coal has 1%–1.5%; cattle manure and chicken waste contain high amounts of N (Table 2).

Heating Value (ASTM D3286)

The gross or higher heating value (HHV) of a fuel is the amount of heat released when a unit (mass or volume) of the fuel is burned. The HHV of solid fuel is determined using ASTM D3286 with an isothermal jacket bomb calorimeter. For rations fed to animals and animal waste fuels, the HHV for DAF roughly remains constant at about 19,500 kJ/kg (8400 Btu/lb),[4] irrespective of stage of decomposition of animal waste. The HHV can also be estimated using the ultimate analysis of the fuel and the following empirical relation from Boie[5]:

$$\mathrm{HHV_{fuel}}\left(\mathrm{kJ/kg\ fuel}\right) = 35,160\ Y_c + 116,225\ Y_H - 11,090\ Y_O + 6280\ Y_N + 10465\ Y_S \tag{1}$$

$$\mathrm{HHV_{fuel}}\left(\mathrm{BTU/lb\ fuel}\right) = 15,199\ Y_C + 49,965\ Y_H - 4768\ Y_O + 2700\ Y_N + 4499\ Y_S, \tag{2}$$

where Y denotes the mass fraction of an element C, H, O, N, or S in the fuel. The higher the oxygen content, the lower the HV, as seen in biomass fuels.

Annamalai et al. used the Boie equation for 62 kinds of biosolids with good agreement.[6] For most biomass fuels and alcohols, the HHV in kilojoules per unit mass of stoichiometric oxygen is constant at 14,360–14,730 kJ/kg of O_2 (6165–6320 Btu/lb of O_2).[7]

TABLE 2 Coal, Advanced Feedlot Biomass (FB) and Litter Biomass (LB)

Parameter	Wyoming Coal	Cattle Manure (FB)	Chicken Manure (LB)a	Advanced Feedlot Biomass (AFB)b	High-Ash Feedlot Biomass (HFB)b
Dry loss (DL)	22.8	6.8	7.5	10.88	7.57
Ash	5.4	42.3	43.8	14.83	43.88
FC	37.25	40.4	8.4	17.33	10.28
VM	34.5	10.5	40.3	56.97	38.2
C	54.1	23.9	39.1	50.08	49.27
H	3.4	3.6	6.7	5.98	6.13
N	0.81	2.3	4.7	38.49	38.7
O	13.1	20.3	48.3	4.58	4.76
S	0.39	0.9	1.2	0.87	0.99
Cl	<0.01%	1.2			
HHV-as received (kJ/kg)	21385	9560	9250	14983	9353
T adiab, Equilc	2200 K (3500°F)	2012 K (3161°F)			
DAF formula	$CH_{0.76}O_{0.18}N_{0.013}S_{0.0027}$	$CH_{1.78}O_{.64}N_{.083}S_{.014}$	$CH_{2.04}O_{0.93}N_{0.10}S_{0.012}$	$CH_{1.4184}O_{0.5764}N_{0.078}S_{0.0056}$	$CH_{1.4775}O_{0.5892}N_{0.083}S_{0.0076}$
HHV-DAF (kJ/kg)	29785	18785	18995	20168	19265
CO_2, g/GJ					
N, g/GJ					
S, g/GJ					

a Priyadarsan et al.

b Priyadarsan et al.[37]

c Equilibrium temperature for stoichiometric mixture from THERMOLAB Spreadsheet software for any given fuel of known composition (Annamalai and Puri.[36] website http://www.crcpress.com/e_products/downloads/download.asp?cat_no=2553)

TABLE 3 Coal Composition (DAF basis)

ASTM Rank	State (U.S.A.)	Ash, % (dry)	C	H	N	S*	O**	HHV Est kj/kg	CO_2 kg/GJ	N kg/GJ	S kg/GJ
Lignite	ND	11.6	63.3	4.7	0.48	0.98	30.5	24,469	94.8	0.196	0.401
Lignite	MT	7.7	70.7	4.9	0.8	4.9	22.3	28,643	90.4	0.279	1.711
Lignite	ND	8.2	71.2	5.3	0.56	0.46	22.5	28,782	90.7	0.195	0.160
Lignite	TX	9.4	71.7	5.2	1.3	0.72	21.1	29,070	90.4	0.447	0.248
Lignite	TX	10.3	74.3	5	0.37	0.51	19.8	29,816	91.3	0.124	0.171
Sbb. A	WY	8.4	74.3	5.8	1.2	1.1	17.7	31,092	87.6	0.386	0.354
Sbb. C	WY	6.1	74.8	5.1	0.89	0.3	18.9	30,218	90.7	0.295	0.099
HVB	IL	10.8	77.3	5.6	1.1	2.3	13.6	32,489	87.2	0.339	0.708
HVC	IL	10.1	78.8	5.8	1.6	1.8	12.1	33,394	86.5	0.479	0.539
HVB	IL	11.8	80.1	5.5	1.1	2.3	11.1	33,634	87.3	0.327	0.684
HVB	UT	4.8	80.4	6.1	1.3	0.38	11.9	34,160	86.2	0.381	0.111
HVA	wv	7.6	82.3	5.7	1.4	1.8	8.9	34,851	86.5	0.402	0.516
HVA	KY	2.1	83.8	5.8	1.6	0.66	8.2	35,465	86.6	0.451	0.186
MV	AL	7.1	87	4.8	1.5	0.81	5.9	35,693	89.3	0.420	0.227
LV	PA	9.8	88.2	4.8	1.2	0.62	5.2	36,153	89.4	0.332	0.171
Anthracite	PA	7.8	91.9	2.6	0.78	0.54	4.2	34,974	96.3	0.223	0.154
Anthracite	PA	4.3	93.5	2.7	0.24	0.64	2.9	35,773	95.8	0.067	0.179

Notes: HHV_{est}: Boie Equation. CO_2 in g/MJ or kg/GJ = C content in % ×36645/{HHV in kJ/kg}. CO_2 in lb per mmBtu = Multiply CO_2 in (g /MJ) or kg/GJ by 2.32. N in g/MJ or kg/GJ = N% × 10000/{HHV in kJ/kg}. For NO_x estimation, multiply N content in g/MJ by 1.15 to get NO_x in g/MJ which assumes 35% conversion of fuel N; For SO_2 estimation, multiply S content in g/MJ by 2 to get SO_2 in g/MJ assuming 100% conversion of fuel S (Multiply HHV in kJ/kg by 0.430 to get Btu/lb); *Organic sulfur; **by difference.

TABLE 4 Ultimate Analyses and Heating Values of Biomass Fuels

Biomass	C	H	O	N	S	Residue	Measured HHVm	[a]Estimated HHV	CO_2 g/MJ	N, g/MJ	S, g/MJ
Field crops											
Alfalfa seed straw	46.76	5.40	40.72	1.00	0.02	6.07	18.45	18.27	92.9	0.542	0.011
Bean straw	42.97	5.59	44.93	0.83	0.01	5.54	17.46	16.68	90.2	0.475	0.006
Com cobs	46.58	5.87	45.46	0.47	0.01	1.40	18.77	18.19	90.9	0.250	0.005
Com stover	43.65	5.56	43.31	0.61	0.01	6.26	17.65	17.05	90.6	0.346	0.006
Cotton stalks	39.47	5.07	39.14	1.20	0.02	15.10	15.83	15.51	91.4	0.758	0.013
Rice straw (fall)	41.78	4.63	36.57	0.70	0.08	15.90	16.28	16.07	94.0	0.430	0.049
Rice straw (weathered)	34.60	3.93	35.38	0.93	0.16	25.00	14.56	12.89	87.1	0.639	0.110
Wheat straw	43.20	5.00	39.40	0.61	0.11	11.40	17.51	16.68	90.4	0.348	0.063
Switchgrass[b]	42.02	6.30	46.10	0.77	0.18	4.61	15.99	15.97	96.3	0.482	0.113
Orchard prunings											
Almond prunings	51.30	5.29	40.90	0.66	0.01	1.80	20.01	19.69	93.9	0.330	0.005
Black Walnut	49.80	5.82	43.25	0.22	0.01	0.85	19.83	19.50	92.0	0.111	0.005
English Walnut	49.72	5.63	43.14	0.37	0.01	1.07	19.63	19.27	92.8	0.188	0.005
Vineyard prunings											
Cabernet Sauvignon	46.59	5.85	43.90	0.83	0.04	2.71	19.03	18.37	89.7	0.436	0.021

(*Continued*)

TABLE 4 (*Continued*) Ultimate Analyses and Heating Values of Biomass Fuels

Biomass	C	H	O	N	S	Residue	Measured HHVm	[a]Estimated HHV	CO_2 g/MJ	N, g/MJ	S, g/MJ
Chenin Blanc	48.02	5.89	41.93	0.86	0.07	3.13	19.13	19.14	92.0	0.450	0.037
Pinot Noir	47.14	5.82	43.03	0.86	0.01	3.01	19.05	18.62	90.7	0.451	0.005
Thompson seedless	47.35	5.77	43.32	0.77	0.01	2.71	19.35	18.60	89.7	0.398	0.005
Tokay	47.77	5.82	42.63	0.75	0.03	2.93	19.31	18.88	90.7	0.388	0.016
Energy Crops											
Eucalyptus											
Camaldulensis	49.00	5.87	43.97	0.30	0.01	0.72	19.42	19.19	92.5	0.154	0.005
Globulus	48.18	5.92	44.18	0.39	0.01	1.12	19.23	18.95	91.8	0.203	0.005
Grandis	48.33	5.89	45.13	0.15	0.01	0.41	19.35	18.84	91.5	0.078	0.005
Casuarina	48.61	5.83	43.36	0.59	0.02	1.43	19.44	19.10	91.6	0.303	0.010
Cattails	42.99	5.25	42.47	0.74	0.04	8.13	17.81	16.56	88.5	0.415	0.022
Popular	48.45	5.85	43.69	0.47	0.01	1.43	19.38	19.02	91.6	0.243	0.005
Sudan grass	44.58	5.35	39.18	1.21	0.08	9.47	17.39	17.63	93.9	0.696	0.046
Forest residue											
Black Locust	50.73	5.71	41.93	0.57	0.01	0.97	19.71	19.86	94.3	0.289	0.005
Chaparral	46.9	5.08	40.17	0.54	0.03	7.26	18.61	17.98	92.3	0.290	0.016
Madrone	48	5.96	44.95	0.06	0.02	1	19.41	18.82	90.6	0.031	0.010
Manzanita	48.18	5.94	44.68	0.17	0.02	1	19.3	18.9	91.5	0.088	0.010
Ponderosa Pine	49.25	5.99	44.36	0.06	0.03	0.3	20.02	19.37	90.1	0.030	0.015
Ten Oak	47.81	5.93	44.12	0.12	0.01	2	18.93	18.82	92.6	0.063	0.005
Redwood	50.64	5.98	42.88	0.05	0.03	0.4	20.72	20.01	89.6	0.024	0.014
White Fur	49	5.98	44.75	0.05	0.01	0.2	19.95	19.22	90.0	0.025	0.005
Food and fiber processing wastes											
Almond hulls	45.79	5.36	40.6	0.96	0.01	7.2	18.22	17.89	92.1	0.527	0.005
Almond shells	44.98	5.97	42.27	1.16	0.02	5.6	19.38	18.14	85.0	0.599	0.010
Babassu husks	50.31	5.37	42.29	0.26	0.04	1.73	19.92	19.26	92.5	0.131	0.020
Sugarcane bagasse	44.8	5.35	39.55	0.38	0.01	9.79	17.33	17.61	94.7	0.219	0.006
Coconut fiber dust	50.29	5.05	39.63	0.45	0.16	4.14	20.05	19.2	91.9	0.224	0.080
Cocoa hulls	48.23	5.23	33.09	2.98	0.12	10.25	19.04	19.56	92.8	1.565	0.063
Cotton gin trash	39.59	5.26	36.33	2.09		16.68	16.42	16.13	88.4	1.273	0.000
Macadamia shells	54.41	4.99	39.69	0.36	0.01	0.56	21.01	20.55	94.9	0.171	0.005
Olive pits	48.81	6.23	43.48	0.36	0.02	1.1	21.39	19.61	83.6	0.168	0.009
Peach pits	53	5.9	39.14	0.32	0.05	1.59	20.82	21.18	93.3	0.154	0.024
Peanut hulls	45.77	5.46	39.56	1.63	0.12	7.46	18.64	18.82	90.0	0.874	0.064
Pistachio shells	48.79	5.91	43.41	0.56	0.01	1.28	19.26	19.25	92.8	0.291	0.005
Rice hulls	40.96	4.3	35.86	0.4	0.02	18.34	16.14	15.45	93.0	0.248	0.012
Walnut shells	49.98	5.71	43.35	0.21	0.01	0.71	20.18	19.45	90.8	0.104	0.005
Wheat dust	41.38	5.1	35.19	3.04	0.19	15.1	16.2	16.78	93.6	1.877	0.117

[a] HHV based on Boie equation.

[b] Aerts et al[20]; [Adapted from Ebeling and Jenkins[3] and AnnamalaiJ[17] See foot note of Table 3.2 for conversions to English units and estimation of NO_x and SO_2 emissions.

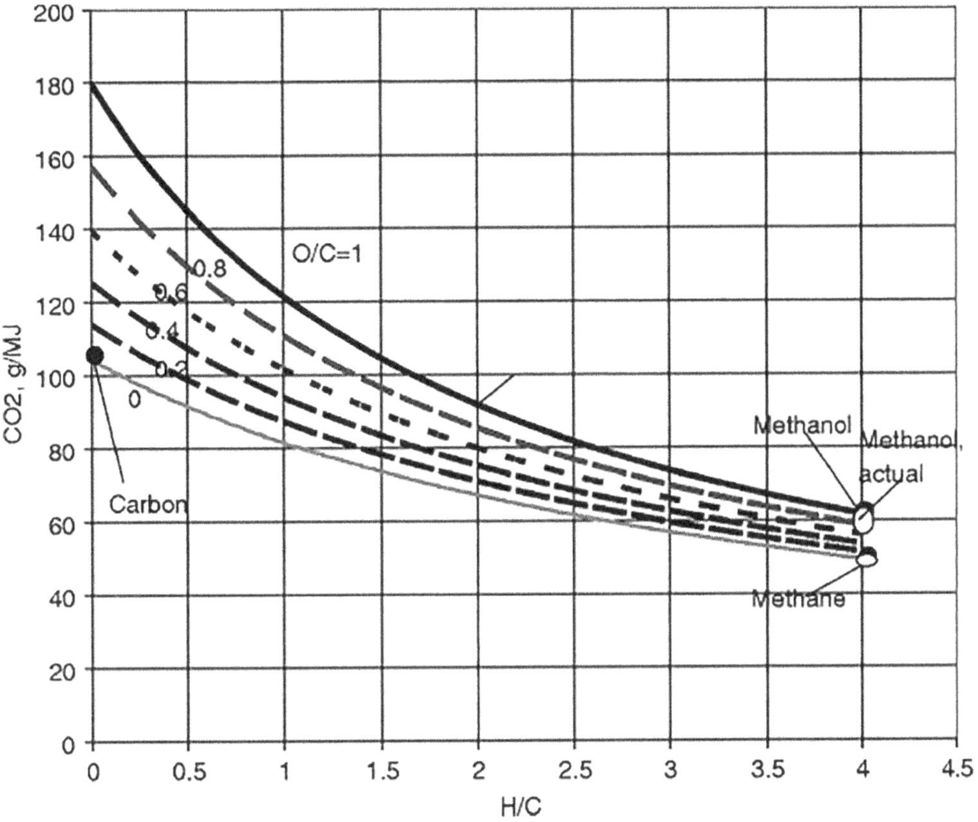

FIGURE 3 Emission of CO_2 as a function of H/C and O/C atom ratios in hydrocarbon fuels.
Source: Adapted from Annamalai and Puri.[8]

Estimate of CO_2 Emission

Using the Boie-based HVs for any fuel of known elemental composition, one can plot the CO_2 emission in g/MJ (Figure 3) as a function of H/C and O/C ratios.[8] Comparisons for selected fuels with known experimental HVs are also shown in the same figure. Coal, with H/C ratio $\approx$ 0.5, releases the highest CO_2, while natural gas (mainly CH_4) emits the lowest CO_2. Because the United States uses fossil fuels for 86% of its energy needs (100 quads), the estimated CO_2 emission is 6350 million ton/year, assuming that the average CO_2 emission from fossil fuels is 70 kg/GJ (methane: 50 kg/GJ vs coal: 90 kg/GJ). Figure 1 seems to confirm such estimation within a 10% error.

Flame Temperature

Figure 4 shows a plot of maximum possible flame temperature vs moisture percentage with combustion for biomass fuels. The result can be correlated as follows[4]:

$$T(\mathrm{K}) = 2290 - 1.89\mathrm{H_2O} + 5.06\ \mathrm{Ash} - 0.309\ \mathrm{H_2O}\ Ash - 0.180\ \mathrm{H_2O_2} - 0.108\ \mathrm{Ash}^2 \quad (3)$$

$$T(°\mathrm{F}) = 3650 - 3.40\ \mathrm{H_2O} + 9.10\ \mathrm{Ash} - 0.556\ \mathrm{H_2O\ Ash} - 0.324\ \mathrm{H_2O_2} - 0.194\ \mathrm{Ash}^2 \quad (4)$$

The adiabatic flame temperature decreases if the ash and moisture contents increase.

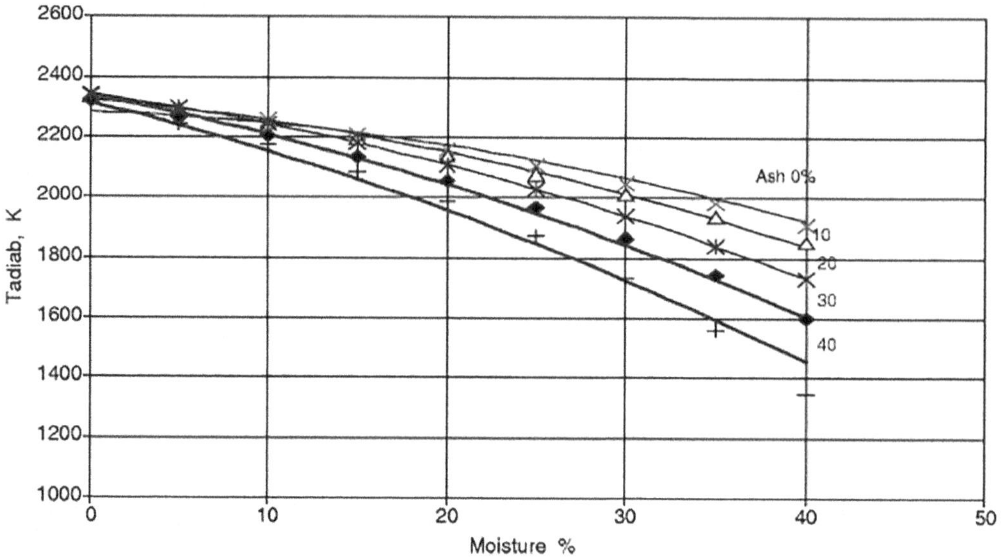

FIGURE 4 Correlation of adiabatic flame temperature with moisture and ash contents.

Flue Gas Volume

The flue gas volume for C–H–O is almost independent of O/C ratios. The fit at 6% O_2 in products gives the following empirical equation for flue gas volume (m^3/GJ) at SATP[8]:

$$\text{Flue gas}_{\text{vol}}\left(m^3/GJ\right) = 4.96\left(\frac{H}{C}\right)^2 - 38.628\left(\frac{H}{C}\right) + 389.72 \quad (5)$$

$$\text{Flue gas}_{\text{vol}}\left(ft^3/\text{mmBtu}\right) = 184.68\left(\frac{H}{C}\right)^2 - 1439.28\left(\frac{H}{C}\right) + 14520.95 \quad (6)$$

Liquid Fuels

Liquid fuels, used mainly in the transportation sector, are derived from crude oil, which occurs naturally as a free-flowing liquid with a density of $\rho \approx 780\,kg/m^3$–$1000\,kg/m^3$, containing 0.1% ash and 0.15%–0.5% nitrogen. Crude oil normally contains a mixture of hydrocarbons, and as such, the "boiling" temperature keeps increasing as the oil is distilled. Most fuel oils contain 83%–88% carbon and 6%–12% (by mass) hydrogen.

Gaseous Fuels

The gaseous fuels are cleaner-burning fuels than liquid and solid fuels. They are a mixture of HC but dominated by highly volatile CH_4 with very little S and N. Natural gas is transported as liquefied natural gas (LNG) and compressed natural gas (CNG), typically at 150–250 bars. Liquefied petroleum gas (LPG) is a byproduct of petroleum refining, and it consists mainly of 90% propane. A low-Btu gas contains 0–7400 kJ/SCM (Standard Cubic Meter, 0–200 Btu/SCF, standards defined in Table 1); a medium-Btu gas. 7400–14,800 kJ/SCM (200–400 Btu/SCF); and a high-Btu gas, above 14,800 kJ/SCM (more than 400 Btu/ SCF). Hydrogen is another gaseous fuel, with a heat value of 11,525 kJ/SCM (310 Btu/SCF). Because the fuel quality (heat value) may change when fuel is switched, the thermal output rate at a fixed gas-line pressure changes when fuels are changed.

Coal and Biomass Pyrolysis, Gasification, and Combustion

Typically, coal densities range from 1100 kg/m^3 for low-rank coals to 2330 kg/m^3 for high-density pyrolytic graphite, while for biomass, density ranges from 100 kg/ m^3 for straw to 500 kg/m^3 for forest wood.[9] The bulk density of cattle FB as harvested is 737 kg/m^3 (CF) for high ash (HA-FB) and 32 lbs/CF for low ash (LA-FB).[10] The processes during heating and combustion of coal are illustrated in Figure 5, and they are similar for biomass except for high VM. The process of release of gases from solid fuels in the absence of oxygen is called pyrolysis, while the combined process of pyrolysis and partial oxidation of fuel in the presence of oxygen is known as gasification. If all combustible gases and solid carbon are oxidized to CO_2 and H_2O, the process is known as combustion.

Pyrolysis

Solid fuels, like coal and biomass, can be pyrolyzed (thermally decomposed) in inert atmospheres to yield combustible gases or VM. While biomass typically releases about 70%–80% of its mass as VM (mainly from cellulose and hemicellulose) with the remainder being char, mainly from lignin content of biomass, coal releases 10%–50% of its mass as VM, depending upon its age or rank. Typically, a medium-rank coal consists of 40% VM and 60% FC, while a high-rank coal has about 10% VM. Bituminous coal pyrolyzes at about 700 K (with 1% mass loss for heating rates <100°C/s), as in the case of most plastics. Pyrolytic products range from lighter volatiles like CH_4, C_2H_4, C_2H_6, CO, CO_2, H_2, and H_2O to heavier molecular mass tars. Apart from volatiles, nitrogen is also evolved from the fuel during pyrolysis in the form of HCN, NH_3, and other compounds or, more generally, XN.

Sweeten et al. performed the thermogravimetric analysis (TGA) of feedlot manure.[4] The results are shown in Figure 6. In the case of manure, drying occurred between 50 and 100°C, pyrolysis was initiated around 185°C–200°C for a heating rate of 80°C/min, and the minimum ignition temperature was approximately 528°C. The gases produced during biomass pyrolysis can also be converted into transportation fuels like biodiesel, methanol, and ethanol, which may be used either alone or blended with gasoline.

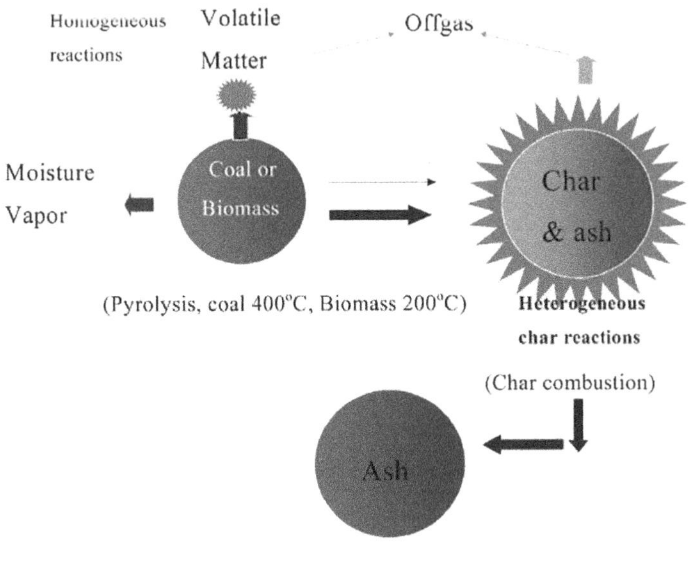

FIGURE 5 Processes during coal pyrolysis, gasification, and combustion.

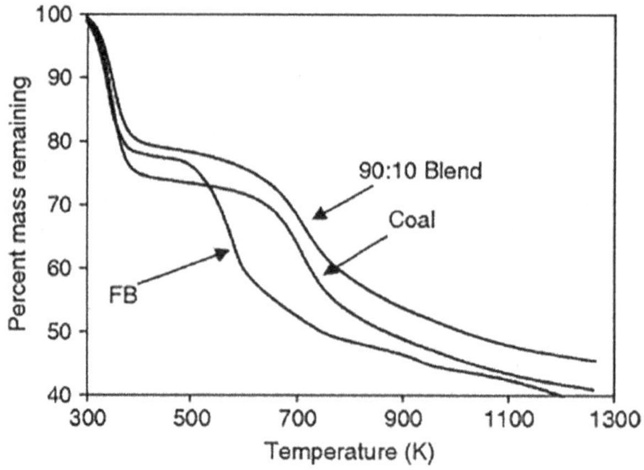

FIGURE 6 Thermo-gravimetric analyses of Feedlot Biomass (FB or cattle manure), coal, and 90:10 coal: FB blends.
Source: Sweeten et al.[4]

Volatile Oxidation

Once released, volatiles (HC, CO, H_2, etc.,) undergo oxidation within a thin gas film surrounding the solid fuel particle. The oxidation for each HC involves several steps. The enveloping flame, due to volatile combustion, acts like a shroud by preventing oxygen from reaching the particle surface for heterogeneous oxidation of char. Following Dryer,[11] the one-step global oxidation of a given species can be written as

$$\text{Fuel} + \upsilon_{CO_2} O_2 \xrightarrow{\text{oxidation}} \upsilon_{CO_2} CO_2 + \upsilon_{H_2O} H_2O \tag{7}$$

$$-\frac{d[\text{Fuel}]}{dt}, \frac{\text{kg}}{\text{m}^3\ \text{sec}} = \text{A}\exp\left(\frac{-E}{RT}\right)[Y_{\text{fuel}}]^a [Y_{O_2}]^b \tag{8}$$

where [ ] represents the concentration of species in kg/m^3, Y the mass fraction, A the pre-exponential factor, E the activation energy in kJ/kmole, and a and b the order of reaction; they are tabulated in Bartok and Sarofim for alkanes, ethanol, methanol, benzene and toluene.[11]

Char Reactions

The skeletal char, essentially FC, undergoes heterogeneous reactions with gaseous species. The heterogeneous combustion of carbon or char occurs primarily via one or more of the following reactions:

Reaction I. $C + \frac{1}{2} O_2 \rightarrow CO$

Reaction II. $C + O_2 \rightarrow CO_2$

Reaction III. $C + CO_2 \rightarrow 2\,CO$

Reaction IV. $C + H_2O \rightarrow CO + H_2$

Assuming a first-order reaction for scheme I, the oxygen consumption rate is given as

$$\dot{m}_{O2} \approx \pi d_P^2 B_I T^n \exp\left(-\frac{E}{R_V T_P}\right) \rho_\infty Y_{O2,w.} \tag{9}$$

The dominant oxygen transfer mechanism at high temperatures is via reaction I with an E/R (a ratio of activation energy to universal gas constant) of about 26,200 K, where $B_I = 2.3 \times 10^7$ m/s and $n = 0.5$ to 1. Reaction II has an E/R of 20,000 K, and $B_{II} = 1.6 \times 10^5$ m/s. Reaction III, the Boudouard reduction reaction, proceeds with an E/R of about 40,000 K. The reduction reactions, III and IV, may become significant, especially at high temperatures for combustion in boiler burners. Reaction with steam is found to be 50 times faster than CO_2 at temperatures up to 1800°C at 1 bar for 75–100 micron-sized Montana Rosebud char.[12] The combustible gases CO and H_2 undergo gas phase oxidation, producing CO_2 and H_2O.

Ignition and Combustion

Recently, Essenhigh et al. have reviewed the ignition of coal.[13] Volatiles from lignite are known to ignite at T > 950 K in fluidized beds. Coal may ignite homogeneously or heterogeneously depending upon size and volatile content.[14,15] A correlation for heterogeneous char ignition temperature is presented by Du and Annamalai, 1994.

Once ignited, the combustion of high volatile coal proceeds in two stages: combustion of VM and combustion of FC. Combustion of VM is similar to the combustion of vapors from an oil drop. The typical total combustion time of 100-micron solid coal particle is on the order of 1 s in boilers and is dominated by the time required for heterogeneous combustion of the residual char particle, while the pyrolysis time ($t_{pyr} = 10^6$ (s/m^2) d_p^2) is on the order of 1/10th–1/100th of the total burning time. Since bio-solid contains 70%–80% VM (coal contains 10% VM), most of the combustion of volatiles occurs within a short time (about 0.10 s).

For liquid drops and plastics of density ρ_c, simple relations exist for evaluating the combustion rates and times. If the transfer number B is defined as

$$B = \frac{cp\{T_\infty - T_w\}}{L} + \frac{Y_{O_2,\infty}}{\nu_{O_2}} \frac{h_c}{L}, \tag{10}$$

where $T_w \approx$ TBP for liquid fuels; $T_w = T_g$, the temperature of gasification for plastics; L is the latent heat for liquid fuel and $L = q_g$, heat of gasification for plastics; $Y_{O_2\infty}$ is the stoichiometric oxygen mass per unit mass of fuel (typically 3.5 for liquid fuels); and h_c is the lower heating value of fuel; then the burn rate ($\dot{m}$) and time (t_b) for spherical condensates (liquid drops and spherical particles of diameter d_p and density ρ_c) are given by the following expressions:

$$\dot{m} \approx 2\pi \frac{\lambda}{c_p} d_p \ln(1+B) \tag{11}$$

$$t_b = \frac{d_0^2}{\alpha_c}, \tag{12}$$

Where

$$\alpha_c = 8 \frac{\lambda}{c_p} \frac{\ln(1+B)}{\rho_c} \tag{13}$$

and c_p and λ are the specific heat and thermal conductivity of gas mixture evaluated at a mean temperature (approximately 50% of the adiabatic flame temperature).

The higher the B value, the higher the mass loss rate, and the burn time will be lower. The value of B is about 1–2 for plastics (polymers), 2–3 for alcohols, and 6–8 for pentane to octane. The burn time of plastic waste particles will be about 3–4 times longer than single liquid drops of pentane to octane (≈ gasoline) of similar diameter.

Combustion in Practical Systems

The time scales for combustion are on the order of 1000, 10, and 1 ms for coal burnt in boilers and liquid fuels burnt in gas turbines and diesel engines. Coal is burnt on grates in lumped form (larger-sized particles, 2.5 cm or greater with a volumetric intensity on the order of 500 kW/m^3), medium-sized particles in fluidized beds (1 cm or less, 500 kW/m^3), or as suspensions or pulverized fuel (pf; 75 micron or less, 200 kW/m^3) in boilers.

Apart from pyrolysis, gasification, and combustion, another option for energy conversion (particularly if solid fuel is in slurry firm, such as flushed dairy manure), is the anaerobic digestion (in absence of oxygen) to CH_4 (60%) and CO_2 (40%) using psychrophilic (ambient temperature), mesophilic (95°F) and thermophilic (135°F) bacteria in digesters.[10] Typical options of energy conversion, indicated in Figure 7, include anaerobic digestion (path 1, the biological gasification process), thermal gasification with air to produce CO, HC, CO_2 (path 2A) or with steam to produce CO_2 and H_2 (path 2B), cofiring (path 3), reburn (path 4; see "Reburn with Bio-Solids"), and direct combustion (path 5).

Suspension Firing

In suspension-fired boilers, solid fuel is pulverized into smaller particles (d_p = 75 µm or less) so that more surface area per unit mass is exposed to the oxidant, resulting in improved ignition and combustion characteristics. Typical boiler burners use swirl burners for atomized oil and pulverized coal firing, while a gas turbine uses a swirl atomizer in highly swirling turbulent flow fields. A swirl burner for pf firing is shown in Figure 8. The air is divided into a primary air stream which transports the coal (10%–20% of the total air, heated to 70°C–100°C to prevent condensation of vapors and injected at about 20 m/s to prevent settling of the dust, loading dust and gas at a ratio of 1:2) and a secondary air stream (250°C at 60–80 m/s) which is sent through swirl vanes, supplying the remaining oxygen for combustion and imparting a tangential momentum to the air. In wall-fired boilers, burners are stacked above each other on the wall; while in tangential-fired boilers, the burners are mounted at the corners of rectangular furnaces.

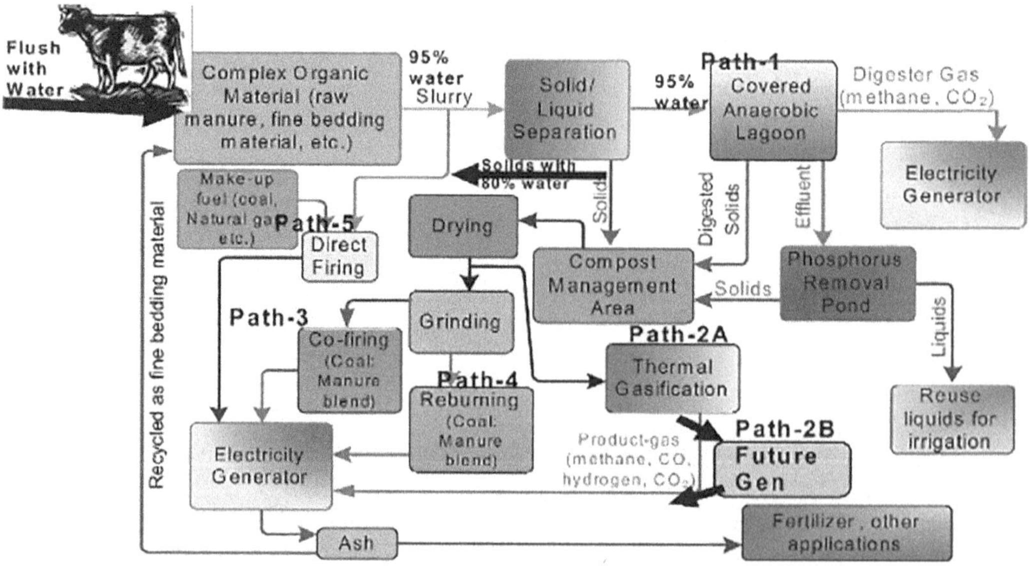

FIGURE 7 Flow chart showing several energy conversion options for a typical dairy or cattle operation.

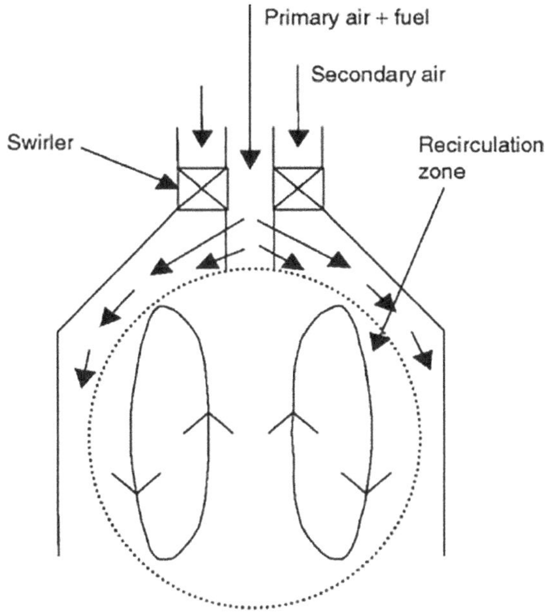

FIGURE 8 Pulverized Fuel (pf) fired swirl burner.

Stoker Firing

The uncrushed fuel [fusion temperature <1093°C (2000°F); volatile content >20%; sizes in equal proportions of 19 mm × 12.5 mm (3/4 in. × ½ in.), 6.3 mm × 3.2 mm (½ in. × ¼ in.), 3.2 mm × 3.2 mm (¼ in × ¼ in.)][16] is fed onto a traveling chain grate below to which primary air is supplied (Figure 9), which may be preheated to 177°C (350°F) if moisture exceeds 25%. The differential pressure is on the order of 5–8 mm (2–3 in.). The combustible gases are carried into an over-fire region into which secondary air (almost 35% of total air at three levels for low emissions) is fired to complete combustion. The over-fire region acts like a perfectly stirred reactor (PSR). It is apparent that solid fuels need not be ground to finer size.

Fixed Bed Combustor

The bed contains uncrushed solid fuels, inert materials (including ash), and processing materials (e.g., limestone to remove SO_2 from gases as sulfates). It is fed with air moving against gravity for complete combustion, but the velocity is low enough that materials are not entrained into the gas streams. Large solid particles can be used.

Fluidized Bed Combustor

When air velocity (V) in fixed bed combustor (FXBC) is increased gradually to a velocity called minimum fluidization velocity V_{mf}, the upward drag force is almost equal to the weight of the particle, so that solids float upward. The bed behaves like a fluid (like liquid water in a tank), i.e., it becomes fluidized. If $V > V_{mf}$, then air escapes as bubbles and is called a bubbling fluidized bed combustor (BFBC). The bed has two phases: the bubble phase, containing gases (mostly oxygen), and the emulsion phase (dense phase, oxygen deficient), containing particles and gas. Many times gas velocity is so high that gaseous combustibles produced within the bed burn above the bed (called free board region), while solids (e.g., char and carbon) burn within the bed. Fluidized Bed Combustor (FBC) is suitable for fuels which are difficult to combust in pf-fired boilers.

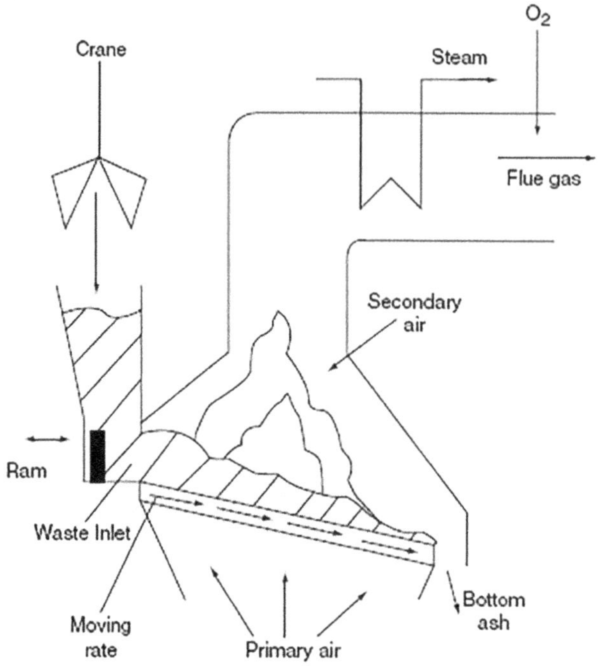

FIGURE 9 Schematic of stoker firing.
Source: Loo and Kessel.[39]

Circulating Fluidized Bed Combustor (CFBC)

When air velocity in FBC is increased at velocity $V >> V_{mf}$, particles are entrained into the gas stream. Since the residence time available to particles for combustion is shorter, unburned particles are captured using cyclones located downstream of the combustor and circulated back to the bed.

The residence time (t_{res}) varies from a low value for pf-fired burners to a long residence time for fixed-bed combustors. The reaction time (t_{reac}) should be shorter than t_{res} so that combustion is complete. The reaction time includes time to heat up to ignition temperature and combustion. The previous section on fuel properties and the homogenous (e.g., CH_4, CO oxidation) and heterogeneous (e.g., carbon oxidation) reaction kinetics can be used to predict t_{reac} or burn time t_b.

Coal and Bio-Solids Cofiring

General Schemes of Conversion

Most of the previously reviewed combustion systems typically use pure coal, oil, or gas. The same systems require redesign for use with pure biomass fuels. A few of the technologies, which utilize bio-solids as an energy source, are summarized in Annamalai et al.[17] These technologies include direct combustion (fluidized beds), circulating fluidized beds, liquefaction (mostly pyrolysis), onsite gasification for producing low to medium Btu gases, anaerobic digestion (bacterial conversion), and hydrolysis for fermentation to liquid fuels like ethanol.[18,19]

Cofiring

Although some bio-solids have been fired directly in industrial burners as sole-source fuels, limitations arose due to variable moisture and ash contents in bio-solid fuels, causing ignition and combustion problems for direct combustion technologies. To circumvent such problems, these fuels have been fired along with the primary fuels (cofiring) either by directly mixing with coal and firing (2%–15% of heat input basis) or by firing them in between coal-fired burners.[20–24]

Cofiring has the following advantages: improvement of flame stability problems, greater potential for commercialization, low capital costs, flexibility of adaptation of biomass fuels and cost effective power generation, mitigated NO_x emissions from coal-fired boilers, and reduced CO_2 emissions. However, a lower melting point of biomass ash could cause fouling and slagging problems.

Some of the bio-solid fuels used in cofiring with coal are cattle manure,[25,26] sawdust and sewage sludge,[21] switch grass,[20] wood chips,[24,27] straw,[22,28] and refuse-derived fuel (RDF).[21] See Sami et al. for a review of literature on cofiring.[7]

Coal and Agricultural Residues

Sampson et al. reported test burns of three different types of wood chips (20%, HHV from 8320 to 8420 Btu/lb) mixed with coal (10,600 Btu/lb) at a stoker (traveling grate) fired steam plant.[24] The particulate emission in grams per SCF ranged from 0.05 to 0.09. An economic study, conducted for the 125,000 lb/h steam power plant, concluded that energy derived from wood would be competitive with that from coal if more than 30,000 tons of wood chips were produced per year with hauling distances less than 60 mi. Aerts et al. carried out their experiments on cofiring switch grass with coal in a 50-MW, radiant, wall-fired, pulverized coal boiler with a capacity of 180 tons of steam at 85 bar and 510°C (Figure 10). The NO_x emissions decreased by 20%, since switchgrass contains lesser nitrogen (Table 4).[20] It is the author's hypothesis that a higher VM content of bio-solids results in a porous char, thus accelerating the char combustion process. This is validated by the data from Fahlstedt et al. on the cofiring of wood chips, olive pips and palm nut shells with coal at the ABB Carbon 1 MW Process Test Facility; they found that blend combustion has a slightly higher efficiency than coal-only combustion.[27]

Coal and RDF

Municipal solid waste includes residential, commercial, and industrial wastes which could be used as fuel for production of steam and electric power. MSW is inherently a blended fuel, and its major components are paper (43%); yard waste, including grass clippings (10%); food (10%); glass and ceramics (9%); ferrous materials (6%); and plastics and rubber (5%). Refer to Tables 5 and 6 for analyses. When

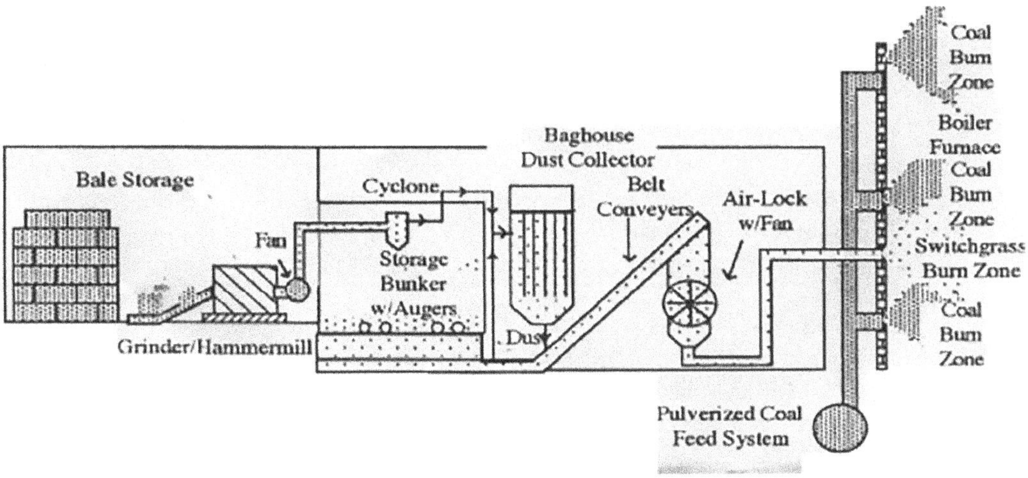

FIGURE 10 A cofiring scheme for coal and biomass (Alternate Fuel Handling Facility at Blount St. Generating Station).
Source: Aerts et al.[20]

TABLE 5 Chemical Composition of Solid Waste

Proximate Analysis	Percent Range					Percent Typical
Volatile matter (VM)	30–60					50
Fixed carbon (FC)	5–10					8
Moisture	10–45					25
Ash	10–30					25

	Percent by Mass (dry basis)					
Ultimate Analysis	C	H	O	N	S	Ash
Yard wastes	48	6	38	3	0.3	4.7
Wood	50	6	43	0.2	0.1	0.7
Food wastes	50	6	38	3	0.4	2.6
Paper	44	6	44	0.3	0.2	5.5
Cardboard	44	6	44	0.3	0.2	5.5
Plastics	60	7	23			10
Textiles	56	7	30	5	0.2	1.8
Rubber	76	10		2		12
Leather	60	9	12	10	0.4	8.6
Misc. organics	49	6	38	2	0.3	4.7
Dirt, ashes, etc.	25	3	1	0.5	0.2	70.3

TABLE 6 Heat of Combustions of Municipal Solid Waste Components

	Inerts (%)		Heating Values (kJ/kg)	
Component	Range	Typical	Range	Typical
Yard wastes	2–5	4	2,000–19,000	7,000
Wood	0.5–2	2	17,000–20,000	19,000
Food wastes	1–7	6	3,000–6,000	5,000
Paper	3–8	6	12,000–19,000	17,000
Cardboard	3–8	6	12,000–19,000	17,000
Plastics	5–20	10	30,000–37,000	33,000
Textiles	2–4	3	15,000–19,000	17,000
Rubber	5–20	10	20,000–28,000	23,000
Leather	8–20	10	15,000–20,000	17,000
Misc. organics	2–8	6	11,000–26,000	18,000
Glass	96–99	98	100–250	150
Tin cans	96–99	98	250–1,200	700
Nonferrous	90–99	96		
Ferrous metals	94–99	98	250–1,200	700
Dirt, ashes, etc.	60–80	70	2,000–11,600	7,000

raw waste is processed to remove non-combustibles like glass and metals, it is called RDF. MSW can decompose in two ways, aerobic and anaerobic. Aerobic decomposition (or composting) occurs when O_2 is present. The composting produces CO_2 and water, but no usable energy products. The anaerobic decomposition occurs in the absence of O_2. It produces landfill gas of 55% CH_4 and 45% CO_2.

Coal and Manure

Frazzitta et al. and Annamalai et al. evaluated the performance of a small-scale pf-fired boiler burner facility (100,000 Btu/h) while using coal and premixed coal manure blends with 20% manure. Three types of feedlot manure were used: raw, partially composted, and fully composted. The burnt fraction was recorded to be 97% for both coal and coal-manure blends.[25,26]

NO_x Emissions

During combustion, the nitrogen evolved from fuel undergoes oxidation to NO_x; and this is called fuel NO_x to distinguish it from thermal NO_x, which is produced by oxidation of atmospheric nitrogen. Unlike coal, most of the agricultural biomass being burned is very low in nitrogen content (i.e., wood or crops), but manure has a higher N content than fossil fuels. A less precise correlation exists between cofiring levels on a Btu basis and percent NO reduction under cofiring. The following Eq. (valid between 3 and 22% mass basis cofiring) describes NO_x reduction as a function of cofiring level on a heat input basis:

$$NO_x \text{ Reduction}(\%) = 0.0008(\text{COF}\%)^2 + 0.0006\ \text{COF}\% + 0.0752, \tag{14}$$

where COF% is the percentage of co-firing on a heat input basis. The mechanisms used to reduce NO_x emissions by cofiring vary between cyclone firing and PC firing.

Figure 11 shows the percentage reduction in NO with percentage cofiring of low-N agricultural biomass fuels. This relationship does not apply to high-N biofuels such as animal manure.

Fouling in Cofiring

Hansen et al. investigated the ash deposition problem in a multi-circulating fluidized bed combustor (MCFBC) fired with fuel blends of coal and wood straw.[25] The Na and K lower the melting point of ash. For ash fusion characteristics see Table 7. Rasmussen and Clausen evaluated the performance of an 80-MW co-generation power plant at Grenaa, Denmark, fired with hard coal and bio-solids (surplus straw from farming). Large amounts of Na and K in straw caused superheater corrosion and

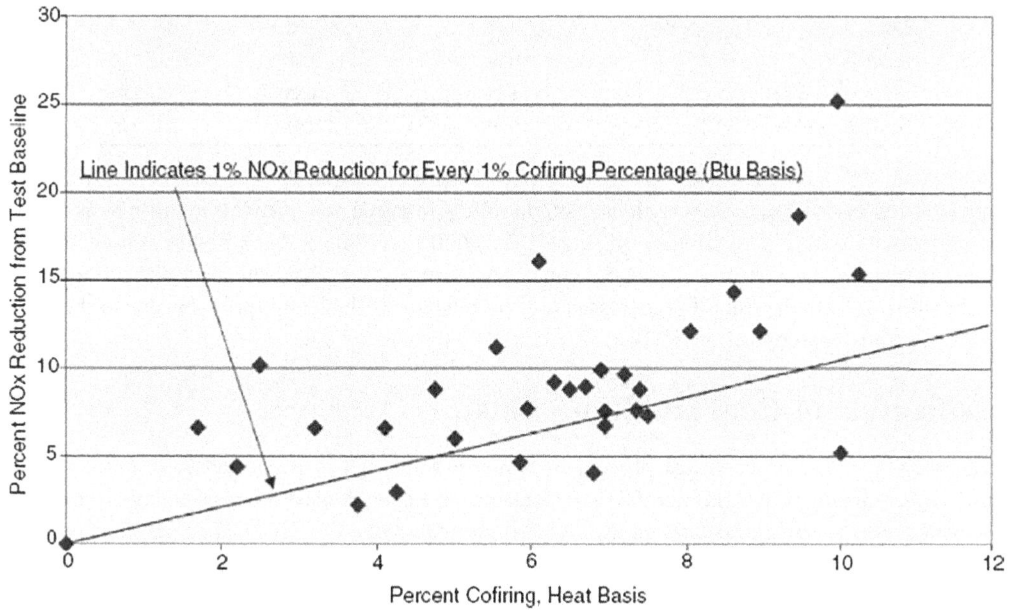

FIGURE 11 NO_x due to cofiring with low-N agricultural residues.
Source: Grabowski.[40]

TABLE 7 Ash Fusion Behavior and Ash Composition, Fusion Data: ASTM D-188

	FB	PRB Coal	Blend
Ash Fusion, (reducing) Initial deformation, IT, °C (°F)	1140 (2090)	1130 (2060)	NA
Softening, °C (°F)	1190 (2170)	1150(2110)	NA
Hemispherical, HT, °C (°F)	1210(2210)	1170(2130)	NA
Fluid, °C (°F)	1230 (2240)	1200 (2190)	NA
Ash fusion, (oxidizing) Initial deformation, IT, °C (°F)	1170(2130)	1190(2180)	NA
Softening, °C (°F)	1190 (2180)	1200 (2190)	NA
Hemispherical, HT, °C (°F)	1220(2230)	1210(2210)	NA
Fluid, °C (°F)	1240(2270)	1280 (2330)	NA
Slagging Index, Rs, °C (°F)	1160(2120)	1140 (2090)	
Slagging classification	High	Severe	
Ash composition (wt%) SiO_2	53.63	36.45	43.56
Al_2O_3	5.08	18.36	12.87
Fe_2O_3	1.86	6.43	4.54
TiO_2	0.29	1.29	0.88
CaO^+	14.60	19.37	17.40
MgO^+	3.05	3.63	3.39
Na_2O^+	3.84	1.37	2.39
K_2O^+	7.76	0.63	3.58
P_2O_5	4.94	0.98	2.62
SO_3	3.71	10.50	7.69
MnO_2	0.09	0.09	0.09
Sum	98.84	99.11	99.00
Volatile Oxides	30.77	28.25	
Basic oxides	32.73	35.51	
Silica ratio	0.73	0.53	
Na_2O+K_2O	11.60	2.00	5.97
Inherent Ca/S Ratio	6.71	1.86	2.48
kg alkali (Na_2O+K_2O)/GJ	5.37	0.06	0.29

combustor fouling.[29] Annamalai et al. evaluated fouling potential when feedlot manure biomass (FB) was cofired with coal under suspension firing.[30] The 90:10 Coal:FB blend resulted in almost twice the ash output compared to coal and ash deposits on heat exchanger tubes that were more difficult to remove than baseline coal ash deposits. The increased fouling behavior with blend is probably due to the higher ash loading and ash composition of FB.

Gasification of Coal and Bio-Solids

Gasification is a thermo-chemical process in which a solid fuel is converted into a gaseous fuel (primarily consisting of HC, H_2 and CO_2) with air or pure oxygen used for partial oxidation of FCs. The main products during gasification are CO and H_2, with some CO_2, N_2, CH_4, H_2O, char particles, and tar (heavy hydrocarbons). The oxidizers used for the gasification processes are oxygen, steam, or air. However, for air, the gasification yields a low-Btu gas, primarily caused by nitrogen dilution present in the supply air. Syngas (CO + H_2) is produced by reaction of biomass with steam. The combustible product, gas, can be used as fuel burned directly or with a gas turbine to produce electricity; or used to make chemical feedstock (petroleum refineries). However, gas needs to be cleaned to remove tar, NH_3,

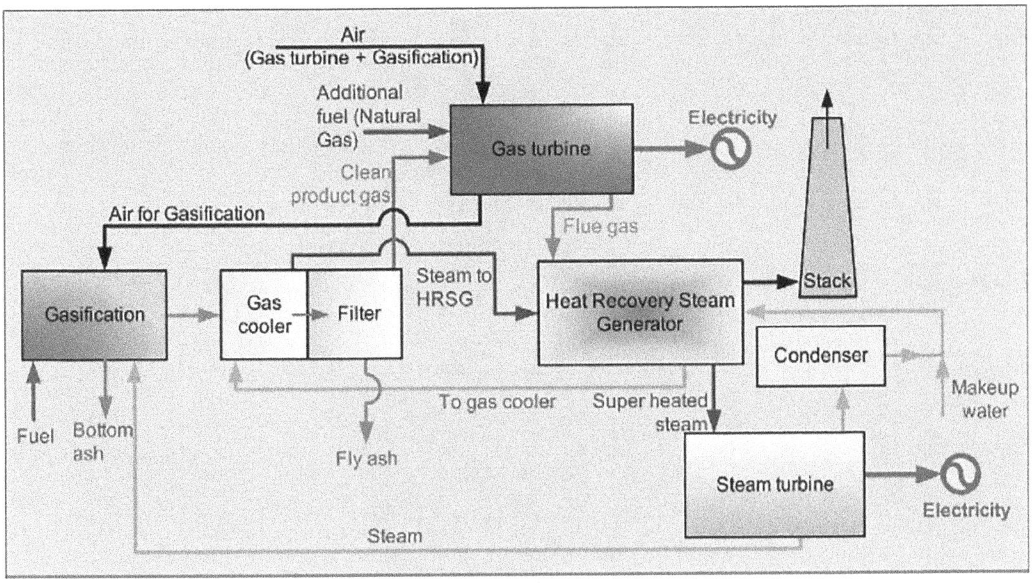

FIGURE 12 Fluidized-bed gasification for Integrated Gasification Combined Cycle (IGCC) Process.

and sulfur compounds. The integrated gasification combined cycle (IGCC) (Figure 12), for combined heat and power (CHP), and traditional boilers use combustible gases from gasifiers for generation of electric power.

Typically in combined cycles, gaseous or liquid fuel is burnt in gas turbine combustors. High-temperature products are expanded in a gas turbine for producing electrical power; a low-temperature (but still hot) exhaust is then used as heat input in a boiler to produce low-temperature steam, which then drives a steam turbine for electrical power. Therefore, one may use gas as a topping cycle medium, while steam is used as fluid for the bottoming cycle. The efficiency of a combined cycle is on the order of 60%, while a conventional gas turbine cycle has an efficiency of 42%.[31] Commercial operations include a 250-MW IGCC plant at Tampa, Florida, operating since 1996; a 340-MW plant at Negishi, Japan, since 2003; and a 1200-MW GE-Bechtel plant under construction in Ohio for American Electric Power, to start in 2010.[31]

There are three basic gasification reactor types: (i) fixed-bed gasifiers (Figure 13); (ii) fluidized-bed gasifiers, including circulating-bed (CFB) or bubbling-bed; and (iii) entrained-flow gasifiers. The principles of operation are similar to those of combustors except that the air supplied is much below stoichiometric amounts, and instead of a combination of steam, air and CO_2, air can also be used. The oxidant source could also include gases other than air, such as air combined with steam in Blasiak et al.[32]

Futuregen

FutureGen is a new U.S. initiative to build the world's first integrated CO_2 sequestration and H_2 production research power plant using coal as fuel. The technology shown in Figure 14 employs modern coal gasification technology using pure oxygen, resulting in CO, C_nH_m (a hydrocarbon), H_2, HCN, NH_3, N_2, H_2S, SO_2, and other combinations which are further reacted with steam (reforming reactions) to produce CO_2 and H_2. The bed materials capture most of the harmful N and S compounds followed by gas-cleaning systems; the CO_2 is then sequestered and H_2 is used as fuel, using either combined cycle or fuel cells for electricity generation or sold as clean transportation fuel. With partial oxidation of gasification products and char supplying heat for pyrolysis and other endothermic reactions (i.e., net zero external

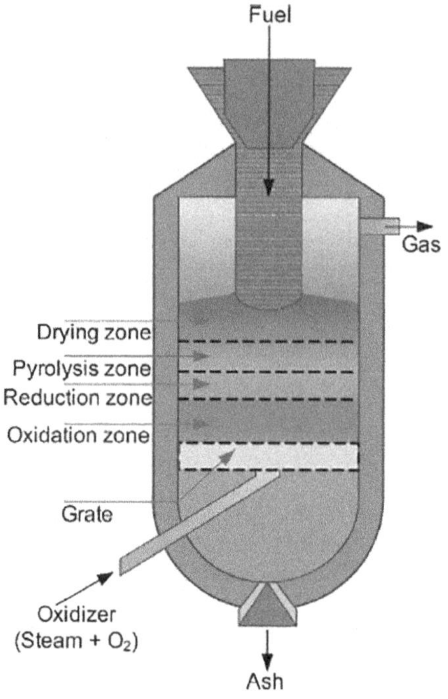

FIGURE 13 Updraft fixed-bed gasifier.

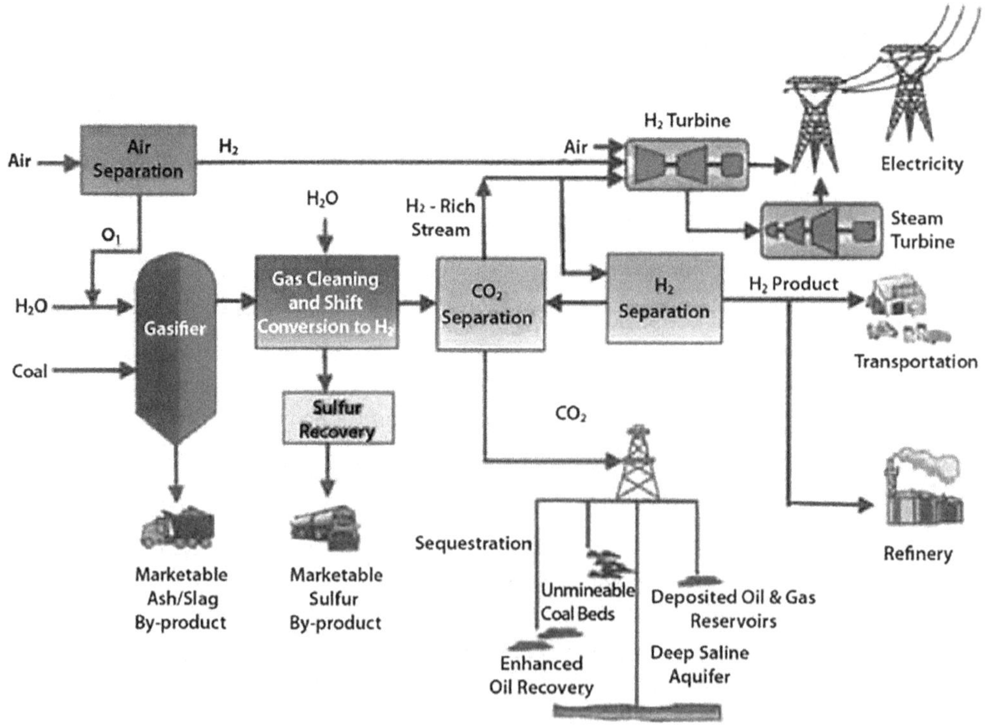

FIGURE 14 FutureGen layout. Source: http://www.fe.doe.gov.

heat supply in gasifier), the overall gasification reaction can be represented as follows for 100 kg of DAF Wyoming coal:

$$\text{Reaction V} \quad C_{6.3}H_{4.4}N_{0.0076}O_{1.15}S_{0.014} + 8.14\ H_2O(1) + 1.67\ O_2 \rightarrow 6.3\ CO_2 + 10.34\ H_2 + 0.038\ N_2 + 0.014\ SO_2$$

It is apparent that the FutureGen process results in enhanced production of H_2, using coal as an energy source to strip H_2 from water. For C–H–O fuels, it can be shown that theoretical H_2 production (N_{H2}) in moles for an empirical fuel CH_hO_o is given as $\{0.4115\,h - 0.6204\,o + 1.4776\}$ under the above conditions. For example, if glucose $C_6H_{12}O_6$ is the fuel, then empirical formulae is CH_2O; thus, with h = 2, o = 1, $N_{H_2} = 1.68$ kmol, using atom balance, $CH_2O + 0.68\ H_2O(\ell) + 0.16\ O_2\ CO_2 + 1.68\ H_2$.

Reburn with Bio-Solids

NO_x is produced when fuel is burned with air. The N in NO_x can come both from the nitrogen-containing fuel compounds (e.g., coal, biomass, plant residue, animal waste) and from the N in the air. The NO_x generated from fuel N is called fuel NO_x, and NO_x formed from the air is called thermal NO_x. Typically, 75% of NO_x in boiler burners is from fuel N. It is mandated that NO_x, a precursor of smog, be reduced to 0.40–0.46 lb/mmBtu for wall and tangentially fired units under the Clean Air Act Amendments (CAAA). The current technologies developed for reducing NO_x include combustion controls (e.g., staged combustion or low NO_x burners (LNB), reburn) and post-combustion controls (e.g., Selective Non-Catalytic Reduction, SNCR using urea).

In reburning, additional fuel (typically natural gas) is injected downstream from the primary combustion zone to create a fuel rich zone (optimum reburn stoichiometric ratio (SR), usually between SR 0.7 and 0.9), where NO_x is reduced up to 60% through reactions with hydrocarbons when reburn heat input with CH_4 is about 10%–20%. Downstream of the reburn zone, additional air is injected in the burnout zone to complete the combustion process. A diagram of the entire process with the different combustion zones is shown in Figure 15. There have been numerous studies on reburn technology found in literature, with experiments conducted, and the important results summarized elsewhere.[33] Table 8 shows the percentages of reduction and emission obtained with coal or gas reburn in coal-fired installations and demonstration units.

The low cost of biomass and its availability make it an ideal source of pyrolysis gas, which is a more effective reburn fuel than the main source fuel, which is typically coal. Recently, animal manure has been tested as a reburn fuel in laboratory scale experiments. A reduction of a maximum of 80% was achieved for pure biomass, while the coal experienced a reduction of between 10 and 40%, depending on the equivalence ratio.[34] It is believed that the greater effectiveness of the feedlot biomass is due to its greater volatile content on a DAF basis and its release of fuel nitrogen in the form of NH_3, instead of HCN.[35]

Acknowledgments

Most of this work was supported in part by the U.S. Department of Energy of Pittsburgh, PA, the DOE of Golden, CO, the USDA, the Texas Advanced Technology Program, and the Texas Commission on Environmental Quality (TCEQ) of Austin, TX, through the Texas Engineering Experiment Station (TEES), Texas A&M University, and College Station, TX.

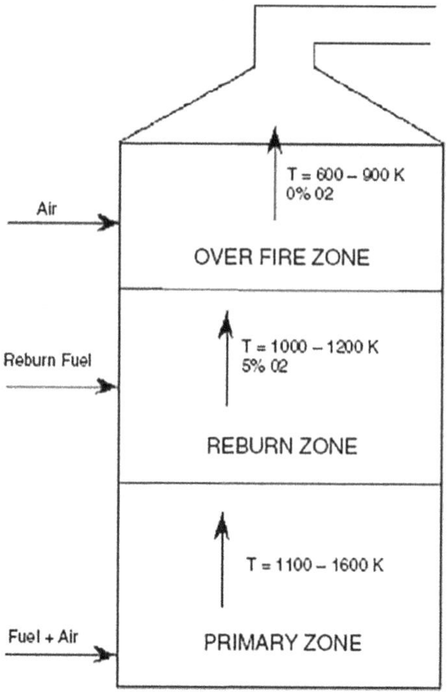

FIGURE 15 Schematic of reburn process.

TABLE 8 Percentage Reduction in NO_x: Demonstration and/or Operating Reburn Installations on Coal-Fired Boilers in the United States.

Type of Burner	% Reburn Heat in	% Reduction	NOx with Reburn lb/mmBtu[a]
Gas reburning			
Tangential	18	50–67	0.25
Cyclone	20–23	58–60	0.39–0.56
Wall without LNB	18	63	0.27
Coal reburn			
Cyclone (micronized)	30 (17)	52 (57)	0.39 (0.59)
Tangential (micron) with LNB	14	28	0.25

Note: LNB: Low NO_x Burners.
[a]*1 lb per mmBtu = 0.430 kg/GJ.*
Source: DOE.[38]

References

1. U.S. DOE, Energy Information Administration, available at http://www.eia.doe.gov/ (accessed April 12, 2005).
2. Priyadarsan, S.; Annamalai, K.; Sweeten, J.M.; Holtzapple, M.T.; Mukhtar, S. *Co-gasification of blended coal with feedlot and chicken litter biomass,* Proceedings of the 30th Symposium (International) on Combustion; The Combustion Institute: Pittsburgh, PA, 2005; Vol. 30, 2973–2980.
3. Ebeling, J.M.; Jenkins, B.M. Physical and chemical properties of biomass fuels. Trans. ASAE **1985**, *28* (3), 898–902.

4. Sweeten, J.M.; Annamalai, K.; Thien, B.; McDonald, L. Co-firing of coal and cattle feedlot biomass (FB) fuels, part I: feedlot biomass (cattle manure) fuel quality and characteristics. Fuel **2003**, *82* (10), 1167–1182.
5. Boie, W. Wiss z Tech. Hochsch. Dresden 1952/1953, 2, 687.
6. Annamalai, K.; Sweeten, J.M.; Ramalingam, S.C. Estimation of the gross HVs of biomass fuels. Trans. Soc. Agric. Eng. **1987**, *30*, 1205–1208.
7. Sami, M.; Annamalai, K.; Wooldridge, M. Co-firing of coal and biomass fuel blends. Prog. Energy Combust. Sci. **2001**, *27*, 171–214.
8. Annamalai, K.; Puri, I.K. 2006, *Combustion Science and Engineering*, Taylor and Francis: Orlando, FL, 2006.
9. Tillman, D.A.; Rossi, A.J.; Kitto, W.D. *Wood Combustion: Principles: Processes and Economics*; Academic Press: New York, 1981.
10. Sweeten, J.M.; Heflin, K.; Annamalai, K.; Auvermann, B.; Collum, Mc; Parker, D.B. *Combustion-Fuel Properties of Manure Compost from Paved vs Un-paved Cattle feedlots*; ASABE 06-4143: Portland, OR, July 9–12, 2006.
11. Bartok, W., Sarofim, A.F. 1991, *Fossil Fuel Combustion*, chapter 3: FL Dryer, John Wiley, and Hobokan, NJ. pp. 121–214.
12. Howard, J.B.; Sarofim, A.F. Gasification of coal char with CO_2 and steam at 1200–1800°C. *Energy Lab Report;* Chemical Engineering; MIT: MA, 1978.
13. Essenhigh, R.H.; Misra, M.K.; Shaw, D.W. Ignition of coal particles: a review. Combust. Flame **1989**, *77*, 3–30.
14. Annamalai, K.; Durbetaki, P. A theory on transition of ignition phase of coal particles. Combust. Flame **1977**, *29*, 193–208.
15. Du, X.; Annamalai, K. The transient ignition of isolated coal particles. Combust. Flame **1994**, *97*, 339–354.
16. Johnson, N. *Fundamentals of Stoker Fired Boiler Design and Operation*, CIBO Emission Controls Technology Conference, July 15–17, 2002.
17. Annamalai, K.; Ibrahim, Y.M.; Sweeten, J.M. Experimental studies on combustion of cattle manure in a fluidized bed combustor. Trans. ASME, J. Energy Res. Technol. **1987**, *109*, 49–57.
18. Walawender, W.P.; Fan, L.T.; Engler, C.R.; Erickson, L.E. Feedlot manure and other agricultural wastes as future material and energy resources: II. Process descriptions. *Contrib.30,* Deptartment of Chemical Engineering, Kansas Agricultural Experiment Station: Manhattan, KS, 30, 1973.
19. Raman, K.P.; Walawander, W.P.; Fan, L.T. Gasification of feedlot manure in a fluidized bed: effect of temperature. Ind. Eng. Chem. Proc. Des. Dev. **1980**, *10*, 623–629.
20. Aerts, D.J.; Bryden, K.M.; Hoerning, J.M.; Ragland, K.W. *Co-firing Switchgrass in a 50 MW Pulverized Coal Boiler,* Proceedings of the 59th Annual American Power Conference, Chicago, IL, **1997**, *50* (2), 1180–1185.
21. Abbas, T.; Costen, P.; Kandamby, N.H.; Lockwood, F.C.; Ou, J.J. The influence of burner injection mode on pulverized coal and biosolid co-fired flames. Combust. Flame **1994**, *99*, 617–625.
22. Siegel, V.; Schweitzer, B.; Spliethoff, H.; Hein, K.R.G. Preparation and co-combustion of cereals with hard coal in a 500 kW pulverized-fuel test unit. Biomass for energy and the environment, *Proceedings of the 9th European Bioenergy Conference,* Copenhagen, DK, June 24–27, 1996; 2, 1027–1032.
23. Hansen, L.A.; Michelsen, H.P.; Dam-Johansen, K. Alkali metals in a coal and biosolid fired CFBC—measurements and thermodynamic modeling, *Proceedings of the 13th International Conference on Fluidized Bed Combustion*, Orlando, FL, May 7–10, 1995; 1, 39–48.
24. Sampson, G.R.; Richmond, A.P.; Brewster, G.A.; Gasbarro, A.F. Co-firing of wood chips with coal in interior Alaska. Forest Prod. J. **1991**, *41* (5), 53–56.
25. Frazzitta, S.; Annamalai, K.; Sweeten, J. Performance of a burner with coal and coal: feedlot manure blends. J. Propulsion Power *1999*, *15* (2), 181–186.

26. Annamalai, K.; Thien, B.; Sweeten, J.M. Co-firing of coal and cattle feedlot biomass (FB) fuels part II: performance results from 100,000 Btu/h laboratory scale boiler burner. Fuel **2003**, *82* (10), 1183–1193.
27. Fahlstdedt, I.; Lindman, E.; Lindberg, T.; Anderson, J. Cofiring of biomass and coal in a pressurized fluidized bed combined cycle. Results of pilot plant studies, *Proceedings of the 14th International conference on Fluidized Bed Combustion*, Vancouver, Canada, May 11–14, 1997; 1, 295–299.
28. Van Doom, J., Bruyn, P., Vermeij, P. Combined combustion of biomass, fluidized sewage sludge and coal in an atmospheric Fluidized bed installation, Biomass for energy and the environment, *Proceedings of the 9th European Bioenergy Conference*, Copenhagen, DK, June 24–27, 1996; 1007–1012.
29. Rasmussen, I.; Clausen, J.C. ELSAM strategy of firing biosolid in CFB power plants, *Proceedings of the 13th International Conference on Fluidized Bed Combustion*, Orlando, FL, May 7–10, 1, 1995; 557–563.
30. Annamalai, K.; Sweeten, J.; Freeman, M.; Mathur, M.; O'Dowd, W.; Walbert, G.; Jones, S. Co-firing of coal and cattle feedlot biomass (FB) fuels, part III: fouling results from a 500,000 Btu/hr pilot plant scale boiler burner. Fuel **2003**, *82* (10), 1195–1200.
31. Langston, L.S. 2005 New Horizons. *Power and Energy* vol 2, June 2, 2005.
32. Blasiak, W.; Szewczyk, D.; Lucas, C.; Mochida, S. *Gasification of Biomass Wastes with High Temperature Air and Steam*, Twenty-First International Conference on Incineration and Thermal Treatment Technologies, New Orleans, LA, May 13–17, 2002.
33. Thien, B.; Annamalai, K. *National Combustion Conference,* Oakland, CA, March 25–27, 2001.
34. Arumugam, S.; Annamalai, K.; Thien, B.; Sweeten, J. Feedlot biomass co-firing: a renewable energy alternative for coal-fired utilities, Int. Natl J. Green Energy **2005**, *2* (4), 409–419.
35. Zhou, J.; Masutani, S.; Ishimura, D.; Turn, S.; Kinoshita, C. Release of fuel-bound nitrogen during biomass gasification. Ind. Eng. Chem. Res. **2000**, *39*, 626–634.
36. Annamalai, K.; Puri, I.K. *Advanced Thermodynamics;* CRC Press: Boca Raton, FL, 2001.
37. Priyadarsan, S.; Annamalai, K.; Sweeten, J.M.; Holtzapple, M.T.; Mukhtar, S. Waste to energy: fixed bed gasification of feedlot and chicken litter biomass. Trans. ASAE **2004**, *47* (5), 1689–1696.
38. DOE, Reburning technologies for the control of nitrogen oxides emissions from coal-fired boilers Topical Report, No 14, U.S. Department of Energy, May, 1999.
39. Loo, S.V.; Kessel, R., 2003, Optimization of biomass fired grate stroker systems, EPRI/IEA Bioenergy Workshop, Salt Lake City, Utah, February 19, 2003.
40. Grabowski, P. *Biomass Cofiring, Office of the Biomass Program*, Technical Advisory Committee, Washington, DC, March 11, 2004.

18

Energy Demand: From Individual Behavioral Changes to Climate Change Mitigation

Leila Niamir and
Felix Creutzig

Introduction

Decarbonization of the economy requires massive worldwide efforts and a strong involvement of regions, cities, businesses, and individuals in addition to commitments at national levels (1). Human consumption, in combination with a growing population, contributes to climate change by increasing the rate of green house gas (GHG) emissions (2,3). Over the last decade, instigated by the Paris Agreement, the efforts to limit global warming have been expanding. However, significant attention is being devoted to new energy technologies on both the production and consumption sides, while changes in individual behavior and management practices as part of the mitigation strategy are often neglected (4–8).

Although demand-side solutions are promising, they are not given the same level of attention as technological supply-side solutions in assessments, like the IPCC's AR5, and in integrated assessment models (IAMs) in general, or in the popular media, as mirrored in the observation that directed innovation perversely privileges energy-supply technologies over efficient end-use technologies (10). This potentially has two main reasons: (a) demand-side solutions are often embedded in a complex network of social institutions and practices (11), and thus less prone to quantitative analysis and clear-cut implementation; (b) demand-side solutions also often involve explicit normative positions, or values, making those solutions subject to value-laden discourses (12). Specifically, as demand-side solutions presuppose modified behavior, they are less compatible with the revealed-preference framework prevalent in economics. Both reasons apply less to supply-side solutions: people consume electricity via plug-in, and the original supply of energy for electricity is irrelevant for actual consumption.

Table 1 shows the potential contribution to climate change mitigation from three sectors: buildings, transport, and food under two main perspectives: (a) hard infrastructures, such as the built environment, and (b) soft infrastructures, such as habits and norms. On the one hand, hard infrastructures, epitomized by the urban built infrastructure, provide a physical setting for shaping preferences, practices, and opportunity spaces. On the other hand, changing norms, practices, and nudges modify the opportunity space and can induce direct behavioral change (13).

In building sector, the behavioral change potential can be as high as 50% over long periods of time (Table 1). There is a range in the energy savings achievable in buildings due to behavioral changes, depending on the type of end use. For example, savings from heating loads of 10–30% are possible for changes in the thermostat setting. Similarly, cooling savings of 50–67% are recorded with measures such as substituting air-conditioning with fans in moderately hot climates with tolerable, brief heat exposures. Meanwhile, subsequent literature has provided quantitative estimates of the importance of urban transport and behavioral options (14). Overall, behavioral and infrastructural measures in cities can potentially reduce GHG emissions from urban passenger transport by 20–50% until 2050 (15). This may be achieved via three routes, technological change, modal shift, and reduced travel demand, and

TABLE 1 Climate Mitigation Options and Their Quantitative Reduction Potential

End-User Sector	Mitigation Option	Reduction Potential	Reference(s)
Urban areas and spatial planning	Modifying the emerging urbanization	20–25% reduction of future urban energy use until 2050	(17)
Transportation	Optimal pricing	15–20% reduction in automobile transport	(17)
	Compactness	United States: 10% reduction in distance traveled	(18)
		Europe: 5% reduction in distance traveled	(19)
	Behavioral measures (marketing, information provision, and tailored services)	10% reduction in transport demand	(20,21)
Tourism	Transport modal shifts and increases in average length of stay	44% emission reductions worldwide until 2035	(22)
Buildings	Developed context: short-term behavioral change potential	>20%	(2)
	Developed context: long-term behavioral change potential	50%	(23,24)
	Developing context: behavioral change potential	Relevant but lower potential	(25,26)
	Heating: adjusting thermostat setting	10–30%	(27)
	Cooling: using substitutes for air-conditioning, increasing thermostat setting, changing dress codes	50–67%	(27,28)
	Clothes washing and drying behavior (e.g., by operation at full load versus one-third to one-half load)	10–100%	(29)
	Water-heating and cooking energy savings (e.g., by shorter showers, using different cooking practices)	50%	(29)
	Lighting energy savings (e.g., by turning off unneeded lights)	70%	(29)
	Refrigerator energy savings (e.g., by smaller refrigerators)	30–50%	(29)
	Dishwasher energy savings (e.g., by full-load operation)	75%	(29)
Agriculture and other land use	Technical potential of demand-side options for 2050	70%	(30,31)
	Healthy diet: meat, fish, and egg consumption of 90 g/(person/day)	36% greenhouse gas savings	(32)

Source: Ref. (13).

information technology plays a key role in all three areas of innovation (16). Behavioral options could be fostered by information campaigns that facilitate social learning, active choice, and frame alternative mobility choices.

Reshaping urban forms and the urban environment provides ample opportunity to make private car travel obsolete and save high amounts of energy in buildings. Both transport and buildings also offer significant opportunities for reduced energy demand by soft measures, such as providing targeted information on sustainable mobility for specific groups of people. In agriculture, demand-side action, in particular, dietary shift, could reduce emissions by more than 70% compared to the trend in 2055, thereby surpassing the potential of technological options.

Demand-Side Solutions

Demand-side solutions for mitigating climate change include strategies targeting technology choices, consumption, behavior, lifestyles, coupled production–consumption infrastructures and systems, service provision, and associated sociotechnical transitions. Disciplines vary in their approaches and in the research questions that they ask on demand-side issues (4). For example, psychologists and behavioral economists focus on emotional factors and cognitive biases in the decision-making process (33); economists elaborate on how, under rational decision-making, carbon pricing and other fiscal instruments can trigger change in demand (34); sociologists emphasize everyday practices, structural issues, and socioeconomic inequality (35); anthropologists address the role of culture in energy consumption (36); and studies in technological innovation consider sociotechnical transitions and the norms, rules, and pace of adoption that support dominant technologies (37).

Figure 1 proposed a demand-side assessment framework and discuss key topics that need to be addressed: the characterization of demand; policy instruments and how they would affect demand; techno-economic evaluation; implications for well-being; mitigation pathways; and the sustainable development context.

The starting point for a demand-side assessment seeks to characterize patterns of demand for energy, mobility, food and shelter, and the associated GHG emissions. Hence, the first question to ask is: what norms, values, preferences, and structural factors shape energy demand and GHG emissions (Figure 1a)? Policy instruments can spur demand-side solutions, in ways that depend on the specific energy service and socioeconomic context. Different disciplines have provided important pieces of this big jigsaw, but much remain to be done to put the assessment of policy instruments together in a truly interdisciplinary effort and address the questions posed (Figure 1b).

Beyond specific technologies, research should take a wider scope and ask for the efficient and reliable provision of end-use services, rather than efficient technology design alone (Figure 1c). Technological studies contribute to an understanding of dynamic systems, describing cost reductions and strategies to overcome barriers on the path from research and development of a technology to market-scale deployment and uptake. Another question is: how do demand-side mitigation measures affect well-being (Figure 1d)? Reducing energy use or GHG emissions need to be balanced with the goal of enhancing human well-being (39).

Sketched approaches such as transition theory, study of behavioral tipping points and social norms, and political economy insights into policy sequencing have the potential for laying out short-term and action-oriented mitigation pathways. Such approaches, together with bottom-up assessments from technological studies, can be soft-coupled and combined with IAMs and similar economic models that assess system-wide potentials, reflecting the interaction between sectors and mitigation options (Figure 1e). The linkage between sustainable development and climate change is also articulated in the 'nationally determined' language of the Paris Agreement, which promotes climate mitigation that coincides with nationally determined development outcomes. A demand-side assessment should also be able to inform sustainable development pathways (Figure 1f).

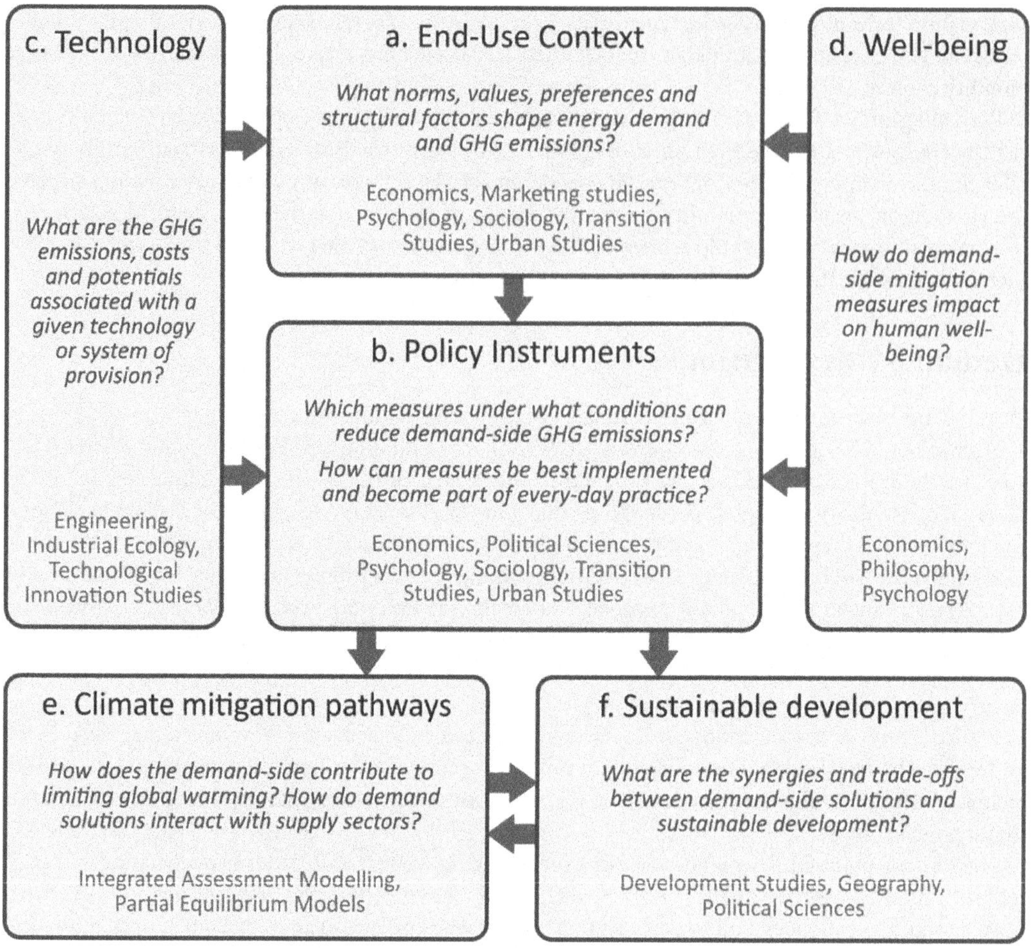

FIGURE 1 Key research questions and contributing disciplines for assessing demand-side solutions. (From Ref. 38.)

Modeling Individuals Energy Behavioral Changes

The understanding of how bottom-up processes can impact climate mitigation guides us to effective development and implementation of policies (9,40). In the last decade, a variety of macroeconomic models and assessment tools emerged (41–43) and were predominately used to support climate-energy policy decisions (44) range from the assessment of macroeconomic and cross-sectoral impacts (45–47), to detailed micro-simulation models for a specific technology (5,47,48). Much can be done to make the assumptions in macroeconomic and integrated assessment models more realistic concerning the scale and nature of damage (5). These models usually assume that economic agents form a representative group(s), have perfect access to information and adapt instantly and rationally to new situations, maximizing their long-run personal advantage (49–52).

Unlike other approaches, agent-based model (ABM) is not limited to perfectly rational agents or to abstract microdetails in aggregate system-level equations. Instead, ABM can represent the behavior of energy consumers – such as individual households – using a range of behavioral theories (9,50,53). In addition, ABM has the ability to examine how interactions of heterogeneous agents at micro-level give rise to the emergence of macro-outcomes, including those relevant for climate mitigation such as an adoption of low-carbon behavioral strategies and technologies over space and time (54). The ABM approach simulates complex and nonlinear behavior that is intractable in equilibrium models (52).

Individual energy behavior, especially when amplified through social context, shapes energy demand and, consequently, carbon emissions. By changing their behaviors, individuals can play an essential role in the transformation process towards a low-carbon society and global emissions reduction (9). However, explaining and affecting human behavior is a difficult task since human nature is complex and heterogeneous. As a result, quantitative tools to assess cumulative household emissions, given the diversity of behavior and a variety of psychological and social factors influencing it beyond purely economic considerations, are scarce (6,40,55).

BENCH Agent-Based Model

To assess the impact of individual behavior on carbon emissions, we went beyond classical economic models and the stylized representation of a perfectly informed optimizer. An agent-based simulation model was designed and developed to quantify the cumulative impacts of individual behavioral changes on regional dynamics of saved energy and CO_2 emissions (40). The BENCH (Behavioral change in ENergy Consumption of Households) model[1] builds up on the advances in agent-based modeling applied in the energy domain, and adds theoretically and empirically grounded individual behavioral rules that drive households' energy-related choices (40,52).

Driven by the empirical evidence from environmental behavioral studies (56–61), the *BENCH* model assumes that a decision regarding energy action is driven by psychological and social factors in addition to the standard economic drivers such as prices relative to incomes (40,52,55). Behavioral factors including personal norms and awareness may either amplify the economic logic behind a decision-making or impede it, serving either as a trigger or a barrier (Figure 2).

Household agents in *BENCH* are heterogeneous in socioeconomic characteristics, preferences, and awareness of environment and climate change, so they can pursue various energy choices and actions. Figure 3 illustrates our conceptual framework that represents household energy behavioral change as a dynamic process unfolding in stages. Knowledge and awareness can have an important role in triggering individual behavior change (4,15,62–67). If individuals have enough knowledge and awareness about climate, environment, and energy issues, a feeling of guilt develops and activates motivational factors, which may lead to energy-related behavior change. Motivation is enhanced by personal and social norms (58,67), which can lead to a feeling of responsibility and provoke an individual to change their behavior. When intentions for the latter are high, individuals do a formal feasibility assessment according their income, dwelling conditions, and own perceived behavioral control. Individuals compare their current energy-use habits with alternatives, and if things can be improved, the intention to pursue an alternative rises and may lead to a behavior change. This conceptual framework combines some behavioral constructs that are common between TPB (in red) and NAT (in green).

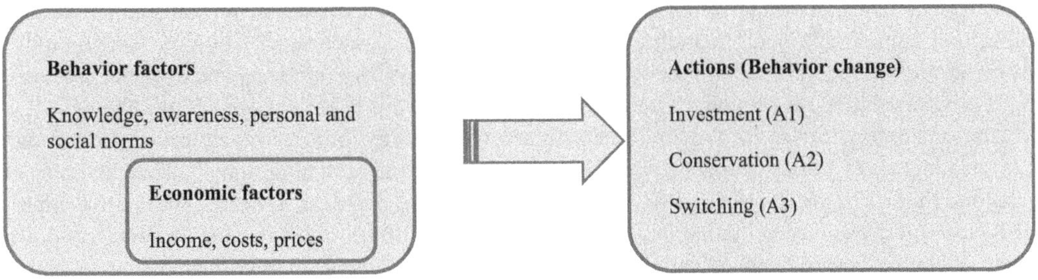

FIGURE 2 Factors affecting household decision-making regarding their energy use. (From Ref. 52.)

[1] **B**ehavioral change in **EN**ergy **C**onsumption of **H**ouseholds Model.

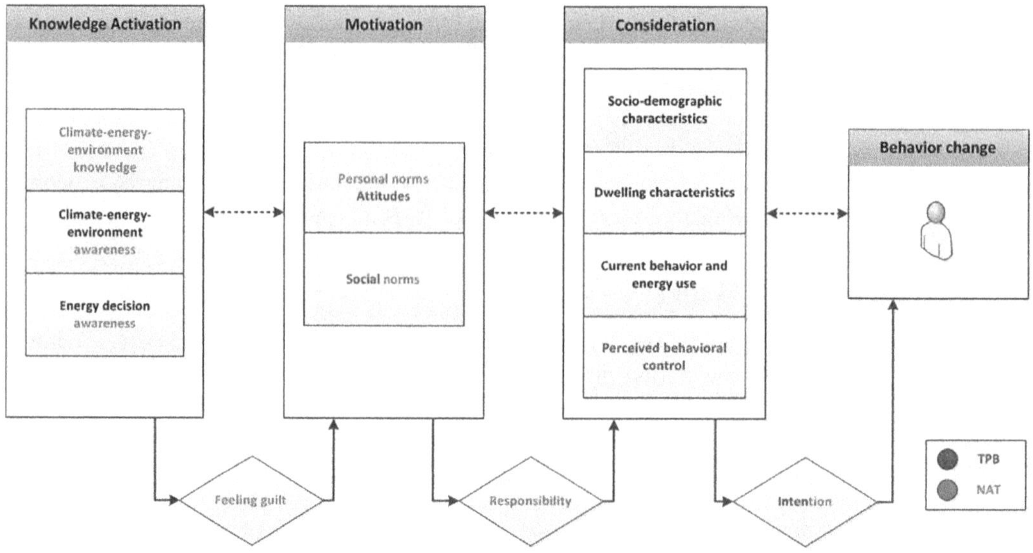

FIGURE 3 conceptual representation of multi-stage household behavioral change. (From Ref. 68.)

Empirical Data

Household agents in *BENCH* are heterogeneous in socioeconomic characteristics, preferences, and awareness of environment and climate change, so they can pursue various energy choices and actions. In order to quantify which factors – socioeconomic (e.g., income, age), behavioral (e.g., personal and social norms, knowledge and awareness about the environment, social influence), and structural (e.g., size and type of house) – trigger or attenuate a transition to a lower and greener energy footprint at the household level, the comprehensive survey (N=1,790 households) was designed and conducted in two provinces in Europe.

Our data analysis demonstrates that awareness and personal and social norms are equally as important as monetary factors when it comes to individual energy actions. Education and structural dwelling factors, e.g., size and type of dwelling appear to be very significant in bottom-up actions contributing to the reduction of the regional CO_2 footprint from the residential sector (68) (Figure 4).

End-User Scenarios

The *BENCH* calculated changes in electricity consumption annually and implied carbon emissions based on the primary source of energy by simulating individuals' behavior under different end-user behavioral and climate scenarios. The results indicate that accounting for demand-side heterogeneity provides a better insight into possible transition pathways to a low-carbon economy and climate change mitigation. Namely, the model including household heterogeneity, as represented by socio-demographic, dwelling, and behavioral factors, shows rich dynamics and provides a more realistic image of socioeconomics by simulating the economy through the social interactions of heterogeneous households.

Four end-user scenarios are designed and analyzed, which varied from the baseline scenario through the introduction of agent heterogeneity, the intensity of social interactions among households (slow or fast), and the lack or presence of carbon price (52). By comparing end-user scenarios, the relative impact of bottom-up drivers (social dynamics and learning on the diffusion of information) and top-down market policies (carbon price) on carbon emission reduction are estimated (Figure 5).

The impact of household attribute heterogeneity and social dynamics brought about a 5–9% CO_2 emission reduction by 2030. Adding carbon price cuts CO_2 emission down to 55% compared to the baseline scenario, which mimics the traditional economic setup of a representative rational, fully informed household making the optimal decision.

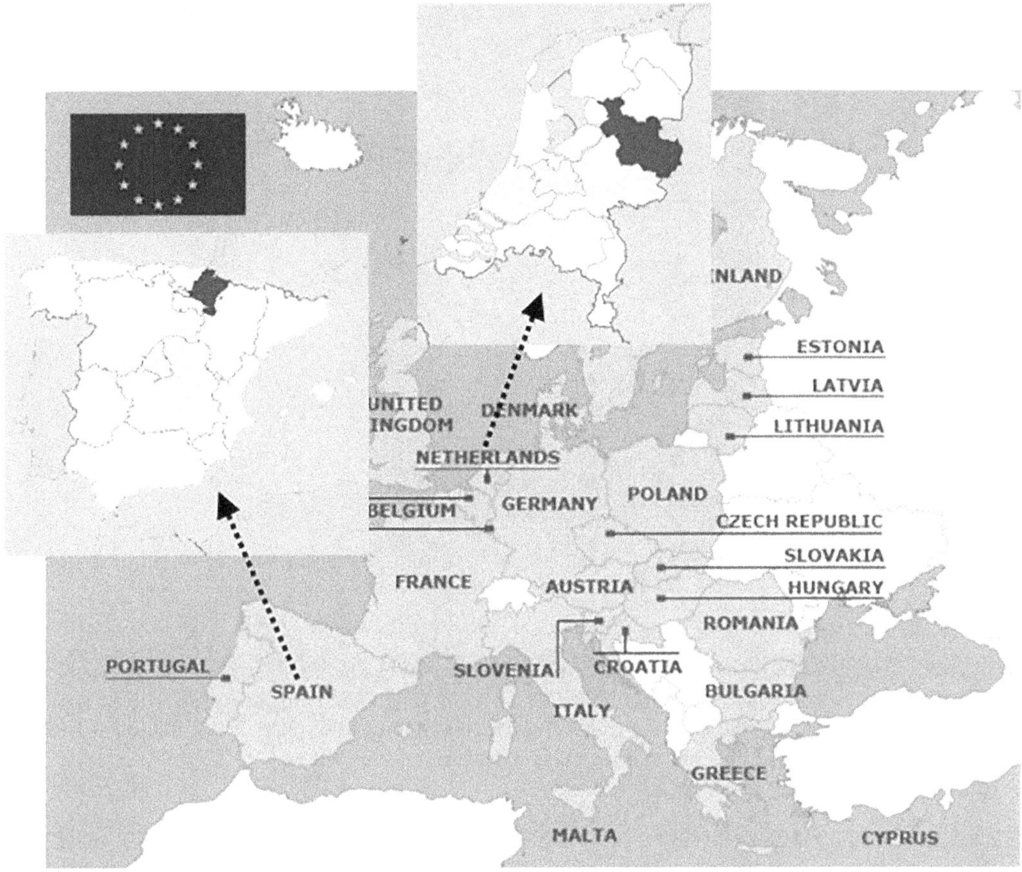

FIGURE 4 Survey case studies: the Overijssel province in the Netherlands and the Navarre province in Spain. (From Ref. 68.)

These insights have also implications for more standard policymaking. As preferences are malleable, and change with infrastructures and policy environments (13,69), policies themselves need to consider social norms and malleable preferences to avoid inefficiencies, or to even crowd out desirable outcomes (70,71).

Conclusion

Avoiding and shifting consumption plays a key role to achieve Sustainable Development Goals (SDGs) and for climate change mitigation. We can achieve the environmental targets (e.g., SDGs) through controlled consumption. Demand-side management should hence be better represented in energy systems scenarios. Demand-side management covers a broad range from technology interventions and diffusion to energy efficiency solutions and behavioral changes. Designing energy systems scenario requires an urgent understanding of which human activities are more culpable, what causes them, and how we can effectively change them. Since the nature of human is complex and heterogeneous, explaining and affecting human behavior is a difficult task. Therefore, this requires quantitative tools to assess cumulative individual emissions, given the diversity of behavior and variety of psychological and social factors influencing it beyond purely economics considerations.

The potential of individual behavioral changes in reducing carbon emissions attracts considerable attention as one of the climate change mitigation strategies (13,52). The theoretically and empirically

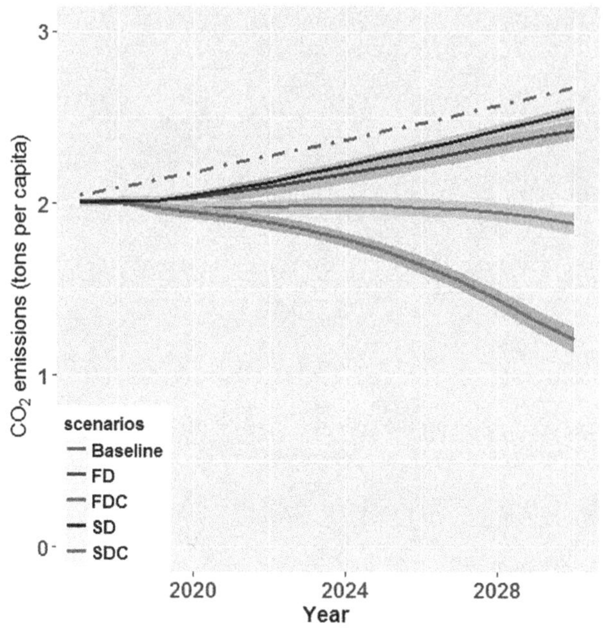

FIGURE 5 Macroimpact of heterogeneous households behavioral change combing (bottom-up) strategy and carbon price pressure (top-down) strategy on CO_2 emissions over time. Combining behavioral-climate scenarios: combination of carbon price and slow and fast social dynamics (SDC, FDC) (2017–2030). The shaded bounds around the curves indicate the uncertainty intervals across 100 runs. (From Ref. 52.)

grounded modeling tools such as the BENCH model can serve as useful instruments to quantify regional impacts of qualitative and untraceable individual behavioral aspects. The model can serve as a simulation platform to support the engagement of stakeholders. The results of this novel model indicate that accounting for the demand-side heterogeneity provides better insights into possible transition pathways to a low-carbon economy and into potential of behavioral changes as a climate change mitigation strategy (72). In order to facilitate this transition, the broader view on the social environment, cultural practices, public knowledge, producers technologies and services, and the facilities used by consumers are needed to design implementable and politically feasible policy options. It offers policymakers ways to explore various policy mixes combining price instruments (subsidies and taxes) with various targeted information policies to amplify the positive effect of individual behavioral changes regarding energy use. The reflected consideration have important implications for management practices and professionals. They point to the importance of seeing avoid and shift measures as part of management in business, administrations, and among professions.

The key message is the individuals are more than just consumers in climate change mitigation. Therefore, climate mitigation policies should go beyond economic cost-benefit incentives (e.g., subsidies and taxes). First, the social environment, cultural practices, public knowledge, producer technologies and services, and facilities used by consumers – all should be considered when designing implementable and politically feasible policy options. Second, various financial, social, and other instruments in the policy mix should be designed as a coherent set to reinforce each other, optimizing the joint effectiveness. In particular, policies such as the provision of targeted information, social advertisements, and power of celebrities for the broader public in combination with education can be used to create more knowledge and awareness in the longer run and could accompany and reinforce the effectiveness of other stimuli such as subsidies. These types of policies (soft policies) may prove to be more effective in promoting green energy solutions implemented by households compared to fiscal policy measures alone.

References

1. Grubler A, Wilson C, Bento N, Boza-Kiss B, Krey V, McCollum DL, et al. A low energy demand scenario for meeting the 1.5°C target and sustainable development goals without negative emission technologies. *Nat Energy*. 2018;3(6):515. Available from: https://www.nature.com/articles/s41560-018-0172-6
2. Dietz T, Gardner GT, Gilligan J, Stern PC, Vandenbergh MP. Household actions can provide a behavioral wedge to rapidly reduce US carbon emissions. *Proc Natl Acad Sci U S A*. 2009 Nov 3 [cited 2019 Jun 3];106(44):18452–6. Available from: http://www.ncbi.nlm.nih.gov/pubmed/19858494
3. Dietz T, Rosa EA. Effects of population and affluence on CO_2 emissions. *Proc Natl Acad Sci U S A*. 1997 Jan 7 [cited 2019 Jun 3];94(1):175–9. Available from: http://www.ncbi.nlm.nih.gov/pubmed/8990181
4. Creutzig F, Roy J, Lamb WF, Azevedo IML, Bruine de Bruin W, Dalkmann H, et al. Towards demand-side solutions for mitigating climate change. *Nat Clim Chang.* 2018 Apr 3 [cited 2019 May 27];8(4):260–3. Available from: http://www.nature.com/articles/s41558-018-0121-1
5. Stern N. Economics: Current climate models are grossly misleading. *Nature*. 2016 Feb 24 [cited 2019 May 27];530(7591):407–9. Available from: http://www.nature.com/articles/530407a
6. Farmer JD, Foley D. The economy needs agent-based modelling. *Nature*. 2009 Aug 5;460(7256):685–6.
7. Niamir L. Behavioural mitigation from individual energy choices to demand-side potential. [cited 2019 May 10]. Available from: https://ris.utwente.nl/ws/portalfiles/portal/86376675/Dissertation_LeilaNiamir_BehaviouralClimateChangeMitigation.pdf
8. Niamir L, Filatova T. From climate change awareness to energy efficient behaviour. *Int Congr Environ Model Softw*. 2016 Jul 13 [cited 2019 Jun 3]. Available from: https://scholarsarchive.byu.edu/iemssconference/2016/Stream-A/74
9. Niamir L. *Behavioural Climate Change Mitigation*. Enschede, The Netherlands: University of Twente; 2019 [cited 2019 May 27]. Available from: http://purl.org/utwente/doi/10.3990/1.9789036547123
10. Wilson C, Grubler A, Gallagher KS, Nemet GF. Marginalization of end-use technologies in energy innovation for climate protection. *Nat Clim Chang.* 2012;2(11):780–8. Available from: http://www.nature.com/nclimate/journal/v2/n11/abs/nclimate1576.html
11. Spaargaren G. Theories of practices: Agency, technology, and culture: Exploring the relevance of practice theories for the governance of sustainable consumption practices in the new world-order. *Glob Environ Chang.* 2011 Aug 1 [cited 2019 Jun 12];21(3):813–22. Available from: https://www.sciencedirect.com/science/article/abs/pii/S0959378011000379?via%3Dihub
12. Tribe LH, Schelling CS, Voss J. *When Values Conflict: Essays on Environmental Analysis, Disco urse, and Decision*. Cambridge, MA: Ballinger Pub. Co; 1976 [cited 2019 Jun 12]. 178 p. Available from: https://www.tib.eu/en/search/id/TIBKAT%3A022211993/When-values-conflict-essays-on-environmental-analysis/
13. Creutzig F, Fernandez B, Haberl H, Khosla R, Mulugetta Y, Seto KC. Beyond Technology: Demand-Side Solutions for Climate Change Mitigation. 2016 [cited 2019 May 27]. Available from: www.annualreviews.org
14. Creutzig F, Jochem P, Edelenbosch OY, Mattauch L, van Vuuren DP, McCollum D, et al. Transport: A roadblock to climate change mitigation? *Science*. 2015;350(6263):911–2. Available from: http://www.sciencemag.org/content/350/6263/911
15. Creutzig F. Evolving narratives of low-carbon futures in transportation. *Transport Rev.* 2015 [cited 2019 Jun 3];36:341–60. Available from: http://dx.doi.org/10.1080/01441647.2015.1079277
16. Nykvist B, Whitmarsh L. A multi-level analysis of sustainable mobility transitions: Niche development in the UK and Sweden. *Technol Forecast Soc Chang.* 2008 [cited 2019 Jun 12];75:1373–87. Available from: http://www.matisse-project.net

17. Creutzig F, Baiocchi G, Bierkandt R, Pichler P-P, Seto KC. Global typology of urban energy use and potentials for an urbanization mitigation wedge. *Proc Natl Acad Sci.* 2015;112(20):6283–8. Available from: http://www.pnas.org/content/112/20/6283
18. Ewing R, Bartholomew K, Winkelman S, Walters J, Chen D. Growing cooler: the evidence on urban development and climate change. *Renew Resour J.* 2009;25(4):6–13.
19. Echenique MH, Hargreaves AJ, Mitchell G, Namdeo A. Growing cities sustainably. *J Am Plan Assoc.* 2012 Apr [cited 2019 Jun 19];78(2):121–37. Available from: http://www.tandfonline.com/doi/abs/10.1080/01944363.2012.666731
20. Salon D, Boarnet MG, Handy S, Spears S, Tal G. How do local actions affect VMT? A critical review of the empirical evidence. *Transp Res Part D Transp Environ.* 2012 Oct 1 [cited 2019 Jun 19];17(7):495–508. Available from: https://www.sciencedirect.com/science/article/pii/S136192091200051X
21. Cairns S, Sloman L, Newson C, Anable J, Kirkbride A, Goodwin P. Smarter choices: Assessing the potential to achieve traffic reduction using 'soft measures.' *Transp Rev.* 2008 Sep [cited 2019 Jun 19];28(5):593–618. Available from: http://www.tandfonline.com/doi/abs/10.1080/01441640801892504
22. UNWTO (UNWorld Tour.Organ.), UNEP (UNEnviron. Progr.). Climate Change and Tourism: Responding to Global Challenges. 2008 [cited 2019 Jun 19]. Available from: www.unep.fr
23. Fujino J, Hibino G, Ehara T, Matsuoka Y, Masui T, Kainuma M. Back-casting analysis for 70% emission reduction in Japan by 2050. *Clim Policy.* 2008 Jan [cited 2019 Jun 19];8(supl):S108–24. Available from: http://www.tandfonline.com/doi/abs/10.3763/cpol.2007.0491
24. Skea J, Ekins P, Winskel M. Energy 2050: Making the transition to a secure low carbon energy system. *Earthscan.* 2011 [cited 2019 Jun 19]. 381 p. Available from: https://www.ingentaconnect.com/content/rout/324pkr/2011/00000001/00000001/art00010;jsessionid=6olit2pe2ffh0.x-ic-live-02
25. Wei Y-M, Liu L-C, Fan Y, Wu G. The impact of lifestyle on energy use and CO_2 emission: An empirical analysis of China's residents. *Energy Policy.* 2007 Jan 1 [cited 2019 Jun 19];35(1):247–57. Available from: https://www.sciencedirect.com/science/article/pii/S0301421505003022
26. Shukla PR, Dhar S, Mahapatra D. Low-carbon society scenarios for India. *Clim Policy.* 2008 Jan [cited 2019 Jun 19];8(supl):S156–76. Available from: http://www.tandfonline.com/doi/abs/10.3763/cpol.2007.0498
27. Jaboyedoff P, Roulet C-A, Dorer V, Weber A, Pfeiffer A. Energy in air-handling units—results of the AIRLESS European Project. *Energy Build.* 2004 Apr 1 [cited 2019 Jun 19];36(4):391–9. Available from: https://www.sciencedirect.com/science/article/pii/S0378778804000477
28. Lin Z, Deng S. A study on the characteristics of nighttime bedroom cooling load in tropics and subtropics. *Build Environ.* 2004 Sep 1 [cited 2019 Jun 20];39(9):1101–14. Available from: https://www.sciencedirect.com/science/article/pii/S0360132304000472
29. Goswami DY, Kreith F. *Energy Efficiency and Renewable Energy Handbook.*
30. Smith P, Haberl H, Popp A, Erb K, Lauk C, Harper R, et al. How much land-based greenhouse gas mitigation can be achieved without compromising food security and environmental goals? *Glob Chang Biol.* 2013 Aug 1 [cited 2019 Jun 20];19(8):2285–302. Available from: http://doi.wiley.com/10.1111/gcb.12160
31. Edenhofer O, Pichs-Madruga R, Sokona Y, Farahani E, Kadner S, Seyboth K, et al., editors. *IPCC, 2014: Climate Change 2014: Mitigation of Climate Change. Contribution of Working Group III to the Fifth Assessment Report of the Intergovernmental Panel on Climate Change.* Cambridge University Press.
32. Stehfest E, Bouwman L, van Vuuren DP, den Elzen MGJ, Eickhout B, Kabat P. Climate benefits of changing diet. *Clim Change.* 2009 Jul 4 [cited 2019 May 27];95(1–2):83–102. Available from: http://link.springer.com/10.1007/s10584-008-9534-6
33. Vandenbergh MP, Nielsen KS. From myths to action. *Nat Clim Chang.* 2019 Jan 17 [cited 2019 Jun 12];9(1):8–9. Available from: http://www.nature.com/articles/s41558-018-0357-9
34. Hendrickson CY. Global carbon pricing: The path to climate cooperation. *Glob Environ Polit.* 2018 Jun 28;18(3):159–61. Available from: https://doi.org/10.1162/glep_r_00475

35. Shove E, Watson M, Spurling N. Conceptualizing connections. *Eur J Soc Theory.* 2015 Aug 22 [cited 2019 Jun 12];18(3):274–87. Available from: http://journals.sagepub.com/doi/10.1177/1368431015579964
36. Crate SA, Nuttall M. *Anthropology and Climate Change: From Actions to Transformations.* Routledge; 2019 [cited 2019 Jun 12]. 450 p. Available from: https://www.routledge.com/Anthropology-and-Climate-Change-From-Actions-to-Transformations-2nd-Edition/Crate-Nuttall/p/book/9781629580012
37. Nemet GF. *How Solar Energy Became Cheap: A Model for Low-carbon Innovation.* Routledge; 2019 [cited 2019 Jun 12]. Available from: https://www.routledge.com/How-Solar-Energy-Became-Cheap-A-Model-for-Low-Carbon-Innovation/Nemet/p/book/9780367136598
38. Creutzig F, Roy J, Lamb WF, Azevedo IML, Bruine De Bruin W, Dalkmann H, et al. Towards demand-side solutions for mitigating climate change. *Nat Clim Change.* 2018 [cited 2019 May 27];8:260–3. Available from: www.nature.com/natureclimatechange
39. Steinberger JK, Roberts JT, Peters GP, Baiocchi G. Pathways of human development and carbon emissions embodied in trade. *Nat Clim Change.* 2012 Feb 22 [cited 2019 Jun 12];2(2):81–5. Available from: http://www.nature.com/articles/nclimate1371
40. Niamir L, Filatova T, Voinov A, Bressers H. Transition to low-carbon economy: Assessing cumulative impacts of individual behavioral changes. *Energy Policy.* 2018;118:325–45.
41. Riahi K, van Vuuren DP, Kriegler E, Edmonds J, O'Neill BC, Fujimori S, et al. The Shared Socioeconomic Pathways and their energy, land use, and greenhouse gas emissions implications: An overview. *Glob Environ Change.* 2017 Jan 1 [cited 2019 Jun 12];42:153–68. Available from: https://www.sciencedirect.com/science/article/pii/S0959378016300681
42. Kriegler E, Weyant JP, Blanford GJ, Krey V, Clarke L, Edmonds J, et al. The role of technology for achieving climate policy objectives: overview of the EMF 27 study on global technology and climate policy strategies. *Clim Change.* 2014 Apr 28 [cited 2019 Jun 12];123(3–4):353–67. Available from: http://link.springer.com/10.1007/s10584-013-0953-7
43. O'Neill BC, Kriegler E, Riahi K, Ebi KL, Hallegatte S, Carter TR, et al. A new scenario framework for climate change research: the concept of shared socioeconomic pathways. *Clim Change* [Internet]. 2014 Feb 15 [cited 2019 Jun 12];122(3):387–400. Available from: http://link.springer.com/10.1007/s10584-013-0905-2
44. Babatunde KA, Begum RA, Said FF. Application of computable general equilibrium (CGE) to climate change mitigation policy: A systematic review. *Renew Sustain Energy Rev.* 2017 Oct 1 [cited 2019 Jun 12];78:61–71. Available from: https://www.sciencedirect.com/science/article/pii/S1364032117305713
45. Kancs A. Predicting European enlargement impacts: A framework of interregional general equilibrium. *East Europ Econ.* 2001 Sep 8 [cited 2019 Jun 5];39(5):31–63. Available from: https://www.tandfonline.com/doi/full/10.1080/00128775.2001.11041001
46. Siagian U, Yuwono B, Fujimori S, Masui T, Siagian UWR, Yuwono BB, et al. Low-carbon energy development in indonesia in alignment with intended nationally determined contribution (INDC) by 2030. *Energies.* 2017 Jan 5 [cited 2019 Jun 5];10(1):52. Available from: http://www.mdpi.com/1996-1073/10/1/52
47. Hunt LC, Evans J. *International Handbook on the Economics of Energy.* Edward Elgar; 2009. 831 p.
48. Bhattacharyya SC. *Energy Economics: Concepts, Issues, Markets and Governance.* Springer; 2011. 721 p.
49. Niamir L, Filatova T. Linking agent-based energy market with computable general equilibrium model: An integrated approach to climate-economy-energy system. 2015 [cited 2019 Jun 3]; Available from: https://ris.utwente.nl/ws/portalfiles/portal/17707456/linking.pdf
50. Filatova T, Niamir L. Changing climate – changing behavior: empirical agent based computational models for climate change economics. In: Lamperti F, Monasterolo I, Roventini A, editors. *Complexity and Climate Change: An Evolutionary Political Economy Approach.* Taylor & Francis; 2019.

51. Lebaron B, Tesfatsion LS, Lebaron B, Tesfatsion L. Modeling macroeconomies as open-ended dynamic systems of interacting agents. *The American Economic Review* 2008 Aug 19 [cited 2019 Jun 12];98:246–250. Available from: https://econpapers.repec.org/paper/isugenres/12973.htm
52. Niamir L, Kiesewetter G, Wagner F, Schöpp W, Filatova T, Voinov A, et al. Assessing the macroeconomic impacts of individual behavioral changes on carbon emissions. *Clim Change*. 2019.
53. Rai V, Robinson SA. Agent-based modeling of energy technology adoption: Empirical integration of social, behavioral, economic, and environmental factors. *Environ Model Softw*. 2015 Aug 1 [cited 2019 Jun 12];70:163–77. Available from: https://www.sciencedirect.com/science/article/pii/S1364815215001231?via%3Dihub
54. Rai V, Henry AD. Agent-based modelling of consumer energy choices. *Nat Clim Chang*. 2016 Jun 9 [cited 2019 Jun 12];6(6):556–62. Available from: http://www.nature.com/articles/nclimate2967
55. Niamir L, Filatova T. Transition to low-carbon economy: Simulating nonlinearities in the electricity market, Navarre region-Spain. Vol. 528, Advances in Intelligent Systems and Computing. 2017.
56. Abrahamse W, Steg L. Factors related to household energy use and intention to reduce it: The role of psychological and socio-demographic variables. *Human Ecol Rev*. 2011 [cited 2019 Jun 12];18:30–40. Available from: https://www.jstor.org/stable/24707684
57. Bamberg S, Rees J, Seebauer S. Collective climate action: Determinants of participation intention in community-based pro-environmental initiatives. *J Environ Psychol*. 2015 Sep 1 [cited 2019 Jun 12];43:155–65. Available from: https://www.sciencedirect.com/science/article/abs/pii/S0272494415300190?via%3Dihub
58. *Bamberg* S, Hunecke M, Blöbaum A. Social context, personal norms and the use of public transportation: Two field studies. *J Environ Psychol*. 2007 Sep [cited 2019 Jun 12];27(3):190–203. Available from: https://linkinghub.elsevier.com/retrieve/pii/S0272494407000357
59. *Steg* L. Values, norms, and intrinsic motivation to act proenvironmentally. *Annu Rev Environ Resour*. 2016 Nov 20 [cited 2019 Jun 12];41(1):277–92. Available from: http://www.annualreviews.org/doi/10.1146/annurev-environ-110615-085947
60. Steg L, Vlek C. Encouraging pro-environmental behaviour: An integrative review and research agenda. *J Environ Psychol*. 2009 Sep 1 [cited 2019 Jun 12];29(3):309–17. Available from: https://www.sciencedirect.com/science/article/abs/pii/S0272494408000959
61. Mills B, Schleich J. Residential energy-efficient technology adoption, energy conservation, knowledge, and attitudes: An analysis of European countries. *Energy Policy*. 2012 Oct 1 [cited 2019 Jun 12];49:616–28. Available from: https://www.sciencedirect.com/science/article/pii/S0301421512005897
62. Matsui K, Ochiai H, Yamagata Y. Feedback on electricity usage for home energy management: A social experiment in a local village of cold region. *Appl Energy*. 2014 May 1 [cited 2019 Jun 12];120:159–68. Available from: https://www.sciencedirect.com/science/article/pii/S0306261914000683
63. Long C, Mills BF, Schleich J. Characteristics or culture? Determinants of household energy use behavior in Germany and the USA. *Energy Effic*. 2018 Mar 8 [cited 2019 Jun 12];11(3):777–98. Available from: http://link.springer.com/10.1007/s12053-017-9596-2
64. Kobus CBA, Klaassen EAM, Mugge R, Schoormans JPL. A real-life assessment on the effect of smart appliances for shifting households' electricity demand. *Appl Energy*. 2015 Jun 1 [cited 2019 Jun 12];147:335–43. Available from: https://www.sciencedirect.com/science/article/pii/S0306261915001099
65. Donaldson SI, Grant-Vallone EJ. Understanding self-report bias in organizational behavior research. *J Bus Psychol*. 2002 [cited 2019 Jun 12];17(2):245–60. Available from: http://link.springer.com/10.1023/A:1019637632584
66. Ajzen I, Fishbein M. *Understanding Attitudes and Predicting Social Behavior*. Prentice-Hall; 1980. 278 p.

67. Abrahamse W, Steg L. How do socio-demographic and psychological factors relate to households' direct and indirect energy use and savings? *J Econ Psychol.* 2009 Oct 1 [cited 2019 Jun 12];30(5):711–20. Available from: https://www.sciencedirect.com/science/article/abs/pii/S0167487009000579
68. Niamir L, Ivanova O, Filatova T, Voinov A, Bressers H. Demand-side solutions for climate change mitigation: bottom-up drivers of household energy behavior changestle. *Energy Res Soc Sci.* 2019.
69. Mattauch L, Ridgway M, Creutzig F. Happy or liberal? Making sense of behavior in transport policy design. *Transp Res Part D Transp Environ.* 2016 Jun 1 [cited 2019 Jun 12];45:64–83. Available from: https://www.sciencedirect.com/science/article/pii/S136192091500111X
70. van den Bijgaart I. Too slow a change? Deep habits, consumption shifts and transitory tax policy. 2018 [cited 2019 Jun 12]. CESifo Working Paper Series No. 6958. Available from: https://papers.ssrn.com/sol3/papers.cfm?abstract_id=3185965
71. Mattauch L, Hepburn C, Stern N. Pigou pushes preferences: Decarbonisation and endogenous values. 2018 [cited 2019 Jun 12]. Available from: https://papers.ssrn.com/sol3/papers.cfm?abstract_id=3338758
72. Bressers H, Ligteringen J. Political-administrative policies for sustainable household behavior. 2001 [cited 2019 Jun 12]. Available from: https://research.utwente.nl/en/publications/political-administrative-policies-for-sustainable-household-behav

19

Wind Farms: Noise

Daniel Shepherd,
Chris Hanning,
and Bob Thorne

Introduction

Planning authorities, environmental agencies, and policy makers in many parts of the world are seeking information on possible links between wind turbine noise and health in order to legislate permissible noise levels or setback distances. Concurrently, larger and noisier wind turbines are emerging, and consent is being sought for progressively larger windfarms to be placed even closer to human habitats. While noise standards can effectively and fairly facilitate decision-making processes if developed properly, the current standards on offer suffer severe conceptual difficulties. Specifically, noise metrics considered by many in the industry as best practice may in fact relate little to health outcome variables such as annoyance or sleep disruption. In this entry, we describe the physical characteristics of wind turbine noise, review the impact of such noise on humans, and critique current approaches to mitigation.

Industrial Wind Turbines

Industrial wind turbines transform kinetic energy from the wind into electricity, a practice dating back over 100 years. Structurally, wind turbines can be decomposed into three key components (Figure 1). First, wind turbines possess a rotor, consisting of one or more blades designed to rotate when exposed to wind. The rotor can be thought of as a type of sail, catching wind in order to induce movement. Depending on the axis of blade rotation, wind turbines can be categorized as either horizontal-axis (the most common) or vertical-axis turbines. The second major component is the generator or "dynamo." The generator component includes a gearbox to regulate the speed of the dynamo and components to change blade pitch and plane of rotation with respect to wind direction. The dynamo can be used as a motor to maintain rotation at very low wind speeds. Third, there is a tower supporting the rotor and, typically, the generator. The size of a wind turbine can be specified either as a dimension (e.g., tower

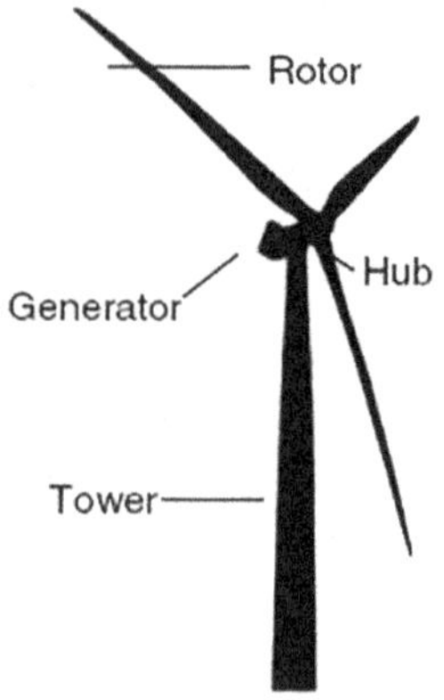

FIGURE 1 Components of a typical horizontal-axis wind turbine.

height measured from the ground to the top of a blade at its highest point) or as an electrical output (e.g., watts). Currently, turbines range from approximately 2 to 200 m high and from approximately 50 W to 6 MW.

Wind turbines can be erected in isolation or in sets and be located either onshore (i.e., terrestrial) or offshore (i.e., marine), though the latter is associated with higher construction costs. Industrial-scale wind energy generation, involving the saturation of an optimum number of wind turbines in a fixed area of land, gives rise to the concept of the "windfarm" or "wind park." Wind energy developers seek areas that have good consistent wind flow and close access to energy grids. The proliferation in the number of windfarms established globally in the past decade has been largely driven by environmental concerns such as climate change, renewableness and sustainability, and strategic energy considerations relating to the depletion of fossil fuels.[1] However, in the absence of large-scale electricity storage devices (i.e., batteries), the contribution of wind energy to a nation's electricity needs is likely to be peripheral. Another barrier is social acceptance, with reviewed social surveys indicating citizens supporting renewable energy in principle but opposed to having windfarms in their immediate vicinity due to visual impacts on the landscape, shadow flicker from the blades, and fears of noise-induced annoyance and sleep disruption.

Acoustic Profile of Wind Turbine Noise

The sound generated from a windfarm is qualitatively different from any sound source commonly met in the environment, can rapidly switch from being stationary to nonstationary, and can vary by as much as 20 dB within a single minute. When it interferes with human activities, wind turbine sound becomes a type of noise. Analysis of windfarm noise poses distinct challenges, including the identification of acoustic energy that can be directly attributed to the turbines and the detection of special audible characteristics, including distinct tonal complexes and modulation effects. Windfarm noise is often a broadband low-amplitude noise constantly shifting in character ("waves on beach," "rumble-thump," "plane never landing." etc.). In this respect, windfarm noise is not like, for example, traffic noise or the continuous hum from plant and machinery. When assessed, wind turbine noise is often related to either wind speed (m/s) or electrical output (watts) and typically increases with both.

When the wind reaches a blade, it flows both over and under the blade. The part of the airflow with momentum great enough to break away forms trailing vortexes and turbulence behind the blade, producing a set of sound sources. The power of each sound source depends on the strength of the turbulence, which in turn depends on the speed of airflow; the compressibility and viscosity of the air; the design and surface texture (roughness) of the blade; the wind speed; and the velocity of the blade at that point. The faster the blade rotates, the earlier the breakup in the boundary vortexes and the greater the interaction between the vortexes emanating by adjacent wind turbines. An amplification of potential noise occurs when two or more turbines are, or nearly are, synchronous, such that the blade passing pulses coincide and then go out of phase again.[2] With exact synchronicity, there is a fixed interference pattern; with near synchronicity, concurrent arrival of pulses will change over time and place.

Noise emissions from modern wind turbines are primarily due to turbulent flow and trailing edge sound, blade characteristics, blade/tower interaction, and to a lesser degree, mechanical processes. The most commonly used description of wind turbine noise is the A-weighted sound pressure level, which is expressed in decibels (notated dBA). The most commonly used noise compliance assessment methods for windfarms involve the "time average" sound level $L_{A_{eq}}$ or the background sound level L_{A90}. These levels are quite different as the time-averaged ambient sound level includes all noises from near and far. The difference between these levels, and other common levels, is illustrated in Figure 2. The chart shows that sound levels change over time and that any derived sound level index is a summary of fluctuating levels in that time period. In a relatively short time period, such as 10 minutes, the unique noise events such as bangs or thuds from turbines shifting in the wind may be captured. If the time period is relatively long, for example, an hour, then evidence of unique short-term noise events is reduced because the sound energy is "averaged" over the whole hour, and the single-value A-weighted level will not represent short-term variations in sound character. If extraneous noise (e.g., insect noise) is included in the wind turbine measurement, its contribution to the overall level must be determined, though how this is undertaken remains a challenge.[3]

The A-frequency-weighted sound pressure level or "sound level" is the most common sound descriptor and is reputedly analogous to our hearing at medium sound levels. This is not strictly true, and the A-weighting has a significant restriction in that it does not permit measurement or assessment of low-frequency sound (i.e., 20 to 250 Hz). For more complex situations where dominant tonal components are significant (i.e., special audible characteristics), a procedure for determining tonal adjustment requiring one-third octave band frequency or narrow-band analysis is needed. These assessment procedures require the "C" weighting for low frequency or the unweighted (also known as "Z") response to measure both low-frequency and infrasonic sound. Whereas the dBC metric is able to include low-frequency sounds such as the audible rumble and thump from wind turbines, the dBZ response is more

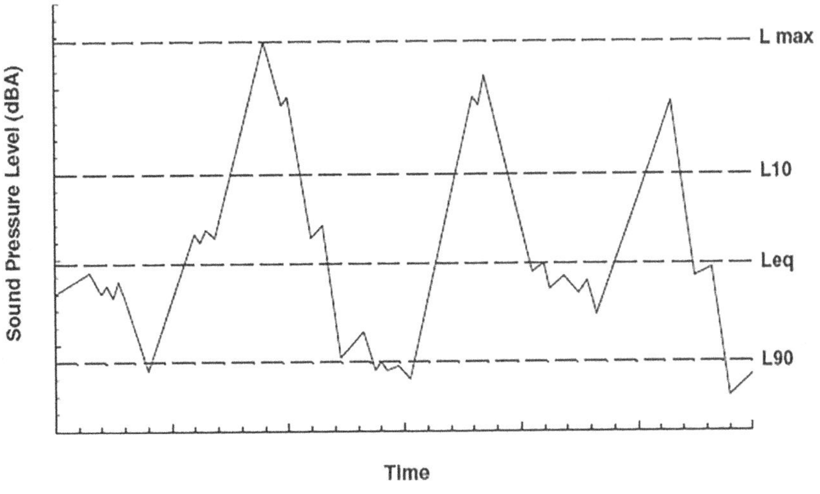

FIGURE 2 Chart illustrating different noise descriptors. L10 is the level exceeded 10% of the time, while L90 is the level exceeded 90% of the time. The time-average (equivalent continuous) sound pressure level, Leq, represents the average acoustic energy across a defined measurement epoch.

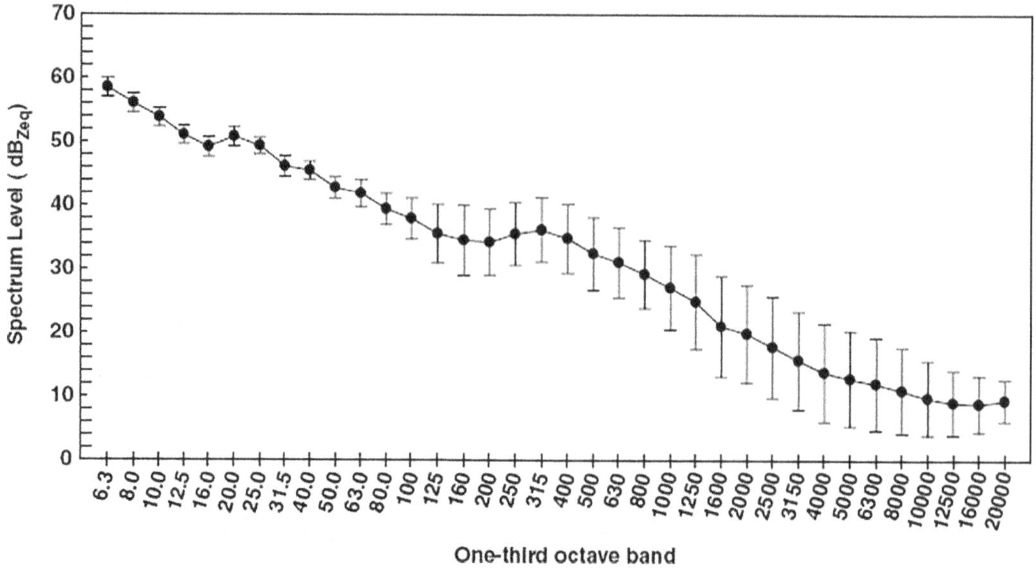

FIGURE 3 One-third octave band analysis of time-average unweighted sound pressure level (dB_{Zeq}) for wind turbine sound measured from 7:00 p.m. to 1:00 a.m. outside of a residence.

suitable for infrasound measurements (i.e., typically inaudible energy below 20 Hz). Figure 3 presents a third octave band analysis of outdoor wind turbine noise recorded over a 6-hr period. Other measures include assessments for tonality or low- frequency sound referenced to third octave bands and the "G" weighting for infrasound. Aside from physical measures of amplitude (e.g., dBA), wind turbine noise can be quantified with a variety of other acoustical and objective psychoacoustic measures, including amplitude modulation (for example, 100 msec samples of peak, time-average, or fast response), sound quality (including audibility, dissonance, roughness, fluctuation strength, sharpness, tonality), loudness (for steady, time-varying, and impulsive sounds), and unbiased annoyance.[4]

Certification of wind turbine noise is undertaken in accordance with the International Standard IEC 61400-11:2002.[5] Emission levels are to be reported as A- weighted time-averaged (L_{Aeq}) sound levels in one-third octave bands. Audibility is calculated by reference to tones. An informative entry in IEC 61400-11 states the following: "In addition to those characteristics of wind turbine noise described in the main text, this emission may also possess some, or all of the following: infrasound; low-frequency noise; impulsivity; low-frequency modulation of broad band or tonal noise; other, such as a whine, hiss, screech, or hum, etc., distinct pulses in the noise, such as bangs, clatters, clicks or thumps, etc." Unfortunately, many of these parameters are not reported by the turbine manufacturer and cannot be predicted with the simple calculation methods currently available. The prediction of windfarm sound levels is most often referenced to national or international standards that have been based on ISO 9613-2.[6] The propagation method is calculated with the receivers being downwind from the noise source(s). All prediction models have uncertainty to their accuracy of prediction. Table 5 of the ISO 9613-2 standard gives an estimated accuracy for broadband noise of ±3 dB at between 100 and 1000 m. This is due to the inherent nature of the calculation algorithms that go into the design of the model, the assumptions made in the implementation of the model, and the availability of good source sound power data. The ISO 9613-2 method holds for wind speeds of between approximately 1 and 5 m/s, measured at a height of 3 to 11 m above the ground. However, wind turbines are sound sources that operate at higher wind speeds than allowed for under the standard, and an accuracy of ±7 dB can be expected. [3] Ultimately, the received noise levels at residences will vary subject to varying meteorological conditions in the locality (e.g., wind speed and direction, wind shear, temperature, humidity, inversions),

TABLE 1 Factors Affecting the Prediction of Wind Farm Noise Levels at a Receiver[a]

• The true sound power level of the turbine(s) at the specified wind speed
• The reduction in sound level due to ground effects
• The increase or reduction in sound level due to atmospheric (meteorological) variations and wind direction
• The variation due to modulation effects from wind velocity gradient
• Increase and reduction in sound levels due to wake and turbulence modulation effects due to turbine placement and wind direction
• Increased sound levels due to synchronicity effects of turbines in phase due to turbine placement and wind direction
• Building resonance effects for residents inside a dwelling

[a] A conservative set of noise predictions should take all factors into account.

among other factors (see Table 1), all of which must be accounted for when measuring or modeling wind turbine noise levels.

Human Impacts of Wind Turbine Noise

A Psychological Description of Wind Turbine Noise

At the psychological level of description, wind turbine noise is most frequently characterized as a swishing or lashing sound or less commonly as thump/throb, low- frequency rumble, or a rustling sound.[7,8] Wind turbines produce noise with an impulsive character[9] and while the actual cause of the swishing or thumping has not yet been fully elucidated, it has been demonstrated that the swishing or thumping pattern is common with larger turbines[10] and may result from a fluctuating angle of attack between the trailing edge of the rotor blade and wind, or wind speed inequalities across the area being swept by the rotor blades.[11] It is thought that the swishing sound may be linked to activity in the 2000 to 4000 Hz band, with the pace of the rotor blades determining the degree of amplitude modulation.[12] Unfortunately, such amplitude-modulated sounds are generally attenuated poorly by background noise, especially so in rural areas.[13] Further, because human sensory systems behave as contrast analyzers, fluctuations in the incoming stimulus field tend to direct attention and so are more easily detected. Thus, amplitude-modulated sounds such as wind turbine noise are readily perceived and difficult to filter out, making them especially intrusive.[14] The loudness of a wind turbine depends on a number of factors, including wind speed, sound-attenuating materials between the turbines and the receiver, other masking sounds, the season, and time of day. The loudness of a modern 2 to 3 MW wind turbine can be compared to a car on a motorway, autobahn, or freeway,[15] with a sound power level of 94 to 104 dBA at a windspeed of 8 m/s.[16] Wind turbine noise is perceived louder at night and during the summer months and when the wind is blowing from the direction of the turbines toward the receiver.[7,8]

Quantifying the Health Impacts of Wind Turbine Noise

Elucidating a causal mechanism between an environmental event and health is a complicated undertaking, and noise effects are commonly "indirect" as opposed to "direct." According to the biomedical model of health (Figure 4a), a direct health effect implies a direct pathological relationship between an environmental parameter (e.g., noise level) and a target organ. An alternative approach (Figure 4b) distinguishes between direct health effects and psychosomatic illness, the latter indicting that any physiological illness coinciding with the onset of wind turbine noise is caused by a negative psychological response to the noise and not the noise *per se*. Thus, anxiety or anger in the presence of wind turbine noise induces stress and strain that, if maintained, can eventually lead to adverse health effects. A counter argument to this approach is that some individuals are simply more susceptible to noise

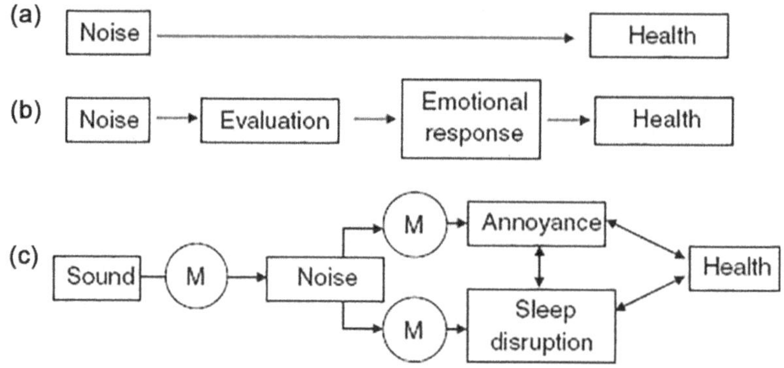

FIGURE 4 Three models representing the relationship between noise and health: the biomedical model (a) stipulating a direct causal relationship and indirect models (b and c) containing moderators and mediators.

than other individuals, which fits with the general concept of biological and physical variation. In the field of epidemiology, the differential susceptibilities of individuals are known as risk factors or vulnerabilities, with noise sensitivity being one risk factor related to negative responses to intrusive noise. A second challenge to the psychosomatic approach comes from documented instances of individuals who initially welcomed wind turbines into the community but who later campaigned to have them removed due to undesirable noise exposure.[17] Lastly, the veracity of psychosomatic arguments lessens in the face of feasible biological mechanisms describing the relationship between health and noise.[18]

An alternative and more accepted approach would be to adopt the World Health Organization's (WHO's) definition of health:[19] "A state of complete physical, mental and social well-being and not merely the absence of disease or infirmity." The forerunner of the biopsychosocial model, the WHO's definition states that optimal human functioning is determined by the interplay of biological, environmental, psychological, and social factors. Figure 4c displays a model consistent with the WHO's approach, in which the impact of noise is moderated by environmental, psychological, and social factors. A context-relevant model proposed by van den Berg and colleagues,[8] based on previous wind turbine literature, takes a similar shape to that presented in Figure 4c. They dichotomize moderators (denoted "M" in Figure 4c) into environmental moderators (e.g., degree of urbanization, house type, and ambient sound level) or psychological and demographic moderators (e.g., age, gender, education, employment status, attitudes to wind energy, noise sensitivity, and whether the individual receives a monetary return from the turbines). Other models linking wind turbine sound and health have been proposed[20] but can be considered extensions of that presented in Figure 4c.

As a new source of noise, the impact of wind turbine noise is understandably understudied relative to aviation and road traffic noise. Consequently, little data exist with which to assess the impacts of wind turbine noise on health, a state of affairs compounded by rapid development of wind turbine technology, in which data collected for smaller and less powerful turbines are not generalizable to larger, more modern turbines.[9,21] As of 2011, there have been two approaches to collecting wind turbine noise impact data, either epidemiological studies relying on masked surveys or direct clinical case studies.[22] Both approaches typically focus on the emotional impacts of noise (i.e., annoyance), upon sleep disruption, and/or the degradation of well-being and increases in stress that arise from sleep disturbance and annoyance. Irrespective of approach, however, case studies,[23–25] and epidemiological studies[7,8,20] have provided evidence that, like road traffic and aviation noise, wind turbine noise can be associated with negative health outcomes.

Wind Turbine Noise and Annoyance

People generally respond more negatively to man-made noise than to natural sounds,[26] and this generalization holds true for wind turbine noise.[16] From a psychological perspective, chronic exposure to community noise can impact health through information overload, overarousal, loss of coping strategies, loss of privacy, and loss of perceived control. These mechanisms give rise to a number of subjective responses to noise, of which the most common is annoyance. As a psychological stressor,[27] noise annoyance can express itself through malaise, fear, threat, uncertainty, restricted liberty, excitability, or defenseless- ness.[28] Furthermore, annoyance may be accompanied by intense anger, especially if one believes that they are being harmed unnecessarily. Thus, the term "annoyance" is often misinterpreted by the layperson as a feeling brought about by the presence of a minor irritant. The medical usage, in contrast, exists as a precise technical term and defines annoyance as a mental state capable of degrading health and well-being,[29,30] and it is classified as an adverse health effect by the WHO.[31]

There have been few studies estimating the health impacts of windfarms, with a series of studies undertaken in Scandinavia contributing the most to current knowledge. A seminal Swedish study undertaken by Pedersen and Persson Waye[7] sought to document the prevalence of wind turbine–induced annoyance and, further, to generate dose–response relationships between the two. Respondents were located between 150 and 1200 meters from the nearest wind turbine and were classified into noise exposure categories (see Figure 5). A significant relationship between dose (dBA) and annoyance was reported, but the variability in annoyance scores explained by noise level was small (adjusted $R^2 = 0.13$). Those reporting annoyance indicated a daily or nearly-everyday intrusion of windfarm noise. Those describing the noise as "swishing" were more likely to report noise annoyance, a finding replicated in a subsequent study reporting a high correlation ($r = 0.664$) between the swishing sound and annoyance.[14] Among those who noticed the noise, 11.2% reported being annoyed when indoors. A small but significant correlation was found between noise annoyance and noise sensitivity, with approximately 50% of the rural-dwelling respondents describing themselves as noise sensitive. Those making negative appraisals of the wind turbines, for example, as visually incongruent with the landscape, were at higher risk of an annoyance response. On the basis of their data, the authors undertook follow-up studies[14–16,22] supporting their conclusion that wind turbine noise maybe more potent than other categories of environmental noise (e.g., road or aviation) and appealed for further studies to determine why this might be. In a later report, Pedersen[22] suggests that coping strategy may moderate the relationship between wind turbine noise and stress.

Van den Berg et al.[8] analyzed data from 725 Dutch nationals residing within 2.1 km of a wind turbine and who were exposed to calculated outdoor noise levels between 24 and 54 dB(A). Approximately 60% of the sample could hear the turbines outdoors, while 33% reported that they could hear the wind turbines indoors. Of the 45% ($n = 231$) who noticed the sound of the rotor blades, 24.7% were not annoyed, 25.8% were slightly annoyed, 19.5% were rather annoyed, and 29.9% were very annoyed. The sound level explained approximately 25% of the variability in annoyance scores, and those who compared the noise to an amplitude modulation (i.e., swishing or lashing) were more likely to be annoyed, though this is not a novel finding.[14,32,33] Figure 5 plots the data from van den Berg et al., presenting proportions of detection and elicited annoyance as a function of noise level, for their entire dataset (Figure 5, circles) and for those receiving no economic benefit (Figure 5, squares). Note that, for those receiving no economic benefit, a monotonic relationship is evident, while a nonmonotonic function occurs when individuals benefiting financially from the turbines are included. Van den Berg[8] reports that this depreciation in annoyance of those benefiting economically can be explained by the control they have over the wind turbines, such that they can impede their operation if noise levels increase. Finally, it was reported that annoyance was positively correlated with stress scores, though a causal relationship could not be inferred.

It is accepted that both the physical parameters of the noise and the psychological characteristics of the listener combine to produce noise annoyance.[34] On the physical side, the relatively high annoyance

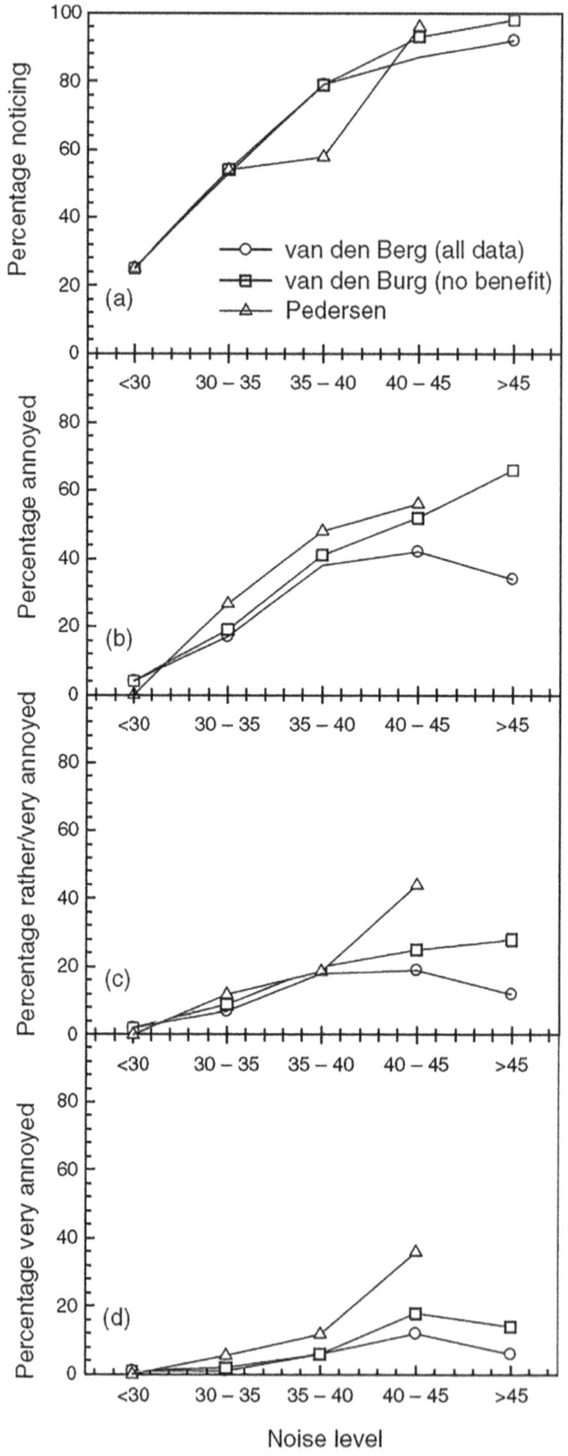

FIGURE 5 Perception of wind turbine noise as a function of noise level for three sets of data: Tables 7.25 (complete data set) and 7.26 (no economic benefit of turbines) from van den Berg et al.[8] and Pedersen and Persson Waye's[7] Table V. Plot (a) is percentage noticing the noise, while plots (b) to (d) are for annoyance. Plot (b) includes data from plots (c) and (d), and plot (c) includes data from plot (d).

levels elicited by wind turbine noises (e.g., swishing or thumping) may be explained by the increased fluctuation of the sound, up to 4 to 6 dB for a single turbine operating in a stable atmosphere.[11] Individuals are also highly sensitive to changes in frequency modulation variations of approximately 4 Hz or greater.[4] Noting that amplitude-modulated sound is known to be more annoying than unmodulated sound, Lee et al.,[34] in a laboratory setting, demonstrated that amplitude-modulated wind turbine noise was consistently judged to be more annoying than its unmodulated counterpart. Thus, the dominant acoustic driver of annoyance is likely to be noise dynamics rather than noise level. Other physical parameters linked to annoyance include terrain complexity, with rural terrain associated with greater annoyance than urban areas, possibility due to more complicated terrain exhibiting various focusing or defocusing effects and greater ground reflection.

While there is a strong correlation between the sound pressure level (i.e., amplitude) of a sound wave and the perceived loudness of a sound, there is no one-to-one mapping between sound pressure level and the psychological responses that individuals have to a sound.[35] Many non-acoustical factors determine how annoyed one will become toward a source of noise.[36–38] Thus, the response of the individual to the sound is just as important as the parameters of the acoustic wave, and the "people" side of noise should not be omitted from acoustical reports. Table 2 lists, in no particular order, non-acoustical factors found to influence levels of noise annoyance.[39] In relation to windfarms, the personal factors listed in Table 2 have been found to strongly influence how exposed individuals perceive the noise.[16] In addition, perceptions of amenity, individuals seeking refuge from urban noise, or the lower ambient sound levels typical of the rural environment may explain why annoyance responses are higher in rural as opposed to urban settings.[13,16]

When considering wind turbine noise and annoyance data emerging from the literature, a number of risk factors are evident, including an effect of age and educational status but not gender.[8] Employment status was also linked to wind turbine noise–induced annoyance in one study, possibly due to impeded restoration,[16] but to date, there are no data meaningfully comparing ethnicity or national groups (but see Pedersen et al.[40]). The general public view wind turbines as necessary but ugly,[14] and it is possible that the visual impact of a windfarm can interact with noise level to cause moderate annoyance. This amplification of annoyance is possibly due to a violation of the landscape soundscape continuum constructed by those who choose to live in areas that later contain windfarms,[41] or alternatively, multisensory engagement may enhance detection and identification of wind turbine noise.[42] The degree of influence of the visual aspects of windfarms has yet to be determined, with laboratory studies suggesting that it is wind turbine noise and not the visual impact that underlies the annoyance response,[41] while epidemiological studies suggest that the visual effects are nontrivial. [40]

TABLE 2 Non-Acoustical Factors Influencing the Degree of Annoyance to Noise

- Perceived predictability of the noise level changing
- Perceived control, either by the individual or others
- Trust and recognition of those managing the noise source
- Voice, the extent to which concerns are listened to
- General attitudes, fear of accidents, and awareness of benefits
- Personal benefits, how one benefits from the noise source
- Compensation, how one is compensated due to noise exposure
- Noise sensitivity
- Home ownership, concern about plummeting house values
- Accessibility to information relating to the noise source

Source: Flindell and Stallen.[39]

Wind Turbine Noise and Sleep

The deleterious effects of noise on sleep and the consequences of sleep loss are well documented and are a major concern for governments.[43] In comparison with road, rail, and aircraft noise, there is little research on the effects of wind turbine noise on sleep. However, there is no doubt that wind turbine noise can and does disturb the sleep of those living nearby. Sleep disruption is the predominant symptom in the thousands of anecdotal cases reported in the press and on the Internet and is confirmed by more structured surveys.[25] The quantity, consistency, and ubiquity of complaints has been taken as *prima facie* epidemiological evidence of a causal link between wind turbine noise, sleep disruption, and ill health.[44]

Early investigations into wind turbine noise and sleep are difficult to interpret as researchers used imprecise outcome measures, generally relying on recalled sleep disturbances such as difficulty in initiating or returning to sleep, which tends to underestimate the magnitude of the noise impact and its consequences.[45] One of the earliest studies ($n = 128$) reported that approximately 16% of respondents living at calculated outdoor turbine noise exposures exceeding 35 dB L_{Aeq} stated that wind turbine noise disturbed their sleep.[7] A New Zealand study of 604 households within 3.5 km of a windfarm found that 42 reported occasional and 26 frequent sleep disturbance.[46] The largest wind turbine noise study to date, "Project WINDFARM- perception,"[8] concluded that turbine noise was more of an annoyance at night and that interrupted sleep and difficulty in returning to sleep increased with both indoor and outdoor calculated noise levels. Even at the lowest noise levels, 20% of 725 respondents reported disturbed sleep at least one night per month. In a meta-analysis[40,47] of three European datasets ($n = 1764$),[7,8,16] there was a clear increase in levels of sleep disturbance with dB L_{Aeq} in two of the three studies. In one study, an increment in self-report sleep disturbance occurred between 35 and 40 dBA, while in the other, it occurred between 40 and 45 dBA.

More recent research into wind turbine noise and sleep includes two studies reported by Nissenbaum, Aramini, and Hanning.[48] In the first, a pilot study, a structured questionnaire was administered to 22 subjects living 370 to 1100 m from 28 1.5 mW turbines and a control group ($n = 28$) living at least 4.5 km from the nearest turbine. The study group had clinically and statistically worse sleep disturbance, headache, vestibular symptoms, and psychiatric symptomatology. The second study, using validated questionnaires, administered the Pittsburgh Sleep Quality Index (PSQI), Epworth Sleepiness Score (ESS), and Short- form health survey (SF36) to 79 subjects living between 375 and 6600 m from two windfarms. Those living within 375–1400 m reported worse sleep, were sleepier, and had worse SF36 mental summary scores than those between 3 and 6.6 km from a turbine. Psychiatric symptom scores (irritability, stress, anger, hopelessness, and anxiety) were significantly greater, as was a composite mental health score. They were also more likely to report headaches, nausea (31.6% vs. 12.2%), and a willingness to move away. Modeled dose–response curves of both sleep and health scores against distance from nearest turbine (Figs. 6–8) were significantly related after controlling for gender, age, and household clustering. There was a sharp increase in effects between 1 and 2 km. This study is the first to use appropriate sleep outcome measures[45] and to use a control group. While the sample size is modest ($n = 78$), it is convincing evidence that wind turbine noise adversely affects sleep and health for those living within 1.5 km of turbines.

Mechanisms explaining the effects of wind turbine noise on sleep have been considered, but would benefit from further empirical support.[45] Noise of any description can interfere with sleep by preventing the onset of sleep either at sleep initiation or at the return to sleep after a spontaneous or induced awakening. The amplitude, character, and associations of the noise are all important as is the noise sensitivity of the individual and the psychological response to the noise. In this respect, wind turbine noise seems to be particularly annoying, possessing an impulsive nature with short bursts of low-frequency sound, making it audible 10–15 dBA below background level.[38,49] Nocturnal atmospheric stability ensures that wind turbine noise is maintained while ground level ambient noise diminishes. Indoor noise levels for most noise sources can be reduced by closing windows; however, the low-frequency

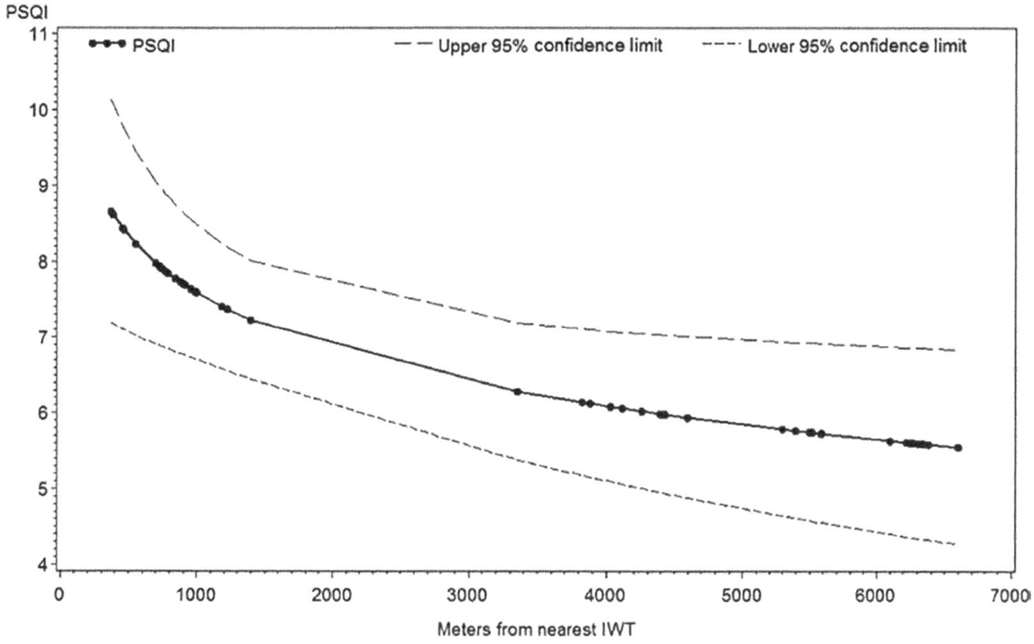

FIGURE 6 Mean Pittsburgh Sleep Quality Index (PSQI) scores as a function of setback distance. The dashed lines are 95% confidence intervals.
Source: Nissenbaum, Aramini, and Hanning.[48]

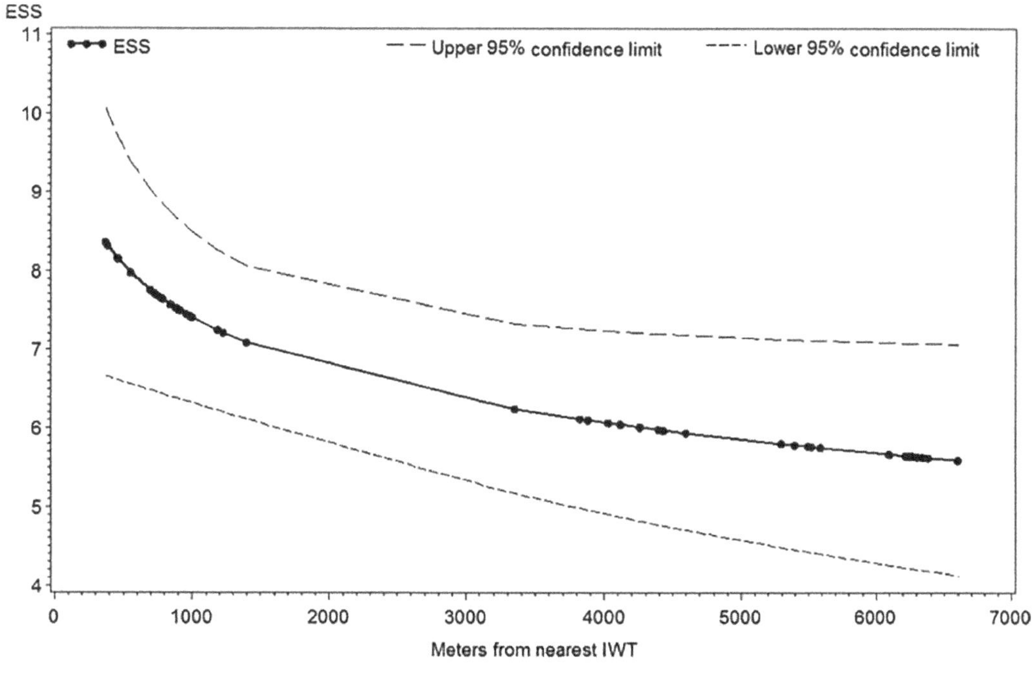

FIGURE 7 Mean Epworth Sleepiness Scale (ESS) scores as a function of setback distance. The dashed lines are 95% confidence intervals.
Source: Nissenbaum, Aramini, and Hanning.[48]

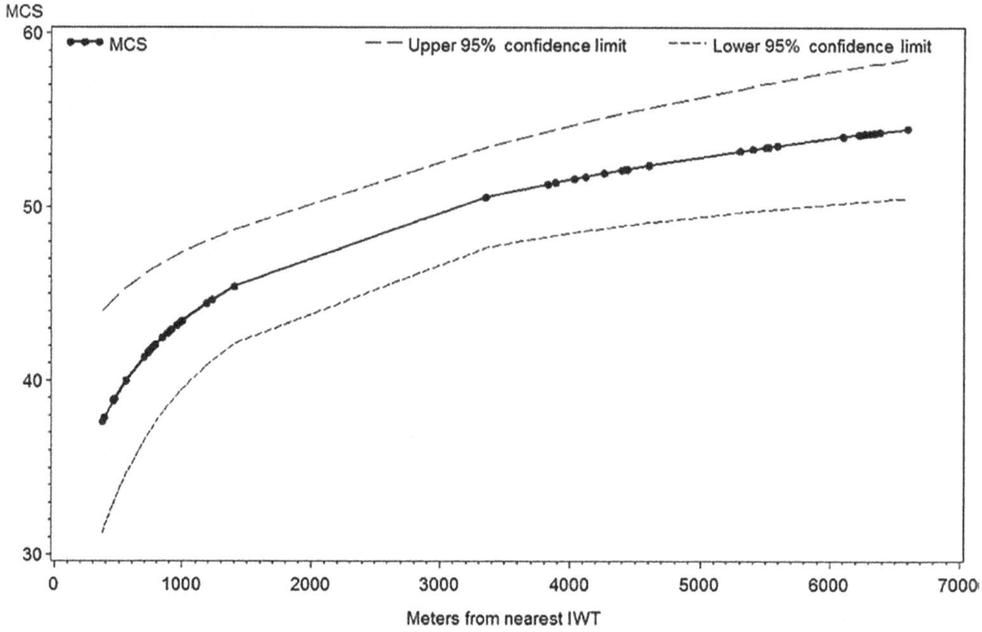

FIGURE 8 Mean SF36 mental component score (MCS) as a function of setback distance. The dashed lines are 95% confidence intervals.
Source: Nissenbaum, Aramini, and Hanning.[48]

content of wind turbine noise means that it may be more audible indoors than outdoors. Additionally, during warmer months, windows are more likely to stay open to control thermal parameters, whence the inability to control or modify wind turbine noise will contribute to the annoyance and, presumably, the effect on sleep onset.[16]

Noise may also cause awakenings and arousals. Arousal is a brief lightening of sleep that is not recalled. Sleep becomes fragmented and, if enough arousals occur, induces the same consequences as reduction of total sleep time. Awakenings are arousals of sufficient degree for wakefulness to be reached and long enough (greater than 10 sec) to be recalled. Arousals are more likely than awakenings, and thus, relying on reported awakenings underestimates the magnitude of the noise effects. The likelihood of an arousal depends upon the volume, character, and duration of the noise as well as the sleep stage and individual propensity (i.e., noise sensitivity). In an investigation into hospital noise, dose–response curves were created for different noises in different sleep stages.[50] Noises with characteristics designed to alert (e.g., telephone, alarms) were more likely to arouse. These noises tend to be impulsive in character, as does wind turbine noise. Noises that were classified as continuous broadband noises (e.g., traffic noise) were less likely to arouse. Another study[51] has shown that subjects with fewer sleep spindles (electrophysiological markers characteristic of stage II sleep) are more easily aroused by noise (Figure 9). Sleep spindles are taken as a marker of sleep stability and may provide a physiological marker of sleep quality.

To date, there are no electrophysiological studies of wind turbine noise on sleep. However, it is reasonable to expect that, in common with road, rail, and aircraft noise, it will induce arousals, fragmenting sleep, as well as preventing the onset of and return to sleep. The sleep measures used in the study by Nissenbaum, Aramini, and Hanning[48] (i.e., ESS and PSQI) are average scores, determining sleepiness and sleep quality, respectively, over a period of weeks. Thus, occasional sleep disturbance would not alter scores as the sleep loss would have been compensated quickly over one or two nights. The study results imply strongly that sleep was being disturbed to some degree on sufficient nights to prevent compensation occurring, thus leading to persistent daytime symptoms.

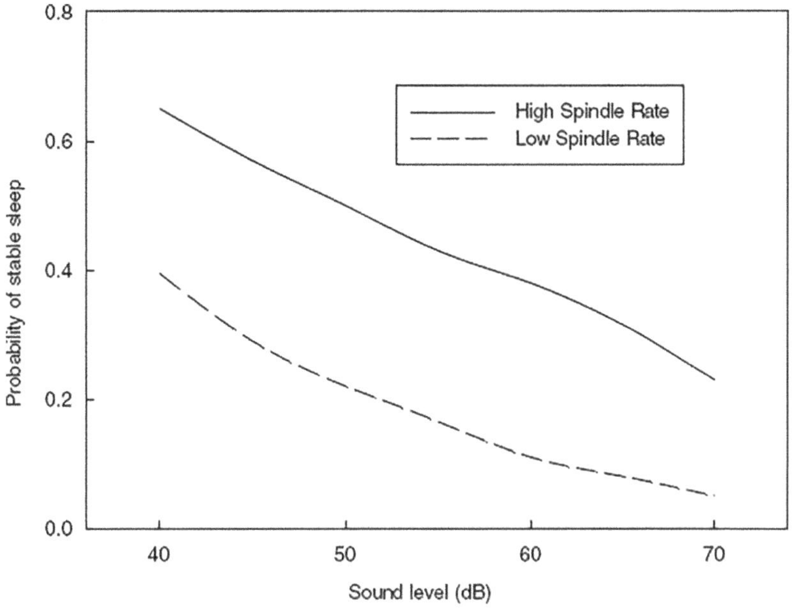

FIGURE 9 Sleep stability as a function of sound level for noise-resistant (high-spindle) and noise-sensitive (low-spindle) groupings.
Source: Estimated from Dang-Vu et al.[51]

Wind Turbine Syndrome

Wind turbine syndrome refers to a cluster of symptoms, which Pierpont,[24] who coined the phrase, claims are associated with exposure to wind turbine noise. Using direct clinical case studies, Pierpont describes the following symptoms to be characteristic of many individuals residing in close vicinity of wind turbines: insomnia, headaches, dizziness, unsteadiness, nausea, exhaustion, anxiety, anger, irritability, depression, memory loss, eye problems, problems with concentration and learning, and tinnitus. Pierpont hypothesizes that wind turbines may affect the vestibular system, that part of the inner ear that plays an important role in the maintenance of balance and stable visual perception. Wind turbines may compromise this system in two ways: first, by the visual disturbance of the moving blades and shadows (i.e., the flicker), and second, by direct vibration of the vestibular system. Such a model would explain why some residents in the close proximity of wind turbines (i.e., less than a kilometer) complain of vertigo, dizziness with nausea, and migraines. Wind turbine syndrome awaits further validation from the medical and scientific establishments, specifically the confirmation of a cause-and-effect relationship between wind turbine noise and vestibular function.

Wind Turbine Noise and Low-Frequency/Infrasound Components

Recent enquiry has focused on the impacts of low-frequency (20–200 Hz) and infrasonic frequencies (typically taken as below 20 Hz) being emitted by wind turbines. Infrasound is characterized by fluctuating pressure sensations at the eardrum, is atonal and countable, and is of a level proportional to wind speed.[21] Low-frequency acoustic waves emitted by wind turbines may be amplified by ground reflection and originate from varying lift forces as the rotors travel through spaces differing in wind speed and density.[21] Compared with medium (i.e., 250 to 4000 Hz) and high frequencies (above 4000 Hz), low-frequency energy decays slowly with distance, is less attenuated by conventionally designed structures, causes certain building materials to vibrate, and can sometimes resonate within rooms and undergo amplification. The effect of air absorption must also be taken into account, in which higher frequencies are attenuated at a greater rate as a function of distance, resulting in a shifting of the spectrum toward lower frequencies. The relationship between low-frequency wind turbine noise and building type creates an interesting proposition in which the low-frequency sound may be louder inside a dwelling than out,[21,52] and the assumption that walls and windows attenuate sound by 15 dB may not be applicable to frequencies below 200 Hz.

Research has shown that low-frequency noise increases cortisol levels in those who are sensitive to noise[12] and disturbs rest and sleep at levels below noise otherwise free from lower-frequency components.[31] Low-frequency noise and infrasound are known disturbers of sleep; however, the contribution, if any, of the low-frequency noise emissions of wind turbines to the sleep disturbances they induce remains to be scientifically determined. Beyond infrasound, the phenomenon of vibroacoustic disease is worthy of note. Humans chronically exposed to infrasound may exhibit elevated cortisol levels and generalized cell damage: a condition known as vibroacoustic disease.[53] A number of human and animal models explaining how infrasound can lead to cardiovascular and respiratory disease have been proposed[54] and applied to wind turbine noise.[55] The phenomenon of vibroacoustic disease is supported by correlational evidence coupled with a thoroughly detailed mechanism. However, further research is required to establish the veracity of this approach to human health within and beyond the wind turbine context.

Mitigation

There are multiple ways in which to reduce the impacts of audible and inaudible wind turbine noise. The first, and often the most effective, method is to control audible noise at the sound source. Thus, mechanical solutions invite technologies designed to attenuate wind turbine noise or to shift its spectral character in order to eliminate salient tonal characteristics. To safeguard health is more difficult, however, because wind turbine noise is largely aerodynamic in origin,[7] and it is not possible to obtain solutions that completely attenuate the noise at its source. Having minimized the noise through the implementation of technology, other approaches are often required, normally involving the application of noise standards to limit exposure levels or the determination of "safe" setback distances to mitigate noise impact. Still other approaches involve the positioning of wind turbines around preexisting noise generators,[15] in remote areas away from human habitations, or using social processes to determine wind turbine location.[27,56]

Regulating Permissible Noise Level

Permissible or safe exposure levels are often set in national noise standards, which may or may not be specific to wind turbine noise. These standards may serve one of two purposes, or sometimes both, with noise compliance guidelines naturally emerging from the two. The first purpose relates to methodologies for the physical quantification of the noise. This may involve standardized procedures for measuring noise from preexisting windfarms or detailing accepted mathematical models affording noise predictions of a planned windfarm. The second purpose is to determine what exposure levels can be considered safe and to clearly state criteria to this effect. However, there are a number of flaws inherent in wind turbine noise standards, including the metrics used to represent the noise, oversimplified modeling approaches that yield unrealistically low predictions of noise levels representing "best case" conditions,[5] or stimulus-oriented approaches that fail to account for human factors.[3,57]

There exists, in respect to levels-based noise standards, disagreement as to the relevance of physical measures such as dBA to human response,[58] not only for windfarm noise[14] but also for traffic and aviation noise. Of the few parametric studies that have been published,[7,8] only marginal dose–response relationships between wind turbine noise intensity and health measures have emerged. For example, Pedersen[22] noted that stress was not related to wind turbine noise level but rather noise annoyance.

Persson Waye and Öhtrsöm[12] reported that annoyance ratings varied for five distinct recordings of wind turbine noise, even though all five had equivalent noise levels. Others note that both laboratory and field studies have consistently found that the equivalent dBA measure fails to account for the relationship between wind turbine noise and annoyance.[14]

To some degree, then, it must be accepted that there is an uncoupling between wind turbine noise level and human response. A hitherto rarely measured characteristic of wind turbine noise is amplitude modulation, whereby noise levels fluctuate periodically as a function of blade passing frequency. Lee et al.[34] recommend that standardized metrics based on the modulation depth spectrum be developed and used in conjunction with sound levels. Other approaches to measuring amplitude modulation have existed for some time[4,59] but have yet to be seriously applied to the wind turbine noise context. However, the inability to account for amplitude modulation arises primarily due to the time-averaged dBA levels applied by noise standards, and arguably, smaller sampling epochs of around 100 msec should be adopted as best practice in order to record the amplitude modulation inherent in turbine noise.[60,61] The New Zealand Standard[62] applies a penalty for amplitude modulation, but does not describe an objective assessment. Furthermore, using aggregated metrics that average noise level over long periods underestimates the effect of peak levels and crest factors, important when considering sleep disturbance.

For the most part, the acceptable noise limits recommended by noise standards are derived from WHO guide-lines.[31,63,64] However, as Figure 10 demonstrates, using recommended noise levels from guidelines based on transport data risks exposing the population to unacceptable levels of noise. It follows that the Ldn (the "day–night" level in the United States) or Lden (the "day–evening– night" level in Europe) measures, derived from the measured LAeq sound level can be used in a wind farm context, but with caution.[65] Inspection of Figure 10 suggests that, relative to transport guidelines, at least a 10 dBA penalty should be placed on wind turbine noise. The differences in annoyance ratings between wind turbine noise and transport noise maybe accounted for by amplitude modulation, the typical location

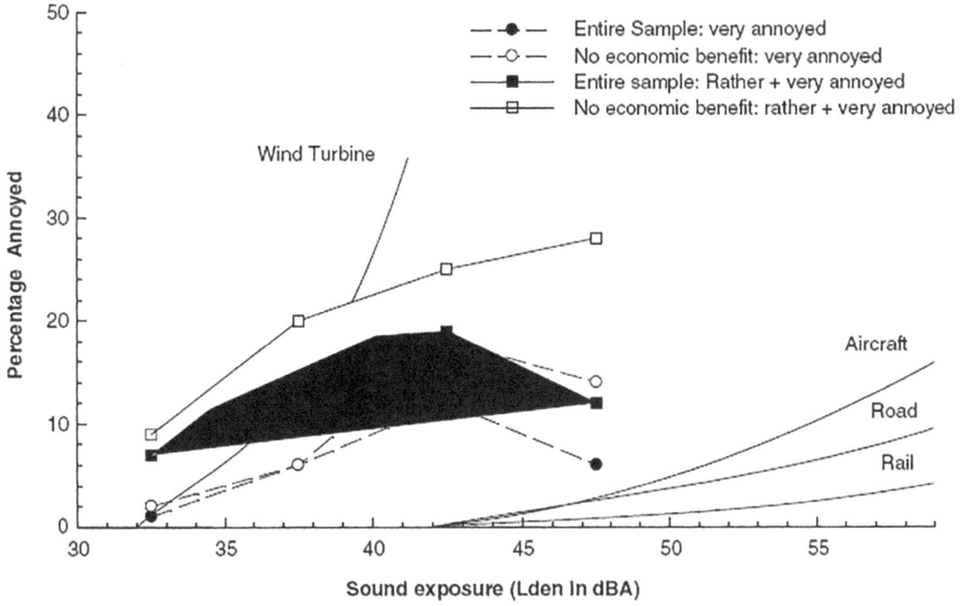

FIGURE 10 Annoyance plotted as a function of noise level for four theoretical models (rail, road, and air parameters: Miedema and Oudshoorm;[66] wind turbine parameters: Pedersen et al.;[7] and four sets of data obtained from Tables 7.24–7.26 of van den Berg et al.[8]). For the data, closed symbols are for the entire sample, while open symbols are for those who identified that they had no economic interest. Circles represent the percentage of "very annoyed" responses, while squares represent the sum of "very annoyed" and "rather annoyed" responses.

of windfarms (e.g., rural areas), or the over-representation of noise-sensitive individuals. A recent meta-analysis of three epidemiological studies revealed a consistent trend in wind turbine noise exposure and both annoyance and sleep disruption.[22] On the basis of her analysis, Pederson recommends that outdoor levels should not exceed 40 dBA, though this level could be more-or-less depending on situational factors, that is, ambient noise levels or the building's construction materials. When noise is continuous, the WHO[31] stipulates an indoor limit of 30 dBA, though for noises containing lower frequencies (e.g., wind turbine noise), a lower limit still is recommended. Thus, careful examination of the lower end of the frequency spectrum is important when judging appropriate exposure to wind turbine noise, and the use of dBC or spectral analysis in one-third octave bands or narrow bands is necessary.

In the comparison of global wind turbine noise level standards, there exist two chief methodologies, namely, sound levels not to be exceeded (usually in dBA) or a not- to-be-exceeded limit derived from the sum of the preconstruction ambient limit and a constant (e.g., L_{A90}+10 dBA). Critique of both these approaches can be found in Thorne.[3] The fact that noise limits differ between, and even within, a country is testament to the impoverished research database guiding their development or the political sensitivities around wind turbine placement. Examples of noise limits are presented in Table 3, and the variability in guidelines is evident. Based on the authors' collective experience, an interim guideline, providing a conservative noise limit capable of protecting the health of the public and susceptible individuals, would be a sound level of L_{Aeq} 35 dBA outside the residence and below the individual's threshold of hearing inside a residence. More specific guidelines are presented in Appendix A of this document.

Regulating Setback Distances

A setback distance is defined as the minimum distance between a dwelling and the closest wind turbine required to protect the health of the inhabitants. One difficulty is whether such setback distances can be standardized, as they will differ depending on a number of factors, including turbine type, terrain, and climate. Lee et al.[34] report that the perception of amplitude-modulated noise decreases with distances beyond a kilometer, though others claim that amplitude-modulated turbine noise can be heard up to 4km away from the source.[67] Setback distances maybe based on noise level, which, as

TABLE 3 A Comparison of Wind Turbine Noise Guidelines Taken from Nine Countries

Country	State	Limit (dBA)	Background Plus Constant
Australia	Victoria	LA90 35 or 40	LA90+5 dBA
	South Australia	LAeq 35 or 40	LA90+5 dBA
Australia	Queensland	LAeq 30 indoors	Health and well-being criteria
Canada	Ontario	LAeq 40 to 51	
Denmark		40	
France			Day: LA90+5 dBA Night: LA90+3 dBA
Netherlands		40	
New Zealand		LA90 35, 40	LA90+5 dBA
United Kingdom		Day: 40 Night: 43	LA90+5 dBA
United States	Illinois	Day: 50 Night: 46	
	Michigan	55	
	Oregon	35	

discussed in the preceding section, maybe an invalid approach. Instead, a better approach may be to link setbacks to turbine type. MOller and Pedersen,[21] investigating the detection and annoyance of lower-frequency sound emitted from wind turbines, suggest that, for flat terrain, the minimum setback distance for modern turbines (2 to 3.6 MW) should be between 600 and 1200 metres. Other approaches rely on the establishment of dose–response curves relating a health outcome variable (e.g., annoyance or disturbed sleep) and distance (e.g., Figure 6). Medical professionals have proposed setback distances of 2.4 km[23,24] or 1.5 km.[45] Other research recommends a minimum of 2 km if wind turbines are sited in rough terrain.[3,20]

Conclusion

Windfarms have significant potential for sleep disruption and annoyance due to the intermittent nature and amplitude modulation of their sound emissions, even though exposure may be of low amplitude. The interactions between ambient levels, amplitude modulation, and the tonal character of windfarm noise overlaid within a soundscape are complex and difficult to measure and assess in terms of health and individual amenity. Additionally, currently employed sound level measurement and prediction approaches for complex noise sources of this nature are only partially relevant to environmental risk assessment. Aside from acoustic parameters, other factors such as noise sensitivity or amenity expectations may also predict the human response to wind turbine noise. Unfortunately then, for policymakers, there appears to be no proportional relationship between wind turbine noise levels and health, as these outcome factors will be influenced by characteristics associated with both the noise and the listener.[39]

As a relatively new source of intrusive noise, there is little research to draw upon when judging if a proposed windfarm constitutes a health threat to the exposed public. A liberal approach to assessing health impact will involve the application of previous knowledge obtained from other noise sources (e.g., road, aviation). A conservative approach, consistent with the precautionary principle, will consider wind turbine noise more potent than these other harmful noise sources. Thus, at this time, a constellation of acoustic and social metrics should be taken at preexisting wind farms in order to assess potential threat. Peak and crest noise levels, level metrics assessing low-frequency contributions (e.g., dBC), and amplitude modulation indices constitute the acoustic measures of importance. It should also be remembered that predicted levels derived from computer models represent estimates and not precise values, are constrained by numerous assumptions, contain substantial uncertainty, and as such should not constitute the sole criteria for wind turbine positioning. What form the social measures will take is yet to be elucidated, but research suggests that noise sensitivity[67] and procedural fairness[27] are the best approaches to minimize the health impacts and facilitate social acceptance of windfarms.

Acknowledgments

The authors thank Rick James of E-Coustic Solutions for sharing his acoustical expertise on the matter of wind turbine noise guidelines.

Appendix A

'Proposed Wind Turbine Siting Sound Limits', a revision by Thorne, R, of the Kamperman James criteria (2008) to include updates to ISO 1996-2 and U.K. Court of Appeal (Hulme re: Den Brook).

1. Audible Sound Limit
    a. No wind turbine or group of turbines shall be located so as to cause an exceedance of the preconstruction/operation background sound levels by more than 5 dBA. The background sound levels shall be the L_{A90} sound descriptor measured during a pre-construction noise study during the quietest time of evening or night. All data recording shall be a series of

contiguous ten (10) minute measurements. L_{A90} results are valid when L_{A10} results are no more than 15 dBA above L_{A90} for the same time period. Noise sensitive sites are to be selected based on wind development's predicted worst-case sound emissions in L_{Aeq} and L_{Ceq} which are to be provided by the developer.

b. Test sites are to be located along the property line(s) of the receiving non-participating property(s).
c. A 5 dB penalty is applied for tones as defined in IEC 61400-11 at the turbine and ISO1996-2 at any affected residence.
d. A 5 dB penalty is applied for amplitude modulation as defined following. When noise from the wind farm has perceptible or audible characteristics that are perceived by the complainant as being cause for complaint, or greater than expected, the measured sound level of the source shall have a 5 dB penalty added. Audible characteristics include tonal character measured as amplitude or frequency modulation (or both); and tonality (where the tonal character/tonality of noise is described as noise with perceptible and definite pitch or tone). Amplitude modulation is the modulation of the level of broadband noise emitted by a turbine at blade passing frequency. Amplitude modulation will be deemed greater than expected if the following characteristics apply:
   i. A change in the measured L_{Aeq}, 125 ms turbine noise level of more than 3 dB (represented as a rise and fall in sound energy levels each of more than 3 dB) occurring within a 2 second period.
   ii. The change identified in (i) above shall not occur less than 5 times in any one minute period provided the L_{Aeq}, 1 minute turbine sound energy for that minute is not below 28 dB.
   iii. The changes identified in (i) and (ii) above shall not occur for fewer than 6 minutes in any hour.

Noise emissions are measured outside a complainant's dwelling and shall be measured not further than 35 metres from the relevant building, and not closer than within 3.5 metres of any reflective building or surface, or within 1.2 metres of the ground.

2. Low Frequency Sound Limit
   a. The L_{Ceq} and L_{C90} sound levels from the wind turbine at the receiving property shall not exceed the lower of either:
      i. L_{Ceq} -L_{A90} greater than 20 dB outside any occupied structure, or
      ii. A maximum not-to-exceed sound level of 50 dBC measured as the background sound level (L_{C90}) from the wind turbines without other ambient sounds for properties located at one mile or more from state highways or other major roads or measured as the background sound level (L_{C90}) for properties closer than one mile.
      iii. These limits shall be assessed using the same night-time and wind/weather conditions required in 1(a). Turbine operating sound emissions (L_{Aeq} and L_{Ceq}) shall represent worst case sound emissions for stable night-time conditions with low winds at ground level and winds sufficient for full operating capacity at the hub.
3. General Clause
   a. Sound levels from the activity of any wind turbine or combination of turbines shall not exceed L_{Aeq} 35 dB within 100 feet of any noise sensitive premises.
   b. The monitoring shall include all the sound levels as required by these noise conditions and shall include monitoring for the characteristics described in Annex A of IEC 61400-11 including infrasound, low-frequency noise, impulsivity, low-frequency modulation of broad-band or tonal noise, and other audible characteristics. Wind speed and wind direction shall be measured at the same location as the noise monitoring location.
4. Requirements
   a. All instruments must meet ANSI or IEC Class 1 integrating sound level meter performance specifications.
   b. Procedures must meet ANSI S12.9, IEC61400-11 and ISO1996-2

c. Procedures should meet ANSI, IEC and ISO standards applicable to the measurement of sound or its characteristics.
d. Measurements must be made when ground level winds are 2m/s (4.5 mph) or less. Wind shear in the evening and night often results in low ground level wind speed and nominal operating wind speeds at wind turbine hub heights.
e. IEC 61400-11 procedures are not suitable for enforcement of these requirements except for the presence of tones near the turbine.

5. Definitions

ANSI S12.9 Quantities and Procedures for Description and Measurement of Environmental Sound, Parts 1 to 6.

IEC 61400-11 Wind turbine generator systems—Part 11: Acoustic noise measurement techniques.

ISO 1996-2 Acoustics—Description, measurement and assessment of environmental noise—Part 2: Determination of environmental noise levels.

L_{A90}, L_{A10} Statistical measures calculated under ANSI S12.9.

L_{Aeq}, L_{Ceq} Time average levels calculated under ANSI S12.9 or ISO 1996-2.

Noise sensitive premises includes a residence, hotel, hostel or residential accommodation premises of any type.

6. References

ANSI S12.9-2008, Quantities and Procedures for Description and Measurement of Environmental Sound—Part 6: Methods for Estimation of Awakenings Associated with Outdoor Noise Events Heard in Homes, 2008.

Hulme V Secretary of State (2011). Hulme v Secretary of State for Communities and Local Government and Anor. Approved Judgment in the England and Wales Court of Appeal (Civil Division) on Appeal from the High Court of Justice, Case No: C1/2010/2166/ QBACF. Neutral Citation Number [2011] EWCA Civ 638; Available from: http://www.bailii.org/ew/cases/EWCA/Civ/2011/636.html.

IEC 61400-11Wind Turbine Generator Systems—Part 11: Acoustic Noise Measurement Techniques, 2nd ed. (International Technical Commission, Geneva, 2002 plus Amendment 1 2006).

International Standards Organization, (2007). ISO 1996-2 second edition, Acoustics—description, assessment and measurement of environmental noise- part 2: Determination of environmental noise levels. International Standards Organization.

Kamperman, G. W.; James, R. R. In: Proceedings of the 30th Annual Conference of ASME's Noise Control and Acoustics Division, 28–31 July 2008, Dearborn, Michigan, USA.

References

1. Warren, C.R.; Lumsden, C.; O'Dowd, S.; Birnie, R.V. 'Green on green': Public perceptions of wind power in Scotland and Ireland. J. Environ. Plann. Manage. 2010, *48* (6), 853–875.
2. Van Den Berg, G.P. Wind-induced noise in a screened microphone. J. Acoust. Soc. Am. 2006, *119* (2), 824–833.
3. Thorne, R. The problems with "noise numbers" for wind farm noise assessment. Bull. Sci., Technol. Soc. 2011, *31*, 262–290.
4. Zwicker, E.; Fastl, H. *Psychoacoustics: Facts and Models*; 2nd Edition, Springer-Verlag, Berlin Heidelberg, New York, 1999.
5. International Electrotechnical Commission. *IEC 61400-11 Wind Turbine Generator Systems—Part 11: Acoustic Noise Measurement Techniques, 2nd Ed.*; International Technical Commission: Geneva, 2002 plus Amendment 1, 2006.
6. International Standards Organization. ISO 9613-2 Acoustics—Attenuation of Sound during Propagation Outdoors— Part 2: General Method of Calculation; International Organization for Standardization: Geneva, 1996.

7. Pedersen, E.; Persson Waye, K. Perception and annoyance due to wind turbine noise—A dose-response relationship. J. Acoust. Soc. Am. 2004, *116*, 3460–3470.
8. Van den Berg, G.P.; Pedersen, E.; Bouma, J.; Bakker, R. *Project WINDFARM perception. Visual and Acoustic Impact of Wind Turbine Farms on Residents.* FP6-2005- Science-and-Society-20. Specific Support Action Project no. 044628. Final report; University of Groningen: Holland, 2008.
9. Van Den Berg, G.P. Effects of the wind profile at night on wind turbine sound. J. Sound Vib. 2004, *277* (4–5), 955–970.
10. Stigwood, M. Large wind turbines—are they too big for ETSU-R-97? *Wind Turbine Noise*; Institute of Acoustics: Bristol, 2009.
11. Van Den Berg, G.P. The beat is getting stronger: The effect of atmospheric stability on low frequency modulated sound of wind turbines. J. Low Freq. Noise Vib. Active Control 2005, *24* (1), 1–24.
12. Persson Waye, K.; Öhtrsöm, E. Psycho-acoustic characters of relevance for annoyance of wind turbine noise. J. Sound Vib. 2002, *250* (1), 65–73.
13. Arlinger, S.; Gustafsson, H.A. How a Broadband Noise with Constant Sound Pressure Level Masks a Broadband with Periodically Varying Sound Pressure Levels; Department of Technical Audiology, Linkoping University: Sweden, 1988.
14. Pedersen, E.; Waye, K.P. Wind turbines—Low level noise sources interfering with restoration? Environ. Res. Lett. 2008, *3* (1), 1–5.
15. Pedersen, E.; van den Berg, F.; Bakker, R.; Bouma, J. Can road traffic mask sound from wind turbines? Response to wind turbine sound at different levels of road traffic sound. Energy Policy 2010, *38* (5), 2520–2527.
16. Pedersen, E.; Persson Waye, K. Wind turbine noise, annoyance and self-reported health and well-being in different living environments. Occup. Environ. Med. 2007, *64* (7), 480–486.
17. Martin, M. (2008). Statement of Evidence in Chief of Mr Murray Martin In the Matter of an Appeal Under Sections 120 and 121 of that Act Between Motorimu Wind Farm Ltd (Appellant) and Palmerston North City Council, Environment Court, ENV-2007-WLG-000098.
18. Lercher, P. Environmental noise and health: An integrated research perspective. Environ. Int. 1996, *22*, 117–129.
19. World Health Organization. Preamble to the Constitution of the World Health Organization as Adopted by the International Health Conference; WHO: New York, 1948.
20. Shepherd, D.; McBride, D.; Welch, D.; Dirks, K. N.; Hill, E. M. Evaluating the impact of wind turbine noise on health-related quality of life. Noise and Health, 2011, *13* (54), 333–339.
21. Møller, H.; Pedersen, C.S. Low-frequency noise from large wind turbines. J. Acoust. Soc. Am. 2011, *129* (6), 3727–3744.
22. Pedersen, E. Health aspects associated with wind turbine noise—Results from three field studies. Noise Control Eng. J. 2011, *59* (1), 47–53.
23. Harry, A. *Wind Turbines, Noise and Health*; 2007, available at http://www.flat-group.co.uk/pdf/wtnoise_health_2007_a_barry.pdf (accessed July 2011).
24. Pierpont N. *Wind Turbine Syndrome: A Report on a Natural Experiment*; K Selected Publications: Santa Fe, New Mexico, 2009.
25. Krogh, C.M.E.; Lorrie Gillis, L.; Kouwen, N.; Aramini, J. WindVOiCe, a self-reporting survey: Adverse health effects, industrial wind turbines, and the need for vigilance. Bull. Sci., Technol. Soc. 2011, *31*, 334–345.
26. Alvarsson, J.J.; Wiens, S.; Nilsson, M.E. Stress recovery during exposure to nature sound and environmental noise. Int. J. Environ. Res. Public Health 2010, *7* (3), 1036–1046.
27. Maris, E.; Stallen, P.J.; Vermunt, R.; Steensma, H. Noise within the social context: Annoyance reduction through fair procedures. J. Acoust. Soc. Am. 2007, *121* (4), 2000–2010.
28. Niemann, H.; Maschke, C. *WHO LARES: Report on Noise Effects and Morbidity*; World Health Organization: Geneva, 2004.

29. European Union. *Noise Policy of the European Union—Year* 2; Luxembourg, 2000, http://www.ec.europa.eu/environment/noise/pdf/noise_expert_network.pdf (accessed August 2011).
30. European Commission. *Relating to the Assessment and Management of Environmental Noise*, European Parliament and Council Directive 2002/49/EC; 2002, http://eur-lex.europa.eu/LexUriServ/LexUriServ.do?uri=OJ:L:2002:189:0012:0025:EN:PDF (accessed August 2011).
31. Berglund, B.; Lindvall, T.; Schwela, D.H. *Guidelines for Community Noise*; The World Health Organization: Geneva, 1999.
32. Hayes, M.; McKenzie, A. *The measurement of Low Frequency Noise at Three U.K. Windfarms;* Hayes Mckenzie Partnership Ltd, report to the Department of Trade and Industry: London, England, 2006.
33. Berglund, B.; Nilsson, M. Intrusiveness and dominant source identification for combined community noises. In *Fechner Day 1997,* Proceedings of the 13th Annual Meeting of the International Society for Psychophysics, Poznan, Poland, Aug 16–19, 1997; Preis, A., Hornowski, T., Eds.; The International Society for Psychophysics, 1997.
34. Lee, S.; Kim, K.; Choi, W.; Lee, S. Annoyance caused by amplitude modulation of wind turbine noise. Noise Control Eng. J. 2011, *59* (1), 38–46.
35. Shepherd, D. Wind turbine noise and health in the New Zealand context. In: *Sound, Noise, Flicker and the Human Perception of Wind Farm Activity;* Rapley, B., Bakker, H., Eds.; Atkinson and Rapley: Palmerston North, 2010; ISBN 978-0-473-16558-1.
36. Kuwano, S.; Seiichiro, N. Continuous judgment of loudness and annoyance. In *Fechner Day 90,* Proceedings of the Sixth Annual Meeting of the International Society of Psychophysics, Wurzburg, Germany, Aug 18–22, 1990; Muller, F., Ed.; International Society of Psychophysics.
37. Johansson, M.; Laike, T. Intention to respond to local wind turbines: The role of attitudes and visual perception. Wind Energy 2007, *10,* 435–45.
38. Nelson, D. 2007. Perceived loudness of wind turbine noise in the presence of ambient sound. In Proceedings of the Fourth International Meeting on Wind Turbine Noise, Rome, Italy, Aprl2–14, 2011; INCE Europe; ISBN: 97888-88942-33-9.
39. Flindell, I.; Stallen, P.J. Non-acoustical factors in environmental noise. Noise Health 1999, *3,* 11–16.
40. Pedersen, E.; van den Berg, F.; Bakker, R.; Bouma, J. Response to noise from modern wind farms in the Netherlands. J. Acoust. Soc. America 2009, *126,* 634–643.
41. Pheasant, R.J.; Fisher, M.N.; Watts, G.R.; Whitaker, D.J.; Horoshenkov, K.V. The importance of auditory–visual interaction in the construction of 'tranquil space.' J. Environ. Psychol. 2010, *30,* 501–509.
42. Calvert, G.A. Crossmodal processing in the human brain: Insights from functional neuroimaging studies. Cereb. Cortex 2001, *11* (12), 1110–1123.
43. World Health Organization. *WHO Technical Meeting on Sleep and Health*; World Health Organization, Regional Office for Europe: Bonn, 2004.
44. Phillips, C.V. Properly interpreting the epidemiologic evidence about the health effects of industrial wind turbines on nearby residents. Bull. Sci., Technol. Soc. 2011, *31,* 303–315.
45. Hanning, C.D.; Nissenbaum, M. Selection of outcome measures in assessing sleep disturbance from wind turbine noise. In Proceedings of the Fourth International Meeting on Wind Turbine Noise, Rome, Italy, Apr 12–4, 2011; INCE Europe; ISBN: 978–88–88942–33–9.
46. Phipps, R. A.; Amati, M.; McCoard, S.; Fisher R.M. (2007). Visual and Noise Effects Reported by Residents Living Close to Manawatu Wind Farms: Preliminary Survey Results. New Zealand Planners Institute Conference, Palmerston North, March 27–30, 2007.
47. Pedersen, E. Effects of wind turbine noise on humans. In Proceedings of the Third International Meeting on Wind Turbine Noise, Aalborg, Denmark, June 17–19, 2009; INCE Europe.
48. Nissenbaum, M.; Aramini, J.; Hanning, C. Adverse health effects of industrial wind turbines: A preliminary report. In Proceedings of 10th International Congress on Noise as a Public Health Problem (ICBEN), London, July 24–28, 2011; Foxwoods, C.T., Ed; International Commission on Biological Effects of Noise.

49. Bolin, K. *Wind Turbine Noise and Natural Sounds-Masking, Propagation and Modelling*; Doctoral Thesis; Royal Institute of Technology: Stockholm, 2009.
50. Solet, J.M.; Buxton, O.M.; Ellenbogen, J.M.; Wang, W.; Carballiera, A. *Evidence-Based Design Meets Evidence- Based Medicine: The Sound Sleep Study*; the Center for Health Design: Concord, California, 2010.
51. Dang-Vu, T.T.; McKinney, S.M.; Buxton, O.M.; Solet, J.M.; Ellenbogen, J.M. Spontaneous brain rhythms predict sleep stability in the face of noise. Curr. Biol. 2010, *20*, R626–7.
52. Casella, S. *Low Frequency Noise Update*; DEFRA Noise Programme, Department of the Environment, Northern Ireland Scottish National Assembly for Wales, 2001; 1–11.
53. Castelo-Branco, N.A.A.; Alves-Pereira, M. The clinical stages of vibroacoustic disease. Aviat., Space Environ. Med. 1999, *70* (3), A32–9.
54. Castelo-Branco, N.A.A.; Alves-Pereira, M. Vibroacoustic disease. Noise Health 2004, *6* (23), 3–20.
55. Alves-Pereira, M.; Castelo-Branco, N.A.A. In-home wind turbine noise is conducive to vibro-acoustic disease. In Proceedings of the Second International Meeting on Wind Turbine noise, Lyon, France, Sept 20–21, 2007.
56. Gross, C. Community perspectives of wind energy in Australia: The application of a justice and community fairness framework to increase social acceptance. Energy Policy 2007, *35*, 2727–2736.
57. Shepherd, D.; Billington, R. Mitigating the acoustic impacts of modern technologies: Acoustic, health and psychosocial factors informing wind farm placement. Bull. Sci., Technol. Soc. 2011, doi:10.1177/0270467611417841.
58. Fidell, S. The Schultz curve 25 years later: A research perspective. J. Acoust. Soc. of Am. 2003, *114* (6), 3007–3015.
59. Ando, Y.; Pompoli, R. Factors to be measured of environmental noise and its subjective responses based on the model of auditory-brain system. J. Temporal Des. Archit. Environ. 2002, *2* (1), 2–12.
60. Lundmark, G. Measurement and analysis of swish noise: A new method. In Proceedings of the Fourth International Meeting on Wind Turbine Noise, Rome, Italy, Apr 12–4, 2011; INCE Europe; ISBN: 978-88-88942-33-9.
61. Hulme. Approved Judgment in the Court of Appeal (Civil Division) on Appeal from the High Court of Justice, Case No: C1/2010/2166/QBACF, Neutral Citation Number [2011] EWCA Civ 638, http://www.bailii.org/ew/cases/EWCA/Civ/2011/636.html (accessed August 2011).
62. New Zealand Standards. *NZS6808:2010 Acoustics—Wind farm noise*; Standards New Zealand: Wellington, 2010.
63. World Health Organization, Regional Office for Europe. *Night Noise Guidelines for Europe*; World Health Organization: Geneva, Switzerland, 2009, http://www.euro.who.int/document/e92845.pdf (accessed August 2011).
64. World Health Organization. *Burden of Disease from Environmental Noise*; Bonn: World Health Organization: Geneva, 2011, http://www.euro.who.int/_data/assets/pdf_file/0008/136466/e94888.pdf (accessed August 2011).
65. Pedersen T.H, 2007, The "Genlyd" Noise Annoyance Model Dose-Response Relationships Modelled by Logistic Functions. Delta Acoustics and Electronics on behalf of the Ministry of Science, Technology and Innovation, Denmark.
66. Di Napoli, C. Long distance amplitude modulation of wind turbine noise. In Proceedings of the Fourth International Meeting on Wind Turbine Noise, Rome, Italy, Apr 1–14, 2011; INCE Europe; ISBN: 978-88-88942-33-9.
67. Miedema, H.M.E.; Oudshoorm, C.G.M. Annoyance from transportation noise: Relationship with exposure metrics DNL and DENL. Environ. Health Perspect. 2001, *109* (4), 409–416.

IV

DIA: Diagnostic Tools: Monitoring, Ecological Modeling, Ecological Indicators, and Ecological Services

20

Exergy: Analysis

Marc A. Rosen

Introduction

Energy analysis is based on the first law of thermodynamics, which embodies the principle of conservation of energy and is the traditional method used to assess the performance and efficiency of energy systems and processes.

Exergy analysis is a thermodynamic analysis technique for systems and processes that is based on the second law of thermodynamics. Exergy analysis has been increasingly applied over the last several decades, in large part because of its advantages over energy analysis:

- More meaningful efficiencies are evaluated with exergy analysis because exergy efficiencies are always a measure of the approach to the ideal.
- Inefficiencies in a process are better pinpointed with exergy analysis because the types, causes, and locations of the losses are identified and quantified.

In this entry, the role of exergy analysis in the assessment and improvement of energy systems is examined. First, exergy and its use as an analysis technique are briefly described. Second, the ranges of energy systems that have been assessed with exergy analysis are surveyed. Third, several example applications of exergy analysis are presented, ranging from simple devices to large and complex systems.

Exergy

Exergy can be regarded as a measure of the usefulness or quality of energy. Technically, exergy is defined as the maximum amount of work that can be produced by a stream of energy or matter, or from a system, as it is brought into equilibrium with a reference environment. Unlike energy, exergy is consumed during real processes due to irreversibilities and conserved during ideal processes. Exergy and related concepts have been recognized for more than a century.[1]

Exergy analysis is a methodology that uses the first and second laws of thermodynamics for the analysis, design, and improvement of energy and other systems.[2–14] The exergy method is useful for improving the efficiency of energy-resource use, for it quantifies the locations, types, and magnitudes of wastes and losses. In general, more meaningful efficiencies are evaluated with exergy analysis rather than energy analysis because exergy efficiencies are always a measure of the approach to the ideal. Therefore, exergy analysis accurately identifies the margin available to design more efficient energy systems by reducing inefficiencies.

In evaluating exergy, the characteristics of the reference environment must be specified,[2–15] usually by specifying the temperature, pressure, and chemical composition of the reference environment. The results of exergy analyses, consequently, are relative to the specified reference environment, which in most applications is modeled after the actual local environment. The exergy of a system is zero when it is in equilibrium with the reference environment. The tie between exergy and the environment has implications regarding environmental impact.[7,8,16,17]

The theory and the applications of exergy have been described in specialized books, e.g.,[2–8] general thermodynamics texts, e.g.,[9,10], and journal entries, e.g.,[11–14]. Many applications of exergy analysis have been reported in fields ranging from power generation,[18] hydrogen energy,[18] and cogeneration[9,19] to district energy,[19] thermal processes,[20,21] and thermal energy storage[21–23] and on to systems as large as countries[24] and the world.[25]

Exergy and the Reference Environment

Exergy quantities are evaluated with respect to a reference environment. The intensive properties of the reference environment in part determine the exergy of a stream or system. The reference environment is in stable equilibrium, with all parts at rest relative to one another and with no chemical reactions occurring between the environmental components. The reference environment acts as an infinite system and is a sink and source for heat and materials. It experiences only internally reversible processes in which its intensive state remains unaltered (i.e., its temperature T_0, pressure P_0, and the chemical potentials μ_{i00} for each of the i components present remain constant). The exergy of the reference environment is zero. More information on reference-environment models can be found in this encyclopedia in an entry by the present author entitled "Exergy: Environmental Impact Assessment."

Exergy Balances

Energy and exergy balances can be written for a general process or system.

Since energy is conserved, an energy balance for a system may be written as

$$\text{Energy input} - \text{Energy output} = \text{Energy accumulation} \tag{1}$$

Energy input and output refer, respectively, to energy entering and exiting through system boundaries. Energy accumulation refers to build-up (either positive or negative) of the quantity within the system.

By contrast, an exergy balance can be written as

$$\text{Exergy input} - \text{Exergy output} - \text{Exergy accumulation} = \text{Exergy accumulation} \tag{2}$$

This expression can be obtained by combining the principles of energy conservation and entropy non-conservation, the latter of which states that entropy is created during a process due to irreversibilities. Exergy is consumed due to irreversibilities, with exergy consumption proportional to entropy creation.

Equations 1 and 2 demonstrate an important difference between energy and exergy—energy is conserved while exergy, a measure of energy quality or work potential, can be consumed.

Definitions

It is helpful to define some terms related to exergy for readers. The following are exergy quantities:

Exergy: A general term for the maximum work potential of a system, a stream of matter, or a heat interaction in relation to the reference environment (see definition below) as the datum state; or the maximum amount of shaft work obtainable when a steady stream of matter is brought from its initial state to the dead state (see definition below) by means of processes involving interactions only with the reference environment.

Physical exergy: The maximum amount of shaft work obtainable from a substance when it is brought from its initial state to the environmental state (see definition below) by means of physical processes involving interaction only with the environment.

Chemical exergy: The maximum work obtainable from a substance when it is brought from the environmental state to the dead state by means of processes involving interaction only with the environment.

Thermal exergy: The maximum amount of shaft work obtainable from a given heat interaction using the environment as a thermal energy reservoir.

Exergy consumption: The exergy consumed during a process due to irreversibilities within the system boundaries.

The following terms relate to the reference environment and its state:

Reference environment: An idealization of the natural environment, which is characterized by a perfect state of equilibrium, i.e., absence of any gradients or differences involving pressure, temperature, chemical potential, kinetic energy, and potential energy. The reference environment constitutes a natural reference medium with respect to which the exergy of different systems is evaluated.

Dead state: The state of a system when it is in thermal, mechanical, and chemical equilibrium with a conceptual reference environment, which is characterized by a fixed pressure, temperature, and chemical potential for each of the reference substances in their respective dead states.

Environmental state: The state of a system when it is in thermal and mechanical equilibrium with the reference environment, i.e., at the pressure and temperature of the reference environment.

Reference state: A state with respect to which values of exergy are evaluated. Several reference states are used, including environmental state, dead state, standard environmental state, and standard dead state.

Exergy Analysis

Exergy analysis involves the application of exergy concepts, balances, and efficiencies to evaluate and improve energy and other systems. Many engineers and scientists suggest that devices are best evaluated and improved upon using exergy analysis in addition to or in place of energy analysis.

A journal devoted to exergy matters entitled *The International Journal of Exergy* was established by Inderscience. Some extensive bibliographies have been compiled, including one by Goran Wall (see the website http://exergy.se).

A simple procedure for performing energy and exergy analyses involves the following steps:

- Subdivide the process under consideration into as many sections as desired, depending on the depth of detail and the understanding desired from the analysis.
- Perform conventional mass and energy balances on the process, and determine all basic quantities (e.g., work, heat) and properties (e.g., temperature, pressure).
- Based on the nature of the process, the acceptable degree of analysis complexity and accuracy, and the questions for which answers are sought, select a reference-environment model.

- Evaluate energy and exergy values relative to the selected reference-environment model.
- Perform exergy balances, including the determination of exergy consumptions.
- Select efficiency definitions depending on the measures of merit desired, and evaluate the efficiencies.
- Interpret the results and draw appropriate conclusions and recommendations relating to such issues as design changes and retrofit plant modifications.

Exergy Analysis and Efficiency

Increases in efficiency are subject to two constraints, which are often poorly understood:

- Theoretical limitations, which establish the maximum efficiency theoretically attainable for a process by virtue of the laws of thermodynamics
- Practical limitations, which further limit increases in efficiency.

First, consider practical limitations on efficiency. In practice, the goal when selecting energy sources and utilization processes is not to achieve maximum efficiency, but rather to achieve an optimal trade-off between efficiency and such factors as economics, sustainability, environmental impact, safety, and societal and political acceptability. This optimum is dependent on many factors controllable by society. Furthermore, these factors can be altered to favor increased efficiency (e.g., governments can offer financial incentives that render high-efficiency technologies economically attractive or provide disincentives for low-efficiency alternatives through special taxes and regulations).

Next, consider theoretical limitations on efficiency, which must be clearly understood to assess the potential for increased efficiency. Lack of clarity on this issue in the past has often led to confusion, in part because energy efficiencies generally are not measures of how nearly the performance of a process or device approaches the theoretical ideal. The consequences of such confusion can be significant. For example, extensive resources have at times been directed towards increasing the energy efficiencies of devices that in reality were efficient and had little potential for improvement. Conversely, devices at other times have not been targeted for improved efficiency even though the difference between the actual and maximum theoretical efficiencies, which represents the potential for improvement, has been large.

The difficulties inherent in energy analysis are also attributed to the fact that it only considers quantities of energy and ignores energy quality, which is continually degraded during real processes. Exergy analysis overcomes many of the problems associated with energy analysis.

Overview of Exergy Analysis Applications

Exergy analysis has been applied to a wide range of processes and systems, including those that are mechanical, thermal, electrical, and chemical. The types of applications of exergy methods that have been reported over the last several decades include

- Electricity generation using both conventional devices such as fossil and nuclear power plants as well as alternative devices such as fuel cells and solar energy systems
- Energy storage systems such as batteries, pumped storages, and thermal energy storages
- Combustion technologies and systems and engines of various types
- Transportation systems for land, air, and water transport
- Heating and cooling systems for building systems and industrial applications
- Cogeneration systems for producing heating and electrical needs simultaneously
- Chemical processes such as sulfuric acid production, distillation, and water desalination, as well as petrochemical processing and synthetic fuels production
- Metallurgical processes such as lead smelting.

Examples of Exergy Analysis Applications

Three examples of differing complexity of applications of exergy analysis are presented:

- An electrical resistance space heater (a simple component)
- A thermal energy storage system (a simple system containing a number of components)
- A coal-fired electrical generating station (a complex system).

Electrical Resistance Space Heater

An electrical resistance space heater converts electricity to heat at a temperature suitable for room comfort and is illustrated in Figure 1a.

The energy efficiency of electric resistance space heating often exceeds 99%, implying that the maximum possible energy efficiency for electric resistance heating is 100%, corresponding to the most efficient device possible.

This understanding is erroneous; however, energy analysis ignores the fact that in this process, high-quality energy (electricity) is used to produce a relatively low-quality product (warm air). Exergy analysis recognizes this difference in energy qualities and indicates the exergy of the heat delivered to the room to be about 5% of the exergy entering the heater. Thus, the exergy efficiency of electric resistance space heating is about 5%.

The exergy results are useful. Since thermodynamically ideal space heating has an exergy efficiency of 100%, the same space heating can in theory be achieved using as little as 5% of the electricity used in conventional electric resistance space heating. In practical terms, one can achieve space heating with a greatly reduced electricity input using an electric heat pump (see Figure 1b), using 15% of the electricity that electric resistance heating would require, for a heat pump with a "coefficient of performance" of 7.

Thermal Storage System

A thermal energy storage system receives thermal energy and holds it until it is required. Thermal storages can store energy at temperatures above or below the environment temperature, and they come in many types (e.g., tanks, aquifers, ponds, caverns).

The evaluation of a thermal energy storage system requires a measure of performance which is rational, meaningful, and practical. The conventional energy storage efficiency is inadequate. A more perceptive basis is needed if the true usefulness of thermal storages is to be assessed and their economic benefit optimized, and exergy efficiencies provide such performance measures.

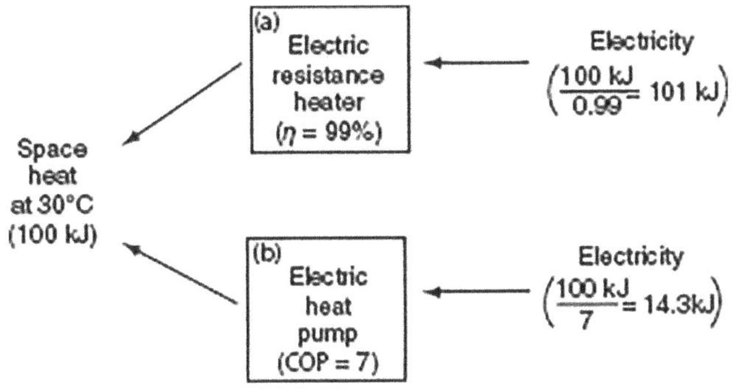

FIGURE 1 Comparison of the quantity of electricity required to provide 100 kJ of space heat using two different heating devices: (a) an electric resistance heater and (b) an electric heat pump. Here, η denotes the energy efficiency and COP the coefficient of performance.

The notion that energy efficiency is an inappropriate measure of thermal storage performance can be illustrated. Consider a perfectly insulated thermal storage containing 1,000 kg of water, initially at 40°C. The ambient temperature is 20°C, and the specific heat of water is taken to be constant at 4.2 kJ/kg K. A quantity of 4,200 kJ of heat is transferred to the storage through a heat exchanger from an external body of 100 kg of water cooling from 100°C to 90°C. This heat addition raises the storage temperature 1.0°C to a value of 41°C. After a period of storage, 4200 kJ of heat is recovered from the storage through a heat exchanger, which delivers it to an external body of 100 kg of water, raising the temperature of that water from 20°C to 30°C. The storage is returned to its initial state at 40°C.

For this storage cycle, the energy efficiency—the ratio of heat recovered from the storage to heat injected—is 4,200kJ/4,200 kJ = 1, or 100%. But the recovered heat is at only 30°C and is of little use, having been degraded even though the storage energy efficiency was 100%. The exergy recovered in this example is 70 kJ and the exergy supplied 856 kJ. Thus, the exergy efficiency, the ratio of the thermal exergy recovered from storage to that injected, is 70/856 = 0.082 or 8.2%, a much more meaningful expression of the achieved performance.

Consequently, a device which appears to be ideal on an energy basis is correctly shown to be far from ideal on an exergy basis, clearly demonstrating the benefits of using exergy analysis for evaluating thermal storage.

Coal-Fired Electrical Generating Station

Energy and exergy analyses are applied to the former Nanticoke coal-fired electrical generating station in Ontario, Canada, which has a net unit electrical output of approximately 500 MWe and is operated by the provincial electrical utility, Ontario Power Generation (formerly Ontario Hydro). This example illustrates how exergy analysis allows process inefficiencies to be better pinpointed than an energy analysis does and how efficiencies are to be more rationally evaluated.

A detailed flow diagram for a single unit of the station is shown in Figure 2. The symbols identifying the streams are described in a–c for material, thermal, and electrical flows, respectively, with corresponding data. Figure 2 has four main sections:

- *Steam Generation.* Eight pulverized-coal-fired natural circulation steam generators each produce 453.6 kg/s steam at 16.89 MPa and 538°C and 411.3 kg/s of reheat steam at 4.00 MPa and 538°C. Air is supplied to the furnace by two 1,080 kW 600-rpm motor-driven forced draft fans. Regenerative air preheaters are used. The flue gas passes through an electrostatic precipitator rated at 99.5% collection efficiency and exits the plant through two multiflued, 198 m high chimneys.
- *Power Production:* The steam passes through a series of turbine generators linked to a transformer. Extraction steam from several points on the turbines preheats feedwater in several low- and high-pressure heat exchangers and one spray-type open deaerating heat exchanger. The low-pressure turbines exhaust to the condenser at 5 kPa. Each station unit has a 3,600-rpm, tandem-compound, impulse-reaction turbine generator containing one single-flow high-pressure cylinder, one double-flow intermediate-pressure cylinder, and two double-flow low-pressure cylinders. Steam exhausted from the high-pressure cylinder is reheated in the combustor.
- *Condensation:* Cooling water from Lake Erie condenses the steam exhausted from the turbines. The cooling-water flow rate is adjusted to achieve a specified cooling-water temperature rise across the condenser.
- *Preheating:* The temperature and pressure of the feedwater are increased in a series of pumps and feedwater-heater heat exchangers.

The reference-environment model used here has a temperature of 15°C (the approximate mean temperature of the lake cooling water), a pressure of 1 atm, and a chemical composition consisting of air saturated with water vapor, and the following condensed phases at 15°C and 1 atm: water (H_2O),

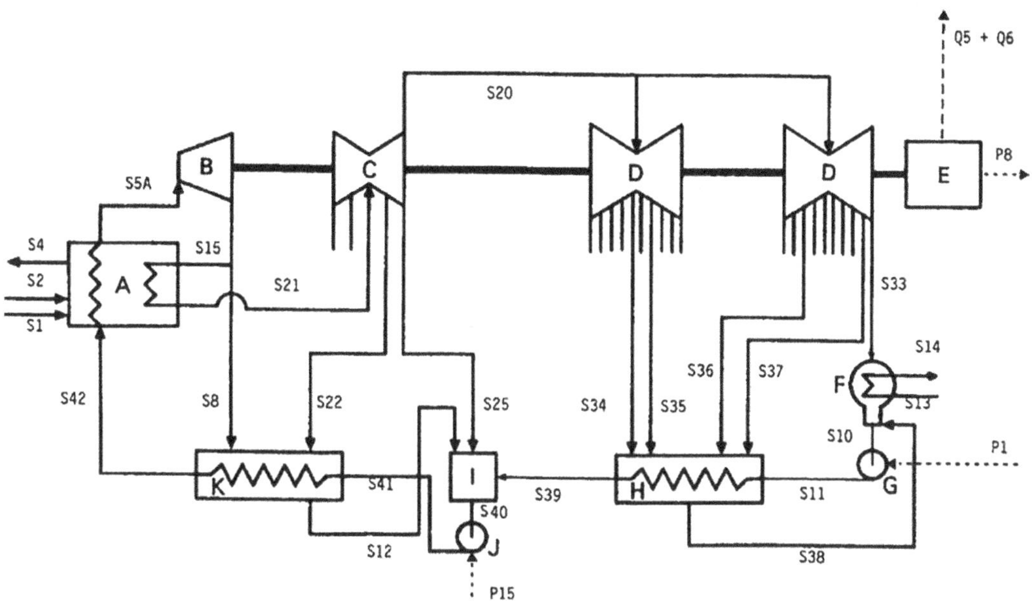

FIGURE 2 A unit of the coal-fired electrical generating station. Lines exiting the turbines represent extraction steam. The station has four main sections: *Steam Generation* (Device A), *Power Production* (B–E), *Condensation* (F), and *Preheating* (G–K). The external inputs for Device A are coal and air, and the output is stack gas and solid waste. The external outputs for Device E are electricity and waste heat. Electricity is input to Device G and Device J, and cooling water enters and exits Device F. A: steam generator and reheater, B: high- pressure turbine, C: intermediate-pressure turbine, D: low-pressure turbines, E: generator and transformer, F: condenser, G: hot well pump, H: low-pressure heat exchangers, I: open deaerating heat exchanger, J: boiler feed pump, and K: high-pressure heat exchangers.

gypsum ($CaSO_4 \bullet 2H_2O$), and limestone ($CaCO_3$). For simplicity, heat losses from external surfaces are assumed to occur at the reference-environment temperature of 15°C.

Energy and exergy values for the streams identified in Figure 2 are summarized in Tables 1a–1c. Exergy-consumption values for the devices are listed, according to process section, in Table 2. Figure 3a and b illustrates the net energy and exergy flows and exergy consumptions for the four main process sections.

Overall energy and exergy efficiencies are evaluated as

$$\text{Energy Efficiency} = (\text{Net energy output with electricity})/(\text{Energy input}) \tag{3}$$

and

$$\text{Exergy Efficiency} = (\text{Net exergy output with electricity})/(\text{Exergy input}) \tag{4}$$

Coal is the only input source of energy or exergy, and the energy and exergy efficiencies are 37% and 36%, respectively. The small difference in the efficiencies is due to the fact that the specific chemical exergy of coal is slightly greater than its energy. Although the station energy and exergy efficiencies are similar, these efficiencies differ markedly for many station sections.

In the *Steam Generation* section, exergy consumptions are substantial, accounting for 659 MW (or 72%) of the 916 MW station exergy loss. Of this 659, 444 MW is consumed with combustion and 215 MW with heat transfer. The energy and exergy efficiencies for the *Steam Generation* section, considering the increase in energy or exergy of the water as the product, are 95% and 49%, respectively.

TABLE 1A Data for Material Flows for a Unit of the Coal-Fired Electrical Generating Station

Stream	Mass Flow Rate (kg/s)[a]	Temperature (°C)	Pressure (N/m^2)	Vapor Frac.[b]	Energy Flow Rate (MW)	Exergy Flow Rate (MW)
S1	41.74	15.00	1.01×10^5	solid	1367.58	1426.73
S2	668.41	15.00	1.01×10^5	1.0	0.00	0.00
S3[c]	710.15	1673.59	1.01×10^5	1.0	1368.00	982.85
S4	710.15	119.44	1.01×10^5	1.0	74.39	62.27
S5A	453.59	538.00	1.62×10^7	1.0	1585.28	718.74
S8	42.84	323.36	3.65×10^6	1.0	135.44	51.81
S10	367.85	35.63	4.50×10^3	0.0	36.52	1.20
S11	367.85	35.73	1.00×10^6	0.0	37.09	1.70
S12	58.82	188.33	1.21×10^6	0.0	50.28	11.11
S13	18,636.00	15.00	1.01×10^5	0.0	0.00	0.00
S14	18,636.00	23.30	1.01×10^5	0.0	745.95	10.54
S15	410.75	323.36	3.65×10^6	1.0	1298.59	496.81
S20	367.85	360.50	1.03×10^6	1.0	1211.05	411.16
S21	410.75	538.00	4.00×10^6	1.0	1494.16	616.42
S22	15.98	423.23	1.72×10^6	1.0	54.54	20.02
S25	26.92	360.50	1.03×10^6	1.0	88.64	30.09
S33	309.62	35.63	4.50×10^3	0.93	774.70	54.07
S34	10.47	253.22	3.79×10^5	1.0	32.31	9.24
S35	23.88	209.93	2.41×10^5	1.0	71.73	18.82
S36	12.72	108.32	6.89×10^4	1.0	35.77	7.12
S37	11.16	60.47	3.45×10^4	1.0	30.40	5.03
S38	58.23	55.56	1.33×10^4	0.0	11.37	0.73
S39	367.85	124.86	1.00×10^6	0.0	195.94	30.41
S40	453.59	165.86	1.00×10^6	0.0	334.86	66.52
S41	453.59	169.28	1.62×10^7	0.0	347.05	77.57
S42	453.59	228.24	1.62×10^7	0.0	486.75	131.93

[a] The composition of all streams is 100% H_2O, except that, on a volume basis, the composition of S1 is 100% carbon, of S2 is 79% N_2 and 21% O_2, and of both S3 and S4 is 79% N_2, 6% O_2, and 15% CO_2.
[b] Vapor fraction is listed as 0.0 for liquids and 1.0 for superheated vapors.
[c] Stream S3 (not shown in Figure 2) represents the hot product gases for adiabatic combustion.

TABLE 1B Data for Principal Thermal Flows for a Unit of the Coal-Fired Electrical Generating Station

Stream	Energy Flow Rate (MW)	Exergy Flow Rate (MW)
Q5	5.34	0.00
Q6	5.29	0.00

TABLE 1C Data for Principal Electrical Flows for a Unit of the Coal-Fired Electrical Generating Station

Stream	Energy (and Exergy) Flow Rate (MW)
P1	0.57
P8	523.68
P15	12.19

TABLE 2 Breakdown of Exergy Consumption Rates for a Unit of the Coal-Fired Electrical Generating Station

Section/Device	Exergy Consumption Rate (MW)	
Steam generation		
Steam generator (including combustor)	659.0	
		659.0
Power production		
High-pressure turbine	26.4	
Intermediate-pressure turbine	22.3	
Low-pressure turbines	59.2	
Generator	5.3	
Transformer	5.3	
		118.5
Condensation		
Condenser	43.1	
		43.1
Preheat		
Low-pressure heat exchangers	10.7	
Deaerating heat exchanger	5.1	
High-pressure heat exchangers	6.4	
Hot well pumps	0.1	
Boiler feed pumps	1.1	
		23.4
Total		844.0

The *Steam Generation* section thus appears significantly more efficient on an energy basis than on an exergy basis. Physically, this discrepancy implies that although 95% of the input energy is transferred to the preheated water, the energy is degraded as it is transferred. Exergy analysis highlights this degradation.

In the condensers, a large quantity of energy enters (775 MW for each unit), of which close to 100% is rejected, while a small quantity of exergy enters (54 MW for each unit), of which about 25% is rejected and 75% is internally consumed. Thus, energy analysis leads to the erroneous conclusion that almost all losses in electricity-generation potential for the station are associated with the heat rejected by the condensers, while exergy analysis demonstrates quantitatively and directly that the condensers are responsible for little of these losses (see Figure 3b). This discrepancy arises because heat is rejected by the condensers at a temperature very near that of the environment.

In the *Power Production* and *Preheating* sections, energy losses are small (less than 10 MW) and exergy losses moderately small (118 MW in *Power Production* and 23 MW in *Preheating*). The exergy losses are almost completely associated with internal consumptions.

In assessing the thermodynamic characteristics of a coal-fired electrical generating station, several illuminating insights into performance are acquired:

- Although energy and exergy efficiencies are similar for the station, energy analysis does not identify the location and cause of process inefficiencies, while exergy analysis does. Energy losses are associated with emissions (mainly heat rejected by condensers), and exergy losses are primarily associated with consumptions (mainly in the combustors).
- Because devices with the largest thermodynamic losses have the largest margins for efficiency improvement, efforts to increase the efficiencies of coal-fired electrical generating stations should focus on the combustors. For instance, technologies capable of producing electricity without

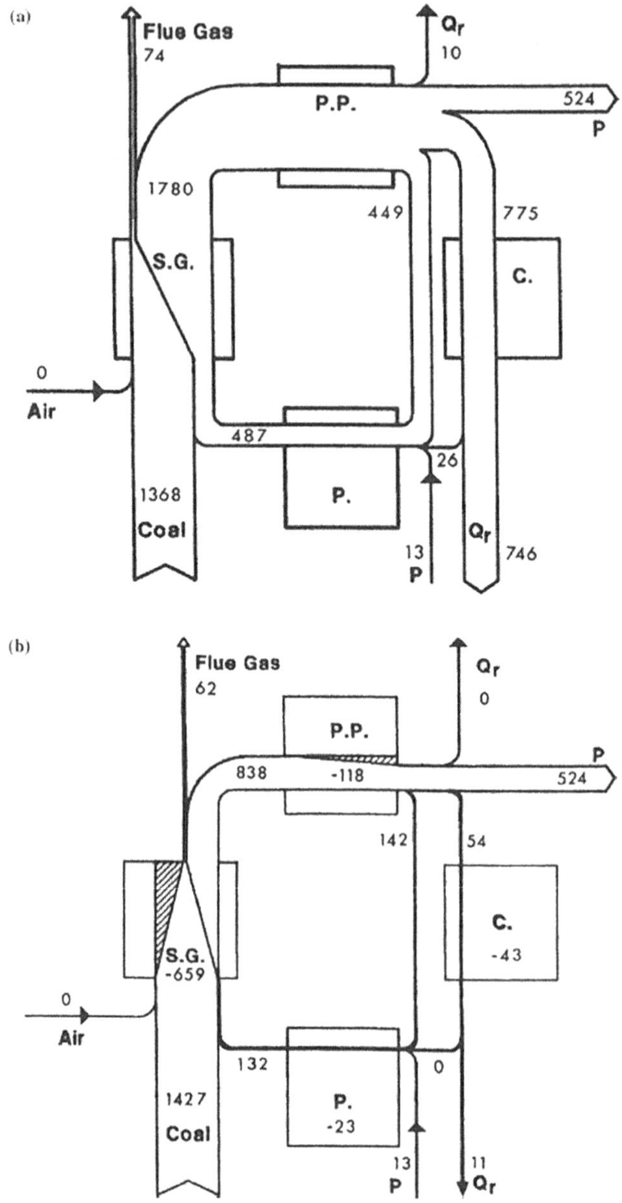

FIGURE 3 (a) Diagram for a coal-fired electrical generating station unit indicating net energy flow rates (MW) for streams. Stream widths are proportional to energy flow rates. Station sections shown are *Steam Generation* (S.G.), *Power Production* (P.P.), *Condensation* (C.), and *Preheating* (P.). Streams shown are electrical power (P), heat input (Q), and heat rejected (Q_r). (b) Diagram for a coal-fired electrical generating station unit indicating net exergy flow rates for streams and consumption rates (negative values) for devices (in MW). Stream widths are proportional to exergy flow rates and shaded regions to exergy consumption rates. Other details are as in (a).

combustion (e.g., fuel cells) or utilizing heat at high temperatures could increase efficiencies significantly. This suggestion is, of course, overly simplistic, as such decisions must also account for other technical and economic factors.

- The use of heat rejected by condensers only increases the exergy efficiencies by a few percent.

Applications beyond Thermodynamics

Exergy concepts can be applied beyond thermodynamics in such fields as environmental impact assessment,[7,8,16,17] economics,[5,17,20,26] and policy.[27,28]

Exergy and Environment

Many suggest that the impact of energy utilization on the environment is best addressed by considering exergy. Although the exergy of an energy form or a substance is a measure of its usefulness, exergy is also a measure of its potential to cause change. The latter point suggests that exergy may be or may provide the basis for an effective measure of the potential of a substance or energy form to impact the environment. The relation between exergy and the environment is discussed in this encyclopedia in an entry entitled "Exergy: Environmental Impact Assessment."

Exergy and Economics

Another area in which applications of exergy are increasing is that of economics. In the analysis and design of energy systems, techniques are often used that combine scientific disciplines like thermodynamics with economics to achieve optimum designs. For energy systems, costs are conventionally based on energy. Many researchers, however, have recommended that costs are better distributed among outputs based on exergy. Methods of performing exergy-based economic analyses have evolved (e.g., thermoeconomics, second-law costing, and exergoeconomics). These analysis techniques recognize that exergy, not energy, is the commodity of value in a system, and assign costs and prices to exergy-related variables. These techniques usually help in appropriately allocating economic resources so as to optimize the design and operation of a system and its economic feasibility and profitability (by obtaining actual costs of products and their appropriate prices).

Conclusion

Exergy analysis provides information that influences design, improvement, and application decisions, and it is likely to be increasingly applied. Exergy also provides insights into the "best" directions for research, where "best" is loosely considered most promising for significant efficiency gains. There are two main reasons for this conclusion:

- Unlike energy losses, exergy losses represent true losses of the potential to generate the desired product from the given driving input. Focusing on exergy losses permits research to aim at reducing the losses that degrade efficiency.
- Unlike energy efficiencies, exergy efficiencies always provide a measure of how closely the operation of a system approaches the ideal. By focusing research on plant sections or processes with the lowest exergy efficiencies, effort is directed to those areas that inherently have the largest margins for efficiency improvement. By focusing on energy efficiencies, on the other hand, research can inadvertently be expended on areas for which little margins for improvement exist, even theoretically.

Exergy analysis results typically suggest that improvement efforts should concentrate more on internal rather than external exergy losses based on thermodynamic considerations, with a higher priority for the processes that have larger exergy losses. Of course, effort should still be devoted to processes having low exergy losses when cost- effective ways to increase efficiency can be identified.

Energy-related decisions should not be based exclusively on the results of energy and exergy analyses even though these results provide useful information to assist in such decision-making. Other factors must also be considered, such as economics, environmental impact, safety, and social and political implications.

Glossary

COP	coefficient of performance
i	component
P	pressure
P_0	reference-environment pressure
T	temperature
T_0	reference-environment temperature
η	energy efficiency
μ	chemical potential
$\mu 00$	reference-environment chemical potential

Acknowledgments

The author is grateful for the support provided for this work by the Natural Sciences and Engineering Research Council of Canada.

References

1. Rezac, P.; Metghalchi, H. A brief note on the historical evolution and present state of exergy analysis. *Int. J. Exergy* **2004**, *1*, 426–437.
2. Kotas, T.J. *The Exergy Method of Thermal Plant Analysis*; Krieger: Malabar, FL, Reprint ed.; 1995.
3. Moran, M.J. *Availability Analysis: A Guide to Efficient Energy Use*; American Society of Mechanical Engineers: New York, Revised ed.; 1989.
4. Brodyanski, V.M.; Sorin, M.V.; Le Goff, P. *The Efficiency of Industrial Processes: Exergy Analysis and Optimization*; Elsevier: London, 1994.
5. Szargut, J.; Morris, D.R.; Steward, F.R. *Exergy Analysis of Thermal, Chemical and Metallurgical Processes*; Hemisphere: New York, 1988.
6. Yantovskii, E.I. *Energy and Exergy Currents (An Introduction to Exergonomics)*; Nova Science Publishers: New York, 1994.
7. Szargut, J. *Exergy Method: Technical and Ecological Applications*; WIT Press: Southampton, 2005.
8. Dincer, I.; Rosen, M.A. *Exergy: Energy, Environment and Sustainable Development*, 2nd ed.; Elsevier: Oxford, 2013.
9. Moran, M.J.; Shapiro, H.N.; Boettner, D.D.; Bailey, M.B. *Fundamentals of Engineering Thermodynamics*, 9th ed.; Wiley: New York, 2018.
10. Sato, N. *Chemical Energy and Exergy: An Introduction to Chemical Thermodynamics for Engineers*; Elsevier: Oxford, 2005.
11. Moran, M.J.; Sciubba, E. Exergy analysis: principles and practice. *J. Eng. Gas Turbines Power* **1994**, *116*, 285–290.
12. Aghbashlo, M.; Rosen, M.A. Consolidating exergoeconomic and exergoenvironmental analyses using the emergy concept for better understanding energy conversion systems. *J. Clean. Prod.* **2018**, *172*, 696–708.
13. Milia, D.; Sciubba, E. Exergy-based lumped simulation of complex systems: an interactive analysis tool. *Energy* **2006**, *31*, 100–111.
14. Frangopoulos, C.A.; von Spakovsky, M.R.; Sciubba, E. A brief review of methods for the design and synthesis optimization of energy systems. *Int. J. Appl. Thermodyn.* **2002**, *5* (4), 151–160.
15. Rosen, M.A.; Dincer, I. Effect of varying dead-state properties on energy and exergy analyses of thermal systems. *Int. J. Therm. Sci.* **2004**, *43* (2), 121–133.
16. Jorgensen, S.E.; Svirezhev, Y.M. *Towards a Thermodynamic Theory for Ecological Systems*; Elsevier: Oxford, 2004.

17. Rosen, M.A. *Environment, Ecology and Exergy: Enhanced Approaches to Environmental and Ecological Management*; Nova Science Publishers: Hauppauge, NY, 2012.
18. Al-Zareer, M.; Dincer, I.; Rosen, M.A. Multi-objective optimization of an integrated gasification combined cycle for hydrogen and electricity production. *Comput. Chem. Eng.* **2018**, *117*, 256–267.
19. Rosen, M.A.; Koohi-Fayegh, S. *Cogeneration and District Energy Systems: Modelling, Analysis and Optimization*; Institution of Engineering and Technology: London, 2016.
20. Mahmoudi, S.M.S.; Salehi, S.; Yari, M.; Rosen, M.A. Exergoeconomic performance comparison and optimization of single-stage absorption heat transformers. *Energies* **2017**, *10*(4), 532.
21. Rad, F.M.; Fung, A.S.; Rosen, M.A. An integrated model for designing a solar community heating system with borehole thermal storage. *Energy Sustainable Dev.* **2017**, *36C*, 6–15.
22. Dincer, I.; Rosen, M.A. Energy and exergy analyses of thermal energy storage systems. In *Thermal Energy Storage: Systems and Applications*, 2nd ed.; Wiley: London, 2011; 233–334.
23. Rezaie, B.; Reddy, B.V.; Rosen, M.A. Thermodynamic analysis and the design of sensible thermal energy storages. *Int. J. Energy Res.* **2017**, *41*(1), 39–48.
24. Rosen, M.A.; Dincer, I. Sectoral energy and exergy modeling Turkey. *ASME J. Energy Resour. Technol.* **1997**, *119*, 200–204.
25. Rosen, M.A. Assessing global resource utilization efficiency in the industrial sector. *Sci. Total Environ.* **2013**, *461–462*, 804–807.
26. Jawad, H.; Jaber, M.Y.; Bonney, M.; Rosen, M.A. Deriving an exergetic economic production quantity model for better sustainability. *Appl. Math. Modell.* **2016**, *40*(11–12), 6026–6039.
27. Rosen, M.A. Benefits of exergy and needs for increased education and public understanding and applications in industry and policy. Part I: Benefits. *Int. J. Exergy* **2006**, *3*(2), 202–218.
28. Rosen, M.A. Benefits of exergy and needs for increased education and public understanding and applications in industry and policy. Part II: Needs. *Int. J. Exergy* **2006**, *3*(2), 219–229.

V

ENT: Environmental Management Using Environmental Technologies

21

Air Pollution: Monitoring

Waldemar Wardencki

Introduction

Concern about air quality is not new. The first reports of air pollution problems appear to have been made by writers in ancient Rome who were aware of its adverse effects on human health. Air pollution and its consequences had originally been considered to be relatively local phenomena associated with urban and industrial centers. Complaints were recorded in the 13th century when coal was first used in London. Now, it has become apparent that pollutants may be transported long distances in the air, causing adverse effects in environments far removed from the source emission. Scientific research, conducted over 200 years, has evidently shown that polluted air has a negative influence on health and, in some cases, may lead to death. The World Health Organization (WHO) appraises that air pollution causes approximately 2 million premature deaths worldwide per year. The levels of pollutants, which have a negative influence on life, are nowadays well defined. Because current thresholds set by national or global air quality guideline values are frequently exceeded, further reductions of emissions are necessary.

The first essential step in controlling and mitigating air pollution is to quantify the emissions of air pollutants. Most countries entail controlling a range of key pollutants at their point of discharge. The most important tool in environmental protection is monitoring. Environmental monitoring is the general term for systematic observations of what is going on in the environment. In the broadest context, environmental monitoring is defined as a system of measurements, evaluations, and forecasts of environmental states, and the collecting, processing, and spreading of information on the environment.

Air pollution and its control are a global issue demanding international cooperation. Monitoring of air pollution is a very important source of data. However, measurement of air pollutant concentrations, in comparison to monitoring of other elements of the environment, is the most difficult. This is related to the dynamics of the atmosphere, making it the main route of pollutant transport between the remaining environmental compartments. Unlike the case of water and soil pollution, environmental pollution is not geographically restricted, as a result of which large human populations can be exposed to it. Another problem is low concentration of air pollutants and their interaction with other gases.

This entry reviews the issues in the field of air pollution monitoring. At the beginning, the general objectives of air monitoring, ambient air quality standards for so-called criteria pollutants, and their sources are presented. In the next part, both analytical methods and instruments for monitoring of ambient air and stack gases are briefly presented. Additionally, other approaches applied in air pollution monitoring, such as biomonitoring, geographical information system (GIS), or remote monitoring, are also briefly characterized.

Objectives of Air Monitoring

Collecting information on the presence and concentration of pollutants in the environment, both naturally occurring or from anthropogenic sources, may be achieved by measurements of such substances or phenomenon of interest. For realistic assessment, temporal and spatial variations of concentrations in the particular environmental compartment, repeated measurements rather than single ones, are made.

The general aim of monitoring is to provide information about the actual levels of harmful or potentially harmful pollutants to indicate areas in which the quality of air does not fulfill proper standards. The main objectives of air monitoring are as follows:

- To measure pollutant mixing ratios and their interactions, patterns, and fate in the environment.
- To carry out ecotoxicological studies and assessment of the effects of pollution on man and the environment, to identify possible cause-and-effect relationship between pollutant concentration and health effects.
- To assess emission sources and the need for legislative controls on emissions of pollutants and to ensure compliance with emission standards.
- To activate emergency procedures in areas prone to acute pollution episodes.
- To obtain a historical record of air quality to provide a database for future use.

The area of applications of air monitoring data is presented in Figure 1.

When the objectives of monitoring are clearly defined, several decisions should be made to generate suitable data for the intended use. Decisions on what to monitor, when and where to monitor, and how to monitor are usually undertaken at the beginning. More difficult are next decisions, e.g., establishing the number and location of sampling sites, the duration of the survey, and the time resolution of sampling. All the steps in the design of a monitoring program are presented in Figure 2.

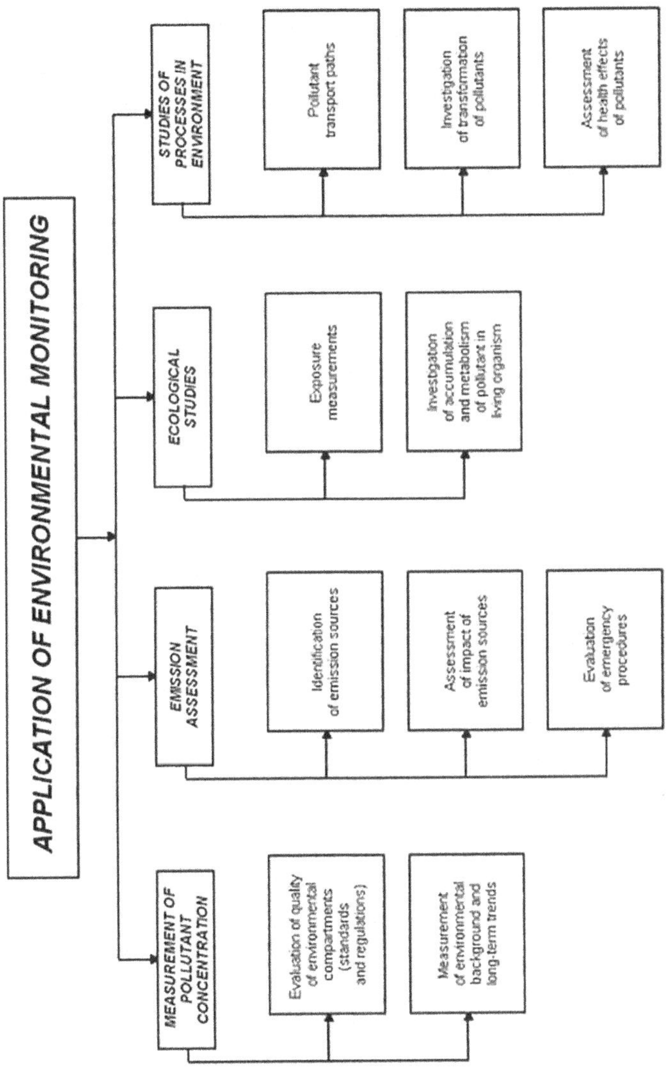

FIGURE 1 The detailed goals of activities in air monitoring.

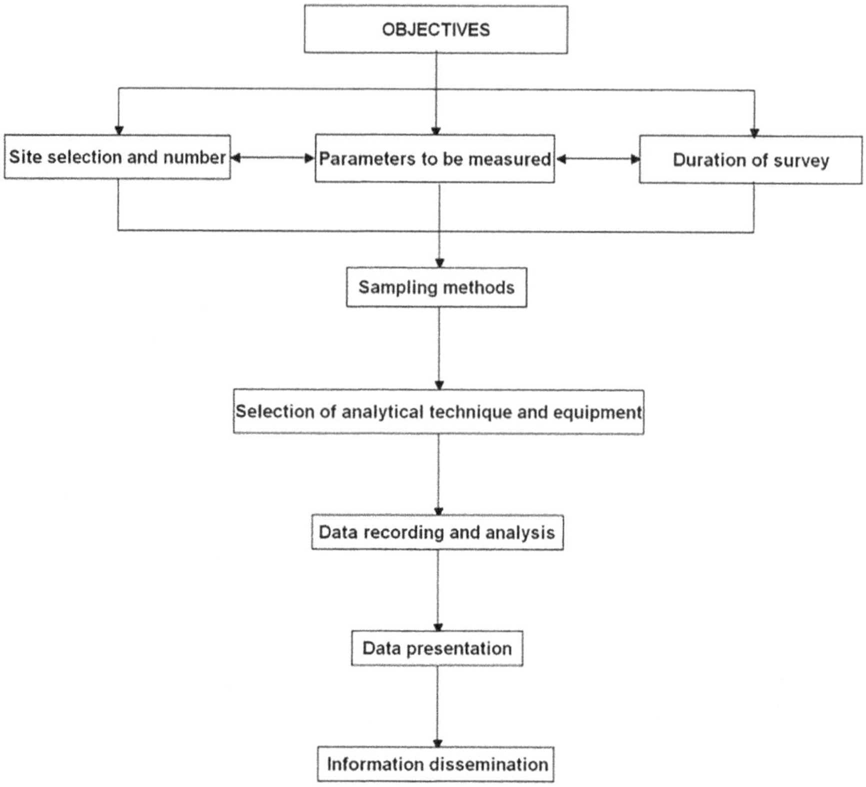

FIGURE 2 Steps in the design of a monitoring program.

History of Air Pollution Legislation

A growing concern over the influence of different air pollutants on human health was the main driving force to develop and implement air quality criteria and standards. Impetus was given to the development of air quality standards in 1958 when it was realized that photochemical problems could not be resolved without control of motor vehicle emission.

Air Quality Standards

Efforts to regulate air quality by law were discussed and undertaken in the 1960s. One of the first proposition was presented by Atkisson and Gaines[1] in 1970. The initial regulations were set in California in response to concerns over human health.[2] After that, similar air quality programs were soon adopted nationally. In 1967, the U.S. Congress enacted the Air Quality Acts, the first modern environmental law.

The Clean Air Act,[3] which was last amended in 1990, requires the United States Environmental Protection Agency (US EPA) to set National Ambient Air Quality Standards (NAAQS) for pollutants considered harmful to public health and the environment. The Clean Air Act established two types of national air quality standards. *Primary standards* set limits to protect public health, including the health of "sensitive" populations such as asthmatics, children, and the elderly. *Secondary standards* set limits to protect public welfare, including protection against decreased visibility and damage to animals, crops, vegetation, and buildings. EPA has established NAAQS for six principal pollutants, which are called criteria pollutants: sulfur dioxide, particulate matter (PM), nitrogen oxide, carbon monoxide, ozone, and lead. These standards are threshold concentrations based on a detailed review of scientific information related to effects.

In Europe, the first international air quality standards were introduced by the European Commission in 1980 for SO_2 and suspended particulates, mainly aimed at protecting human health. A few years later, the WHO, recognizing ecological damage as being relevant to human health, introduced air quality guidelines for Europe, which include the former as well as the latter revision in 2000.[4] The newest directive on ambient air quality and cleaner air of the European Union (EU) entered into force in June 2008.[5] It merges four earlier directives and one Council decision into a single directive on air quality. The new directive of the EU on air quality takes into account concerns from latest WHO air quality guidelines[6] on fine particles. Reflecting the latest WHO air quality guidelines that identify fine particles (PM2.5) as one of the most dangerous pollutants for human health, the new EU directive sets objectives and target dates for reducing population exposure to PM2.5. It also maintains limits for concentration of coarser particles known as PM10 and other main pollutants already subject to legislation.

Table 1 presents examples of air quality standards issued by the EPA, the WHO, and some states.

TABLE 1 Comparison of Limit Values (pm/m^3) for a Given Averaging Time and the Number of Exceedances per Year Issued by Different Countries and Organizations

Pollutant	Averaging Time	WHO	EPA	EU	UK	France	Germany	Poland
SO_2	10–15 min	500	–	–C	266 (not more than 35 times)	–	–	–
	30 min	–	–	–	–		–	–
	1 hr	–	–	350 (not more than 24 times)	350 (not more than 24 times)	350 (not more than 24 times)	350	350 (not more than 24 times)
	3 hr	–	1,300	–	–	–	–	–
	24 hr	125	365	125 (not more than 3 times)	125 (not more than 3 times)	125 (not more than 3 times)	125	125 (not more than 3 times)
	Year	50	80	20	20	20	20	20
NO_2	30 min	–	–	–	–	–	200	–
	1 hr	200	–	200 (not more than 18 times)	200 (not more than 18 times)	230 (not longer than 0.2 % of time)	–	200 (not more than 18 times)
	24 hr	–	–	–	–	–	100	–
	Year	40	100	40	40	46	–	40
PM10	30 min	–	–	–	–	–	–	–
	24 hr	20	150	50 (not more than 35 times)	50 (not more than 35 times)	50 (not more than 35 times)	50	50 (not more than 35 times)
	Year	50	–	40	40	40	40	40
CO	10–15 min	100,000	–	–	–	–	–	–
	30 min	60,000	–	–	–	–	–	–
	1 hr	30,000	4,000	–	–	–	–	–
	8 hr	10,000	1,000	1,000	1,000	1,000	1,000	1,000
	24 hr	–	–	–	–	–	–	–
	Year	–	–	–	–	–	–	–
O_3	30 min	–	–	–	–	–	–	–
	1 hr	–	235	–	–	–	–	–
	8 hr	100	157	120	100 (not more than 10 times)	120	120	120 (not more than 25 days)
Pb	24 hr	–	–	–	–	–	5	–
	3 months	–	1.5	–	0.5	–	–	–
	Year	0.5	–	0.5	–	0.5	0.5	0.5
Benzene	Year	–	–	5	16.25	8	–	5

Source: WHO.[6]

Air Quality Index

The Air Quality Index (AQI), also known as the Air Pollution Index (API) or Pollutant Standard Index (PSI), is a number used by different government agencies to characterize the quality of the air at a given location. The index aims to help the public easily understand air quality level and protect the health of people from air pollution. As the AQI increases, an increasingly large percentage of the population is likely to experience increasingly severe adverse health effects. Computing the AQI requires an air pollutant concentration from a monitor or model. The function used to convert air pollutant concentration to AQI varies by pollutant and is different among countries. AQI values expressed in different values (most frequently from 0 or 1 to 10, 100, or to 500) are divided into ranges (from 4 to 10), and each range is assigned a descriptor and a color code. Standardized public health advisories are associated with each AQI range. An agency might also encourage members of the public to take public transportation or work from home when AQI levels are high.

Not all air pollutants are characterized by AQI. Many countries monitor only some pollutants, e.g., ground-level ozone, sulfur dioxide, carbon monoxide, and nitrogen dioxide, and calculate AQIs for these pollutants.[7]

The EPA in the United States measures air quality in all parts of the country and publishes a daily AQI based on the data obtained. The following six priority pollutants are measured regularly in order to generate the AQI: carbon monoxide, nitrogen dioxide, particulates, sulfur dioxide, ozone, and lead. EPA has assigned a specific color to each AQI category to make it easier for people to understand quickly whether air pollution is reaching unhealthy levels in their communities. For example, the color orange means that conditions are "unhealthy for sensitive groups," while red means that conditions may be "unhealthy for everyone," and so on.

In Canada, API has values of 0 up to 100+ and is divided into four categories (from good to very poor).

CITEAIR (Common Information to European Air) has developed the first AQIs in Europe.[8] An important feature of the indices is that they differentiate between traffic and city background conditions. The Common Air Quality Index (CAQI) is designed to present and compare air quality in near real time on an hourly or daily basis. The CAQI has five levels, using a scale from 0 (very low) to >100 (very high) and the matching colors range from light green to dark red. The Year Average Common Air Quality Index (YACAQI) uses a different approach adopting the difference to target's principle. If the index is higher than 1.0, it means that for one or more pollutants, the limit values are not met. If the index is below 1, it means that on average the limit values are met. Both indices are practically implemented on a Common Operational Webpage (COW).[9] The project CITEAIR II will further develop the AQIs.[10]

The EU's Sixth Environment Action Programme (EAP), "Environment 2010: Our Future, Our Choice," includes Environment and Health as one of the four main target areas requiring greater effort—air pollution is one of the issues highlighted in this area. The Sixth EAP aims to achieve levels of air quality that do not result in unacceptable impacts on, and risks to, human health and the environment.

The EU is acting at many levels to reduce exposure to air pollution: through EC legislation, through work at the international level to reduce cross-border pollution, through cooperation with sectors responsible for air pollution, through national and regional authorities and NGOs, and through research. The Clean Air for Europe (CAFE) initiative has led to a thematic strategy setting out the objectives and measures for the next phase of European air quality policy.[11]

Regulated Air Pollutants

The contaminants in ambient air that are of concern are basically categorized as criteria and non-criteria pollutants.[12]

Criteria air pollutants are those air contaminants for which numerical concentration limits have been set as the dividing line between acceptable air quality and poor or unhealthy air quality. Criteria pollutants include five gases/vapors and two solids: nitrogen oxides (NO_x), sulfur dioxide (SO_2), carbon monoxide (CO), ozone (O_3), benzene, PM10, and lead (Pb). Non-criteria pollutants are those contaminants designated as toxic or hazardous by legislation or regulation. They fall in two further subcategories, depending on the legislation that defines them. In general, hazardous air pollutants may pose a variety of health effects, whereas toxic ones focus on one physiological response.

Main Sources of Air Pollutants

Air pollution may be defined as a situation in which substances change the qualitative composition of air in relation to average composition of troposphere sufficiently high above their normal ambient air levels to produce a measurable effect on humans, animals, vegetation, or material.[13–17]

Pollutants (both organic and inorganic) may be present in different forms such as gases, aerosols (liquid, solid), and sorbates and have a very broad range of concentration.

The concentrations of ambient air pollutants are expressed in terms of either a mass per unit volume ratio, such as μg/m³, or a volumetric ratio (i.e., volumes of contaminant per million or billion volumes of air). The conversion between mass units and volumetric ratios at standard temperature or pressure is:[12]

$$\mu g / m^3 = ppm \times MW/0.02445 = ppb \times MW/24.45,$$

where:

μg/m³—micrograms per cubic meter
ppm—parts per million by volume ($1{:}10^6$)
ppb—parts per billion by volume ($1{:}10^9$).
MW—molecular weight of the contaminant

Some of the most important atmospheric pollutants, their sources, and impacts on the environment and human health are presented in Table 2.

TABLE 2 Atmospheric Pollutants and Their Sources and Effects

Pollutant	Sources	Impact
Sulfur dioxide (SO_2)	Power generation, industry	Acid deposition, smog formation, threat to human health, smog formation
Nitrogen oxides (NO, N_2O, NO_2)	Transport, power generation, industry	Acid deposition, smog formation, O_3 precursor, threat to human health
Carbon dioxide (CO_2)	Combustion processes, power generation, transport, landfills	Global warming
Carbon monoxide (CO)	Combustion processes, power generation	Toxic to humans
Particulate matter (PM)	Power generation, industry, transport	Threat to human health, reduced visibility
Volatile organic compounds (VOCs)	Transport, industry	Photochemical smog, O_3 precursor, global warming
Ozone (O_3)	Photochemical reactions between VOCs and NO_x	Photochemical smogs, respiratory irritant, crop damage
Methane (CH_4)	Landfills, agriculture, gas industry	Global warming
Benzene, 1,3-butadiene	Transport industry	Carcinogenic
Ammonia (NH_3)	Industry, farming, refrigeration, power plant	Toxic to humans and wildlife
Heavy metals	Industry, transport	Toxic to humans and wildlife
Dioxins and furans	Incineration, electrical equipment	Toxic to humans

Source: Bogue.[18]

Urban traffic has become the most important cause of air pollution in the cities.[15] Road traffic is responsible for emission of several air pollutants; the most important of which are nitrogen oxides (NO and NO_2), sulfur dioxide (SO_2), PM, carbon monoxide (CO), and volatile organic compounds (VOCs), all of which can pose a health hazard.

Characteristics of Criteria Air Pollutants

Air pollutants arise from a wide variety of sources although they are mainly a result of the combustion process.[15] The largest sources include power generation, motor vehicles, and industries. The emissions of pollutants to the atmosphere badly influence vegetation, human and animal life, agriculture, and climate. Emissions of carbon monoxide (CO), nitrogen oxides (NO_x), and hydrocarbons are controlled by catalytic converters on new gasoline-driven cars. Emissions of sulfur oxides are being reduced through a lower sulfur content in gasoline. However, emissions of PM are not decreasing. Any successful strategies for controlling or countering these problems must be based on reliable air quality monitoring data for management, to make informed decisions on air pollution control.

Volatile organic compounds is a collective name for a very large number of different chemical species, including hydrocarbons, halocarbons, and oxygenates that have different physicochemical properties and are directly emitted from both anthropogenic and natural sources, and which can contribute to the formation of secondary pollutants with different efficiencies. For vehicular emissions, the list of compounds is long and variable depending on fuel, type of engine, and operating conditions. Hydrocarbons such as ethane, ethyne, higher aliphatic hydrocarbons, benzene, toluene, and xylenes are typical emissions in most cases. Each of these compounds can be released unreacted or can undergo oxidation reactions. One of them, benzene, is found in highest concentrations. Ambient concentrations are typically between 1 and 50 ppb, but close to major emissions can be as high as several hundred parts per billion. In the unreacted state, it has undesirable ecotoxicological properties. Besides causing annoying physiological reactions such as dizziness and membrane irritation, it is known to be a human carcinogen.

The two nitrogen oxides, NO and NO_2 (together called NO_x), from anthropogenic sources are present as a consequence of various combustion processes from both stationary sources, i.e., power generation (21%), and mobile sources, i.e., transport (44%). These species have very short atmospheric lifetimes, around 5 days, and have been ultimately converted to nitric acid and removed in rainfall. However, nitrogen oxide is important because it is a precursor to tropospheric ozone. Whereas NO does not affect climate, ozone does. A typical sea-level mixing ratio of NO is 5 ppt (parts per trillion, $1{:}10^{12}$), but in urban regions, NO mixing ratios reach 0.1 ppm in the early morning, but it decreases to zero by midmorning due to reaction with ozone. A major source of NO_2 is oxidation of NO, with NO_2 being intermediary between NO emission and O_3 formation. Nitrogen dioxide is one of the six criteria air pollutants for which ambient standards are set by the US EPA under CAAA70 (Clean Air Act Amendments of 1970). In the urban regions, the mixing ratio of NO_2 ranges from 0.1 to 0.25 ppm. It is more prevalent during midmorning than during midday or afternoon because sunlight breaks down most NO_2 past midmorning. Exposure to high concentrations of NO_2 harms the lungs and increases respiratory infections. It may trigger asthma by damaging or irritating and sensitizing the lungs, making people more susceptible to allergens. At higher concentrations, it can result in acute bronchitis or death.

Sulfur dioxide (SO_2) is a strong-smelling, colorless gas that is formed by the combustion of fossil fuels, smelting, manufacture of sulfuric acid, conversion of wood pulp to paper, incineration of refuse, and production of elemental sulfur. Power plants, which may use coal or oil high in sulfur content, can be major sources of SO_2, accounting for about 50% of annual global emissions. SO_2 and other sulfur oxides contribute to the problem of acid deposition and can be major contributors to smog. Natural background levels of SO_2 are about 2 ppb. Hourly peak values can reach 750 ppb on infrequent occasions. Sulfur dioxide can lead to lung diseases. SO_2 is a criteria air pollutant.

Ozone (O_3) is not directly emitted from both anthropogenic and natural sources. Its only source into air is chemical reaction. In the urban air, ozone mixing ratios range from less than 0.01 ppm at night to 0.5 ppm (during afternoons in the most polluted cities worldwide), with typical values of 0.15 ppm during moderately polluted afternoons. Ozone causes headaches at concentrations greater than 0.15 ppmv, chest pains at mixing ratios greater than 0.25 ppm, and sore throat and cough at mixing ratios greater than 0.30 ppm. Exceeding the level 0.30 ppm, it decreases lung functions. Symptoms of a respiratory condition include coughing and breathing discomfort. Ozone can also accelerate the aging of lung tissue. It also interferes with the growth of plants and deteriorates organic materials, such as rubber, textiles, and some paints and coatings. Furthermore, ozone increases plant and tree stress and their susceptibility to disease, infestation, and death.

PM, frequently described simply as particle pollution, in the atmosphere arise from natural sources, such as wind-borne dust, sea spray, and volcanoes, and from anthropogenic activities, such as combustion of fuels. Particle pollution in the air includes a mixture of solids and liquid droplets and come in a wide range of sizes. Those less than 10 micrometers (μm) in diameter (PM10) are so small that they can get into the lungs, potentially causing serious health problems. Particles less than 2.5 pm in diameter are called fine particles. These particles are so small that they can be detected only with an electron microscope. Sources of fine particles include all types of combustion, including motor vehicles, power plants, residential wood burning, forest fires, agricultural burning, and some industrial processes. Particles between 2.5 and 10 μm in diameter are referred to as coarse. Sources of coarse particles include crushing or grinding operations and dust stirred up by vehicles traveling on roads. After releasing into air, particles can change their size and composition by condensation of vapor species or by evaporation, by coagulating with other particles, or by chemical reaction. Particles with a diameter smaller than 1 pm generally have atmospheric concentrations in the range from around 10 to several thousands per cubic meter; those exceeding 1 pm diameter are usually found at concentration less than 1 cm^{-3}.

Carbon monoxide (CO) is a colorless, odorless gas that is produced by the incomplete burning of carbon-based fuels including petrol, diesel, and wood. It is also produced from the combustion of natural and synthetic products such as cigarettes. Natural background levels of CO fall in the range of 10–200 ppb. Levels in urban areas are highly variable, depending upon weather conditions and traffic density. Eight-hour mean values are generally less than 10 ppm, but sometimes, they can be as high as 500 ppm. Carbon monoxide lowers the amount of oxygen that enters the blood. It can slow human reflexes and make people confused and sleepy.

Air Quality Monitoring

Design of Monitoring Networks for Air Pollution

For the purpose of monitoring and reporting air pollution, most industrialized countries have been divided into regions or zones and urban areas or agglomeration, e.g., in Europe, in accordance with EC Directive 96/62/EC.[19] This Directive sets a framework for ways how to monitor and report ambient levels of air pollutants. Other directives set ambient air limit values for particular pollutants:

- Directive 99/30/EC for nitrogen dioxide and oxides of nitrogen, sulfur dioxide, lead, and PM.
- Directive 2000/69/EC for benzene and carbon monoxide.
- Directive 2002/3/EC for ozone.

The monitoring sites are organized into automatic and non-automatic networks (regional and national) that gather a particular kind of information using a particular method. For example, across the U.K., there are more than 1500 monitoring sites that monitor air quality, and these are organized into networks (automatic and non-automatic) that gather a particular kind of information, using a particular method. The pollutants measured and method used by each network depend on the reason for setting

FIGURE 3 Stationary monitoring station of the ARMAAG network.

FIGURE 4 General view of an interior of a monitoring station.

up the network, and what the data are to be used for. In Poland, monitoring of air quality has been performed systematically since 1992, mainly by using automatic air monitoring stations. Air quality monitoring data are used at national, voivodship (provincial), and local scale. An exemplary air automated monitoring station, together with a general view of analyzers situated in them, is shown in Figures 3 and 4. This station belongs to the Agency of Regional Air Monitoring of the Gdansk Agglomeration (ARMAAG), which is Poland's first local government-owned air monitoring network. Foundation ARMAAG provides information of air condition in Gdansk agglomeration in real time from the automatic measurement network.

The obtained data from air monitoring are used in air quality inventories and bulletins.

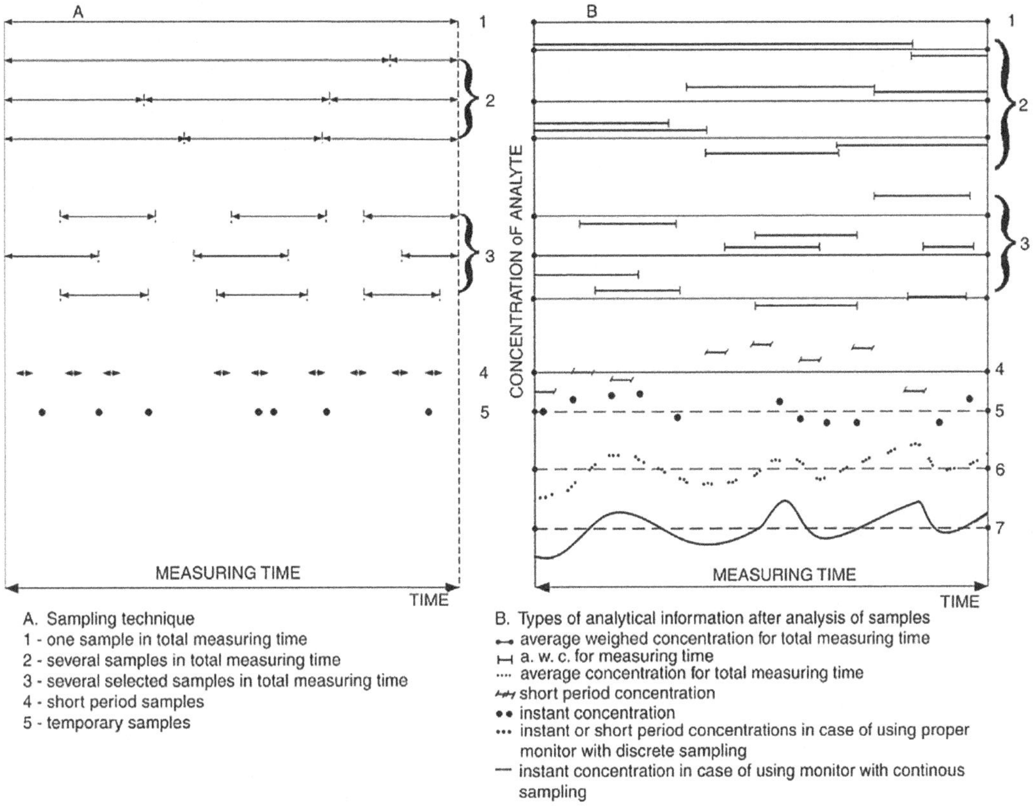

FIGURE 5 Schematic diagram of different sampling techniques used for getting information on concentration of pollutants in determined measuring time.

Types of Information Obtained from Air Monitoring

The obtained information concerns different types of concentration of investigated pollutants depending on applied sampling techniques and measuring period. The results of measurements may be referred to real time (instantaneous concentrations) or to a selected period of time (e.g., 1 hr, 8 hr, 24 hr, 1month, 1year). Final measurements represent averaged concentrations.

Figure 5 presents schematically different forms of concentration obtained as a function of sampling time.[20]

Considering the frequency of sampling, discrete, periodic, and instantaneous measurements are distinguished. Taking into account space, parameter measurements are divided to a point, averaged along a defined part of space, and averaged on the selected area. Point monitoring is inadequate to measure poorly mixed gases such as fugitive emissions over large areas. If the point instrument is wrongly placed, measurement results are not representative. Final measurements enable determination of weighted average concentrations over the sampling period.

General Requirements of the Instruments Used in Monitoring

Monitoring of air pollution is a prerequisite of air quality control and is carried out by a wide variety of analytical methods employing different measuring instruments (analyzers) that have different sensitivities and specificities.[21]

The basic requirement of the analyzers for air monitoring is high measurement sensitivity, i.e., the low limit of detection (LOD) and the low limit of quantitation (LOQ). It gives a chance to detect the pollutants at required levels. The instruments that should acquire analytical data in real time or only with a small time delay have to possess the following additional capabilities:

- Providing high data resolution (characterized by low response time).
- Providing automatic calibration and zeroing.
- Long functioning without service.

The last demand means that they should be equipped with an independent power supply and be able to automatically regenerate or exchange worn-out filters and, depending on the type of detector used (sensor), fulfill special demands, e.g., for electrochemical sensors, exchange or supplement the working solution and reagents, and in devices with Flame Ionisation Detector (FID) or Flame Photometric Detector (FPD) detection, protect against flame extinguishing.

Depending on the number of analytes that an instrument can determine in a single sample, they can be single-parameter (single gas) or multiparameter (multigas) instruments.

Based on sampling frequency, analyzers can be discrete (for single measurement), periodic (for measurements at preset intervals), or continuous (for permanent monitoring).

Classification of the Instruments Used for the Detection and Monitoring of Air Pollutants

The analytical instruments currently used for the detection and determination of atmospheric pollutants can be classified according to various criteria.

Recently, measuring techniques based on a physical (or physicochemical) principle are more frequently used in the assessment of air quality. Such methods involve direct determination of a physical property of a pollutant, sometimes after its interaction with another compound. In this approach, better stability, sensitivity, and reliability may be easily achieved. Furthermore, the practical application requires less maintenance. Instruments based on this principle can be easily automated, which enables their use in providing continuous measurements needed for up-to-date assessments of air quality. It is especially relevant to environmental monitoring because many existing standards refer to specified period of time, i.e., 1 hr, 24 hr, or 1 year.

According to the location where measurements are taken, instruments can be stationary or *on-site*. In the first case, analysis is performed in the laboratory and sophisticated instruments are applied, such as mass, electron mobility, or x-ray fluorescence spectrometers. *On-site* systems enable measuring of pollution levels in the field. Since access to a sophisticated laboratory is not required, the devices (usually uncomplicated, relatively cheap, and portable) hold great promise for use in remote locations. The main advantage of on-site analysis is the potential for rapid assessment and response to a particular problem.

All monitoring systems can be classified as mobile or stationary. Most existing systems monitoring gaseous pollutants of atmospheric air and ambient aerosols, both automatic and manual, usually perform stationary measurements; i.e., they are directly linked to a specific point or space in the vicinity of that point. Basing on the data obtained from single monitoring sites, it is not possible to assess spatial and temporal variations of air pollutants.

Mobile refers to a continuous-monitoring instrument that is portable or transportable. They are usually designed to perform analytical measurements without preliminary operations. Portable refers to self-contained, battery-operated, or worn or carried by the person using it, or may require the use of special vehicles for placement in a specific area to be monitored. Transportable gas monitors can be mounted on a vehicle such as a car, plane, balloon, ship, or space shuttle, but not to a mining machine or industrial truck.

For mobile systems, the registered values of pollutants have to be correlated with information about the geographical site and actual meteorological conditions (temperature and humidity).

Portable systems for field measurements should meet the following requirements:

- Compactness and robustness
- Ease of handling
- Adaptability to on-site measurements
- Automated operation with a long-lasting power supply
- Stability under aggressive environmental conditions

Several contributions published during the last decade have proven the advantages of mobile systems in getting information concerning the spatial and temporal distribution of atmospheric trace gases, without the need of a dense network of stationary stations. Most of the proposed systems are based on application of mobile laboratories,[21–23] equipped with appropriate monitors. There are also systems that allow to measure pollutants in a stream of vehicles but the measuring unit is installed on any vehicle[24–27] rather than attached to a dedicated van.[28,29]

The general trend in the field of creating instruments for air quality assessment is combining several instruments into one system and forming so-called hybrid multisensor systems, controlled by a microprocessor capable of transferring the obtained data to a central station, frequently using a wireless system. In the central station, the data are collected both from single objects (houses, plants) and from large areas. Many systems are equipped with devices for testing the sensors and for providing diagnosis of the whole instrument. Frequently, they have alarms that warn the user of any dangerous situation due to the breaching of some value limit. Such systems are battery powered and able to work continuously for several days or months.

The environments in which analyzers are used differ from the relative calm of the laboratory. Analyzers have to withstand wide ambient temperatures, fluctuations, and vibrations. Due to this, many systems are completely sealed so as to operate independent of outside conditions and be able to withstand the onslaught of monkey-wrench mechanics.

General Characteristics of the Methods and Analytical Instruments for Air Monitoring

Due to the complexity of environmental problems and the variety of pollutants and their different concentrations (typically parts per million or percentage levels in stack gases and parts per billion in air), there is a wide range of methods and instruments used in measuring ambient air quality.

Based on physicochemical principles, the monitoring instruments involve direct determination of the different physical properties of the pollutant or following its interaction with another compound. These methods allow determination of air pollutants in a continuous and automatic way. Such approach requires extremely sensitive instrumentation. Therefore, the most advanced techniques, comprising chemistry, physics, and microelectronics, should be used. As a result, the instruments are combinations of many different devices giving one measuring system. In developing such a system, it should be remembered that it will be exposed to environmental impact, such as changes in temperature, dustiness, humidity of air, aggressive components of air, vibrations, and transportation stress.

The typical instruments used for atmospheric ambient monitoring are based on optical, electrochemical, and semiconductor principles. Among spectroscopic techniques, chemiluminescence, infrared (IR), and fluorescence are the most frequently applied.

The range of typical measuring methods and techniques used in air monitoring is shown in Table 3.

Among the many different optical spectroscopic methods, differential optical absorption spectroscopy (DOAS) has found wide use in atmospheric research and air quality monitoring. The technique is based on the measurements of absorption features of gas molecules along a path of known length in the open atmosphere. The DOAS systems, due to the calibration-free absolute measurements and the unequivocal identification of many trace pollutants, such as CO, SO, NO, and VOCs, and highly

TABLE 3 Air Quality Monitoring Techniques Used in Air Monitoring

Pollutant	Emissions	Ambient Air
CO_2	FTIR, NDIR, TDLAS	NDIR, DOAS
CO	FTIR, NDIR, TDLAS, DOAS	NDIR (gas filter correlation variant)
SO_2		UV fluorescence
NO_x		Chemiluminescence
PM	Triboelectric, opacity, beta ray attenuation	Beta ray attenuation, oscillating microbalance, gravimetric
VOCs	FID, GC	FID, GC
CH_4	NDIR, FTIR,TDLAS, FID	FID, GC
O_3		UV absorption spectroscopy, DOAS, electrochemical sensors
NH_3	FTIR,TDLAS, chemiluminescence GC	Chemiluminescence, DOAS GC
Benzene, 1,3-butadiene	In situ GC-MS with continuous sampling	MS, GC-MS
Dioxins and furans	XRF, LIBS, cold vapor AFS, atomic	Sampling plus MS or GC-MS
Metals	Emission spectrometry	Sampling plus ICP-MS, DOAS (for Hg)

Source: Bogo.[29]

FTIR, Fourier transform infrared absorption spectroscopy; NDIR, non-dispersive infrared absorption; TDLAS, tunable laser diode absorption spectroscopy; DOAS, differential optical absorption spectroscopy; FID, flame ionization detector; GC, gas chromatography; MS, mass spectrometry; XRF, x-ray fluorescence; LIBS, laser-induced breakdown spectroscopy; ICP, inductively coupled plasma; AES, atomic emission spectrometry; AFS, atomic fluorescence spectrometry.

reactive radicals, e.g., OH, NO, and halogen oxide radicals, are exploited by air quality monitoring agencies around Europe and in the United States. The physical and chemical principles, the current state of this measurement method, and details for users are broadly presented in a recently published book.[30]

In Europe, standard/reference methods (EN) are provided by the European standardization body (CEN). The standard methods use the following principles:

Chemiluminescence for NO EN 14211

Nitrogen oxide reacts with ozone, generated within the instrument, produces an excited molecule of nitrogen dioxide, which emits light returning to its original state. A photomultiplier tube measures the emitted light that, if the volumes of sample gas and excess ozone are carefully controlled, is proportional to the concentration of NO in the gas sample.[31]

The chemiluminescent method used for nitrogen oxides is based on the following reactions:

$$NO + O_3 \rightarrow NO_2^{*} + O_3$$

$$NO_2^{*} \rightarrow NO_2 + h\nu$$

The chemiluminescence technique may be used to measure total oxides of nitrogen (NO_x) by passing the sample over a heated catalyst to reduce all oxides of nitrogen to NO. This is done within the instrument, just prior to the reaction chamber. Some instruments can perform the automatic switching of the catalyst in and out of the sample path so that the resulting signals may be compared to indirectly measure NO_2.

Ultraviolet Fluorescence for SO_2 EN 14212

Sulfur dioxide is measured without chemical pretreatment by gas-phase fluorescence spectrometry in the UV region. Molecules of SO_2 are excited by UV radiation (200–220 nm) into unstable forms, which return to a basic state, emitting radiation in the range of 240–420 nm according to following reactions:[32]

$$SO_2 + h\nu \rightarrow SO_2^*$$

$$SO_2^* \rightarrow SO_2 + (UV)$$

The intensity of fluorescence radiation is proportional to sulfur dioxide in the sample.

Non-Dispersive Infrared for CO EN 14626

The analytical principle is based on absorption of IR light by the CO molecule. NDIR-GFC (non-dispersive infrared-gas filter correlation) analyzers operate on the principle that CO has a sufficiently characteristic IR absorption spectrum such that the absorption of IR by the CO molecule can be used as a measure of CO concentration in the presence of other gases. CO absorbs IR maximally at 2.3 and 4.6 pm. Because many other molecules also absorb IR radiation in practice, different technical designs are proposed. The following approaches are typically applied:[33]

- Measurement of IR absorption at specific wave for CO (2.3 or 4.6 pm)
- Analyzers with two cells, one of which is filled with pure air (compensation of drift)
- Analyzers with turning circle (GFC)

Ultraviolet Photometry for O_3 EN 14625

Upon exposure to UV light, ozone absorbs some of the light and the intensity difference is directly proportional to the concentration of ozone. Frequently, the UV light source is a 254 nm emission line from a mercury discharge lamp.[34]

Online Gas Chromatography for Benzene EN 14662-Part 3

A measured volume of sample air is drawn or forced through a sorbent tube. Provided suitable sorbents are chosen, benzene is retained by the sorbent tube and thus is removed from the flowing air stream. The collected benzene (on each tube) is desorbed by heat and is transferred by inert carrier gas into a gas chromatograph equipped with a capillary column and a flame ionization detector or another suitable detector, where it is analyzed. Prior to entering the column, the sample is concentrated either on a cryo trap, which is heated to release the sample into the column, or on a pre-column, where higher boiling hydrocarbons are removed from the pre-column by back flush. Two general types of instruments are used. One is equipped with a single sampling trap and the other is equipped with two or more traps. The single-trap instrument samples for only part of the time in each cycle, whereas the multitrap instrument samples continuously. Typical cycling times are between 15 min and 1 hr.[35]

PM is usually determined using active filter method by gravimetry as the reference method.[36] In this method, the air is passed through a filter that stops particles above 10 pm (PM10) or 2.5 pm (PM2.5). Measurements are made over a period of 24 hr or longer. The filters are collected and the adsorbed particles are measured in the laboratory. Other methods use beta ray absorption or tapered element oscillating microbalance (TEOM) of PM. In beta gauge instruments, which are used for real-time measurements of particulate emissions from stationary sources, the mass of the sample deposited on the filter tape is automatically measured by beta ray attenuation. The measurement is made first on a blank, then on the particulate-laden filter. The range is 2–4000 mg/m^3 without interference or effect from color, size, or atomic mass of the dust.

An interesting review on *online* analyzers for monitoring of VOCs was recently published.[37]

Other Monitoring Approaches

Biomonitoring Using Plants

Modern air instruments cannot measure the effects air pollution has on living cells and are limited to measuring the present conditions. Biological materials can be an excellent basis for establishing a biomonitoring network on large areas for a long time. Biomonitoring, as a continuous observation of an area with the help of bioindicators, can allow a qualitative survey and quantitative estimation of the pollutants in the environment. Since the 19th century, biological monitoring as a rather simple observation has turned into a serious alternative if not a useful complement to the traditional methods of assessing contamination, from both natural and anthropogenic sources. In the case of airborne pollution, its heyday really began after World War II. The expensive growth of biomonitoring research works has gained momentum mainly from lower organisms such as lichens, bryophytes, and, to a lesser degree, fungi. The use of cosmopolite organisms for assessing pollution has developed notably during the last two decades.

Bioindicators are organisms or organs of such organisms that respond to a certain level of pollution by the change in their life cycle or accumulation of the particular pollutant. Bioindicators, in contrast to direct analysis, reflect complex effects of harmful substances, as such organisms show not only the synergistic effects of a sum of parameters but also a time-integrated picture of the history of their life span. Another advantage is the selective uptake of such substances, as an organism exposed to an environmental pollutant, either through air or via direct uptake, absorbs the bioavailability fraction only. They readily reflect the proportion of the pollutants, which may be dangerous to human beings as well.

In air monitoring, two organisms, i.e., lichens and mosses, have become the most popular bioindicators.

Lichens are unusual organisms because they consist of fungal threads and microscopic green alga living together and functioning as a single organism. Lichens grow on rocks, soil, trees, or artificial structures in unpolluted habitats. Lichens act like sponges, taking in everything that is dissolved in the rainwater and retaining it. Different species of lichens vary in sensitivity to air pollution. Lichens are commonly used as air quality indicators since some species are more pollution tolerant than others. The most sensitive lichens are shrubby and leafy, whereas the most tolerant lichens are crustose. In city centers, lichens may be entirely absent. If the air is clean, shrubby, hairy, and leafy lichens colonize every available surface.

Lichens may be used as bioindicators in two different ways:

- By mapping all species present in a specific area.
- Through the individual sampling of lichen species and measurement of the accumulated pollutants or by transplanting lichens from a clean environment to a contaminated one.

Lichens are used as biomonitors in the examination of the level of pollutants such as sulfur, nitrogen, and phosphorus compounds, as well as ozone, fluorides, chlorides, and heavy metals. Several biomonitoring methods using lichens have been described since the 1970s when Hawksworth and Rose[38,39] proposed a method based on the determination of zones (from 0—strong pollution to 10—clear air) with selected epiphytic lichens (on two different kinds of tree bark) that relate to levels of sulfur dioxide pollution. This method was widely adopted (both in the original scale and in relation to the real concentrations of sulfur dioxide) in many countries mainly due to its simplicity. Over the last decade, new techniques like the European method for mapping lichen diversity (LDV), as an indicator of environmental stress/quality, have been proposed. The procedure is based on the fact that epiphytic lichen diversity is impaired by air pollution and environmental stress. It provides a rapid, low-cost method to define zones of different environmental quality. In addition to information on the long-term effects of air pollutants, data on eutrophication, anthropization, and climatic change on sensitive organisms may likewise be obtained. Data quality depends on the uniformity of growth conditions, and usually standardization in sampling procedure is necessary.

The relative ease of sampling, the absence of any need for complicated and expensive equipment, and the accumulative and time-integrative behavior of the monitor organisms that make biomonitoring of atmospheric pollutants possible could be continued for the foreseeable future, especially in large-scale surveys.[40–45]

GIS in Air Quality Monitoring

Reliable information on air quality is needed not only on temporal trends in air pollution (as, for example, provided by data from fixed-site monitoring stations) but also on geographical variations. Maps are needed, for example, to identify so-called hot spots, to define at-risk groups, to show changes in spatial patterns of pollution, and to provide improved estimates of exposure for epidemiological studies. The development of GIS techniques offers considerable potential to mapping air pollution. GIS technology enables obtaining statistical and spatial data on air quality by estimation of environmental levels of regulated contaminants.[46]

Remote Monitoring Techniques

Remote monitoring techniques enable the measurement of atmospheric pollutants in remote, poorly accessible, and dangerous regions.[47–49] Remote sensing is especially recommended for the detection of diffuse emissions that are hard to quantify with typical ground-point measurements, but its use is restricted to specialized monitoring demands due to very considerable cost of the equipment.

The cheapest and most widely used methods are those of aerial photography, including IR sensing and optical spectroscopy. Different spectrometers, whether they are ground based in a mobile laboratory or airborne, are the most common instruments used in air pollution for determination of SO_2 and NO_2 concentrations in plumes from tall stacks. The results provide reliable data for studying the transport and dispersion of a plume.

Typical gaseous pollutants such as NO_x, SO_2, CO, and O_3 may be monitored using different types of lasers, which, due to long-path absorption measurements, enable the determination of pollutants at very low concentrations. In laser absorption methods, a detector is used to monitor absorption of specific wavelengths in light paths. Lidar (light detection and ranging) transmits light out to a target and part of this light is reflected/scattered back to the instrument where it is analyzed. The time for the light to travel out to the target and back to the lidar is used to determine the range to the target, allowing spatial resolution of pollutant concentration data within the light path; by monitoring back-scatter intensity at two close wavelengths, the species concentration as well as its spatial distribution may be inferred. This method has been successfully used for measurements of SO_2 up to a range of 2 km.

There are three basic generic types of lidars: range finder lidars, differential absorption lidars (dial), and Doppler lidars.

Range finder lidars are used to measure the distance from the lidar instrument to a solid or hard target. Dial is used to measure pollutant concentration in the atmosphere on the basis of the difference of the two return signals having two different wavelengths (one is absorbed by the molecule of interest while the other is not absorbed). The Doppler lidar is used to measure the velocity of a target. The target can be either a hard target or an atmospheric target.

Another type of device used for remote monitoring is sodar (sound detection and ranging). The Doppler sodar sends out sound pulses of several frequencies in slightly different directions. The acoustic signals are back-scattered by inhomogeneities in the atmosphere.

Further significant developments of laser techniques use the Raman back-scatter, which is highly characteristic of the scattering molecule.

Monitoring of Flue and Exhaust Gas Emissions

Different industrial branches, e.g., coal-fired power plants, chemical plants, petrochemical plants, oil refineries, PVC factories, heavy industries, and incinerators, are principal stationary pollution sources emitting, usually by chimneys, stack (flue) gases, containing different pollutants. Monitoring of such emission sources needs continuous, automatically acting systems that are usually a multielement, integrated, and cooperative set of measuring devices, auxiliary equipment, and calibration appliance.[50,51]

Systems for Continuous Monitoring of Stack Gases

The monitoring systems for continuous monitoring can be classified on the basis of different criteria. Depending on the way in which measurement is made, and especially on the applied sampling mode, extractive and in situ systems are distinguished.[52]

In extractive systems, as the name implies, the sample is extracted continuously from a duct or stack from a representative volume of stack gases and transported by transfer line to analyzers (one or more single-component analyzers or one multicomponent analyzer). However, in most cases, some conditioning of the sample is required to remove water vapor and PM. The two main types of extractive systems are fully extractive (sometimes called simply "extractive") and dilution extractive (also known as "dilution").

A typical extractive system (Figure 6) has a stainless steel probe, with a filter to remove coarse particulates. After filtration, a heated, unchanged sample is transferred to a sample conditioner located in the system enclosure. Calibration gas is delivered from the enclosure to the probe and back through the sample tubing to calibrate the system. The simplest sampling conditioning method is cooling the sample and allowing the moisture to condense and drain out of the system.

Instruments based on spectroscopic techniques (mainly UV and IR), paramagnetic properties, and solid electrolytes (zirconium dioxide) for oxygen determination are frequently used to monitor CO, CO_2, NO_x, and SO_2. For low concentrations of NO_x and SO_2, a chemiluminescent method can be applied. Fully extractive systems can be sometimes used without moisture removal, especially when the sample contains components that are easily soluble in water.

Fully extractive systems are recommended for monitoring of pollutants in stack gases with different physicochemical parameters of compounds. Another advantage of such systems is the possibility of monitoring several locations using one analyzer (time-share systems).

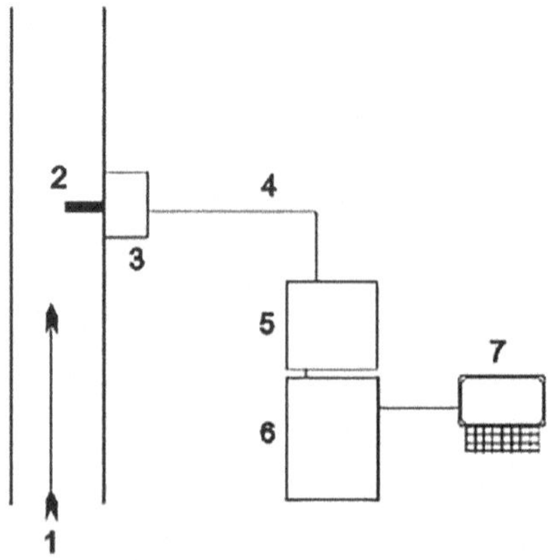

FIGURE 6 Schematic diagram of a fully extractive system for continuous emission monitoring of stack gases: 1, stack gases; 2, probe; 3, filter; 4, heated sampling line; 5, moisture removal; 6, analyzers; 7, storage unit.

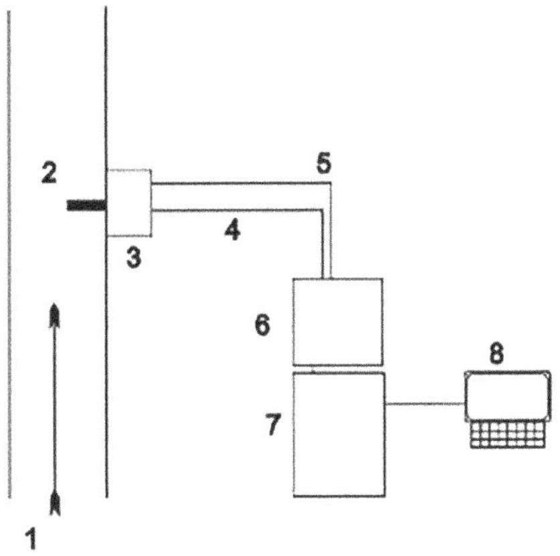

FIGURE 7 Schematic diagram of a dilution-extractive system for continuous emission monitoring of stack gases: 1, stack gases; 2, probe; 3, filter; 4, clean and dry dilution air; 5, diluted sample to analyzers; 6, dilution control unit; 7, analyzers; 8, storage unit.

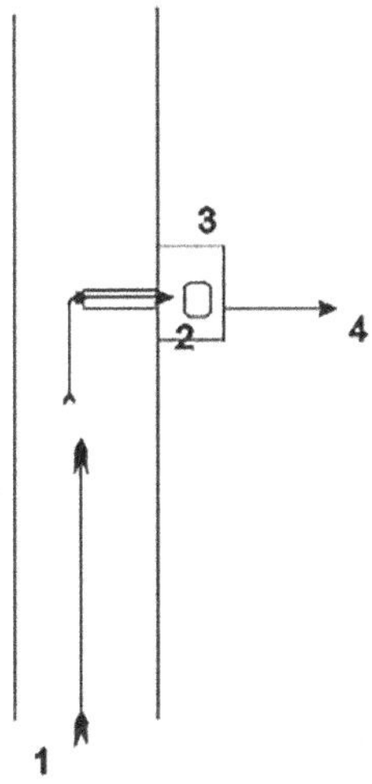

FIGURE 8 Schematic diagram for a point-type in situ system for continuous emission monitoring of stack gases (sensor mounted in the box with the sensor electronics): 1, stack gases; 2, sensor; 3, electronics in enclosure; 4, signal.

Dilution of the sample gas (Figure 7) with clean, dry air to the sample (usually from 50:1 to 250:1) considerably facilitates sample handling and reduces the dew point of the sample gas so that the sampling line can be unheated. Furthermore, the diluted sample is similar in respect of pollutant concentration to ambient air, enabling the use of ambient analyzers. Relatively small amounts of sampled gases increase the time between cleaning the filters. Because most dilution-extractive systems are affected by changes in temperature and barometric pressure, it is recommended that temperature and pressure sensors be installed at the sampling location to compensate these effects. The dilution systems are recommended for plants fueled with carbons when high levels of particulates are present in stack gases (0.1 g/m^3) and corrosive substances (e.g., HCl or SO_3).

In in situ systems (Figure 8), a gas probe is inserted into the wall. It allows monitoring the sample without removing it from the source and does not require sample conditioning or transport of the sample gas, thus minimizing the measurement errors during sampling, transferring, and conditioning

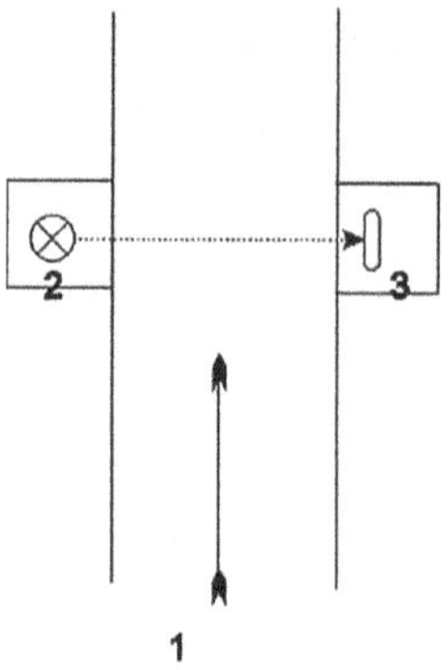

FIGURE 9 Schematic diagram of a single-path-type in situ system for continuous emission monitoring of stack gases: 1, stack gases; 2, light source; 3, receiver.

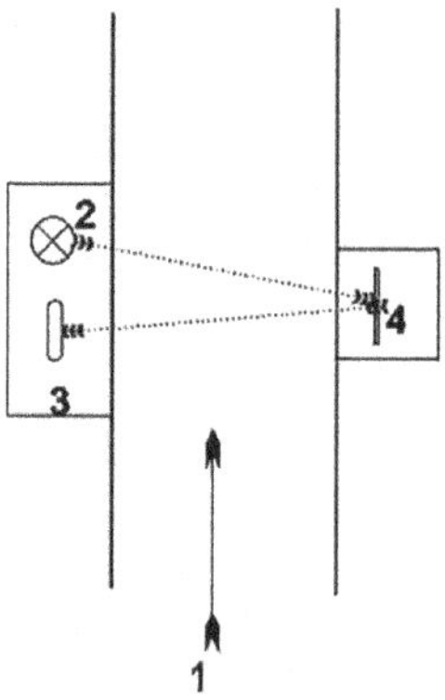

FIGURE 10 Schematic diagram of a dual-path-type in situ system for continuous emission monitoring of stack gases: 1, stack gases; 2, light source; 3, receiver; 4, reflector.

the sample. An optical beam is contained within the probe. This optical beam represents the absorption path that enables the analysis. The sample is drawn into the probe but remains under the conditions found in the stack. The sample is never removed or extracted from the stack.

In practice, two types of in situ systems are used: point and path monitoring systems (Figures 9 and 10). In point monitoring systems, a sample probe and analyzer are installed inside the stack. They are also called in-stack monitors and measure gas at a single point. Therefore, it is important to choose a location that is representative in terms of the components of interest. As analyzers in such systems, spectroscopic instruments are used (based on absorption of UV and IR radiation) as well as electrochemical devices. In situ monitoring systems are recommended for locations with easy stack access and for measuring SO_2 and O_2 in combustion sources because point monitors are very cost-effective for measuring only one or two components.

Path monitoring systems minimize errors that can arise when the location of a measuring point is not representative and due to the disturbances in the flow of stack gases. They measure gas concentration along a path, usually across the diameter of the stack or duct. A light source is mounted on one side of the stack and a beam is passed through to the other side. A single-pass system measures the light that reaches the other side of the stack, whereas a double-pass system uses a reflector and passes the light back across the stack before performing the measurements. Two parameters are limited in these systems: the length of the measuring path (no less than 0.5 m, no more than 8–10 m) and the temperature of stack gases (no more than 300°C).

In situ systems are usually mounted in the ducts after electrostatic precipitators or in chimney ducts. The systems for stack sampling can be easily adapted to process monitoring or even ambient (closed or open path) monitoring.

Conclusions

Air quality has unquestionably adverse effects on human health. For example, air pollution is increasingly being cited as the main cause of lung conditions such as asthma—twice as many people suffer from asthma today compared to 20 years ago.

This is the main reason that the issue of air quality is now a major concern for many countries that have been working to improve air quality by controlling emissions of harmful substances into the atmosphere, improving fuel quality, and integrating environmental protection requirements into the transport and energy sectors. Despite these improvements in air quality over the last few years, the problem of air pollution still remains. Therefore, more needs to be done at the local, national, and international level. For example, a wide interest is observed to establish common criteria (e.g. AQIs-Air Quality Criteria (AQC)) to compare the state of the air for different countries that follow different directives.

During the last decade, different types of AQC were proposed in literature[53,54] and/or adopted by governments. An interesting review on AQI published recently online by the Italian Group of Environmental Statistics (GRASPA) (http://www.graspa.org) shows the lack of a common strategy to compare the state of the air for cities that follow different directives.[55] The major differences between the indices in the literature are found in the number of index classes (and their associated color) and relative descriptive terms (e.g., considered pollutants, averaging time, frequency). Also, the guidelines themselves are sometimes consistently different from state to state, not only in indicating the pollutants to be monitored but also in setting the threshold values and the number of exceedances per year. Furthermore, the way air quality is interpreted on the basis of a country-or city-specific AQI differs considerably.

Monitoring of air pollution is a prerequisite of air quality control and is carried out by a wide variety of analytical methods employing different measuring instruments that have different sensitivities and specificities. Monitoring plays a critical role in protecting the environment and is a key element of all actions related with management and protection of ambient air.

Every developed country has legislation to control or limit emissions of atmospheric pollution. The air quality standards that could be regulated by law and achieved were established. Concentrations of selected pollutants would have to be determined, and reliable analytical methods would be required to measure the levels of the pollutants. Monitoring actions are based on using stationary networks of measuring stations and/ or mobile laboratories equipped with proper instruments.

References

1. Atkisson, A.; Gaines R.S., Eds.; *Development of Air Quality Standards*; Charles E. Merril Publ. Co.: Columbus, Ohio, 1970.
2. California State Department of Public Health, Health and Safety Code, Section 426.1,1959.
3. Available at http://www.epa.gov/air/caa/ (accessed May 15, 2012).
4. World Health Organisation. *Air Quality Guidelines for Europe*, 2nd Ed.; WHO Regional Publications European Series No.91: Copenhagen, 2000.
5. Directive 2008/50/EC of the European Parliament and of the Council of 21 May 2008 on ambient air quality and cleaner air for Europe. Official Journal of the European Union 11.6.2008, L. 152/1–152/44.
6. WHO air quality guidelines for particulate matter, ozone, nitrogen dioxide and sulfur dioxide—Global update 2005— Summary of risk assessment.
7. Available at http://armaag.gda.pl/normy.html (accessed May 15, 2012).

8. Available at http://citeair.rec.org/products.html (accessed May 15, 2012).
9. Available at http://www.airqualitynow.eu (accessed May 15, 2012).
10. Available at http://www.citeair.eu (accessed May 15, 2012).
11. Available at http://europa.eu/legislation_summaries/environment/air_pollution/l28026_en.htm (accessed May 15, 2012).
12. Griffin, R.D. *Principles of Air Quality Management;* Taylor & Francis Group: Boca Raton, 2007.
13. Jacobsen, M.Z. *Atmospheric Pollution. History, Science and Regulation*; Cambridge University Press: Cambridge, U.K., 2002.
14. Hobbs, P.V. *Introduction to Atmospheric Chemistry;* Cambridge University Press: Cambridge, U.K., 2000.
15. Friedrich, R.; Reis, S., Eds. *Emission of Air Pollutants;* Springer-Verlag: Berlin Heidelberg, 2004.
16. Seinfeld, J.H.; Pandis, S.N., Eds. *Atmospheric Chemistry and Physics. From Air Pollution to Climate Change*, 2nd Ed.; John Wiley and Sons, Inc.: New York, 2006.
17. Bell, J.N.B.; Treshow, M., Eds. *Air Pollution and Plant Life*; John Wiley & Sons, Ltd.: New York, 2002.
18. Bogue, R. Environmental sensing: Strategies, technologies and applications. Sens. Rev. **2008**, *28* (4), 275–282.
19. Council Directive 96/62/EC of 27 September 1996 on ambient air quality assessment and management. Official Journal L *296*, 21/11/1996 P, 0055–0063.
20. Michulec, M.; Wardencki, W.; Partyka, M.; Namiesnik, J. Analytical techniques used in monitoring of atmospheric air pollutants. Crit. Rev. Anal. Chem. **2005**, *35*, 1–17.
21. Wardencki, W.; Katulski, R.; Stefanski, J.; Namiesnik, J. The state of the art in the field of non-stationary instruments for the determination and monitoring of atmospheric pollutants. Crit. Rev. Anal. Chem. 2008, *38*, 1–10.
22. Bukowiecki, N.; Dommen, J.; Prevot, A.S.H.; Richter, R.; Weingartner, E.; Baltensperger, U. A mobile pollutant measurement laboratory—Measuring gas phase and aerosol ambient concentrations with high spatial and temporal resolution. Atmos. Environ. **2002**, *36*, 5569–5579.
23. Gouriou, F.; Morin, J.-P.; Weill, M.-E. On-road measurements of particle number concentrations and size distributions in urban and tunnel environments. Atmos. Environ. **2004**, *38* (18), 2831–2840.
24. Pirjola, L.; Parviainen, H.; Hussein, T.; Valli, A.; Hameri, K.; Aaalto, P.; Virtanen, A.; Keskinen, J.; Pakkanen, T.A.; Makela, T.; Hillamo, R.E. "Sniffer"—A novel tool for chasing vehicles and measuring traffic pollutants. Atmos. Environ. **2004**, *38*, 3625–3635.
25. Katulski, R.; Stefanski, J.; Wardencki, W.; Zurek, J. Concept of the mobile monitoring system for chemical agents control in the air, Proceedings of the IEEE Conference on Technologies for Homeland Security—Enhancing Transportation Security and Efficiency, Boston, USA, June 7–8, 2006; 181–184.
26. Katulski, R.; Stefanski, J.; Wardencki, W.; Zurek J.The Mobile Monitoring System (MMS)—A useful tool for assessing air pollution in cities, Proceedings of the Pittsburgh Conference on Analytical Chemistry and Applied Spectroscopy PITTCON'2007, February 25–March 2, 2007, Chicago, USA, 2007.
27. Katulski, R.; Namiesnik, J.; Sadowski, J.; Stefanski, J.; Szymanska, K.; Wardencki, W. Mobile system for on-road measurements of air pollutants. Rev. Sci. Instrum. **2010**, *81*, 045104.
28. Seakins, P.W.; Lansley, D.L.; Hodgson, A.; Huntley, A.; Pope, F. New directions: Mobile laboratory reveals new issues in urban air quality. Atmos. Environ. **2002**, *36*, 1247–1248.
29. Bogo, H.; Negri, R.M.; San Roman, E. Continuous measurement of gaseous pollutants in Buenos Aires city. Atmos. Environ. **1999**, *33*, 2587–2598.
30. Platt, U.; Stutz, J. *Differential Optical Absorption Spectroscopy. Principles and Applications*; Springer-Verlag: Berlin, Heidelberg, 2008.
31. EN 14211. *Ambient air quality.* Standard method for the measurement of the concentration of nitrogen oxide by chemilumiscence, 2005.

32. EN 14212. *Ambient air quality.* Standard method for the measurement of the concentration of sulphur dioxide by ultraviolet fluorescence, 2005.
33. EN 14626. *Ambient air quality.* Standard method for the measurement of the concentration of carbon monoxide by nondispersive infrared spectroscopy, 2005.
34. EN 14625. *Ambient air quality.* Standard method for the measurement of the concentration of ozone by ultraviolet photometry, 2005.
35. EN 14662-3. *Ambient air quality.* Standard method for measurement of benzene concentrations—Part 3: Automated pumped sampling with in situ gas chromatography, 2005.
36. Larssen, S.; de Leew, F. *PM10 and PM2.5 concentration in Europe as assessed from monitoring data reported to AirBase*, ETC/ACC; 2007.
37. Król, S.; Zabiegala, B.; Namiesnik, J. Monitoring of volatile organic compounds (VOCs) in atmospheric air. Part I. On-line gas analyzers. Trends Anal. Chem. **2010**, *29* (9), 1092–1100.
38. Hawksworth, D.L.; Rose F. Quantitative scale for estimating sulphur dioxide air pollution in England and Wales using epiphytic lichens. Nature, **1970**, *227,* 145–148.
39. Nash, T.H.; Gries C. Lichens as bioindicators of sulphur dioxide. Symbiosis **2002**, *33* (1), 1–21.
40. Svoboda, D. Evaluation of the European method for mapping lichen diversity (LDV) as an indicator of environmental stress in the Czech Republic. Biologia **2007**, *62* (4), 424–431.
41. Agraval, M.; Sing, B.; Rajput, M.; Marshall, F.; Bell, J.N.B. Effect of air pollution on peri-urban agriculture: A case study. Environ. Pollut. **2003**, *126,* 323–329.
42. Orendovici, T.; Skelly, J.M.; Ferdinad, J.A.; Savage, J.E.; Sanz, M.-J.; Smith, G.C. Response of native plants of northeastern United States and southern Spain to ozone exposures; determining exposure/response relationships. Environ. Pollut. **2003**, *125* (1), 31–40.
43. Wolseley, P.A. Using lichens on twigs to assess changes in ambient atmospheric conditions. In *Monitoring with lichens—Monitoring Lichens;* Nimis, P.L., Scheidegger, C., Wolseley, P.A., Eds.; NATO Science Series IV; Kluwer: Dordrecht, 1999; Vol. 7, 291–294.
44. Van Haluwyn, C.; van Hert, C.M. Bioindication: the community approach. In *Monitoring with lichens—Monitoring Lichens;* Nimis, P.L., Scheidegger, C., Wolseley, P.A., Eds.; NATO Science Series IV; Kluwer: Dordrecht, 2002; Vol. 7, 39–64.
45. Asta, J.; Erhard, W.; Ferretti, M.; Fornasier, F.; Kirschbaum, U.; Nimis, P.L.; Purvis, O.W.; Pirintsos, S.; Scheidegger, C.; van Haluwyn, C.; Wirth, V. Mapping lichens diversity as an indicator of environmental quality. In *Monitoring with Lichens—Monitoring Lichens;* Nimis, P.L., Scheidegger, C., Wolseley, P.A., Eds; NATO Science Series IV; Kluwer: Dordrecht, 2002; Vol. 7, 273–279.
46. Korte, G.B. *The GIS Book. How to Implement, Manage and Asses the Value of Geographic Information System*; Onward Press: Albany, NY, 2000.
47. Schroter, M.; Obermeier, A.; Bruggemann, D.; Plech-schmidt, M.; Klemm, O. Remote monitoring of air pollutant emissions from point sources by a mobile Lidar/Sodar System. J. Air Waste Manage. Assoc. **2003**, *53*, 716–723.
48. Available at http://www.ghcc.msfc.nasa.gov/sparcle_tutorial.html (accessed May 15, 2012.)
49. Weibring, P.; Andersson, M.; Edner, H.; Svanberg, S. Remote monitoring of industrial emissions by combination of lidar and plume velocity measurements. Appl. Phys. B, **1998**, *66*, 383–388.
50. Jahnke, J.A. *Continuous Emissions Monitoring*; Van Nostrand Reinhold: New York, 1993.
51. White, J.R. Technologies for enhanced monitoring. Pollut. Eng. **1995**, 27 (6), 46–50.
52. Walker, K. Select a continuous emissions monitoring system. Chem. Eng. Progress. **1996**, *92* (2), 28–34.
53. Mayer, H.; Holst, J.; Schindler, D.; Ahrens, D. Evolution of the air pollution in SW Germany evaluated by the longterm air quality index LAQx. Atmos. Environ. **2008**, *42,* 5071–5078.
54. Shooter, D.; Brimblecombe, P. Air quality indexing. Int. J. Environ. Pollut. **2009**, *36* (1–3), 19–29.
55. Plaia, A.; Rugierri, M. *Air quality indices: A review;* GRASPA Working paper no. 39; June 2010.

22

Air Pollution: Technology

Sven Erik Jørgensen

Sources of Particulate Pollution

When considering particulate pollution, the source should be categorized with regard to the contaminant type. Inert particulates are distinctly different from active solids in the nature and type of their potentially harmful human health effects. Inert particulates comprise solid airborne material, which does not react readily with the environment and does not exhibit any morphological changes as a result of combustion or any other process. Active solid matter is defined as particulate material that can be further oxidized or react chemically with the environment or the receptor. Any solid material in this category can, depending on its composition and size, be considerably more harmful than inert matter of similar size.

A closely related group of emissions come from aerosols of liquid droplets, generally below 5 μm. They can be oil or other liquid pollutants (e.g., freon) or may be formed by condensation in the atmosphere. Fumes are condensed metals, metal oxides, or metal halides, formed by industrial activities, predominantly as a result of pyro-metallurgical processes: melting, casting, or extruding operations. Products of incomplete combustion are often emitted in the form of particulate matter. The most harmful components in this group are particulate polycyclic organic matter (PPOM), which are mainly derivatives of benz[a]pyrene.

Natural sources of particulate pollution are sandstorms, forest fires, and volcanic activity. The major sources in towns are vehicles, combustion of fossil fuel for heating and electricity production, and industrial activity.

The total global emission of particulate matter is in the order of 10^7 t per year. Deposition of particles may occur by three processes:

1. Sedimentation (Stokes law may be applied, particles >20 μm)
2. Impaction (determined by differences in concentrations by use of Fick's law, particles between 5 and 20 μm)
3. Diffusion (particles <5μm)

Particles <20 μm are identified as suspended particulate matter. Particles >20 μm may be denoted dust, which will be deposited close to the source due to the high sedimentation rate. Dry deposition consists of gases or dry particles. Wet deposition is raindrops containing gases and particles. Particles may consist of minor concentrations of dissolved salts in water drops, crystals, or a combination of the two.

Particulate Pollution Problem

Particulate pollution is an important health factor, most crucially the toxicity and size distribution. Many particles are highly toxic, such as asbestos and those of metals such as beryllium, lead, chromium, mercury, nickel, and manganese. In addition, particulate matter is able to absorb gases, which enhances the effects of these components. In this context, the particle size distribution is of particular importance, as particles greater than 10 μm are trapped in the human upper respiratory passage and the specific surface (expressed as square meter per gram of particulate matter) increases with l/d, where d is the particle size. The adsorption capacity of particulate matter, expressed as grams adsorbed per gram of particulate matter, will generally be proportional to the surface area. Table 1 lists some typical particle size ranges. However, size as well as shape and density must be considered. Furthermore, particle size is determined by two parameters: the mass median diameter, which is the size that divides the particulate sample into two groups of equal mass, i.e., the 50% point on a cumulative frequency versus particle size plot, and the geometric standard deviation.

Control Methods Applied to Particulate Pollution

Particulate pollutants have the ability to adsorb gases including sulfur dioxide, nitrogen oxides, carbon monoxide, and so on. The inhalation of these toxic gases is frequently associated with this adsorption, as the gases otherwise would be dissolved in mouthwash and spittle before entering the lungs. Particulate pollution may be controlled by modifying the distribution pattern. This method is described in detail below. In principle, it represents an obsolete philosophy of pollution abatement, dilution, but it is still widely used to reduce the concentration of pollutants at ground level and thereby minimizes the effect of air pollution. Particulate control technology can offer a wide range of methods aimed at the removal of particulate matter from gas. These methods are settling chambers, cyclones, filters, electrostatic precipitators, wet scrubbers, and the modification of particulate characteristics.

Modifying the Distribution Patterns

Although emissions, gaseous or particulate, may be controlled by various sorption processes or mechanical collection, the effluent from the control device must still be dispersed into the atmosphere. Atmospheric dispersion depends primarily on horizontal and vertical transport. The horizontal transport depends on the turbulent structure of the wind field. As the wind velocity increases, so does the degree of dispersion and there is a corresponding decrease in the ground-level concentration of the contaminant at the receptor site.

TABLE 1 Typical Practical Size Ranges

	μm
Tobacco smoke	0.01–1
Oil smoke	0.05–1
Ash	1–500
Ammonium chloride smoke	0.1–4
Powdered activated carbon	3–700
Sulfuric acid aerosols	0.5–5

Source: Jørgensen[10]

The emissions are mixed into larger volume of air, and the diluted emission is carried out into essentially unoccupied terrain away from any receptors. Depending on the wind direction, the diluted effluent may be funneled down a river valley or between mountain ranges. Horizontal transport is sometimes prevented by surrounding hills that form a natural pocket for locally generated pollutants. This particular topographical situation occurs for instance in the Los Angeles area, which suffers heavily from air pollution.

The vertical transport depends on the rate of change of ambient temperature with altitude. The dry adiabatic lapse rate is defined as a decrease in air temperature of 1°C per 100 m. This is the rate at which, under natural conditions, a rising parcel of unpolluted air will decrease in temperature with elevation into the troposphere up to approximatelyl0,000 m. Under so-called isothermal conditions, the temperature does not change with elevation. Vertical transport can be hindered under stable atmospheric conditions, which occur when the actual environmental lapse rate is less than the dry adiabatic lapse rate. A negative lapse rate is an increase in air temperature with latitude. This effectively prevents vertical mixing and is known as inversion.

The dispersion from a point source (a chimney for instance) may be calculated from the Gaussian plume model (see, for instance, Reible.[1])

These different atmospheric conditions are illustrated in Figure 1 where stack gas behavior under the various conditions is shown. Further explanations are given in Table 2. The distribution of particulate material is more effective the higher the stack. The maximum concentration, C_{max}, at ground level can be shown to be approximately proportional to the emission and to follow approximately this expression:

$$C_{max} = kQ/H^2 \tag{1}$$

where Q is the emission (expressed as grams per particulate matter per unit of time), H is the effective stack height, and k is a constant. The effective height is slightly higher than the physical height and can be calculated from information about the temperature, the stack exit velocity, and the stack inside diameter.

These equations explain why a lower ground-level concentration is obtained when many small stacks are replaced by one very high stack. In addition to this effect, it is always easier to reduce and control one large emission than many small emissions, and it is more feasible to install and apply the necessary environmental technology in one big installation.

Particulate Pollution Control Equipment

Environmental technology offers several solutions to the problem of particulate matter removal. The methods and their optimum particle size and efficiency are compared in Table 3. The cost of the various installations varies of course from country to country and is dependent on several factors (material applied, standard size or not standard size, automatized, and so on). Generally, electrostatic precipitators are the most expensive solution and are mainly applied for large quantities of air. Wet scrubbers also belong among the more expensive installations, while settling chamber and centrifuges are the most cost-effective solutions.

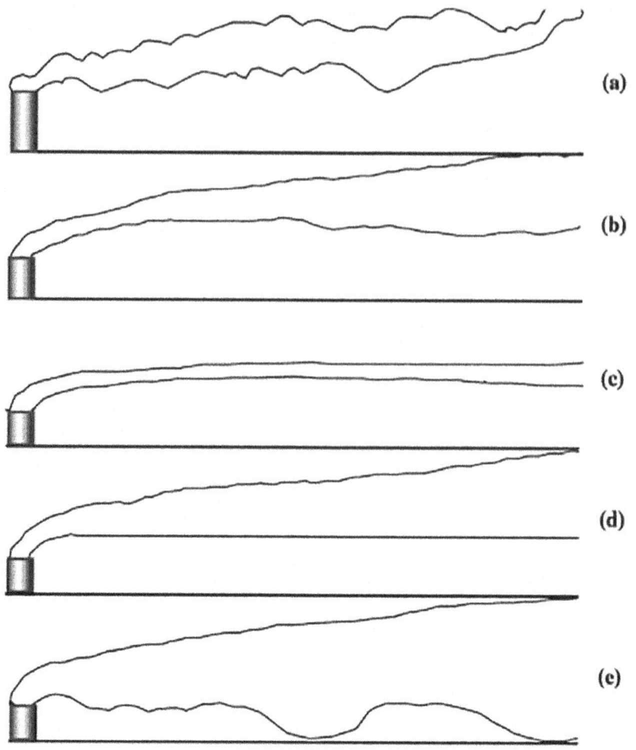

FIGURE 1 Stack gas behavior under various conditions. (a) Strong lapse (looping); (b) weak lapse (coning); (c) inversion (fanning); (d) inversion below, lapse aloft (lofting); (e) lapse below, inversion aloft (fumigation).

TABLE 2 Various Atmospheric Conditions

Strong lapse (looping)	Environmental lapse rate > adiabatic lapse rate
Weak lapse (coning)	Environmental lapse rate < adiabatic lapse rate
Inversion (fanning)	Increasing temperature with height
Inversion below, lapse aloft (lofting)	Increasing temperature below, App. adiabatic lapse rate aloft
Lapse below, inversion aloft (fumigation)	App. adiabatic lapse rate below, increasing temperature aloft

Source: Jørgensen.[10]

TABLE 3 Characteristics of Particulate Pollution Control Equipment

Device	Optimum Particle Size (μm)	Optimum Concentration(g–3)	Temperature Limitations (°C)	Air Resistance (mm H_2O)	Efficiency (% by Weight)
Settling chambers	>50	>100	–30 to 350	<25	<50
Centrifuges	>10	>30	–30 to 350	<50–100	<70
Multiple centrifuges	>5	>30	–30 to 350	<50–100	<90
Filters	>0.3	>3	–30 to 250	>15–100	>99
Electrostatic precipitators	>0.3	>3	–30 to 500	<20	<99
Wet scrubbers	>2–10	>3–30	0 to 350	>5–25	<95–99

Source: Jørgensen.[10]

Settling Chambers

Simple gravity settling chambers depend on gravity or inertia for the collection of particles. Both forces increase in direct proportion to the square of the particle diameter, and the performance limit of these devices is strictly governed by the particle settling velocity. The pressure drop in mechanical collectors is low to moderate, 1–25 cm water in most cases. Most of these systems operate dry, but if water is added, it performs the secondary function of keeping the surface of the collector clean and washed free of particles.

The settling or terminal velocity can be described by the following expression, which has general applicability:

$$V_t = (\partial_p - \partial) g \frac{d_p^2}{17\mu} \tag{2}$$

where V_t is terminal velocity, ∂_p is particle density, ∂ is gas density, d_p is particle diameter, and μ is gas viscosity.

This is the Stokes' law and is applicable to $N_{Re} < 1.9$,

Where

$$N_{Re} = d_p^* V_t^* \frac{\partial}{\mu} \tag{3}$$

The intermediate equation for settling can be expressed as:

$$V_t = \frac{0.153 * g^{0.71} * d_p^{1.14} (\partial_p - \partial)^{0.71}}{\partial^{0.29} * \mu^{0.43}} \tag{4}$$

This equation is valid for Reynolds numbers between 1.9 and 500, while the following equation can be applied for $N_{Re} > 500$ and up to 200,000:

$$V_t = 1.74 \left(d_p^* g \frac{(\partial_p - \partial)^{1/2}}{\partial} \right) \tag{5}$$

The settling velocity in these chambers is often in the range 0.3–3 m/sec. This implies that for large volumes of emission, the settling velocity chamber must be very large in order to provide an adequate residence time for the particles to settle. Therefore, the gravity settling chambers are not generally used to remove particles smaller than 100 µm (= 0.1 mm). For particles measuring 2–5 µm, the collection efficiency will most probably be as low as 1% –2%. A variation of the simple gravity chamber is the baffled separation chamber. The baffles produce a shorter settling distance, which means a shorter retention time. The shown equations can be used to design a settling chamber.

Cyclones

Cyclones separate particulate matter from a gas stream by transforming the inlet gas stream into a confined vortex. The mechanism involved in cyclones is the continuous use of inertia to produce a tangential motion of the particles towards the collector walls. The particles enter the boundary layer close to the cyclone wall and lose kinetic energy by mechanical friction. The forces involved are the centrifugal force imparted by the rotation of the gas stream and a drag force, which is dependent on the particle density, diameter, shape, etc. A hopper is built at the bottom. If the cyclone is too short, the maximum force will not be exerted on some of the particles, depending on their size and the corresponding drag forces.[2] If, however, the cyclone is too long, the gas stream might reverse its direction and spiral up the center.

It is therefore important to design the cyclone properly. The hopper must be deep enough to keep the dust level low. The efficiency of a cyclone is described by a graph similar to Figure 2, which shows the efficiency versus the relative particle diameter, i.e., the actual particle diameter divided by D_{50}, which is defined as the diameter corresponding to 50% efficiency. D_{50} can be found from the following equation:

$$D_{50} = K * \left(\frac{\mu Dc}{V_c * \partial_p} \right)^{1/2} \tag{6}$$

where D_c is the diameter of cyclone, V_c is inlet velocity, ∂p is the density of particles, μ is gas viscosity, and K is a constant dependent on cyclone performance.

If the distribution of the particle diameter is known, it is possible to calculate the total efficiency from a graph such as Figure 2:

$$\text{eff} = \sum m_i^* \text{eff}_i \tag{7}$$

where m_i is the weight fraction in the ith particle size range and eff_i is the corresponding efficiency.

The pressure drop for cyclones can be found from:

$$\Delta p = N^* \frac{V_c^2}{2_g} \tag{8}$$

From these equations, it can be concluded that higher efficiency is obtained without increased pressure drop if D_c can be decreased while velocity V_c is maintained. This implies that a battery of parallel coupled small cyclones will work more effectively than one big cyclone. Such cyclone batteries are available as blocks and are known as multiple cyclones. Compared with settling chambers, cyclones offer higher removal efficiency for particles below 50 μm and above 2–10 μm, but involve a greater pressure drop.

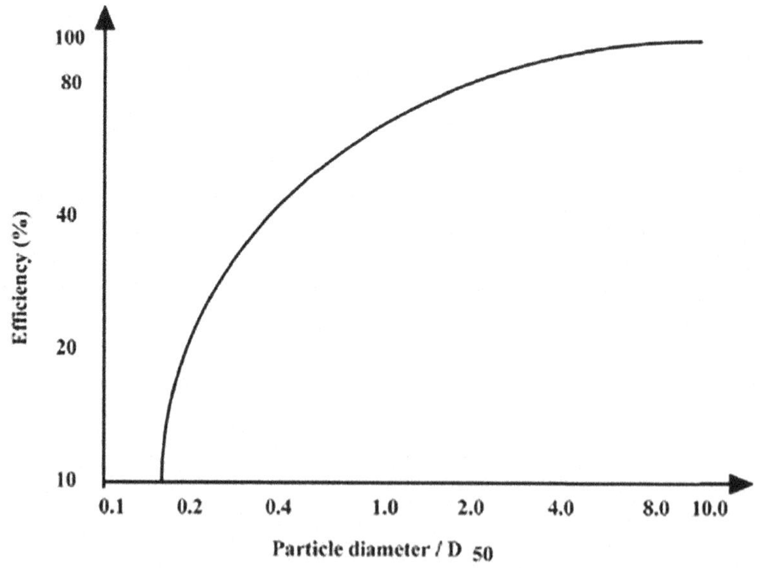

FIGURE 2 Efficiency plotted against relative particle diameter. Notice that it is a log–log plot.

Filters

Particulate materials are collected by filters by the following three mechanisms:

Impaction, where the particles have so much inertia that they cannot follow the streamline around the fiber and thus impact on its surface.

Direct interception, where the particles have less inertia and can barely follow the streamlines around the obstruction.

Diffusion, where the particles are so small (below 1 μm) that their individual motion is affected by collisions on a molecular or atomic level. This implies that the collection of these fine particles is a result of random motion.

Different flow patterns can be used, as demonstrated in Figure 3. The types of fibers used in fabric filters range from natural fibers, such as cotton and wool, to synthetics (mainly polyesters and nylon), glass, and stainless steel.

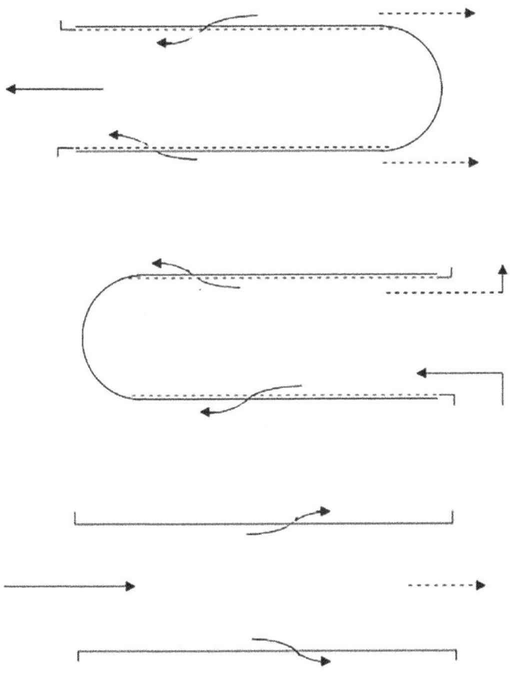

FIGURE 3 Flow pattern of filters.

TABLE 4 Properties of Fibers

Fabric	Acid Resistance	Alkali	Fluoride Strength	Tensile	Abrasion Resistance
Cotton	Poor	Good	Poor	Medium	Very good
Wool	Good	Poor	Poor	Poor	Fair
Nylon	Poor	Good	Poor	Good	Excellent
Acrylic	Good	Fair	Poor	Medium	Good
Polypropylene	Good	Fair	Poor	Very good	Good
Orlon	Good	Good	Fair	Medium	Good
Dacron	Good	Good	Fair	Good	Very good
Teflon	Excellent	Excellent	Good	Good	Fair

Some properties of common fibers are summarized in Table 4. As can be seen, cotton and wool have a low temperature limit and poor alkali and acid resistance, but they are relatively inexpensive. The selection of filter medium must be based on the answer to several questions:[3,4]

What is the expected operating temperature?

Is there a humidity problem that necessitates the use of a hydrophobic material, such as, e.g., nylon?

How much tensile strength and fabric permeability are required?

How much abrasion resistance is required?

Permeability is defined as the volume of air that can pass through 1 m^2 of the filter medium with a pressure drop of no more than 1 cm of water.

The filter capacity is usually expressed as cubic meter of air per square meter of filter per minute. A typical capacity ranges between 1 and 5 $m^3/m^2/min$.

The pressure drop is generally larger than for cyclones and will in most cases be 10–30 cm of water, depending on the nature of the dust, the cleaning frequency, and the type of cloth.

There are several specific methods of filter cleaning. The simplest is backwash, where dust is removed from the bags merely by allowing them to collapse. This is done by reverting the airflow through the entire compartment. The method is remarkable for its low consumption of energy. Shaking is another low-energy filter-cleaning process, but it cannot be used for sticky dust. The top of the bag is held still and the entire tube sheath at the bottom is shaken. The application of blow rings involves reversing the airflow without bag collapse. A ring surrounds the bag; it is hollow and supplied with compressed air to direct a constant stream of air into the bag from the outside.

The pulse and improved jet cleaning mechanism involves the use of a high-velocity, high-pressure air jet to create a low pressure inside the bag and induce an outward airflow, cleaning the bag by sudden expansion and reversal of flow.

In some cases, as a result of electrostatic forces, moisture on the surface of the bag, and a slight degree of hygroscopicity of the dust itself, the material forms cakes that adhere tightly to the bag. In this case, the material must be kept drier, and a higher temperature on the incoming dirty airstream is required. Filters are highly efficient even for smaller particles (0.1–2 m), which explains their wide use as particle collection devices.

Electrostatic Precipitators

The electrostatic precipitator consists of four major components:

1. *A gas-tight shell with hoppers* to receive the collected dust, inlet and outlet, and an inlet gas distributor
2. *Discharge electrodes*
3. *Collecting electrodes*
4. *Insulators*

The principles of electrostatic precipitators are illustrated in Figure 4. The dirty airstream enters a filter, where a high, 20–70 kV, usually negative voltage exists between discharge electrodes. The particles accept a negative charge and migrate towards the collecting electrode. The efficiency is usually expressed by use of Deutsch's equation (see the discussion including correction of this equation in the work of Gooch and Francis[5]). This equation entails that the relationship between migration velocity and particle diameter has a minimum between 0.1 and 1.0 μm.

The operation of an electrostatic precipitator can be divided into three steps:

1. The particles *accept a negative charge*.
2. The charged particles *migrate towards the collecting electrode* due to the electrostatic field.
3. The collected dust *is removed from the collecting electrode* by shaking or vibration, and is collected in the hopper.

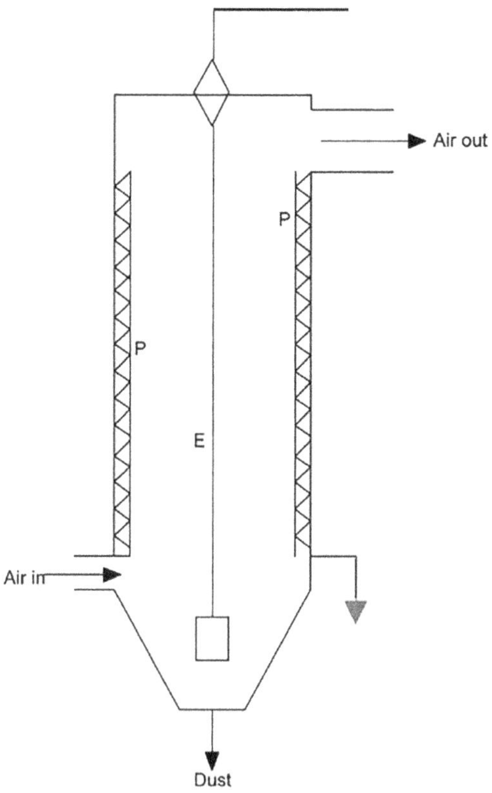

FIGURE 4 The dust is precipitated on the electrode P. E has a high, usually negative voltage and emits a great number of electrons that give the dust particles a negative charge. The dust particles will therefore be attracted to P.

r, the specific electrical resistance, measured in ohm meter, determines the ability of a particle to accept a charge. The practical specific resistance can cover a wide range of about four orders of magnitude, in which varying degrees of collection efficiencies exist for different types of particles. The specific resistance depends on the chemical nature of the dust, the temperature, and the humidity.

Electrostatic precipitators have found wide application in industry. As the cost is relatively high, the airflow should be at least 20,000 m^3/hr; volumes as large as 1,500,000 m^3/hr have been treated in one electrostatic precipitator.

Very high efficiencies are generally achieved in electrostatic precipitators and emissions as low as 25 mg/m^3 are quite common. The pressure drop is usually low compared with other devices—25 mm water at the most. The energy consumption is generally 0.15–0.45 Wh/m^3/hr.

Wet Scrubbers

A scrubbing liquid, usually water, is used to assist separation of particles, or a liquid aerosol from the gas phase. The operational range for particle removal includes material less than 0.2 μm in diameter up to the largest particles that can be suspended in air. Gases soluble in water are also removed by this process. Four major steps are involved in the collection of particles by wet scrubbing. First, the particles are moved to the vicinity of the water droplets, which are 10–1000 times larger. Then, the particles must collide with the droplets. In this step, the relative velocity of the gas and the liquid phases is very important: If the particles have an excessively high velocity in relation to liquid, they cannot be retained by the droplets unless they can be wetted and thus incorporated into the droplets. The last step is the removal of the droplets containing dust particles from the bulk gas phase. Scrubbers are generally very flexible. They are able to operate under peak loads or reduced volumes and within a wide temperature range.[6] They are smaller and less expensive than dry particulate removal devices, but the operating costs are higher. Another disadvantage is that the pollutants are not collected but transferred into water, which means that the related water pollution problem must also be solved.[7]

Several types of wet scrubbers are available and their principles are outlined below:

1. *Chamber scrubbers* are spray towers and spray chambers that can be either round or rectangular. Water is injected under pressure through nozzles into the gas phase.
2. *Baffle scrubbers* are similar to a spray chamber but have internal baffles that provide additional impingement surfaces. The dirty gas is forced to make many turns to prevent the particles from following the airstream.
3. *Cyclonic scrubbers* are a cross between a spray chamber and a cyclone. The dirty gas enters tangentially to wet the particles by forcing its way through a swirling water film onto the walls. There, the particles are captured by impaction and are washed down the walls to the sump. The saturated gas rises through directional vanes, which are used solely to impart rotational motion to the gas phase. As a result of this motion, the gas goes out through a demister for the removal of any included droplets.
4. *Submerged orifice scrubbers* are also called gas induced scrubbers. The dirty gas is accelerated over anaerodynamic foil to a high velocity and directed into a pool of liquid. The high velocity impact causes the large particles to be removed into the pool and creates a tremendous number of spray droplets with a high amount of turbulence. These effects provide intensive mixing of gas and liquid and thereby a very high interfacial area. As a result, reactive gas absorption can be combined with particle removal.
5. The *ejector scrubber* is a water jet pump (see Figure 5). The water is pumped through a uniform nozzle and the dirty gas is accelerated by the action of the jet gas. The result is aspiration of the gas into the water by the Bernoulli principle and, accordingly, a lowered pressure. The ejector scrubber can be used to collect soluble gases as well as particulates.
6. The *venturi scrubber* involves the acceleration of the dirty gas to 75–300 m/min through a mechanical constriction. This high velocity causes any water injected just upstream of or in the venturi throat to be sheared off the walls or nozzles and atomized. The droplets are usually 5–20 μm in size and form

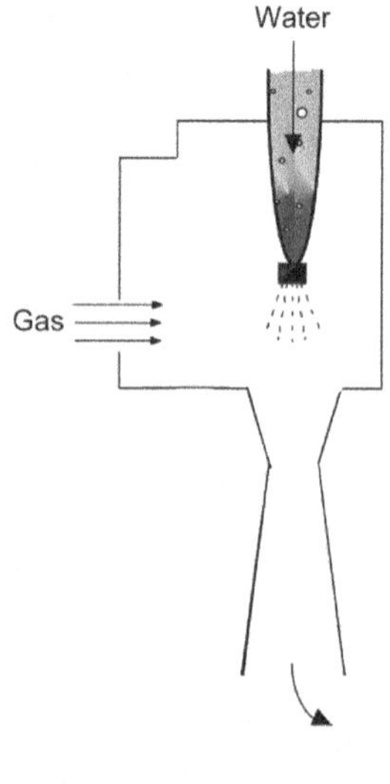

FIGURE 5 Principle of ejector scrubber.

into clouds from 150 to 300 μm in diameter, depending on the gas velocity. The scrubber construction is similar to that of the ejector scrubber, but the jet pump is replaced by a venturi constriction.

7. *Mechanical scrubbers* have internal rotating that which break up the scrubbing liquid into small droplets and simultaneously create turbulence.
8. *Charged-droplet scrubbers* have a high voltage ionization section where the corona discharge produces air ions (as in electrostatic precipitators). Water droplets are introduced into the chamber by the use of spray nozzles or similar devices. The additional collection mechanism provided by the induction of water droplets increases the collection efficiency.
9. *Packed-bed scrubbers* have a bottom support grid, and a top retaining grid (see Figure 6). The fluid (often water or a solution of alkali or acid) is distributed as shown in the figure over the top of the packed section, while the gas enters below the packing.

The flow is normally counter current. Packed-bed scrubbers offer the possibility of combining gas absorption with the removal of particulate material. The pressure drop is often in the order of 3 cm water per meter of packing. If the packing consists of expanded fiber, the bed scrubber is known as a fiber-bed scrubber. The packed-bed scrubber has a tendency to clog under high particulate loading, which is its major disadvantage.

Common packing includes saddles, rings, etc., like those used in absorption towers. Some important parameters for various scrubbers are plotted in Figure 7, which demonstrates the relationship between pressure drop, energy consumption, and D_{50} (the diameter of the particles removed at 50% efficiency).

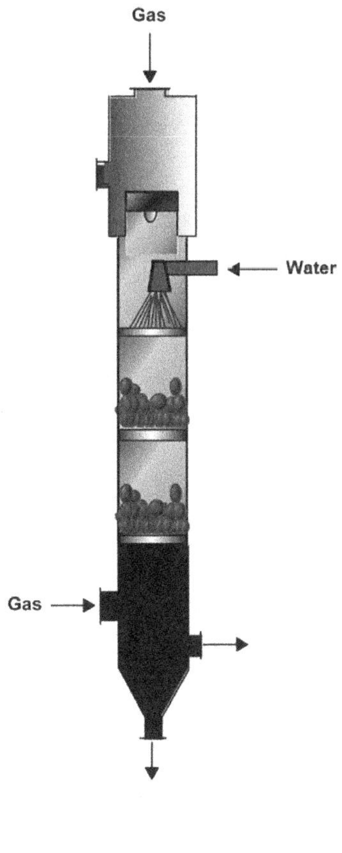

FIGURE 6 Packed-bed scrubber.

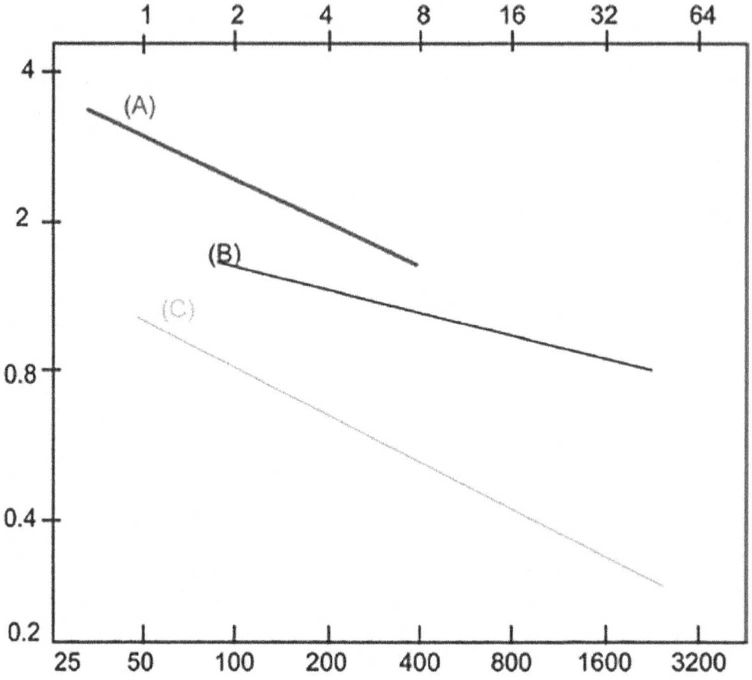

FIGURE 7 Relationship between D_{50} (μm), pressure drop (mm H_2O, lower axis), and energy consumption (kW/m^3/sec; upper axis). A: Packed-bed scrubber. B: Baffled scrubber. C: Venturi scrubber.

Air Pollution Problems of Carbon Hydrides and Carbon Monoxide: Sources of Pollutants

All types of fossil fuel will produce carbon dioxide on combustion, which is used in the photosynthetic production of carbohydrates. As such, carbon dioxide is harmless and has no toxic effect, whatever the concentration levels, but see the entries about the greenhouse effect of carbon dioxide. An increased carbon dioxide concentration in the atmosphere will increase absorption of infrared radiation and the heat balance of the earth will be changed.

Carbon hydrides are the major components of oil and gas, and incomplete combustion will always involve their emission. Partly oxidized carbon hydrides, such as aldehydes and organic acids, might also be present.

The major source of carbon hydrides pollution is motor vehicles.

In reaction with nitrogen oxides and ozone, they form so-called photochemical smog, which consists of several rather oxidative compounds, such as peroxyacyl nitrates and aldehydes. In areas where solar radiation is strong and the atmospheric circulation is weak, the possibility of smog formation increases as the processes are initiated by ultraviolet radiation.

Incomplete combustion produces carbon monoxide. By regulation of the ratio of oxygen to fuel, more complete combustion can be obtained, but the emission of carbon monoxide cannot be totally avoided.

Motor vehicles are also a major source of carbon monoxide pollution. On average, 1 L of gasoline (petrol) will produce 200 L of carbon monoxide, while it is possible to minimize the production of this pollutant by using diesel instead of gasoline.

The annual production of carbon monoxide is more than 200 million tons, of which 50% is produced by the United States alone.

In most industrial countries, more than 75% of this pollutant originates from motor vehicles.

Pollution Problem of Carbon Hydrides and Carbon Monoxide

Carbon hydrides, partly oxidized carbon hydrides, and the compounds of photochemical smog are all more or less toxic to man, animals, and plants. Photochemical smog reduces visibility, irritates the eyes, and causes damage to plants, with immense economic consequences, for example, for fruit and tobacco plantations. It is also able to decompose rubber and textiles.

Carbon monoxide is strongly toxic as it reacts with hemoglobin and thereby reduces the blood's capacity to take up and transport oxygen. Ten percent of the hemoglobin occupied by carbon monoxide will produce symptoms such as headache and vomiting. It should be mentioned here that smoking also causes a higher carboxyhemoglobin concentration. An examination of policemen in Stockholm has shown that non-smokers had 1.2% carboxyhemoglobin, while smokers had 3.5%.

Control Methods Applied to Carbon Dioxide, Carbon Hydrides, and Carbon Monoxide Pollution

Carbon dioxide pollution is inevitable with the use of fossil fuels. Therefore, it can only be solved by the use of other sources of energy.

Legislation plays a major role in controlling the emission of carbon hydrides and carbon monoxide. As motor vehicles are the major source of these pollutants, control methods should obviously focus on the possibilities of reducing vehicle emission. The methods available today are as follows:

1. Motor technical methods
2. Afterburners
3. Alternative energy sources

The first method is based upon a motor adjustment according to the relationship between the composition of the exhaust gas and the air/fuel ratio. A higher air/fuel ratio results in a decrease in the carbon hydrides and carbon monoxide concentrations, but to achieve this, a better distribution of the fuel in the cylinder is required, which is only possible through the construction of another gasification system. This method may be considered cleaner technology.

At present, two types of afterburners are in use—*thermal* and *catalytic afterburners*. In the former type, the combustible material is raised above its autoignition temperature and held there long enough for complete oxidation of carbon hydrides and carbon monoxide to occur. This method is used on an industrial scale[8,9] when low-cost purchased or diverted fuel is available; in vehicles, a manifold air injection system is used.

Catalytic oxidation occurs when the contaminant-laden gas stream is passed through a catalyst bed, which initiates and promotes oxidation of the combustible matter at a lower temperature than would be possible in thermal oxidation. The method is used on an industrial scale for the destruction of trace solvents in the chemical coating industry. Vegetable and animal oils can be oxidized at 250–370°C by catalytic oxidation. The exhaust fumes from chemical processes, such as ethylene oxide, methyl methacrylate, propylene, formaldehyde, and carbon monoxide can easily be catalytically incinerated at even lower temperatures. The application of catalytic afterburners in motor vehicles presents some difficulties due to poisoning of the catalyst with lead. With the decreasing lead concentration in gasoline, it is becoming easier to solve that problem, and the so-called double catalyst system is now finding wide application. This system is able to reduce nitrogen oxides and oxidize carbon monoxide and carbon hydrides simultaneously. New catalysts are currently coming on the market and offer a higher efficiency.

Lead in gasoline has been replaced by various organic compounds to increase the octane number. Benzene has been applied, but it is toxic and causes air pollution problems because of its high vapor pressure. MTBE (methyl tertiary butyl ether) is another possible compound for increasing the octane number. It is, however, very soluble and has been found as a groundwater contaminant close to gasoline stations.

Application of alternative energy sources is still at a preliminary stage. The so-called Sterling motor is one alternative, as it gives more complete combustion of the fuel, but there have been several improvements of the efficiency of the motor during the last years due to the increasing cost of fossil fuel. Much interest has, however, been devoted to electric and hybrid vehicles.

Air Pollution of Sulfur Dioxide: Sources, Problems, and Solutions

Fossil fuel contains approximately 2%–5% sulfur, which is oxidized by combustion to sulfur dioxide. Although fossil fuel is the major source, several industrial processes produce emissions containing sulfur dioxide, for example, mining, the treatment of sulfur containing ores, and the production of paper from pulp. The total global emission of sulfur dioxide has been decreasing during the last 25 years due to the installation of pollution abatement equipment, particularly in North America, the European Union, and Japan. The concentration of sulfur dioxide in the air is relatively easy to measure, and sulfur dioxide has been used as an indicator component. High values recorded by inversion are typical.

Sulfur dioxide is oxidized in the atmosphere to sulfur trioxide, which forms sulfuric acid in water. Since sulfuric acid is a strong acid, it is easy to understand that sulfur dioxide pollution indirectly causes the corrosion of iron and other metals and is able to acidify aquatic ecosystems.[10]

The health aspects of sulfur dioxide pollution are closely related to those of particulate pollution. The gas is strongly adsorbed onto particulate matter, which transports the pollutant to the bronchi and lungs. There is a clear relationship between concentration, effect, and exposure time, which is reflected in the emission standards for sulfur dioxide (see Table 5).

Clean Air Acts were introduced in all industrialized countries during the 1970s and 1980s. Table 5 illustrates some typical sulfur dioxide emission standards, although these may vary slightly from country to country.

The approaches used to meet the requirements of the acts as embodied in the standards can be summarized as follows:

1. Fuel switching from high to low sulfur fuels.
2. Modification of the distribution pattern—use of tall stacks.
3. Abandonment of very old power plants that have a particular high emission.
4. Flue gas cleaning.

Desulfurization of liquid and gaseous fuel is a well-known chemical engineering operation. In gaseous and liquid fuels, sulfur either occurs as hydrogen sulfide or reacts with hydrogen to form hydrogen sulfide. The hydrogen sulfide is usually removed by absorption in a solution of alkanol- amine and then converted to elemental sulfur. The process in general use for this conversion is the so-called Claus process. The hydrogen sulfide gas is fired in a combustion chamber in such a manner that one-third of the volume of hydrogen sulfide is converted to sulfur dioxide. The products of combustion are cooled and then passed through a catalyst-packed converter, in which the following reaction occurs:

$$2H_2S + SO_2 = 3S + 2H_2O \tag{9}$$

TABLE 5 SO_2 Emission Standards

Duration	Concentration (ppm)	Comments
Month	0.05	
24 hr	0.10	Might be exceeded once a month
30 min	0.25	Might be exceeded 15 times/month

Source: Jorgensen.[10]

The elemental sulfur has commercial value and is mainly used for the production of sulfuric acid.

Sulfur occurs in coal both as pyritic sulfur and organic sulfur. Pyritic sulfur is found in small discrete particles within the coal and can be removed by mechanical means, e.g., gravity separation methods. However, 20%–70% of the sulfur content of coal is present as organic sulfur, which can hardly be removed today on an economical basis. Since sulfur recovery from gaseous and liquid fuels is much easier than that from solid fuel (which has other disadvantages as well), much research has been and is being devoted to *the gasification or liquefaction of coal.* It is expected that this research will lead to an alternative technology that will solve most of the problems related to the application of coal, including sulfur dioxide emission. Approach (2) listed above has been mentioned earlier in this entry, while approach (3) needs no further discussion. The next subsection is devoted to (4) flue gas cleaning.

Flue Gas Cleaning of Sulfur Dioxide

When sulfur is not or cannot be economically removed from fuel oil or coal prior to combustion, removal of sulfur oxides from combustion gases will become necessary for compliance with the stricter air pollution control laws.

The chemistry of sulfur dioxide recovery presents a variety of choices and five methods should be considered:

1. Adsorption of sulfur dioxide on active metal oxides with regeneration to produce sulfur.
2. Catalytic oxidation of sulfur dioxide to produce sulfuric acid.
3. Adsorption of sulfur dioxide on charcoal with regeneration to produce concentrated sulfur dioxide.
4. Reaction of dolomite or limestone with sulfur dioxide by direct injection into the combustion chamber. A lime slurry is injected into the flue gas beyond the boilers.
5. Fluidized bed combustion of granular coal in a bed of finely divided limestone or dolomite maintained in a fluid-like condition by air injection. Calcium sulfite is formed as a result of these processes.

In particular, the two latter methods have found wide application, particularly to large industrial installations. It is possible to recover the sulfur dioxide or elemental sulfur from these processes, making it possible to recycle the spent sorbing material.

Air Pollution Problems of Nitrogenous Gases: Sources, Problems, and Control

Seven different compounds of oxygen and nitrogen are known: N_2O, NO, NO_2, NO_3, N_2O_3, N_2O_4, and N_2O_5— often summarized as NO_x. From the point of view of air pollution, it is mainly NO (nitrogen oxide) and NO_2 (nitrogen dioxide) that are of interest. Nitrogen oxide is colorless and is formed from the elements at high temperatures. It can react further with oxygen to form nitrogen dioxide, which is a brown gas. The major sources of the two gases are combustion of gasoline and oil (nitrogen oxide) and combustion of oil, including diesel oil (nitrogen dioxide). The production of NO is favored by high temperature. In addition, a relatively small emission of nitrogenous gases originates from the chemical industry. The total global emission is approximately 10 million tons per year. This pollution has only local or regional interest, as the natural global formation of nitrogenous gases in the upper atmosphere by the influence of solar radiation is far more significant than the man-controlled emission.

As mentioned above, nitrogen oxide is oxidized to nitrogen dioxide, although the reaction rate is slow—in the order of 0.007/hr. However, it can be accelerated by solar radiation. Nitrogenous gases take part in the formation of smog, as the nitrogen in peroxyacyl nitrate originates from nitrogen oxides. They are highly toxic but as their contribution to global pollution is insignificant, local and regional problems can partially be solved by changing the distribution pattern (see the section on *Control Methods Applied to Particulate Pollution).*

The emission from motor vehicles can be reduced by the same methods as mentioned for carbon hydrides and carbon monoxide. The air/fuel ratio determines the concentration of pollutants in exhaust gas. An increase in the ratio will reduce the emission of carbon hydrides and carbon monoxide, but unfortunately will increase the concentration of nitrogenous gases. Consequently, the selected air/ fuel ratio will be a compromise. A double catalytic afterburner is applied today, and it is able to reduce nitrogenous gases and simultaneously oxidize carbon hydrides and carbon monoxide. The application of alternative energy sources will, as for carbon hydrides and carbon monoxide, be a very useful control method for nitrogenous gases at a later stage.

Between 0.1 and 1.5 ppm of nitrogenous gases, of which 10%–15% consists of nitrogen dioxide, are measured in urban areas with heavy traffic. On average, the emission of nitrogenous gases is approximately 15 g per liter of gasoline and 25 g per liter of diesel oil.

Nitrogenous gases in reaction with water form nitrates that are washed away by rainwater. In some cases, this can be a significant source of eutrophication. For a shallow lake, for example, the increase in nitrogen concentration due to the nitrogen input from rainwater will be rather significant. In a lake with a depth of 1.7 m and an annual precipitation of 600 mm, which is normal in many temperate regions, the annual input will be as much as 0.3 mg/L.

The methods used for control of industrial emission of nitrogenous gases, including ammonia, will be discussed in the next section that discusses industrial air pollution, but as pointed out above, industrial emission is of less importance, even though it might play a significant role locally. The emission of nitrogenous gases by combustion of oil for heating and the production of electricity can hardly be reduced.

Industrial Air Pollution, Overview, and Control Methods

The rapid growth in industrial production during recent decades has exacerbated the industrial air pollution problem, but due to increased application of continuous processes, recovery methods, air pollution control, use of closed systems, and other technological developments, industrial air pollution has, in general, not increased in proportion to production.

Industry displays a wide range of air pollution problems related to a large number of chemical compounds in a wide range of concentrations.

It is not possible in this context to discuss all industrial air pollution problems; instead, we shall touch on the most important problems and give an overview of the control methods applied today. Only the problems related to the environment will be dealt with in this context.

A distinction should be made between air quality standards, which indicate that the concentration of a pollutant in the atmosphere at the point of measurement should not be greater than a given amount, and emission standards, which require that the amount of pollutant emitted from a specific source should not be greater than a specifically indicated amount.

The standards reflect, to a certain extent, not only the toxicity of the particular component but also the possibility for its uptake.

Here, the distribution coefficient for air/water (blood) plays a role. The more soluble the component is in water, the greater the possibility for uptake. For example, the air quality standard for acetic acid, which is very soluble in water, is relatively lower than the toxicity of aniline, which is almost insoluble in water.

Since industrial air pollution covers a wide range of problems, it is not surprising that all three classes of pollution control methods mentioned previously have found application: modification of the distribution pattern, alternative (cleaner) production methods, and particulate and gas/vapor control technology.

All the methods mentioned in the section on *Control Methods Applied to Particulate Pollution* also apply for industrial air pollution control.

In gas and vapor technology, a distinction has to be made between condensable and non-condensable gaseous pollutants. The latter must usually be destroyed by incineration, while the condensable gases can be removed from industrial effluents by absorption, adsorption, condensation, or combustion.

Recovery is feasible by the first three methods.

Gas Absorption

Absorption is a diffusion process that involves the mass transfer of molecules from the gas state to the liquid state along a concentration gradient between the two phases. Absorption is a unit operation that is enhanced by all the factors generally affecting mass transfer, i.e., high interfacial area, high solubility, high diffusion coefficient, low liquid viscosity, increased residence time, turbulent contact between the two phases, and possibilities for reaction of the gas in the liquid phase. This last factor is often very significant and almost 100% removal of the contaminant is the result of such a reaction. Acidic components can easily be removed from gaseous effluents by absorption in alkaline solutions, and, correspondingly, alkaline gases can easily be removed from effluent by absorption in acidic solutions (Table 6).

Carbon dioxide, phenol, and hydrogen sulfide are readily absorbed in alkaline solutions in accordance with the following processes:

$$+CO_2 + 2N_aOH \rightarrow 2N_a^+ + CO_3^{2-} \quad (10)$$

$$H_2S + 2NaOH \rightarrow 2N_a^+ + S_{2-} + 2H_2O \quad (11)$$

$$C_6H_5OH + NaOH \rightarrow C_6H_5O^- + N_a^+ + H_2O \quad (12)$$

Ammonia is readily absorbed in acidic solutions:

$$NH_3 + H_2SO_4 \rightarrow 2NH_4 + SO_4^{2-} \quad (13)$$

Gas Adsorption

Adsorption is the capture and retention of a component (adsorbate) from the gas phase by the total surface of the adsorbing solid (adsorbent). In principle, the process is the same as when dealing with wastewater treatment; the theory is equally valid for gas adsorption.

Adsorption is used to concentrate (often 20 to 100 times) or store contaminants until they can be recovered or destroyed in the most economical way. Figure 8 illustrates some adsorption isotherms applicable to practical gas adsorption problems. These are often described as either Langmuir's or Freundlich's adsorption isotherms. Adsorption is dependent on temperature: increased temperature

TABLE 6 Absorber Reagents

Reagents	Applications
$KMnO_4$	Rendering, polycyclic organic matter
NaOCl	Protein adhesives
Cl_2	Phenolics, rendering
Na_2SO_3	Aldehydes
NaOH	CO_2, H_2S, phenol, Cl_2, pesticides
Ca(OH)2	Paper sizing and finishing
H_2SO_4	NH_3, nitrogen bases

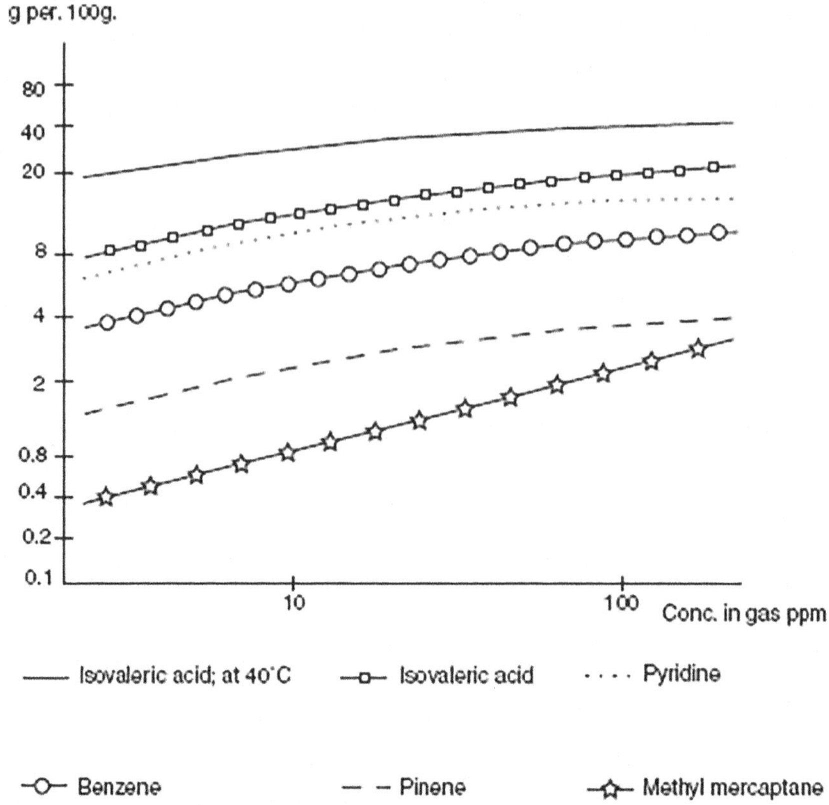

FIGURE 8 Adsorption isotherms, at 20°C.

means that the molecules move faster and therefore it is more difficult to adsorb them. There are four major types of gas adsorbents, the most important of which is activated carbon, but aluminum oxide (activated aluminum), silica gel, and zeolites are used as well.

The selection of adsorbent is made according to the following criteria:

1. High selectivity for the component of interest.
2. Easy and economical to regenerate.
3. Availability of the necessary quantity at a reasonable price.
4. High capacity for the particular application so that the unit size will be economical. Factors affecting capacity include total surface area involved, molecular weight, polarity activity, size, shape, and concentration.
5. Pressure drop, which is dependent on the superficial velocity.
6. Mechanical stability in the resistance of the adsorbent particles to attrition. Any wear and abrasion during use or regeneration will lead to an increase in bed pressure drop.
7. Microstructure of the adsorbent should, if at all possible, be matched to the pollutant that has to be collected.
8. The temperature, which has a profound influence on the adsorption process, as already mentioned.

Regeneration of the adsorbents is an important part of the total process. A few procedures are available for regeneration:

1. *Stripping* by use of steam or hot air.
2. *Thermal desorption* by raising the temperature high enough to boil off all the adsorbed material.

3. *Vacuum desorption* by reducing the pressure enough to boil off all the adsorbed material.
4. *Purge gas stripping* by using a non-adsorbed gas to reverse the concentration gradient. The purge gas may be condensable or non-condensable. In the latter case, it might be recycled, while the use of a condensable gas has the advantage that it can be removed in a liquid state.
5. *In situ oxidation* based on the oxidation of the adsorbate on the surface of the adsorbent.
6. *Displacement* by use of a preferentially adsorbed gas for the desorption of the adsorbate. The component now adsorbed must, of course, also be removed from the adsorbent, but its removal might be easier than that of the originally adsorbed gas, for instance, because it has a lower boiling point.

Although the regeneration is 100%, the capacity of the adsorbent may be reduced 10%–25% after several regeneration cycles, due to the presence of fine particulates and/or high molecular weight substances that cannot be removed in the regeneration step. A flowchart of solvent recovery using activated carbon as an adsorbent is shown in Figure 9 as an illustration of a plant design.

Combustion

Combustion is defined as rapid, high-temperature gas-phase oxidation. The goal is the complete oxidation of the contaminants to carbon dioxide and water, sulfur dioxide, and nitrogen dioxide.

The process is often applied to control odors in rendering plants, paint and varnish factories, rubber tire curing, and petrochemical factories. It is also used to reduce or prevent an explosion hazard by burning any highly flammable gases for which no ultimate use is feasible. The efficiency of the process is highly dependent not only on temperature and reaction time but also on turbulence or the mechanically induced mixing of oxygen and combustible material. The relationship between the reaction rate, r, and the temperature can be expressed by Arrhenius' equation:

$$r = A^{*}e - \frac{E}{RT} \tag{14}$$

where A is a constant, E is the activation energy, R is the gas constant, and T is the absolute temperature. A distinction is made between combustion, thermal oxidation, and catalytic oxidation, the latter two being the same in principle as the vehicle afterburners.

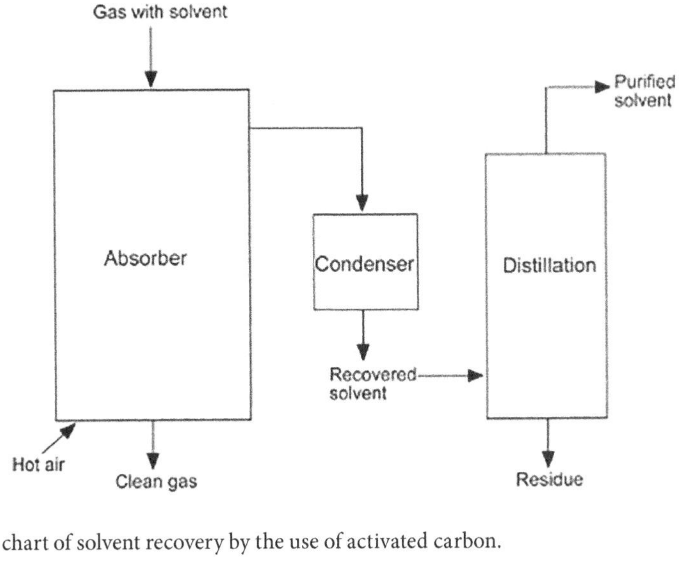

FIGURE 9 Flow chart of solvent recovery by the use of activated carbon.

Heavy Metals as Air Pollutants

Heavy metals, which may be defined as the metals with a specific gravity >5.00 kg/L, comprise 70 elements. Most of them are, however, only rarely found as pollutants. The heavy metals of environmental interest form very heavy soluble compounds with sulfide and phosphate and form very stable complexes with many ligands present in the environment. It means, fortunately, that most of the heavy metals are not very bioavailable in most environments (see also *Bioremediation*, p. 408).

A number of enzymes activated by metal ions and metalloenzymes are known. Members of the first mentioned group, comprising iron, cobalt, chromium, vanadium, selenium, copper, zinc, iron, cobalt, and molybdenum, are able with a stronger bond to form metalloenzymes: metalloproteins, metalloporphyrins, and metalloflavins.

Heavy metals are emitted to the atmosphere by energy production and a number of technological processes (see Table 7). It makes the atmospheric deposition of heavy metals originating from human activities the dominant pollution source for the vegetation of natural ecosystems— forests, wetlands, peat lands, and so on. The heavy metal content in sludge and fertilizers plays a more important role for agricultural land where also the inputs of heavy metals by irrigation, natural fertilizers, and application of chemicals including pesticides may add to the overall pollution level. The atmosphere and hydrosphere both have a well- developed ability for "self-purification"—for heavy metals by removal processes, for instance, sedimentation. The lithosphere has a high buffer capacity toward the effects of most pollutants and also has an ability to self-purify, for instance, by runoff and uptake by plants, although the rates usually are much lower than in the two other spheres. Table 8 illustrates the removal rates.

Heavy metals are bound to clay particles due to their ion-exchange capacity and to hydratized metal oxides, such as iron sesquioxide (As, Cr, Mo, P, Se, and V) and manganese sesquioxides (Co, Ba, Ni, and lanthanides). Calcium phosphate is further better able to bind As, Ba, Cd, and Pb in alkaline soil. Fulvic acids (molecular weight about 1000) and humic acid (molecular weight about 150,000) are able to form complexes with a number of heavy metals, Hg(II), Cu(II), Pb(II), and Sn(II). The mobility of heavy metals is dependent on a number of factors. The soil pore water contains soluble organic compounds (acetic acid, citric acid, oxalic acid, and other organic acids), partly excreted by the roots. These small

TABLE 7 Important Atmospheric Pollution Sources of Heavy Metals

Source	Heavy Metals
Incineration of oil	V, Ni
Incineration of coal	Hg, V, Cr, Zn, As
Gasoline	Pb (leaded gasoline)
Metal industry	Fe, Cu, Mn, Zn, Cr, Pb, Ni, Cd, and others
Application of pesticides	Hg, Cr, Cu, As
Incineration of solid waste	Hg, Zn, Cd, and others

Source: Jørgensen.[10]

TABLE 8 Removal of Heavy Metals by Runoff and Drainage from a Typical Cultivate Clay Soil

Metal	Removal (mg/m^2/yr)	Removal % of Pool 1–3
Pb	0.5	1–3
Cu	1.2	2–3
Zn	15.9	30–50
Cd	0.07	15–30

Source: Waid.[8]

organic molecules form chelated, soluble compounds with metal ions such as Al, Fe, and Cu. Activity of living organisms in soil may also enhance the mobility of heavy metal ions. Fungi and bacteria may utilize phosphate and thereby release cations. Formation of insoluble metal sulfide under anaerobic conditions from sulfate implies a reduced mobility. The lower oxidation stages of heavy metals are generally more soluble than the higher oxidation stages, implying increased mobility.

The many possibilities of binding heavy metals in soil explain the long residence time. Cadmium, calcium, magnesium, and sodium have the most mobile metal ions with a residence time of about 100 years. Mercury has a residence time of about 750 years, while copper, lead, nickel, arsenic, selenium, and zinc have residence times of more than 2000 years under temperate conditions. Tropic residence times are typically lower (for all heavy metals, about 40 years).

The biological effect of heavy metal pollution occurs in accordance with Sections 4.4 and 4.5 on two levels: on an organism level and on the higher level—the ecosystem level.

Plant toxicity is very dependent on the presence of other metal ions. For instance, Rb and Sr are very toxic to many plants, but the presence of the biochemically more useful K and Ca is able to reduce or eliminate toxicity. The toxicity of arsenate and selenate can be reduced in the same manner by sulfate and phosphate.

Formation of complexes by reaction with organic ligands also reduces toxicity due to reduced bioavailability. The plant toxicity of heavy metals in soil is consequently also correlated with the concentration of heavy metal ions in the soil solution.

The heavy metals that are most toxic to plants are silver, beryllium, copper, mercury, tin, cobalt, nickel, lead, and chromium. With the exception of silver and chromium, the divalent form is most toxic. For silver, it is Ag^+, and for chromium, it is chromate and dichromate that are most toxic. Silver and mercury ions are very toxic to fungus spores, and copper and tin ions are very toxic to green algae; lethal concentrations may be as low as 0.002–0.01 mg/L.

One of the key processes on ecosystem level is the mineralization process, because it determines the cycling of nutrients. Heavy metals can inhibit the mineralization due to the blocking of enzymes. The effect is known not only for the enzymes produced in the organisms but also for extracellular enzymes—exoenzymes—originated from dead cells or excreted from roots and living microorganisms. As the various processes forming the cycling of nutrients are coupled, the entire mineralization cycle is disturbed if only one process is reduced. It is therefore possible to determine the change of the mineralization cycle by measuring the respiration, the transformation of nitrogen, and the release of phosphorus. As low a concentration of copper as 3–4 times the background concentration may imply a reduced soil respiration. A few hundred milligrams of copper per kilogram of soil is furthermore able to diminish the nitrogen release rate by one half.

The most sensitive mineralization process is phosphorus cycling. Biological material binds phosphorus as esters of phosphoric acid. The phosphate is released by the hydrolysis of the ester bond, a process catalyzed by phosphatase. This process is inhibited by the presence of heavy metals. The inhibition is decreasing in the following sequence: molybdenum (VI) > wolframate (VI) > vanadate (V) > nickel (II) > cadmium > mercury (II) > copper (II) > chromate (VI) > arsenate (V) > lead (II) > chromium (III).

The inhibition of exo-enzymes by heavy metals does not form a clear pattern. It is therefore difficult to generalize. Most experiments, however, give a clear picture of the influence of heavy metals on mineralization: the rate of mineralization may be reduced significantly with a consequent reduction of the productivity of the entire ecosystem.

In Denmark (a country with relatively little heavy industry and good pollution control), atmospheric deposition causes an average annual increase of the total content of heavy metal in soil between 0.4% and 0.6%, but it varies very much from location to location. In accordance with the many possibilities for side reactions of heavy metals in soil, including adsorption to the soil particles, the amount of heavy metal ions that are available to plants is only a fraction of the total content. If only the bioavailable heavy metals are used as the basis, the annual percentage increase in the soil concentration due to atmospheric deposition is probably higher.

Most lead in soil is not mobile and cannot be transported via the root system to the leaves and stems. This is in contrast to cadmium, which is very mobile. About 50% of the cadmium in soil will be found in the plants after the growth season, although the concentration may be very different in different parts of the plants. The cadmium in grains for instance has not increased parallel to the increased atmospheric deposition of cadmium.

The heavy metal pollution of soil is one of the major challenges in environmental management in industrialized countries. Due to the many diffuse sources of heavy metal pollution, the solution of the problem requires a wide spectrum of methods, the first of which is application of cleaner technology (see the section on *Industrial Air Pollution, Overview, and Control Methods*). It is in other words necessary to reduce the total emission of heavy metals. Dilution (for instance, higher chimneys) is not an applicable solution. Moreover, as pollution, particularly air pollution, has no borders, it is necessary to take international initiatives and agree on international standards, particularly for the most problematic heavy metals, i.e., cadmium, mercury, nickel, chromium, and vanadium. A three-point program must be adopted:

- A nationally and internationally accepted environmental strategy.
- Agreed international standards and long-term goals.
- A monitoring program to assess the pollution level and compare the measured concentrations with standards.

References

1. Reible, D.D. *Fundamentals of Environmental Engineering;*; Lewis Publ.: Boca Raton, New York, Washington, DC, 1998; 526 pp.
2. Leith, D.; Licht, W. The collection of cyclone type particles—A new theoretical approach. AICHE Symp. Ser. **1975**, *68*, 196–206.
3. Pring, R.T. Speciation considerations for fabric collectors. Pollut. Eng. **1972**, *4*, 22–24.
4. Rullman, D.H. Backhouse technology: A perspective. J. Air Pollut. Control Assoc. **1976**, *26*, 16–18.
5. Gooch, N.P.; Francis, N.L. A theoretically based mathematical model for calculations of electrostatic precipitators performance. J. Air Pollut. Control Assoc. **1975**, *25*, 106–113.
6. Onnen, J.H. Wet scrubbers tackle pollution. Environ. Sci. Technol. **1972**, 6, 994–998.
7. Hanf, E.B. A guide to a scrubber selection. Environ. Sci. Technol. **1970**, *4*, 110–115.
8. Waid, D.E. Controlling pollutants via thermal incineration. Chem. Eng. Prog. **1972**, *68*, 57–58.
9. Waid, D.E. *Thermal Oxidation or Incineration*; Pollut. Control Assoc.: Pittsburgh, PA, 1974; 62–79.
10. Jørgensen, S.E. *Principles of Pollution Abatement*; Elsevier: Amsterdam, 2000; 520 pp.

23

Alternative Energy: Hydropower

Andrea Micangeli,
Sara Evangelisti, and
Danilo Sbordone

Introduction

Approximately 70% of the earth's surface is covered with water, a resource that has been exploited for many centuries. Hydropower is currently the most common and the most important renewable energy source: throughout the world, it produces 3288 TWh, just over 17% of global production and the 84% of energy produced by renewable energy sources[1] from an installed capacity of about 850 GW.[2]

The International Energy Agency (IEA) has developed a number of scenarios that describe the efforts needed to reduce carbon dioxide emissions. The "business-as-usual" baseline scenario foreshadows the situation in the absence of policy change and major supply constraints leading to increases in oil demand and CO_2 emissions. The "**BLUE**" scenario is the most ambitious, bringing emissions at 50% of the 2005 level by 2050. This implies of course higher investment costs, as well as greater needs in technological and policy developments. In *Energy Technology Perspectives 2010*, it states that hydro could produce up to 6000 TWh in 2050.

The main characteristics that make hydropower a successful energy source are its plant storage capacity and fast responses to meet sudden fluctuations in electricity demand. Global hydropower generation has increased by 50% since 1990, with the highest absolute growth in China, as shown in Figure 1.

IEA estimates the *global technically exploitable hydropower potential* (the *technically exploitable potential* is the annual energy potential of all natural water flows that can be exploited within the limits of current technologies[3]) at more than 16,400 TWh per year.[1] However, hydroelectric plants of big dimensions, with million cubic meter water basins, have negatively affected the natural and social environment of the territories. Small plants are characterized by a different management, distributed on the territory, managed in small communities, integrated in the multiple and balanced use of the water resources.

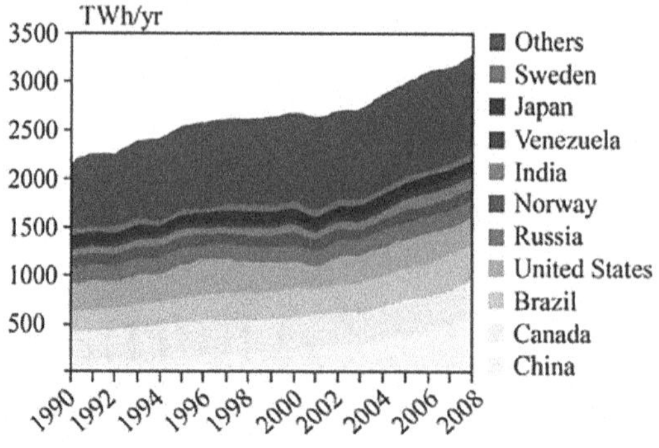

FIGURE 1 Evolution of global hydropower generation, 1990–2008.
Source: International Energy Agency.[1]

The contribution of *small hydropower (SHP) plants*, defined as those with installed capacity of up to 10 MW, to the worldwide electrical supply is about 1%–2% of the total one, amounting to about 61 GW.[2] Europe with about 13 GW installed capacity has the second biggest contribution to the world's installed capacity, just behind Asia. Moreover, the SHP potential is estimated in 180,000 MW.

SHP has a key role to play in the development of renewable energy resources and an even greater role in developing countries. In the face of increasing electricity demand, international agreements to reduce greenhouse gases, environmental degradation from fossil fuel extraction and use, and the fact that, in many countries, large hydropower sites have been mostly exploited, there is an increasing interest in developing SHP. Indeed, SHP has a huge, as yet largely untapped potential, which will enable it to make a significant contribution to future energy needs, offering a very good alternative to conventional sources of electricity, not only in the developed world but also in developing countries.

A hydropower sector technological maturity has already been reached during the last century, but only big plants have received all the benefits from technological development, while those of smaller dimension have been neglected. Nowadays the economy of scale, social and environmental implications, suggest this solution due to their economical feasibility and environmental respectful, allowing sustainable distributed production with an easy installation and great applicability in developing countries.

This entry is organized as follows: first, a classification of the hydropower plants is given. Then, the basics of the technology of mini and micro-hydro plants are illustrated, together with a description of the main civil works that occurred in a hydropower scheme. Finally, a conclusion on the potential and shortcomings of the hydropower technology is drawn.

Classification of the Hydroelectrical Plants

Hydropower plants can be generally classified in terms of power outputs:

Micro-hydro plant, with a nominal power lower than 100 kW, subdivided into *low-head plants*, when the vertical drop is lower than 50 m, and *low-flow rate plants*, when the water flow is lower than 10 m^3/sec.

Mini-hydro plant, with a nominal power between 100 kW and 1000 kW, subdivided into *mini-head plants*, when the vertical drop is between 50 and 250 m, and *mini-flow rate plants*, if the water flow is between 10 and 100 m^3/sec.

Small hydro plants, with a nominal power between 100 kW and 10 MW, subdivided into *medium-head plants*, when the drop is between 250 and 1000 m, and *medium flow-rate plants*, when the water flow is between 100 and 1000 m^3/sec.

Big hydro plants, as shown in Figure 2, with a nominal power of more than 10 MW, defined as *high-head plants* if the drop is higher than 1000 m and as *high-flow rate plants* with a water flow of more than 1000 m^3/sec.

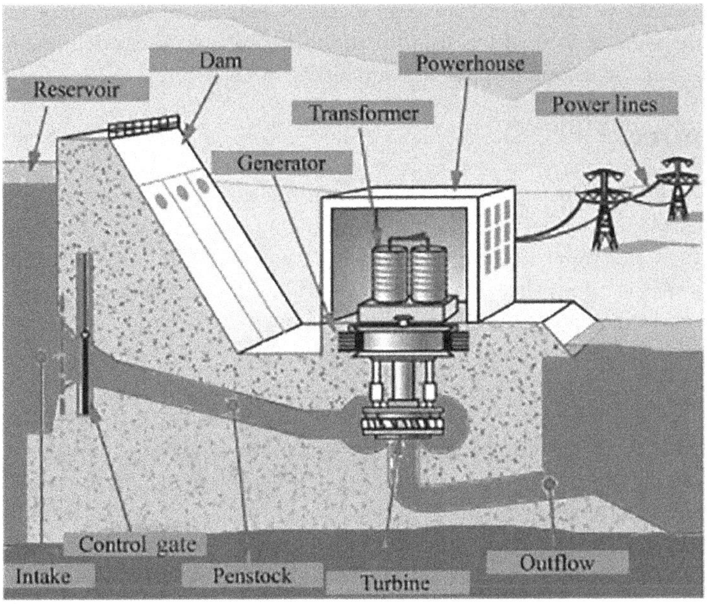

FIGURE 2 Big hydro storage scheme.

Another important way to classify hydroelectric plants is on the basis of their typology. In particular, they can be classified as follows:

Run-of-river scheme: or fluent water plants, they take a portion of a river through a canal or penstock. They do not require the use of a dam or catch basin. Because of that, they aim to affect upstream water levels and downstream stream flow less than any other power plants. Electricity generation from these plants could change in the amount of water flowing in the river.

Storage scheme: an impound water behind a dam, as a reservoir. Water is released through turbine generators to produce electricity. The water storage and release cycles can be relatively short, for instance, storing water at night for daytime power generation, or the cycles can be long, storing spring runoff for generation in the summer, when air conditioner use increases power demand. Some projects operate on multiyear cycles carrying over water in a wet year to offset the effects of dry years.

Pumped-storage scheme: these plants use off-peak electricity to pump water from a lower reservoir to an upper reservoir. During periods of high electrical demand, the water is released back to the lower reservoir to generate electricity.

The following sections focus on mini and micro-hydropower plants and run-of river scheme.

Mini-Hydro Plants

Mini-hydropower (MHP) plants have an installed power lower than 1 MW actually in Europe is 3 MW but in many country (in particular USA 5 MW) it may be more. Generally, these plants need less civil works, consistently reducing the costs connected to the realization of the plant and justifying their realization also under an economic point of view. If the plants are well planned and placed, their environmental impact is reduced for their limited dimensions. The simplicity of construction allows them to be introduced in contexts where the technology of the sector is not yet developed and there is a strong need for mechanical or electrical power. Also in those cases, MHP plants can be operated and maintained locally, even with less-specialized technicians. Changing hydropower plant size or typology, many things are the same, such as a turbine installation. This entry focuses on mini-hydro due to its low environmental impact and its opportunities of developing in the future.

Water Resource

Hydraulic energy, as almost every forms of energy on the earth, comes from the sun, which is the "engine" of the hydrological cycle. The sun, irradiating and warming up the atmosphere, makes seas and lakes evaporate; the water vapor rises up and thickens the clouds that move because of the wind, also generated by the sun; the clouds then produce precipitations in the form of snow, hail, and rain. When the rainfall ends up in the natural basin situated at a higher level, energy is transformed to *potential energy*. This energy is naturally stocked in rivers and in creeks that flow into the sea, closing in this way the hydrological cycle (see Figure 3).

The amount of available energy, which the water basin can produce at a given height, comes from the water level reached at the end of the cycle. In other words, to know the potential energy of a basin, it is necessary to evaluate the available rise, depending on the orography of the territory and on specific water works such as dams or small barriers.

The amount of water available is defined as the mass of water flowing per time unit (*flow rate*). In general, the potential power from a reservoir can be calculated by the following equation:

$$P_0 = \rho \cdot g \cdot Q \cdot H_0 \qquad (1)$$

where

P_0 = theoretical power (W)
ρ = water density (1000 kg/m^3)
g = gravitational acceleration (9.8 m/sec^2)
Q = flow rate (m^3/sec)
H_0 = net head (m)

The output power of the plant is a percentage of it, due to mechanical, electrical, and friction losses.

Hydrology and Rain Measurements

In order to exploit the energy of water for power purposes, a *hydro geological analysis* of the territory is needed, particularly for catch basins. The analysis is based on the evaluation of the supply of the basin and its outflow. To obtain a balanced catch basin, it has to take into consideration the *meteoric flow rate,*

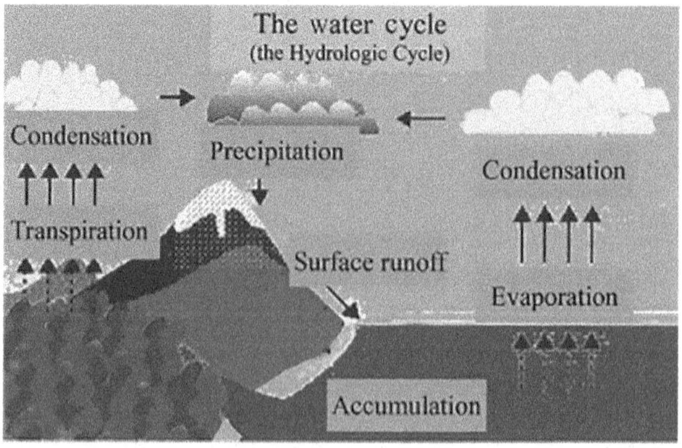

FIGURE 3 Hydrological cycle. (Source: Harvey et al.[4])

evaporation, and both *superficial* and *underground circulation*. The quantity of water from the rainfall to the basin, the *meteoric supply*, must be evaluated not only on the water surface of the basin itself but also on the whole area in which the rainfall is collected as well as the streams towards the basin (see Equation 2). The flows, which depend on soil permeability, are essentially of two types: *superficial* and *underground*, i.e., when the water filters through the soil and supplies underground basins and water-bearing stratum.

It is possible to evaluate the *meteoric supply*—superficial flow, as:

$$P = E + D + (I - C) \tag{2}$$

where P is the *meteoric supply* to the basin, E is the contribution given to the evaporation, D is the outflows, and I and C are the increase and decrease of the basins, respectively.

The term E is the amount of the following different contributions:

- Evaporation of water from the soil
- Transpiration of plants
- Evaporation of water intercepted by vegetation
- Evaporation of internal basins

Similarly, the term in relation to draining—underground flow—can be subdivided into the following:

- Natural water draining underground toward the external (*groundwater*)
- Artificial draining water toward the external (*inversion*)
- Natural superficial water inflow from the external (*water flows*)
- Underground natural water inflow (*water-bearing stratum*)
- Artificial inflow from the external (*adduction*)

Joining of different rivers has to be considered as well. The evaluation of the meteoric intake is usually performed through specialized devices, such as *rain gauges*, very common all over the world (see Figure 4).

The intensity of the rainfall flow rate is not constant through time and it has to be referred to different periods of the year (usually a multiyear). Not all the water from the rainfall ends up in the catch basin as

FIGURE 4 Example of a rain gauge.

shown before. Generally, the phenomenon is estimated by introducing a *coefficient of draining*, depending on the waterfall and the water collected into the catch basin, as shown in Equation 3:

$$C = \frac{V}{V_0} \tag{3}$$

where V is the real caught volume and V_0 is the waterfall.

Once V is determined, it is possible to calculate directly the energy exploitable from the plant in a given site:

$$E = 0.00273 \cdot \eta \cdot H_0 \cdot V \tag{4}$$

where H_0 indicates the *net drop*, in other words, the available drop minus the losses in the work of adduction with η output efficiency of the turbine.

Within a natural *hydrological basin*, it is necessary to analyze the head and the flow rate available along the whole river bed, through the *hydrodynamic curve* (see Figure 5). It shows the surfaces of the catch basin on the horizontal axes and the height of the water flow on the vertical one. Through the hydrodynamic curve, it is possible to optimize the entire use of a catch basin, while for the realization of a single plant, without having the intention to optimize the use of the resource along the whole river bed, it is enough to measure the flow rate of the river and the consequent evaluation of the quantity of water that can be taken from the basin or from the available drop.

Measure of the Water Flow

The determination of the water flow of a catch basin can be done by using specific devices, but it has to be undertaken only in absence of historical data of the course along the years. Different methods exist.[4] However, for each of them, it would be necessary to repeat the measures along a period of time to obtain the variation that occur throughout the year.

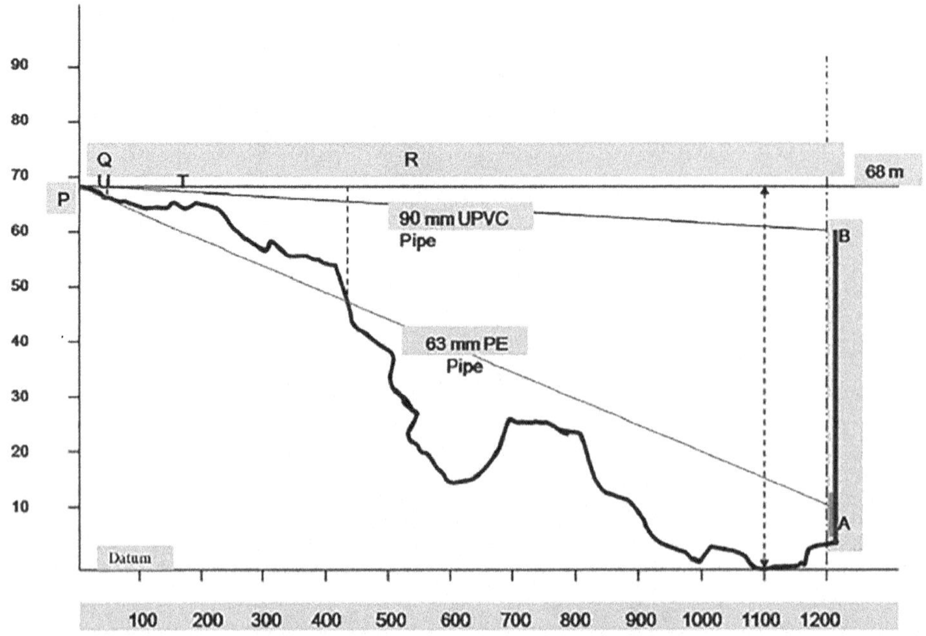

FIGURE 5 Example of a hydrodynamic curve.

One of the simplest methods that can be used is to force the flow to get into a container of known dimensions, measuring the necessary time to fill it up. This method is known for its simplicity; however, it is limited in that it can be applied only in rivers with small water flows.

A second method (Figure 6) consists of the realization of a weir of known dimensions, in which the river is forced to get into—the *weir method*. This method can be used to bring up to around 1 m^3/sec. To measure how much is carried, the second level reached by the water is taken into consideration as illustrated in Figure 7.

The water flow Q can be calculated using:

$$Q = 0.41 \cdot B \cdot H \cdot \sqrt{2 \cdot g \cdot H} \tag{5}$$

In this formula, H can be calculated as the difference between H_2 and H_1 (see Figure 7), B is the width of the weir, and g is the gravitational acceleration.

Another method of measuring water flow involves the evaluation of water velocity on the cross-sectional area. Velocity measurements can be made through a very simple method: with a floating, not too light in order to avoid the friction with the air, in a place where the river is pretty regular and flat, measuring the time that the floating takes to cover a specific distance. The measurement must be repeated more than once and the sought value taken into consideration must be the average between the distance covered and the time spent to cover it. The final velocity must be corrected with a factor between 0.75 and 0.85, given by the losses due to the friction with the sides of the canal.

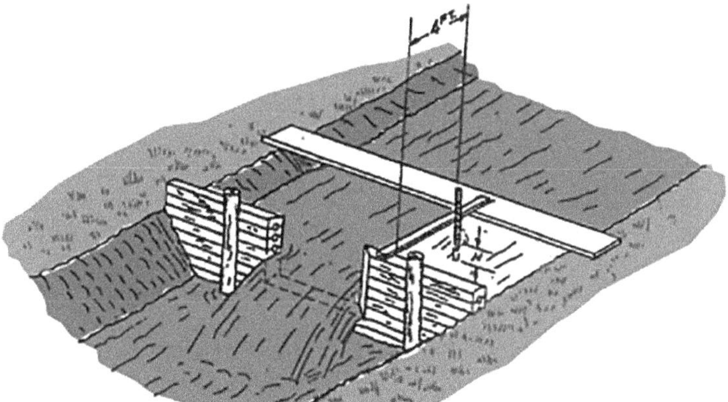

FIGURE 6 Weir method.

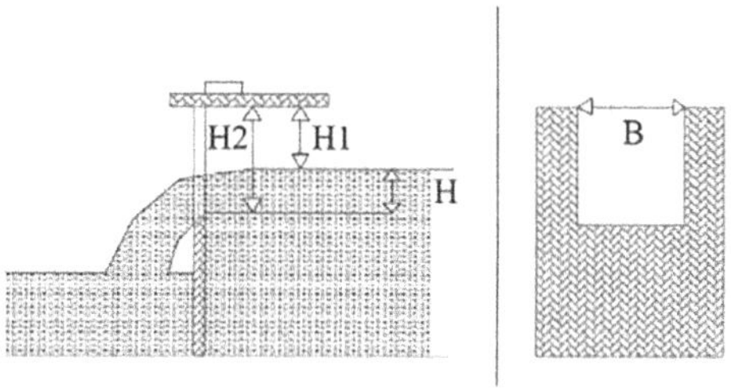

FIGURE 7 Weir realization.

TABLE 1 Value of the *Factor of Manning* (n) for Different River Bed Typologies

River Bed Typology	n
Regular river bed with a minimum annual flow	0.030
Stable flow condition	0.035
River with stagnant water, aquatic vegetation, and meanders	0.045
River with stones and shrubs with shallow pools and lush vegetation	0.060

The velocity can be obtained also through different methods, such as the use of "titled solutions," taking into account the variations on the electricity conductivity of the river when it flows with a known quantity of salt inside.

Once the cross section is evaluated, the water flow can be then calculated through the *formula of Manning*:[5]

$$Q = \frac{A \cdot R^{\frac{2}{3}} \cdot S^{\frac{1}{2}}}{n} \tag{6}$$

where A represents the cross section, R is the hydraulic radius, and S is the slope of the water surface. The value of n can be obtained from Table 1.

Flow Duration Curve

Water flow measurements are always referred to a specific period, as the water of the river changes during the year, passing several times from huge quantities of water to smaller ones. In general, the curve is uneven as it reports the water flow rate throughout the year as shown in Figure 8.

To organize the collected measurement data, it is possible to use another graphic that puts them all together. Indeed, in Figure 8, it is possible to note that there are two evident points of absolute maximum and absolute minimum, corresponding to the maximum quantity of water that occurs for a very short period and the minimum quantity of water that is the quantity available all year. Moreover, Figure 9 shows a *flow duration curve*, which presents the duration of each amount of the flow rate.

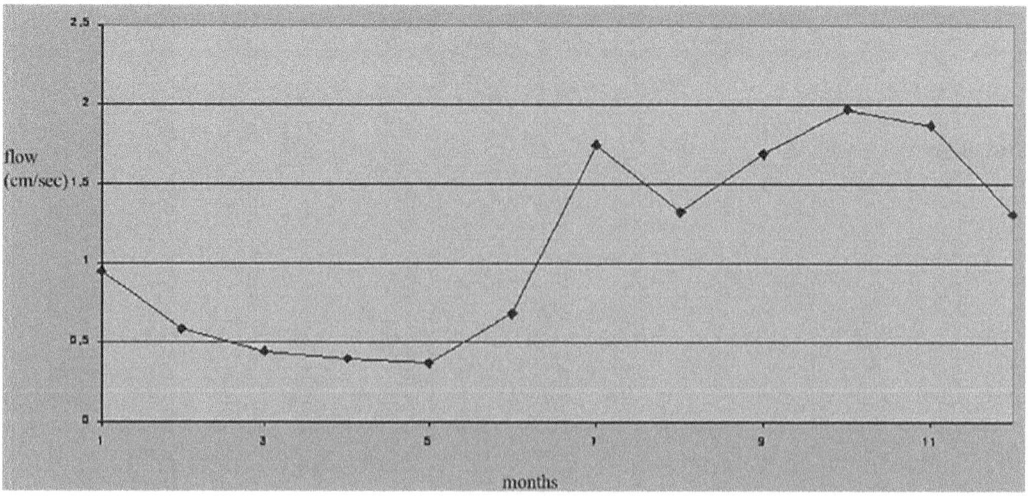

FIGURE 8 Example of a daily flow curve.

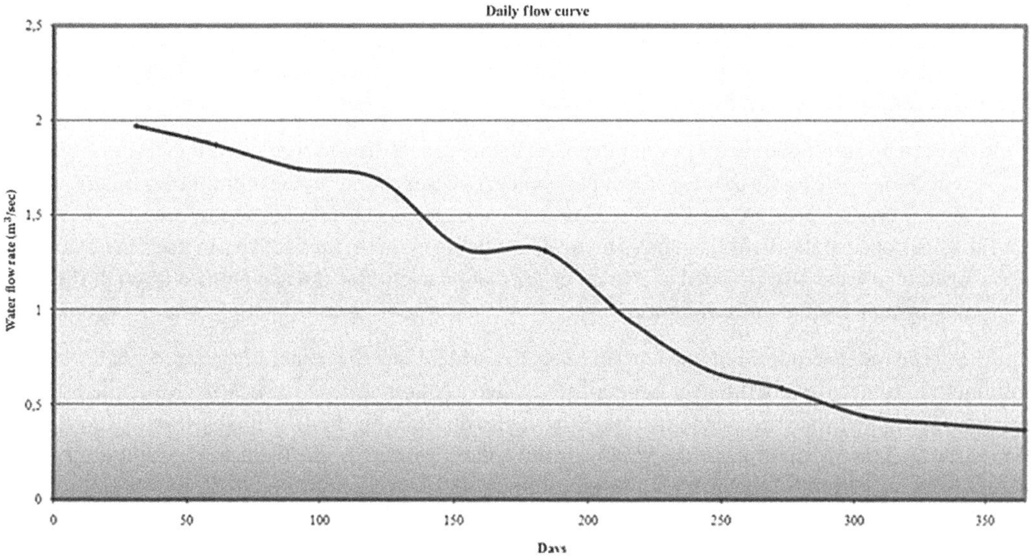

FIGURE 9 Example of a flow duration curve.

Measurements of the Gross Head

To measure the height difference between two points, it is necessary to utilize a level and to follow the scheme in Figure 10. The operator must simply read the values of each ruler to come out with the height by computing the difference between the two values. This procedure can be repeated until the final point is reached. If a level is not available, it is possible to use a table with a carpenter level—even if it requires a lot of patience—or to proceed with a plastic transparent pipe filled up with water, which fulfills the same characteristics of a level. While measuring from the available head, it is wise to also calculate the length of the *forced penstock* as distance from the hold point to the arrival of the penstock itself. This is essential both for the choice of the material of the penstock and for the evaluation of the pressure drop.

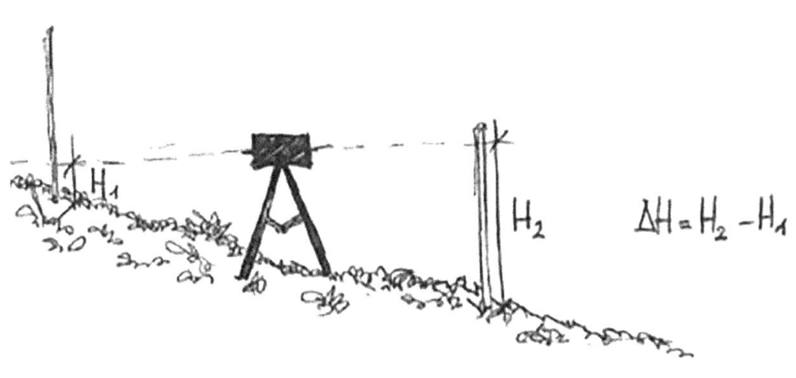

FIGURE 10 Example of measurement of the gross head.

Instream Flow and Environmental Impact

The balance of any catch basin is connected not only to the water balance but also to the real possibility of exploitation, characterized by other aspects:

Rivers can be used not only for power purposes. Before proceeding to the derivation of the outflow, it is necessary to be sure that it will not have a negative impact on further communities. It is necessary to verify all the aspects connected with the multiple uses of the water resource.

The subsistence of the natural balances involved in the river and in the catch basin itself. An example, to adduce water to a riverbed of a catch basin, can be useful for the fish fauna, although the flow rate is high and it comes from nearby.

The use of the water resource of a catch basin refers to a wider issue that generally is approached letting a minimum natural course of the river, generally indicated as *minimum instream flow* (*IF*), defined as "the minimum height of water needed to maintain the values of the basin at an acceptable level." The calculation of the IF is essential: in fact, if the minimum flow rate is lower than the IF during the planning stage, the no-working periods of the plant can be estimated.

To guarantee a minimum flow means to preserve the biological balance and the need of the use of a civil work as a caption downriver. The derivation established on the basis of this context must be lower than the limit beyond which it may influence the river ecosystem and may cause the entering in crisis however the natural water regime must be guaranteed.

The river regime model must therefore take into account the following aspects:

Biological species that would suffer from the uncontrolled derivation of water.

Hydrologic characterization for the protection of water balance (equilibrium) and the defense of soil.

The use of the water resource represented by the social and economic wardship of needs.

Two different relations are commonly used for the calculation of the minimum IF:

$$Q_{IF} = \frac{15 \cdot \alpha \cdot Q_{media}}{\left[\ln(\alpha \cdot Q_{media})\right]^2} \tag{7}$$

where α is the *coefficient of perpetuity* given by the relation between low intake and the medium IF values, expressed in liters per second:

$$IF_{hydrol.} = 6.IF_{microhabitat} \tag{8}$$

The equation shows that conserving the IF from a hydrological point of view also means, with a big margin, that the conservation of IF from a biological point of view is connected with the microhabitat. If the value of the low intake is unknown and only the average is known, it is possible to assume that the coefficient of perpetuity α is equal to 0.24, showing the outflow of hydropower plant that guarantees the preservation of the fish fauna and hydrology. Figure 11 shows a typical example of the realization of a canal to guarantee the conservation of hydrology and of the fish fauna in a hydropower plant.

Civil Works in MHP Plants

The *civil works* for *hydropower facilities* have the functions to capture, to exploit, and to return the water downriver. In small rivers, the realization of those works is done by deflecting the water flow for a short period of time, to operate in dry conditions. To deflect the river, a provisional river bed has to be built or the plant must be realized during the dry season, if feasible.

FIGURE 11 Example of a system to conserve fish fauna.

Dams and Weirs

The choice of the typology of the dam must be made taking into consideration the orography of the territory and the water resource to exploit.[6] Generally, hydraulic works must create a good drop between mountain and valley, to have the possibility to exploit the potential energy from the mass of water. The structure can be *dams* (Figure 12) or *catch basins* (Figure 13).

A *dam* is a "lung" of water available that is able to compensate, partially and totally depending on the volume created, the variation of the flow rate of the river during the year. Although the creation of a dam needs a big investment, this makes it practically not usable in MHP.[7]

A type of barrier—often installed in small plants—is the *derivation weir*. The derivation weir is smaller and more cost-effective than a traditional dam and generally the water overflows the traverse, reversing in the natural path of the river. If the plant is big enough, the derivation weir turns into a small dam that basically does not function like a water accumulator, but it aims at raising the level of the water flow.

FIGURE 12 Example of a dam.

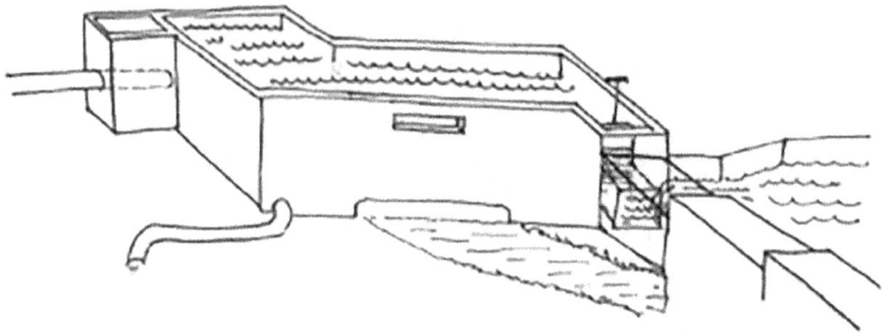

FIGURE 13 Example of a catch basin.

The derivation weir can be realized with several different materials and, when possible, it is suggested that locally available material, such as rocky materials, be used. Otherwise, it is possible to use blocks of flat rocks and soil kept altogether by metal nets to constitute the *barrage*.

It is important that the barrage is realized with a central waterproof nucleus, made with clay, and supported with soil or minerals. If there sand or gravel is available, a barrier made of concrete can be also taken into consideration.

If the regimes of full flow are huge, it is necessary that the intake must be drained through ad hoc dischargers not easy to realize when the barrage has been made in the river. Instead, in places with much seismic activities or in very cold climates, it is better to avoid rigid structures, so floor barriers are preferred.

Stability of the Dam

In small plants, the stability of the dam generally depends on the weight of the dam itself. The strains that the dam applies to the ground through its own weight can be higher than those that the ground can take. It is necessary to minimize the infiltration of water underneath the dam with the use of spillway drain and erosion or diaphragms.

Also, the stability to the overturning of the dam must be guaranteed in all the conditions of charge concerning the hydrostatic push related to the dragging of solid material or to seismic waves. In order to guarantee that, it is enough to ensure that all the strains against the ground are not negative, in order to avoid dangerous situations that can bring the structure to overturn. This entails designing a dam where all the horizontal and vertical strains fall internally in the central part of its own basement (see Figure 14).

As the hydrostatic strains generally go from upstream to downstream, the side of the dam facing downstream is more sloping as compared to the perpendicular of the basement; thus, the profile of the dam upstream has to be more vertical than the downstream.

System of Elevating the Free Water Surface

To control the water flow, *floodgates* are generally used. Given that most of these devices are used to control the free water surface elevation being stored or routed, they are also known as *crest gates*.

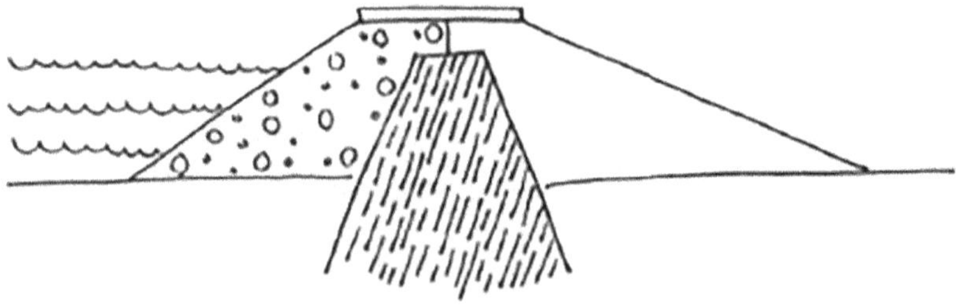

FIGURE 14 Stability of the dam.

A removable type of *floodgates* is *flashbooks gates*— wooden panels usually installed on the cap of the barrier that allow the increase of the water surface and that can be removed during floods, avoiding the inundation of the upstream fields.[8]

Generally, to avoid the manual intervention and to check upstream flooding, it is possible to install a gate that can be opened progressively during full intakes. Another solution is to install *fusegates*, concrete crates that flip over when the water level is reached (Figure 15). Finally, another type is the *dinghy*, anchored to the cap of the barrier that blows up during full intakes of water, only to deflate again during the rest of the year.

Spillway

If the flow rate varies during the year, it could be that the floodgates systems are not enough to guarantee the integrity of the plant. In this case, it is necessary to foresee a system that takes the surplus flow rate to downstream— the *spillway*. Except during flood periods, water does not normally flow over a spillway. The surplus water usually flows at a high velocity and often it is necessary to insert a system to reduce its kinetic energy. In small plants, the introduction of a drainage, which allows emptying the loading tank to help in the maintenance of the plant, is always considered.

The spillway is also used in times of emergency, i.e., shutting down the plant.

Intake

The *water intake* (see Figure 16) is a structure in which water is adduced in order to bring it to the *forebay tank*. It must be able to address in the penstock or in the drainage canal the amount of water estimated in the project. It is pleonastic to underline how the civil work has to be studied and realized in a way to minimize friction losses and the impact on the environment. It must also be designed to minimize maintenance and to reduce the costs. Practical aspects of the project related to civil work therefore must follow the following criterion: Hydraulics and structural works have to guarantee the resistance of the pipes to minimize the waste of energy and to be cost-effective, to avoid transportation of solid material inside the pipes to the powerhouse and for easier maintenance, and to reduce the passage of fishes and to not compromise the ecosystem of the area in which it has been realized.

FIGURE 15 Example of fusegates: "Sant'Antonio 1" IdroPower Station (Italy).

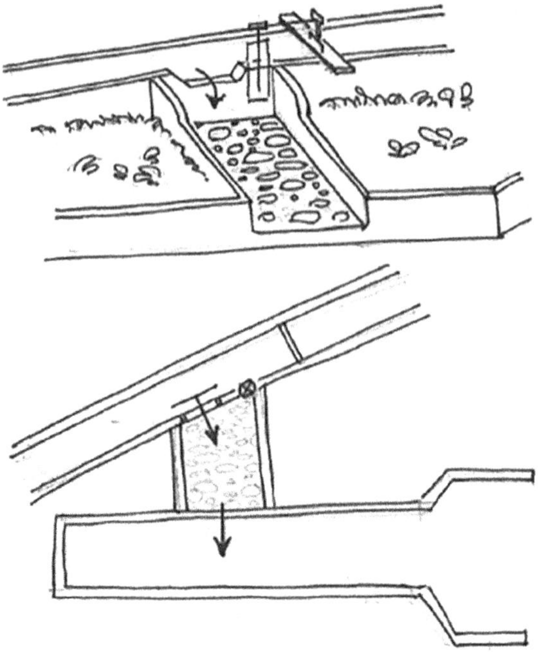

FIGURE 16 Example of intake works.

The water intake works can be realized in different ways, depending on the peculiarity of the orography of the territory. Often it is a channel that brings the water to the point where the forced pipe is. In this case, the channel is realized in a way that the water flows slowly with a contained slope of the pipe in order to contain the losses and the erosion of the walls and to preserve the jump.

The choice of the dimension of the intake channel and the water velocity are results of a compromise to avoid frequent maintenance due to the deposits of sand and slime, to avoid losses and erosion against the wall. Generally, the water flows along the channel with a velocity between 0.1 and 0.4 m/sec. If the quantity of solid materials to be transported is huge, then a bigger tank, where the deposits end up due to the reduction of the speed of the water as a result of a bigger section, is needed, based on the fundamental hypothesis of the continuity of the intake.

The *sedimentation tank* is necessary if the channel has not an open surface, due to the costs connected to the maintenance of the closed channel. In small plants, the drainage tank usually works as a sedimentation tank as well; however it must be cleaned more frequently and it will have a bigger dimension due to the sedimentation process. The transportation of solid material is very deleterious not only for turbine performance but also for the life of the device itself.[5]

The *orientation* of the *intake* is crucial when choosing a project, as it can reduce the accumulation of material over the grid itself and the frequency of the intervention of maintenance. The best position is parallel to the flow letting the full flows the task of removing the material stuck in front of the grid. Anyway, it does not have to be located in areas of stagnant water as the whirlpool and the parasite flows tend to accumulate solid material in front of the grill. If there is a discharger, it is good to place the grill next to it to simplify maintenance as the deposits can be pushed to the discharger as well.

Forebay Tank

At the end of the channel or coincident with the intake, replacing the channel of charge, a little tank, known as *forebay tank* or *basin of charge,* has to be realized. Its function is to guarantee the presence of upstream water in the penstock, in order to avoid the entrance of air along the pipe and the formation of whirlpools.

Channel

If the plant needs it, the sampling of the flow to adduce in the central, is done by using a channel. The channel can be realized both as open channel and as under pressure pipe. In the small plant, the technical solution is oriented towards an open surface channel, the sizing of which is done considering first the intake to derive. The intake is the function of the section of the channel as well as of the slope and the roughness that depends on the material used and on the degree of finishing of the wall. The channel can be made from different materials such as soil, wood, and concrete.

Generally, for small plants where the banks have an inclination of 45° with the base, if the width is L, the width of water surface is $2L$ and the height is $L/2$. Concerning concrete structures, using one that is rectangular shaped, which helps in the cleaning of the channel, is usually preferred.

Penstock and Pressure Drop

The *penstock* (see Figure 17) takes the water from the load tank and pushes it to the turbine. It can be realized with the use of different materials. One should take into account the cost, weight, type of joints, and the conditions of the ground when choosing the type of material to be used. The *penstock* is also characterized by the diameter of the conduit itself that must contain the loss of load.

The choice of the pipe diameter has to be made as a compromise between three needs:

1. Keeping the costs down and therefore realizing a small-diameter pipe.
2. Containing load losses.
3. Realizing a bigger pipe to increase the energy.

The first head measure can be seen as a gross head, keeping into consideration the losses inside the conduits and all the other works of adduction.

The real *head* exploited by the turbine is lower than the first value above. Indeed, the definition of drop goes together with the *net head* that identifies the usable jump by the turbine; thus, the gross head minus the losses of the adduction works. Such a definition allows dividing the losses of the hydraulic parts from those related to the turbine. The value of the net head obviously depends on the pressure

FIGURE 17 Example of a penstock.

drops occurring inside the penstock. Through the definition of *net head*, it is also possible to define the output of the section of the hydraulics work simply as:

$$\eta_{idr} = \frac{H_n}{H_0} \tag{9}$$

where n and 0 refer to the net head and the gross head, respectively. The bigger is the hydraulic output of the plant, the better will be the exploitation of the water resource as higher power can be obtained with the same load or the same power can be obtained with less load. Moreover, if the entity of the load losses compared to the available drop is modest in high- and very-high-fall plants, in the low-fall plant with 6 m of available drop, a load loss of 1 m is almost 20% of the produced power. Thus, the hydraulic works of an adduction channel have to be realized with focus on MHP plants.

The penstock is the part of the plant in which the water flows faster. Considering that the losses are proportional to the square of the speed, the realization of the penstock is very important in terms of hydraulic performances of work of adduction.

The amount of water that flows inside the penstock is functional to the section of the pipe, its diameter and the water velocity. Once the flow is designed, a relation between speed flooding and penstock diameter is needed.

The problem can be solved with a dimensional analysis that puts into evidence how the *Fanning factor* is a function of the *Reynolds number* and *relative roughness*, known from fluid dynamic theory. The Fanning factor is connected to load losses and it represents their adimensionalization; the Reynolds number comes from the relation between inertia forces, viscosity, and velocity. Relative roughness is connected to the choice of material and the level of superficial finishing. Generally speaking, it is verified that, depending on the fluid regime, the Fanning factor tends to depend only on one of the variables.[8]

Walls

The thickness of the walls and veins are subject to the pressure of the impulse load, which also includes a water hammer. Nevertheless, in the case of a water hammer, plastic pipelines react better than iron ones, because the elasticity of the plastic tends to absorb overpressure better than other materials.

Once the ideal material is selected, the formula of the thickness can be found using the *Mariotte's formula*:[9]

$$t = \frac{P \cdot D}{2\sigma_f} \tag{10}$$

where t is the thickness of the pipe, P stands for the hydrostatic pressure, D is the diameter, and σ is the allowable stress.

Equation 10 is only valid for stationary systems, where both capacity reductions and closure operations are not verified. Moreover, it does not take into account the problems that occur in iron pipes. Therefore, Equation 11 should be amended, and, taking into account the types of joints, it becomes:

$$t = \frac{P \cdot D}{2\sigma_f \cdot k_f} + t \cdot s \tag{11}$$

where k_f is the efficiency of the welding and $t \times s$ represents the overpressure due to corrosion. k_f can be derived using Table 2.

In general, the value obtained as the thickness is always corrected when it is too low to take into account other factors: the tube must have achieved a sufficient rigidity to be moved without deformation. If the plant has a high fall, then a conduct with variable thicknesses (based on the pressure) can

TABLE 2 Value of k_f for Different Types of Joints

Type of Joint	k_f
Without welding	1
Welding checked with x-ray	0.9
Welding checked with x-ray and subjected to a relaxation	1

be used in order to reduce the cost of the materials. In addition to resistance to pressure increases, a conduct has also to withstand internal depressions to avoid collapsing:

$$P_c = 882.500 \cdot \left(\frac{t}{D}\right)^3 \quad (12)$$

where P_c is the pressure of collapse.

The depressions can be avoided through an *aerophore* with a minimum diameter:

$$d = 7.47\sqrt{\frac{Q}{\sqrt{P_c}}} \quad (13)$$

where d is the diameter of the aerophore.

Finally, to conclude the calculation of the wall thickness, a water hammer has to be considered. The *Allievi-Michaud formula* can be modified if the pressure is expressed in water column, as:

$$\Delta P = c\frac{\Delta V}{g} \quad (14)$$

where c is the propagation speed in the middle of the pressure wave that depends on the water density and the elasticity of the material:

$$c = \sqrt{\frac{k}{\left(1+\frac{kD}{Et}\right)}} \quad (15)$$

where k is the water cubic compression module (2.1×10^9 MPa); E is Young's modulus of the conducting material; t and D are the thickness and diameter of the tube, respectively; and ρ is the water density. By applying the relationships (Equations 14 and 15) to PVC and iron pipes, it is possible to calculate for an instant closure (d of 400 mm, PVC thickness of 14 mm, and iron thickness of 4 mm):

$$c_{\text{pvc}} = 305 \text{ m/sec}$$

$$c_{\text{acciaio}} = 1023 \text{ m/sec}$$

$$\Delta P_{\text{pvc}} = 123 \text{ m}$$

$$\Delta P_{\text{acciaio}} = 417 \text{ m}$$

This provides a quantitative demonstration of the previously described nature of the two materials. If the operating time increases, the pressure is drastically reduced. Indeed, the maneuvering speed plays a crucial role in the generation of the overpressure. In large systems, it is common not to install pipes that can withstand overpressure, or water hammer, but rather to install mechanisms for the exclusion of the load to prevent the turbine from going into "overspeed."

Another device that serves to absorb the pressure waves that can occur within the pipes is the *piezometric borehole.* To evaluate if its installation is needed, the following formula can be considered:

$$I = \frac{V \cdot L}{gH} \tag{17}$$

where I is the constant acceleration. If I is less than 3, it can be assumed that the piezometric borehole is not necessary.[5]

The penstocks are also anchored to the ground, or supported on special works such as anchor blocks or saddles. The distance between two saddles or between two anchor blocks has to be as much as to make the pipe's arrow acceptable when it is full.

Water Turbines

A turbine converts energy in the form of falling water into rotary shaft power. The selection of the best turbine depends on several factors: the net head of the plant, the nominal flow, the power rating, and the shape of the turbine. The most installed turbine models, mainly used in big hydro plants, are three different typologies: Pelton, Francis, and Kaplan. In MHP, the turbine choice is made on different considerations, not only economical. Indeed, its construction and operational simplicity become essential, especially in developing contexts. Often, it is possible to install simpler versions of big hydro plant turbines—not for the Pelton model, which is the simplest one yet, for example, changing the blade edges.

As stated before, the choice of the turbine depends on the net head and the flow rate, as well as on the available water resources and the plant typology. A good criterion to select the turbine is resumed in the following well-known diagram, where they have a range in head and flow (Figure 18).

It should be stressed that the fields of employment are not very narrow and are only suggestions for the best choice. In fact, a Pelton turbine could be installed in a low head-high flow rate plant, even if the turbine efficiency will be strongly penalized. This results in areas of the diagram where different typologies of turbines can be used at the same time. The final choice has to be taken considering also other factors, such as the operation of the plant.

Conclusion

Hydropower has been used as far back as the Roman empire and through history has been used to power water mills, textile machines, sawmills, and irrigation systems. In the early 1800s, however, people started to see that the use of water to power small factories and machines is but a minor application of its potential. As early as the 19th century, waterpower was being used as a source of electricity. Though primitive hydropower technology only consisted of wheels, buckets, and river flow, it was from this point on that waterpower's potential as one of the most efficient and abundant sources of renewable energy became apparent.

Mini-hydropower is probably the least common of the three readily used renewable energy sources (i.e., water, sun, and wind), but it has the potential to produce the most power, more reliably than solar or wind power if you have the right site. Small-scale hydro is in most cases run-of-river, without dam or water storage, and is one of the most cost-effective and environmentally benign energy technologies for developing countries and further hydro development in Europe.[17]

In this entry, a summary of the main advantages and shortcomings of small-scale hydropower has been presented. The hydro resource is a much more concentrated energy resource than either wind or solar power and the energy available is readily predictable. Moreover, no fuel and only limited maintenance

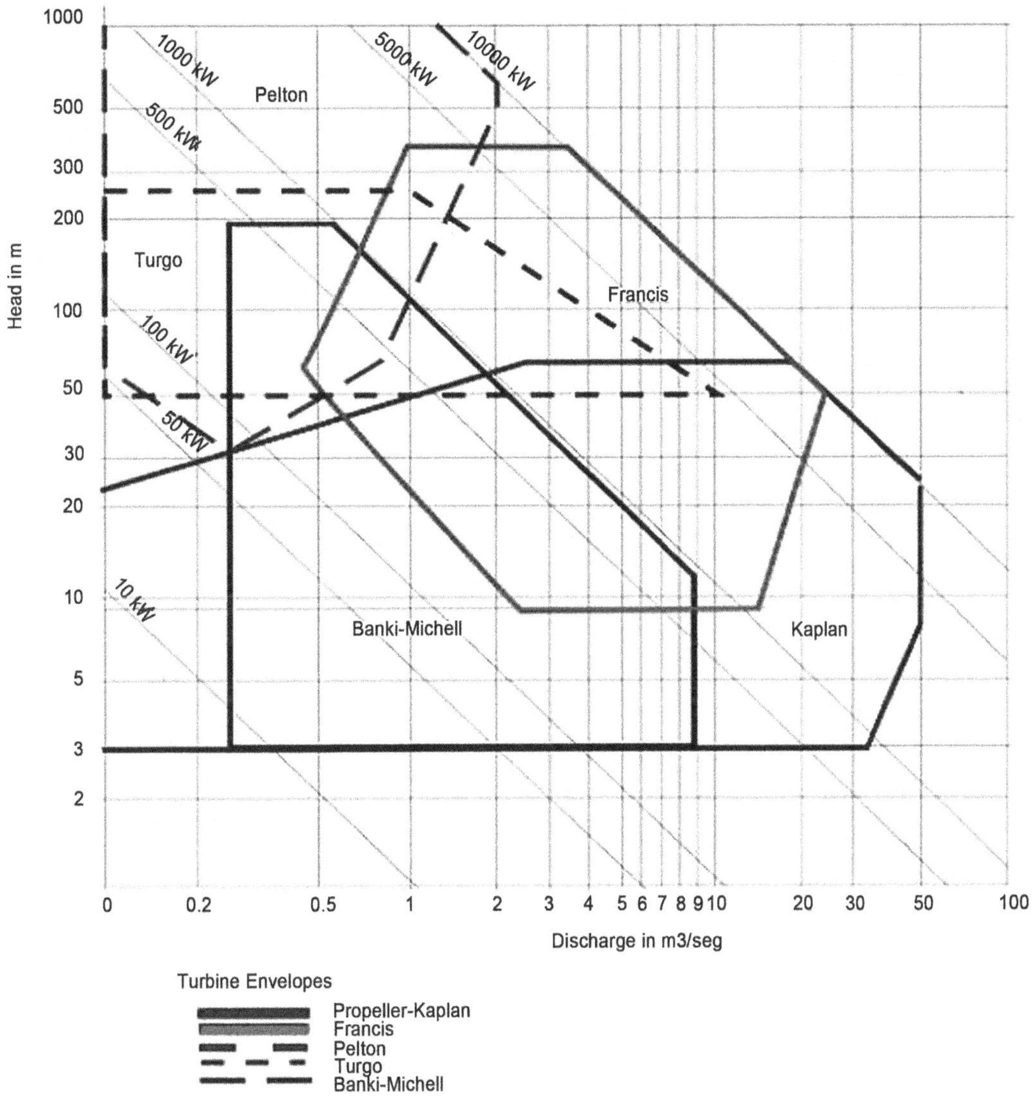

FIGURE 18 Values of flow and height for each field of application of different turbines.

are required and, if well designed, it has almost no environmental impact. On the other hand, it has to be considered that it is a site-specific technology and no general consideration has to be taken in the design of the plant; otherwise, environmental, social, or economical problems could occur, such as conflicts with fisheries interests on low-head plants and with irrigation needs on high-head plants. Furthermore, river flows often vary considerably with the seasons, especially where there are monsoon-type climates, and this can limit the firm power output to quite a small fraction of the possible peak output.

However, where a hydropower resource exists, experience has shown that there is no more cost-effective, reliable, and environmentally sound means of providing power than a hydropower system. Even with the various advantages of hydropower, it is still an underused alternative energy source. As of 2008, only 6% of the United States' electricity production came from hydropower, while nearly 50% came from the non-renewable source that is coal.[12] Due to a lack of economic speculation, a vast amount of potential for renewable hydropower remains untapped. Third world countries and underdeveloped areas have many areas that would be highly conducive to hydropower. The construction and use of hydropower facilities in these countries/areas, along with an increase in hydropower in the United States, could result in a great increase in renewable, affordable, and non-polluting energy.

However, if the prospected potential has to be realized, significant challenges have to be faced, in terms of *decision-making process,* establishing an equitable, credible, and effective environmental assessment procedure that takes into account both environmental and social concern and that takes into consideration the share of the benefits with local communities, both in the short term and in the long term. Finally, increasing efficiency, developing high-tech turbines, and reducing the costs of very low-head schemes, along with proper technology transfer of appropriate turbines to local manufacturers and technical support to the developers, will help realize our long-term objectives.

References

1. International Energy Agency. *Renewable Energy Essentials: Hydropower*; OECD/IEA, 2010.
2. *State of the art of small hydro power in the EU 25*; Thematic Network of Small Hydropower Project, European Small Hydro Power Association, 2005; 1–20.
3. *2007 Survey of Energy Resources*; World Energy Council, ISBN: 0 946121 26 5, 2007; 1–600.
4. Harvey, A., et al. *Micro-Hydro Design Manual*; IT Publications Ltd.: London, 1993.
5. Caputo, *C. Gli impianti convertitori di energia;* Casa edi-trice Ambrosiana, ristampa, 2008.
6. *International Journal of Hydropower and Dams, World Atlas;* Aquamedia Publications: Sutton, 2000.
7. Khennas, S.; Barnett, A. Best practices for sustainable development of micro-hydro in developing countries.
8. Available at http://www.friulanacostruzioni.it/pages/ita/prodotti/paratoie/paratoia-a-tenuta-su-tre-lati-manuale.php.
9. Churchill, S.W. Friction factor equations spans all fluid-flow ranges. Chem. Eng. **1977**, 91.
10. Arrighetti, C. *Dispense del corso di Macchine II; 2007.*
11. Paish, O. Micro-Hydro Power: Status and Prospects. Journal of Power and Energy, Professional Engineering Publishing, 2002.
12. Energy Technology Perspective 2010, IEA, Paris.

24

Alternative Energy: Photovoltaic Solar Cells

Ewa Klugmann-Radziemska

Introduction

Nowadays, fossil fuels are the main sources of energy from which electricity is obtained, but these sources will not last forever, so in due course renewable energies will have to replace them in this role. One of these new sources is solar energy. Each year, the Earth receives around 1×10^{18} kWh of solar energy, which is more than 1000 times the current global energy demand. This is therefore a vast source of energy that can be tapped to satisfy human energy requirements. To generate electricity from sunlight, solar cells (photovoltaic cells) are used. These devices are based on the photovoltaic effect, in which a p–n semiconductor is exposed to light, and photons are absorbed by electrons, providing an electric current. The electrons that are set free are pulled through the electric field and into the n-area. The holes produced move in the other direction, into the p-area.

The use of solar energy releases no CO_2, SO_2, or NO_2 gases and does not contribute to global warming. Photo-voltaics is now a proven technology that is inherently safe, as opposed to some dangerous electricity-generating technologies. Over its estimated life, a photovoltaic module will produce much more electricity than was used in its production. A 100 W module will prevent the emission of more than 2 tons of CO2. Photovoltaic systems make no noise and cause no pollution while in operation.

Photovoltaic Effect

The solar cells now in use are the practical application of fundamental physical phenomena observed already in the 19th century (see Table 1).

The absorption of light in semiconductors takes place when electrons are released from interatomic chemical bonds. In order to produce a free electron in a given semiconductor material, a certain quantity of energy must be supplied, equal at least to that of the energy band gap, which in the case of silicon at a temperature of 300 K is $E_g = 1.12\,\text{eV}$. The liberated electron leaves behind it a hole that can move about by diffusion or drift under the influence of an electrical field owing to its being positively charged (Figure 1).

The introduction of other atoms in place of the parent atoms (or at interstitial positions) of a pure intrinsic semiconductor considerably improves its electrical conductivity. Energy levels of elements with one valence electron more than the semiconductor atoms form donor levels in the neighborhood

TABLE 1 The Beginnings of Photovoltaics Worldwide

Year	Achievement
1839	Alexander Edmund Becquerel observed the photovoltaic effect in a circuit of two illuminated electrodes immersed in an electrolyte.
1843	Fritts produced the first tin–selenium solar cell.
1879	Adams and Day observed the photovoltaic effect at the interface between two solid bodies (selenium–platinum).
1930	The first copper/copper oxide solar cell.
1941	Ohl patented the silicon cell (monocrystalline silicon doped during its growth).
1954	At the Bell laboratory (United States), Chapin, Fuller, and Pearson developed a cell on monocrystalline silicon with a diffusion p–n junction of 6% efficiency, which was subsequently manufactured by two companies.
1954	Lindmayer and Allison obtained a cell with an efficiency of 16% (radiation intensity, 1000 w/m^2).
1954	Reynolds produced the first multiple-junction Cu_2S/ CdS cell.
1958	Monocrystalline solar cells were used for the first time in the Vanguard I satellite, where modules of six cells supplying 5 W of electrical power were installed; the traditional batteries ran out after a few months, but the photovoltaic panels powered the satellite's transmitter for another 6 years.
1962	The first thin-layer Cu_2S/CdS photocell was obtained.

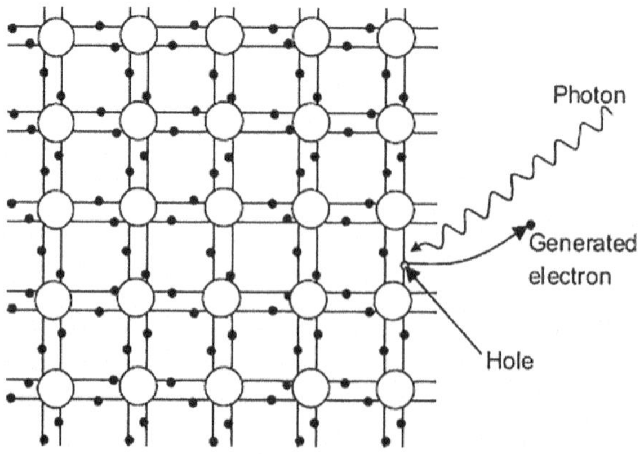

FIGURE 1 The generation of an electron–hole pair by a photon of energy $h\nu > E_g$.

of the conduction band (n-type). Energy levels of elements with one valence electron less than the semiconductor atoms form acceptor levels in the vicinity of the valence band (p-type).

If a p –n junction is formed from the p- and n-type areas of the semiconductor, then the charge carriers move around in such a way that the Fermi level will be identical throughout the crystal (Figure 2b). At room temperature (300 K), practically all donor and acceptor dopants are ionized; hence, the concentrations of majority carriers (electrons in the n-type area and holes in the p-type area) are approximately equal to the concentrations of the relevant dopants (Figure 2a).

At the instant these two areas are brought into intimate contact, a very large concentration gradient of electrons and holes across the boundary between them comes into existence. This gradient causes electrons to diffuse from the n-type area to the p-type area and the holes to move in the opposite direction. As a result of this diffusion, a space charge region comes into being near the junction: on the n-type side, this is positive, since electrons have left this area, while the uncompensated positive charges of immobile donor ions remain along with the holes newly arrived from the p-type area; on the p-type side, it is negative, because in the same way carrier diffusion has given rise to an area of negative charge consisting of immobile acceptor ions and electrons newly arrived from the n-type area. In this way, a dipole space charge layer is formed in the area around the p–n junction (Figure 3). A potential barrier and electric field are formed within this layer that counteract further diffusion and restrict the diffusion

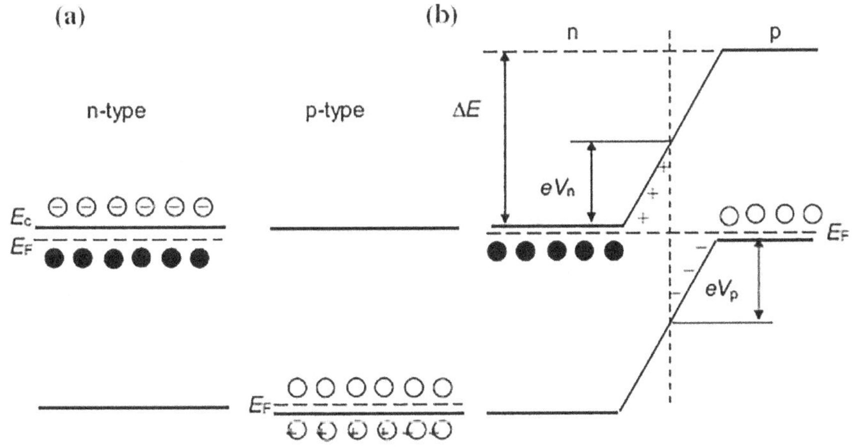

FIGURE 2 Formation of an abrupt p–n junction (b) as a result of the juxtaposition of n- and p-type areas (a), $\Delta E = e(V_n + V_p)$.

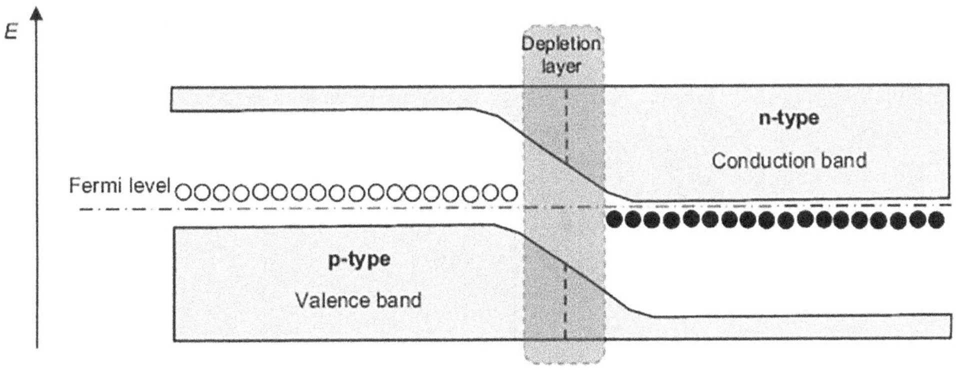

FIGURE 3 Equilibrium in the p–n junction region.

current. Apart from majority carriers, there are minority carriers in the two areas on either side of the junction, which come about as a result of the thermal generation of electron–hole pairs. The potential barrier formed as a result of majority carrier diffusion favors the outflow of minority carriers from both areas. The movement of these carriers creates a dark current, which flows in the opposite direction to that of the diffusion current.

If the p–n junction is illuminated by photons with an energy equal to or greater than the band gap width E_g $(h\nu \geq E_g)$, then electron–hole pairs form on either side of the junction, as in the case of thermal generation (Figure 4).

Carriers forming no farther from the potential barrier than the diffusion length of minority carriers will diffuse towards the potential barrier and will be distributed there by the electric field due to the presence of the junction (the diffusion length is the mean distance that minority carriers have to move before they recombine with majority carriers). This field causes the carriers to move in opposite directions—electrons to the n-type area and holes to the p-type area. If an electron–hole pair forms on the p-type side of the junction, the electron reaches the junction before it has any chance of recombining with the hole (if recombination does occur, the resultant energy is emitted in the form of heat, and the effect is entirely useless as far as the photovoltaic effect is concerned), and the hole in this pair stays on the p-type side since it is repelled by the barrier in the junction. There is no danger of recombination here as there is an excess of holes in this region. The same thing happens when an electron–hole pair is generated by light on the n-type side of the junction. Then, the liberated electron remains on the n-type side, as it is repelled by the barrier. On the p-type side, however, we now have very few free electrons capable of recombining. This causes an increase in negative charge on the n-type side and of positive charge on the p-type side, which leads to a charge imbalance in the cell. This charge separation gives rise to a potential difference across the junction. As a result, a photoelectric current I_f comes into being in a closed circuit, regardless of the height of the potential barrier.

The generation of a photoelectric current I_f by a stream of photons in a solar cell can be demonstrated using the model of a current generator connected in parallel with a diode representing the p–n junction of the cell. As Figure 5 shows, the output current I flowing through the series resistance r_s of the cell and the load resistance is equal to the difference between the generated photoelectric current I_f and the diode current I_d.

$$I = I_f - I_d = I_f - I_s\left(\exp\frac{eU}{mkT} - 1\right),$$

where I_s is the saturation current and m is the diode ideality factor.

It emerges from the above equation that for cell operated at open circuit ($I = 0$):

$$I_d = I_f \text{ and } U = U_{oc} = \frac{mkT}{e}\ln\left(\frac{I_f}{I_s} + 1\right).$$

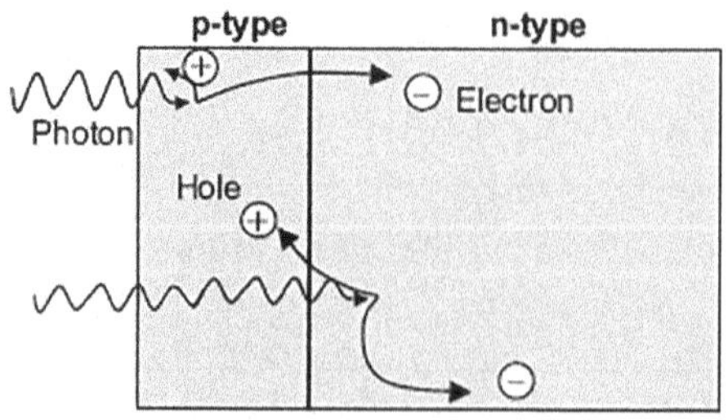

FIGURE 4 The potential barrier in a solar cell distributes the charge carriers generated by light.

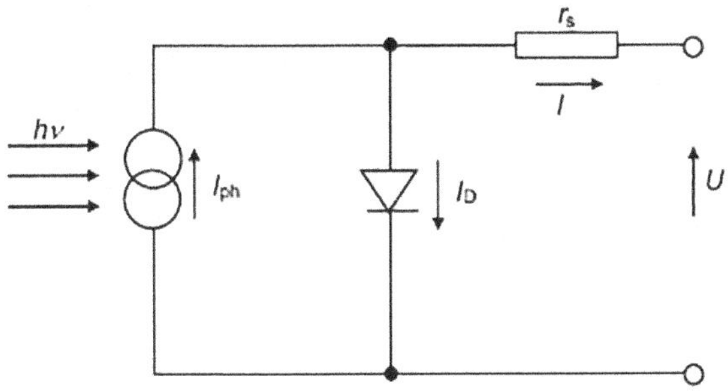

FIGURE 5 Electrical model of a solar cell.

From this last relationship, we obtain:

$$I_f = I_s\left(\exp\frac{eU_{oc}}{mkT} - 1\right) \text{ and}$$

$$\text{finally}: I = I_s\left(\exp\frac{eU_{oc}}{mkT} - \exp\frac{eU}{mkT}\right).$$

For an exact description, however, we replace the single-diode electrical model with a two-diode equivalent circuit, which has two resistors: r_s is the series resistance of the cell, which consists of a number of components, and r_p is the effect of all defects in the crystal in the p–n junction area and is a shunt resistor (Figure 6).

In this model, the current generated is described as a function of the cell voltage as follows:[2]

$$I = I_{s1}\left(\exp\frac{e(U - Ir_s)}{m_1kT} - 1\right) + I_{s2}\left(\exp\frac{e(U - Ir_s)}{m_2kT} - 1\right) + \frac{U - Ir_s}{r_p} - I_{ph}$$

where I_{ph} is the photoelectric current, I_{s1} and I_{s2} are saturation currents, and m_1 and m_2 are non-ideality factors of the characteristics of the two diodes.

The parameters of the model are defined in such a way as to ensure that the above equation gives a good description of the real characteristic of a photovoltaic cell. The first exponential term in the characteristic equation represents the diffusion current, whereas the second one represents the recombination currents in the entire cell, particularly in the space charge region. The characteristics, which also enable the parameters of the two-diode model to be determined, are measured when the cell is polarized in the

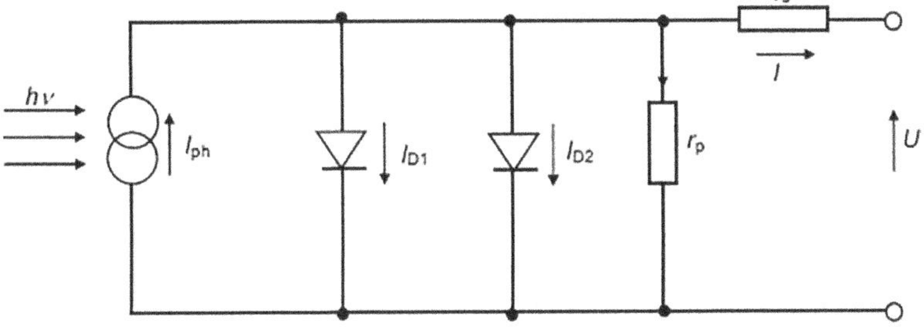

FIGURE 6 Equivalent circuit of a two-diode model of a solar cell.
Source: Stutenbaeumer and Masfin.[1]

forward direction and in the complete absence of any illumination, the dark current being measured as a function of the external voltage.

Solar Cell Characteristics

The usable voltage from solar cells depends on the semiconductor material. In silicon cells, it amounts to approximately 0.6 V.

Under illumination, the fourth quadrant of the light *I*–*U* is the region of interest (Figure 7), and the figures of merit for the device are the following:

1. The open-circuit voltage (U_{oc}) is the maximum voltage obtainable under open-circuit conditions (Figure 8).
2. The short-circuit current (I_{sc}) is the maximum current through the load under short-circuit conditions (Figure 8).
3. Fill factor (*FF*).

The output voltage of the photovoltaic cell is only slightly dependent on irradiance, while the current intensity increases with intensity of insolation. The working point of the solar cell therefore depends on load and insolation. In addition, the output voltage of a solar cell is temperature dependent. A higher cell working temperature leads to lower output and, hence, to lower efficiency (Figure 9)

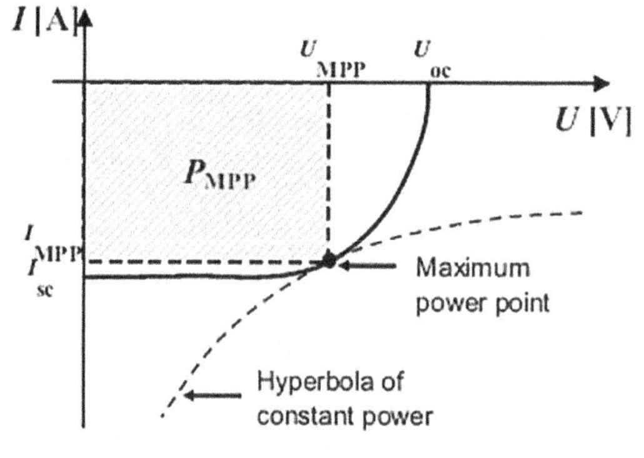

FIGURE 7 Current–voltage characteristic of illuminated photovoltaic solar cell.

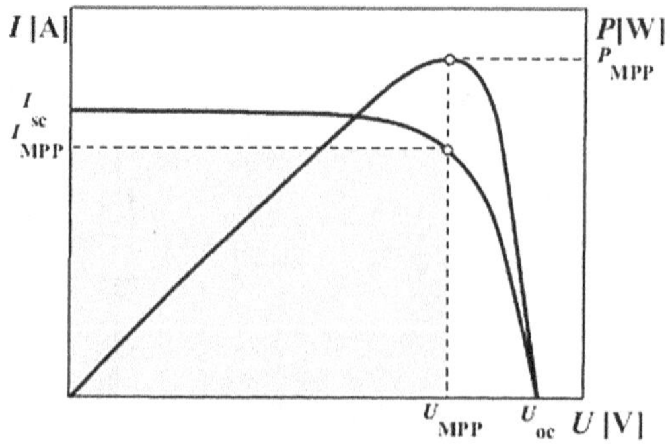

FIGURE 8 Current–voltage and power–voltage characteristics of the solar cell.

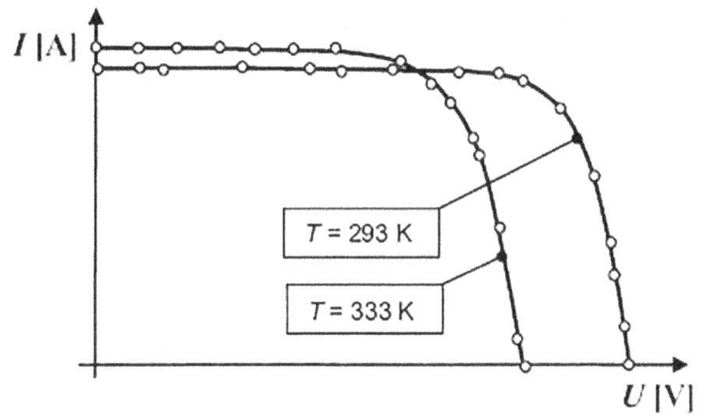

FIGURE 9 Temperature dependence of the solar cell characteristic.
Source: Radziemska and Klugmann.[3]

An important parameter as regards the application of a PV module in photovoltaics is the peak power obtainable from the module at the load resistance R_{opt}, where the rectangle under the characteristic $I(U)$ has the maximum area equal to the maximum power $P_{mpp} = I_{MPP}\ U_{MPP}$, and the point of intersection with the curve of $I(U)$ is in this case the maximum power point (MPP). The load resistance R in the cell circuit or PV module should be chosen such that the power it generates takes the maximum value, i.e., $P = P_{Mpp}$.

The MPP is the point at which the coordinates I_{MPP} and U_{MPP} form a rectangle with the largest possible area under the $I(U)$ curve.

The level of efficiency indicates how much of the radiated quantity of light is converted into usable electrical energy. The photovoltaic conversion efficiency of the cell η_{PV} is calculated from the maximum output power point (MPP) in the $I(U)$ curve:

$$\eta_{pv} = \frac{I_{MPP} I_{MPP}}{E.S_C}.100\%,$$

where S_C is the total surface of solar cell and E is the irradiance (W/m^2).

To describe solar cell quality, a special parameter—the fill factor (*FF*)—is used. It can be calculated from the following equation:

$$FF = \frac{I_{MPP} U_{MPP}}{I_{sc}.U_{oc}},$$

where I_{MPP} is the MPP current, U_{MPP} is the MPP voltage, I_{sc} is the short-circuit current, and U_{oc} is the open-circuit voltage.

For the ideal solar cell, the fill factor is a function of open-circuit parameters and can be calculated as follows:

$$FF \approx \frac{v_{oc-\ln(v_{oc}+0.72)}}{v_{oc}+1},$$

where v_{oc} is the voltage, calculated from the equation:

$$v_{oc} = U_{oc} \frac{e}{mkT},$$

where k is the Boltzmann constant = 1.38×10^{-23} J K^{-1}, T is the temperature (K), e is the charge on an electron = 1.6×10^{-19} C, and m is the diode ideality factor (–).

Solar Cell Materials: Production and Features

Silicon

The most important material for solar cell production is silicon. At the present time, it is almost the only material used for the mass production of solar cells. Being the most often used semiconductor material, it has some important advantages.

In nature, it is readily found in large quantities. Silicon dioxide forms one-third of the Earth's crust. It is environmentally friendly and not poisonous, and its waste does not cause any problems. It is easily melted and handled and it is fairly easy cast into its monocrystalline form. Its electrical properties, which remain unchanged up to temperatures of 125°C, allow the use of silicon semiconductor devices even in the harshest environments and applications.

In technology, pure silicon is the only widely used chemical element produced at such a high level of purity. The percentage of pure silicon in "pure silicon" is at least 99.9999999%. The concentration of silicon is 5×10^{22} atoms/cm^3, which means 5×10^{13} impure atoms/cm^3. Quantities of impure atoms are measured using sophisticated physical methods like mass spectrometry.

Pure silicon is produced from sand (silicon dioxide—SiO_2) by reduction at carbon electrodes at 1800°C in specially designed furnaces. The final material contains 98%–99% pure silicon. The complete reaction is:

$$SiO_2 + C \rightarrow Si + CO_2.$$

Such silicon is the raw material for the production of pure silicon. It is also used in steel and aluminium production as a supplementary material. The most important producers of raw silicon are Canada, Norway, and Brazil. Fifteen to twenty-five kilowatt-hours of electrical energy is needed to produce 1kg of silicon. Silicon tetrachloride (tetrachlorosilane) gas is obtained by the chlorination of finely ground metallurgical-grade silicon in a special reactor. Additions or impurities are eliminated in the form of chlorine salt.

$$Si + 2CI \rightarrow SiCI_4.$$

The following reaction produces trichlorosilane gas:

$$SiCI_2 + HCI \rightarrow SiHCI_3 \quad .$$

This gas is then further purified with the removal of any remaining tetrachlorosilane and other silanes. The purification is followed by reduction in a hydrogen atmosphere at 950°C:

$$4SiHCl_3 + H_2 \rightarrow 2Si + SiCl_4 + SiCl_2 + 6HCl.$$

Besides pure silicon, the procedure yields a number of gaseous by-products, which condense outside the reactor. Tetrachlorosilane is one of these by-products. At 1200°C, it can be converted into trichlorosilane using the following reaction:

$$SiCl_4 + H_2 \rightarrow SiHCl_3 + HCl$$

This example illustrates one possible way of producing pure silicon. There are other procedures using different chemical reactions, but the end product is the same—pure silicon.

Crystalline Silicon Solar Cells

Polycrystalline as well as monocrystalline solar cells belong to this group. The basic form for crystalline solar cell production is the silicon ingot (see the description of the production procedure above). The ingot (block of silicon), cut with a diamond saw into thin wafers, is the basis of solar cell production. One-millimeter-thick wafers sawn accurate to 1/10 mm are placed between two plane-parallel metal plates rotating in opposite directions. This procedure enables the wafer thickness to be adjusted to within 1/1000 mm. The subsequent procedure for solar cell production consists of the following steps:

- Doped wafers are first etched some micrometers deep. The procedure removes crystal structure irregularities caused by sawing and cleans the wafer. During the extraction of pure silicon, the material is doped either as powdered polycrystalline silicon or by the addition of a suitable gas. This is then followed by diffusion. Phosphorus, supplied inside the material in gaseous form, diffuses at 800°C. The n-doped layer and the p-rich oxide layer form on top of the wafers as a result of reaction with oxygen.
- The wafers are then folded to form a cube and etched in oxygen plasma, which removes the n-doped layer from the edges. Wet chemical etching then removes the oxide layers from the top of the wafer.
- At the rear, the contact surface is produced from silver containing 1% aluminium. Special procedures enable silver to be printed over mask on cell surface. The pressed cells are then sintered at high temperatures. A similar procedure is used to print the contacts on the front cell surface, and the anti-reflex layer is applied likewise. In this case, titanium paste is used, which, on sintering, forms titanium dioxide (TiO_2) or silicon nitride (Si_3N_4).

Polycrystalline Silicon Production

The extraction of pure polycrystalline silicon from trichlorosilane can be carried out in special furnaces, such as those developed by Siemens. The furnaces are heated by electric current, which, in most cases, flows through silicon electrodes. These 2 m long electrodes are 8 mm in diameter. The current flowing through the electrodes can be as much as 6000 A. The furnace walls are cooled to prevent the formation of unwanted reactions producing gaseous by-products. The procedure yields pure polycrystalline silicon, which is used as a raw material for solar cell production. Polycrystalline silicon can be extracted from silicon by heating it up to 1500°C and then cooling it down to 1412°C, which is just above the melting point of the material. As material cools, a 40 × 40 × 30 cm ingot of fibrous polycrystalline silicon forms.

Monocrystalline Silicon Production

Two different technological procedures are used to produce monocrystalline silicon from pure silicon.

Czochralski's Method

In 1918, the Polish scientist Jan Czochralski discovered a method for producing monocrystalline silicon, from which monocrystalline solar cells could be manufactured. The first monocrystalline silicon solar cell was constructed in 1941. In Czochralski's method, silicon is extracted from the melt in a graphite-lined induction oven at a temperature of 1415°C. A silicon crystal with a set orientation is placed on a rod. Spinning the rod in the melt makes the crystal grow. The rod spins at 10 to 40 revolutions per minute and grows in length at a rate of between 1 μm and 1 mm per second. This allows the production of rods measuring 30 cm in diameter and several meters in length. The whole process takes place in an inert atmosphere. Possible impurities are burnt or eliminated in the melt.

Float Zone Method

With this method, monocrystalline silicon is produced from polycrystalline silicon. The main advantage of this procedure over the previous one is the better yield of pure silicon.

The silicon rods produced measure 1 m in length and 10 cm in diameter. This procedure, in which an induction heater travels along the rod melting the silicon, also takes place in an inert atmosphere. Monocrystalline silicon is produced during the cooling stage. Monocrystalline or polycrystalline silicon ingots are then sawn and the wafers are worked upon until they can serve as a foundation for solar cell production. Sawing causes approximately 50% of the material to be wasted.

Amorphous Solar Cells

Amorphous silicon is produced in high-frequency furnaces under partial vacuum. In the presence of a high-frequency electrical field, gases like silane, B_2H_6, or PH_3 are blown through the furnaces, supplying silicon with boron and phosphorus.

Amorphous solar cells are produced with technologies similar to those used in the manufacture of integrated circuits. Due to this procedure, these modules are also known as thin-film solar cells (thin-film modules). Here is a brief summary of amorphous solar cell production:

- The glass substrate is cleaned thoroughly.
- The lower contact layer is applied.
- The surface is then structured—it is divided into bands.
- The amorphous silicon layer is applied under vacuum and in the presence of a high-frequency electric field.
- The surface is rebanded.
- The upper metal electrodes are fixed.

Other Solar Cells

The less frequently used solar cell types include solar cells produced by the EFG (edge-defined film-fed growth) method, as well as Apex solar cells made from silicon, cadmium telluride (CdTe) solar cells, and copper-indium selenide (CIS) solar cells. EFG monocrystalline solar cells are produced directly from the silicon melt, which eliminates wafer sawing; production costs are thus lower and material is saved since there is no "sawdust." In the EFG procedure, an octagonal tube of silicon, several meters long, is extruded from the silicon melt. The flat sides of this tube are then laser sawn into separate solar cells. Most solar cells are square in shape and 100 × 100 mm in size. Consequently, the module power is greater with a smaller surface compared to crystal modules of square cells with truncated sides. Contacts take the form of copper bands. The separate cells are then combined in the same way as other cell types.

EFG cells are produced by Schott Solar. In contrast to EFG cells, Apex cells are polycrystalline. Their production procedure is protected by patent. The production procedure was developed by Astropower Inc.

Cadmium Telluride

Thin-film material produced by deposition or by sputtering is a promising low-cost foundation for photovoltaic applications in the future. The disadvantage of this procedure, however, is that the materials used in production are toxic. The efficiency of solar cells in the laboratory is as high as 16%, but that of commercial types is only 8%.

Copper-Indium-Diselenide ($CuInSe_2$, or CIS)

$CuInSe_2$ is a thin-film material with an efficiency of up to 17%. The material is promising, but not yet widely used owing to production-specific problems. CdTe and CIS cells have so far been used mostly in laboratory research. Commercial modules made from these materials are still hard to find.

Gallium Arsenide (GaAs)

GaAs is used in the production of high-efficiency solar cells. It is often utilized in concentrated PV systems and space applications. Their efficiency is as good as 25%, and even 28% at concentrated solar radiation. Special types have an efficiency of more than 30%.

Structure and Manufacture of Photovoltaic Cells from Crystalline Silicon

In principle, a photovoltaic cell consists of the following elements (see Figure 10):

- A mono-or polycrystalline silicon wafer in which a p–n junction has been formed.
- Contacts, i.e., the front and rear electrodes; the front one should be shaped in such a way that the maximum amount of incident radiation can reach the junction region, the depth of which is limited by the permeability of silicon to radiation.
- An antireflective coating (ARC) on the front side of the cell.

The manufacture of a crystalline cell takes place in the following stages.

Surface Preparation

The surfaces of silicon wafers, cut from monocrystalline ingots, are degreased, cleaned, polished (mechanically or chemically), and etched in an aqueous solution of sodium hydroxide (40% NaOH) at a temperature of 383 K. A pyramidal surface structure is thereby obtained, which is then rinsed in hydrochloric or nitric acid. Dry (plasma) etching is also possible.

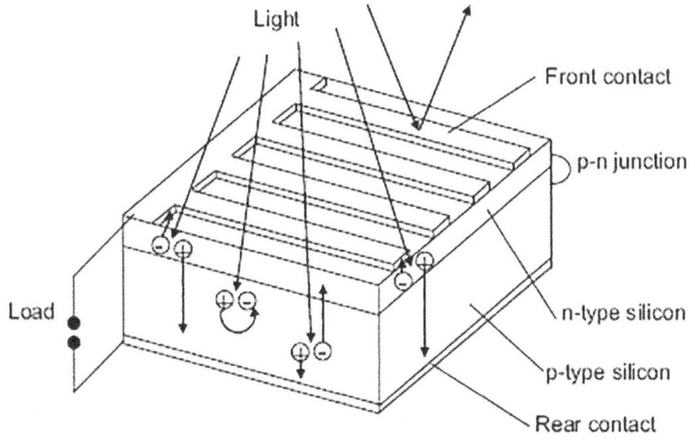

FIGURE 10 The structure and functioning of a photovoltaic cell (not to scale).

Diffusion Formation of a p–n Junction

A dopant, usually an acceptor, is added to the silicon base during crystal growth, whereas the n–p junction is produced by the diffusion of a dopant (usually a donor) to the p-type base wafer across one of its surfaces. If the donor concentration in the subsurface layer of the silicon (initially p-type) is greater than the acceptor concentration, this layer then becomes an n-type semiconductor.

The source of the dopant may be a solid or a gas. There are a number of techniques involving diffusion from the solid phase:

- Vacuum deposition of a thin layer of dopant
- Doping a vacuum-deposited layer of SiO_2
- Doping a mechanically deposited or screen-printed layer of SiO_2
- Coating the silicon wafer with a material containing phosphorus and silicon dioxide

If the diffusion of phosphorus to silicon is carried out at a temperature of 1220 K, 10 min are sufficient to obtain an n-type layer 0.25 μm in thickness.

Formation of an p–n Junction by Ion Implantation

Ion implantation is a technique for obtaining shallow, abrupt n–p junctions; it allows precise control of junction depth and dopant concentration. It is based on the implantation of dopant ions into silicon when its surface is bombarded with a beam of ions (of the order of 10^{16} ions/cm^2) of energy 5–300 keV.

The depth of the junction depends on the energy of the ions. A drawback of this method it that its gives rise to a large number of structural defects in the silicon; these can be removed after implantation by either heating the silicon wafer or irradiating it with an electron or laser beam. Ion implantation is quite a costly method—the necessary equipment (ion source, ion separator and accelerator, vacuum) is expensive.

Passivation of the Silicon Surface

Both the quantum efficiency for blue light and the open circuit voltage can be considerably enhanced if the front surface of the silicon is passivated. This is easily done with a layer of silicon dioxide, which forms on the surface when this oxidizes. Such a thin layer of SiO_2 of controlled thickness can be obtained by heating the silicon wafer in a stream of neutral gas like nitrogen containing a small quantity of dry oxygen. In the cells designed by Green, there is also a very thin layer of SiO_2 (ca. 2 nm thick) underneath the metal front electrode.

Metal Contacts

Since cells are an integral part of an electrical installation, metal contacts are made on either side of the cell. The front electrode is a fine grid, so as to reduce shading of the light-sensitive surface to a minimum. The "fingers" are ca. 0.1 mm wide, and the bus bar is from 1.5 to 2 mm wide. During the module's construction, the bus bar is connected to the rear electrode of the adjacent cell by means of copper strips soldered to it.

The metallic layers in a cell should be in ohmic contact with the silicon and have a low contact resistance, good adhesion to the silicon surface, and good soldering properties. Various techniques are used worldwide for producing metal contacts fulfilling these requirements.

Unlike the front electrode, the rear one can cover the entire area of the wafer. To improve efficiency, the rear electrode is made from a layer of aluminium vacuum deposited on the silicon surface between silver contacts in the form of strips or squares 2.5–6 mm wide. When this electrode is heated at temperatures from 770 to 1070 K, aluminium diffuses to the silicon forming a thin p^+ layer. To obtain a p–p^+

junction, the p^+ layer must be far thicker (0.2 μm); this is achieved by heating the cell to a temperature of 970–1070 K.

A very much cheaper method of making the two electrodes is the chemical deposition of nickel, or screen printing using a paste containing silver, aluminium, copper, or some other metal. Screen printing is a method that was used to produce lettering with a stencil over a thousand years ago. In 1975, it was first used in silicon cell technology to deposit the front and rear electrodes, thus replacing the costly vacuum deposition technique.

Silver paste is used for producing the front electrode, while aluminium paste combined with a small quantity of silver is usually used for the rear one. The silver pastes used for screen printing metal contacts on a silicon surface consist of a conducting phase (powdered silver of grain size 1–3 pm), an auxiliary phase (assisting the sintering of solid-phase grains—an enamel formed from the melting of a mixture of inorganic oxides), and an organic carrier facilitating the screen-printing process, which is burnt off when the layer is fired. The silver layers are usually sintered at 850°C.

About 86% of the silicon cells produced in 2006 had screen-printed metal contacts. At present, standard silicon cells with screen-printed electrodes achieve an efficiency of around 15% if they are polycrystalline and ca. 18% if they are monocrystalline (produced by Czochralski's method). During the screen printing of metal contacts, the mesh must be placed at a constant distance from the front side of the wafer. Silver paste is applied to the mesh and then imprinted using two squeegees. To ensure that the paste properly fills all the openings in the mesh, one squeegee moves along it, spreading the paste down its whole length. The other squeegee then applies just enough pressure to force the paste out of the mesh openings onto the wafer surface. After drying at 120°C for ca. 60 min, the printed layer consists of an aggregation of loose grains 1–2 μm in size; this must now be fired in order to impart stability to the layer.[4]

Cells with Rear Contacts

In cells of this type, both the positive and negative contacts are made on the rear surface of the cell. In this way, the whole of the front light-sensitive surface can be used to harvest light and the space between the cells can be minimized. The SunPower company produces commercial cells and modules of this type from n-type monocrystalline silicon—the efficiency of the cells is 21.5%, and that of the modules is 18.6%. The p–n junction is produced in the lower layer in the form of bands. This means that the photogenerated charges have to cover quite a long distance to reach the junction region, so only high-quality silicon is suitable for this type of cell.

Deposition of the ARC

One of the most significant parameters affecting the efficiency of a cell is the coefficient of light reflection from its surface—in the case of silicon, this is 33%–54%. This can be minimized by applying a transparent ARC to the cell's active surface.

ARCs can be applied in various ways: chemical vapor deposition (CVD), spraying, spin-on, and screen printing. The spin-on technique is the simplest one, as it does not require expensive equipment and is very efficient, but it can only be usefully applied to silicon wafers with a smooth, polished surface. If the wafer surface is textured, the ARC obtained in this way will not be of uniform thickness. Plasma-enhanced chemical vapor deposition (PECVD) produces ARCs with very good refractive index, photonic band gap, homogeneity, chemical composition, and controlled thickness. The surfaces of silicon wafers are usually coated with one or two antireflective layers.

The presence of an ARC is responsible for the color of the cells: polycrystalline cells are blue and monocrystalline ones are dark blue to black. By optimizing the thickness of the ARC, it is now possible to produce cells that are green, gold, brown, and violet in color, but this is only at the expense of their efficiency. One can, of course, do without the ARC and apply the cells in their original silver (polycrystalline) or dark gray (monocrystalline) colors; depending on architectural requirements, solar panels without an ARC can be integrated, for example, into the façades of buildings.

Dye-Sensitized Solar Cells

Dye-sensitized cells (DSCs) imitate the way that plants and certain algae convert sunlight into energy. The cells are inexpensive, easy to produce, and can withstand long exposure to light and heat compared with traditional silicon-based solar cells. This is a relatively new class of low-cost solar cell. It is based on a semiconductor formed between a photosensitized anode and an electrolyte, a photoelectrochemical system.

The fruits, flowers, and leaves of plants are tiny factories in which sunlight converts carbon dioxide gas and water into carbohydrates and oxygen. Although not very efficient, they are very effective over a wide range of sunlight conditions. Despite the low efficiency and the fact that the leaves must be replaced, the process has worked for hundreds of millions of years and forms the primary energy source for all life on earth. On the basis of this principle, there were early attempts to cover crystals of semiconductor titanium dioxide with a layer of chlorophyll. Unfortunately, the efficiency of the first solar cells sensitized in this way was about 0.01%. In 1991, Michael Grätzel and Brian O'Regan at the École Polytechnique Fédérale de Lausanne (Switzerland) used a sponge of small particles, each about 20 nm in diameter, coated with an extremely thin layer of pigment to obtain the effective surface area available for absorbing light.[5] These dye-sensitized solar cells (DSSCs or DSCs) are also known as Grätzel cells. Following much academic research, the energy conversion efficiency of laboratory cells made on glass substrates with liquid electrolytes has steadily increased to around 10% at air mass 1.5, 1 Sun conditions (for testing, the cells, regardless of design and active material, are typically insolated at a constant density of roughly 1000 W/m^2, which is defined as the standard 1 Sun value).

Dye-sensitized photoelectrochemical solar cells differ from conventional photovoltaic solar cells in that they separate the function of light absorption from charge carrier transport. The cells are made up of a porous film of tiny (nanometer sized) white pigment particles of titanium dioxide. These are covered with a layer of dye that is in contact with an electrolyte solution. Photoexcitation of the dye results in the injection of an electron into the conduction band of the oxide. The original state of the dye is subsequently restored by electron donation from a redox system, such as the iodide/tri-iodide couple. This process results in the conversion of sunlight into electrical energy.

In the case of the original Grätzel design, the cell has three primary parts. On the top is a transparent anode made of fluorine-doped tin dioxide (SnO_2:F) deposited on the back of a (typically glass) plate. On the back of the conductive plate, a thin layer of titanium dioxide (TiO_2) is deposited, which forms into a highly porous structure with an extremely high surface area. The plate is then immersed in a mixture of a photosensitive ruthenium-polypyridine dye and a solvent. After the film has been soaked in the dye solution, a thin layer of the dye is left covalently bonded to the surface of the TiO_2. A separate backing is made with a thin layer of the iodide electrolyte spread over a conductive sheet, typically platinum metal. The front and back parts are then joined and sealed together to prevent the electrolyte from leaking. The construction is simple enough that there are hobby kits available for hand-constructing them. Although they use a number of advanced materials, these are inexpensive compared to the silicon needed for normal cells because they require no expensive manufacturing steps. TiO_2, for instance, is already widely used as a paint base.

Principles

DSSCs separate the two functions typical of a traditional silicon cell design. In the crystalline silicon solar cells, the silicon acts as the source of photoelectrons and provides the electric field to separate the charges and produce the current.

In the DSSC, the bulk of the semiconductor is used solely for charge transport, while the photoelectrons are provided from a separate photosensitive dye. Charge separation occurs at the semiconductor/dye/electrolyte interface.

Although photoelectrochemical cells can operate without an organic dye, the efficiency of such cells is very low due to the low light-harvesting ability of inorganic n-conductors, which normally absorb light only from the high-energy ultraviolet region of the solar spectrum. The introduction of an organic dye makes for a significant increase in the absorption ability of the cells that extends across almost the entire solar spectrum.

DSC Structure

The structure of currently produced DSSCs is similar to those produced by Grätzel and O'Regan: on the top, there is a transparent anode made of indium tin oxide (ITO), deposited on the back of a glass plate. ITO (or tin-doped indium oxide) is a solid solution of indium (III) oxide (In_2O_3) and tin (IV) oxide (SnO_2), typically 90% In_2O_3 + 10% SnO_2 by weight. In thin layers, it is transparent and colorless, but in bulk form, it is yellowish to gray. In the infrared region of the spectrum, it is a metal-like mirror. ITO's main feature is the combination of electrical conductivity and optical transparency. A compromise has to be reached during film deposition, as a high concentration of charge carriers will increase the material's conductivity but decrease its transparency. Thin films of ITO are most commonly deposited on surfaces by electron beam evaporation or a range of sputter deposition techniques.

The thin oxide coating on one side of the glass makes the glass surface electrically conducting. On the back of the conductive plate, a thin layer of titanium dioxide (TiO_2) is deposited, which forms into a highly porous structure with an extremely high surface area—a 10 μm layer of randomly stacked nanoparticles (ca. 20 nm in diameter) (Figure 11).

The plate is then immersed in a mixture of a photosensitive ruthenium–polypyridine dye and a solvent. After the film has been soaked in the dye solution, a thin layer of the dye is left covalently bonded to the surface of the TiO_2. A separate backing is made with a thin layer of the iodide electrolyte spread over a conductive sheet, typically platinum metal. The front and back parts are then joined and sealed together to prevent the electrolyte from leaking. Sunlight enters the cell through the transparent top contact, striking the dye on the surface of the TiO_2.

Titanium Dioxide

Titanium dioxide, also known as titanium(IV) oxide or titania, is the naturally occurring oxide of titanium TiO_2. When used as a pigment, it is called titanium white.

Titanium dioxide occurs in nature as the well-known, naturally occurring minerals rutile (the most common and stable form), anatase, and brookite. Crude titanium dioxide is purified via titanium tetrachloride in the chloride process. In this process, the crude ore (containing at least 90% TiO_2) is reduced with carbon, oxidized with chlorine to give titanium tetrachloride.

This titanium tetrachloride is distilled and then reoxidized with oxygen to give pure titanium dioxide. Another widely used process utilizes ilmenite as the titanium dioxide source, which is digested in sulfuric acid (as in Millennium Inorganic Chemicals). The by-product, iron(II) sulfate, is crystallized

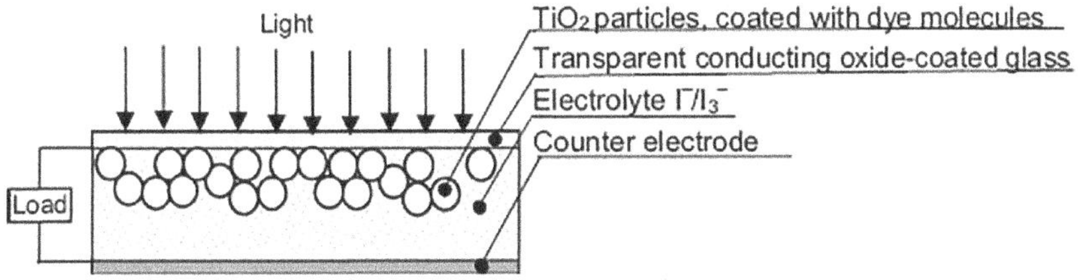

FIGURE 11 DSC structure.

and filtered off to yield only the titanium salt in the digestion solution, which is processed further to yield pure titanium dioxide.

The TiO_2 semiconductor has three functions in the DSSC:

1. It provides the surface for the dye adsorption.
2. It functions as electron acceptor for the excited dye.
3. It serves as electron conductor.

Colloid preparation and layer deposition have been developed to optimize the TiO_2 for these functions. Most important for the performance of desensitized solar cells was the development of a mesoporous semiconductor structure. This becomes evident considering the limited light capture of a dye monolayer on a flat surface.

The conductivity of nanophase TiO_2 films in a vacuum has been found to be very low, ~10^{-9} $(\Omega\ \text{cm})^{-1}$ at room temperature.[6] However, on exposure to UV light, the conductivity is much increased, indicating that the low conductivity in the dark is due to the low electron concentration in the conduction band rather than to poor electrical contacts between the particles.

Untreated TiO_2 is an insulator that becomes "photodoped" and therefore conductive following electron injection of the adsorbed dye. Electronic contact between the nanoparticles is established by sintering the nanoparticles together, which enables the entire surface-adsorbed molecular layer to be accessed electronically. The interconnection of the nanoparticles by the sintering process allows the deposition of a mechanically stable, transparent film, typically a few microns thick. It is not necessary to increase the free electron concentration in the dark; indeed, this may even be detrimental to the photoelectrochemical behavior of the TiO_2. Among several semiconductors studied for photoelectrochemical applications, TiO_2 is by far the most commonly used, because of its energetic properties, its stability, and the ability to attach dyes. It is, furthermore, a low-cost material that is widely available. TiO_2 is used in its low-temperature stable form anatase (pyramid-like crystals), as rutile shows non-negligible absorption in the near-UV region (350–400 nm). This excitation within the band gap leads to the generation of holes, which are strong oxidants and cause long-term instability issues in the solar cell.

Mechanism of Operation

Photons striking the dye with enough energy to be absorbed will create an excited state in the dye, from which an electron can be injected directly into the conduction band of the TiO_2, and from where it moves by diffusion (as a result of an electron concentration gradient) to the clear anode on top (Figure 12).

Meanwhile, the dye molecule has lost an electron and the molecule will decompose if another electron is not provided. The dye strips one from the iodide in the electrolyte below the TiO_2, oxidizing it to tri-iodide. This reaction occurs quite quickly compared to the time that it takes for the injected electron to recombine with the oxidized dye molecule, thus preventing the recombination reaction that would

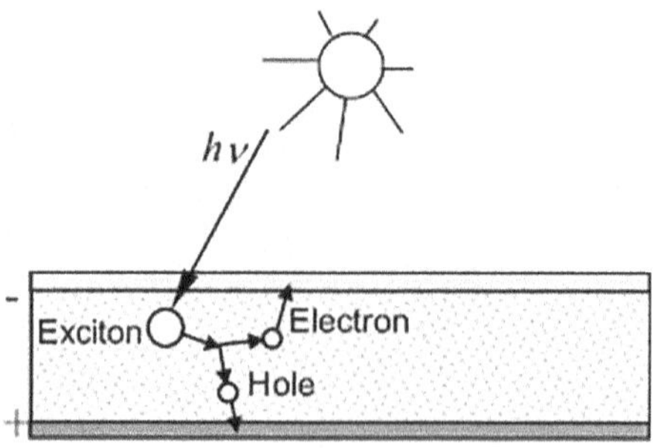

FIGURE 12 Photogeneration of charge carriers in DSC.

effectively short-circuit the solar cell. The tri-iodide then recovers its missing electron by mechanically diffusing to the bottom of the cell, where the counter electrode reintroduces the electrons flowing through the external circuit.

The dye-sensitized oxide is usually deposited on a highly doped transparent conducting oxide (TCO), which allows light transmission while providing sufficient conductivity for current collection. Recently, high-conductivity organic polymers deposited onto plastic foil have found increasing application as substrates for flexible devices. The conductivities of metal foils are superior to those of TCOs and polymers. Because of their opacity, the illumination of the cell has to be established through the counter electrode. The surface of TCO should make good mechanical and electrical contact with the porous TiO_2 film.

To reduce dark current losses due to the short-circuiting of electrons in the substrate with holes in the hole conductor, a thin underlayer of TiO_2 is introduced between the SnO_2 layer and the nanocrystalline TiO_2 layer. This thin compact layer improves the mechanical adhesion of the porous TiO_2 film to the substrates, especially to SnO_2 layers of low haze, i.e., layers with less surface roughness and thus less contact area. Figure 13 illustrates a model of charge carrier separation and charge transport in a nanocrystalline film.[7] The electrolyte is in contact with the individual nanocrystallites. Illumination produces an electron–hole pair in one crystallite. The hole transfers to the electrolyte and the electron traverses several crystallites before reaching the substrate.

The photogenerated hole always has a short distance (roughly equal to the particle radius) to cover before reaching the semiconductor–electrolyte interface whenever an electron–hole pair is created in the nanoporous film. However, the probability that the electron will recombine depends on the distance between the photoexcited particle and the back contact.

The mechanism for converting solar energy into electrical energy in a DSSC is a five-step process:

1. Solar energy (photons of light *hv*) causes electrons in the molecular orbitals within the adsorbed dye sensitizer (S) molecules to become photoexcited (S*):

   S (adsorbed on TiO_2) + $hv \rightarrow$ S* (adsorbed on TiO_2)

   The trapping of solar energy by a sensitizer molecule is analogous to the light-absorbing chlorophyll molecule found in nature, which converts carbon dioxide and water to glucose and oxygen.

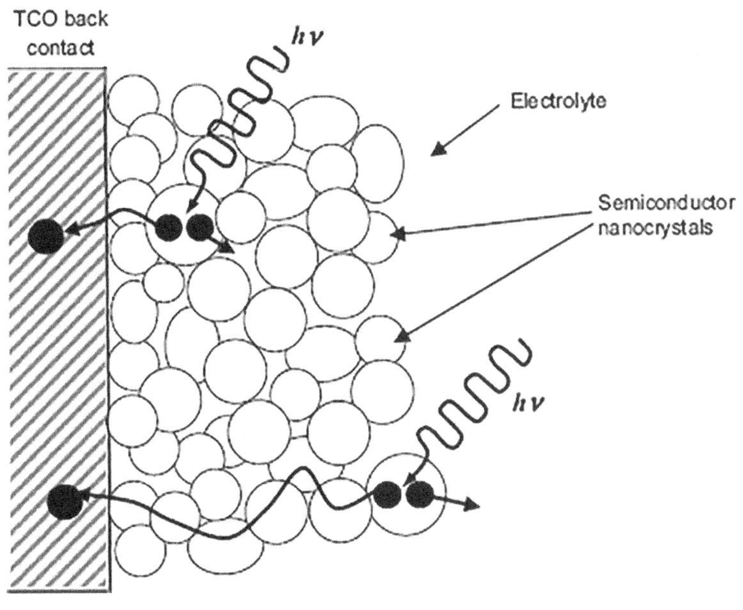

FIGURE 13 Qualitative model of photocurrent generation in nanocrystalline films.

2. The excited electrons escape from the dye molecules: S* (adsorbed on TiO_2) → S^+ (adsorbed on TiO_2) + e^-
3. The free electrons then move through the conduction band of TiO_2, gather at the anode (the dyed TiO_2 plate), and then start to flow as an electric current through the external load to the counter electrode.
4. The oxidized dye (S^+) is reduced to the original form (S) by regaining electrons from the organic electrolyte solution that contains the iodide/tri-iodide redox system, with the iodide ions being oxidized (loss of electrons) to tri-iodide ions:
   S^+ (adsorbed on TiO_2) + 3/2 I^- → S (adsorbed on TiO_2) + 1/2 I_3^-
5. To restore the iodide ions, free electrons at the counter (graphite) electrode (which have travelled around the circuit) reduce the tri-iodide molecules back to their iodide state. The dye molecules are then ready for the next excitation/oxidation/reduction cycle.

Initially, the solutions of iodine–iodide mixtures in volatile solvents, usually acetonitrile, were used as redox electrolytes:

Electrolyte:	I2 + I− ↔ I3−
Anode (Dye):	2 Dye+ + 3 I− → 2 Dye0 + I3−
Cathode:	I3− + 2e− → 3 I−

On the basis of different measurements, it is possible to indicate the orders of magnitude for the rate of the reaction steps. Upon illumination, the sensitizer is photoexcited in a few femtoseconds and electron injection is ultrafast from S* to TiO_2 on the subpicosecond time scale, where they are rapidly (less than 10 fs) thermalized by lattice collisions and phonon emissions. The nanosecond-ranged relaxation of S* is rather slow compared to injection, which ensures that the injection efficiency is unity. The ground state of the sensitizer is then recuperated by I^- in the microsecond domain, effectively annihilating S^+ and intercepting the recombination of electrons in TiO_2 with S^+, which happens in the millisecond time range. This is followed by the two most important processes—electron percolation across the nanocrystalline film and redox capture of the electron by the oxidized relay, I_3^-, within milliseconds or even seconds.[8]

Recently, solvent-free redox electrolytes, prepared from ionic liquids (liquid ionic organic compounds) or from ionically conducting polymer–nanocrystal blends were found to be very efficient. Figure 14 shows the processes taking place during the conversion of light into electrons in a DSSC.

Upon excitation, the dye injects electrons into the conduction band of the titanium oxide. The photo-induced electrons diffuse through the porous TiO_2 network and are extracted at the SnO_2 substrate. The dye itself is regenerated by the electrolyte containing the I^-/I_3^- redox pair. In the most efficient DSCs, the dye is regenerated by I^-, present in an external electrolyte at high (~0.45 M) concentrations.[10] The electronic circuit is closed by the reduction of the iodide couple at the platinized SnO_2 counter electrode.

Titania Solar Cells: Manufacturing Process

Manufacturing processes utilize the production methodologies and equipment already in use in the manufacture of electronic components. Production of cells needs a relatively low capital investment, the equipment does not rely on highly skilled operators, and—most importantly—the technology is environmentally friendly. A number of these manufacturing processes are protected by patents.

During production, a piece of glass is coated with fluorine-doped tin oxide and then sputtered onto a 500 nm thick layer of titanium. The layer is then anodized by placing it in an acidic bath with a mild electric current, and the titanium dioxide nanotube arrays grow to about 360 nm. The tubes are then

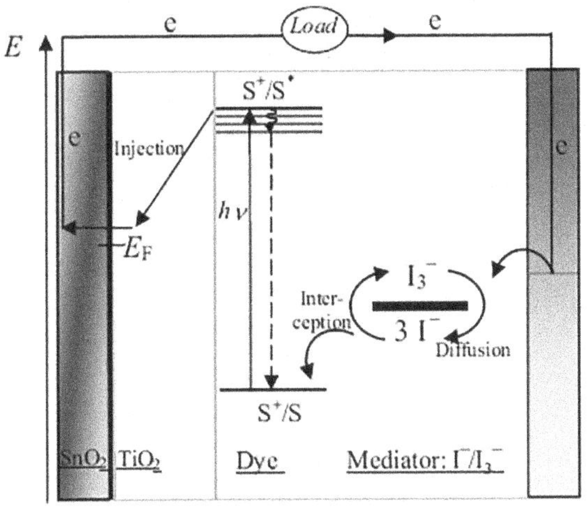

FIGURE 14 The electron processes taking place in an illuminated DSSC under open-circuit conditions.
Source: Anandan.[9]

heated in oxygen so that they crystallize. The process turns the opaque coating of titanium into a transparent coating of nanotubes.

In the next step, the nanotube array is coated with a commercially available dye—this becomes the negative electrode. The cell is sealed with a positive electrode, which contains an iodized electrolyte.

The most important advantages of the DSC manufacturing processes are as follows:

- The materials can be produced cheaply in large quantities.
- They use standard processing and assembly equipment.
- They are not very energy intensive (about 32 kWh/m^2).
- They can be automated.
- The equipment consists of a number of processing stations, each easily reprogrammable to adopt a wide range of DSC designs.

DSSCs can be connected in series or parallel and integrated into solar panels (modules). The tiles are normally ochre, but other colors such as gray, green, and blue are to be introduced shortly.

Advantages of DSC in Comparison with Other Solar Cells

The important features that distinguish a DSC from a conventional silicon crystalline solar cell are as follows:

- It is a photoelectrochemical cell: charge separation occurs at the interface between a wide band gap semiconductor and an electrolyte.
- It is a nanoparticulate titania cell: it is not a dense film like amorphous silicon, but a "light sponge."
- In a DSC, the dye monolayer chemically absorbed on the semiconductor is a primary absorber of sunlight; free charge carriers are generated by electron injections from a dye molecule, excited by visible radiation.
- In a DSC, light absorption and charge carrier transport are separate, whereas in a conventional cell, both processes are performed by the semiconductor.
- An electric field is necessary for charge separation in the p–n junction cell. Nanoparticles in the DSC are too small to sustain a built-in field; accordingly, charge transport occurs mainly via diffusion.

- Inside a p–n junction, minority and majority charge carriers coexist in the same bulk volume. This makes conventional solar cells sensitive to bulk recombination and demands the absence of any recombination centres such as trace impurities. DSCs are majority charge carrier devices in which the electron transport occurs in the TiO_2 and the hole transport in the electrolyte. Recombination processes can therefore only occur in the form of surface recombination at the interface.
- The maximum voltage generated by DSC in theory is the difference between the Fermi level of the TiO_2 and the redox potential of the electrolyte, equal to about 0.7 V (V_{oc})—it is slightly higher than V_{oc} for silicon, which has an open cell voltage equal to 0.6 V. However, the most important differences are dominated by current production.
- Compared to other solar cells, the titania solar cell has the following advantages:
- It is much less sensitive to the angle of incidence of radiation—it can therefore utilize refracted and reflected light.
- It performs over a much wider range of light conditions owing to the high internal surface of the titania (light sponge)—it can thus be designed for operation under very poor light conditions.
- It can be designed to operate optimally over a wide range of temperatures; unlike silicon solar cells, whose performance declines with increasing temperature, DSSC devices are only negligibly influenced when the operating temperature increases from ambient to 60°C.
- There is an option for transparent modules—they can be used for daylighting, roof lighting, and displays.
- DSC production needs only commonly available (nonvacuum) processing equipment, making it appreciably cheaper to set up facilities.
- DSC has a significantly lower embodied energy than all other forms of solar cell.
- Crystalline silicon PV modules are suited to full sun applications, particularly sun tracking systems and roofs, and are best suited to cold climates and clear sky conditions.
- In contrast, the DSC is particularly suited to target markets in temperate and tropical climates, because of its good temperature stability and excellent performance under indirect radiation, during cloudy conditions, and when temporarily or permanently partially shaded.
- The DSC is 7also particularly suited to indoor applications that require stability of voltage and power output over a wide range of low-light conditions.

Comparison of Different Types of Solar Cell

Table 2 compares the different solar cell types.

Conclusion

In many countries, the decentralization of power production, the increasing proportion of energy generated from renewable sources, and the development of cogenerative systems for producing heat and electricity are viewed as the paths to be taken by energy production in the future. A secure, long-term plan for energy development that takes environmental conservation requirements into account should therefore fulfill two conditions:

- The supply of power from conventional sources should decrease.
- The proportion of energy generated from renewable sources should increase.

Electricity consumption is increasing by 1% per annum in developed countries and by ca. 5% per annum in developing countries.[11] This implies the necessity to look for sources of electrical energy other than the traditional ones. Photovoltaics is one such non-traditional source, which satisfies all the criteria now required of energy sources: solar energy is universally available, and photovoltaic cells and modules are some of the environmentally safest devices for energy conversion.

TABLE 2 Comparison of Different Types of Solar Cell (copyright: ©pvresources.com)

Material	Thickness	Efficiency (%)	Color	Features
Monocrystalline Si solar cells	0.3 mm	15–18	Dark blue, black with AR coating; gray, without AR coating	Lengthy production procedure, wafer sawing necessary. Best researched solar cell material—highest power/area ratio
Polycrystalline Si solar cells	0.3 mm	13–15	Blue, with AR coating; silver-gray, without AR coating	Wafer sawing necessary. Most important production procedure, at least for the next 10 years
Polycrystalline transparent Si solar cells	0.3 mm	10	Blue, with AR coating; silver-gray, without AR coating	Lower efficiency than monocrystalline solar cells. Attractive solar cells for different BIPV (Building Integrated Photovoltaics) applications
EFG	0.28 mm	14	Blue, with AR coating	Limited use of this production procedure. Very fast crystal growth; no wafer sawing necessary
Polycrystalline ribbon Si solar cells	0.3 mm	12	Blue, with AR coating; silver-gray, without AR coating	Limited use of this production procedure; no wafer sawing necessary. Decrease in production costs expected in the future
Apex (polycrystalline Si) solar cells	0.03 to 0.1 mm + ceramic substrate	9.5	Blue, with AR coating; silver-gray, without AR coating	Production procedure used only by one producer; no wafer sawing necessary; production in band form possible. Significant decrease in production costs expected in the future
Monocrystalline dendritic web Si solar cells	0.13 mm including contacts	13	Blue, with AR coating	Limited use of this production procedure; no wafer sawing necessary; production in band form possible
Amorphous silicon	0.0001 mm + 1 to 3 mm substrate	5–8	Red-blue, black	Lower efficiency; shorter life span. No sawing necessary; possible production in band form
CdTe	0.008 mm + 3 mm glass substrate	6 –9 (module)	Dark green, black	Toxic raw materials; significant decrease in production costs expected in the future
CIS	0.003 mm + 3 mm glass substrate	7.5 –9.5 (module)	Black	Limited supply of indium in nature. Significant decrease in production costs possible in the future
Hybrid silicon (HIT) solar cell	0.02 mm	18	Dark blue, black	Limited use of this production procedure; higher efficiency; better temperature coefficient and lower thickness

References

1. Stutenbaeumer, U.; Masfin, B. Equivalent model of monocrystalline, polycrystalline and amorphous silicon solar cells. Renewable Energy **1999**, *18*, 501.
2. Abdel Rassoul, R.A. Analysis of anomalous current–voltage characteristics of silicon solar cells. Renewable Energy **2001**, *23*, 409.

3. Radziemska, E.; Klugmann, E. Photovoltaic maximum power point varying with illumination and temperature. J. Sol. Energy Eng. ASME **2006**, *128/1*, 34–39.
4. Panek, P.; Lipinski, M.; Beltowska-Lehman, E.; Drabczyk, K.; Ciach, R. Industrial technology of multicrystalline silicon solar cells. Opto-Electron Rev. **2003**, *11* (4), 269–275.
5. Lund, J.W.; Freeston, D. World-wide direct uses of geothermal energy 2000. Geothermics **2001**, *30*, 29–68.
6. Hagfeldt, A.; Grätzel, M. Light redox reactions in nanocrystalline systems. Chem. Rev. **1995**, *95*, 49–68.
7. O'Regan, B.; Grätzel, M. A low-cost, high-efficiency solar cell based on dye-sensitized colloidal TiOz films. Nature **1991**, *353*, 737–740.
8. Zhang, Z. PhD Thesis N°4066, École Polytechnique Fédérale de Lausanne: Suisse, 2008.
9. Anandan, S. Recent improvements and arising challenges in dye-sensitized solar cells. Sol. Energy Mater. Sol. Cells **2007**, *91*, 843–846.
10. Marton, C.; Clark, C.C.; Srinivasan, R.; Freundlich, R.E.; Narducci Sarjeant, A.A.; Meyer, G.J. Static and dynamic quenching of Ru(II) polypyridyl excited states by iodide. Inorg. Chem. **2006**, *45*, 362–369.
11. Muneer, T.; Asif, M.; Munawwar, S. Sustainable production of solar electricity with particular reference to the Indian economy. Renewable Sustainable Energy Rev. **2005**, *9*, 444.

25

Alternative Energy: Solar Thermal Energy

Andrea Micangeli,
Sara Evangelisti, and
Danilo Sbordone

Introduction

The sun produces energy through nuclear fusion at its core, where tremendous amounts of energy are released by the fusion of nuclei into more massive nuclei under extreme conditions such as extreme pressures and temperatures. In this way, the conversion of hydrogen nuclei into the much heavier helium causes neutrinos and photons to be discharged. This energy travels through space by radiation that is a form of energy transmitted in electromagnetic waves. The energy is transferred from the core to the photosphere through the convection zone. The photosphere, or the radiating surface of the sun, is where the energy will be radiated into space. Electromagnetic radiation is mostly emitted here, though sometimes small amounts of microwave, radio, and x-ray emissions can also be emitted. A typical photon journey lasts about 100,000 years from the core of the sun to its surface, while a photon will take only 8 min from the sun's surface to reach the Earth.

Solar intensity can be measured by the Inverse Square Law, where the intensity of radiation hitting or striking objects in space, such as planets, asteroids, or dust, can be quantitatively assessed. The total energy given off by the sun is very high, owing through the tremendous energy released by nuclear fusion. Scientists estimate the energy output at 63,000,000 W/m^2 (watts per square meter).

Given the distance of the Earth from the sun, and due to the intercepting atmosphere, this figure is definitely much lower. Radiation in the outer atmosphere amounts to approximately 1367 W/m^2. Of these, only about 40% will reach the surface of the Earth as shown in Figure 1.

The solar energy that reaches the surface of the Earth can be used by a photovoltaic system to produce electricity or by a thermal system to produce heat. When the solar energy is used to produce heat, it is called *solar thermal energy* technology. In this case, the solar radiation is used to warm up a fluid (water, air, solutions appropriated to each system) that can circulate, mainly:

- In heat exchangers at the beginning of the circuit that will use the returned available heat.
- In pipes and radiant objects put in place to warm up.
- In the refrigeration cycles to evaporate the volatile substances that are used in the condensation phase.

Through the surface, rarely adjustable, the solar energy must be accumulated and transformed into thermal energy, so it is possible:

- To concentrate it through mirrors or lenses that reflect the radiation towards panels and a boiler to be able to use directly to warm water or to the production of vapor in the pipe and into a turbine.
- To collect it from applied panels or integrated panels in closure of buildings (walls, roofs, parapet, etc.)

The most important application that can be used by everyone is a *low-temperature solar thermal technology*. It consists of a system that uses a solar collector to warm up a liquid. The purpose is to capture and to transfer solar energy to produce *domestic hot water* (DHW) or to control the temperature inside a

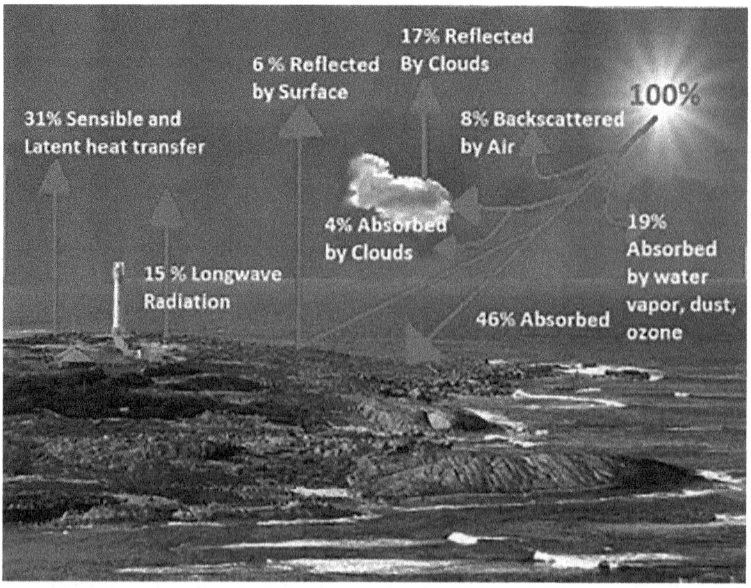

FIGURE 1 Earth's energy budget.

building. The term "low temperature" refers to the temperature that a working fluid can reach, generally up to 100–120°C.

This technology is ideal for application on a small scale, because it is cost-effective and it is simple to install and to manage. Low-temperature solar collectors can be classified as follows:

- *Flat plate collectors,* the most common type that consists of a dark flat-plate absorber of solar energy, a transparent cover that allows solar energy to pass through but reduces heat losses, a heat-transport fluid (air, antifreeze, or water) to remove heat from the absorber, and a heat-insulating backing.
- *Unglazed collectors* that are realized with tubes in plastic materials such as propylene, neoprene, synthetic rubber, and PVC. These are cheap because there is no insulation or transparent coverage. They have good performance only during summertime. They are recommended only if thermal energy for open swimming pools (and the like) is requested for.
- *Integrated collector storage,* recommended only for temperate climates, where the collector itself is the storage.
- *Evacuated tube panels,* where it is possible to eliminate the air between the capturing plate (reduced to a strip) and the evolved transparent sheet in a glass cylinder to resist the pressure difference.

The principal elements that are used by solar collectors are as follows:

- *The capturing plate*
- *The insulated material*
- *The transparent coverage*
- *The external casing*
- *The working fluid*

Solar Collector Technology

Flat-Plate Solar Collector

The *capturing plate* is realized with copper or steel and it is treated with satin and dark paint to reduce the reflection losses to bost the absorption capability to the wavelength of solar radiations with a low emissivity in the infrared radiations.

Normally, the canalizations on the plate are built to resist a pressure of about 6–7 bar; some collectors guarantee the resistance to a pressure of up to 10 bar.

The *insulated material* is a barrier against the conduction losses of the plate toward the external part of the collector. The materials used are always characterized by a porous or alveolar conformation in order to create microscopic motionless air spaces (that constitute a perfect barrier to the heat transmission). Polyurethane, polyester wool, glass wool, and rock wool are the most used materials. An enemy of the insulating material is humidity, which can appear for many reasons inside the collector (moisture, rain due to the gasket caused by little leakage in the pipes); often, the above-mentioned materials are covered with a thin aluminum layer that acts as a barrier to humidity and, at the same time, reflects towards the absorption plate.

The *transparent coverage* has the dual function of limiting the loss of energy towards the outside of the collector and facilitating the penetration of the radiation inside the collector. To satisfy this request, the coverage should be the most transparent as possible to the wavelength typical of the solar radiation (approximately between 0.2 and 2.5 mm) and at the same time it should be matt to the infrared radiation coming from the pipe-table while their temperature rises. The material that meets these qualifications is glass, above all if treated to acquire more transparency. Sometimes, due to some adverse elements, such as fragility and weight, plates made of plastic materials (polycarbonate) are preferred to glass plates.

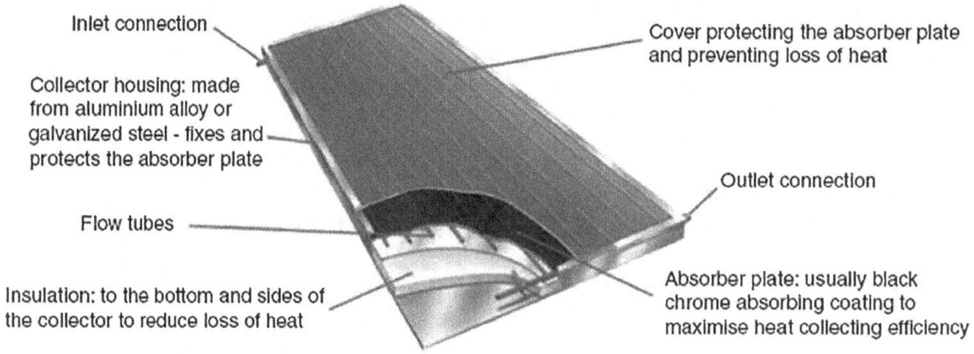

FIGURE 2 Flat plate solar collector scheme.

The *external casing* has the dual function of contributing consistency and mechanical solidity to the collector and protecting the internal elements from dirt and atmospheric agents. It can be made of stainless steel (zinced) or aluminium.

The *working fluid* that flows along the pipe system must take the largest quantity as much as possible. The fluid should also have a high density even at high temperature. It is important that it does not have a corrosive effect along the wall of the circuit; it must be inert and stable at temperatures below 100°C, and it should also have limited hardness to avoid limestone deposits. The hardness of the water refers to the quantity of magnesium and calcium salts in the water. The fluid should have a low freezing point. The option used among the producers of solar panels is a water solution of propylene glycol (not toxic and has a good anti-freezing action). In Figure 2, a section of a flat plate collector with all components is shown.

Unglazed or Open Collectors

An *unglazed collector* is a simple form of flat-plate collector without a transparent cover. Typically, polypropylene, Ethylene-Propylene Diene Monomer (EPDM) rubber, or silicone rubber is used as an absorber as shown in the left panel of Figure 3. Used for pool heating, they can work quite well when the desired output temperature is near the ambient temperature (i.e., when it is warm outside); moreover, they are cost-effective. As the ambient temperature gets cooler, these collectors become less effective. They can be used as preheat make-up ventilation air in commercial, industrial, and institutional buildings with a high ventilation load. They are called "transpired solar panels," and they employ a painted perforated metal solar heat absorber that also serves as the exterior wall surface of the building. Heat conducts from the absorber surface to the 1 mm thick thermal boundary layer of air on the outside of the absorber and to air that passes behind the absorber. The boundary layer of air is drawn into a nearby perforation before the heat can escape by convection to the outside air. The heated air is then drawn from behind the absorber plate into the building's ventilation system (Figure 3, right panel).

Integrated Collector Storage

An *integrated collector* storage system is constituted by a unique element that assumes the role of capturing plate, absorber, and external accumulation. In this type of solar collector, the storage is located in the collector itself, and it is exposed to a slow heating process.

The water is placed inside the insulated tank and in this case the heat losses due to the exposure surface cannot be ignored. The integrated collector storage works without pumps or electrical devices: the panel absorbs the solar energy and the water inside the collector rises up for conduction, moving towards the outlet pipe and reaching the domestic network when hot water is required (see Figure 4).

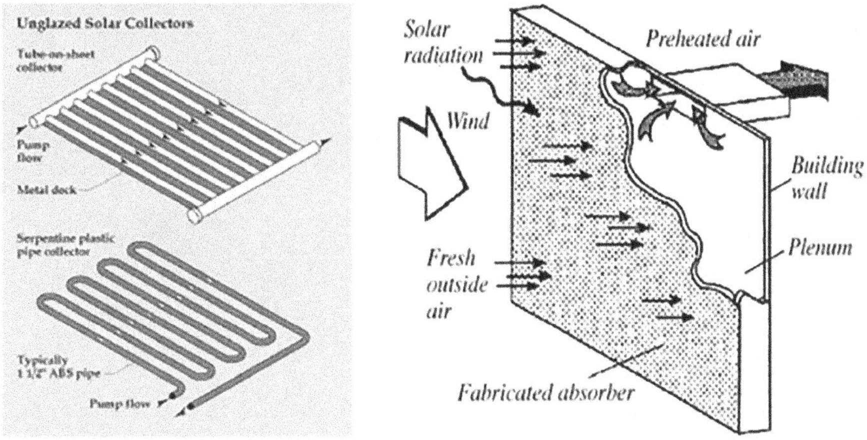

FIGURE 3 Unglazed solar collector scheme.

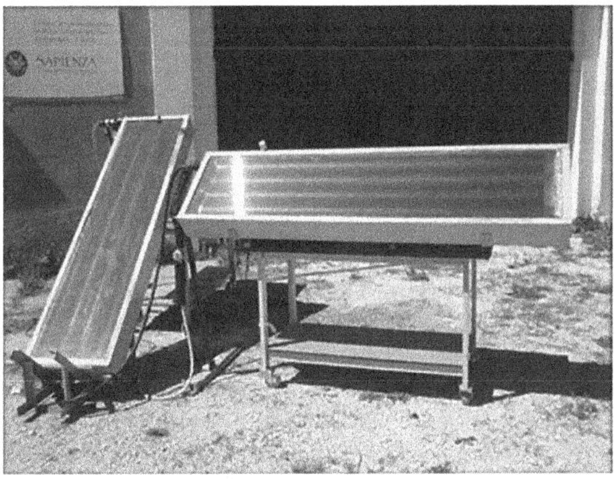

FIGURE 4 Example of an integrated collector storage system.

The performance of such equipment is not fully satisfactory, because during the discharge phase, the water temperature decreases rapidly, reducing the overall usability of the collector.

Evacuated Tube Collectors

The *evacuated tube collectors* are obtained, reducing the presence of air in the space between the plate and the transparent cover, thus avoiding losses caused by convective movements. In spite of their higher cost, these collectors are able to perform well even when the environment temperature is low. Among the possible technological solutions that can be used to build these collectors, the *heat pipe* technology does not limit the exchange of heat between liquids but in the case the fluid flows in thin tube pipes, also between vapor and liquid, taking advantage of the heat for the condensation along its way, resulting more efficient although more complex. Taking advantage of this type of heat exchange, the pipe system where the thermal vector liquid flows should be depressurized to decrease the evaporation temperature. A collector is built by the plate in long metal cylinders (copper) superficially covered with black and selected paint; this tube is in a second glass tube, in a way that surface of the first one is perfectly tallying with the internal of the second one; it is concentrically inserted in a second glass pipe. Figure 5 shows a section of the collector.

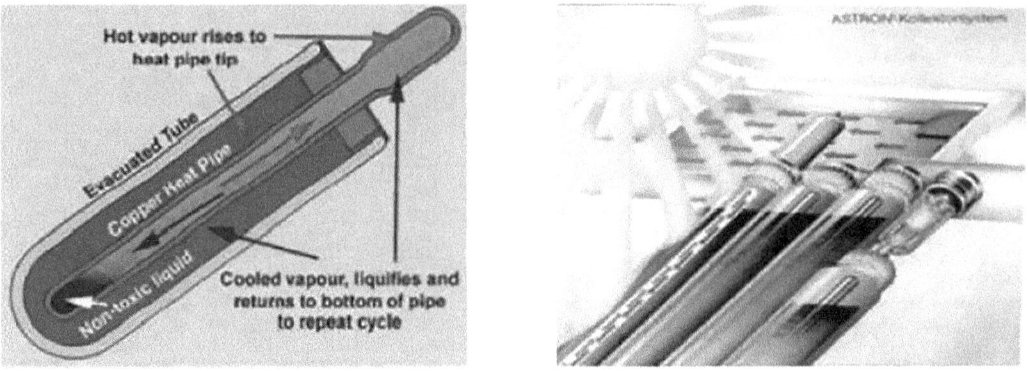

FIGURE 5 Section of an evacuated tube collector.

The air between the two tubes is vacuumed out, until a pressure of $P = 5 \times 10^{-3}$ *Pa* is reached. A small tube goes lengthwise through the copper cylinder following a U path, inside the thermal vector fluid flows reaching a temperature close to 100°C. In order to maximize the use of the heat pipe, a bigger pipe can be used in a concentric position; the heating of the fluid in the pipe will increase because the captured radiation increases. Heater exchangers are placed at the ends of the pipes, in order to transfer heat to the users. The evacuated tube collectors are divided into two main types:

1. *Evacuated pack collectors* with direct circulation of thermal vector fluid.
2. *Heat pipe evacuated pack collectors:* the fluid inside the pipe system evaporates along the way and give its heat, due to the condensation process.

In the first type, the plate is divided in long metal cylinders superficially treated with selected black paint, with each of these tubes inserted in a glass tube, which is also inserted in a larger glass tube, and then vacuum packed. A small tube goes through the copper cylinder following a U path.

In the heat pipe evacuated pack collectors, the little tube under pressure (heat pipe) receives heat from the capturing plate. The tube contains water or alcohol that evaporates at around 25°C under pressure. The vapor goes up until the head, where it exchanges heat through the condensation phase giving heat to another external fluid.

Energy Balance of a Collector

The phenomena that interact within a manifold are multiple and interconnected: the main energy exchanges between solar radiation and the various elements of the system are here described.

As shown in Figure 6, the solar radiation (E_0) hits the glass cover. A small amount of radiation (E_1) is reflected and absorbed by the transparent cover. The copper absorber does not absorb all the remaining radiation into useful heat and partly reflects and dissipates heat (E_2) by convection, conduction, and radiation to the outside. Treatment with selective coatings, as mentioned above, reduces leakages. On one side, the transparent cover prevents the reflection of the solar radiation from the plate to disperse outwards favoring the greenhouse effect inside the collector, and on the other side, it limits the heat convection dispersion (Q_1). If a good thermal insulation in the back and sides of the collector is designed, with standard insulating materials such as rock wool or polyurethane foam, the energy losses by thermal conduction are reduced to a minimum (Q_2).

Only a portion of the incident solar energy (E_0) is transferred to the fluid as useful heat (Q_3) due to the different energy losses (E_1, E_2, Q_1, and Q_2).

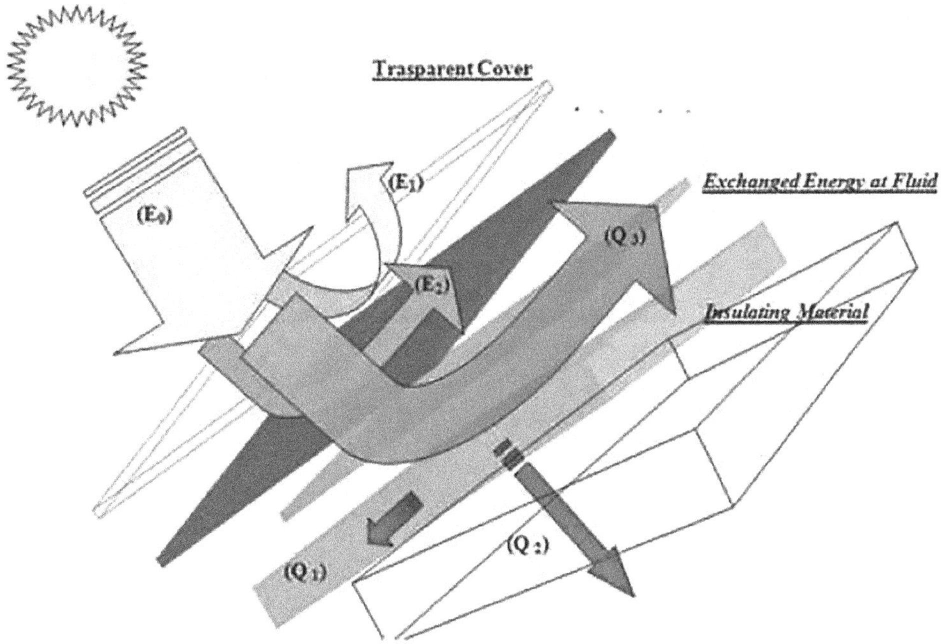

FIGURE 6 Energy balance of a solar thermal collector.

Collector Efficiency Curves

The thermal energy transferred to the fluid per time unit is calculated as the difference between solar radiation captured by the plate and converted into heat, and the heat losses by convection, conduction, and radiation.

$$Q_n = E_c - Q_p \tag{1}$$

E_c takes into account the absorbance and transmittance of the glass plate, and it is calculated as the product of the irradiance E, the transmittance τ, and absorbance of the plate α, as follows:

$$E_c = E^* \tau^* \alpha \tag{2}$$

The heat loss depends on the temperature difference between the plate and the environment. As a first approximation (for low temperatures of the plate), this relationship is linear and it can be described through the coefficient of total losses:

$$Q_p = k^* \Delta T \tag{3}$$

where $\Delta T = T_p - T_a$, T_p is the average temperature of the plate, and T_a is the ambient temperature.

By substituting these relations in the efficiency of the collector, the following are obtained:

$$\eta = \left(E^* \tau^* \alpha - k^* \Delta T\right)/E, \tag{4}$$

$$\eta = \left(E^* \tau^* \alpha / E\right) - \left(k^* \Delta T/E\right), \tag{5}$$

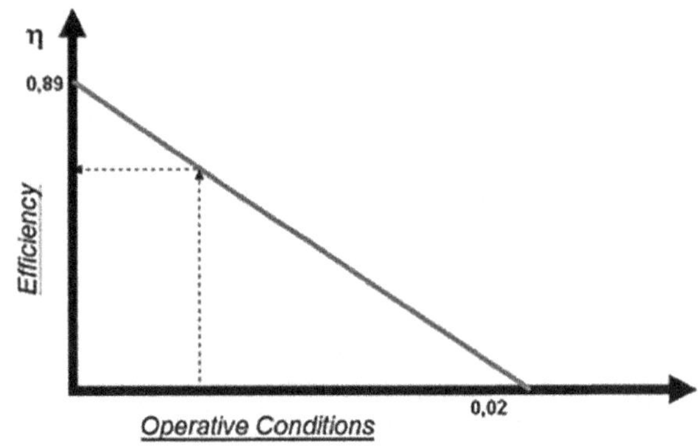

FIGURE 7 Efficiency of a solar thermal collector at different operative conditions.

and

$$\eta = \tau^{*}\alpha - \left(k^{*}\Delta T / E\right), \tag{6}$$

Whereas

$$\eta_0 = \tau^{*}\alpha. \tag{7}$$

It follows that the efficiency of the collector is equal to:

$$\eta = \eta_0 - \left(k^{*}\Delta T / E\right). \tag{8}$$

Conventionally, it is defined:

$$\Delta T^{*} = \left(T_{\mathrm{mf}} - T_a\right) / E. \tag{9}$$

Finally, the efficiency of a collector can be expressed as:

$$\eta = F'^{*}\eta_0 - K\Delta T^{*}. \tag{10}$$

In Figure 7, the efficiency curve of a collector is shown.

Instantaneous Efficiency of a Collector

The instantaneous efficiency depends on the optical losses (E_1 and E_2) and on the temperature (Q_1 and Q_2).

The overall losses of the total heat occurring in the collector by conduction, convection, and radiation can be expressed as the coefficient of total loss in K (W/m^2 * °C).

The graph shows that, with constant irradiance, the higher is the difference between the average temperature of the fluid and the ambient temperature, the heat losses also increase and, consequently, the efficiency of the collector decreases.

Comparison between Different Types of Collectors

By comparing the two efficiency curves, the principal characteristics of the collectors can be classified:

- *No glass collector* is the one with the best possibility to absorb the incidental radiations; its efficiency though decreases fast until zero in situations where the other collectors still have valid performances.
- *Flat-plate collector* with a selective plate has a better performance than the open one, practically in every working condition.
- *The evacuated collector* has a more stable efficiency curve, and it guarantees good performance even during bad weather conditions.

From Figure 8, which shows the efficiency curve of different collector types, it is possible to observe that the unglazed collectors have better optics than others. In fact, the absence of covering helps eliminate the untransparency losses and reflection that often occur with the glass covering.

Natural Circulation Systems

Once the differences between an open and a closed circuit have been examined, it is possible to analyze the circulation of the thermal vector fluid of the plant.

Natural circulation and closed circulation systems are based on the convective movements flown from the thermal vector fluid caused by a difference of temperature in the fluid itself. Indeed, the fluid warms up inside the serpentine of the capturing plate and it naturally goes up to the top of the collector. It will need an inclination compared to the floor to maximize the quantity of transferred energy.

Over the collector, a "storage tank" is placed with the heat exchanger inside. In the tank for the accumulation, there are two separate flows: the closed circuit of the collector with the thermal vector fluid and the water net of the running system designated to the final users.

The water, in contact with the heat flow, has a lower temperature, with higher heat absorption capacity, thanks to its stratification. While liberating heat to the running water, the fluid cools off moving to the lower part of the capturing plate; meanwhile, the part of the fluid that was at a colder temperature, being at this point at a higher temperature, tends to go up to the accumulation tank, cooling off while transferring heat to the running water. The quantity of fluid inside the solar collector remains constant and does not need any regulation pump for the circulation of the water, because a self-regulation natural mechanism works thanks to the trigger of the convective movement (see Figure 9).

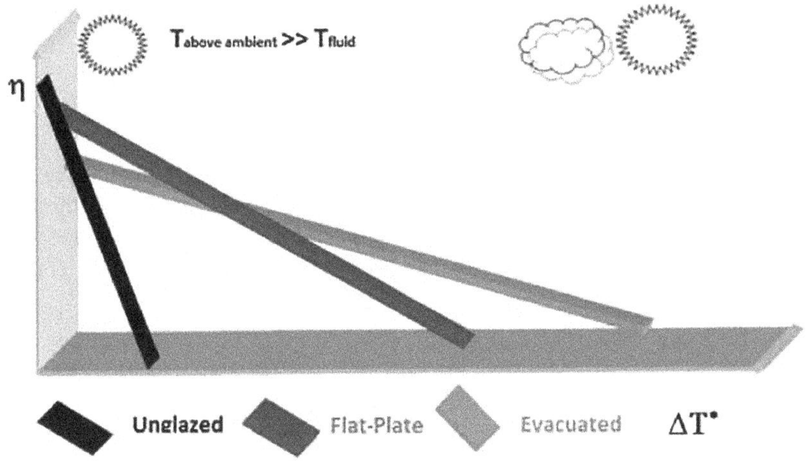

FIGURE 8 Efficiency curves of different collector systems.

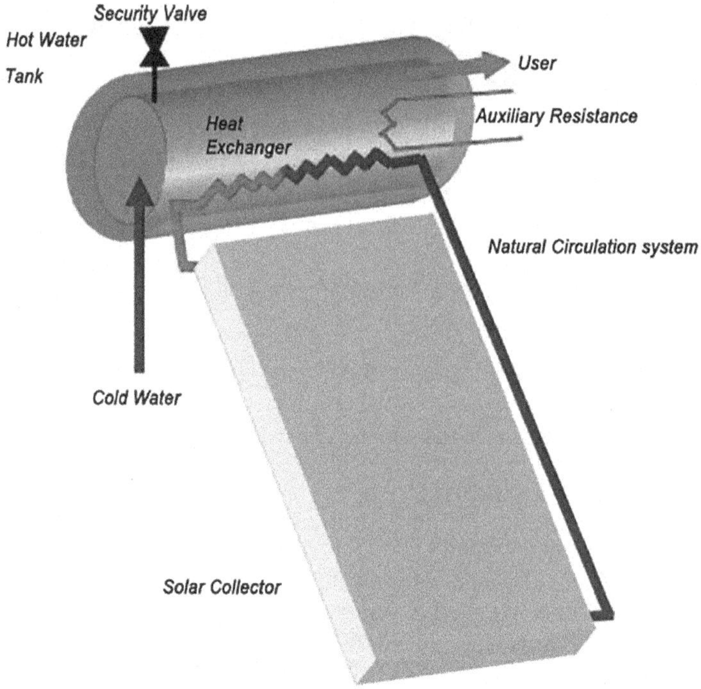

FIGURE 9 Natural circulation system.

Forced Circulation Systems

In the system where the circulation is forced, the presence of the circulation pump, driven by a differential thermostat, allows the fluid to circulate inside the pipes. After a selected difference between the water temperature and the fluid, it activates automatically. The tube in which the thermal vector fluid circulates represents a primary circuit. In the highest part of the accumulation tank, there is an exchange of heat coming from an auxiliary traditional circuit that starts functioning when the water temperature designated to the users does not arrive to the requested one. The fluid that liberates its heat inside the tank of accumulation cools off through a pump of circulation that is sent to the lower part of the collector usually sited on the roof. See Figure 10 for the scheme.

Open Circuit Systems

Open circuit systems have some advantages, such as the simplicity of the hydraulic circuit realization and the lowest thermal losses that always occur when heat goes from one fluid to another. There are two problems that put limitations to this type of plant: the possible freezing of the water in case the temperature reaches values below 0°C and the calcium deposits along the tube system of the collector. In both cases, the collectors could go out of service.

Closed Circuit Systems

The closed circuit is the most common solution. In this case, there are two different hydraulic systems: the main one, in which the thermal vector fluid circulates, and the secondary one, where the water coming from the hydro net is used (a third circuit is foreseen in the large-scale plans for the heat storage).

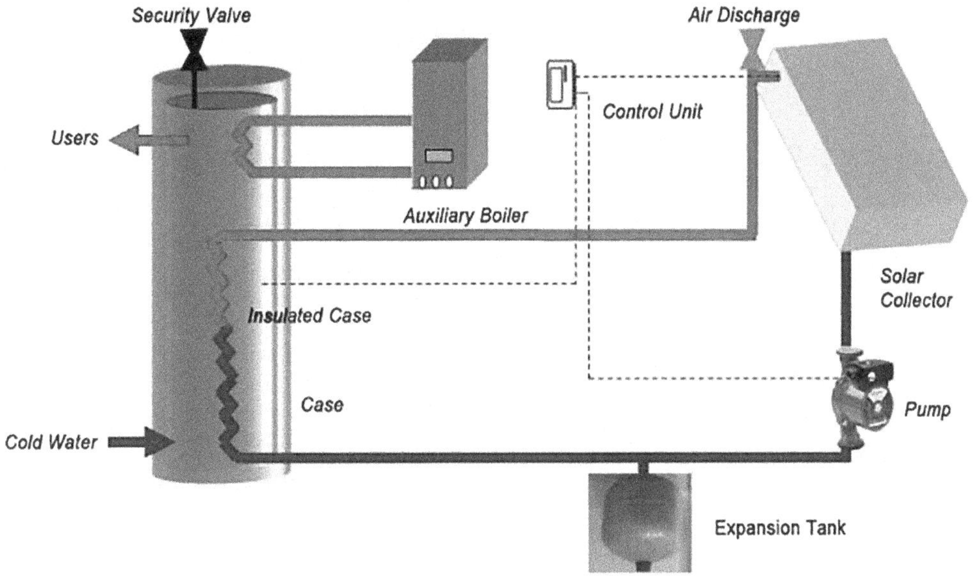

FIGURE 10 Forced circulation system.

The thermal vector fluid must respond to specific functions:

1. To increase density and specific heat capacity to be able to use the smaller pipe systems.
2. To avoid limestone deposits due to hardness of the water.
3. To reduce freezing points and viscosity.
4. To be not toxic (in the case of plant for sanitary warm water).
5. To be chemically inert, stable, and not corrosive.

The option adopted by most of the producers is a solution of water and polyethylene glycol (usually 25%–45% of glycol).

The purpose of the thermal vector fluid is to take thermal energy captured by plane and transfer it to the water to heat up. In this energy transfer, the fluid can give some heat to the cold water through the exchanger in a proportional measure depending on the difference of temperature between the two fluids. In Figure 11, the plate exchanger of heat is shown together with the tank with a double exchanger and a serpentine plunged. The larger the interface, the greater the amount of energy exchanged. The differences in temperature are of the utmost importance. To unify the need of big exchange surfaces with compact exchangers, the immersion serpentines are used.

Solar Cooling

The physical principle of generating solar cooling power is almost similar to the operating principle of conventional air-conditioning systems (condenser-compressor type) air-conditioning. Both systems rely on systems full vacuum single glass heat pipe collector picture liquid-to-vapor phase change energy of the refrigerant to attract heat (i.e., to produce cooling effect). The way the two systems achieve this is quite different because the condenser machine achieves a cooling effect by expanding compressed refrigerant into a low-pressure chamber, while the solar cooling machine relies on the absorbing action of the absorbent to create near-vacuum inside its chamber. In near-vacuum, the refrigerant will evaporate at a very low temperature, removing latent heat from the refrigerant (i.e., producing a cooling effect). This happens at a temperature significantly lower than the refrigerant's evaporation temperature

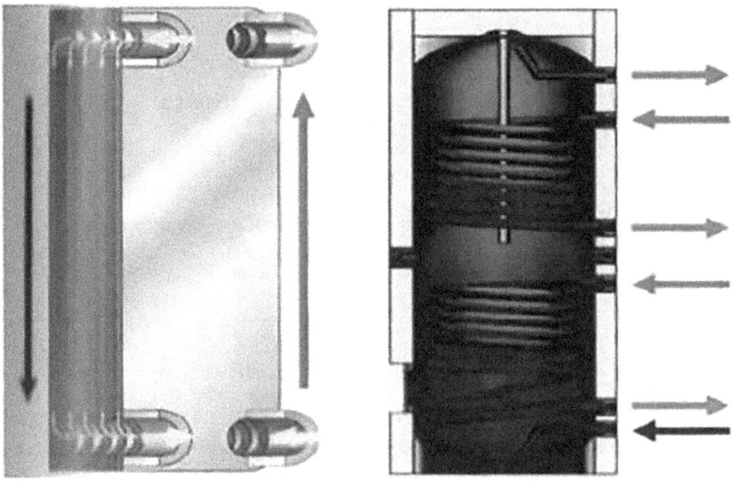

FIGURE 11 The plate exchanger of heat is shown together with a tank with a double exchanger and a serpentine plunged.

at atmospheric pressure. A cooling effect is thus achieved at usefully low refrigerant temperatures, making this principle practical for commercial use.

Heat supply from a field solar collector to the solar air-conditioning system is required not to directly provide the cooling action directly but to maintain the absorbent concentration. This ensures that low chamber pressure and low evaporation temperature of the refrigerant are maintained.

Preliminary Analysis and Solar Thermal Plant Design

Due to the abundance and benefits of solar energy, solar thermal plants can be very useful especially if there is a willingness to invest some capital into them, but the choice should be made taking into account the quality of the project and the devices.

The covered need can be high (60%–70% or more), for various reasons such as the randomness of the sunshine and the urban or market situations; it will not be able to meet the demand of the users. Most of the time, the plant will limit the use of fuel for heat production.

In the supply of DHW (domestic home water) through this type of plant, it must take into consideration that the weather conditions will not always be able to provide a sufficient quantity of energy to satisfy the demand and the complete comfort of the users. Demand and supply of heat will not be equal at the same time, and so to obtain independent plants in an annual scale, the capturing surface should be overestimated for summer (to supply the energy needed during winter).

Nowadays, investing in large seasonal storage is still in an experimental phase and involves very interesting cases. Figure 12 shows the offset of the thermal charge compared to the availability of the energetic source as the solar radiation incident. The latter is present in a solar thermal plant.

Matching Energy Availability and Thermal Energy Need

The way to estimate the energy needed to consume hot water is simple, but it is important to pay attention to the details involved.

Previously collected data can be used, but it is important to verify who collected them and what the original purpose was. If purposes are not reliable, it is necessary to go through monitoring to carry out data collection using a manual or digital device (calorie counter or a simple flow meter), gathering the data daily for at least 1 mo. In the sizing of the solar plants, the most important element is the energy

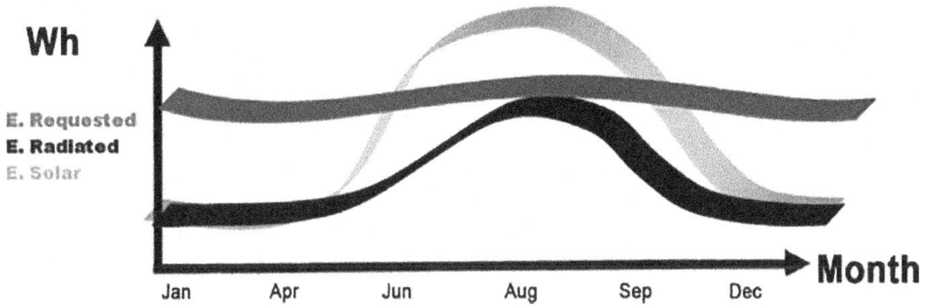

FIGURE 12 Offset of the thermal charge compared to the availability of the energetic source as the solar radiation incident.

need that will be accumulated and not the power need, which will be obtained at the right moment, from the mix of the solar heat storage and the auxiliary heater. Moreover, it is worth to invest on a thermal solar plant, as in all the renewable energies, only after a strong initiative to reduce the energy consumption, for example the aerator to be put at the final tap, can reduce up to 50% of the consume of hot water.

The Design Phase

The main design activities for a thermal solar system will be described here. The parameters and the different components that must be chosen for the plant and its dimensioning will be analyzed by the on-site procedure. It is important to note that in order to obtain the estimation of the costs of a solar plant, there is no need to proceed with detailed dimensioning, but it is sufficient to make an estimation of the following data:

- The surface of the collectors
- The volume of the tanks
- The thermal energy needs of the user

The phases of the project can be organized in the following way:

1. On-site investigation
2. Selection of the typology of plant and collectors
3. Dimensioning of the capturing surface
4. Dimensioning of the different components

On-Site Investigation

This is the first step in the design of thermal solar plants that includes three main targets:

- A detailed analysis of user energy consumption to estimate energy need (in joules, kilowatt-hours, or calories).
- A set of possible solutions to reduce the consumptions through energy saving.
- A check on possible solutions considering the logistic realization of the plant, with particular reference to the collectors' colocation.

Analysis of the Users' Consumptions

The first step toward a correct project is to estimate the exact heat consumptions; this allows the dimensioning of the solar system to satisfy the user energy needs through solar energy. There are two rapid assessment methods through which the evaluation of the users on thermal energy consumption is made:

1. The study of energy bills of the previous years
2. The study of the consumption habit of the users

Saving Energy Interventions

The dimensioning of a solar system cannot be based exclusively on the previous analysis; it needs to examine the opportunity to minimize each and every cost, to utilize clean energy after dispersion and waste reduction. Saving energy represents an objective of main importance, which can be realized in two ways:

1. Working on the demand level, boosting the users to modify their habits (this is called energy sacrifice), not always accepted but free of cost.
2. Working on the level of the offer, promoting the substitution of old and common devices with high efficiency products having the same performance but lower consumption.

The realization of a recovery intervention implies almost all the time an economic investment, the convenience of which must be analyzed before comparing the saved energy that would be obtained. It has to be found out *how much is saved.*

In the first case, the person who is taking care of the project must underline the waste, applying devices of energy saving. For example, in the case of DHW, both flow reducers and an air–water mixer limit the flow of hot water. In this way, users have the impression that the jet of water is the same although the quantity is much lower.

Logistic Aspects

The purpose of the on-site visit is to analyze some logistic aspects because it is very important to evaluate the feasibility of the solar plant. This phase is often ignored but very important. The first examination should verify if any historical constraint, related to the buildings or to the landscape in the area, exists. There could be situations where it is not possible at all to install a solar plant because it should be a waste of time to proceed. Under a technical point of view, it is necessary to find out a free area in which the solar panel could be installed, on the surface of which it is necessary to measure:

- Typology and material of the surface (characteristic of the roof, for instance)
- Gross area available
- Obstacles (antenna, chimney, others)
- Shade elements nearby or far away (buildings, trees, etc.)
- Accessibility for the installation and for the following maintenance operations
- Azimuth (orientation compared to the south) and inclination compared to the horizon

With reference to the specific problem of the shade, it can be useful in areas that are growing, to evaluate the local urbanization policies on the short and medium terms and to avoid that some years after the installation, the panels might fall in the shade of a new building.

Choice of Solar Plant Type

From an initial analysis, it is possible to distinguish the solar thermal plants on the basis of the hydraulic circuit. In an open circuit, the thermal vector fluid coincides with the thermal vector fluid used by the final users. A closed circuit consists of three hydraulic circuits divided into

1. Thermal vector fluid
2. Accumulation fluid
3. Water from the user

Analyzing the system circulation two categories of plants can be underlined: systems of natural circulation and systems of forced circulation. In most cases, the plant is designed with a closed circuit because there is the possibility of using a thermal vector fluid different from simple water.

The use of water as a fluid has some disadvantages like the presence of calcium and a low freezing point on one hand, and the open circuit is more economic and rapid to be installed on the other hand.

Estimation of the DHW Need

The production of DHW represents one optimum solution for the use of solar energy because it allows using solar radiation also when it is at its maximum power. A reason that has increased their new diffusion is the economic return in terms of capital investment due to the costs relatively contained for the installation and the good reliability of the plant. The evaluation of the theoretical thermal need in the systems of DHW can be quantified through the following expression:

$$Q_{ac} = V^* \left(t_{ut} - t_{al}\right)^* C_s \tag{11}$$

where Q_{ac} is the needed daily thermal energy (kcal/g), V is the requested water volume (L/g), t_{al} is the temperature of water source (°C), t_{ut} is the temperature of water output (°C), and C_s is the specific heat (kcal/kg * °C).

The supplied temperature depends on the place of the plant installation and on the water net from which it is taken, as it is possible to have differences of temperature depending on the period of the year. The temperature of warm water supply depends on the user considering that the maximum temperature of the water to the thermal generation should be $t_{ut} = 48 + 5$°C, where the second term indicates the maximum tolerance. To be able to guarantee this temperature until the water reaches the user, it is necessary that the distribution system must be suitably insulated. For the buildings that have no isolated distribution system, the temperature of supply would be higher than 50–55°C with peaks of 60°C.

The effective calculation takes into consideration the thermal losses that occur in the distribution network that brings the water from the production point to where it should be used. These losses make one think of the necessity of an effective thermal load higher than the theoretical one, to be able to satisfy the users' needs.

The total output of the distribution network depends on the grade of the insulation of the network itself. The average of the final output can have values in the following ranges:

- 0.85 to 0.90 in the case of recent constructions with insulated pipe system and with the recirculation.
- 0.75 to 0.85 for the constructions with plants without recirculation or with a cycle that works only during the day.
- 0.65 to 0.705 for the construction of plants with a working recycle during full time.

Sizing the Collector Field Surface

The first step to designing a solar plant system is to evaluate the surface needed for the plant as a compromise between the technical need and the economic need: unless a solar roof is used, it is necessary to think about the tilt angle and its right positioning. Basically, the exposure should not face north and the surface should be enough to have all the collectors needed. That surface of collectors must avoid getting over economic convenience, even only on log term. The dimension of the capturing also depends on the type of chosen collectors, particularly on their performance and on the orientation of the tilt in the roof.

The quantity of energy on the surface of the collectors varies in function of two angles: the azimuth angle θ and the tilt angle φ. The azimuth angle θ is the projection on the horizon plane surface and the tilt angle φ defines the inclination of the collector to the surface as shown in Figure 13.

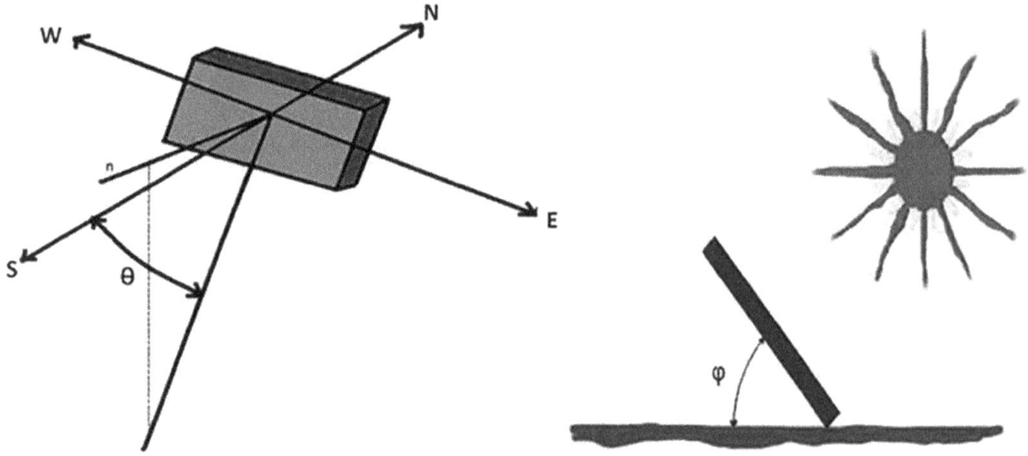

FIGURE 13 Azimuth and tilt angles.

TABLE 1 Increment of the Surface of the Collectors under Suboptimal Conditions

	Tilt Angle						
	0°	**15°**	**30°**	**45°**	**60°**	**75°**	**90°**
South	12%	3%	0%	1%	8%	20%	45%
South/East or South/West	12%	6%	3%	5%	11%	23%	43%
East and west	12%	14%	15%	20%	28%	41%	61%

The tilt angle depends on the building on which the plant is installed and on the orientation of the building itself as well. Since it is often not economically and aesthetically convenient, they are fixed directly on the top of the roof to create an ad hoc structure to hold the collectors.

About the typology of collectors, the choice depends on the conditions of the performance that will influence the output, and so from

- The internal temperature: the temperature to which the water is heated up.
- The external temperature: it depends on the period of the year in which it will be used.

The most important data (for the output) are the difference between the temperatures of the collector and the external environment because it characterizes the thermal losses. If data on consumptions are not yet available (for example, the building has just been built), an estimation on the energy needed (referring to average values) must be done. The corrections match the increases of the surface in bad orientation conditions, as seen in Table 1.

For combined plants, used for house heating and for the production of DHW would be appropriate to increase the value of the inclination to reduce the difference between the summer and winter production as underlined in Figure 14.

Heat Storage System

The energy needed for a large number of applications depends on the time, but often in a different way from solar energy. As a consequence, it is necessary that the accumulation of thermal energy produced with solar panels firstly in those cases in which the solar energy must cover an important fraction of consumes. The optimal skill of the heat storage system depends on

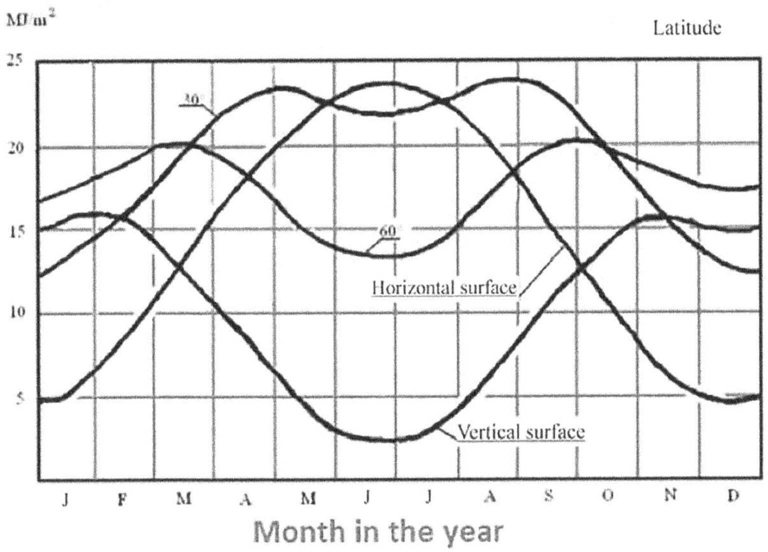

FIGURE 14 Monthly average solar radiation during a year at 40° latitude.

1. The availability of solar energy.
2. The nature of the loads.
3. The level of reliability needed in supply processes.
4. The economic analysis that sets the optimal percentage of traditional means.

Normally, a thermally isolated tank and a boiler are used. The boiler should store the solar energy when available and give it back when requested.

The most common energy discrepancy include alternation of day and night, sunny days as against cloudy days, and the summer season compared to the winter season. The latter is very important under a technical point of view because it reduces the technical problems connected to the summer overproduction as well.

A typical situation is a plant with accumulation. Paste Figs. 9 and 10:

Heat Storage Systems Dimensioning

Considering the time course of thermal loads and solar energy available:

- G_c represents the incident solar power.
- q_u represents the power delivered to the fluid.
- L is the power required by the loads that vary over time.

The areas above the red line and below the green one represent the time intervals that exceed the energy needs and must be accumulated. In contrast, the areas above the line in red and green show the time intervals in which the heat must be supplied to the loads from the storage system as in Figure 15.

The storage systems can be divided into two broad categories:

- Sensible heat storage systems
- Latent heat storage systems

In the sensitive system, energy is stored by raising the temperature of a suitable material. Latent heat storage systems take advantage of the latent energy at the phase change (usually the liquefaction) of a substance, and in this case, the process takes place at constant temperature.

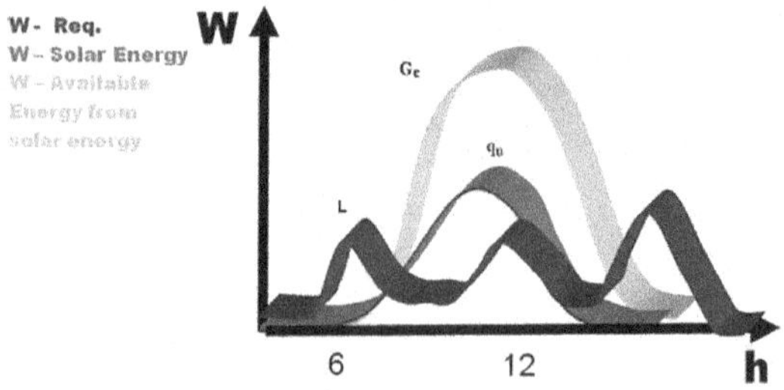

FIGURE 15 Energies during the day for a typical plant with storage.

In the first type, energy is taken by decreasing the temperature of the substance in storage while in the second type, stored energy is made available, causing a phase change.

In the build-sensitive systems, adopted in most cases, the energy is proportional to the temperature change of the substance contained in the batteries themselves:

$$Q_s = Mc_p\Delta T_s \tag{12}$$

where Q_s represents the accumulated energy, M is the mass storage, c_p is the specific heat of substance accumulation, and ΔT_s is the temperature variation.

Often a single temperature does not characterize the accumulation, because it cannot be considered perfectly mixed; it is actually layered. The top tank is the hottest part and the bottom is colder. The stratification is beneficial because of the collectors; reducing the average temperature of the absorber plate improves the efficiency to capture solar energy.

The best storage material in liquid systems is certainly water, since it has a low cost, has high specific heat, and is not toxic, and its boiling temperature at atmospheric pressure is high enough. The size of the storage system depends on the absorbing surface. Recommended values for solar systems for hot water are 100 L/m^2. The most important aspect to consider is, as already mentioned, the stratification of the water tank. As an index to assess the extent of the stratification of the water in a reservoir, the extraction efficiency is defined as:

$$\eta = \left(Qt^{*}\right)/V \tag{13}$$

where Q is the volumetric tank drain, V is the volume of accumulation, and t^* is the time required for starting from a completely mixed storage; the temperature difference of input-output has fallen to 90% of initial value.

Since the tank has a considerable cost, mainly due to heat insulation, sizing is necessary to make technical evaluations of economics. The storage tanks can be recharged in a sunny day or in an entire season. The first are those most commonly used and consist of an insulated tank to maintain hot water temperature. The degree of isolation of the accumulation should be as high as possible if the tank is installed outside, as often happens in systems with forced circulation. The storage tanks must also possess other important characteristics:

- They must be suitable for containment of potable water and must have had internal anticorrosion treatment.
- They must be resistant to high pressure (6 bar).
- They must be equipped with safety devices such as air vent, expansion vessel, and safety valve.
- They must be equipped with the following measuring devices: temperature gauge and pressure gauge for measuring pressure.

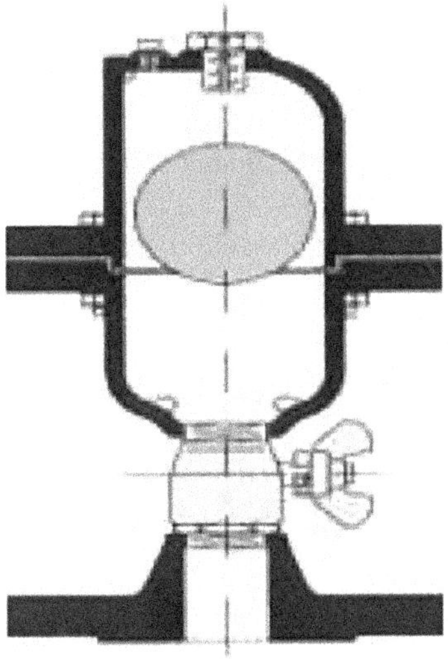

FIGURE 16 Example of an expansion tank.

When the heat demand is roughly constant during the day, usually about 20 L/m^2 of collector is sufficient, while this figure varies between 50 and 100 L/m^2 of collector daily batch loads such as residential use, where consumption of water for showers are concentrated at certain times of day (morning or evening).

An important phenomenon is the *stagnation* of the system. In summer, the energy produced is often greater than the amount of thermal energy users and, if the accumulation is too small or absent, the excess energy increases the temperature inside the collector to allow evaporation of the fluid heat transfer. This leads to the thermal gradients and overpressure under serious problem the hydraulic components of the system, and it can compromise the integrity of the plant. Also, bear in mind that the plant also has air release valves to eliminate the air that enters the piping system itself, inevitably compromising the thermal exchanges and thus the operation. If it is used, an automatic valve (Jolly) is strongly discouraged, due to the stagnation of the primary because the emptying of the liquid transfer medium such as air comes out unwanted in automatic air vents. The sizing of the accumulation is performed for the reasons outlined above, depending on the surface of the manifold: in practice, it takes an average of 50 to 90 L of storage per square meter of collector area (see Figure 16).

Size of the Auxiliary

The sizes of the auxiliary are dependent on the diameter pipes that are obviously related to flow values (Table 2).

Another very important element, which has a major impact on system performance, is the pressure drop. The volume flow can be calculated by the following equation:

TABLE 2 The Size of the Pipes according to the Scale

Flow (L/hr)	External Diameter per Thickness (mm)
<240	15 × 1
240–410	18 × 1
410–570	22 × 1
570–880	28 × 1.5
880–1450	35 × 1.5

$$P_v = Q/\left(c_g \Delta T\, m_v\right) \tag{14}$$

where Q is the thermal power made available by the solar collector (W/m^2); c_g = 1.3 Wh/(kg °C), the specific heat of the fluid; m_v is the density of the fluid (1 kg/L,); and ΔT is the temperature difference between inlet and outlet collectors (10°C).

The section of pipe can be calculated from the volumetric flow rate and velocity of the fluid P_v.

Prevention and Control of Legionella Exposure Risk

A problem that needs to be focused on in water systems for large facilities (hotels, prisons, hospitals, etc.,) is the presence of *Legionella.*

During the design phase of some facilities, a plant configuration to eventually prevent *Legionella bacteria* has already been furnished. Legionnaires' disease is contracted by breathing, by inhalation, or by aerosol microaspiration that contains the bacterium. The aerosol is formed by the droplets generated by water spray or the impact of water on solid surfaces. Most droplets are small; the more dangerous water droplets are those with a diameter less than 5 microns as they can more easily reach the lower respiratory tract. To ensure a reduction in the risk of legionellosis, the following preventive measures should be followed:

- Always keep hot water at a temperature above 55°C.
- Slide the water (either hot or cold) taps and showers that are not used for a few minutes at least once a week.
- Keep showers, jet showers, and speaker of the taps clean and free of fouling; replace as needed.
- Clean and disinfect all water filters regularly every 1–3 mo.
- Ensure that any changes made to the system, including new installations, do not create dead arms or pipes with no water or stream flows intermittently.
- Adhering to a point, the aim is to ensure a continuous thermal disinfection of hot water.
- Heating the water to 55°C results to, in fact, short-term elimination of *Legionella bacteria,* as evidenced by the Hodgson-Casey diagram taken as a point of reference for setting temperature and the time needed to implement thermal disinfection against *Legionella.*

However, in order to properly set up a continuous thermal disinfection, it is necessary that the water plant is at least maintained at 55°C.

It should be considered, in fact, that if the thermal head high facilities are not well insulated and well balanced between supply points and certain areas of the circulation thermal head high can be determined, as shown in the diagram below (Figure 17). So, even if hot water is supplied at a temperature below 50°C, there may be circulating in the networks of temperatures and can encourage the development of Legionella.

Large Systems

Solar systems are designed to be large to produce hot water for space heating. Possible applications include the following (see Figure 18): hospitals, home care for seniors, military barracks, hotels, gyms, prisons, and residential complexes—situations where the demand for heat is constant both during the day and at different seasons of the year.

They are called large systems when the surface of the solar field is greater than 100 m^2. The investments made to realize these plants are vital because of the current costs of the collectors; thus, before spending a lot of money, an in-depth energy audit to evaluate the requests of the users and assess the economic breakeven point should be undertaken. These systems, of course, are not able to meet the total energy demand, so they must be integrated into a conventional source.

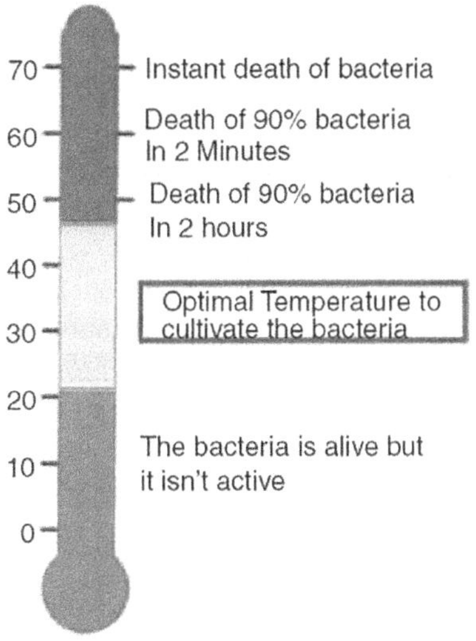

FIGURE 17 Behavior of the *Legionella* bacteria, depending on different temperatures.

FIGURE 18 Abruzzo region post-earthquake emergency camps (2009).

In support of these investments often involved in campaign financing from the state, such as loans to grants and tax deductions on the cost of the material. In recent years, the problem of pollution and continuous increases in the cost of traditional energy sources have caused people to engage in new ways, i.e., try to change their old habits, often met by indiscriminate consumption of energy.

Solar Integrated Collector Storage System Innovations

The use of a solar integrated collector storage (ICS) system (see Figure 19) represents a well-established technology for heat storage in civil and industrial applications.

An innovative solar thermal device has been used as an integrated collector storage providing DHW (up to 50–80°C). Here, the collector also acts as a storage unit, without requiring an external vessel. This device was successfully used in several circumstances, especially in extreme situations such as in the post-earthquake tent cities or to feed remote users in developing countries.

The efficiency of this device is strongly related to the draw-off curve. In fact, it is strongly desirable that during the draw-off, the water temperature remains as high as possible. Presently, this aspect is not optimal, leading to a strong reduction of the water temperature (more than 50% when 50% of the hot water filling the collector was discharged).

The collector was designed for use in emergency situations or for feeding remote users. Then, very simple configuration was adopted.

A series of eight J-type thermocouples have been placed to measure the temperature during the thermal energy storage phase and during discharge. The accuracy of the thermocouples was ±1.5°C and the acquisition frequency was 0.88 Hz. Solar radiation fluctuations have been followed by a pyranometer system with an accuracy of ±10 μV/(W m^2). The eight thermocouples (seven have been used for monitoring the pipe and one for the ambient temperature) and the pyranometer have been connected through a thermocouple and RTD modules, with a National Instrument field point with Ethernet connection, and connected to a data processing system through LabVIEW, as shown in Figure 19.

During the charge phase, the pipe has been placed in vertical position in order to be completely filled with water and to eliminate the air. Then, it has been placed on its supports, with a tilt of 42°.

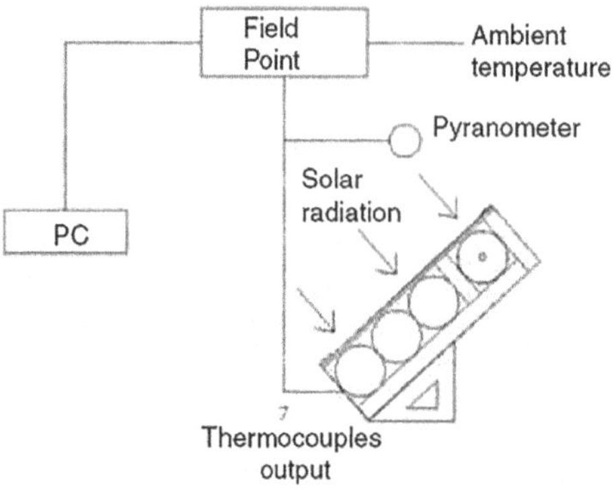

FIGURE 19 Experimental device used for the ICS prototype.

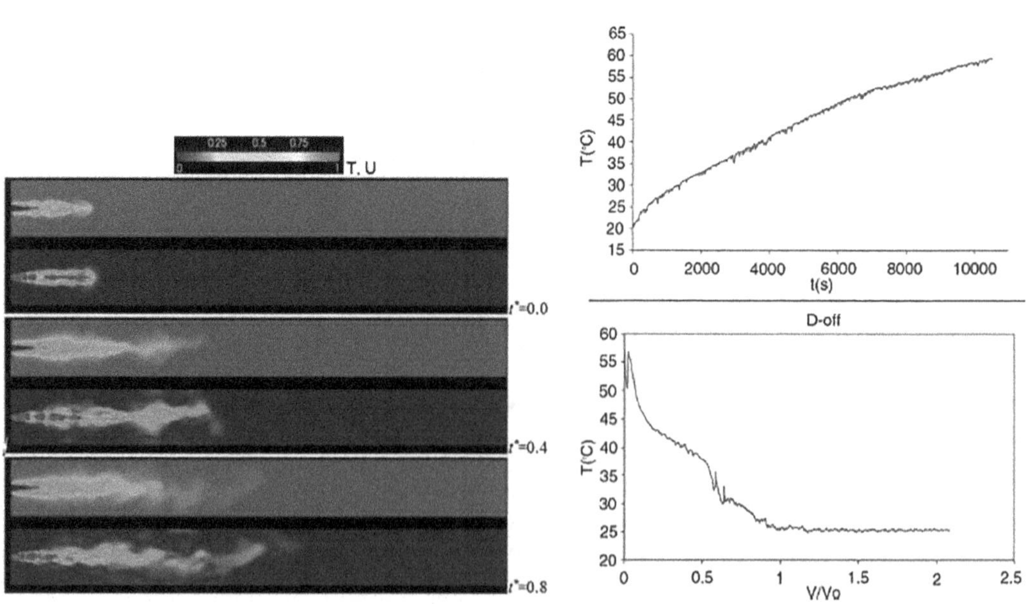

FIGURE 20 Longitudinal view of the temperature (top) and velocity magnitude (bottom) fields in three successive time steps.

The experimental analysis demonstrates that in the present configuration, the ISC has a reasonable performance in supplying hot water during the discharge phase.

Conclusions

The investments made to realize the plants are important because of the current costs of the collectors; therefore, before spending such quantity of money, it is better to undertake an in-depth energy audit to evaluate not only the requests of the users but also the breakeven point.

These plants are not able to supply the total energy demand; therefore, they should be integrated with a traditional source.

To support such investments, often there are financial aids from the state, such a free grant loan or allowances on the V.A.T. on the costs of the material.

In recent years, the pollution problem and the constant rise in the cost of traditional energy sources have resulted to users trying to find new solutions, changing old habits, often entailing excessive energy consumption.

References

1. Weiss, W.; Biermayr, P. Potential of solar thermal in Europe. European Solar Thermal Industry Federation (ESTIF), RESTMAC 6th Framework EU founded project, 2009, available at http://www.solarthermalworld.org/node/878.
2. Kalogirou, S.A. Solar thermal collectors and applications. Progress Energy Combustion Sci. **2004**, *30*, 231–295.
3. Hazami, M.; Kooli, S.; Lazaar, M.; Farhat, A.; Belghith, A. Performance of a solar storage collector. Desalination **2005**, *183*, 167–172.
4. Battisti, R.; Corrado, A. Environmental assessment of a solar thermal collectors with integrated water storage. J. Cleaner Prod. **2005**, *13*, 1295–1300.
5. Micangeli, A. *Design and implementation of innovative systems for the exploitation of renewable sources in social conflict areas: Desalination with heat recovery and hydroelectric energy production.* Sapienza University Doctoral Thesis, 2000.
6. Lesieur, M.; Metais, O.; Comte, P. *Large Eddy Simulations of Turbulence;* Cambridge University Press: New York, 2005; ISBN 0-521-78124-8.
7. Borello; Corsini; Delibra; Evangelisti; Micangeli. *Experimental and computational investigation of a new solar integrated collector storage system*, Third International Conference on Applied Energy, May 16–18, 2011, Perugia, Italy.

26

Alternative Energy: Wind Power Technology and Economy

K.E. Ohrn

Introduction (Web Preview)

This entry is a broad overview of wind power. It covers a range of topics, each of which could be expanded considerably. It is intended as an introductory reference for engineers, students, policy-makers, and the lay public.

Wind power is a small but growing source of electrical energy. Its economics are well known; there are several large and competent manufacturers, and technical problems are steadily being addressed. Wind power can now be considered as a financially and operationally viable alternative when planning additional electrical capacity. However, as shown below, wind power is only a minor component of present energy sources.

World final energy consumption (2002)[1]: 100%
World final electrical consumption (2002)[2]: 16.1%
World final wind power consumption (2004)[3]: 0.15%

Wind power's main deficiency as a power source is variability. Since wind velocity cannot be controlled or predicted with pinpoint accuracy, alternatives must be available to meet demand fluctuations.

Wind power carries few environmental penalties and makes use of a renewable resource. It has the potential to become a major but not dominant part of the future energy equation.

History

People have used wind to move boats, grind grains, and pump water for thousands of years. Wind-powered flour mills were common in Europe in the 12th century. In the 1700s, the Dutch added technical sophistication to their windmills with improved blades and a method to follow the prevailing wind. Isolated farms in the last century used windmills to generate electricity until the availability of the electrical power grid became widespread.

Past interest in wind power has tended to rise and fall with fuel prices for the predominant method of electrical production—thermal plants burning oil, natural gas, and coal.

Current

Electrical Production

Although small, wind power is a fast-growing part of the energy picture. Since 1990, worldwide installed capacity has grown about 27% (Table 1).[3]

Business is good for the leading manufacturers of wind power devices. Sales have increased; the technology is stable and predictable, with low maintenance and high availability.

Geographical Distribution (Countries)

The European Union had around 72% of installed capacity in 2004, and Germany, Spain, and the United States accounted for 66.1%. Denmark, Spain, and Germany had by far the largest 2004 capacity in terms of MW per million populations, and 10.6% of the world's population had 81.9% of its wind power capacity. In 2004, Denmark produced about 20% of its electrical power from wind power in 2004 (Table 2).[3]

TABLE 1 Capacity Growth

Year	Capacity (MW)	Growth Rate Year-over-Year (%)
1990	1,743	13.8
1991	1,983	17.0
1992	2,321	20.7
1993	2,801	26.1
1994	3,531	36.5
1995	4,821	26.6
1996	6,104	25.1
1997	7,636	33.0
1998	10,153	33.9
1999	13,594	27.7
2000	17,357	66.3
2001	28,857	7.9
2002	31,128	26.9
2003	39,500	20.3
2004	47,500	13.8
	Average Growth Rate	26.7

Source: Reprinted with permission from European Wind Energy Association.[3]

TABLE 2 Capacity Distribution

Country	Wind Power Installed Capacity (MW) 2001	2002	2003	2004	One-Year (%)	Three-Year (%)	Population (millions)	Capacity (MW/Million)	Percent of World Capacity
Denmark	2,456	2,880	3,076	3,083	0.2	7.9	5.4	570.9	6.4
Spain	3,550	5,043	6,420	8,263	28.7	32.5	40.3	205.0	17.2
Germany	8,734	11,968	14,612	16,649	13.9	24.0	82.4	202.1	34.7
Netherlands	523	727	938	1,081	15.2	27.4	16.4	65.9	2.3
USA	4,245	4,674	6,361	6,750	6.1	16.7	295.8	22.8	14.1
Italy	700	806	922	1,261	36.8	21.7	58.1	21.7	2.6
UK	525	570	759	889	17.1	19.2	60.4	14.7	1.9
Japan	357	486	761	991	30.2	40.5	127.4	7.8	2.1
India	1,456	1,702	2,125	3,000	41.2	27.2	1080.4	2.8	6.3
China	406	473	571	769	34.7	23.7	1306.4	0.6	1.6
Total	22,952	29,329	36,545	42,736	16.9	23.0	3073		89.2

Manufacturing capacity in 2002 was largely confined to this group of countries, with five big vendors accounting for 76% of sales. European Union vendors accounted for 85% of manufacturing market share.[3]

Economics

Cost per kWh

Wind power is a viable method of producing electricity that is capital intensive with low operating costs. The cost of production compares favorably with traditional fossil fuel or nuclear plant costs.

The major cost elements of a modern wind power installation are as follows[3,5,9]:

Capital

Onshore: 1200–1500 USD/kW
Offshore: 1700–2200 USD/kW

Operating: usually about 1.5%–2.0% of capital cost per year.[12]

Capital costs include wind capacity survey and analysis, land surveying, permits, roads, foundations, towers and turbines, sensors and communications systems, cabling to transformers and substations, maintenance facilities, testing, and commissioning. By far the largest individual capital cost is the turbine (up to 75%).

Operating and maintenance costs include management fees, insurance, property taxes, rent, and both scheduled and unscheduled maintenance.

Financing costs are a major portion of energy production costs, making them very sensitive to interest rates, incentives and subsidization.

Energy production cost estimates vary considerably. Optimists in the industry, such as the British Wind Energy Association, quote a low of 4.8 USD cents per kWh for an onshore plant in an optimal location. Pessimists, like the Royal Academy of Engineering in the U.K., quote up to 13.2 USD cents per kWh for an offshore plant, partly by including a controversial 3.1 USD cents per kWh cost for "standby capacity" required to supply demand when wind power is not available (Table 3).

Capacity Factor

The power generated by a wind turbine depends on the speed of the wind, and on how often it is available. At any given site, this is measured by the capacity factor, or the ratio of actual generated energy

TABLE 3 Windower Costs

	Cents (US) per kWh			
Wind Power Costs	Wind Onshore	Wind Offshore	Coal CFB	Gas CCGT
RAE [15]	6.8	9.9	4.8	4.0
RAE 2[a]	10.1	13.2	9.2	6.0
EWEA [3]	5.5	8.5		
AWEA (5)	6.6			
AWEA 2[b]	5.4			
BWEA (minimum)[4]	4.8	6.9	4.8	4.8
BWEA (maximum)[c]	6.8	9.1	6.8	5.9
Euro to USD conversion (2004)				1.22
GBP to USD conversion (2004)				1.83

[a] Adds 3.1 cents per kWh for wind power backup capacity and 1.9–4.6 cents/kWh for coal and gas carbon capture.
[b] AWEA figures adjusted to delete 1.8 cents US/kWh production tax credit and are for onshore sites with different average wind speeds.
[c] BWEA figures for a range of site types in November 2004 and 1–2 cents/kWh for carbon capture.
Source: RAE, Royal Academy of Engineering[15]; BWEA, British Wind Energy Association[4]; EWEA, European Wind Energy Association[3]; AWEA, American Wind Energy Association.[5]

to the theoretical maximum. Wind power turbine electrical output rises as the cube of the wind speed. When wind speed doubles, energy output increases eightfold. A typical turbine begins to turn when wind speed is at 9 MPH and will cut out at 56 MPH for safety reasons.

Capacity factors vary by site but are typically in the range of 20%–30% with occasional very good offshore sites reaching 40%. The yearly energy output from a wind farm is given by the following formula:

Site

Power production costs, site size, site design, and energy output and variability will depend mainly on details about the wind. These details include wind speed, wind direction, and the geographical distribution of favorable wind profiles. During analysis of potential sites, most planners use high (60 m plus) anemometer towers—often several of them—to gather at least one year's data per site. These data are usually correlated with national meteorological observations, if these are available and suitable. If not available, it would be prudent to gather site data for a longer period of up to three years.

Investors and regulators are increasingly aware of the crucial nature of wind data in estimating the quantity and timing of potential power production at a specific site. This research is crucial to the financial analysis of a potential wind power venture.

Other site analysis factors are accessibility via road for heavy equipment, electrical grid proximity and capacity, land ownership, and environmental impact.

Location

Favored Geography

Many countries have developed wind charts of broad areas based on meteorological data gathered for weather and aviation purposes. These charts show potential areas for investigation, where wind strength is high and constant over long periods of the year. Once potential sites, and their extent, have been identified, on-site data measurements provide the basis for analysis and modeling of potential energy production for a specific site.

After wind modeling, the site's geographical, environmental, financing, and ownership issues can be explored in detail.

Generally, sites are either onshore or offshore. Onshore sites are cheaper to construct, but have lower capacity factors due to wind turbulence from nearby hills, trees, and buildings. Offshore sites can have more potential energy available due to higher wind speeds and lower turbulence, which also reduces turbine component wear. Good offshore sites can be near high-demand load areas such as coastal cities, which also increases transmission options. Aesthetic and noise concerns are often fewer offshore, and sea-bed environmental concerns can be lower than landuse concerns for an onshore site.

Sizing a Location

Wind farm towers are usually spread over a large area in order to minimize wake losses. A spacing of five rotor diameters is often recommended. In a typical wind farm, the land physically occupied by tower foundations, buildings, and roads is often less than 2% of overall land area.[6] The remaining land is quite suitable for agriculture and other uses.

Limits to Maximum Production

How much capacity exists to generate electricity from wind? Is it possible that we will require more energy than wind can provide? After surveying wind patterns in the United States and applying energy density and extraction calculations, Elliott and Schwartz[14] concluded in 1993 that 6% of the available U.S. land mass could provide 150% of then-current U.S. electrical consumption. Furthermore, the needed land would be sparsely affected by the wind farm installation, with the vast majority of it (95%–98%) unoccupied by tower foundations, roads, or ancillary equipment and suitable for farming, ranching, and other uses. This study excluded land that is environmentally or otherwise unsuitable, such as cities, forests, parks, wildlife refuges, and environmental exclusion areas.

In the European Union, potential wind power capacity is also larger than current electrical consumption.

Future

Projected Growth

Thanks to increasing concern over the environmental effects of greenhouse gas emissions, the rising cost of fossil fuels, and the impending decrease in availability of oil and natural gas, wind power has a bright future.

Current 25%–30% growth rates are likely not sustainable, due to equipment production volume constraints and limits to perceived need for further capacity. Given the Eurocentric, highly clustered nature of current installed capacity, there is significant potential for high-rate growth elsewhere. However, even in European countries like Germany and Denmark, steady growth will be driven by predicted cost reductions in the 10%–20% range and by regulatory and governmental initiatives aimed at reducing emissions from electrical energy production and transitioning to renewable resources.

Projected Cost

Wind power technology is well down the cost improvement curve, with costs having fallen to present levels, below ten cents USD per kWh, from over $1.00 U.S. in 1978. Costs for a medium-sized turbine have dropped 50% since the mid-1980s, reflecting increasing maturity in the market. Cost projections range from a further 9%–17% drop as installed capacity doubles in the near-term future.

Projected Production

With increasing governmental policy support and commitment, growth rates of 15%–20% appear achievable in five to ten years. But there is likely an upper limit to the amount of electrical energy that can be produced from the wind.

Reaching Maximum Production

Production limits for wind power are based on its variable nature. Other types of electrical production capacity will be needed to provide base-load electrical capacity in the event that there is little wind available. Wind power will then become one player in a mix of generating technologies.

Hydrogen Economy

As wind power becomes a larger portion of electrical supply, occasionally its supply will exceed demand. Rather than simply curtailing wind plant production, it is attractive to think of using this excess electrical power capacity to generate hydrogen via electrolysis. This has the effect of storing wind energy that would otherwise not be harvested. This energy, in the form of hydrogen, can be used directly as a non-polluting fuel or as an input source to fuel cells to produce electricity at a later time.

When there is a significant hydrogen economy, with transmission lines, storage, and fuel cell capacity, this use for wind power will become a very attractive scenario.

Other Issues for the Future

Learning more about wind and forecasting—predicting the best locations, wind farm output, gusts, and directional shear.[10] This will help reduce financing costs when wind power plant output and impact on the grid are better understood and more predictable.

- Improving the control of demand through incentives around end-user load shedding, rescheduling and simple conservation methods. This could be used to offset wind power production shortfalls as an alternative to other forms of generation.
- Advancing aerodynamics specific to wind turbine blades and control systems.
- Designing extremely large wind tunnels to study wake effects minimization, structural load prediction, and energy output maximization at lower wind speeds.
- Enhancing power system capacity planning models to include wind farm components.
- Re-planting, or upgrading older mechanical and electrical components at existing wind farms.
- Wind farm siting further offshore and on floating platforms.
- Combining wind power and hydroelectric capacity by using surplus wind power to pump water behind dams and so store power that might otherwise be wasted.
- Determining how and whether to allocate full costs of environmental impact to fossil and nuclear plants.

Strengths and Weaknesses

Strengths

Environment

Wind power installations do not emit air pollution in the form of carbon dioxide, sulfur dioxide, nitrogen oxides, or other particulate matter such as heavy metal air toxins. Wind power installations do not use water or discharge any hazardous waste or heat into water. Conventional coal, oil, and gas electric power plants produce significant emissions of all kinds. Nuclear power plants produce dangerous and long-lasting radioactive waste. Greater use of wind power means less impact on health and the environment, particularly regarding climate change due to greenhouse gas emissions.

Renewable

Wind power produces energy from a resource that is constantly renewed. The energy in wind is derived from the sun, which heats different parts of the earth at different rates during the day and over the seasons. Unlike fossil and nuclear plants, the source of energy is essentially inexhaustible.

Costs

Wind power's costs are well known and are dropping to the point at which this technology is very competitive with other means of production. Fuel costs are nil, meaning that fuel costs have no uncertainty. Wind power costs should be more stable and predictable over the lifetime of the plant than power costs for fossil fuel plants.

Local and Diverse

Wind power plants provide energy source diversity and reduce the need to find, develop, and secure sources of fossil or nuclear fuel. This reduces foreign dependencies in energy supply, and reduces the chances of a political problem or natural disaster interfering with and diminishing the supply of electricity.

Quick to Build, Easy to Expand

Wind power plants of significant capacity can be constructed and installed within a year, a much shorter time than conventional plants. The planning time horizon is similar to conventional plants, given the need to accurately survey site wind characteristics and deal with normal environmental and related site issues. This means that capacity can be increased in closer step with demand than with conventional plants. With the right site and design, a wind power plant can be incrementally expanded very quickly.

Weaknesses

Natural Variability

A single wind farm produces variable amounts of energy, and its output is not yet as predictable as a traditional plant. As the geographical distribution and number of wind plants increases, and as research into predicting wind continues, these problems should be minimized, allowing cost-effective and orderly scheduling and dispatch of total grid capacity sources—but it is difficult to see traditional power sources disappearing entirely.

Connection to Grid

As the amount of electrical power supplied by wind power plants increases, concern increases over its effects on the electrical grid.

In order to maintain a reliable supply of electricity that matches demand, utility operators maintain emergency reserve capacity in order to deal with plant outages (failures) and unexpected demand across their entire system. This reserve is in the form of purchased power, unused capacity at conventional plants running below their maximum, or quick-start plants such as gas-fired turbines. Often, conventional plants on the grid are allocated a cost to cover this reserve based on their capacity (large plant, large reserve) and reliability (more outages, more reserve).

The industry is working on ways to determine and allocate this reserve cost for wind power plants. Yet to be agreed upon is the statistical basis for calculating such wind power plant reserve costs. Improvements in day-ahead wind forecasting will greatly reduce the uncertainty around wind plant output, and so decrease the cost burden to provide this reserve.

Several current estimates prepared for U.S. utilities show this reserve cost burden (or ancillary services cost) to increase with the amount of capacity provided by wind power, and to be in the range of 0.1–0.5 cents USD per kWh for penetrations between 3.5 and 29%. In no case was it thought necessary to allocate a reserve equal to 100% of the wind power capacity.[11] German experience is similar,[8] with no additional reserve capacity required for the 14% wind energy share of the national electrical consumption forecast for 2015.

When wind power supplies less than 20% of electrical consumption, these problems are not severe. At larger penetrations, reserves become a major issue. Interestingly, wind power plants may be subject to shutdown or voluntary power reductions in the event of coincident high wind, low demand situations. This is occasionally the case today in Denmark and Spain.

In some cases, wind power sites are situated far from the location of high electrical power demand, placing strain or potential overload on existing transmission facilities. In these cases, there are often cost, ownership and responsibility issues yet to be resolved.

Power quality problems around power factor, harmonic distortions, and frequency and voltage fluctuations are being successfully addressed in modern large production wind farms.[8]

This is one of the most difficult sets of issues facing the future of wind power as it matures from small-scale and local to large-scale penetration.

Local Resource Shortage

In a few places, high-quality wind power sites are not available or are already in production, leaving these places to import electrical power or use traditional sources.

Noise

Noise levels have decreased and are now confined to blade noise in modern units. Generator and related mechanical noise has been effectively eliminated. Noise, however, will always be a significant factor. Blade noise is described as a "whoosh, whoosh" sound, and is in the 45–50 decibel range at a distance of 200–300 m. This noise level is consistent with many national noise level regulations. However, this noise buffer zone adds to the overall land requirement for a wind power plant and so increases costs.

Visual Impact

Onshore wind farms are highly visible due to the height of towers and the size of the blades and generator. The impact of this varies with each person. Each wind plant operator needs to determine the levels of support and opposition from those who live and work within sight of the plant. Offshore plants attract fewer detractors than onshore plants—one of the reasons for their increasing popularity.

Offshore wind plants are less likely to cause unwanted noise since they are far from human habitation. This reduces turbine and blade design constraints and can lead to higher capacity factors.

Bird Impact

Bird deaths are a regrettable reality. The bird death rate at a specific wind farm project is quite variable. Several early wind farms (Altamont Pass, California, and La Tarifa, Spain) caused concern over death rates. The California Energy Commission estimates the death rate at Altamont (5400 turbines) to be 0.33–0.87 bird deaths per turbine per year.[16] The overall recent U.S. national average[13] is 2.3 bird deaths per turbine per year. Prudently located sites are off migration routes and not in nesting, over-wintering, or feeding areas. Their tower designs do not offer nesting or even roosting places. In such locations, death rates are lower, and overall impact is much lower than that caused by other types of human activity.

Since climate change is a very serious environmental problem faced by bird populations, wind power and other renewables are an important part of the solution.

Technology

Overview

Wind turbine design has three major components, and there are large economies of scale in design.

- *Tower height:* Wind turbine energy output is proportional to the cube of wind speed. Since moving air (wind) is subject to drag and turbulence from its contact with the earth and the objects on the earth, wind speed increases with height (vertical shear). The higher the tower, the more advantage there is for power generation. The tradeoff is between tower costs and increase in power generation. Typically, tower heights are rising, and are currently in the 100 m range. Off shore, vertical wind shear is generally less than onshore, so towers can be shorter, with wave height clearance being the factor that determines tower height (Figure 1).
- *Blade diameter:* The power capacity (watts) of a wind turbine varies with the square of its blade diameter, because a blade with a larger diameter has a larger area available for harvesting the wind energy passing through it. The coefficient of performance defines the actual power capacity compared to the maximum—how much energy can be extracted from the wind compared to the available energy. Modern wind turbines can achieve a coefficient of performance approaching 0.5, very close to the theoretical maximum of 0.59 derived by Betz.[3,18] This maximum is derived from the concept that if 100% of wind energy were extracted, the wind exiting the turbine would be at zero speed, so no new air could enter the turbine. Larger capacity turbines benefit significantly from economies of scale in foundation and support costs as well as swept area (Figure 2).
- *Controls and generating equipment:* The turbine's hub (or nacelle) is the most costly component and contains the generator, gear boxes, yaw controls, brakes, cooling mechanism, computer controls, anemometer, and wind directional vane.

Generators

As the blades turn, they drive a generator to produce electricity. Generating capacity ranges from a few hundred kW to over 3 MW. In older designs, there is usually a 40:1 gearbox to match low, fixed rotor speeds (~30 RPM) to required generator speed (1200 RPM for a 60 Hz output, 6-pole generator).

FIGURE 1 Typical large wind turbine. Note entrance steps and utility pole at base for scale.
Source: Photo courtesy of Suncor Energy, Inc.

FIGURE 2 Site assembly of large wind turbine nacelles and blades.
Source: Photo courtesy of Suncor Energy, Inc.

The gearbox often incorporates brakes as a part of the overall wind turbine control system. Generators that operate at low RPM are available and are called direct drive generators. These would eliminate the gearbox.

In more modern designs, rotor speed is variable and controlled to optimize power extraction from the available wind. Generator output is converted to d.c. and then back to a.c. at the required grid frequency and voltage. The conversion equipment is sometimes located at a central part of the wind power plant.

This is an active area for ongoing technical innovation.

Blades

In order to maximize power capacity through size, blades are very long, up to 50 m. To minimize noise, they must turn slowly so as to reduce tip speed, the primary blade noise source. Typical rotation speeds are in the 10–30 RPM range. Blades are increasingly made from composites (carbon fibre reinforced epoxy resins).

Rotor blade aerodynamics[19] have much in common with the aerodynamics of a propeller or a helicopter blade, but they are sufficiently different that the aerodynamics of wind turbine blades is an evolving field. The difference is that wind turbine airflows are unsteady due to gusting, turbulence, vertical shear, turbine tower upstream shadow, yaw correction lag, and the effects of rotation on flow development. For example, at present it can be difficult to predict rotor torque (and therefore power output) accurately for normal turbine operating conditions. Further development of theory and modeling tools should allow the industry to improve rotor strength, weight, power predictability, power output, and plant longevity while controlling cost and structural life.

For a given site wind velocity, the rotor blade's tip has very different air flow than its root, requiring the blade to be designed in a careful twist. The outer third of the blade generally produces two-thirds of the rotor's power. The third nearest the hub provides mechanical strength to support the tip, and also provides starting torque in startup situations.

Each blade generally has lightning protection in the form of a metallic piece on the tip and a conductor running to the hub.

Some manufacturers place Whitcomb winglets at the blade tips to reduce induced drag and rotor noise, in common with aircraft wing design.

In order to control blades during high wind speeds, some are designed with a fixed pitch that will progressively stall in high wind speeds. Others incorporate active pitch control mechanisms at the hub. Such control systems use hydraulic actuators or electric stepper motors and must act very quickly to be effective.

Wind Sensors

Wind turbines incorporate an anemometer to measure wind speed and one or more vanes to measure direction. These are primary inputs to the control mechanism and data gathering systems usually incorporated into a wind turbine.

Control Mechanisms (Computer Systems)

Control systems are used to yaw the wind turbine to face into the wind, and in some designs to control blade pitch angle or activate brakes when wind gets too strong. In sophisticated cases, the controllers are redundant closed-loop systems that operate pumps, valves and motors to achieve optimum wind turbine performance. They also monitor and collect data about wind strength and direction; electrical voltage, frequency and current; nacelle and bearing temperatures; hydraulic pressure levels; and rotor speeds, vibration, yaw, fluid levels, and blade pitch angle. Some designs provide warnings and alarms to central site operators via landline or radio. Manufacturers do not release much detail about these systems, since they are a critical contributor to a wind turbine's overall effectiveness, safety, and mechanical longevity.

Role of Governments and Regulators

Governments play a large part in determining the role and scale of wind energy in our future mix of energy production capabilities.

Subsidies, Tax Incentives

As part of programs to encourage wind power production, the following are used in varying ways[7]:

- Outright subsidies, grants and no-interest loans.
- Tax incentives such as accelerated depreciation.
- Fixed prices paid for produced electrical power.
- Renewable energy quantity targets imposed on power utility operators.

Grid Interconnection and Regulatory Issues

Since many power utility operators are owned by governments, and most are regulated heavily, governments have a role to play in encouraging solutions to grid interconnection issues. There must be a political will to address issues, find solutions, and develop practices and different management strategies that will allow greater penetration of wind power into the electrical supply.

Improving Wind Information

Climate and environmental information is most often collected and supplied by national governments in support of weather and aviation services. Wind atlases are an invaluable resource to the wind plant planning process. National efforts to improve long-and short-term wind forecasting, atmospheric modeling tools, and techniques will benefit wind power projects' ability to forecast power output for long-term and short-term planning purposes.[10]

Environmental Regulation

In this controversial area, government can tighten its regulation of air quality, carbon emissions and other environmental areas. This would have the effect of increasing the apparent cost of conventional thermal electrical power, which is responsible for significant emissions. It is often argued that wind power would already be cost competitive if environmental and health costs were to be fully allocated to conventional oil, gas, and particularly coal-powered plants, or if such plants were required to make investments to significantly reduce emissions.

Conclusion

Western societies depend on a steady supply of energy, much of it in the form of electricity. Most of that supply comes from thermal plants that burn oil, natural gas, and coal, or from nuclear plants. Where will our electrical energy come from in the future? How will we keep our environment livable and healthy?

One part of this answer lies in wind power. Its costs are within reason; the technology has matured with some gains yet to be realized; it carries little environmental penalty; and the source of its energy is renewable. As long as the sun heats the earth, wind power will be available.

Wind power will not likely be the complete answer; it is an intermittent source because the wind doesn't always blow. But there is a very large amount of it available for us to harvest. As wind power moves quickly from small-scale to large-scale, its future path depends on governments and regulators as much as it does on technical innovators and manufacturers.

References

1. Annual Energy Review 2003, DOE/EIA-0384(2003), Energy Information Administration, U.S. Department of Energy, Washington, DC, September 2004.
2. Key World Energy Statistics -2004 Edition, International Energy Agency, OECD/IEA -2, rue André-Pascal, 75775 Paris Cedex 16, France or 9, rue de la Fédération, 75739 Paris Cedex 15, France, 2004.
3. Wind Energy, The Facts, Analysis of Wind Energy In the EU-25, European Wind Energy Association, Brussels, Belgium, 2004.
4. BWEA Briefing Sheet, Wind and the U.K.'s 10% Target, The British Wind Energy Association, http://www.bwea.com/energy/10percent.pdf (accessed July 2005).
5. The Economics of Wind Energy, American Wind Energy Association, http://www.awea.org/pubs/factsheets/Economics OfWind-Feb2005.pdf (accessed July 2005).
6. Saddler, H., Dr.; Diesendorf, M., Dr.; Denniss, R. A Clean Energy Future For Australia, A Study By Energy Strategies for the Clean Energy Future Group, Australia, March 2004.
7. Policies to Promote Non-hydro Renewable Energy in the United States and Selected Countries, Energy Information Administration, Office of Coal, Nuclear, Electric and Alternate Fuels, United States Department of Energy, Washington, DC, February 2005.
8. BRIEFING May 10th, 2005. German Energy Agency DENA study demonstrates that large scale integration of wind energy in the electricity system is technically and economically feasible. http://www.ewea.org/documents/0510_EWEA_BWE_VDMA_dena_briefing.pdf (accessed July 2005).
9. Offshore Wind Experiences, International Energy Agency, 9, rue de la Fédération, Paris Cedex 15, 2004.
10. Milborrow, D. Forecasting for Scheduled Delivery, Windpower Monthly, December 2003.
11. Utility Wind Interest Group—Wind Power Impacts On Electric-Power-System Operating Costs, November 2003.
12. Danish Wind Industry Association, http://www.windpower. org/en/tour.htm (accessed July 2005).
13. Wind Turbine Interactions With Birds and Bats: A Summary of Research Results and Remaining Questions, Fact Sheet, Second Edition, National Wind Coordinating Committee, Washington, DC, November 2004. http://www.nationalwind.org/publications/avian/wildlife_factsheet.pdf (accessed July 2005).
14. Elliott, D.L.; Schwartz, M.N., Wind Energy Potential In the United States, Pacific Northwest Laboratory, PNL-SA-23109, Richland, Washington, USA, September 1993.

15. The Cost of Generating Electricity, Royal Academy of Engineering, http://www.raeng.org.uk/news/publications/list/reports/Cost_of_Generating_Electricity.pdf (accessed July 2005).
16. Developing Methods to Reduce Bird Mortality in the Al- tamont Pass Wind Resource Area, August 2004, http://www.energy.ca.gov/reports/500-04-052/500-04-052_00_ EXEC_SUM.PDF (accessed July 2005).
17. Wind Force 12, Global Wind Energy Council, June 2005, http://www.ewea.org/03publications/WindForce12.htm (accessed July 2005).
18. Proof of Betz Theorem, Danish Wind Industry Association, http://www.windpower.org/en/stat/betzpro.htm (accessed July 2005).
19. National Wind Technology Center, http://www.nrel.gov/wind/about_aerodynamics.html (accessed July 2005).

27

Electric Power: Microgrids

Ryan Hanna

Introduction

Microgrids are insular power systems embedded within the bulk power grid that use distributed energy resources (DERs), such as fossil fuel generators, renewable generators, and energy storage, to provide energy locally near the point of consumption (Figure 1). They may be small or large, for example, a single building or a campus with dozens of buildings, and they may be simple or complex, for example, consisting of a single backup generator or numerous DERs working in concert to provide primary power. In any case, their defining attribute is their ability to disconnect, or "island", from the larger grid during power outages, thereby shielding customers within from disruptions and improving the reliability of electric service. (Some taxonomies include remote or "off-grid" systems—those not embedded within a bulk grid.) Although microgrids vary widely in size, from tens of kilowatts to tens of megawatts, those powering a small commercial building (approximately hundreds of kilowatts) are indeed "micro" or on the order of 10^{-6} the capacity of the bulk U.S. power grid.

Since their inception, power systems have been built to provide electricity that is reliable and economic—two competing design criteria that require balancing (Billinton and Allan 1996). Today, a third criterion—that power systems provide carbon-free electricity in accordance with long-term climate stabilization goals—is now widely accepted. The climate criterion presents new and interesting design trade-offs for microgrids. Historically, the fossil fuel pathway—i.e., the use of fossil fuel generators, first diesel but more recently natural gas—has provided high levels of reliability at low cost. Given future projections for very low gas prices, it is likely to remain the status quo absent new carbon-minded policy. Some policies are emerging—typically in the form of restrictions on diesel use—while governments have variously supported low-carbon pathways such as combined heat and power (CHP), renewables with energy storage, biofuels, and hydrogen. Although some options remain far-off, consideration is important today because each pathway offers a fundamentally different vision for how microgrids interact with a decarbonized energy system. Pathways can differ in use of fuels, generators, load electrification, outside requirements for energy infrastructure, and, importantly, outcomes for carbon emissions and pollution.

Recent support for microgrids has stemmed from concern about potential impacts to power systems from an increasing frequency of severe weather, including wildfires, to which climate change is contributing. The conventional wisdom is that microgrids can increase grid reliability and resiliency while also

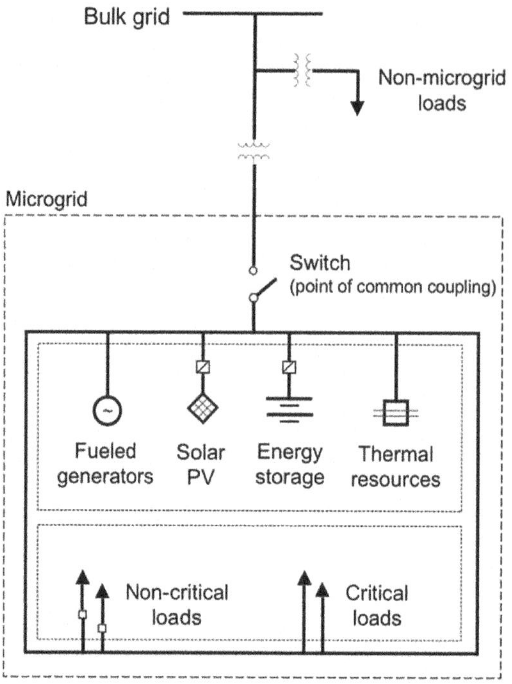

FIGURE 1 A simple representation of a grid-connected microgrid. Distributed energy resources (DERs) and loads comprise the microgrid and lie behind a single point of common coupling to the bulk grid. The microgrid can disconnect from and reconnect to the bulk grid and operate autonomously. (Fossil fuel and thermal energy networks are present but, for clarity, not pictured.)

helping to address climate change. An appraisal of decarbonization pathways, however, shows that this is not always so (Section "Decarbonization Pathways for Microgrids"), and comparison of cost, emissions, and performance helps explain where co-benefits and trade-offs exist. A picture of historical deployments and present-day activities around microgrids provides further context (Section "Microgrids, Past and Present"). Although the focus here is the U.S.A, deployments are on the rise globally, particularly in Asia Pacific (e.g., Australia, India, China, and Japan) where forecasted growth is in fact greatest and driven by deployment of remote systems in developing areas that lack grid infrastructure (Willette and Asmus 2019).

Microgrids, Past and Present

Microgrids are not a new phenomenon. Today, we recognize the earliest power systems—pioneered by Thomas Edison during the 1880s that provided lighting for hundreds of customers across a few New York City blocks—as microgrids. These grids, and hundreds in other cities, would first experience decades of expansion, as industry pursued larger customer bases and scale economies and governments mandated rural electrification in the 1930s–1940s to combat poverty (Tuttle et al. 2016), and later interconnection to increase redundancy and power plant utilization rates.

Centralized grids now underpin nearly all of the industrialized world, limiting the role for microgrids to niche applications. The high cost of microgrids has been justified where interconnection to the bulk grid is impractical or uneconomic, such as with island villages, military bases, and remote industrial facilities. In Alaska, diesel and hydropower microgrids power over 200 remote communities. It has also been justified where services within the bulk grid are deemed critical. Microgrids provide backup power for the health care industry, where hospitals must comply with emergency generator standards and maintain onsite a 96-hour supply of diesel fuel, as well as for the telecommunications industry's enormous network of cell towers, where the use of small (<10 kW) diesel generators and lead-acid batteries has dominated. More recently, incentives have supported the use of fuel cells—over 3,000 have been deployed since 2007 (Ma et al. 2017). Microgrids have also been economic in campus settings that have space and the institutional capacity for ownership and operation. Prominent examples include those at New York University, Princeton University, the Illinois Institute of Technology, the University of California San Diego, and the University of California Irvine, among others. Natural gas cogeneration forms the physical and economic backbone of many, providing significant cost savings on electricity, heating, and cooling.

Today, the role for microgrids is expanding as governments begin to take a proactive approach in guiding their development. State governments in the northeast U.S.A are seeking to create a role for "community" microgrids that embed pockets of resiliency into the grid and serve the public interest. Following severe weather in 2012 that caused widespread blackouts, Connecticut in 2013, Massachusetts and New Jersey in 2014, New York in 2015, and Maryland in 2019 each enacted a community microgrid grant program to increase resiliency at critical infrastructure such as hospitals, police and fire stations, wastewater treatment plants, public shelters, and municipal commercial areas. Recent power outages at the Atlanta, Las Vegas, Washington D.C., Los Angeles, Hamburg, and Amsterdam international airports have spurred funding for microgrids that support critical airport services.

Other jurisdictions are pursuing a market-based approach with the development of commercialization pathways that tie in to broader efforts to decentralize electricity generation. California is leading, having recently awarded $25.7 million for seven microgrids in 2014 that demonstrate "low-carbon" and "renewables-based" configurations and $45 million for nine microgrids in 2018 that demonstrate "repeatable, commercial-scale" configurations that support the state's emissions reduction goals. Hawaii passed legislation in 2018 to establish a tariff that defines the value of services exchanged between a microgrid and the bulk grid. In the Caribbean, Puerto Rico's post-Hurricane Maria reconstruction plans aim to decentralize generation on the island, and part of that vision includes a new regulatory framework for microgrids that encourages deployment by third-party (non-utility) providers. The island's goals are to increase reliability, and reduce fossil fuel consumption and carbon dioxide emissions, among others. The U.S. Virgin Islands is also rebuilding through microgrids, deploying two on the island of St. John comprised of diesel generators, solar photovoltaics (PV), and battery energy storage.

To what extent are these microgrids low-carbon? In all historical use cases, fossil fuel generators have been the preferred technology. This holds true today in the emerging community microgrid paradigm, despite falling prices for renewable energy and energy storage. For the programs in Connecticut (Microgrid Pilot Program Round 1 and 2) and New York (NY Prize Stage 2), for example, for which good data on installed or planned DERs exists, natural gas generators comprise 78% of total capacity, of which 61% is CHP, while PV comprises 4%, which is on par with diesel generators (also 4%). Battery storage, on a power capacity basis, comprises 3%. Five of the six microgrids supported by the New Jersey Energy Resilience Bank are based entirely on natural gas CHP. By contrast, California has set out to support renewable microgrids and its 2014 program produced 73% solar PV, 11% battery storage, and 15% fossil fuel generation by capacity. Programs in other jurisdictions remain too nascent to draw conclusions.

Each of these jurisdictions pushing microgrids has enacted climate policies that aim to decarbonize the power sector through mandated cuts in greenhouse gas emissions, procurement targets for renewable energy, or both. Puerto Rico and Hawaii have passed into law 100% renewable energy portfolio standards (RPS), for instance, while California has passed a 100% clean energy standard. The five northeast states with microgrid grant programs have RPS targets ranging from 25% to 50% by 2030. Accordingly, some have acted to limit fossil fuel use in microgrids. Puerto Rico's 2018 rules, for example, mandate renewable energy and restrict fossil fuel use in non-CHP generators in three ways: to 25% of the total microgrid energy output on an annual basis, by generator heat rate (not greater than 13,000 btu/kWh), and by the total microgrid energy input–output ratio (not greater than 2,500 btu/kWh). CHP generators qualify but must adhere to efficiency standards (useful thermal energy must be greater than 50%

of the total energy output) and generator heat rate (not greater than 7,000 btu/kWh). California's 2018 microgrid legislation (SB 1339) prohibits compensation for all diesel use and natural gas generators that fail to meet strict emission standards, effectively precluding non-CHP fossil fuel generators. By contrast, Hawaii, which is in the process of establishing a microgrid services tariff, has not set any prohibitions on fossil fuel use.

Decarbonization Pathways for Microgrids

Although the merits of particular choices for technology and fuel in microgrids are case-specific (Hanna et al. 2017; Flores and Brouwer 2018), an assessment of pathways generally should address the three power system design criteria—reliability, cost, and emissions. Table 1 summarizes that assessment qualitatively for the present day, with characteristics scored from very poor (—) to very good (++). It also includes the practicability of deployment, or dependency on outside energy infrastructure, as well as scalability, or the capability of the pathway to scale up to meet a significant portion of society's electricity needs. Power density, a watt per square meter (W/m^2) measure of the concentration of energy generation or consumption, is a useful metric with which to assess the capabilities of distributed generators that, unlike rural power plants, are confined by limited space in built environments. Here, it is a proxy for reliability.

The diesel pathway, or use of diesel reciprocating engines, is the historical pathway and taken in this assessment as the base case against which other pathways are scored. Diesel generators are technologically mature, highly versatile, and have high power densities and low cost. Overuse can significantly degrade local air quality, however, and many jurisdictions limit their application to emergency operation only, typically a few hundred hours per year. Thus, they are neither scalable nor desirable options for widespread decentralization; other solutions are needed. Figure 2 shows the physical features for technology choice and fuel that distinguish pathways.

Combined Heat and Power

Over 2,000 CHP systems across the U.S.A provide nearly 2.3 GW of capacity (Darrow et al. 2017). These systems are fossil fuel generators like reciprocating engines and turbogenerators that capture and use engine waste heat to supply building heating and cooling needs. In making use of heat that is otherwise dissipated to the environment, CHP provides two benefits: it increases the efficiency of fuel use, from 25–45% in standalone (i.e., non-CHP) generators to 65–80% in generators with heat recovery, and it reduces (and can eliminate) direct gas combustion for heating as well as electricity for cooling, thereby reducing greenhouse gas emissions in the commercial and residential sectors. CHP is successful in reducing emissions today because the combustion of gas to meet electrical and thermal energy demand in concert results in a lower carbon intensity (gCO_2/MJ) than the separate production of electricity from the bulk grid and heat from a gas boiler or furnace.

TABLE 1 Microgrid Pathways—Characteristic Technology Choice, Fuel Use, and Assessment[a]

Pathway	Technology	Fuel	Power Density	CO_2 Emissions	Pollutant Emissions	Cost	Practicability	Scalability
Diesel	ICE	Diesel	++	—	—	++	++	—
CHP	ICE-CHP	NG	++	-	-	+	++	+
All-electric (solar-plus-storage)	PV, BS, HP	None	-	++	++	-	++	+
Biomethane	ICE-CHP	Biomethane	++	+	-	-	+	—
Hydrogen	FC-CHP	Hydrogen	++	++	++	—	—	++

[a] *Abbreviations:* ICE (internal combustion engine), CHP (combined heat and power), PV (photovoltaics), BS (battery storage), HP (heat pumps), FC (fuel cell), and NG (natural gas).

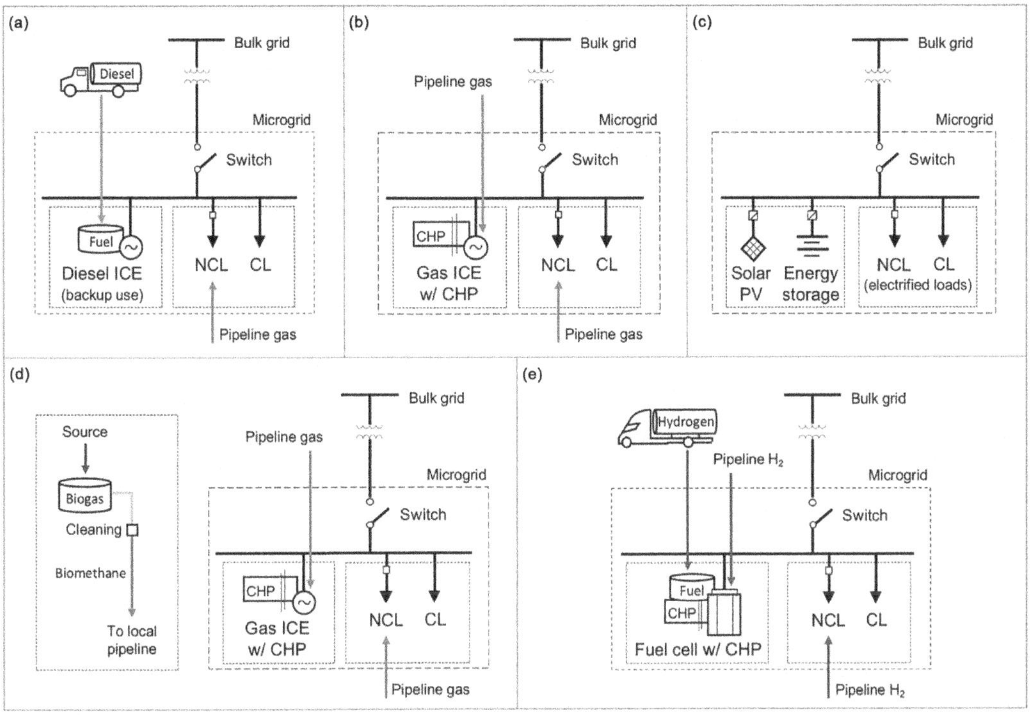

FIGURE 2 Physical features for technology choice and fuel that distinguish pathways from one another. The five pathways are (a) a diesel genset, (b) gas-fired genset, (c) solar-plus-storage (all-electric), (d) biomethane-fired genset, and (e) hydrogen fuel cell. All fueled gensets (less the diesel genset; b–d) make use of CHP. NCL denotes non-critical load, CL critical load, ICE internal combustion engine, CHP combined heat and power (i.e., infrastructure for capturing and using waste heat), and H_2 hydrogen.

Governments have supported CHP for decades as non-utility generators. For example, the Obama Administration in 2012 set a nationwide target of 40 GW of new capacity by 2020, while in 2010 the state of California set a target of 6.5 GW by 2030. Governments are now coupling support of CHP to microgrids. California and Puerto Rico, through new microgrid rules, are discouraging (or banning outright) the use of standalone generators while encouraging CHP use, while the community microgrid programs emerging in the northeast U.S.A are largely supporting natural gas CHP.

The central issue with CHP is its reliance on fossil fuels. The CHP pathway is therefore at most a low-carbon pathway. For this reason, it also lacks scalability, as local air quality regulations limit criteria pollutant emissions. Its ability to reduce carbon emissions depends on the carbon intensity of grid electricity, which is expected to decarbonize over time, as well as the presence of gas combustion in buildings, which are expected to electrify over time. The fossil CHP pathway is therefore limited in efficacy: it locks in long-lived fossil generation that is effective in reducing carbon emissions in the short term but whose effectiveness declines as the bulk grid decarbonizes. (Fuel switching from fossil gas to renewable gas could make CHP carbon-neutral—but that is a distinct pathway considered separately.)

All-Electric: Solar-plus-Storage

One alternative to the use of fossil fuels is to discard them entirely via a fuel-free "all-electric" pathway that pairs solar PV with electric storage such as lithium-ion batteries and that might further include electrified loads and heat pumps, deep energy efficiency, and demand response. PV and storage provide power and substitute for gas generators, electrified loads obviate gas combusted for heating, and responsive demand provides operational flexibility. Combinations of these technologies offer a legitimate path to zero carbon and zero pollution in microgrids.

Of all pathways, the solar-plus-storage option is perhaps the most touted by advocates of decentralization. It maps technologically to other transitions taking shape in the power sector, where utility-scale PV and batteries are emergent and electrification of load and responsive demand are expected to assume important roles in decarbonization. Like CHP, solar-storage microgrids are not dependent on technological breakthrough nor new energy infrastructure; they are proven solutions with numerous operational today, including three funded in 2014 by the State of California as well as many financed privately (Asmus et al. 2017).

Yet the solar-storage pathway faces fundamental challenges related to performance and applicability, while high costs remain an issue. Fundamentally, the use case for solar-storage microgrids is limited by the high power demand of modern cities and low power density of PV (Smil 2019). For example, an analysis of energy use for a selection of the most common types of buildings in the U.S.A shows power densities of 10–250 Wm^{-2}, with warehouses (5–10 Wm^{-2}), retail spaces (15–20 Wm^{-2}), schools (15–60 Wm^{-2}), and smaller office buildings (15–55 Wm^{-2}) constituting the lower end and hotels (150–165 Wm^{-2}), large offices (190–230 Wm^{-2}), and hospitals (235–260 Wm^{-2}) constituting the higher. Variation derives from thermal loads that differ by climate zone—in this example bounded by temperate coastal and hot inland conditions in southern California. In comparison, average annual solar irradiance in coastal and inland California is approximately 210 Wm^{-2} and 240 Wm^{-2}, respectively. With 19% efficient PV cells, PV power densities are 40–46 Wm^{-2}. Thus, only for a subset of these buildings—warehouses, retail spaces, schools, and small office buildings—is PV able to provide a surplus of energy on an annual basis. For the remaining buildings, 100–600% more space is needed for PV installations to achieve parity between annual energy demand and generation. In all cases, substantial energy storage is required to shift excess PV generation to the nighttime when operating in island mode. Demand response can reduce building power density during times of stress by turning off or delaying non-critical load, whereas building electrification increases building power density because it electrifies loads otherwise met with gas.

The solar-storage pathway also faces challenges of cost and performance. Compared to solar-storage microgrids, microgrids that further include fossil fuel generators can provide higher levels of reliability at lower cost because fossil fuel generators are inexpensive by comparison and with fuel reserves are able to supply power day and night. Because PV power is weather-dependent, microgrids depend heavily on storage during times of stress, possibly requiring significant oversizing of storage to ensure islanded operation. As such, microgrids with fossil generators outperform solar-storage microgrids on a reliability-per-cost basis.

Ultimately, low power density and lack of space for PV installations in urban environments limit scalability. Barring radical breakthroughs in cell efficiency or cost decreases for high-efficiency cells, distributed PV will remain unable to meet the power demand of much of modern civilization. A restriction on fossil generators achieves climate benefits but limits applicability because fossil generators are high-power-density DERs that are best suited to serving high-density urban areas. Absent new carbon-minded policy, such as carbon pricing, addressing resiliency and climate change simultaneously with solar-storage microgrids, requires sacrificing cost, performance, or both.

Biomethane

A second alternative to the use of fossil fuels is to substitute them with biomethane while keeping the generator. Biomethane, or renewable natural gas, is biogas that has been cleaned of impurities, upgraded to pipeline quality, compressed, and injected into the natural gas network. Biogas can be sourced from landfills, wastewater treatment plants, municipal solid waste, and manure, among others. Although its emissions when combusted are similar to those of fossil natural gas, biomethane is considered carbon-neutral because it displaces fossil gas in the power sector and when sourced from waste streams reduces emissions from the source sector (either industry or agriculture).

Although in some cases microgrids are collocated with a source of biogas—the MCAS Miramar microgrid in San Diego, California, which uses landfill gas, is one such example—production and consumption need not be physically linked. With such "directed" biomethane, a source facility captures and treats previously vented or flared biogas and injects it into the local gas network, displacing an equal volume of fossil gas as a result. The consuming generator, located perhaps tens or hundreds of kilometers distant, combusts pipeline gas locally but through reporting methodologies is credited with the combustion of the injected, carbon-neutral biomethane. Allowing for directed biomethane is practical because the load centers that microgrids serve may be located far from contracted biogas sources. It also enables market forces to identify the cheapest sources of supply. The University of California, as part of its 2025 carbon neutrality initiative, has contracted for directed biomethane to permit continued use of its CHP gas turbines (Meier et al. 2018).

The biomethane pathway is a carbon-neutral pathway and allows microgrids to retain use of fossil fuel generators. Microgrids can therefore achieve high levels of reliability and are not limited in applicability due to issues of power density. Nevertheless, issues of cost, scalability, and pollutant emissions constrain and limit its potential. In the short term, biomethane commands a price premium relative to fossil gas: prices for pipeline-ready biomethane are $10–24/mmbtu depending on the source (Sheehy and Rosenfeld 2017), while U.S. fossil gas prices stand near $3/mmbtu. Whether biomethane microgrids are cost competitive with fossil CHP microgrids depends on the contracted price for biomethane and prevailing carbon pricing; whether they are cost competitive with solar-storage microgrids requires granular analysis of performance and cost that considers local conditions such as electric tariff structures and rules limiting air pollution or pricing carbon. Although biomethane prices exceed those of fossil gas, governments have created policies and incentives to encourage the development of biogas resources, such as methane emission reduction targets and direct financial support for projects. However, the bulk of these incentives, notably the federal Renewable Fuel Standard (RFS) and Low-Carbon Fuel Standard (LCFS) in California, place emphasis on the use of biogas as an alternative transportation fuel rather than as a fuel for electricity generation. Competition with the transportation sector may constrict supply or increase price.

A second major issue is that long-term biogas resources are limited. The most recent estimate of U.S. biogas availability is 331 billion cubic feet per year or about 1–1.5% of U.S. fossil gas consumption (National Renewable Energy Laboratory 2013). Biogas may therefore provide for early adopters, but it lacks scalability to support a large customer base. As with fossil CHP microgrids, scalability might also become constrained jurisdictionally by air quality regulations. Developers could instead use fuel cells, which would emit only water vapor and carbon dioxide, but, at present, these command a significant cost premium compared to internal combustion engines and turbines.

Hydrogen

A fourth pathway, and the final discussed here, is the hydrogen pathway—the use of hydrogen fuel with engines and fuel cells. Although long-term visions for transitioning to hydrogen (Staffell et al. 2019) remain distant, the hydrogen pathway merits inclusion because it is the most promising in theory: it offers a path to reliable, scalable, pollution-free, carbon-free electricity. It is similar technologically to the CHP and biomethane pathways in its use of fueled generators and can further make use of fuel cells to provide pollution-free electricity. In either case, it is a zero-carbon pathway. Although today the majority of hydrogen is produced via steam methane reforming of natural gas, an underlying assumption of a future hydrogen system is production from renewable energy and electrolysis.

Hydrogen can be combusted in conventional spark-ignition engines and gas turbines with only small modification and efficiency losses and with emissions limited to water vapor and nitrous oxides (no carbon dioxide or carbon monoxide). A more efficient, cleaner, and ideal use of hydrogen is via electrochemical oxidation in fuel cells, where water vapor is the sole product. Hydrogen microgrids can provide high levels of reliability (i.e., they are not limited by low power densities) and are carbon-free.

They therefore match CHP and biomethane microgrids for provision of reliability and exceed them for emissions reductions, while the reverse is true in comparison with solar-storage microgrids.

Though the most promising, the hydrogen pathway faces the largest hurdles to implementation. Cost is a major issue in the short term, as hydrogen fuel commands a cost premium relative to fossil gas and the cheapest form of production remains steam methane reforming, which is carbon-intensive. Fuels cells also remain expensive relative to engines for prime power generation applications, although smaller deployments do exist today—notably over 3,000 hydrogen fuel cells that provide backup power for the telecommunications sector (Ma et al. 2019).

In the long term, the use of hydrogen places the largest demand of any pathway on new infrastructure outside the microgrid. Although hydrogen can be blended into the natural gas network in small amounts (concentrations larger than a few percent on a volume basis may embrittle pipeline materials), blending limits are a constraint and require that a future system of centralized hydrogen production use a dedicated transmission network. The development of hydrogen microgrids is therefore reliant on substantial build-out of infrastructure, including pipelines, production facilities, and storage sites—little of which exists today—and would require coordinated buy-in from the public, private industry, and government. Centralized hydrogen production also requires high penetrations of utility-scale renewable energy.

A ubiquitous hydrogen network for production and distribution, as envisioned in a future hydrogen economy, would make hydrogen microgrids fully scalable. In its absence, however, incremental scale-up is a challenge and the need today for distributed refueling networks limits applicability geographically. Experimentation with distributed refueling networks is ongoing, for example, with the U.S. Department of Energy's support for trigeneration fuel cells (U.S. Department of Energy 2016).

Conclusion

In the context of long-term climate stabilization goals, sustainable management and use of microgrids requires zero-carbon pathways for technology and fuel use. Competing objectives for reducing cost and emissions while increasing reliability preclude a single "best" pathway, while applicability and scalability should also be considered when planning long-term support for pathways. With respect to climate goals, the all-electric and hydrogen pathways are ideal because they make use of carbon- and pollution-free generation. The biomethane pathway is also ideal because it incents reductions in industrial and agriculture sector emissions, which must be brought to near-zero. Today, these pathways incur high costs, however, while low-cost options (diesel and CHP) are the least ideal. Aligning microgrid deployments with climate goals therefore requires action by policymakers. Carbon and pollution taxes, bans, or mandates for technology choice, and technology runtime limits are all policy options that can help encourage low-carbon investment choices. It is not the intent here to provide a ranking, however, and further work is needed to understand which options are best and why, or whether such policies are in fact sensible at this time given the nascency of the microgrid industry.

That microgrids be decarbonized to support climate stabilization goals assumes microgrids will play a non-negligible role in the future power system. Their future role remains uncertain, however, and will be guided by policy and market forces. There is another distinct possibility that microgrid deployments will remain limited in application, for example, to only the most critical of infrastructure. In such a scenario, it is possible that negative emissions in other sectors of the economy permit the continued use of fossil fuels in a relatively small number of strategic microgrids.

References

Asmus, P., Forni, A., and L. Vogel. 2017. Microgrid Analysis and Case Study Report. Navigant Consulting, Inc., Report No. CEC-500-2018-022.

Billinton, R., and R.N. Allan. 1996. *Reliability Evaluation of Power Systems*. Springer US.

Darrow, K., Tidball, R., Wang, J., and A. Hampson. 2017. Catalog of CHP Technologies. Environmental Protection Agency.

Flores, R. J., and J. Brouwer. 2018. Optimal design of a distributed energy resource system that economically reduces carbon emissions. *Applied Energy* 232:119–138.

Hanna, R., Ghonima, M., Kleissl, J., Tynan, G., and D.G. Victor. 2017. Evaluating business models for microgrids: Interactions of technology and policy. *Energy Policy* 103:47–61.

Ma, Z., Eichman, J., and J. Kurtz. 2017. Fuel Cell Backup Power Unit Configuration and Electricity Market Participation: A Feasibility Study. National Renewable Energy Laboratory, Report No. NREL/TP-5400-67408.

Ma, Z., Eichman, J., and J. Kurtz. 2019. Fuel cell backup power system for grid service and microgrid in telecommunication applications. *Journal of Energy Resources Technology* 141(6):062002.

Meier, A., S.J. Davis, D.G. Victor, K. Brown, L. McNeilly, M. Modera, R.Z. Pass, J. Sager, D. Weil, D. Auston, A. Abdulla, F. Bockmiller, W. Brase, J. Brouwer, C. Diamond, E. Dowey, J. Elliott, R. Eng, S. Kaffka, C.V. Kappel, M. Kloss, I Mezić, J. Morejohn, D. Phillips, E. Ritzinger, S. Weissman, J. Williams. 2018. University of California Strategies for Decarbonization: Replacing Natural Gas. UC TomKat Carbon Neutrality Project.

National Renewable Energy Laboratory. 2013. Biogas Potential in the United States. Report No. NREL/FS-6A20-60178.

Sheehy, P., and J. Rosenfeld. 2017. Potential to Develop Biomethane, Biogas, and Renewable Gas to Produce Electricity and Transportation Fuels in California. 2017 Joint Agency IEPR Workshop on Renewable Gas.

Smil, V., 2019. Distributed Generation and Megacities: Are Renewables the Answer? *IEEE Power and Energy Magazine* 17(2):37–41.

Staffell, I., Scamman, D., Abad, A.V., Balcombe, P., Dodds, P.E., Ekins, P., Shah, N., and K.R. Ward, 2019. The role of hydrogen and fuel cells in the global energy system. *Energy & Environmental Science* 12(2):463–491.

Tuttle, D.P., Gülen, G., Hebner, R., King, C.W., Spence, D.B., Andrade, J., Wible, J.A., Baldwick, R., and R. Duncan. 2016. The History and Evolution of the U.S. Electricity Industry. White Paper UTEI/2016-05-2. http://energy.utexas.edu/the-full-cost-of-electricity-fce.

U.S. Department of Energy Fuel Cell Technologies Office. 2016. Tri-Generation Success Story: World's First Tri-Gen Energy Station–Fountain Valley.

Willette, S., and P. Asmus. 2019. Microgrid Deployment Tracker 2Q19: Project and Trends in the Global Microgrid Market by Region, Segment, Business Model, and Top Vendors. Navigant Consulting, Inc.

28

Energy Conservation: Benefits

Eric A. Woodroof,
Wayne C. Turner,
and Steven D. Heinz

Executive Summary

"It's not the age ... it's the mileage" ...

It is logical that a car driven 25% less each year will last longer. The same is true for most energy-consuming equipment, such as lights, motors, and even digital equipment. By turning "off "energy-consuming equipment when it is not needed, an organization can find a financial jackpot, which extends beyond the utility budget. *It doesn't matter how energy-efficient an organization is, there are savings from turning equipment "off" when it is not needed.* Listed here are some "secret" benefits of energy conservation and these are benefits that can be attained without a negative impact on productivity.

Budgetary Improvements

1. Efficient Net Income: When energy is conserved, utility budgets are reduced. This is no secret, but what is noteworthy is that conservation savings impact a bottom line far more efficiently than many other investment initiatives. *For example: an energy conservation program that saves $100,000 in operating costs is equivalent to generating $1,000,000 in new revenue (assuming the organization has a 10% profit margin). It is more difficult to generate $1,000,000 in new revenue, and would require more marketing, infrastructure, etc. Thus, the energy conservation/efficiency program is an investment with less risk and quickly improves cash flow.*

2. Extended Equipment Lives: If assets are lasting longer (owing to reduced operation per year), replacements are less frequent, thereby reducing capital budget requirements. *For example, if a lighting system is operating 30% fewer hours per year, it could last up to 30% longer. A 15-year replacement policy could be changed to 20 years.* Further savings could result from considering that if equipment lasts longer, then staff/engineering/project management time is reduced for reviewing new equipment proposals, evaluating competing bids, overseeing installation efforts, coordinating invoices and payables with accounting.
3. Reduced Maintenance Costs: When equipment runs fewer hours per year, maintenance material/labor requirements are reduced. *For example, if maintenance on a motor is done on a "run hour" basis and there are less "run hours" per year, there should be fewer maintenance visits. Further, if the motor is part of a ventilation system, air filter replacements would occur less often, reducing material and labor costs. Predictive maintenance technologies can also assist in this strategy and reduce the cost per horsepower by 50%. (Ameritech)*
4. Reduced Risk to Energy Supply Price Spikes: *For example, if less energy is consumed; the operational budget is less vulnerable when electric/gas/heating oil prices hit their seasonal spikes. The avoided costs can be worth millions to a large organization.*

Beyond the large financial benefits mentioned here, there are many strategic benefits of energy conservation, which can significantly add to your organization's "jackpot."

Long-Term Strategic Benefits

1. Ability to Sell "Carbon Credits": Organizations can claim emissions reductions from energy conservation. When energy is saved, power plants do not have to produce as much electricity, thereby reducing "smoke stack" emissions. *Emission benefits from energy conservation can be expressed in terms of "equivalent trees planted," or "equivalent barrels of oil not consumed."* There are environmental markets where "emissions credits" (from energy conservation) can be sold, generating revenue for an organization. These markets are already liquid in Europe (and are motivated by carbon-related legislation). California and other states already require emissions reporting and reductions, and federal regulations are in process that will open the door to a similar trading environment in the United States.[1]
2. Enhanced Public Image: Organizations that conserve/manage energy (thereby reducing emissions) can differentiate themselves as "environmentally friendly" and "good" members of a community. This can have tremendous political, strategic, competitive, and morale-building value for organizational leaders. Many benefits (such as attracting and retaining better employees, faculty, students, clients, suppliers, etc.,) result from being the "leader" in your field. A recent study showed that 92% of young professionals want to work for an organization that is environmentally friendly.[2] Even stock prices of corporations have been proven to improve dramatically (21.33% on average) when energy management programs are announced.[3]
3. Reduced Risk to Environmental/Legal Costs: If assets are replaced less frequently, an organization will generate less waste and be less vulnerable to environmental regulations governing disposal. (Disposal of batteries and fluorescent lamps is already regulated in most states) Greater environmental regulations are inevitable and unforeseen legal costs can pose a significant expense and political risk.[4,5]

As will be shown in the example on the next page, benefits 2–7 represent a significant improvement (18%–50%) to the original savings estimates.

Climate Change and Its Effect on Energy Conservation Approaches

"The Writing Is on the Wall..."

The glaciers are melting and climate change is here. Climate change is a result of changes in the Earth's atmosphere. The growth of "greenhouse gases" between 10,000 years ago and the 1800s was approximately 1% for that period. Since the 1800s, greenhouse gases have increased 33%. Thus, it is logical that this growth is due to human-caused activities with the dawn of the industrial age.

— *Time Magazine-Special Report, December 2007. Also quoted by the UN Intergovernmental Panel on Climate Change-February, 2007 as well as the US EPA website: http://www.epa.gov.*

The data is compelling and creating change in consumer choices.[6] Consumers are becoming more "green-minded" in their purchases, especially young people and college students. Studies show that more consumers are choosing to reduce their "carbon footprint," and thereby are choosing products, companies, and colleges that are more environmentally friendly. Carbon offset trading growth is greater than 200%: "in June 2007, the Chicago Climate Exchange reported that in the past 6 months, it had already traded 11.8 $MtCO_2e$—more than had been traded in the entire year of 2006"[7] and 69% of consumers shop for brands aligned with a social cause.[8] Federal and state governments are introducing legislation that will mandate carbon emissions reporting and management.[9] In September 2006, Governor Schwarzenegger signed the California Global Warming Solutions Act, which mandates a 25% cut in emissions by 2020 and an 80% cut by 2050. In summary, the need for a "carbon diet" is driving activity in the energy-conservation industry.

The "Good" News ...

Companies, colleges, and governments are responding to this growing "green" consumer market and competitors are innovating to be the "environmental leaders" of their fields. Hewlett-Packard says that in 2004, $6 billion of new business depended on answers to customer questions about the company's environmental record—a 660% growth from 2002.[10] "Sustainability efforts protect our license to grow" said Wal-Mart CEO Lee Scott in 2005.[11] Energy efficiency/conservation is ranked by corporate executives as the no. 1 way to reduce emissions in a cost-effective manner.[12] Because buildings contribute approximately 43% of the carbon emissions in the USA, an opportunity exists to reduce a large part of these emissions and become "environmental heros."[13] In addition, organizations perceived as more "environmentally friendly" can recruit better faculty, students, suppliers, and employees.[14] Finally, the "secret benefits" (discussed in the Executive Summary) are increasing in value and importance. *An energy conservation program is more valuable today because the material, waste, labor, emissions, and risk savings are more valuable in today's economy.*

A Simple Example to Demonstrate the "Secret Benefits"

A lighting conservation measure will serve as the example, although similar calculations could be applied towards motor systems. Motors and lights consume the majority of electricity in a typical building.[15] Computers and other digital equipment are also worth mentioning, because they can consume considerable amounts of energy, and it is noted that "plug loads" (computers, printers, and other digital equipment) have increased significantly during the past 20 years.

For this example, consider a large school with 10,000 light fixtures. Through a variety of energy conservation measures, it is common to reduce consumption by 25%.[16] First, we will calculate the dollar savings from electricity conservation. Then, we will show the "secret benefits," which have impacts beyond the utility budgets. A spreadsheet will illustrate the total savings/benefits.

Benefit 1: Reduced Utility Budget from Lighting Conservation

Assume the fluorescent lights are relatively new and consume 60 watts per 2-lamp fixture and operate 5000 hr/yr. (This example uses a standard T-8 lighting system, although the energy conservation savings would be even greater with a less efficient lighting system, such as a T-12.) Our baseline energy consumption is:

$$= (5000\ \text{hr/yr})(0.060\ \text{kW/fix})(10{,}000\ \text{fix})$$

$$= 3{,}000{,}000\ \text{kWh/yr}$$

If the school pays approximately $.08/kwh, then the dollars spent on electricity for this lighting system:

$$= \$240{,}000/\text{yr}$$

Thus, a 25% reduction from the baseline usage would equal: 750,000 kWh/yr, or $60,000/yr in savings, *which goes immediately to the bottom line* and improves cash flow. Note that we will not count demand (kW) savings as the electrical load reduction would most likely occur during non-occupied hours (off-peak electrical rates). *However, it is not unusual for conservation programs to reduce both kWh and kW.* In addition to "direct dollar savings," there are tax rebates and credits available that can further improve the financial results from energy conservation/ efficiency programs/projects.

Benefit 2: The Value of Extended Equipment Lives (Reducing Capital Budgets)

If lights are used 25% less, the lighting system (ballasts) should last about 25% longer. Fluorescent fixture and wiring replacement costs are not included, as these components typically last longer than the ballast. We will address lamp life as a part of "maintenance costs" in Benefit 3. A lighting ballast is rated for 60,000 hr of operation. If the school operates the lights 5000 hr/yr, they would need to replace the ballasts at the 12th year and dispose of the old ballasts. If there are 5000 ballasts, each costing $25-$55 (material, installation, and disposal costs vary by geographic location), the replacement cost (minimum) at the 12th year would be:

$$= (\$25/\text{ballast})(5000\ \text{ballasts})$$

$$= \$125{,}000$$

Annualized replacement cost would be:

$$= (\$125{,}000)(1/12\ \text{yr})$$

$$= \$10{,}417/\text{yr}$$

With a use rate of only 3750 hr/yr (a 25% reduction), the ballasts should last 16 years. If replacement occurs at failure or based on run time, these savings automatically occur. If replacements are planned in advance, planners should adjust their schedules to insure that savings are captured from extended equipment lives (not replacing assets pre-maturely). This would reduce the annualized replacement cost to:

$$= (\$125{,}000)(1/16\ \text{yr})$$

$$= \$7813/\text{yr}$$

Thus, the Annualized Savings (calculated as the difference between the original replacement cost minus the reduced replacement cost) are:

$$= \$10,417/\text{yr} - \$7813/\text{yr}$$

$$= \$2604/\text{yr} \ (\text{at } \$25 \text{ per ballast})$$

Using the same equations, at $55/ballast, the annualized savings, (from replacing at 16 years instead of 12 years) would be:

$$= \$5729 \text{ per year}$$

Thus, due to extended equipment life, we have reduced the annualized replacement cost by a minimum of $2604/ yr to a maximum of $5729/yr.

Benefit 3: The Value of Reduced Maintenance Costs (Operating Expenses, Not Capital Replacements)

If the lights are used 25% less, the lamps should last about 25% longer. (Note that if lamps are turned "on" and "off " frequently at less than 3 hr intervals, the lamp's expected life will be reduced by approximately 25%, which would erode the savings in this category. Lamp life is rated at the factory by turning lamps on and off every 3 hr until they burn out. If the frequency of on/off cycling is less than 3 hr, lamp lives will decline by 25% on average. Therefore, turning a lamp off for longer periods is better than shorter periods. For example, it is better to find locations where you can turn off lamps for 5 hr out of 15 hr, instead of 1 min out of every 3 min, although the % time off is the same.) A typical fluorescent lamp life is 20,000 hr. With a use rate of 5000hr/yr, the school would need to replace lamps at the 4th year. If there are 10,000 lamps, each costing $3–$5 (material, installation, and disposal costs vary by location), the replacement cost (minimum) at the 4th year would be:

$$= (\$3/\text{lamp})(10,000 \text{ lamp}) = \$30,000$$

Annualized replacement cost would be $30,000/4 = $7500.

With a use rate of only 3750 hr, the lamps should last 5.3 years, thereby reducing the annualized replacement cost to:

$$= \$30,000/5.3 \text{ yr}$$

$$= \$5660/\text{yr}$$

Thus, Annualized Savings are:

$$= \$7500 - \$5660/\text{yr}$$

$$= \$1840/\text{yr} \ (\text{at } \$3/\text{lamp})$$

Using the same equations, at $5/lamp, the re-lamping cost would be $50,000 and the annualized savings from replacing at 5.3 years instead of at 4 years would be = $3066/yr.

Thus, due to extended lamp life, we have reduced the annualized maintenance cost by a minimum of $1840/yr to a maximum of $3066/yr.

Benefit 4: The Value of Reduced Risk to Energy Supply Price Spikes

Assume that on average, for one-quarter of the year, energy prices are 25%–50% higher ($.02–$.04 more per kWh) due to seasonal/supply spikes. Similar calculations could be used for systems that use natural gas, owing to its seasonal volatility.

If we are using less energy, we will pay less of a premium for the price spike. The avoided price spike premium is equal to:

$$=(\text{price premium})(\text{kWh saved})(\text{premium period})$$

$$=(\$.02/\text{kWh})(750{,}000\ \text{kWh/yr})(1/4)$$

$$=\$3750/\text{yr}$$

Using the same equations, a 50% price spike would represent an avoided premium worth:

$$=(\text{price premium})(\text{kWh saved})(\text{premium period})$$

$$=(\$.04/\text{kWh})(750{,}000\ \text{kWh/yr})(1/4)$$

$$=\$7500/\text{yr}$$

Thus, owing to reduced risk from price spikes, the avoided premiums are $3750–$7500/yr.

Benefit 5: The Value of Carbon Credits

According to the Environmental Protection Agency,[17] 1.37 lbs of CO_2 are created for every kWh burned. So, if we are saving 750,000 kWh/yr, the avoided power plant emissions would be equivalent to:

$$=(750{,}000\ \text{kWh saved})(1.37\ \text{Ibs of}\ CO_2/\text{kWh})$$

$$=1{,}027{,}500\ \text{Ibs of}\ CO_2\ \text{saved per year F}$$

Translating lbs to Metric Tons:

$$=(1{,}027{,}500\ CO_2)(.000454\ \text{Metric Tons/Ib})$$

$$=466.5\ \text{Metric Tons of}\ CO_2\ \text{saved per year}$$

These avoided power plant emissions could be claimed as "carbon credits" and sold to another party who wants to buy "carbon credits." An "aggregator" may be required to trade carbon credits in small quantities. Note that as of this printing, European prices for carbon credits are well over five times the price of carbon credits in the USA. The US carbon market is expected to follow Europe's lead as US regulations begin to take effect. Therefore, it is logical that the US prices will approach the European prices, which are currently at $34/metric ton.

Assuming for now a market price of $6 per metric ton, the additional revenue generated by selling the carbon credits would be:

$$=(466\ \text{Metric Tons of}\ CO_2/\text{yr})(\$6/\text{M}-\text{Ton})$$

$$=\$2799/\text{yr}$$

Using the same equations, at $30 per metric ton, the additional revenue generated by selling the carbon credits would be:

$$=\left(466 \text{ Metric Tons of } CO_2/\text{yr}\right)\left(\$30/\text{M}-\text{Ton}\right)$$

$$=\$13{,}980 \text{ per year}$$

Thus, due to the new carbon market, there is a possible additional revenue stream worth a minimum of $2799 to a maximum of $13,980 per year from selling carbon credits. In addition, as carbon prices go higher ... so does the value of this new revenue stream.

Benefit 6: The Value of Enhanced Public Image

Although calculation of this value is difficult and is not generalized here, it can be far greater than any of the benefits mentioned above. In today's "green-minded" economy, many organizations such as Patagonia, Google, GE, and Home Depot, have used "green" programs as a very effective marketing tool to differentiate themselves from the competition, achieve business objectives, secure and retain talent, improve productivity, and capture a greater market share.

The green-shaded area of Figure 1 shows the "equivalent environmental benefits" from avoided power plant emissions. These reductions/benefits can be published in various places to improve the organization's green image with employees, clients, students, suppliers, distributors, shareholders, and other groups relevant to the success of an organization.

Thus, due to energy conservation program, the school can claim environmental benefits equivalent to removing 1008 cars off the road, thereby improving the school's public image. Although not calculated here, the benefits of attracting better faculty, students, employees, etc., could far out-weigh all the benefit estimates in this entry.

See Figure 1 for additional expressions of environmental benefits.

PROFITABLEGREENSOLUTIONS

Complete Emissions Calculator

INSTRUCTIONS: Type in the kWh savings and see the emissions-environmental benefits in green-shaded areas. Insert your own $$ values for the Strategic Benefits in blue text.

Type the amount of electricity your program will save ⇨ 750,000 kWh/year

Emissions Reductions:	Annual Reductions	Reductions over 10 years
Conversion Factor: 1 kWh is worth 1.37 lbs of CO2 (Source: EPA 2006)		
GreenHouse Gas Reduction (in pounds of CO2)	1,027,500 lbs	10,275,000 lbs
or when converted to Metric Tons of CO2 >>>	466.5 Metric Tons	4,665 Metric Tons

Equivalent Environmental Benefits (mutually-exclusive):	Annual Reductions	Reductions over 10 years
Acid Rain Emission Reduction	5,625.0 lbs of SOx	56,250 lbs of SOx
Smog Emission Reductions	2,700.0 lbs of NOx	27,000 lbs of NOx
Barrels of Oil Not Consumed	1,085.0 Barrels	10,850 Barrels
Cars off the Road	100.8 Cars	1,008 Cars
Gallons of Gas not Consumed	53,130.3 Gallons	531,303 Gallons
Acres of pine trees reducing carbon	388.6 Acres	3,886 Acres

Strategic Benefits (quantifiable at site-specific level)	Annual Benefits	Benefits over 10 years
Annual Report to Shareholders,	?	?
Community Morale & "Green Image",	?	?
Productivity Improvements, Cost-Competitiveness	?	?
Avoided Future Capital Outlay	?	?
LEED Points, White Certificates, RECs	?	?
FREE Public Press (GREAT), Political/Strategic	?	?
Legal Risk Reduction, Avoided Penalties	?	?

FIGURE 1 Complete emissions calculator.

Example: The "Secret Benefits" of Energy Conservation

	Additional Benefits Estimates Min	Additional Benefits Estimates Max
Assumptions: Baseline Electricity Expenses from the Lighting System = $240,000 per year. A 25% savings via basic energy conservation measures would yield $60,000 in savings/year	$/Year	$/Year
Value of "Secret Benefits" (most exist outside the utility budget)		
Benefit #2: Extended Equipment Lives (Avoided Annual Capital Costs)	$2,604	$5,729
Benefit #3: Reduced Maintenance Costs (Avoided Operational Expenses)	$1,840	$3,066
Benefit #4: Reduced Risk to Energy Price Spikes (Avoided Premium Costs)	$3,750	$7,500
Benefit #5: Selling Carbon Credits (emissions reductions via energy conservation)	$2,799	$13,980
Total Additional Value from Quantifiable "Secret Benefits">>>	**$10,993**	**$30,275**
% Savings of Baseline Electricity Expenses ($240,000/year) of the Lighting System	***4.6%***	***12.6%***
% Savings Improvement from Original Estimate of $60,000/year in Savings	***18.3%***	***50.5%***

FIGURE 2 Secret benefits of energy conservation. (Note: Estimates are Conservative because Dollar Values for Benefits #6 and #7 were not included here).

Benefit 7: The Value of Reduced Risk of Environmental/Legal Costs

Although calculation of this value is also difficult and is not generalized here, it can be very significant. The risk is real, but unknown. This is demonstrated by the following environmental disasters that significantly crippled or destroyed the organizations deemed responsible:

- The Union Carbide accident in Bhopal
- Love Canal's hazardous waste
- Mercury poisoning at Alamogordo, NM

It is also interesting to note that Exxon's penalties and fees were four times the actual clean-up costs for the Valdez oil spill.

More relevant to this entry is that emissions regulations are quite likely to become a standard in the United States. Organizations that are implementing energy conservation programs will have a regulatory advantage over those that do not. Inaction could pose legal risks.

Thus, due to its energy conservation program, the school in this example can reduce its risk from unknown environmental and legal risks that may arise in the future.

Fi@@g. 2 summarizes the dollar value from the benefits mentioned in this entry. The approach and calculations for these benefits could be used as a guide to identify the "secret benefits" of other energy consuming systems, such as HVAC and motors, etc.

Conclusion

This entry has presented additional benefits from energy conservation. The example described an energy conservation project that was achieving a 25% reduction in electrical consumption from the lighting system. Beyond obvious energy savings, the "secret benefits" 2–5 yield additional value worth $10,993 to $30,275 per year. *In other words, if energy conservation saves 25% of a utility budget, the "secret benefits" are worth an additional 4.6%–12.6%.*

Looking at this in a different way, the "secret benefits" contribute additional value worth a minimum 18% improvement from the original estimated savings of $60,000 per year. *In other words, if we value the secret benefits as worth only an additional $10,993, this represents a minimum improvement of 18% to our energy savings of $60,000. In addition, there is a $4660 value improvement for each $10 rise in US carbon prices.*

Finally, all estimates in this entry only included the quantifiable "secret benefits" (benefits 2–5). Actual values could be much higher when accounting for enhanced public image and a reduction in legal and environmental risks (benefits 6 and 7).

We hope that this entry motivates additional action for energy conservation, dollar savings, and environmental benefits.

Acknowledgments

This entry originally appeared as "Overcoming Barriers to Approval for Energy and Green Projects" in *Strategic Planning for Energy and Environment,* Vol. 27, No. 3, 2008. Reprinted with permission from AEE/Fairmont Press.

References

1. Senate pases cap and trade legislation. New York Times, December 6, 2007.
2. How going green draws talent, cuts costs. Wall Street Journal, November 13, 2007.
3. Wingender, J.; Woodroof, E. When firms publicize energy management projects: Their stocks prices go up. Strategic Planning for Energy and the Environment 1997 (Summer Issue).
4. McCain-Lieberman Senate Proposal, 110th Congress, 8/2/2007.
5. Mayor Bloomberg calls for carbon tax. New York Times Articles, November 7, 2007.
6. If the entire world lived like North Americans, it would take three planet Earths to support the present world population (in 2006)—Source: Harvard Business Review on Business and the Environment.
7. State of the voluntary carbon market 2007, The Chicago Climate Exchange, July 18, 2007.
8. Survey data from United States Green Building Council.
9. Bingaman-Specter, Kerry-Snowe, Standers-Boxer Senate Proposals, 110th Congress, 8/2/2007.
10. Daniel, E. Green to Gold; Yale University Press, 2006.
11. Harvard Green Campus Initiative, UCSB Sustainability Program.
12. Getting Ahead of the Curve: Corporate Strategies That Address Climate Change. Pew Center on Global Climate Change, 2006.
13. Pew Center on Global Climate Change. The U.S. Electric Power Sector and Climate Change Mitigation and Towards a Climate Friendly Built Environment.
14. How going green draws talent, cuts costs. Wall Street Journal, November 13, 2007.
15. Association of Energy Engineers. Certified Energy Manager Program Workbook, 2008.
16. Gregerson, J. Cost effectiveness of commissioning 44 existing buildings, 5th National Conference on Building Commissioning.
17. EPA. Avoided power plant emissions calculations (updates October 2006. Available at http://www.ProfitableGreenSolutions.com ("Resources" tab).

29

Energy Conservation: Industrial Processes

Harvey E. Diamond

Introduction and Scope

Energy conservation is a broad subject with many applications in governmental, institutional, commercial, and industrial facilities, especially because energy costs have risen so high in the last few years and continue to rise even higher. Energy conservation in industrial processes may well be the most important application—not only due to the magnitude of the amount of potential energy and associated costs that can be saved, but also due to the potential positive environmental effects such as the reduction of greenhouse gases associated with many industrial processes and also due to the potential of the continued economic success of all of the industries that provide jobs for many people.

This entry will focus on energy conservation in industrial processes—where energy is used to manufacture products by performing work to alter feedstocks into finished products. The feedstocks may be agricultural, forest products, minerals, chemicals, petroleum, metals, plastics, glass, or parts from other industries. The finished products may be food, beverages, paper, wood building products, refined minerals, refined metals, sophisticated chemicals, gasoline, oil, refined petroleum products, metal products, plastic products, glass products, and assembled products of any kind.

This entry will distinguish industrial processes and the characteristics that differentiate them in order to provide insight into how to most effectively apply energy conservation within industries. The level of applied technology, the large amount of energy required in many cases to accomplish production, the extreme conditions (e.g., temperature, pressure, etc.,) that are frequently required, and the level of controls that are utilized in most cases to maintain process control will be addressed in this entry.

This entry will outline the analytical procedures needed to address energy conservation within industrial processes and will comment on general analytical techniques that will be helpful in analyzing energy consumption in industrial facilities.

Many of the main energy intensive processes, systems, and equipment used in industries to manufacture products will be identified and discussed in this entry and some common ways to save energy will be provided.

This entry will cover main energy intensive processes, systems, and equipment in a general format. If more in-depth instruction is needed for explanation of a particular industrial process, system, or type of equipment or regarding the analytical procedures required for a specific process, then the reader should refer to the many other articles included in this Encyclopedia of Energy Engineering and Technology, to references at the Association of Energy Engineers, to the references contained in this entry, and if further detail is still needed, then the reader should contact an applicable source of engineering or an equipment vendor who can provide in-depth technical assistance with a specific process, system, or type of equipment.

In addition to the analytical methods of energy conservation, managerial methods of energy conservation will be briefly discussed. The aspects of capital projects versus managerial and procedural projects will be discussed. The justification of managerial efforts in industrial processes will be presented.

Industrial Processes—Differentiation

Industrial processes require large amounts of energy, sometimes the highest level of technology, and often require very accurate process controls for process specifications, safety, and environmental considerations.

Industrial processes utilize an enormous amount of energy in order to produce the tons of production that are being produced within industrial facilities. Industrial processes utilize over one-third of the total energy consumed in America.[1] Consider the amount of energy that is required to melt all of the metals being manufactured, to vaporize all of the oil and gasoline being refined, to dry all of the finished products that are made wet, to heat all of the chemicals that react at a certain temperature, to vaporize all of the chemicals that must be distilled for purity, to vaporize all of the steam that is used to heat industrial processes, to mechanically form all of the metal objects that we use, etc.,—this list is too long to be fully included in this entry. This is an enormous amount of energy that produces all of the things that humans need and use—food, clothes, homes, appliances, cars, municipal facilities, buildings, roads, etc.

The level of technology required by current industrial processes is the highest in many cases and it is always at a high level in most industrial processes. Most industrial processes are utilizing technology that has been developed in the last 100 years or so, and consequently it has been further improved in the most recent years. Industrial processes most often utilize aspects of chemistry and physics in a precise manner in order to produce the sophisticated products that benefit people in our culture today. Very often, industrial processes require a very high or low temperature or pressure. Often they require a very precise and sophisticated chemistry and commonly they require highly technical designed mechanical processes. The application of electrical equipment and facilities in industrial processes is the highest level of technology for electrical power systems.

Industrial processes often require the highest level and accuracy of controls in order to produce products that meet product specifications, keep processes operating in a safe manner, and maintain environmental constraints. Due to each of these requirements or due to a combination of these requirements, the process controls for the processes within industrial facilities are often real-time Distributed Control Systems (DCSs), that are of the most sophisticated nature. A typical DCS for industrial processes functions to control process variables instantaneously on a real-time basis, whereby each process variable is being measured constantly during every increment of time and a control signal is being sent to the control element constantly on a real-time basis. The accuracy of a DCS in an industrial facility today is comparable to that of the guidance systems that took the first men to the moon

Most industrial facilities with DCS controls also utilize a process historian to store the value of most process variables within the facility for a certain increment or period of time. The stored values of these process variables are used for accounting purposes and technical studies to determine optimum operating conditions and maintenance activities.

Energy Conservation Analyses for Industrial Processes

In any industrial facility, the first analysis that should be performed for the purpose of energy conservation should be that of determining a balance of the energy consumed for each form of energy. This balance is used to determine how much energy is consumed by each unit, area, or division of the plant or possibly by major items of equipment that are consuming major portions of the energy consumed by the plant. This balance should be determined for each form of energy, whether it is for natural gas, electricity, coal, fuel oil, steam, etc., (see Table 1 for an example of an energy balance). It might be best that this determination not be called a balance (in that the numbers might not exactly come to a precise balance) but that it sufficiently quantifies the amount of energy consumed by each unit, area, or division of the plant. A better term for this determination might be an "Energy Consumption Allocation." The term balance is more usually applied to chemical and thermodynamic processes where heat and material balances are worked together mathematically to determine a calculated variable and the numbers have to exactly balance in order to arrive at the correct mathematical solution.

Once the amount of energy that is being actually consumed by each part of a plant has been determined, an energy consumption analysis should be performed for each item of energy consuming equipment and each major energy consuming facility in order to determine how much energy should realistically be consumed by each part of the plant. Notice that these calculations are called "realistic" as opposed to just theoretical because the object of these calculations is to determine as closely as possible how much actual energy each item of equipment or part of the plant should be consuming. By comparing these calculations with the actual energy consumption allocations mentioned above, it should be possible to obtain at least an initial indication of where energy is being wasted in a plant (see Table 2 for an example of an energy consumption analysis for a process feed heater). During the course of obtaining the values of process variables that are required to make these energy consumption analyses, it is

TABLE 1 Energy Balance

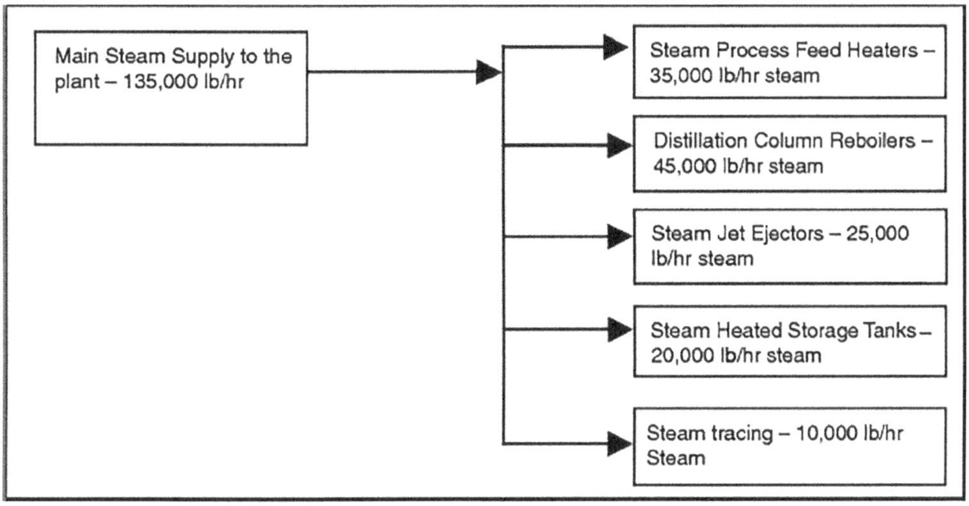

TABLE 2 Example of Energy Consumption Analysis: Process Feed HTR

• A process feed heater heats 1,199,520 lb/day of liquid feed material with a specific heat of 0.92 Btu/lb-°F, from 67 to 190°F. A realistic heater efficiency for this type of heater is determined to be 88%. The amount of realistic heat required for this heater is calculated to be: Q = 1,199,520 lb/day × 0.92 Btu/lb-°F × (190°F − 67°F) ÷ 0.88 = 154,247,367.3 Btu/day of realistic heat consumption.
• It is observed that this feed heater is consuming 186,143,720 Btu/day.
• This feed heater is being operated in a wasteful way and is wasting over 20% of its heat.

possible that indications will be observed of energy wastage due to the presence of an inordinate value of some process variable, such as too high or low of a temperature or pressure. When this type of indication is discovered, it usually also provides insight into what is operating in the wrong way to waste energy. There are numerous instances of energy wastage that can be discovered during these analyses, such as the inordinate manual control of a process, loose operational control of a process variable, or simply not shutting down a piece of equipment when it is not needed.

The analyses discussed in the above paragraph encompass all technical engineering science subjects, such as chemistry, thermal heat transfer, thermodynamics, fluid mechanics, mechanical mechanisms, and electrical engineering.

The next set of energy conservation analyses that should be performed are used to calculate the efficiencies of each item of equipment or facility to which efficiency calculations would be applicable, such as boilers, fired heaters, furnaces, dryers, calciners, and all other thermodynamic processes (efficiency calculations for boilers and other combustion equipment is available in the *Energy Management Handbook* by Wayne C. Turner and Steve Doty[2] and in the *Guide to Energy Management* by Barney L. Capehart, Wayne C. Turner, and William J. Kennedy)[3] and for electrical and mechanical equipment such as motors, pumps, compressors, vacuum pumps, and mechanical machinery. Once the actual efficiencies of any of the above have been determined, these numbers can be compared to the realistic efficiency for the type of equipment or facility that is prevalent throughout industry. These calculations and comparisons will also reveal wastage of energy and will frequently identify the causes of energy wastage and the possible issues to be corrected.

The next level of energy conservation analysis that may be performed is process analysis that can be conducted on a particular chemical, thermodynamic, thermal, fluid flow, mechanical, or electrical process. These analyses are usually performed by experienced engineers to examine the process itself and the process variables to determine if the process is being operated in the most effective and efficient manner. Here again, an indication will be provided as to whether or not energy is being wasted in the actual operation of the process. Chemical, thermo-dynamic, thermal, fluid flow, and other processes, as well as combinations of any of these processes can often require process simulation software such as PROMAX by Bryan Research and Engineering, Inc.,[4] in order to properly analyze these processes. The analysis of distillation columns, evaporators, and dryers can fall into this category. A good example would be the process analysis of a distillation column to determine if an effective and efficient level of reflux to the column and reboiler duty is being used.

Another analysis that has been very useful in the past few years in identifying energy conservation projects is Pinch analysis. This analysis is performed on thermodynamic and thermal processes in order to identify sources of energy within existing processes that can be used to supply heat for these processes instead of having to add additional heat to the entire process. The net effect is to reduce the amount of energy required for the overall process. The performance of a Pinch analysis on a particular process or facility will usually identify capital projects where revisions to the facility can be made to decrease the total amount of energy required. These are very often waste heat recovery projects. See "Use Pinch Analysis to Knock Down Capital Costs and Emissions" by Bodo Linnhoff, Chemical Engineering, August 1994[5] and "Pinch Technology: Basics for the Beginners."[6]

Main Industrial Energy Processes, Systems, and Equipment

This section provides an overview and a list of the more common energy intensive industrial processes that are used to manufacture products in industrial facilities. Most energy intensive industrial processes can be classified into about eight general process categories—process heating, melting, chemical reactions, distillation-fractionation, drying, cooling, mechanical processes, and electrical processes. These processes are intended to be the main general energy intensive processes that are most commonly used and to which variations are made by different industries in order to make a specific product. In this regard, this is an overview—these processes are often not the specific process but a general category to which variations can be made to achieve the specific process.

In the following paragraphs, each process will be discussed by addressing its description, what systems it utilizes, what products are generally made, how it uses energy, and frequent ways that energy can be saved.

Common energy consuming systems and equipment that work to manufacture products in industrial facilities are also listed below and discussed in the same manner as the main industrial processes, as they are also common to industrial facilities and are related to these processes.

Process Heating

- *Description.* The addition of heat to a target in order to raise its temperature. Temperatures can range from the hundreds to the thousands in industrial process heating.
- *Energy form.* Heat must be generated and transferred to the intended object or medium.
- *Energy unit.* Btu, calorie, joule, therm or watt-hour.
- *Examples.* The application of heat in order to heat feed materials, to heat chemical processes, to heat metals for forming, to heat materials for drying, to heat materials in a kiln or calciner, to heat minerals and metals for melting.
- *Applied systems.* Combustion systems, steam systems, thermal systems, hot oil systems, heating medium systems such as Dowtherm[7] or Paracymene,[8] and electrical resistance or induction heating systems.
- *Common equipment.* Boilers, furnaces, fired heaters, kilns, calciners, heat exchangers, waste heat recovery exchangers, preheaters, electrical resistance heaters, and electrical induction heaters.
- *Common energy conservation issues.* Keeping the heat targeted at the objective—proper insulation, seals on enclosures, eliminating leakage, and eliminating unwanted air infiltration. Control issues—maintaining sufficient control of the heating process, temperatures, and other process variables to avoid waste of heat. Management issues—shutting down and starting up heating processes at the proper times in order to avoid waste of heat and management of important process variables to reduce the amount of heat required to accomplish the proper process. Application of Pinch Technology—identify process areas where heat can be recovered, transferred, and utilized to reduce the overall process heat requirement. Waste heat recovery.

Melting and Fusion

- *Description.* The addition of heat or electrical arc energy at a high temperature in order to melt metals, minerals, or glass. The melting process involves more than just process heating, it involves fluid motion, fluid density equilibrium, chemical equilibrium, cohesion, and sometimes electromagnetic inductance. Reference: "The study showed that the fluid equations and the electromagnetic equations cannot be decoupled. This suggests that arc fluctuations are due to a combination of the interactions of the fluid and the electromagnetics, as well as the rapid change of the boundary conditions."[9]

- *Energy forms.* Heat at high temperatures or electrical arc energy in the form of high voltage and high current flowing in an arc.
- *Energy units.* Heat, Btu, calorie, joule, or therm.
- *Electrical arc. kWhrs.*
- *Examples.* Melting of ores in order to refine metals such as iron, aluminum, zinc, lead, copper, silver, etc. Melting of minerals in order to refine minerals such as silica compounds, glass, calcium compounds, potassium alum, etc.
- *Applied systems.* Combustion systems, chemical reactions, and electrical systems.
- *Common equipment.* Blast furnaces, arc furnaces, electrical resistance heaters, and electrical induction heaters.
- *Common energy conservation issues.* Pre-condition of feed material—moisture content, temperature, etc. Feed method—efficiency of melting process effected by the feed method, feed combinations, and feed timing. Control of electromagnetics during melting and use of magnetic fields during separation. Over-heating can waste energy without yielding positive process results. Heat losses are due to poor insulation, the failure of seals, or lack of shielding or enclosure.

Chemical Reactions

- *Description.* Chemicals react to form a desired chemical, to remove an undesired chemical, or to break out a desired chemical. The chemical reaction can involve heat, electrolysis, catalysts, and fluid flow energy.
- *Energy forms.* Heat, electrolysis, and fluid flow.
- *Energy units.* Heat, Btu, calorie, joule, or therm.
- *Electrolysis. kWhrs.*
- *Fluid flow.* ft-lbs or kg-m.
- *Examples.* Reaction of chemical feed stocks into complex chemicals, petrochemical monomers into polymers, the oxidation of chemicals for removal, dissolving of chemicals to remove them, reaction of chemicals with other chemicals to remove them, the reaction of lignin with reactants in order to remove it from cellulose, the electroplating of metals out of solution to refine them.
- *Applied systems.* Feed systems, catalysts systems, heating systems, cooling systems, vacuum systems, run-down systems, separation systems, filtering systems, and electroplating systems.
- *Common equipment.* Reactors, digesters, kilns, calciners, smelters, roasters, feed heaters, chillers, pressure vessels, tanks, agitators, mixers, filters, electrolytic cells.
- *Common energy conservation issues.* Close control of heating and cooling for chemical reactions. Close control of all reaction process variables—balance of all constituents, amount of catalyst, proper timing on addition of all components. Management of feedstocks, catalysts, and run-down systems for proper timing and correct balance for highest efficiency. Pinch analysis of feed heating, run-down products, cooling system, etc. Conservation of heating and cooling—proper insulation, sealing, and air infiltration. Waste heat recovery.

Distillation-Fractionation

- *Description.* A thermo-dynamic and fluid flow equilibrium process where components of a mixture can be separated from the mix due to the fact that each component possesses a different flash point.
- *Energy form.* Heat and fluid flow.
- *Energy units.* Heat, Btu, calorie, joule, or therm.
- *Fluid flow.* Ft-Lbs or KG-M.
- *Examples.* Distillation-fractionation of hydrocarbons in oil and gas refineries and chemical plants. Distillation of heavy hydrocarbons in gas processing plants where natural gas is processed to remove water and heavy hydrocarbons.

- *Applied systems.* Feed heating systems, over-head condensing systems, reflux systems, reboil systems, vacuum systems.
- *Common equipment.* Distillation columns or towers, over-head condenser heat exchangers and accumulators—vessels, reflux pumps, reboiler heat exchangers, feed pumps, feed—effluent heat exchangers, vacuum steam jet ejectors.
- *Common energy conservation issues.* Feed temperatures, reflux ratios, reboiler duty. Close control on pressures, temperatures, feed rates, reflux rates, and reboil duty. Management of overall operation timing—running only when producing properly. Concurrent use of vacuum systems—only when needed. Pinch analysis for feed and effluent streams and any process cooling systems. Proper insulation and elimination of lost heat for fired heater reboilers.

Drying

- *Description.* The use of heat and fluid flow to remove water or other chemical components in order to form a more solid product.
- *Energy forms.* Heat and fluid flow.
- *Energy units.* Heat, Btu, calorie, joule, or therm.
- *Fluid flow.* Ft-Lbs or KG-M.
- *Examples.* Spray dryers that dry foods, sugar, fertilizers, minerals, solid components, and chemical products. Rotary dryers that dry various loose materials. Line dryers that dry boards, tiles, paper products, fiberglass products, etc. Other dryers that dry all kinds of products by flowing heated air over finished products in an enclosure.
- *Applied systems.* Combustion systems, steam systems, thermal heating systems, cyclone systems, air filter systems, incinerator systems, Regenerative Thermal Oxidizer (RTO) systems.
- *Common equipment.* Spray dyers, spray nozzles, natural gas heaters, steam heaters, electrical heaters, blowers, fans, conveyors, belts, ducts, dampers.
- *Common energy conservation issues.* Efficient drying process for the components being eliminated. Proper amount of air flowing through dryer for drying. Proper insulation, seals, and elimination of lost heat due to infiltration. Waste heat recovery.

Process Cooling

- *Description.* The removal of heat by a cooling medium such as cooling water, chilled water, ambient air, or direct refrigerant expansion.
- *Energy form. Heat.*
- *Energy unit.* Btu, calorie, joule, or therm.
- *Examples.* Cooling water or chilled water circulated through cooling heat exchanger, an air cooled heat exchanger, or a direct expansion evaporator that cools air for process use.
- *Applied systems.* Cooling water systems, chilled water systems, refrigerant systems, thermal systems.
- *Common equipment.* Cooling towers, pumps, chillers, refrigeration compressors, condensers, evaporators, heat exchangers.
- *Common energy conservation issues.* Use evaporative cooling as much as possible. Keep chillers properly loaded. Restrict chilled water flow rates to where 10°F temperature difference is maintained for chilled water. Limit cooling water pumps to the proper level of flow and operation. Apply Pinch analysis to achieve most efficient overall cooling configuration. Proper insulation, seals, and elimination of air infiltration.

Mechanical Processes

- *Description.* Physical activities that involve force and motion that produce finished products. Physical activities can be discrete or can be by virtue of fluid motion.
- *Energy form.* Physical work.
- *Energy unit.* Ft-Lbs or KG-M.
- *Examples.* Machining of metals, plastics, wood, etc.; forming or rolling or pressing of metals, minerals, plastics, etc.; assembly of parts into products; pumping of slurries thru screens or filters for separation; cyclone separation of solids from fluids; pneumatic conveyance systems that remove and convey materials or products and separate out solids with screens or filters.
- *Applied systems.* Machinery, electrical motors, hydraulic systems, compressed air systems, forced draft or induced draft conveyance systems, steam systems, fluid flow systems.
- *Common equipment.* Motors, engines, turbines, belts, chains, mechanical shafts, bearings, conveyors, pumps, compressors, blowers, fans, dampers, agitators, mixers, presses, moulds, rolls, pistons, centrifuges, cyclones, screens, filters, filter presses, etc.
- *Common energy conservation issues.* Equipment efficiencies, lubrication, belt slippage, hydraulic system efficiency, compressed air system efficiency. Control of process variables. Application of variable speed drives and variable frequency drives. Management of system and equipment run times.

Electrical Processes

- *Description.* The application of voltage, current, and electromagnetic fields in order to produce products.
- *Energy form.* Voltage-current over time; electromagnetic fields under motion over time.
- *Energy units. KWh.*
- *Examples.* Arc welding, arc melting, electrolytic deposition, electrolytic fission, induction heating.
- *Applied systems.* Power generator systems, power transmission systems, amplifier systems, rectifier systems, inverter systems, battery systems, magnetic systems, electrolytic systems, electronic systems.
- *Common equipment.* Generators, transformers, relays, switches, breakers, fuses, plates, electrolytic cells, motors, capacitors, coils, rectifiers, inverters, batteries.
- *Common energy conservation issues.* Proper voltage and current levels, time intervals for processes, electromagnetic interference, hysteresis, power factor, phase balance, proper insulation, grounding. Infrared scanning of all switchgear and inter-connections.

Combustion Systems

Combustion systems are found in almost all industries in boilers, furnaces, fired heaters, kilns, calciners, roasters, etc. Combustion efficiency is most usually a prime source of energy savings.

Boilers and Steam Systems

Boilers and steam systems may well be the most widely applied system for supplying process heat to industrial processes. "Over 45% of all the fuel burned by U.S. manufacturers is consumed to raise steam."[10] Boiler efficiencies, boiler balances (when more than one boiler is used), and steam system issues are usually a prime source of energy savings in industrial facilities.

Flare Systems and Incinerator Systems

Flare and incinerator systems are used in many industrial facilities to dispose of organic chemicals and to oxidize volatile organic compounds. Proper control of flares and incinerators is an issue that should always be reviewed for energy savings.

Vacuum Systems

Vacuum systems are used to evaporate water or other solvents from products and for pulling water from products in a mechanical fashion. Vacuum systems are also used to evacuate hydrocarbon components in the petroleum refining process. Vacuum systems are frequently used in the chemical industry to evacuate chemical solvents or other components from a chemical process. Steam jet ejectors and liquid ring vacuum pumps are commonly used to pull vacuums within these systems. The efficiencies of the ejectors and the liquid ring vacuum pumps can be a source of energy savings as well as the management of vacuum system application to production processes. Pneumatic conveyance systems that utilize a fan or blower to create a low-level vacuum are sometimes used to withdraw materials or products from a process and separate the matter within a screen or filter. For large conveyance systems, the efficiencies of the equipment and the management of their operation can be a source of energy savings.

Furnaces, Fired Heaters, Kilns, Calciners

The above comments on combustion systems are applicable to these equipment items and additional energy savings issues can be found relative to them.

Centrifugal Pumps

Centrifugal pumps are used widely in industries. The flow rate being pumped is a primary determining factor for the amount of power being consumed and it is sometimes higher than required. Good control of the pumping rate is an important factor in saving energy in centrifugal pumps. The application of variable frequency drives to the motor drivers can be a good energy saving solution for this issue.

Fans and Blowers

The flow rate for fans and blowers is analogous to the pumping rate above for centrifugal pumps. Good control of the flow rate and the possible application of Variable Frequency Drive (VFD) apply here as well for fans and blowers.

Centrifugal Compressors

Compressors are used widely in industry. The above discussions of flow rates, control of flow rates, and application of VFDs apply here as well. Centrifugal compressors frequently will have a recycle flow that is controlled in order to prevent the compressor from surging. Close control of this recycle flow at its minimum level is very important for compressor efficiency.

Liquid Ring Vacuum Pumps

As mentioned above in several places, liquid ring vacuum pumps are used widely in industry. The amount of sealing liquid that is recycled to the pump and the temperature of the sealing liquid are important determinates of the efficiency of the Liquid Ring Vacuum Pump (LRVP).

The above overview and list of industrial processes, systems, and equipment has been general in nature due to the limitations of this entry. Greater and more specific familiarity with each of these industrial energy intensive processes, systems, and equipment will yield greater applicable and effective insight into ways to save energy related to each of these items.

Capital Projects versus Improved Procedures

Energy conservation effort applied in industrial facilities can identify capital projects whereby the facilities can be changed in order to achieve greater overall energy efficiency or the efforts can identify changes to in day-to-day operating and maintenance procedures that can reduce waste of energy and also improve the overall efficiency of the facility. Frequently, energy-saving procedural changes to day-to-day operations and maintenance activities within an industrial plant can be identified by taking and recording operating data once the processes, systems, and equipment have been studied and analyzed for energy consumption. Procedural changes to operations and maintenance within an industrial plant can often amount to low costs or possibly no costs to the facility. This aspect of energy conservation is often overlooked by highly technical personnel that have worked hard to design industrial facilities because they have technically designed the facility very well for energy consumption considerations and the more mundane activities related to day-to-day operation and maintenance tend to not register in their highly technical perspective. None-the-less, a considerable amount of energy can usually be saved within most industrial processes, systems, and equipment due to changes in the way they are operated and maintained. A general tendency within industrial plants is that operations will often operate the processes and systems at a point that provides a comfortable separation between an operating variable and its limitation in order to understandably ensure no upsets occur within the process or system. However, with the cost of energy being what it is today, it is frequently found that a significant amount of energy can be saved by operating processes and systems more tightly and efficiently, even though it may require more attention, increased control, and the monitoring of process variables.

Effective Energy Management Systems

Another aspect of energy conservation that can be very productive in saving energy within industrial processes is that of an effective energy management system. An effective energy management system is comprised of operational and maintenance managers functioning in conjunction with an accurate and concurrent data collection system in order to eliminate waste and improve overall efficiency of industrial processes. It is not possible to manage any activity unless the activity is being properly monitored and measured with key performance metrics (KPMs). The data collection system part of an effective energy management system within any industrial facility provides the accurate and concurrent measurement data (KPMs) that is required in order to identify actions that are needed to eliminate waste of energy and improve overall efficiency of the facility. An effective energy management system is first built upon acquiring total knowledge of the facility down into every level of operation and maintenance of the facility. Such a level of thoroughness and complete analysis of energy consumption within a facility is sometimes referred to as *Total Quality Energy* management.[11] Once an effective energy management system has been established and is effectively controlling energy consumption of an industrial facility, it should be maintained, in effect, so that it will continue to monitor KPMs to maintain energy conservation for the facility. An effective energy management system within an industrial manufacturing facility can eliminate as much as two to three percent of the energy costs by eliminating waste of energy on a day-to-day operational and maintenance basis. In most industrial facilities, this level of cost reduction is significant and will justify an effective energy management system.

Current Need for Greater Energy Conservation in Industry

With the present cost of all forms of energy today, it would certainly seem logical that all of industry would be seeking greater energy conservation efforts within their facilities. Unfortunately, many corporate industrial managers are not aware of the true potential of conserving energy within their processes and facilities. Greater awareness of the ability to conserve energy on the basis of increased efficiencies of processes, systems, and equipment is needed; and also due to the application of an effective energy management system. For the good of society and environment, corporate industrial managers should be more open to the possibility of the improvement of industry that will work to sustain their business and improve the world that we live in. This is in opposition to corporate political thinking, which does not want to consider making changes and wants no one to interfere with their present activities. Human beings should be willing to examine themselves and make changes that will make things better. The same outlook should be applied to businesses and industry in order to make things better. Greater management support is needed in industry today to accomplish greater and very much needed increased energy conservation.

Current Application of Increased Energy Management

With the recent technological advancements that have been made in digital computer and communications systems, data collection systems can be implemented in industrial facilities in a much more cost effective manner. Wireless communication systems for metering and data collection systems have advanced dramatically in the last few years and network-based computer communication has enabled whole new systems for measurement and control. With all of these new fields of configuration for data collection systems, with the increased technology, and with the lower costs to accomplish data collection systems, it is now possible to apply energy management systems to industry today with much greater applicability. Hopefully this will be recognized and result in greater applications of effective energy management systems.

From recent observations, it appears that most of industry today is a candidate for improved and more effective energy management systems. In conjunction with the increased technology and lower cost potentials, it seems that there is a definite match between supply and need for the application of increased energy management systems.

Summary

Industrial processes have commonality in processes, systems, and equipment. There are logical and systematic analyses that can be performed in industrial processes that can identify ways to save energy. Effective energy management systems are needed in industry today and there are great possibilities to save energy in industrial processes. Energy can be conserved in industrial processes by analyses that will improve efficiencies, by implementation of procedures that eliminate waste, and by application of an effective energy management system.

References

1. U.S. Department of Energy–Energy Efficiency and Renewable Energy, available at http://www.eere.energy.gov/EE/industry.html (accessed on 2006).
2. Turner, W.C.; Doty, S. *Energy Management Handbook,* 6th Ed.; Fairmont Press: Lilburn, GA, 2005.
3. Capehart, B.L.; Turner, W.C.; Kennedy, W.J. *Guide to Energy Management,* 5th Ed.; Fairmont Press: Lilburn, GA, 2004.
4. Bryan Research and Engineering, Inc. PROMAX; BRE, Bryan, TX; available at http://www.bre.com (accessed on 2006).

5. Linnhoff, B. *Use pinch analysis to knock down capital costs and emissions.* Chemical Engineering August 1994 http://www.che.com.
6. Solar Places Technology. *Pinch Technology: Basics for the Beginners,* available at http://www.solarplaces.org/pinchtech.pdf (accessed on 2006).
7. Dowtherm, Dow Chemical http://www.dow.com/heattrans/index.html (accessed on 2006).
8. Paracymene, Orcas International, Flanders, NJ 07836 available at http://www.orcas-intl.com (accessed on 2006).
9. King, P.E. *Magnetohydrodynamics in Electric Arc Furnace Steelmaking.* Report of Investigations 9320; United States Department of the Interior, Bureau of Mines, available at http://www.doi.gov/pfm/ar4bom.html (accessed on 2006).
10. U.S. Department of Energy-Energy Efficiency and Renewable Energy, available at http://www.eere.energy.gov/EE/industry.html the common ways (accessed on 2006).
11. Energy Management International, Inc. Total Quality Energy, available at http://www.wesaveenergy.com (accessed on 2006).

30

Energy Master Planning

Fredric S. Goldner

Introduction

Developing an energy management program or more broadly energy master planning (EMP) is the process of transitioning an organization's culture from the traditional "fixed cost, line item" view of energy to one in which energy is recognized as the opportunity and risk that it has become.

An EMP can guide an organization in longer-range planning of energy cost reduction and control as part of their facility maintenance, management, and design. An EMP can even lead the energy budget to be recognized as a potential profit center and source of opportunity rather than just another business expense. An EMP moves beyond the confines of traditional engineering to include energy procurement, energy-related equipment purchasing, measurement and verification (M&V), staffing and training, communications, and setting energy consumption targets and tracking/feedback loop systems. The long-term perspective goes beyond simply cutting last year's energy use. It makes energy awareness part of the everyday operation and "mindset" of the organization.

If you're thinking this doesn't apply to your firm or clients because you're too small, think again. This approach works for an organization as small as a single site to an owner with half a dozen small buildings to Fortune 100 companies. The effort and level of detail vary, respectively, but the approach is basically the same. The good news is that there are resources available to assist professionals and the organizations they serve to understand the EMP process and get started on this path.

To be successful an organization must treat energy in the same business-like manner that they do all other major expenses, such as labor and materials. "If you can measure it, you can manage it"[1] is the catch phrase of Paul Allen, the energy manager at Disney World, who has been instrumental in implementing one of the most successful EMPs in the country. "Energy is a competitive opportunity

… Winners manage it effectively!" is the driving force of another highly successful program at Owens Corning. Many other organizations throughout this country and across the globe have recognized that to achieve significant and sustained energy cost control, organization needs to make energy management an integrated part of their business/operations. There are various ways to pursue the process, but a key requirement is an interdisciplinary mix of engineering/technical, behavioral, and organizational or management components. The EMP must be integrated into the basic business operations.

Unexpected Benefits

One of the most potent driving factors in many organizations' efforts to address energy issues is increased profitability that can be realized through reduced/optimized energy expenditures. Beyond the "bottom line" impacts, an EMP can also provide an organization with a more secure energy supply, reduced downtime of systems, improved equipment availability, reduction in maintenance costs/ premature system replacement expenditures, and overall productivity gains. Additional benefits that have been documented include quality-of-life improvements, enhanced product quality, better operational safety, reduced raw material waste in industrial plants, and increased rentability in commercial facilities. An often overlooked outcome of an EMP is the reduction of emissions, among other environmental impacts that help organizations become perceived as better corporate citizens.

As more companies move toward an integrated corporate strategy that links environmental, economic, and social considerations, the results of an EMP can be used to considerable public relations advantage. Ratings in one of the sustainability indices and publication of an annual sustainability report (using, for example, the Global Reporting Initiative guidelines) can give an organization a higher standing in the business community and can result in a higher level of trust by stockholders.

A Process for Optimization

Though energy prices are volatile, and energy security is often far from reliable, facilities now face leaner operating funds and increased directives to do more with less. Optimizing a facility's operations budget frequently means cutting energy costs. But how do you do so without cutting occupant comfort or productivity? How do you know where to start and what steps to take? And how do you persuade upper management that energy costs can be controlled?

While the general goal of an EMP is the same as that of conventional energy management, the two disciplines are far from identical. Traditional energy management, which is technically oriented, is essentially centered around the boiler or mechanical room. Energy master planning, on the other hand, is a business management procedure for commercial, institutional, and industrial operations. With this approach, it is not enough simply to manage installations. The process involves:

- Developing strategies
- Creating processes to fulfill those strategies
- Identifying barriers and finding procedures to overcome them
- Creating accountability
- Providing feedback loops to monitor and report progress

Clarification of the terminology for these disciplines is important, as the terms mean different things to different people. In other English-speaking nations, for example, the defining feature of "energy management" is the emphasis on integration with business practices to analyze, manage, and control energy. In the U.S., however, "energy management" has traditionally referred to developing technical and operational measures involving equipment handled by facility managers—not processes such as energy procurement and business planning usually handled by purchasing agents, production personnel, and corporate economists. A typical U.S. energy manager's responsibility rarely extends much beyond utility bill analysis, an occasional energy audit, or managing installation of system upgrades.

Though an EMP encompasses traditional efforts to cut energy costs, it also includes many steps not usually taken under standard energy management. Rather than being just equipment-oriented, an EMP starts long before a comprehensive energy audit and extends beyond commissioning of new systems. An EMP may be thought of as a road map to savings that starts before and continues after energy- efficiency measures are involved. Why do you need such a map? Because you can't get there if you don't know where "there" is. How many of us are willing to undertake a trip that will have costs and risks (like any business or personal decision) if we don't know where we are going? A map makes clear not just your final destination, but also how to get there—and it leaves no question about the starting point.

An EMP is a process to organize and improve your existing energy-related resources and capabilities. Resources, in this case, include standard operating procedures, institutional memory, and actual records (such as energy bills, plans and blueprints, and energy contracts). Capabilities include facilities staff familiar with mechanical room equipment, consultants for energy costs or usage, energy cost accounting and management systems, and meters and software that monitor them. Once organized and integrated, these resources and capabilities become powerful tools for managing energy and producing savings.

Today's Practices Found Wanting

Current thinking about managing energy often falls short of the EMP perspective. Today's energy management frequently reflects a short-term crisis mentality: A facility manager or energy manager concentrates on whatever immediate 'fires' have to be put out at his facility. By contrast, with an EMP mindset, the energy manager might first look to increase the efficiency of systems already in place and then move ahead to lay a solid foundation for improving performance via tight energy specs and training.

Current thinking is also often characterized by piecemeal 'solutions' that lead to short-sighted component replacement. When equipment breaks down, it gets replaced without anyone's asking whether this is the best option, long-term. Typically, business thinks in a quarterly mindset because of the short budget cycles of our economic system. That kind of "right now and right here" viewpoint creates situations in which life-cycle thinking is not possible because potentially higher initial costs are visible, but potential benefits tend to be invisible. As a result, when first cost becomes the main criterion for purchasing, such a focus distorts planning and decision-making. Too often, facilities choose the cheapest solution, based on the current quarter's budget, without realizing it could cost them more later.

A critical difference between energy master planning and conventional energy management is an orientation toward the future. With an EMP, you don't just look to increase the efficiency of systems already in place. Instead, you plan for new or changed loads based on your detailed information about the facility's long-term business strategy and projected growth.

Energy master planning deals with a longer timeframe than just simple payback periods. For example, typical financial constraints for a commercial building upgrade often dictate a 2- to 3-year timeframe. Energy master planning, however, looks deeper than simple payback and goes beyond merely reducing energy use. Therefore, an energy professional with an EMP considers life-cycle costs and views long-range planning of energy cost minimization/ optimization as part of overall facility maintenance, management, and design. To sustain the savings over time, an EMP calls for hiring an energy manager/ coordinator and setting up an energy team. But who is this energy leader? Not simply the facility manager wearing yet another hat, but rather, a highly trained and, often, certified specialist with sufficient acumen and expertise to understand, handle, and maintain whatever new energy systems and practices are to be put in place. Willingness to identify (from within the organization) or hire an individual with the required qualifications and to define his or her responsibility as managing energy rather than managing the facility is a requisite indication of senior management's true commitment to energy master planning. The policy guidance needs to come from senior management levels to convey "buy-in" throughout the management structure of the organization.

Improving Business as Usual

Energy master planning is one way of creating a new norm of "business as usual," taking its cues from time-tested business management practices. For example, rather than carrying out upgrade projects with a defined start and end point, an energy manager uses processes that continue to turn up new sources of savings. This requires a level of creativity for identifying and capturing new opportunities. If a chiller replacement is needed in Building A, for instance, a better approach might be to expand the capacity of the existing central chilled water plant in Building B and run piping from Building B to Building A.

Without an EMP, the Purchasing department often buys equipment and energy, rather than the Facilities department. And Facilities is so busy handling emergencies that they have little interaction with other departments in the organization, let alone industry groups or other end users. An EMP avoids either departmental isolation and turf wars by including representatives from such departments on the energy team. The team consists of more than an energy manager, facilities personnel, and design and construction specialists. To be effective, it needs to incorporate representatives from every department in the organization impacted by energy. This may mean including purchasing, accounting, engineering, environmental affairs, maintenance, legal, health and safety, corporate relations, human resources and training, public relations and marketing, and members from the rank and file (hourly employees).

Energy master planning is a significant challenge. It often rejects the status quo and may question existing components of an organizational culture that do not support energy master planning. If you have always done something one way, you don't necessarily have to perpetuate what could be a costly mistake. For example, using outdated specs (e.g., calling for T12 lamps instead of T8) allows inefficiency to continue.

Origins of the American EMP Approach

A common set of energy master planning definitions and processes has taken root in English-speaking countries other than the U.S. In the United Kingdom, the energy master planning concept has been practiced for well over a decade and vigorously promoted by the government's Action Energy program. It has been so successful that Canada, New Zealand, and Australia have each adapted the process to their own conditions.

In the U.S., there are a few recent models cover some, but not all, aspects of energy master planning management system for energy (MSE) 2000 is a specialized quality improvement standard (like ISO 14000), developed by Georgia Tech. It's an American National Standards Institute (ANSI)-approved management system for energy that covers all sectors, not only buildings. Georgia Tech offers a certificate program to train energy professionals in this standard. The Association of Energy Engineers delivers the Developing an Energy Management Master Plan and Creating a Sustainable Energy Plan Workshops to both 'real' and 'virtual' end users. Live presentations and online seminars present an energy master planning approach with strong emphasis on integration with business strategy.

Steps to an Energy Master Plan

To fill this gap in the U.S., the Energy Master Planning Institute (EMPI) was established and has developed a set of steps that lay out the process for applying energy master planning to a commercial, institutional, or industrial facility (See Figure 1). This model, which builds on accepted international approaches, offers US organizations a broad and integrated business approach for managing energy that is both strategic and sustainable.

The steps presented in Figure 1 appear sequential, but built into the energy master planning process are series of feedback loops, evident in Figure 1. These should not be overlooked, as they guarantee the viability of the process and offer many points for input from internal and external stakeholders, from the Chief Executive Officer (CEO) to boiler room personnel, and from the local community to organizational peers across the country.

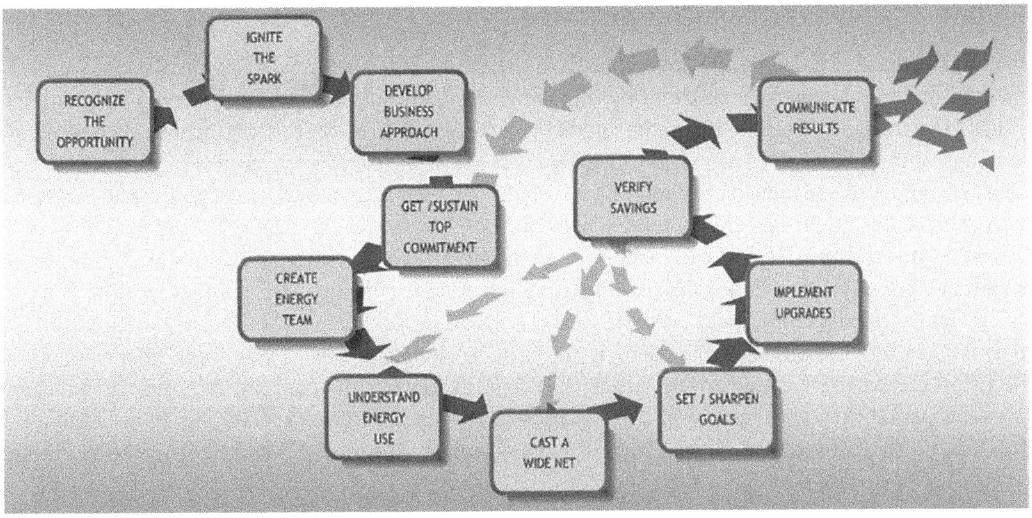

FIGURE 1 Energy Master Planning Institute (EMPI) steps to an energy master plan (EMP). Copyright 2004. Energy Master Planning Institute. All rights reserved.

Recognize the Opportunity

This is where the process starts—becoming aware of the facility's major energy-related opportunities and challenges. Whether it's the facility manager or the energy manager, whoever takes the initial action must define the opportunities and challenges succinctly so they can be clearly communicated. When the leader creates and implements an EMP, that plan can actually generate a revenue stream and gain recognition for the leader's contribution to the company's bottom line. By integrating energy concerns into the overall corporate business strategy, the energy budget will come to be seen as a potential profit center and source of opportunity—and not just an uncontrollable expense.

Ignite the Spark

Since top executives in an organization are the ultimate decision-makers, particularly concerning funding, it's critical to spark their attention early in the process. It's unlikely that the facility manager (or the energy manager, if there is one), has direct access to upper management, so the right individual to make the pitch to the CEO or Chief Financial Officer CFO must be identified. This could be someone along your management chain, or a consultant or board member or a senior officer in another department who can sell an idea at the top. To find the right person, it's important to understand the decision-making structure of your organization as well as the vision, mission, and long-term business plan.

Develop a Business Approach

To get access to senior executives and persuade them to listen to a new idea, you must speak to them in their own language—dollars/ft^2/year savings, not kWh. Since most executives have not been introduced to the bottom-line value of energy master planning, your task is to change management culture so executives no longer view energy as an uncontrollable expense. Part of the marketing message is that managing energy is no different from tracking, controlling, and accounting for the costs of raw materials, IT, personnel, safety, or the corporate fleet.

Obtain and Sustain Top Commitment

Without question, this step is the most difficult in the energy master planning process. Serious commitment from senior management means providing ongoing financial resources and personnel with appropriate credentials. A one-time memo of support is not effective. To secure top executive commitment, you must show how an EMP can support key business goals, such as growth, customer satisfaction, or a sharper competitive advantage. You might, for instance, explain that lower operating costs and increased energy efficiency can bring a higher level of occupant comfort, which, in turn, can mean a lower worker absentee rate–and, possibly, a greater employee retention rate. To gain top executive commitment, make sure that CEO and CFO see the dollar savings highlighted in the energy team's regular reports.

Top management commitment is the single most important goal in an effective and lasting EMP. Not only is it crucial to have this commitment, it must also be obvious to everyone throughout the organization. Top management should participate in the program start-up and continue to reinforce that commitment periodically with both words and actions. Such organizations as Walt Disney have achieved this top-level buy-in and gained significant and lasting bottom-line results from energy master planning.

Create an Energy Team

A middle manager in your organization may declare that energy master planning is only the responsibility of the Facilities department. However, when an energy team represents the company's broad interests, energy concerns can be successfully integrated into the overall business plan. Every department that's impacted by the organization's energy use should be invited to participate on the team. This includes

- Facilities
- Construction/engineering
- Purchasing
- Accounting
- Inventory
- Environmental, health, safety
- Legal
- Public affairs
- Property/asset management
- Leasing/real estate
- Risk management
- Security
- Financial service

This should result in the creation of an Energy Committee. Depending on the size of the organization, there may be separate Technical and Steering Energy Committees. In addition to representatives from the departments listed above it is important to include members of the rank and file (hourly employees or line workers). By including such folks in the planning process it allows not only those individuals, but their peers as well, to become aligned with the EMP objectives and particular initiatives during the early stages. Having these employees on the team helps the rest of the rank and file staff (who will be needed to carry out many of the activities) see the energy program as something other than 'just another management flavor of the month.' It also increases success, as feedback from these employees often helps address many of the nuts-and- bolts "bugs" in advance.

The energy team should be headed by an energy manager or coordinator. The energy manager needs to have a mix of technical, people, and communications skills, and must be enthusiastic. He or she should thoroughly understand the organization's operations and should report to someone as close to the top of the organization as possible, so he or she has the clout need to get things accomplished.

Understand the Organization's Energy Use

Conduct disciplined information analysis. Collect and use metered–not just billed–energy data, and analyze it with software designed to manage energy costs. "Fully understanding current energy use practices–the when, where and how much of energy consumption–in detailed qualitative terms is an essential precondition to energy management master planning."[2] For example, to determine exactly where the energy is going, specify and install sub-metering and a data acquisition system. Such tools will help you monitor and collect, as well as analyze, the meter data for electricity, fossil fuels, steam and condensate, and water. For loads of several hundred kW, it's advantageous to use interval meters that allow a close look at 15-minute interval or hour-by-hour use of energy, so you can see where peaks and valleys in usage occur. This kind of metering is also essential for internal billing. Besides establishing points of excessive usage, it can also help manage loads and pinpoint efficiency opportunities.

Cast a Wide Net

Casting a wide net means looking beyond the central plant for savings opportunities, such as lighting, office equipment, elevators, and localized process loads. It means making construction and equipment specs energy-conscious and creating ways to "enforce" such specs. For example, specifying T8 lamps for efficiency upgrades is not sufficient. The architecture, design, and construction staff should build T8s into their specs for non-energy-related upgrades, such as converting a library to training rooms. Casting a wide net also means revising inventory and purchasing practices to support energy efficiency. As in the example above, to ensure that the new installation continues to function properly, the purchasing specs need to list T8s, not T12s.

Set or Sharpen Goals

Once the organization's energy use is determined, strategic thinking linked to the long-term growth objectives will determine how much energy you should aim to save, where the savings should come from, and when those savings should occur. The energy manager along with the energy team, proposes a phase-one timetable with quantifiable goals to reduce energy use and operating and energy costs. Senior management, however, must mandate the goals and schedule or they carry no weight. Ideally, the CEO, CFO, or the Board issues a position calling for measurable reductions at the end of a 3-year fiscal cycle, based on current use: a percentage reduction in peak electrical demand across the entire facility, a reduction in annual kWh consumption of electricity, and a reduction in overall British Thermal Unit (BTU) for fuel per gross square foot. Lesser annual goals will help keep progress on track. A key is to set achievable goals, and then work to meet or beat them.

Implement Upgrades

Because energy managers generally have experience with efficiency upgrades, it is critical not to fall into old patterns of sporadic efforts and a piecemeal approach, with an eye on the quickest payback. To carry out an EMP, start the upgrade process with a comprehensive energy audit—not just a walk-through—to determine which upgrades will give the best results, not just the best rate-of-return. For example, in a commercial building operation, to avoid a rush upgrade for a new tenant, pre-audit all your buildings so you know what work needs to be done before that tenant signs their lease. Or, if you're considering recommissioning, look at it from the longer-term energy master planning perspective. Recommissioning may not be sufficient, as it means only the existing systems would operate more efficiently. But what if they need to be ripped out as part of the upgrade?

Don't fall into the common trap of focusing just on the high-profile, capital-intensive, projects. While a new chiller or micro-turbine cogeneration system may be a good photo opportunity, focusing on the less visible details can often provide the lion's share of savings through lower-cost measures that improve operations and maintenance (O&M) or maintenance and management practices.

Verify Savings

Once the upgrade process is underway, you must first validate the savings with a recognized technique such as the International Performance Measurement and Verification Protocol (IPMVP or MVP). However, instead of following the M&V protocol after the fact, build measurement and verification into the design—that is, install it as part of the upgrade—so you can identify the savings as soon as the device is turned on. Then, set up an energy accounting system that tracks usage and savings, thereby providing objective accountability. All too often energy management activities are not perceived as being of value because the energy team did not plan ahead and position themselves and theirs efforts for recognizable success. A North American floor coverings manufacturer reported how "in the past we would complete a project to find that we did not have the baseline or operational data to judge whether the project was successful."[3] How willing will management be to fund the next energy related project or initiative if the energy team cannot prove that prior efforts actually saved what was projected?

With an energy accounting system, everyone on the energy team will be working from the same data, and you can prepare regular progress reports, plus quarterly reports to the CFO and an annual report to the Board. This is critical: The energy team must account for the savings—or lack of savings-to senior management, using a feedback process so goals and targets can be reset if they aren't met. Use this process to verify energy use and to help identify any new opportunities for savings, and then to fine-tune your goals. If you fall short of your targets, you need to figure out why. Or perhaps you did meet your targets, but parts of the data were wrong. Perhaps the metering is off, or the energy accounting system needs readjustment. Regularly scheduled feedback from these reports will keep the energy master planning process up-to-date and realistic and will ensure accountability. One approach that has been successfully used in many organizations is the use of "energy report cards" and "intra-organizational listings/scorecards." Produced by the energy manager and sent out to all facilities/ operating groups/ departments on a monthly basis, these tools inform as well as motivate, based on the certainty that no one wants to be on the bottom of the list.

Communicate Results

Successful implementation of all stages of the EMP can be a useful vehicle for departmental—and personal—recognition. With a representative of the corporate PR office already on the energy team—enlist that individual's expertise for internal and external communications. Inform the organization's Board about the financial savings and improved asset value. Let the staff and community know about the environmental benefits. Apply for energy awards for national recognition in your sector. And use the good will generated by the success to keep the energy master planning process moving forward.

Tips for Success

From these steps to an EMP, two points emerge as the most critical to success.

The first is ensuring buy-in from the top, with a long term commitment and binding, formal statement of energy policy for the organization. Commitment to action also means that senior executives support the people in the middle—delegating authority to the energy manager or facility manager. If the EMP lacks commitment from the top down, no amount of effort by the energy manager will make it succeed. Once you've caught executives' attention, keep energy master planning on their radar screen by having the energy team build it into the organization's business plan. Such a commitment must also be apparent. Top management must continue to be seen (by all levels of staff) reaffirming their commitment, lest the EMP be perceived as just another short-term initiative of the organization.

The second critical component is line-management accountability, making specific individuals accountable for sustaining the savings. On the management side, executives need to mandate quantifiable goals and targets, strengthened by obligatory deadlines. Such requirements should even be built into job descriptions and evaluated as part of annual performance standards reviews. Likewise, incentives can be offered through these same personnel standards for individuals who meet their energy goals. Employee teams with day-to-day knowledge of the facility's operation can also be organized to identify additional opportunities for savings. At the upper level of the organization, corporate accountability for successful energy management can be communicated (and made public) through such mechanisms as the Global Reporting Initiative, especially in firms with sustainability principles that follow a "triple bottom-line."

Conclusion

Energy master planning is an effective and long-term shift in organizational cultures to address and adapt to the impact energy resources can have on their competitive posture and economic success. Energy is a universal raw material and is essential to the operation of almost every commercial, institutional, governmental, and even nonprofit organization. Significant changes to the basic structure of the energy supply chain and significant energy price volatility, driven by a wide range of political and physical events (e.g., climate change, weather), have made energy both a competitive opportunity and a risk requiring active long-term planning to manage. Energy master planning requires buy-in from top levels of an organization down through the rank and file, selection/hiring of a dedicated energy manager, and creation of an energy team with membership from across the organization. This team must be given the resources and top-level access to develop and implement long-term planning for procurement and operations throughout the organization that places a premium on lasting energy-use reductions and cost optimization, rather than "burst" efforts that provide quick and quickly forgotten energy management efforts.

Successful organizations do not exist on a quarterly basis. They plan for the long-term in all aspects of their operations. Experience has shown that organization-wide adoption of EMP is effective in optimizing energy costs and needs to ensure competitive posture and success in the long-term business reality in which organizations exist. This is the new bottom line for energy.

Acknowledgments

Permission has kindly been granted by the Energy Master Planning Institute for use of materials on its Web site and for materials published and presented by Coriolana Simon, founder of the EMPI.

References

1. Allen, P.J.; Kivler, W.B. *Walt Disney World's Utility Efficiency Awards and Environmental Circles of Excellence,* Proceedings of WEEC 1995; Association of Energy Engineers: Atlanta, GA, 1995.
2. Tripp, D.E.; Dixon, S. *Making Energy Management "Business as Usual": Identifying and Responding to the Organizational Barriers,* Proceedings of the 2003 World Energy Engineering Congress; Association of Energy Engineers: Atlanta, GA, 2003.
3. Key, G.T.; Benson, K.E. *Collins and Aikman Floorcoverings: A Strategic Approach to Energy Management,* Proceedings of the 2003 World Energy Engineering Congress; Association of Energy Engineers: Atlanta, GA, 2003.

31

Energy: Solid Waste Advanced Thermal Technology

Alex E.S. Green and Andrew R. Zimmerman

Solid Waste, Solid Fuels, and Their Properties

In 2011 the United States was heavily (~50%) reliant on foreign sources for its liquid fuels and somewhat (~10%) dependent upon imports for its gaseous fuels. Our country is now expending "blood and treasure" in its efforts to stabilize regions of the globe that supply these premium fuels. Yet the United States is well endowed with solid fuels in the form of coal, oil shale, and substantial quantities of renewable but wasted solids. As part of a continuing long search for alternatives to oil,[1–10] this entry is focused on converting our solid waste to energy by advanced thermal technologies (SWEATTs) while mitigating environmental and economic problems. Table 1 is a list of United States' abundant supply of solid waste (SW) whose organic matter can be converted into gaseous and liquid fuels as well as charcoal. The value society places on a specific fuel or energy type is very sensitive to its physical form as indicated in Table 2[11] which gives prices of various forms of energy in the United States at the beginning of 2010. The large carbon dioxide neutral (neither net producing nor consuming CO_2) plant matter components in Table 1 can help in greenhouse heating mitigation. The great diversity of physical and chemical characteristics of fuel wastes (feedstock) in Table 1 implies that the world now needs "omnivorous feedstock converters" (OFCs) to change these solid fuels into much more usable liquid or gaseous fuels or better solid fuels. Figure 1 is a conceptual illustration of an OFC adapted from a number of prior papers[8–10] in which a SW pyrolyzer–gasifier–liquifier–carbonizer is coutilized with a natural gas-fired combined cycle (NGCC) system, as will be discussed below.

Table 3 shows major ranks of coals as well as of peat, wood, and cellulose and their ultimate and proximate analyses as measured by industry for over a century. The numbers listed in columns labeled C, H,

TABLE 1 Potential Sources of Useful Non-Conventional Fuels

Waste Type	MDT[a]
1. Agricultural residues	1000
2. Forest under-story and forestry residues	400
3. Hurricane debris	40
4. Construction and deconstruction debris	20
5. Refuse-derived fuels	10
6. Urban yard waste	20
7. Food-serving and food-processing waste	80
8. Used newspaper and paper towels	30
9. Used tires	60
10. Energy crops on underutilized lands	50
11. Ethanol production waste	20
12. Anaerobic digestion waste	10
13. Bio-oil production waste	10
14. Waste plastics	40
15. Infested trees (beetles, canker, spores)	20
16. Invasive species (cogon-grass, melaleuca, cat-tail)	50
17. Plastics mined when restoring landfills	30
18. Biosolids (dried sewage sludge)b	40
19. Poultry and pig farm wasteb	20
20. Water plant remediators (algae, hydrilla)b	10
21. Muck pumped to shoreb	10
22. Manure from cattle feed lots	10
23. Plants for phyto-remediation of toxic sites	10
24. Treated wood past its useful life	10
Total	2000

[a] MDT = million dry tonne.
[b] Denote water remediation-related items.

and O (wt% of carbon, hydrogen, and oxygen, respectively) essentially apply to ideal carbon, hydrogen, and oxygen (CHO) materials by correcting measurements to their dry-, ash-, sulfur-, and nitrogen-free (DASNF) form. Then [C]+[H]+[O] = 100 and any two of the three variables fixes the third. In this work we mostly focus on the variables [O] and [H] which then essentially specifies [C]. The column labeled higher heating values (HHV) gives typical HHV in millions of joules per kilogram (MJ/kg) as measured with standard bomb calorimeters after allowing for the minor components.

Figure 2a is mainly a plot of [H] (solid diamonds with values read on the left scale) vs. [O] on the top scale for 185 representative DANSF CHO materials taken from ultimate analysis data available in the technical literature. The trend can be represented by [H] = 6 (1 − exp([O]/2)). The bottom scales give conventional coal ranks, some potential names for the biomass region, and some names that might foster more friendly discussions between the coal and biomass sectors. This [H] vs. [O] coalification plot shows that apart from the anthracite region, all natural DANSF feed-stock have [H] values that are close to 6%. The [H] and [O] coordinates of the three main components of all plant matter are lignin-6.1, 32.6; cellulose-6.2, 49.4; and hemi-cellulose-6.7, 53.3. Materials present in SW can depart substantially above and below the coalification curve. For example, the coordinates of polyethylene and polypropylene are 14.2 and 0, respectively.

TABLE 2 Market and Energy Prices, December 2010[11]

Fossil Fuels	Market Price	$/MMBtu
Crude oil	$84.93 $/Barrel	$14.64
Gasoline	$2.865 $/Gallon	$22.92
Diesel Fuel	$3.116 $/Gallon	$24.21
Natural gas	$3.56 $/MMBtu	$3.56
Liquid Propane (Gulf)	$1.11 $/Gallon	$12.19
Heating oil	$2.084 $/Gallon	$15.10
Electricity retail, resid.	12.02 c/kWh	$35.23
Coal	$47.25 $/ton	$2.00
Liquid Fuels		
Ethanol (Iowa)	$2.42 $/Gallon	$31.78
biodiesel (Iowa)	$4.13 $/Gallon	$34.96
Soybean oil (Central IL)	51.68 c/Lb	$30.40
No 2, Yellow grease	$32.88 $/cwt	$21.35
Solid Fuels		
Fuel pellets	$206.60 $/Ton	$12.91
Shelled corn	$5.37 $/Bushel	$11.76
Compost	$25.00 $/cu. yard	$3.63
Wheat straw	$80.00 $/Ton	$5.41
Grass hay (lg md bale)	$50.00 $/Ton	$3.33
DDGS	$156.00 $/Ton	$8.30

Source: Adapted from Jenner.[11]

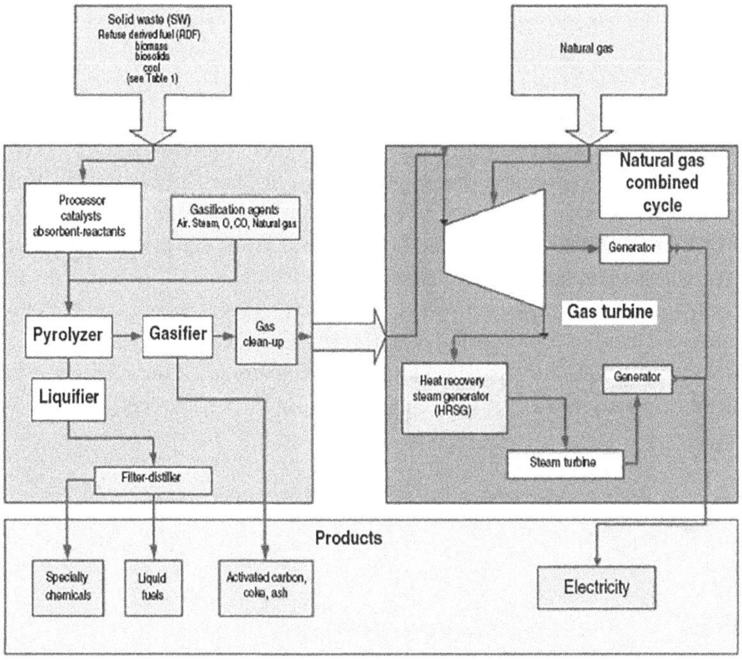

FIGURE 1 Diagram of the Omnivorous Feedstock Converter (OFC) illustrating the addition of a solid waste system to an existing NGCC plant to create an effective SWCC system.
Source: Pyrolysis in Waste to Energy Conversion (WEC).[10]

TABLE 3 Properties of Fuels along Nature's Coalification Path

	Ultimate Analysis			Proximate Analysis			Other Properties		
Name	C	H	O	HHV	VT	FC	Dens	E/vol	charR
Anthracite	94	3	3	36	7	93	1.6	58	1.5
Bituminous	85	5	10	35	33	67	1.4	49	5
Sub-Bituminous	75	5	20	30	51	49	1.2	36	16
Lignite	70	5	25	27	58	42	1	27	50
Peat	60	6	34	23	69	31	0.8	18	150
Wood	49	7	44	18	81	19	0.6	11	500
Cellulose	44	6	50	10	88	12	0.4	9	1600

C, H, and O are wt% of carbon, hydrogen, and oxygen, HHV = higher heating value (millions of joules per kilogram: MJ/kg), V_T = weight percent volatiles, FC = fixed carbon weight percent, Dens = g cm^{-3}, E/vol = relative energy density, charR = relative char reactivity.

The [C] vs. [O] data calculated with [C] = 100−[O]−[H] for DASNF feedstock are also shown in Figure 2A. When the smooth [H] vs. [O] formula is used one gets the smooth upper curve in relation to the data. This figure provides strong reasons for regarding peat and biomass simply as lower rank coals. The diagram suggests that coalification is a natural geophysical deoxygenating process. Much of this treatise on SWEATT will be devoted to attempting to bring some order to the confused literature on artificial pyrolysis, deoxygenating or carbonizing processes and their gaseous, liquid, and solid products. For many purposes, natural solid fuels could be ranked simply by [O] to replace the different ranking systems of various countries (a Tower of Babel!). For example, using 34-O for peat, called "turf' in Ireland, might help temper the "turfwars" in fuel sector competitions and in energy vs. environmental confrontations on the use of our available fuels.

HHV of various fuels measured with calorimeters are often reported along with proximate analyses. Representative values for the various coal ranks are given in the column labeled HHV in Table 3. The column labeled V_T gives representative "total volatiles," V_T, as determined by an American Standard Test Measurement Method. A solid sample is heated (pyrolyzed) in a platinum crucible at 950°C for 7 minutes. The weight percent loss due to the escaping volatiles is designated as the total volatile yield (V_T). The balance from 100% then represents the weight percent of the fixed carbon (FC) plus ash. The ash wt% is the weight percent remaining after combustion in full atmosphere at 750°C for 6 hours.

The columns of Table 3 labeled Dens and E/vol give the physical density (in g/cc) and relative energy density of the various natural solid fuels. These are important factors in determining handling and transportation costs. The column labeled charR gives some relative measures of the reactivity of the chars that are produced by the pyrolysis of these natural feedstock.

Figure 2b displays HHV data for the compilation of 185 materials after correction to DASNF cases. Most points within this scattered HHV data can be fit within a few percent by a two variable form of Dulong's formula:

$$\text{HHV in MJ/kg} = 34.9 - 0.453[\text{O}] + 0.829[\text{H}] \tag{1a}$$

Or

$$\text{HHV in MBtu/1b} = 15.00 - 0.194[\text{O}] + 0.356[\text{H}] \tag{1b}$$

The first form is simplified from the six-variable DuLong formula found by Channiwala and Parikh[12] who fit a large body of HHV measurements of biomass and other fuels. The smooth curve in Figure 2b shows the trend of the HHV vs. [O] when [H] = 6 (1 − exp([O]/2)) is used. When measured [H] values are

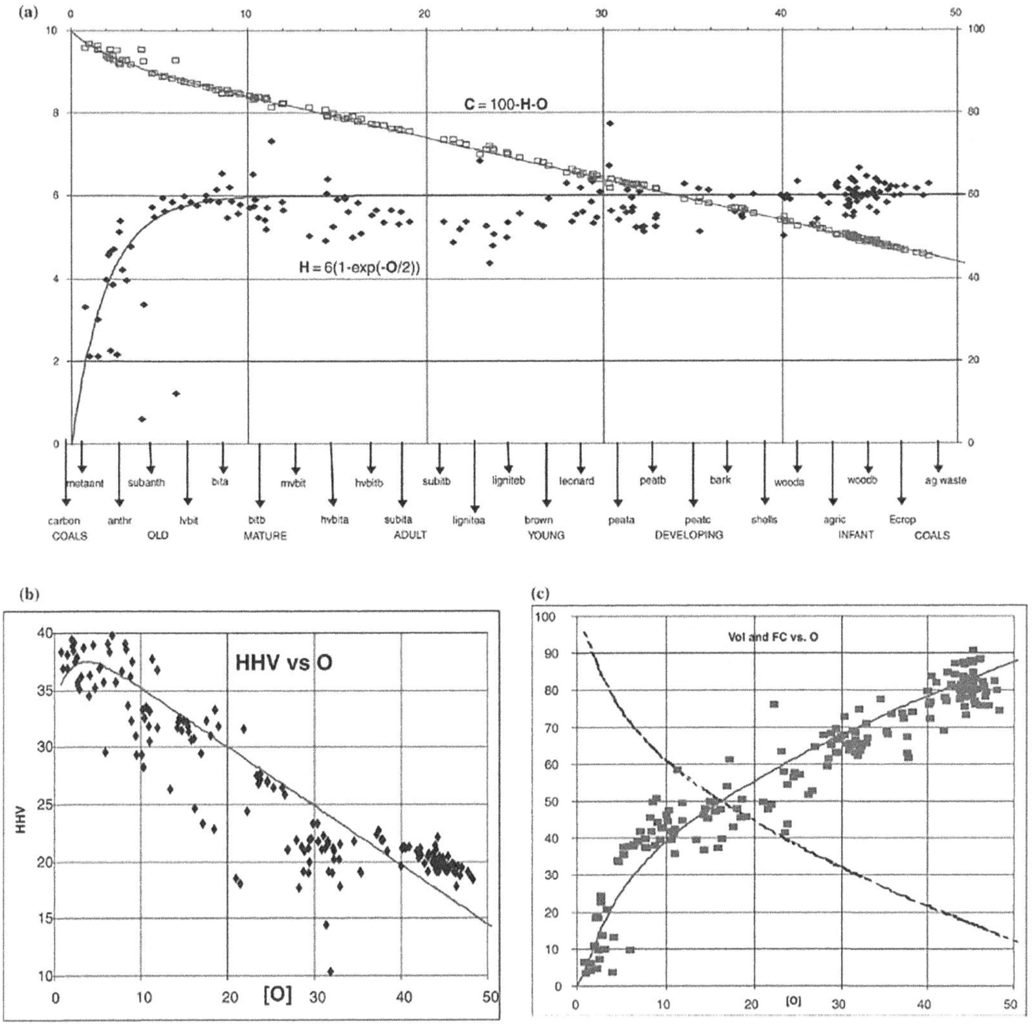

FIGURE 2 (a) Weight percentages of hydrogen [H, left] vs. [O, top] for 185 DANSF carbonaceous materials (diamonds) and additional solid waste materials (red circles). Classification labels are given at the bottom axis. (b) Higher heating values (HHV) of 185 carbonaceous materials (corrected to DANSF) vs. [O]. The smoothed curve represents: Eq. 1a when [H] = 6(1- exp([O]/2)) is used. (c) Total volatile weight percentages (left) vs. [O] for 185 DASNF carbonaceous materials (squares) from proximate analysis. The curve through the data points satisfies: $V_T = 62([H]/6)([O]/25)^{1/2}$. The analytic fixed carbon (FC) vs. [O] is shown as a dashed line.
Source: Green.[4]

used the HHV formula given fits within a few percent. From a HHV standpoint deoxygenating biomass by pyrolysis endows the char progressively with some properties of the higher ranks of coal except that chars tend to be more porous.

The general trends of total volatiles along nature's coalification curve can approximately be represented by the empirical formula $V_T = 62([H]/6)([O]/25)^{1/2}$. Note the rapidly increasing trend in V_T from low [O] materials to high [O] materials (Figure 2c). Because of the large production of volatiles by high [O] materials, pyrolysis of these materials is substantially equivalent to gasification. The [H] dimension is also important and small deviations of [H] from the smooth coalification path have a large impact on the volatile release.

The three diagrams in Figure 2 all indicate the importance of [O] in determining the fuel and carbonization or pyrolysis properties of organic materials. Coalification might be called nature's carbonization or deoxidization process whereas pyrolysis is an artificial process for carbonization or deoxidation of organic feedstock.

Global and U.S. Primary Energy Supplies

Figure 3a presents an overview of the world total primary energy supply (TPES) in 2004 (see International Energy Agency website). Among the major sources of energy, combustible renewables and waste (CRW, mostly biomass) need only be doubled to be together with coal and natural gas, and tripled to be competitive with petroleum. Note that the global CRW is currently about twice as large as nuclear. On the other hand, wind and solar must grow by factors of over 100 to become major global energy supplies.

This global TPES picture is not representative of the industrial world, particularly the United States today. Figure 3b shows the percentage subdivisions of the US TPES in 2007 when the total consumption

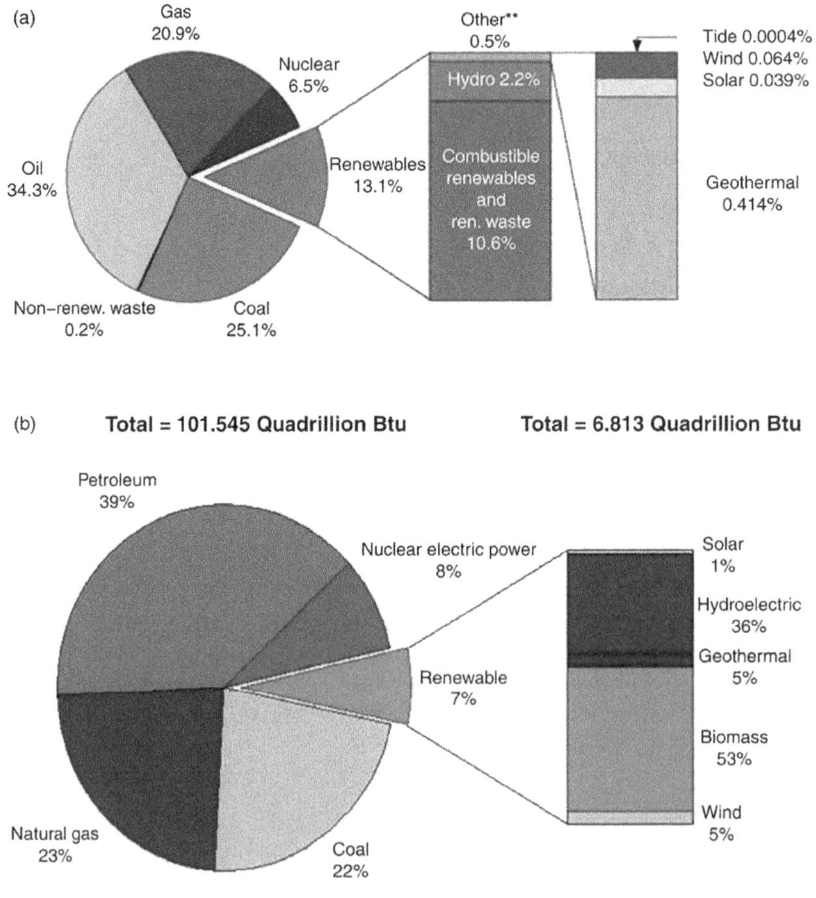

FIGURE 3 (a) Total primary energy supply (TPES) for the globe at 2004 (IEA Website). (b) TPES for the United States in 2007 (EIAWebsite). *TPES is calculated using the IEA conventions (physical energy content methodology). It includes international marine bunkers and excludes electricity/heat trade. The figures include both commercial and non-commercial energy. **Geothermal, solar, wind, tide/ wave/ocean. Totals in graph might not add up due to rounding.
Source: *IEA Energy* Statistics.

was over 101.5 quads, (quadrillion British thermal units (Btu)/annum or quads). It is seen that about 39% of our energy consumption is in the form of petroleum that is mainly consumed in our transportation sector. As Figure 3c illustrates renewables only constitute 7% of the U.S. TPES. The percentage subdivisions of these in 2007 are shown in the figure, and it is seen that biomass was 53% of the 7%. The major thrust of this work is that the solid wastes listed in Table 1, consisting mostly of biomass, were only a minor component (~3.5%) of the U.S. annual TPES, and could in the near term become a major component comparable to coal and natural gas, both still at about 23%. The more popular renewables, geothermal wind and solar, have much further to go than solid waste before becoming a major primary energy source in the United States. Since SWEATT is based on locally available solid waste, it would also create good non-exportable local industries and jobs while mitigating serious U.S. energy import and waste disposal problems. An Oak Ridge National Laboratory study[13] estimates the sustainable supply of the first few biomass categories in Table 1 at about 1.4 billion dry tons. The remaining categories should readily bring the total sustainable U.S. solid waste available to over 2 billion dry tons. Assuming a conservative HHV of 7500 Btu/lb, a simple calculation shows that with SWEATT technologies similar to those that are now in place in Japan, U.S. solid waste contribution to its primary energy supply could reach the 25% level.

Without a doubt, the biggest energy problem faced by the United States today is the need to find alternatives to oil.[1–3] In the 1970s and early 1980s, the United States focused heavily on alternatives to oil in the utility sector. The alternatives first were pulverized coal plants and in the late 1980s and 1990s, on NGCC systems. The U.S. and the globe have been slow in generating electricity by Solid Waste with Advanced Thermal Technologies (SWEATT) and flexible systems such as those illustrated in Figure 1.

It is important to differentiate secondary energy supplies (SESs) from the primary energy supplies (PESs) shown in Figure 3. Secondary energy supplies include steam, syngas, reactive chemicals, hydrogen, charges in batteries, fuel cells, and other energy sources that draw their energy from PESs. If a SES is converted to another type of energy, say mechanical energy via a steam turbine, the mechanical energy becomes a tertiary energy supply (TES). This TES can be converted to electrical energy using magnetic generators, in which case the electricity is a quaternary (QES) supply. In the case of electricity, the many conversions are usually justified since electricity can readily be distributed by wire and has so many uses as a source of energy for highly efficient electric motors, illumination systems, home appliances, computers, etc. Table 2 points to the high cost/value placed on electricity and on liquid fuels.

In many communities debates are underway as to whether increasing electricity needs should be met with the solid fuel coal, MSW, biomass via conventional steam and steam turbine generator systems, or via conversion to a gaseous fuel and using integrated gasifier combined cycle (IGCC) systems. Granting that the steam turbine route has had many advances over the last century, converting the solid fuel to gaseous fuel is increasingly being accepted as the ATT route of the future. The ATT route is not only driven by environmentally acceptable waste disposal needs and increased needs for electricity, but also by the need for liquid and gaseous fuels. A number of petroleum resource experts have recently advanced the date that the globe's supply of oil and natural gas will run out. The prices of oil and to a lesser extent natural gas now reflect this drawdown and are already high enough that conversion of organic matter in solid waste to liquid and gaseous fuels makes economic sense. We should recognize, however, that for the most part, cartels govern fuel prices not free markets. Thus we should not abandon alternative fuels efforts whenever cartels, for their interests, lower prices.

Advanced Thermal Technologies

The largest solid waste to energy systems in operation today are direct combustion municipal solid waste (MSW) incinerators with capacities in the range of 1000–3000 tons SW per day. In such mass burn systems, the organic constituents of the solid waste are combusted into the gaseous products CO_2 and H_2O. These have no fuel value but can be carriers of the heat of combustion as in coal and biomass boiler-furnace systems. Along with the flame radiation, these gases may be used to transfer heat to pressurized water to produced pressurized steam that drives a steam turbine-driven electric generator. The steam can also serve as a valuable SES to distribute heat for heating buildings, industrial processes, etc. The production and use of steam, along with the steam engine, launched the industrial age.

Instead of using the heat released to raise steam, in SWEATT systems the solid waste is first converted into gaseous or liquid fuels and, in pyrolysis systems, partly to char. The volatiles, gases, and vaporized liquids fuel then serve as a SES that can be used in efficient internal combustion engines (ICEs), combustion turbines or, in the future, in fuel cells, none of which can directly use solid fuels. Over the past century automotive and aircraft developments have pushed ICEs and gas turbines (GT) to very high levels of efficiency. Furthermore, with the use of modern high temperature GTs in NGCC systems, the heat of the exhaust gases can be used with a heat recovery steam generator (HRSG) to drive a steam turbine. Alternatively, the HRSG can provide steam for heating buildings or industrial applications of steam. These combined heat and power (CHP) system at this time make the most efficient use of the original solid fuel energy.

If one considers the United States' heavy dependence on foreign sources of liquid and gaseous fuels, the most challenging technical problem facing the United States today should be recognized as the development and implementation of efficient ways of converting our abundant domestic solid fuels into more useful liquid and gaseous fuels. In view of the diversity of feedstock represented in agricultural, municipal, or institutional solid waste, the United States and the world need an omnivorous feedstock converter such as is illustrated in Figure 1. Here, the right block represents a typical NGCC system, whereas the left block represents a conceptual Omnivorous Conversion System that can convert any organic material into a gaseous or liquid fuel.

We will first consider the gross nature of the output gas from biomass or cellulosic type material, the major organic components of most solid waste streams. Apart from minor constituents such as sulfur and nitrogen, the cellulosic feed types are complex combinations of carbon, hydrogen, and oxygen such as ($C_6H_{10}O_5$) that might serve as the representative cellulosic monomer.

ATT systems used to produce output can be divided into 1) air blown partial combustion (ABPC) gasifiers; 2) oxygen blown partial combustion (OBPC) gasifiers; and 3) pyrolysis (PYRO) systems. The three types of systems for converting waste into a gaseous fuel have many separate technical forms depending upon the detailed arrangements for applying heat to the incoming feed and the source of heat used to change the solid into a gas or liquid fuel. We use "producer gas" as a generic name for gases developed by partial combustion of the feedstock with air as in many traditional ABPC gasifiers that go back to Clayton's coal gasifier of 1694. We will use "syngas" for gases developed by partial combustion of the feedstock with oxygen as in OBPC gasifiers, which are mainly a development of the twentieth century. We will use "pyrogas" for gases produced by oxygen-free heating of the feedstock such as in indirectly heated (pyro)gasifiers. The objective is to replace the natural gas, that is, fossil fuel gas, that has a HHV, ~1000 Btu/cft = 1 MBtu/cft (with Btu $M = 1000$) with a biomass-generated fuel gas having similar energy and combustion qualities.

When an ABPC gasifier is used with cellulosic materials (cardboard, paper, wood chips, bagasse, etc.), the HHV of biomass producer gas is very low, 100–200 Btu/cft. Essentially, the useful product of partial combustion of biomass is CO that only has a HHV of 322 Btu/cft. Unfortunately, considerable CO_2 and H_2O are produced during partial combustion and together with the air-nitrogen these inerts substantially dilute the output gas. The "syngas" obtained from biomass with OBPC gasifiers is better, ~320 Btu/cft, since it is not diluted by the atmospheric nitrogen. However, because of the partial combustion it is still somewhat lower than the energy contained within the feedstock molecules. Additionally, the oxygen separator is a major capital cost component of an OBPC gasifier. With a Pyro system, the original cellulosic polymer is broken by the applied indirect heat to its monomers and then to the major pyro-products CO, CO_2, and H_2O as well as hundreds of hydrocarbons (HCs) and carbohydrates (HCOs), each with yields that depend upon the applied temperature, heating time, and particular processing arrangement. Cellulosic pyrogas can have heating values in excess of 400 Btu/cft.

Among the pyro-volatiles coming from pyrolysis systems are the paraffins (CH_4, C_2H_6, C_3H_8, ...), olefins (C_2H_4, C_3H_6, ...), acetylenes (C_2H_2, C_3H_3, ...), and various carbohydrates, carbonyls, alcohols, ethers, aldehydes and phenols, and other oxygen-containing gaseous products. Attempting to find some patterns or regularities in the literature on products of pyrolysis from various natural and man-made fuels has been the goal of multiyear effort.[4–26] Table 4 is a list of the families of molecules that have been detected in pyrolysis volatiles and the rules that connect the family member, labeled by $j = 1, 2, 3, \ldots$ (see "The ASEM and Organization of Pyrolysis Products").

HC plastics such as polyethylene and polyolefins are heavily represented in many solid waste streams. Thus one might use (C_2H_4) as representative of the monomers in the plastic component of MSW or refuse-derived fuels (RDFs). Polyethylene pyrolysis products are dominated by C2–C4 olefins, acetylenes, and other HCs and at higher temperatures by H_2 as well as aromatics (Ar) and polynuclear aromatics (PNAs) identified in Table 4. On a per unit weight basis, all but H_2 have gross heating values in the range 18–23 MBtu/lb, similar to oil, whereas H_2 has a gross heating value of 61 MBtu/lb. On a per unit volume basis, polyethylene pyrolysis products have gross heating values ranging from 1 to 5 MBtu/cft whereas H_2 is 0.325 MBtu/ cft = 325 Btu/cft. Because natural gas is typically about MBtu/cft, we would expect the pyrogas from polyethylene to have a gross heating value comparable or greater than that of natural gas and much greater than cellulosic pyrogas.

In summary since cellulosic feedstock is already oxygenated compared with pure HC plastics, its pyrogas, syngas, and producer gas will all have considerably lower heating values than the corresponding gases from HC feedstock. From the viewpoint of maximizing the HHV of SW derived gas, Pyro gasification scores better than OBPC gasification, both of which score much better than ABPC gasification. Pyrolysis also leaves more of a solid residue in char-ash form than ABPC gasification or OBPC gasification. Figure 4 illustrates a typical pattern of evolution of the solid in an indirectly heated slow pyrolysis system.[27] For DASNF materials the asymptote of the solid (char) curve would represent the FC and the balance from 100% would represent the total volatiles. Figure 4 also shows a typical pattern of evolution of the tar and gas from wood feedstock in the pyrolysis of a small particle of wood when its temperature is raised at a slow rate such as 10°C/min. These curves are representative of results from the analytical semiempirical model (ASEM).

TABLE 4 Organization of Functional Groups by Families

Families	a	B	c
Paraffins	j	$2a + 2$	0
Olefins	$j+1$	$2a$	0
Acetylenes	$j+1$	$2a - 2$	0
Aromatics	$5+j$	$4+2j$	0
Polynuclear	$6+4j$	$6+2j$	0
Aldehydes	$j+1$	$2a$	1
Carbonyls	J	$2a$	1
Alcohols	J	$2a + 2$	1
Ethers	$j+1$	$2a + 2$	1
Phenols	$5+j$	$4+2j$	1
Formic acids	J	$2a$	2
Guaiacols	$6+j$	$6+2j$	2
Syringols 1	$7+j$	$8+2j$	3
Syringols 2	$8+j$	$10+2j$	4
Sugars 1	$4+j$	10	5
Sugars 2	$5+j$	$10+2j$	5

a, b, and c are the subscripts in $C_aH_bO_c$, $j = 1, 2, 3\ldots$

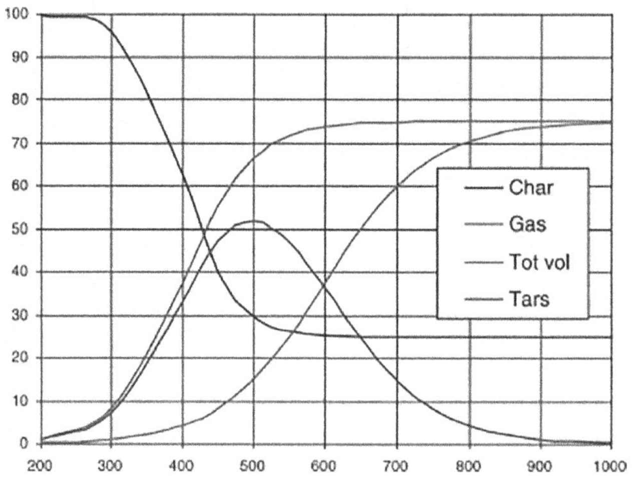

FIGURE 4 Typical pattern of evolution of char, tar, and gas from wood feedstock as temperature is raised at a slowly (10°C/min).

Studies of the evolution of chars, tars (volatiles condensable at standard temperature), gas (not condensable at standard temperature), and total volatiles have been carried out at heating rates (r) from 1°C/min to 1000°C/sec. Usually the heat rate is coupled to the temperature via a linear relationship such as $T = rt + T_0$ (where t is time and T_0 is an initial temperature). The dependence of these products upon temperature then changes dramatically from that shown in Figure 4 in ways that are difficult to track via standard kinetic modeling. However, they can be relatively easily represented via formulas used in the ASEM by letting the parameters of the model be simply dependent upon the heating rate. This approach is often dismissed as "just curve fitting" in academic circles where the search for models that depend upon fundamentals physical variables has become traditional. Unfortunately, in pyrolysis studies, because of the complexity of products released at various temperatures, this quest is still far from realization. Accordingly, it appears to some investigators that after the experimental assembly of reproducible scientific data (the first step of the scientific method), organizing the results in some robust analytical form (the usual second step of the scientific method) cannot only be useful for applications, but could help in achieving a fundamental model.

The ASEM and Organization of Pyrolysis Products

Proximate analyses of coal and biomass measured for over a century provide extensive data on total volatile content. However, quantitative data as to the molecular constituents in these volatiles have only been reported in recent years and a predictive method for identifying these molecules is still not available. This despite the fact that such knowledge could provide a more fundamental understanding of humankind's oldest technology (the use of fire). For control and application of a pyrolysis system it would be useful to have at least an engineering-type knowledge of the expected yields of the main products from various feedstock subjected to oxygen-free thermal treatment (pyrolysis).

In most attempts to describe the systematic of pyrolysis yields of organic materials such as coal and biomass, including the initial CCTL studies,[14–19] it has been customary to characterize the feedstock by its atomic ratios $y = H/C$ and $x = O/C$. In recent studies,[20–26] it has been found more advantageous to work with the weight percentages [C], [H], and [O] of the feedstock after correcting to DASNF conditions (i.e., pure CHO materials). Focusing on weight percentages appears to facilitate easier connections between the great complexity of compounds that evolve from pyrolysis and the gas, liquid (tar), and solid products of pyrolysis.

The ASEM is a phenomenological attempt to find some underlying order in the pyrolysis yields of any product $C_aH_bO_c$ vs. the [O] and [H] of the DASNF feedstock and the temperature (T) and time (t) of exposure. The ASEM was developed so as to be useful for a number of applications of pyrolysis.[18–26] Some progress has been made in including the time dimension but much more work remains on that front. When the time dimension is not an important factor, as in many cases of slow pyrolysis, the yield Y(T) as a function of temperature of each product for slow pyrolysis (or fast pyrolysis at a fixed time) is represented by

$$Y(T) = W\left[L(T:T_0,D)\right]^P\left[F(T:T_0,D)\right]^q \tag{2}$$

where

$$L(T:T_0,D) = \frac{1}{1+\exp\left((T-T_0)/D\right)} \tag{3}$$

and

$$F(T:T_0,D) = 1 - L(T) = \frac{1}{1+\exp\left((T-T_0)/D\right)} \tag{4}$$

Here $L(T)$ is the well-known logistic function that is often called the "learning curve." Its complement, $F(T) = 1 - L(T)$ thus might be called the "forgetting curve." For engineering applications this "curve fitting" approach provides a more robust and convenient means for organizing pyrolysis data than traditional methods that use conventional Arrhenius reaction rate formulas.[27] In the ASEM each product is assigned five parameters (W, T_0, D_0, p, q) to represent its yield vs. temperature profile. The objective has been to find how these parameters depend on the [H] and [O] of the feedstock and the a, b, c of the $C_aH_bO_c$ products for the data from particular types of pyrolyzers. Studies by Xu and Tomita (XT)[28] that gave data on 15 products from 17 coals at 6 temperatures have been particularly helpful in revealing trends of the parameters with [O] and [H]. In applying the ASEM to the CCTL data collection, the XT collection, and several other collections, a reasonable working formula was found for the yield of any *abc* product for any [O], [H] feedstock given by

$$Y(C_aH_bO_c) = W_{abc}z^{\alpha}h^{\beta}x^{\gamma}\ \left[L(T:T_0,D)\right]^P\left[F(T:T_0,D)\right]^q \tag{5}$$

where z = [C]/69, h = [H]/6, and x = [O]/25 and the parameters α, β, and γ. T_0, D, p, and q were found to have simple relationships to the feedstock and product defining parameters [H], [O], a, b, and c. The final ASEM formulas that fit the data could then be used to extrapolate or interpolate the XT results to any [H], [O] feedstock and temperature. Figure 5 gives an overview of the interpolated and extrapolated outputs $Y(T)$ for a selection of products for six representative feedstock along nature's coalification path.

Several hundreds, even thousands, of organic products of pyrolysis have been identified in the literature. Thus, to bring order from chaos will require some comprehensive organization of these products. Toward this goal, the ASEM approach groups products into the families as summarized in Table 4, which gives rules for the a, b, cs that connect these groups. This list can be subdivided into pure HCs, i.e., (C_aH_b), and the oxygenates (C_aH_bO, $C_aH_bO_2$, $C_aH_bO_3$, …, etc.). Isomers (groups with identical a, b, and c) can differ in detailed pyrolysis properties and hence parameters. We use j = 1, 2, 3, … to denote the first, second, third, etc., members of each group or the carbon number (n). In the most recent ASEM studies[20–26] of specific feedstock, pyrolysis formulas have been proposed and tested for the dependence of the W, T_0, D_0, p, and q parameters upon the carbon number of the product within each group. This makes it possible to compact a very large body of data with simple formulas and a table of parameters.

The case of polyethylene is an example of such a study. It is not shown on Figure 2a, as it is far removed from the coalification curve having the position [H] = 14.2 on the [O] = 0 axis. Without oxygen in the

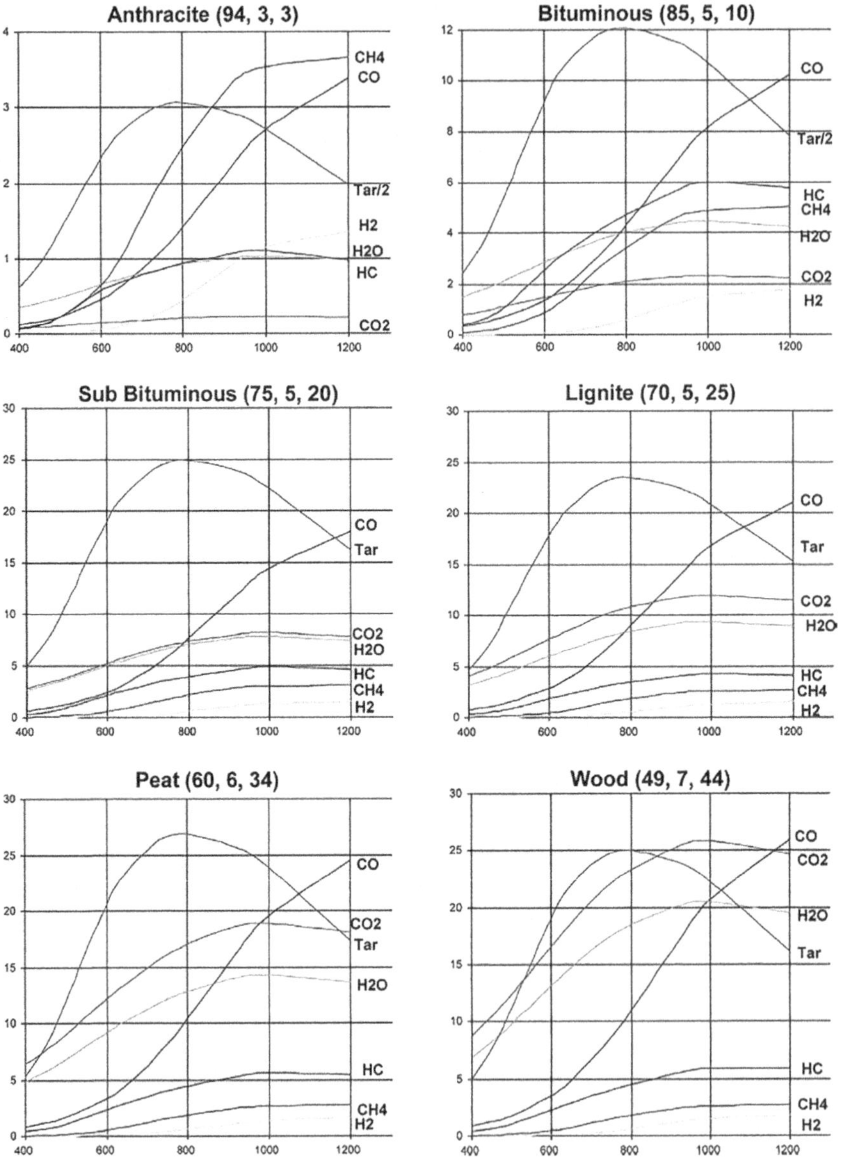

FIGURE 5 Weight percent yields (left axis) vs. temperature (in °C, bottom axis) of pyrolysis products of anthracite, bituminous coal, subbituminous coal, lignite, peat, and wood of ([C], [H], [O]) composition as shown. HC represents C2 and C3 gasses and aromatics.
Source: Green and Bell.[10]

feedstock, the pyrolysis products are much fewer and the ASEM is much simpler to use than with carbohydrates. Thus, only the first 6 rows of Table 4 are needed to cover the main functional groups involved in organizing the pyrolysis products of polyethylene. Figure 31.6 gives an ASEM-type summary of the product yields vs. temperature for polyethylene based on fits to the experimental data of Mastral et al.[29,30] at five temperatures that were constrained to approximately satisfy mass, [C], and [H] balances. Once the parameter systematic is identified, the ASEM representation can be used to estimate the pyrolysis product of polyethylene pyrolysis at any intermediate temperature or at reasonable extrapolated

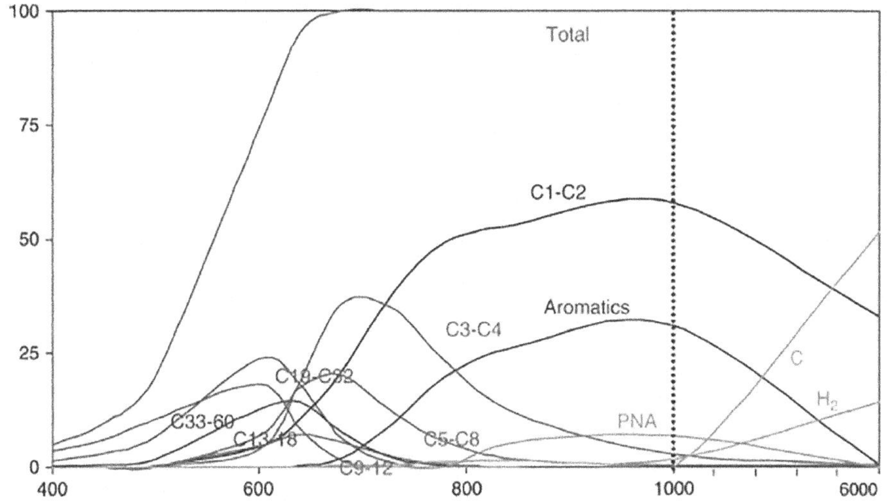

FIGURE 6 Weight percent yields (left axis) vs. temperature (in °C, bottom axis) of pyrolysis products of polyethylene in various hydrocarbon groups.
Source: Green and Sadrameli.[22]

temperatures. The experimental data was only available up to 850°C but the extrapolations to 1000°C were constrained in detail to conform to mass, [C], and [H] balances. Figure 6 also shows extrapolations to 6000°C that might be of interest if one goes to very high temperatures, for example, by plasma torch heating. Here we incorporate a conjecture that at the highest temperatures H_2 and C emerge among the products at the expense of the C1-C2 compounds as well as Ar and PNA components. While we have already found that an ASEM can begin to bring some order into pyrolysis yields, clearly there is a long way to go. When the time dimension is important, the overall search is for a reasonable function of seven variables [H], [O], *a*, *b*, *c*, *T*, and *t*. In comparison, Einstein's special theory of relativity only dealt with four variables *x*, *y*, *z*, and *t*.

Analytical Cost Estimation and SW-IGCC vs. NGCC

Before World War II (WWII) almost every town had its own gas works, mainly using coal as a feedstock. After WWII, cheap natural gas became available and became a major PES for home heating and cooking as well as for industrial purposes. In the 1980s, factory produced natural gas combined cycle (NGCC) systems became available and natural gas became a base load fuel source for many electric utilities. This hastened the drawdown of U.S. conventional domestic supplies and natural gas prices rose to high levels. However, natural gas prices have proven very volatile so that pursuing combinations of SWEATT with NGCC provides good flexibility in facing high energy price swings. For most biomass and plastic feedstock, pyrolysis is substantially equivalent to gasification.

Most comparative economic analyses use detailed life cycle analysis (LCA) or other forms of cost-benefit (C/B) approaches. However, it must be recognized that in recent years fuel costs, an important component of LCA, or C/B approaches have become so volatile that long-term projections based on assumed fuel cost can be grossly inaccurate. Fortunately, the economic feasibility of using a gasifier in front of a gas-fired system can be examined with simple arithmetic and algebra using an analytical cost estimation (ACE) method.[8–10] ACE takes advantage of the almost linear relationship between the cost of electricity (COE = *Y*) vs. cost of fuel (COF = *X*) observed in utility practice and in many LCA for many technologies, i.e.,

$$Y(X) = K + SX \tag{6}$$

In Eq. 6, Y is in ct/kWh (cents per kilowatt hour), X is in \$/MMBtu, and S is the slope of the $Y(X)$ line in ct/kWh/\$/ MMBtu or 10,000 Btu/kWh. S relates to the net plant heat rate (NPHR, see Chapter 37 in Stultz and Kitto[31]) via

$$S = \frac{\text{NPHR}}{10{,}000} \text{ or efficiency via } S = \frac{34.12}{\text{Eff}} \tag{7}$$

A slope $S_{ng} = 0.7$ is now a reasonable assignment for a NGCC system reflecting the high efficiency of recent gas and steam turbines.

In Eq. 6 the parameter K is obviously the COE if the fuel comes to the utility without cost (i.e., $X = 0$). In previous studies,[8–10] $K_{ng} = 2$ was used as a reasonable zero fuel cost parameter for a 100-MW NGCC system.[32,33] This low number reflects the low capital costs of factory-produced gas and steam turbines in NGCC systems. In addition the contribution to the intercept K from operations and maintenance costs are reasonable. The K_{sw} for a solid waste-integrated gasification-combined cycle (SW-IGCC) system is higher than K_{ng} because the capital costs and operating cost must include the gasifier and gas cleanup system. The value of S_{sw} is also higher than S_{ng} because we must first make a SES producer gas, syngas, or pyrogas which involves conversion losses. A study of the literature[33–37] suggests that $S_{sw} = 1$ is a reasonable estimated slope for a SW-IGCC system. The X_{sw} for a SW-IGCC system that would compete with a NGCC system at various X_{ng} thus must satisfy

$$K_{SW} + X_{SW}S_{SW} = K_{ng} + X_{ng}S_{ng} \tag{8}$$

It follows that the solid waste fuel cost X_{sw} that would enable a SWCC system to deliver electricity at the same cost as a NGCC system paying X_{ng} is given by

$$X_{SW} = \frac{K_{ng} - K_{SW}}{S_{SW}} + X_{ng}\left(\frac{S_{ng}}{S_{SW}}\right) \tag{9}$$

In the following, all X numbers are in \$/MMBtu and all Y and K numbers are in ct/kWh. Let us use Eq. 9 with $K_{kg} = 2$, $S_{ng} = 0.7$, $S_{sw} = 1$, and $K_{sw} = 4$ as reasonable estimates based on several SWCC analyses.[6–10] Then the first term in Eq. 9 is –2. Now when the $X_{ng} = 2$ to generate SWCC electricity at the same cost the solid waste provider must deliver the fuel at a negative price, i.e., pay the tipping fee of –0.7. However, if X_{ng} is near 6 as it has been several times between 2004 and 2010 the SWCC utility could pay up to 2.4 for the SW fuel. If the X_{ng} is at 12, the SWCC facility could pay 7.1 to the SW supplier. This X_{sw} price is much higher than that of coal whose delivered price (X_c) these days usually is in the 2 range. It is also much higher than a pulp and paper mill would pay for waste wood. This simple cost comparison is illustrated in Figure 7 that shows the opportunities for SWCC systems when natural gas prices are above say \$5/MMBtu. The results are slightly less favorable if the K_{sw} were higher say at $K = 5$. However, the conclusions that at high natural gas prices SWCC electricity becomes competitive with NGCC electricity would be similar. It is conceivable that Ksw could be held as low as 2 ct/kWh by retrofitting a NGCC system stranded by high natural gas prices. In this case the first term in Eq. 9 vanishes and the competitive $K_{sw} = (S_{ng}/S_{sw})X_{ng}$. This illustrates the main point that at high natural gas prices, with an ATT system, SW can be a valuable PES. Indeed, this simple algebraic–arithmetic exercise establishes the feasibility of a new paradigm in which solid waste (mostly biomass) can become potentially valuable marketable assets.

As described above, the values of K and S are the key factors in determining the COF_{sw} that can be used in a SW-integrated pyrolysis combine cycle (IPCC) to have the COE equal or less than the COE with a NGCC system at the available COF_{ng}. The ACE method can be extended to the use of SW or biomass with other technologies if we can identify the K and S for each technology. Where actual facilities

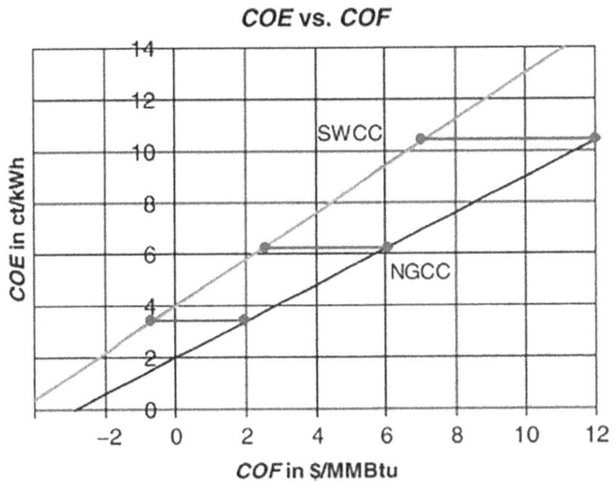

FIGURE 7 Cost of electricity (COE) vs. the cost of fuel (COF) for an solid waste combined cycle (SWCC) system and for a natural gas combined cycle (NGCC) system. COF comparisons of 2, 6, 12 dollar/MMBtu are indicated in red lines.
Source: Green and Bell.[10]

have been built and placed in operation, the K_s and S_s can be assigned on the basis of actual plant experience which is the case for many fossil fuel technologies. However, for new technologies proposed for the renewable age, some accepted form of C/B analysis is needed to provide COE vs. COF relationships. The ACE method cannot serve in this role. However, if a detailed C/B analysis or LCA is available at a particular power level, its *Y(X)* results can generally be cast into the Eq. 6 form which is generally more useful and transparent than tables of numbers. A major advantage of doing so is the possibility of then making reasonable *Y(X,P)* extrapolation to other power levels *(P)* on the basis of many years of economy of scale experience with a wide range of technologies.

The ACE method has been applied to reformulations of a large body of COE vs. COF calculations on biomass use presented in an Antares Group Inc. report (AGIR)[34] for a number of technologies. The technologies investigated in the AGIR were for systems in which 100 tons per day forest thinning were available in wild land–urban interface areas. This assumption limited the power level (*P*) for that technology quite severely. The technologies in the AGIR included a biomass-integrated gasifier combined cycle (B-IGCC) system, a B-IG simple cycle (B-IGSC) system, a B-IG internal combustion (B-IGIC) system, a biomass–gasification–coal co-firing B-IGCC system, a direct cofiring of biomass and coal in a coal steam boiler BCoSt, a direct use of biomass in a feed water heat recovery arrangement (FWHR), direct use of biomass in a Stoker fire boiler steam turbine (SFST) system, and direct firing in a CHP plant (CHP) with a steam market at \$6/MMBtu. For each technology, it was possible to approximately represent the tabulated COE vs. COF results of their detailed economic analysis by Eq. 6 and to evaluate *K* and *S* for that technology at that power level.

The most interesting result of this ACE digest of the massive tables of the AGIR was that by slight extrapolations to higher power levels[9] the competitive results in several important cases were opposite to those for the power levels limited by the 100 tons biomass per day assumption.

Several other detailed economic COE vs. COF analyses have been used to refine ACE and generalize the ACE methodology. In particular, *K* has been broken into components $K = K_c + K_{om} + K_{ne}$, where *c* stands for capital costs, om for operating and maintenance costs, and *en* for environmental costs. Establishing the magnitudes of these components for various technologies and power levels is still at the cutting edge of utility economic analyses, and there are large disagreements particularly on K_{en}. In one generalized component form of ACE (CACE) Eq. 6 is replaced by

$$Y(X,P) = K_{cr}\left(\frac{P_r}{P}\right)^{\alpha} + K_{omr}\left(\frac{P_r}{P}\right)^{\beta} + K_{er}\left(\frac{P_r}{P}\right)^{\gamma} + XS_r\left(\frac{P_r}{P}\right)^{\delta} \tag{10}$$

where K_{cr}, K_{omr}, K_{er}, and S_r are established on the basis of a detailed analysis at a reference power level P_r and α, β, γ and 5 are scaling parameters intended to reflect the tendency of per energy unit cost to go down as the power goes up (economy of scale). Table 5 lists CACE parameters extracted from a detailed analysis "Options for Meeting the Electrical Supply Needs of Gainesville" prepared by ICF Consulting.[35] Here the final COE is given in 2003 ct/kWh. The third and fourth cases NGCCc and NGCC-d have been added to better reflect the high volatility of natural gas prices that have ranged from \$1/MMBtu to \$16/MMBtu over the past 25 years.

The final column shows that at the reference power levels without the NGCCc case the IGCC scores the lowest COE as was concluded in the ICF report (ICFR). The value of the ACE analysis is that with a bit of algebra anyone can easily consider other fuel cost projections and other power levels (with the assigned values of α, β, γ and δ). Based on prior exploratory work and economy of scale investigations it was estimated that for costly field erected facilities $\alpha = \beta = 0.3$ are reasonable choices. However, with factory fabrication of gas and steam turbines these parameters might not follow the usual economy of scale pattern and that their α may be somewhat smaller in magnitude. Assigning a value for γ is a wide open question since environmental costs and methods of incorporating them into the COE are still highly debated issues.[36] Reasonable values for δ are also somewhat difficult to find. For NGCCs the author tentatively assigns close to zero or a very small value (~0.1) perhaps because the development of highly efficient aero-derivative turbines has proceeded on a wide range of power levels.

A Somewhat simper generalized formula for $Y(X,P)$ has been developed in the form [37]

$$Y(X,P) = \alpha C_{cr}\left\{\left(\frac{P_r}{P}\right)^{\gamma}\right\}(1+f_{om}) + SX(1+f_e) \tag{11}$$

Here, $\alpha C_{cr} = K_{cr}$ at a reference power level where C_{cr} is the specific capital cost for that facility in dollar per watt (\$/W or \$1000/kW) for that technology at the reference power level. Based on tabular data contained in the RA study a value $\alpha = 1.34$ was identified as the coefficient the approximately relates specific capital costs (in \$/W) to K_{cr} (in ct/kWh). This essentially is the COE when the fuel cost is free and OM can be ignored. As for economies of scale, based on an extensive literature survey. Green et al.[37] fond

TABLE 5 Analytical Cost Estimation (ACE) Results from ICFR for Five Technologies—Four Natural Gas Prices Are Assumed for the NGCC Technology

Technology	P_r	K_o	K_{om}	K_{en}	S_o	COF	COE
NGCC-a	220	0.598	0.234	−0.17	0.68	11.34	8.37
NGCC-b	220	0.598	0.234	−0.17	0.68	6.1	4.81
NGCC-c	220	0.598	0.234	−0.17	0.68	5	4.06
NGCC-d	220	0.598	0.234	−0.17	0.68	4	3.38
SCPC	800	1.491	0.299	0.714	0.93	1.91	4.28
CFB-CB	220	2.531	0.261	0.618	1.05	1.41	4.89
CFB-B	75	2.845	0.261	0.039	1.39	1.67	5.47
IGCC	220	2.2	0.196	0.407	0.86	1.41	4.02

COE = cost of electricity, COF = cost of fuel.

the different technologies had scaling parameters in the range $0.13<\gamma<0.33$. When no information is available one might use some intermediate gamma between these extremes or let $\gamma = 0$.

Typical dimensionless values for f_{om} can be identified for various technologies from the Antares report on 6 technologies, ICER on 5 technologies, and the RA study of 14 widely ranging technologies. In most cases they are less than unity and can be assigned within reasonable bounds on the basis of experience.

In Eq. 11 f_e is an added dimensionless "correction" to the delivered COF that reflects environmental costs not included in the cost charged by the utility but paid for by the public in other ways (reduced visibility, added coughs, higher cancer rates, etc.). The landmark RA[36] study that incorporated externalities into a levelized cost of energy analysis for 14 electric generating plants provided a basis for estimating the dimensionless externality correction f_e to the price of fuel. Table 6 translates RA's results for the 14 utilities using their minimum externality cost estimates into the generalized analytic cost estimation (GACE) analytical form of Eq. 11. As one sees these f_e are substantially larger than 1 for fossil fuel technologies but small (0.15) for a biomass system. Wind turbines, photo voltaic, and landfill gas systems do not have a fuel cost hence f_e cannot be assigned. However, these technologies are directly assigned externality C_{ex} by RA. The major uncertainty in the future COE is probably represented by the variable X, the externality parameter f_e, and possibly the power level of the facility. Having an explicit formula with these variables and parameters can help in bringing transparency to important policy decisions.

The GACE approach to reaching decisions in the face of large uncertainties might be viewed as application of the operations analysis methods used by one of the authors in WWII.[38] It will be interesting to compare this approach with recent European Union operations analysis effort for incorporating externalities in electricity-generating technology evaluations.[39]

Going back to the 2007 SWEATT study it mainly considered the competition between NG-fueled technologies and coal-steam-generated electricity. These included supercritical coal-burning units that reached efficiencies as high as 40%. However, when coal-steam turbine's expensive scrubber costs are included in the K_c and environmental costs in their K_{en} this technology did not compete compared with the IGCC. Thus a major conclusion of the 2007 SWEATT study was that the age of making gas has returned. At the same time the most favorable EEE position of natural gas among the fossil fuels is recognized.

TABLE 6 Roth-Ambs Externality Impacts on Cost of Electricity for 14 Technologies Using Low RA Estimates, Impacts, and Derived GACE Parameters

Technology	C/W	K_c	f_{om}	S	C_{of}	C_{ex}	f_e	COE	COE_e
Coal Boiler	1.80	2.81	0.36	0.995	1.06	4.45	4.20	4.86	9.29
Adv Fld Bed	2.20	3.52	0.49	0.975	1.04	2.86	2.75	6.26	9.05
IGCC (coal)	2.10	3.28	0.28	0.889	0.95	2.64	2.78	5.05	7.40
Oil Boiler	1.30	2.15	0.19	0.943	3.22	6.03	1.87	5.59	11.27
Gas Turb SC	0.70	10.1	0.12	1.15	3.47	4.62	1.33	15.28	20.59
Gas T Adv	0.40	0.82	0.51	1.09	3.29	4.45	1.35	4.83	9.68
NGCC	0.60	0.91	0.34	0.683	2.11	3.46	1.64	2.66	5.02
MSW Inc.	5.70	9.63	0.44	1.687	5.16	7.7	1.49	22.55	35.54
LFG	1.50	3.3	0.3	1.215	0	0.7		4.29	5.14
SOFC	1.60	2.42	2.71	0.758	2.29	2.75	1.20	10.71	12.79
Wind Turb	1.00	5.74	0.29	0	0	0.7		7.40	7.40
PV Utility	4.70	49.5	0.02	0	0	0.25		50.53	50.53
Hybrid solar	3.70	20.3	0.15	0.346	1.07	2.38	2.22	23.64	24.46
Biomass	2.40	3.54	0.73	1.431	2.75	0.41	0.15	10.07	10.65

COE_e are Roth-Amb's total COE with low externalities.

It should be noted that in an effort to minimize a major environmental externality, global warming, the U.S. Department of Energy is now investing a $4 billion plus effort in carbon capture and storage (CCS) technologies that first starts with oxygen-blown coal gasification. DOE funds will be matched by about $7 billion from the coal and utility industry. This effort could bring up to ten commercial demonstration projects online by 2016. The goal of the program is to provide the information needed to evaluate whether such CCS technologies are commercially deployable.

Mercury emission control has recently become mandatory for coal fired plants. Injecting activated carbon as a sorbent to capture flue gas mercury has shown the most promise as a near-term mercury control technology. The process is still in its early stages and its effectiveness under varied conditions (e.g., fuel properties, flue gas temperatures, and trace-gas constituents) is still being investigated.

Bioenergy and Biochar

During the first decade of the 21st century, the most widely pursued sources of renewable biofuels were fermentation of corn or corn stover leading to ethanol, anaerobic digestion of animal waste yielding methane, and compression of plant seeds to extract bio-oil. Since this Solid Waste to Energy by Advanced Thermal Technologies (SWEATT) work is focused on advanced thermal technologies (ATTs), we refer the reader to the literature on these non-thermal conversion methods. This section is largely devoted to the economic, environmental, and energy (EEE) impacts on SWEATT that the production of "pyro-char" or "black carbon" may have. These terms can be used to include many solid pyrolysis products that might serve as charcoal, biochar, or activated carbon. Charcoal and activated carbon are well documented in the technical literature. However, the literature on biochar is just developing. Biochar applications have recently inspired an International Biochar Initiative, a community of scientists and enthusiasts that envision large-scale conversion of waste biomass into biochar while generating energy at the same time.[40,41] This biochar can be applied to soils, both enhancing soil fertility and mitigating climate change by sequestering CO_2 drawn from the atmosphere.

Charcoal was used as early as 5000 BCE in the smelting of copper, and 2000 years later, it became commonplace in the smelting of iron and bronze. Charcoal was burned to produce temperatures in excess of 1000°C that are needed to produce these alloys. Other uses include blacksmith forges and household cooking, in which maximal heat with minimal smoke is desirable, and the filtering and removal of impurities such as in the spirit or sugar processing industries. Charcoal was commonly made in covered conical piles of wood, sometimes covered with earth, constructed so as to exclude air, thus attaining greater yields of charcoal. Its large-scale production is presumed to have led to widespread deforestation of Europe and Eastern North America in the 18th century, until coal supplanted charcoal as an industrial fuel.

In an extensive review, Antal and Gronli[42] have summarized knowledge of the production and properties of charcoal that has been accumulated over the past 38,000 years. They point out that biomass carbonization can be carried out, leading to high char yields (~30%) by the manipulation of pressure, moisture content, and gas flow involved in the process. The review also provides a good summary of measurements of the heat of pyrolysis from various plant feedstock that range from +0.7 MJ/kg (exothermic) to −0.3 MJ/kg (endothermic). When viewed in the light of the fact that the higher heating values (HHV) for most dry biomass are about 17 MJ/kg and pure carbon is 32 MJ/kg, it should be clear that, in any case, the heat cost of pyrolysis is small compared to the heat content of the feedstock. However, in various practical biomass pyrolysis arrangements, component system losses need to be kept as small as possible so that acceptable conversion efficiencies from the feedstock to the desired form of energy are achieved.

Conventional combustion technologies, or even advanced combustion systems, when applied to biomass/solid waste (SW), leave little carbon in the fly or bottom ash. Thus, it might be difficult to adapt these technologies to the useful production of pyro-char products. Air-blown partial combustion and oxygen-blown partial combustion systems might be adapted since they essentially take advantage of substoichiometric combustion to produce CO fuel rather than CO_2 and, depending upon the oxygen content of the feedstock, could leave a substantial carbon residue. The solid char residues produced during forest fires are an example of natural combustion under limited oxygen conditions. The charred woods and plastics remaining after building fires are further examples of limited oxygen combustion. An extensive technical literature is available on fire and fire protection technology that could be drawn upon if carbonization again becomes a widespread technology. Pyro-char or black carbon, when produced intentionally in partial combustion gasifiers or more efficiently in indirectly heated pyrolyzers, is now being referred to as "biochar." Some uses of biochar are quite ancient, while others are quite recent.[40–47]

It is only fairly recently that another ancient use of biochar has come to the wider attention of environmental scientists. *Terra preta* are small plots (20 ha, on average) of highly fertile Amazonian soils, enriched in organic carbon and nutrients, that are surrounded by Oxisols, typical of tropical soils, that are extremely depleted in nutrients and organic matter. Because *terra preta* are associated with high concentrations of charcoal and ceramic fragments, and can be dated to have formed between 800 BCE and 500 ACE, they are presumed to be anthropogenic, made either intentionally by some method of slash and char forestry for agricultural purposes or accidentally through the dumping of kitchen fire wastes over long time periods.[41] This discovery has inspired the current "biochar movement" and interest in identifying optimum methods of producing and using biochar. It is recognized that the optimum characteristics of biochar are still uncertain and are the subject of research.

The unique properties of some biochar, particularly its high adsorption capacity, can be attributed to its high surface-specific surface area (SA) as well as its surficial functional group content. Although charcoal can be "activated," that is, altered with physical or chemical treatment or by "carbonization," heating above 800°C, to produce extremely high SAs or oxidized surfaces, even non-activated biochars can possess some of these features. Biochar SA tends to increase with pyrolysis temperature but starting biomass type and pyrolysis atmosphere and duration of heating will also play a role. For example, Zimmerman[48] has reported N_2-BET SA of less than 13 m^2/g for a variety of biomass types pyrolyzed at 400°C (3 hr), including grasses as well as softwoods and hardwoods. At 525°C, SA ranged from 31 to 501 m^2/g, and at 650°C, 220–550 m^2/g. Between 800°C and 1000°C, SA of between 400 and 1000 m^2/g are commonly recorded by Downie et al.[44] These measurements, however, include only pores larger than a few nanometers (nm) in diameter. Surface present within micropores (pores smaller than about 1.5 nm in diameter), measured using CO_2 sorptometry, have yielded SA in the range of 160 to 650 m^2/g and have been found to be more strongly related to the ability of a biochar to sorb low molecular weight organic compounds and cations.[40,50]

One can envision a number of possible ways in which the sorbent properties of biochar could be utilized. First, much as activated carbons have been used for many centuries in a wide variety of industrial process that require the adsorption of noxious, odorous, or colored substances from gases or liquids, biochar could be used as a low-cost alternative, especially in circumstances where large volumes of material are required. Much like activated carbon,[51] biochar can be powdered to increase SA or granulated for use in fixed bed filtration systems. Although somewhat lower in SA and, thus, sorption capacity compared to activated carbons, its characteristics can be tuned via production conditions, for sorption of specific components. For example, biochar made from anaerobically digested bagasse has been shown to be a superior sorbent of metal including lead.[52] Its most cost-effective industrial use is likely to be in the areas of primary or secondary water treatment or in contaminant remediation as reactive media for surficial or subsurface permeable barriers such as trenches, wall barriers, funnels and gates, or landfill bottom linings. In all these cases, both the biochar C and the adsorbed C may be sequestered from the atmosphere and, thus, may be considered an additional C sink, or at least an avoided C source. With the 2010 EPA (Environmental Protection Agency) limitations on mercury emissions from coal plants and municipal waste incinerators, one might anticipate a large market increase for mercury-adsorbing activated or nonactivated carbons.

Much as black-carbon-enriched soil such as *terra preta* in the Amazon has been prized for centuries for its ability to produce sustained enhanced crop yields, it is presumed that biochar amendments to soil, if carried out properly, can increase soil fertility in both the United States and perhaps more critically in the third world where soil depletion is reaching critical levels. Some biochars have high cation exchange capacity, lending it the ability to adsorb and retain such essential plant nutrients as nitrate, ammonium, calcium, and potassium.[43,46] Biochars have also been shown to adsorb the critical anionic nutrient phosphate, though the chemical mechanism for this is unclear. Other positive agricultural effects may include better soil moisture retention and the encouragement of unique microbial populations that may be beneficial to plant growth.[53] Thus, while not yet shown on a large-scale basis, biochar amendment may reduce a farmer's costs for fertilizer and irrigation, while reducing runoff of environmentally damaging nutrients into surrounding surface waters and groundwater (cultural eutrophication) and reliance on inorganic fertilizers made using energy from fossil fuels (another CO_2 source).

A recent life-cycle assessment (LCA) study assessed the energy and carbon impacts of four biochar-cropping systems.[47] They found that, for late and early corn stover, switch grass, and yard waste as biomass feedstock sources, the net energy generated was +4116, +3044, +4899, and +4043 MJ t^{-1} dry feedstock, respectively. Most of the energy consumed was in either agrochemicals or feedstock drying, and most of the energy yield was in syngas heat. Net greenhouse gas emissions were negative for both stover types and yard waste, with the majority of the total reductions, 62%–66%, realized from C sequestration by the biochar. For switch grass, however, land use change and field emissions were high enough to drive net emissions to positive. The main conclusion of this LCA analysis was that the energy and carbon impact of small-scale use of pyrolysis systems using yard waste is the most economically favorable at this time. However, many numbers used in these calculations were, by necessity, broad estimates. Much more research is required to improve the inputs to these types of models.

Because of biochar's environmental stability, conversion of biomass to biochar represents a long-term transfer from a C pool rapidly cycling between biomass and the atmosphere to a pool held sequestered within soils or even aquatic sediments. Conversion of 1% of all biomass to biochar each year could reduce the atmosphere CO_2 by 10% in only 14 years (assuming 50% conversion efficiency and no biochar C degradation). These figures are certain not realistic, however. First, it has been shown that biochars degrade abiotically as well as microbially at rates ranging from C half-lives of a few 100 years (for lower-temperature chars, particularly those made from grasses) to 10^5 years for higher-temperature chars with additional losses to be expected from leaching. Second, the amount of biomass that could be reasonably used as feedstock without using major quantities of fossil fuels in the process of gathering and transportation, and without endangering soil stocks, habitat, or human food resource security (i.e., without land-use conversion), is likely in the range of 2.27 Pg C $^{yr-1}$.[55] Aside from C sequestration and enhanced crop growth, further reduction in greenhouse gas concentration and associated climate change may be obtained via reductions in methane (CH_4) production associated with waste land filling and suppression of nitrous oxide (N_2O) production when biochar is added to soils[43,46] and energy extraction. Using this estimate of maximum sustainable feedstock generation and accounting for all possible benefits, biochar production could potentially offset a maximum of 12% of current anthropogenic CO_2-C equivalent emissions each year.[54] Another interesting finding of this study was that the greatest environmental benefits are to be had by the biochar approach in regions of infertile soils or where water resources are scarce. However, where soils are already fertile, and particularly in regions where coal emissions can be offset, bioenergy (see next section) may be a better approach.

It is hoped that production of biochars or application of biochar to soils may soon qualify as a "carbon offset" or be traded on the open market should a "C cap and trade" policy be implemented. The biochar concept has received formal political support in the U.S. and globally. The U.S. 2008 Farm Bill established the first federal-level policy in support of biochar production and utilization programs nationally and biochar has been included in the United Nations Framework Convention on Climate Change (UN-FCCC in Dec. 2009).

Bio-Liquid Fuels

Intensive use of liquid petroleum products, particularly diesel and gasoline, by automobiles, trucks, airplanes, trains, and ships in the 20th century drew down national and global reserves to the point that energy security has become a major concern of the United States, other industrial countries, and the globe in general. During this same period, CO_2, the major product of hydrocarbon combustion, has further increased from its preindustrial level of 280 ppm to its current level of 385 ppm. Global warming is now emerging as the biggest environmental problem of the 21st century and "What to do about CO_2" is the biggest environmental question.[56–59]

Producing liquid fuels from plants could potentially mitigate both security and environmental problems in countries that have land available that is not in food production. Plants use the sun's energy, the atmosphere's CO_2, and the soil's H_2O to make carbohydrates such as cellulosic matter and lignin. When biomass, or a converted form of it, is combusted (oxidized), the CO_2 is returned to the atmosphere as a part of a short-term cycle that can be considered *carbon neutral*. On the other hand, combustion of coal and petroleum fossil fuels adds CO_2 that had been extracted by plants from ancient atmospheres to today's atmosphere.

In contrast to solid and gaseous fuels, the convenience of energy storage and transfer makes liquid fuels much more useful in the transportation sector. Because ethanol, pyrolysis oil, vegetable oil, and biodiesel are biomass-derived liquid fuels that are closer to carbon neutral, they are now under rapidly increasing consideration as replacements of or supplements to conventional diesel and gasoline. The high monetary value of liquid fuels is illustrated in Table 2. Table 7 lists typical properties of these liquid fuels.

Technologies to convert plant simple sugars to ethanol go back to the beginnings of the wine, beer, and liquor industries thousands of years ago. In recent years, genetic manipulation[60] has led to microbes that can convert cellulose to ethanol as well. We refer the reader to the extensive biochemical literature for such recent developments. As compared to ethanol, vegetable oil, biodiesel and diesel in Table 7 have about twice the energy/volume, important in transportation applications.

The properties of pyrolysis oils vary over wide ranges depending upon the feedstock, the rate of heating, the temperature reached, catalysts used if any, the speed of quenching after the polymeric bonds are broken, and other specifics of the thermal processing. Fluidized bed systems with fast heat transfer followed by rapid quenching produce bio-oil yields ranging from 50% to 75% of feedstock weight, with pyro-gas and pyro-char representing most of the remainder material. Extensive R&D efforts are now underway to upgrade pyro-oils into more energy dense, water-free, and oxygen-reduced liquid fuels.[61]

Since the 1973 oil embargo, vegetable oils such as corn, soybean, canola, rapeseed, sunflower, palm, and coconut have been given serious consideration for liquid fuels.[62] In effect, this would be a return to what Rudolf Diesel used with his first compression ignition engine (CIE). Used vegetable oil from fast-food restaurants has become a favorite inexpensive source of such feedstock. The high viscosity of most vegetable oils presents CIE problems, but these problems can be overcome by suitable preheaters. Since the supply of used vegetable oil is limited, there has been rising interest in high-yield, non-food vegetable oils, such as *Jatropha, Camelina,* flax, and algae that can be grown on marginal lands.

TABLE 7 Some Key Properties of Biomass-Derived Liquid Fuels and Diesel

Fuel	C (wt%)	H (wt%)	O (wt%)	HV (MJ/kg)	Density (kg/m^3)	E/vol (GJ/m^3)	Visc. cs
Ethanol	52.2	13.0	34.8	22.6	790	17.9	1.1
Pyro-oil	~45	~8	~47	~18	~1000	~18	var.
Vegetable oil	74.5	10.6	10.8	40.4	906	36.6	46.7
Biodiesel	79.0	12.9	8.0	41.2	920	37.9	4.7
Diesel	87	13	0	45.3	852	38.6	3.2

The trend in bio-oil production is now towards converting vegetable oil to low-viscosity biodiesel by mixing it with an alcohol and a catalyst. In this esterification process, the oil's glycerin is replaced by the alcohol, making a mono-alkyl ester, which greatly reduces the viscosity and slightly increases the HHV. Biodiesel can be used in unmodified diesel engines as a sole fuel or in mixtures with diesel. Emissions of sulfur oxides and other regulated pollutants, apart from NO_x, are generally lower from biodiesel than from conventional diesel. Since biodiesel is derived from plants that were made with solar energy, it is considered approximately carbon neutral.

Several states are initiating non-food vegetable oil programs to meet transportation needs beginning with the fuel needs of agriculture and particularly the needs of the individual farmer. In the Pacific Northwest, canola (rapeseed) has[63] and is being studied but not yet adopted on a large scale because revenues from growing canola are generally somewhat lower than that from wheat. The differential is currently being subsidized in the form of a federal blender's credit of $1.00 per gallon to jump-start this new industry.

In Texas, flax, an annual plant and prolific biomass producer, is now under consideration as a source of vegetable oil for transportation applications. Flaxseed has long been used as the source of linseed oil, which is used as a component of many wood-finishing and other industrial products and as a nutritional supplement. In a description of recent developments by a Texas A&M researcher,[64] "It's kind of like [Texas] is coming full circle. Flax was grown on about 400,000 acres in the 1950s and Texas AgriLife Research at A&M had an active flax breeding program. Those varieties were known nationwide for having good cold tolerance. That's what we needed, a flax variety that was something you could plant in the fall, survive the winter, avoid late freezes, and produce seed in the spring. Now we're evaluating this as a possible biodiesel product or (one which) could be used in the vegetable oil industry."[64]

Florida is developing a bio-oil production program utilizing *Camelina sativa* planted on non-food-producing land.[65] When *Camelina* seed is cold pressed, 30% (by weight) oil is obtainable. The resulting pressed cake called meal has oil content between 10% and 12% (by weight) that can also be extracted with organic solvents, but this process is expensive. For biodiesel production to be economically competitive, it is essential to minimize production costs and maximize all potential revenue streams. The use of the meal as animal feed is one such stream. However, if non-food vegetable oil ever reaches its full transportation potential, the seed meal and plant residue would far exceed what could be consumed by existing cattle herds. With current budget deficits, the prospects for adequate subsidies in the future are not favorable. Thus, additional services or commodities must be developed to generate additional revenues.

In the spirit of this overall study of SWEATT, the use of on-site SWEATT systems to extract additional liquid fuel, energy, biochar, and specialty chemicals from the waste generated in a farmer's overall bio-oil production cycle warrants careful consideration. The meal produced by pressing the canola, flax, or *Camelina* seed and the plant stover could be dried and continuously fed into an on-site pyrolyzer yielding thermal energy, pyro-oil, biochar, and possibly activated carbon or other valuable products. The biochar then could be used together with local biosolids from wastewater sewage to restore or improve the farmer's soil. The pyro-oil could be collected and distilled off-site into additional transportation fuel and chemicals.[61] The thermal energy can be used on-site for heating, drying, hot water, and steam and other farm applications. Along with biochar and a productive outlet for sewage sludge (sometimes called biosolids) and its otherwise unusable water, such a comprehensive strategic program could provide cost-competitive renewable liquid fuel alternatives to diesel from imported petroleum.

The production of biodiesel in the United States increased from 75 million gallons in 2005 to 250 million gallons in 2006 and 450 million gallons in 2007, with an expected capacity of well over 1000 million gallons in the next few years.[66] This is still a small rate compared to over 60,000 million gallons of petroleum-based diesel consumed in the United States in 2009.

Sewage sludge represents a potentially large source of fatty acids for biodiesel production. After completion of the treatment cycle inside a sewage plant, these biosolids can be chemically processed to extract a biodiesel.[66,67] The waste residual solid and liquid can still serve for soil fertility enhancement thanks to its nitrogen, phosphate, and other mineral components.

The displacement of petroleum diesel can be further hastened and increased substantially with rendered animal fats, using thermal animal fat rendering technology, advanced greatly in 1811 by the French chemist Chevault. This technology has been further advanced by recent research and development.[68]

Recycling and SWEATT

Waste to energy is viewed by some groups as a threat to recycling, but the opposite is more likely the case. Municipal or institutional recycling programs in which the household sorts the SW for collective pickup can actually help maximize the return on waste components and minimize environmental problems. For example, if, at a given time, waste newspaper has no recycling value but, instead represents a disposal cost, preprocessing it for use as dry high-energy feedstock for SWEATT systems could be the optimum response. In effect, the marketplace could decide whether to recycle via the material recovery route or via an energy or fuel recovery route. Separation at the source generally requires simpler, less expensive preprocessing technologies than if all the waste sorting was carried out at the SWEATT plant. Sorting at the source also lends itself to a front-end pollution prevention program in which problem materials are separated and directed to hazardous waste facilities.[69,70]

Recycling with sorting at the source also facilitates application of beneficial feedstock-blending strategies. The analytical semi-empirical model (ASEM) study shows that polyethylene can serve as a rich source of energetic hydrogenic compounds. Thus, blending this plastic with biomass in a SWEATT pyrolysis system is expected to yield more energetic pyro-gas or pyro-liquid fuels and higher heat outputs than simple biomass. This has been demonstrated in high-temperature (1000°C) intermediate (between slow and fast) pyrolysis processing of Meals Ready to Eat (MRE) waste in which the heating value of the pyro-gas measured about 900 Btu/ft^3, quite close to the 980 Btu/ft^3 of natural gas. Food waste pyrolysis generally is expected to yield a low HHV pyro-gas (<150 Btu/ft^3). The high HHV of MRE pyrolysis is due to the large percentage of the ethylene monomer (C_2H_4) contributed by polyethylene plastic packaging. Plastics are used extensively in modern agriculture[71] and, after a growing season, present either a disposal problem or a good SWEATT feedstock blending opportunity. Alternatively, since plastics melt at relatively low temperatures, a sorting–recycling program together with very low temperature pyrolysis can relatively easily restore hydrocarbon plastics to liquid fuel forms.[72]

A Biomass Alliance with Natural Gas is a promising fuel blending strategy. Partial combustion gasifiers using biomass feedstock produce a low heating gas (~150 Btu/ ft^3) that will result in de-rating a natural gas-designed turbine-generator system. Coutilizing the biomass pyro-gas with natural gas can insure that the input energy requirement matches the output needs at least until the maximum rating of the generator is required. In a SW/biomass alliance with natural gas, an additional option becomes available when the SW comes from a recycling community. Then, the utility as a means of getting a richer gas to follow peak loads without calling upon the full use of natural gas might prepare and store high-energy plastics for increased use during times of high electricity demand. Another type of co-use strategy might be profitable if natural gas prices are low (≤$4/MMBtu) as they were at the end of 2011 and biodiesel is high (≥ $35/MMBtu). Then, the SWEATT system should be configured and operated and use the pyro gas yield supplemented by cheap natural gas to maximize the most valued pyro-liquid and pyro-chemical yields.[73] Many other opportunities for efficient co-use of domestic resources are described in the Proceedings of an International Conference on Co-utilization of Domestic Fuels.[6]

SWEATT: Summary and Conclusions

The acronym SWEATT was inspired by the senior author's (A.G.) recollection of Winston Churchill's historic call to arms in the darkest hours of World War II while admitting "I have nothing to offer but blood, toil, tears, and sweat." Today, the energy, environmental, economic, and security (EEES) problems of many nations are such as to bring to mind the problems of the World War II era and the need for bold imaginative solutions. The sustainable supplies of now wasted solids particularly in the United States that have an annual energy potential comparable to the total primary energy supply (TPES) contributed by coal as well as the TPES contributed by natural gas is a resource that should be a component of these solutions. Converting this SWEATT could multiply its current contribution to the U.S. TPES by a factor of about 10.

With low-energy density feedstock, minimizing transportations costs is very important. Thus, in addition to robust thermal technologies that can handle municipal solid waste (MSW), small on-site SWEATT systems have a particularly important role to play. Such systems should soon be available to harvest the energy and valuable chemicals in many of the SW categories listed in Table 1, most of which are greenhouse gas neutral and could gainfully be processed on-site, Table 1 does not list oil shale or tar sands in the United States that could substantially increase the available "solid waste" tonnage to address the U.S. need for liquid transportation fuels. A 2005 Rand study[74] shows that low temperature pyrolysis could be used to extract oils from domestic oil shale. The same can be said about hydraulic fracturing for the production of both oil and natural gas. Indeed this "fracking" might evolve as a big game changer if environmental concerns can be mitigated.

Marginal agricultural lands not in food production that could grow hardy, high-yield, fast-growing vegetable oil crops or can be improved to do so represent another form of waste that is gaining attention to solve energy security and climate change problems and particularly the need for biodiesel. Demonstration programs to date look favorable but the biodiesel price comes out higher than conventional diesel (see Table 2). Thus, it is important in such operations to generate additional revenue streams or to lower the costs of production. On-site SWEATT systems that convert the pressed seeds (meal) and plant residual (stover) into valuable pyro-oil, biochar, and thermal energy and reduce waste disposal costs could potentially make biodiesels more price competitive with diesel distilled from petroleum. It must be recognized, however, that imported petroleum is priced by cartels rather than a free market, Hence, countries poorly endowed with petroleum resources should not abandon biodiesel RD&D efforts or subsidies whenever this cartel to regain market share drastically lowers its price as in 1986.

Japan, a country with an outstanding sustainability record, has established the technical and environmental feasibility of large-scale conversion of SW to energy with thermal pyrolysis and gasification systems.[75] This should allay the concerns of environmentalists and risk-averse utility decision makers in other countries. A recent comprehensive report on the environmental performance of thermal conversion technologies throughout the world[76] identified more than 100 facilities that are using conversion technologies to convert MSW (mainly biomass) for energy production. The study used independently validated emissions data from operating facilities in five nations and found that pyrolysis and gasification facilities currently operating throughout the world with waste feedstocks meet each of their air quality emission limits. With few exceptions, most meet all of the current emission limits mandated in California, the United States, the European Union, and Japan. In the case of toxic air contaminants (dioxins/furans and mercury), every process evaluated met the most stringent emission standards worldwide. Facilities with advanced environmental controls are very likely to meet regulatory requirements in California. Thus, the report concludes that thermochemical conversion technologies possess unique characteristics that as a part of an integrated waste management system can generate useful energy and substantially reduce the amount of material that must be landfilled.

Finally, the following are our main conclusions:

The United States and most industrial nations are excessively reliant on imported oil for their liquid fuels and imported natural gas for their gaseous fuels.

The United States is well endowed with solid coal and oil shale, which can be converted into useful liquid and gaseous fuels by ATTs to mitigate its energy security problem, albeit not its problem of excessive CO_2 emissions.

The United States is also well endowed (~2 billion dry tons) with many forms of carbon-neutral forms of SW that can be converted to natural gas supplements, bio-liquid fuels, thermal energy, and pyro-char by on-site SWEATT systems.

ATTs are the fastest and most efficient SW to energy conversion methods. The high volatile content of biomass makes high-temperature pyrolysis of the biomass component of SW a direct form of gasification.

- If environmental and security externalities are included, renewable energy often compete with fossil fuel energy.

Approximate analytic representations of cost of electricity vs. the important controlling variables can be helpful in making SWEATT policy decisions in the face of highly volatile natural gas prices and uncertainties as to values assigned to externalities.

- Co-use of volatiles from the pyrolysis of SW with natural gas can be useful in overcoming a number of EEE problems.

SWEATT generally results in lower harmful emissions than traditional high-temperature waste to energy systems (incinerators).

- Given a free biomass source such as SW, SWEATT becomes cost competitive with NGCC energy production systems when NG fuel prices are above about $4/ MMBtu.
- Blending feedstock and including inexpensive catalysts can enhance production of energy and high-valued products.

Many areas of engineering research will be needed to optimize SWEATT systems.

The solid residuals of SWEATT, biochar, have potential value as soil amendments to boost fertility and as industrial and environmental adsorbents.

- The biochar dimension offers promise of achieving systems with carbon negativity. Thus, the economic profitability of pyrolysis SWEATT systems could be hastened if and when "C credits" or "C trading" becomes available.[40,41]
- Optimizing the production of pyro-char products and its characteristics for specific needs could enhance the economic and environmental benefits of SWEATT systems without greatly reducing its energy benefits.
- SWEATT can be used in parallel with other fuel-generating enterprises such as biodiesel production and presorted waste recycling operations to increase cost- competitiveness and maximize environmental benefit.
- At present, biochar might only deliver climate change mitigation benefits and be environmentally and financially viable as a distributed system using waste biomass, i.e., small-scale on-site SWEATT.[47]

Finally, the authors conclude that applications of SWEATT can play a significant role in the variety of solutions that will be needed to address today's EEES problems.

Acknowledgments

This work was partially supported by a grant from NSF- EAR #0819706, Geobiology and Low Temperature Geochemistry Program to A. Zimmerman. Alex Green would like to thank his many sponsors and collaborators in his quest for alternatives to oil that began in 1973 and in his work on On-Site SWEATT that began with Bruce Green in 1986. In this update effort, he acknowledges the help of Professor Nicholas Comerford, Dr. Deborah Green, and Julio Castro Vazquez.

Nomenclature and Acronyms Used in Chapter

ABPC	Air Blown Partial Combustion
ACE	Analytical Cost Estimation
AGIR	Antares Group Inc. Report
Ar	Aromatics
ASEM	Analytical Semi-Empirical Model
ATT	Advanced Thermal Technologies
BTU	British Thermal Units
CCTL	Clean Combustion Technology Lab
CHP	Combined Heat and Power
CIE	Compressed Ignition Engines
DANSF	Dry Ash, Nitrogen and Sulfur Free
EU	European Union
FC	Fixed Carbon
GACE	Generalized Analytic Cost Estimation
HHV	Higher Heating Values
HRSG	Heat Recovery Steam Generator
ICE	Internal Combustion Engines
IGCC	Integrated Gasifier Combined Cycle
MSW	Municipal Solid Waste
NGCC	Natural Gas-Fired Combined Cycle
NPHR	Net Plant Heat Rate
OBPC	Oxygen Blown Partial Combustion
OFC	Omnivorous Feedstock Converter
PES	Primary Energy Supplies
PNA	Polynuclear Aromatics
PYRO	Pyrolysis Systems
QES	Quaternary Energy Supply quads Quadrillion BTUs
RDF	Refuse Derived Fuels
SES	Secondary Energy Supplies
SW	Solid Waste
SWANG	Solid Waste Alliance with Natural Gas
SWEATT	Solid Waste to Energies by Advanced Thermal Technologies
TES	Tertiary Energy Supply
TPES	Total Primary Energy Supply
V_T	Volatiles
WEC	Waste to Energy Conversion

References

1. Green, A., Ed. *An Alternative to Oil, Burning Coal with Gas,* University of Presses of Florida: Gainesville, FL, 1981.
2. Green, A., et al. *Coal-Water-Gas, An All American Fuel for Oil Boilers,* Proceedings of the Eleventh International Conference on Slurry Technology, Hilton Head, SC, 1986.
3. Green, A., Ed. *Solid Fuel Conversion for the Transportation Sector,* FACT-Vol 12 ASME New York, NY. Proceedings of Special Session at International Joint Power Generation Conference San Diego, 1991.
4. Green, A. *A Green Alliance of Biomass and Coal (GABC),* Appendix F, National Coal Council Report May 2002: Proceedings of 27th Clearwater Conference, Clearwater, FL, March 2003, 2002.

5. Green, A.; Feng, J. A Green Alliance of Biomass and Natural Gas for a Utility Services Total Emission Reduction (GANGBUSTER), Final Report to School of Natural Resources and the Environment, 2003.
6. Green, A.; Hughes, E.; (EPRI), Kandiyoti, R.; (Imperial College London) conference organizers. Proceedings of the first international conference on co-utilization of domestic fuels. Int. J. Power Energy Syst. **2004**, *24*, 3.
7. Green, A.; Smith, W.; Hermansen-Baez, A.; Hodges, A.; Feng, J.; Rockwood, D.; Langholtz, M.; Najafi, F.; Toros, U.; Multidisciplinary Academic Demonstration of a Biomass Alliance with Natural Gas (MADBANG). Proceedings of the International Conference on Engineering Education, University of Florida, Conference Center, Gainesville, FL, Oct 2004.
8. Green, A.; Swansong, G.; Najafi, F. Co-utilization of Domestic Fuels Biomass Gas/Natural Gas, GT2004-54194, IGTI meeting in Vienna June 14–17, **2004**.
9. Green, A.; Feng, J. Assessment of Technologies for Biomass Conversion to Electricity at the Wildland Urban Interface Proceedings of ASME Turbo Expo 2005: Reno-Tahoe, 2005.
10. Green, A.; Bell, S. Pyrolysis in Waste to Energy Conversion (WEC), Proceedings of NAWTEC 14 Conference Tampa, FL, May, 2006.
11. Jenner, M. Biomass Rules. LLC Cited in Natural Gas Futures NYMEX Jan 10th, 2010.
12. Channiwala, S.; Parikh, P. A unified correlation for estimating HHV of solid liquid and gaseous fuels. Fuel **2002**, *81*, 1051–1064.
13. Perlack, R.; Stokes, B.; Erbach, D. Biomass as Feedstock for a Bioenergy and Bioproducts Industry: The Technical Feasibility of a Billion-Ton Annual Supply, Oak Ridge National Laboratory, ORNL/TM-2005/66, U.S. Department of Energy, 2005.
14. Green, A.; Peres, S.; Mullin, J.; Xue, H. *Co-gasification of Domestic Fuels*, Proceedings of IJPGC. Minneapolis, MN, ASME-NY, NY, 1996.
15. Green, A.; Zanardi, M.; Mullin, J. Phenomenological models of cellulose pyrolysis. Biomass Bioenergy **1997**, *13*, 15–24.
16. Green, A.; Zanardi, M. Cellulose pyrolysis and quantum chemistry. Int. J. Quantum Chem. **1998**, *66*, 219–227.
17. Green, A.; Mullin, J. Feedstock blending studies with laboratory indirectly heated gasifiers J. Eng. Gas Turbines Power; **1999**, *121*, 1–7.
18. Green, A.;Mullin, J.; Schaefer, G.; Chancy, N.A.; Zhang, W. *Life Support Applications of TCM-FC Technology*, 31st ICES Conference, Orlando, FL, Jul 2001.
19. Green, A.; Venkatachalam, P.; Sankar, M.S. Feedstock Blending of Domestic Fuels in Gasifier/Liquifiers, TURBO EXPO 2001, Amsterdam, GT, 2001.
20. Green, A.; Chaube, R.; Pyrolysis Systematics for Coutilization Applications. TURBO EXPO 2003, Jun 2003, Atlanta, GA, 2003.
21. Green, A.; Chaube, R.A. A unified biomass-coal pyrolysis reaction model. Int. J. Power Energy Syst. **2004**, *24*(3), 215–223.
22. Green, A.; Sadrameli, S. M. Analytical represntations of experimental polyethylene pyrolysis yields. J. Anal. Appl. Pyrolysis **2004**, 72, 329–335.
23. Sadrameli, S.; Green, A. Systematics and modeling representations of naphtha thermal cracking for olefin production. J. Anal. Appl. Pyrolysis, **2005**, *73*, 305–313.
24. Green, A.; Feng, J. Systematics of corn stover pyrolysis yields and comparisons of analytical and kinetic representations. J. Anal. Appl. Pyrolysis **2006**, *76*, 60–69.
25. Feng, J.; Green, A. Peat pyrolysis and the analytical semiemperical model. Energy Sources, Part A. **2007**, *29*, 1049–1059.
26. Feng, J.; YuHong, Q.; Green, A., Analytical model of corn cob pyroprobe-FTIR data. J. Biomass Bioenergy **2006**, *30*, 486–492.
27. Gaur, S.; Reed, T. *Thermal Data for Natural and Synthetic Fuels*; Marcel Dekker: New York, NY, 1998.

28. Xu, W.C.; Tomita, A. Effects of coal type on the flash pyrolysis of various coals. Fuel **1987b**, 66, 627–636.
29. Mastral, F.J.; Esperanza, E.; Garcia, P.; Juste, M. Pyrolysis of high-density polyethylene in a fluidised bed reactor. Influence of the temperature and residence time J. Anal. Appl. Pyrolysis **2002**, *63*, 1–15.
30. Mastral, F.J.; Esperanza, E.; Berruco, C.; Juste, M. Fluidized bed thermal degradation products of HDPE in an inert atmosphere and in air-nitrogen mixtures. J. Analy. Appl. Pyrolysis **2003**, 70, 1–17.
31. Stultz, S.; Kitto, J., Ed., *Steam* 40th Ed.; Babcock & Wilcox: Barberton, OH, 1992.
32. Liscinsky, D.; Robson, R.; Foyt, A.; Sangiovanni, J.; Tuthill, R.; Swanson, M. *Advanced Technology Biomass-Fueled Combined Cycle,* Proceedings of ASME Turbo Expo 2003, Power for Land, Sea and Air, Atlanta, GA, GT2003–38295, 2003.
33. Phillips, B.; Hassett, S., *Technical and Economic Evaluation of a 79 MWe (Emery) Biomass IGCC,* Gasification Technologies Conference, San Francisco, 2003.
34. Antares Group, Inc. Assessment of Power Production at Rural Utilities Using Forest Thinnings and Commercially Available Biomass Power Technologies. Landover, MD, Sept 2003.
35. ICF Consulting. *Options for Meeting the Electrical Supply Needs of Gainseville* Report March 6, 2006.
36. Roth, I.; Ambs, L. Incorporation externalities into a full cost approach to electric power generation life-cycle costing. Energy **2004,** 129(12–15), P2125–P2144.
37. Green, A.; Palacio, J.; Bell, S.; Avery, M. Sustainable Energy for Communities and Institutions, Proceedings of Sustainability Conference, University of Florida 2006.
38. Green, A. Operations Analysis and the 20th Air Force Slide Rules of WW II, J. Oughtred Soc. **2000**, 9, 212.
39. Mirasgedis, S.; Diakoulaki, D. Multicriteria analysis vs. externalities assessment for the comparative evaluation of electricity generation systems. Eur. J. Oper. Res. **1997**, 102(2), 364–379.
40. Lehmann, J.; Joseph, S. Eds., *Biochar for Environmental Management: Science and Technology.* Proceedings of Newcastle U.K. Conference, Earthscan Publishers Ltd., 2009.
41. Woods, W.; Teixeira, W.; Lehmann, J.; Steiner, C.; Winkler Prins, A.; Rebellato, L. Eds. *Amazonian Dark Earths: Wim Sombroek's Vision;* Springer Publishers: Dordrecht., 2009, 504 p.
42. Antal, M.J.; Gronli, M. The art, science, and technology of charcoal production. Ind. Eng. Chem. Res. **2003**, *42,* 1619–1640.
43. Falcão, N.; Comerford, N.; Lehmann, J. Determining nutrient bioavailability of amazonian dark earth soils—methodological challenges. In *Amazonian Dark Earths: Origin, Properties, and Management;* Lehmann, J., Kern, D., Glaser, B., Woods, W., Eds.; Kluwer Academic Publishers: The Netherlands, 2003, 255–270.
44. Downie, A.; Crosky, A.; Munroe, P. Physical properties of biochar. In *Biochar for Environmental Management: Science and Technology;* Lehmann, J.; Joseph, S., Eds.; Earth- scan: London, 2009; pp. 183–205.
45. Glaser, B. Prehistorically modified soils of central Amazonia: A model for sustainable agriculture in the twenty- first century. Phil. Trans. R. Soc. B Biol. Sci. **2007**, *362,* 187–196.
46. Lehmann, J. Bio-energy in the black. Front. Ecol. Environ. **2007**, 3, 381–387.
47. Roberts, K.; Gloy, B.A.; Joseph, S.; Scott, N.R.; Lehmann J. Life cycle assessment of biochar systems: Estimating the energetic, economic, and climate change potential. Environ. Sci. Technol. **2010**, *44,* 827–833.
48. Zimmerman, A. Abiotic and microbial oxidation of laboratory-produced black carbon (biochar). Environ. Sci. Tech- nol. **2010**, 44(4), 1295–1301, DOI: 10.1021/es903140c.
49. Kasozi, G.N.; Zimmerman, A.R.; Nkedi-Kizza, P.; Gao, B. Catechol and humic acid sorption onto a range of laboratory- produced black carbons (biochars). Environ. Sci. Technol. **2010**, 44, 6189–6195.

50. Mukherjee, A.; Zimmerman, A.R.; Harris, W. Surface chemistry variations among a series of laboratory-produced biochars. Geoderma. **2011**, *163*, 247–255.
51. Bansal, R.; Goyal, M. *Activated Carbon Adsorption;* Taylor and Francis, 2005; 497 pp.
52. Inyang, M.; Gao, B.; Pullammanappallil, P.; Ding, W.; Zimmerman, A.R.; Cao, X. Enhanced lead removal by biologically activated biochar from sugarcane bagasse. Separation Sci. Technol. **2011**, *46*, 1950–1956.
53. Khodadad, C.L.M.; Zimmerman, A.R.; Uthandi, S.; Foster, J.S. Changes in microbial community composition in soils amended with pyrogenic carbon. Soil Biol. Biochem. **2010**, *43*, 385–392.
54. Woolf, D.; Amonette, J.E.; Street-Perott, A.; Lehmann, J.; Joseph, S. Sustainable biochar to mitigate global climate change. Nat. Commun. **2010**, 1, 1–9.
55. Inyang, M.; Gao, B.; Pullammanappallil, P.; Ding, W.; Zimmerman, A.R. Biochar from anaerobically digested sugarcane bagasse. Bioresour. Technol. **2010**, *101*, 8868–8872.
56. Green, A. Transportation. In *Encyclopedia of Climate and Weather;* Schneider, S.H., Root, T.H. and Mastrandrea, M.D. Eds.; Oxford University Press: Oxford, N.Y., 2011, 184–187.
57. Green, A.; Schaefer, G. What to Do with CO_2. Proceedings Turbo Expo 2001, New Orleans paper IGTI 2001-GT-1.
58. Hansen, J.; Mki. Sato, P.; Kharecha, D.; Beerling, R.; Berner, V.; Masson-Delmotte, M.; Pagani, M.; Raymo, D.L.; Royer, and J.C. Zachos, 2008: Target atmospheric CO_2: Where should humanity aim? *Open Atmos. Sci. J,* 2, 217–231, doi:10.2174/1874282300802010217.
59. Princiotta, F. Global climate change and the mitigation challenge. J. Air Waste Manage. Assoc. **2009**, 39(10), 1194–1211.
60. Doran-Peterson, J.; Amruta Jangid, S. K.; Brandon, E.; De- Crescenzo-Henriksen, E.; Ingram, L.O. Simultaneous saccharification and fermentation and partial saccharification and co-fermentation of lignocellulosic biomass for ethanol production. Biofuels Methods Mol. Biol. **2009**, *381*, 263–280.
61. Holmgren, J.; Marinangeli, R.; Elliott, D.; Bain, R. Consider upgrading pyrolysis oils into renewable fuels. Hydrocarbon Process. **2008**, *Sept*, 95–103.
62. Peterson, C.; Ault, D. Technical overview of vegetable oil as a transportation fuel. ASME Fact **1991**, *12*, 45–54.
63. Brown, S.; Henry, C. Natural Selection Farms (Nov. 29, 2010), available at http://wwwkingcounty-gov/environment/wastewater/Biosolids/BiosolidsRecyclingProjects/ NaturalSelectionFarms.aspx (accessed Nov. 30, 2010).
64. Morgan, G. *Flax Finding New Life as Biodiesel Stock— Domestic Fuel.* Agra Life Communications April 4, 2008, College Station, Texas.
65. Vasden, B. *Commercial Biomass and Biofuel Feedstock Farming in Florida;* 2010; available at http://www.usqocom(accessed Dec. 1, 2010).
66. Kargbo, D.M. Biodiesel production from municipal sewage sludges. Energy Fuels **2010**, *24* (5), 2791–2794.
67. Dufreche, S.; Hernandez, R.; French, T.; Sparks, D.; Zappi, M.; Alley, E. Extraction of lipids from municipal wastewater plant. Microorganisms for production of biodiesel. J. Am. Oil Chem. Soc. **2007**, *84*, 181–187.
68. Dias J., et al. Production of biodiesel from acid waste lard. Bioresour. Technol. **2009**, *100*, 6355–6361.
69. Green, A., Ed. *Pollution Prevention and Medical Waste Incineration;* Van Nostrand Reinhold: New York, 1992
70. Wisconsin Department of Natural Resources Agricultural Plastics (Sept. 1, 2010) available at http://dnrwigov/org/aw/wm/recycle/newpages/agplastics.htm (accessed Nov. 30, 2010).
71. Smith, C. Plastic to Oil Fantastic (Apr. 4, 2009), available at http://www.rexresearch.com/ito/ito.htm (accessed Nov. 30, 2010).
72. Akinori Ito Invents a Machine Converting Waste Plastic into Oil (Aug. 23, 2010), available at http://www.neverthelessnation.com/2010/08/akinori-ito-invents-machine-converting.html (accessed Nov. 30, 2010).

73. Vispute, T.P.; Zhang, H.; Sanna, A.; Xiao, R.; Huber, G.W. Renewable chemical commodity feedstocks from integrated catalytic processing of pyrolysis oils. Science **2010**, 330(6008), 1222–1227.
74. Bartis, J.; LaTourrette, T., et al. *Oil shale development in the United States: Prospects and policy issues*; RAND Corp. ISE Division, NETL, US DoE: Pittsburgh, PA, 2005.
75. California Integrated Waste Management Board. Conversion Technologies Report to the Legislature; 2005, available at http://www.ciwmb.ca.gov/Organics/Conversion/ Events Agra Life Communications, College Station, Texas.
76. Bioenergy Producers Association. UC-R Issues Report on Thermal Conversion Technologies (Jan.19, 2010), available at http://www.bioenergyproducers.org/index.php/ucr-report(accessed Nov. 31, 2010).

32

Energy: Walls and Windows

Therese Stovall

Introduction

The building envelope protects you from the weather, separating the indoor air that you have paid to heat or cool from the outdoor air. The walls, roofs, and windows form the major surfaces of this envelope. Energy travels through these surfaces along many paths and in many forms, such as unintended air leakage through a wall or sunlight streaming through a window. We can improve the energy efficiency of the building envelope if we carefully consider all of these energy pathways.

Here, we are focusing on energy efficiency, but keep in mind that each part of the building envelope performs multiple jobs under challenging conditions. The roof keeps out rain, hail, and snow; bakes in the hot summer sun; freezes in the cold winter night; and must be sturdy enough to survive a workman's boots. The walls must repel the rain, hold up the roof, stop the wind, and provide a rigid support for windows and doors. The windows have to let in the light, allow ventilation when they are open, and keep out drafts when they are shut. Any change we make to the building envelope to conserve energy must account for these multiple functions and the complex interactions between the envelope components.

Resistance to heat transfer is often expressed as an *R*-value. For a complete description of R-values and their use, please refer to the DOE Insulation Fact Sheet (http://www.ornl.gov/roofs+walls/insulation).

Building Types

Buildings fall into two main classes: high-rise and low-rise. The high-rise buildings are typically custom engineered with structural steel frames. Low-rise buildings are typically divided into commercial, low-rise multifamily, and single-family buildings. The high-rise, commercial, and low-rise multifamily buildings are more likely to have low-slope (often mislabeled as "flat") roofs, whereas the single-family houses are more likely to have pitched or steep-sloped roofs.

Construction methods can be roughly grouped according to the portion of the assembly performed on site and the portion performed at a factory. With today's engineered wood products and premade trusses, few buildings are strictly built on site; but we still refer to a stick-built building as the one where the greatest part of the assembly takes part on the construction site. At the other end of the spectrum are manufactured buildings that can be moved from one site to another with relatively little effort. In the middle are factory-built modular buildings and panelized construction. A factory-built modular building typically includes one or more modules that are placed on a permanent foundation. In some modular buildings, the windows are installed after the modules have been installed. A panelized building is closer to the site-built model, but will have major wall, floor, or foundation sections prebuilt and delivered to the site.

Every building must conform to local building codes, which can limit the material choices or construction methods. Many local building codes now include energy conservation clauses or incorporate the Model Energy Code or the International Energy Conservation Code.[1]

Walls

The walls make up most of the exterior surface area of many buildings and therefore the energy transported through these surfaces is very important. Wall issues vary according to the building type. A framed building is constructed with a wood or metal skeleton that provides structural support to both the building and all the other wall components. A framed wall is characterized by numerous parallel heat paths and multiple layers of different materials. A non-framed building uses bulk material, such as masonry or adobe, to provide the structural support. This type of building is characterized by a more homogenous heat path and relatively few layers of materials.

For either type of building, the connections between the wall and the roof and between the wall and the foundation are important construction details from an energy conservation standpoint. These connections can provide unintended air passageways and may be overlooked in the overall insulation scheme. An air drywall approach is an effective way to limit air leakage from walls, roofs, and windows. Here, a rubber gasket is fitted along the perimeter of the window frame and along the exterior wall's base board and ceiling plate to compensate for openings that occur as the wood changes shape with time. The gasket, once placed, seals against the drywall gypsum board and makes an airtight barrier.

Whether the wall is built on a frame or constructed from masonry, there is a wide selection of exterior siding choices. These include brick; wood, fiber cement, or vinyl siding; or an exterior insulation finish system (EIFS). For all of these facades, repelling rain and wind is often more complex than it looks. For example, a brick wall looks like a solid surface, but the mortar joints provide capillary paths for moisture, especially when subjected to wind-driven rain. Similar pathways exist for other cladding materials. For this reason, air gaps are often provided behind the outermost wall layer. Depending on the vent/drain arrangement, this may provide a true pressure-equalized rain screen or a simple break in the capillary path. Old-fashioned wood lap siding is a good example of a simple rain screen, as shown in Figure 1. The small air gap behind each wood layer is well vented and drained to the outside, so that the air pressure within the air gap is equal to the air pressure outside the wall. This pressure equalization reduces the moisture moving into the air space, and therefore reduces the amount of moisture available to penetrate the rest of the wall.[2,3]

Proper moisture management is important for energy conservation for two reasons. First, unmanaged moisture must be removed by additional ventilation, which entails the energy load needed to heat and cool the additional air mass. Second, moist building materials will always have a higher thermal conductivity than dry materials. When wet, some insulation materials become matted and lose the greater part of their insulating value.

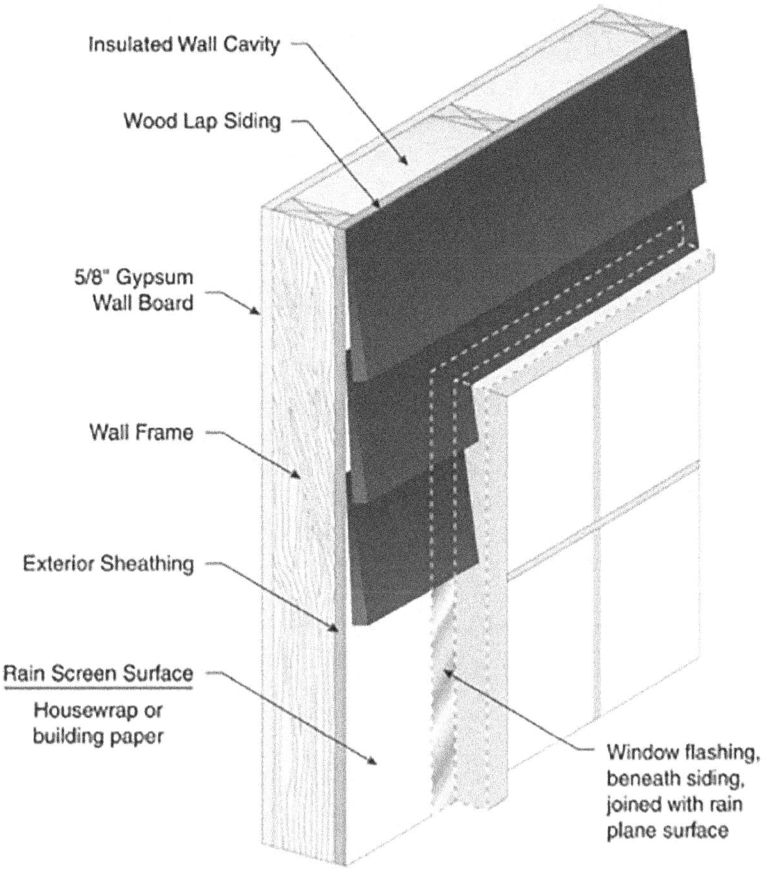

FIGURE 1 Simple rain screen within a typical wood lap siding wall.

Every wall system stores energy. This storage quality is called "thermal mass." The thermal mass can reduce the amount of heat lost or gained through the wall whenever the outdoor temperature varies above and below the indoor temperature on a daily basis. This occurs during the spring, summer, and fall for most of the United States, and during the winter as well for the southern regions. The relative benefit of thermal mass is therefore determined by both the wall properties and the climate. A wood-framed wall has very little thermal mass compared with a masonry wall. One study compared the energy consumed by a house with traditional wood-framed walls with a house constructed with masonry walls for six cities. Depending on the location and wall thickness, the more massive wall reduced the household energy use by an amount equal to increasing the traditional wall's thermal resistance by 10%–50%, with the savings greatest in Denver and Phoenix and least in Miami.[4]

Although we typically think of the exterior walls when we think about energy losses, the interior walls are also important. Air enters the wall cavity through a number of penetrations, some visible and some not. Holes made in the drywall to accommodate electrical outlets and plumbing connections also allow uncontrolled airflow. Other gaps are often present at the floor–wall and floor–ceiling connections. Therefore, it is important to provide a continuous top plate above every wall cavity. This top plate separates the interior wall cavity from the attic space, thus preventing a free flow of conditioned air from your house into the attic.

In addition to the energy used to heat and cool a building, energy is also embodied in the building materials and expended in the construction process. Among low-rise residential buildings, studies have shown that many of the building elements, including gypsum drywall, roofing materials, and carpeting, are common to all the wall types. But the wall type still makes a significant difference in the overall energy embodied in the building, with a wood-framed house containing about 15% less embodied energy than either a metal-framed or concrete house.[5,6]

Wood-Framed Walls

Wall construction methods and materials vary somewhat according to the local climate and natural resources, but most walls in the United States are made from wood framing with insulation between the studs, drywall on the inner surface, and exterior sheathing layer(s) (Figure 2). The wall studs used are either nominal 2 × 4, with a 3.5-in. cavity depth, or nominal 2 × 6, with a 5.5-in. cavity depth. Cavity insulation options for new construction include batts, a blown-in mixture of a foam binder and loose-fill insulation, blown-in loose-fill insulation secured by nets, or blown-in foam insulation.[8] In commercial buildings, high-density batts are sometimes used to provide both thermal insulation and acoustical buffering. In retrofit situations, professional installers can blow loose-fill insulation into the wall cavities by drilling a series of holes through the interior or exterior facade between each pair of adjacent studs.

Energy flows through the wood studs more easily than through the surrounding insulation. Most walls contain much more framing than you would think, as shown in Figure 3. In addition to the studs, there is additional wood framing around each window, around each door, at each building corner, where the wall sits on the foundation, where the roof sits on the wall, where an interior wall meets an

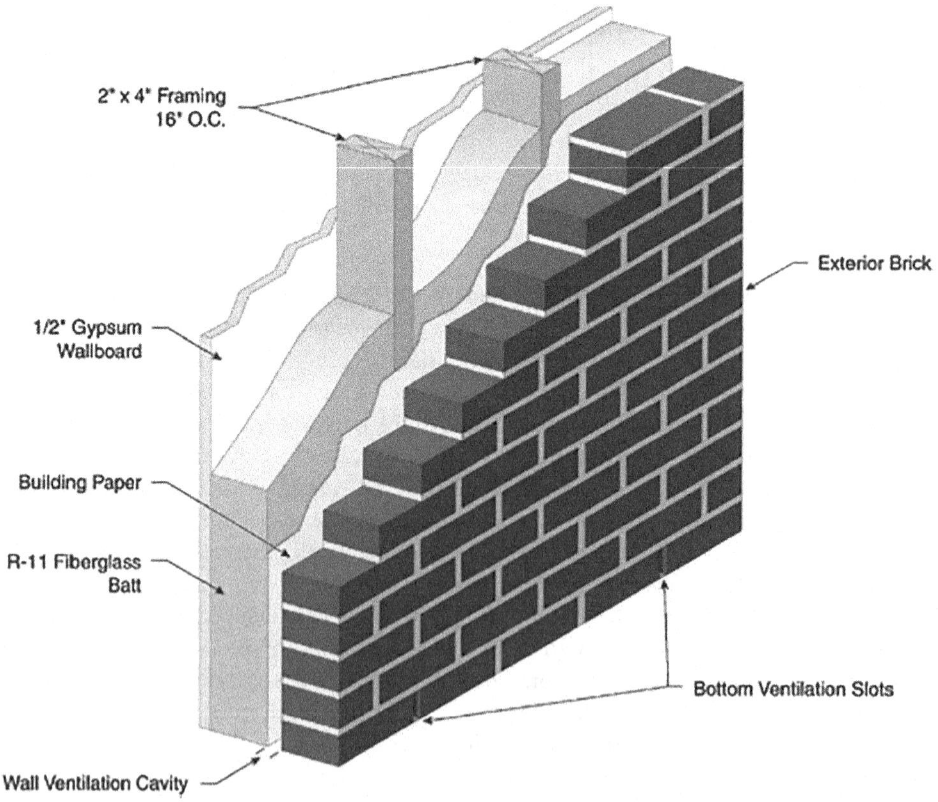

FIGURE 2 Typical wall structure for a wood-framed wall with brick cladding.
Source: ASHRAE Special Publications (see Christian et al.[7])

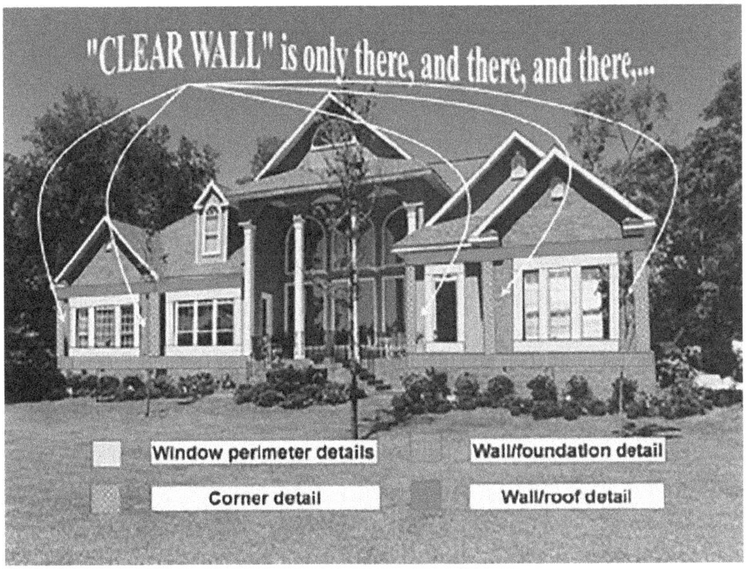

FIGURE 3 Framing lies behind a significant portion of the wall area in many houses.
Source: ASHRAE Special Publications (see Joint Center for Housing Studies of Harvard University, Remodeling, Measuring the Benefits of Home Remodeling[9]).

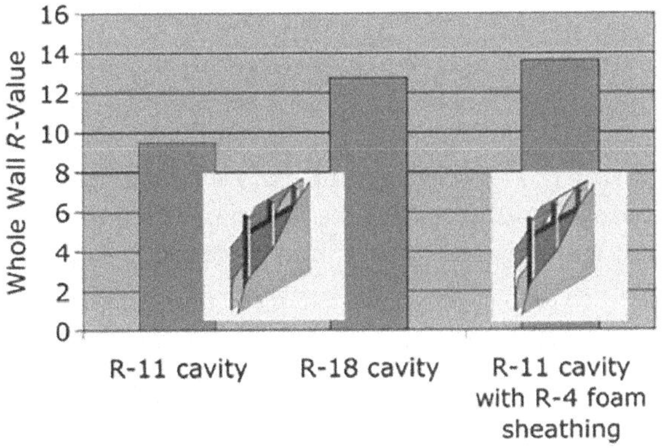

FIGURE 4 Whole wall *R*-values for wood-framed (nominal 2 × 4) walls on 16-in. centers with interior gypsum and exterior wood siding in a one-story ranch house.

exterior wall, and between floors in a multistory building. When you combine all of these thermal "short circuits," a wall (in a one-story ranch style house) filled with R11 insulation provides an overall performance of only R9–R10. If you replace that insulation with foam in the wall cavity, thereby increasing the cavity insulation from R11 up to about R18, you get an overall performance of R13 or an increase that is about half of the increased insulation value. One way to improve the thermal performance of a wall is therefore to place insulation between the studs and the exterior surface of the wall. For example, adding only 1 in. (R5) of foam sheathing to the R10 wall brings its overall R-value up to R14 (Figure 4).

Advanced framing techniques are available which reduce the amount of lumber required. These methods reduce the energy losses through the framing and allow more room for insulation. Many of these techniques also provide for improved air sealing.[10]

The wall sheathing provides a flat uniform surface to support the exterior air barrier, vapor retarder, and siding. If a wood product is used for the sheathing, it will also provide the structural stiffness needed at the building corners. If a foam insulation product is used instead, some form of additional bracing will be necessary at the corners. Sometimes a layer of foam is placed on top of a layer of wood product sheathing. This greatly improves the wall's thermal resistance because that layer covers the thermal short circuits provided by the wood frame. However, this configuration can require extra care during the finishing process, with longer nails needed to fasten siding materials. Also, specialty brick ties will be needed if a brick decor is selected for finishing the wall.

Steel-Framed Walls

Steel-framed walls share many of the characteristics of a wood-framed wall, but the steel components themselves have a very high thermal conductivity. Therefore, most steel-framed walls are built with a layer of foam sheathing to break that thermal pathway. There is also research underway to produce complex steel shapes to provide the structural support needed while providing a longer heat transfer pathway, and therefore greater thermal resistance.[11] Some metal-framed products also include an integral foam insulation element for the same reason. Connections between the walls and a steel-framed roof can be problematic from an energy point of view, especially if the steel framing extends out in the eaves. Such arrangements act like the fins on a heat exchanger and can cause excessive energy consumption if appropriate insulation arrangements are not included in the design. Steel-framed walls are most attractive where the heating loads are modest and where insect damage is more challenging.

Masonry Walls

Masonry walls can refer to walls with a masonry veneer, such as brick or stone, or to a wall where the masonry also provides the structural support, such as a poured concrete or concrete block wall. Such buildings are relatively resistant to the corrosive environment common near the ocean. Masonry also provides thermal mass that can both save energy and improve the interior comfort level in a hot climate with daily temperature swings. Aside from this thermal mass effect, masonry veneer walls share the same energy characteristics as other wood- or steel-framed walls.

Concrete Block and Poured Concrete

Full masonry walls are often used for foundations or basements. Whole houses built from masonry are popular in the warmer climates where termites and other insects are more populous. A full masonry wall can be built from concrete blocks or by pouring concrete into forms. For a load-bearing wall, steel reinforcement rods, or rebar, are used with either method to add strength. For a reinforced concrete block wall, reinforcing rods are placed vertically in the block cavities which are then filled with mortar. Steel wires or mesh are laid in the horizontal mortar joints to resist shear stress. In a poured concrete wall, the steel reinforcing rods are positioned both vertically and horizontally within the forms before the concrete is poured. Because these steel rods tend to be perpendicular to the heat transfer direction, they have little effect on the overall wall thermal resistance.

The thermal characteristics of a masonry wall vary, depending on the whether blocks or poured concrete is used and on the density of the concrete. But in general, the thermal resistance of the masonry portion of such walls will be very small, ranging from R1 to R2 for an 8-in. thick wall, even if the cores of a hollow concrete block are filled with perlite or vermiculite.[12] Foam board insulation, with a thermal resistance in the range of 5*R*/in. is often used to increase the total thermal resistance of a masonry wall and can be placed on the inside and/or the outside surface. The foam board must be covered by some material with an appropriate fire rating, such as gypsum board. One study has shown that the thermal mass is more effective when the concrete is in good thermal contact with the interior building, i.e., when the insulation is placed on the outside of the wall.[13]

Autoclaved aerated concrete can be used in place of ordinary concrete for low-rise buildings. This type of concrete is much lighter than standard concrete, available in a variety of sizes and shapes, and may be reinforced. The autoclaved aerated concrete has a much higher *R*-value (1.25*R*/in.) than the standard concrete (0.05*R*/in.).[3] The overall *R*-value of a wall built with this material will depend on the shape and thickness of the concrete and on the thermal resistance of other wall components, such as air cavities. Several walls tested with aerated concrete (no facings applied), both in the traditional hollow concrete block and solid block forms, had *R*-values between 6 and 9.[14]

Precast Concrete

Walls can be made from reinforced concrete slabs that have been precast into the desired shape. Unless insulation is applied, these walls will have approximately the same thermal resistance as a site-poured concrete wall, i.e., about R1–R2 for an 8-in. thick wall. This construction method is more often used for larger buildings, such as apartment buildings or hospitals, but is also used for residential buildings. In the smaller buildings, the precast panels were first used for basement walls, but precast panels can be used for all exterior walls. The panels are precast and cured in the factory, thus avoiding weather limitations associated with pouring and curing concrete at the building site. The factory environment also allows the production of concrete that is stronger and more water resistant than site-poured concrete. Some of the panels are cast against foam insulation to improve the wall's *R*-value. In addition to the steel reinforcing, these walls can be produced with cavities for electrical wiring and rough openings. Some of the panels have been designed to provide the appearance of bricks and limestone, so that a new building will blend into an existing urban neighborhood. A crane is usually used to place the precast panels on top of a bed of crushed stone. The panels are then connected in place using weld joints or bolts and sealants. Installation for a typical residential unit can be completed in a single day.[15,16]

Insulated Concrete Forms

An insulated concrete form (ICF) wall is composed of a set of joined polystyrene or polyurethane forms that are filled with reinforced concrete, as shown in Figure 5. The joining methods vary from one manufacturer to another. Some have interlocking foam panels; others are linked with ties made from polypropylene or steel. The foam panels become a permanent part of the wall, providing a continuous layer of insulation on both the inner and outer wall surfaces. The concrete portion of the wall is reinforced with rebar positioned inside the forms before the concrete is poured to provide needed strength. The ICF walls can be covered on the outside with light-weight stucco, brick, or wood or vinyl siding. Gypsum board fastens onto the interior side. The attachment methods for these facing materials vary from one manufacturer to another. In the laboratory, the simple steady-state *R*-value of unfinished ICF walls varied from 12 to 18.[14] The time-varying effective energy performance of these walls is also determined by the temperature profile within the wall, which is improved by the walls' well-insulated thermal mass. Therefore, the ICF wall design provides both greater thermal mass and a higher thermal resistance than a typical 2×4 frame wall. Although the airtightness of any wall system tends to vary depending upon the expertise of the construction crew, an ICF wall is generally more airtight than a wood-framed wall.

A variant of the ICF wall uses forms made from a polystyrene–cement composite. The composite walls may have a greater thermal mass, but provide a significantly lower steady-state thermal resistance (about R8 for one 10-in. thick configuration).[14]

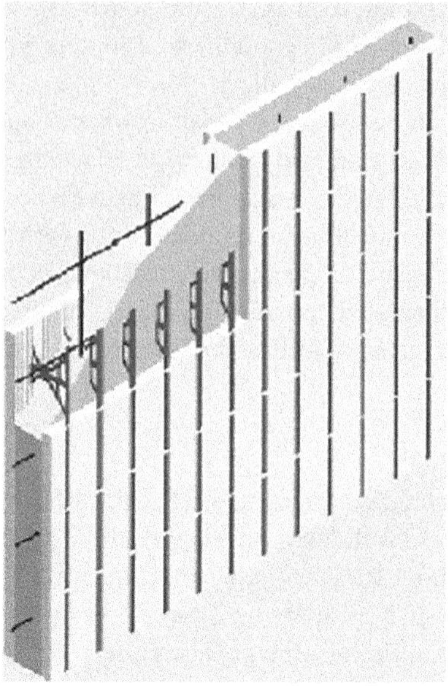

FIGURE 5 Insulated concrete form (ICF) wall.

Adobe

In parts of the southwest, adobe walls are popular because they absorb and store the daytime heat until it can be released to the cooler night air. Traditional adobe bricks are sun-dried, not fired, mixtures of clay, sand, gravel, water, and straw or grass. (For fired or stabilized bricks that are made to look like adobe, see the previous section on concrete masonry). The traditional production method produces a brick that swells and shrinks depending on its fluctuating water content. Adobe walls are relatively fragile and must be sealed with a protective covering.

The low compressive strength of the adobe bricks leads to the use of very thick (10–30 in.) walls that are seldom more than two stories high. These thick solid walls provide significant thermal mass. The thermal mass effect is especially important because the steady-state thermal resistance of these walls is not very great. A 14-in. thick wall would have a thermal resistance of from R2 to R10, depending on the density and water content of the wall. Insulation can of course be added to the interior or exterior face of the wall if covered with an appropriate coating.[17]

Exterior Insulation Finish Systems

The EIFS can be placed on a wood- or steel-framed wall or a masonry wall. In this system, a layer of polystyrene board insulation, one or more inches thick, is applied to the wall which is then covered with multiple coatings that produce the finished appearance of stucco. This wall system provides a continuous cover of insulation that breaks all the thermal short circuits associated with the framing materials and has the thermal advantage previously shown in Figure 4 for foam sheathing. The EIFS system has been and continues to be one of the most popular exterior claddings for commercial and institutional buildings. However, residential construction jobs often do not have the same high level of quality control and job oversight. This difference led to moisture-related problems in residential EIFS walls where moisture seeped into inadequately sealed window openings and became trapped within the walls.

Subsequent building failures led to the development of two classes of EIFS: barrier and drainable. Both classes are used in the commercial building class, but only the drainable system is allowed for residential construction in many locations. In the barrier class, the outer finish layer is designed to be the one and only weather-resistive barrier on the wall. In the drainable class, some form of spacer is placed between the polystyrene board insulation and a second weather-resistive barrier is located atop the wall's structural sheathing layer. This drain plane allows any moisture that seeps into the wall to drain safely out of the wall.[3]

Structural Insulated Panels (SIPS)

The SIPS walls are made by sandwiching foam insulation, typically 4–6 in. thick, between two sheets of a wood product, thus providing structural support, insulation, and exterior sheathing in a single panel. Each manufacturer specifies the proper method and materials to use when joining adjacent panels. Because the panels themselves have such a high thermal resistance, these joints are critical in maintaining a high thermal resistance for the whole wall. Walls have been measured with *R*-values of about R14 for a wall with a 3.5-in. thick foam core and about R22 for a wall with a 5.5-in. thick foam core.[14] The wall sections are relatively light and the exterior walls of a building can often be completed in a day. Two variations of the system include the use of metal sheets for the exterior skin and the use of alternative insulation materials.

Straw Bale Walls

Exterior walls can be made from stacked straw bales. The straw is a natural insulation material, but must be protected from the weather by an exterior surface, often a stucco-type finish applied over a wire mesh. Gypsum board can be used on the interior surface using a number of methods.[3] Experiments have shown that it is very important not to leave any air space between the straw and the surfacing material. Such air gaps work in concert with the hollow straw tubes to set up convection loops that may cut the overall thermal resistance from around R50 down to around R16.[18]

Windows

Transparent glass was first used for windows during Roman times. Technology has gradually advanced, improving the smoothness, strength, and clarity, and increasing the maximum size of manufactured glass. More recently, double-pane windows became popular as energy costs rose and the use of air conditioning became more common. By the mid-1990s, nearly 90% of all residential windows sold had two or more layers of glass.[19] The windows in older homes are being upgraded to double-pane windows as well; from 2000 to 2001, about nine million homeowners spent $15 billion on window and door replacements.[20] From 1990 to 2003, the replacement window market made up from 38% to 52% of the total window market.[21] However, in 2004, two-thirds of residential buildings and about half of the nonresidential building stock still had single-pane windows. The potential energy savings in this population is great, especially with the new selective coating techniques for glass.

Energy Transport

The primary function of windows is to let light and air into a building; so it is not surprising that the windows can be very challenging from an energy conservation point of view. Energy travels through a window by all three energy transfer phenomena, as shown in Figure 6.

The radiation portion of the energy transport includes short-wave radiation, or ultraviolet, that tends to fade the colors in fabric and paint; visible radiation, or light, that we desire; and long-wave radiation or heat. Long-wave radiation is a normal part of the solar spectrum. It helps to heat our buildings during the winter but increases our air conditioning load during the summer. Some values for the solar heat gain through a few prototypical windows are shown in Figure 7. Long-wave radiation also occurs between any two surfaces, traveling from the warmer surface to the cooler surface. Because the outside environment is warmer than the inside surfaces in the summer and the reverse in the winter, this

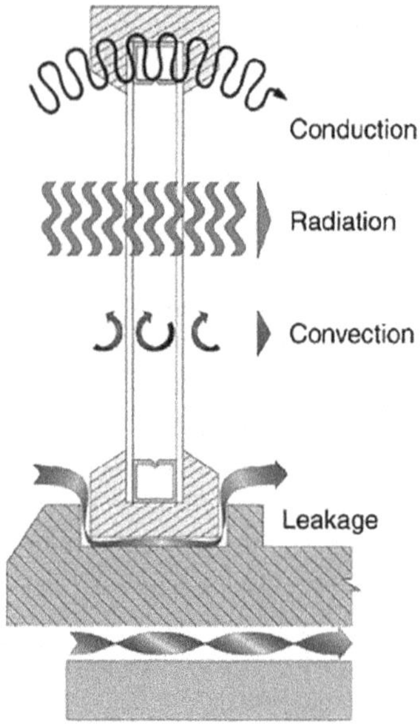

FIGURE 6 Energy traveling through windows.

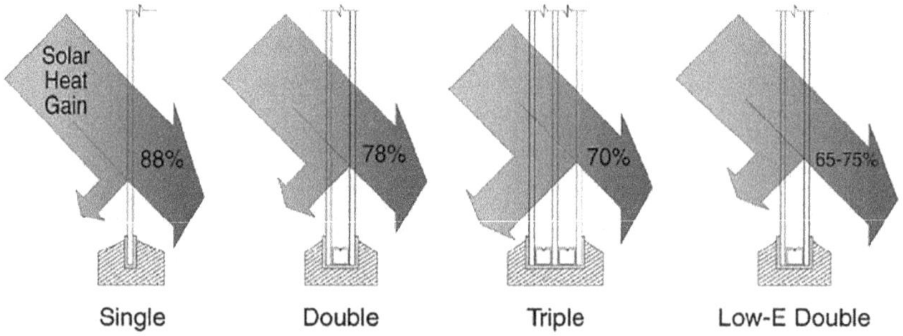

FIGURE 7 Solar heat gain through various windows.

long-wave radiation travels through the window and increases both our winter heating and summer cooling energy use. Newer windows have a "low-e" coating to reduce long-wave radiation and thus reduce the energy losses.

The conduction portion of the energy transport includes the heat that travels through the window frame and heat conducted through the glass pane(s) and through any gas between the panes. The energy that travels through the window frame and sashes is a complex function of the frame material(s), the shape (including any hollow cavities), and the exposed surface area. Each material used in windows has positive and negative qualities. In general, wood has a lower thermal conductivity than metal or plastic. However, wood is more likely to change in shape during its lifetime, so that sealing air leakage out of a wood window over a long time period may be more problematic. Also, wood is more susceptible to moisture damage and must be kept painted or varnished. Metal frames will conduct more heat than wood, but the exposed area can be reduced because the metal is a stronger material. Some of the best performance is found in windows that combine multiple materials. For example, a metal- or vinyl-clad window frame will require less maintenance than a wood window, but will conduct less heat than a solid metal or solid plastic frame. Some special gases, such as argon, have a lower thermal conductivity than air and are sometimes used to fill the gap between panes in multiple-pane windows.

The convection portion of the energy transport includes exterior air movement, or wind, across the glass surface; gas movement between the panes in a multilayer window; and air movement across the interior face of the window. The gas movement within a multilayer window is determined by the thickness of the gap, the height of the window, the gas temperature, and the temperature difference between the two panes of glass. Closed draperies or shades can reduce the air flow across the inside pane of glass.

The energy carried by air leakage falls into two major categories. As Figure 6 shows, some air leaks around the window frame itself. This leakage path should be sealed when the window is installed, although caulking around the frame of an existing window can also reduce the air leakage. Air also leaks through any moving joint in a window. These joints must be sealed by weather stripping or special gaskets that are built into the windows themselves.

Storm windows have been used for a long time and were very popular in northern climates before the introduction of multiple-pane windows. Storm windows require annual installation and storage and there are more glass surfaces to clean. Storm windows can be mounted on either the inside or outside of the window frame.[22] Tests have shown that the energy savings can be substantial when storm windows are added to an existing single-pane window. However, replacing an existing single-pane window with a modern double-pane window will save more energy than the addition of a storm window.[23,24]

Window Rating Systems

Considering the complexities of energy transport through windows, it can be difficult for consumers to compare one window with another. Fortunately, there are two important tools available to help with window selection. The National Fenestration Rating Council (NFRC) has developed a rating system that includes a standard label.[25] The U.S. Departments of Environmental Protection and Energy have cooperated in the production of an Energy Star label for windows.[26]

The NFRC label (Figure 8) shows four values: the *U*-factor, the solar heat gain coefficient (SHGC), the visible transmittance, and air leakage. Manufacturers may also choose to show a value for condensation resistance. The two most important factors used to rate the energy efficiency of a window are the

FIGURE 8 National Fenestration Rating Council (NFRC) label.

U-factor and the solar heat gain coefficient. The *U*-factor is the inverse of the *R*-value (the label that is quoted for insulation), so that a lower *U*-factor indicates a slower rate of heat transfer for any given temperature difference.[27] The NFRC *U*-factor ratings for windows sold nowadays range from 0.2 to 1.2.[28] The SHGC ranges from 0 to 1 and measures how much heat from the sunlight incident upon a window will enter the building. Windows with lower SGHC ratings do a good job of blocking this heat.

The Energy Star label is available to windows that have been certified and labeled by NFRC and that meet special standards. The standards vary among the four NFRC climatic regions because of the trade-off between desirable winter heating and undesirable summer heating. The four regions used by the Energy Star program for windows are shown in Figure 9 and Table 1 shows the required *U*-factors and SHGCs for each region.[29]

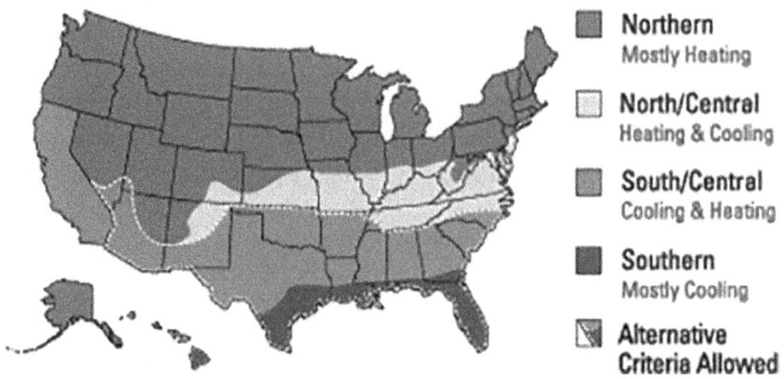

FIGURE 9 Energy Star's four regions used for window rating program.

TABLE 1 Energy Star Window Criteria

Climate Zone	U-Factor	Solar Heat Gain Coefficient G-Factor (SHGC)	
Northern	≤0.35	Any	
North/Central	≤0.40	≤0.55	
South/Central	≤0.40	≤0.40	Prescriptive
	≤0.41	≤0.36	Equivalent performance (excluding CA)
	≤0.42	≤0.31	
	≤0.43	≤0.24	
Southern	≤0.65	≤0.40	Prescriptive
	≤0.66	≤0.39	Equivalent performance
	≤0.67		
	≤0.68	≤0.38	
	≤0.69	≤0.037	
	≤0.70		
	≤0.71	≤0.36	
	≤0.72	≤0.35	
	≤0.73		
	≤0.74	≤0.34	
	≤0.75	≤0.33	

Future Improvements

Researchers are working on a portfolio of window designs with automatic energy-saving features. Some forms of "smart" windows with switchable glazing have become commercially available, albeit at a relatively high cost. These windows vary the amount of light and heat transmitted based upon an electric current, which is usually programmed to respond to either the temperature or the amount of sunlight hitting the window. The electrochromic window uses a multilayer electrically conductive film where ions are moved from one layer to another by a short electrical signal. In one layer, the ions allow only 5% of the sunlight through the window; when the ions move to the other layer, 80% of the sunlight is transmitted. This system has the advantage that once the change from one state to another has been made, no electrical energy is required to maintain that state. Another switchable glazing, the suspended particle display (SPD) places a solution containing suspended particles between two glass panes. When an electrical charge is applied, the particles align and light is transmitted through the window. Without an electrical charge, the particles move about randomly, blocking up to 90% of the light. Other switchable windows are designed to provide privacy by changing from transparent to translucent, but are not effective at saving energy.[30,31] Another proposed window design includes sensors that automatically raise or lower a blind enclosed between two panes of glass based upon the outdoor temperature and solar radiation. This design admits solar radiation when it will help heat the house, but lowers the blind to block solar radiation when it will increase the air conditioning load.[32]

Conclusion

Walls and windows are often selected to achieve a desired appearance. Considering today's emphasis on energy conservation and overall sustainability, it is important to consider their thermal characteristics as well. Selecting more energy-efficient walls and windows will permit the building designer to specify a smaller heating and cooling system, so that often the total building cost is little more than that of a standard building. When you consider the reduced cost of heating and cooling the building during its lifetime, any added investment during construction is returned many times over.

Other Useful Guides

Graphic Guide to Frame Construction: Walls, by *Rob Thallon,* Published by Taunton Press, 2000, ISBN 1-56158-3537, # 070470, (http://www.taunton.com/finehomebuilding/pdf/Framing%20Walls.pdf) and http://www.taunton.com/finehomebuilding/pdf/Grading%20and%20Drainage.pdf.

References

1. Available at http://www.energycodes.gov/ (accessed February 12, 2007).
2. Ontario Association of Architects, *The Rain Screen Wall,* available at http://www.cmhc-schl.gc.ca/en/inpr/bude/himu/coedar/loader.ffm?url=/commonspot/security/getfilecfm&PageID=70139, other related publications from the Canadian Mortgage Housing C. available at http://www.-cmhc-schl.gc.ca/en/ (accessed February 12, 2007).
3. Toolbase Services, a consortium including the U.S. Department of Housing and Urban Development and the National Association of Home Builders (NAHB) provides a wealth of information about current and future building materials and housing systems, NAHB Research Center, Maryland, 2005, available at http://www.toolbase.org (accessed February 13, 2007).
4. Kosny, J.; Kossecka, E.; Desjarlais, A.O.; Christian, J.E. In *Dynamic Thermal Performance of Concrete and Masonry Walls,* Thermal Envelopes VII, Conference Proceedings, Clearwater Beach, FL, December 1998; ASHRAE Special Publications, 1998.

5. Lippke, B.; Wilson, J.; Perez-Garcia, J.; Bowyer, J.; Meil, J. CORRIM: life-cycle environmental performance of renewable building materials. Forest Prod. J. **2004**, *54* (6), 8–19, available at http://www.corrim.org/reports/pdfs/FPJ_Sept2004.pdf, (accessed February 13, 2007).
6. Brown, M.A.; Southworth, F.; Stovall, T.K. *Towards a Climate-Friendly BuildtEnvironment*; Pew Center on Global Climate Change: Arlington, VA, 2005; available at http://www.pewclimate.org/document.cfm?documentIDZ469 (accessed February 13, 2007).
7. Stovall, T. K., and Karagiozis, A., Airflow in the Ventilation Space Behind a Rain Screen Wall, Thermal Envelopes IX, Conference Proceedings, Clearwater Beach, FL, December 2004; ASHRAE Special Publications, 2004.
8. U.S. Department of Energy, Technology Fact Sheet: Wall Insulation, October, 2000, http://www.eere.energy.gov/buildings/info/documents/pdfs/26451.pdf (accessed February 13, 2007).
9. Kosny, J.; Mohiuddin, S. A. Interactive Internet-Based Building Envelope Materials Database for Whole-building Energy Simulation Programs, Thermal Envelopes IX, Conference Proceedings, Clearwater Beach, FL, December 2004; ASHRAE Special Publications, 2004.
10. Toolbase Services, a consortium including the U.S. Department of Housing and Urban Development and the National Association of Home Builders (NAHB) provides a wealth of information about current and future building materials and housing systems, NAHB Research Center, Maryland, 2005, found at www.toolbase.org.
11. Kosny, J. Steel Stud Walls; Breaking a Thermal Bridge. Home Energy July 2001, http://www.homeenergy.org/article_preview.php?id=249&article_title=Steel_Stud_Walls:_Breaking_the_Thermal_Bridge (accessed February 13, 2007).
12. ASHRAE Handbook of Fundamentals, 2005 Chapter 25.
13. Kosny, J.; Petrie, T.; Gawin, D.; Childs, P.; Desjarlais, A.; Christian, J. Thermal Mass—Energy Savings Potential in Residential Buildings, August 2001, http://www.ornl.gov/sci/roofsCwalls/research/detailed_papers/thermal/index.html (accessed February 13, 2007).
14. Advanced Wall Systems Hotbox Test R-value Database, Oak Ridge National Laboratory, http://www.ornl.gov/sci/roofsCwalls/AWT/Ref/TechHome.htm.
15. NAHB Research Center, ToolBase Services, PATH Technology Inventory, Precast Concrete Foundation and Wall Panels, Upper Marlboro, MD, 2005, http://www.toolbase.org/Technology-Inventory/Foundations/precastconcrete-panels (accessed February 13, 2007).
16. Precast/Prestressed Concrete Institute, Single Family Housing, http://www.pci.org/markets/projects/single.cfm (accessed February 13, 2007).
17. Thermal conductivity estimated considering the reported thermal conductivity of unfired clay bricks found at http://www.constructionresources.com/products/ and the measured thermal conductivity of a wall built from rammed earth bricks reported at http://www.ornl.gov/sci/roofsCwalls/AWT/Ref/TechHome.htm (accessed February 13, 2007).
18. Christian, J.E.; Desjarlais, A.O.; Stovall, T.K. In *Straw Bale Wall Hot Box Test Results and Analysis*, Conference Proceedings, Clearwater Beach, FL, December, 1999; Thermal Performance of the Exterior Envelopes of Buildings VII, 1999; 275–285.
19. Carmody, J.; Selkowitz, S.; Heschong, L. *Residential Windows, A Guide to New Technologies and Energy Performance;* W.W. Norton and Company: New York, 1996.
20. Joint Center for Housing Studies of Harvard University, *Remodeling, Measuring the Benefits of Home Remodeling;* Harvard University: Cambridge, MA, 2003.
21. U.S. Department of Energy, 2004 Buildings Energy Databook, available at http://buildingsdatabook.eren.doe.gov/default.asp
22. U.S. Department of Energy, A Consumer's Guide to Energy Efficiency and Renewable Energy, Storm Windows, available at http://www.eere.energy.gov/consumer/your_home/windows_doors_skylights/index.cfm/mytopicZ13490.
23. Klems, J.H. In *Measured Winter Performance of Storm Windows*, ASHRAE 2003 Meeting, Kansas City, MO, June 28–July 2, 2003; LBNL-51453.

24. Turrell, C. Storm windows Save Energy, Home Energy Magazine Online, July/August 2000, available at http://homeenergy.org/archive/hem.dis.anl.gov/eehem/00/000711.html.
25. NFRC. Various informative fact sheets regarding window issues. Available at http://www.nfrc.org/factsheets.aspx (accessed February 13, 2007).
26. Available at http://www.energystar.gov/index.cfm?cZwindows_doors.pr_windows.
27. The *U-factors* used by the NFRC and the Energy Star program have units of Btu/h-ft^2-°F.
28. Available at http://www.nfrc.org/label.aspx (accessed February 13, 2007).
29. Available at http://www.energystar.gov/index.cfm?cZwindows_doors.win_elig.
30. Fanjoy, R. Smart Windows Ready to Graduate, HGTV Pro. com, available at http://www.hgtvpro.com/hpro/dj_construction/article/0,2619,HPRO_20156_3679648,00.html (accessed February 13, 2007).
31. Toolbase Services, a consortium including the U.S. Department of Housing and Urban Development and the National Association of Home Builders (NAHB) provides a wealth of information about current and future building materials and housing systems, NAHB Research Center, Maryland, 2005, Available at http://www.toolbase.org.
32. Kohler, C.; Goudey, H.; Arasteh, D. A First-Generation Prototype Dynamic Residential Window, LBNL-56075, Lawrence Berkeley National Laboratory, Berkeley, CL, October 2004, available at http://www-library.lbl.gov/docs/LBNL/560/75/PDFZLBNL-56075.pdf (accessed February 13, 2007).

33

Energy: Waste Heat Recovery

Martin A. Mozzo, Jr.

Introduction

In many industrial and commercial energy applications, only a portion of the energy input is used in the process. The remainder of the useful energy is rejected to the environment. This rejected energy may potentially be recaptured as useful energy through waste heat recovery. Not all rejected energy can be recovered due to quality, usefulness in a host's load profile, and/or economical reasons that may make its recovery infeasible. This entry will serve as a guide to waste heat recovery in order to provide a framework for the energy engineer to develop waste heat recovery projects. It will discuss the concept of quality vs quantity, engineering concerns in waste heat recovery, sample calculations for waste heat recovery, and the types of waste heat recovery equipment that can be used.

Concept of Quality vs. Quantity

While at first glance, a waste heat recovery project may appear feasible, this may not always be the case. The concept of quality vs. quantity plays an important role. For the purposes of this entry, there are three classifications of waste heat. These are: (1) high-grade waste heat, generally 1000°F and above; (2) medium-grade waste heat, generally in the range of 400°F–1000°F; and (3) low-grade waste heat, generally below 400°F. Typically, the higher the grade of waste heat, the better the application for a successful and economical waste heat recovery project. This is referred to as the "quality" of the waste heat. It is better to have marginal amounts of high quality waste heat than large quantities of lower-grade waste heat.

There are numerous reasons why higher quality waste heat recovery provides a better application. One reason is that the waste heat stream has a much higher temperature than the recovery medium, resulting in higher temperature differences, and, thus, waste heat recovery equipment sizes can be smaller. This is a result of the higher heat transfer efficiency between the waste heat stream at a higher temperature to a lower temperature heat recovery medium. Second, the recovery medium can take on many forms, namely air, steam, or hot water, whichever is best suited for the host facility, whereas, with lower quality heat recovery, the form may be very limited, especially with low- grade waste heat. For example, if you have a waste heat stream that is at 400°F, steam is not a viable heat recovery medium, as the temperature difference will be too low. Finally, cost and savings become important issues in regard to the quality of the waste heat recovery system. Generally, the higher the quality of waste heat, the lower the capital costs for waste heat recovery equipment. This is because better heat transfer efficiencies will be realized with heat recovery equipment. Additionally, the higher the quality of the waste heat stream, the higher the savings usually are. Thus, with lower costs and higher savings, simple payback periods will be shortened, a result which is generally well- received by management.

A word of caution should be offered regarding the quality of waste heat. There are vendors who will reduce the grade of the available waste heat so that their equipment can be used in a heat recovery application. One such vendor manufactured only low temperature heat transfer equipment, which he was trying to sell as waste heat recovery equipment. His proposals included the dilution of high-grade waste heat streams by mixing the high temperature waste heat stream with ambient air in order to bring the temperature down to levels that his equipment could handle. His argument was that the total Btu content of the waste heat stream was not changed, just the temperature. Unfortunately, in reducing the temperature and quality of the waste heat stream, the size of the heat recovery equipment had to be increased to recover the waste energy, and the type of waste heat recovery medium was limited. This results in higher costs, lower savings, and longer-term payback periods.

There is one instance where the dilution of a waste heat stream may be justified. Steel is usually used in the manufacture of waste heat equipment, and there is an upper temperature limit for the waste stream where steel can be used. A substitution of other materials, for example, ceramics, can be made; however, the capital costs may become too prohibitive. In this instance, the dilution of the waste heat stream may be justified to bring the waste stream temperature down for the safe operation of the waste heat recovery equipment. It should, however, never be diluted simply to enable the use of a particular piece of heat recovery equipment. The primary reason is that by diluting the waste stream, energy savings may decrease and equipment costs will definitely increase; thus, payback periods will increase, perhaps significantly.

Engineering Considerations

There are several engineering factors that must be evaluated when considering and designing a waste heat recovery system. In terms of steps, these are: (1) quantifying the waste heat stream; (2) determining the value of the waste heat stream; (3) evaluating the best form of heat recovery for the host facility; (4) determining the host site heat load profile; (5) determining the grade of waste heat; (6) determining the cleanliness and quality of the waste stream; and (7) selecting the proper waste heat recovery equipment by considering size, location, and maintainability.

The first step that should be executed is to quantify the waste heat stream by determining how many Btu/h are in the waste stream. The equation to calculate this is as follows:

$$Q = M \times \text{specific heat} \times \text{delta temp} \tag{1}$$

where Q = total heat flow rate of waste stream in Btu/h; M = mass flow rate in Lb/h; specific heat = for air, 0.24 (Btu/Lb/°F); delta temp = ($T_{\text{upper}} - T_{\text{lower}}$) in °F.

The specific heat (Cp) changes as the temperature of the air rises. For example, at 1000°F, Cp is 0.26. Since this value is higher than that of Eq. 1, using the lower value, rather than correcting for temperature, provides a conservative calculation of the total heat rate in the waste stream.

The mass flow rate (M) is calculated as follows:

$$M = \rho \times V \tag{2}$$

where M = mass flow rate in Lb/h; ρ = density of the waste stream in Lb/ft^3; V = volumetric flow rate in ft^3/h (in standard cubic feet).

Note that the value of Q is not the total amount of waste heat that will be recovered, but, rather, the total amount of waste heat that is ideally available for recovery. Not all of this waste heat will be recovered, or even can be recovered. The total amount that will be recovered will be determined by numerous other factors, such as the cleanliness of the waste stream and the form of recovery (i.e., high pressure, superheated steam, saturated steam, or hot water). This step is necessary in determining if there are sufficient volumes available for waste heat recovery.

One common mistake committed when quantifying the amount of waste heat available is assuming values for flow and/or temperatures without making measurements. Too often, actual conditions vary from assumed conditions, a variation that can often cause disastrous results in a waste heat recovery project. The cost of making actual field measurements is a small price to pay to obtain reliable data.

After the quantity of the waste stream is correctly determined, step two is to find the dollar value of the waste heat stream to determine how much capital cost a potential project can bear, or if it even makes economical sense. In all waste heat recovery projects, the heat recovered will displace a medium, such as steam, which would have to be generated using another piece of equipment, such as a boiler. This equipment likewise has a related efficiency, and the heat output is always less than the heat input. To determine the dollar value of the waste heat stream, use Eqs. 3 and 4 below:

$$\text{Value} = Q \times \text{unit cost} \tag{3}$$

where value = the dollar value of the waste heat stream, per hour; Q = total heat flow rate of waste stream in Btu/h as calculated from Eq. 1; unit cost = unit cost of the waste stream in dollars/Btu.

$$\text{Unit cost} = \frac{\text{fuel cost}}{\text{efficiency}} \tag{4}$$

where unit cost = dollars/Btu; fuel cost = cost for fuel displaced in dollars/Btu; efficiency = efficiency of unused equipment. For example, a steam boiler @ 75%.

The third through fifth steps tend to overlap and be interrelated, and are discussed together. In order for waste heat recovery to be acceptable to a host facility, its use must be consistent with current energy usage at the facility. To best apply waste heat recovery, the engineer needs to examine current energy usage patterns, heat loads for the site, and the quality of the waste heat that is available. The load factor of the waste heat recovery stream is the first thing to determine. For example, the waste heat recovery stream may be exhaust air flows from a process furnace that is periodic in nature, rather than a continuous operation. In this case, the best potential would be to return the waste heat to the process in some manner that allows the usage to follow the waste heat stream generation. Examples include utilization in a drying operation or as preheated combustion air for the process furnace. In another situation, the process furnace may run continuously and be of a high enough grade to generate steam; however, if the host facility only uses steam for heating in the winter, the loads do not match and, thus, steam is not a good recovery medium. On the other hand, there may be a need for large quantities of hot water in the facility, and the loads may match up. A thorough evaluation of both the source and operation of the waste heat and potential uses to recover the waste heat must be performed.

The sixth step, determining the cleanliness and quality of the waste stream, is important. If the waste heat stream is the product of the combustion of a natural gas-fired or fuel oil-fired operation, then the gases should be relatively safe for waste heat recovery. The important issue here would be the condensation of acidic liquids in the products of combustion gas stream, especially if fuel oil is used. This condensation would happen if the temperature of the waste stream is allowed to drop to low temperatures, typically below 300°F. The waste stream could also contain a burn-off from the product itself, which could condense on the waste heat recovery equipment and cause blockages. Here, it is best to maintain the exit temperature significantly above the condensation temperature to avoid the formation of acids or other potential hazards from the waste stream.

If the waste heat stream contains the products of combustion and the products of the process, great care must be exercised when utilizing the waste heat. The selection and design of waste heat recovery equipment could lower the waste heat exhaust temperature below acceptable levels, resulting in condensation or blockage problems with the heat recovery equipment. When unsure of the condensation temperatures and the effect on the heat recovery equipment, simple tests must be conducted to obtain this information.

Some actual examples wherein the waste heat stream was dirty and created problems follow. The first was an air-to-air heat recuperator used in a process operation. The products of combustion gases were dirty, creating a buildup on the heat exchanger. A soot blower was used to remove the buildup on a periodic basis; however, this was not totally successful. During actual heat recovery operations, the soot blowers could not remove the buildup adequately. A solution to this problem occurred accidentally. The operation was a 24 h/five days a week, and by mistake one weekend, the soot blower was left on, while the waste heat recovery equipment was shut off. Because the buildup and the heat exchanger were composed of different materials, the heat exchanger and the buildup thermal contraction rates were different. As a result, the soot blower was able to break up the buildup on the heat exchanger surfaces. This method was then used to clean the heat exchanger each week. The efficiency of the heat exchanger did decrease somewhat during the week, though not significantly. During the weekend, the buildup was blasted off the heat exchanger surface, thus increasing the heat exchanger's efficiency at the beginning of another week.

The second example did not turn out as well. In this scenario, the quality of the waste heat stream was low, and the products of combustion, combined with product burn-off, created a dirty waste stream. As a result, a serious problem quickly developed. With the waste heat recovery system in operation, gas products in the waste heat stream condensed on the heat exchanger surfaces, and within a very short period of time, the heat exchanger was completely blocked. The only corrective action was to remove the waste heat recovery equipment from service and steam clean it, a very costly operation that eventually resulted in the failure of the project.

The seventh step in the design of a heat recovery system is the selection of the waste heat recovery equipment. Descriptions of several different types of equipment are given in another entry in this encyclopedia. Selection of the proper piece of equipment will be based on the quality of available waste heat; the recovery fluid that can be generated, i.e., steam, hot water, or air; the location of the equipment; and the maintenance capabilities of the host facility. While the first two criteria should be obvious, the last two are sometimes neglected.

The location where the equipment is to be placed is important. It obviously should be close to the waste heat stream, yet should not interfere with the other equipment involved in the generation of waste heat, for example, a process furnace. The equipment also must not be squeezed into an area where maintainability is an issue. Sufficient access for maintenance and overhaul must be included since these pieces of equipment will require periodic maintenance. Accessibility to perform maintenance is not just desirable, but mandatory.

Sample Calculations

This section describes some sample calculations that are used in the design of waste heat recovery systems. The first step is to calculate the amount of waste heat that is available for recovery. This is done using Eq. 1.

Example—Waste heat is available in the form of hot products of combustion from a natural gas-fired furnace. Assume that there are no contaminants in the waste heat stream. Measurements indicate that there are 14,000 available actual cubic feet per minute (ACFM) flowing from a process at 1000°F. The actual volumetric flow must first be converted to standard cubic feet per minute (SCFM) at 60°F, using the following equation:

$$\text{SCFM} = \text{ACFM} \times (T_{\text{absolute}} + 60)/(T_{\text{absolute}} + T_{\text{actual}}) \quad (5)$$

where SCFM = standard cubic feet per minute; ACFM = actual cubic feet per minute; $T_{absolute} = 460°F$; T_{actual} = actual gas temperature.

Using the data available here,

$$\text{SCFM} = 14{,}000 \times \left(460 + 60\right)/460 + 1000\right) = 4{,}986\ ft^3/\text{min}$$

Since the product of combustion is essentially air, a good approximation of the density of the gases is similar to that of air at standard conditions, or $\rho = 0.074$ Lb/ft³.

Using Eq. 2, we can calculate the mass flow rate for the waste stream as follows:

$$M = \rho \times V = 0.074 Lb/ft^3 \times 4{,}986 ft^3/\text{min} \times 60\text{min}/h = 22{,}139\ Lb/h$$

The next step is to calculate how much waste heat is available for transfer to the heat recovery system. The upper temperature has been given as 1000°F. The lower temperature will be determined by the medium used for heat recovery, that is, steam, hot water, or air, as well as the cleanliness and condensation temperature of the waste stream.

To continue the sample calculations, we will use three different applications of heat recovery mediums: 125 PSIG steam, the hot water required at 350°F, and pre-heated combustion air at whatever final temperature is available. Each of these applications is analyzed separately.

The first heat recovery application medium is saturated steam at 125 PSIG (140 PSIA). From steam tables, we find the enthalpy for saturated liquid (h_f) to be 324.82 Btu/Lb. and for saturated vapor (h_g) 1193.0 Btu/Lb. Thus, we need 868.18 Btu for every pound of steam (1193.0–324.82). The temperature of steam at this pressure is 353°F. For this temperature of steam, we can take the waste stream down to approximately 400°F without the risk of condensation (since condensation usually does not occur above about 325°F). Using Eq. 1, we can calculate the available waste heat as follows:

$$Q = 22{,}139 Lb/h \times 0.24\ Btu/Lb/°F \times \left(1000 - 400\right) = 3{,}188{,}041\ \text{Btu/h} = 3.188041\text{MMB}tu/h$$

This waste heat stream's value can be calculated using Eqs. 3 and 4. Assuming we have natural gas priced at \$8.00/MMBtu in a steam boiler with an efficiency of 75%, the value becomes:

$$\text{Unit cost} = \$8.00/\text{MMBtu}/0.75 = \$10.67/\text{MMBtu}$$

$$\text{Value} = \$10.67/\text{MMBtu} \times 3.188{,}042\ \text{MMBtu/h} = \$34.00/\text{h}$$

To obtain steam mass flow, we must use Eq. 6 below:

$$M_s = \frac{Q}{\left(h_g - h_f\right)} = \frac{3{,}188{,}041\ \text{Btu/h}}{\left(1193.0 - 324.82\right)\text{Btu/Lb}} = 3{,}672\ Lb/h \qquad (6)$$

The second heat recovery application medium will be hot water leaving the heat recovery equipment at 250°F, using the same waste stream. Water will be supplied to the heat recovery equipment at 200°F. With the temperatures involved, we can take the waste stream down to approximately 300°F, provided this low temperature does not cause any condensation issues. For the purposes of this example, we have determined that condensation occurs below 325°F. We will use 350°F as our lower limit. Using Eq. 1, we can calculate the available heat as:

$$Q = 22{,}139\ Lb/h \times 0.24\ Btu/Lb/°F \times \left(1000 - 350\right)$$

$$= 3{,}453{,}711 Btu/h$$

In this case, we need to determine the mass flow rate, expressed as gallons per minute (GPM), of hot water we can heat from 200 to 250°F. Eq. 7 below provides the methodology:

$$Q' = 500 \times V \times \left(T_{\text{upper}} - T_{\text{lower}}\right) \quad (7)$$

where Q' = total heat recovered in Btu/h; V = volumetric flow rate of water in GPM. 500 is the constant used to convert GPM to Lb of water per hour.

To calculate the volume of water required for this system, we calculate:

$$V = \frac{Q'}{500 \times \left(T_{\text{upper}} - T_{\text{lower}}\right)}$$

$$= \frac{3{,}453{,}711\ Btu/h}{500 \times (250 - 200)} = 138 \text{ GPM}$$

Thus, our waste heat recovery system will generate approximately 138 GPM of hot water raised from 200 to 250°F.

The third heat recovery application medium is an air- to-air recuperation system. The waste stream is the same as in the previous example, providing 3,453,711 Btu/h (assuming that 350°F will be the lower limit that we can go to without condensation issues). The recovery medium will be air applied as pre-heated combustion air with an input temperature of 75°F. In this system, we need to provide a set volume of combustion air to the waste heat recovery equipment, so we are interested in finding the final temperature of the pre-heated combustion air.

Using equation

$$Q = M \times 0.24 \times \left(T_{\text{upper}} - T_{\text{lower}}\right)$$

$$3{,}453{,}711\ Btu/h = 50{,}000 Lb/h \times 0.24 Btu/Lb - {}^{\circ}F \times \left(T_{\text{upper}} - 75\right)$$

Solving for T_{upper} we get 362.8°F

Heat Recovery Equipment

There is a variety of equipment manufactured and/or sold as heat recovery equipment. The following is a list of some more common types of equipment.

1. Waste heat steam recovery—This piece of equipment can be either a water-tube or fire-tube boiler that uses the hot waste heat gas stream to heat boiler feed water to generate either low-pressure or high-pressure steam. Typically, this piece of equipment is called a heat recovery steam generator (HRSG). The unit can be a heat recovery-specific piece of equipment; however, some boiler manufacturers will sell their boilers without a burner package and call them HRSGs. This type of boiler can lose some of its effectiveness as a steam generator; however, it still is an acceptable piece of equipment. The reason for the decrease in recovery effectiveness is solely based on the possibility that the waste heat stream temperature will be somewhat lower than the burner flame temperature.
2. Recuperator—This piece of equipment is generally an air-to-air heat exchanger that transfers heat from the waste heat gas stream to air on the recovery side of the heat exchanger. This air can be used as pre-heated combustion air, or host make-up air in the facility.
3. Shell and tube heat exchanger—This piece of equipment consists of a bundle of tubes within a steel shell. It is usually used for water-to-water heat recovery. One of its uses may be to recover

heat from a boiler condensate blow down on the shell side to heat up boiler feed water on the tube side.

4. Fin-tube heat exchanger—This type of heat exchanger uses air (usually the waste heat stream) blowing across finned coils that contain water. A typical application for this type of heat exchanger is using boiler products of combustion gases to preheat boiler feed water. Plugging of the finned coils could be a problem if the fins are closely spaced, the exhaust gases are dirty, or the condensation temperature (of the gases) is approached.
5. Plate and frame heat exchanger—This piece of equipment consists of two frames sandwiching thin plates. It is usually used with fluids of relatively low temperature, which flow through alternate plates. Since the plates are thin, heat transfer is usually good; however, both fluid streams need to be relatively clean or the exchanger will plug up.
6. Heat wheels—These are typically used in low temperature applications, such as the exhaust of environmental (space) air coupled with the introduction of outside air. The two ducts, outside and exhaust air, must be adjacent to one another so that the wheel turns through both ducts. The wheel will turn at low revolutions per minute (RPM) collecting the waste heat or cooling, depending on what season it is, and exchange the heat or cooling to the outside air being introduced to the facility.
7. Heat pipes—This equipment consists of a pipe heat exchanger with the interior containing a coolant. The coolant is alternately vaporizing and condensing between an exhaust and outside air stream, exchanging cooling and heating between the two air streams. The exhaust air and outside air ducts must reside together in the same manner as a heat transfer wheel.
8. Run-around coils—These are similar to heat pipes in operation. They cool or heat exhaust air and outside air streams. Their typical construction is a finned water coil with air blown across the coil. The advantage of the run-around coil is that the exhaust air and outside air ducts can be physically separated. The disadvantage is that a pump, with pumping power, is required. Additionally, if the coils are subject to freezing conditions, then a water/glycol solution must be used. This solution will decrease the effectiveness of the heat transfer.

Summary

The purpose of this entry has been to provide the energy engineer with some general guidelines to applying waste heat recovery. There are numerous applications in the industrial, institutional, and commercial sectors where waste heat recovery can be used cost effectively. The best applications are situations wherein the waste heat stream is of high quality, produced for many hours of the year, preferably 24/7, with a heat load that matches the waste heat availability. Care must be taken by the engineer to ensure that the waste heat stream will not block or destroy the waste heat recovery equipment.

There are also numerous pieces of equipment that can be used for waste heat recovery. These include steam boilers, hot water boilers, recuperators, coil heat exchangers, shell and tube heat exchangers, plate and frame heat exchangers, heat transfer wheels, heat pipes, and run-around coil systems.

Bibliography

Thumann, A.; Metha, D.P. *Handbook of Energy Engineering*; The Fairmont Press, Inc.: Lilburn, GA, 1997.
Turner, W.C. *Energy Management Handbook*, 3rd Ed.; The Fairmont Press, Inc.: Lilburn, GA, 1997.

34

Fuel Cells: Intermediate and High Temperature

Xianguo Li, Gholamreza Karimi, and Kui Jiao

Introduction

The three major types of low-temperature fuel cells are discussed in a separate entry entitled "Fuel Cells: Low Temperature." This entry will elaborate on the intermediate- and high-temperature fuel cells including high-temperature proton exchange membrane fuel cells (HT-PEMFCs), phosphoric acid fuel cells (PAFCs), molten carbonate fuel cells (MCFCs), and solid oxide fuel cells (SOFCs). The HT-PEMFCs can be considered as the PEMFCs with membranes and other cell components that can tolerate temperatures above 100°C. The operating temperature range of HT-PEMFC is about between 100°C and 200°C. It is still at the laboratory research level mainly due to the durability issue of the membrane at high temperatures and is expected to be more suitable for stationary rather than automotive applications due to its slower start-up characteristics (higher operating temperature to be reached) than the conventional PEMFC. The PAFC is the most commercially developed fuel cell. It is being used in applications such as hospitals, hotels, offices, and schools. It can also be used in larger vehicles such as buses. The operating temperature range of PAFC is between 160°C and 220°C, and its efficiency is about 40%. The high-temperature MCFCs and SOFCs have advantages over conventional energy-generating systems in terms of reliability, fuel flexibility, modularity, low emission of NO_x and SO_x pollutants, and environmental friendliness. They also show relatively high tolerance to trace levels of impurities in the gas stream. In addition, due to their high operating temperatures typically in the range of 600°C to 1000°C, hydrocarbon fuels such as methane and natural gas can be reformed within the stack, eliminating the need for expensive external reformer systems. The high operating temperatures requires that most applications for these fuel cells be limited to large, stationary power plants. The high-quality heat produced can be used in cogenerated hybrid power systems such as space heating, industrial processing, or even in steam turbines to generate more electricity, improving the system efficiencies to high levels (up to 90%).

High-Temperature Proton Exchange Membrane Fuel Cells

The HT-PEMFCs with operating temperatures higher than 100°C have attracted growing interests in the past decade. By comparing with the conventional PEMFCs operating at around 80°C, the HT-PEMFCs with elevated operating temperatures feature faster electrochemical kinetics, simpler water management (presence of liquid water can be neglected), higher carbon monoxide tolerance, and easier cell cooling and waste heat recovery. The operating temperature range of HT-PEMFC is about between 100°C and 200°C.

Although HT-PEMFCs have many attractive features, technical challenges still remain and are mostly related to the PEM. The durability issue of the PEM at high temperatures is the reason that the HT-PEMFC remains at the laboratory research level. The conventional PEMs (e.g., Nafion membrane) widely used in PEMFCs suffer significant decrement in mechanical strength at the high operating temperature of HT-PEMFCs, and the much lower relative humidity (RH) in HT-PEMFCs than in conventional PEMFCs due to the significantly increased vapor saturation pressure with temperature also results in severe reduction of the proton conductivity of the conventional PEMs. Therefore, developing PEMs with high mechanical strength at temperatures higher than 100°C and with high proton conductivity in anhydrous environments becomes the major challenge, and most of the previous HT-PEMFC-related researches focused on this important issue.

Among the different high-temperature PEMs being developed, polybenzimidazole (PBI) membranes have been recognized as promising PEMs when doped with a strong oxo-acid (e.g., phosphoric acid or sulfuric acid) for HT-PEMFCs. At present, the phosphoric-acid-doped PBI membrane perhaps offers the best combination of durability and proton conductivity for HT-PEMFCs. The durability of HT-PEMFC with this type of PEM is over 6000 hr under continuous operation, and the durability with frequent start-stop cycles is still questionable. The durability issue and the previously mentioned slow start-up make the HT-PEMFC more favorable for stationary rather than automotive applications. Since most of the cell components are similar between PEMFC and HT- PEMFC, it is expected that the cost of HT-PEMFC is similar to conventional PEMFC.

Phosphoric Acid Fuel Cells

The PAFC is the most advanced type of fuel cells and is considered to be "technically mature" and ready for commercialization after nearly 30 years of RD&D and over half a billion dollars expenditure. Therefore, the PAFC has been referred to as the first-generation fuel cell technology. Unlike the alkaline fuel cell systems that were primarily developed for space applications, the PAFC was targeted initially for terrestrial applications with the carbon-dioxide- containing air as the oxidant gas and hydrocarbon-reformed gas as the fuel for electrochemical reactions and electric power generation.

The basic components of a PAFC are the electrodes consisting of finely dispersed platinum catalyst or carbon paper, silicon carbide matrix holding the phosphoric acid, and a bipolar graphite plate with flow channels for fuel and oxidant. The operating temperature ranges between 160°C and 220°C, and it can use either hydrogen or hydrogen produced from hydrocarbons (typically natural gas), ethanol, or methanol as the anodic reactant. In the case of hydrogen produced from a reformer with air as the anodic reactant, a temperature of 200°C and a pressure of as high as 8 atm are required for better performance. PAFCs are advantageous from a thermal management point of view. The rejection of waste heat and product water is very efficient in this system and the waste heat at about 200°C can be used efficiently for the endothermic steam-reforming reaction. The waste heat can also be used as a cogeneration for space heating and hot water supply.

However, the PAFC cannot tolerate the presence of carbon monoxide and H_2S, which are commonly present in the reformed fuels. These contaminants poison the catalyst and decrease its electrochemical catalytic activity. A major challenge for using natural gas reformed fuel, therefore, lies in the removal of carbon monoxide to a level of less than 200–300 ppm. Carbon monoxide tolerance is better at the operating temperature of above 180°C. However, removal of sulfur is still essential. Further, the PAFC has a lower performance, primarily due to the slow oxygen reaction rate at the cathode. Therefore, PAFC is typically operated at higher temperature (near 200°C) for better electrochemical reactivity and for smaller internal resistance, which is mainly due to the phosphoric acid electrolyte. As a result, PAFC exhibits the problems of both high- and low-temperature fuel cells, but possibly none of the advantages of either option.

The PAFC system is the most advanced fuel cell system for terrestrial applications. Its major use is in on-site integrated energy systems to provide electrical power such as in apartments, shopping centers, office buildings, hotels, and hospitals. These fuel cells are commercially available in the range from 24 V, 250 W portable units to 200 kW on-site generators. PAFC systems of 0.5–1.0 MW are being developed for use in stationary power plants of 1–11 MW capacity. The power density of PAFC system is about 200 mW/cm^2, and the power density for 36 kW brassboard PAFC fuel cell stack has been reported to be 0.12 kW/kg and 0.16 kW/L. One representing PAFC system is the PC-25 from the International Fuel Cells in Connecticut (the United States). It costs about $3000/kW, while the conventional thermal power generation system costs only about $1000/kW. In fact, present PAFC systems can offer reasonable durability and performance, and the cost is the primary issue that hinders the commercialization of PAFC.

Molten Carbonate Fuel Cells

Introduction

The MCFC is often referred to as the second-generation fuel cell because its commercialization is normally expected after the PAFC. It is believed that the development and technical maturity of the MCFC is about 5–7 yr behind the PAFC. At present, the MCFC has reached the early demonstration stage of precommercial stacks, marking the transition from fundamental and applied R&D towards product development. MCFCs are being targeted to operate on coal-derived fuel gases or natural gas. This contrasts with the PAFCs, as discussed earlier, which prefer natural gas as primary fuel.

The MCFC operates at higher temperature than all the fuel cells described so far. The operating temperature of the MCFC is generally around 600–700°C, typically 650°C. Such high temperature produces high-grade waste heat that is suitable for fuel processing, cogeneration, or combined cycle operation, leading to higher electric efficiency. It also yields the possibility of utilizing carbonaceous fuels (especially natural gas) directly, through internal reforming to produce the fuel (hydrogen) ultimately used by the fuel cell electrochemical reactions. This results in simpler MCFC systems (i.e., without external reforming or fuel processing subsystem), less parasitic load, and less cooling power requirements, hence higher overall system efficiency as well. The high operating temperature reduces voltage losses due to reduced activation and ohmic and mass transfer polarization. The activation polarization is reduced to such an extent that it does not require expensive catalysts as low-temperature fuel cells do, such as PAFCs and PEMFCs. It also offers great flexibility in the use of available fuels, say, through in situ reforming of fuels. It has been estimated that the MCFC can achieve an energy conversion efficiency of 52%–60% (from chemical energy to electrical energy) with internal reforming and natural gas as the primary fuel. Some studies have indicated that the MCFC efficiency of methane to electricity conversion is the highest attainable by any fuel cell or other single pass/simple cycle generation scheme.

Basic Operating Principle

A schematic of a MCFC is illustrated in Figure 1. A MCFC consists of two porous gas-diffusion electrodes (anode and cathode) and a carbonate electrolyte in liquid form. The electrochemical reaction occurring at the anode and the cathode is

$$\text{Cathode}: \frac{1}{2}O_2 + CO_2 + e^- = CO_3^{2-}$$

and the net cell reaction is

$$H_2 + \frac{1}{2}O_2 = H_2O$$

Beside the hydrogen oxidation reaction at the anode, other fuel gases such as carbon monoxide, methane, and higher hydrocarbons are also oxidized by conversion to hydrogen. Although direct electrochemical oxidation of carbon monoxide is possible, it occurs very slowly compared to that of hydrogen. Therefore, the oxidation of carbon monoxide is mainly via the water–gas shift reaction

$$CO + H_2O = CO_2 + H_2,$$

which, at the operation temperature of the MCFC, equilibrates very rapidly at catalysts such as nickel. Therefore, carbon monoxide becomes a fuel, instead of a contaminant as in the previously described low-temperature fuel cells. Direct electrochemical reaction of methane appears to be negligible. Hence, methane and other hydrocarbons must be steam reformed, which can be done either in a separate reformer (external reforming) or in the MCFC itself (the so-called internal reforming).

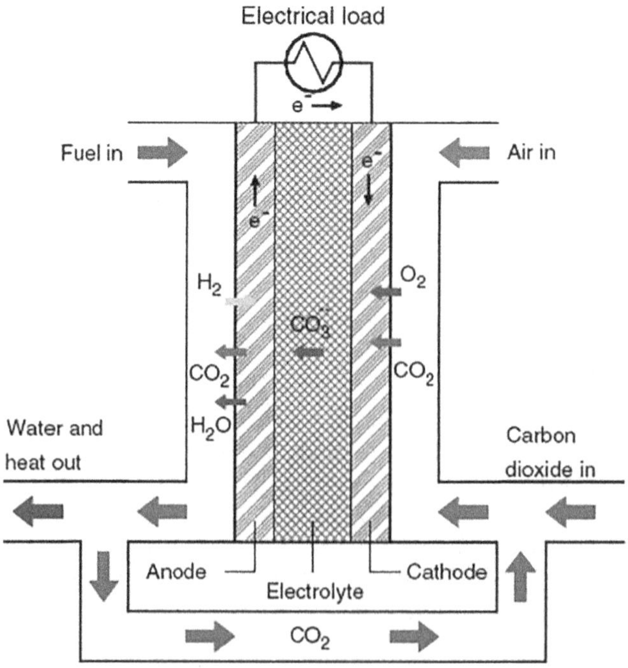

FIGURE 1 Schematic of a MCFC.

As a result, water and carbon dioxide are important components of the feed gases to the MCFCs. Water, produced by the main anode reaction, helps to shift the equilibrium reactions to produce more hydrogen for the anodic electrochemical reaction. Water must also be present in the feed gas, especially in low-Btu (i.e., high CO content) fuel mixtures, to avoid carbon deposition in the fuel gas flow channels supplying the cell, or even inside the cell itself. Carbon dioxide, from the fuel exhaust gas, is usually recycled to the cathode as it is required for the reduction of oxygen.

The MCFCs use a molten alkali carbonate mixture as the electrolyte, which is immobilized in a porous lithium aluminate matrix. The conducting species is carbonate ions. Lithiated nickel oxide is the material of the current choice for the cathode, and nickel, cobalt, and copper are currently used as anode materials, often in the form of powdered alloys and composites with oxides. As a porous metal structure, it is subject to sintering and creeping under the compressive force necessary for stack operation. Additives such as chromium or aluminum form dispersed oxides and thereby increase the long-term stability of the anode with respect to sintering and creeping. MCFCs normally have about 75%–80% fuel (hydrogen) utilization.

Acceptable Contamination Levels

MCFCs do not suffer from carbon monoxide poisoning, and in fact, they can utilize carbon monoxide in the anode gas as the fuel. However, they are extremely sensitive to the presence of sulfur (<1 ppm) in the reformed fuel (as hydrogen sulfide, H_2S) and oxidant gas stream (SO_2 in the recycled anode exhaust). The presence of HCl, HF, and HBr causes corrosion, while trace metals can spoil the electrodes. The presence of particulates of coal/fine ash in the reformed fuel can clog the gas passages.

Major Technological Problems

The main research efforts for the MCFCs are focused on increasing the lifetime and endurance and reducing the long-term performance decay. The main determining factors for the MCFC are electrolyte loss, cathode dissolution, electrode creepage and sintering, separator plate corrosion, and catalyst poisoning for internal reforming.

Electrolyte loss results in increased ohmic resistance and activation polarization, and it is the most important and continuously active factor in causing the long-term performance degradation. It is primarily a result of electrolyte consumption by the corrosion/dissolution processes of cell components, electric potential-driven electrolyte migration, and electrolyte vaporization. Electrolyte evaporation (usually Li_2CO_3 and/or K_2CO_3) occurs either directly as carbonate or indirectly as hydroxide.

The cathode consists of NiO, which slowly dissolves in the electrolyte during operation. It is then transported towards the anode and precipitates in the electrolyte matrix as Ni. These processes lead to a gradual degradation of cathode performance and the shorting of the electrolyte matrix. The time at which shorting occurs depends not only, via NiO solubility, on the CO_2 partial pressure and the cell temperature but also on the matrix structure, i.e., on the porosity, pore size, and, in particular, thickness of the matrix. Experience indicates that this cell shorting mechanism tends to limit stack life to about 30,000 hr under the atmospheric reference gas conditions, and much shorter for real operating conditions.

Electrode (especially anode) creepage and sintering (i.e., a coarsening and compression of electrode particles) result in increased ohmic resistance and electrode polarization. NiO cathodes have quite satisfactory sinter and creepage resistance. Creep resistance of electrodes has an important effect on maintaining low contact resistance of the cells and stacks. The corrosion of the separator plate depends on many factors, such as the substrate, possible protective layers, composition of the electrolyte, local potential and gas composition, and the oxidizing and reducing atmospheres at the cathode and anode, respectively. Poisoning of the reforming catalyst occurs for direct internal-reforming MCFCs. It is caused by the evaporation of electrolyte from the cell components and condensation on the catalyst, which is the coldest spot in the cell, and by liquid creep within the cell.

Technological Status

MCFC technology is in the first demonstration phase and under the product development with full-scale systems at the 250 kW to 2 MW range. The short-term goal is to reach a lifetime of 40,000 hr. It is estimated that the capital cost is about $1000–1600/kW for the MCFC power systems. The cost breakdown is, at full-scale production levels, about one-third for the stack, and two-thirds for the balance of the plant. It is also generally accepted that the cost of raw materials will constitute about 80% of total stack costs. Although substantial development efforts supported by fundamental research are still needed, the available knowledge and number of alternatives will probably make it possible to produce precommercial units in the earlier part of the coming decade at a capital cost of $2000–4000/kW. Precompetitive commercial units may be expected some years later by which time further cost reduction to full competitiveness will be guided by extensive operating experience and increased volume production.

Applications

The MCFC is being developed for their potential as baseload utility generators. However, their best application is in distributed power generation and cogeneration (i.e., for capacities less than 20 MW in size), and in this size range, MCFCs are 50% to 100% more efficient than turbines—the conventional power generator. Other applications have been foreseen, such as pipeline compressor stations, commercial buildings, and industrial sites in the near term and repowering applications in the longer term. Due to its high operation temperature, it only has very limited potential for transportation applications. This is because of its relatively low power density and long start-up times. However, it may be suitable as a powertrain for large surface ships and trains.

Solid Oxide Fuel Cells

Introduction

SOFCs have emerged as a serious alternative high- temperature fuel cell, and they have been often referred to as the third-generation fuel cell technology because their commercialization is expected after the PAFCs (the first generation) and MCFCs (the second generation).

SOFC is an all-solid-state power system, including the electrolyte, and it is operated at high temperature of around 1000°C for adequate ionic and electronic conductivity of various cell components. The all-solid-state cell composition makes the SOFC system simpler in concept, design, and construction; two-phase (gas–solid) contact for the reaction zone reduces corrosion and eliminates all the problems associated with the liquid electrolyte management. The high-temperature operation results in fast electrochemical kinetics (i.e., low activation polarization), without the need for noble metal catalysts. The fuel may be gaseous hydrogen, H_2/CO mixture, or hydrocarbons because the high-temperature operation makes the internal in situ reforming of hydrocarbons with water vapor possible. It is specially noticed that CO is no longer a contaminant; rather, it becomes a fuel in SOFCs. Even with external reforming, the SOFC fuel feedstock stream does not require the extensive steam reforming with shift conversion as it does for the low-temperature fuel cell systems. More important, the SOFC provides high-quality waste heat that can be utilized for cogeneration applications or combined cycle operation for additional electric power generation. The SOFC operating condition is also compatible with the coal gasification process, which makes the SOFC systems highly efficient when using coal as the primary fuel. It has been estimated that the chemical-to-electrical energy conversion efficiency is 50% to 60%, even though some estimates go as high as 70% to 80%. Also, nitrogen oxides are not produced and the amount of carbon dioxide released per kilowatt hour is around 50% less than for power sources based on combustion because of the high efficiency.

Basic Operating Principle

As mentioned earlier, both hydrogen and carbon monoxide can be oxidized in the SOFCs directly. Hence, if hydrogen or hydrogen-rich gas mixture is used as fuel, and oxygen (or air) is used as oxidant, the half cell reaction becomes (Figure 2)

$$\text{Anode}: H_2(g) + O^{2-} = H_2O(g) + 2e^-$$

$$\text{Cathode}: 2e^- + \frac{1}{2}O_2(g) = O^{2-}$$

$$H_2(g) + \frac{1}{2}O_2(g) = H_2O(g).$$

and the overall cell reaction becomes

Note that g represents gaseous phase.

However, if carbon monoxide is provided to the anode instead of hydrogen, the anode reaction becomes

$$\text{Anode}: CO(g) + O^{2-} = CO_2(g) + 2e^-$$

with the cathode reaction remaining the same, the cell reaction becomes

$$CO(g) + \frac{1}{2}O_2(g) = CO_2(g)$$

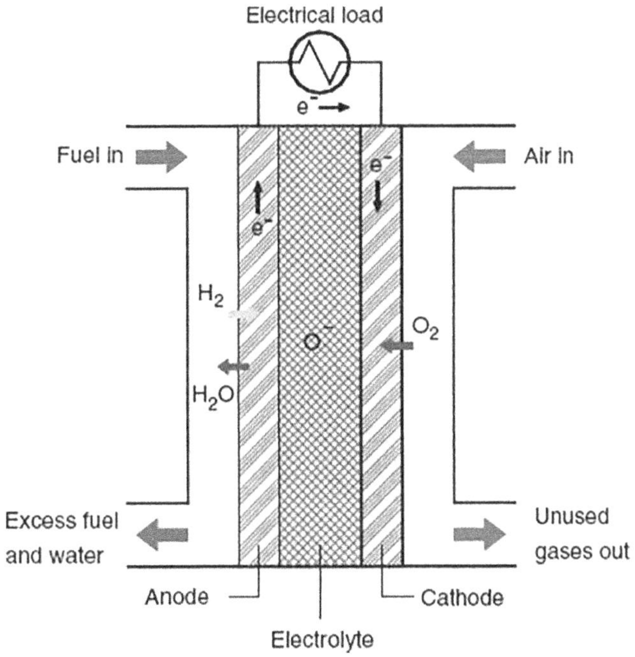

FIGURE 2 Schematic of SOFC.

If the fuel stream contains both hydrogen and carbon monoxide as is the case for hydrocarbon reformed gas mixture, especially from the gasification of coal, the oxidation of hydrogen and carbon monoxide occurs simultaneously at the anode, and the combined anode reaction becomes

$$\text{Anode}: a\,H_2(g) + b\,CO + (a+b)O^{2-} = a\,H_2(g) + b\,CO_2 + 2\,(a+b)e^-$$

Consequently, the corresponding cathode and overall cell reaction become

$$\text{Cathode}: \frac{1}{2}(a+b)O_2 + 2(a+b)e^- = (a+b)O^{2-}$$

$$\text{Cell}: \frac{1}{2}(a+b)O_2 + a\,H_2 + b\,CO = a\,H_2O + b\,CO_2$$

The solid electrolyte in SOFCs is usually yttria-stabilized zirconia (YSZ); thus, a high operating temperature of around 1000°C is required to ensure adequate ionic conductivity and low ohmic resistance. This is especially important because the cell open-circuit voltage is low, compared with low-temperature fuel cells, typically around 0.9–1 V under the typical working conditions of the SOFCs. The high-temperature operation of the SOFCs makes the activation polarization very small, resulting in the design operation in the range dominated by the ohmic polarization. The conventional material for the anode is nickel-YSZ-cermet, and the cathode is usually made of lanthanum-strontium-manganite. Metallic current collector plates of a high-temperature corrosion-resistant chromium-based alloy are typically used.

Acceptable Contamination Levels

Because of high temperature, the SOFCs can better tolerate impurities in the incoming fuel stream. They can operate equally well on dry or humidified hydrogen or carbon monoxide fuel or on mixtures of them. However, hydrogen sulfide (H_2S), hydrogen chloride (HCl), and ammonia (NH_3) are impurities typically found in coal gasified products, and each of these substances is potentially harmful to the performance of SOFCs. The main poisoning factor for SOFCs is H_2S. Though the sulfur tolerance level is approximately two orders of magnitude greater than other fuel cells, the level is below 80 ppm. However, studies have shown that the effect of hydrogen sulfide (H_2S) is reversible, meaning that the cell performance will recover if hydrogen sulfide is removed from the fuel stream or clean fuel is provided after the contaminant poison has occurred.

Major Technological Problems

The high-temperature operation of the SOFCs places stringent requirements on materials used for cell construction, and appropriate materials for cell components are very scarce. Therefore, the key technical challenges are the development of suitable materials and the fabrication techniques. Of the material requirements, the most important consideration is the matching of the thermal expansion coefficients of electrode materials with that of the electrolyte to prevent cracking or delamination of SOFC components either during high-temperature operation or heating/cooling cycles. One of the remedies for the thermal expansion mismatch is to increase the mechanical toughness of the cell materials by either developing new materials or doping the existing materials with SrO and CaO.

The electrode voltage losses are reduced when the electrode material possesses both ionic and electronic conductivities (the so-called mixed conduction), for which the electrochemical reactions occur throughout the entire surface of the electrode rather than only at the three-phase interface of, e.g., the cathode, the air (gas phase), and the electrolyte. Therefore, it is important for performance enhancement to develop mixed-conduction materials for both the cathode and anode, which have good thermal expansion match with the electrolyte used and good electrical conductivity to reduce the ohmic polarization that dominates the SOFC voltage losses.

Another focus of the current development is the intermediate-temperature SOFCs operating at around 800°C for better matching with the bottoming turbine cycles and lessening requirements for the cell component materials. Again, appropriate materials with adequate electrical conductivity are the key areas of the development effort, and thermal expansion matching among the cell components is still necessary.

Technological Status

There are three major configurations for SOFCs: tubular, flat plate, and monolithic. Even though SOFC technology is in the developmental stage, the tubular design has gone through development at Westinghouse Electric Corporation since the late 1950s and has been demonstrated at user sites in a complete operating fuel cell power unit of nominal 25 kW (40 kW maximum) capacity. The flat plate and the monolithic designs are at a much earlier development status typified by subscale, single cell, and short stack development (up to 40 cells). The present estimated capital cost is $1500/kW, but it is expected to be reduced with improvements in technology. Therefore, the SOFCs may become very competitive with the existing technology for electric power generation. However, it is believed that the SOFC technology is at least 5 to 10 years away from commercialization.

Applications

SOFCs are very attractive in electrical utility and industrial applications. The high operating temperature allows them to use hydrogen and carbon monoxide from natural gas steam reformers and coal gasification plants, a major advantage as far as fuel selection is concerned. SOFCs are being developed for the large (>10 MW, especially 100 to 300 MW) baseload stationary power plants with coal as the primary fuel. This is one of the most lucrative markets for this type of fuel cells.

A promising field for SOFCs is the decentralized power supply in the megawatt range, where the SOFC gains interest due to its capability to convert natural gas without external reforming. In the range of one to some tenths of a megawatt, the predicted benefits in electrical efficiency of SOFC-based power plants over conventional methods of electricity generation from natural gas can only be achieved by an internal-reforming SOFC. Thus, internal reforming is a major target of present worldwide SOFC development.

Concluding Remarks

A summary of the preceding discussion, including the operational characteristics and technological status of the intermediate- and high-temperature fuel cells, is given in Table 1. Note that the costs for PAFC, MCFC, and SOFC are for stationary power systems.

HT-PEMFC is still at the laboratory research level mainly due to the durability issue of the membrane at high temperatures and is expected to be more suitable for stationary rather than automotive applications due to its slower start-up characteristics (higher operating temperature to be reached) than the conventional PEMFC. PAFC is the most commercially developed fuel cell operating at intermediate temperatures. PAFCs are being used in applications such as hospitals, hotels, offices, and schools with relative high conversion efficiency. The high-temperature fuel cells like MCFCs and SOFCs may be most appropriate for cogeneration and combined cycle systems (with gas or steam turbine as the bottoming cycle). The MCFCs have the highest energy efficiency attainable from methane to electricity conversion in the size range of 250 kW to 20 MW, whereas the SOFCs are best suited for baseload utility application

TABLE 1 Operational Characteristics and Technological Status of Intermediate- and High-Temperature Fuel Cells

Type of Fuel Cells	Operating Temperature (°C)	Power Density (mW/cm^2), (present) Projected	Projected Rated Power Level (kW)	Fuel Efficiency	Lifetime Projected (hr)	Capital Cost Projected ($/kW)	Application Areas
HT-PEMFC	100–200	(600) >900	1–1000	45–70	>40,000	35	Transportation, space
PAFC	160–220	(200) 250	100–5000	55	>40,000	3000	Dispersed and distributed power
MCFC	600–700	(100) >200	1000–100,000	60–65	>40,000	1000	Distributed power generation
SOFC	800–1000	(240) 300	100–100,000	55–65	>40,000	1500	Baseload power generation

operating on coal-derived gases. It is estimated that the MCFC technology is about 5 to 10 years away from commercialization, and the SOFCs are probably years afterwards. In addition, due to the high operating temperatures and fuel reforming of MCFC and SOFC, the management of the exhaust gases (e.g., NO_x and SO_2) is important due to the environmental concerns.

Bibliography

1. Blomen, L.J.M.J.; Mugerwa, M.N. *Fuel Cell Systems;* Plenum Press: New York, 1993.
2. Larminie, J.; Dicks, A. *Fuel Cell Systems Explained,* 2nd Ed.; Wiley, Chichester, 2003.
3. Kordesch, K.; Simader, G. *Fuel Cells and Their Applications;* VCH Press: New York, 1996.
4. Appleby, A.J.; Foulkes, F.R. *Fuel Cell Handbook*; Van Nostrand Reinhold Company, New York, 1989.
5. Hoogers, G. *Fuel Cell Technology Handbook*; CRC Press, New York, 2002.
6. Vielstich, W.; Gasteiger, H.; Lamm, A. *Handbook of Fuel Cells—Fundamentals, Technology,* Applications, 4-volume set, Wiley, Chichester, 2003.
7. Li, X. *Principles of Fuel Cells*; Taylor and Francis, New York 2005.

35

Fuel Cells: Low Temperature

Xianguo Li
and Kui Jiao

Introduction

Fuel cell technology has been developed and improved dramatically in the past two decades, and this has once again captured public attention as well as the industry's attention concerning the prospect of fuel cells as practical power sources for terrestrial applications. Fuel cell is a highly energy-efficient and environmentally friendly technology for power generation that is also compatible with alternative fuels and renewable energy sources and carriers for sustainable development.

Fuel cell offers additional advantages for both mobile and stationary power generation applications. Fuel cell, as an electrochemical device, has no moving components except for peripheral equipment. As a result, its operation is very quiet, virtually without vibration and noise, thus capable of on-site cogeneration with no need of long distance power transmission lines that consume approximately 10% of electric energy delivered to consumers in North America. Its inherent modularity allows for simple construction and operation with possible applications for baseload electricity generation, dispersed, distributed, and portable power generation, because it may be made in any size from a few watts up to a megawatt-scale plant with equal efficiency. Its fast response to the changing load condition while maintaining high efficiency makes it ideally suited to load following applications. Its high efficiency represents less chemical, thermal, and carbon dioxide emissions for a given amount of power generation.

At present, fuel cell technology is being routinely used in many specific areas, notably in space explorations, where fuel cell operates on pure hydrogen and oxygen with over 70% efficiency and drinkable water as the only by-product. Fuel cell technology is also being widely used for terrestrial applications; impressive technical progress has been achieved in terms of higher power density and better durability as well as reduced capital and maintenance and operation cost, and is driving the development of competitively priced fuel-cell-based power generation systems with advanced features for terrestrial use, such as utility power plants and zero-emission vehicles. In light of decreasing fossil fuel reserves and increasing energy demands worldwide, fuel cell will probably become one of the major energy technologies with fiercest international competition in the 21st century.

The major terrestrial commercial applications of fuel cells are electric power generation in the utility industry and as a zero-emission powertrain in the transportation sector. For these practical applications, the efficiencies of fuel cells range somewhere from 40% to 65% based on the lower heating value (LHV) of hydrogen. Typically, the cell electric potential is only about 1V across a single cell, and it decreases due to various loss mechanisms under the operational conditions. Thus, multiple cells are required to be connected together in electrical series in order to achieve a useful voltage for practical purposes, and these connected cells are often referred to as a fuel cell stack. A fuel cell system consists of one or multiple fuel cell stacks connected in series and/or parallel and the necessary auxiliaries whose composition depends on the type of fuel cells and the kind of primary fuels used. The major accessories include thermal management (or cooling) subsystem, fuel supply, storage and processing subsystem, and oxidant (typically air) supply and conditioning subsystem.

In light of the recent advancement in fuel cell technology and its potential impact on municipal utilities, the energy delivery marketplace, and transportation sector, an assessment of the current state of fuel cell technology has been conducted. The result of this study is described in the first and second parts of this report, which summarizes the state-of-the-art technology for the seven major types of fuel cells, including

- Alkaline fuel cells (AFCs)
- Proton exchange membrane fuel cells (PEMFCs)
- High-temperature PEMFCs (HT-PEMFCs)
- Direct methanol fuel cells (DMFCs)
- Phosphoric acid fuel cells (PAFCs)
- Molten carbonate fuel cells (MCFCs)
- Solid oxide fuel cells (SOFCs)

A critical parameter that determines the potential applications of each type of fuel cell is the operating temperature. For instance, AFCs, DMFCs, and PEMFCs have potential applications in transportation because they do not produce much heat (which otherwise would have to be eliminated by some cooling device) and, as a result of this, have a very short start-up period (a few minutes). These types of fuel cells are discussed in this entry.

On the other hand, HT-PEMFCs, PAFCs, MCFCs, and SOFCs producing high-temperature heat are more complex to run and are better fit for stationary applications like combined heat power generation or CHP. The fuel cells under this category will be discussed in a separate entry entitled "Fuel Cells: Intermediate and High Temperature."

The study was based on a literature review and search of technical databases, and recent advancement of fuel cell technology has been identified and delineated for each of the seven major types of fuel cells listed above. The following sections describe each of the three low- temperature fuel cells. Their technological status is presented with respect to their basic operating principle, acceptable contamination levels, the state-of-the-art technology, major technical barriers to commercialization, their economics and suitability for utility, and transportation applications.

Alkaline Fuel Cells

AFCs give the best performance among all the fuel cell types under the same or similar operating conditions when running on pure hydrogen and oxygen. Hence, they are among the first fuel cells to have been studied and taken into development for practical applications, and they are the first type of fuel cells to have reached successful routine applications, mainly in space programs such as space shuttle missions in the United States and similar space exploration endeavors in China and Europe, where pure hydrogen and oxygen are used as reactants. Because of their success in space programs, AFCs are also the type of fuel cells on which probably the largest number of fuel cell development programs have begun in the world in an effort to bring them down to terrestrial applications, particularly in Europe, and almost all the AFC development programs have come to an end. In recent years, AFC with solid electrolyte membrane has attracted people's attention once again. Because with solid electrolyte membrane, AFC can be expected to offer similar or better power density and durability when compared with PEMFC. In addition to its most efficient energy conversion mechanism among all the fuel cell types, it is becoming a promising candidate for energy conversion in automotive applications.

The AFCs have the highest energy conversion efficiency among all types of fuel cells under the same operating condition if pure hydrogen and pure oxygen are used as reactants. That was one of the important reasons that AFC was selected for the U.S. space shuttle missions. The AFCs used in the shuttle missions are operated at about 200°C for better performance (i.e., high energy conversion efficiency of over 70% and high power density, which is critical for space applications), and the alkaline electrolyte is potassium hydroxide (KOH) solution immobilized in the asbestos matrix. As a result, the AFCs operate at high pressure in order to prevent the boiling and depletion of the liquid electrolyte. Consequently, these severe operating conditions of high temperature and high pressure dictate extremely strict requirement for cell component materials that must withstand the extreme corrosive oxidizing and reducing environment of the cathode and anode. To meet these requirements, precious metals such as platinum, gold, and silver are used for the construction of the electrodes, although these precious metals are not necessary for the electrochemical reactions leading to electric power generation. Each shuttle flight contains 36 kW AFC power system. Its purchase price is about $28.5 million, and it costs NASA additional $12–19 million annually for operation and maintenance. Although the manufacturer claims about 2400 hr of lifetime, NASA's experience indicates that the real lifetime is only about 1200 hr. With sufficient technology development, 10,000 hr are expected as the life potential (or upper limit) for the AFC system. This belief is based on the nature of the AFC systems and the data accumulated on both stacks and single cells.

The typical working temperature of AFC power systems aimed at commercial and terrestrial applications ranges from 20°C to 90°C, and the electrolyte is KOH solution (30%–35%) or solid electrolyte membrane. There are five different cell types investigated:

1. Cell with a free liquid electrolyte between two porous electrodes
2. ELOFLUX cell with liquid KOH in the pore systems.
3. Matrix cell where the electrolyte is fixed in the electrode matrix
4. The falling film cell
5. Cell with solid electrolyte membrane

Research groups are working on the development of technical AFC systems; the group name and the current status of the work are listed below:

- Siemens, Erlangen, Germany: stopped working on the AFCs
- VARTA AG, Kelkheim, Germany: terminated working on preparation of technical electrodes and technical fuel stacks in 1993
- GH, Kassel, Germany: stopped working on preparation of technical electrodes and technical fuel cell stacks in 1994
- ISET, Kassel, Germany: stopped working on technical AFC systems in 1994
- DLR-ITT (German Aerospace Research Establishment), Stuttgart: stopped working on the investigation of the degradation of technical electrodes, the development of new catalysts for AFC, and the theoretical simulation of stacks and systems in 1994
- ELENCO, Antwerpen, Belgium: stopped working on electrodes, stacks, and systems, as well as bus demonstration program (it went bankrupt in 1995)

- Royal Institute of Technology, Stockholm: working on the field of stationary fuel cells powered by biofuels
- Hoechst AG, Frankfurt: terminated working on AFC electrodes and stopped development of the falling film cell
- Technical University, Graz, Austria: planning an investigation of degradation effects
- Zevco with headquarters in London: demonstration of AFCs for mobile applications, particularly for bus with ELENCO's technology
- University of California, Riverside, California: development of solid electrolyte membrane for AFC

The following is a list of some selected technical applications and demonstration projects in Europe in recent years:

- Space applications: the space applications are the development of an AFC system for the European space shuttle HERMES (Siemens, ELENCO, VARTA/GHK) and the bipolar matrix AFC system Photon for space applications (Ural Electrochemical Integrated plant).
- Vehicles: applications of AFCs in vehicles are as follows:
  1. Forklift truck of VARTA AFC
  2. VW Van, 14 kW ELENCO AFC and battery
  3. VW van, 17.5 kW Siemens AFC
  4. Submarine, 100 kW Siemens AFC
  5. EUREKA-Bus, 80 kW ELENCO AFC and battery
- Decentralized energy supply: these supplies comprise the following:
  1. A meteorological station, 5 W, VARTA AFC for long-term operation
  2. A TV-transmitting installation Ruppertshain, 100 W VARTA AFC
  3. A mobile current supply unit for the Belgian Geological Service, 40 kW ELENCO AFC
- Energy storage: the solar hydrogen demonstration plant in Neunburg vorm Wald, 6.5 kW Siemens AFC, and the solar hydrogen system at Fachhochschule Wiesbaden, 1.2 kW ELENCO AFC, are typical examples. Many problems concerning AFCs are described in the literature. The most important problems are the following:
  - Preparation method of the electrodes: the electrodes consist of porous material that is covered with a layer of catalyst. Generally, it is very difficult to distribute the catalyst at the surface and to produce a defined pore system for the transportation of the reactants.
  - Costs of the electrode, stacks, and fuel cell systems: the preparation of electrodes with noble metal catalysts is very expensive. In general, the electrodes are manufactured in a small-scale production with high overhead costs.
  - Lifetime of the electrode/degradation: the electrolyte is very corrosive and the catalyst materials are sensitive to high polarization. Using nickel and silver as catalysts, in order to reduce the costs of the fuel cell, leads to a high degradation of these catalysts. This problem is expected to be solved or abated with solid electrolyte membrane.
  - Diaphragm made of asbestos: the diaphragm of low- temperature AFCs is made of asbestos. This material is hazardous for health, and in some countries, its use is even banned. Therefore, new diaphragms should be developed, but it is difficult to find a material with a similar behavior in alkaline electrolyte. This issue is solved when solid electrolyte membrane is used.
  - Carbon-dioxide-contaminated fuel and oxidant streams (carbonating of electrolyte and electrodes): the electrolyte intolerance of carbon dioxide is the most important disadvantage of air-breathing AFCs with reformate gases from primary fossil fuels. This problem is expected to be solved or abated with solid electrolyte membrane.

Other problems associated with the AFC power systems are the concern on the safety and reliability of AFC power systems, mainly related to AFCs with liquid electrolyte. For example, the liquid KOH electrolyte contained in an asbestos matrix can only withstand a 5 psi limit of pressure differential between the anode and cathode reactant gases. This dictates the need for sophisticated pinpoint pressure control during the operation including transient, start-up, and shutdown process. It is also a safety issue because of its greater likelihood of mixing of the reactants occurring in the AFC system with the possibility of a serious fire breaking out. In terms of general safety considerations, the use of the corrosive potassium hydroxide electrolyte in the AFCs represents the need for hazardous handling, and the handling of asbestos matrix poses potential hazard to one's health. With flowing reactant gases, the potential for the gradual loss of the liquid electrolyte, drying of the electrolyte matrix, reactant crossover of the matrix, and ensuing life-limiting reactant mixing (or actual AFC stack failure due to fire) is very real in the AFC system. Moreover, the liquid electrolytes, like potassium or sodium hydroxide, do not reject carbon dioxide; even the 300–350 ppm of carbon dioxide in the atmospheric air is not tolerated (carbon dioxide concentration in both cathode and anode gases must be less than 10–100 ppm by volume), while terrestrial applications almost invariably require the use of atmospheric air as oxidant due to technical and economic considerations. For municipal electric applications, hydrocarbon fuels, especially natural gas, are expected to be the primary fuel, and their reformation into hydrogen-rich gases invariably contains a significant amount of carbon dioxide, e.g., steam reforming of the natural gas results in the reformate gas consisting approximately of 80% hydrogen, 20% carbon dioxide, and a trace amount of other components such as carbon monoxide. Carbonaceous products of aging and corrosion shorten AFC life; they degrade the alkaline electrolyte. Whether originating as impurities in the gaseous reactants or from some fuel cell materials, oxides of carbon will chemically react with the alkaline electrolyte and produce irreversible decay, which will decrease performance and shorten life. Consequently, AFCs with liquid electrolyte have been restricted to specialized applications where pure hydrogen and oxygen are utilized.

The revival of this technology seems to have started in recent years, mainly due to the development of solid alkaline electrolyte membrane. For AFCs with solid electrolyte membrane, most of the above-mentioned problems can be solved or abated. The solid electrolyte membrane makes the AFC technology comparable with PEMFC in terms of safety, cost, power density, and durability. In addition, since AFC has the most efficient energy conversion mechanism among all the fuel cell types, it therefore also has higher possibility to use no-precious catalyst than PEMFC. Nevertheless, AFC with solid electrolyte membrane is still at its early stage of development; the cost, durability, and performance are expected to be further examined and improved, and the transport phenomena inside the cell need to be investigated as well.

Proton Exchange Membrane Fuel Cells

Introduction

The PEMFC is also called solid polymer (electrolyte) fuel cell. It is perhaps the most elegant of all fuel cell systems in terms of design and mode of operation. It was the first type of fuel cell that was put into practical application (in Gemini space missions from 1962 to 1966). It consists of a solid polymeric membrane acting as the electrolyte. The solid membrane is an excellent proton conductor, sandwiched between two platinum-catalyzed porous carbon electrodes. It has fast start capability and yields the highest output power density among all types of the fuel cells. Because of the solid membrane as the electrolyte, there is no corrosive fluid spillage hazard, and there is lower sensitivity to orientation. It has no volatile electrolyte and has minimal corrosion concerns. It has truly zero pollutant emissions with potable liquid product water when hydrogen is used as fuel. As a result, the PEMFC is particularly suited for vehicular power application, although it is also being considered for stationary power application, albeit to a lesser degree.

The proton-conducting polymer membrane belongs to a class of materials called ionomers or polyelectrolytes, which contain functional groups that will dissociate in the presence of water. The dissociation produces ions fixed to the polymer and simple counterions that can freely exchange with ions of the same sign from the solution. The current available polyelectrolytes have cation as the counterion. In the case of hydrogen, the cation is proton. Therefore, the membrane must be fully hydrated in order to have adequate ion conductivity. As a result, the fuel cell must be operated under conditions where the product water does not evaporate faster than it is produced, and the reactant gases, both hydrogen and oxygen, need to be humidified. Therefore, water and thermal management in the membrane become critical for efficient cell performance, are fairly complex, and require dynamic control to match the varying operating conditions of the fuel cell. Because of the limitation imposed by the membrane and problems with water balance, the operating temperature of PEMFCs is usually less than 100°C, typically at 80°C. This rather low operating temperature requires the use of noble metals as catalysts in both the anode and cathode side with generally higher catalyst loadings than those used in PAFCs.

Currently, the polymer electrolyte used is made of per-fluorinated sulfonic acid membrane, or it is essentially acid, though in solid polymeric form. Hence, PEMFCs are essentially acid electrolyte fuel cells, with its operational principle essentially the same as PAFCs. As a result, most of PEMFC design, material selection, component fabrication, etc., are similar to those of PAFCs. The only difference is the humidification of reactant gases dictated by the membrane performance. Reactant humidification is often achieved by a number of techniques, e.g., by passing gas stream through a water column, by using in-stack humidification section of cell and membrane arrangement, and by spraying water into the reactant streams. In the early stage of the PEMFC development, the membranes were based on polystyrene, but since 1968, a Teflon-based product named "Nafion" by DuPont is used. This offers high stability, high oxygen solubility, and high mechanical strength.

Basic Operating Principle

The schematic of a single PEMFC is illustrated in Figure 1. The PEMFC requires hydrogen gas as the fuel and oxygen (typically air) as the oxidant. The half-cell reactions are

$$\text{Anode}: H_2 = 2H^+ + 2e^-$$

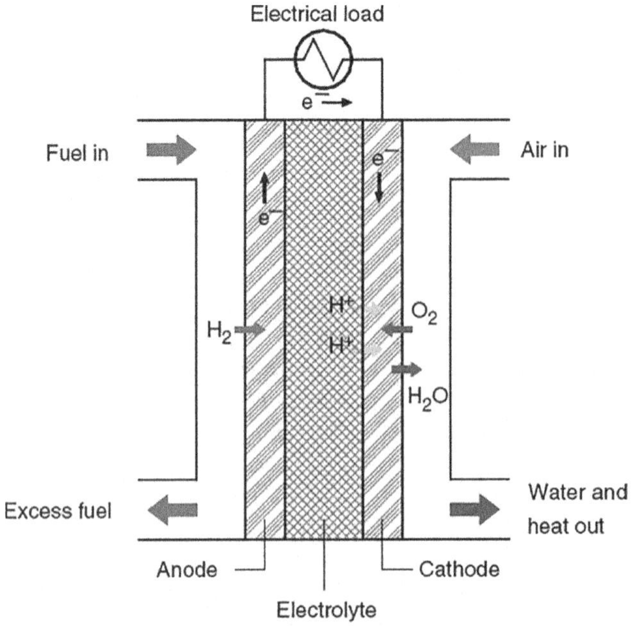

FIGURE 1 Schematic of a PEMFC.

$$\text{Cathode}: 1/2O_2 + 2H^+ + 2e^- = H_2O$$

and the overall cell reaction is

$$H_2 + 1/2O_2 = H_2O + \text{Heat Generated} + \text{Electric Energy}.$$

The current PEMFCs use perfluorinated sulfonic acid membrane as the proton-conducting electrolyte; carbon paper or cloth as the anode and cathode backing layers; and platinum or its alloys, often supported on carbon black, as the catalyst. The bipolar plate with the reactant gas flow fields is often made of graphite plate. The stoichiometry is around 1.1 to 1.2 for the fuel and 2 for the oxidant (oxygen). The PEMFCs usually operate at about 80°C and 1–8 atm pressure. The pressures, in general, are maintained equal on either side of the membrane. Operation at high pressure is necessary to attain high power densities, particularly when air is chosen as the cathodic reactant.

To prevent the membrane dryout leading to local hot spot (and crack) formation, performance degradation, and lifetime reduction, both fuel and oxidant streams are often fully humidified, and the operating temperature is limited by the saturation temperature of water corresponding to the operating pressure. The product liquid water formed at the cathode can dissolve in the electrolyte membrane or be removed from the cell by the excessive oxidant gas stream. The accumulation of liquid water in the cathode backing layer blocks the oxygen transfer to the catalytic sites, thus resulting in the phenomenon called "water flooding," causing performance reduction. Local hot and cold spots will cause the evaporation and condensation of water. Thus, an integrated approach to thermal and water management is critical to PEMFCs' operation and performance, and a proper design must be implemented.

Cold Start

In winter conditions, it is unavoidable for vehicles driving below the freezing point of water (0°C); therefore, for successful commercialization of PEMFC in automotive applications, rapid start-up from subzero temperatures must be achieved, which is referred to as "cold start" of PEMFC. The major problem of PEMFC cold start is that the product water freezes when the temperature inside the PEMFC is lower than the freezing point of water. If the catalyst layer is fully covered by ice before the cell temperature rises above the freezing point, the electrochemical reaction may be stopped due to the blockage of the reaction site. In addition, ice formation may also result in serious damage to the structure of the membrane electrode assembly. Most of the present PEMFCs employ various assisted cold start methods such as resistance heating using DC from batteries, coolant heating, hot air blowing, and catalytic hydrogen/oxygen reaction inside the PEMFC. The cold start performance can also be improved by design optimizations. Membrane thickness, ionomer content in catalyst layer, total heat capacity of the cell, and cell insulation are all important design factors to improve the cold start performance.

Acceptable Contamination Levels

As an acid electrolyte fuel cell operating at low temperature, the PEMFC is primarily vulnerable to carbon monoxide poisoning. Even a trace amount of CO drastically reduces the performance levels, although CO poisoning effect is reversible and does not cause permanent damages to the PEMFC system. Further, the performance reduction due to CO poisoning takes a long time (on the order of 2 hr) to reach steady state. This transient effect may have profound implication for transportation applications. Therefore, the PEMFC requires the use of a fuel virtually free of CO (must be less than a few parts per million). Also, high-quality water free of metal ions should be used for the cell cooling and reactant humidification to avoid the contamination of the membrane electrolyte. This requirement has a severe implication on the materials that can be used for cell components. On the other hand, carbon dioxide does not affect PEMFC operation and performance except through the effect of reactant dilution (the Nernst loss).

Major Technological Problems

For practical applications, PEMFC performance in terms of energy efficiency, power density (both size and weight), durability, and capital cost must be further improved. This can be accomplished by systematic research in the following:

i. New oxygen reduction electrocatalysts: This includes the development of non-precious catalyst and the reduction of precious metal platinum and its alloys loading from 4 mg/cm^2 to 0.4 mg/cm^2 or lower without affecting the long-term performance and the lifetime, as well as the development of CO-tolerant catalysts.
ii. New types of polymer electrolyte with higher oxygen solubility, thermal stability, long life, and low cost. A self-humidified membrane or a polymer without the need of humidification will be ideal for PEMFC operation and performance enhancement with significant simplification of system complexities and reduction of the cost.
iii. Profound changes in oxygen (air) diffusion electrode structure to minimize all transport-related losses. The minimization of all transport losses is the most promising direction for PEMFC performance improvement.
iv. Optimal thermal and water management throughout the individual cells and the whole stack to avoid local hot and dry spot formation and to avoid water flooding of the electrode.
v. Optimal design from single-cell component to system level to enhance the cold start performance.

Figure 2 illustrates the schematic of a PEMFC stack and its major components. In addition to the above issues, the development of low-cost lightweight materials for construction of reactant gas flow fields and bipolar plates is one of the major barriers to PEMFCs' large-scale commercialization. The successful solution to this problem will further increase the output power density, and it includes an optimal design of flow fields with the operating conditions, as well as an appropriate selection of materials and fabrication techniques. It has been reported that as much as over 20% improvement in the performance of PEMFC stacks can be obtained just by appropriate design of flow channels alone. The current leading technologies for bipolar plate design include the following: injection-molded carbon-polymer composites, injection-molded and carbonized amorphous carbon, assembled three-piece metallic, and stamped unitized metallic.

Technological Status

PEMFCs have achieved a high power density of over 2.2 kW/L, perhaps the highest among all types of the fuel cells currently under development. It is also projected that the power density may be further improved with unitized metallic (stainless steel) bipolar plates. The U.S. Department of Energy reported that the capital cost of 80 kW automotive fuel cell system costs in volume production (projected to 500,000 units per year) is $61/kW in 2009, and the future target is $35/kW. It is expected that PEMFC technology is about 5 to 10 years from commercialization, and precommercial demonstration for buses and passenger vehicles is under way with increasing intensity. The first demonstration for residential combined heat and power application has began at the end of 1999, and application of PEMFCs in powering portable and mobile electronics such as laptops has already been started.

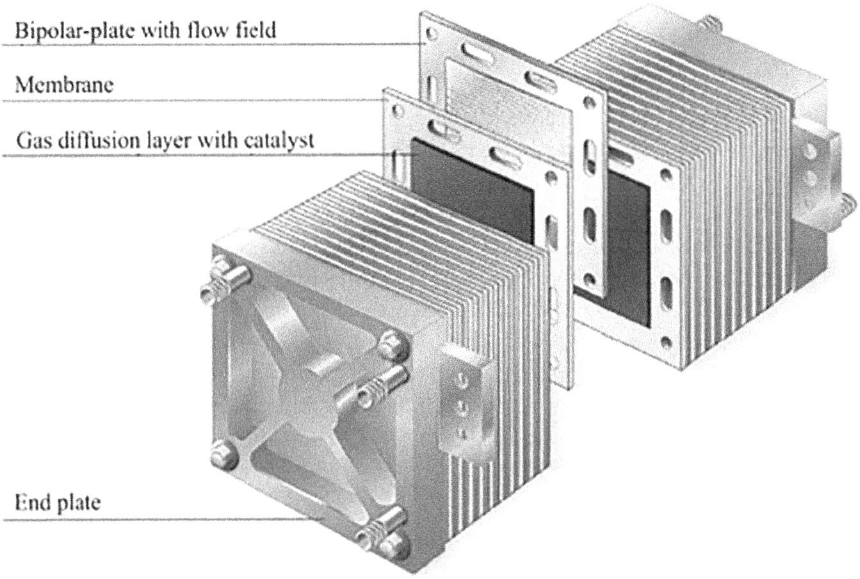

FIGURE 2 Schematic of a PEMFC stack and its components. (Source: Ballard Power Systems: http://www.ballard.com.)

Applications

PEMFCs have a high power density, a variable power output, and a short start-up time due to low operating temperature; the solid polymer electrolyte is virtually corrosion free and can withstand a large pressure differential (as high as 750 psi reported by NASA) between the anode and cathode reactant gas streams. Hence, PEMFCs are suitable for use in the transportation sector. Currently, they are considered the best choice for zero-emission vehicles as far as present-day available fuel cell technologies are concerned. Their high power density and small size make them primary candidates for light-duty vehicles, though they are also used for heavy-duty vehicles. For high-profile automobile application, pure hydrogen and air are used as reactants at the present. However, conventional gasoline and diesel engines are extremely cheap, estimated to cost about \$30–\$50/kW. Therefore, the cost of PEMFC systems must be lowered in order to be competitive with the conventional heat engines in the transportation arena.

For electricity generation from the hydrocarbon fuels, a reformer with carbon monoxide and sulfur cleaning is necessary. It is estimated that the cost of the reforming system is about the same as the fuel cell stack itself, which is also the same as the cost of other ancillary systems. Apart from the high cost, the optimal chemical-to-electric conversion efficiency is around 40%–45%, and the low operating temperature makes the utilization of the waste heat difficult, if possible at all; for the reforming of hydrocarbon fuels, cogeneration of heat and combined cycles can be applied. On the other hand, conventional thermal power plants with combined gas and steam turbines have energy efficiency approaching 60% with a very low capital cost of \$1000/kW. Therefore, the best possible application of the PEMFC systems interesting to utility industry is the use of PEMFCs in the size of tens to hundreds of kilowatts for remote region, as well as a possibility for residential combined heat and power application.

In addition, NASA has been conducting a feasibility study of using the PEMFC power systems for its space programs (mainly space shuttle missions) in place of its current three 12 kW AFC power modules. As discussed in the "Alkaline Fuel Cells" section, NASA is motivated by the extremely high cost, low lifetime, and maintenance difficulty associated with its current AFC systems.

Direct Methanol Fuel Cells

Introduction

All the fuel cells reviewed above for commercial applications require the use of gaseous hydrogen directly or liquid/solid hydrocarbon fuels, e.g., methanol, reformed to hydrogen as the fuel. Pure oxygen or oxygen in air is used as the oxidant. Hence, these fuel cells are often referred to as hydrogen-oxygen or hydrogen-air types of fuel cells. The use of gaseous hydrogen as a fuel presents a number of practical problems, such as storage system weight and volume as well as handling and safety issues especially for consumer and transportation applications. Although liquid hydrogen has the highest energy density, the liquefaction of hydrogen needs roughly one-third of the specific energy, and the thermal insulation required increases the volume of the reservoir significantly. The use of metal hydrides decreases the specific energy density, and the weight of the reservoir becomes excessive. The size and weight of a power system are extremely important for transportation applications, as they directly affect the fuel economy and vehicle capacity, although they are less critical for stationary applications. The low volumetric energy density of hydrogen also limits the distance between vehicle refueling.

Methanol as a fuel offers ease of handling and storage, and potential infrastructure capability for distribution. Methanol also has a higher theoretical energy density than hydrogen (~5 kWh/L compared with 2.6 kWh/L for liquid hydrogen). Easy refueling is another advantage for methanol. However, in the conventional hydrogen-air or hydrogen-oxygen fuel cells, a reformer is needed, which adds complexity and cost as well as production of undesirable pollutants such as carbon monoxide. The addition of a reformer also increases response time.

Therefore, direct oxidation of methanol is an attractive alternative in view of its simplicity from a system point of view. The DMFCs utilizing proton exchange membrane (PEM) have the capability of efficient heat removal and thermal control through the circulating liquid, as well as elimination of humidification required to avoid membrane dryout. These two characteristics have to be accounted for in the direct and indirect hydrogen systems that impact their volume and weight and, consequently, the output power density.

Basic Operating Principle

Figure 3 illustrates the basic operating principle of a DMFC. The DMFC allows the direct use of an aqueous, low-concentration (3%), liquid methanol solution as the fuel. Air is the oxidant. The methanol and water react directly in the anode chamber of the fuel cell to produce carbon dioxide and protons that permeate the PEM and react with the oxygen at the cathode. The half-cell reactions are as follows:

$$\text{Anode}: CH_3OH(1) + H_2O(1) = CO_2(g) + 6H^+ + 6e^-$$

$$\text{Cathode}: 6H^+ + 6e^- + 3/2O_2(g) = 3H_2O(1)$$

and the net cell reaction is

$$CH_3OH(1) + 3/2O_2(g) = CO_2(g) + 2H_2O(1).$$

Because the PEM (e.g., Nafion membrane) is used as the electrolyte, the cell operating temperature must be less than the water boiling temperature to prevent the dryout of the membrane. Typically, the operating temperature is around 90°C, and the operating pressure ranges from 1 atm to several atmospheres.

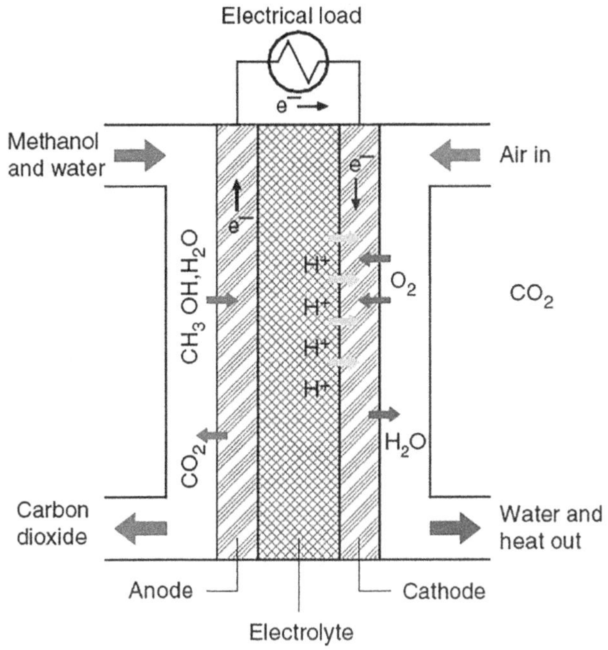

FIGURE 3 Schematic of DMFC.

Acceptable Contamination Levels

The system is extremely sensitive to carbon monoxide (CO) and hydrogen sulfide (H_2S). Carbon monoxide may exist as one of the reaction intermediaries and can poison the catalyst used. There are arguments whether CO is present in the anode during the reaction. Sulfur may be present if methanol is made of petroleum oils and needs to be removed.

Major Technological Problems

The PEM used in the DMFCs is the same as that employed in the PEMFCs. Since the electrolyte in DMFCs is essentially acid, expensive precious metals (typically platinum or its alloys) are used as the catalyst. However, the most serious problem is the so-called "methanol crossover." This phenomenon is caused by the electro-osmotic effect. When the protons migrate through the electrolyte membrane, a number of water molecules are dragged along with each proton, and because methanol is dissolved in liquid water on the anode side, methanol is dragged through the membrane electrolyte to reach the cathode side together with the protons and water. Fortunately, the methanol at the cathode is oxidized into carbon dioxide and water at the cathode catalyst sites, producing no safety hazards. However, the methanol oxidation in cathode does not produce useful electric energy. The development of a new membrane with low methanol crossover is a key to the success of DMFCs.

Such a low methanol crossover membrane has a number of advantages. First, it reduces the methanol crossover, enhancing fuel utilization and, hence, energy efficiency. Second, it reduces the amount of water produced at the cathode, leading to a lower activation and concentration polarization, thus allowing higher cell voltage at the same operating current. Third, it allows higher methanol concentration in the fuel stream, resulting in better performance.

Technological Status

Although methanol itself has simpler storage requirements than hydrogen and is simpler to make and transport, its electrochemical activity is much slower than that of hydrogen; that is, its oxidation rate is about four orders of magnitude smaller than that of hydrogen. Also, the conversion takes place at low temperature (about 80–90°C), and the contaminant problem is one of the serious issues.

The state-of-the-art performance is an energy conversion efficiency of 34% (from methanol to electricity) at 90°C using 20 psig air, that is, a cell voltage of 0.5 V (corresponding to a voltage efficiency of 42%) together with the methanol crossover accounting for 20% of the current produced (equivalent to a fuel efficiency of 80%). This 20% methanol crossover occurs when the fuel stream used is an aqueous solution containing only 3% methanol. It has been projected that with the better membrane developed by the University of Southern California (USC) and improvement of membrane electrode assembly, a cell voltage of 0.6 V can be achieved with only 5% methanol crossover. This is equivalent to 50% voltage efficiency and 95% fuel efficiency, resulting in an overall stack efficiency of 47% (from methanol to electricity). The DMFC system efficiency will be lower due to running the necessary auxiliary systems. Since the current DMFCs basically use the same cell components, materials, construction, and fabrication techniques as the PEMFCs, it is therefore expected that the system and component costs will be similar to that of the PEMFCs.

Applications

DMFCs have low power density due to the sluggish reaction kinetics of methanol, but high volumetric energy capacity since the volumetric energy density of methanol is high. These characteristics make DMFCs unlikely suitable for high power density applications such as powering large vehicles, but ideal for smaller applications such as laptops, cellular phones, digital cameras, battery chargers, or other electronic devices. Current DMFC systems are available with power outputs from 25 W to 5 kW with durations up to 100 hr between refueling operations. DMFCs therefore feature longer working period and shorter refueling/recharging time than conventional lithium batteries. In addition, since methanol can be made from agriculture products, the use of methanol is also compatible with renewable energy sources to allow for sustainable development.

Concluding Remarks

A summary of the preceding discussion, including the operational characteristics and technological status of the three major types of low-temperature fuel cells, is given in Table 1.

For stationary power generation in the utility industry, the economic fuels of today and in the near future are the fossil fuels. Although AFCs have the best performance when operating on pure hydrogen and oxygen, the liquid electrolyte's intolerance of carbon dioxide eliminates its role for utility

TABLE 1 Operational Characteristics and Technological Status of Various Low-Temperature Fuel Cells

Type of Fuel Cells	Operating Temperature (°C)	Power Density (mW/cm²), (Present) Projected	Projected Rated Power Level (kW)	Fuel Efficiency	Lifetime Projected (hr)	Capital Cost Projected ($/kW)	Application Areas
AFC	<90	(100–400) >300	10–100	4070	>10,000	35	Space, transportation
PEMFC	50–90	(600) >900	1–1000	45–70	>40,000	35	Transportation, space
DMFC	90	(230)	0.025–5	34	>10,000	35	Portable electronic devices, transportation

applications. However, the development of solid electrolyte membrane for AFC lifts the curtain of revival of this technology. The cost of PEMFCs has been significantly reduced in recent years, and it is expected to be comparable with internal combustion engines. PEMFCs are believed to be the most promising candidate for transportation application because of their high power density, fast start-up, high efficiency, and easy and safe handling. DMFCs are most suitable for portable applications such as laptops, cellular phones, digital cameras, battery chargers, or other electronic devices, because they have low power density but high energy capacity.

Bibliography

1. Blomen, L.J.M.J.; Mugerwa, M.N. *Fuel Cell Systems;* Plenum Press: New York, 1993.
2. Larminie, J.; Dicks, A. *Fuel Cell Systems Explained,* 2nd Ed.; Wiley, Chichester, 2003.
3. Kordesch, K.; Simader, G. *Fuel Cells and Their Applications;* VCH Press: New York, 1996.
4. Appleby, A.J.; Foulkes, F.R. *Fuel Cell Handbook;* Van Nostrand Reinhold Company: New York, 1989.
5. Hoogers, G. *Fuel Cell Technology Handbook;* CRC Press: New York, 2002.
6. Vielstich, W.; Gasteiger, H.; Lamm, A. *Handbook of Fuel Cells—Fundamentals;* Technology, Applications, 4-volume set; Wiley, Chichester, 2003.
7. Li, X. *Principles of Fuel Cells;* Taylor and Francis: New York, 2005.

36

Global Climate Change: Gasoline, Hybrid-Electric, and Hydrogen-Fueled Vehicles

Robert E. Uhrig

Introduction

It is theoretically possible for plug-in hybrid-electric light transportation vehicles to utilize electricity provided by electric utilities to displace almost 75% of the energy of gasoline used for light transportation vehicles (automobiles, SUVs, pickup trucks, minivans, etc.). It was also shown that replacing this gasoline energy with electricity would require 200–250 GW of new electrical generating capacity. Calculations show that about 930 GW of new electrical generating capacity would be required to produce hydrogen by electrolysis to replace all hydrocarbon fuels used in all U.S. transportation systems (including heavy trucks, aircraft, military vehicles, etc.). Hence, the choice of fuels for of these new electrical generating plants could have a large environmental impact if a significant fraction of gasoline used for light and heavy vehicle transportation was replaced by electricity or hydrogen.

Greenhouse Gases

Sunlight enters the atmosphere striking the earth, and it is reflected back towards space as infrared radiation (heat energy). Greenhouse gases absorb this infrared radiation, trapping the heat in the atmosphere. Until about 150 years ago, the amount of energy absorbed was about the same as the amount of energy radiated back into space. At that time, this equilibrium was upset, and the concentration of greenhouse gases—especially CO_2 (carbon dioxide)—began to increase. The concentration is continuing to increase at a rate that seems to parallel the increase in production of CO_2. During the same period, the average temperature of the atmosphere increased with subsequent climate changes in a manner that is thought by many to be a cause–effect relationship.[1]

The principal greenhouse gases are water vapor (60%–65%) and CO_2 (20%–25%). Water vapor stays in the atmosphere for a relatively short time—a matter of hours or days—whereas CO_2 has an average residence period of about a century. As a result, the amount of water vapor remains relatively constant related to the rate at which it is produced by weather phenomena, while CO_2 tends to accumulate as the amount emitted increases. All other greenhouse gases (10%–20%) such as methane, ozone, nitrous oxide, and carbon monoxide tend to have less effects on the atmosphere over time because of their small quantities or short residence times. Hence, the only greenhouse gas considered in this analysis is CO_2.

In contrast with the "well to wheels" approach that includes the emissions produced by processing the original fuel, this analysis starts with the fuel used to generate electricity or to produce hydrogen. This simplification does not materially change the relationship between the emissions of greenhouse gases by the various processes analyzed.

Models Used for Analyses

The models to quantitatively evaluate the greenhouse gas emissions, specifically carbon dioxide, when electrical energy or hydrogen is used to replace gasoline as an automotive fuel are those used in the author's previous publications.[2,3] These models were based on information provided by or extrapolated from data provided by the Department of Energy (DOE) Energy Information Administration (EIA).[4]

These models utilize the following data:

1. There were 225 million light transportation vehicles in the United States in 2004, including 133 million automobiles and 92 million light truck-based vehicles (sport utility vehicles, vans, pickup trucks, passenger minivans, and delivery vans).
2. On any given day, 50% of these vehicles traveled less than 32.2 km (20 mi).
3. The average distance traveled by each vehicle in the United States was 19,749 km/year (12,264 mi/year). This distance traveled by each average vehicle is derived from the 9 million barrels of oil per day used to make gasoline for the 225 million light vehicles having an average fuel consumption of 8.51 km/l (20 mi/gal), the approximate average gasoline mileage reported by EIA.[4]

4. The quantitative index used for comparing electricity and hydrogen produced by different methods using different source fuels is kilograms of CO_2 per vehicle-year (pounds per vehicle-year) for the average distance traveled by each vehicle in the United States—19,749 km/year (12,264 mi/year). Actual emissions of specific vehicles may be greater or less than the values presented here depending upon whether the fuel consumption of the particular vehicle is greater or less than the average value of 8.51 km/l (20 mi/gal).

CO_2 Emissions for Gasoline-Fueled Light Vehicles

TerraPass, a vendor dealing in carbon dioxide emission credits, has developed a "carbon dioxide calculator" that gives the emission of carbon dioxide for virtually all American automobiles for the past 20 years.[5] Experimenting with this calculator shows that the quantity of carbon dioxide, the most important greenhouse gas from combustion of gasoline in internal combustion engines, emitted per unit of time is a direct function of the amount of gasoline used in that time. This calculator utilized the fact that the combustion of 1 gal of gasoline emits 2.35 kg of CO_2 per liter (19.56 lbs of CO_2 per gallon) of gasoline. Because the average gas mileage in our model is 8.51 km/l (20 mi/gal), the annual emission of carbon dioxide is

$$\frac{19749 \text{ km/year} \times \text{kg } CO_2 \text{ per liter}}{8.51 \text{ km/liter}} = 5458 \text{ km/vehicle} - \text{year}$$

$$\left(11{,}997 \; lb \; CO_2 \; \text{ per vehicle-year}\right)$$

Clearly, the amount of CO_2 for smaller vehicles is less than for larger vehicles because the gasoline mileage is greater. However, the model for this analysis deals with average vehicles that achieve 8.51 km/l (20 mi/gal)—the reference performance used for all comparisons.

CO_2 Emissions for Micro, Mild, and Full Hybrid Vehicles

Because of the energy to propel all traditional types of hybrid vehicles is provided by gasoline, we will utilize the increased fuel mileages assigned in the author's earlier publication[3] and reduce the total CO_2 emissions for traditional hybrid-electric vehicles accordingly, as shown in Table 1.

These values are for hybrid vehicles of a size corresponding to a traditional vehicle that attains 8.51 km/l (20 mi/gal). A larger or smaller vehicle would have higher or lower emissions, respectively.

CO_2 Emissions for Plug-In Hybrid-Electric Vehicles

The model utilized in the author's hybrid article[3] has half the vehicles traveling 24.2 km (15 mi) per day on electricity, or 8816 km (5475 mi) per year. The other half of the vehicles travel 56.4 km (35 mi) per day, or 20,572 km (12,775 mi) per year on electricity. These assumptions result in an average of 14,694 km (9125 mi) per year on electricity for all vehicles.

TABLE 1 Emissions for Hybrid-Electric Vehicles

	Gasoline Mileage		CO_2 per Vehicle-Year	
Type Hybrid	km/l	mi/gal	Kg	lb
Micro hybrid	9.35	22	4,955	10,906
Mild hybrid	10.62	25	4,362	9,598
Full hybrid	12.32	29	3,761	8,274

The second half of these vehicles must also travel an additional 5055 km (3139 mi) per year as a full hybrid using gasoline at 12.32 km/l (29 mi/gal). Because the average total distance traveled by each car per year is 19,749 km (12,264 mi), the distance traveled using gasoline is 25.6% of the average distance each vehicle travels. The other 74.4% of this distance is traveled using electricity generated by a utility using nuclear fuels, fossil fuel, solar energy (wind, hydro, or photovoltaic systems), or renewable energy systems. Such a substitution of electricity for gasoline would save 6.7 million of the 9.0 million barrels of oil per day used in the United States for light vehicle transportation.

Electricity Generated Using Nuclear, Solar, and Renewable Fuels

Plug-in hybrid vehicles have two sources of CO_2—the CO_2 emitted due to operation in the full-hybrid mode and the CO_2 emitted in generating the electrical energy used for the rest of the time. Because nuclear, solar, and renewable fuels generate a net of zero CO_2, the only CO_2 generated is by operation in the full-hybrid mode for 25.6% of the distance traveled. Hence, the emission is 25.6% of the value in Table 1 for a full-hybrid, 964 kg (2122 lb) per vehicle-year.

Electricity Generated Using Fossil Fuels (Natural Gas, Oil, and Coal)

When the electricity is supplied by utilities using fuels that emit carbon dioxide during combustion, this CO_2 from generating electricity must be added to the CO_2 of the full-hybrid operation discussed above. Data on actual emissions from fossil power plants in the United States for 1999 (the last year for which complete data is available), provided by EPA and (DOE) (DOE-EIA 1999) show the emission rates in Table 2.

A reasonable average of 0.374 kWh/km (0.603 kWh/ mi) for the expenditure of electrical energy for hybrid vehicles that would be comparable to a vehicle getting 8.51 km/l (20 mi/gal) of gasoline using an internal combustion engine. Previously in this entry, it was shown that the average distance traveled by all plug-in hybrid vehicles while operating on electricity alone was 14,694 km/year (9125 mi/year). Hence, the electricity used is

$$\left(14{,}694\ \text{km/year}\right) \times \left(0.374\ \text{kWh/km}\right) = 5{,}496\ \text{kWh/year}.$$

If we multiply this value by the amount of CO_2 emission per kilowatt-hour for the three fossil fuels given in Table 2 and add the 964 kg/vehicle-year (2122 lb/vehicle-year) of CO_2 for operation in the hybrid mode, we get the results for fossil fuels given in Table 3.

TABLE 2 Emission Rates for Fossil Fuels

		CO_2 Emission Rate	
Fuel to Generate Electricity	% of Total Generation	kg/kwh	lb/kwh
Coal	51.0	0.952	2.095
Oil	3.2	0.895	1.969
Natural gas	15.2	0.600	1.321
Renewables	0.6	0 (net)	0 (net)
Non-fossil[a]	30.0	0	0

[a]Nuclear, Solar, Wind, and Hydro.
Source: Energy Information Administration (see DOE-EIA[6]).

TABLE 3 Emissions for Fossil Fuels

	Emissions of CO_2	
Fuel to Generate Electricity	kg/Vehicle-Year	lb/Vehicle-Year
Coal	6,205	13,650
Oil	5,889	12,956
Natural gas	4,271	9,396

Source: Energy Information Administration (see DOE–EIA[6]).

Hydrogen-Fueled Vehicles

The emission of carbon dioxide from vehicles utilizing electricity generated with fuel cells operating on hydrogen is negligible except for the emissions from the various processes used to produce the hydrogen. These emissions of carbon dioxide from the production processes must be taken into account to give a valid comparison with emissions of greenhouse gases from the other vehicle configurations and their potential impact upon climate and weather modification.

Steam Methane Reforming

Some 95% of hydrogen used today is produced by steam methane reforming (SMR) from natural gas (~98% methane—CH_4). The two steps of SMR are

$$CO_4 + H_2O + \text{Heat}(206\ \text{kJ/mol}) \xrightarrow{\text{Catalyst}} CO + 3H_2$$

$$(\text{Steam Reforming Reaction})$$

$$CO + H_2O \xrightarrow{\text{Catalyst}} CO_2 + H_2 + \text{Heat}(41\ \text{kJ/mol})$$

$$(\text{Water–Gas Shift Reaction})$$

The hydrogen comes from the methane and the steam. The steam reforming reaction is endothermic with the required 206 kJ/mol heat energy normally produced by combustion of some of the methane. The water-gas shift reaction is exothermic, providing 41 kJ/mol heat energy that, if recovered, can reduce the amount of methane burned. Then the methane burned has to provide only the net 165 kJ/mol that represent ~17% of the total methane energy. About 3.3 mol of hydrogen (~83% of the theoretical maximum of 4 mol of hydrogen from CH_4) are produced for each mol of methane. A well-designed SMR plant will yield hydrogen having about 80% of the energy of the methane supplied. Unfortunately, SMR produces CO_2 in both the methane combustion and in the water–gas shift reaction.

Emissions of Vehicles Using Hydrogen Produced by SMR

In an earlier publication,[3] it was established that 0.228 kWh/km (0.367 kWh/mi) of mechanical energy at the tire- pavement interface corresponded to 8.51 km/l (20 mi/gal). If we use the following efficiencies:

- 70% efficiency for the electric motor drive (electricity to mechanical energy at the tire-pavement interface), the same used in the above reference,
- 60% efficiency (hydrogen to electricity) for the fuel cell, and
- 85% efficiency is distributing and dispensing the hydrogen to the fuel cell, the needed energy of the hydrogen input to the fuel cell is

$$\frac{0.228\ kWh/km}{0.70 \times 0.60 \times 0.85}$$

$$= 0.638\ \text{kWh/km}\ (1.028\text{kWh}/\text{mi})$$

of hydrogen energy. Because the average total distance traveled per vehicle is 19,749 km/year (12,264 mi/year), the total kilowatt-hour per vehicle-year is

$$(19749\ \text{km}/\text{year}) \times (0.638\ \text{kWh of H}_2\ \text{energy}/\text{km})$$

$$= 12607\ \text{kWh of H}_2\ \text{energy}/\text{vehicleyear}.$$

The energy of hydrogen (lower heating value) is 119.9 MJ/kg (51,600 Btu/lb), so the amount of H_2 used per year is

$$\frac{12607\ \text{kWh per vehicle} - \text{year}}{119.9\ \text{MJ}/\text{kg} \times 0.278\ \text{kWh}/\text{MJ}}$$

$$= 379\text{kg H}_2\ \text{per vehicle} - \text{year}$$

$$(834\ lb\ H_2\ \text{per vehicle-year})$$

Goswami indicates that SMR produces a net of 0.43 mol of CO_2 for each mol of H_2 produced using SMR.[7] Because the molecular weights of CO_2 and H_2 are 44 and 2, respectively, the specific CO_2 emission can be calculated by

$$(0.43\ \text{mol}\ CO_2\ /mol\ H_2) \times \frac{(44\ gm/\text{mol}\ CO_2)}{(2\ gm/\text{mol}\ H_2)}$$

$$= 9.46\ \text{gm CO}_2\ /\ \text{gm H}_2 = 9.46\ \text{kg CO}_2\ /\ \text{kg H}_2$$

and the total emission of CO_2 per vehicle-year is

$$(379\ \text{kg H}_2\text{per vehicle} - \text{year})$$

$$\times (9.46\ kg\ CO_2\ \text{per}\ kg\ H_2)$$

$$= 3{,}585\ \text{kg CO}_2\ \text{per vehicle} - \text{year}$$

$$(= 7{,}888\ \text{lb CO}_2\ \text{per vehicle} - \text{year}).$$

Emission for Vehicles Using Hydrogen Produced by SMR with Heat Provided by a High-Temperature Nuclear Reactor

Recent work in Japan has demonstrated the feasibility of substituting high-temperature heat from a gas-cooled nuclear reactor to replace the heat supplied by the combustion of methane. This increases the amount of hydrogen produced to 4 moles per mole of methane and eliminates the CO_2 produced by combustion of methane, but not the CO_2 produced by the water–gas shift reaction. The overall reaction of the steam reforming and water shift reactions is the production of 1 mole of CO_2 and 4 mol of hydrogen for each mole of CH_4. Hence, multiplying this ratio by the molecular weights and the amount of hydrogen used per year gives

$$\frac{1 \text{ mol } CO_2}{4 \text{ mol } H_2} \times \frac{44}{2} \times 379 \text{ kg } H_2 \text{ per vehicle} - \text{year}$$

$$= 2{,}084 \text{ kg } CO_2 \text{ per vehicle} - \text{year}$$

$$\left(4{,}586 \ lb\ CO_2 \text{ per vehicle-year}\right).$$

Emissions for Vehicles Using H_2 Produced by Electrolysis

Conventional electrolysis of water to produce hydrogen is a well-developed technology and production units as large as 10 MWe are commercially available today. However, the typical overall efficiency of hydrogen production using electrolysis based on the thermal content of the generating plant fuel is about 25% today, consisting of two components—about 33% efficiency in converting fossil or nuclear fuel to electricity and about 75% in using electricity to separate water into hydrogen and oxygen. If a high-temperature gas-cooled reactor or a modern high-efficiency gas-fired combined cycle plant is used, the overall efficiency in producing hydrogen could approach 45%.

A leading manufacturer of electrolysis equipment indicated that 1 MW of electricity can generate 0.52 ton (1040 lb) of hydrogen per day or 0.473 kg H_2/kW day (1.04 lb H_2/kW day).[8] In the case of SMR, it was calculated that the hydrogen required per year for a vehicle using hydrogen to travel 12,260 mi/year was 379 kg of H_2/ vehicle-year (832 lb/vehicle-year). Hence, the amount of electricity required per year is

$$\frac{\left(379 \text{ kg } H_2 / \text{year}\right) \times \left(24 \text{ h/day}\right)}{0.473 \text{ kg}H_2 / \text{kW day}} = 19{,}230 \text{ kWh/year}.$$

The carbon dioxide emitted for hydrogen-fueled vehicles is the carbon dioxide emitted in producing the electricity required to produce the hydrogen. The choices of fuel for generating electricity are coal, oil, natural gas, nuclear energy, and solar energy (wind, photovoltaics, and hydro). Hence, the emissions of carbon dioxide are the product of the kilowatt-hour per year and the appropriate emission per kilowatt-hour from Table 2 for the fuel used.

Emissions for Vehicles Using H_2 Produced by Electrolysis Using Solar, Nuclear, and Renewable Energy

Because none of these energy sources emit carbon dioxide when generating electricity, there are no emissions of greenhouse gases associated with these arrangements.

Emissions for Vehicles Using H_2 Produced by Electrolysis Using Fossil Energy (Coal, Oil, and Natural Gas)

The emissions of carbon dioxide for these fossil fuels are the products of the 19,230 kWh/year and the appropriate emission per kilowatt-hour for the fuel as given in Table 2. The results are shown in Table 4.

TABLE 4 Emissions for Fossil Electrolysis

Fuel	kg CO_2/Year	lb CO_2/Year
Natural gas	11,538	25,384
Oil	17,211	37,963
Coal	18,307	40,276

Emissions for Vehicles Using H_2 Produced by Thermochemical Processes

The overall efficiency of the thermochemical process of producing hydrogen, such as the sulfur–iodine process, approaches 50% at 900°C, while the overall efficiency of the electrolysis process is about 25%. Hence, for the same fuel, the carbon dioxide emissions will be half as much for thermochemical processes as for electrolysis.

Emissions for Vehicles Using Hydrogen Produced by Thermochemical Processes Using Solar, Nuclear, and Renewable Energy

Because none of these energy sources emit carbon dioxide when generating electricity, there are no emissions of carbon dioxide associated with these arrangements.

Emissions for Vehicles Using Hydrogen Produced by Thermochemical Processes Using Fossil Energy (Natural Gas, Oil, and Coal)

The emissions of carbon dioxide for these fossil fuels are half of those for electrolysis, as shown in Table 4. The results are shown in Table 5.

Results and Conclusions

The results of this analysis are presented in Tables 6 and 7, which provide emissions of carbon dioxide for the various hybrid and hydrogen-fueled vehicle arrangements for the reference case of 8.50 km/l (20 mi/gal) for the average vehicle being driven 19,749 km/year (12,264 mi/year).

TABLE 5 Emissions for Hydrogen Produced by Thermochemical Processes by Fossil Fuels

Fuel	kg CO_2year	lb CO_2year
Natural gas	5,769	12,692
Oil	8,605	18,932
Coal	9,153	20,138

TABLE 6 Comparison of Emissions of Carbon Dioxide for Hybrid Vehicles

Vehicle Emissions of Carbon Dioxide	kg CO_2 Vehicle-Year	lb CO_2 Vehicle-Year
Reference gasoline vehicle average 8.51 km/l (20 mi/gal)	5,500	12,000
Micro-hybrid vehicle 9.35 km/l (22 mi/gal)	4,950	10,900
Mild-hybrid vehicle 10.62 km/l (25 mi/gal)	4,350	9,600
Full-hybrid vehicle 12.32 km/l (29 mi/gal)	3,750	8,300
Plug-in hybrid vehicle		
Electricity generated with nuclear	950	2,100
Electricity generated with solar energy	950	2,100
Electricity generated with renewable	950	2,100
Electricity generated with natural gas	4,250	9,400
Electricity generated with oil	5,900	13,000
Electricity generated with coal	6,200	13,700

Vehicle travels 19,749 km (12,264 mi) per year.

TABLE 7 Comparison of Emissions of Carbon Dioxide for Hydrogen-Fueled Fuel Cell Vehicles

Vehicle Emissions of Carbon Dioxide	kg CO_2/Vehicle-Year	lb CO_2/Vehicle-Year
Reference gasoline vehicle utilized an average 8.50 km/l (20 mi/gal)	5,500	12,100
Vehicle uses hydrogen produced using SMR	3,600	7,900
Vehicle using hydrogen produced using SMR with heat supplied by		
High temperature gas cooled reactor	2,100	4,600
Vehicle using hydrogen produced using electrolysis with electricity		
Produced using natural gas	11,550	25,400
Produced using oil	17,200	37,900
Produced using coal	18,300	40,300
Produced using nuclear energy	0	0
Produced using solar energy (wind, hydro, and photovoltaic)	0	0
Produced using renewables	0	0
Vehicle using hydrogen generated using a thermochemical process with heat		
Generated using natural gas	5,750	12,700
Generated using oil	8,600	18,900
Generated using coal	9,150	20,100
Generated using nuclear energy	0	0
Generated using solar energy (wind, hydro, and photovoltaic)	0	0
Generated using renewables	0	0

Vehicle travels 19,749 km (12,264 mi) per year.

Plug-In Hybrid-Electric Vehicles

In Table 6, the vehicle emissions of CO_2 for all traditional hybrid-electric vehicles are inversely related to the gas mileage. The gasoline mileage for micro, mild, and full hybrids are assumed to be 9.35, 10.62, and 12.32 km/l (22, 25, and 29 mi/gal), respectively. These are reasonable overall values for the newer, larger hybrid vehicles that have larger gasoline engines and electric motors such as the Toyota Highlander and Honda Accord hybrids. The rounded emissions in Table 6 for electricity generated using oil and coal are greater than for the reference case of gasoline. However, the emission of a plug-in hybrid vehicle where the electricity is generated by nuclear or solar energy is less than 20% of the reference gasoline case. Hence, the only way to significantly improve the greenhouse gas situation if plug-in hybrids are implemented on a large scale is to use only nuclear energy, solar (wind, hydro, or photovoltaic) energy, or renewables to generate the needed electricity. Because hydro power is virtually out of the question due to limited sites and environmental concerns, renewables have large land requirements, and both photovoltaic and wind are intermittent in nature with availabilities of less than about 30%, nuclear power would appear to be the primary choice.

Hydrogen-Fueled Fuel Cell Vehicles

The vehicle emissions of CO_2 for hydrogen-fueled fuel-cell-driven vehicles are presented in Table 7, where the emissions have been rounded off. Because fuel cells do not emit CO_2, the only emissions are from the processes used to produce the hydrogen. The carbon dioxide emissions for hydrogen produced by steam methane reforming, electrolysis, and thermochemical methodologies are given.

Perhaps the most important observation is that, contrary to the widespread belief that hydrogen-fueled vehicles produce no greenhouse gases, the emissions for hydrogen- fueled vehicles vary widely depending upon the method used to generate the electricity or heat to produce the hydrogen. In the case of electrolysis, in which the electricity is generated using nuclear, solar, and renewable energy, the average annual emissions per vehicle-year are negligible. However, the emissions for fossil fuels (natural gas, oil, and coal) are about 11,500, 17,200, and 18,300 kg/year (25,300, 37,800, and 40,200 lb/year), respectively—almost three times those for plug-in hybrid vehicles. These are extremely large emissions compared to almost any other method investigated here.

The emissions for producing hydrogen for a hydrogen- fueled vehicles using SMR produces CO_2 by two methods: combustion of some of the methane to drive the reaction and the water–gas shift reaction. These reactions produce about 3600 kg (7900 lb) CO_2 per vehicle-year, about the same as a full-hybrid vehicle.

If the heat required to drive the reaction in steam methane reforming can be supplied by an outside source such as a high temperature gas cooled reactor, then the emission of CO_2 is reduced to about 2100 kg (4600 lb) per vehicle- year.

The conclusion is that using hydrogen in fuel cells to propel vehicles is a complex process with many steps (converting thermal energy to electricity to hydrogen to electricity to propelling the vehicle), each consuming considerable energy. Hence, the resulting CO_2 emissions when fossil fuels are used to generate the electricity for electrolysis or thermochemical processes are correspondingly high. Given the concerns about the influence of CO_2 on weather and climate, it seems compelling that fossil fuels not be used in producing electricity for electrolysis or thermochemical processes to produce hydrogen for transportation.

References

1. DOE–EIA, Energy Information Administration, *Greenhouse Gases, Climate Change, and Energy*, 2005; at http://www.eia.doe.gov/oiaf/1605/ggccebro/chapter1.html (accessed May 2006).
2. Uhrig, R.E. Engineering challenges of the hydrogen economy. The BENT of Tau Beta Pi, 2004; XCL (2), 10–19.
3. Uhrig, R.E. Using plug-in hybrid vehicles to drastically reduce petroleum-based fuel consumption and emissions. The BENT of Tau Beta Pi, 2005; XCLI (2), 13–19.
4. DOE–EIA, Energy Information Administration, Petroleum Flow 2002. *Annual Energy Review*, 2003; available at http://www.eia.doe.gov/emeulaer/diagram2.html (accessed January 2006).
5. TerraPass, *Carbon Dioxide Calculator*, Terrapass.com/ carboncalc.html (accessed January 2005).
6. DOE–EIA, Department of Energy, Energy Information Administration (DOE and EPA), *Carbon Dioxide Emissions from the Generation of Electric Power in the United States*, July 2000.
7. Goswami, D.Y. *Hydrogen Supply Technologies: Introduction to Issues*, 2005; available at http://www.ases.org/hydrogen_forum03/Goswami.pdf (accessed June 2005).
8. Stuart, S. *Nuclear-Electric Synergies: For Neat Hydrogen Production and Cleaner Fossil Reserves Harvesting;* ANS President's Special Session: Hydrogen Systems, ANS Meeting: Reno, NV, 2001.

37

Heat Pumps

Lu Aye

Introduction

This entry provides foundation knowledge about heat pumps for both professional and non-professional readers. A brief history of heat pumps, basic terms used in the heat pump industry and research and the fundamentals of heat pumps are presented. Working principles of the thermoelectric heat pump, the absorption heat pump, the gas compression heat pump and the vapor compression heat pump are explained by using schematic diagrams. Performance parameters of the heat pump systems—coefficient of performance (COP), energy efficiency ratio (EER), primary energy ratio (PER), and ambient energy fraction (AEF)—are also discussed.

Major advances have been made in heat pump technologies over the last 35 years. "Heat pump systems offer economical alternatives of recovering heat from different sources for use in various industrial, commercial and residential applications."[1] Heat pumps in their various and diverse forms allow one to harness, in a cost-effective manner, solar energy that has already been stored in the atmosphere and biosphere. To make a fair systematic comparison with other systems that provide the same thermal output, AEF may be used. The concept and the method of estimating the AEF are discussed. The relationships between the AEF, the heating coefficient of performance (HCOP) and PER are also presented. The short-term and long-term potential use of the heat pumps and corresponding greenhouse gas (GHG) emissions are discussed.

Brief History

Although most people are familiar with the common household refrigerator, the concept of using such a device to provide heating rather than cooling is less widely understood. The basic operational principle of this machine called the heat pump, however, was laid down in thermodynamic terms by Lord Kelvin, the first professor of natural philosophy at the University of Glasgow, Scotland, in the middle of the 19th century. Whereas the thermodynamic principle of the heat pump was found in 1852, it was not until 1855 that it was realized for producing heat by means of an open-cycle mechanical vapor recompression unit in Ebensee, Austria.[2] Much later, the closed vapor compression process was used for generating useful heat. After World War II, heat pump units for air-conditioning homes and individual rooms became common. Now, heat pump technology is well known as one of the energy conservation technologies.

Fundamentals

A heat pump is a thermodynamic system whose function is to heat, at the required temperature, with the aid of heat extracted from a source at lower temperature.[3] Heat pumps are devices designed to utilize low-temperature sources of energy that exist in atmospheric air, in lake or river water, and in the earth. These sources are referred to as "ambient energies." Heat pumps can also be operated using waste heat from commercial and industrial processes, thereby upgrading this to the required temperature level for some thermal process operations.[4]

The heat pump can be considered as a heat engine in reverse.[5] A heat engine removes heat from a high-temperature source and discharges heat to a low temperature and in doing so can deliver work (see Figure 1). Heat pumps capture heat energy from low-grade sources and deliver at a higher temperature. They require external work (W) to upgrade the heat absorbed (Q_L) (see Figure 2). Note that the heat energy flows from a low-temperature heat source to a higher application temperature. This is the

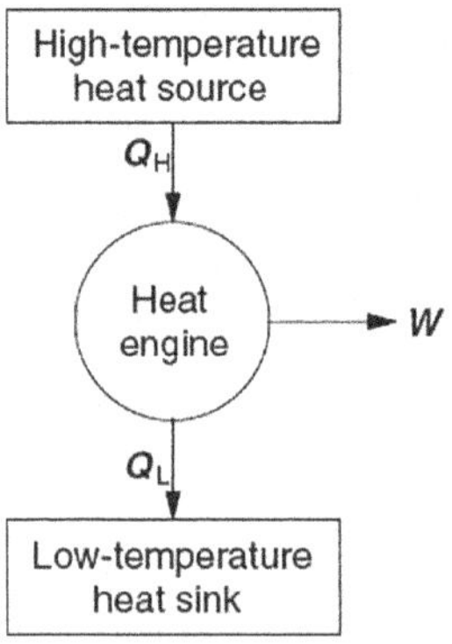

FIGURE 1 Thermodynamic model of a heat engine.

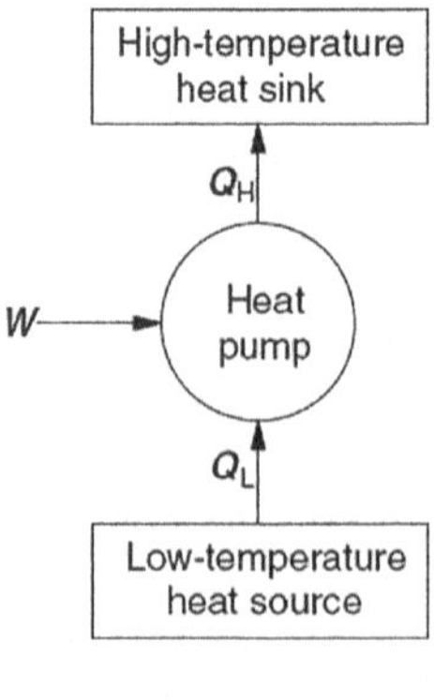

FIGURE 2 Energy flow of a heat pump.

TABLE 1 Commonly Used Heat Sources

Heat Source	Temperature Range (°C)
Ambient air	–10 to 15
Building exhaust air	15–25
Groundwater	4–10
Lake water	0–10
River water	0–10
Seawater	3–8
Rock	0–5
Ground	0–10
Wastewater and effluent	>10

reverse of the natural flow direction of heat from a higher temperature to a lower temperature. For ideal conditions without heat losses, the energy balance of a heat pump provides (Eq. 1):

$$Q_H = Q_L + W \tag{1}$$

where Q_H is the useful heat delivered, Q_L is the heat absorbed and W is the work input.

The useful heat delivered is always greater than the work input to the heat pump for ideal conditions.

The technical and economic performance of a heat pump is closely related to the characteristics of the heat source. Table 1 lists current commonly used heat sources. Ambient and exhaust air, soil and groundwater are practical heat sources for small heat pump systems, whereas sea/lake/river water, rock (geothermal) and wastewater are generally used for large heat pump systems.[6]

The fundamental theories of heat pumps and refrigerators are the same. In engineering practice, a distinction is made between heat pumps and refrigerators. The principal purpose of the latter device is to remove heat from a low-temperature heat source, and the purpose of the former device is to supply heat to a high-temperature sink. In many practical devices, the heating and cooling elements are interchangeable. The term *reverse cycle air conditioner* is frequently applied to heat pumps with interchangeable heating and cooling elements. In Japan and in the United States, reversible air-conditioning units are called heat pumps, and in Europe, the term *heat pump* is used for heating-only units.[2]

Carnot Coefficient of Performance (Theoretical Cop)

The heating performance of a heat pump is measured by the HCOP. The HCOP is the ratio of the quantity of heat transferred to the high-temperature sink to the quantity of energy driving the heat pump (see Figure 2).

$$\text{HCOP} = \frac{Q_H}{W} \tag{2}$$

In the Carnot ideal heat pump cycle, the heat is delivered isothermally at T_H and received isothermally at T_L. By using the laws of thermodynamics and definition of entropy, it can be shown that the Carnot HCOP is given by Eq. 3.

$$\text{HCOP} = \frac{T_H}{T_H - T_L} = \frac{T_L}{T_H - T_L} + 1 \tag{3}$$

No practical heat pump constructed can have a better performance than this theoretical ideal HCOP. The best that our practical heat pump cycles can do is struggle towards achieving this ideal performance.[5]

Facts about Heating Relevant to Heat Pumps

It is argued that heat pumps are very energy efficient and therefore environmentally benign. The International Energy Agency (IEA) Heat Pump Centre provides the basic facts about heat supply and discusses the value of heat pumps.[7] The basic facts about heating explained are as follows:

- Direct combustion to generate heat is never the most efficient use of fuel.
- Heat pumps are more efficient because they use renewable energy in the form of low-temperature heat.
- If the fuel used by conventional boilers were redirected to supply power for electric heat pumps with HCOP of 3.5 to 4.5, about 35–50% less fuel would be needed, resulting in 35–50% less emissions.
- Around 50% savings are made when electric heat pumps with HCOP of 4.5 are driven by combined heat and power (CHP) or cogeneration systems.
- Whether a fossil fuel, nuclear energy or renewable energy is used to generate electricity, electric heat pumps make far better use of these resources than do resistance heaters.
- The fuel consumption, and consequently the emissions rate of an absorption or gas-engine heat pump is about 35–50% less than that of a conventional boiler.

Because heat pumps consume less primary energy than conventional heating systems, they are an important technology for reducing the unwanted air pollutants such as respirable particulate matters (PMs), carbon monoxide (CO), nitrogen oxides (NO_x), sulfur dioxide (SO_2), and carbon dioxide (CO_2) that harm the human environment.

Heat Pump Types

Five major types of heat pump may be identified in the literature.[8,9] These are the following:

- the magnetic heat pump
- the thermoelectric heat pump
- the absorption heat pump
- the gas compression heat pump
- the vapor compression heat pump

The Magnetic Heat Pump

The magnetic heat pump requires a solid magnetic material and it is based on the magnetocaloric effect. The magnetocaloric effect is the heating and cooling of a magnetic material in response to the application and removal of a sufficiently high external magnetic field.[10] The principle of the magnetic heat pump is shown in Figure 3.[11] With thermal Switch 1 off and thermal Switch 2 on, a magnetic field is applied to the solid magnetic material. The solid magnetic material is heated up above the heat sink temperature and heat will flow to the heat sink. Next, thermal switch 1 is on while thermal switch 2 remains on, and the magnetic field is partially removed; this cooled down the magnetic material to the heat source temperature. Thermal switch 2 is then off while the magnetic field is totally removed, completing the cooling of the magnetic material, and the magnetic material gains the heat from the heat source. Thermal switch 2 is then on and a magnetic field is applied to heat the magnetic material. Then, thermal switch 1 is off and the cycle is repeated to transfer heat from the low-temperature heat source to the higher-temperature heat sink.

The magnetic heat pumps do not use ozone-depleting refrigerants, and they have compact configuration, low noise, high efficiency, high stability, and longevity. They have great applicable prospects.[12,13]

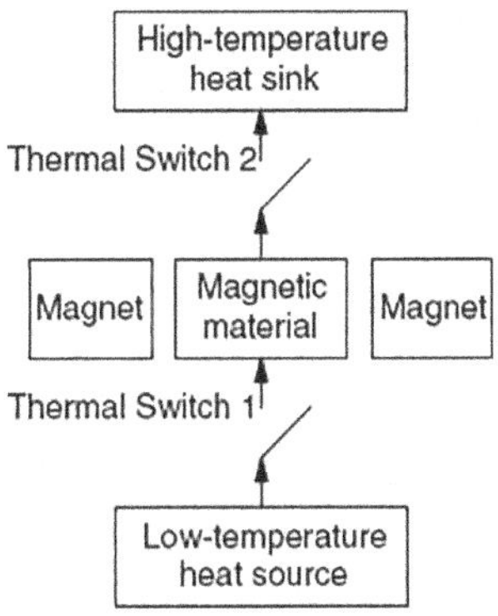

FIGURE 3 Schematic of a magnetic heat pump.

The Thermoelectric Heat Pump

The thermoelectric heat pump uses the Peltier effect. Two materials with different thermoelectric properties (*p* type and *n* type) are arranged to form hot and cold junctions and are connected in series with a direct current voltage source (Figure 4). These devices do not have moving parts and the cooling and heating elements may be reversed simply by changing the direction of the electric current. Current commercially available thermoelectric heat pumps have efficiencies that are well below those of vapor compression heat pumps. They are generally used in conditions where the solid-state nature (no moving parts, maintenance-free) outweighs efficiency, e.g., cooling of electronic components and small instruments.

The Absorption Heat Pump

An absorption cycle is a heat-activated thermal cycle. The absorption heat pump can be considered as a heat engine driving a heat pump (see Figure 5). All absorption cycles include at least three thermal exchanges with their surroundings (i.e., energy exchange at three different temperatures). For the system to operate, the generator temperature (T_g) must be greater than the condenser temperature (T_c) and the absorber temperature (T_a), must in turn be greater than the evaporator temperature (T_e).

The components of a simple single-effect absorption system are as follows:

- generator (desorber)
- condenser
- refrigerant expansion valve
- evaporator

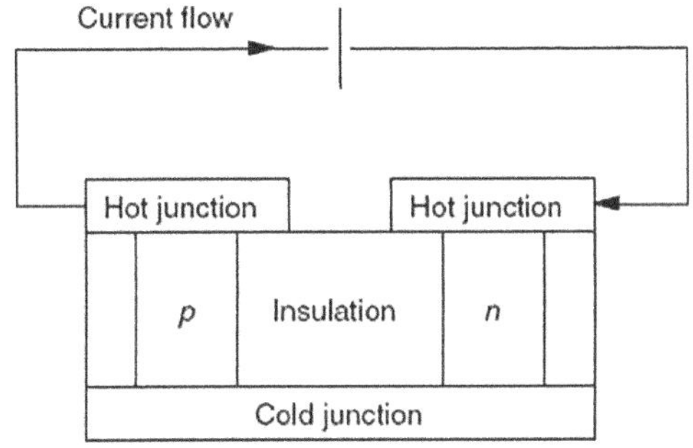

FIGURE 4 Schematic of a thermoelectric heat pump.

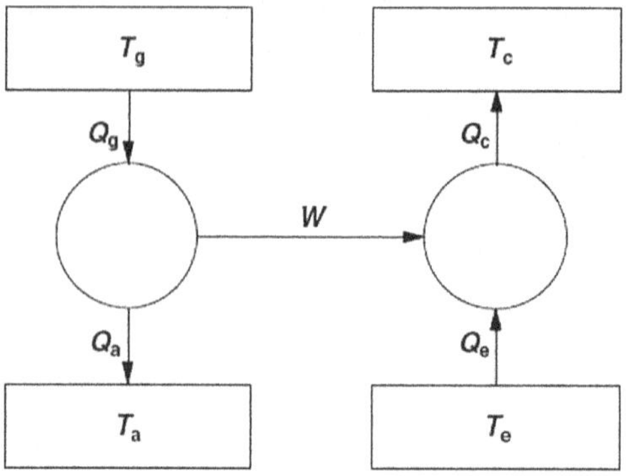

FIGURE 5 Thermodynamic model of an absorption heat pump.

- absorber
- pump
- solution heat exchanger
- solution pressure reducer
- connecting pipes

The absorption system utilizes a sorbent–refrigerant pair, lithium bromide and water, and water and ammonia, being the most widely used. A schematic of the continuous absorption heat pump system is shown in Figure 6. The major exchanges with its surroundings are thermal energy, plus a small amount of mechanical work at the pump to transport the working fluids.

A high-temperature heat source supplies heat to the generator to drive off the refrigerant vapor from solution. The condenser receives the refrigerant vapor from the generator. In the condenser, the refrigerant changes phase from vapor to liquid and heat is rejected at a medium temperature. The refrigerant expansion valve reduces the pressure of the refrigerant from the condensing pressure to

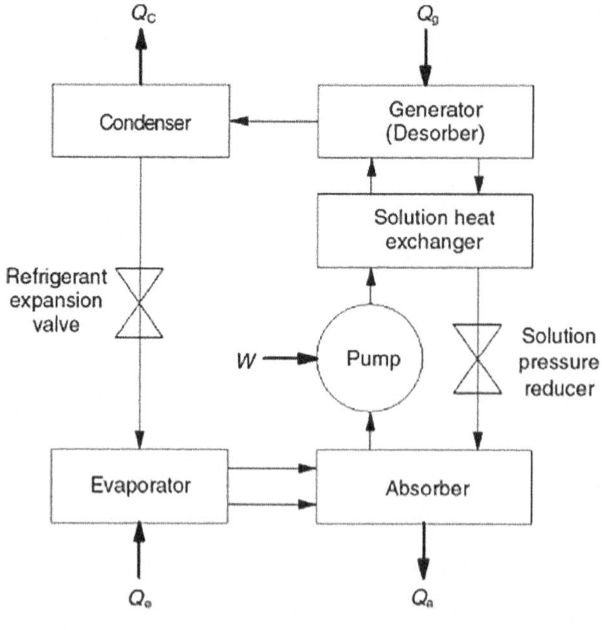

FIGURE 6 Simple absorption cycle.

the evaporating pressure. In the evaporator, the liquid refrigerant collects the low-temperature heat and changes to vapor phase. In other words, the working fluid (refrigerant) provides the cooling effect at the evaporator and the vaporized refrigerant enters the absorber. The strong-concentration sorbent absorbs the refrigerant vapor and becomes a weak solution. Note the heat rejected at the absorber. The low-pressure weak solution (refrigerant and sorbent) is pumped to the solution heat exchanger to exchange heat with the strong-concentration sorbent. The solution pressure reducer accepts the high-pressure strong solution from the generator and delivers low-pressure strong solution to the absorber. The generator receives the high-pressure weak solution from the solution heat exchanger and separates the refrigerant and sorbent.

The Gas Compression Heat Pump

Gas compression heat pumps use a gas as the working fluid. Phase changes do not occur during the cycle of operation. Major components of a gas compression heat pump system are as follows:

- gas compressor
- high-pressure heat exchanger
- gas expander (or turbine)
- drive motor
- low-pressure heat exchanger

A schematic of a gas compression heat pump is shown in Figure 7. The gas is drawn in to the compressor and compressed. The hot high-pressure gas delivers heat at the high-pressure heat exchanger. At the expander (turbine), the high-pressure gas does expansion work that assists to drive the compressor. In general, the external work input to the compressor-expander system is provided by a drive motor.

The Vapor Compression Heat Pump

Almost all heat pumps currently in operation are based on either a vapor compression or an absorption cycle.[14] Heat pumps and refrigerators of the vapor compression type utilize a working fluid that undergoes phase changes during operation. The phase changes occur principally during the heat collection and rejection stages of the cycles. The phase-changing processes occurring in the vapor compression heat pumps are accompanied by extremely favorable heat transfer conditions. Boiling agitation in evaporators and dropwise condensation in condensers make the heat transfer rates very high, which helps to improve the system efficiency. To carry out evaporation and condensation, any practical heat pump uses four separate components as illustrated in Figure 8:

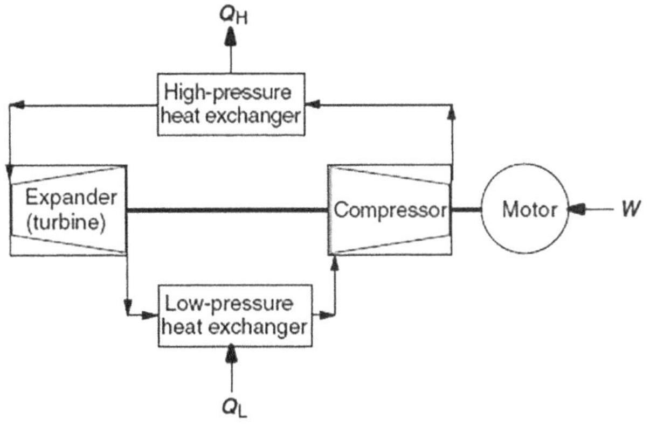

FIGURE 7 Schematic of a gas compression heat pump.

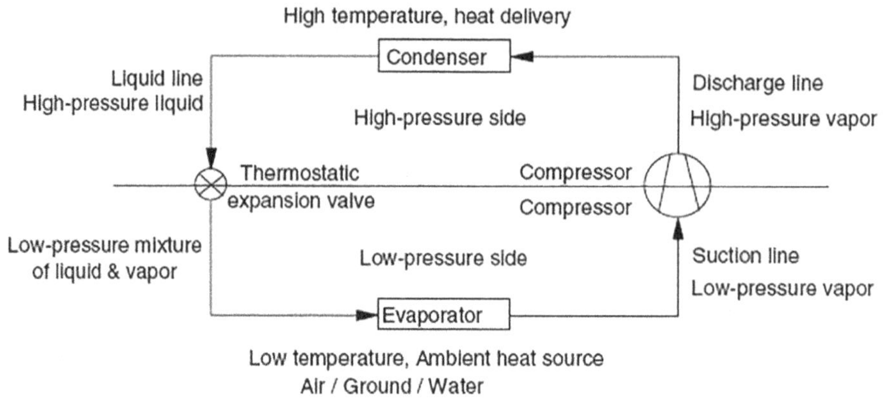

FIGURE 8 Schematic diagram of a vapor compression heat pump.

- compressor
- condenser
- expansion device
- evaporator

The compressor is the heart of the vapor compression heat pump system. By using mechanical power, the compressor increases the pressure of the working fluid vapor received from the evaporator and delivers it to the condenser. The working fluid condenses and provides useful heat at the condenser. The pressure of the liquid working fluid coming out of the condenser is reduced to the evaporating pressure by an expansion device. The low-grade heat is collected at the evaporator and the phase of the working fluid changes from liquid to vapor. The working fluid enters the compressor in vapor form to repeat the cyclic flow.[15]

Compressors used in practical units are generally of the positive displacement type: either reciprocating or rotary vanes. In practice, nearly all vapor compression heat pumps use thermostatic expansion (T-X) valves because of the availability of the valve to handle a wide range of operating conditions and because of the fact that the pressure reduction is essentially isenthalpic. These valves, which are relatively low-cost devices, control the liquid working fluid flow to the evaporator by sensing the superheat condition of the working fluid leaving the evaporator. Such control ensures that nearly all of the available evaporator surface is covered with a forced convection nucleate boiling film with consequential excellent heat transfer characteristics in the evaporation process.

Reversible circuit heat pumps for building cooling and heating are equipped with a four-way exchange valve that can be operated automatically or manually. This valve enables the normal evaporator to become the condenser and the normal condenser to become the evaporator so that the machine can be made to operate in reverse fashion. This arrangement is often incorrectly called reverse-cycle operation because of the end effect achieved—i.e., the interchange of roles for the cooling and heating parts of the circuit. It would be better to call this reverse-flow operation as the basic thermodynamic cycle remains the same for each mode of operation.

In general, vapor compression heat pumps are classified in many different ways. One method commonly used distinguishes among air, water and ground sources of low-temperature energy, and also between the high-temperature energy delivery medium (i.e., air or water). In such a classification, the most common types of heat pump units would include air-to-air units and water-to-water units. Solar energy could also be captured at the evaporator as a heat source. These types of heat pumps are called solar-assisted and solar-boosted heat pumps.[16]

Changes in condensing and evaporating temperatures affect the work input required at the compressor and consequently the COP. Decreasing the condenser or increasing the evaporator temperatures will

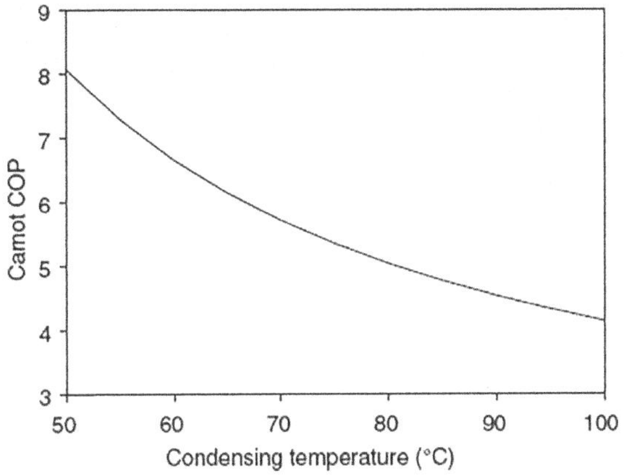

FIGURE 9 Effect of condensing temperature on ideal COP (evaporating temperature = 10°C).

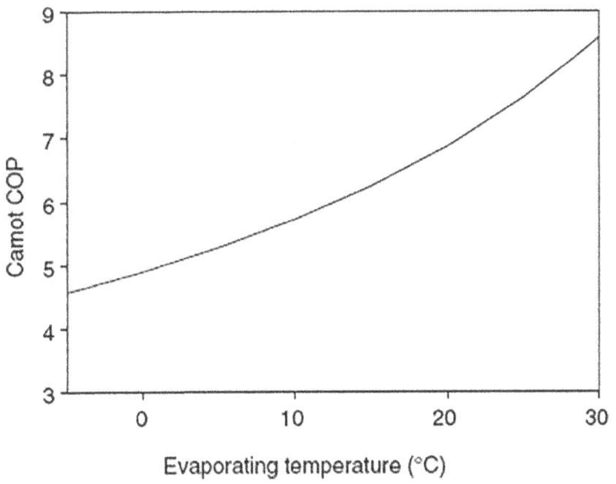

FIGURE 10 Effect of evaporating temperature on ideal COP (condensing temperature = 70°C).

decrease the compressor work and increase the COP. Figure 9 shows the effect of condensing temperature on COP for a fixed evaporator temperature. The effect of the evaporating temperature on the COP for a fixed condenser load temperature is shown in Figure 10. For simplicity, these curves have been obtained assuming operation on the ideal heat pump cycle—i.e., the Carnot cycle of two isothermal and two isentropic (reversible adiabatic) processes not allowing any practical deviations. Although the COP values for real machines will be substantially lower, the trend will be the same.

Performance Parameters

This section defines and explains some performance parameters that are used to compare various heat pumps for a particular application and also presents the relationships among them.

SCOP and EER

System coefficient of performance (SCOP) is defined as output heating capacity per unit of power input to the system (see Eq. 4).

$$\text{SCOP} = \frac{\text{Heating capacity (W)}}{\text{Power input (W)}} \tag{4}$$

SCOP is a dimensionless number that measures the performance of a heat pump. If the heating capacity is expressed in units other than watts, it is called heating energy efficiency ratio (HEER). Heat pumps sold in the United States are often stated in terms of HEER (see Eq. 5).

$$\text{HEER} = \frac{\text{Heating capacity (Btu/hr)}}{\text{Power input (W)}} \tag{5}$$

Seasonal heating energy efficiency ratio (SHEER) is the time average HEER value for the heating season, and it is also known as seasonal performance factor (SPF). In the equation described, HEER has a unit of "Btu/hr per W." It should be noted that 1 Btu/hr = 0.2928 W. Therefore, the relationship between HEER and SCOP can be expressed as in Eq. 6.

$$\text{SCOP} = \text{HEER} \times 0.2928 \tag{6}$$

Primary Energy Ratio

The HCOP provides a measure of the usefulness of the heat pump system in producing heat from work. It does not express the fact that energy available as work is normally more useful than energy available as heat. To assess different heat pump systems using compressor drives from different fuel or energy sources, the PER is applied. The PER takes into account not only the heat pump COP but also the efficiency of conversion of the primary fuel into the work that drives the compressor. PER is defined as in Eq. 7.[5] This can be also expressed as in Eq. 8.

$$\text{PER} = \frac{\text{Useful heat delivered by heat pump}}{\text{Primary energy consumed}} \tag{7}$$

$$\text{PER} = \frac{Q_H}{E_{pe}} = \frac{Q_H}{W} \cdot \frac{W}{E_{pe}} = \text{HCOP} \cdot \eta_{pp} \tag{8}$$

where Q_H is the load, E_{pe} is the primary energy used by the heat pump system, and η_{pp} is the power plant efficiency.

The drive energy of heat pumps is most commonly electricity. "Ideally a heat pump where free work is available should be contemplated e.g., wind or water power."[5] Consider an electric heat pump powered by a conventional power plant fueled by a non-renewable energy (see Figure 11). The power plant efficiency, η_{pp}, is up to 58% for oil- or gas-fired combined-cycle power plants currently available on the market. The PER is equal to the HCOP for direct power generation from renewable ambient energy sources such as solar and wind.[2] This concept is illustrated in Figure 12.

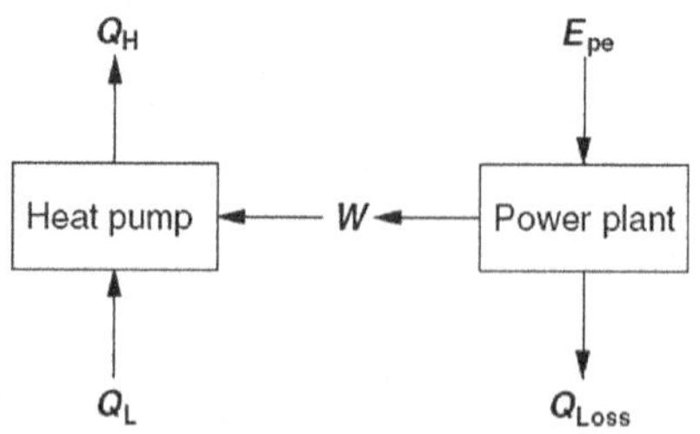

FIGURE 11 Energy flow of a conventional electric heat pump.

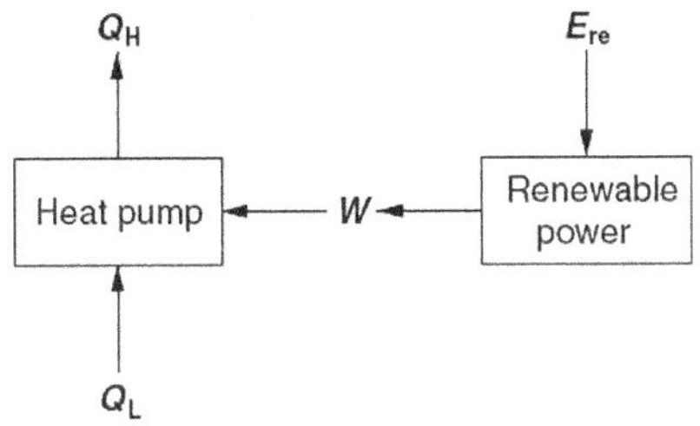

FIGURE 12 Energy flow of a renewable electric heat pump.

The amount of renewable or ambient energy spent (used up) to produce work for the heat pump is equal to the amount of work (i.e., $E_{re} = W$). For this case, by definition, $\eta_{pp} = 1$.[2] It should be noted that unlike the losses in a fossil fuel power plant, the unused ambient energy passing through the renewable power plant is still in the form of ambient energy and it is available to be used.

Ambient Energy Fraction

To make a fair systematic comparison of heat pumps and other systems that provide the same heat output, the term *AEF* was developed by Aye et al.[17] The term *solar fraction* is widely understood, accepted and used in the solar energy field. It is defined as the fractional reduction of purchased energy when a solar energy system is used.[18] It is the fraction of the load contributed by the solar energy, which can be calculated by Eq. 9.

$$f = \frac{L - E}{L} = \frac{L_s}{L} \tag{9}$$

where L is the load, E is the auxiliary energy supplied to the solar energy system, and L_s is the solar energy delivered.

Similar to the solar fraction of a solar system, the term AEF of a heat pump can be defined as the fraction of the load contributed by the ambient energy, which may be calculated as in Eq. 10.

$$\text{AEF} = \frac{Q_H - E_{pe}}{Q_H} = 1 - \frac{E_{pe}}{Q_H} = 1 - \frac{1}{\text{PER}} = 1 - \frac{1}{\text{HCOP} \times \eta_{pp}} \tag{10}$$

Figure 13 illustrates the relationship between the AEF and the PER.

Table 2 shows the AEFs of a heat pump, which has a HCOP of 3.5, for various electric power plants. It can be clearly seen from Table 2 that for the same heat pump, the AEF may vary from 1% to 71% depending on how electricity is generated for driving the heat pump compressor. The AEF for the renewable

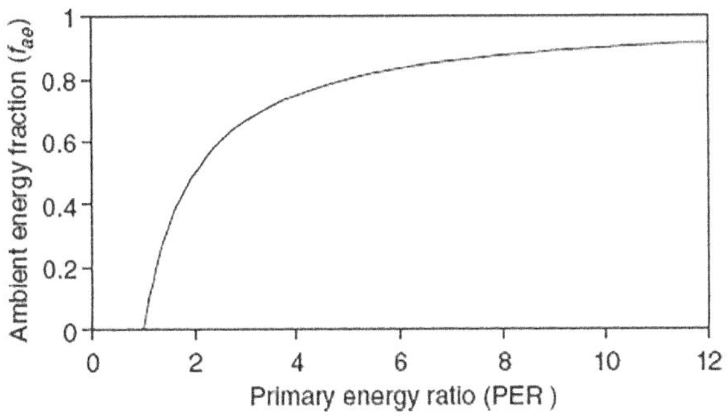

FIGURE 13 The AEF versus the PER.

TABLE 2 AEFs of a Heat Pump Powered by Typical Power Plants

Power Plant	Brown Coal	Nuclear	Black Coal	Gas-Fired Combined Cycle	Renewable
Power plant efficiency, η_{pp}	0.29a	0.33b	0.35a	0.58b	1.00
HCOP of heat pump	3.50	3.50	3.50	3.50	3.50
PER	1.02	1.16	1.23	2.03	3.50
AEF	0.01	0.13	0.18	0.51	0.71

[a] Typical Australian data.
Source: Taylor and Labson.[19]
[b] Typical European data.
Source: Gac et al.[3]

TABLE 3 Effect of HCOP on AEF of a Heat Pump Powered by a Renewable Source

HCOP of heat pump	1.50	2.50	3.50	4.50
Renewable power plant efficiency, η_{pp}	1.00	1.00	1.00	1.00
PER	1.50	2.50	3.50	4.50
AEF	0.33	0.60	0.71	0.78

electricity-driven heat pump is the highest (71%). The renewable energy used is only 29% (i.e., 100% – 71%) of the total thermal load. The energy use of the brown coal electricity-driven heat pump is 99% of the total thermal load (i.e., 0.99 MJ of brown coal energy is required for 1 MJ of thermal load). Table 3 shows the effect of heat pump COP on the AEF of a heat pump powered by a renewable source.

Heat Pump Applications

Heat pumps have been used for domestic, commercial, and industrial applications.

Domestic applications are as follows:

- provision of space heating
- provision of hot water
- swimming pool heating

Commercial and industrial applications are as follows:

- space heating
- water heating
- swimming pool heating
- drying and dehumidification
- evaporation and boiling
- desalination

Potential for the Use of Heat Pumps

In 2000, the primary energy used in buildings was 149 EJ, and the world total primary energy used was 387 EJ.[20] The building sector represented 38% of the total primary energy use and 34% of the total GHG emissions in 2000. Price et al.[20] reported that, in 2004, the emissions from the building sector, including the electricity consumed, were 8.6 Gt CO_2, 0.1 Gt $CO_{2\text{-e}}$ N_2O, 0.4 Gt $CO_{2\text{-e}}$ CH_4, and 1.5 Gt $CO_{2\text{-e}}$ halocarbons (including CFCs and HCFCs). These emissions arise from the following:

TABLE 4 GHG Saving Potential of Heat Pumps (Space Heating, Space Cooling, and Water Heating Applications) for the Building Sector

	2004	2010	2020	2030
GHG emissions from buildings (GtCO2-e)	6.5	7.5	9.6	12.4
Heat pump market share (%)	3	5	14	36
HCOP	3.5	3.7	4.1	4.5
Average power plant efficiency (%)	30	36	48	65
Global average PER	1.05	1.33	1.98	2.93
GHG saving (Mt CO2-e)	9	100	657	2911

- direct combustion of fossil fuels in residential and commercial buildings, amounting to 3.3 $GtCO_2$ (almost 1.7 $GtCO_2$ from combustion of oil, around 1.3 $GtCO_2$ from gas, and about 0.3 $GtCO_2$ from coal)[21]
- indirect or upstream CO_2 emissions from the demand of electricity and district heat were about 5.4 $GtCO_2$[21]
- combustion of biomass produces N_2O and CH_4 equivalent to 0.5 $GtCO_{2\text{-e}}$
- refrigerant or working fluid (halocarbons) leakages accounting for 1.5 $GtCO_{2\text{-e}}$

In 2004, about 60% of total GHG emissions generated by the building sector (6.5 $GtCO_{2\text{-e}}$ out of 10.7 $GtCO_{2\text{-e}}$) is estimated to be due to space heating, space cooling, and water heating. Heat pumps can meet these requirements in all types of buildings. Many low-temperature industrial heating requirements can also be met by heat pump technology. It is apparent that heat pumps have a large potential for saving energy and GHG emissions due to buildings. Energy and GHG saving potential of heat pumps can be estimated by using PER, one of the heat pump performance parameters presented. Table 4 shows the estimated GHG saving potential. The following assumptions were made for these estimations until 2030:

- annual GHG emissions growth rate for the building sector is 2.5%
- annual market growth rate for vapor compression heat pump is 10%
- HCOP improvement is 1% per annum
- annual global average power plant efficiency improvement is 3%

The GHG reduction potential of about 3 $GtCO_{2\text{-e}}$ in 2030 was estimated for the building sector alone. The IEA Heat Pump Centre reported a minimum of 0.2 $GtCO_{2\text{-e}}$ saving potential by industrial heat pumps in 1997, an estimation based on a study by project Annex 21. The total GHG reduction potential of 1.2 $GtCO_{2\text{-e}}$ was estimated in the year 2010 by IEA HPC. "This is one of the largest that a single technology can offer and this technology is already available in the market place."[22]

Conclusions

The fundamentals of heat pumps have been presented together with the working principles of the thermoelectric heat pump, the absorption heat pump, the gas compression heat pump and the vapor compression heat pump. It should be noted that vapor compression heat pumps driven by electricity dominate the current market. Heat pumps are very energy-efficient and therefore environmentally benign compared to other available heating technologies. The technical and economic performance of a heat pump is closely related to the characteristics of the heat source.

Various performance parameters are available for comparing heat pumps; HCOP and HEER are the most widely used. PER and AEF can be used to compare various heat pump systems systematically and fairly. The AEF of an electric heat pump depends on the HCOP and the power plant efficiency based on the primary energy used. The AEF is highly dependent on the type of power plant used to generate the electricity that drives the heat pump compressor. Heat pumps driven by renewable electricity offer the possibility of reducing energy consumption significantly. In the future, the AEF may be used widely as a performance parameter of heat pumps since energy resource issues are becoming more important.

Acknowledgments

I wish to thank Emeritus Professor William W. S. Charters, former Dean of Engineering at the University of Melbourne, who introduced me to heat pump technology and for supporting my career and research in heat pumps.

References

1. Chua, K.J.; Chou, S.K.; Yang, W.M. Advances in heat pump systems: A review. Appl. Energy **2010**, *87* (12), 3611–3624.
2. Halozan, H.; Rieberer, R. Energy-efficient heating and cooling systems for buildings. IIR Bulletin, **2004**, *LXXXI* (2004-6), 6–22.
3. Gac, A.; Vrinat, G.; Blaise, J.-C.; Camous, J.-P.; Fleury, M. *Guide for the Design and Operation of Average and Large Capacity Electric Heat Pumps*; International Institute of Refrigeration: Paris, 1988.
4. McMullen, J.T.; Morgan, R. *Heat Pumps;* Adam Hilger Ltd: Bristol, U.K., 1981.
5. Reay, D.A.; Macmichael, D.B.A. *Heat Pumps Design and Applications: A Practical Handbook for Plant Managers, Engineers, Architects and Designers*; Pergamon Press: Oxford, 1979.
6. IEAHPC (International Energy Agency Heat Pump Centre). Heat Sources. Available at http://www.heatpumpcentre.org/About_heat_pumps/Heat_sources.asp (accessed September 2005).
7. IEAHPC (International Energy Agency Heat Pump Centre). How Heat Pumps Achieve Energy Savings and CO_2 Emissions Reduction—An Introduction. Available at http://www.heatpumpcentre.org/About_heat_pumps/Energy_and_CO$_2$.asp (accessed September 2005).
8. Brown, G.V. Magnetic heat pumping near room temperature. J. Appl. Phys. **1976**, *47* (7), 3673–3680.
9. Taylor, L.E. *The Design, Construction, and Evaluation of a Solar Boosted Heat Pump;* Master of Engineering Science thesis, Department of Mechanical Engineering, The University of Melbourne, 1978.
10. Phan, M-H.; Yu, S.-C. Review of the magnetocaloric effect in manganite materials. J. Magn. Magn. Mater. **2007**, *308* (2), 325–340.
11. Steyert, W.A. Stirling-cycle rotating magnetic refrigerators and heat engines for use near room temperature. J. Appl. Phys. **1978**, *49* (3), 1216–1226.
12. Yu, B.; Liu, M.; Egolf, P.W.; Kitanovski, A. A review of magnetic refrigerator and heat pump prototypes built before the year 2010. Int. J. Refrig. **2010**, *33* (6), 1029–1060.
13. Yu, B.F; Gao, Q.; Zhang, B.; Meng, X.Z.; Chen, Z. Review on research of room temperature magnetic refrigeration. Int. J. Refrig. **2003**, *26* (6), 622–636.
14. IEAHPC (International Energy Agency Heat Pump Centre). Heat Pump Technology. Available at http://www.heatpumpcentre.org/Aboutheatpumps/HP_technology.asp (accessed September 2005).
15. Aye, L.; Charters, W.W.S. Electrical and engine driven heat pumps for effective utilisation of renewable energy resources. Appl. Therm. Eng. **2003**, *23* (10), 1295–1300.
16. Aye, L.; Charters, W.W.S; Chaichana, C. Solar boosted heat pump. *Progress in Solar Energy Research;* Nova Science Publishers, Inc.: NY, 2006.

17. Aye, L.; Fuller, R.J.; Charters, W.W.S. Ambient energy fraction of a heat pump. In *Solar 2005, Renewable Energy for a Sustainable Future—A challenge for a post carbon world,* Proceedings of the 43rd Annual Conference of Australian and New Zealand Solar Energy Society, November 28–30, 2005, Dunedin, 2006.
18. Duffie, J.A.; Beckman W.A. *Solar Engineering of Thermal Processes,* 2nd Ed.; A Wiley-Interscience Publication: New York, 1991.
19. Taylor, M.; Labson, S. *Profiting from Cogeneration.* DPIE: Commonwealth Department of Primary Industries and Energy/Australian Cogeneration Association: Canberra, Australia, 1997.
20. Price, L.; de la Rue du Can, S.; Sinton, J.; Worrell, E.; Nan, Z.; Sathaye, J.; Levine, M. *Sectoral Trends in Global Energy Use and Greenhouse Gas Emissions*; Lawrence Berkeley National Laboratory: Berkeley, CA. LBNL-56144, 2006.
21. Stern, N.H. *The Economics of Climate Change: The Stern Review*; Cambridge University Press: Cambridge, 2007.
22. IEAHPC (International Energy Agency Heat Pump Centre). How Heat Pumps Achieve Energy Savings and CO_2 Emissions Reduction: An Introduction, 2010. Available at http://www.heatpumpcentre.org/en/aboutheatpumps/howheatpumpsachieveenergysavings/Sidor/default.aspx (accessed August 2010).

38

Hydroelectricity: Pumped Storage

Jill S. Tietjen

Introduction

The dynamics of electric utility system operation are such that integrating large amounts of intermittent or variable renewable energy resources (such as wind and solar) will require energy storage mechanisms. Wind and solar resources are weather dependent—they generate power when the wind blows or the sun shines—they are often unavailable at times when the electric utility's requirement for power is at or near peak levels. Wind generation is often at its peak output when the requirement for electricity is at or near its lowest levels. Pumped storage hydro is a commercially available technology that has been in use for more than 100 years to provide generation when it is most beneficial. The operational benefits of pumped storage hydro enhance the ability of the power system to accommodate renewable generation while meeting reliability requirements. These benefits will allow renewable generation to become a more significant portion of electric utility generation portfolios around the world.

This entry describes the dynamics of utility system operation. This is followed by information on wind and solar renewable energy resources and the historical experience with how and when the wind blows and the sun shines, enabling these resources to produce electricity. An overview of various energy storage mechanisms is provided, followed by a more in-depth discussion of pumped storage hydro. Information on how electric utilities use pumped storage hydro is followed by a discussion of how pumped storage hydro can be integrated with and provide benefits for renewable energy resources.

Hour	MW
1	54.19282
2	51.58654
3	49.39094
4	49.98949
5	52.04352
6	54.05509
7	60.46665
8	68.17764
9	74.70331
10	81.64417
11	90.04101
12	95.16006
13	98.23288
14	98.29861
15	100
16	98.22293
17	97.93742
18	94.14076
19	91.889
20	87.04675
21	80.462
22	75.42743
23	67.80997
24	59.64637

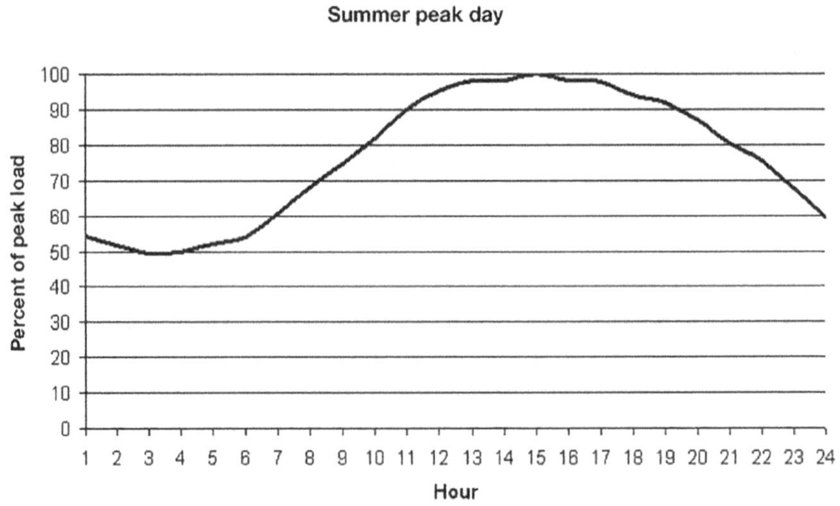

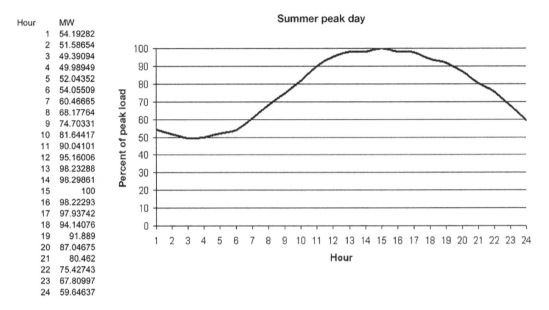

FIGURE 1 Electric customer demand—summer peak day.

Dynamics of Utility System Operation

Demand for electricity is constantly changing—with the hour of the day, the season—and it is dependent on customer behavior. Heating and cooling needs of consumers comprise a significant portion of the demand for electricity. Electric system operators balance the demand for electricity with its supply as electricity has to be used as soon as it has been generated. Generally, there is a greater demand for electricity in the summer (hot temperatures) and winter (cold temperatures) than in the more moderate temperatures associated with the spring and fall. In the aggregate, customers use more electricity during the day and less at night (see Figure 1 where 1 A.M. is hour 1, noon is hour 12, and midnight is Hour 24). Demand patterns change from minute to minute and from one hour to the next. Power system operators must manage the changes that occur with each of the demand fluctuations to balance the system and keep the lights on.[1,2]

Hour-by-hour changes in demand are normally managed by increasing or decreasing the set of resources that need to be operating to meet demand in that hour (unit commitment). Minute-by-minute fluctuations in demand are managed by directing the more flexible resources to increase or decrease their output. Every few seconds, system operators change the generation on specific generating units up or down to keep the system in balance.[1] By balancing fluctuations between supply and demand, system operators provide an ancillary service referred to as regulation. Some generating units, such as combustion turbines and conventional hydroelectric power, are particularly suited for providing regulation as they can operate across a wide range of generating levels. Baseload coal-fired units and nuclear units generally cannot operate with this level of flexibility. Highly variable generation resources such as wind and solar require a significant increase in the regulation capability of on-line generating resources in order to establish the required real-time balance between generation and demand.[3]

Wind and Solar Resources

Renewable energy resources other than hydroelectric are appearing in more and more utility generation portfolios in significant numbers. This trend is being driven by efforts to reduce carbon emissions and other greenhouse gases worldwide and by specific regulatory or other mandates to increase the percentage of capacity and the energy generated by these renewable resources. Due to advances in technology, tax incentives, and cost reductions, wind resources and solar resources are becoming ever more common.

However, these two types of technology present a major challenge to the operators of utility systems: they only generate electricity when the wind blows or the sun shines but electric load and electric generation must be balanced at each instant from on-line generation. The system cannot absorb or provide extra energy beyond the generating resources operating at any given time. Thus, when the wind dies down or the sun goes behind a cloud, the electric utility must have another generation resource ready and able to provide backup electricity almost instantaneously. Due to the variability of the electric generation from these resources, solar and wind resources are often referred to as intermittent resources. Solar resources generate electricity during the daylight hours. However, the sun is most intense around noon whereas the peak demand for electricity generally occurs in the evening hours in the summer and in a bimodal manner (early morning/evening) in the winter. These patterns mean that the peak availability of solar energy does not occur at the same time as the utility peak; the term used for this by the industry is *non-coincident.*

Wind resources tend to be much more non-coincident with a utility's peak demand than solar resources (see Figure 2).[4] In fact, wind energy resources often peak at night when the electricity demand is at its lowest levels.[5]

Energy storage technologies offer a way to continue the encouragement of the development of wind and solar resources to maximize the value of the energy they provide and to enable the electric utility system to operate reliably once those resources have been added.

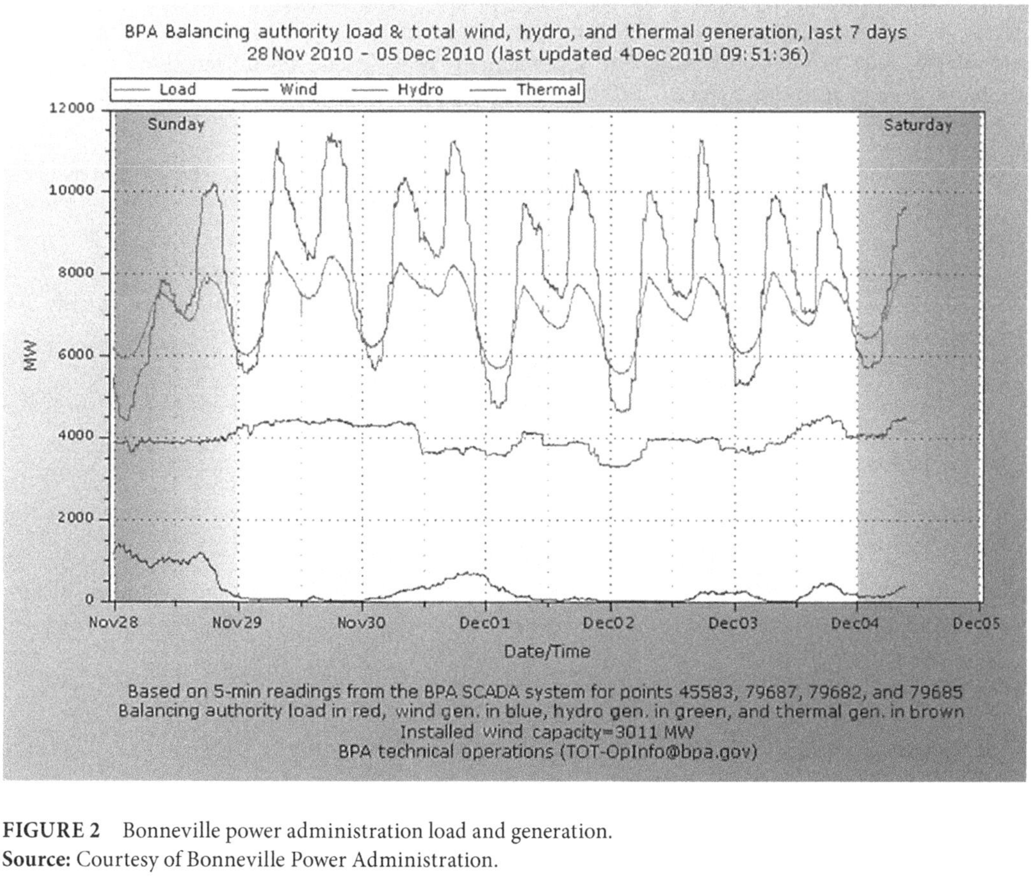

FIGURE 2 Bonneville power administration load and generation.
Source: Courtesy of Bonneville Power Administration.

Storage Technologies

Energy storage technologies are expected to play an important role in the integration of intermittent resources into the electricity grid worldwide. Energy storage involves a device that accepts electrical energy from the grid, converts it into an energy form suitable for storage, and then converts it back into electricity, which then, minus efficiency losses, is returned to the grid. Energy storage enhances the performance and economics of the power system, as it allows the capture and storage of renewable energy resources for later use, thus preserving resources. Energy storage can also be a valuable instrument for providing operational flexibility.[2,6]

Technologies that provide the ability to store intermittently generated power can smooth out the variations in the hour-by-hour, minute-by-minute, and second-by-second availability experienced by wind and solar resources. Energy storage can increase the usefulness of wind power by absorbing excess wind generated overnight and supplying that power to the grid during peak daytime hours. It can also help grid operators deal with the second-by-second variability of wind and solar by providing additional regulation service.[5]

Energy storage technologies also can help utilities provide the power quality and reliability required by increasingly complex and sensitive equipment. Energy storage devices improve system responsiveness, reliability, and flexibility, while reducing capital and operating costs for both suppliers and customers. Suppliers can use energy storage for transmission line stabilization, spinning reserve, and voltage control, which means customers would receive improved power quality and reliability.[7]

Many generating resources cannot be shut down during the off-peak period and must be kept in operation at minimum levels. For some utilities, a particular challenge is having more resources that must be kept operational during off-peak hours than the level of load demanded from customers. Energy storage technologies provide a sink for this minimum load energy. In addition, some energy storage technologies, such as pumped storage hydro, can actually provide load when needed to meet system minimum generation requirements.

In total, benefits of energy storage technologies include the following:[8]

- Enhancement to the value of intermittent renewable energy resources on the power grid by firming their energy.
- Improvement in power quality by providing ancillary services such as voltage regulation, spinning resources, and so forth.
- Ability to store low-value, excess energy when power supplies exceed demand until the energy can be economically used to meet load.
- Enhancement of the flexibility of the existing transmission grid.
- Relief of transmission congestion to defer capital expenditures on system upgrades.
- Conversion of less costly off-peak energy into higher- value on-peak power.
- Reduction of problems associated with minimum generation requirements.

The types of energy storage technologies in service and being evaluated worldwide include the following:[9–11]

- Batteries: sodium-sulfur (NaS), vanadium redox flow (VRB), lithium-ion, lead-acid, nickel-cadmium, nickel-metal hydride, zinc-bromine
- Pumped storage hydro
- Compressed Air Energy Storage (CAES)
- Electric double-layer capacitors
- Superconducting Magnetic Energy Storage (SMES)
- Flywheel systems
- Thermal

TABLE 1 Energy Storage Technology Development Status

Commercial	Precommercial	Demonstration Phase	Developmental
Pumped storage hydro		Electrochemical capacitor Flywheel	Lithium ion (grid applications)
Flywheels (local power quality)	Flywheel (grid device)	Hydrogen loop	SMES (grid applications)
CAES		Zinc-bromine battery	
Lead-acid battery (distribution level)		Vanadium redox battery	
Nickel-cadmium battery (distribution level)			
NaS battery (distribution level)			

- Regenerative fuel cells
- Hydrogen

The commercialization status of each of the energy storage technologies is shown on Table 1.[12,13]

The technology that presents the largest scale for utility application (up to 1000 MW), is able to handle large quantities of energy, and is already in the most widespread use is pumped storage hydro.[14,15]

Pumped Storage Hydro

Pumped storage hydro is a form of hydroelectric power generation for electric utilities that incorporates an energy storage feature.[16] The water, the source of potential and kinetic energy, moves between two reservoirs—an upper and a lower one—with a significant vertical separation (see Figure 3).[17] Water is stored in the upper reservoir until such time as the utility determines it is economic to use it to produce electricity for the system. The water in the upper reservoir is stored gravitational energy.[18] When the water is released, the force of that water spins the blades of a turbine that connects to a generator that produces electricity.[19]

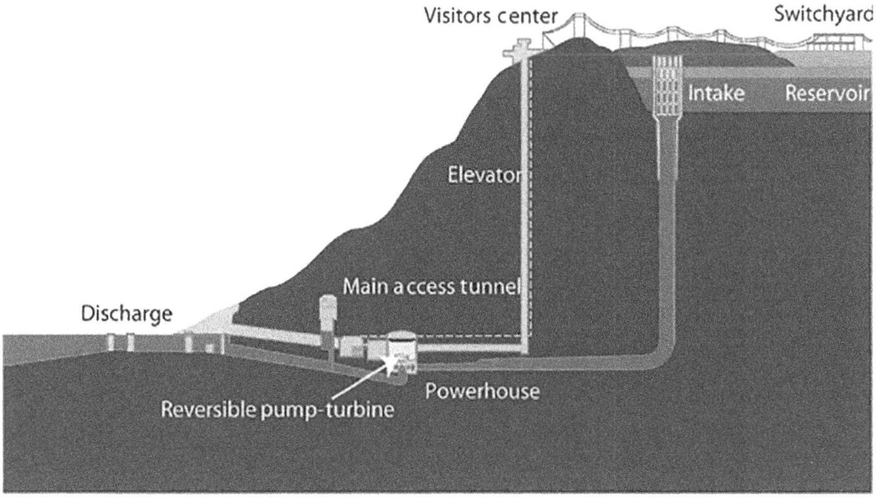

FIGURE 3 Schematic diagram of pumped storage hydro.
Source: Courtesy of the Tennessee Valley Authority.

Water is generally released from the upper reservoir to produce electricity during peak hours or during times of system need. After passing through the turbines, the water is discharged into the lower reservoir. At night and on weekends or during light load or off-peak hours, water is pumped up from the lower reservoir into the upper reservoir by the turbines that have now been reversed to work as electric-motor-driven pumps. Figure 4 shows the upper and lower reservoirs of the Rocky Mountain Hydroelectric Plant in Georgia, United States.

In the upper reservoir, the water is essentially stored energy. Water can be stored for a long time or a short time in the upper reservoir, depending on the needs of the utility (see Figure 5 for a view of the upper reservoir of the Taum Sauk Project in Missouri, United States). A vertical separation of at least 100 m (328 ft) is generally necessary to make a pumped hydro facility economic.[20] This height difference between the upper and lower reservoirs is called the head. The amount of potential energy in the water is directly proportional to the head, so the greater the height, the more energy that can be stored in a reservoir of a given size.[16,21]

FIGURE 4 Rocky mountain hydroelectric plant.
Source: Courtesy of Oglethorpe Power Corporation.

FIGURE 5 Taum Sauk Upper Reservoir.
Source: Courtesy of Ameren Missouri.

Pumped storage hydro is the most widespread energy storage system in use on power networks. Its main applications are for energy management, frequency control, and provision of reserve. Pumped hydro storage may make use of any generating resource on the system. It can be used to absorb temporary generating surpluses, to capture economic energy resources that would otherwise not be utilized, and to allow other generating resources to stay on-line by meeting their minimum generating requirements.[22]

Innovations in variable speed motors have helped these plants operate at partial capacity and greatly reduced equipment vibrations, increasing plant life.[23] Efficiency of any specific pumped storage facility, which is primarily dependent on the height between the upper and lower reservoirs, ranges from 70% to 85%. Many of the proposed pumped storage projects in the United States are off-stream from main stem rivers and either use existing reservoirs or are completely closed-loop projects. As such, the environmental impacts resulting from new pumped storage projects are significantly less than traditional large conventional hydroelectric power projects.[24]

Pumped storage hydro was first used in Italy and Switzerland in the 1890s to enable those utilities to store surplus nighttime output from run-of-river hydro stations for use in meeting their peak power demand requirements the following day. In 1929, the first major pumped storage hydroelectric plant, Rocky River, was built in the United States in New Milford, Connecticut. By 1933, reversible pump turbines with motor generators had become available. The turbines could operate both as turbine generators and, in reverse, as electric-motor-driven pumps.[24–26]

About 3% of total global generation capacity, over 127 GW, is pumped storage capacity. In 2010, the European Union had nearly 40 GW of pumped storage capacity in operation. In 2010, 42 projects provided more than 23 GW of pumped storage capacity in the United States. The largest pumped storage facility in the United States is in Bath County, Virginia, which has a capacity of 2100 MW. Pumped storage plants are characterized by long construction times and high capital expenditures.[27–30]

Technology Description

In conventional hydroelectric generation, hydraulic turbines rotate due to the force of moving water (its kinetic energy) as it flows from a higher to a lower elevation. This water can be flowing naturally in streams or rivers or it can be contained in man-made facilities such as canals, reservoirs, or pipelines. Dams raise the water level of a stream or river to a height sufficient to create an adequate head (height differential) for electricity generation.[31]

If the dam stops the flow of the river, then water pools behind the dam to form a reservoir or artificial lake. As hydroelectric generation is needed by the electric utility, the water is released to flow through the dam and powerhouse. In other cases, the dam is simply built across the river and the water in the river moves through the power plant or powerhouse inside the dam on its way downstream.[32] This latter type of hydroelectric facility is referred to as run-of-river.

In either case, as the water actually moves through the dam, the water pushes against the blades of a turbine causing the blades to turn. The turbine converts the energy in the form of falling water into rotating shaft power to turn a generator to produce electricity. The mechanical efficiency of hydroelectricity is high, about 95%. Availability of the generators is about 90%.[32–34]

The selection of the best turbine for any specific hydroelectric site is primarily dependent on the head (the vertical distance through which the water falls) and the water flow (measured as volume per unit of time) available. Generally, a high-head plant needs less water flow than a low- head plant to produce the same amount of electricity.[33,35]

The power available in a stream of water is

$$P = \eta^* \rho^* g^* h^* \dot{V},$$

Where η is turbine efficiency, ρ is power (joules per second or watts), g is acceleration of gravity (9.81 m/sec^2), h is head (meters, this is the difference in height between the inlet and outlet water surfaces), and $\dot{V}$ is flow rate (cubic meters per second).[36]

This equation can be roughly approximated as:

$$\text{POWER}(\text{kW}) = 5.9 \times \text{FLOW} \times \text{HEAD},$$

where FLOW is measured in cubic meters and HEAD is measured in meters.[37] In general terms, 1 gal/sec falling 100 ft can generate 1 kW of electrical energy.[38]

Hydroelectric power is generally found in mountainous areas where there are lakes and reservoirs and along rivers. Hydroelectric power currently provides about 10% of all of the electricity produced in the United States. Hydroelectricity provides about one-fifth of the world's electricity. Worldwide capacity of all hydroelectric capacity as of 2008 was approximately 850,000 MW.[32,38,39]

Producing electricity from hydropower is so economical because once the dam is built and the equipment has been installed, the flowing water has no cost. In addition, the dams are very robust structures and the equipment is relatively mechanically simple. Hydro plants are dependable and long lived, and their maintenance costs are low compared to most other forms of electricity generation, including fossil-fired and nuclear generation.[35] Traditional hydroelectric power once comprised almost 40% of U.S. electricity production. Due to land use issues in developed areas, the potential to build additional traditional hydroelectric facilities is limited and their relative contribution to the generation portfolio in the United States will continue to decrease. There is a significant effort in the United States to explore the utilization of existing dams and infrastructure to add power at dams that currently do not have associated electricity generation. At present, there are approximately 80,000 dams in the United States, of which about 2400 have powerhouses. Pumped storage hydro generation facilities capture many of the system benefits associated with traditional hydroelectric power.

Pumped storage hydro generation is a specific kind of hydroelectric generation requiring an upper and lower reservoir and special equipment that can both generate power as water flows downhill and then reverse and serve as a pump to move the water back uphill. This requirement for location in suitable topological areas limits the geographic flexibility of pumped storage hydro and, at least in Europe, means that most cost-effective sites have already been developed. Some existing conventional hydropower dams are, however, being upgraded with pumped storage capacity. In addition, seawater pumped storage facilities are being examined and a pilot project in Japan has been developed to research the feasibility of using seawater as opposed to freshwater.[40] Modular pumped storage systems, discussed in the next section, may provide additional siting opportunities.

Facility Description

A pumped storage hydropower plant typically has three major components: the upper reservoir, the lower reservoir, and the pumping/generating facilities (see Fig. 3). The water from the upper reservoir used for electricity generation is allowed into the intake shaft through the opening of the headgates. Water moves through the high-pressure shaft and steel-lined power tunnel until it reaches the turbines in the powerhouse.[16] The water turns the turbines that then drive the generators to produce electricity. The water then moves through the tailrace tunnel until it is discharged into the lower reservoir. Water discharge capacity in the upper reservoir can require from several hours to several days.[24]

Pumped storage facilities can be categorized as "pure" or "combined." Pure pumped storage plants, also referred to as closed loop, continually shift water between an upper and a lower reservoir.[29] Combined pumped storage plants also generate their own electricity like conventional hydroelectric plants through natural steam flow. Closed-loop pumped storage systems use closed water systems that are artificially created instead of natural waterways or watersheds and can be as large as traditional large hydroelectric power stations. The water for closed-loop storage usually is only put into the system when it begins operation, either from groundwater or possibly from municipal wastewater.[31,41] Old iron ore quarries and other types of old mines are being examined for use as reservoirs for closed-loop systems worldwide.[42,43]

The efficiency of pumped storage plants generally ranges from 70% to 85%. This means that 70–85% of the electrical energy used to pump the water into the upper reservoir is actually generated when water flows back down through the turbines to the lower reservoir. The losses of energy are primarily related to the mechanical efficiency losses during conversion from flowing water to electricity. Such losses are higher with older designed equipment.[41]

Today, there are three kinds of pumped storage hydro plants available. The most commonly used is a basic version utilizing single or synchronous speed machines that cannot be regulated in pump mode. The pumps can only be run at full capacity or not run at all. Some of these plants use a combined pump and turbine unit instead of two separate units. The compact size of the equipment is beneficial when the plant is built underground.[44]

The second type of pumped storage hydro facility is called "hydraulic short circuit," which allows more flexible operation. Pumps and turbines can be operated simultaneously. The pump can still only operate at full capacity. When large amounts of excess electricity are available, the turbine reduces capacity and more water is stored.[44]

The third type of plant uses variable speed pump technology in order to quickly adapt to fluctuating grid frequency. The operator is able to control how much power is consumed by the pumps. This type of pumped storage plant can experience a round-trip efficiency of 85%.[45]

Environmental Issues

Issues related to permitting of conventional hydroelectric and on-steam pumped storage hydroelectric facilities include factors relating to establishing both the reservoirs themselves and the operation of the facilities. Establishing the reservoirs may involve flooding of land currently in use for other purposes. During operation, the level of the water in both the upper and lower reservoirs fluctuates, which can lead to environmental concerns as well:[46]

- Water resource impacts—stream flows, reservoir surface area, groundwater recharge, water temperature, turbidity, oxygen content
- Biological impacts—displacement of terrestrial habitat, alteration of fish migration patterns, other impacts due to changes in water quality and quantity
- Potential damage to archaeological, cultural, or historic sites
- Visual quality changes
- Loss of scenic or wilderness resources
- Increased risks of landslides and erosion
- Navigation impacts
- Gain in recreational resources

These concerns are much less significant for closed-loop pumped storage plants that are not associated with natural waterways and watersheds. Usually, closed-loop pumped storage plants are specifically not located near existing rivers, lakes, streams, and other sensitive environmental areas to avoid the regulatory lag time and complexity associated with combined pumped storage hydroelectric facilities.[31,46]

In addition to closed-loop systems, underground powerhouses and lines are being used around the world to mitigate the environmental effects of pumped storage projects and reduce visual surface disturbances.[26]

TABLE 2 Countries with Pumped Storage Installed or under Construction

Argentina	India	Slovakia
Australia	Ireland	South Africa
Austria	Italy	South Korea
Belgium	Japan	Spain
Bulgaria	Lithuania	Sweden
Canada	Luxembourg	Switzerland
China	Norway	Taiwan
Croatia	Philippines	Ukraine
Czech Republic	Poland	United Kingdom
France	Portugal	United States
Germany	Russia	
Greece	Serbia	

Installed Facilities Worldwide

Pumped storage facilities are currently installed in many countries around the world (see Table 2) and numerous new facilities are under construction.[27,47]

Electric Utility Usage of Pumped Storage Hydro

Pumped storage facilities use power from available generating resources to pump water into the upper reservoir when it is cost-effective to do so (often at night). This allows the utility to have the water to use during higher load periods the next day to generate electricity at the pumped storage facility. Figs. 6 and 7

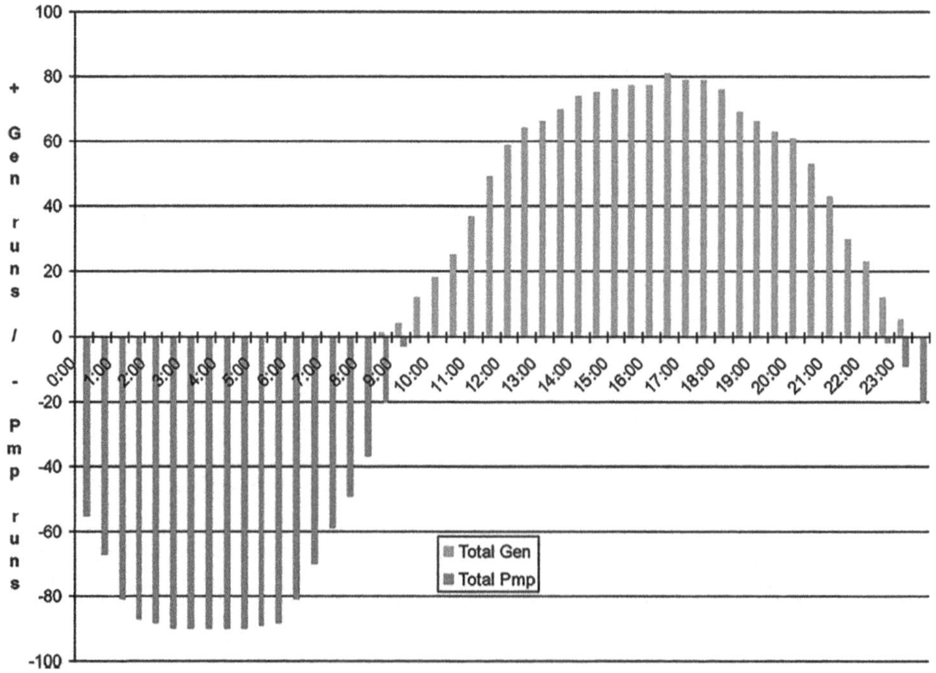

FIGURE 6 Pumped storage unit—typical summer operation.
Source: Courtesy of Oglethorpe Power Corporation.

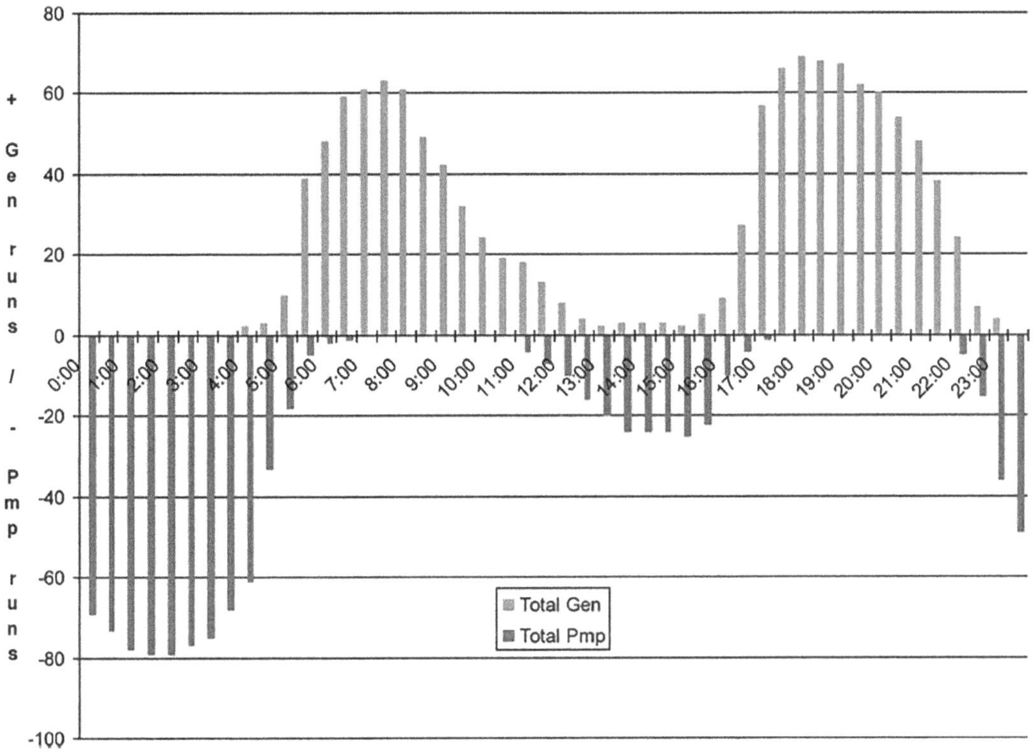

FIGURE 7 Pumped storage unit—typical winter operation.
Source: Courtesy of Oglethorpe Power Corporation.

show the typical operational patterns for pumped storage usage by a utility in each of the summer and winter periods. Power is generated by the pumped storage facility during the higher load hours across all seasons and pumping occurs in the off-peak hours and lower load hours.

Conventional hydroelectric power and pumped storage are actively being used as energy storage for Danish wind generation. Electricity from wind turbines in Denmark is stored in Norway's and Sweden's large, conventional storage reservoirs and then is sent back to Denmark during periods of low wind generation.[48–50] Pumped storage hydro projects in Germany are also critical to maintaining grid reliability and stable power flows within Denmark. Denmark has no native load balancing, so those balancing services are provided via its interconnections with Germany, Norway, and Sweden.[11] The curves in Fig. 8 demonstrate the non-coincident behavior of the wind generation versus the load demand from January 2007.[51]

How Pumped Storage Hydro Can Benefit Renewable Energy Resources

Pumped storage hydro is one of the best available energy storage mechanisms for use on electric utility systems. Pumped storage hydro allows utilities to integrate renewable energy resources such as wind and solar and helps to optimize the operation of the utility grid. Pumped storage hydro helps utilities achieve maximum efficiency for the power-generating system through load shifting. The ability for fast ramping provides a myriad of benefits from serving as spinning reserve to maintaining power system stability to providing voltage control.[52] Pumped storage hydro units can perform black starts and

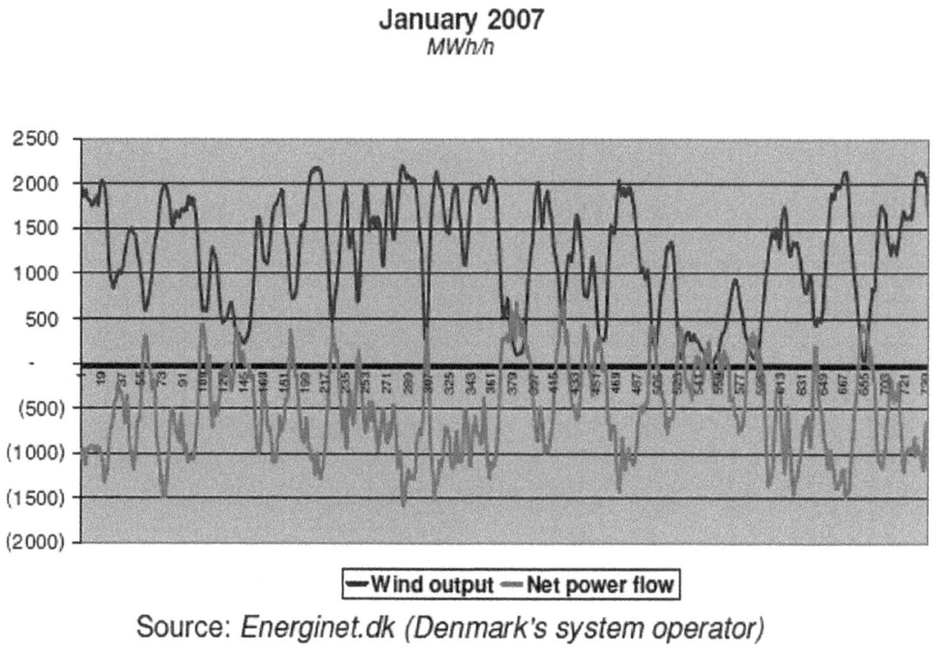

FIGURE 8 Western Denmark—wind output and net electricity flows.

deliver electricity within a few seconds. Modern pumped storage plants have the highest efficiency of all energy storage systems— 85%.[3,44]

Because of their ability to both pump and generate, pumped storage hydro units provide a load-shifting device for system operators. Pumped storage hydro can firm intermittent resources by absorbing excess generation at times of high output and low demand and releasing that stored energy during peak demand periods. Pumped storage hydro units can help systems avoid minimum generation problems and serve as critical backup facility during periods of excessive demand on the grid system.[3,30,53–56] Pumped storage hydro provides system operators a variety of ways to help balance portfolios from rapid starts and stops to generation in peak load hours.

Pumped storage hydro can move from idle position to full load very quickly, often in seconds, if so designed.[53,57]

This capability allows pumped storage units to provide system frequency control, regulation, and voltage control and to compensate promptly for the loss of other generating units or transmission lines. It allows pumped storage units to count for spinning reserve and standby reserve/ black start units. Typical ramping characteristics of modern pumped storage projects are as follows:[58]

• Shutdown to on-line	60–90 sec
• On-line to full-load generating	5–15 sec
• Spinning to full-load generating	5–15 sec
• Shutdown to normal pumping	6 min
• Spinning to normal pumping	60 sec

Rapid ramping is particularly desirable in the case of a unit becoming unavailable or forced out of service or on utility systems with high amounts of intermittent resources such as solar and wind. For example, the Dinorwig pumped storage facility in North Wales, United Kingdom, can go from 0 MW to full capacity of 1320 MW in 12 sec and can usually maintain this level until other generating units on the utility's system can be brought on-line. This makes pumped storage plants the "race cars" among power-generating facilities.[16,20,41,59] Pumped storage hydro projects can charge or discharge over a 2, 6, or even 12 hr period, as necessary for the utility. The Dinorwig station in North Wales, for example, has an energy storage capacity equivalent to about 5 hr of operation at full capacity. Most of Japan's modern pumped storage projects have energy storage capacity equivalent to 6 to 8 hr of operation at full capacity. The upper reservoirs for many of Japan's pumped storage projects are designed for long-term seasonal storage and drawdown.[26,53]

The dynamic, rapid response capabilities make pumped storage hydro highly valuable for keeping electrical grids stable and reliable with higher percentages of renewable resources, especially as pumped storage hydro can provide both real and reactive power.[59] The ability of pumped storage hydro to produce watts (real) and vars (reactive) power enables system operators to better ride out any system disturbances such as the loss of a major generating unit or transmission line or the significant instantaneous increase or decrease in available energy from renewable resources.

Pumped storage hydro contributes no greenhouse gas emissions and thus assists utilities, and countries, achieve a lower carbon footprint. Utility systems that utilize pumped storage units are also able to operate their thermal units more efficiently—with fewer stops and starts—also contributing to overall reduced emissions.[53,60]

Depending on its location, pumped storage hydro may help optimize transmission line loadings and defer the need for additional transmission assets. Pumped storage units near load centers can operate in condenser mode to generate or absorb reactive power as may be required for system voltage regulation.[53,60]

Conclusion

Energy storage devices in the form of pumped storage hydro facilities offer many benefits to utilities including the integration of the intermittent renewable resources—wind and solar. Due to the weather-dependent nature of these renewable resources and the fact that much wind generation is non-coincident with utility peak demands, pumped storage hydro provides a way for utilities to utilize the renewable resources more efficiently and more cost-effectively. Where topologically feasible, the reliability benefits associated with the rapid ramping capabilities of pumped storage hydro, in conjunction with the many other positive operational capabilities, will drive development of this energy storage technology worldwide for many years to come.

Acknowledgments

The author wishes to acknowledge Russ Schussler, Vice President, System Planning, Georgia Transmission Corporation, and William G. Tharp, Plant Manager, Rocky Mountain Hydroelectric Plan, Oglethorpe Power Corporation, for their assistance.

References

1. *Energy Storage in the New York Electricity Markets*; A New York Independent System Operator White Paper, March 2010; 2.
2. Lohiya, A.K. *Electrical Energy Storage—Large Scale*; August 2009; 1 available at http://www.scribd.com/doc/25835943/Pumped-Storage-Hydroelectricity (accessed December 2010).
3. Montero, F.P.; Pérez, J.J. Wind–hydro integration: Pumped storage to support wind. Hydro Rev. Worldwide **2009**, *17* (3).
4. Available at http://transmission.bpa.gov/Business/Operations/Wind/baltwg.aspx (accessed December 4, 2010).

5. *Energy Storage in the New York Electricity Markets*; A New York Independent System Operator White Paper, March 2010; 4.
6. Cailliau, M.; Foresti, M.; Villar, C.M. Winds of change. IEEE Power Energy Mag. **2010**, 59.
7. Schaber, C.; Mazza, P.; Hammerschlag, R. Utility-scale storage of renewable energy. Electr. J. **2004**, 21–22.
8. Gilbert, K.; Lee, H.; Manwaring, M.; MWH America, Inc. What's so hard about licensing a pumped storage project? Renewable Energy World North Am. **2010**, 76.
9. Inage, S.-I. *Prospects for Large-Scale Energy Storage in Decarbonised Power Grids;* International Energy Agency Working Paper, Paris, France, 2009.
10. Schaber, C.; Mazza, P.; Hammerschlag, R. Utility-scale storage of renewable energy. Electr. J. **2004**, 23–28.
11. Miller, R. Wind Integration Utilizing Pumped Storage. Platts 2nd Annual Power Storage, February 8–9, 2010.
12. Slowe, J. Emerging electricity storage technologies. Cogeneration and On-Site Power Production, **2008**; 59–71.
13. Delta Research Brief. Electric Storage Technologies—Time for Utility Engagement? available at http://www.deKa-ee.com (accessed June 2008).
14. Schaber, C.; Mazza, P.; Hammerschlag, R. Utility-scale storage of renewable energy. Electr. J. July 2004, 25.
15. Bandyk, M. PG&E explores 1,200-MW pumped water storage plant to back up renewables. SNL Energy Electric Utility Report, August 30, 2010; 35.
16. Pumped Storage Reservoirs—Storing Energy to Cope with Big Demands, available at home.clara.net/darvill/altenerg/pumped.htm (accessed March 2006).
17. Courtesy of the Tennessee Valley Authority, available at http://www.tva.gov/power/pumpstorart.htm (accessed December 22, 2010).
18. U.S. Department of Energy—Energy Efficiency and Renewable Energy, Distributed Energy Program. *Pumped Hydro*, available at http://www.eere.energy.gov/de/pumped_hydro.html (accessed March 2006).
19. Hydroelectric Power, available at http://www.tva.gov/power/hydro.htm (accessed March 2006).
20. Available at http://www.uaf.edu/energyin/webpage/webpage-withframes/pumpedhydro.text.htm (accessed January 2006).
21. Hydroelectricity, from Wikipedia, available at http://en.wikipedia.org/wiki/Hydroelectricity (accessed March 2006).
22. Special Report: Accommodating High Levels of Variable Generation; North American Electric Reliability Corporation, April 2009; 50.
23. Schaber, C.; Mazza, P.; Hammerschlag, R. Utility-scale storage of renewable energy. Electr. J. July 2004, 26.
24. Electricity Storage Association. Technologies and Applications, Technologies, Pumped Hydro Storage, available at http://www.electricitystorage.org/tech/technologies_tech-nologies_pumpedhydro.htm (accessed March 2006).
25. U.S. Bureau of Reclamation. *The History of Hydropower Development in the United States,* available at http://www.usbr.gov/power/edu/history.html (accessed April 2006).
26. Lohiya, A.K. *Electrical Energy Storage—Large Scale;* August 2009; 2, available at http://www.scribd.com/doc/25835943/Pumped-Storage-Hydroelectricity (accessed December 2010).
27. Ingram, E.A. Development Activity Snapshots. Hydro Rev. Worldwide **2009** *17,* (6).
28. Robb, D. Could CAES answer wind reliability concerns? Power **2010**, 59.
29. Gilbert, K.; Lee, H.; Manwaring, M.; MWH America, Inc. What's so hard about licensing a pumped storage project? Renewable Energy World North Am. **2010**; 78.
30. Miller, R. Opportunities in pumped storage hydropower. Power Eng. **2010**, 50.

31. California Energy Commission. *Hydroelectric Power in California,* available at http://www.energy.ca.gov/electricity/hydro.html (accessed April 2006).
32. Energy Story: Chapter 12—Hydro Power, available at http://www.energyquest.ca.gov/story/chapter12.html (accessed April 2006).
33. Micro Hydropower Basics: Turbines, available at http://www.microhydropower.net/turbines.html (accessed April 2006).
34. Blankinship, S. Hydroelectricity: The versatile renewable. Power Eng. **2009**, available at http://www.powergenworldwide.com/index/display/articledisplay/364609/articles/power-engineering/volume-113/issue-6/features/hydroelectricity-the-versatile-renewable.html (accessed November 2010).
35. Learn About Energy.org, "Part 11" Hydropower, available at http://www.learnaboutenergy.org/focus/part11.htm (accessed April 2006).
36. Water turbine, from Wikipedia, available at http://en.wikipedia.org/wiki/Water_turbine (accessed April 2006).
37. Hydro-Electric Power, available at http://www.ecology.com/archived-links/hydroelectric-energy/ (accessed April 2006).
38. Union of Concerned Scientists. *How Hydroelectric Energy Works*, available at http://www.ucsusa.org/clean_energy/re-newable_energy_basics/how-hydroelectric-energy-works_html (accessed April 2006).
39. *U.S. Energy Information Administration—Independent Statistics and Analysis*; International Energy Statistics, available at http://tonto.eia.doe.gov/cfapps/ipdbproject/IEDIndex3.cfm (accessed December 2010).
40. Geschler, T. *Survey of Energy Storage Options in Europe*; London Research International, March 2010; 5–6, available at http://www.londonresearchinternational.com/Pu13_energy_storage.pdf (accessed December 2010).
41. Pumped-storage hydroelectricity, from Wikipedia, available at http://en.wikipedia.org/wiki/Pumped-storage_hydroelectricity (accessed March 2006).
42. Fodstad; Audun, L. *Pumped Storage,* 13, available at http://www.risoe.dtu.dk/en/Conferences/Workshop_Sustainable_Energies/~/media/Risoe_dk/Conferences/Energyconf/Documents/Storage/lars_fodstad_tilladelse.ashx (accessed December 2010).
43. Mandel, J. Renewable Energy: DOE Promotes Pumped Hydro as Option for Grid Storage, Greenwire 10/15/2010, available at http://www.eenews.net/public/Greenwire/2010/10/15/10 (accessed December 2010).
44. Geschler, T. *Survey of Energy Storage Options in Europe,* London Research International, March 2010; 3, available at http://www.londonresearchinternational.com/Pu13_energy_storage.pdf (accessed December 2010).
45. Geschler, T. *Survey of Energy Storage Options in Europe,* London Research International, March 2010; 3–4, available at http://www.londonresearchinternational.com/Pu13_energy_storage.pdf (accessed December 2010).
46. Fodstad; Audun, L. *Pumped Storage,* 15, available at http://www.risoe.dtu.dk/en/Conferences/Workshop_Sustainable_Energies/~/media/Risoe_dk/Conferences/Energyconf/Documents/Storage/lars_fodstad_tilladelse.ashx (accessed December 2010).
47. List of pumped-storage hydroelectric power stations, available at http://en.wikipedia.org/wiki/List_of_pumped-storage_hydroelectric_power_stations (accessed December 2010).
48. van Loon, J. Energy: Where wind power is blowing away profits. Bloomberg Businessweek **2010**, 22.
49. Miller, R. Opportunities in Pumped Storage Hydropower. Power Eng. **2010**, 51.
50. Robb, D. Could CAES answer wind reliability concerns? Power **2010**, 58.
51. Wind Energy: The Case of Denmark, CEPOS—Center for Politiske Studier, September 2009; 11.

52. Lohiya, A.K. *Electrical Energy Storage—Large Scale,* August 2009; 4, available at http://www.scribd.com/doc/25835943/Pumped-Storage-Hydroelectricity (accessed December 2010).
53. Makansi, J. Bulk storage could optimize renewable energy. Power **2010**; 68.
54. Miller, R. Johnson, C.Y. Efforts turn to storage for renewable energy. The Boston Globe September 1, 2008, available at http://www.boston.com/news/science/articles/2008/09/01/efforts_turn_to_storage_for_renewable_energy/?page=1 and http://www.boston.com/news/science/articles/2008/09/01/efforts_turn_to_storage_for_renewable_energy/?page=12 (accessed February 2010).
55. Collinson, A. The costs and benefits of electrical energy storage, S684/003/99, as published in Renewable Energy Storage: Its Role in Renewable and Future Electricity Markets, IMechE Seminar Publication 2000-7, Professional Engineering Publishing Limited: London, 26.
56. Fodstad, L.A. *Pumped Storage,* 19, available at http://www.risoe.dtu.dk/en/Conferences/Workshop_Sustainable_Energies/~/media/Risoe_dk/Conferences/Energyconf/Documents/Storage/lars_fodstad_tilladelse.ashx (accessed December 2010).
57. Fodstad, L.A. *Pumped Storage,* 9, available at http://www.risoe.dtu.dk/en/Conferences/Workshop_Sustainable_Energies/~/media/Risoe_dk/Conferences/Energyconf/Documents/Storage/lars_fodstad_tilladelse.ashx (accessed December 2010).
58. Lohiya, A.K. *Electrical Energy Storage—Large Scale,* August 2009; 3–4, available at http://www.scribd.com/doc/25835943/Pumped-Storage-Hydroelectricity (accessed December 2010).
59. Vansant, C. Viewpoint: Pumped storage: An idea whose time has come...again. Hydro Rev. Worldwide **2008,** *16* (6).
60. Lohiya, A.K. *Electrical Energy Storage—Large Scale*; August 2009; 5, available at http://www.scribd.com/doc/25835943/Pumped-Storage-Hydroelectricity (accessed December 2010).

39

Integrated Energy Systems

Leslie A. Solmes and Sven Erik Jørgensen

Introduction

On August 6, 2003, the Board of Governors of the Electric Power Research Institute reported, "... there is a growing concern that the electricity sector's aging infrastructure, workforce, and institutions are losing touch with the needs and opportunities of the 21st Century. The investment gap—inadvertently reinforced by the regulatory uncertainty of electricity restructuring—is exacting a significant reliability cost that is seen as just the tip of the iceberg in terms of the electricity infrastructure's growing vulnerability to capacity, reliability, security, and service challenges."[1]

On August 14, 2003, the northeast United States experienced power failure that impacted 50 million consumers in the United States and Canada. The front page of the *Financial Times* stated, "Public to pay for power upgrade. Energy Secretary (Spencer Abraham) warns that much of the $50bn (billion) bill will be passed on to customers." The headline that followed on page 3 was, "Failure of the power came as little surprise to experts. Wake-up call was just latest alarm."[2] Energy supply investments are needed, but the public cost for energy should decline.

What does not get front-page headlines are the billions of dollars needed to upgrade unreliable, old, customer energy supply production and distribution systems like heat, air conditioning, and water. One example is APPA: The Association of Higher Education Facilities Officers. APPA reported in 1995 that the backlog of deferred maintenance for colleges and universities was $26 billion.[3] The major portion of this backlog was needed for utility systems upgrades. The problem has not gone away.

Investments in electricity generation and transmission systems are essential for social and economic well-being. Upgrade and expansion of customer production and distribution systems are equally essential and costly. These needs are not problems but opportunities—*huge investment opportunities.*

Electric utilities are large industrial customers who use massive amounts of energy—more than any other industry. The second law of thermodynamics states that every time one form of energy is converted into another form, some energy is lost. On average, two-thirds of the British thermal unit (BTU) value of the fuels used to produce electricity is thrown away as waste heat. More BTUs are lost in converting and transporting electricity to consumers. Consumers, in turn, convert electricity to heat energy to heat water, heat buildings, and cook. Consumers also burn fuels to produce heating, cooling, and process steam and distribute it to the end use. Every conversion and transportation event causes BTU loss.

The opportunity to save energy in producing and transporting electricity, heat, steam, and chilled water is enormous. However, it requires utilities and customers to plan and invest together. Investing in reliable electricity that is efficient, competitive, and environmentally friendly, and that offers the greatest financial benefits to utilities and their customers is directly related to how utilities and customers select, implement, and manage investments in electric, heat, and cooling capacity and dispatching of all forms of energy.

A comprehensive, integrated energy supply investment assessment at Illinois State University (ISU) offers a case study of the financial opportunities and energy savings for both ISU and the electric utility. Investment implementation roadblocks related to cultural practices and risk aversion are discussed. Through use of an integrated, energy supply system investment methodology adapted from development of independent power plants and a Web-based business plan software tool, incentives to overcome cultural inertia and manage life-cycle project risks are detailed. Examples of dynamic Web-based reports serve as a communication medium among the project team members and allow them to conduct real-time asset management. Finally, profits from electric utility and customer collaboration are highlighted.

Integrated Solution

Imagine if electric utilities could reduce the number of new plants and transmission facilities and instead invest in customer heating, cooling, process steam, or other heat energy needs with on-site electricity generation and load shifting technologies. Imagine that both utilities and consumers could benefit financially by integrating their energy supply investments and reducing the amount of energy consumed.

Culturally, electric utilities and customers have diametrically opposed business goals. Utilities want to sell the most energy at the highest price possible. Customers want to purchase the least amount of energy to reliably meet their operating requirements at the lowest cost. Investments in integrated energy supply systems not only create energy efficiency and emission reduction but also result in economic benefits to both utilities and consumers. Let us use a case study from ISU to demonstrate how to create valuable investment for both utilities and customers.

In 1999, ISU determined that they could fix their facilities operating budget at their 1997 level for 20 years by investing $37.56 million in utility and building systems infrastructure upgrades. Annual budget savings were conservatively estimated to be $3.45 million per year with debt service at $3 million. Net present value ranged from $8.7 to $13 million.

Five megawatts of on-site electricity generation was planned. Using the waste heat and load shifting technologies, electricity load was leveled and reduced by 3.2 MW. The electrical capacity addition to the grid was 8.2 MW. More than half of the waste heat from electrical generation became useful energy. Owing to ISU's level load, the utility was able to gain greater use of its existing power plants by supplying the remaining electricity required by ISU on a 24/7 basis.

History

For nearly 30 years, the development of energy business plans, using an integrated systems methodology with investments sized to meet the business goals of the end users, has demonstrated the ability to:

- Finance millions of dollars of infrastructure upgrades from operating budget savings
- Level electricity load requirements
- Reduce BTU consumption and unit costs
- Improve energy reliability and operating productivity

So why were so few of these opportunities implemented? Customers are not in the utility business. Instead, they are focused on investments that provide products and services. As a result, no energy savings or investment occurs.

More importantly, the problem has been, "How to communicate and manage risks?" How to insure risks associated with energy price escalation, building the project on time and budget, costs of financing and ownership, equipment performance, and changes in business goals can cause major headaches. What business manager in his or her right mind would risk his or her career on a performance-based investment that cannot be measured or managed?

At the same time, as customers were struggling to update their heat/steam and cooling systems, electric utility generation investments almost stopped. The risks associated with building central electricity generation plants on time and budget, the rising cost of financing, the uncertainty of the regulatory world, and competition from independent power producers strangled central power plant development.

On both sides of the meter, risks associated with energy systems investments caused inaction. Investing in efficient energy supply systems requires integration. Integrated investment in a dynamic marketplace demands that all stakeholders have the ability to measure, manage, communicate, and report project status at any point in time.

Stakeholders

An energy supply investment team or stakeholders include the following: chief executive; financial, budget, facilities, and utilities managers; engineering, procurement, construction, capital planning, maintenance, and information technology personnel; contractors; facility occupants; financiers; and investment partners. Every member of the team contributes to the assumptions that go into energy supply systems planning and implementation.

Once energy master plan assumptions are adopted, each team member now knows the complete game plan and his or her performance requirements. As the plan is implemented and measured over time, each team member is responsible for communicating changes that will occur. Instantaneously, the impact of these changes on the total project performance and each team member's area of responsibility must be calculated, compared with the original plan, and reported to all team members. Team members can, in turn, adjust and manage their area of responsibility.

Web-Based, Real-Time Communication and Reports

To conduct the investment communication necessary to manage energy investments, it is essential to conduct energy planning, investment, and asset management using a dynamic, Web-based tool. The tool should not only compare project investment options but also provide timely updates of business plan assumptions based on market changes so that all project team members can communicate and manage change. Businesses and organizations require adoption and commitment to an operating software system that is their master business planning and asset management standard. Web-based workflow connects all the elements of integrated energy systems needed to manage market and business interactivity and create the lowest energy unit cost at the highest reliability.

Using the ISU project, let us look at the kinds of reports that team members will need to see. Presented here are five examples: Executive Summary, Pro Forma, Cogeneration, Steam, and Chilled Water Operating Calculations. Note that although the ISU project includes cogeneration, each consumer is encouraged to assess all on-site generation options that apply to the facility/site situation.

Table 1 and Figure 1 are life-cycle examples of a preliminary engineering, implementation, and financial assessment of an infrastructure investment strategy for ISU. This project analysis was conservatively conducted and included all campus buildings and utility systems. In summary, the

TABLE 1 ISU Executive Summary

Executive Summary					
Illinois State University - Test Case 3					
Current utility budget					$7,833,897
Energy cost savings					
Facility efficiency recommendations					$1,264,034
Electrical generation					$1,880,989
Steam production					$392,913
Chilled water production					$(89,549)
Water					$0
	Total utility budget savings			$3,448,386	
	Percent of current utility			**44.02%**	
Project costs					
Facility efficiency recommendation					Electrical generation $3,595,110
	$2,625,000				
Steam systems					$4,260,000
Chilled water systems					$10,707,100
Infrastructure					$7,100,000
Engineering/contingencies					$5,657,442
Additional project costs					$2,069,200
	Total EPC cost			$36,013,851	
	Annual debt service			$3,049,409	
	Energy budget savings: Debt service			**113.08%**	
Unit energy costs					
Average cost for purchased electricity					$0,046 /kWh
Average cost for cogenerated electricity					$0,043 /kWh
Average cost for steam					$6,994 /KLbs
Average cost for chilled water					$0,312 /ton-hr
Financial					
Capitalization			NPV		
Term debt		$36,013,851			
			Before Tax @ 8%		$12,927,353
Debt service reserve		$1,550,863	10%		$10,547,035

(*Continued*)

TABLE 1 (*Continued*) ISU Executive Summary

		Executive Summary			
		Illinois State University - Test Case 3			
Total Debt		$37,564,714		After Tax @ 12%	8%
	$8,718,363	$12,927,353			
Equity		$0		10%	$10,547,035
Total capital		$37,564,714		12%	$8,718,363
Proforma assumptions					
			Debt coverage		
Term	Interest Rate		20	5 85% A	Average 1.55
Equity	100%		Minimum	1.29	
Partner equity	Effective Tax Rate	0%	0.00%		
Interest on reserves	0.00%		Maximum	**1.69**	

Proforma

(Numbers in thousands of dollars)

ILLINOISSTATEUNIVER-TESTCASE2

	2001	2002	2003	2004	2005	2006	2007	2008	2009	2010
INCOME STATEMENT										
Revenue										
Capacity Payments	0	0	0	0	0	0	0	0	0	0
Power Sales to Utility	0	0	0	0	0	0	0	0	0	0
Electrical to Host	0	0	0	0	0	0	0	0	0	0
Chilled Water to Host	0	0	0	0	0	0	0	0	0	0
Steam to Host	0	0	0	0	0	0	0	0	0	0
Development Fees	0	0	0	0	0	0	0	0	0	0
Service Revenues	0	0	0	0	0	0	0	0	0	0
Performance Revenues	0	0	0	0	0	0	0	0	0	0
Budget as Revenue Base	9,669	9,862	10,059	10,210	10,363	10,519	10,676	10,837	10,999	11,164
Total Revenue	9,669	9,862	10,059	10,210	10,363	10,519	10,676	10,837	10,999	11,164
Operating Expense										
Electricity	1,538	1,536	1,538	1,492	1,447	1,404	1,361	1,321	1,360	1,401
Natural Gas	3,416	3,468	3,520	3,573	3,626	3,681	3,736	3,792	3,849	3,906
Fuel Oil	0	0	0	0	0	0	0	0	0	0
Coal	0	0	0	0	0	0	0	0	0	0
Water	800	824	849	874	900	927	955	984	1,013	1,044
Operations & Maintenance	(13)	(14)	(14)	(15)	(15)	(16)	(16)	(17)	(17)	(18)
Property Taxes	0	0	0	0	0	0	0	0	0	0
Total Operating Expense	5,741	5,816	5,892	5,924	5,958	5,996	6,036	6,080	6,205	6,333
Operating Income	**3,928**	**4,046**	**4,167**	**4,287**	**4,405**	**4,523**	**4,640**	**4,757**	**4,794**	**4,831**
Debt Service / Lease	(3,049)	(3,049)	(3,049)	(3,049)	(3,049)	(3,049)	(3,049)	(3,049)	(3,049)	(3,049)
Debt Reserve Withdrawals(Additions)	0	0	0	0	0	0	0	0	0	0
Pre-tax Cash Flow to Equity Owners	879	997	1,118	1,237	1,356	1,474	1,591	1,708	1,745	1,781
Interest Payments	2,289	2,245	2,198	2,148	2,095	2,039	1,980	1,917	1,851	1,781
Debt Reserve Additions(Withdrawals)	0	0	0	0	0	0	0	0	0	0
Book Depreciation	849	849	849	849	849	849	849	849	849	849
Book Income Before Taxes	790	953	1,121	1,290	1,461	1,635	1,812	1,991	2,094	2,201
Book Taxes	0	0	0	0	0	0	0	0	0	0
Book Income	790	953	1,121	1,290	1,461	1,635	1,812	1,991	2,094	2,201
Income Tax Calculations										
Operating Income	3,928	4,046	4,167	4,287	4,405	4,523	4,640	4,757	4,794	4,831
Interest	2,289	2,245	2,198	2,148	2,095	2,039	1,980	1,917	1,851	1,781
Tax Depreciation / Amortization	3,394	5,431	3,259	1,955	1,995	978	0	0	0	0
Taxable Income	(1,756)	(3,629)	(1,289)	184	355	1,506	2,660	2,840	2,943	3,049
Income Tax	0	0	0	0	0	0	0	0	0	0
CASH FLOW STATEMENT										
Operating Income	3,928	4,046	4,167	4,287	4,405	4,523	4,640	4,757	4,794	4,831
Return on DSR	0	0	0	0	0	0	0	0	0	0
Return on Working Capital	16	16	16	16	16	16	16	16	16	16
Debt Reserve Withdrawals(Additions)	0	0	0	0	0	0	0	0	0	0
Use of Funds										
Interest	2,289	2,245	2,198	2,148	2,095	2,039	1,980	1,917	1,851	1,781
Principal	760	805	852	902	954	1,010	1,069	1,132	1,198	1,268
Project Pre-Tax Cash Flow	895	1,013	1,134	1,253	1,372	1,490	1,607	1,724	1,761	1,797
Income Tax	0	0	0	0	0	0	0	0	0	0
Project After-Tax Cash Flow	895	1,013	1,134	1,253	1,372	1,490	1,607	1,724	1,761	1,797

	2011	2012	2013	2014	2015	2016	2017	2018	2019	2020
INCOME STATEMENT										
Revenue										
Capacity Payments	0	0	0	0	0	0	0	0	0	0
Power Sales to Utility	0	0	0	0	0	0	0	0	0	0
Electrical to Host	0	0	0	0	0	0	0	0	0	0
Chilled Water to Host	0	0	0	0	0	0	0	0	0	0
Steam to Host	0	0	0	0	0	0	0	0	0	0
Development Fees	0	0	0	0	0	0	0	0	0	0
Service Revenues	0	0	0	0	0	0	0	0	0	0
Performance Revenues	0	0	0	0	0	0	0	0	0	0
Budget as Revenue Base	11,332	11,502	11,674	11,849	12,027	12,207	12,390	12,576	12,765	12,956
Total Revenue	11,332	11,502	11,674	11,849	12,027	12,207	12,390	12,576	12,765	12,956
Operating Expense										
Electricity	1,443	1,486	1,531	1,577	1,624	1,673	1,723	1,775	1,828	1,883
Natural Gas	3,965	4,024	4,085	4,146	4,208	4,271	4,335	4,400	4,466	4,533
Fuel Oil	0	0	0	0	0	0	0	0	0	0
Coal	0	0	0	0	0	0	0	0	0	0
Water	1,075	1,107	1,140	1,175	1,210	1,246	1,284	1,322	1,362	1,403
Operations & Maintenance	(18)	(19)	(19)	(20)	(20)	(21)	(22)	(22)	(23)	(24)
Property Taxes	0	0	0	0	0	0	0	0	0	0
Total Operating Expense	6,465	6,599	6,737	6,878	7,022	7,169	7,321	7,475	7,633	7,795
Operating Income	**4,867**	**4,902**	**4,937**	**4,971**	**5,005**	**5,038**	**5,070**	**5,101**	**5,132**	**5,161**
Debt Service / Lease	(3,049)	(3,049)	(3,049)	(3,049)	(3,049)	(3,049)	(3,049)	(3,049)	(3,049)	(3,049)
Debt Reserve Withdrawals(Additions)	0	0	0	0	0	0	0	0	0	0
Pre-tax Cash Flow to Equity Owners	1,817	1,853	1,888	1,922	1,956	1,989	2,021	2,052	2,082	2,112
Interest Payments	1,707	1,628	1,545	1,457	1,364	1,266	1,161	1,051	934	813
Debt Reserve Additions(Withdrawals)	0	0	0	0	0	0	0	0	0	0
Book Depreciation	849	849	849	849	849	849	849	849	849	849
Book Income Before Taxes	2,311	2,425	2,543	2,665	2,792	2,924	3,060	3,202	3,349	3,502
Book Taxes	0	0	0	0	0	0	0	0	0	0
Book Income	2,311	2,425	2,543	2,665	2,792	2,924	3,060	3,202	3,349	3,502
Income Tax Calculations										
Operating Income	4,867	4,902	4,937	4,971	5,005	5,038	5,070	5,101	5,132	5,161
Interest	1,707	1,628	1,545	1,457	1,364	1,266	1,161	1,051	934	813
Tax Depreciation / Amortization	0	0	0	0	0	0	0	0	0	0
Taxable Income	3,160	3,274	3,392	3,514	3,641	3,772	3,909	4,050	4,198	4,351
Income Tax	0	0	0	0	0	0	0	0	0	0
CASH FLOW STATEMENT										
Operating Income	4,867	4,902	4,937	4,971	5,005	5,038	5,070	5,101	5,132	5,161
Return on DSR	0	0	0	0	0	0	0	0	0	0
Return on Working Capital	16	16	16	16	16	16	16	16	16	16
Debt Reserve Withdrawals(Additions)	0	0	0	0	0	0	0	0	0	0
Use of Funds										
Interest	1,707	1,628	1,545	1,457	1,364	1,266	1,161	1,051	934	813
Principal	1,342	1,421	1,504	1,592	1,685	1,784	1,888	1,999	2,116	2,239
Project Pre-Tax Cash Flow	1,833	1,869	1,904	1,938	1,972	2,005	2,037	2,068	2,098	2,128
Income Tax	0	0	0	0	0	0	0	0	0	0
Project After-Tax Cash Flow	1,833	1,869	1,904	1,938	1,972	2,005	2,037	2,068	2,098	2,128

FIGURE 1 ISU pro forma.

university could fix its 1997 utility budget at $7.8 million for 20 years and finance $37.5 million in facility and utility systems upgrades. The cash flow benefit came from simultaneously installing highly efficiency equipment, efficiently sizing and dispatching equipment, providing for fuel switching, reducing the amount of BTUs needed to provide reliable service, and lowering the BTU unit costs needed to serve the university.

The investment reflects a capital investment, which included a 5 MW gas turbine, new boiler, boiler plant upgrades, a new boiler structure, steam and condensate lines, electric chillers, thermal energy storage system, new chiller structure, chilled water building loop and distribution system upgrades, lighting conversions, air-side HVAC upgrades, new equipment control system, system recommissioning, electrical distribution systems upgrades, a new substation, energy information management system, and $5 million in asbestos abatement. Total capital costs were estimated to be more than $28 million. Engineering, procurement, construction, and contingency costs were $5.6 million with additional costs for permitting, legal fees, start-up, and numerous other project costs estimated at more than $2 million. The financing term is 20 years with an interest rate of $5.85 and a debt service reserve of $1.5 million.

Note the area called *Unit energy costs*. Unit costs including capital, O&M, and energy are averaged by system including costs for purchase of electricity. Change in the technology solutions to any system impacts the BTU unit costs for the other systems. For example, if ISU were to purchase steam from another provider, the unit costs for electricity and chilled water would be impacted. This Executive Summary represents one plan option. To gain the greatest BTU value at the lowest combined BTU unit costs, numerous project options should be compared.

Figure 1 presents the annual cash flow of the master energy plan for the project life. Additional reports shown later detail the operating calculation assumptions associated with each of the project utility systems' financial reports. Cogeneration, steam, and chilled water operating calculation reports are shown in Tables 2–4.

Once a capital investment strategy is identified, decision-makers will want to see an assessment of what might happen if the assumptions change. Risks in an infrastructure investment largely include changes in assumptions related to energy growth and cost escalation, getting the project built on time and within budget, financing costs, and long-term operating and maintenance costs. Table 5 shows a range of cost impacts to the project if the initial assumptions were to change by –20%, –10%, or –5%. Notice that the project shown stays healthy at –5% but loses money at –10%. Its project analysis should analyze worse- and best-case scenarios and be prepared to modify the project or find a way to mitigate the risks.

An executive Summary, supplemented by systems operating assumptions, a life-cycle financial pro forma, and sensitivity analysis are essential to gain the approval from top management to proceed with the project. Although this assessment demonstrated that the opportunity for upgrading the university infrastructure was significant, the project was not implemented.

Planning Methodology

The OA methodology slide presented in Figure 2 is a summary of the myriad assumptions that are needed to create a life-cycle business plan for any facility.

Starting with the first step or lowest level of the OA methodology pyramid, the planning process begins with identifying the facility, its owner, and the business goals associated with the use of the facility over the life of the proposed debt. Infrastructure goals can range from environmental improvements, to business expansion, comfort needs, new operating modifications, financing requirements, and code compliance. For ISU, the university wanted to fix its energy budget and finance all infrastructure capital backlog.

Using a life-cycle approach, the growth assumptions should reflect the energy systems impacts associated with the facility use business goals. For ISU, these included growth of electricity, steam, and chilled water for 20 years on a year-by-year basis. Cost escalation estimates by year were also inputted individually for electricity, natural gas, other fuels, budget growth, O&M, and general inflation. Detail of base year budget and tax status is the last group of assumptions usually required to determine the business situation for the facility. Obviously, ISU is a tax-exempt entity.

TABLE 2 ISU Cogeneration Operating Calculations

		Operational Calculations					
		Cogeneration Systems					
		Illinois State University - Test Case 3					
Gas turbine/engine	Capacity	Fuel rate		Cap factor	Availability O&M cost		Production (kWh)
Fuel Generator #1	5,000	56		92.00%	88.00%	$0.00	35,460,480
Steam turbines	Capacity	Inlet enthalpy	Outlet enthalpy	Cap factor	Availability O&M cost		Production (kWh)
Customer annual usage		68,892,296 kWh			Retail rate		0.0000 $/kWh
Wheeling		0 kWh					
		Gas turbine/engine					**Steam Turbines**
Generation capacity		4,400 kW					0 kW
Turbine output		35,460,480 kWh/Yr.					0 kWh/Yr.
Annual fuel usage		397,157 MMBtu					0 MMBtu
Annual fuel cost		$1,247,074					$0
Annual O&M cost		$0					$0
Annual steam output		0 KLbs.					
Fuel chargeable to power		0 Btu/kWh					
							Annual
Cost breakdown:							
Fuel		0.0352 $/kWh					$1,247,074
O&M		0.0000 $/kWh					$0
Debt service & returns		0.0077 $/kWh					$271,297
Total		**0.0428** $/kWh					
Current avg. cost		0.0558 $/kWh					
Projected avg. Cost		0.0460 $/kWh					
Weighted avg. Cost		0.0444 $/kWh					
Customer Savings from Cogeneration:							$1,609,693
Saving Without Debt Service							$1,880,989

Determining the assumptions for ISU's business situation (the first step of the pyramid) begins to highlight the risks associated with a final business investment. Just estimating the costs for purchase of natural gas or electricity for the next 20 years is a major guess. As planning continues up the pyramid, the stability of the entire project can be compromised by changing assumptions at the base. Having the capability to automatically update and integrate the impacts of changes in cost and growth assumptions or any other baseline assumption greatly improves the team's ability to manage and report these impacts.

Step 2 of the pyramid, first offers the ability to determine the base year energy consumption, loads, and costs associated with purchasing and producing steam/heat, cooling, and electricity. Further data input is needed to describe each system's equipment, age, operating characteristics, maintenance requirements, and other data to calculate each system's BTU unit costs.

At this stage, a status quo business situation pro forma should be run to determine the life-cycle costs of what the business entity will spend if no investments are made. The planner should also chart the facility energy load requirements for electricity, steam/heat, and chilled water energy over time when compared with the equipment capability for each system to provide reliable service. For example,

TABLE 3 ISU Steam Operating Calculations

			Operational Calculations			
			Steam System			
			Illinois State University - Test Case 3			
Assumptions						
Cogen steam output				0 KLbs.		
Fuel used	From GT			0 MMBtu		
	Supplemental			0 MMBtu		
Cost				$0		
Projected steam requirements				532,619 KLbs. per Year		
Steam required beyond cogen supply				**532,619** KLbs. per Year		
Projected steam quality		*Enthalpy*	*Temperature*	*Pressure*		
	High Pressure Intermediate Pressure 1,203		1,189 Btu/Lb.	Btu/Lb. 0 400	deg. F	deg. F
	0	125 psig	psig			
	Low Pressure	203 Btu/Lb.	200 deg. F	50 psig		
	Hot Water	Condensate 0	5 Btu/Lb.	Btu/Lb. 0 85	deg. F	deg. F
	0	220 psig	psig			
Steam Production:	**Capacity**	**Fuel**	**Efficiency**	**Pressure**	**Production**	
Standard boiler#10	35,000	Primary	81%	Intermediate	54,622	
Standard boiler #6	85,000	Primary	78%	Intermediate	81,933	
Standard boiler#7	85,000	Primary	78%	Intermediate	191,178	
Standard boiler#8	85,000	Primary	81%	Intermediate	136,556	
Standard boiler #9	65,000	Primary	81%	Intermediate	81,933	
Summary						
Boiler plant:						
Total steam capacity	355,000 Lbs./Hr.			Operation & Maintenance 1.50/KLbs.		
Annual steam output	546,222 KLbs.			MMBTU		
Consumption	815,501					
Cost	$2,560,673					
Cost breakdown:						
Boiler plant				**Annual**		
O&M		Fuel 1.50		4.69 /KLbs. /kLbs.		
$819,333	$2,560,673					
Debt service & returns		0.81 /KLbs.	$440,276			
		6.99 /KLbs.				
Steam from Cogen				**Annual**		
Fuel		0.00 /KLbs.		$0		
O&M	Debt Service & Returns 0.00			0.00 /kLbs. /kLbs.		
$103,351		$0				
		0.00 /KLbs.				
	Weighted average cost per klbs			$6.99		
	Current cost per Klbs			$6.91		
Customer savings from new steam production				**$(150,714)**		
Saving without debt service				$392,913		

TABLE 4 ISU Chilled Water Operating Calculations

Operational Calculations				
Chilled Water System				
Illinois State University - Test Case 3				
Assumptions				
Chiller	**Type**	**Capacity**	**COP**	**Production**
Existing units	Electric	1,500	5.10	1,577,513
New 1000(2)	Steam	2,000	0.90	3,155,025
New 2000(2)	Electric	5,000	6.00	5,784,213
				10,516,751
Projected avg. electrical cost	$0.0444 $/kWh			
Projected fuel cost	$3.14 $/MMBTU			
(Delivered to burner tip)				
Operation & maintenance	$0.01 /ton			
Summary				
Total chilled water capacity			8,500 tons	
Annual production			10,516,751 ton-hrs.	
		Electric	*Gas*	*Steam*
Consumption		42,766,369 kWh	0 MMBTU	42,664 MMBtu
Cost		$1,897,116	$0	$167,254
Cost breakdown:			**Annual**	
Fuel		$0.1963 /ton-hr.	$2,064,370	
O&M		$0.0100/ton-hr.	$105,168	
Debt services & returns		$0.1052 /ton-hr.	$1,106,590	
		$0.3115 /ton-hr.		
	Equal to:	$25.96 /MMBTU		
Current cost per ton of chilled water			$0,177 **/ton-hr.**	
Saving on total new chilled water production				$(1,196,140)
Saving without debt service				$(89,549)

TABLE 5 Sensitivity Analysis

Project cost	–20%	–10%	–5%	estimate
	$51.9 M	$47.6 M	$45.4 M	$43.3 M
Interest rate	–20%	–15%	10%	
	7.02%	6.73%	6.44%	5.85%
Savings	–20%	–15%	10%	
	$2.3 M	$2.5M	$2.6M	$2.9M
Electrical cost	–15%	–10%	–5%	
	4.270	4.090	3.900¢	3.720¢
Natural gas cost	–15%	–10%	–5%	
	$3.45	$3.30	$3.15	$3.00
12% NPV	–$8.5 M	–$2.6 M	$2.3 M	$6.5 M

Figure 3 shows the baseline demand (the diamond line) to exceed the central plant capacity (the triangle line) within a few years.

Step 3 begins data input to come up with integrated energy supply and production systems technology solutions to reliably and efficiently meet business energy needs over time. First, before addressing the supply purchase, production, and distribution technologies, investments are needed to make the end

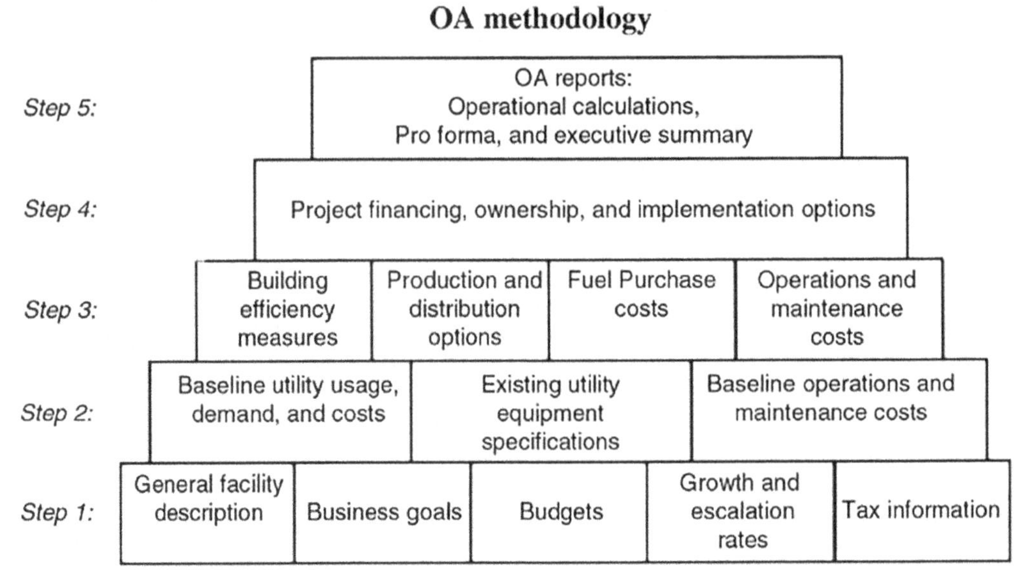

FIGURE 2 OA methodology.

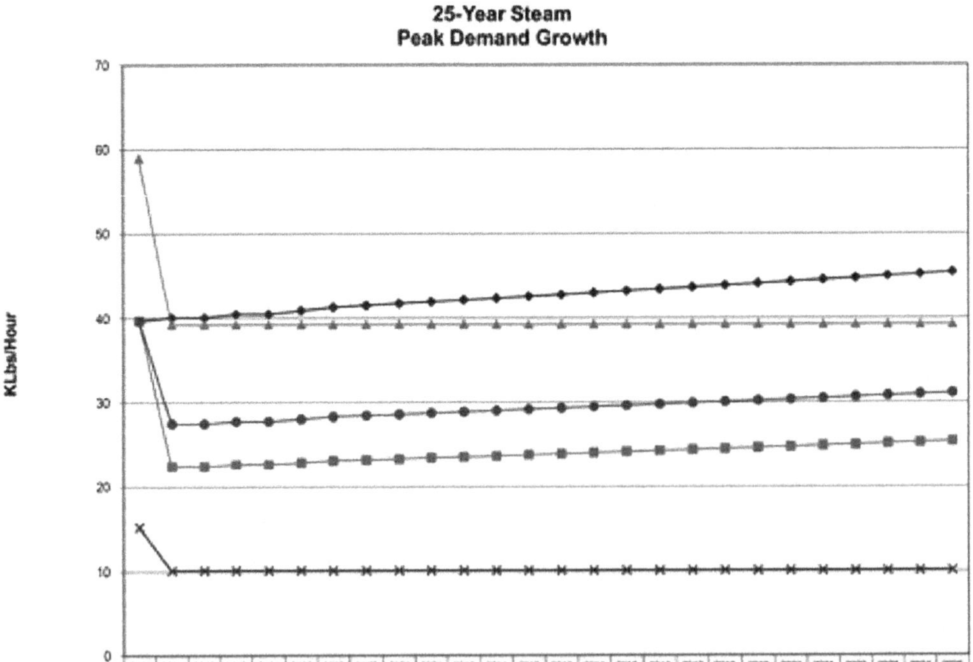

	FY98	2001	2002	2003	2004	2005	2006	2007	2008	2009	2010	2011	2012	2013	2014	2015	2016	2017	2018	2019	2020	2021	2022	2023	2024	2025
Baseline Demand	39.6	40.0	40.0	40.4	40.4	40.8	41.2	41.4	41.7	41.9	42.1	42.3	42.5	42.7	42.9	43.1	43.3	43.6	43.8	44.0	44.2	44.4	44.7	44.9	45.1	45.3
Adjusted Baseline Demand	39.6	22.4	22.4	22.6	22.6	22.8	23.0	23.2	23.3	23.4	23.5	23.6	23.7	23.9	24.0	24.1	24.2	24.3	24.5	24.6	24.7	24.8	25	25.1	25.2	25.3
Base Case Demand	39.6	27.4	27.4	27.7	27.7	28.0	28.2	28.4	28.5	28.7	28.8	29	29.1	29.2	29.4	29.5	29.7	29.8	30.0	30.1	30.3	30.4	30.6	30.7	30.9	31.1
Heat Recovery Boiler Capacity	15.2	10.1	10.1	10.1	10.1	10.1	10.1	10.1	10.1	10.1	10.1	10.1	10.1	10.1	10.1	10.1	10.1	10.1	10.1	10.1	10.1	10.1	10.1	10.1	10.1	10.1
Total Central Plant Capacity	58.9	39.2	39.2	39.2	39.2	39.2	39.2	39.2	39.2	39.2	39.2	39.2	39.2	39.2	39.2	39.2	39.2	39.2	39.2	39.2	39.2	39.2	39.2	39.2	39.2	39.2

Year

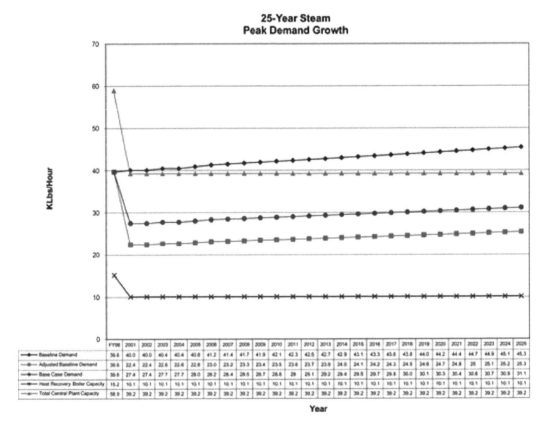

FIGURE 3 Twenty-five-year steam peak demand.

use reliable and efficient. Lighting, controls, commissioning, air-side HVAC improvements, and other end-use thermal and electrical improvements need to be identified. For example, if lighting retrofits are implemented, the load demand for heating and cooling will change. A new peak demand line or adjusted baseline (square line) is added to the peak demand growth chart. Now, the planner is ready to select and assess the benefits of new energy supply systems' equipment solutions. The Base case demand (circle line) shows the reduction in steam energy or thousand pounds per hour demand that resulted from the improvement in equipment efficiency. The old baseline assumptions are modified to incorporate the new project technology solution assumptions.

Before running a new pro forma, the planner must input assumptions for how the project will be implemented (Step 4). What are the schedule; financing structure; costs for engineering, procurement, construction, contingency, permitting, start-up, and project management; legal fees; and the myriad other costs associated with implementing the project? Who will own the project? How will it be financed? What are the annual principal and interest payments? Are there tax benefits? All these assumptions are inputted so a full life-cycle cost pro forma can be run to compare the new business plan with the status quo or baseline. ISU Executive Summary and Pro Forma show the comparison of ISU's new business plan with its status quo.

We have now reached the top of the pyramid. Step 5 is the business plan. Look at the assumptions that enter into each business investment and asset management decision. The very bottom of the pyramid requires the planner to begin by guessing the cost escalation for primary fuels, electricity, labor, maintenance, inflation, and budgets over the life of equipment. Debt terms could be 20 years. Now, factor in the business growth planned for that same period of time and try to project the increases or decrease in business growth for electricity, heat, and cooling energy. Planners have not begun to deal with equipment selection and performance or the assumptions associated with implementation to get a project financed, built on time and within budget, and other investment assumptions that impact investment risks.

Weakness in each level of assumptions could make the entire investment crash. The dynamic nature of changing energy prices coupled with business changes is already enough to discourage anyone from pursuing integrated energy investments by using the old-fashioned paper plan approach. Assumptions in a Web-based plan enable constant updates and integrated, real-time impact reporting.

Asset Management

Let us put all those assumptions into a Web model and automate the update of critical assumptions like time-of-day energy pricing, equipment performance, and load changes. In addition, let us approach the investment in a manner that allows the user to turn on and turn off equipment or conduct fuel switching based on unit costs. Further, let us integrate price signals into equipment controls to automate the activity. Let us look at what new capital planning does to the investment. Let us evaluate proposals from potential business partners to determine the value of laying of risk and measuring results.

Software abounds for the following:

- Metering, accounting, billing, tracking, profiling, and verification
- Equipment energy management, measurement, optimization, and automated dispatching
- Integrated energy engineering analysis, calculating technology applications for energy and water consumption, load, and cost savings
- Enterprise energy management for energy systems analysis and optimization
- Conversion of hand-drawn prints to an AutoCAD database
- Work order management

The software listed can be accessed, reported, and analyzed for timely decision-making using Web-based information management software.

What does not abound are software systems that provide for developing, managing, and reporting integrated energy supply investments. Nobody ties supply side energy efficiency investment altogether into a financial pro forma, yet all the energy information software described above can be used to update and manage the investment.

Let us create a communication and reporting platform where the entire business investment team can view real time reports. Reports show the financial pro forma and operating calculation for each and all facilities that need to be reviewed. The reports can be delivered from anywhere, disaggregated at any level of granularity, evaluated compared with any time sequence, rearranged, and put back together.

All team members are required to work together to create each project's assumptions and present the project for management approval. This includes risk assessment and mitigation options. Each team member will know his or her job requirements associated with each project and can be evaluated based on meeting his or her portion of the investment plan performance. In addition, each team member is responsible for communicating market and business changes in the business plan assumptions with the rest of the team.

Now let us connect and integrate consumer energy information systems to electric utility supply systems. What if the utility was able to coordinate dispatching of its generation and transmission options with consumer energy supply systems dispatching? The optimum use of all investments in energy supply systems can result. Not only does efficiency and emissions control occur, but also fewer new electric central plants and transportation investments will be needed. Everyone gains greater productivity through modernization, efficiency, and communication. Less energy is consumed.

Risk Management

For decades, independent power producers have set up limited liability corporations (LLCs) and financed generation project based on the strength of performance contracts for sale of energy, purchase of primary fuels and standby power, financing, engineering/procurement/construction (EPC), and operation/maintenance.

All of these risks can be mitigated contractually. The question is, What is it worth to have business relationships or partners who will take the risks compared to the facility owner assuming the risk? The fun part of having a Web- based business plan is that it is easy to assess the value of laying off risks onto other business entities. Comparison of proposals can be integrated into the Web-based energy plan and compared with the cost to the facility owner. During the course of EPC, the plan can track the actual costs and schedule. Constant monitoring and updating of the Web plan strengthen the ability of both owner and contractor to make decisions and communicate project development.

Risks associated with purchase of fuels or electricity are often most difficult for building owners to control. In addition, regulatory changes and changes in business goals can significantly impact long-term investments. With the economic and political strength of electric utilities, the fact that building and operating energy supply systems are their core competency, and the direct links electric companies have to fuel supply opportunities, both consumers and utilities can benefit from partnering.

Energy service companies, worldwide, are in business to earn profits by taking these risks. However, supply system investments often are not integrated. Methods of asset management and reporting are varied and often subject to dispute.

Profits

Profits to electric companies from partnering with consumers can be expansive. For example, institutions especially often have EPC rules, and situations where the project cost for EPC can be very high—as much as a 100% increase in capital costs. This situation leaves a lot of room for having a business arrangement where another company is paid a profit to build the project and demonstrate equipment performance.

Profits can result:

- By adding value to customer energy systems investments and sharing the savings
- Through implementation of EPC contracts
- From operating and maintaining customer energy supply systems
- From sale of primary fuels
- From sale of heat, chilled water, water, and wastewater
- By investing in customer infrastructure
- From tax benefits
- By gaining emissions credits
- Through access to customer information to dispatch integrated systems
- From advanced growth management information
- From firm power sales contracts
- Through optimization of current supply systems' fixed capital
- From reduced financing costs

In turn, the customer is able to upgrade his or her infrastructure, improve the reliability of energy systems, control facility operating budget, and gain an experienced partner who can manage the short- and long-term investment risks. Emissions are reduced, and productivity is increased, often with no customer debt on his or her balance sheet.

Any and all of these benefits are possible depending on the specific site and partnering options. Another paper could be written describing the structure and value of integrated, energy supply systems business opportunities. The benefits to all suppliers and customers include the ability to:

- Finance millions of dollars of infrastructure upgrades from operating budget savings
- Level electricity load requirements
- Reduce BTU consumption and unit costs
- Improve energy reliability and operating productivity
- Communicate and manage energy supply systems risks

Conclusion: "Big Collaboration, It Is Now and the Subject Is Energy" [4]

Thomas Friedman in his book, *The World Is Flat,* recognizes, "One of the unintended consequences of the flat world is that it puts different societies and cultures in much greater direct contact with one another."[51] Promoting electric utility and customer partnering seeks a change in different cultures. Friedman's vision of a radical new approach to energy use and conservation is not just among countries to jointly develop clean alternative energies.[61] Part of the radical new approach is closer to home—be- tween electric utilities and their customers. Inherent in the solution to energy resource efficiency is a radical new approach to earning profits from value-added. To achieve this, new relationships must be developed. Trust is built from sharing a commitment to save energy and communicate and manage risk.

Traditionally, energy supplier generation, transmission, consumer production and distribution systems, and building end use have been dealt with separately—an approach that has proven to be costly and inefficient. Rather than losing money and energy through the traditional approach, right now, we can build systems as integrated investments, save money, make money, and stop wasting energy.

Inherent to integrated energy investments is commitment to using a methodology and dynamic Web-based tool on which to plan, finance, implement, and sustain joint energy supplier and customer economic and environmental opportunities.

A Note on Sources

Energy gained discussions of energy infrastructure investment practices and solutions rely largely on experience and information drawn from publicly available sources: public-sector studies, Web sites, industry groups, professional associations, and U.S. government sources. The ISU data resulted from a study that my staff and I conducted with the help of the University staff. Extensive reference information can be found at the following Web sites: http://www.acee.org, http://www.aeecenter.org, http://www.appa.org, http://www.chpa.co.uk, http://www.cogeneration.net, http://www.eei.org, http://www.eia.doe.gov, http://www.ieadhc.org, http://www.epri.com, http://www.naruc.org, and http://www.naesco.org. Numerous other Web sites exist.

Addendum

Questions

It has been asked, "Would it be possible to give an estimation of how much energy could be saved from investment and use of integrated energy systems?" Putting a simple number to answer this question is not easy and is complicated by the many ways that experts define and measure energy. The answer varies based on two primary factors: 1) the age, location, pollution costs, and inefficiency of current energy systems and 2) our level of commitment to invest in and operate highly efficient, reliable, clean energy supply systems.

Our goals are to gain the greatest amount of energy output from every unit of energy mined, transported, and converted to electricity, heat, and cooling and to size production and distributions systems to meet reliable, efficient end-use facilities and equipment. (The complexity of this topic cannot be fully explained in this short narrative. Further elaboration and definition of terms can be found in my textbook, *Energy Efficiency: Real Time Energy Infrastructure Investment and Risk Management.*) The goal also includes how to dispatch and operate the energy commodities listed above to achieve efficient, reliable, and nonpolluting water and wastewater systems investments at the lowest production and distribution unit costs.

Although it is important to answer to "how much energy could be saved?," the more important questions are as follows:

- How do we implement and earn profits from integrated resource efficiency investments?
- How do we evaluate alternative fuels; technologies; operating and maintenance options; engineering, procurement, and construction progress; financing and ownership structures; life cycle operation and maintenance maximization; and regulatory and technology changes; and follow market demand changes and numerous other risks to close deals and earn the best returns on investments over time?

Savings Potential

In addition to the coal, natural gas, and petroleum used to produce electricity and heat, the U.S. Department of Energy, Energy Information Administration, reports in its World Consumption of Primary Energy by Energy Type and Selected Country Groups, Table 1, the "net" generation of electricity from numerous renewable fuels and nuclear. The efficiency or net output of useful electricity compared to the total energy value contained in the primary fuel during the 20th century is reported to have a sustained average in the United States of about 33%. The amount of energy consumed is embedded in the conversion factors to change the primary fuel to secondary commodities like electricity and heat. Each primary fuel has energy content. When a primary fuel is converted to electric energy in the United States, on average, two-thirds of the original energy content is thrown away.

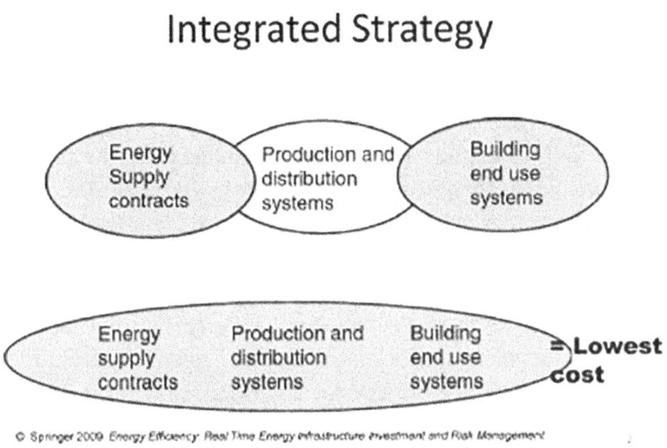

FIGURE A1 Integrated strategy.

Further energy losses also result from decisions about selection of sites to build electric, heat, and cooling production and transportation systems as well as energy use efficiencies, choices of equipment, time of use, fuel switching, storage, and other infrastructure investment efficiency options.

Investment in efficient energy systems requires a comprehensive, integrated approach solving for gaining the greatest output and reliability from every energy unit consumed at the lowest energy unit cost. Figure A1—Integrated Strategy (Solmes, Leslie, *Energy Efficiency: Real Time Energy Infrastructure Investment and Risk Management*, Springer, 2009, p. 20)—first reflects the traditional way of investing in energy systems as separate parts: primary fuels; production and distribution of electricity, heat, and cooling systems; and building end-use systems.

In order to create the greatest production and transmission/distribution efficiency, investment analyses begin with updating end-use facilities and equipment based on defined business goals and projections of end-use growth and cost escalation for primary fuels, labor, emissions, budgets, and other long-term cost considerations that will impact the returns of investment over its life cycle. An assessment of end-use efficiency and system upgrades will result in adjusting baselines for electricity, heat, and cooling requirements. Selection and sizing of new production and distribution equipment will be affected and often reduced.

Add up the energy and load savings that result from comprehensive, integrated energy systems' investments that include the following:

1. Updating of all aged, inefficient, unreliable, and polluting end-use facilities and equipment—*savings of 10%–50%*. A July 2009 McKinsey report on energy efficiency projects that there exists an investment opportunity of $520 billion that would save $1.5 trillion through 2020, reduce U.S. consumption by 23%, and more than offset U.S. expected growth in energy use (http://green.blogs.nytimes.com/2009/07/29/mckinsey-report-cites-12-trillion-in-potential-savings- from-energy-efficiency/). Success stories abound.
2. Sizing production and transmission/distribution systems to meet new, efficient end-use loads—*savings based on load reductions from end-use investments, often 20%–30%*. End-use efficiency will change the size and number of electric, heat, and cooling infrastructure, thus lowering the capital costs for replacement and/or new capacity additions and gaining greater optimization and efficiency of new production equipment and reducing the need for distribution systems expansions. For example, just one building retrofit at a major California university showed a 30% reduction in chilled water requirements due to end-use efficiency savings. The capital cost, even with 15% redundancy, was lowered by 22%.

3. Sizing on-site generation of electricity to meet the thermal load resulting in the greatest energy output from every unit burned—*savings from doubling energy output of primary fuel to produce electricity and dramatic reduction in consumption of primary fuels to produce heat energy could be 40%.* The U.S. Clean Heat and Power Association reports that combined heat and power investment get up to 2–3 times the useful energy products from the fuel and can effectively cut fuel costs by up to two-thirds (http://www.uschpa.org/14a/pages/index.cfm?pageid=3297).
4. Using recycled heat energy to provide heat and cooling also reduces the amount of new equipment needed—*same as above including reduction in new equipment needed to produce heat energy.*
5. Leveling the electrical load by using recycled heat energy for cooling—*savings in electric consumption and need for investments in electric peaking capacity.* Optimal equipment sizing for electricity production usually means that some purchase of electricity from the grid will be needed. However, use of waste heat for heating and air conditioning will result in flattening the electric load and reducing the need for building electric generation capacity to meet peak loads. It will also enable a high percentage of supplemental electricity to be purchased 24/7 thus improving the efficiency of the energy supplier's generation station. (Solmes 2009)
6. Incorporating multiple production equipment solutions to efficiently address shoulder months where heating and cooling vary greatest—*savings from turning off equipment operation during shoulder months and operating equipment at optimum efficiency.* For example, multiple small boilers can be run at full load during low heat demand compared to operating a large boiler for limited hours or partially loaded.
7. Incorporating district heating and cooling systems efficiencies—*it can result in as much as 50% energy savings in industrial, campus, military, hospital, dense city, and similar environments.* Placing boilers and chillers in each building means buying equipment sized to meet the higher heating and cooling loads for each building. Most of the year, this equipment runs partially loaded. A central plant with multiple generation units can run equipment to meet the aggregate load of multiple buildings, thus optimizing equipment performance efficiency and needing to invest in less equipment.
8. Increasing load and efficiency of electricity purchases from utilities, thus improving existing generation efficiency and capital utilization—*improved 24/7 use of electric generation capacity.* See (5).
9. Reducing investment requirements and losses associated with electricity transport systems—*savings from elimination of electric transportation losses and reduction in grid requirements.*
10. Improving load requirements on equipment and systems used to provide primary fuels—*savings from 12-month purchase and/or use of primary fuels.* Greater contracting cost leverage can result for longer-term contracts.

No doubt, more efficiency items can be listed as a part of integrated energy systems investments. Factor in the energy and capital efficiency that can be gained by the combination of all 10 items. Now, include the efficiency that results from financing a single larger project rather than numerous small ones. Add the savings that result from emissions reductions and reduced need to build new electricity generation and transmission systems. This cascading of energy and capital savings by integrating energy systems investments can fund the investment debt as well as earn significant returns on investment.

Thomas Casten in his book *Turning Off the Heat: Why America Must Double Energy Efficiency to Save Money and Reduce Global Warming* provides an answer to "how much could be saved from integrated energy systems investment?"—at least half of current consumption. The opportunity is that due to age and inefficiency, use of the integrated approach will result in funding investments, greater productivity, less pollution, more jobs, and a stronger economy.

New Energy Supply Systems Should Reduce Costs to Consumers

The savings reflected by McKinsey is appropriately titled "end-use" energy savings. Although some waste heat recovery is reflected in the table, the more than half a billion dollars in energy infrastructure investment does not address the investment opportunity that exists in investing in highly efficient, integrated electricity, heat, and cooling production and distribution systems. When Casten wrote about doubling our energy efficiency, he was citing the doubling effect of committing to supply-side resource efficiency—conversion of primary fuels to electricity, heat, and cooling production and distribution combined with efficient end use. This formula will result in meeting the goals of resource efficiency at the lowest cost to the consumer, doubling emissions reductions and tremendous opportunities to earn profits from value-added.

Tracking Investment Returns and Managing Risks

Numerous international measurement and verification guidelines and protocols have been developed and adopted to determine the energy savings from energy efficiency measures and renewable fuels. The oversight of this effort is now done by a non-profit organization called the Energy Valuation Organization or EVO. According to EVO's website, "EVO is the only non-profit organization in the world solely dedicated to creating measurement and verification (M&V) tools to allow efficiency to flourish. Our vision is a global market that properly values the efficiency resource, enabling and assisting the optimal investment in these opportunities."

The limitation to the work accomplished to date is that Web-based energy systems business investment plans, real time tracking, and calculation of electric, heat, and cooling unit costs as a means to achieve the lowest combined energy unit cost are not available. A software application that communicates, tracks, and reports all the assumptions that go into creating a baseline, comparing the benefits of a combination of efficient investment options and investment strategies, and automating the update of these assumptions over the term of the debt does not exist.

To gain the greatest value from integrated energy investments, stakeholders need to communicate and see the financial impact of market and business changes in real time. Even going through the investment stages of baseline development to preliminary assessment to detailed engineering, procurement, and construction entails very big financial risks if continuous automated tracking and data comparison using internet technology are not instituted to address critical time factors for decision making. An energy infrastructure investment and risk management software application that provides the answers to achieving resource efficiency is the means to successfully accomplish the efficiency investments described above and to encourage electric utilities, service companies, consumers, and investors to earn returns from integrated efficiency investments. An initiative to create this application standard is how to answer the question, "How much energy could be saved from investment and use of integrated energy systems?"

References

1. Electric Power Research Institute for the Board of Directors, Ed. Electricity Sector Framework for the Future, Summary Report; EPRI: Palo Alto, CA, 2003; 2—August 6.
2. ed.;—Financial Times August 18, **2003**, 1 and 3.
3. APPA, CRDM: *Capital Renewal/Deferred Maintenance,* http://www.appa.org.
4. Friedman, T.L., Ed. *The World Is Flat: A Brief History of the Twenty-First Century;* 2005; 413.
5. Friedman, T.L., Ed. *The World Is Flat: A Brief History of the Twenty-First Century;* 2005; 391.
6. Friedman, T.L., Ed. *The World Is Flat: A Brief History of the Twenty-First Century;* 2005; 412–413.

Bibliography

1. Diamond, J., Ed. *Collapse: How Societies Choose to Fail or Succeed;* The Penguin Group: New York, 2005.
2. Friedman, T.L., Ed. *The World Is Flat: A Brief History of the Twenty-First Century*; Farrar, Straus and Giroux: New York, 2005.
3. Greenspan, A., Ed. *The Age of Turbulence: Adventures in a New World;* The Penguin Press: New York, 2007.
4. Vaitheeswaran, V.V., Ed. *Power to the People: How the Coming Energy Revolution Will Transform an Industry, Change Our Lives, and Maybe Even Save the Planet*; Farrar, Straus and Giroux: New York, 2003.

40

Bioreactors for Waste Gas Treatment

Sarina J. Ergas

- Low solubility compounds.
- Cometabolism of chlorinated organic compounds.
- Compounds requiring specialized microbial populations or conditions.
- Indoor air applications.

Introduction

Successful biofiltration applications have been limited to control of relatively soluble volatile organic compounds (VOCs) at low loading rates. Biofiltration is also of limited use for compounds that produce acidic or toxic metabolites or are degraded via cometabolism. To overcome the limitations described above, gas-phase biological treatment systems must be developed which (1) incorporate mass transfer of gas-phase pollutants across a media with high specific surface area and low diffusion length, (2) incorporate high biomass concentrations to maintain high biodegradation rates, (3) provide a method for wasting biomass to prevent clogging, and (4) provide a method for addition of pH buffers, nutrients, cometabolites, and/or other amendments to support the microbial population and neutralize acidic metabolites.

Hollow fiber membrane bioreactor (HFMB) systems have been under investigation which meet the above requirements and have been shown to achieve high VOC removal efficiencies in small reactor volumes. A schematic of a typical HFMB system is shown in Figure 1. The hollow fiber membranes serve as a support for the microbial population and provide a large surface area for VOC and oxygen mass transfer. Waste gases containing VOCs are passed through the lumen of the hollow fibers. Soluble compounds in the gas phase are transferred through the membrane pores and partition into

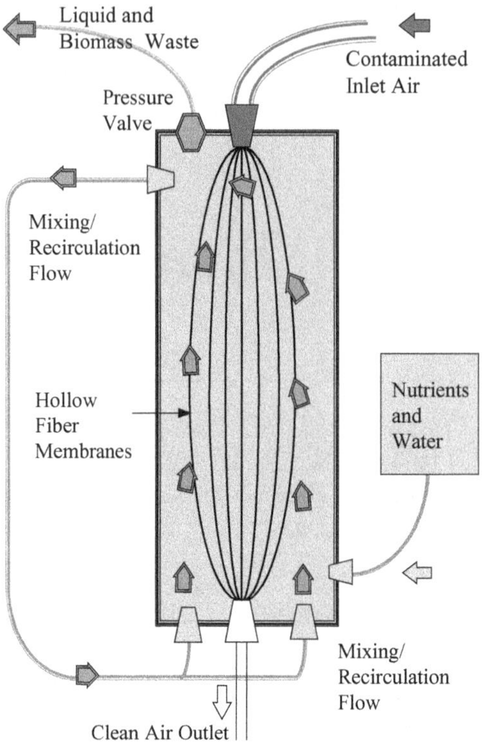

FIGURE 1 Schematic of a hollow fiber membrane bioreactor for biological air pollution control. (From Shumway, 1997.)

a VOC-degrading biofilm surrounded by a circulating nutrient media. Compounds in the biofilm are available for biodegradation. Potential advantages of HFMBs for waste gas treatment include

- Ability to continuously remove biomass to prevent clogging at high VOC loading rates.
- Ability to remove degradation by-products and add pH buffers, nutrients, and/or co-substrates to support the microbial population and neutralize acidic metabolites.
- Ability to treat low solubility compounds such as nitric oxide (NO) and methane.
- Separation of the microbial process from the gas being treated. This may be useful in indoor air applications where carryover of biomass into the ambient air is a concern.
- Optimal humidification of the gas stream.
- Low pressure drop.
- Gas and liquid flow rates can be varied independently without flooding or foaming.
- Modular design, no moving parts.

Disadvantages of HFMBs for waste gas treatment include high capital costs and that the technology has not been demonstrated at full scale.

Membrane Bioreactor Applications for Pollution Control

Hollow fiber membrane bioreactors have been used in a number of pollution control applications including

- Separation and retention of biosolids.
- Bubble-free aeration of bioreactors.

- Providing hydrogen and other low solubility gases to bioreactors.
- Extractive membrane bioreactors for controlled transfer of pollutants from industrial wastewaters.
- Biological air pollution control.

Membrane bioreactors are most often used for separation of biomass. The ultrafiltration of microfiltration membranes is used as a substitute for sedimentation. These systems are used to produce high-quality effluent and/or provide high solid retention times without washout of biomass. These applications are well documented in the literature and are outside the scope of this discussion (Stephenson et al., 2000).

Gas-to-liquid transfer membrane systems were first introduced for use in blood oxygenation and have been studied extensively for this application (Lund et al., 1996). Pollution control research into membrane aeration HFMBs has been focused on the enhancement of oxygen transfer for in high oxygen demanding applications such as the treatment of high-strength industrial wastewaters (Yamagiwa et al., 1998) and nitrification (Brindle and Stephenson, 1996; Brindle et al., 1998). These systems have also been used for bioreactors degrading VOCs, since the membranes allow for reactor aeration without stripping volatile compounds to the atmosphere (Pressman et al., 1999). Due to the low solubility of oxygen, oxygen transfer often limits aerobic degradation of high oxygen demanding wastewaters. Many facilities switch from atmospheric air to pure oxygen to enhance oxygen mass transfer; however, conventional oxygenation systems have high power requirements and are incompatible with attached growth reactors (Stephenson et al., 2000). Hollow fiber membranes provide a very high surface area for transfer of oxygen directly to the biofilm. Hydrophobic hollow fibers have been developed with sealed hydrophilic ends that enable 100% utilization efficiency in gas-to-liquid mass transfer applications (Ahmed and Semmens, 1992a,b).

A novel application of HFMB gas-transfer technology, hydrogenotrophic denitrification of drinking water, has been investigated by several researchers (Gantzer, 1995; Lee and Rittmann, 2000; Ergas and Reuss, 2001). A number of common genera of bacteria can use hydrogen as an electron donor and nitrate as an electron acceptor under anoxic conditions. These organisms have been shown to denitrify nitrate contaminated drinking water to acceptable levels. Disadvantages of hydrogenotrophic denitrification include lower denitrification rates and the difficulty in dissolving sufficient quantities of hydrogen into the water due to its low solubility. Use of HFMBs has been shown to support high biomass densities and improve hydrogen mass transfer rates while preventing the waste of excess hydrogen and avoiding accumulation of explosive gasses in a confined space.

Extractive membrane bioreactors have been used to biologically treat industrial wastewaters in the presence of high concentration of acids, bases, and salts that can inhibit degradation (Livingston, 1994; Livingston et al., 1998). Extractive membrane bioreactors utilize dense silicone membranes that selectively extract organic pollutants such as chloroethanes, chlorobenzenes, and toluene, from polar and ionic compounds. A VOC-degrading biofilm grows on the surface of the membranes creating a driving force for mass transfer. Due to the selectivity of the membranes, the biofilm is isolated from the harsh conditions in the wastewater to ensure high biodegradation rates.

Membrane Fundamentals

A membrane separates two distinct bulk phases of a system while allowing the transport of compounds from one phase to the other. In waste gas treatment applications, gases are most often blown through the lumen (inside) of tubes made from membrane materials. Pollutants from the gas phase diffuse through the membranes to a liquid phase on the shell side (outside) of the membranes. The membranes also serve as a support for the microbial population. Once in the liquid, compounds are biodegraded creating a concentration gradient, which serves as a driving force for mass transfer.

Membranes are available in a wide variety of materials, porosities, and pore sizes. For successful application, membrane materials must strike a balance between reasonable mechanical strength, high permeability, and high selectivity (Stephenson et al., 2000). For microporous membranes, high selectivity requires a membrane material with a narrow range of pore sizes, and high permeability requires a membrane material with a high porosity.

Dense and porous membranes are fundamentally different from each other. Dense membranes rely on physical–chemical interactions between the permeating compounds and the membrane materials. The mass transfer rate through a dense membrane depends on the solubility and the diffusivity of the permeating compound in the dense matrix (Reij et al., 1998). Dense membranes are limited to polymeric materials, such as latex and silicone, and they can be operated at high gas pressures, and are resistant to chemical and mechanical abrasion (Fitch et al., 2000; Stephenson et al., 2000). They have also been shown to be more resistant to biofouling than porous membranes (Coté et al., 1988, 1989), possibly because the hydrophobic nature of silicone resists attachment of microorganisms.

Microporous hydrophobic membranes are most often used in gas-transfer applications because they provide a high gas permeability, while not allowing transport of water across the membrane. The membrane pores remain gas filled and compounds transfer from the gas stream through the membrane pores by gaseous diffusion. At excess liquid side pressures above the critical pressure, ΔP_{cr}, water enters the pores of the membranes, dramatically decreasing mass transfer rates (Ergas and Reuss, 2001). Gas side pressures greater than the bubble point result in bubble formation in the liquid phase (Semmens et al., 1999). Thus, over the excess pressure range of 0 to ΔP_{cr}, the gas/liquid interface is immobilized at the mouth of the membrane pore on the liquid side.

Microporous hydrophobic membrane materials include polytetrafluorethylene (PTFE), polypropylene, Teflon™, Gore-Tex™ (PTFE/nylon), and other composites. Hydrophobic microporous membranes coated with an extremely thin layer of silicone have also been investigated (Sikar, 1992). The thin silicone layer increases mass transfer resistance but also decrease biofouling. Microporous hydrophobic membranes are available with pore diameters between 0.1 and 1.0 μm (Stephenson et al., 2000). The membranes are manufactured as small-diameter (200–400 μm ID) hollow fiber bundles that provide surface area-to-volume ratios as high as 30–100 cm^{-1} (Sirkar, 1992). This is an order of magnitude greater than equivalent sized packed towers.

Gas-to-liquid transfer can also be carried out using microporous hydrophilic membranes such as polysulfone and cellulose membranes. In these applications, gas pressure must be higher than liquid side pressures. The gas/liquid interface is thus immobilized at the pore mouth on the gas side of the membrane.

Control of biomass thickness has been shown to be a key operational consideration in continuously operated HFMBs for aeration of wastewaters (Stephenson et al., 2000) and biological waste gas treatment (Ergas and McGrath, 1997). Decreased HFMB performance has been observed in bioreactors after the development of a thick biofilm due to substrate mass transfer limitations, membrane fiber plugging, decreased biomass activity, and/or metabolite accumulation (Brindle and Stevenson, 1996; Freitas dos Santos et al., 1997). Hollow fiber bundles also tend to clump together when biofilm growth is high, resulting in fiber tangling and reduction of available membrane surface area. Several operational strategies have been used to maintain film thickness at an optimum level including the use of crossflow membrane configurations (Ahmed and Semmens, 1996) and periodic shearing of biomass from the membranes using high liquid velocities combined with scouring with gas bubbles (Pankhania et al., 1994; Dolasa and Ergas, 2000).

Research Overview

A summary of the laboratory-scale investigations of HFMBs for air pollution control is shown in Table 1. In an early study by Hartmans et al. (1992), a HFMB was used to control air emissions of toluene and dichloromethane. Mass transfer coefficients were determined for a number of different membrane materials. Using the experimentally determined mass transfer coefficients, dichloromethane removal was simulated. Results of simulations suggested that a significantly lower reactor volume would be required

TABLE 1 Summary of HFMB for Waste Gas Treatment Research

Reactor Type	Compound	Membrane Materials	Fiber ID (mm)	Pore Size (μm)	Membrane area (m^2)	References
–	Toluene, dichloromethane	Polypropylene	–	0.10	0.004	Hartmans et al., 1992
Flat sheet	Propene	Polypropylene	NA	0.1	0.0040	Reij et al., 1995
Hollow fiber	Propene	Polypropylene	1.8	0.2	0.10	Reij and Hartsmans, 1996
Hollow fiber	Toluene	Polysulfone	1.1	–	0.028	Parvatiyar et al., 1996a
Hollow fiber	Trichloroethene	Polysulfone	1.1	–	0.028	Parvatiyar et al., 1996b
Flat sheet	Propene	Polypropylene	NA	0.1	0.0040	Reij et al., 1997
Hollow fiber	Toluene	Polyethylene	0.28	–	0.23	Ergas and McGrath, 1997
Hollow fiber	Toluene	Polypropylene	0.20	0.05	0.37	Ergas et al., 1999
Hollow fiber	Toluene/TCE	Polypropylene	0.20	0.05	0.37	Dolasa and Ergas, 2000
Hollow fiber	Ammonia	Polyolefin multilayer	0.20	–	0.063	Keskiner and Ergas, 2001
Hollow fiber	Benzene	Latex rubber	9.5	NA	0.012	Fitch et al., 2000
Hollow fiber	Benzene	Silicone rubber	9.5	NA	0.012	Fitch et al., 2000
Hollow fiber	Benzene	Polypropylene	0.20	0.2	0.30	Fitch et al., 2000
Hollow fiber	Butanol	Polysulfone	1.1, 1.9, 2.7	0.05	0.030, 0.022, 0.013	Fitch et al., 2000
Hollow fiber	Trichloroethene	Polypropylene	0.24	0.03	0.70	Pressman et al., 2000

for a HFMB than for a biotrickling filter. Greater than 95% removal of toluene and dichloromethane were observed in experiments with a flat sheet membrane bioreactor.

Parvatiyar et al. (1996a) used a two module-in-series polysulfone HFMB to investigate toluene removal from a contaminated airstream. Toluene removal reached 84% with a 16-second gas residence time and an inlet concentration of 600 ppm. A similar experimental system was used by the authors to study degradation of trichloroethylene (TCE) (Parvatiyar et al., 1996b). The biofilm was initially acclimated to toluene and then gradually weaned from a toluene/TCE mixture to 100% TCE. A 30% TCE removal efficiency was achieved with a 36-second gas residence time.

Fitch et al. (2000) compared mass transfer and biodegradation rates for benzene and butanol contaminated gases in HFMBs that utilized dense (latex and silicone), microporous hydrophobic (polypropylene), and microporous hydrophilic (polysulfone) membranes. The highest overall pressure drops were observed with the polypropylene membranes due to the smaller diameter of these fibers compared with the other fibers. Significant sorption of benzene was observed in initial tests with the polysulfone membranes; therefore, butanol was used in subsequent experiments with these membranes. Removal efficiencies for butanol of up to 99% were obtained at an inlet concentration of 200 ppm. Dense high permeability latex and silicone membranes were found to have high benzene mass flux rates, possibly because the greater solubility of benzene in the polymers than in air creates a greater effective concentration gradient than observed in the air phase. The low total surface area of the dense membrane tubes limited overall removal; however, and the polypropylene membrane unit was the most effective on the basis of removal per total unit volume of reactor.

In my own laboratory, we have conducted a number of studies using toluene as a model VOC. In our first set of experiments (Ergas and McGrath, 1997), a laboratory-scale HFMB was constructed and operated with toluene at varying loading rates. The gasses passed from an inlet manifold to a membrane distributor made from 336 polypropylene hollow fibers (280 μm ID, 63% porosity, and an active fiber length of 1.1 m). A plot of removal efficiency vs. gas flow rate over the experimental period is shown in Figure 2. Toluene removal efficiencies of greater than 97% were achieved with an inlet toluene concentration of 100 ppm and gas flow rates less than 1.0 L/min (1.4 second residence time). When the gas flow rate was increased above 1.2 L/min (1.1-second residence time), a significant decrease in removal

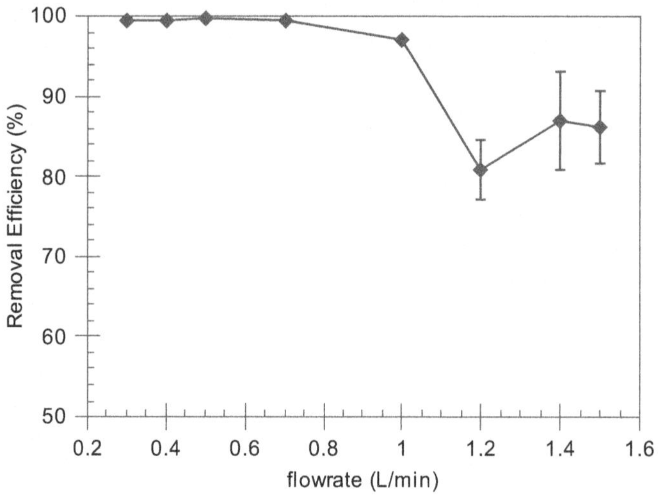

FIGURE 2 Removal efficiency in a HFMB gas flow rate. (From Ergas and McGrath, 1997.) Removal efficiency was determined after an acclimation period at each flow rate. The inlet toluene concentration was maintained at 100 ppm$_v$, and the liquid recirculation rate was 4 L/min.

efficiency was observed. Removal efficiency was found to decrease over the four-month operational period due to clogging of the bioreactor with microbial biomass, possibly due to the growth of nitrifying bacteria.

In subsequent experiments (Ergas et al., 1999), we investigated the effects of toluene loading rate, gas residence time, and liquid-phase turbulence on toluene removal in a laboratory-scale HFMB. Nitrate was used as a nitrogen source to discourage the growth of nitrifying bacteria. Initial acclimation of the microbial culture to toluene occurred over a period of nine days, after which a 70% removal efficiency was achieved at an inlet toluene concentration of 200 ppm and a gas residence time of 1.8 seconds (elimination capacity of 20 g/m^3/min). At higher toluene loading rates, a maximum elimination capacity of 42 g/m^3/min was observed. Liquid-phase recirculation rate had no effect on toluene removal in the HFMB.

We have also investigated the use of a HFMB to control ammonia using nitrifying bacteria (Keskiner and Ergas, 2001). The reactor utilized polyolefin multilayer membrane bundles consisting of 200 fibers, with a length of 50 cm, an inner diameter of 200 μm, and a porosity of 42%. Greater than 92% removal efficiency was obtained at an inlet ammonia concentration of 60 ppm and a gas residence time of less than 0.4 seconds. Ammonia mass transfer rates were found to increase at higher recirculation rates. Biomass adhesion and accumulation was not a problem in this system, possibly due to the slow growth rates of nitrifying bacteria. Experiments are currently being conducted with this reactor using nitrifying bacteria to remove nitric oxide from combustion gas streams.

Treatment of Low Solubility Compounds

In biological air pollution control systems, compounds must partition from the gas phase into the moist biofilm before they can be degraded. Pollutant concentrations at the gas/biofilm interface can be described by Henry's law:

$$S_L = C_g / H \tag{1}$$

where H is the Henry's law coefficient, C_g is the gas-phase concentration, and S_L is the liquid phase concentration. Conventional biofilters have therefore been limited to the control of relatively soluble compounds. A number of environmentally relevant compounds such as nitric oxide (NO), hexane, and methane are biodegradable but have high Henry's law constants. For these compounds, mass transfer from the gas phase to the biofilm limits removal unless very large reactor volumes are used. Due to their higher mass transfer rates, HFMBs may make biological treatment of these compounds more economically feasible.

A number of studies have been carried out using propene as a model VOC because its low solubility makes it difficult to remove in conventional biofilters (air/water partition coefficient at 25°C of 8.6). A flat sheet microporous polypropylene HFMB inoculated with *Xanthobacter Py2* was investigated by Reij et al. (1995). After five days with an inlet propene concentration of 2,300 ppm, the biofilm acclimated and 58% propene removal was maintained for the duration of the 30-day test. Because of propene's poor solubility, all mass transfer resistance was found to be in the liquid phase. For more soluble compounds, the authors determined that membrane-phase resistance could approach the same order of magnitude as liquid-phase resistance. A similar reactor system and bacterial culture was investigated by Reij et al. (1997) to degrade propene at concentrations varying from 10 to 1,000 ppm. Once a biofilm was established, propene flux to the membranes was stable, even at low concentrations (9–30 ppm) when mass transfer limitations should be greatest.

Reij and Hartsmans (1996) investigated a HFMB that utilized 40 polypropylene hollow fibers with a length of 500 mm, an inner diameter of 1.8 mm and a pore size of 0.2 μm. Propene was again used as a model compound due to its low solubility. Maximum propene removal rates were 70–110 $g/m^3/h$. A gas residence time of 80 seconds was required for 95% removal at an inlet propene concentration of 480 ppm. The reactor was changed from ammonia as a nitrogen source to nitrate to discourage the growth of nitrifying bacteria. Increasing the shell side velocity was found to alleviate clogging of the fibers with biomass. A gradual decrease in propene degradation was observed, possibly due to aging of the biofilm.

Cometabolism

A number of chlorinated organic compounds, such as TCE, can only be degraded aerobically under cometabolic conditions. Cometabolism is defined as the transformation of a compound by a microorganism that is unable to use the substrate as a source of energy or as an essential nutrient element. A second substrate (primary substrate) is used to support growth and induce the enzymes necessary to cometabolize the target compound. The primary substrates often inhibit metabolism of the target compound due to enzyme competition (Alvarez-Cohen and McCarty, 1991). In plug-flow reactors, such as biofilters, the induced bacteria are located near the inlet of the bioreactor where the primary substrate is readily degraded. In these areas, competitive inhibition occurs between the two compounds. In the remaining sections, the bacteria do not get sufficiently induced with the low primary substrate concentrations remaining to cometabolize the target compound. Microbial activity can also be inhibited by the toxicity of the target compound or its metabolites. The result is low removal efficiencies for TCE in conventional biofilters (Speitel and McLay, 1993).

Several researchers have studied HFMBs for cometabolism in TCE in wastewaters. Aziz et al. (1995) investigated a HFMB with an external semi-batch reactor for TCE cometabolism. Wastewater contaminated with TCE flowed inside the hollow fibers, with a liquid media containing an active methanotrophic culture circulating around the fibers. At residence times of 5–9 minutes in the fiber lumen, TCE conversions of 80–95% were observed. Pressman et al. (1999) presented a follow-up study with a similar system and methanol as the primary substrate. Over 93% of the transferred TCE was biodegraded.

Pressman et al. (2000) used a HFMB for cometabolism of TCE-contaminated gases. A pure culture of *Methylosinus trichosporium* OB3b PP358 was grown in a continuous-flow chemostat and circulated through the fiber lumen of a HFMB, while TCE-contaminated air was circulated on the shell side of the reactor. *M. trichosporium* OB3b PP358 is a methanotrophic bacterium that has the ability to rapidly cometabolize chlorinated solvents when grown on either methane or methanol. Between 54% and 84% TCE transfer was observed, and 92–96% of the transferred TCE was cometabolized at gas residence times of 1.6–5.0 minutes. Biomass clogging did not occur in this system, possibly because the biomass was pumped through the lumen rather than the shell side of the fibers.

In my own laboratory, we have investigated TCE cometabolism in HFMBs using toluene as a primary substrate (Dolasa and Ergas, 2000). A TCE cometabolizing culture enriched from a wastewater seed was inoculated into the HFMB. Initially, toluene was supplied to the reactor to build a sufficient biomass density on the fibers. After steady-state toluene removal was achieved, TCE was added to gas phase of the reactor. Toluene was added in three different configurations: (1) as a mixture with TCE in the gas phase, (2) by pulsing into the gas phase, or (3) to the liquid phase. Addition of a toluene/TCE mixture through the fibers resulted in an initial decrease in toluene removal followed by complete recovery within 5 days and a maximum TCE removal efficiency of approximately 30%. Pulsing of toluene and TCE through the membranes did not result in significant TCE removal. Adding TCE in the gas phase and toluene in the liquid phase resulted in a maximum TCE removal efficiency of 23%; however, results were highly variable and appeared to be related to liquid-phase toluene utilization rates and biofilm thickness.

Theoretical Models

A number of researchers have presented mathematical models of mass transfer and biodegradation of substrates by biofilms growing on the surfaces of gas-transfer membranes (Livingston, 1993; Essila et al., 2000). The general biofilm model reported here was developed in collaboration with Dr. Mark W. Fitch of the University of Missouri, Rolla, and tested using experimental data from HFMB studies conducted using toluene as a model VOC (Ergas et al., 1999). The model uses an inert surface to establish a boundary condition, with substrate entering the biofilm from the gas/liquid interface. The model was derived for a single lumen and related to the total removal by the number of fibers. A conceptual model of this system is shown in Figure 3. Model assumptions include steady-state operation; Monod biodegradation kinetics; and constant biomass density, ρ_b. Because concentration varies both axially and radially, no analytical solution exists for a single lumen. Therefore, the lumen is divided along the axis into n sections, each with an axial length, Δz. The influent gas concentration to the nth section is the concentration exiting the previous section, C_{n-1}. The concentration exiting the nth section, C_n, is equal to the influent less the removal in the section due to the mass transfer and biodegradation.

Membrane Mass Transfer

The flux, J_n, of substrate through the membrane can be expressed as

$$J_n = k_m A_m \left(C_{n,m} - C_{n-1}\right) \tag{2}$$

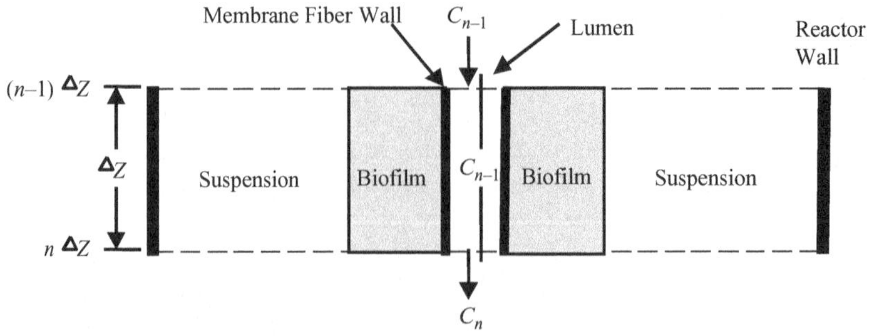

FIGURE 3 Conceptual model of the phases of the hollow fiber membrane reactor. (From Ergas et al., 1999.)

where k_m is the membrane mass transfer coefficient, A_m is the area of the membrane in the section, and $C_{n,m}$ is the gas-phase concentration on the outer face of the membrane. A number of authors have assumed that gas and membrane resistances are negligible compared to liquid-phase resistance (Ergas and McGrath, 1997; Yang and Cussler, 1986; Coté et al., 1989) and therefore $C_{n,m}$ is approximately equal to the concentration in the gas stream, C_{n-1}. Since the membrane is surrounded by biofilm, the liquid-phase concentration at the biofilm interface, $S_{n,0}$, can be related to C_{n-1} using Henry's law:

$$S_{n,0} = C_{n-1}/H \tag{3}$$

Equation 3 sets the inner surface (left-hand) boundary condition for the biofilm model.

Suspension Mass Transfer and Degradation

The suspension (liquid volume) was treated as a continuous-flow stirred tank reactor (CFSTR). The mass flux of the substrate, J_b, from the biofilm to the liquid can be described by

$$J_b = k_L A_B \left(S_{n,i} - S_L\right) \tag{4}$$

where k_L is the liquid mass transfer coefficient, A_b is the outer surface area of the biofilm, $S_{n,i}$ is the VOC concentration at the outer surface of the biofilm, and S_L is the bulk liquid VOC concentration. Assuming Monod biodegradation kinetics, a mass balance on the liquid volume yields:

$$\frac{\rho_L}{Y}\frac{\mu_{\max} S_L}{K_S + S_L} V = QS_o - QS_L + J_b \tag{5}$$

where V is the liquid volume, ρ_L is the biomass density in the liquid, $\mu_{\max}$ is the maximum specific growth rate, Y is the yield coefficient, K_S is the half saturation coefficient, Q is the liquid flow rate, and S_O is the influent VOC concentration to the suspension. Substituting Eq. (4) into Eq. (5) and solving for $S_{n,i}$ yields

$$S_{n,i} = \left(1 + \frac{V\rho_L\mu_{\max}}{Yk_L A_b (K_s + S_L)}\right)S - \frac{Q}{k_L A_b}(S_O - S_L) \tag{6}$$

This relationship sets the exterior surface (right-hand) boundary condition for the biofilm model.

Biofilm Mass Transfer and Degradation

Assuming no advection in the biofilm, steady-state ($\partial S/\partial t = 0$), no concentration gradient in the z direction, and Monod substrate utilization kinetics, the continuity equation for the biofilm in cylindrical coordinates (r, z, and θ) is

$$D_S \frac{1}{r}\left(\frac{1}{r}\frac{\partial^2 S}{\partial r^2} + \frac{\partial S}{\partial r}\right) = -\frac{\rho_b}{Y}\frac{\mu_{\max} S}{K_S + S} \tag{7}$$

where S is the substrate concentration in the biofilm and D_S is the VOC diffusion coefficient in the biofilm. There is no analytical solution for Eq. (7); therefore, a numerical solution was generated.

To fit the models, the biofilm density was varied until the predicted removal matched the observed removal (Ergas et al., 1999). The biofilm biomass density was calibrated at 29,000 mg/L, slightly higher than that reported by Characklis and Marshall (1990). The numerical model predicted the observed trend that shell side liquid flow rate had little effect on observed removal once a biofilm was established. The model slightly underpredicted the effect that substrate loading rate had on removal in the system.

The only case of poor prediction by the model was the effect of gas residence time on removal. Observed removals were higher than predicted removals for these runs. If the pore space of the membranes became water-filled during the course of experimentation, as discussed above, an added resistance to mass transfer would be expected, and mass transfer to the biofilm could become dependent upon the gas flow rate (gas-to-liquid transfer). Sensitivity analysis indicated that removal was a strong function of the biofilm phase biomass density and also of the biofilm diffusion coefficient, with diffusion rates below $(10)^{-9}\,m^2/s$ resulting in decreased removal rates.

Conclusions

Hollow fiber membrane bioreactors are a promising technology for the treatment of biodegradable gas phase pollutants. It has been shown to be effective for aerobic degradation of a range of compounds including ammonia, benzene, butanol, dichloromethane, propene, TCE, and toluene. Advantages of HFMBs include high mass transfer rates, low pressure drops, and small reactor volume requirements. In addition, the ability to separate the microbial population from the gases being treated allows for independent optimization of each phase of the system. In the opinion of this author, the most promising areas of air pollution research for HFMBs include

- Low solubility compounds.
- Cometabolism of chlorinated organic compounds.
- Compounds requiring specialized microbial populations or conditions.
- Indoor air applications.

The greatest disadvantages for the technology are high capital costs and that it has yet to be demonstrated at full scale. Due to their modular design; however, HFMBs should be relatively easy to scale up. EnviroGen, Inc. and the Medical University of South Carolina have developed a prototype with funding from the Department of Energy that they hope to field test in 2001 (Togna, 2000). Long-term studies have been conducted with membrane aeration bioreactors (Stephenson et al., 2000), which are similar in concept and operation. Membrane reactors have also been in use for several decades for blood oxygenation (Sikar, 1992) and separation and retention of biosolids (Brindle and Stephenson, 1996). In common with conventional biofilters, HFMBs have problems with excess biofilm growth and long-term stability of VOC-degrading biofilms. It is also unknown what effect long periods of association of membrane materials and biomass will have on mass transfer rates and mechanical strength in these systems.

Acknowledgments

I would like to thank my students Ayesha Dolasa, Carolyn Gendron, Yenner Keskiner, Michael McGrath, Fereshteh Mehmandoust, Andreas Reuss, and Leslee Shumway for their work on these projects. This material is based on work supported by the National Science Foundation (NSF). Any opinions, findings, conclusions, or recommendations expressed in this material are those of the authors and do not necessarily reflect the views of NSF.

References

Ahmed, T., Semmens, M.J. 1992a. Use of sealed end hollow fibers for bubbleless membrane aeration: experimental studies. *J. Membr. Sci.*, 69:1–10.

Ahmed, T., Semmens, M.J. 1992b. The use of independently sealed microporous hollow fiber membranes for oxygenation of water: model development. *J. Membr. Sci.*, 69:11–20.

Ahmed, T., Semmens, M.J. 1996. The use of transverse hollow fibers for bubbleless membrane aeration. *Water Res.*, 30(2):440–446.

Alvarez-Cohen, L., McCarty, P. 1991. Effects of toxicity, aeration, and reductant supply on trichloroethylene transformation by a mixed methanotrophic culture. *Appl. Envir. Microbiol.*, 57:228–235.

Aziz, C.E., Fitch, M.W., Linquist, L.K., Pressman, J.G., Georgiou, G., Speitel, G.E. 1995. Methanotrophic biodegradation of trichloroethylene in a hollow fiber membrane bioreactor. *Envir. Sci. Technol.*, 47:2574–2583.

Brindle, K., Stephenson, P. 1996. Mini review: the application of membrane biological reactors for the treatment of wastewaters. *Biotechnol. Bioengrg.*, 49(6):601–610.

Brindle, K., Stevenson, P. 1996. Nitrification in a bubbleless oxygen mass transfer membrane bioreactor. *Wat. Sci. Tech.*, 34(9):261–267.

Brindle, K., Stevenson, P., Semmens, M.J. 1998. Nitrification and oxygen utilization in a membrane aeration bioreactor. *J. Membr. Sci.*, 144:197–209.

Characklis, W.G., Marshall, K.C. 1990. *Biofilms*. Wiley, New York.

Coté, P., Bersillion, J.L., Huyard, A. 1989. Bubble-free aeration using membranes: mass transfer analysis. *J. Membr. Sci.*, 47:91–106.

Coté, P., Bersillon, J.L., Huyard, A., Faup, G. 1988. Bubble-free aeration using membranes: process analysis. *J. Water Pollution Control Fed.*, 60:1986–1992.

Dolasa, A.R., Ergas, S.J. 2000. Membrane bioreactor for cometabolism of trichloroethene air emissions. *J. Environ. Engr. ASCE.*, 126(10):969–973.

Ergas, S.J., McGrath, M.S. 1997. Membrane bioreactor for control of volatile organic compound emissions. *J. Envir. Engrg., ASCE*, 123:593–598.

Ergas, S.J., Reuss, A. 2001. Hydrogenotrophic denitrification of drinking water using a hollow fiber membrane bioreactor. *J. Water Supply: Res. Technol.-Aqua*, 50(3):161.

Ergas, S.J., Shumway, L., Fitch, M.W., Neeman J.J. 1999. Membrane processes for biological treatment of contaminated airstreams. *Biotechnol. Bioengrg.*, 63:431–441.

Essila, N.J., Semmens, M.J., Voller, V.R. 2000. Modeling biofilms on gas permeable supports: concentration and activity profiles. *J. Environ. Engr. ASCE.*, 126(3):250–257.

Fitch, M., Sauer, S., Zhang, B. 2000. Membrane biofilters: material choices and diurnal loading effects. *Proceedings USC-TRG Conference on Biofiltration*, University of Southern California, October 19–20, 83–90.

Freitas dos Santos, L.M., Pavasant, P., Strachan, L.F., Pistikopoulos, E.N., Livingston, A.G. 1997. Membrane attached biofilms for waste treatment – Fundamentals and applications. *Pure Appl. Chem.*, 69(11):2459–2469.

Gantzer, C. J. 1995. Membrane dissolution of hydrogen for biological nitrate removal. *68th Annual Conference and Exhibition, Water Environment Federation, Research Symposium*, Miami Beach, FL, 49–60.

Hartmans, S., Leenen, E.J.T.M., Voskuilen, G.T.H. 1992. Membrane bioreactor with porous hydrophobic membranes for waste-gas treatment. In *Biotechniques for Air Pollution Abatement and Odour Control Policies*; A.J. Dragt and J. van Ham, Eds. Elsevier Science Pub, B.V., Amsterdam.

Keskiner, Y., Ergas, S.J. 2001. Control of ammonia and NOx emissions using a nitrifying membrane bioreactor. *Proceedings 94th Annual Meeting of the Air & Waste Management Association*, Orlando, FL.

Lee, K.C., Rittmann, B.E. 2000. A novel hollow-fibre membrane biofilm reactor for autohydrogenotrophic denitrification of drinking water. *Water Sci. Technol.*, 41(4–5):219–226.

Livingston, A.G. 1993. A novel membrane bioreactor for detoxifying industrial wastewater: I. Biodegradation of phenol in a synthetically concocted wastewater. *Biotechnol. Bioengrg.*, 41(10):915–926.

Livingston, A.G. 1994. Extractive membrane bioreactors – a new process technology for detoxifying chemical-industry wastewaters. *J. Chemical Technol. Biotechnol.*, 60(2):117–124.

Livingston, A.G., Arcangeli, J.P., Boam, A.T., Zhang, S.F., Marangon, M., dos Santos, L.M.F. 1998. Extractive membrane bioreactors for detoxification of chemical industry wastes: process development. *J. Membr. Sci.*, 151(1):29–44.

Lund, L.W., Federspiel, W.J., Hattler, B.G. 1996. Gas permeability of hollow fiber membranes in a gas-liquid system. *J. Membr. Sci.*, 117:207–219.

Pankhania, M., Stephenson, T, Semmens, M.J. 1994. Hollow fibre bioreactor for wastewater treatment using bubbleless membrane aeration. *Water Res.*, 28(10):2233–2236.

Parvatiyar, M.G., Govind, R., Bishop, D.F. 1996a. Biodegradation of toluene in a membrane biofilter. *J. Membr. Sci.*, 119:17–24.

Parvatiyar, M.G., Govind, R., Bishop, D.F. 1996b. Treatment of trichloroethylene (TCE) in a membrane biofilter. *Biotechnol. Bioengr.*, 50:57–64.

Pressman, J.G., Georgiou, G, Speitel, G.E. 1999. Demonstration of efficient trichloroethylene biodegradation in a hollow-fiber membrane bioreactor. *Biotechnol. Bioengrg.*, 62:681–692.

Pressman, J.G., Georgiou, G., Speitel, G.E. 2000. A hollow-fiber membrane bioreactor for the removal of trichloroethylene from the vapor phase. *Biotechnol. Bioengr.*, 68(5):548–556.

Reij, M. W., de Gooijer, K.D., de Bont, J.A.M., Hartmans, S. 1995. Membrane bioreactor with a porous hydrophobic membrane as a gas-liquid contactor for waste gas treatment. *Biotechnol. Bioengr.*, 45:107–115.

Reij, M.W., Hamann, E.K., Hartmans, S. 1997. Biofiltration of air containing low concentrations of propene using a membrane bioreactor. *Biotechnol. Prog.*, 13:380–386.

Reij, M.W. and Hartsmans, S. 1996. Propene removal from synthetic waste gas using a hollow fiber membrane bioreactor. *J. Appl. Microbiol. Biotechnol.*, 45:730–736.

Reij, W.M., Keurentjes, J.T.F., Hartmans, S. 1998. Membrane bioreactors for waste gas treatment. *J. Biotechnol.*, 59:155–167.

Semmens, M.J., Gulliver, J.S., Anderson, A. 1999. An analysis of bubble formation using microporous hollow fiber membranes. *Water Environ. Res.*, 71(3):307–315

Shumway, L. 1997. A membrane process for biological treatment of contaminated airstreams. M.S. Thesis, Department of Civil and Environmental Engineering, University of Massachusetts, Amherst.

Sirkar, K.K. 1992. Other new membrane processes. In *Membrane Handbook*; W.S.W. Ho and K.K. Sirkar, Eds. Chapman & Hall, New York.

Speitel, G.E.; McLay, D.S. 1993. Biofilm reactors for treatment of gas streams containing chlorinated solvents. *J. Envir. Engrg., ASCE*, 119, 658–678.

Stephenson, T., Judd. S., Jefferson, B., Brindle, K. 2000. *Membrane Bioreactors for Wastewater Treatment*. IWA Publishing, London.

Togna, P. 2000. Personal communication. EnviroGen Inc. Lawrenceville NJ.

Yamagiwa, K., Yoshida, M., Ito, A., Ohkawa, A. 1998. A new oxygen supply method for simultaneous organic carbon removal and nitrification by a one-stage biofilm process. *Wat. Sci. Tech.*, 37(4–5):117–124.

Yang, M-C, Cussler, E.L. 1986. Designing hollow fiber contactors. *AIChE J.*, 32, 1910.

41

Review of Fine-Scale Air Quality Modeling for Carbon and Health Co-Benefits Assessments in Cities

Andrew Fang and
Anu Ramaswami

Introduction

Globally, cities exist as centers of population, economic activity, and innovation. Currently, cities around the world are grappling with serious air quality challenges that require action from local and national policymakers to protect the well being of urban residents. Many cities around the world have annual air pollution levels exceeding the WHO standards (Health Effects Institute, 2019), and globally, this excess ambient air pollution contributed to 2.9 million premature mortalities in 2017 (Stanaway et al., 2018). As policymakers attempt to reduce urban air pollution exposure, they are challenged by air pollution emission sources which exist outside the city (and which they may not be able to manage) and inequities in intracity air pollution levels due to pollution "hotspots".

Cities are also important centers of climate change action as evidenced by the over 9,000 cities that have committed to the Global Covenant of Mayors for Climate and Energy and various organizations (e.g., ICLEI and C40) which support cities in their efforts to mitigate and adapt to climate change. Because cities are embedded within larger infrastructure and trade networks, various tools have been developed to assess current community-wide emissions, project future greenhouse gas (GHG) emissions, and assess the impact of climate mitigation policies.

Because cities are concurrently trying to manage carbon and air pollution emissions, urban environmental policies should take into account the co-benefits of carbon mitigation, particularly as they relate to reducing local air pollution levels. Recent literature (Mayrhofer and Gupta, 2016; Nemet, Holloway, and Meier, 2010) has documented the challenges of quantifying co-benefits, but the health benefits of reducing air pollution are too large to ignore. The health benefits of the reduction in air pollutants (PM2.5) from both clean energy and carbon mitigation policies may outweigh the costs of carbon mitigation in the United States (Fowlie, 2018) and globally (Markandya et al., 2018). Further, we know these benefits will be heterogeneous (Muller and Mendelsohn, 2009; Goodkind et al., 2019), and we know they will be concentrated in cities (West et al., 2013). The heterogeneity and size of these health benefits means that they also have significant equity implications (Anderson et al., 2018; Cushing et al., 2018). It is therefore important to develop accurate estimates of these benefits at the urban scale, so that the size and distribution of health co-benefits of carbon mitigation are accurately reflected in environmental policies.

For cities grappling with air quality challenges, the first question is how can we reduce annual average air pollution levels and the number of acute pollution episodes? The answer to this question varies from city-to-city depending on the economic structure of the city and various physical and meteorological characteristics. In particular, cities must identify which sectors (industrial, transportation, residential, agriculture, etc.) contribute to their local air quality challenges and whether the sources of pollution are located within or outside the city. Once policies are identified that will prove beneficial to reducing exposure to local air pollution, the next question becomes how do we make sure these benefits are distributed fairly? Potential approaches to distributing these health benefits equitably are complicated by economic (e.g., Where are marginal benefits highest? Where are marginal costs lowest?) and political concerns (e.g., Which neighborhoods historically have high levels of pollution? Does a given policy reduce this burden?). Policymakers seek to address both sets of questions as they develop strategies for urban air quality management; therefore, co-benefit models need to account for emissions inside and outside the city, while assessing changes in air quality within the city.

Separately, cities have also been developing carbon emission inventories for local climate action planning and for reporting to the Global Covenant of Mayors in Climate and Energy. Carbon accounting of community-wide emissions for cities has become more standardized over the past few years as cities have begun to report emissions through the Global Protocol for Community-Scale Greenhouse Gas Emission Inventories (GPC) (Fong et al., 2014). Community-wide carbon emissions are reported across multiple spatial scales grouped by three categories:

- Scope 1 emissions: Emissions that physically occur within the city
- Scope 2 emissions: Energy-related emissions from the use of electricity, steam, and/or heating/cooling supplied by grids which may or may not cross city boundaries
- Scope 3 emissions: Emissions that physically occur outside the city but are driven by activities taking place within the city's boundaries.

Historically, cities have used production-based approaches to quantify Scope 1 and 2 emissions. Production-based approaches have historically been ill-defined, but many have used the term in reference to territorial emissions or emissions that are occurring geographically within the city boundary (Chen et al., 2019). Recent studies (Lin et al., 2015) have re-defined the term to account for the goods and services imported into the city that are used by industries/manufacturers to better account for the impact of industrial symbiosis and other interventions improving the efficiency of urban-industrial supply chains. But cities may account for different Scope 3 emissions depending on the methodology they are following. For example, many cities have used a consumption-based approaches (C40, 2018; Lin et al., 2017) which account for Scope 1, 2, and 3 emissions due to the production of goods and services that are consumed by residents of the city, but exclude emissions from exported goods and services that are produced by the city and consumed elsewhere. The production-based approach likely underestimates emissions in cities with high consumption activity (such as cities with tourist-based economies), while the consumption-based approach has limited application in cities with high industrial or manufacturing activity (Chavez and Ramaswami, 2013). None of these approaches is perfect, but all seek to better account for carbon emissions associated with city activity that are outside the geographic boundary, and it is therefore critical that there is transparency in the methods used to quantify these emissions.

For cities using consumption-based approaches, there may be a spatial mismatch between the emissions accounted for in the carbon emissions inventory and the local/regional emissions required for air pollution modeling.

Many models and methods have been developed to manage urban air pollution and carbon emissions, but few models have been able to address the nexus of carbon, air pollution, and health co-benefits within cities. Assessing co-benefits at the city scale requires sophisticated models that are able to account for the transboundary nature of the supply chains supporting city activity and the potential for non-local emissions to impact local air quality, while also having the spatial resolution to accurately assess intracity air pollution concentrations. This review assesses the state-of-the-science tools that have been developed to assess co-benefits at the city scale.

Overview of City-Relevant Co-Benefit Models

Models which are able to assess carbon, air pollution, and health co-benefits are summarized below with a focus on nested models. This list is not meant to be exhaustive but illustrates how carbon and air pollution modeling methods can be coupled. Table 1 demonstrates how the authors' view the scope and spatial scale of models that have the ability to assess co-benefits of carbon and air pollution emission reductions. The first type of models develop carbon and air pollution co-pollutant inventories but assign air pollution/health benefits directly to each unit of emission without using air pollution and transport modeling to assess changes in air pollution concentration and exposure (e.g., UrbanFootprint, 2013). A second type of model (e.g., AERMOD) utilizes similar community-scale emission inventories to mechanistically estimate how changes in local emissions impact local air pollution concentration and exposure levels based upon non-linear dynamics of air pollution transport.

Nested models, a third type of model, are most applicable to addressing city-scale co-benefits because they acknowledge the fact that city activity induces emissions outside the territorial boundary of the city. While the location of carbon emissions is not as important due to its global impacts, the location of air pollution emissions is critical to determining to associated impacts on air quality and pollution exposure. For example, SIM-Air (Guttikunda, Nishadh, and Jawahar, 2019; Guttikunda and Jawahar, 2012), GAINS (Wagner et al., 2018), COBRA (USEPA, 2018), and AP3 (National Research Council, 2010) are well-known models which utilize gridded emission inventories and regional estimates of air pollution transport and formation to predict changes in air quality/pollution exposure resulting from changes in emissions. Models like the Carbon Footprinting and Air Pollution Dispersion (CFAD) (Ramaswami et al., 2017) utilize community-wide carbon and air pollution inventories coupled with air pollution models (e.g., AERMOD) to distinguish between how changes in local and regional emissions may impact urban air quality and exposure. Lastly, nested-reduced form models, such as InMap (Tessum et al., 2019; Hill et al., 2019), have the potential to couple carbon and air pollution inventories at multiple scales (national, state, city) with reduced form air pollution models in order to estimate changing air quality and exposure at multiple scales resulting from changes in consumption of goods and services.

Air Pollution Models

In order to assess carbon and air pollution co-benefits, air pollution models are required to connect spatially explicit carbon and air pollutant inventories with air pollution formation and transport in the atmosphere. This is critical because of the non-linear relationship between air pollution emissions and air pollution concentrations (e.g., each unit of air pollutant emission may have a varied effect on concentration). Air pollution concentrations then become the unit of analysis for air quality standards, public health, and exposure estimates. The relationship between emissions and concentration may be governed by physical characteristics of the pollution source (e.g., stack height and toxicity of pollutants), non-linear dynamics of physical processes governing pollution in the atmosphere (e.g., transport, reaction, and deposition), and geographic characteristics of the exposure site (e.g., topography and wind patterns).

TABLE 1 Boundary of Carbon Emission Accounting and Air Pollution Models for City-Relevant Co-Benefit Models

		Boundary of Carbon Emission Accounting		
		Geographic (community-wide, Scope 1+2)	Geographic (community-wide, Scope 1+2+3)	Consumption-based (Scope 1+2+3)
Boundary of Air Pollution Modeling	Air pollution emission inventories	Urban Footprint		
	Community-scale air quality models	AERMOD		
	Nested local-regional-national models (source–receptor)	GAINS, SIM-Air, COBRA, AP3		
	Nested local-regional-national models (dispersion)		CFAD	
	Nested local-regional-national models (reduced form)			InMap

Note: Co-benefit models with the ability to couple across spatial scales are organized based upon the geographic boundaries of the carbon emission accounting and air pollution exposure.

To accurately assess the impact of these factors, air pollution models are needed to determine the dynamics of each unit of emissions once they enter the atmosphere. For city-level co-benefits modeling, these tools need to have the potential utilize urban and regional emissions inventories while evaluating air pollution concentration at a fine-grained spatial resolution so that intercity and intracity air pollution/health benefits can be estimated. This means that the air pollution models that are most applicable at the city scale need to balance data availability and accessibility with the computational complexity necessary to model air pollution transport and formation.

Given the uncertainty in modeling and the need to estimate air quality across varying spatial and temporal resolution, various methods exist for estimating air quality. Monitors are generally used to track air quality over time and to ensure compliance with air quality regulations; they also provide data validation for modeling results. Land use regression techniques use the historical data from monitors in comparison with land use variables (e.g., density) to predict air pollution levels with high spatial resolution. Recently, aerosol optical dispersion techniques (Di et al., 2017; van Donkelaar et al., 2016) have advanced considerably, using satellite data with increasingly high spatial resolution to estimate air pollution concentrations over time in locations lacking monitoring data.

While these methods are extremely sophisticated, they do not connect emissions reductions directly to concentration reductions. Therefore, these techniques are not able to directly inform policy design and implementation but are useful for empirical and historical air quality studies; for example, many studies have explored how health outcomes change before and after a policy change (e.g., Ebenstein et al., 2017). The models described below estimate air pollution concentrations and changes in exposure based on spatially resolved emissions inventories. This is critical to developing prospective scenarios that determine a priori the benefits of policies and interventions which reduce air pollution.

The challenge for air pollution models is to balance the complexity of modeling atmospheric processes that induce primary and secondary formation of pollutants with the spatial and temporal resolution necessary to answer policy-relevant questions, for example, where are high air pollution concentrations located or how can cities improve region-wide air quality? Urban air pollution concentration will be a function of the emissions (local and non-local) and the meteorological conditions (wind speed, wind direction, mixing height, etc.) that govern pollution formation at any given time. In practice, modelers must make the choice between model processing time and spatial resolution (Gilmore et al., 2019; Tessum, Hill, and Marshall, 2017). This tradeoff is very important for determining which models can assess the benefits of policy-induced emissions reduction at the city scale.

Application of Fine-Scale Air Pollution Models to Co-Benefits in Cities

Air pollution and transport models are tailored to specific spatial and temporal scales, which means they can estimate air pollutant concentrations across multiple spatial scales ranging from global to between buildings. Figure 1 illustrates the spatial dimensions of carbon and air pollution emission inventories compared to the scale at which air pollution models operate. The discussion below will focus on those tools that have potential to be applied to urban-scale co-benefits evaluation and policy scenarios.

Models that are able to address co-benefits at the urban scale need to be designed to utilize emissions data that distinguishes between local and non-local emissions, estimate changes in air pollution exposure within cities, and be customizable to policies that are specific to the jurisdiction in question. Not all air pollution models will be well suited to evaluate co-benefits at the city-wide scale, because many may be designed for alternate purposes such as regulatory compliance. The air pollution models which are most relevant need to differentiate between in-boundary and transboundary emissions so that urban and non-urban pollution control strategies can be evaluated. They must also account for the variation of inter- and intra-city air pollution exposure, so that the distribution of air pollution and health benefits can be assessed to determine which areas and populations will benefit most from air pollution control policies being considered. To systematically evaluate the air pollution and transport models that have the ability to estimate city-scale co-benefits, the following air pollution model characteristics need to be considered:

1. Primary/Secondary PM2.5 formation – Primary PM2.5 drives the local emissions component of urban PM2.5 concentrations because it consists of the particles directly emitted during energy combustion, industrial, and agricultural processes. In contrast, secondary PM2.5 is formed in the atmosphere by reactions between SOx, NOx, NH_3, etc. that may occur over larger provincial and regional areas.

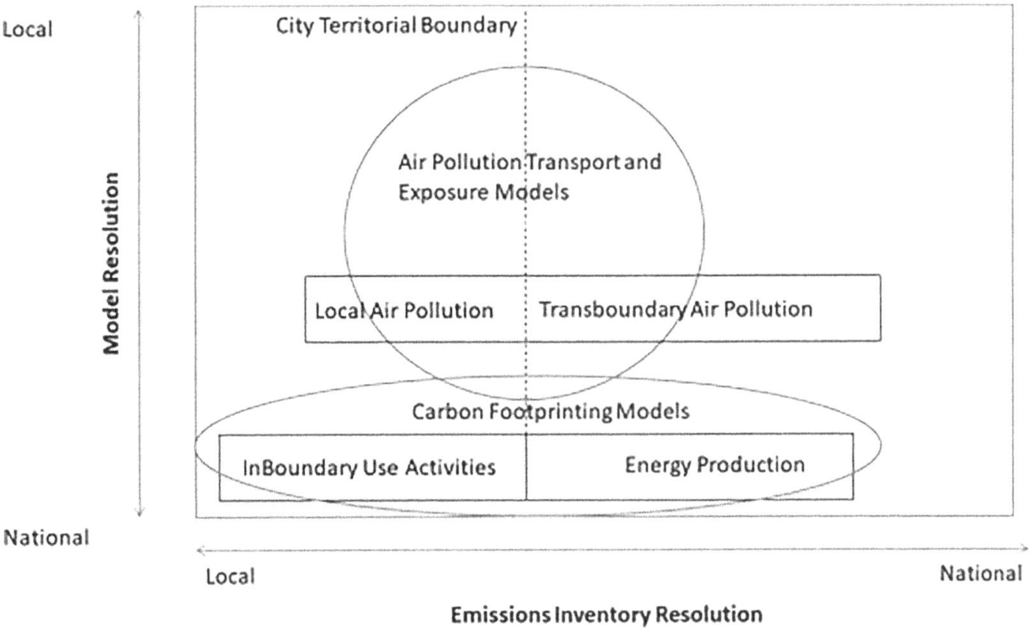

FIGURE 1 Spatial Resolution of Carbon Footprinting and Air Pollution Models – Geographic Scale of Carbon/Air Pollution Models shown as black ovals and geographic scale of emissions inventories shown as black boxes. Although carbon and air pollution emissions are aligned at the city scale, carbon and air pollution models evaluate outcomes at different scales because air pollution damages occur on a local-regional scale while climate damages occur on a global scale.

2. Spatial Resolution – Model inputs (emission inventories) and outputs (concentration estimates) will be set to a standardized or variable grid ranging from 0.1 to 50 km. The spatial resolution of these grids has implications for data availability and the model's ability to assess air pollution exposure within the city
3. Ground-Truthing and Relevance – Some models are comparable to real-time monitored air pollution levels, while others are intended to predict changes in air pollution levels based on changes in emissions. Recently, a number of reduced form tools (AP3, EASIUR, InMap) have been developed to quantify marginal benefits in pollution reduction. Critically, these reduced form tools do not claim to be more accurate than previous models, nor is their intention to be used for regulatory compliance. Instead, they focus on developing assessments of changes in pollution exposure due to policies that reduce local, regional, or national emissions.
4. Application – Certain air pollution models may be better suited for different policy-relevant questions. Models may be tailored to assess the health impacts of changes in emissions and exposure to air pollution, while others may be better suited for the identification of concentrated areas of pollution.

Although a variety of air pollution models exist, the following section is focused on the subset of models that can be readily applied to city-scale co-benefits. It is critical to note that the most state-of-the-art models may not be widely used by local policymakers due to data, computational, and technical capacity limitations. Reduced form models have been developed to make air pollution models more accessible and more policy-relevant by reducing computational effort and allowing for broader evaluation tailored to location-specific policies/interventions. The following three sections evaluate three types of air pollution models using the criteria above to determine which air pollution and transport models are best suited for different questions surrounding co-benefits at the urban scale.

Dispersion Models

- *Description*: Gaussian plume dispersion models (e.g., AERMOD) estimate air pollution levels that are downwind of individual sources or source groups. They are useful for predicting pollution impacts within cities due to nearby sources but are not recommended for predictions of long-range pollution transport (>50 km) (Cimorelli et al., 2005; USEPA, 2015).
- *PM2.5 Formation*: Primary PM2.5 only; assumes that increased concentration in cities is mainly due to local sources.
- *Spatial Resolution*: Fine-scale resolution (<1 km) enables assessment of intracity air pollution exposure.
- *Application*: Used for regulatory purposes to determine near-source impacts of large emitters, but generally cannot estimate secondary formation of PM2.5 or predict long-range pollution transport.
- *Co-Benefits Examples*: Applied at the local to regional scale, can be nested with carbon footprinting models (e.g., CFAD) (Ramaswami et al., 2017).

Chemical Transport Models

- *Description*: Chemical transport models (CTMs) are three-dimensional mechanistic models that predict ambient concentrations of pollutants using mass balance principles accounting for emissions, transport, dispersion by winds, chemical transformations, and atmospheric removal processes. CTMs are the most scientifically sophisticated and rigorous tools available for linking emissions to ambient concentrations (e.g., CMAQ, WRF-Chem) (Byun and Ching, 1999; Grell et al., 2005).
- *PM2.5 Formation*: Primary and Secondary PM2.5 including gridded emission inventory of all PM2.5 precursors.
- *Spatial Resolution*: Coarse (>4 km) due to model complexity, computational processing time, and availability of nationwide fine-scale emissions inventory data (US-based models utilize the triennial National Emissions Inventory).

TABLE 2 Summary of Air Pollution Models for Urban Application

	Increasing Complexity		
	Reduced Form Models	Air Pollution Dispersion	Chemical Transport Models
Example	InMap, AP3	AERMOD	CMAQ, WRF-Chem
Contribution	Less computationally intensive, predictions of marginal damages	High spatial resolution, intracity variability (identification of pollution hotspots)	State-of-the-art physical and chemical transport process model, most complete spatial/temporal coverage
Limitations	Limited temporal coverage, limited comparability to observed concentrations	Only primary PM2.5, not recommended for long-range transport	Computationally intensive, coarse spatial scale`

Note: In order of increasing computational complexity, this table provides examples, contribution, and limitations of air pollution models that can be applied to urban co-benefits.

- *Application*: Can be compared to real-time air pollution monitoring and used for regulatory impact assessment, but limited spatial and temporal resolution.
- *Co-Benefits Examples*: Applied at the National or Regional Scale (Thompson et al., 2016; Zhang et al., 2016).

Reduced Form Models

- *Description*: To improve the availability and accessibility of state-of-the-science CTM air quality modeling and cost estimates, the air quality research community has recently created a new set of simplified models, known as reduced form or reduced-complexity models, to estimate the marginal social costs of air pollution in terms of monetized health damages per pollutant (Gilmore et al., 2019). These models estimate marginal social costs with nested local-regional-national emission inventories.
- *PM2.5 Formation*: Variable – AP3 (Primary PM2.5 only), InMap (Primary and Secondary PM2.5 Formation).
- *Spatial Resolution*: InMap (1–48 km) but coarser than dispersion models.
- *Application*: Able to model prospective policy scenarios and sensitivity to model/scenario uncertainties due to lower computational intensity. These models are meant to estimate marginal air quality and health benefits based upon the changes in air quality that can be attributed to changes in emissions; may be limited in their comparability to real-time pollution data.
- *Co-Benefits Examples*: Applied at multiple scales – national (Markandya et al., 2018), regional (Muller and Mendelsohn, 2009), and local (Goodkind et al., 2019; Paolella et al., 2018) (Table 2).

In the US context, air pollution dispersion and CTMs have historically been used by state and national regulators to ensure regulatory compliance and perform regulatory impact assessments. These models on their own have robust applications to urban air quality management strategies, but are not necessarily aligned with local analyses that seek to address carbon, air pollution, and health co-benefits. Dispersion and reduced form models have already been integrated with city-scale carbon accounting methods due to their ability to incorporate nested emission inventories at multiple spatial scales. Local policymakers and practitioners should use these tools or tools with similar functionality to quantify the magnitude and distribution of carbon, air pollution, and health co-benefits

Conclusion

Reducing air pollution has significant benefits both globally and locally. Environmental policies that are designed and implemented at the local-regional scale must take into account both the magnitude and distribution of these health benefits because of their implications on cost-benefit analyses and developing political support for local environmental policies. Models developed to date have been limited in their ability to measure urban carbon and air pollution co-benefits. First, it has been challenging to acquire data on activity and emissions delineated between local sources and regional/national sources of pollution (>50 km from city). Second, the complexity of state-of-the-art models leads to limitations in terms of the software/equipment, computational time, and technical capacity needed to apply these tools towards the development of local air pollution and carbon policy. Further, these state-of-the-art models have been designed for state/national level assessments and may not have the spatial resolution necessary to estimate urban air pollution/health benefits.

A variety of commercially available models have been developed to assess how carbon emissions will respond to changes in land use, technology, and policy/behavior changes at the urban scale. Certain models also attempt to measure air pollution benefits of these changes using linear marginal benefits estimates (UrbanFootprint, 2013) or neural networks to predict air quality changes based on weather, traffic, and air pollution sensor data (Siemens, 2019). While these estimates may have some predictive value, these models have limited ability to inform the design and implementation of local policies that impact air pollution. Models that inform policy should estimate both the change in criteria air pollutant emissions *and* the changes in local air pollution concentrations due to the non-linear relationship between air pollution emissions reduction and air pollution exposure reduction. This is critical because of the heterogeneity that exists in the marginal benefits of reducing one unit of air pollution and because these benefits will be higher within cities but distributed unequally throughout the city. These marginal benefits need to be estimated at high spatial resolution (e.g., sub-county scale) because there may be an order of magnitude difference within counties (Goodkind et al., 2019).

The recent development of reduced-complexity models is a promising alternative, recognizing the importance of estimating the marginal benefits of reductions in air pollution with high spatial resolution and limited computational complexity (Gilmore et al., 2019). These models combine the ability to estimate the impact of both local/long-range pollution sources on air quality levels within cities with the granularity to estimate variations in air pollution exposure at the sub-county level. As these models develop and become used in discussions around equity (Tessum et al., 2019; Tessum, Hill, and Marshall, 2017), it is important to assess the relationship between mechanistic and empirical models.

The spatial resolution of relevant co-benefits models is a critical variable to assess, particularly when considering equity implications of the distribution of air pollution/health benefits (Paolella et al., 2018). Because carbon and air pollution emission inventories also take place across nested spatial scales (city, region, national), models addressing co-benefits must also have the ability to differentiate between urban and non-urban emission reductions. Models which can couple spatially explicit emission inventories with regional air pollution transport methods are best suited for assessing co-benefits at the city scale. If these models can combine regional air pollution transport with high spatial resolution at the sub-county scale, then they have the ability to inform environmental policy design so that the health benefits of reducing air pollution are both optimal and distributed more equitably.

Given the uncertainties in underlying local emissions, exposure, and demographic data, policymakers must acknowledge two key truths. First, not all emissions reductions are created equal; improving air quality across urban areas requires an understanding of the implications of emitter location and physical characteristics of local and non-local emitters. Second, the marginal health benefits of reducing local air pollution will not be distributed uniformly across the city, so policy needs to be designed to account for the presence of concentrated areas of pollution, particularly when addressing multi-pollutant policies (e.g., policies and interventions which reduce multiple pollutants, such as carbon and criteria air pollutant emissions). The complex nature of air pollution and atmospheric transport has resulted in multiple models that may be able to serve different co-benefits-related questions. Moving forward, it is critical that the modeling community develops tools that are suited to the questions that policymakers are asking so that environmental policy design is being informed by the best available science. These tools need to have the ability to utilize local and regional emissions data across multiple spatial scales spanning the city boundary, while also measuring air pollution exposure variation within the city. The existing suite of tools is not perfect but can be used to evaluate the magnitude and distribution of carbon, air pollution, and health co-benefits in cities. Policymakers can use these tools for prospective analyses to determine which carbon and air pollution policy options may have the most health benefits at the city scale and then utilize other empirical methods involving satellite-derived and air quality monitor data to evaluate how these policies change air quality and health outcomes over time.

References

Anderson, C. M., Kissel, K. A., Field, C. B., and K. J. Mach. 2018. "Climate Change Mitigation, Air Pollution, and Environmental Justice in California." *Environmental Science & Technology* 52 (18): 10829–38. https://doi.org/10.1021/acs.est.8b00908.

Byun, D. W., and J. K. S. Ching. 1999. "Science Algorithms of the EPA Models-3 Community Multiscale Air Quality (CMAQ) Modeling System." *United States Environmental Protection Agency* 44 (6): 1765–78. https://nepis.epa.gov/Exe/ZyPDF.cgi/30003R9Y.PDF?Dockey=30003R9Y.PDF.

C40. 2018. "Consumption-Based GHG Emissions of C40 Cities." https://www.c40.org/researches/consumption-based-emissions.

Chavez, A., and A. Ramaswami. 2013. "Articulating a Trans-Boundary Infrastructure Supply Chain Greenhouse Gas Emission Footprint for Cities: Mathematical Relationships and Policy Relevance." Energy Policy 54: 376–84. https://doi.org/10.1016/j.enpol.2012.10.037.

Chen, G., Shan, Y., Hu, Y., Tong, K., Wiedmann, T., Ramaswami, A., Guan, D., Shi, L. and Y. Wang. 2019. "Review on City-Level Carbon Accounting." *Environmental Science & Technology* 53 (10): 5545–58. https://doi.org/10.1021/acs.est.8b07071.

Cimorelli, A. J., Perry, S. G., Venkatram, A., Weil, J. C., Paine, R. J., Wilson, R. B., … R. W. Brode. 2005. "AERMOD: A Dispersion Model for Industrial Source Applications. Part I: General Model Formulation and Boundary Layer Characterization." *Journal of Applied Meteorology* 44 (5): 682–93. https://doi.org/10.1175/JAM2227.1.

Cushing, L., Blaustein-Rejto, D., Wander, M., Pastor, M., Sadd, J., Zhu, A., and R. Morello-Frosch. 2018. "Carbon Trading, Co-Pollutants, and Environmental Equity: Evidence from California's Cap-and-Trade Program (2011–2015)." *PLOS Medicine* 15 (7): e1002604. https://doi.org/10.1371/journal.pmed.1002604.

Di, Q., Wang, Y., Zanobetti, A., Wang, Y., Koutrakis, P., Choirat, C., Dominici, F., and J.D. Schwartz. 2017. "Air Pollution and Mortality in the Medicare Population." *New England Journal of Medicine* 376 (26): 2513–22. https://doi.org/10.1056/NEJMoa1702747.

Ebenstein, A., Fan, M., Greenstone, M., He, G., and M. Zhou. 2017. "New Evidence on the Impact of Sustained Exposure to Air Pollution on Life Expectancy from China's Huai River Policy." *Proceedings of the National Academy of Sciences* 114 (39): 10384–89. https://doi.org/10.1073/pnas.1616784114.

Fong, W. K., Sotos, M., Doust, M., Schultz, S., Marques, A., and C. Deng-Beck. 2014. "Global Protocol for Community-Scale Greenhouse Gas Emission Inventories." http://www.iclei.org/fileadmin/user_upload/ICLEI_WS/Documents/Climate/GPC_12-8-14_1_.pdf.

Fowlie, M. 2018. "Air Pollution Co-Benefits Matter." Energy Institute Blog. 2018. https://energyathaas.wordpress.com/2018/11/13/air-quality-matters/.

Gilmore, E. A., Heo, J., Muller, N. Z., Tessum, C. W., Hill, J. D., Marshall, J. D., and P.J. Adams. 2019. "An Inter-Comparison of Air Quality Social Cost Estimates from Reduced-Complexity Models." Environmental Research Letters. http://iopscience.iop.org/10.1088/1748-9326/ab1ab5.

Goodkind, A. L., Tessum, C. W., Coggins, J. S., Hill, J. D., and J. D. Marshall. 2019. "Fine-Scale Damage Estimates of Particulate Matter Air Pollution Reveal Opportunities for Location-Specific Mitigation of Emissions." *Proceedings of the National Academy of Sciences* 116 (18): 8775–80. https://doi.org/10.1073/pnas.1816102116.

Grell, G. A., Peckham, S. E., Schmitz, R., McKeen, S. A., Frost, G., Skamarock, W. C., and B. Eder. 2005. "Fully Coupled 'Online' Chemistry within the WRF Model." Atmospheric Environment 39 (37): 6957–75. https://doi.org/10.1016/j.atmosenv.2005.04.027.

Guttikunda, S. K., and P. Jawahar. 2012. "Application of SIM-Air Modeling Tools to Assess Air Quality in Indian Cities." Atmospheric Environment 62: 551–61. https://doi.org/10.1016/j.atmosenv.2012.08.074.

Guttikunda, S. K., Nishadh, K. A., and P. Jawahar. 2019. "Air Pollution Knowledge Assessments (APnA) for 20 Indian Cities." *Urban Climate* 27: 124–41. https://doi.org/10.1016/j.uclim.2018.11.005.

Health Effects Institute. 2019. "State of Global Air 2019." http://www.stateofglobalair.org.

Hill, J. D., Goodkind, A., Tessum, C. W., Thakrar, S., Tilman, D., Polasky, S., … J.D. Marshall. 2019. "Air-Quality-Related Health Damages of Maize." *Nature Sustainability* 2 (5): 397–403. https://doi.org/10.1038/s41893-019-0261-y.

Lin, J., Hu, Y., Cui, S., Kang, J., and A. Ramaswami. 2015. "Tracking Urban Carbon Footprints from Production and Consumption Perspectives." *Environmental Research Letters* 10 (5): 1–12. https://doi.org/10.1088/1748-9326/10/5/054001.

Lin, J., Hu, Y., Zhao, X., Shi, L. and J. Kang. 2017. "Developing a City-Centric Global Multiregional Input-Output Model (CCG-MRIO) to Evaluate Urban Carbon Footprints." Energy Policy 108: 460–66. https://doi.org/10.1016/j.enpol.2017.06.008.

Markandya, A., Sampedro, J., Smith, S. J., Van Dingenen, R., … M. González-Eguino. 2018. "Health Co-Benefits from Air Pollution and Mitigation Costs of the Paris Agreement: A Modelling Study." *The Lancet Planetary Health* 2 (3): e126–33. https://doi.org/10.1016/S2542-5196(18)30029-9.

Mayrhofer, J. P., and J. Gupta. 2016. "The Science and Politics of Co-Benefits in Climate Policy." Environmental Science and Policy 57: 22–30. https://doi.org/10.1016/j.envsci.2015.11.005.

Muller, N. Z., and R. Mendelsohn. 2009. "Efficient Pollution Regulation: Getting the Prices Right." *American Economic Review* 99 (5): 1714–39. https://doi.org/10.1257/aer.99.5.1714.

National Research Council. 2010. Hidden Costs of Energy: Unpriced Consequences of Energy Production and Use - Appendix C. Environmental Forum. Vol. 27. Washington, DC: National Academies Press. https://doi.org/10.17226/12794.

Nemet, G. F., Holloway, T., and P. Meier. 2010. "Implications of Incorporating Air-Quality Co-Benefits into Climate Change Policymaking." *Environmental Research Letters* 5 (1): 014007. https://doi.org/10.1088/1748-9326/5/1/014007.

Paolella, D. A., Tessum, C. W., Adams, P. J., Apte, J. S., Chambliss, S., Hill, J., … J. D. Marshall. 2018. "Effect of Model Spatial Resolution on Estimates of Fine Particulate Matter Exposure and Exposure Disparities in the United States." *Environmental Science & Technology Letters* 5 (7): 436–41. https://doi.org/10.1021/acs.estlett.8b00279.

Ramaswami, A., Tong, K., Fang, A., Lal, R. M., Nagpure, A. S., Li, Y., … S. Wang. 2017. "Urban Cross-Sector Actions for Carbon Mitigation with Local Health Co-Benefits in China." Nature Climate Change 7 (10): 736–42. http://dx.doi.org/10.1038/nclimate3373.

Siemens. 2019. "City Air Management Tool." 2019. https://new.siemens.com/global/en/company/topic-areas/intelligent-infrastructure/city-performance-tool.html.

Stanaway, J. D., Afshin, A., Gakidou, E., Lim, S. S., Abate, D., Abate, K. H., … C. J. L. Murray. 2018. "Global, Regional, and National Comparative Risk Assessment of 84 Behavioural, Environmental and Occupational, and Metabolic Risks or Clusters of Risks for 195 Countries and Territories, 1990–2013: A Systematic Analysis for the Global Burden of Disease Study 2017." *The Lancet* 392 (10159): 1923–94. https://doi.org/10.1016/S0140-6736(18)32225-6.

Tessum, C. W., Apte, J. S., Goodkind, A. L., Muller, N. Z., Mullins, K. A., Paolella, D. A., … J. D. Hill. 2019. "Inequity in Consumption of Goods and Services Adds to Racial–Ethnic Disparities in Air Pollution Exposure." *Proceedings of the National Academy of Sciences* 116 (13): 6001–6. https://doi.org/10.1073/PNAS.1818859116.

Tessum, C. W., Hill, J. D., and J. D. Marshall. 2017. "InMAP: A Model for Air Pollution Interventions." *PLoS ONE* 12 (4): e0176131. https://doi.org/10.1371/journal.pone.0176131.

Thompson, T. M., Rausch, S., Saari, R. K., and N. E. Selin. 2016. "Air Quality Co-Benefits of Subnational Carbon Policies." *Journal of the Air & Waste Management Association* 66 (10): 988–1002. https://doi.org/10.1080/10962247.2016.1192071.

UrbanFootprint. 2013. "Technical Documentation Model Version 1.0." https://urbanfootprint.com/.

USEPA. 2015. Revision to the Guideline on Air Quality Models: Enhancements to the AERMOD Dispersion Modeling System and Incorporation of Approaches to Address Ozone and Fine Particulate Matter. https://www3.epa.gov/ttn/scram/11thmodconf/9930-11-OAR_AppendixW_Proposal.pdf.

USEPA. 2018. "User's Manual for the Co-Benefits Risk Assessment (COBRA) Screening Model." https://www.epa.gov/statelocalenergy/users-manual-co-benefits-risk-assessment-cobra-screening-model.

van Donkelaar, A., Martin, R. V., Brauer, M., Hsu, N. C., Kahn, R. A., Levy, R. C., ... D.M. Winker. 2016. "Global Estimates of Fine Particulate Matter Using a Combined Geophysical-Statistical Method with Information from Satellites, Models, and Monitors." *Environmental Science & Technology* 50 (7): 3762–72. https://doi.org/10.1021/acs.est.5b05833.

Wagner, F., Borken-Kleefeld, J., Kiesewetter, G., Klimont, Z., Nguyen, B., Rafaj, P., and W. Schopp. 2018. "The GAINS PMEH-Methodology - Version 2.0."

West, J. J., Smith, S. J., Silva, R. A., Naik, V., Zhang, Y., Fry, M. M., Anenberg, S., ... J.-F. Lamarque. 2013. "Co-Benefits of Mitigating Global Greenhouse Gas Emissions for Future Air Quality and Human Health." *Nature Clim. Change* 3 (10): 885–89. https://doi.org/10.1038/NCLIMATE2009.

Zhang, Y., Bowden, J. H., Adelman, Z., Naik, V., Horowitz, L. W., Smith, S. J., and J. J. West. 2016. "Co-Benefits of Global and Regional Greenhouse Gas Mitigation for US Air Quality in 2050." *Atmospheric Chemistry and Physics*. https://doi.org/10.5194/acp-16-9533-2016.

42

Thermal Energy: Solar Technologies

Muhammad Asif
and Tariq Muneer

Introduction

The sun is a source of energy on Earth, and in a sense, it is a source of life, as energy is the most important commodity for life besides providing light and heat to the planet. The reaction between the sun's energy and Earth's atmosphere determines weather patterns and rainfall, and Earth's tilt towards the sun creates the seasons. Its role in photosynthesis helps plants to grow and in biodegradation complete the natural cycle of ecosystems. The sun also sources several other forms of energy on the planet: wind power depends on the sun's impact on atmospheric movement as it creates wind patterns; through photosynthesis, sun contributes to bioenergy (wood and other organic materials); and fossil fuels indirectly owe their creation millions of years ago to solar energy.[1]

Solar energy is one of the most promising renewable technologies. It is abundant, inexhaustible, environmentally friendly, and widely available. Solar energy has the potential not only to play a very important role in providing most of the heating, cooling, and electricity needs of the world, but also to solve global environmental problems. Solar energy can be exploited through solar thermal and solar photovoltaic (SPV) routes for various applications. While SPV technology enables direct conversion of sunlight into electricity through semiconductor devices called solar cells, solar thermal technologies utilize the heat energy from the sun for a wide range of purposes.

Solar thermal technologies are quite diverse in terms of their operational characteristics and applications—they include fairly simple technologies such as solar space heating and solar cooking as well as complex and sophisticated ones like solar air conditioning and solar thermal power generation. Solar thermal technologies have also a broad bandwidth in terms of their economical standing. Solar water heating and solar space heating, for example, are very cost-effective and are regarded among the most economical renewable energy technologies, while high-temperature technologies such as solar thermal power generation and solar air conditioning are on the higher economic bandwidth. Solar thermal technologies on the basis of their working temperature can be classified into the following three types:

- Low-temperature technologies (working temperature <70°C)—solar space heating, solar pond, solar water heating, and solar crop drying.
- Medium-temperature technologies (70°C< working temperature <200°C)—solar distillation, solar cooling, and solar cooking.
- High-temperature technologies (working temperature >200°C)—solar thermal power generation technologies such as parabolic trough, solar tower, and parabolic dish.

In the coming sections, this entry provides an overview of these technologies, briefly highlighting their technology fundamental and operating principles.

Solar Space Heating

Solar energy can be used to accomplish heat for comfort in buildings. The application, referred to as solar space heating, is used to optimize the reduction of auxiliary energy consumption in such a way that minimum overall cost is obtained. In combination with conventional heating equipment, solar heating provides the same levels of comfort, temperature stability, and reliability as conventional systems. A building that includes some arrangement to admit, absorb, store, and distribute solar energy as an integral part is also referred to as a solar house. A solar space-heating system can consist of a passive system, an active system, or a hybrid of the two.

Passive Space Heating

In passive space heating, the buildings are designed or modified so that they independently capture, store, and distribute solar heat throughout the building without using any electrical or mechanical equipment. Inherently flexible passive solar design principles typically accrue energy benefits with low maintenance risks over the life of the building. The design does not need to be complex, but it does involve knowledge of solar geometry, window technology, and local climate. Passive solar heating techniques generally fall into three categories: direct solar gain, indirect solar gain, and isolated solar gain, as shown in Figure 1.

Direct Solar Gain Design

Direct gain passive designs typically have large windows with predominately equatorial aspects. In this design, solar radiation directly penetrates and is stored in the building's inherent thermal mass, in materials such as concrete, stone floor slabs, or masonry partitions that hold and slowly release heat.

Indirect Solar Gain Design

In indirect solar passive design, a glazed heat collector, also referred to as Trombe wall, collects and stores solar radiation during the day. A Trombe wall consists of an 8–16″ thick masonry wall coated with a dark, heat-absorbing material and covered by a single or double layer of glass, placed from about

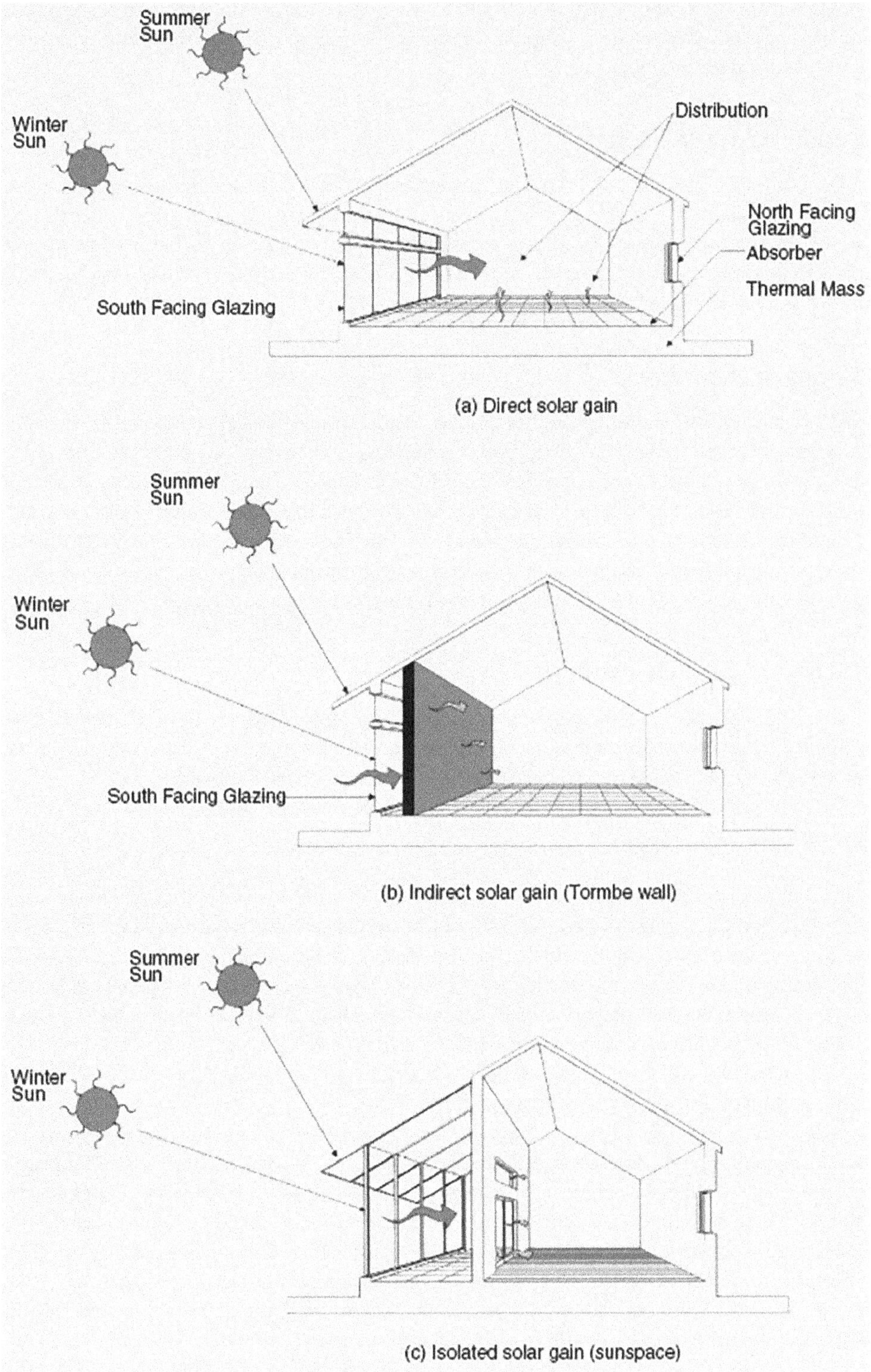

FIGURE 1 Passive solar space-heating principles. a-c captions adequately describe the relevant techniques, same is the case with Figure 6.

¾″ to 6″ away from the masonry wall. Heat from the sun is stored in the air space between the glass and the dark material, and conducted slowly to the interior of the building through the masonry through the conduction and convection mechanisms.

Isolated Solar Gain Design

In isolated solar passive systems, an extra highly glazed unheated room—a sun-space or conservatory—is added to the south side of the house. Solar gains always make sun-spaces warmer than the outside air, and this reduces heat losses from the house and warms any ventilation air that passes through the sun-space. When solar gains raise the sun-space above house temperature, the heat collected can be let into the house by opening communicating doors and windows.[2]

Active Space Heating

In active space heating of buildings, additional electrical and mechanical equipment is incorporated to circulate solar heated water or air. The main components of an active system are the heat collectors, storage tanks or pebble bed storage, heat exchangers, heat emitters, fans/pumps, connecting pipes or ducts, and controls. Active solar heating systems can be designed to provide the same levels of control of condition in the heated spaces as conventional systems. With indoor temperature essentially fixed at or little above a minimum, load estimations can be done by conventional methods. Passively heated buildings in many cases are not controlled within the same narrow temperature ranges.[3]

Hybrid Solar Space Heating

Solar space-heating system can also be of hybrid nature, combining both the passive and active modes. For example, in a hybrid system, a roof-space collector accomplishes passive collection of solar energy that can be actively distributed in the house using a fan and associated ductwork.

Solar Water Heating

Water heating is an essential feature of energy requirements in industrial and commercial sectors in general and in domestic sector in particular. A solar water heater consists of two main elements—the collector and the water storage tank, which, respectively, have the functions of absorbing solar radiation and transferring it to the water, and storing the water for usage. The collectors in solar water heaters can be broadly classified into two categories—flat plate and evacuated tube. A flat plate collector consists of an absorber plate that absorbs solar energy, while a glazing above it is used to reduce convective heat loss. An evacuated tube collector consists of tubes with vacuum maintained between the tubes and glazing for better protection against convective heat loss.

Solar water heaters come in three main types: thermosyphon, built-in-storage, and forced circulation. There are two operating principles for solar water heaters: passive system, which relies on natural circulation of water (such as thermosyphon, built-in-storage types); and active system, which uses an external element such as an electric pump to circulate the water (such as the forced circulation type). Another criterion that distinguishes solar water heaters is the way they transfer heat to water. Again, there are two types: direct system, in which the collector itself transfers heat to water; and indirect system, in which a heat-transfer fluid, circulating in collector in a closed loop, transfers heat to water through a heat exchanger. Figure 2 shows an indirect active solar water heater.

The efficiency of a solar water heater depends upon its design and the available solar radiation. In this entry, solar water heating has been classified as a low-temperature thermal technology because most of its application is in residential sector where it operates at a temperature of ≤70°C. Also, in industrial applications, solar water heaters are used as preheaters and to hold supply water at almost the

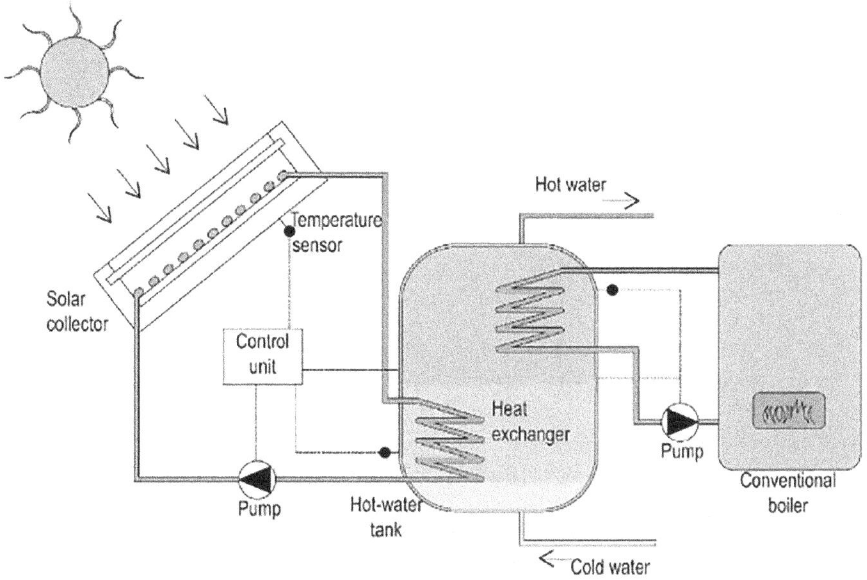

FIGURE 2 Indirect active solar water heating system.

same temperature for further heating by conventional means. Solar water heaters are a cost-effective technology—the payback period of solar water heaters can be as low as 3 years while having a service life of more than 20 years. Owing to its technical and economical viability across the world, solar water heating is one of the most established and efficient application of solar energy. Among solar thermal technologies, solar water heating holds the greatest market share and the highest market growth rate.

Solar Ponds

Solar ponds are naturally occurring salt gradient lakes that collect and store solar energy. A solar pond contains salt water with increasing concentrations of salt, hence the density of the solution. When solar radiation is absorbed, the density gradient prevents heat in the lower layers from moving upward by convection and leaving the pond. This results in an increased temperature at the bottom of the pond and a near atmospheric temperature at the top of the pond. The phenomenon of solar ponds was first discovered in 1902 by von Kalecsinsky, who reported that the Medve Lake in Transylvania, containing nearly saturated NaCl solution at a few meters depth with almost fresh water at its surface, had a bottom temperature of 70°C.

A solar pond has three distinctive zones. The top layer is the surface zone that has a low salt content and is at atmospheric temperature. It is also called the upper convective zone (UCZ), as shown in Figure 3. The bottom layer has a very high salt content and is at a high temperature, 70°C–90°C. This is the zone that collects and stores solar energy in the form of heat, and it is called the lower convective zone (LCZ). There is an intermediate insulating zone with a salt gradient. It establishes a density gradient that prevents heat exchange by natural convection, and hence, it is called the nonconvective zone (NCZ). In this zone, salt content increases with depth, creating salinity.

Solar ponds can be broadly classified into two main types: nonconvective and convective. In nonconvective solar ponds, the heat loss to environment is reduced by suppressing natural convection normally by using salt stratification. While in convective ponds, heat loss to environment is reduced by covering the pond surface with a transparent material. The heat trapped in the solar ponds can be used for many different purposes, such as industrial process heating, the heating of buildings, desalination, and to drive a turbine for generating electricity.

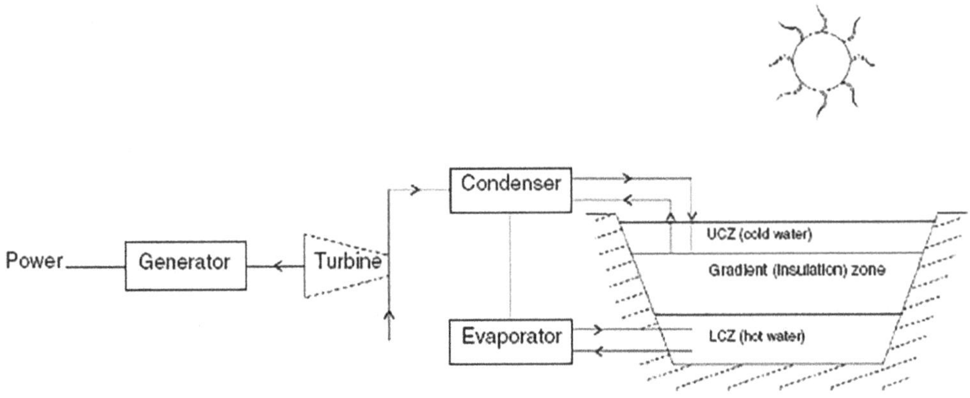

FIGURE 3 Schematic of solar pond power generation system.

The first artificial solar pond was developed in Israel in 1958. Since then, many countries such as Australia, the United States, China, India, Iran, Italy, and Mexico have constructed solar ponds, mostly for research and development purposes. During the last decade, significant success in operational practices and applications of solar pond technologies has been achieved.[4]

Solar Crop Drying

Drying is the oldest technique used to preserve food. Until around the end of the 18th century when canning was developed, drying was virtually the only method of food preservation. Solar energy is the main driving force that utilizes warm air to dry food. In drying, the moisture from the food is reduced to a certain level—as low as 5–25% depending on the type of food-to prevent decay and spoilage in an environment free of contaminations such as dust and insects. Successful drying depends on[5]

- Enough heat to draw out moisture, without cooking the food
- Dry air to absorb the released moisture
- Adequate air circulation to carry off the moisture.

Solar drying can be carried out in open air under the sun by simply spreading the material on a clean surface or in particularly designed solar dryers. Solar dryers, however, exhibit many advantages over open air drying. First, solar dryers are more efficient because they require lesser drying time and area. Second, the product is protected from rain, insects, animals, and dust, which may contain fecal material. Third, faster drying reduces the likelihood of mold growth. Fourth, higher drying temperatures mean that more complete drying is possible, and this may allow much longer storage times (only if rehumidification is prevented in storage). Finally, more complex types of solar driers allow some control over drying rates. Solar dryers can be made in many different designs depending upon various factors, i.e., the type of produce, scale of operation, and local economical and environmental conditions. In terms of their operational mode, solar dryers can be broadly classified into two main types, active and passive dryers, which can both be further subclassified into direct (in which the produce is directly heated from sun) and indirect types (in which the produce is not directly exposed to sun).

Almost all types of food—for example, vegetables, fruits, milk, herbs, spices, meat, and fish—can be dried by solar energy. The advantages of solar food drying are numerous. Dried foods, for example, are tasty, nutritious, lightweight, easy to prepare, and easy to store and use. The energy input is less than what is needed to freeze or can, and the storage space is minimal compared with that needed for canning jars and freezer containers.

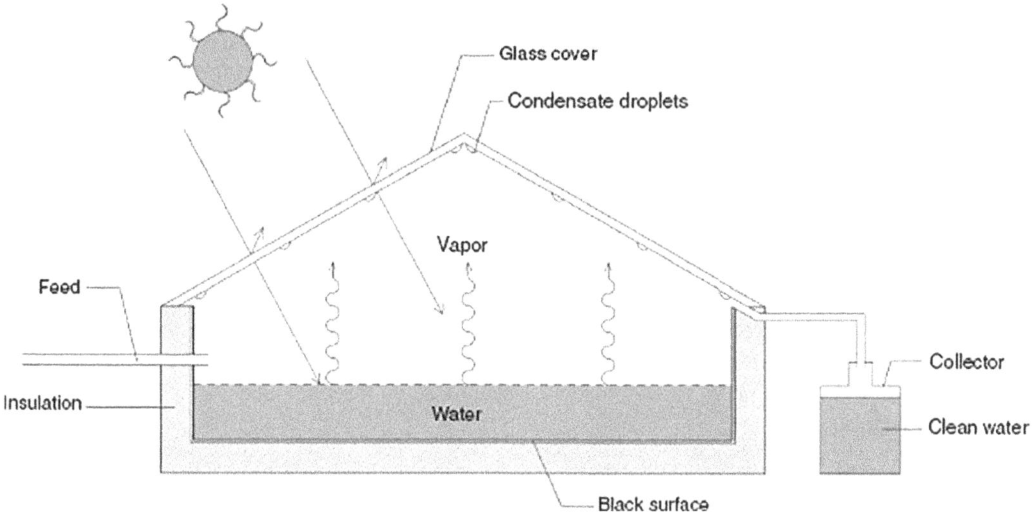

FIGURE 4 Schematic diagrams of a solar still.

Solar Distillation

Solar distillation is a process that utilizes solar energy to purify water through evaporation and condensation processes. The process is also referred to as water desalination when solar energy is used to purify water from saline water. Solar water distillation is a solar technology with a very long history. Installations were built over 2,000 years ago, although they were to produce salt rather than drinking water. Documented use of solar stills (the distillation unit) began in the 16th century. An early large-scale solar still was built in 1872 that spread over an area of 4,600 m^2 capable of producing 23,000 L of drinking water for a mining community in Chile. Mass production occurred for the first time during the World War II when 200,000 inflatable plastic stills were made to be kept in life-crafts for the US Navy.[6] In addition to their use in obtaining drinking water, solar stills are also suitable for the production of distilled water if there is appreciable demand for it in industry, laboratories, and medical facilities or to fill lead acid batteries.

Solar stills come in different designs; however, the main features of operation are the same for all of them. In its simple form, water can be placed in an airtight basin that has a sloped transparent cover normally made of glass or plastics, although glass is preferred for its high transparency. The basin is coated with a black lining to maximize absorption of solar radiation. The incident solar radiation is transmitted through the glass cover and is absorbed as heat by the black surface in contact with the water to be distilled. The water is thus heated and gives off water vapor. The vapor condenses on the glass cover, which is at a lower temperature because it is in contact with the ambient air, and runs down into a tray where it is fed to a storage tank, as shown in Figure 4. The economic viability of solar stills is determined to a critical degree by the design, the construction, the materials employed, and the local market conditions.

Solar Cooking

A solar cooker or solar oven harnesses solar energy to cook food. The solar cooker was first developed by a Swiss scientist Horace de Saussure in 1767.[7] Solar cookers are now being used in many countries across the world, especially in remote areas of poor countries. Solar cookers accomplish free cooking with environment friendliness as they only capitalize solar energy. Solar cooking can be very helpful in reducing the deforestation and pollution that originate from consumption of wood, and animal and agricultural residues for cooking in remote areas that lack access to electricity and gas. Solar cookers are capable of performing various types of cooking phenomena, i.e., frying, baking, and boiling. The maximum achievable temperature depends on the intensity of the available solar radiation and the design and size of the solar cooker. Solar cookers come in a wide range of designs, which can be categorized under the following three major types.

Solar Box Cookers

A solar box cooker consists of an insulated box with a transparent top and a reflective lid. It is designed to capture solar radiation and make use of the greenhouse effect to cause heat to accumulate inside. The top is removable to allow food pots to be placed inside. Temperatures in a typical box cooker can reach above 200°C, but the temperatures achieved obviously depend on the size and design parameters of the cooker and the location of use.

Solar Panel Cookers

The solar panel cooker is the simplest solar cooker, and it consists of multiple simple reflectors arranged to focus solar radiation onto a covered black pot enclosed in a clear heat-resistant plastic bag or other transparent enclosure, such as glass bowel.

Solar Parabolic Cookers

Parabolic solar cookers, also called concentrated cookers, consist of a concave disk that focuses the light onto the bottom of a pot that is arranged at the focal length of the disk, as shown in Figure 5. These are the most efficient types of solar cookers.

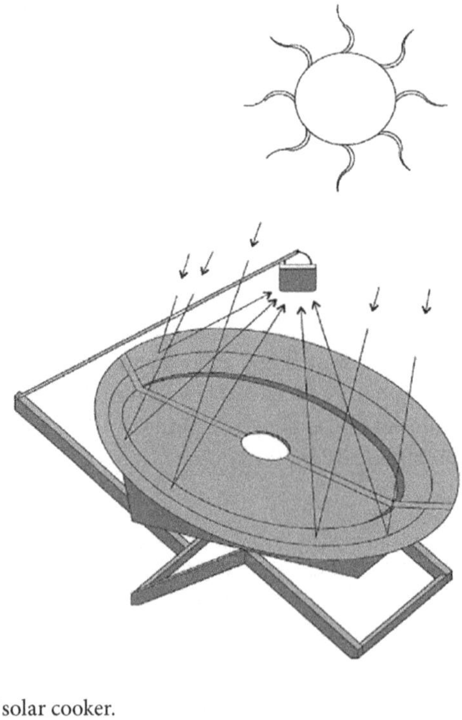

FIGURE 5 Parabolic type of solar cooker.

TABLE 1 Overview of Processes for Thermally Powered Cooling and Air Conditioning

Solar Thermal Cooling System Design	Adsorption Refrigeration	Absorption Refrigeration	Desiccant Air Conditioning
Solar collector	Vacuum tube collector, flat plate collector	Vacuum tube collector	Flat plate collector, solar air collector
Coolant circulation process	Closed refrigerant circulation systems	Closed refrigerant circulation systems	Open refrigerant circulation systems (in contact with the atmosphere)
Process basic principle	Cold water production	Cold water production	Air dehumidification and evaporative cooling
Sorbent type	Solid	Liquid	Solid
Refrigerant/sorbent	Water–silica gel ammonia–salt	Water–water–lithium bromide, ammonia–water	Water–silica gel water–lithium chloride–cellulose
Typical operating temp.	60°C–95°C	80°C–110°C (one step) 130°C–160°C (two step)	45°C–95°C

Solar Cooling/Air Conditioning

Solar thermal energy can be used for cooling and dehumidification. Collectors play a critical role in extracting the energy from solar radiation to operate the cooling device. The collectors used in solar thermal cooling could be of various types, such as low-temperature flat plates and high-temperature evacuated tubes and concentrators. The basic principle behind solar thermal cooling is the thermo-chemical process of sorption—a liquid or gaseous substance is either attached to a solid, porous material (adsorption) or taken in by a liquid or solid material (absorption). The heat-transfer fluid is heated in the solar collectors to a temperature well above ambient and used to power a cooling device—a type of heat-actuated pump. The heat-transfer fluid may be air, water, or another fluid; it can also be stored in a hot state for use during times of no sunshine. Heat extracted by the cooling device from the conditioned space and from the solar energy source is rejected to the environment using ambient air or water from a cooling tower.[8]

The solar thermal cooling process can be broadly classified under open cycle systems and closed cycle systems. Open cycle systems are those in which the refrigerant is in direct contact with atmosphere and is discarded from the system after providing the cooling effect and new refrigerant is supplied in an open-ended loop. In closed systems, on the other hand, the refrigerant is not in direct contact with the atmospheric air. Open and closed cycle systems can further be distinguished according to the type of sorbent used, which can be in a liquid or a solid form. The three main designs of solar thermal cooling technologies that have gained the most attraction include solar adsorption, solar absorption, and solar desiccant. The key features of these designs are provided in Table 1.[9]

Solar Thermal Power Generation

Solar thermal power generation systems start with capturing heat from solar radiation. Direct solar radiation can be concentrated and collected by a range of concentrating solar power technologies to provide medium- to high-temperature heat. This heat then operates a conventional power cycle—for example, through a steam turbine or a Stirling engine to generate electricity. Solar thermal power plants can be designed for solar-only or hybrid operation, where some fossil fuel is used in case of lower radiation intensity to secure reliable peak-load supply. Five distinct solar thermal power generation concepts are available:

- Solar pond
- Solar chimney

TABLE 2 Characteristics of Typical Concentrated Solar Collectors

Solar Collector Technology	Typical Operating Temperature (°C)	Concentration Ratio	Tracking	Maximum Conversion Efficiency (Carnot) (%)
Solar Fresnel reflector technology	260–400	8–80	One-axis	56
Parabolic trough collectors	260–400	8–80	One-axis	56
Heliostat field + central receiver	500–800	600–1000	Two-axis	73
Paraboloidal dish concentrators	500–1200	800–8000	Two-axis	80

- Solar parabolic trough
- Solar central receiver or solar tower
- Solar parabolic dish.

Solar pond and solar chimney are nonconcentrated types of technology. In this section, the three concentrated types of solar thermal technologies—solar parabolic trough, solar central receiver, and solar parabolic dish—are discussed, as they have received the greater degree of attention over the years due to their favorable technical and commercial characteristics. These technologies can be used to generate electricity for a variety of applications, ranging from remote power systems as small as a few kilowatts (kW) up to grid-connected applications of 200–350 megawatts (MW) or more.[10]

Solar thermal power generation systems have three essential elements needed to produce electricity: a concentrator (to collect and focus solar radiation), a receiver (to convert concentrated solar radiation into heat), and an engine cycle (to generate electricity). Some systems also involve a transport or storage system. Solar collectors have a crucial role to play in the whole system and can be mainly classified into two types: concentrating and nonconcentrating. They are further categorized on the basis of their concentrator optical properties and the operating temperature that can be obtained at the receiver. Most of the techniques for generating electricity from heat need high temperatures to achieve reasonable efficiencies. Concentrating systems are hence used to produce higher temperatures. Table 2 shows the operational characteristics of concentrated collectors.[11]

Parabolic Trough

The parabolic trough systems consist of large curved mirrors or troughs that concentrate sunlight by a factor of 80 or more onto thermally efficient receiver tubes placed in the trough's focal line, as shown in Figure 6a. A thermal transfer fluid, such as synthetic thermal oil, is circulated in the tubes at focal length. Heated to approximately 400°C by the concentrated sun's rays, this oil is then pumped through a series of heat exchangers to produce superheated steam.[12] The steam is converted to electrical energy in a conventional steam turbine generator, which can either be part of a conventional steam cycle or integrated into a combined steam and gas turbine cycle, as shown in Figure 7. Parabolic trough power plants are the only type of solar thermal power plant technology with existing commercial operating systems.

It is also possible to produce superheated steam directly using solar collectors. This makes the thermal oil unnecessary and also reduces costs because the relatively expensive thermo oil and the heat exchangers are no longer needed. However, direct solar steam generation is still in the prototype stage.

Central Receiver or Solar Tower

In solar thermal tower power plants, hundreds or even thousands of heliostats (large individually tracking mirrors) are used to concentrate sunlight onto a central receiver mounted at the top of a tower, as indicated in Figure 6b. A heat-transfer medium in this central receiver absorbs the highly concentrated radiation reflected by the heliostats and converts it into thermal energy to be used for the subsequent

(a) Parabolic trough

(b) Central receiver or solar tower

(c) Parabolic dish (Stirling engine)

FIGURE 6 Solar thermal power generation technologies.

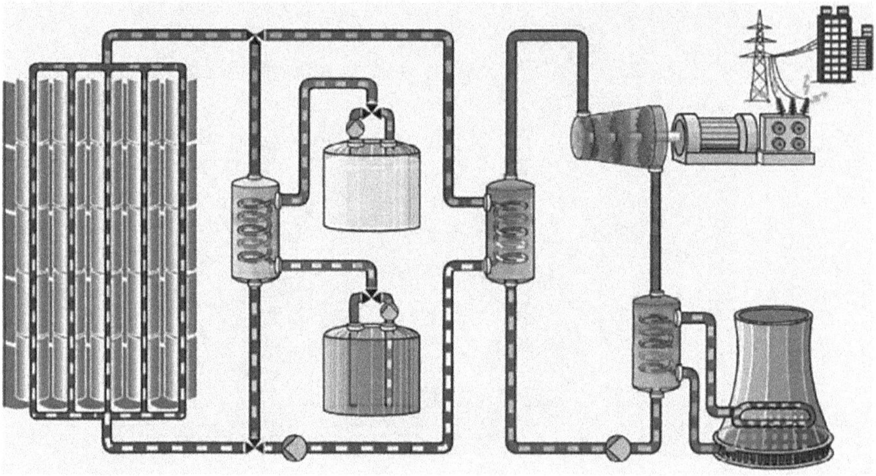

FIGURE 7 Schematic of solar parabolic trough power plant.

generation of superheated steam for turbine operation. To date, the heat-transfer media demonstrated include water or steam, molten salts, liquid sodium, and air. If pressurized gas or air is used at very high temperatures of about 1000°C or more as the heat-transfer medium, it can even be used to directly replace natural gas in a gas turbine, thus making use of the excellent cycle of modern gas and steam combined cycles.

Parabolic Dish

A parabolic dish system uses a parabolic concave mirror to concentrate sunlight onto a receiver located at the focal point of the mirror, as highlighted in Figure 6c. The concentrated beam radiation is absorbed into the receiver to heat a fluid or gas (air) to approximately 750°C. This fluid or gas is then used to generate electricity in a small piston, Stirling engine, or a microturbine attached to the receiver. These systems stand alone, and they are normally used to generate electricity in the kilowatts range.[13]

Solar Thermal Technologies—Market Growth and Trends

Solar thermal technologies, like other renewables, are experiencing a rapid growth. Between 2007 and 2017, for example, the global installed capacity of solar water heaters increased from 145 GW_{th} to 472 GW_{th}. The year 2017 saw an addition of 35 GW_{th}. China alone accounts for over 71% of the world's total solar water heating installed capacity. Other leading countries include Turkey, India, Brazil, the USA, and Germany. In terms of collector type, while glazed flat plate collectors are the preferred choice around the world, in China and India glazed evacuated tube collectors account for over two-thirds of the market. Of the new installations in 2017, vacuum tube collectors had a share of 73%. Flat plate collectors and unglazed collectors accounted for respective shares of 23% and with 4%.[14]

Over the past decade, concentrated solar power (CSP) has experienced a rapid growth —between 2007 and 2017 the global installed capacity increased from 0.45 to 4.9 GW, with around 2 GW of projects being under construction. Spain has emerged as the global leader over this period installing 2.3 GW of CSP projects. Over 80% of the world's total installed capacity is in Spain and the USA, the two countries having respective figures of 2.3 and 1.7 GW. In terms of the ongoing developments, China is having 20 projects—including parabolic trough, tower, and Fresnel facilities—at various stages of construction, with a combined capacity of 1 GW. The Middle East and North Africa (MENA) region, being rich in the direct solar radiation, has shown a significant interest in the technology. United Arab Emirates (UAE), having already installed 100 MW of CSP, has awarded tender for 700 MW (600 MW parabolic basin complex and a 100 MW solar tower) of projects. Morocco is having two projects with total capacity of 350 MW at advanced stages of completion. Projects are also underway in Kuwait, Israel, and Saudi Arabia with respective capacities of 50, 121, and 93 MW. Projects in Chile and Australia of respective capacity 110 and 100 MW are also under construction. Driven by factors like competition and technology cost reduction, the year 2017 has also witnessed record-low tariffs: AUD78/MWh in Australia, USD50/MWh in Chile, and USD73/MWh in UAE.[14]

Solar thermal technologies are regarded to be playing an important role in the future energy scenarios. A scenario of what could be achieved by the year 2025 was prepared by Greenpeace International, the European Solar Thermal Industry Association, and International Energy Agency (IEA) SolarPACES projects. It suggested that by 2025, the total installed capacity of solar thermal power around the world will reach over 36 GW. It is also projected that by 2040, more than 5% of the world's electricity demand may be satisfied by solar thermal power.[12]

Conclusions

Solar thermal technologies operate by converting solar radiation into heat, which can be either directly utilized in various applications such as solar space heating, solar water heating, and solar air conditioning, or can be transformed into electricity to serve any purpose similar to conventional electricity. The key element in all solar thermal technologies is the collector, whose function is to gather the heat of solar radiation. Collectors normally come in three different types: flat plate, evacuated tube, and concentrated, and they operate in a wide range of temperatures, i.e., from less than 50°C to more than 1200°C. Solar thermal technologies normally operate in passive or active modes. Different types of solar thermal technologies are gaining huge attention across the world depending upon their technical and economic viability. Solar thermal power generation is also expected to grow at a healthy rate in coming years, as it is projected that by 2040, more than 5% of the world's electricity demand could be satisfied by solar thermal power.

References

1. Asif, M. *Energy Crisis in Pakistan: Origins, Challenges and Sustainable Solutions*; Karachi: Oxford University Press, 2011.
2. Muneer, T. Solar energy. KEMPS Engineering Year Book; John H Stephens (Ed.) Miller Freeman: New York, 2000.
3. Duffie, J.; Beckman, W. Solar Engineering of Thermal Processes, 2nd Ed.; Wiley: New York, 1991.
4. Akbarzadeh, A.; Andrews, J.; Golding, P. Solar pond technology: A review and future directions. Advances in Solar Energy; Yogi Goswami (Ed.) Earthscan: London, 2005.
5. Whitfield, D.E. Solar drying. *International Conference on Solar Cooking*, South Africa, November 26–29, 2000.
6. Solar distillation. Technical Brief, Intermediate Technology Development Group. http://www.itdg.org/?idZtechnical_briefs (accessed on March 2006).
7. History of Solar Cooking, https://www.solarcooker-at-cantinawest.com/solarcooking-history.html (accessed on 15 May, 2000).
8. Grossman, G. Solar cooling, dehumidification, and airconditioning, solar thermal power generation. In Encyclopedia of Energy, Cutler Cleveland (Ed.) Vol. 5; Elsevier: Amsterdam, 2004.
9. Asif, M., and Muneer, T. Thermal Energy: Solar Technologies, Encyclopaedia of Environmental Management; Taylor & Francis: New York, 2013; Vol. IV, 2526-2535.
10. Concentrating Solar Power: Energy from Mirrors, report Produced by NREL 2001. Available at: http://www.nrel.gov/docs/fy01osti/28751.pdf.

11. Luzzi, A.; Lovegrove, K. Solar thermal power generation. In Encyclopedia of Energy, Vol. 5; Cutler Cleveland (Ed.) Elsevier: Amsterdam, 2004.
12. Concentrated solar thermal power—now! Report by European Solar Thermal Industry Association and Greenpeace, 2005 http://www.greenpeace.org/raw/content/international/press/reports/Concentrated-Solar-Thermal-Power.pdf.
13. Quaschning, V. Solar thermal power plants. Renewable Energy World, June 2003.
14. Renewables 2018 Global Status Report, Renewable Energy Policy Network for the 21st Century, Paris.

VI

PRO: Basic Environmental Processes

43

Acid Rain

Umesh Kulshrestha

Introduction

Any form of precipitation (rain, snow, or hail) having high acidity is known as acid rain. The term "acid rain" was first used by Robert Angus Smith in his book *Air and Rain: The Beginnings of a Chemical Climatology*, published in England in 1872.[1] He had chemically analyzed the rainwater near Manchester and observed three types of rain composition—"that with carbonate of ammonia in the fields of distance, that with sulfate of ammonia in the suburbs and that with sulfuric acid or acid sulfate, in the town."

In broader perspectives, acid rain refers to wet deposition (rain, snow, hail, cloud water, fog, dew, or sleet) and dry deposition (absorption of SO_2, NO_x, other acidic gases and particles) of acidic compounds. High acidity is generally caused by higher levels of sulfuric and nitric acids. These acids are contributed by their precursor gases (SO_2 and NO_2), which are emitted by natural as well as anthropogenic sources. Natural sources include volcanoes, vegetation decay, various biological processes on the land, and oceans, while major anthropogenic sources of these gases are fossil fuel combustion and smelting of metal ores. In regions of North America, the rates of anthropogenic emissions of these two gases have gone up to 100 times more than the natural rates, adding to higher atmospheric acidity.[2] Acid rain has caused severe damage in Europe, North America, parts of China and Japan through acidification of lakes and other water bodies, decline of forests, acidification of soils, and corrosion of building materials.

How Acid Rain Happens

Pure water (H_2O) is neutral in nature, having a pH value of 7. Any aqueous solution having a pH higher than 7 is said to be alkaline, while one having a pH lower than 7 is known as an acidic solution. Rainwater in remote and unpolluted atmospheres has a slightly acidic pH of around 5.6 due to the presence of carbonic acid formed at equilibrium due to dissolution of atmospheric carbon dioxide in cloud water:[3]

$$CO_2 + H_2O \leftrightarrow H_2CO_3 \quad (1)$$

In water, carbonic acid is dissociated, forming bicarbonate ion:

$$H_2CO_3 \leftrightarrow HCO_3 - + H^+ \quad (2)$$

Rainwater pH is further depressed to about 5.2 in unpolluted regions by organic acids. However, anthropogenic acid rain arises due to oxidation of SO_2 and NO_2 in the atmosphere to form sulfuric and nitric acids. There are a number of probable reactions for the oxidation of these gases involving both homogeneous and heterogeneous oxidation.[4] The gas-phase oxidation of these gases is initiated by reaction with hydroxyl radicals:

$$SO_2 + OH \rightarrow HOSO_2 \quad (3)$$

$$HOSO_2 + O_2 \rightarrow HO_2 + SO_3 \quad (4)$$

$$SO_3 + H_2O \rightarrow H_2SO_4 \quad (5)$$

Homogeneous aqueous-phase oxidation of SO_2 takes place by its dissolution and dissociation in water, forming equilibrium similar to CO_2:

$$SO_2 + H_2O \leftrightarrow SO_2H_2O \quad (6)$$

$$SO_2.H_2O \leftrightarrow HSO_3 - + H^+ \quad (7)$$

$$HSO_3 - \leftrightarrow SO_3^{2-} + H^+ \quad (8)$$

Gas-phase oxidation of NO_2 is faster than SO_2 by one order of magnitude:

$$NO_2 + OH \rightarrow HNO_3 \quad (9)$$

In addition, significant formation of nitric acid takes place through ozone and NO_3 radical reactions.

During daytime, NO_3 radical is formed as follows:

$$NO_2 + O_3 \rightarrow NO_3 + O_2 \quad (10)$$

NO_3 radical so formed reacts with NO_2 at nighttime, finally resulting in the formation of HNO_3:

$$NO_3 + NO_2 \leftrightarrow N_2O_5 \quad (11)$$

$$N_2O_5 + H_2O \rightarrow 2HNO_3 \quad (12)$$

At ambient levels of NO, its aqueous-phase oxidation is very slow due to its low solubility in water and also the dependence on NO_2 concentrations. It can be faster at higher NO_2 levels. However, the reaction follows the path

$$2NO_2 + H_2O \leftrightarrow 2H^+ + NO_3 - + NO_2 - \quad (13)$$

Heterogeneous oxidation of SO_2 and NO_2 involves gas- particle reactions. In the liquid phase, SO_2 is rapidly converted into sulfate by H_2O_2. SO_2 is also converted into sulfate on freshly emitted soot particles, but subsequently, the rate of oxidation is retarded due to saturation of soot particle surface. Preferable oxidation of SO_2 onto soil dust particles is reported in dusty regions where formation of calcium sulfate takes place instead of free sulfuric acid. NO_2 is also oxidized onto particles—for example, it reacts with NaCl of sea salt, forming $NaNO_3$ on the surface. However, over time, the surface is saturated, and the rate of oxidation becomes lower.

History of Acid Rain

The major cause of acid rain is the increased combustion of fossil fuels, which has been practiced at larger scale after the industrial revolution. The presence of sulfur compounds in the air of Sweden and England was realized in the 18th century. In fact, Robert Boyle, in 1692, mentioned in his book *A General History of the Air* the "nitrous and salino-sulphureous spirits" in the air.[4] The term "acid rain" was first used in 1872 by Robert Smith, who discovered acid rain in 1852 in the area surrounding Manchester. He referred to this term in a treatise on the chemistry of rain published in England in 1872. Robert Smith mentioned various factors such as coal combustion and the amount of rain affecting the precipitation. Unfortunately, this wonderful publication was overlooked until it was revisited and critiqued by Gorham in 1981.[1]

Acid rain attracted attention of the scientific community and society when Odén[6,7] reported that large-scale acidification of surface waters in Sweden could be attributed to pollution from the United Kingdom and central Europe. The worst-hit areas of acid rain were Scandinavia and Central and Southern Germany. In Europe, the rain pH was observed to be as low as 3.97 in Germany. Drastic loss of fish population was seen in lakes of Sweden and parts of southwest Norway. European data show that most of the acidity was intensified during 1955–1970, with a sudden increase in the mid-1960s. In parts of Germany and other European countries, forest damage and loss of needles from pine and spruce trees were noticed due to acid rain. Acid smog killed almost 4000 people in London in 1952.

Sources of Acidity

As mentioned earlier, in natural conditions, atmospheric CO_2 when dissolved in water forms carbonic acid (H_2CO_3), which brings down the pH of water. Other gases such as SO_2 and NO_2 also form acids, viz., sulfuric (H_2SO_4) and nitric acid (HNO3), respectively. The main cause of acid rain is excess contribution of H_2SO_4 and HNO_3 in precipitation due to anthropogenic sources, especially through combustion of coal and petroleum. Sulfur is present in significant amounts in fossil fuel (coal and petroleum), which is the major source of SO_2. Oxidation of SO_2 is accelerated by higher concentrations of H_2O_2 and O_3 found in polluted air. Martin and Barber[8] noticed that acidity of precipitation at several sites in England was the highest during spring, when O_3 concentrations were higher. During past century, huge consumption of fossil fuel in North America and Europe resulted in high SO_2 emissions.[9,10]

Natural Acidity Contributed by Organic Acids

Apart from sulfuric and nitric acids, acidity in rainwater is also contributed by organic acids. Formic acid (HCOOH) and acetic acid (CH3COOH) are the major species reported in rainwater, contributing around two-thirds of the total acidity at remote sites.[11] Generally, formic acid is found to dominate over acetic acid. A relatively higher contribution of organic acids is observed at tropical sites than at temperate ones. These organic acids are produced in gas phase by the oxidation of isoprene and terpenes emitted by the vegetation,[12,13] which are then scavenged by the rain. These are also formed through aqueous-phase oxidation of aldehydes. Sometimes, these acids are emitted by soils.[14] It is to be noted that organic acids may be important for pH in cloud and rainwater in some areas, but their contribution to acidification of soils and surface waters is small because these are quickly consumed by microorganisms.

Spread and Monitoring of Acid Rain

Considering the degree of damage caused by acid rain, several efforts are made by European countries to monitor and control it. The European Air Chemistry Network was started in the early 1950s by Stockholm University in collaboration with the Swedish University of Agricultural Sciences. Both institutes served as centers for the network for the chemical analysis of samples. Originally, the purpose of this network was to study the depositions of plant nutrients to forest and agriculture systems. Under this network, continuous data related to chemical composition of precipitation have been available since 1955.[15] Later on, this network became part of the Swedish National Monitoring Programme. A Norwegian program called SNSF, "Acid Precipitation: Effects on Forests and Fish" was run up to 1980.[16] Immediately after the Stockholm conference in 1972, the European Organization for Economic Cooperation and Development (OECD), in 1978, established a network to monitor long-distance transport of pollutants and the impacts of European countries on their neighbors, known as the Cooperative Program for Monitoring and Evaluation of Long-Range Transmission of Air Pollutants in Europe (EMEP). Further, in 1983, the Convention on Long-Range Transboundary Air Pollution (CLRTAP) was signed by more than 30 countries, including the United States, Canada, and the European Union, to deal with transboundary air pollution.

The discovery of acid rain in Europe attracted attention of scientific community in North America too. Odén's study was followed up by the United Nations (UN).[17] This also led to the first international conference on acid rain in Columbus, Ohio, United States, in 1975. Later on, under the Acid Precipitation Act of 1980, U.S. Congress formed a national network called National Acid Precipitation Assessment Program (NAPAP), which supported the expansion of National Atmospheric Deposition Program (NADP) to monitor the trends in long-term precipitation chemistry and deposition. Further, the NADP was changed to the NADP National Trends Network, which has around 250 monitoring sites.

Similar to Europe and United States, Canada also experienced acid rains. Environment Canada has developed its program called the Canadian Air and Precipitation Monitoring Network (CAPMoN) to monitor the regional patterns and trends of atmospheric pollutants. including acid rain, smog, particulate matter, etc., Canada started the Canadian Network for Sampling Precipitation in 1978, which was renamed as the Air and Precipitation Network (APN). CAP- MoN is the new name of APN (changed in 1983).

Later on, the spread of acid rain was also noticed in East Asia. After successful implementation of CLRTAP in Europe, the UN Conference on Environmental Development adopted to continue and share the experience gained from acid rain programs in Europe and North America and established the Acid Deposition Monitoring Network in East Asia (EANET) in 1993. This network includes Japan, Russia, China, the Republic of Korea, Mongolia, Thailand, Singapore, Cambodia, Lao People's Democratic Republic, Myanmar, Vietnam, the Philippines, Malaysia, and Indonesia.

Measurements through the long-term acid rain network have not been carried out extensively in other parts of the world such as India, Africa, and Latin America. However, programs such as the Composition of Atmospheric Aerosols and Precipitation in India and Nepal and the Composition of Asian Deposition (CAD), as part of the Regional Air Pollution in Developing Countries (RAPIDC) program funded by the Swedish International Development Cooperation Agency), were very effective in providing a summarized picture of the acid rain scenario in the Indian region.[18,19] The RAPIDC program was coordinated by the Stockholm Environment Institute, which facilitated international cooperation on air pollution issues to develop relevant knowledge to support decision making in Asia and Africa. The CAD program was a part of the International Global Atmospheric Chemistry/Deposition of Biogeochemically Important Trace Species (IGAC/DEBITS) activities of the International Geosphere-Biosphere Programme, which focused on good-quality measurements at rural sites in Asia to produce high-quality data so as to understand the Asian wet deposition scenario. In the African region, the IGAC/DEBITS-Africa program has its network of 10 stations for the measurement of wet and dry depositions at selected sites.

In Australia, acid rain studies have been carried out under the Commonwealth Scientific and Industrial Research Organisation (CSIRO) network of sites.[20] Globally, the 1989 initiative of the World Meteorological Organization, under the Global Atmospheric Watch (GAW), is carrying out precipitation measurements at around 80 stations in different countries. Earlier, GAW used to be known as the Background Air Pollution Monitoring Network (BAPMoN).

Regional Acidity of Precipitation

United States

As described by Gibson,[10] northeastern United States and southeastern Canada were the most affected areas by acid rain. Likens[21] was the first who evaluated the 1955–1956 and 1972–1973 data and found a significant increase in acidity in northeastern United States and southeastern Canada during the two decades. He also noticed a significant increase in the spread of acid rain in the areas of southeastern and Midwestern United States.[22]

Canada

Long-term measurements of acid deposition in Canada showed that other than local sources, acidity was also contributed by the long-range transport of oxides of sulfur and nitrogen from sources located southerly in the United States. According to Environment Canada, more than half of the acid deposition in eastern Canada is originated from the United States. Studies by the APN showed higher acidity in southern Ontario, having a pH of around 4.2.[23] Regionally representative sites Long Point and Chalk River experienced that, most of the time, wind parcels came from southerly source areas. These sites are the receptor sites to the major sources in Ontario and the lower Great Lakes region. Estimates of the year 1995 showed that 3.5–4.2 Tg per year of SO_2 was transported from the United States to Canada. Acid deposition in Canada can be reduced by the joint measures of the United States and Canada. Collaborative efforts in this direction are already in progress.

Europe

Areas affected by acid rain in Europe include northern and western Europe, southeast England, Germany, the Netherlands, and parts of Denmark. Table 1 shows a drastic decrease of precipitation pH in western Europe during the 1950s and 1970s.[24]

TABLE 1 Change in pH of Precipitation in Western Europe during the 1950s and 1970s

Country	pH in 1950s	pH in 1970s
Southern Norway	5.0–5.5	4.7
Northern Sweden	5.5–6.0	4.3
Southern Sweden	5.5–6.0	4.3
Southeast England	4.5–5.0	4.2

Source: Environmental Resources Ltd. Pearce.[24]

OECD 1977 estimates indicate that anthropogenic emissions of sulfur in Europe increased by 50% during 1955–1970. However, later on (1972–1982), many European countries, viz., the United Kingdom, West Germany, the Netherlands, Sweden, Norway, and Denmark reduced their sulfur emissions. These reduction measures improved the situation in Europe.[15]

Asia

In the Asian region, much of the acid rain problem prevails in East Asia, covering China, Japan, North Korea, and Thailand. Among these, China is the biggest polluter. According to estimates, China's sulfur emissions will triple between 1990 and 2020.[25] After the United States and Europe, China is the biggest consumer of fossil fuel. Rapid increase in SO_2 and NO_x emissions from 2000 onward is a major reason for the spread of acid rain in China.[26] The area most affected by acid rain in China is south of the Yangtze River, where average pH is recorded to be less than 4.5.

Acidity levels in precipitation in Japan show seasonality. During the summer season, most of the acidity is observed to be due to local sources of sulfur oxides, whereas during the winter season, increased level of acidity is due to long-range transport from the Asian continent, which results in higher acidity of precipitation at the sites in western Japan.[27] Similarly, Thailand also experiences acid rain. Around 50% of rain events are reported acidic due to high concentration of sulfate and nitrate. Of these oxides, 70%–80% are contributed by Thai sources.[28]

In the Indian subcontinent, precipitation is reported to have relatively higher pH (>5.6).[19,29] The pH of rainwater at some of the continental sites in India is as high as 8.3 (Table 2) due to interference of soil dust (rich in calcium carbonate) suspended in the atmosphere. Abundance of soil dust in air is a

TABLE 2 Average pH of Rainwater at Various Sites in India

Site	Nature of Site	pH	Reference
Calcutta	Urban	6.8	Das[31]
Nainital	High altitude	6.2	Hegde et al.[32]
Iqbalpur	Rural	7.1	Jain et al.[33]
Mumbai (Colaba)	Urban	5.9	Khemani et al.[34]
Darjeeling	High altitude semiurban	6.4	Kulshrestha[35]
Haflong	High altitude rural	7.3	Kulshrestha[35]
Delhi	Urban	5.7	Kulshrestha et al.[36]
Hyderabad	Urban	6.4	Kulshrestha et al.[37]
Jorhat	Rural	5.8	Kulshrestha et al.[38]
Hudegadde	Reserve forest	6.0	Kulshrestha et al.[38]
Agra (Dayalbagh)	Semiurban	7.1	Kumar et al.[39]
Malikadevi	Remote	6.4	Mahadevan et al.[40]
Allahabad[a]	Urban	7.1	Mukhopadhyay et al.[41]
Jodhpur[a]	Rural	8.3	Mukhopadhyay et al.[41]
Kodaikanal[a]	Rural	6.1	Mukhopadhyay et al.[41]
Mohanbari[a]	Rural	6.4	Mukhopadhyay et al.[41]
Nagpur[a]	Rural	6.3	Mukhopadhyay et al.[41]
Srinagar[a]	Rural	7.0	Mukhopadhyay et al.[41]
Pune	Urban	6.3	Pillai et al.[42]
Sinhagad	High altitude rural	6.2	Pillai et al.[42]
Silent Valley	Reserve forest	5.3	Rao et al.[43]
Agra (Tajganj)	Semiurban	7.0	Saxena et al.[44]
Indian Ocean (during Jan–Mar)	Northern and central	Below 5.6	Kulshrestha et al.[45]

[a] BAPMoN sites.

TABLE 3 Average pH and Major Ions (μeq/L) in Precipitation at Banizaumbou during 1994–2005

Parameter	Value
pH	6.05
SO_4^{2-}	9.4
NO^{3-}	11.6
Ca^{2+}	27.3
NH^{4+}	18.1

Source: Galy-Lacaux et al.[47]

common feature of the Indian atmospheric environment. The pH of most soils of India is very high as compared with the pH of soils in acidified regions of the world. Generally, in India, the pH of rainwater is the mirror image of the pH of soil in the region. The acidity generated by the oxidation of gases like SO_2 and NO_x is buffered by soil-derived particles. Acidity of SO_2 is buffered by $CaCO_3$ of soil dust forming calcium sulfate, which is removed by below-cloud scavenging (Figure 1). Due to this, the spread of acid rain at continental sites in India is controlled by the continuous suspension of loose soil during prevailing dry weather conditions. However, a bigger number of hot spots of higher wet deposition of non-sea salt sulfate (nss SO_4) are reported in urban and industrial areas than in rural areas. Several of these larger hot spots lie in the Indo-Gangetic region (Figure 2).

Although rainwater pH higher than 5.6 is more frequently recorded in India, sometimes, occurrence of acid rain (pH <5.6) is also reported. Figure 3 shows the frequency of acid rain reported from various sites in India.[46] Rainwater is noticed to be acidic in India if any of following applies:

1. Rain continues for a long time, washing off soil dust from the atmosphere. A similar situation prevails over the Indian Ocean, where soil dust interference is at a minimum.
2. In the areas where a large part of ground is covered with vegetation.
3. In the areas where soil itself is acidic (northeast, east, and southwest India).
4. Near heavy sources of SO_2 (e.g., thermal power plants).

In the African region, the pH of precipitation is reported to be nearly similar to that in the Indian region. Longterm data (1994–2005) showed that the acidity of precipitation at Banizaumbou, a regional

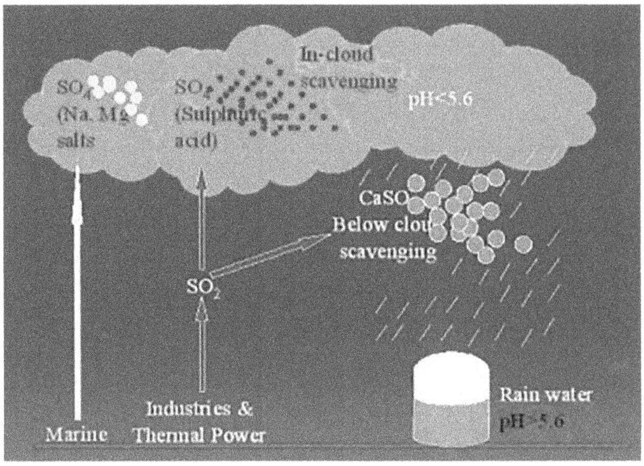

FIGURE 1 Schematic diagram showing alkaline rains by removal of soil dust during below-cloud scavenging process in India.
Source: Kulshrestha.[30]

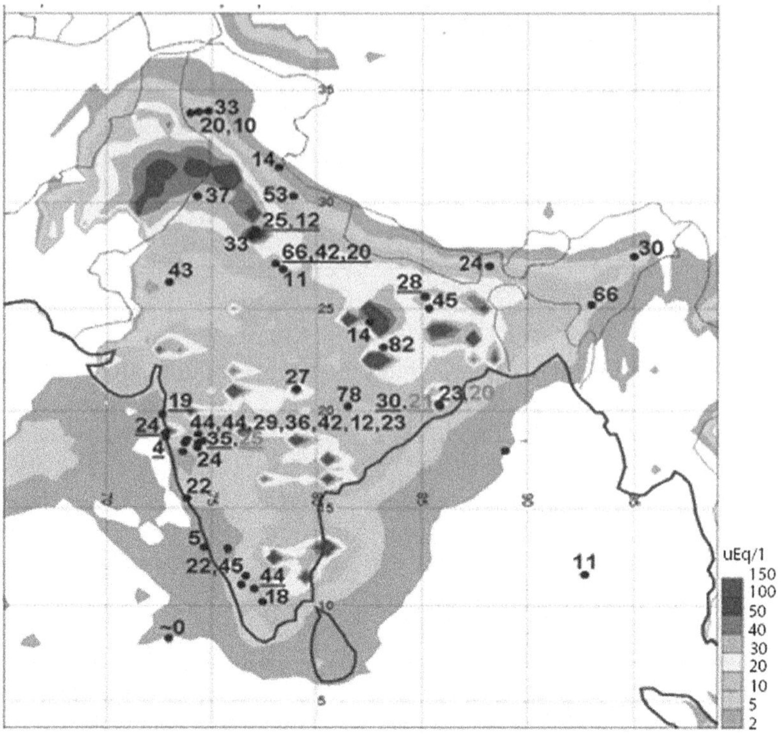

FIGURE 2 Concentration of nss SO_4 in rainwater. Data from measurements at rural and suburban (underlined) sites obtained with bulk (black) and wet (red) collectors only (scaled to year 2000) compared with the concentration field obtained with the Multiscale Atmospheric Transport and Chemistry Model (MATCH) for the year 2000.
Source: Kulshrestha et al.[19]

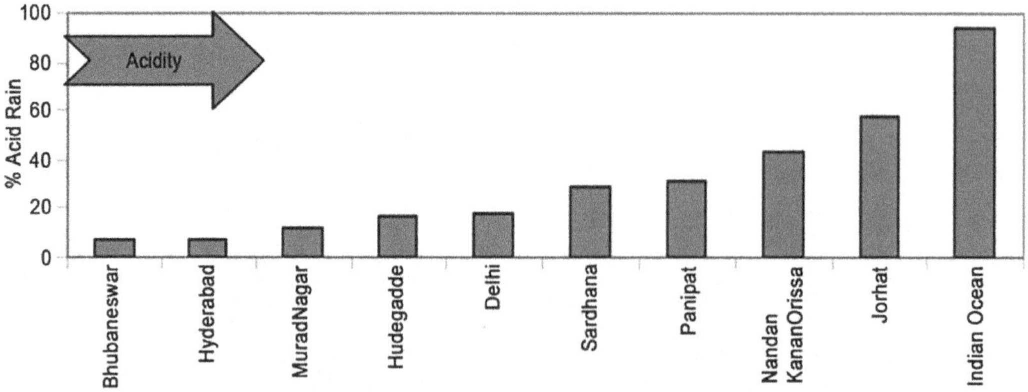

FIGURE 3 Percent frequency of acid rain reported in Indian region.
Source: Kulshrestha et al.[46]

representative site in the semiarid savanna region, has high interference of soil dust, resulting in higher pH.[47] In addition, neutralization by ammonium ion is also partly responsible for elevated pH in Africa.

At a glance, the model based global distribution of pH of rain water is shown in Figure 4.[48] The distribution of pH shows high acidity in Europe, eastern North America, and East Asia. To some extent, acidity is seen in the west coast of South America and Africa also. High acidity in these areas is due to free acidity

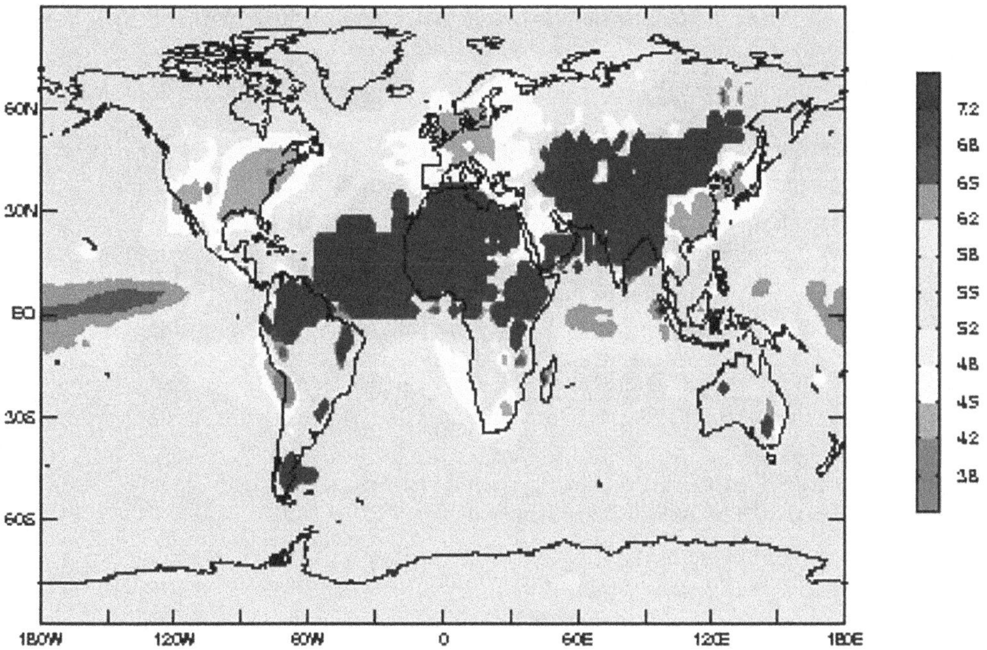

FIGURE 4 Estimated global distribution of pH of precipitation.
Source: Rodhe et al.[48]

contributed by sulfuric acid. However, in areas such as north of South America, northern Africa, South Asia, and part of China, the acidity is lower which is due to neutralization by ammonia and soil dust.

Trends in Acidity

United States and Europe

After 1980, significant reduction in SO_2 emissions has improved the situation of acidity in Europe.[49] Similarly, in Canada and the United States, effective steps of reduction in SO_2 emissions have contributed to improved acidity levels. In the United States, the average reduction in SO_2 emissions from 1980 to 2008 was around 54%, as shown in Figure 5. Overall, the reduction measures have resulted in a decrease in H^+ in precipitation over these regions. Trends of acidity in the United States show a significant improvement after implementation of the 1995 Clean Air Act Amendment, which forced them to reduce SO2 emissions. Figure 6 is an example of trends of acidity of precipitation in North Carolina during 1985–2005,[52] which shows around 50% reduction in H^+ during two decades. In Europe, sulfate concentrations increased by approximately 50% between the 1950s and the late 1960s but have been declining since the mid-1970s. In Sweden and Norway, an average of 20% reduction in SO_4 levels has been achieved since the 1970s,[15] followed by higher reductions in more recent decades.

Asia

China

In the developed countries, efforts have been made to control SO_2 emissions, but in the developing countries, SO_2 emissions continue to be high. In Asia, China is the biggest SO_2 emitter. Total Chinese SO_2 emission increased from 21.7 Tg to 33.2 Tg (53% increase) from 2000 to 2006, showing an annual growth rate of 7.3% per year.[53]

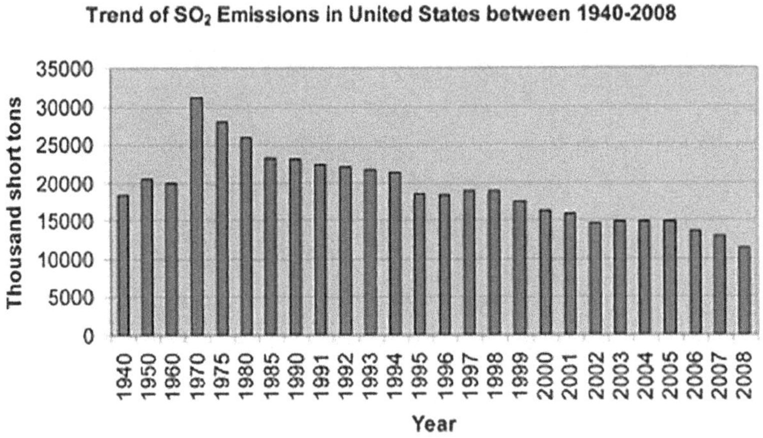

FIGURE 5 Trend of SO_2 Emissions in the United States from 1940 to 2008.
Source: Adapted from the USEPA Web site[50] and Stensland.[51]

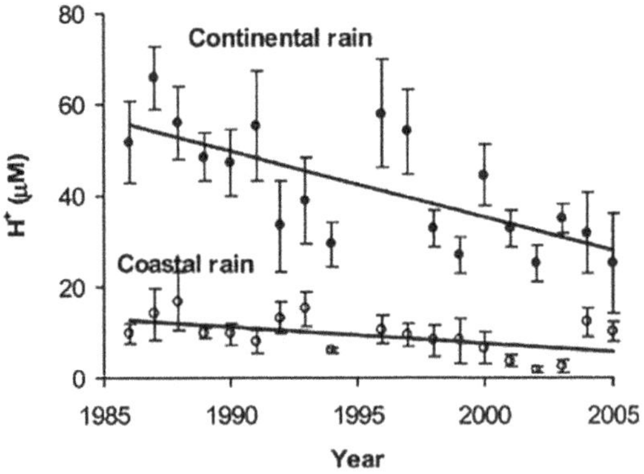

FIGURE 6 Annual volume-weighted average H^+ concentration in Wilmington, North Carolina, continental (filled circles) and coastal (open circles) precipitation from 1985 to 2005.
Source: Willey et al.[52]

Long-term acid precipitation observations show that the temporal and spatial distribution of rain acidity in China has changed remarkably since 2000.[26] Future estimates of acidification potential, using a dynamic soil acidification model, indicate that sensitive soils in south China and Southeast Asia may reach a critical threshold within a few decades.[54] Model-based estimates (Figures 7a and b) indicate higher levels of total depositions of S-SO_x and N-N_r in East Asia.

India

SO_2 emissions in India are relatively less, but the rate of increase is almost doubled from 1985 to 2005. In 2000, Indian SO_2 emission was estimated to be 4.26 Mt.[53] Precipitation studies in India lack long-term measurements. Most of the studies were carried out by individual scientists/groups. A few sites under the GAW network are in operation. In addition, one study from Pune[56] reports that during 1984–2002, there was significant increase in SO_4 and NO_3 concentrations (Figure 8), which resulted in decrease in pH of rainwater from 6.9 in 1982 to 6.5 in 2002. These changes are due to increase in industrial and vehicular activities in the region. Another long-term network study in a rural area of Nandankanan (Orissa state in east India) reported 57% frequency of acid rain events during 1997–1998,[57] which was reduced

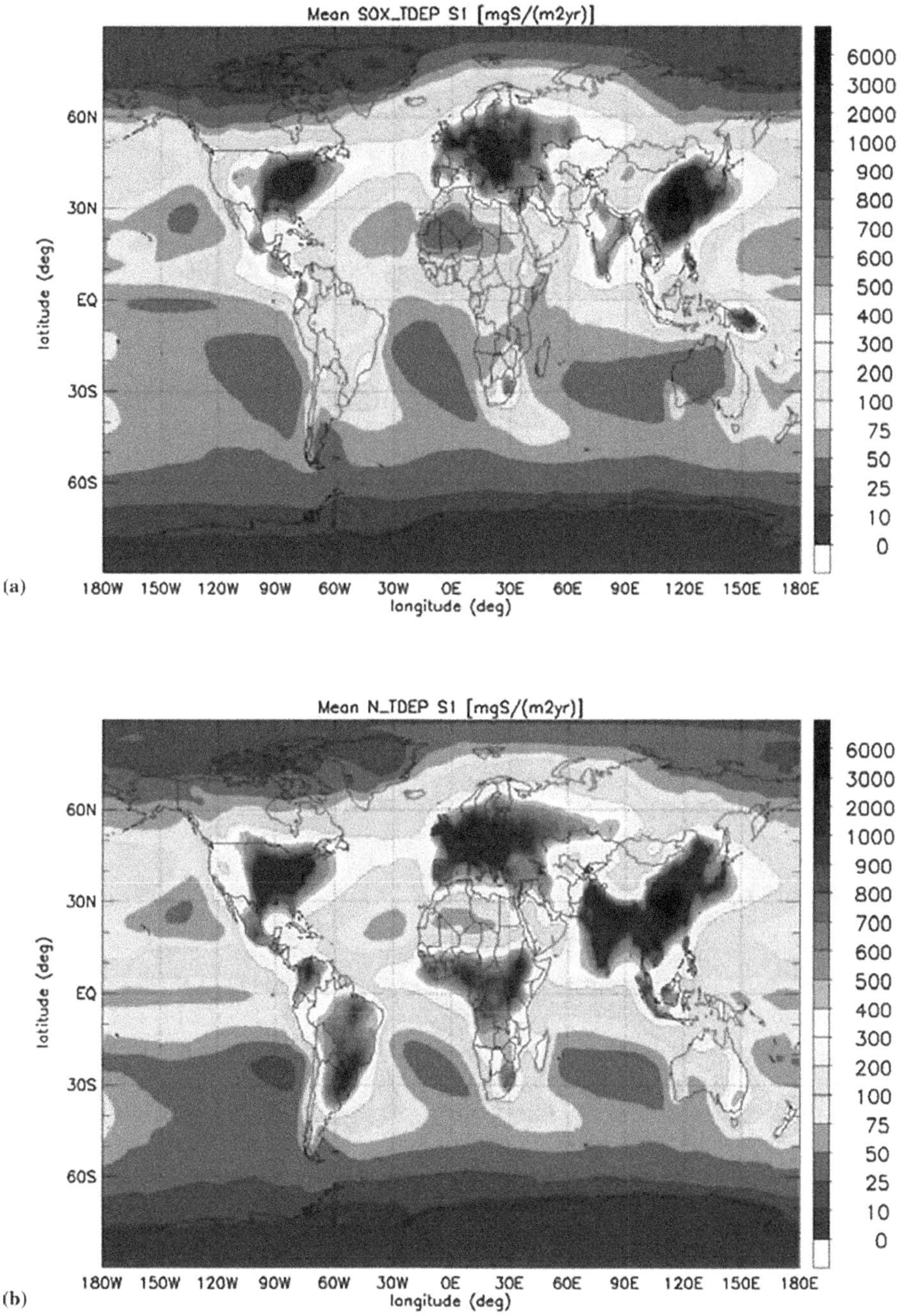

FIGURE 7 Estimated total deposition of (a) S (mg m^{-2} yr^{-1}) of SO_x and (b) N (mg m^{-2} yr^{-1}) of reactive nitrogen for year 2000.
Source: Dentener et al.[55]

to 40% during 2005–2007.[58] SO_4/Na ratios were also reduced drastically from 1.58 in 1997–1998 to 0.519 during 2005–2007. In a review compiled by Kulshrestha and coworkers,[19] it is reported that most of the Indian precipitation measurements lack quality assurance (QA) and quality control (QC) in sampling, storage, and analysis of samples. Hence, this region really needs quality-controlled measurements of

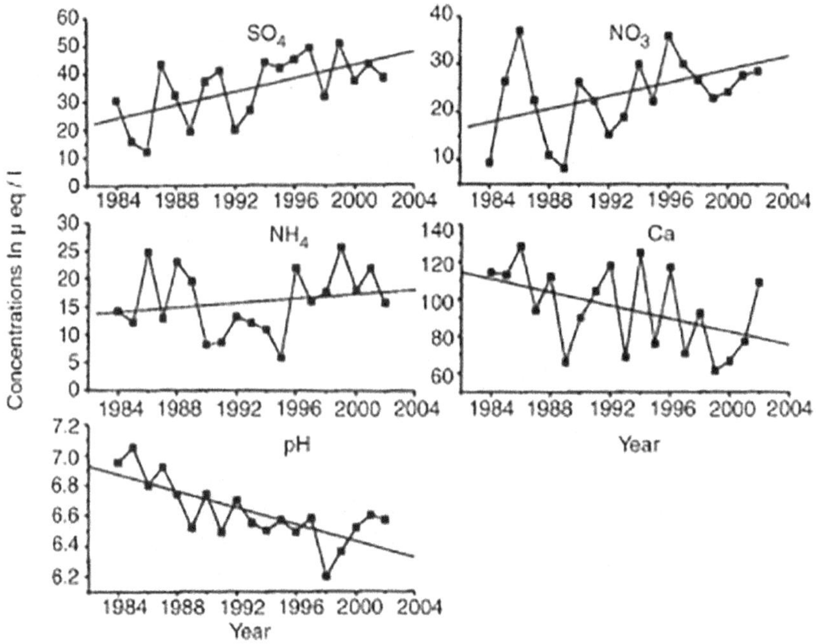

FIGURE 8 Trends of pH, SO_4, NO_3, Ca, and NH_4 in rainwater at Pune during 1984–2000.
Source: Rao et al.[56]

wet and dry depositions at a few selected sites in order to get an idea about the trends of acidity with the growing emissions of oxides of S and N.

Acidification of Oceans

Oceans are the biggest sinks for atmospheric CO_2. Increasing emissions of CO_2 due to anthropogenic sources will lead to ocean acidification through excess dissolution of CO_2 in seawater. Since the industrial revolution, the acidity of the ocean has increased by 30% (from a pH of 8.2 to 8.1). Future projections show that under a business-as-usual scenario, surface ocean pH will be lowered by 0.4 pH units by the end of the century. Acidification of the ocean affects the nitrification process, which further affects marine biota. Apart from CO_2 rise, acid deposition can add to the acidification of oceans. Precipitation having very low pH contributes a significant amount of hydrogen ions in seawater, which in the long term may alter the pH of seawater. Results from Indian Ocean Experiment showed the pH of rainwater to be between 3.8 and 5.6 over the Indian Ocean.[45] The acidic nature of rainwater over the Indian ocean is due to insignificant influence of soil dust and the dominance of anthropogenic sulfate contributed by long-range transport.[59,60] This aspect needs to be investigated in the future in order to protect the marine ecosystem.

Global Sensitivity toward Acidification

Global precipitation acidity and mapping of soil sensitivity to acid deposition suggests three main problematic areas. These are North America, Europe, and southern China (Figure 9), where acid rain control is necessary. Already, in North America and Europe, steps have been taken to reduce SO_2 emissions. Other parts of the world also need to take appropriate steps to reduce sulfur emissions.[62]

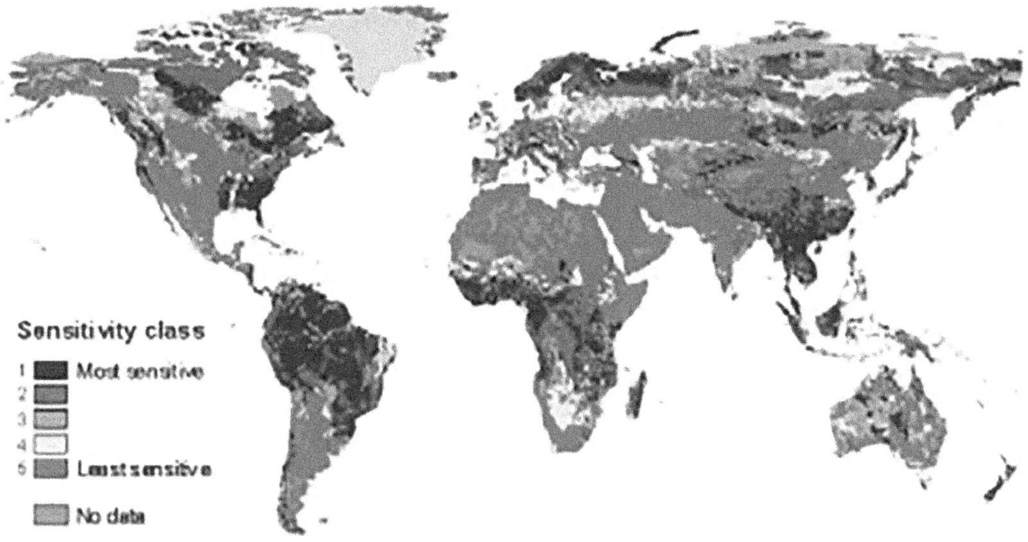

FIGURE 9 Global sensitivity toward acidification.
Source: Kuylenstierna et al.[63]

Global Scenario: Future Projections through Modeling

Recently, the Regional Air Pollution Information and Simulation model has been developed by International Institute for Applied Systems Analysis (IIASA) as a tool for the integrated assessment of alternative strategies to reduce acid deposition in Europe and Asia.[63] Dentener and coworkers[55] have attempted simulation of the global future scenario (up to 2030) of deposition of oxides of nitrogen and sulfur by using 23 atmospheric chemistry transport models. The stud focused mainly upon three emission scenarios: 1) current legislation (CLE); 2) case of the maximum emission reductions (MFR); and 3) pessimistic IPCC SRES A2 scenario. The model output showed a good agreement with observations in Europe and North America primarily because of quality-controlled measurements reported from these regions. The study suggested that in the future, deposition fluxes are going to be controlled mainly by the changes in emissions, with atmospheric chemistry and climate having a very limited role.

Regional Comparison of Precipitation Scenario

Sulfate in the Atmosphere

Normally, free acidity is contributed by the acids of SO_2 and NO_2. In case of free acidity of H_2SO_4, pH value decreases with increasing concentration of SO_4 ions. A comparison of pH and SO4 in rainwater at different sites in the United States, Sweden, and India is shown in Figs. 10–12. The United States and Sweden are examples of developed and acidified countries, whereas India is a developing country having high pH of rainwater. From Figures. 10 and 11, pH decreases with increase of SO_4 concentration, while at Indian sites (Figure 12), even at higher SO_4 levels, higher pH of rainwater is observed. This indicates that in the United States and Sweden, the SO_4 is present as H_2SO_4, which gives free H^+ in rain, but in India, it is present in a different form that does not contribute free H^+ in rainwater. The Indian dusty atmosphere is rich in $CaCO_3$, which allows SO_2 to form calcium sulfate, due to which pH of rainwater in India and other dusty regions is observed to be higher. In the Indian region, the possible

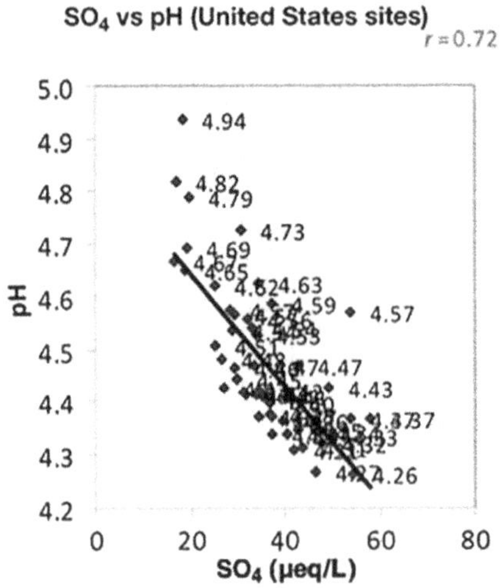

FIGURE 10 Variation of SO_4 and pH in rainwater at the sites in the United States.
Source: Trends in Precipitation Chemistry in the United States, 1983–94: An Analysis of the Effects in 1995 of Phase I of the Clean Air Act Amendments of 1990, Title IV.[61]

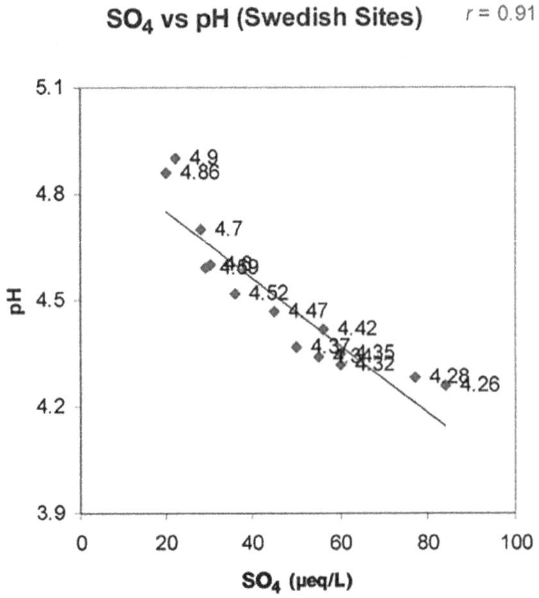

FIGURE 11 Variation of SO_4 and pH in rainwater at Swedish sites.
Source: Granat.[65]

mechanism of SO_2 removal is the adsorption of SO_2 onto the $CaCO_3$-dominated dust particles forming calcium sulfate.[62]

$$CaCO_3 + SO_2 + 1/2O_2 + 2H_2O \rightarrow CaSO_4.2H_2O + CO_2 \quad (14)$$

A comparison of typical composition of precipitation in an acidified region and a dusty region has been reported by Rodhe and coworkers,[48] establishing such differences very clearly.

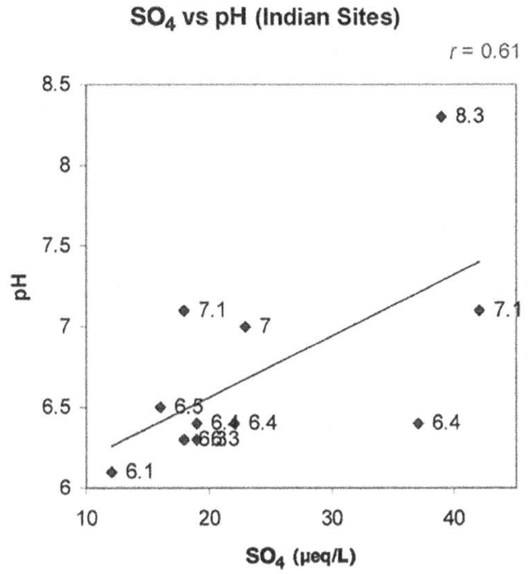

FIGURE 12 Variation of SO_4 with pH at Indian sites.
Source: Granat.[65]

Control of Acid Rain

Acid Rain Control Policy of the United States

Due to pressure from the public, the federal government of the United States adopted the Clean Air Act in 1970. Under this act, emission standards were set for SO_2 and NO_x, and states were directed to compliance with the National Ambient Air Quality Standards. Congress formed a 10 years program, the NAPAP, and mandated it to conduct scientific, technological, and economic study of the acid rain.

In 1990, under Clean Air Act Amendments, the National Deposition Control Program was implemented. This legislation was to control adverse effects of acidic deposition through reductions in emissions of SO_2 and NO_x. It was targeted to achieve a 50% reduction in annual SO_2 emissions by the year 2000. In 2001, SO_2 emissions from utilities subject to the provisions of the acid rain program were 39% below their 1980 levels, and total emissions from all sources were 50% less than their 1980 levels.[67] These implementation steps will help in environment protection, in particular, to check deterioration of historic buildings, to reduce fine particulate matter (sulfates, nitrates) and ground-level ozone (smog), and to improve public health.[68]

Since Canada is affected by transboundary pollution, an air quality accord was signed between the United States and Canada in 1991 under the framework of the UN Economic Commission for Europe. This bilateral accord facilitates the United States' and Canada's meeting their emissions targets for SO_2 and NO_x and putting coordinated efforts into atmospheric modeling and monitoring the effects of transboundary air pollution.[69]

European Policy to Control Acid Rain

Among European countries, most of the scientific research on acid rain effects was conducted in Norway and Sweden. In the beginning, acidification of lakes through transboundary pollution was the primary issue in Scandinavian countries. In 1972, at the UN Conference on the Human Environment in Stockholm, Sweden's case study on the effects of long-range transport of sulfur compounds was presented, which emphasized the need for international agreement to reduce damage from acid deposition. Soon after the Stockholm conference, in 1978, OECD initiated EMEP to monitor long-range pollution.

The 1985 Helsinki Protocol was the first binding commitment on the reduction of sulfur emissions or their transboundary fluxes by at least 30%. Later, a group of 12 countries decided to sign a declaration to reduce by 30% NO_x emissions by 1998 as compared with 1986 (base year). The latest agreement is the 1999 Gothenburg Protocol, which deals with SO_2, NO_x, NH_3, and non-methane volatile organic compounds (VOCs), aiming to mitigate the problem of acidification, eutrophication, and ground-level ozone. The Gothenburg Protocol targets the reduction of Europe's sulfur emissions by at least 63%, NO_x emissions by 41%, VOC emissions by 40%, and NH_3 emissions by 17% by the year 2010 from their 1990 levels.[70]

Effects of Acid Rain

Acid rain is very harmful to the environment as it damages many living and non-living things over a period of time. Acid rain affects both terrestrial as well as aquatic life. There are many inevitable impacts of acid rain, which affect natural as well as anthropogenic environments.

Effects on Aquatic Ecosystem

The aquatic ecosystem is visibly affected by acid rain as it directly falls into water bodies like lakes, streams, and rivers. Also, the extra acidic rainwater from other terrestrial places like forests and roads flows into nearby water bodies. Although the acidic effect of the rain may get diluted after it is mixed into the water bodies, it may lower the average pH of the aquatic system over the period, and in case the water body has low base cation supply or buffering capacity, it can become acidic faster. Charles and Norton[71] have reviewed the situation in lakes in United States and Canada and found that around the 1920s–1950s onward, weakly buffered lakes in some regions became more acidic. There are estimated to be around 50,000 lakes in the United States and Canada with a pH below 5.3. Out of these, hundreds of lakes have very low pH and are unable to support aquatic life, eliminating many existing insect and fish species. At pH lower than 5, the life of aquatic animals is threatened as they are unable to absorb oxygen from water. At pH lower than 4.8, fishes, frogs, and aquatic insects experience increased mortality. However, constructive steps, such as SO_2 emission reduction and liming of lakes, have significant potential for reversibility.[72,73]

Effects on Vegetation and Soil

In addition to the aquatic ecosystem, acid rain significantly affects trees, plants, forests, and other vegetation. Acid rain can damage the leaves and stems of the trees and affect their growth by getting absorbed through roots via soil. Acid rain reacts with leaves and stem wax coatings and allows acidic water to enter the leaves, thereby damaging the trees and plants. Experimental studies have established that acidic deposition causes some physiological effects in plants.[74] Especially at high altitude, the forests are surrounded by clouds carrying acidic water. The moisture of the clouds passing through the forests leads to severe damaging effects on the forests. In addition to individual effects on trees and forests, there are also effects on the soil, which contains the necessary nutrients and microorganisms for the healthy growth of the trees, plants, forests, and other vegetation. Virtual effects of acids rain have been observed all over the world, especially in Europe and eastern North America.[75,76] Hedin and coworkers[77] have reported evidence of steep decline of base cations in precipitation, which is based on long-term quality-controlled measurements, in Europe and North America. These measurements support that decline in base cations in precipitation might have resulted in increased sensitivity of a weakly buffered ecosystem, affecting forests and vegetation. Nitrification of ammonia (NH_3) and ammonium (NH_4^+) leads to the acidification of soil, which also adds to nutrient leaching:

$$NH_3 + O_2 \rightarrow NO_2 - +3H^+ + 2e- \tag{15}$$

$$NO_2 - +H_2O \rightarrow NO_3 - +2H^+ + 2e- \tag{16}$$

Effects on Buildings and Monuments

Acid rain can damage buildings and historic monuments as well by reacting with their paints and construction material. The damaged walls of the buildings and monuments leave a rough surface along with the moisture, which is a favorable place for the growth of microorganisms. Acid rain can affect sculpture and architecture adversely by corrosion. Acid rain can even corrode railway tracks, paints of cars, and joints of bridges and flyovers. A BERG report[78] gives more details of evidence of damage to stone and other materials by acid deposition. Under CLRTAP, the International Cooperative Programme on Effects on Materials, Including Historic and Cultural Monuments, has been set up, with 39 test sites, three of which are in the United States and Canada. According to a report, the corrosion rates of carbon steel, paint on steel, limestone, and bronze have decreased to about 60% from 1987 values in Europe due to decrease in SO_2 levels.[79]

Conclusion

Acid rain, which was a problem of North America and Europe, is now spreading in East Asia due to increased emissions of SO_2 from fossil fuel combustion. Experiences of the global community show that acid rain is spread through transboundary pollution. Scandinavia, Canada, and Japan are such examples. Acid rain damages water bodies, forests, vegetation, buildings, human health, etc., costing a huge loss to the economy. Although appreciable steps are taken by the United States, Canada, and Europe to control SO_2 and NO_x emissions, the considerable increase in SO_2 emission rates in China is of concern in the Asian region. During the past two decades, Chinese SO_2 emissions increased tremendously. Data show that spread of acid rain is very much controlled by suspended atmospheric soil dust in Indian and African regions. At a glance, the following are concluded:

1. Decreasing trends of SO_2 emissions in Europe and North America will be helpful in improving the pH of precipitation in coming decades.
2. Increasing trends of acidity in China in Asia may result in more acid rains in the region.
3. Systematic monitoring networks are established in North America, Europe, and East Asia to monitor acid deposition.
4. There is a strong need for long-term studies on acid deposition, including wet and dry depositions at selected sites in South Asia, Africa, and South America, through extensive networking.

References

1. Cowling, E.B. Acid precipitation in historical perspectives. Environ. Sci. Technol. **1982**, *16*, 110A.
2. Galloway, J.N. Acidification of the world: Natural and anthropogenic. Water Air Soil Pollut. **2001**, *130*, 17–24.
3. Charlson, R.J.; Rodhe, H. Factors controlling the acidity of natural rainwater. Nature **1982**, *295*, 683–685.
4. Seinfeld, J.; Pandis, S. *Atmospheric Chemistry and Physics*, 2nd Ed.; Wiley: New Jersey, 2006.
5. Brimblecombe, P. Interest in air pollution among early fellows of the Royal Society. Notes Rec. R. Soc. **1978**, *32*, 123.
6. Odén, S. Dagens Nyheter (Sweden), October 24, 1967.
7. Odén, S. The acidification of air precipitation and its consequences in natural environment. Ecology Committee Bulletin No. 1; Swedish Sciences Research Council, Stockholm. Translation Consultants Ltd.: Arlington, VA, 1968.
8. Martin, A.; Barber, F.R. Acid gases and acid in rain monitored for over 5 years in rural east-central England. Atmos. Environ. **1984**, *18*, 1715–1724.

9. Vermeulen, A.J. The acidic precipitation phenomenon. A study of this phenomenon and of a relationship between the acid content of precipitation and the emission of sulphur dioxide and nitrogen oxides in the Netherlands. In *Polluted Rain;* Toribara, T.Y., Miller, M.W., Morrow, P.E., Eds.; Plenum Press: New York, 1979; 7–60.
10. Gibson, J.H. Evaluation of wet chemical deposition in north America. In *Deposition Both Wet and Dry,* Acid Precipitation Series; Hicks, B.B., Ed.; Butterworth Publishers: London, 1984; Vol. 4, 1–13.
11. Keene, W.C.; Galloway, J.N.; Holden, J.D. Measurement of weak organic acidity in precipitation from remote areas of the world. J. Geophys. Res. **1983**, 88, 5122.
12. Jacob, D.J.; Wofsy, S.C. Photochemistry of biogenic emissions over the Amazon forest. J. Geophys. Res. **1988**, *93,* 1477–1486.
13. Chameides, W.L.; Davis, D.D. Aqueous phase source for formic acid in clouds. Nature **1983**, *304,* 427–429.
14. Sanhueza, E.; Andreae, M.O. Emission of formic and acetic acids from tropical Savanna soils. Geophys. Res. Lett. **1991**, *18,* 1707–1710.
15. Rodhe, H.; Granat, L. An evaluation of sulfate in European precipitation 1955–1982. Atmos. Environ. **1984**, *18,* 2627–263.
16. Overrein, L.; Seip, H.M.; Tollan, A. *Acid Precipitation— Effects on Forest and Fish,* Final report of the SNSF project 1972–1980. FR 19/80; Norwegian Institute for Water Research: Oslo, Norway, 1980.
17. Bolin, B., Ed. The impact on the environment of sulphur in air and precipitation. Sweden's National Report to the United Nations Conference on the Human Environment; Air Pollution Across Boundaries; Norstadt, Stockholm, 1971.
18. Parashar, D.C.; Granat, L.; Kulshrestha, U.C.; Pillai, A.G.; Naik, M.S.; Momim, G.A.; Prakasa Rao, P.S.; Safai, P.D.; Khemani, L.T.; Naqvi, S.W.A.; Narverkar, P.V.; Thapa, K.B.; Rodhe, H. Chemical composition of precipitation in India and Nepal. A preliminary report on an Indo-Swedish project on atmospheric chemistry. Report CM 90; IMI, Stockholm University: Sweden, 1996.
19. Kulshrestha, U.C.; Granat, L.; Engardt, M.; Rodhe, H. Review of precipitation chemistry studies in India—A search for regional patterns. Atmos. Environ. **2005**, *39,* 7403–7419.
20. Ayers, G.P.; Gillet, R.W. Acidification in Australia. In *Acidification in Tropical Countries,* SCOPE 36; Rodhe, H., Herrera, R., Eds.; John Wiley and Sons: Chichester, England, 1988; 347–400.
21. Likens, G.E.; Acid precipitation. Chem. Eng. News, November 22, 1976; 29–43.
22. Likens, G.E.; Butler, T.J. Recent acidification of precipitation in North America. Atmos. Environ. **1981**, *15,* 1103–1109.
23. Barrie, L.A.; Anlauf, K.; Wiebe, H.A.; Fellin, P. Acidic pollutants in air and precipitation at selected rural locations in Canada. In *Deposition of Both Wet and Dry,* Acid Precipitation Series; Hicks, B.B., Ed.; Butterworth Publishers: Boston, 1984; Vol. 4, 15–36.
24. Environmental Resources Ltd Pearce. *Acid Rain—A Review of the Phenomenon in the EEC and Europe;* Graham and Trotman: London, 1983.
25. Streets, D.G.; Carmichael, G.R.; Amann, M.; Arndt, R.L. Energy consumption and acid deposition in northeast Asia. Ambio **1999**, *28,* 135–143.
26. Jie, T.; Xiao, X.; Bin, B.; Jin, A.; Feng, W.S. Trends of the precipitation acidity over China during 1992–2006. Chin. Sci. Bull. **2010**, 55, 1800–1807. doi: 10.1007/s11434-009-3618-1.
27. Hara, H.; Akimoto, H. National level variations in precipitation chemistry in Japan. In Proceedings of the International Conference on Regional Environment and Climate Changes in East Asia, Taipei. November 30–December 3, 1993.
28. Garivaita, H.; Yoshizumib, K.; Morknoya, D.; Chanatorna, D.; Meepola, J.; Mark-Maia, A. Characterization of wet deposition in suburban area of Bangkok, Thailand. In Proceedings of CAD Workshop, Hyderabad, India, Nov–Dec 2006.
29. Khemani, L.T.; Momin, G.A.; Prakash Rao, P.S.; Safai, P.D.; Singh, G.; Kapoor R.K. Spread of acid rain over India. Atmos. Environ. **1989**, *23,* 757–762.

30. Kulshrestha, U.C. Air quality assessment through atmospheric depositions: A comparison of measurements and model calculations for India. Indian J. Environ. Manage. **2007**, *34*, 51–55.
31. Das, D.K. Chemistry of monsoon rains over Calcutta, West Bengal. Mausam **1988**, *39*, 75–82.
32. Hegde, P.; Kulshrestha, U.C.; Dumka, U.C.; Naza, M.; Pant, P. Chemical characteristics of rainwater and atmospheric aerosols at an elevated site in Himalayan Ranges in India. In Proceedings of CAD Workshop, Hyderabad, India, Nov–Dec 2006.
33. Jain, M.; Kulshrestha, U.C.; Sarkar, A.K.; Parashar, D.C. Influence of crustal aerosols on wet deposition at urban and rural sites in India. Atmos. Environ. **2000**, *34*, 5129–5137.
34. Khemani, L.T.; Momin, G.A.; Rao P.S.P.; Pillai, A.G.; Safai, P.D.; Mohan, K.; Rao, M.G. Atmospheric pollutants and their influence on acidification of rain water at an industrial location on the west coast of India. Atmos. Environ. **1994**, *28*, 3145–3154.
35. Kulshrestha, U.C. Chemistry of atmospheric depositions in India. In Proceedings of Symposium Science at High Altitudes, Darjiling, May 7–11, 1997; 1998. 187–194.
36. Kulshrestha, U.C.; Sarkar, A.K.; Srivastava, S.S.; Parashar, D.C. Investigation into atmospheric deposition through precipitation studies at New Delhi (India), Atmos. Environ. **1996**, *30*, 4149–4154.
37. Kulshrestha, U.C.; Kulshrestha, M.J.; Sekar, R.; Sastry, G.S.R.; Vairamani, M. Chemical characteristics of rain water at an urban site of south-central India. Atmos. Environ. **2003**, 37, 3019–3026.
38. Kulshrestha, M.J.; Reddy, L.A.K.; Satyanarayana, J.; Duarah, R.; Rao, P.G.; Kulshrestha, U.C. Chemical characteristics of rain water at two rural sites of NE and SW India. In Proceedings of CAD Workshop, Hyderabad, India, Nov–Dec 2006.
39. Kumar, N.; Kulshrestha, U.C.; Saxena, A.; Kumari, K.M.; Srivastava, S.S. Formate and acetate in monsoon rain water of Agra. J. Geophys. Res. **1993**, *98*, D3, 5135.
40. Mahadevan, T.N.; Negi, B.S.; Meenakshi, V. Measurements of elemental composition of aerosol matter and precipitation from a remote background site in India. Atmos. Environ. **1989**, *23*, 869–874.
41. Mukhopadhyay, B.; Datar, S.V.; Srivastava, H.N. Precipitation chemistry over the Indian region. Mausam **1992**, *43*, 249–258.
42. Pillai, A.G.; Naik, M.S.; Momin, G.A.; Rao, P.D.; Safai, P.D.; Ali, K.; Rodhe, H.; Granat, L. Studies of wet deposition and dustfall at Pune, India. Water Air Soil Pollut. **2001**, *130*, 475–480.
43. Rao, P.S.P.; Momin, G.A.; Safai, P.D.; Pillai, A.G.; Khemani, L.T. Rain water and throughfall chemistry in the Silent Valley forest in South India. Atmos. Environ. **1995**, *29*, 2025–2029.
44. Saxena, A.; Sharma, S.; Kulshrestha, U.C.; Srivastava, S.S. Factors affecting alkaline nature of rain water in Agra (India). Environ. Pollut. **1991**, *74*, 129–138.
45. Kulshrestha, U.C.; Jain, M.; Mandal, T.R.; Gupta, P.K.; Sarkar, A.K.; Parashar, D.C. Measurements of acid rain over Indian Ocean and surface measurements of atmospheric aerosols at New Delhi during INDOEX pre-campaigns. Curr. Sci. **1999**, *76*, 968–972.
46. Kulshrestha, U.C.; Rodhe, H. Precipitation chemistry metadata from Asia. Presented in the 2nd Steering Committee Meeting of CAD programme, Bangkok, Thailand, Nov 26–27, 2007.
47. Galy-Lacaux, C.; Laouali, D.; Descroix, L.; Gobron, N.; Liousse, C. Long term precipitation chemistry and wet deposition in a remote dry savanna site in Africa (Niger) Atmos. Chem. Phys. **2009**, 9, 1579–1595.
48. Rodhe, H.; Dentener, F.; Schulz, M. The global distribution of acidifying wet deposition. Environ. Sci. Technol. **2002**, *36*, 4382–4388.
49. Zhu, Q. Trends in SO_2 emissions. Report of IEA Clean Coal Center, October 2006. PF 06–09, London.
50. USEPA Web site, available at http://www.epa.gov/air/emissions/so2.htm. (accessed March 2, 2011).
51. Stensland, G.J. Precipitation chemistry trends in the northern United States. In *Polluted Rain;* Toribara, T.Y., Miller, M.W., Morrow, P.E., Eds.; Plenum Press: New York, 1979; 87–108.
52. Willey, J.; Kiber, R.; Avery, G.B., Jr. Changing chemical composition of precipitation in Wilmington, North Carolina, U.S.A.: Implications for the continental U.S.A. Environ. Sci. Technol. **2006**, *40*, 5675–5680.

53. Lu, Z.; Streets, D.G.; Zhang, Q.; Wang, S.; Carmichael, G.R.; Cheng, Y.F.; Wei, C.; Chin, M.; Diehl, T.; Tan, Q. Sulfur dioxide emissions in China and sulfur trends in East Asia since 2000. Atmos. Chem. Phys. Discuss. **2010**, *10*, 8657–8715.
54. Hicks, K.; Kuylenstierna, J.; Owen, A.; Rodhe, H.; Seip, H.; Dentener, F. Assessing the time development of acidification damage in Asian soils at regional scale. In Proceedings of CAD Workshop, Hyderabad, India, Nov–Dec 2006.
55. Dentener, F.; Drevet, J.; Lamarque, J.F.; Bey, I.; Eickhout, B.; Fiore, A.M.; Hauglustaine, D.; Horowitz, L.W.; Krol, M.; Kulshrestha, U.C.; Lawrence, M.; Galy-Lacaux, C.; Rast, S.; Shindell, D.; Stevenson, D.; Van Noije, T.; Atherton, C.; Bell, N.; Bergman, D.; Butler, T.; Cofala, J.; Collins, B.; Doherty, R.; Ellingsen, K.; Galloway, J.; Gauss, M.; Montanaro, V.; Müller, J.F.; Pitari, G.; Rodriguez, J.; Sanderson, M.; Solmon, F.; Strahan, S.; Schultz, M.; Sudo, K.; Szopa, S.; Wild, O. Nitrogen and sulfur deposition on regional and global scales: A multimodel evaluation. Global Biogeochem. Cycles **2006**; *20* (4), GB4003, doi:10.1029/2005GB002672.
56. Rao, P.S.P.; Safai, P.D.; Momin, G.A.; Ali, K.; Chate, D. M.; Praveen, P.S.; Tiwari, S. Precipitation chemistry at different locations in India. In Proceedings of CAD Workshop, Hyderabad, India, Nov–Dec 2006.
57. Das, R.; Das, S.N.; Misra, V.N.; Chemical composition of rainwater and dust fall at Bhubaneswar in the east coast of India. Atmos. Environ. **2005**, *34*, 5908–5916.
58. Das, N.; Das, R.; Chaudhury, G.R.; Das, S.N. Chemical composition of precipitation at background level. Atmos. Res. **2010**, *95*, 108–113.
59. Kulshrestha, U.C.; Jain, M.; Sekar, R.; Vairamani, M.; Sarkar, A.K.; Parashar, D.C. Chemical characteristics and source apportionment of aerosols over Indian Ocean during INDOEX-1999. Curr. Sci. **2001**, *80*, 180–185.
60. Granat, L.; Norman, M.; Leck, C.; Kulshrestha, U.C.; Rodhe, H. Wet scavenging of sulfur and other compounds during INDOEX. J. Geophys. Res. **2002**, *107*, D19, 8025, doi:1011029/2001JD000499.
61. Trends in Precipitation Chemistry in the United States, 198394: An Analysis of the Effects in 1995 of Phase I of the Clean Air Act Ammendments of 1990, Title IV, http://pvbs.usgs.gov/acidrain/index.html#tables.
62. Kulshrestha, M.J.; Kulshrestha, U.C.; Parashar, D.C.; Vairamani, M. Estimation of SO_4 contribution by dry deposition of SO_2 onto the dust particles in India. Atmos. Environ. **2003**, 37, 3057–3063.
63. Kuylenstierna, J.C.I.; Rodhe, H.; Cinderby, S.; Hicks, K. Acidification in developing countries: Ecosystem sensitivity and the critical load approach on a global scale. Ambio **2001**, *30*, 20–28.
64. Alcamo, J.; Shaw, R.; Hordijk, L., Eds. *The RAINS Model of Acidification. Science and Strategies in Europe*. Kluwer Academic Publishers: Dordrecht, Netherlands, 1990.
65. Granat, L. Luft-och nederbordskemiska stationsnatet inom PMK. Report 3942, Meteorological Institute at Stockholm University: Stockholm, 1990.
66. Hara, H. Temporal variation of wet deposition in the EANET region during 2000–2004. In Proceedings of CAD Workshop, Hyderabad, India, Nov–Dec 2006.
67. U.S. Environmental Protection Agency (USEPA). *Acid Rain Program, 2001 Progress report;* U.S. Environmental Protection Agency: Washington, DC, available at http://www.epa.gov/airmarkets/cmprpt/arp01/2001report.pdf.
68. U.S. Environmental Protection Agency (USEPA). *Acid Rain Program: Overview, Environmental Benefits*, available at http://www.epa.gov/airmarkets/arp/overview.html. Retrieved on March 2, 2011.
69. U.S. Environmental Protection Agency (USEPA). Canada- United States agreement, 1994 Progress report, EPA/430/ R-94/013; Washington, DC. Retrieved on March 2, 2011.
70. Menz, F.C.; Seip, F.M. Acid rain in Europe and the United States: An update. Environ. Sci. Policy **2004**, *7*, 253–265.
71. Charles, D.F.; Norton, S.A. Paleolimnological evidence for trends in the atmospheric deposition of acids and heavy metals. In *Atmospheric Deposition: Historic Trends and Spatial Patterns*; Norton, S.A., Ed.; National Academy Press: Washington, DC, USA, 1985; 86–105.

72. Battarbee, R.W.; Flower, R.J.; Stevenson, A.C.; Harriman, R.; Appleby, P.G. Diatom and chemical evidence for reversibility of acidification of Scottish Lochs. Nature (London) **1988**, *332*, 530–532.
73. Jenkins, A.; Whitehead, P.G.; Cosby, B.J.; Birks, H.J.B. Modelling long term acidification: A comparison with diatom reconstructions and the implications for reversibility. Philos. Trans. R. Soc. London **1990**, *B327*, 435–440.
74. Evans, L.S.; Lewin, K.F.; Parri, M.J.; Cunningham, E.A Productivity of field-grown soybeans exposed to simulated acid rain. New Phytol. **1983**, *93*, 377–388.
75. Rehfuess, K.E. On the impact of acid precipitation in forest ecosystems. Forstwiss. Centralbl. **1981**, *100*, 363–370.
76. Cowling, E.B. Regional declines on forests in Europe and North America: The possible role of airborne chemicals. In *Aerosols, Research Risk Assessment and Control Strategies;* Lewis: Chelsea, MI, USA, 1986; 855–864.
77. Hedin, L.O.; Granata, L.; Likens, G.E.; Buishand, T.A.; Galloway, J.N.; Butler, T.J.; Rodhe, H. Steep decline in atmospheric base cations in regions of Europe and north America. Nature **1994**, *367*, 351–367.
78. BERG, Building Effects Review Group. *The Effects of Acid Deposition on Building Materials in U.K.;* Department of Environment, HMSO: London, 1989.
79. Kucera, V. *Atmospheric Corrosion in Urban Air Pollution in Asia and Africa: The Approach of the RAPIDC Programme;* Kuylenstierna, J., Hicks, K., Eds.; Stockholm Environment Institute (SEI): Stockholm, Sweden, 2002.

44

Acid Rain: Nitrogen Deposition

George F. Vance

Introduction

Air pollution has occurred naturally since the formation of the Earth's atmosphere; however, the industrial era has resulted in human activities greatly contributing to global atmospheric pollution.[1,2] One of the more highly publicized and controversial aspects of atmospheric pollution is that of acidic deposition. Acidic deposition includes rainfall, acidic fogs, mists, snowmelt, gases, and dry particulate matter.[3] The primary origin of acidic deposition is the emission of sulfur dioxide (SO_2) and nitrogen oxides (NO_x) from fossil fuel combustion; electric power generating plants contribute approximately two-thirds of the SO_2 emissions and one-third of the NO_x emissions.[4]

Acidic materials can be transported long distances, some as much as hundreds of kilometers. For example, 30%–40% of the S deposition in the northeastern U.S. originates in industrial midwestern U.S. states.[5] After years of debate, U.S. and Canada have agreed to develop strategies that reduce acidic compounds originating from their countries.[5,6] In Europe, the small size of many countries means that emissions in one industrialized area can readily affect forests, lakes, and cities in another country. For example, approximately 17% of the acidic deposition falling on Norway originated in Britain and 20% in Sweden came from eastern Europe.[5]

The U.S. EPA National Acid Precipitation Assessment Program (NAPAP) conducted intensive research during the 1980s and 1990s that resulted in the "Acidic Deposition: State of the Science and Technology" that was mandated by the Acid Precipitation Act of 1980.[6] NAPAP Reports to Congress have been developed in accordance with the 1990 amendment to the 1970 Clean Air Act and present the expected benefits of the Acid Deposition Control Program,[6,7] http://www.nnic.noaa.gov/CENR/NAPAP/. Mandates include an annual 10 million ton or approximately 40% reduction in point-source SO_2 emissions below 1980 levels, with national emissions limit caps of 8.95 million tons from electric utility and 5.6 million tons from point-source industrial emissions. A reduction in NO_x of about 2 million tons from 1980 levels has also been set as a goal; however, while NO_x has been on the decline since 1980, projections estimate a rise in NO_x emissions after the year 2000. In 1980, the U.S. levels of SO_2 and NO_x emissions were 25.7 and 23.0 million tons, respectively.

Acidic deposition can impact buildings, sculptures, and monuments that are constructed using weatherable materials like limestone, marble, bronze, and galvanized steel,[7,8] http://www.nnic.noaa.gov/CENR/NAPAP/. While acid soil conditions are known to influence the growth of plants, agricultural impacts related to acidic deposition are of less concern due to the buffering capacity of these types of ecosystems.[2,5] When acidic substances are deposited in natural ecosystems, a number of adverse environmental effects are believed to occur, including damage to vegetation, particularly forests, and changes in soil and surface water chemistry.[9,10]

Sources and Distribution

Typical sources of acidic deposition include coal- and oil-burning electric power plants, automobiles, and large industrial operations (e.g., smelters). Once S and N gases enter the earth's atmosphere they react very rapidly with moisture in the air to form sulfuric (H_2SO_4) and nitric (HNO_3) acids.[2,3] The pH of natural rainfall in equilibrium with atmospheric CO2 is about 5.6; however, the pH of rainfall is less than 4.5 in many industrialized areas. The nature of acidic deposition is controlled largely by the geographic distribution of the sources of SO_2 and NO_x (Figure 1). In the midwestern and northeastern U.S., H_2SO_4 is the main source of acidity in precipitation because of the coal-burning electric utilities.[2] In the western U.S., HNO_3 is of more concern because utilities and industry burn coal with low S contents and populated areas are high sources of NO_x.[2]

Emissions of SO_2 and NO_x increased in the 20th century due to the accelerated industrialization in developed countries and antiquated processing practices in some undeveloped countries. However, there is some uncertainty as to the actual means by which acidic deposition affects our environment,[11,12] http://nadp.sws.uiuc.edu/isopleths/maps1999/. Chemical and biological evidence, however, indicates that atmospheric deposition of H_2SO_4 caused some New England lakes to decrease in alkalinity.[13,14] Many scientists are reluctant to over-generalize cause and effect relationships in an extremely complex environmental problem. Although, the National Acid Deposition Assessment Program has concluded there were definite consequences due to acidic deposition that warrant

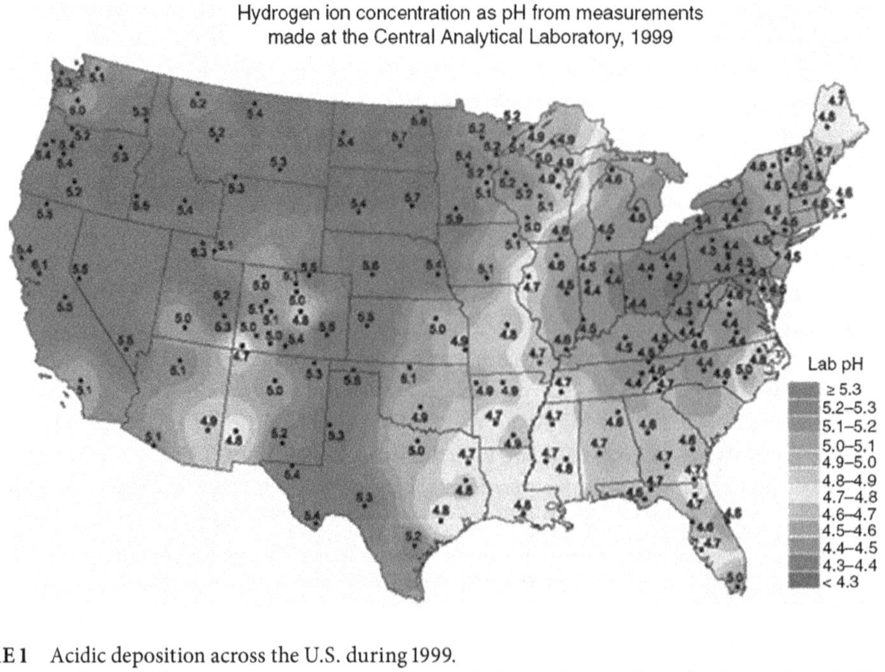

FIGURE 1 Acidic deposition across the U.S. during 1999.
Source: National Atmospheric Deposition Program/National Trends Network http://nadp.sw.uluc.edu.[11]

remediation[6,7] http://www.nnic.noaa.gov/CENR/NAPAP/. Since 1995, when the 1990 Clean Air Act Amendment's Title IV reduction in acidic deposition was implemented, SO_2 and NO_x emissions have, respectively, decreased and remained constant during the late 1990s.[4]

Both H_2SO_4 and HNO_3 are important components of acidic deposition, with volatile organic compounds and inorganic carbon also components of acidic deposition- related emissions. Pure water has a pH of 7.0, natural rainfall about 5.6, and severely acidic deposition less than 4.0. Uncontaminated rainwater should be pH 5.6 due to CO_2 chemistry and the formation of carbonic acid. The pH of most soils ranges from 3.0 to 8.0.[2] When acids are added to soils or waters, the decrease in pH that occurs depends greatly on the system's buffering capacity, the ability of a system to maintain its present pH by neutralizing added acidity. Clays, organic matter, oxides of Al and Fe, and Ca and Mg carbonates (limestones) are the components responsible for pH buffering in most soils. Acidic deposition, therefore, will have a greater impact on sandy, low organic matter soils than those higher in clay, organic matter, and carbonates. In fresh waters, the primary buffering mechanism is the reaction of dissolved bicarbonate ions with H^+ according to the following equation:

$$H^+ + HCO_3^- = H_2O + CO_2 \qquad (1)$$

Human Health Effects

Few direct human health problems have been attributed to acidic deposition. Long-term exposure to acidic deposition precursor pollutants such as ozone (O_3) and NO_x, which are respiratory irritants, can cause pulmonary edema.[5,6] Sulfur dioxide (SO_2) is also a known respiratory irritant, but is generally absorbed high in the respiratory tract.

Indirect human health effects due to acidic deposition are more important. Concerns center around contaminated drinking water supplies and consumption of fish that contain potential toxic metal levels. With increasing acidity (e.g., lower pH levels), metals such as mercury, aluminum, cadmium, lead, zinc, and copper become more bioavailable.[2] The greatest human health impact is due to the consumption of fish that bioaccumulate mercury; freshwater pike and trout have been shown to contain the highest average concentrations of mercury.[5,15] Therefore, the most susceptible individuals are those who live in an industrial area, have respiratory problems, drink water from a cistern, and consume a significant amount of freshwater fish.

A long-term urban concern is the possible impact of acidic deposition on surface-derived drinking water. Many municipalities make extensive use of lead and copper piping, which raises the question concerning human health effects related to the slow dissolution of some metals (lead, copper, zinc) from older plumbing materials when exposed to more acidic waters. Although metal toxicities due to acidic deposition impacts on drinking waters are rare, reductions in S and N fine particles expected by 2010 based on Clean Air Act Amendments will result in annual public health benefits valued at $50 billion with reduced mortality, hospital admissions and emergency room visits.[16]

Structural Impacts

Different types of materials and cultural resources can be impacted by air pollutants. Although the actual corrosion rates for most metals have decreased since the 1930s, data from three U.S. sites indicate that acidic deposition may account for 31%–78% of the dissolution of galvanized steel and copper,[7,8] http://www.nnic.noaa.gov/CENR/NAPAP/. In urban or industrial settings, increases in atmospheric acidity can dissolve carbonates (e.g., limestone, marble) in buildings and other structures. Deterioration of stone products by acidic deposition is caused by: 1) erosion and dissolution of materials and surface details; 2) alterations (blackening of stone surfaces); and 3) spalling (cracking and spalling of stone surfaces due to accumulations of alternation crusts.[8] Painted surfaces can be discolored or etched, and there may also be degradation of organic binders in paints.[8]

Ecosystem Impacts

It is important to examine the nature of acidity in soil, vegetation, and aquatic environments. Damage from acidification is often not directly due to the presence of excessive H^+, but is caused by changes in other elements. Examples include increased solubilization of metal ions such as Al^{3+} and some trace elements (e.g., Mn^{2+}, Pb^{2+}) that can be toxic to plants and animals, more rapid losses of basic cations (e.g., Ca^{2+}, Mg^{2+}), and the creation of unfavorable soil and aquatic environments for different fauna and flora.

Soils

Soil acidification is a natural process that occurs when precipitation exceeds evapotranspiration.[2] "Natural" rainfall is acidic (pH of ~5.6) and continuously adds a weak acid (H_2CO_3) to soils. This acidification results in a gradual leaching of basic cations (Ca^{2+} and Mg^{2+}) from the uppermost soil horizons, leaving Al^{3+} as the dominant cation that can react with water to produce H^+. Most of the acidity in soils between pH 4.0 and 7.5 is due to the hydrolysis of Al^{3+},[17,18] http://www.epa.gov/airmarkets/acidrain/effects/index.html. Other acidifying processes include plant and microbial respiration that produces CO_2, mineralization and nitrification of organic N, and the oxidation of FeS_2 in soils disturbed by mining or drainage.[2] In extremely acidic soils ($pH < 4.0$), strong acids such as H_2SO_4 are a major component.

The degree of accelerated acidification depends both upon the buffering capacity of the soil and the use of the soil. Many of the areas subjected to the greatest amount of acidic deposition are also areas where considerable natural acidification occurs.[19] Forested soils in the northeastern U.S. are developed on highly acidic, sandy parent materials that have undergone tremendous changes in land use in the past 200 years. However, clear-cutting and burning by the first European settlers have been almost completely reversed and many areas are now totally reforested.[5] Soil organic matter that accumulated over time represents a natural source of acidity and buffering. Similarly, greater leaching or depletion of basic cations by plant uptake in increasingly reforested areas balances the significant inputs of these same cations in precipitation.[20,21] Acidic deposition affects forest soils more than agricultural or urban soils because the latter are routinely limed to neutralize acidity. Although it is possible to lime forest soils, which is done frequently in some European countries, the logistics and cost often preclude this except in areas severely impacted by acidic deposition.[5]

Excessively acidic soils are undesirable for several reasons. Direct phytotoxicity from soluble Al^{3+} or Mn^{2+} can occur and seriously injure plant roots, reduce plant growth, and increase plant susceptibility to pathogens.[21] The relationship between Al^{3+} toxicity and soil pH is complicated by the fact that in certain situations organic matter can form complexes with Al^{3+} that reduce its harmful effects on plants.[18] Acid soils are usually less fertile because of a lack of important basic cations such as K^+, Ca^{2+}, and Mg^{2+}. Leguminous plants may fix less N_2 under very acidic conditions due to reduced rhizobial activity and greater soil adsorption of Mo by clays and Al and Fe oxides.[2] Mineralization of N, P, and S can also be reduced because of the lower metabolic activity of bacteria. Many plants and microorganisms have adapted to very acidic conditions (e.g., $pH < 5.0$). Examples include ornamentals such as azaleas and rhododendrons and food crops such as cassava, tea, blueberries, and potatoes.[5,22] In fact, considerable efforts in plant breeding and biotechnology are directed towards developing Al- and Mn-tolerant plants that can survive in highly acidic soils.

Agricultural Ecosystems

Acidic deposition contains N and S that are important plant nutrients. Therefore, foliar applications of acidic deposition at critical growth stages can be beneficial to plant development and reproduction. Generally, controlled experiments require the simulated acid rain to be pH 3.5 or less in order to produce injury to certain plants.[22] The amount of acidity needed to damage some plants is 100 times greater than natural rainfall. Crops that respond negatively in simulated acid rain studies include garden beets, broccoli, carrots, mustard greens, radishes, and pinto beans, with different effects for some cultivars. Positive responses to acid rain have been identified with alfalfa, tomato, green pepper, strawberry, corn, lettuce, and some pasture grass crops.

Agricultural lands are maintained at pH levels that are optimal for crop production. In most cases the ideal pH is around pH 6.0–7.0; however, pH levels of organic soils are usually maintained at closer to pH 5.0. Because agricultural soils are generally well buffered, the amount of acidity derived from atmospheric inputs is not sufficient to significantly alter the overall soil pH.[2] Nitrogen and S soil inputs from acidic deposition are beneficial, and with the reduction in S atmospheric levels mandated by 1990 amendments to the Clean Air Act, the S fertilizer market has grown. The amount of N added to agricultural ecosystems as acidic deposition is rather insignificant in relation to the 100–300 kg N/ha/yr required of most agricultural crops.

Forest Ecosystems

Perhaps the most publicized issue related to acidic deposition has been widespread forest decline. For example, in Europe estimates suggest that as much as 35% of all forests have been affected.[23] Similarly, in the U.S. many important forest ranges such as the Adirondacks of New York, the Green Mountains of Vermont, and the Great Smoky Mountains in North Carolina have experienced sustained decreases in tree growth for several decades.[6] Conclusive evidence that forest decline or dieback is caused solely be acidic deposition is lacking and complicated by interactions with other environmental or biotic factors. However, NAPAP research[6] has confirmed that acidic deposition has contributed to a decline in high-elevation red spruce in the northeastern U.S. In addition, nitrogen saturation of forest ecosystems from atmospheric N deposition is believed to result in increased plant growth, which in turn increases water and nutrient use followed by deficiencies that can cause chlorosis and premature needle-drop as well as increased leaching of base cations from the soil.[24]

Acidic deposition on leaves may enter directly through plant stomates.[1,22] If the deposition is sufficiently acidic (pH ~ 3.0), damage can also occur to the waxy cuticle, increasing the potential for direct injury of exposed leaf mesophyll cells. Foliar lesions are one of the most common symptoms. Gaseous compounds such as SO_2 and SO_3 present in acidic mists or fogs can also enter leaves through the stomates, form H_2SO_4 upon reaction with H_2O in the cytoplasm, and disrupt many metabolic processes. Leaf and needle necrosis occurs when plants are exposed to high levels of SO_2 gas, possibly due to collapsed epidermal cells, eroded cuticles, loss of chloroplast integrity and decreased chlorophyll content, loosening of fibers in cell walls and reduced cell membrane integrity, and changes in osmotic potential that cause a decrease in cell turgor.

Root diseases may also increase in excessively acidic soils. In addition to the damages caused by exposure to H_2SO_4 and HNO_3, roots can be directly injured or their growth rates impaired by increased concentrations of soluble Al^{3+} and Mn^{2+} in the rhizosphere,[2,25] http://nadp.sws.uiuc.edu. Changes in the amount and composition of these exudates can then alter the activity and population diversity of soil-borne pathogens. The general tendency associated with increased root exudation is an enhancement in microbial populations due to an additional supply of carbon (energy). Chronic acidification can also alter nutrient availability and uptake patterns.[8,22]

Long-term studies in New England suggest acidic deposition has caused significant plant and soil leaching of base cations,[1,21] resulting in decreased growth of red spruce trees in the White Mountains.[6] With reduction in about 80% of the airborne base cations, mainly Ca^{2+} but also Mg^{2+}, from 1950 levels, researchers suggest forest growth has slowed because soils are not capable of weathering at a rate that can replenish essential nutrients. In Germany, acidic deposition was implicated in the loss of soil Mg^{2+} as an accompanying cation associated with the downward leaching of SO_4^{2-}, which ultimately resulted in forest decline.[2] Several European countries have used helicopters to fertilize and lime forests.

Aquatic Ecosystems

Ecological damage to aquatic systems has occurred from acidic deposition. As with forests, a number of interrelated factors associated with acidic deposition are responsible for undesirable changes. Acidification of aquatic ecosystems is not new. Studies of lake sediments suggest that increased acidification began in the mid-1800s, although the process has clearly accelerated since the 1940s.[15] Current studies indicate there is significant S mineralization in forest soils impacted by acidic deposition and that the SO_4^{2-} levels in adjacent streams remain high, even though there has been a decrease in the amount of atmospheric-S deposition.[24]

Geology, soil properties, and land use are the main determinants of the effect of acidic deposition on aquatic chemistry and biota. Lakes and streams located in areas with calcareous geology resist acidification more than those in granitic and gneiss materials.[16] Soils developed from calcareous parent materials are generally deeper and more buffered than thin, acidic soils common to granitic areas.[2] Land management decisions also affect freshwater acidity. Forested watersheds tend to contribute more acidity than those dominated by meadows, pastures, and agronomic ecosystems.[8,14,20] Trees and other vegetation in forests are known to "scavenge" acidic compounds in fogs, mists, and atmospheric particulates. These acidic compounds are later deposited in forest soils when rainfall leaches forest vegetation surfaces. Rainfall below forest canopies (e.g., throughfall) is usually more acidic than ambient precipitation. Silvicultural operations that disturb soils in forests can increase acidity by stimulating the oxidization of organic N and S, and reduced S compounds such as FeS_2.[2]

A number of ecological problems arise when aquatic ecosystems are acidified below pH 5.0, and particularly below pH 4.0. Decreases in biodiversity and primary productivity of phytoplankton, zooplankton, and benthic invertebrates commonly occur.[15,16] Decreased rates of biological decomposition of organic matter have occasionally been reported, which can then lead to a reduced supply of nutrients.[20] Microbial communities may also change, with fungi predominating over bacteria. Proposed mechanisms to explain these ecological changes center around physiological stresses caused by exposure of biota to higher concentrations of Al^{3+}, Mn^{2+}, and H^+ and lower amounts of available Ca^{2+}.[15] One specific mechanism suggested involves the disruption of ion uptake and the ability of aquatic plants to regulate Na^+, K^+, and Ca^{2+} export and import from cells.

Acidic deposition is associated with declining aquatic vertebrate populations in acidified lakes and, under conditions of extreme acidity, of fish kills. In general, if the water pH remains above 5.0, few problems are observed; from pH 4.0 to 5.0 many fish are affected, and below pH 3.5 few fish can survive.[23] The major cause of fish kill is due to the direct toxic effect of Al^{3+}, which interferes with the role Ca^{2+} plays in maintaining gill permeability and respiration. Calcium has been shown to mitigate the effects of Al^{3+}, but in many acidic lakes the Ca^{2+} levels are inadequate to overcome Al^{3+} toxicity. Low pH values also disrupt the Na^+ status of blood plasma in fish. Under very acidic conditions, H^+ influx into gill membrane cells both stimulates excessive efflux of Na^+ and reduces influx of Na^+ into the cells. Excessive loss of Na^+ can cause mortality. Other indirect effects include reduced rates of reproduction, high rates of mortality early in life or in reproductive phases of adults, and migration of adults away from acidic areas.[16] Amphibians are affected in much the same manner as fish, although they are somewhat less sensitive to Al^{3+} toxicity. Birds and small mammals often have lower populations and lower reproductive rates in areas adjacent to acidified aquatic ecosystems. This may be due to a shortage of food due to smaller fish and insect populations or to physiological stresses caused by consuming organisms with high Al^{3+} concentrations.

Reducing Acidic Deposition Effects

Damage caused by acidic deposition will be difficult and extremely expensive to correct, which will depend on our ability to reduce S and N emissions. For example, society may have to burn less fossil fuel, use cleaner energy sources and/or design more efficient "scrubbers" to reduce S and N gas entering our atmosphere. Despite the firm conviction of most nations to reduce acidic deposition, it appears that the staggering costs of such actions will delay implementation of this approach for many years. The 1990 amendments to the Clean Air Act are expected to reduce acid-producing air pollutants from electric power plants. The 1990 amendments established emission allowances based on a utilities' historical fuel use and SO_2 emissions, with each allowance representing 1 ton of SO_2 that can bought, sold or banked for future use,[4,6,7] http://www.nnic.noaa.gov/CENR/NAPAP/. Short-term remedial actions for acidic deposition are available and have been successful in some ecosystems. Liming of lakes and some forests (also fertilization with trace elements and Mg^{2+}) has been practiced in European counties for over 50 years.[16,23] Hundreds of Swedish and Norwegian lakes have been successfully limed in the past 25 years. Lakes with short mean residence times for water retention may need annual or biannual liming; others may need to be limed every 5–10 years. Because vegetation in some forested ecosystems has adapted to acidic soils, liming (or over-liming) may result in an unpredictable and undesirable redistribution of plant species.

References

1. Smith, W.H. Acid rain. In *The Wiley Encyclopedia of Environmental Pollution and Cleanup*; Meyers, R.A., Dittrick, D.K., Eds.; Wiley: New York, 1999; 9–15.
2. Pierzynski, G.M.; Sims, J.T.; Vance, G.F. *Soils and Environmental Quality;* CRC Press: Boca Raton, FL, 2000; 459 pp.
3. Wolff, G.T. Air pollution. In *The Wiley Encyclopedia of Environmental Pollution and Cleanup;* Meyers, R.A., Dittrick, D.K., Eds.; Wiley: New York, 1999; 48–65.
4. U.S. Environmental protection agency. In *Progress Report on the EPA Acid Rain Program;* EPA-430-R-99-011; U.S. Government Printing Office: Washington, DC, 1999; 20 pp.
5. Forster, B.A. *The Acid Rain Debate: Science and Special Interests in Policy Formation;* Iowa State University Press: Ames, IA, 1993.
6. *National Acid Precipitation Assessment Program Task Force Report,* National Acid Precipitation Assessment Program 1992 Report to Congress; U.S. Government Printing Office: Pittsburgh, PA, 1992; 130 pp.
7. *National Science and Technology Council,* National Acid Precipitation Assessment Program Biennial Report to Congress: An Integrated Assessment; 1998 (accessed July 2001).
8. Charles, D.F., Ed. The acidic deposition phenomenon and its effects: critical assessment review papers. In *Effects Sciences;* EPA-600/8-83-016B; U.S. Environmental Protection Agency: Washington, DC, 1984; Vol. 2.
9. McKinney, M.L.; Schoch, R.M. *Environmental Science: Systems and Solutions*; Jones and Bartlett Publishers: Sudbury, MA, 1998.
10. United Nations, World band and World resources institute. In *World Resources: People and Ecosystems—The Fraying Web of Life*; Elsevier: New York, 2000.
11. *National Atmospheric Deposition Program (NRSP-3)/National Trends Network.* Isopleth Maps. NADP Program Office, Illinois State Water Survey, 2204 Griffith Dr., Champaign, IL, 61820, 2000 (accessed July 2001).
12. Council on environmental quality. In *Environmental Quality, 18th and 19th Annual Reports;* U.S. Government Printing Office: Washington, DC, 1989.
13. Charles, D.F., Ed.; *Acid Rain Research: Do We Have Enough Answers*? Proceedings of a Speciality Conference. Studies in Environmental Science #64, Elsevier: New York, 1995.
14. Kamari, J. *Impact Models to Assess Regional Acidification;* Kluwer Academic Publishers: London, 1990.
15. Charles, D.F., Ed. *Acidic Deposition and Aquatic Ecosystems;* Springer-Verlag: New York, 1991.
16. Mason, B.J. *Acid Rain: Its Causes and Effects on Inland Waters;* Oxford University Press: New York, 1992.

17. U.S. environmental protection agency. Effects of acid rain: human health. In *EPA Environmental Issues Website*. Update June 26, 2001 (accessed July 2001).
18. Marion, G.M.; Hendricks, D.M.; Dutt, G.R.; Fuller, W.H. Aluminum and silica solubility in soils. Soil Science 1976, 121, 76–82.
19. Kennedy, I.R. *Acid Soil and Acid Rain*; Wiley: New York, 1992.
20. Reuss, J.O.; Johnson, D.W. *Acid Deposition and the Acidification of Soils and Waters*; Springer-Verlag: New York, 1986.
21. Likens, G.E.; Driscoll, C.T.; Buso, D.C. Long-term effects of acid rain: response and recovery of a forest ecosystem. Science 1996, 272, 244–246.
22. Linthurst, R.A. *Direct and Indirect Effects of Acidic Deposition on Vegetation*; Butterworth Publishers: Stoneham, MA, 1984.
23. Bush, M.B. *Ecology of a Changing Planet;* Prentice- Hall: Upper Saddle River, NJ, 1997.
24. Alawell, C.; Mitchell, M.J.; Likens, G.E.; Krouse, H.R. Sources of stream sulfate at the hubbard brook experimental forest: long-term analyses using stable isotopes. Biogeochemistry 1999, 44, 281–299.
25. National Atmospheric Deposition Program. *Nitrogen in the Nation's Rain.* NADP Brochure 2000–01a (accessed July 2001).

45

Carbon Sequestration

Nathan E. Hultman

Introduction

The increasing likelihood of human-caused changes in climate could lead to undesirable impacts on ecosystems, economies, and human health and well-being. These potential impacts have prompted extensive assessment of options to reduce the magnitude and rate of future climate changes. Since climate changes are derived ultimately from increases in the concentrations of greenhouse gases (GHGs) in the atmosphere, such options must target either (a) reductions in the rate of inflow of GHGs to the atmosphere or (b) the removal of GHGs from the atmosphere once they have been emitted. Carbon sequestration refers to techniques from both categories that result in the storage of carbon that would otherwise be in the atmosphere as CO_2.

CO_2 is often targeted among the other GHGs because it constitutes the vast majority of GHG emissions by mass and accounts for three-fifths of the total anthropogenic contribution to climate change. Human emissions of CO_2 come primarily from fossil fuel combustion and cement production (80%), and land-use change (20%) that results in the loss of carbon from biomass or soil.

The rate of inflow of GHGs to the atmosphere can be reduced by a number of complementary options. For CO_2, mitigation options aim to displace carbon emissions by preventing the oxidation of biological or fossil carbon. These options include switching to lower-carbon fossil fuels, renewable energy, or nuclear power; using energy more efficiently; and reducing the rate of deforestation and land-use change. On the other hand, sequestration options that reduce emissions involve the capture and storage of carbon before it is released into the atmosphere.

CO_2 can also be removed directly from the atmosphere. While the idea of a large-scale, economically competitive method of technologically "scrubbing" CO_2 from the atmosphere is enticing, such technology currently does not exist. Policy has therefore focused on the biological process of carbon absorption through photosynthesis, either through expanding forested lands or, perhaps, enhancing photosynthesis in the oceans. This entry describes both the technological and biological approaches to carbon sequestration.

Technological Sequestration: Carbon Capture and Storage

The technological process of sequestering CO_2 requires two steps: first, the CO_2 must be separated from the industrial process that would otherwise emit it into the atmosphere; and second, the CO_2 must be stored in a reservoir that will contain it for a reasonable length of time. This process is therefore often referred to as carbon capture and storage (CCS) to distinguish it from the biological carbon sequestration that is described later.

Sources of Carbon

The best sites for CCS are defined by the efficiency of the capture technique, the cost of transport and sequestration, and the quantity of carbon available. The large capital requirements for CCS also dictate that large, fixed industrial sites provide the best opportunities. Therefore, although fossil-fueled transportation represents about 20% of current global CO_2 emissions, this sector presents no direct options for CCS at this time. The industrial sector, on the other hand, produces approximately 60% of current CO_2 emissions; most of these emissions come from large point sources which are ideal for CCS, such as power stations, oil refineries, petrochemical and gas reprocessing plants, and steel and cement works.[1]

Separation and Capture

Carbon capture requires an industrial source of CO_2; different industrial processes create streams with different CO2 concentrations. The technologies applied to capture the CO_2 will therefore vary according to the specific capture process.[2–4] Capture techniques can target one of three sources:

- Post-combustion flue gases
- Pre-combustion capture from gasification from power generation
- Streams of highly pure CO_2 from various industrial processes

Post-Combustion Capture

Conventional combustion of fossil fuels in air produces CO_2 streams with concentrations ranging from about 4 to 14% by volume. The low concentration of CO_2 in flue gas means that compressing and storing it would be uneconomical; therefore, the CO_2 needs to be concentrated before storage. Currently, the favored process for this task is chemical absorption, also known as chemical solvent scrubbing. Cooled and filtered flue gas is fed into an absorption vessel with a chemical solvent that absorbs the CO_2. The most common solvent for this process is monoethanolamine (MEA). The CO_2-rich solvent is then passed to another reaction vessel called a stripper column. It is then heated with steam to reverse the process, thus regenerating the solvent and releasing a stream of CO_2 with a purity greater than 90%.

Scrubbing with MEA and other amine solvents imposes large costs in energy consumption in the regeneration process; it requires large amounts of solvents since they degrade rapidly; and it imposes high equipment costs since the solvents are corrosive in the presence of O_2. Thus, until solvents are improved in these areas, flue gas separation by this method will remain relatively costly: just the steam and electric load from a coal power plant can increase coal consumption by 40% per net kWh_e. Estimates of the financial and efficiency costs from current technology vary. Plant efficiency is estimated to drop from over 40% to a range between 24 and 37%.[2,5,6] For the least efficient systems, carbon would cost up to \$70/t CO_2 and result in an 80% increase in the cost of electricity.[5] Other studies estimate an increase in the cost of electricity of 25%–75% for natural gas combined cycle and Integrated Gasification Combined Cycle (IGCC), and of 60%–115% for pulverized coal.[4] A small number of facilities currently practice flue gas separation with chemical absorption, using the captured CO_2 for urea production, foam blowing, carbonated beverages, and dry ice production.

In addition, several developments may improve the efficiency of chemical absorption.

Several other processes have been proposed for flue-gas separation. Adsorption techniques use solids with high surface areas, such as activated carbon and zeolites, to capture CO_2. When the materials become saturated, they can be regenerated (releasing CO_2) by lowering pressure, raising temperature, or applying a low-voltage electric current. A membrane can be used to concentrate CO_2, but since a single pass through a membrane cannot achieve a great change in concentration, this process requires multiple passes or multiple membranes. An alternative use for membranes is to use them to increase the efficiency of the chemical absorption. In this case, a membrane separating the flue gas from the absorption solvent allows a greater surface area for the reaction, thus reducing the size and energy requirements of the absorption and stripper columns. *Cryogenic* techniques separate CO_2 from other gases by condensing or freezing it. This process requires significant energy inputs and the removal of water vapor before freezing.

One of the main limitations to flue-gas separation is the low pressure and concentration of CO_2 in the exhaust. An entirely different approach to post-combustion capture is to dramatically increase the concentration of CO_2 in the stream by burning the fuel in highly enriched oxygen rather than air. This process, called oxyfuel combustion, produces streams of CO_2 with a purity greater than 90%. The resulting flue gas will also contain some H_2O that can be condensed and removed, and the remaining high-purity CO_2 can be compressed for storage. Though significantly simpler on the exhaust side, this approach requires a high concentration of oxygen for the intake air. While this process alone may consume 15% of a plant's electric output, the separated N_2, Ar, and other trace gases also can be sold to offset some of the cost. Oxyfuel systems can be retrofitted onto existing boilers and furnaces.

Pre-Combustion Capture

Another approach involves removing the carbon from fossil fuels before combustion. First, the fuel is decomposed in the absence of oxygen to form a hydrogen-rich fuel called synthesis gas. Currently, this process of gasification is already in use in ammonia production and several commercial power plants fed by coal and petroleum byproducts; these plants can use lower-purity fuels and the energy costs of generating synthesis gas are offset by the higher combustion efficiencies of gas turbines; such plants are called IGCC plants. Natural gas can be transformed directly by reacting it with steam, producing H_2 and CO_2. While the principle of gasification is the same for all carbonaceous fuels, oil and coal require intermediate steps to purify the synthesis fuel and convert the byproduct CO into CO_2.

Gasification results in synthesis gas that contains 35%–60% CO_2 (by volume) at high pressure (over 20 bar). While current installations feed this resulting mixture into the gas turbines, the CO_2 can also be separated from the gas before combustion. The higher pressure and concentration give a CO_2 partial pressure of up to 50 times greater than in the post-combustion capture of flue gases, which enables another type of separation technique of physical solvent scrubbing. This technique is well known from ammonia production and involves the binding of CO_2 to solvents that release CO_2 in the stripper under lower pressure. Solvents in this category include cold methanol, polyethelene glycol, propylene carbonate, and sulpholane. The resulting separated CO_2 is, however, near atmospheric pressure and requires compression before storage (some CO_2 can be recovered at elevated pressures, which reduces the compression requirement). With current technologies, the total cost of capture for IGCC is estimated to be greater than \$25 per ton of CO_2; plant efficiency is reduced from 43 to 37%, which raises the cost of electricity by over 25%.[5]

Pre-combustion capture techniques are noteworthy not only for their ability to remove CO_2 from fossil fuels for combustion in turbines, but also because the resulting synthesis gas is primarily H_2. They therefore could be an important element of a hydrogen-mediated energy system that favors the higher efficiency reactions of fuel cells over traditional combustion.[7]

Industrial CO_2 Capture

Many industrial processes release streams of CO_2 that are currently vented into the atmosphere. These streams, currently viewed as simple waste in an economically viable process, could therefore provide capture opportunities. Depending on the purity of the waste stream, these could be among the most economical options for CCS. In particular, natural gas processing, ethanol and hydrogen production, and cement manufacturing produce highly concentrated streams of CO_2. Not surprisingly, the first large-scale carbon sequestration program was run from a previously vented stream of CO_2 from the Sleipner gas-processing platform off the Norwegian coast.

Storage of Captured CO_2

Relatively small amounts of captured CO_2 might be re-used in other industrial processes such as beverage carbonation, mineral carbonates, or commodity materials such as ethanol or paraffins. Yet most captured CO_2 will not be re-used and must be stored in a reservoir. The two main routes for storing captured CO_2 are to inject it into geologic formations or into the ocean. However, all reservoirs have some rate of leakage and this rate is often not well known in advance. While the expected length of storage time is important (with targets usually in the 100–1000 year range), we must therefore also be reasonably confident that the reservoir will not leak more quickly than expected, and have appropriate measures to monitor the reservoir over time. Moreover, transporting CO_2 between the point of capture and the point of storage adds to the overall cost of CCS, so the selection of a storage site must account for this distance as well.

Geologic Sequestration

Geologic reservoirs—in the form of depleted oil and gas reservoirs, unmineable coal seams, and saline formations— comprise one of the primary sinks for captured CO_2. Estimates of total storage capacity in geologic reservoirs could be up to 500% of total emissions to 2050 (Table 1).

Captured CO_2 can be injected into depleted oil and gas reservoirs, or can be used as a means to enhance oil recovery from reservoirs nearing depletion. Because they held their deposits for millions of years before extraction, these reservoirs are expected to provide reliable storage for CO_2. Storage in depleted reservoirs has been practiced for years for a mixture of petroleum mining waste gases called "acid gas."

A petroleum reservoir is never emptied of all its oil; rather, extracting additional oil just becomes too costly to justify at market rates. An economically attractive possibility is therefore using captured

TABLE 1 CO_2 Reservoirs. Carbon Dioxide Storage Capacity Estimates. *E* Is Defined as the Total Global CO_2 Emissions from the Years 2000–2050 in IPCC's Business-as-Usual Scenario IS92A. Capacity Estimates Such as These Are Rough Guidelines Only and Actual Utilization Will Depend on Carbon Economics

	Storage Capacity	
Reservoir Type	Billion Tonnes CO_2	% of E
Coal basins	170	8%
Depleted oil reservoirs	120	6%
Gas basins	700	37%
Saline formations		
Terrestrial		276%
Off-shore		192%
Total	10.490	517%

Source: Dooley and Friedman.[8]

CO_2 to simultaneously increase the yield from a reservoir as it is pumped into the reservoir for storage. This process is called enhanced oil recovery. Standard oil recovery yields only about 30%–40% of the original petroleum stock. Drilling companies have years of experience with using compressed CO_2, a hydrocarbon solvent, to obtain an additional 10%–15% of the petroleum stock. Thus, captured CO_2 can be used to provide a direct economic benefit along with its placement in a reservoir. This benefit can be used to offset capture costs.

Coal deposits that are not economically viable because of their geologic characteristics provide another storage option. CO_2 pumped into these unmineable coal seams will adsorb onto the coal surface. Moreover, since the coal surface prefers to adsorb CO_2 to methane, injecting CO_2 into coal seams will liberate any coal bed methane (CBM) that can then be extracted and sold. This enhanced methane recovery is currently used in U.S. methane production, accounting for about 8% in 2002. Such recovery can be used to offset capture costs. One potential problem with this method is that the coal, as it adsorbs CO_2, tends to swell slightly. This swelling closes pore spaces and thus decreases rock permeability, which restricts both the reservoir for incoming CO_2 and the ability to extract additional CBM.

Saline formations are layers of porous sedimentary rock (e.g., sandstone) saturated with saltwater, and exist both under land and under the ocean. These layers offer potentially large storage capacity representing several hundred years' worth of CO_2 storage. However, experience with such formations is much more limited and thus the uncertainty about their long-term viability remains high. Moreover, unlike EOR or CBM recovery with CO_2, injecting CO_2 into saline formations produces no other commodity or benefit that can offset the cost. On the other hand, their high capacity and relative ubiquity makes them attractive options in some cases. Statoil's Sleipner project, for example, uses a saline aquifer for storage.

Research and experimentation with saline formations is still in early stages. To achieve the largest storage capacities, CO_2 must be injected below 800 m depth, where it will remain in a liquid or supercritical dense phase (supercritical point at 31°C, 71 bar). At these conditions, CO_2 will be buoyant (a density of approximately 600–800 kg/m^3) and will tend to move upward. The saline formations must therefore either be capped by a less porous layer or geologic trap to prevent leakage of the CO_2 and eventual decompression.[9] Over time, the injected CO_2 will dissolve into the brine and this mixture will tend to sink within the aquifer. Also, some saline formations exist in rock that contains Ca-, Mg-, and Fe-containing silicates that can form solid carbonates with the injected CO_2. The resulting storage as rock is highly reliable, though it may also hinder further injection by closing pore spaces. Legal questions may arise when saline formations, which are often geographically extensive, cross national boundaries or onto marine commons.

Ocean Direct Injection

As an alternative to geologic storage, captured CO_2 could be injected directly into the ocean at either intermediate or deep levels. The oceans have a very large potential for storing CO_2, equivalent to that of saline aquifers (~10^3 Gt). While the ocean's surface is close to equilibrium with atmospheric carbon dioxide concentrations, the deep ocean is not because the turnover time of the oceans is much slower (~5000 years) than the observed increases in atmospheric CO_2. Since the ocean will eventually absorb much of the atmospheric perturbation, injecting captured CO_2 into the oceans can therefore be seen as simply bypassing the atmospheric step and avoiding the associated climate consequences. Yet little is known about the process or effects—either ecological or geophysical—of introducing large quantities of CO_2 into oceanic water.

At intermediate depths (between 500 and 3000 m), CO_2 exists as a slightly buoyant liquid. At these depths, a stream of CO_2 could be injected via a pipe affixed either to ship or shore. The CO_2 would form a droplet plume, and these droplets would slowly dissolve into the seawater, disappearing completely before reaching the surface. Depressed pH values are expected to exist for tens of km downcurrent of

TABLE 2 Additional Costs to Power Generation from CCS. Approximate Capture and Storage Costs for Different Approaches to Power Plant Sequestration

	Cost of CCS
Fossil Type	¢ per kWh
Natural gas combined cycle	1–2
Pulverized coal	2–3
Coal IGCC	2–4

Source: Herzog and Golomb,[4] National Energy Technology Laboratory,[5] Dooley et al.,[11] and Freund and Davison.[12]

the injection site, though changing the rate of injection can moderate the degree of perturbation. In addition, pulverized limestone could be added to the injected CO_2 to buffer the acidity.

Below 3000 m, CO_2 becomes denser than seawater and would descend to the seafloor and pool there. Unlike intermediate injection, therefore, this method does not lead to immediate CO_2 dissolution in oceanic water; rather, the CO_2 is expected to dissolve into the ocean at a rate of about 0.1 m/y. Deep injection thus minimizes the rate of leakage to the surface, but could still have severe impacts on bottom-dwelling sea life.

The primary obstacles to oceanic sequestration are not technical but relate rather to this question of environmental impacts.[10] Oceanic carbon storage might affect marine ecosystems through the direct effects of a lower environmental pH; dissolution of carbonates on fauna with calcareous structures and microflora in calcareous sediments; impurities such as sulfur oxides, nitrogen oxides, and metals in the captured CO_2; smothering effects (deep injection only); and changes in speciation of metals and ammonia due to changes in pH. Few of these possibilities have been studied in sufficient detail to allow an informed risk assessment. In addition, the legality of dumping large quantities of CO_2 into the open ocean remains murky.

Overall Costs of CCS

The costs of CCS can be measured either as a cost per tonne of CO_2, or, for power generation, a change in the cost of electricity (Table 2). The total cost depends on the cost of capture, transport, and storage. Capture cost is mainly a function of parasitic energy losses and the capital cost of equipment. Transport cost depends on distance and terrain. Storage costs vary depending on the reservoir but are currently a few dollars per tonne of CO_2. The variety of approaches to CCS and the early stages of development make precise estimates of cost difficult, but current technology spans about \$25–\$85/t CO_2.

Biological Sequestration: Enhancing Natural Carbon Sinks

The previous sections have described processes by which CO_2 could be technologically captured and then stored. Photosynthesis provides an alternate route to capture and store carbon. Enhancing this biological process is therefore an alternative method of achieving lower atmospheric CO_2 concentrations by absorbing it directly from the air.

Terrestrial Carbon Sinks

Carbon sequestration in terrestrial ecosystems involves enhancing the natural sinks for carbon fixed in photosynthesis. This occurs by expanding the extent of ecosystems with a higher steady-state density of carbon per unit of land area. For example, because mature forest ecosystems contain more carbon per hectare than grasslands, expanding forested areas will result in higher terrestrial carbon storage.

Another approach is to encourage the additional storage of carbon in agricultural soils. The essential element in any successful sink enhancement program is to ensure that the fixed carbon remains in pools with long lives.

Afforestation involves planting trees on unforested or deforested land.[13,14] The most likely regions for forest carbon sequestration are Central and South America and Southeast Asia because of relatively high forest growth rates, available land, and inexpensive labor. However, the translation of forestry activities into a policy framework is complex. Monitoring the carbon changes in a forest is difficult over large areas, as it requires not only a survey of the canopy and understory, but also an estimate of the below-ground biomass and soil carbon. Some groups have voiced concern over the potential for disruption of social structures in targeted regions.

Soil carbon sequestration involves increasing soil carbon stocks through changes in agriculture, forestry, and other land use practices. These practices include mulch farming, conservation tillage, agroforestry and diverse cropping, cover crops, and nutrient management that integrates manure, compost, and improved grazing. Such practices, which offer the lowest-cost carbon sequestration, can have other positive effects such as soil and water conservation, improved soil structure, and enhanced soil fauna diversity. Rates of soil carbon sequestration depend on the soil type and local climate, and can be up to 1000 kg of carbon per hectare per year. Management practices can enhance sequestration for 20–50 years, and sequestration rates taper off toward maturity as the soil carbon pool becomes saturated. Widespread application of recommended management practices could offset 0.4 to 1.2 GtC/y, or 5%–15% of current global emissions.[15]

If sinks projects are to receive carbon credits under emissions trading schemes like that in the Kyoto Protocol, they must demonstrate that the project sequestered more carbon than a hypothetical baseline or business-as-usual case. They must also ensure that the carbon will remain in place for a reasonable length of time, and guard against simply displacing the baseline activity to a new location.

Ocean Fertilization

Vast regions of the open ocean have very little photosynthetic activity, though sunlight and major nutrients are abundant. In these regions, phytoplankton are often deprived of trace nutrients such as iron. Seeding the ocean surface with iron, therefore, might produce large phytoplankton blooms that absorb CO_2. As the plankton die, they will slowly sink to the bottom of the ocean, acting to transport the fixed carbon to a permanent burial in the seafloor. While some experimental evidence indicates this process may work on a limited scale, little is known about the ecosystem effects and potential size of the reservoir.[16]

Prospects for Carbon Sequestration

Carbon sequestration techniques—both technological and biological—are elements of a portfolio of options for addressing climate change. Current approaches hold some promise for tapping into the geologic, biologic, and oceanic potential for storing carbon. The costs of some approaches, especially the improved management of agricultural and forest lands, are moderate (Table 3). Yet these opportunities are not infinite and additional options will be necessary to address rising global emissions. Thus, the higher costs of current technological approaches are likely to drop with increasing deployment and changing market rates for carbon.

Possible developments include advanced CO_2 capture techniques focusing on membranes, ionic (organic salt) liquids, and microporous metal organic frameworks. Several alternative, but still experimental, sequestration approaches have also been suggested. Mineralization could convert CO_2 to stable minerals. This approach seeks, therefore, to hasten what in nature is a slow but exothermic weathering process that operates on common minerals like olivine, forsterite, or serpentines (e.g., through selected sonic frequencies). It is possible that CO_2 could be injected in sub-seafloor carbonates. Chemical looping

TABLE 3 Costs of Carbon Sequestration. Estimates for Sequestration Costs Vary Widely. Future Costs Will Depend on Rates of Technological Change

Sequestration Technique	Cost $ per T CO_2
Carbon capture and storage	26–84
Tree planting and agroforestry	10–210
Soil carbon sequestration	6–24

Source: Herzog and Golomb,[4] National Energy Technology Laboratory,[5] Williams[7] Dooley et al.[11] Freund and Davison,[12] Van Kooten,[13] and Richards and Stokes.[14]

describes a method for combusting fuels with oxygen delivered by a redox agent instead of by air or purified oxygen; it promises high efficiencies of energy conversion and a highly enriched CO_2 exhaust stream. Research also continues on microbial CO_2 conversion in which strains of microbes might be created to metabolize CO_2 to produce saleable commodities (succinic, malic, and fumeric acids). In addition, the nascent science of monitoring and verifying the storage of CO_2 will be an important element toward improving technical performance and public acceptance of sequestration techniques.

Acknowledgments

The author thanks the anonymous reviewer for helpful comments on an earlier draft.

References

1. Gale, J. Overview of CO_2 Emission Sources, Potential, Transport, and Geographical Distribution of Storage Possibilities. In *Proceedings of IPCC workshop on carbon dioxide capture and storage* Regina, Canada Nov 18–21, 2002 15–29. http://arch.rivm.nl/env/int/ipcc/pages_media/ ccs2002.html (accessed April 2005). A revised and updated version of the papers presented at the Regina workshop is now available. See, Metz, B.; Davidson, O.; de Coninck, H.; Loos, M.; Meyer, L., Special Report on Carbon Dioxide Capture and Storage. Intergovernmental Panel on Climate Change: Geneva, http://www.ipcc.ch. (accessed May 2006), 2006.
2. Thambimuthu, K.; Davison, J.; Gupta, M. CO_2 Capture and Reuse, In *Proceedings of IPCC workshop on carbon dioxide capture and storage,* Regina, Canada, Nov 18–21, 2002 31–52 http://arch.rivm.nl/env/int/ipcc/pages_media/ccs2002.html (accessed April 2005).
3. Gottlicher, G. *The Energetics of Carbon Dioxide Capture in Power Plants;* U.S. Department of Energy Office of Fossil Energy Washington, DC, 2004; 193
4. Herzog, H.; Golomb, D. Carbon capture and storage from fossil fuel use *Encyclopedia of Energy* Cleveland, C.J. Ed.; Elsevier Science: New York, 2004; 277–287.
5. National Energy Technology Laboratory *Carbon Sequestration: Technology Roadmap and Program Plan* U.S. Department of Energy Washington, DC, 2004.
6. Gibbins, J.R.; Crane, R.I.; Lambropoulos, D.; Booth, C.; Roberts, C.A.; Lord, M. Maximizing the Effectiveness of Post Combustion CO_2 Capture Systems In *Proceedings of the 7th International Conference on Greenhouse Gas Abatement,* Vancouver, Canada, Sept 5–9, 2004 Document E2-2. http://www.ghgt7.ca (accessed June 2004).
7. Williams, R.H. Decarbonized Fossil Energy Carriers and Their Energy Technology Competitors In *Proceedings of IPCC workshop on carbon dioxide capture and storage* Regina, Canada, November, 2002; 119–135, http://arch.rivm.nl/env/int/ipcc/pages_media/ccs2002.html (accessed April 2005).
8. Dooley, J.J.; Friedman, S.J. *A Regionally Disaggregated Global Accounting of CO_2 Storage Capacity: Data and* Assumptions Battelle National Laboratory Washington DC, 2004; 15.

9. Kårsted, O. Geological Storage, Including Costs and Risks, in Saline Aquifers In *Proceedings of IPCC workshop on carbon dioxide capture and storage* Regina, Canada 2002 Nov 1--21 53-60 http://arch.rivm.nl/env/int/ipcc/pages_media/ccs2002.html (accessed April 2005).
10. Johnston, P.; Santillo, D. Carbon Capture and Sequestration: Potential Environmental Impacts, In *Proceedings of IPCC workshop on carbon dioxide capture and storage*, Regina, Canada, Nov 18–21, 2002; 95–110. http://arch.rivm.nl/env/int/ipcc/pages_media/ccs2002.html (accessed April 2005).
11. Dooley, J.J.; Edmonds, J.A.; Dahowski, R.T.; Wise, M.A. Modeling Carbon Capture and Storage Technologies in Energy and Economic Models. In *Proceedings of IPCC workshop on carbon dioxide capture and storage,* Regina, Canada, Nov 18–21, 2002; 161–172. http://arch.rivm.nl/env/int/ipcc/pages_media/ccs2002.html (accessed April 2005).
12. Freund, P.; Davison, J. General Overview of Costs. In *Proceedings of IPCC workshop on carbon dioxide capture and storage,* Regina, Canada, Nov 18–21, 2002 79–93. http:// arch.rivm.nl/env/int/ipcc/pages_media/ccs2002.html (accessed April 2005).
13. Van Kooten, G.C.; Eagle, A.J.; Manley, J.; Smolak, T. How costly are carbon offsets? A meta-analysis of carbon forest sinks. Environ. Sci. Policy **2004,** 7, 239–251.
14. Richards, K.R.; Stokes, C. A review of forest carbon sequestration cost studies: a dozen years of research. Climatic Change **2004,** *68,* 1–48.
15. Lal, R. Soil carbon sequestration impacts on global climate change and food security. Science **2004,** *304,* 1623–1627.
16. Buesseler, K.O.; Andrews, J.E.; Pike, S.M.; Charette, M.A. The effects of iron fertilization on carbon sequestration in the southern ocean. Science **2004,** *304,* 414–417.

46

Energy Conservation

Ibrahim Dincer
and Adnan Midili

Introduction

Civilization began when people found out how to use fire extensively. They burned wood and obtained sufficiently high temperatures for melting metals, extracting chemicals, and converting heat into mechanical power, as well as for cooking and heating. During burning, the carbon in wood combines with O_2 to form carbon dioxide (CO_2), which then is absorbed by plants and converted back to carbon for use as a fuel again. Because wood was unable to meet the fuel demand, the Industrial Revolution began with the use of fossil fuels (e.g., oil, coal, and gas). Using such fuels has increased the CO_2 concentration in the air, leading to the beginning of global warming. Despite several warnings in the past about the risks of greenhouse- gas emissions, significant actions to reduce environmental pollution were not taken, and now many researchers have concluded that global warming is occurring. During the past two decades, the public has became more aware, and researchers and policymakers have focused on this and related issues by considering energy, the environment, and sustainable development.

Energy is considered to be a key catalyst in the generation of wealth and also a significant component in social, industrial, technological, economic, and sustainable development. This makes energy resources and their use extremely significant for every country. In fact, abundant and affordable energy is one of the great boons of modern industrial civilization and the basis of our living standard. It makes people's lives brighter, safer, more comfortable, and more mobile, depending on their energy demand and consumption. In recent years, however, energy use and associated greenhouse-gas emissions and their potential effects on the global climate change have been of worldwide concern.

Problems with energy utilization are related not only to global warming, but also to such environmental concerns as air pollution, acid rain, and stratospheric ozone depletion. These issues must be taken into consideration simultaneously if humanity is to achieve a bright energy future with minimal environmental impact. Because all energy resources lead to some environmental impact, it is reasonable to suggest that some (not all) of these concerns can be overcome in part through energy conservation efforts.

Energy conservation is a key element of energy policy and appears to be one of the most effective ways to improve end-use energy efficiency, and to reduce energy consumption and greenhouse-gas emissions in various sectors (industrial, residential, transportation, etc.). This is why many countries have recently started developing aggressive energy conservation programs to reduce the energy intensity of the their infrastructures, make businesses more competitive, and allow consumers to save money and to live more comfortably. In general, energy conservation programs aim to reduce the need for new generation or transmission capacity, to save energy, and to improve the environment. Furthermore, energy conservation is vital for sustainable development and should be implemented by all possible means, despite the fact that it has its own limitations. This is required not only for us, but for the next generation as well.

Considering these important contributions, the energy conservation phenomenon should be discussed in a comprehensive perspective. Therefore, the main objective of this entry is to present and discuss the world's primary energy consumption and production; major environmental problems; potential solutions to these issues; practical energy conservation aspects; research and development (R&D) in energy conservation, energy conservation, and sustainable development; energy conservation implementation plans; energy conservation measurements; and life-cycle costing (LCC) as an excellent tool in energy conservation. In this regard, this contribution aims to:

- Help explain main concepts and issues about energy conservation
- Develop relations between energy conservation and sustainability
- Encourage energy conservation strategies and policies
- Provide energy conservation methodologies
- Discuss relations between energy conservation and environmental impact
- Present some illustrative examples to state the importance of energy conservation and its practical benefits

In summary, this book contribution highlights the current environmental issues and potential solutions to these issues; identifies the main steps for implementing energy conservation programs and the main barriers to such implementations; and provides assessments for energy conservation potentials for countries, as well as various practical and environmental aspects of energy conservation.

World Energy Resources: Production and Consumption

World energy consumption and production are very important for energy conservation in the future. Economic activity and investment patterns in the global energy sector are still centered on fossil fuels, and fossil-fuel industries and energy-intensive industries generally have been skeptical about warnings of global warming and, in particular, about policies to combat it. The increase of energy consumption and energy demand indicates our dependence on fossil fuels. If the increase of fossil-fuel utilization continues in this manner, it is likely that the world will be affected by many problems due to fossil fuels. It follows from basic scientific laws that increasing amounts of CO_2 and other greenhouse gases will affect the global climate. The informed debate is not about the existence of such effects, but about their magnitudes and seriousness. At present, the concentration of CO_2 is approximately 30% higher than its preindustrial level, and scientists have already been able to observe a discernible human influence on the global climate.[1]

In the past, fossil fuels were a major alternative for overcoming world energy problems. Fossil fuels cannot continue indefinitely as the principal energy sources, however, due to the rapid increase of world energy demand and energy consumption. The utilization distribution of fossil-fuel types has changed significantly over the past 80 years. In 1925, 80% of the required energy was supplied from coal, whereas in the past few decades, 45% came from petroleum, 25% from natural gas, and 30% from coal. Due to world population growth and the advance of technologies that depend on fossil fuels, reserves of those fuels eventually will not be able to meet energy demand. Energy experts point out that reserves are less than 40 years for petroleum, 60 years for natural gas, and 250 years for coal.[2] Thus, fossil-fuel costs are likely to increase in the near future. This will allow the use of renewable energy sources such as solar, wind, and hydrogen. As an example, the actual data[3,4] and projections of world energy production and consumption from 1980 to 2030 are displayed in the following figures, and the curve equations for world energy production and consumption are derived as shown Table 1.

As presented in Figures 1 and 2, and in Table 2, the quantities of world primary energy production and consumption are expected to reach 14,499.2 and 13,466.5 Mtoe, respectively, by 2030. World population is now over six billion, double that of 40 years ago, and it is likely to double again by the middle of

TABLE 1 World Energy Production and Consumption Models through Statistical Analysis

Energy	Production (Mtoe)	Correlation Coefficient Coefficient	Consumption (Mtoe)	Correlation Coefficient
World primary	= 148.70 × Year − 287,369	0.998	= 139.62 × Year − 269,953	0.998
World oil	= 44.47 × Year − 85,374	0.997	= 42.18 ×Year − 80,840	0.998
World coal	= 20.05 × Year − 37,748	0.946	= 23.18 × Year − 43,973	0.968
World NG	= 45.73 × Year − 89,257	0.999	= 46.27 × Year − 90,347	0.999

Note: Mtoe, million tons of oil equivalent.

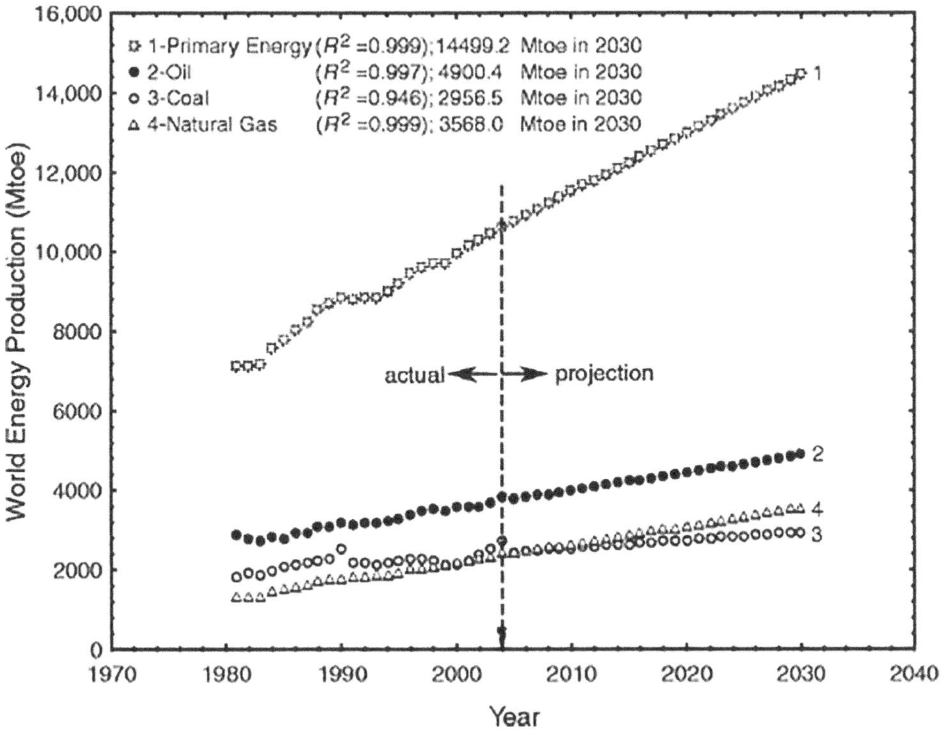

FIGURE 1 Variation of actual data taken from BP[3] and IEE,[4] and projections of annual world energy production. Mtoe, million tons of oil equivalent.
Source: BP[3] and IEE.[4]

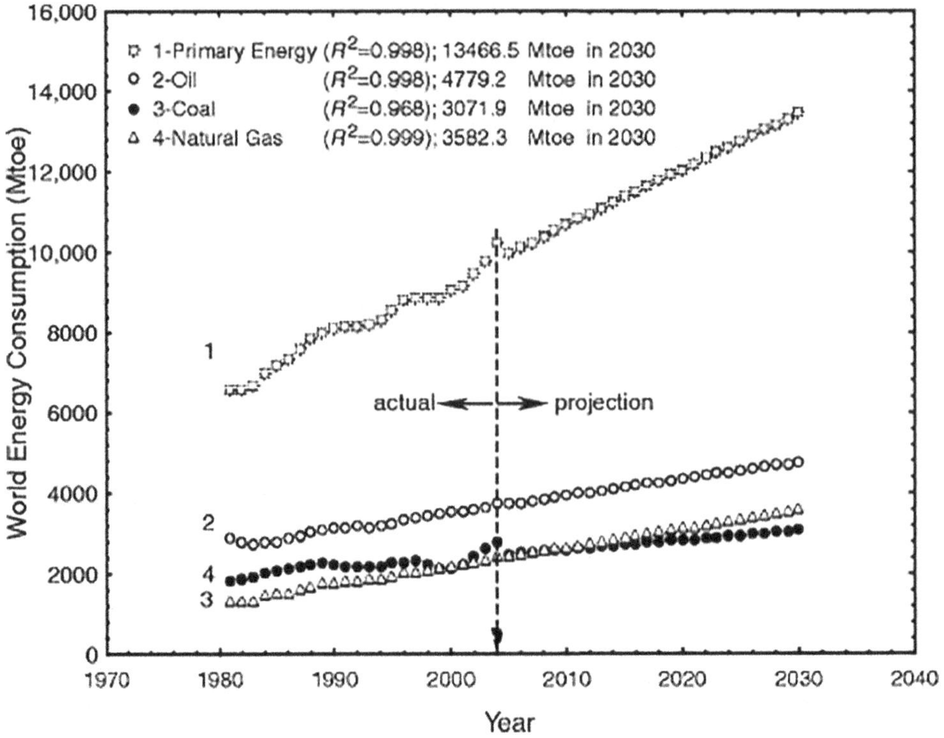

FIGURE 2 Variation of actual data taken from BP[3] and IEE,[4] and projections of annual world energy consumption. Mtoe, million tons of oil equivalent.
Source: BP[3] and IEE.[4]

TABLE 2 Some Extracted Values of World Primary and Fossil Energy Production and Consumption

Year	Primary Energy Production (Mtoe)	Primary Energy Consumption (Mtoe)	Oil Prod. (Mtoe)	Oil Cons. (Mtoe)	Coal Prod. (Mtoe)	Coal Cons. (Mtoe)	NG Prod. (Mtoe)	NG Cons. (Mtoe)
1994	8,996.9	8,310.1	3237.1	3204.4	2178.1	2185.5	1891.2	1876.7
2000	9,981.9	9,079.8	3614.0	3538.7	2112.4	2148.1	2189.9	2194.5
2006	10,930.3	10,115.7	3833.1	3767.0	2475.3	2515.7	2470.5	2471.8
2012	11,822.5	10,953.4	4099.9	4020.0	2595.6	2654.8	2744.9	2749.5
2018	12,714.7	11,791.1	4366.7	4273.1	2715.9	2793.8	3019.2	3027.1
2024	13,606.9	12,128.8	4633.5	4526.1	2836.2	2932.9	3293.6	3304.7
2030	14,499.2	13,466.5	4900.4	4779.2	2956.5	3071.9	3568.0	3582.3

Note: Mtoe, million tons of oil equivalent.

the 21st century. The world's population is expected to rise to about seven billion by 2010. Even if birth rates fall so that the world population becomes stable by 2030, the population still would be about ten billion. The data presented in Figures 1 and 2 are expected to cover current energy needs provided that the population remains constant. Because the population is expected to increase dramatically, however, conventional energy resource shortages are likely to occur, due to insufficient fossil-fuel resources. Therefore, energy conservation will become increasingly important to compensate for shortages of conventional resources.

Major Environmental Problems

One of the most important targets of modern industrial civilizations is to supply sustainable energy sources and to develop the basis of living standards based on these energy sources, as well as implementing energy conservation measures. In fact, affordable and abundant sustainable energy makes our lives brighter, safer, more comfortable, and more mobile because most industrialized and developing societies use various types of energy. Billions of people in undeveloped countries, however, still have limited access to energy. India's per-capita consumption of electricity, for example, is one-twentieth that of the United States. Hundreds of millions of Indians live "off the grid"- that is, without electricity-and cow dung is still a major fuel for household cooking. This continuing reliance on such preindustrial energy sources is also one of the major causes of environmental degradation.[5]

After many decades of using fossil fuels as a main energy source, significant environmental effects of fossil fuels became apparent. The essential pollutants were from greenhouse gases (e.g., CO_2, SO_2, and NO_2). Fossil fuels are used for many applications, including industry, residential, and commercial sectors. Increasing fossil-fuel utilization in transportation vehicles such as automobiles, ships, aircrafts, and spacecrafts has led to increasing pollution. Gas, particulate matter, and dust clouds in the atmosphere absorb a significant portion of the solar radiation directed at Earth and cause a decrease in the oxygen available for the living things. The threat of global warming has been attributed to fossil fuels.[2] In addition, the risk and reality of environmental degradation have become more apparent. Growing evidence of environmental problems is due to a combination of factors.

During the past two decades, environmental degradation has grown dramatically because of the sheer increase of world population, energy consumption, and industrial activities. Throughout the 1970s, most environmental analysis and legal control instruments concentrated on conventional pollutants such as SO_2, NO_x, particulates, and CO. Recently, environmental concern has extended to the control of micro or hazardous air pollutants, which are usually toxic chemical substances and harmful in small doses, as well as to that of globally significant pollutants such as CO_2. Aside from advances in environmental engineering science, developments in industrial processes and structures have led to new environmental problems.[6,7] In the energy sector, for example, major shifts to the road transport of industrial goods and to individual travel by cars has led to an increase in road traffic and, hence, to a shift in attention paid to the effects and sources of NO_x and to the emissions of volatile organic compounds (VOC). In fact, problems with energy supply and use are related not only to global warming, but also to such environmental concerns as air pollution, ozone depletion, forest destruction, and emission of radioactive substances. These issues must be taken into consideration simultaneously if humanity is to achieve a bright energy future with minimal environmental impact. Much evidence exists to suggest that the future will be negatively impacted if humans keep degrading the environment. Therefore, there is an intimate connection among energy conservation, the environment, and sustainable development. A society seeking sustainable development ideally must utilize only energy resources that cause no environmental impact (e.g., that release no emissions to the environment). Because all energy resources lead to some environmental impact, however, it is reasonable to suggest that some (not all) of the concerns regarding the limitations imposed on sustainable development by environmental emissions and their negative impacts can be overcome in part through energy conservation. A strong relation clearly exists between energy conservation and environmental impact, because for the same services or products, less resource utilization and pollution normally are associated with higher-efficiency processes.[8]

Table 3 summarizes the major environmental problems such as acid rain, stratospheric ozone depletion, and global climate change (greenhouse effect)-and their main sources and effects.

As shown in Figure 3, the world total CO_2 production is estimated to be 18,313.13 million tons in 1980, 25,586.7 million tons in 2006, 27,356.43 million tons in 2012, and 29,716.1 million tons in 2020 whereas fossil-fuel consumption is found to be 6092.2 million tons in 1980, 8754.5 million tons in 2006,

TABLE 3 Major Environmental Issues and Their Consequences

Issues	Description	Main Sources	Main Effects
Acid precipitation	Transportation and deposition of acids produced by fossil-fuel combustion (e.g., industrial boilers, transportation vehicles) over great distances through the atmosphere via precipitation on the earth on ecosystems	Emissions of SO_2, NO_x and volatile organic compounds (VOCs) (e.g., residential heating and industrial energy use account for 80% of SO_2 emissions)	Acidification of lakes, streams and ground waters, resulting in damage to fish and aquatic life; damage to forests and agricultural crops; and deterioration of materials, e.g., buildings, structures
Stratospheric ozone depletion	Distortion and regional depletion of stratospheric ozone layer though energy activities (e.g., refrigeration, fertilizers)	Emissions of CFCs, halons (chlorinated and brominated organic compounds) and N_2O (e.g. fossil fuel and biomass combustion account for 65%–75% of N_2O emissions)	Increased levels of damaging ultraviolet radiation reaching the ground, causing increased rates of skin cancer, eye damage and other harm to many biological species
Greenhouse effect	A rise in the earth's temperature as a result of the greenhouse gases	Emissions of carbon dioxide (CO_2), CH_4, CFCs, halons, N_2O, ozone and peroxyacetylnitrate (e.g., CO_2 releases from fossil fuel combustion (w50% from CO_2), CH_4 emissions from increased human activity)	Increased the earth's surface temperature about 0.68C over the last century and as a consequence risen sea level about 20 cm (in the next century by another 28C–48C and a rise between 30 and 60 cm); resulting in flooding of coastal settlements, a displacement of fertile zones for agriculture and food production toward higher latitudes, and a decreasing availability of fresh water for irrigation and other essential uses

Source: Dincer,[9] Dincer,[10] and Dincer.[11]

9424.3 million tons in 2012, and 10,317.2 million tons in 2020. These values show that the CO_2 production will probably increase if we continue utilizing fossil fuel. Therefore, it is suggested that certain energy conversion strategies and technologies should be put into practice immediately to reduce future environmental problems.

The climate technology initiative (CTI) is a cooperative effort by 23 Organization for Economic Cooperation and Development (OECD)/International Energy Agency (IEA) member countries and the European Commission to support the objectives of the united nations framework convention on climate change (UNFCCC). The CTI was launched at the 1995 Berlin Conference of the Parties to the UNFCCC. The CTI seeks to ensure that technologies to address climate change are available and can be deployed efficiently. The CTI includes activities directed at the achievement of seven broad objectives:

- To facilitate cooperative and voluntary actions among governments, quasigovernments, and private entities to help cost-effective technology diffusion and reduce the barriers to an enhanced use of climate-friendly technologies
- To promote the development of technology aspects of national plans and programs prepared under the UNFCCC
- To establish and strengthen the networks among renewable and energy efficiency centers in different regions
- To improve access to and enhance markets for emerging technologies
- To provide appropriate recognition of climate-friendly technologies through the creation of international technology awards

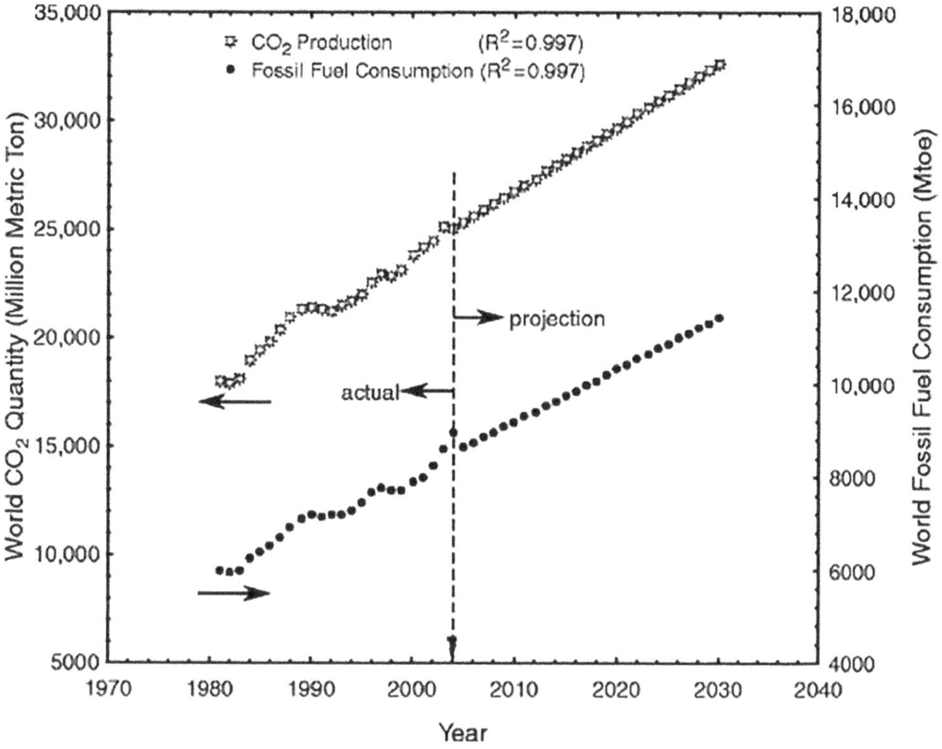

FIGURE 3 Variation of world total fossil-fuel consumption and CO_2 production; actual data from BP[3] and projections. Mtoe, million tons of oil equivalent.
Source: BP.[3]

- To strengthen international collaboration on short-, medium-, and long-term research; development and demonstration; and systematic evaluation of technology options
- To assess the feasibility of developing longer-term technologies to capture, remove, or dispose of greenhouse gases; to produce hydrogen from fossil fuels; and to strengthen relevant basic and applied research

Potential Solutions to Environmental Issues

Although there are a large number of practical solutions to environmental problems, three potential solutions are given priority, as follows[11]:

- Energy conservation technologies (efficient energy utilization)
- Renewable energy technologies
- Cleaner technologies

In these technologies, we pay special attention to energy conservation technologies and their practical aspects and environmental impacts. Each of these technologies is of great importance, and requires careful treatment and program development. In this work, we deal with energy conservation technologies and strategies in depth. Considering the above priorities to environmental solutions, the important technologies shown in Figure 4 should be put into practice.

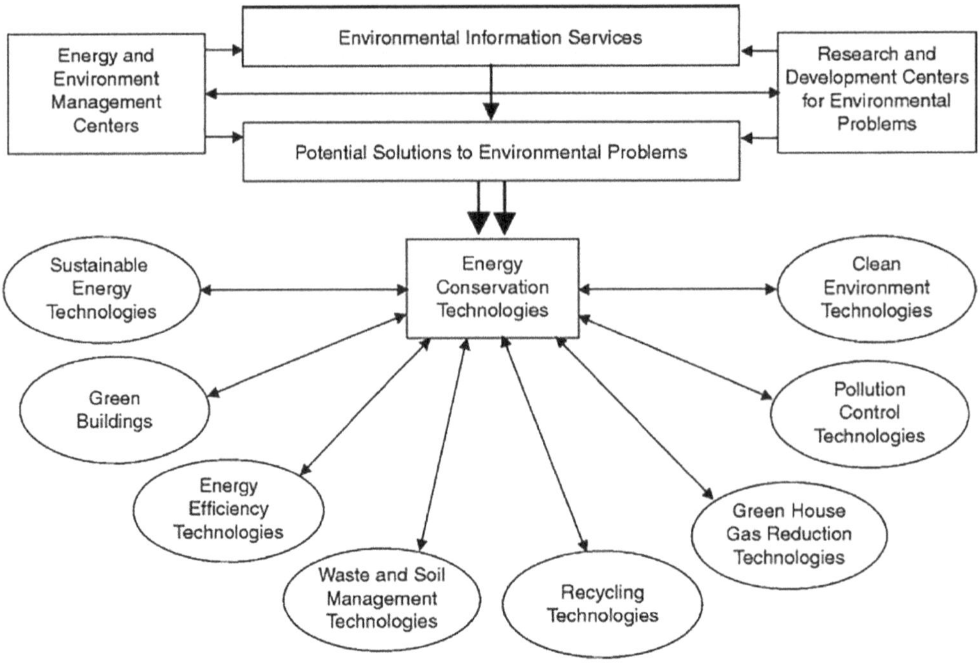

FIGURE 4 Linkages between possible environmental and energy conservation technologies.

Practical Energy Conservation Aspects

The energy-saving result of efficiency improvements is often called energy conservation. The terms efficiency and conservation contrast with curtailment, which decreases output (e.g., turning down the thermostat) or services (e.g., driving less) to curb energy use. That is, energy curtailment occurs when saving energy causes a reduction in services or sacrifice of comfort. Curtailment is often employed as an emergency measure. Energy efficiency is increased when an energy conversion device-such as a household appliance, automobile engine, or steam turbine-undergoes a technical change that enables it to provide the same service (lighting, heating, motor drive, etc.) while using less energy. Energy efficiency is often viewed as a resource option like coal, oil, or natural gas. In contrast to supply options, however, the downward pressure on energy prices created by energy efficiency comes from demand reductions instead of increased supply. As a result, energy efficiency can reduce resource use and environmental impacts;[12]

The quality of a country's energy supply and demand systems is increasingly evaluated today in terms of its environmental sustainability. Fossil-fuel resources will not last indefinitely, and the most convenient, versatile, and inexpensive of them have substantially been used up. The future role of nuclear energy is uncertain, and global environmental concerns call for immediate action. OECD countries account for almost 50% of total world energy consumption: Current use of oil per person averages 4.5 bbl a year worldwide, ranging from 24 bbl in the United States and 12 bbl in western Europe to less than 1 bbl in sub-Saharan Africa. More than 80% of worldwide CO_2 emissions originate in the OECD area. It is clear, then, that OECD countries should play a crucial role in indicating a sustainable pattern and in implementing innovative strategies.[11]

From an economic as well as an environmental perspective, energy conservation holds even greater promise than renewable energy, at least in the near-term future. Energy conservation is indisputably beneficial to the environment, as a unit of energy not consumed equates to a unit of resources saved and a unit of pollution not generated.

Furthermore, some technical limitations on energy conservation are associated with the laws of physics and thermodynamics. Other technical limitations are imposed by practical technical constraints related to the real-world devices that are used. The minimum amount of fuel theoretically needed to produce a specified quantity of electricity, for example, could be determined by considering a Carnot (ideal) heat engine. However, more than this theoretical minimum fuel may be needed due to practical technical matters such as the maximum temperatures and pressures that structures and materials in the power plant can withstand.

As environmental concerns such as pollution, ozone depletion, and global climate change became major issues in the 1980s, interest developed in the link between energy utilization and the environment. Since then, there has been increasing attention to this linkage. Many scientists and engineers suggest that the impact of energy-resource utilization on the environment is best addressed by considering exergy. The exergy of a quantity of energy or a substance is a measure of the usefulness or quality of the energy or substance, or a measure of its potential to cause change. Exergy appears to be an effective measure of the potential of a substance to impact the environment. In practice, the authors feel that a thorough understanding of exergy and of how exergy analysis can provide insights into the efficiency and performance of energy systems is required for the engineer or scientist working in the area of energy systems and the environment.[8] Considering the above explanations, the general aspects of energy conservation can be summarized as shown in Figure 5.

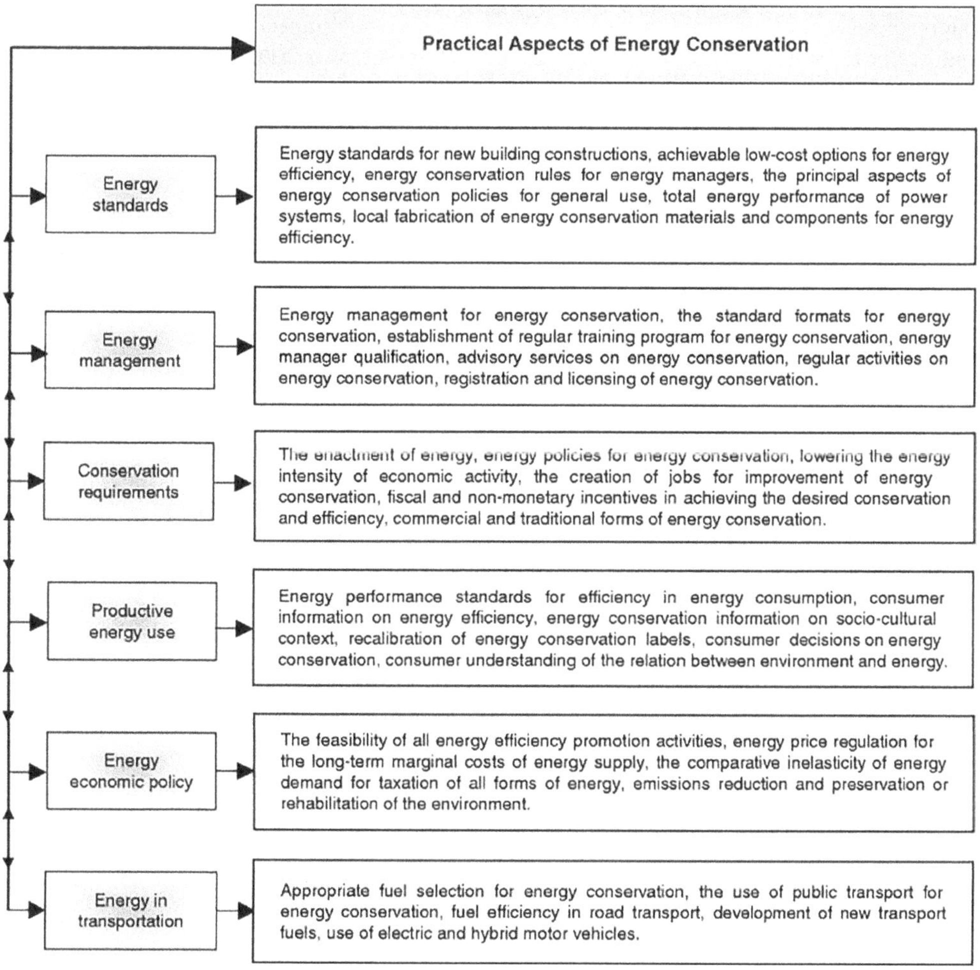

FIGURE 5 A flow chart of practical energy conservation aspects.

Research and Development Status on Energy Conservation

Now we look at R&D expenditures in energy conservation to assess the importance attached to energy conservation in the long range. The share of energy R&D expenditures going into energy conservation, for example, has grown greatly since 1976, from 5.1% in 1976 to 40.1% in 1990 and 68.5% in 2002.[11] This indicates that within energy R&D, research on energy conservation is increasing in importance. When R&D expenditures on energy conservation are compared with expenditures for research leading to protection of the environment in the 2000s, the largest share was spent on environment research. In fact, it is not easy to interpret the current trends in R&D expenditures, because energy conservation is now part of every discipline from engineering to economics. A marked trend has been observed since the mid-1970s, in that expenditures for energy conservation research have grown significantly, both in absolute terms and as a share of total energy R&D. These expenditures also grew more rapidly than those for environmental protection research, surpassing it in the early 1980s. Therefore, if R&D expenditures reflect long-term concern, there seems to be relatively more importance attached to energy conservation as compared with environmental protection.

In addition to the general trends discussed above, consider the industrial sector and how it has tackled energy conservation.

The private sector clearly has an important role to play in providing finance that could be used for energy efficiency investments. In fact, governments can adjust their spending priorities in aid plans and through official support provided to their exporters, but they can influence the vast potential pool of private-sector finance only indirectly. Many of the most important measures to attract foreign investors include reforming macroeconomic policy frameworks, energy market structures and pricing, and banking; creating debt recovery programs; strengthening the commercial and legal framework for investment; and setting up judicial institutions and enforcement mechanisms. These are difficult tasks that often involve lengthy political processes.

Thus, the following important factors, which are adopted from a literature work[13] can contribute to improving energy conservation in real life. Figure 6 presents the improvement factors of energy conservation.

Energy Conservation and Sustainable Development

Energy conservation is vital for sustainable development and should be implemented by all possible means, despite the fact that it has its own limitations. This is required not only for us, but for the next generation as well.

A secure supply of energy resources is generally considered a necessity but not a sufficient requirement for development within a society. Furthermore, sustainable development demands a sustainable supply of energy resources that, in the long term, is readily and sustainably available at reasonable cost and can be utilized for all required tasks without causing negative societal impact. Supplies of such energy resources as fossil fuels (coal, oil, and natural gas) and uranium are generally acknowledged to be finite. Other energy sources (such as sunlight, wind, and falling water) are generally considered to be renewable and, therefore, sustainable over the relatively long term. Wastes (convertible to useful energy forms through, for example, waste- to-energy incineration facilities) and biomass fuels usually also are viewed as being sustainable energy sources. In general, the implications of these statements are numerous and depend on how the term *sustainable* is defined.[14]

Energy resources and their utilization are intimately related to sustainable development. For societies to attain or try to attain sustainable development, much effort must be devoted not only to discovering sustainable energy resources, but also to increasing the energy efficiencies of processes utilizing these resources. Under these circumstances, increasing the efficiency of energy-utilizing devices is important. Due to increased awareness of the benefits of efficiency improvements, many institutes and agencies have started working along these lines. Many energy conservation and efficiency improvement programs have been developed and are being developed to reduce present levels of energy consumption.

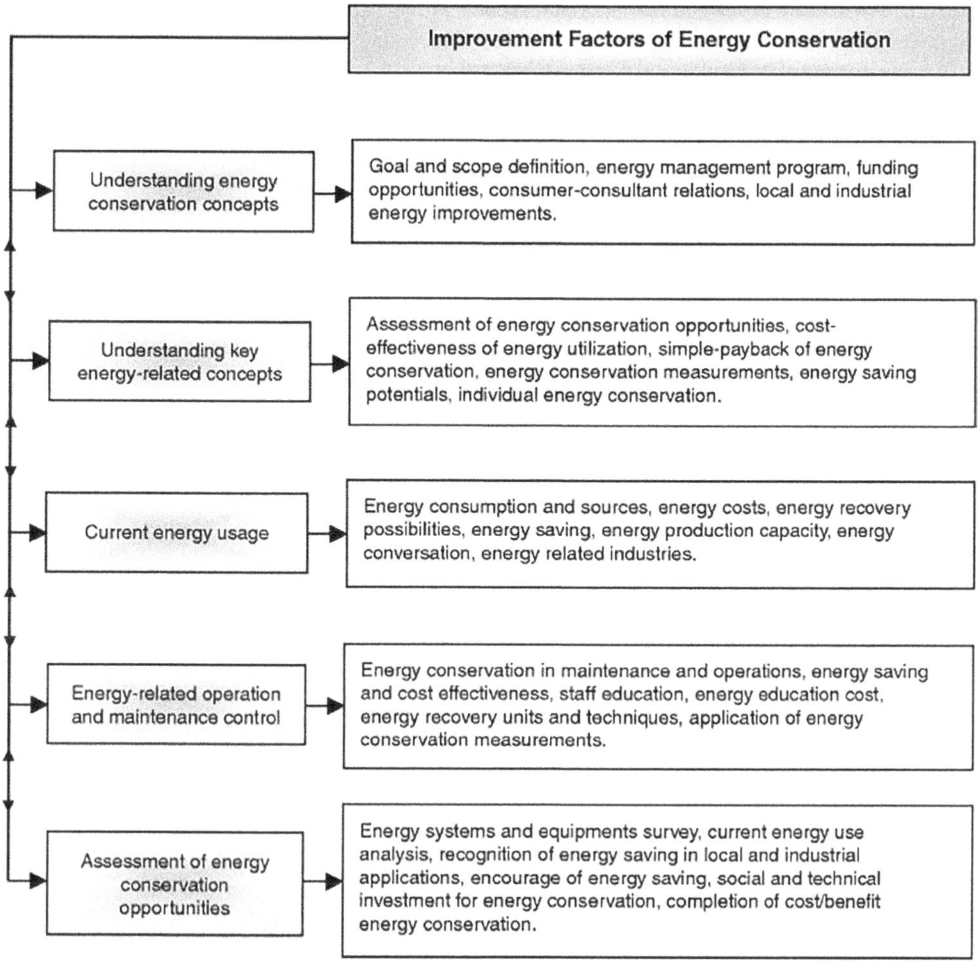

FIGURE 6 Improvement factors of energy conservation.

To implement these programs in a beneficial manner, an understanding is required of the patterns of "energy carrier" consumption-for example, the type of energy carrier used, factors that influence consumption, and types of end uses.[15]

Environmental concerns are an important factor in sustainable development. For a variety of reasons, activities that continually degrade the environment are not sustainable over time-that is, the cumulative impact on the environment of such activities often leads over time to a variety of health, ecological, and other problems. A large portion of the environmental impact in a society is associated with its utilization of energy resources. Ideally, a society seeking sustainable development utilizes only energy resources that cause no environmental impact (e.g., that release no emissions to the environment). Because all energy resources lead to some environmental impact, however, it is reasonable to suggest that some (not all) of the concerns regarding the limitations imposed on sustainable development by environmental emissions and their negative impacts can be overcome in part through increased energy efficiency. Clearly, a strong relationship exists between energy efficiency and environmental impact, because for the same services or products, less resource utilization and pollution normally are associated with increased energy efficiency.

Here, we look at renewable energy resources and compare them with energy conservation. Although not all renewable energy resources are inherently clean, there is such a diversity of choices that a shift to renewables carried out in the context of sustainable development could provide a far cleaner system than would be feasible by tightening controls on conventional energy. Furthermore, being by nature site-specific, they favor power system decentralization and locally applicable solutions more or less independently of the national network. It enables citizens to perceive positive and negative externalities of energy consumption. Consequently, the small scale of the equipment often makes the time required from initial design to operation short, providing greater adaptability in responding to unpredictable growth and/or changes in energy demand.

The exploitation of renewable energy resources and technologies is a key component of sustainable development.[11] There are three significant reasons for it:

- They have much less environmental impact compared with other sources of energy, because there are no energy sources with zero environmental impact. Such a variety of choices is available in practice that a shift to renewables could provide a far cleaner energy system than would be feasible by tightening controls on conventional energy.
- Renewable energy resources cannot be depleted, unlike fossil-fuel and uranium resources. If used wisely in appropriate and efficient applications, they can provide reliable and sustainable supply energy almost indefinitely. By contrast, fossil-fuel and uranium resources are finite and can be diminished by extraction and consumption.
- They favor power system decentralization and locally applicable solutions more or less independently of the national network, thus enhancing the flexibility of the system and the economic power supply to small, isolated settlements. That is why many different renewable energy technologies are potentially available for use in urban areas.

Taking into consideration these important reasons, the relationship between energy conservation and sustainability is finally presented as shown in Figure 7.

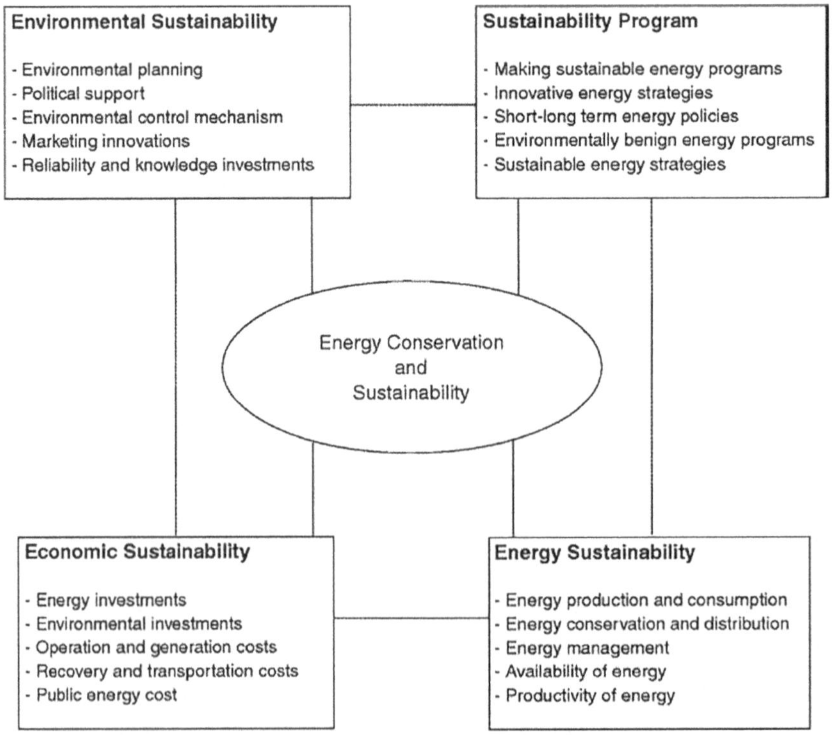

FIGURE 7 Linkages between energy conservation and sustainable development.

Energy Conservation Implementation Plan

The following basic steps are the key points in implementing an energy conservation strategy plan[11]:

1. *Defining the main goals.* It is a systematic way to identify clear goals, leading to a simple goal-setting process. It is one of the crucial concerns and follows an organized framework to define goals, decide priorities, and identify the resources needed to meet those goals.
2. *Identifying the community goals.* It is a significant step to identify priorities and links among energy, energy conservation, the environment, and other primary local issues. Here, it is also important to identify the institutional and financial instruments.
3. *Performing an environmental scan.* The main objective in this step is to develop a clear picture of the community to identify the critical energy-use areas, the size and shape of the resource-related problems facing the city and its electrical and gas utilities, the organizational mechanisms, and the base data for evaluating the program's progress.
4. *Increasing public awareness.* Governments can increase other customers' awareness and acceptance of energy conservation programs by entering into performance contracts for government activities. They can also publicize the results of these programs and projects. In this regard, international workshops to share experiences on the operation would help overcome the initial barrier of unfamiliarity in countries.
5. *Performing information analysis.* This step carries out a wide range of telephone, fax, email, and Internet interviews with local and international financial institutions, project developers, and bilateral aid agencies to capture new initiatives, lessons learned, and viewpoints on problems and potential solutions.
6. *Building community support.* This step covers the participation and support of local industries and communities, and the understanding the nature of conflicts and barriers between given goals and local actors; improving information flows; activating education and advice surfaces; identifying institutional barriers; and involving a broad spectrum of citizen and government agencies, referring to the participation and support of local industrial and public communities.
7. *Analyzing information.* This step includes defining available options and comparing the possible options with various factors (e.g., program implementation costs, funding availability, utility capital deferral, potential for energy efficiency, compatibility with community goals, and environmental benefits).
8. *Adopting policies and strategies.* Priority projects need to be identified through a number of approaches that are best for the community. The decision-making process should evaluate the cost of the options in terms of savings in energy costs; generation of business and tax revenue; and the number of jobs created, as well as their contribution to energy sustainability and their benefit to other community and environmental goals.
9. *Developing the plan.* When a draft plan has been adopted, it is important for the community to review it and comment on it. The public consultation process may vary, but the aim should be a high level of agreement.
10. *Implementing new action programs.* This step involves deciding which programs to concentrate on, with long-term aims being preferred over short-term aims. The option that has the greatest impact should be focused on, and all details should be defined, no matter how difficult the task seems. Financial resources needed to implement the program should be identified.
11. *Evaluating the success.* The final stage is evaluating and assessing how well the plan performed, which helps identify its strengths and weaknesses and to determine who is benefiting from it.

Energy Conservation Measures

For energy conservation measures, the information about the measure's applicability, cost range, maintenance issues, and additional points should be presented. Energy conservation involves efficiency improvements, formulation of pricing policies, good "housekeeping practices," and load management strategies, among other measures. A significant reduction in consumer energy costs can occur if conservation measures are adopted appropriately. The payback period for many conservation programs is less than 2 years.

In spite of the potentially significant benefits of such programs to the economy and their proven successes in several countries, conservation programs have not yet been undertaken on a significant scale in many developed and developing countries. Some reasons for this lack of energy conservation programs relate to the following factors:

- Technical (e.g., lack of availability, reliability, and knowledge of efficient technologies)
- Institutional (e.g., lack of appropriate technical input, financial support, and proper program design and monitoring expertise)
- Financial (e.g., lack of explicit financing mechanisms)
- Managerial (e.g., inappropriate program management practices and staff training)
- Pricing policy (e.g., inappropriate pricing of electricity and other energy commodities)
- Information diffusion (e.g., lack of appropriate information)

Reduced energy consumption through conservation programs can benefit not only consumers and utilities, but society as well. In particular, reduced energy consumption generally leads to reduced emissions of greenhouse gases and other pollutants into the environment.

Accelerated gains in energy efficiency in energy production and use, including those in the transportation sector, can help reduce emissions and promote energy security. Although there is a large technical potential for increased energy efficiency, there exist significant social and economic barriers to its achievement. Priority should be given to market forces in effecting efficiency gains. Reliance on market forces alone, however, is unlikely to overcome these barriers. For this reason, innovative and bold approaches are required by governments, in cooperation with industry, to realize the opportunities for energy efficiency improvements, and to accelerate the deployment of new and more efficient technologies.

Here, we look at energy conservation measures, which may be classified in six elements:

- Sectoral measures
- Energy conservation through systematic use of unused energy
- Energy conservation by changing social behavior
- International cooperation to promote energy conservation to counteract global warming
- Enhancing international and government-industry- university cooperation in developing technologies for energy conservation
- Promoting diffusion of information through publicity and education

The emphasis is on sectoral energy conservation. Table 4 presents some examples of such sectoral energy conservation measures. After determining which energy conservation measures are applicable, you should read the description of each of the applicable energy conservation measures. Information about the savings that can be expected from the measure, maintenance issues related to the measure, and other items to consider is provided for each energy conservation measure.

To evaluate the energy conservation measures, the following parameters should be taken into consideration[13]:

- *Cost estimation.* The first step is to estimate the cost of purchasing and installing the energy conservation measure. Cost estimates should be made for the entire development rather than for a single piece of equipment (e.g., obtain the cost for installing storm windows for an entire

TABLE 4 Sectoral Energy Conservation Measures

Sector	Measures
Industrial	Strengthening of financial and tax measures to enhance adoption and improvement of energy saving technologies through energy conservation equipment investments Re-use of waste energy in factories and/or in surrounding areas Enhancing recycling that reduces primary energy inputs such are iron scraps and used papers, and devising measures to facilitate recycling of manufactured products Retraining of energy managers and diffusion of new energy saving technologies through them Creating database on energy conservation technologies to facilitate diffusion of information
Residential and Commercial	Revising insulation standards provided in the energy conservation law, and introducing financial measures to enhance adoption of better insulation Development of better insulation materials and techniques Developing' energy conservation' model homes and total energy use systems for homes Revising or adopting energy conservation standards for home and office appliances Developing more energy saving appliances Revising guidelines for managing energy use in buildings, and strengthening advisory services to improve energy management in buildings
Transportation	Because 80% of energy consumption of the sector is by automobiles, further improvement in reducing fuel consumption by automobiles is necessary together with improvement in transportation system to facilitate and reduce traffic flow Diffusion of information about energy efficient driving Adopting financial measures to enhance the use of energy saving transportation equipment such as wind powered boats

Source: Adapted from Energy Conservation Policies and Technologies in Japan: A Survey.[16]

development or building, rather than the cost of one storm window). If you are planning to implement the energy conservation measure without the help of an outside contractor, you can obtain cost estimates by calling a vendor or distributor of the product. If, on the other hand, you will be using a contractor to install or implement the energy conservation measure, the contractor should provide estimates that include all labor costs and contract margins.

- *Data survey.* In this step, the questions on fuel consumption and cost should be listed for more than one possible fuel type (e.g., gas, oil, electric, or propane). The appropriate data for each fuel type should be selected and used accordingly for the cost estimation of each fuel.
- *Energy savings.* The amount of energy or fuel used should be estimated.
- *Cost savings.* This step determines the level of savings.
- *Payback period.* The last step in the cost/benefit analysis estimates the simple payback period. The payback period is found by dividing the cost of the measure by the annual cost savings.

Life-Cycle Costing

The term LCC for a project or product is quite broad and encompasses all those techniques that take into account both initial costs and future costs and benefits (savings) of a system or product over some period of time. The techniques differ, however, in their applications, which depend on various purposes of systems or products. Life-cycle costing is sometimes called a cradle-to-grave analysis. A life-cycle cost analysis calculates the cost of a system or product over its entire life span. Life-cycle costing is a process to determine the sum of all the costs associated with an asset or part thereof, including acquisition, installation, operation, maintenance, refurbishment, and disposal costs. Therefore, it is pivotal to the asset management process.

From the energy conservation point of view, LCC appears to be a potential tool in deciding which system or product is more cost effective and more energy efficient. It can provide information about how to evaluate options concerning design, sites, materials, etc., how to select the best energy conservation feature among various options; how much investment should be made in a single energy conservation feature; and which is the most desirable combination of various energy conservation features.

A choice can be made among various options of the energy conservation measure that produces maximum savings in the form of reduction in the life-cycle costs. A choice can be made between double-glazed and triple-glazed windows, for example. Similarly, a life-cycle cost comparison can be made

TABLE 5 An Example of Life-Cycle Costing (LCC) Analysis

Cost of Purchasing Bulbs	Incandescent	Compact Fluorescent
Lifetime of one bulb (hours)	1,000	10,000
Bulb price ($)	0.5	6.0
Number of bulbs for lighting 10,000 h	10	1
Cost for bulbs ($)	10 × 05 = 5.0	1 × 6 = 6
Energy cost		
Equivalent wattage (W)	75	12
Watt-hours (Wh) required for lighting for 10,000 h	75 × 10,000 = 750,000 Wh = 750 kWh	12 × 10,000 = 120,000 Wh = 120 kWh
Cost at 0.05 per kWh	750 kWh × $0.05 = $37.5	120 kWh × $0.05 = $6
Total cost ($)	5 + 37.5 = 42.5	6 + 6 = 12

between a solar heating system and a conventional heating system. The one that maximizes the life-cycle costs of providing a given level of comfort should be chosen. The application of such techniques to energy conservation is related to determining the optimum level of the chosen energy conservation measure. Sometimes, energy conservation measures involve the combination of several features. The best combination can be determined by evaluating the net LCC effects associated with successively increasing amounts of other energy conservation measures. The best combination is found by substituting the choices until each is used to the level at which its additional contribution to energy cost reduction per additional dollar is equal to that for all the other options.

Illustrative Example

Here, we present an illustrative example on LCC to highlight its importance from the energy conservation point of view. This example is a simple LCC analysis of lighting for both incandescent bulbs and compact fluorescent bulbs, comparing their life-cycle costs as detailed in Table 5. We know that incandescents are less expensive (95% to heat and 5% to usable light) and that compact fluorescent bulbs are more expensive but much more energy efficient. So the question is which type of lighting comes out on top in an LCC analysis.

This example clearly shows that LCC analysis helps in energy conservation and that we should make it part of our daily lives.

Conclusion

Energy conservation is a key element in sectoral (e.g., residential, industrial, and commercial) energy utilization and is vital for sustainable development. It should be implemented by all possible means, despite the fact that it sometime has its own limitations. This is required not only for us, but for the next generation as well. A secure supply of energy resources is generally considered a necessary but not a sufficient requirement for development within a society. Furthermore, sustainable development demands a sustainable supply of energy resources that, in the long term, is readily and sustainably available at reasonable cost and can be utilized for all required tasks without causing negative societal impact.

An enhanced understanding of the environmental problems relating to energy conservation presents a high-priority need and an urgent challenge, both to allow the problems to be addressed and to ensure that the solutions are beneficial for the economy and the energy systems.

All policies should be sound and make sense in global terms-that is, become an integral part of the international process of energy system adaptation that will recognize the very strong linkage existing between energy requirements and emissions of pollutants (environmental impact).

In summary this study discusses the current environmental issues and potential solutions to these issues; identifies the main steps for implementing energy conservation programs and the main barriers to such implementations; and provides assessments for energy conservation potentials for countries, as well as various practical and environmental aspects of energy conservation.

References

1. Azar, C.; Rodhe, H. Targets for stabilization of atmospheric CO_2. Science **1997**, *276*, 1818–1819.
2. Midilli, A.; Ay, M.; Dincer, I.; Rosen, M.A. On hydrogen and hydrogen energy strategies: I: Current status and needs. Renew. Sust. Energy Rev. **2005**, *9* (3), 255–271.
3. BP. Workbook 2005. *Statistical Review—Full Report of World Energy*, British Petroleum, 2005, Available at http://www.bp.com/centres/energy (accessed July 25, 2005).
4. IEE. World Energy Outlook 2002. Head of Publication Services, International Energy, 2002, Available at http://www.worldenergyoutlook.org/weo/pubs/we02002/WEO21.pdf.
5. Kazman, S. Global *Warming and Energy Policy;* CEI; 67–86, 2003. Available at http://www.cei.org.
6. Dincer, I. Energy and environmental impacts: Present and future perspectives. Energy Sources **1998**, *20* (4/5), 427–453.
7. Dincer, I. *Renewable Energy, Environment and Sustainable Development.* Proceedings of the World Renewable Energy Congress; September 20–25, 1998; Florence, Italy; 2559–2562.
8. Rosen, M.A.; Dincer, I. On exergy and environmental impact. Int. J. Energy Res. **1997**, *21* (7), 643–654.
9. Dincer, I. Renewable energy and sustainable development: A crucial review. Renew. Sust. Energy Rev. **2000**, *4* (2), 157–175.
10. Dincer, I. *Practical and Environmental Aspects of Energy Conservation Technologies.* Proceedings of Workshop on Energy Conservation in Industrial Applications; Dhahran, Saudi Arabia, February 12–14, 2000; 321–332.
11. Dincer, I. On energy conservation policies and implementation practices. Int. J. Energy Res. **2003**, *27* (7), 687–702.
12. Sissine, F. Energy efficiency: Budget, oil conservation, and electricity conservation issues. *CRS Issue Brief for Congress.* IB10020; 2005.
13. Nolden, S.; Morse, D.; Hebert, S. *Energy Conservation for Housing: A Workbook,* Contract-DU100C000018374; Abt Associates, Inc.: Cambridge, MA, 1998.
14. Dincer, I.; Rosen, M.A. A worldwide perspective on energy, environment and sustainable development. Int. J. Energy Res. **1998**, *22* (15), 1305–1321.
15. Painuly, J.P.; Reddy, B.S. Electricity conservation programs: Barriers to their implications. Energy Sources **1996**, *18*, 257–267.
16. Anon. *Energy Conservation Policies and Technologies in Japan: A Survey,* OECD/GD(94)32, Paris, 1994.

47

Energy Conservation: Lean Manufacturing

Bohdan W. Oppenheim

Introduction

At the time of this writing (2005), the world is experiencing strong contradictory global trends of diminishing conventional energy resources and rapidly increasing global demands for these resources, resulting in substantial upwards pressures in energy prices. Because the energy used by industry represents a significant fraction of the overall national energy use, equal to 33% in the United States in the year 2005, a major national effort is underway to conserve industrial energy.[1] The rising energy prices place escalating demands on industrial plants to reduce energy consumption without reducing production or sales, but by increasing energy density.

Optimization of industrial hardware and its uses, including motors and drives, lights, heating, ventilation and cooling equipment, fuel-burning equipment, and buildings, are well understood, have been practiced for years,[2] and are important in practice. However, they offer only limited energy conservation opportunities, rarely exceeding a few percent of the preoptimization levels. In contrast, the impact of productivity on energy use and energy density offers dramatically higher savings opportunities in energy and in other costs. In the extreme case, when transforming a factory from the traditional "process village" batch-and- queue system to the state-of-the-art, so-called Lean system, the savings in energy can reach 50% or more.

The best organization of production known at this time is called Lean, developed at Toyota in Japan.[3] It is the flow of value-added work through all processes required to convert raw materials to the finished products with minimum waste. Major elements of Lean organization include: steady single-piece flow with minimum inventories and no idle states or backflow; flexible production with flexible equipment and operators and flexible floor layouts ready to execute the order of any size profitably and just-in-time; reliable and robust supplies of raw materials; minimized downtime due to excellent preventive maintenance and quick setups; first-pass quality; clean, uncluttered, and well-organized work space; optimized work procedures; and, most importantly, an excellent workforce–well trained, motivated, team-based and unified for the common goals of having market success, communicating efficiently, and being well-managed. The Lean organization of production is now well understood among productivity professionals, but it is not yet popular among the lower tier suppliers in the United States. Its implementation would save energy and benefit the suppliers in becoming more competitive.

The engineering knowledge of energy conservation by equipment improvements is well understood and can be quantified with engineering accuracy for practically any type of industrial equipment.[2] In contrast, industrial productivity is strongly influenced by intangible and complex human factors such as management, work organization, learning and training, communications, culture, and motivation. These work aspects are difficult to quantify in factory environments. For this reason, the accuracy of productivity gains and the related energy savings are typically much less accurate than the energy savings computed from equipment optimization. Simple quantitative models with a conservative bias are therefore recommended as tools for energy management in plants. This entry includes some examples. They are presented in the form of energy savings or energy cost savings that would result from implementing a given productivity improvement, or eliminating a given productivity waste, or as simple metrics measuring energy density.

It is remarkable that in most cases, these types of energy savings occur as a natural byproduct of productivity improvements, without the need for a direct effort centered on energy. Thus, the management should focus on productivity improvements. In a traditional non-Lean plant intending to transform to Lean production, the first step should be to acquire the knowledge of the Lean system. It is easily available from industrial courses and workshops, books,[3,4] and video training materials.[6] The next step should be the actual transformation of production to Lean. Most of the related energy savings will then occur automatically. Implementation of individual productivity elements such as machine setup time reduction will yield some energy savings, but the result will not be as comprehensive as those yielded by the comprehensive implementation of Lean production.

Traditional vs. Lean Production

The traditional organization of production still used frequently in most factories tends to suffer from the following characteristics:

- Supplier selection is based on minimum cost, resulting in a poor level of mutual trust and partnership, the need for receiving inspection, and often large inventories of raw materials (RM).
- Work-in-progress (WIP) is moving in large batches from process village to process village and staged in idle status in queues in front of each machine, while the machine moves one piece at a time. This work organization is given the nickname "batch-and-queue" (BAQ).[3]
- Finished goods (FG) are scheduled to complex forecasts rather than customer orders, resulting in large inventories.
- The floor is divided into "process villages" populated with large, complex, and fast similar machines selected for minimum unit cost.
- Minimum or no information is displayed at workstations, and the workers produce quotas.
- Work leveling is lacking, which results in a random mix of bottlenecks and idle processes.
- Unscheduled downtime of equipment occurs frequently.
- Quality problems with defects, rework, returns, and customer complaints are frequent.
- Quality assurance in the form of 100% final inspections attempts to compensate for poor production quality.
- The floor space is cluttered, which makes moving around and finding items difficult.
- The workforce has minimum or no training and single skills.
- The management tends to be authoritarian.
- A language barrier exists between the workers and management.
- There is a culture of high-stress troubleshooting rather than creative trouble prevention.

In such plants, the waste of materials, labor, time, space, and energy can be as much as 50%–90%.[3]

The Lean production method developed primarily at Toyota in Japan under the name Just-In-Time (JIT), and generalized in the seminal work[3] is the opposite of the traditional production in almost all respects, as follows:

- Raw materials are bought from reliable supplier– partners and delivered JIT in the amount needed, at the price agreed, and with the consistently perfect quality that obviates incoming inspection.
- Single-piece flow (SPF) of WIP is steadily moving at a common takt time (Takt time is the common rhythm time of the pieces moving from workstation to workstation on the production line. It is the amount of time spent on EACH operation. It precisely synchronizes the rate of all production operations to the rate of sales JIT.), from the first to the last process.
- The FG are produced to actual customer orders JIT resulting in minimum inventories.
- The floor is divided into flexible production lines with small simple machines on casters that can be pushed into position and setup in minutes.
- The labor is multiskilled, well motivated and well trained in optimized procedures.
- Quality and production status are displayed on large visible boards at each workstation, making the entire production transparent for all to see.
- Preventive maintenance assures no unscheduled downtime of equipment.
- All process operators are trained in in-line quality checks and variability reduction.
- No final inspection is needed, except for occasional sampled checks of FG.
- Defects, rework, returns, and customer complaints are practically eliminated.
- The floor space is clean and uncluttered.
- The workforce is trained in company culture and commonality of the plant mission, customer needs, workmanship, and quality.
- The culture promotes teamwork, multiple job skills, supportive mentoring management, and company loyalty.
- The management promotes trouble prevention and "stopping the line" at the first sign of imperfection so that no bad pieces flow downstream.

According to Womack et al. the transformation from traditional to Lean production can reduce overall cost, inventory, defects, lead times by 90%, and space by 50%, and vastly increase plant competitiveness, customer satisfaction, and workforce morale. The resultant energy savings can be equally dramatic. Liker,[4] contains interviews with industry leaders who have succeeded in this transformation.

Impact on Energy

The impact of productivity on plant energy falls into the following two broad categories:

1. Productivity improvements that save infrastructure energy. These improvements reduce the energy consumed by all plant support systems, which tend to be energized regardless of the actual production activities, such as lights, space cooling and heating devices, cooling towers, combustion equipment (boilers, molten metal furnaces), air compressors, forklift battery chargers, conveyors, etc. To the first approximation, the infrastructure energy is reduced in proportion to the production time reductions, which can be huge in the Lean system. In order to perform more detailed estimates of the infrastructure energy savings, the management would have to conduct detailed energy accounting and understand how much energy is used by each support system under different production conditions. This knowledge is rarely available; therefore the former simplistic approach, combined with conservative estimates, offer useful tools.
2. Process energy savings. In this category, the energy savings of process equipment are obtained by improving the process productivity. Examples include the reduction of unscheduled machine downtime or setup time and the elimination of process variability, defects, rework, scrap, excessive labor time, etc.

Single Piece Flow (SPF)

Changing the traditional BAQ production to Lean production is by far the most effective productivity transformation a plant can undertake, creating dramatic savings in the overall throughput time, cost, quality, and energy. The example shown in Figure 1 compares just one aspect of the transformation—a reduction of batch size from five to one, i.e., the SPF. In both cases, four processes of equal one-minute takt time are assumed. The benefits of the SPF alone are dramatic, as follows:

1. In BAQ, the batch is completed in 20 min and in SPF in only 8 min, a 60% reduction.
2. In BAQ, only one machine at a time produces value, while three others are idle. If the idle machines remain energized, as is the case, e.g., with injection molding, three of the four machines (75%) would be wasting energy, and doing it for 16 min each, adding up to 64 min of machine energy

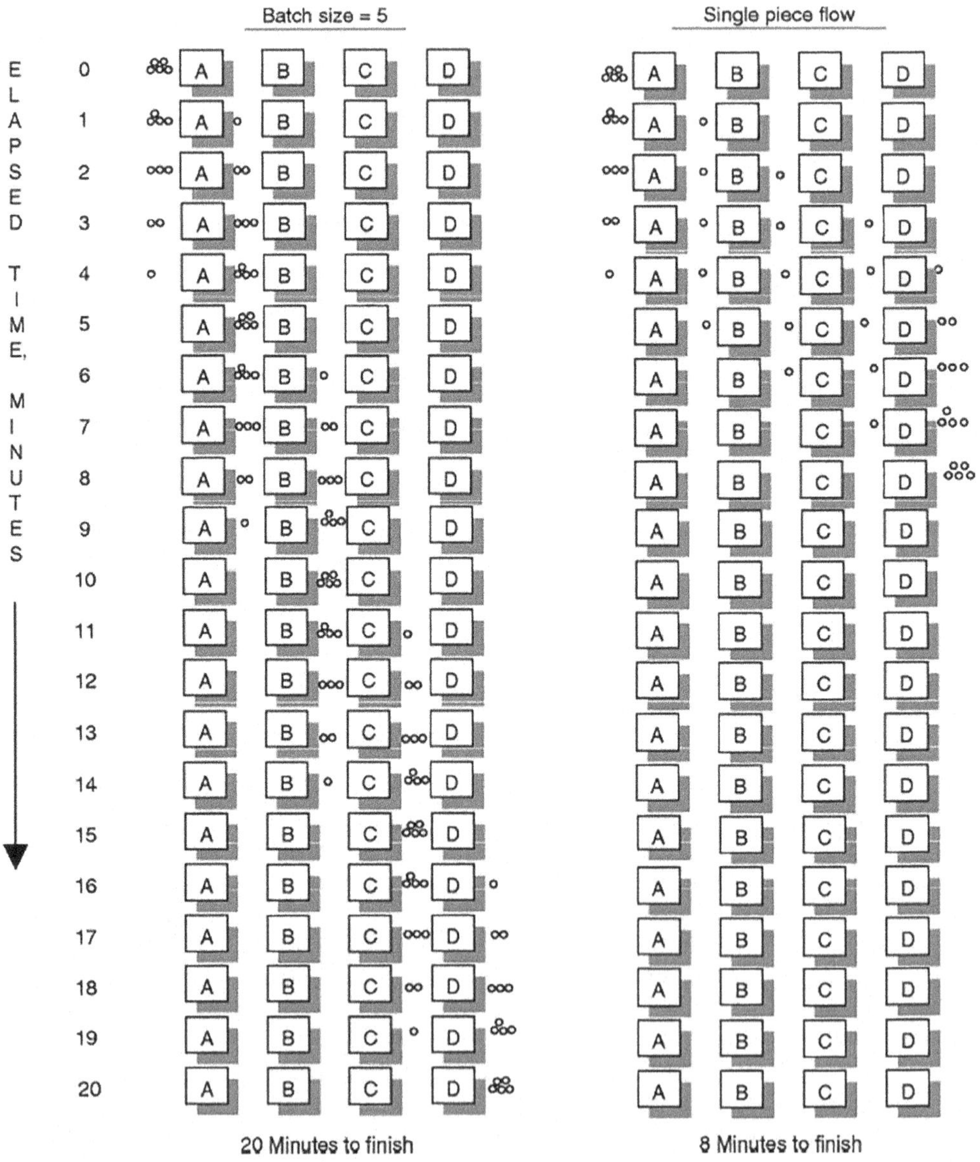

FIGURE 1 BAQ with batch size of five vs SPF.

wasted. In the SPF system, no machine energy is wasted as no machine would be idle, except for the lead and tail of each process of 4 min, adding up to 16 min of machine energy wasted, a savings of 75% from BAQ.

3. Reducing the batch throughput time by 60% reduces the infrastructure energy by the same amount, assuming the production is completed faster and the plant is de-energized. Alternatively, the freed 60% time and energy could be used for additional production and profits.
4. An important additional benefit is that in SPF, a defect can be detected on the first specimen–as soon as it reaches the next process, while in the BAQ, the entire batch may be wasted before the defect is discovered and a corrective action undertaken, with the energy used for making the batch wasted.

This simple example clearly illustrates the dramatic impact of SPF on both overall productivity and energy consumption. Typically, as the factories transform to the Lean system, their sales, production, and profits increase simultaneously and the energy used decreases. A convenient metric to track the overall benefit is the gross energy density, ED_1 or ED_2:

$$ED_1 = \frac{EC_T}{P} \tag{1a}$$

$$ED_2 = \frac{EC_T}{AC} \tag{1b}$$

where EC_T is the overall annual cost of energy in the plant, P is the number of products produced per year and AC is the total annual costs (sales minus net profit). ED_1 should be used if similar products are made most of the time, and ED_2 should be used if the plant has a wide menu of dissimilar products. The ED ratios will decrease as progress is made from BAQ to SPF. If the volume of production remains constant during the transformation, energy savings and energy cost savings alone may be more convenient metrics to track plant energy efficiency.

Inventory Reduction

All inventories, whether in RM, WIP, or FG, beyond the immediate safety buffers, are detrimental. Inventory means that company capital is "frozen" on the floor; cutting into the cash flow; wasting labor for inventory control, storage, and security; wasting infrastructure energy for lights, forklift energy, and possible cooling or heating of the inventory spaces if the goods require temperature or humidity control; wasting space and the associated lease/mortgage fees and taxes; and becoming scrap if not sold (a frequent waste in large inventories). Inventory and inventory space reductions lead to infrastructure energy savings. Process energy can also be saved by not making the FG that end up in inventory, cannot be sold, and become scrap. Womack and Jones[3] and Liker[4] contain case studies for, among others, inventory reductions. A convenient nondimensional metric to track the overall impact of all inventories on energy savings is

$$EC_T \times \frac{I_T}{AC} \tag{2}$$

where I_T is the number of inventory turns per year.

Workmanship, Training, and Quality Assurance

In the ideal Lean system, the processes, equipment, procedures, and training are perfected to the degree that guarantees consistent and robust production with predictable effort, timing, quality, and cost; with no variability, defects, or rework, and with maximum ergonomics and safety. This is accomplished by a

TABLE 1 Energy Waste from Poor Workmanship

A plant with \$20,000,000 in sales and \$2,000,000 in profits spends \$1,000,000 on energy per year. The typical order requires 10 processes of roughly equal energy consumption. The production equipment consumes 60% and the supportive infrastructure consumes 40% of the plant energy. Sequential process #5 has the defect rate of 10%. In order to compensate for the defects, the first 5 processes must produce 10% extra pieces. The annual waste of energy cost (and the energy cost savings, if the defective process is fixed) is then:
(\$1,000,000/yr)(5/10 processes) (60% process energy) (10% defect rate) = \$30,000/yr (3)
The additional production time of 10% waste not only the cost of the process energy computed in (3) but also the infrastructure energy cost of:
(\$1,000,000/yr) (40% infrastructure energy) (10% defects) = \$40,000/yr (4)
Such delays also extend the promised delivery time and reduce customer satisfaction and factory competitiveness. Adding (1) and (2) together. (not counting the direct productivity losses), the wasted energy cost alone of \$70,000/yr represents 3.5% of the annual profits and 7% in annual energy costs. Based on the author's experience,[5] these numbers are not infrequent in industry. Fixing the productivity of process #5 would eliminate these wastes.

consistent long-term strategy of continuous improvement of all the above elements, including intensive initial training of the workforce and subsequent retraining in new procedures. A procedure must be developed for each process until it is robust and predictable and optimized for minimum overall cost, required quality, maximum ergonomics, and safety. Process operators must be trained in the procedures as well as in the process quality assurance, and they must be empowered to stop the process and take corrective action or call for help if unable to avoid a defect. Management culture must be supportive for such activities. Any departure from this ideal leads to costly penalties in quality, rework, delays, overtime or contract penalties, crew frustrations, and customer dissatisfaction. These, in turn, have negative impacts on energy as follows:

1. Defects require rework, which requires additional energy to remake or repair the part. The best metric to use here is the energy or energy cost per part used in the given defective process multiplied by the number of bad parts produced per year.
2. Variability in the process time or delays caused by defects mean that the production takes more time and more infrastructure and process energy for the same amount of value work and profits when compared with the ideal nonvariable process. Table 1 illustrates cases (1) and (2).
3. Defective processes usually require a massive final inspection to sort out the good products. Finding the finished goods defective is the most inefficient means of quality assurance because often the entire batch must then be remade, consuming the associated energy. The inspection space, labor, and energy represent a direct waste and should be replaced with in-line quality assurance (ILQA) that detects the first bad piece (Governmental, medical, etc., orders usually require a 100% final inspection. In the Lean system, this is performed as a formality because everybody in the plant knows that all pieces will be perfect because all imperfections have been removed in real time before the inspection process.) and immediately undertakes a corrective action. Typically, the ILQA can be implemented in few days of operators' training and has the simple payback period measured in days or weeks.[5]

Overage Reduction

Many a plant compensates for its notorious defects by routinely scheduling production in excess of what the customer orders. Some minimum overage is usually justified for machine setups, adjustments, and QA samples. In a Lean plant this rarely exceeds a fraction of one percent. In a traditional plant, the value of 5%–15% is not infrequent. A 5% overage means that the plant spends 105% of the necessary costs. If the profit margin is 5%, the overage alone may consume the entire profit. The overall energy waste (and the opportunity to save energy) is simply proportional to the overage amount. Overage is one of the most wasteful ways of compensating for defective processes. The best remedy is to simply identify the defective process with ILQA, find the root cause (typically the lack of training, excessive work quotas, or bad process or material), and repair it.

Unintentional overage can also be destructive to profits and energy use. Example: A worker is asked to cut only a few small pieces from a large sheet of metal, but instead he cuts the entire sheet, thinking, "my machine is already setup and soon they will ask me to cut the rest of the sheet anyway, so I may as well do it now." The excessive pieces then move through all processes, unknowingly to the management, consuming energy, labor and fixed costs, to end up as excessive FG inventory and, in the worst case, find no buyer and end up as scrap. Uncontrolled and careless overage can easily consume all profits, and, of course, waste energy proportionately to the overage amount.

Downtime

Equipment downtime and idleness may occur due to scheduled maintenance, unscheduled breakdowns, machine setups, and poor process scheduling. The downtime may cause proportional loss of both profits and energy. The downtime may have fourfold impact on energy use, as follows:

1. When a process stops for whatever reason during an active production shift, the plant infrastructure continues to use energy and loosing money, as in Equation 4. A good plant manager should understand what fraction of the infrastructure energy is wasted during the specific equipment downtime. With this knowledge, the energy waste can be estimated as being proportional to the downtime.
2. Some machines continue using energy during maintenance, repair, or setup in proportion to the downtime (e.g., the crucible holding molten metal for a die casting machine remains heated by natural gas while the machine is being setup or repaired). Reducing the setup time or eliminating the repair time saves the gas energy in direct proportion to the downtime saved. In order to calculate energy savings in such situations, it is necessary to understand the energy consumption by the equipment per unit of time multiplied by the downtime reduction.
3. When a particular machine is down, additional equipment upstream or downstream of that machine may also be forced into an idle status but remain energized, thus wasting energy. In an ideal single-piece flow, the entire production line (As in the saying "In Lean either everything works or nothing works") will stop. In order to estimate the energy-saving opportunity from reducing this cumulative downtime, the energy manager must understand which equipment is idled by the downtime of a given machine and how much energy it uses per unit time while being idle.
4. Lastly, energized equipment should be well managed. A high-powered machine may be left energized for hours at a time when not scheduled for production. A good practice is to assign each of these machines to an operator who will have the duty of turning the machine off when not needed for a longer time, if practical, and to turn it back on just in time to be ready for production exactly when it is needed.

Preventive maintenance and setup time reduction have a particularly critical impact on both productivity and related energy use, as follows:

Preventive Maintenance

Practical and routine preventive maintenance should be done during the hours free of scheduled production (e.g., during night shifts, on weekends, or during layover periods). The maintenance should be preventive rather than reactive (The term "preventive" tends to be replaced with "productive" in modern industrial parlance). Well-managed "total" preventive[6] maintenance involves not only oiling and checking the machines per schedule but also ongoing training of the mechanics; developing a comprehensive database containing information on the particular use and needs of various machines; preparing a schedule of part replacement and keeping inventory of frequently used spare parts; and a

TABLE 2 Energy Savings from Setup Time Reduction

A plant operates on two shifts, 260 days per year, performing on average of 20 two-hour setups per day on their electrically heated injection molding machines. Each machine consumes 20 kW when idle bur energized. By a focused continuous improvement system and training, the crew reduces the routine setup time 0.5 h, with few, if any expenses for additional hardware, thus saving:

(260 days/yr) (20 setups/day) (1.5 h saved/setup) = 7800 machine h/yr.

The resultant process energy saved will be:

(7800 h/yr) (20 kW) = 156,000 kWh/yr (5)

In addition, infrastructure energy will be saved because of the reduced downtime. Using the data from Example 1, if the work is done in two shifts for 260 days per year (4160 h/yr.), the plant infrastructure uses 40% of the plant energy, and each machine consumes 2% of the plant infrastructure energy during the set up, the additional energy cost savings due to the setup time reduction will be:

(7,800 hr/yr) (0.02) (0.04) ($1,000,000)/(4160 h/yr) = #15,000 (6)

well-managed ordering system for other parts, including vendor data so that when a part is needed it can be ordered immediately and shipped using the fastest possible means. Industry leaders have demonstrated that affordable preventive maintenance can reduce the unscheduled downtime and associated energy waste to zero. This should be the practical goal of well-run factories.

Setups

Modern market trends push industry towards shorter series and smaller orders, requiring, in turn, more and shorter setups. Industry leaders have perfected routine setups to take no more than a few minutes. In poorly managed plants, routine setups can take as long as several hours. In all competitive modern plants, serious efforts should be devoted to setup time reductions. The effort includes both training and hardware improvements. The training alone, with only minimal additional equipment (such as carts), can yield dramatic setup time reductions (i.e., from hours to minutes). Further gains may require a change of the mounting and adjustment hardware and instrumentation. Some companies organize competitions between teams for developing robust procedures for the setup time reductions. In a plant performing many setups, the opportunity for energy savings may be significant, both in the process and infrastructure energy, as shown in Table 2.

Flexibility

Production flexibility, also called agility, is an important characteristic of competitive plants. A flexible plant prefers small machines (if possible, on casters) that are easy to roll into position and plug into adjustable quick-connect electrical and air lines and that are easy to setup and maintain over the large fixed machines selected with large batches and small unit costs in mind (such machines are called "monuments" in Womack and Jones[3]). Such an ideal plant will also have trained a flexible workforce in multiple skills, including quality assurance skills. This flexibility allows for the setup of new production lines in hours or even minutes, optimizing the flow and floor layout in response to short orders, and delivers the orders JIT. The energy may be saved in two important ways, as follows:

- Small machines processing one piece at a time use only as much energy as needed. In contrast, when excessively large automated machines are used, the typical management choice is between using small batches JIT, thus wasting the large machine energy, or staging the batches for the large machine, which optimizes machine utilization at the expense of throughput time, production flow, production planning effort, and the related infrastructure energy.
- Small machines are conducive to flexible cellular work layout, where 2–4 machines involved in the sequential processing of WIP are arranged into a U-shaped cell with 1–3 workers serving all processes in the cell in sequence, and the last process being quality assurance. This layout can be made very compact—occupying a much smaller footprint in the plant compared to traditional

"process village" plants, roughly a reduction of 50%[3,4]—and is strongly preferred by workers because it saves walking and integrates well the work steps. Such a layout also saves forklift effort and energy and infrastructure energy due to the reduction of the footprint.

Other Productivity Elements

The complete list of productivity elements is beyond the scope of this entry, and all elements have some leverage on energy use and conservation. In the remaining space, only the few most important remaining aspects are mentioned, with their leverage on energy. Descriptive details can be found in Ohno[7] and numerous other texts on Lean production.

- *Visual factory:* Modern factories place an increasing importance on making the entire production as transparent as possible in order to make any problem visible to all, which is motivational for immediate corrective actions and continuous improvements. Ideally, each process should have a white board displaying the short-term data, such as the current production status (quantity completed vs required); the rate of defects or rejects and their causes; control charts and information about the machine condition or maintenance needs; and a brief list and explanation of any issues, all frequently updated. The board should also display long-term information such as process capability history, quality trends, operator training, etc. Such information is most helpful in the optimization of, among other things, process time and quality, which leads to energy savings, as discussed above.
- *"Andon" signals:* The term refers to the visual signals (lights, flags, markers, etc.,) displaying the process condition, as follows: "green=all OK," "yellow=minor problem being corrected," and "red=high alarm, stopped production, and immediate assistance needed." The signals are very useful in identifying the trouble-free and troubled processes, which is conducive to focusing the aid resources to the right places in real time, fixing problems immediately and not allowing defects to flow downstream on the line. These features, in turn, reduce defects, rework, delays, and wasted costs, which improve overall productivity and save energy, as described above. It is also useful to display the estimated downtime (Toyota and other modern plants have large centrally located Andon boards that display the Andon signal, the workstation number, and the estimated downtime.). Knowing the forecasted downtime frees other workers to perform their pending tasks which have waited for such an opportunity rather than wait idle. This leads to better utilization of the plant resources, including infrastructure energy.
- *"5Ss"*: The term comes from five Japanese words that begin with the "s" sound and loosely translate into English as: sorting, simplification, sweeping, standardization, and self-discipline (many other translations of the words are popular in industry); and describes a simple but powerful workplace organization method.[8] The underlying principle of the method is that only the items needed for the immediate task (parts, containers, tools, instructions, materials) are kept at hand where they are needed at the moment, and everything else is kept in easily accessible and well-organized storage in perfect order, easy to locate without searching, and in just the right quantities. All items have their designated place, clearly labeled with signs, labels, part numbers, and possibly bar codes. The minimum and maximum levels of inventory of small parts are predefined and are based on actual consumption rather than the "just-in-case" philosophy. The parts, tools, and materials needed for the next shift of production are prepared by a person in charge of the storage during the previous shift and delivered to the workstation before the shift starts. The floor is uncluttered and marked with designated spaces for all equipment. The entire factory is spotlessly clean and uncluttered. Walls are empty except for the visual boards. In consequence of these changes, the searching for parts, tools, and instructions which can represent a significant waste of labor and time is reduced, and this, in turn, saves energy. Secondary effects are also important. In a well-organized place, fewer mistakes are made; fewer wrong parts are used; less inspection is

FIGURE 2 In this messy plant, the workers waste close to 20% of their time looking for items and scavenging for parts and tools, also wasting the plant energy.

needed; quality, throughput time, and customer satisfaction are increased; and costs and energy are decreased. Figure 2 illustrates a fragment of a messy factory, where the average worker was estimated to waste 20% of his shift time looking for and scavenging for parts and tools. This percentage multiplied by the number of workers yields a significant amount of wasted production time, also wasting plant energy in the same proportion. Sorting, cleaning, and organizing the workplace is one of the simplest and most powerful starting points on the way to improved productivity and energy savings.

Conclusion

Large savings in energy are possible as an inherent byproduct of improving productivity. The state-of-the-art Lean productivity method can yield dramatic improvements in productivity. In the extreme case of converting from the traditional batch-and-queue and "process village" manufacturing system to Lean production, overall costs, lead times, and inventories can be reduced by as much as 50%–90%, floor space and energy by 50%, and energy density can be improved by 50%. The amount of energy that can be saved by productivity improvements often radically exceeds the savings from equipment optimization alone, thus providing a strong incentive to include productivity improvements in energy-reduction efforts.

Productivity strongly depends on human factors such as management, learning, and training, communications, culture, teamwork, etc., which are difficult to quantify, making accurate estimates of the cost, schedule, and quality benefits from various productivity improvements and the related energy savings difficult to estimate with engineering accuracy. For this reason, simple metrics and models are recommended, and some examples have been presented. If applied conservatively, they can become useful tools for energy management in a plant. The prerequisite knowledge includes an understanding of Lean Flow and its various productivity elements and a good accounting of energy use in the plant, including the knowledge of the energy used by individual machines and processes both when in productive use and in the idle but energized state, as well as the energy elements used by the infrastructure (various light combinations, air- compressors, cooling and heating devices, combusting systems, conveyers, forklifts, etc.). In the times of ferocious global competition and rising energy prices, every industrial plant should make every effort to improve both productivity and energy use.

Acknowledgments

This work is a result of the studies of energy conservation using the Lean productivity method performed by the Industrial Assessment Center funded by the U.S. Department of Energy at Loyola Marymount University. The author is grateful to Mr. Rudolf Marloth, Assistant Director of the Center, for his help with various energy estimates included herein and his insightful comments, to the Center students for their enthusiastic work, and to his son Peter W. Oppenheim for his diligent editing.

References

1. U.S. Department of Energy, Energy Efficiency and Renewable Energy, available at http://www.eere.energy.gov/industry/ (accessed on December 2005).
2. U.S. Department of Energy, available at http://eereweb.ee.doe.gov/industry/bestpractices/plant_assessments.html, (accessed on December 2005).
3. Womack, P.J.; Jones, D.T. *Lean Thinking*. 2nd Ed.; Lean Enterprise Institute: Boston, 2005; (http://www.lean.org), ISBN: 0–7432-4927-5.
4. Liker, J. *Becoming Lean, Inside Stories of U.S. Manufacturers;* Productivity Inc.: Portland, OR, 1998; service@produc tivityinc.com.
5. Oppenheim, B.W. *Selected Assessment Recommendations,* Industrial Assessment Center, Loyola Marymount University: Los Angeles, (boppenheim@lmu.edu), unpublished 2004.
6. *Setup Reduction for Just-in-Time,* Video/CD, Society of Manufacturing Engineers, Product ID: VT90PUB2, available at http://www.sme.org/cgi-bin/get-item.pl?VT392&2&SME&1990, (accessed on December 2005).
7. Ohno, T. *Toyota Production System: Beyond Large Scale Production;* Productivity Press: New York, 1988; info@pro ductivityinc.com.
8. Hiroyuki, H. *Five Pillars of the Visual Workplace, the Sourcebook for 5S Implementation;* Productivity Press: New York, 1995; info@productivityinc.com.

48

Global Climate Change: Carbon Sequestration

Sherwood Idso
and Keith E. Idso

Introduction

Concomitant with mankind's growing numbers and the progression of the Industrial Revolution, there has been a significant increase in the burning of fossil fuels (coal, gas, and oil) over the past 200 years, the carbon dioxide emissions from which have led to ever-increasing concentrations of atmospheric CO_2. This "large-scale geophysical experiment," to borrow the words of two of the phenomenon's early investigators,[1] is still ongoing and expected to continue throughout the current century. Furthermore, this enriching of the air with CO_2 is looked upon with great concern, because CO_2 is an important greenhouse gas, the augmentation of which is believed by many to have the potential to produce significant global warming. Therefore, and because of perceived serious consequences, such as the melting of polar ice, rising sea levels, coastal flooding, and more frequent and intense droughts, floods, and storms,[2] a concerted effort is underway to slow the rate at which CO_2 accumulates in the atmosphere, with the goal of stabilizing its concentration at a level that would prevent dangerous anthropogenic interference with the planet's climate system.

Removing Carbon from the Air

Role of Man

There are only two ways to significantly increase the natural flux of carbon from the atmosphere to the biosphere within the time frame required for effective ameliorative action if the ongoing rise in the air's CO_2 content is indeed a bona fide global warming threat: 1) increase the rate of vegetative CO_2 assimilation (photosynthesis) per unit leaf area and/or 2) increase the total plant population of the globe, i.e., leaf area per unit land area. Additionally, these things must be done without increasing the rate at which carbon is lost from the soil.

Man can do certain things to promote both of these phenomena while meeting the latter requirement as well. He can, for example, increase the rate of CO_2 assimilation per unit leaf area in agro-ecosystems by supplying additional nutrients and water to his crops. As has recently been noted, however, there are significant carbon costs associated with the production and application of fertilizers, as well as the transport of irrigation water; and factoring the CO_2 emissions of these activities into the equation often results in little net CO_2 removal from the atmosphere via these intensified agricultural interventions.[3]

Man can also draw more CO_2 out of the air by increasing the acreage of land devoted to growing crops, but this approach simultaneously releases great stores of soil carbon built up over prior centuries. When the plow exposes buried organic matter and it is oxidized, for example, prodigious amounts of CO_2 are produced and released to the atmosphere. But if a transition to less intensive tillage is made on fields that have a long history of conventional management and have thus been largely depleted of carbon, there is a good opportunity for nature to rebuild previously lost stores of soil organic matter.[4]

This approach to carbon sequestration is doubly beneficial for it results in a net removal of CO_2 from the atmosphere at the same time that it enhances a whole host of beneficial soil properties.[4,5] Also, abandoned farmlands will gradually replenish their carbon stores, both above- and below-ground, as native vegetation gradually reestablishes itself upon them. And, of course, the process can be hastened and made even more effective if trees are planted on such lands. Even without trees, it has been estimated that agricultural "best management practices" that employ conservation tillage techniques have the potential to boost the current U.S. farm and rangeland soil carbon sequestration rate of 20 million metric tons of carbon per year to fully 200 million metric tons per year,[6] which is approximately 13% of the country's yearly carbon emissions.[7]

Commercial forests also offer excellent opportunities for CO_2 removal from the air for considerable periods of time, especially when harvested wood is used to produce products that have long lifetimes. In addition, since some species of trees, such as many of those found in tropical rainforests,[8] can live in excess of a thousand years, CO_2 can be removed from the atmosphere and sequestered within their tissues—if man protects the trees from logging—until either long after the Age of Fossil Fuels has run its course or until significant changes in energy systems have reduced our dependence on fossil fuels and the CO_2 content of the air has returned to a level no longer considered problematic. Furthermore, carbon transferred to the soil beneath the trees via root exudation and turnover has the potential to remain sequestered even longer.

Role of Nature

The fact that the biosphere has maintained itself over the eons in the face of a vast array of environmental perturbations (albeit with significant modifications) suggests that earth's plant life has great resiliency and may even be able to exert a restraining influence on climate change.

A particularly important negative feedback of this type is the biosphere's ability to intensify its rate of carbon sequestration in the face of rising atmospheric CO_2 concentrations, as this phenomenon slows the rate of rise of the air's CO_2 content and thereby reduces the degree of intensification of the atmosphere's greenhouse effect. This particular climate-moderating influence of atmospheric CO_2 enrichment was first described in quantitative terms by Idso.[9,10] It begins when the aerial fertilization effect produced by the rising CO_2 content of the atmosphere elicits an increase in plant CO_2 assimilation rate per unit leaf area and when the concomitant plant water use efficiency-enhancing effect of the elevated CO_2 leads to an increase in the total plant population of the globe, due to the ability of more water-use-efficient plants to live and successfully reproduce in areas where it was formerly too dry for them to survive. In fact, these two effects are so powerful, they may actually be able to stabilize the CO_2 content of the atmosphere sometime during the current century, but only if anthropogenic CO_2 emission rates do not rise by an inordinate amount in the interim.[9,10] At the very least, together with the things man can do, they have the potential to "buy time" until other less-CO_2-emitting technologies become available.[11]

Keeping Carbon in the Soil

As more carbon is added to soils via CO_2-enhanced root growth, turnover and exudation, as well as from CO_2- induced increases in leaf litter and other decaying plant parts, the trick of significantly augmenting soil carbon sequestration is to keep at least the same percentage of this carbon in the soil as has historically been the case and to do so in the face of potential global warming.

A number of studies have addressed various aspects of this subject in recent years, with most of them finding that atmospheric CO_2 enrichment has little to no significant effect on plant litter decomposition rates. Furthermore, in nearly all of the cases where elevated CO_2 was observed to impact this phenomenon, the extra CO_2 was found to actually slow the rate of plant decomposition.[12] Much the same results have been obtained when analogous studies have used temperature as the independent variable. Warming has had either no effect on CO_2 evolution from the soil, or it has led to an actual decrease in CO_2 loss to the atmosphere.[13] Hence, the balance of evidence obtained from these studies suggests that the same—or a greater—percentage of plant material produced in a world of elevated atmospheric CO_2 concentration (and possibly higher mean air temperature) would indeed be retained in the soils of the terrestrial biosphere.

Even more compelling are the results of experiments where scientists have made direct measurements of changes in soil carbon storage under conditions of elevated atmospheric CO_2. Nearly every such study has observed increases in soil organic matter. In a Free-Air CO_2 Enrichment (FACE) experiment where portions of a cotton field were exposed to a 50% increase in atmospheric CO_2, for example, Leavitt et al.[14] found that 10% of the organic carbon present in the soil below the CO_2-enriched plants at the conclusion of the three-year experiment came from the extra CO_2 supplied to the FACE plants. In addition, some of the stored carbon had made its way into a very recalcitrant portion of the soil organic matter that had an average soil residence time of 2200 years.

Here, too, most experiments indicate that concomitant increases in temperature do not negate the increased carbon storage produced by atmospheric CO_2 enrichment. In a two-year study of perennial ryegrass grown at ambient and twice-ambient atmospheric CO_2 concentrations, as well as ambient and ambient+3°C temperature levels, for example, Casella and Soussana[15] determined that the elevated CO_2 increased soil carbon storage by 32% and 96% at low and high levels of soil nitrogen supply, respectively, "with no significant increased temperature effect." Hence, as in the case of studies of plant decomposition rates, the balance of evidence obtained from these studies also suggests that the same—or a greater—percentage of plant material produced in a world of elevated atmospheric CO_2 concentration (and possibly higher mean air temperature) would indeed be retained in the soils of the terrestrial biosphere.

Conclusions

As the air's CO_2 content continues to rise, there will almost certainly be a significant upward trend in the yearly production of terrestrial vegetative biomass, due to the growth-enhancing aerial fertilization effect of atmospheric CO_2 enrichment and the concomitant CO_2-induced increase in plant water use efficiency that enables plants to grow where it is currently too dry for them. Experimental evidence further suggests that at least the same percent- age—but in all likelihood more—of this yearly-increasing mass of plant tissue will be sequestered in earth's soils. Consequently, it is almost impossible to conclude that the carbon sequestering prowess of the planet will not be greatly enhanced in the years ahead, even without any overt actions on the part of man. Hence, if the nations of the earth were to implement even a modicum of carbon- conserving measures—such as 1)using minimum tillage techniques wherever possible in agricultural settings; 2) allowing abandoned agricultural land to revert to its natural vegetative state; 3) allowing stands of trees that can grow to very old age to actually do so; and 4) employing wise forestry practices to produce wood for making products that have long lifetimes—it is possible that

TABLE 1 Potential Rates of Carbon Sequestration (Kilograms Carbon per Hectare per Year) due to Land Management Practices That Could Be Employed for This Purpose

Improved rangeland management	50 to 150
Improved pastureland management	
Commercial fertilizer applications	100 to 200
Manure applications	200 to 500
Use of improved plant species	100 to 300
Improved grazing management	300 to 1300
Nitrogen fertilization of mountain meadows	100 to 200
Restoration of eroded soils	50 to 200
Restoration of mined lands	1000 to 3000
Conversion of cropland to pasture	400 to 1200
Conversion of cropland to natural vegetation	600 to 900
Conversion from conventional to conservation tillage	
No till	500
Mulch till	500
Ridge till	500

Source: Adapted from data reported by Follett, R.F.; Kimble, J.M.; Lal, R. The *Potential of U.S. Grazing Lands to Sequester Carbon and Mitigate the Greenhouse Effect;* Lewis Publishers, Boca Raton, FL, 2001; 1–442, and by Lal et al.[4]

the antiwarming feedback produced by the subsequent removal of CO_2 from the atmosphere would be sufficient to keep the risk of potential greenhouse gas-induced global warming at an acceptable level. Estimates of the carbon-sequestering power of some of these "best management practices" are given in Table 1.

References

1. Revelle, R.; Suess, H.E. Carbon dioxide exchange between atmosphere and ocean and the question of an increase of atmospheric CO_2 during the past decades. Tellus **1957**, *9*, 18–27.
2. Intergovernmental panel on climate change. In *Climate Change 2001: The Scientific Basis, Summary for Policy Makers and Technical Summary of the Working Group I Report*; Cambridge University Press: Cambridge, U.K., 2001; 1–98.
3. Schlesinger, W.H. Carbon sequestration in soils: some cautions amidst optimism. Agric. Ecosys. Environ. **2000**, *82*, 121–127.
4. Lal, R.; Kimble, J.M.; Follett, R.F.; Cole, C.V. *The Potential for U.S. Cropland to Sequester Carbon and Mitigate the Greenhouse Effect*; Sleeping Bear Press: Chelsea, MI, 1998; 1–128.
5. Idso, S.B. *Carbon Dioxide and Global Change: Earth in Transition*; IBR Press: Tempe, AZ, 1989; 1–292.
6. Jawson, M.D.; Shafer, S.R. Carbon credits on the Chicago board of trade? Agric. Res. **2001**, *49* (2), 2.
7. Comis, D.; Becker, H.; Stelljes, K.B. Depositing carbon in the bank. Agric. Res. **2001**, *49* (2), 4–7.
8. Chambers, J.Q.; Higuchi, N.; Schimel, J.P. Ancient trees in Amazonia. Nature **1998**, *391*, 135–136.
9. Idso, S.B. The aerial fertilization effect of CO_2 and its implications for global carbon cycling and maximum greenhouse warming. Bull. Amer. Meteorol. Soc. **1991**, *72*, 962–965.
10. Idso, S.B. Reply to comments of L.D. Danny Harvey, Bert Bolin, and P. Lehmann. Bull. Amer. Meteorol. Soc. **1991**, *72*, 1910–1914.
11. Izaurralde, R.C.; Rosenberg, N.J.; Lal, R. Mitigation of climatic change by soil carbon sequestration: issues of science, monitoring, and degraded lands. Adv. Agron. **2001**, *70*, 1–75.

12. Nitschelm, J.J.; Luscher, A.; Hartwig, U.A.; van Kessel, C. Using stable isotopes to determine soil carbon input differences under ambient and elevated atmospheric CO_2 conditions. Global Change Biol. **1997**, *3*, 411–416.
13. van Ginkel, J.H.; Whitmore, A.P.; Gorissen, A. *Lolium perenne* grasslands may function as a sink for atmospheric carbon dioxide. J. Environ. Quality **1999**, *28*, 1580–1584.
14. Leavitt, S.W.; Paul, E.A.; Kimball, B.A.; Hendrey, G.R.; Mauney, J.R.; Rauschkolb, R.; Rogers, H.; Lewin, K.F.; Nagy, J.; Pinter, P.J., Jr.; Johnson, H.B., Jr. Carbon isotope dynamics of free-air CO_2-enriched cotton and soils. Agric. For. Meteorol. **1994**, *70*, 87–101.
15. Casella, E.; Soussana, J.-F. Long-term effects of CO_2 enrichment and temperature increase on the carbon balance of a temperate grass sward. J. Exp. Bot. **1997**, *48*, 1309–1321.

49

Global Climate Change: Earth System Response

Amanda Staudt and
Nathan E. Hultman

Introduction

For the past century, Earth's climate has been changing due to human activities. Observations show that Earth's surface warmed by approximately 0.6°C (1.1°F) on average in the 20th century. Much of this warming has been attributed to increasing abundances of greenhouse gases emitted to the atmosphere by human activities, although it is difficult to quantify this contribution against the backdrop of natural variability and climate-forcing uncertainties. Atmospheric abundances of the major anthropogenic greenhouse gases (carbon dioxide; methane; nitrous oxide; halocarbons manufactured by humans, such as chlorofluorocarbons; and tropospheric ozone) reached their highest recorded levels at the end of the 20th century, and all but methane have continued to rise. Major causes of this rise have been fossil fuel use, agriculture, and land-use change.

The emerging impacts of climate change on natural systems include melting glaciers and ice caps, the rising sea level, extended growing seasons, changes in precipitation regimes, and changes in the geographical distributions of plant and animal species. Additional impacts, to which it may be difficult for human and natural systems to adapt, could arise from events whose triggers are poorly understood. Human-induced global warming will continue during the 21st century and beyond, because many parts of the Earth's system respond slowly to changes in greenhouse gas levels and because altering established energy-use practices is difficult. Uncertainties remain about the magnitude and the impacts of future climate change, largely due to gaps in understanding of climate science and the socioeconomic drivers of climate change.

The Climate System and the Natural Greenhouse Effect

While climate conventionally has been defined as the long-term statistics of the weather (e.g., temperature, cloudiness, precipitation), improved understanding of the atmosphere's interactions with the oceans, the cryosphere (ice-covered regions of the world), and the terrestrial and marine biospheres has led scientists to expand the definition of climate to encompass the oceanic and terrestrial spheres as well as chemical components of the atmosphere (Figure 1). Physical processes within the atmosphere are influenced by ocean circulation, the reflectivity of Earth's surface, the chemical composition of the atmosphere, and vegetation patterns, among other factors.

The Sun provides almost all of Earth's energy. Solar radiation intercepted by Earth first encounters the atmosphere, which allows most of it to pass to Earth's surface. The intensity of radiation at the surface depends on the amount of incident radiation and on the orientation of the surface with respect to that radiation. The surface either reflects or absorbs this incoming radiation. Different surfaces reflect different amounts of sunlight. The fraction of solar energy reflected is defined as a surface's albedo. Albedos range from about 10% for open water, dark soil, and asphalt to about 80% for fresh snow.[1] Earth's average albedo is about 31%.

Although we do not normally think of it as a radiative body, Earth—like all bodies with a nonzero temperature—emits electromagnetic radiation. For Earth's temperature, most of this radiation is in the form of infrared light. In the absence of an atmosphere, all the radiation emitted by Earth would escape to space. The balance of incoming solar radiation and outgoing infrared radiation would result in a global-mean temperature for Earth of 255 K (–18°C/0°F). However, some molecules in Earth's atmosphere absorb some of this outgoing infrared light, thereby increasing their temperature. These greenhouse gases in the atmosphere emit some energy back toward Earth, warming Earth's surface. This natural

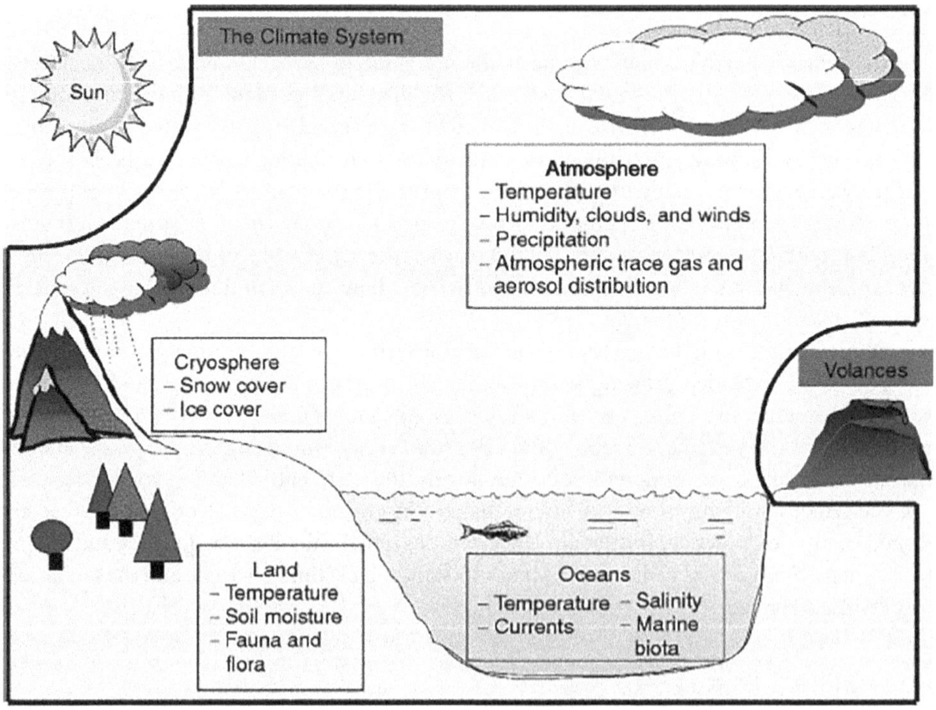

FIGURE 1 The climate system.
Source: National Academies Press (see NRC[5]).

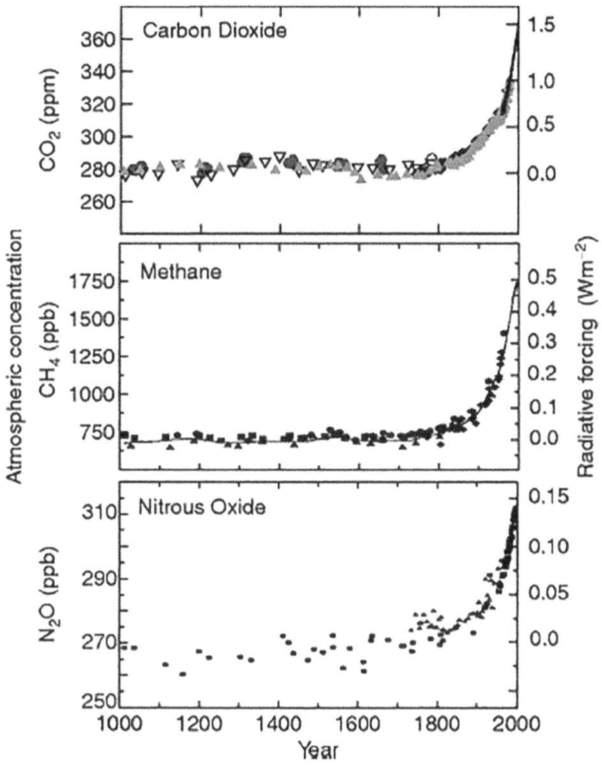

FIGURE 2 Concentrations of major greenhouse gases retrieved from gas bubbles trapped in ice cores from Antarctica and Greenland.
Source: Intergovernmental Panel on Climate Change (see IPCC[10]).

greenhouse effect, which is present in the absence of human activities, raises the global-mean surface temperature from 255 K to a comfortable 288 K (or about 15°C/59°F).[2]

Greenhouse gases that are present naturally in the atmosphere include water vapor (H_2O), carbon dioxide (CO_2), methane (CH_4), nitrous oxide (N_2O), and ozone (O_3). The most common greenhouse gas by quantity and the one exerting the greatest influence on the climate is water vapor; however, because water has a very short lifetime in the atmosphere (~1 week), any human perturbation will dissipate quickly. In most cases, the "greenhouse effect" or "climate change" refers not to this natural phenomenon but to additional, anthropogenic enhancements to the atmosphere's capacity to trap heat. Much higher concentrations of CO_2, CH_4, and N_2O have been observed in the past century than were naturally present for the past 1000 years (Figure 2) and likely much longer.[3] Earth's surface is warmer now on average than it was at any time during the past 400 years, and it is likely warmer now than it was at any time in the past 2000 years.[4]

Climate Forcings and Feedbacks

Factors that affect climate change are usefully separated into forcings and feedbacks. Climate forcings are energy imbalances imposed on the climate system either externally or by human activities.[5] Examples include human- caused emissions of greenhouse gases, as discussed in the preceding section, as well as changes in solar energy input; volcanic emissions; deliberate land modification; or anthropogenic emissions of aerosols, which can absorb and scatter radiation. Climate forcings can be either direct or indirect. Direct radiative forcings are simple changes to the drivers of Earth's radiative balance. For

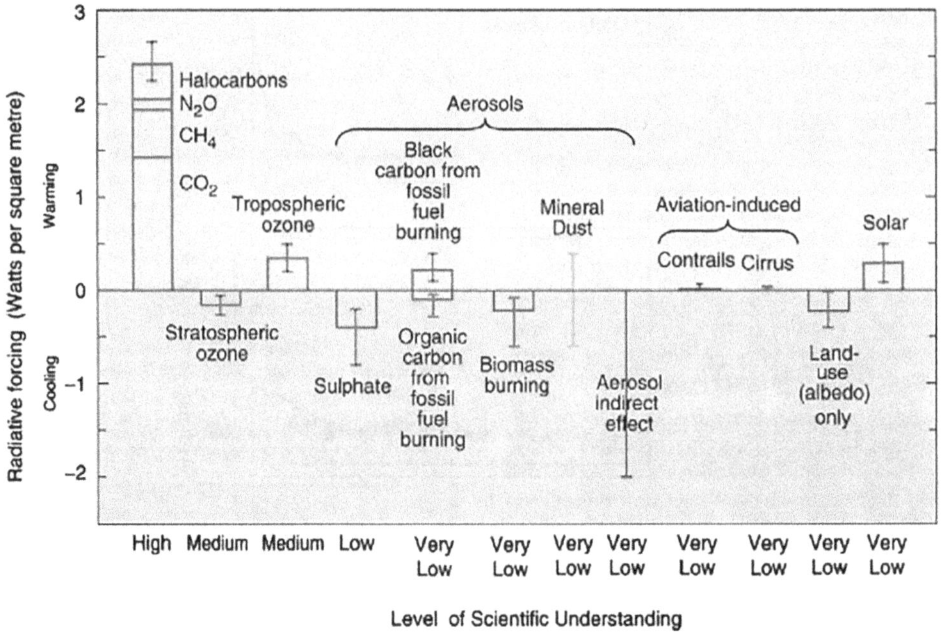

FIGURE 3 Estimated radiative forcings since preindustrial times for Earth and the troposphere system. The height of the rectangular bar denotes a central or best estimate of the forcing, while each vertical line is an estimate of the uncertainty range associated with the forcing, guided by the spread in the published record and physical understanding, and with no statistical connotation. Each forcing agent is associated with a level of scientific understanding, which is based on an assessment of the nature of assumptions involved, the uncertainties prevailing about the processes that govern the forcing, and the resulting confidence in the numerical values of the estimate. On the vertical axis, the direction of expected surface temperature change due to each radiative forcing is indicated by the labels "warming" and "cooling."
Source: Intergovernmental Panel on Climate Change (see IPCC[10]).

example, added CO_2 absorbs and emits infrared radiation. Indirect radiative forcings create a radiative imbalance by first altering climate system components that lead to consequent changes in radiative fluxes; an example is the effect of aerosols on the precipitation efficiency of clouds. Figure 3 provides a summary of the estimated contribution from major climate forcings. Additional information about specific climate forcings is provided in the discussion below.

Climate feedbacks are internal climate processes that amplify or dampen the climate response to an initial forcing.[6] An example is the increase in atmospheric water vapor that is triggered by an initial warming due to rising CO_2 concentrations, which then acts to amplify the warming through the greenhouse properties of water vapor (Figure 4). Other climate feedbacks involve snow and ice cover, biogeochemistry, clouds, and ocean circulation. Some of the uncertainty about how the climate will change in the future stems from unresolved research questions on climate change feedbacks.

Natural Climate Forcings: Solar and Volcanic Variability

Variations in the Sun's activity and in Earth's orbital parameters cause natural forcing of climate. Radiometers on various spacecraft have been measuring the total solar irradiance since the late 1970s. There is an 11-year cycle in total solar irradiance of peak-to-peak amplitude ~1 W m^{-2} (0.1%) in the past three cycles. Allowing for reflection of 30% of this incident energy (Earth's albedo) and

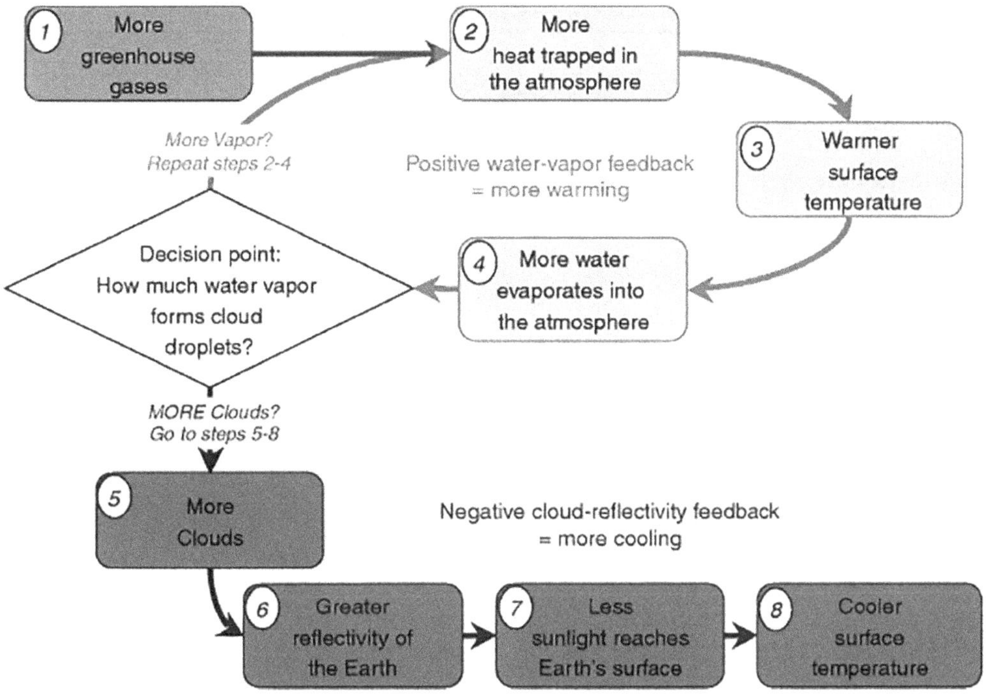

FIGURE 4 This schematic illustrates just two out of the dozens of climate feedbacks identified by scientists. The warming created by emitting more greenhouse gases leads to evaporation of water into the atmosphere. But water itself is a greenhouse gas and can cause even more warming via positive water-vapor feedback. On the other hand, if the water vapor leads to more clouds, the result could be to counteract some of the warming because clouds can reflect incoming sunlight back to space. This chain of events is negative cloud- reflectivity feedback. Trying to understand whether water vapor will create more clouds and what kinds of clouds water vapor will create is a major research objective right now. The answer depends on weather patterns, where the evaporation takes place, and the amount of small soot particles suspended in the air.

averaging over the globe, the corresponding climate forcing is of order 0.2 W m^{-2}, although recent analyses have found little secular trend in solar irradiance over the past 30 years.[7] Knowledge of solar irradiance variations prior to 1979 is less certain, as it relies upon models of how sunspot and facular influences relate to solar irradiance observed since then. These models are used to extrapolate variations back to about 1610, when telescopes were first used to monitor sunspots. The amount of energy Earth receives from the Sun also depends on Earth's distance from the Sun, which does not remain constant. The eccentricity of Earth's orbit (currently 0.0167) and the tilt of its axis relative to the orbital plane result in continual changes to the amount and distribution of energy Earth receives. In modern times, this variation is ±3.5% during the year, with maximum energy and minimum distance in January.

Volcanic forcing has been the dominant source of natural global radiative forcing over the past millennium. Emissions from volcanic eruptions have multiple effects on climate, as listed in Table 1.[8] The greater prevalence of explosive volcanic activity during both the early and the late 20th century and the dearth of eruptions from 1915 to 1960 represent a significant natural radiative forcing of 20th century climate.[9] Similarly, longer-term volcanic radiative forcing has been associated with a significant long-term forced cooling from 1000 to 1900, resulting from a general increase in explosive volcanic activity in later centuries.

TABLE 1 Effects of Large Explosive Volcanoes on Weather and Climate

Effect and Mechanism	Begins	Duration
Reduction of diurnal cycle	Immediately	1–4 days
	Blockage of shortwave and emission of longwave radiation	
Reduced tropical precipitation	1–3 months	3–6 months
	Blockage of shortwave radiation, reduced evaporation	
Summer cooling of Northern Hemisphere tropics and subtropics	1–3 months	1–2 years
	Blockage of shortwave radiation	
Reduced Sahel precipitation	1–3 months	1–2 years
	Blockage of shortwave radiation, reduced land temperature, reduced evaporation	
Stratospheric warming	1–3 months	1–2 years
	Stratospheric absorption of shortwave and longwave radiation	
Winter warming of Northern Hemisphere continents	6–18 months	1 or 2 winters
	Stratospheric absorption of shortwave and longwave radiation, dynamics	
Global cooling	Immediately	1–3 years
	Blockage of shortwave radiation	
Global cooling from multiple eruptions	Immediately	Up to decades
	Blockage of shortwave radiation	
Ozone depletion, enhanced UV radiation	1 day	1–2 years
	Dilution, heterogeneous chemistry on aerosols	

Source: Reproduced/modified by permission of the American Geophysical Union (see Robock[8]).

Greenhouse Gas Forcing

The role of greenhouse gases in the climate system is well understood by scientists because instruments can accurately measure the abundances of these gases in the atmosphere and their radiative properties. The concentrations of CO_2, CH_4, N_20, various halocarbons, and O_3 have increased substantially since preindustrial times, and they are the greatest contributors to total anthropogenic radiative forcing.[10] Many of these greenhouse gases are emitted primarily as a byproduct of fossil fuel combustion.

For a given gas, the total amount of heat-trapping ability depends on the efficiency of heat trapping for a given unit of gas (i.e., radiative forcing), the number of units present in the atmosphere, and the average length of time a given unit spends in the atmosphere. While these three components are enough to characterize a single gas, the large number of gases has prompted the development of an index called the global warming potential (GWP), which represents the relative impact of a particular greenhouse gas on the atmosphere's radiative balance.[10] See Table 2 for some GWP calculations. As the standard reference gas, CO_2 has a GWP of 1, by definition. Over a time horizon of 100 years, CH_4 and N_2O have GWPs of 23 and 296, respectively. In other words, 1 additional kg of CH_4 in the atmosphere absorbs as much radiation as 23 additional kg of CO_2. However, these numbers change if the time horizon shifts.[10] By allowing greenhouse gases to be compared directly, GWPs enable policies that can reduce total climate impact by addressing the least-cost abatement options first.[11,12]

TABLE 2 Radiative Forcing Characteristics of Some Major Greenhouse Gases

	Contribution to Direct Radiative Forcing				Global Warming Potential (GWP) for Different Time Horizons		
	Wm2	%	Concentration in 1998	Lifetime (Years)	20 Years	100 Years	500 Years
Carbon dioxide (CO_2)	1.46	60	365 ppm	5–200a	1	1	1
Methane (CH_4)	0.48	20	1745 ppb	12.0	62	23	7
Nitrous oxide (N_2O)	0.15	6	314 ppb	114	275	296	156
Halocarbons and related compounds	0.34	14	—	0.3–3200	40–15100	12–22200	4–16300

Source: Intergovernmental Panel on Climate Change (see IPCC[10]).

Total direct radiative forcing uncertainty is approximately 10%. The abbreviations parts per million (ppm) and parts per billion (ppb) refer to the ratio of greenhouse gas molecules to molecules of dry air.

[a] No single lifetime can be defined for CO2 because it is affected by multiple removal processes with different uptake rates.

Atmospheric Aerosol Forcing

Aerosols are small particles or liquid droplets suspended in the atmosphere. Aerosols both scatter and absorb radiation, representing a direct radiative forcing. Scattering generally dominates (except for black carbon particles) so that the net effect is of cooling. The average global mean of aerosol-direct forcing from fossil fuel combustion and biomass burning is in the range of –0.2 to –2.0 W m^{-2}.[10] This large range results from uncertainties in aerosol sources, composition, and properties used in different models. Recent advances in modeling and measurements have provided important constraints on the direct effect of aerosols on radiation.[13–15] Aerosols have several indirect effects on climate, all arising from their interaction with clouds— particularly from their roles as cloud condensation nuclei (CCN) and ice nuclei (Table 3).

TABLE 3 Overview of the Different Aerosol Indirect Effects Associated with Clouds

Effect	Cloud Type	Description	Sign of Top of the Atmosphere Radiative Forcing
First indirect aerosol effect (cloud albedo or Twomey effect)	All clouds	For the same cloud water or ice content more but smaller cloud particles reflect more solar radiation	Negative
Second indirect aerosol effect (cloud lifetime or Albrecht effect)	All clouds	Smaller cloud particles decrease the precipitation efficiency, thereby prolonging cloud lifetime	Negative
Semidirect effect	All clouds	Absorption of solar radiation by soot leads to evaporation of cloud particles	Positive
Glaciation indirect effect	Mixed-phase clouds	An increase in ice nuclei increases the precipitation efficiency	Positive
Thermodynamic effect	Mixed-phase clouds	Smaller cloud droplets inhibit freezing, causing super cooled droplets to extend to colder temperatures	Unknown
Surface energy budget effect	All clouds	The aerosol-induced increase in cloud optical thickness decreases the amount of solar radiation reaching the surface, changing the surface energy budget	Negative

Source: National Academies Press (see NRC[5]).

Land-Use Change Forcing

Land-use changes include irrigation, urbanization, deforestation, desertification, reforestation, the grazing of domestic animals, and dryland farming. Each of these alterations in landscape produces changes in radiative forcing, both directly and indirectly.[16,17] Direct effects include the change of albedo and emissivity resulting from the different types of land covers. For example, the development of agriculture in tropical regions typically results in an increase of albedo from a low value of forest canopies (0.05–0.15) to a higher value of agricultural fields, such as pasture (0.15–0.20). The Intergovernmental Panel on Climate Change (IPCC)[10] reports the globally averaged forcing due to albedo change alone as -0.25 ± 0.25 W m^{-2}. Significant uncertainties remain in estimating the effect of land-use change on albedo because of the complexity of land surfaces (e.g., the type of vegetation, phenology, density of coverage, soil color).

Indirect effects of land-cover change on the net radiation include a variety of processes related to (1) the ability of the land cover to use the radiation absorbed at the ground surface for evaporation, transpiration, and sensible heat fluxes (the impact on these heat fluxes caused by changes in land cover is sometimes referred to as thermodynamic forcing); (2) the exchange of greenhouse and other trace gases between the surface and the atmosphere; (3) the emission of aerosols (e.g., from dust); and (4) the distribution and melting of snow and ice.[5] These effects are not yet well characterized or quantified.

Evidence of Human-Induced Climate Change

Because we do not have a "control Earth" against which to compare the effects of our current changing atmosphere, incontrovertibly linking human activities and observed climate change is difficult. Scientists therefore rely on multiple, overlapping evidence of changes and then compare observed patterns of change with what our scientific understanding indicates should happen under anthropogenic climate change. This two-stage concept of discovering changes in climate and linking them to human activity is called detection and attribution. Evidence used to detect climate change is summarized in this section, and the use of climate models for attribution is discussed in the following section.

One piece of evidence of global warming is an increase in surface temperature since the 1900s, with particularly rapid increases since the late 1970s (Figure 5). This dataset caused some controversy when researchers discovered that readings taken near the surface of Earth with thermometers appeared to be higher than readings of the lower atmosphere taken by satellites from above. Subsequent studies concluded that the warming trend in global-mean surface temperature observations during the past 30 years is undoubtedly real and is substantially greater than the average rate of warming during the 20th century.[18] Satellite-and balloon-based observations of middle-troposphere temperatures, after several revisions of the data, now compare reasonably with one another and with observations from surface stations, although some uncertainties remain.[19,20]

The ocean, which represents the largest reservoir of heat in the climate system, has warmed by about 0.118°C (0.212°F), averaged over the layer extending from the surface down to 700 m, from 1955 to 2003 (Figure 6).[21] Approximately 84% of the total heating of Earth's system (oceans, atmosphere, continents, and cryosphere) over the past 40 years has gone into warming the oceans. Recent studies have shown that the observed heat storage in the oceans is what would be expected by a human-enhanced greenhouse effect. Indeed, increased ocean heat content accounts for most of the planetary energy imbalance (i.e., when Earth absorbs more energy from the Sun than it emits back to space) simulated by climate models.[22]

Changes in several other climate indicators have been observed over the past decades, providing a growing body of evidence consistent with a human impact on the climate. For example, reductions

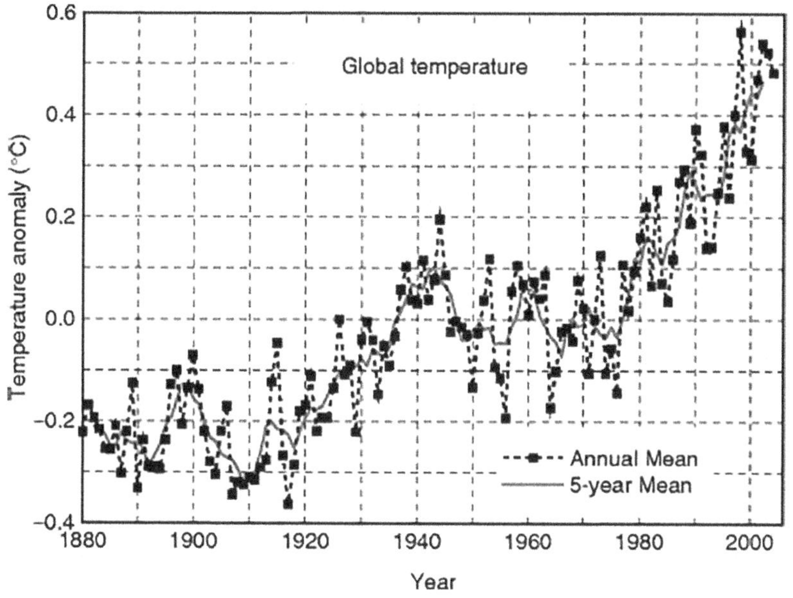

FIGURE 5 Global annual-mean surface air temperature change derived from the meteorological station network. Data and plots are available from the Goddard Institute for Space Sciences (GISS) at http://data.giss.nasa.gov/gistemp/graphs.

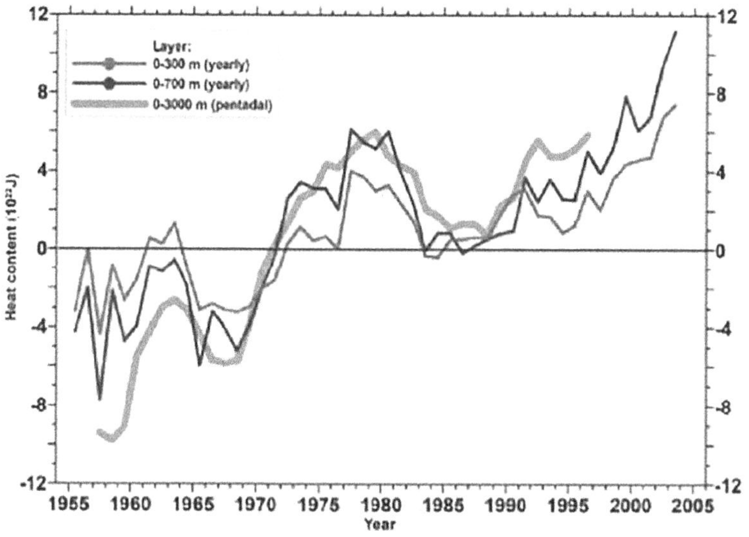

FIGURE 6 Time series of (i) yearly ocean heat content (10^{22}J) for the 0–300 m and 0–700 m layers and (ii) 5-year running averages for 1955–1959 through 1994–1998 for the 0–3000 m layer.
Source: Reproduced/modified by permission of the American Geophysical Union (see Levitus[21])).

in snow and ice cover are one important indicator.[10] Satellite observations indicate that snow cover has decreased by about 10% since the 1960s, while spring and summer sea-ice extent in the Northern Hemisphere has decreased by about 10%–15% since the 1950s. The shrinking of mountain glaciers in many nonpolar regions has also been observed during the 20th century.

Attribution of Observed Climate Change to Human Influence

An important question in global climate change is to what extent the observed changes are caused by the emissions of greenhouse gases and other human activities. Climate models are used to study how the climate operates today, how it may have functioned differently in the past, and how it may evolve in response to forcings. Built using our best scientific knowledge of atmospheric, oceanic, terrestrial, and cryospheric processes, climate models and their components are extensively tested against the full suite of observations of current and past climate to verify that they simulate a realistic version of the climate. Discrepancies between models and observations provide indications that we need to improve understanding of physical processes, model representations of the processes, or in some cases the observations themselves. Hence, climate models contain our accumulated wisdom about the underlying scientific processes and can be no better than our observations of the system and our understanding of the climate.

Figure 7 shows how scientists have used climate models to make the case that human activities have perturbed the climate since preindustrial times. In this experiment, the model is run with three different sets of climate forcings: (a) natural only, (b) anthropogenic only, and (c) natural and anthropogenic. When the natural or anthropogenic forcings are employed separately, the model is unable to reproduce the global-mean variation of temperature anomalies over the simulated time period. Only when both sets of forcings are used does the model capture the nature of the variations, providing evidence that human activities have caused a significant fraction of warming in the past 150 years.[10]

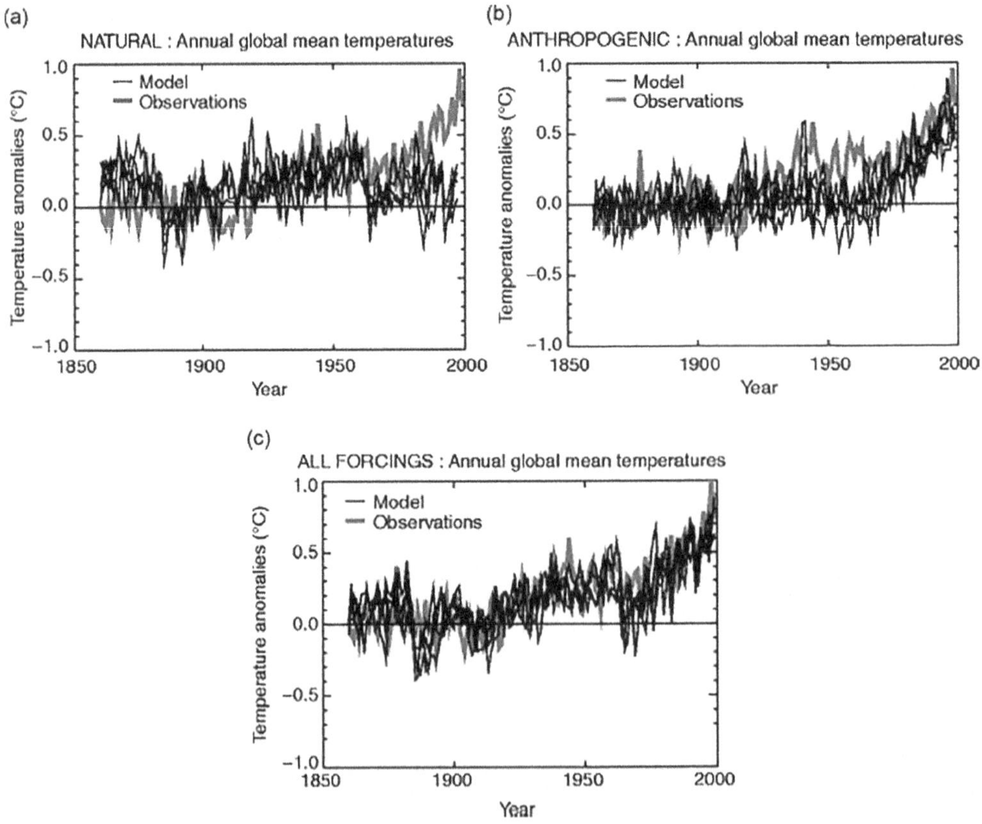

FIGURE 7 Climate model results with (a) solar and volcanic forcings only; (b) anthropogenic forcings only; and (c) all forcings, both natural and anthropogenic.
Source: Intergovernmental Panel on Climate Change (see IPCC[10]).

Projections for Future Climate Change

The IPCC has concluded that by 2100, global surface temperatures will likely be from 1.4 to 5.8°C (2.5°F–10.4°F) above 1990 levels (Figure 8) and that the combined effects of ice melting and seawater expansion from ocean warming will cause the global-mean sea level to rise by between 0.1 and 0.9 m.[10] Uncertainties remain about the magnitude and impacts of future climate change, largely due to gaps in understanding of climate science and the difficulty of predicting societal choices.

Climate changes in the coming century will not be uniformly distributed; some regions will experience more warming than others. There will be winners and losers from the impacts of climate change, even within a single region, but globally, the losses are expected to far outweigh the benefits. A changed climate will increase the likelihood of extreme heat and drought events.[23] High latitudes and polar regions are expected to see comparatively greater increases in average temperatures than lower latitudes, resulting in melting of permafrost and sea ice, which will result in additional costs for residents and in disruption to wildlife and ecosystems.[24] Precipitation changes, which are of great importance to agriculture, may have even more regional variability that is hard to predict.

Finally, several elements of Earth's system seem to be vulnerable to rapid destabilization. For example, the West Antarctic ice sheet and the Greenland ice sheet may be more prone to rapid melting than

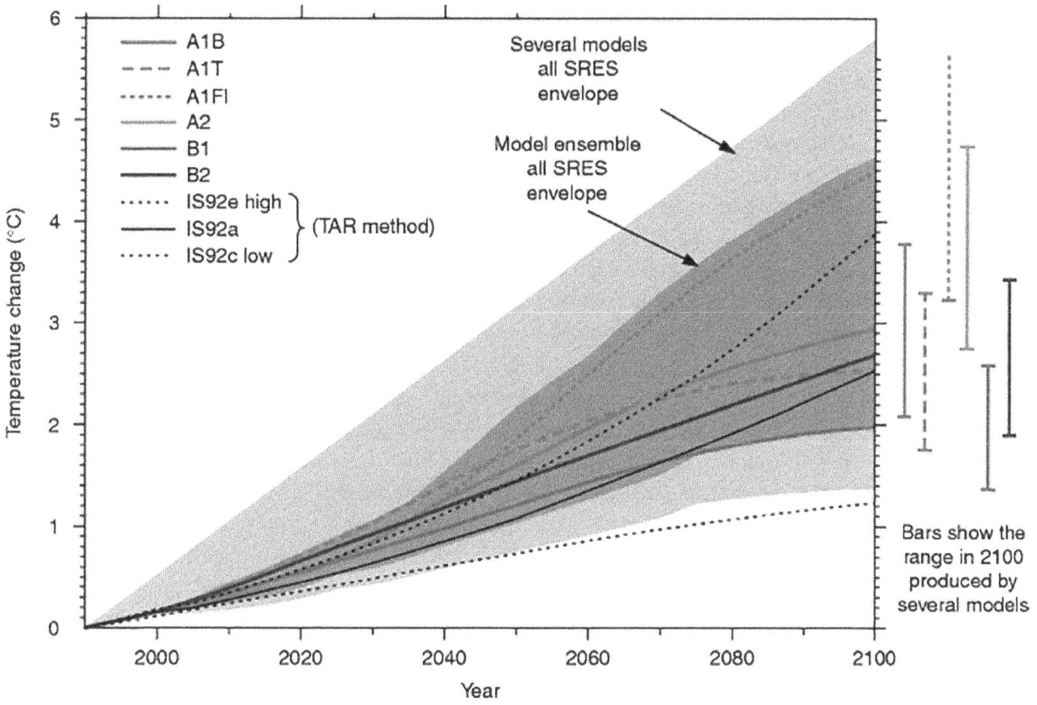

FIGURE 8 Climate models are often used to simulate possible future climates to help inform decisions about policy responses to potential climate changes. This figure shows the range of plausible global-mean temperature change over the next 100 years, simulated by a collection of models. The spread in 2100 temperatures from 1.4 to 5.8°C reflects two factors: (1) Each model was run multiple times using different scenarios (indicated by different line textures) for future climate forcings, and (2) each model makes different assumptions about how the climate responds to those forcings. The scenarios range from those that assume continued acceleration of greenhouse gas emissions to those that assume more moderate growth or leveling off of emissions rates. SRES refers to the collection of scenarios presented in the Special Report on Emissions Scenarios. TAR refers to the IPCC Third Assessment Report.
Source: Intergovernmental Panel on Climate Change (see IPCC[10]).

previously thought, and the loss of either of these would result in a large sea-level rise greater than 5 m. Moreover, the stability of the oceanic circulation that brings heat to Northern Europe has also been questioned. Because of feedback processes and the large uncertainty in system sensitivity, these outcomes are not easy to model and are usually not included in the gradual climate change projections quoted above. Nevertheless, they are nontrivial threats and represent active areas of current research.

Unfortunately, the regions that will be most severely affected are often the regions that are the least able to adapt. Bangladesh, one of the poorest nations in the world, is projected to lose 17.5% of its land if sea level rises about 1 m (40 in.), displacing tens of thousands of people.[10] Several islands throughout the South Pacific and Indian Oceans will be at similar risk for increased flooding and vulnerability to storm surges. Although wetland and coastal areas of many developed nations—including the United States—are also threatened, wealthy countries may be more able to adapt to sea-level rise and threats to agriculture. Solutions could include limiting or changing construction codes in coastal zones and developing new agricultural technologies.

Conclusion

Research conducted to understand how the climate system may be changing—and in turn affecting other natural systems and human society—has led to significant advancement in scientific understanding, but many questions remain. Society faces increasing pressure to decide how best to respond to a changing climate and associated global and regional changes.

One way to address global climate change is to take steps to reduce the amount of greenhouse gases in the atmosphere. Because CO_2 and other greenhouse gases can remain in the atmosphere for many decades, the climate-change impacts from concentrations today will likely continue throughout the 21st century and beyond. Failure to implement significant reductions in net greenhouse gas emissions now will make the job much harder in the future—both in terms of stabilizing CO_2 concentrations and in terms of experiencing more significant impacts. While no single solution can eliminate all future warming, many potentially cost-effective technological options could contribute to stabilizing greenhouse gas concentrations. These options range from personal choices such as driving less to national choices such as regulating emissions or seeking technologies to remove greenhouse gases from the atmosphere to international choices such as sharing energy technologies.

At the same time, it will be necessary to seek ways to adapt to the potential impacts of climate change. Climate is becoming increasingly important to public and private decision-making in various fields such as emergency management, water-quality assurance, insurance, irrigation and power production, and construction. For example, developing practical, "no regrets" strategies that could be used to reduce economic and ecological systems' vulnerabilities to change could provide benefits whether a significant climate change ultimately occurs or not. No-regrets measures could include low-cost steps to improve climate forecasting; to slow biodiversity loss; to improve water, land, and air quality; and to make institutions—such as the health care enterprise, financial markets, and transportation systems—more resilient to major disruptions.

References

1. Houghton, J. *Global Warming: The Complete Briefing,* 3rd Ed.; Cambridge University Press: Cambridge, 2004; 351.
2. Houghton, J. *The Physics of Atmospheres,* 3rd Ed.; Cambridge University Press: Cambridge, U.K., 2002.
3. Petit, J.R.; Jouzel, J.; Raynaud, D.; et al. Climate and atmospheric history of the past 420,000 years from the Vostok ice core, Antarctica. Nature, **1999**, *399*, 429–436.
4. NRC (National Research Council). *Surface Temperature Reconstructions for the Last 2000 Years;* National Academies Press: Washington, DC, 2006.

5. NRC (National Research Council). *Radiative Forcing of Climate Change: Expanding the Concept and Addressing Uncertainties*; National Academies Press: Washington, DC, 2005.
6. NRC (National Research Council). *Understanding Climate Change Feedbacks*; National Academies Press: Washington, DC, 2003.
7. Lean, J.; Rottman, G.; Harder, J.; et al. SORCE contributions to new understanding of global change and solar variability. Solar Phy. **2005**, *230*, 27–53.
8. Robock, A. Volcanic eruptions and climate. Rev. Geophys. **2000**, *38*, 191–219.
9. Crowley, T.J. Causes of climate change over the past 1000 years. Science **2000**, *289*, 270–277.
10. IPCC (Intergovernmental Panel on Climate Change). In *Climate Change 2001: The Scientific Basis;* Houghton, J.T., Ding, Y., Griggs, D.J., Eds.; Contribution of Working Group 1 to the Third Assessment Report of the Intergovernmental Panel on Climate Change, Cambridge University Press: Cambridge, U.K., 2001.
11. Prinn, R.; Jacoby, H.; Sokolov, A.; et al. Integrated global system model for climate policy assessment: Feedbacks and sensitivity studies. Climatic Change **1999**, *41* (3–4), 469–546.
12. Schneider, S.H.; Goulder, L.H. Achieving low-cost emissions targets. Nature **1997**, *389* (6646), 13–14.
13. Ramanathan, V.; Crutzen, P.J.; Kiehl, J.T.; et al. Aerosols, climate and the hydrological cycle. Science **2001**, *294*, 2119–2124.
14. Russell, L.M.; Seinfeld, J.H.; Flagan, R.C.; et al. Aerosol dynamics in ship tracks. J. Geophys. Res. Atmos. **1999**, *104* (D24), 31077–31095.
15. Conant, W.C.; Seinfeld, J.H.; Wang, J.; et al. A model for the radiative forcing during ACE-Asia derived from CIRPAS Twin Otter and R/V Ronald H. Brown data and comparison with observations. J. Geophys. Res. **2003**, *108* (D23), 8661.
16. Pitman, A.J. Review: The evolution of, and revolution in, land surface schemes designed for climate models. Int. J. Climatol. **2003**, *23*, 479–510.
17. Kabat, P., Claussen, M., Dirmeyer, P.A. et al., Eds.; *Vegetation, Water, Humans and the Climate—A New Perspective on an Interactive System*, Berlin, 2004.
18. NRC (National Research Council) *Reconciling Observations of Global Temperature Change*; National Academy Press: Washington, DC, 2000.
19. Mears, C.A.; Wentz, F.J. The effect of diurnal correction on satellite-derived lower tropospheric temperature. Science **2005**, *309*, 1548–1551.
20. Sherwood, S.C.; Lanzante, J.R.; Meyer, C.L. Radiosonde daytime biases and late-20th century warming. Science **2005**, *309*, 1556–1559.
21. Levitus, S.; Antonov, J.; Boyer, T. Warming of the world ocean, 1955–2003. Geophys. Res. Lett. **2005**, *32* (L02604).
22. Hansen, J.; Nazarenko, L.; Ruedy, R.; et al. Earth's energy imbalance: Confirmation and implications. Science **2005**, *308*, 1431–1435.
23. Meehl, G.A.; Tebaldi, C. More intense, more frequent, and longer lasting heat waves in the 21st century. Science **2004**, *305*, 994–997.
24. ACIA (Arctic Climate Impact Assessment) *Impacts of a Warming Arctic: Arctic Climate Impact Assessment*; Cambridge University Press Cambridge, U.K., 2004.

50

Global Climate Change: Gas Fluxes

Pascal Boeckx and Oswald Van Cleemput

Sampling Techniques, Sample Handling, and Analysis

Chamber Techniques

Flux chambers are simple inverted containers, which form an enclosure for gases emitted from the soil surface.[1] Both closed (static) and open (dynamic) chambers can be used. Advantages and disadvantages of chamber techniques are listed in Table 1.

TABLE 1 Advantages and Disadvantages of the Closed Chamber, Open Chamber, and Micrometeorological Methods to Measure Gas Fluxes from Soils

Method	Advantage	Disadvantage
Closed chamber	Simple and low cost	Labor intensive
	Multiple gases can be sampled	Small area is covered
	Small fluxes can be measured	Only a short-term emission event is monitored (1–2 h)
	Manual and automated gas sampling can be used	Disturbance of the emitting surface upon installation
		Altered conditions of temperature and soil atmosphere exchange
		Different functioning of plants in the chamber
Open chamber	Relatively simple	Small area is covered
	Environmental condition close to uncovered field	Disturbance of the emitting surface upon installation

(*Continued*)

TABLE 1 (*Continued*) Advantages and Disadvantages of the Closed Chamber, Open Chamber, and Micrometeorological Methods to Measure Gas Fluxes from Soils

Method	Advantage	Disadvantage
	Continuous long-term monitoring possible	Pressure deficits can cause artificially high fluxes
		Automated sampling is required
Micrometeorological	Useful for diurnal and seasonal variations	Expensive and sophisticated instrumentation needed
	Large areas can be monitored (aggregate flux)	Dependence on a uniform, large surface and constant atmospheric conditions
	Minimal disturbance of the emitting surface	

Closed and Open Chambers

Emissions of CO_2, CH_4, and N_2O are very variable, both spatially and temporally. It has not been determined whether the size of the flux chambers has an influence on this variability. Nevertheless, at least six chambers should be used per campaign. Flux chambers (cylindrical, or square or rectangular and box-like) can either be installed as a complete assembly, eventually for a short period and then removed until the next sampling occasion (Figure 1a), or they can exist out of a basal part, which is installed for the entire duration of the experiment with a gas-tight chamber attached to it for short periods (Figure 1b). This last variant is often used in flooded systems (e.g., paddy soils). The normal procedure for installation is to make a slit in the soil with a metal cutting edge, the same size and shape as the collar of the chamber, and to insert the chamber collar for about 3 cm into the slit.

When a closed chamber is fixed in place, gas samples can be taken from the headspace at different time intervals. The change in concentration in the chamber over time is used to calculate the gas flux. The calculation of the flux goes through a linear regression analysis of the gas concentration increase with time (corrected for eventual temperature changes) and a calculation of the chamber volume and area. Typical expressions of fluxes are g GHG $ha^{-1}d^{-1}$. A minimum of three measurements should be made to check the linearity of concentration increase in the chamber. A non-linear increase could indicate an inadequate sealing of the chamber or an important increase of pressure in the chamber due to temperature increases.[2] However, venting of the closed chambers can create large errors. The chamber cover should be removed once the final sample has been taken to minimize disturbance of the environmental conditions of the area covered by the chamber.

Open chambers can be used as well (Figure 2). In open chambers outside air flows into the chamber via an inlet and is forced to flow over the enclosed soil surface before leaving the chamber via an outlet. The concentration of the respective gas is measured at the in- and outlet sides. The gas flux from the soil surface can be calculated from the concentration difference between the in- and outlet, gas flow rate, and volume and area covered by the chamber.

Manual or Automated Sampling and Analysis

Using closed chambers, the headspace can be sampled by syringe. The gas samples are transferred to the laboratory into sealed containers (evacuated vials fitted with rubber septa) for analysis. To obtain a representative sample of the headspace, a diffusive mixing over 30 min is adequate, unless a mixing fan is used. Gas samples should be taken from the headspace immediately after sealing and at equal time intervals, thereafter, during which, gas concentrations increase linearly.

With automatic sampling, manual gas sampling from the chamber is replaced by a gas flow system providing a periodic sample transfer to a detector. The basic elements of an automatic system consist of closed chambers (equipped with lids that open and close automatically) or open chambers, gas flow

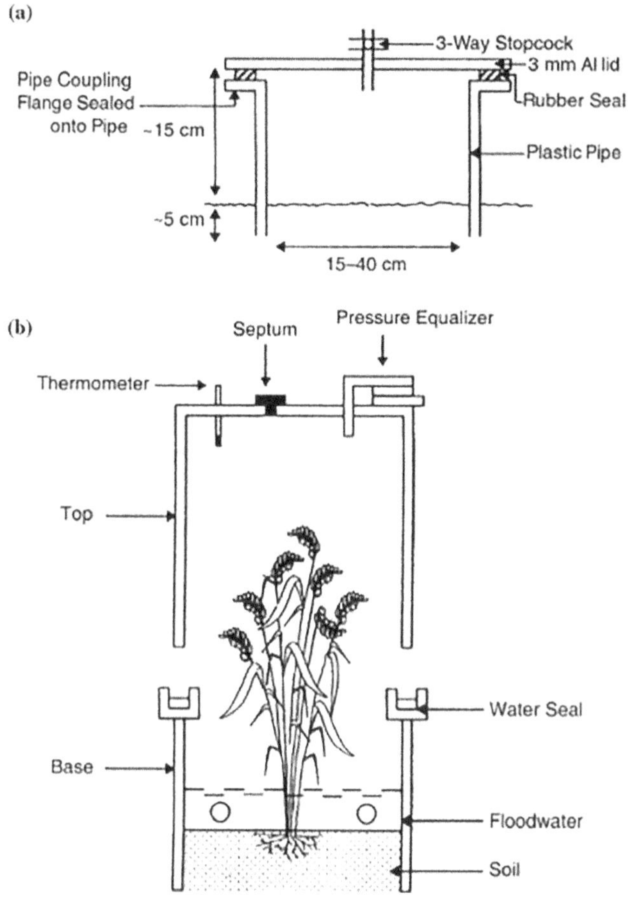

FIGURE 1 (a) Typical closed chamber used for greenhouse gas flux measurements from aerobic soils. **Source:** Adapted from Manual on Measurement of Methane and Nitrous Oxide Emissions from Agriculture.[2] (b) Schematic drawing of closed chamber used for greenhouse gas flux measurements in flooded soil.
Source: Lindau.[3]

FIGURE 2 Field set-up of an open chamber with removable lid to measure NO fluxes from forest soils: In the front, the outlet with the air sampling tube; in the back, the air inlet with the air sampling tube and the NO analyzer with data acquisition system.

systems (tubing and pump); a sampling unit; an analytical unit (detector); a time controller; and a data acquisition system. Automatic sampling is more expensive and it needs to be done in the vicinity of the analysis device. However, this technique is helpful when extensive data sets need to be collected over longer periods of time. It increases the reliability of the emission data obtained, because the number of manipulations is reduced. In most cases open chambers are sampled automatically.

Analytical Aspects

Gas Chromatography

Gas samples are injected into the gas chromatograph (GC) either manually or through the use of a sample loop (automated). Depending on the type of gas to be analyzed, specific GC settings and columns are used. Carbon dioxide is detected using a thermal conductivity detector (TCD). Methane is detected using a flame ionization detector (FID). Nitrous oxide is detected using a ^{63}Ni electron capture detector (ECD). For analysis of N_2O, care must be taken to remove both CO_2 and water vapor. In experiments with acetylene, to block the last step in the denitrification process, one may need to bypass the ECD with the acetylene in the sample after it exits from the chromatographic column. Acetylene alters both the sensitivity and stability of some ECDs.[2]

Photo-Acoustic-Infrared Detector (PAID)

The measuring principle is based on photo-acoustic detection of the absorption of infrared light. This means that any gas that absorbs infrared light of a specific wavelength can be measured. The most common PAID is the *Brüel and Kjær multi-gas monitor.* The PAID is equipped with a pump, which draws air from the flux chamber into an analysis cell inside the gas monitor. This cell can be sealed hermetically. Light from an infrared source is pulsated by a mechanical chopper and then passes through one of the optical filters of the filter carousel. The filter produces an infrared wavelength, which is selectively absorbed by the gas being monitored. Absorption of infrared light by the gas in the closed analysis cell causes the temperature to increase. Because the infrared light is pulsating, the gas temperature increases and decreases. In the closed cell this results in an increase and decrease of the pressure, which can be measured via 2 microphones. This acoustic signal is proportional to the concentration of the gas monitored. The *Brüel and Kjær multi-gas monitor* allows the analysis of 5 different gases and water vapor from 1 sample in approximately 120 sec. The device has the ability to compensate for temperature fluctuations and water vapor interference. Using an automated sampling system, careful calibration of the multi-gas monitor and the sampling unit is required.[4]

Micrometeorological Methods

The basic concept of micrometeorological methods for measuring trace gas fluxes to or from the soil surface is that gas transport is accomplished by the eddying motion of the atmosphere which displaces parcels of air from one level to another. Transport of a gas through the free atmosphere is provided by turbulent diffusion and convection in which the displacement of individual eddies is the basic transport process. Micrometeorological methods are based on the assumption that the flux to or from the soil surface is identical to the vertical flux measured at the reference level some distance above the surface. Therefore, a flat and homogeneous terrain is needed. The flux measured at the reference level provides than the average flux over the upwind area (fetch), provided that sampling point at the reference level is in the height range in which the vertical flux is constant with height (fetch $\approx$ 100H, where H = height of the sampling point). In the simplest of the micrometeorological methods, the flux may be measured by sensing the concentrations and velocities of components of the turbulence. [5,6]

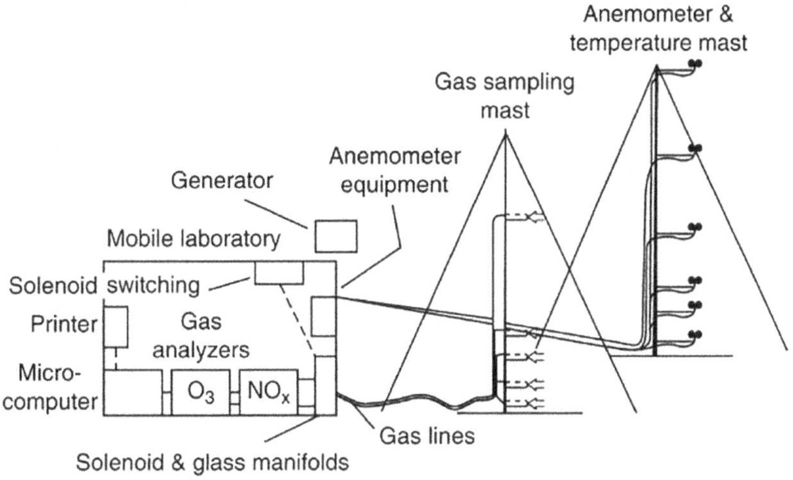

FIGURE 3 Typical instrument for flux gradient measurements of greenhouse gas fluxes in the field.
Source: Adapted from Denmead.[5]

Two general micrometeorological techniques are used to measure trace gas fluxes: the eddy correlation and flux- gradient technique (Figure 3). Application of both approaches is limited to situations in which the air analyzed has passed over a homogeneous exchange surface for a long distance so that profiles of gas concentration in the air are in equilibrium with the local rates of exchange. These methods also require that horizontal concentration gradients are negligible. Eddy correlation methods require a fast response detector. The tuneable diode laser (TDL) technique is based on infrared absorption spectrometry, whereby the absorption depends upon path length, line strength, and absorber concentration. Liquid nitrogen temperature diodes are commercially available to cover the infrared spectrum from about 2–10 μm, the region where most trace gases have absorption spectra.[5,6] Advantages and disadvantages of micrometeorological methods are listed in Table 1.

Nonisotopic Tracer Methods

Tracer methods involve the release of an inert tracer gas, most commonly sulphur hexafluoride (SF_6), from an emitting surface. The tracer gas is released at a known rate in a pattern similar to the release pattern of the GHG, perpendicular to the direction of the prevailing wind. This method can be applied when a definite plume of the GHG can be readily detected in the ambient environment. Under these conditions the plume of the dispersed emission is located based on analyses of upwind and downwind air samples. The flux rate is computed, using the ratio of the plume concentration of the tracer and the GHG and the known release rate of the tracer. The advantage of this technique is that aggregate gas emissions can be collected from heterogeneous areas, such as landfills, circumventing the problem of spatial heterogeneity. However, the high costs, dependence on meteorological conditions, and the potential for interfering sources limit its application.[2]

Ultra-Large Chambers with Long-Path Infrared Spectrometers

Infrared absorption spectrometers are available that can give an average value for the gas concentration over distances of tens or hundreds of meters. They are useful for measurements of average emissions from a whole experimental plot, by covering the plot temporarily with a large canopy to act as a chamber and retain the gas emitted from the soil. Two systems are available: 1) the Fourier Transform Infrared (FTIR) spectrometer with a mirror system, which allows multiple reflections and thus a total path of up to 1 km that is capable of measuring GHG concentration changes down to a fraction of 1 ppb; and 2) a simpler, less-sensitive IR spectrometer with the capacity to detect a concentration change of about 25 ppbv of N_2O and 10 ppbv of CH_4.[2]

References

1. Hutchinson, G.L.; Livingstone, G.P. Use of chamber techniques to measure trace gas fluxes. In *Agricultural Ecosystems on Trace Gases and Global Change*; ASA Special Publication 55; ASA: Madison, 1993; 63–78.
2. Manual on Measurement of Methane and Nitrous Oxide Emissions from Agriculture; IAEA-TECDOC-674; IAEA: Vienna, 1992; 91 pp.
3. Lindau, C.W.; Bollich, P.K.; De Laune, R.D.; Patrick, W.H., Jr.; Law, V.J. Effect of urea fertilizer and environmental factors on CH_4 emissions from a Louisiana USA rice field. Plant and Soil **1991**, *136*, 195–203.
4. De Visscher, A.; Goossens, A.; Van Cleemput, A. Calibration of a multipoint sampling system connected with a photoacoustic detector. Intern. J. Environ. Anal. Chem. **2000**, *76*, 115–133.
5. Denmead, O.T. Micrometeorological methods for measuring gaseous losses of nitrogen in the field. In *Gaseous Loss of Nitrogen from Plant-Soil Systems;* Freney, J.R., Simpson, J.R., Eds.; Nyhoff Junk Publ.: The Hague, 1983; 133–157.
6. Fowler, D.; Duyzer, J. Micrometeorological techniques for the measurement of trace gas exchange. In *Exchange of Trace Gases Between Terrestrial Ecosystems and the Atmosphere;* Andreae, M.O., Schimel, D.S., Eds.; John Wiley and Sons: Chichester, 1989; 189–207.

Index

Page numbers followed by f and t indicate figures and tables, respectively

A

B

C

D

E

F

G

H

J

K

L

M

N

Q

R

S

T

U

V

W

X

Y

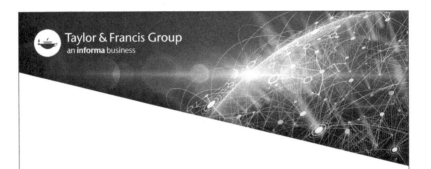